Jane's
C4I
SYSTEMS

Edited by Giles Ebbutt

Sixteenth Edition
2004-2005

Total number of entries $\boxed{997}$ New and updated entries $\boxed{619}$
Total number of images $\boxed{1,279}$ New images $\boxed{88}$

Visit jc4i.janes.com and view the list of latest updates that have been added to the online version of *Jane's C4I Systems* subsequent to this print edition.

Bookmark jc4i.janes.com today!

Jane's C4I Systems online site gives you details of the additional information that is unique to online subscribers and the many benefits to upgrading to an online subscription. Don't delay, visit jc4i.janes.com today and view the list of latest updates to this online service.

ISBN 0 7106 2619 3

"Jane's" is a registered trade mark

Copyright © 2004 by Jane's Information Group Limited, Sentinel House, 163 Brighton Road, Coulsdon, Surrey CR5 2YH, UK

In the USA and its dependencies

Jane's Information Group Inc, 110 N Royal Street, Suite 200, Alexandria, Virginia 22314, USA

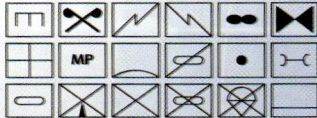

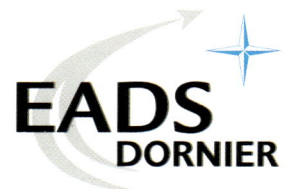

Contents

Jane's C4I website: jc4i.janes.com

Front cover image: The interior of a Bradley Command Vehicle shown at AUSA 2003, equipped with a mission equipment package that provides a mobile tactical operations centre capability. The package consists of networked workstations which host the integrated Army Battle Command System (ABCS) suite of software, plus related combat net radios. Nearly 100 of such packages are planned for both Bradley and Stryker (Patrick Allen/Jane's) *NEW*/1029539

How to use *Jane's C4I Systems*

Jane's C4I Systems is laid out in four main sections, each of which relates to an area of C4I activity: Command Information Systems, Communication Systems, Intelligence Systems and Computing. The first two of these main sections are subdivided into Joint, Land, Maritime and Air subsections, with an additional subsection for C2 Facilities in the CIS section and one for Satellite in the Communications section. The Intelligence Systems section is subdivided into Surveillance and Reconnaissance, Direction-Finding, Signals Intelligence, Imagery Intelligence, and Analysis and Support; and the Computing section into Terminals and Workstations, and Software. Each subsection is subdivided by country alphabetically and within these subdivisions the individual entries are arranged in alphabetical order. Each entry provides descriptions of the system or equipment, development details, operational status and technical specifications, as far as can be determined at the time of writing. Then follows, where appropriate, the name and location of the manufacturer(s) or contractor(s). In some cases where there is a nominated prime contractor, this is indicated. Full details of contractors' addresses and contact details can be found towards the end of this publication and there is also a listing of each manufacturer's products mentioned. If any specific item is required, the name or designation may be found using the main index.

To help users evaluate the data of this edition, the following identifiers have been used:

● **VERIFIED** The editor has made a detailed examination of the entry's content and checked its relevancy and accuracy for publication in the new edition to the best of his ability.

● **UPDATED** During the verification process, significant changes to content have been made to reflect the latest position known to Jane's at the time of publication.

● **NEW ENTRY** Information on new equipment and/or appearing for the first time in the title.

● **NEW** New images are identified as **NEW** and some are followed by a seven digit number for ease of identification by our image library.

A full list of all entries indicating their current status is provided in the index.

Total number of entries 997 New and updated entries 619
Total number of images 1,279 New images 88

Visit jc4i.janes.com and view the list of latest updates that have been added to the online version of *Jane's C4I Systems* subsequent to this print edition.

Copyright enquiries
Contact: Keith Faulkner, Tel/Fax: +44 (0) 1342 305032, e-mail: keith.faulkner@janes.com

British Library Cataloguing-in-Publication Data.
A catalogue record for this book is available from the British Library.

Printed and bound in Great Britain by Biddles Ltd, King's Lynn

EDITORIAL AND ADMINISTRATION

Director: Ian Kay, e-mail: Ian.Kay@janes.com

Managing Editor: Mike Bryant, e-mail: Mike.Bryant@janes.com

Content Services Director: Anita Slade, e-mail: Anita.Slade@janes.com

Content Systems Manager: Jo Agius, e-mail: Jo.Agius@janes.com

Pre-Press Manager: Christopher Morris, e-mail: Christopher.Morris@janes.com

Team Leader: Neil Grace, e-mail: Neil.Grace@janes.com

Content Editor: Elizabeth Glendinning, e-mail: Elizabeth.Glendinning@janes.com

Production Controller: Victoria Powell, e-mail: Victoria.Powell@janes.com

Content Update: Jacqui Beard, Information Collection Co-Ordinator
Tel: +44 (0) 20 8700 3808   Fax: +44 (0) 20 8700 3959
e-mail: yearbook@janes.com

Jane's Information Group Limited, Sentinel House, 163 Brighton Road, Coulsdon, Surrey CR5 2YH, UK
Tel: +44 (0) 20 8700 3700   Fax: +44 (0) 20 8700 3788
e-mail: jc4i@janes.com

SALES OFFICE

Send Europe and Africa enquiries to: *Mike Gwynn – Head of Information Sales*
Jane's Information Group Limited, Sentinel House, 163 Brighton Road, Coulsdon, Surrey CR5 2YH, UK
Tel: +44 (0) 20 8700 3700   Fax: +44 (0) 20 8763 1006
e-mail: info.uk@janes.com

Send USA enquiries to: *Robert Loughman – Sales Director*
Jane's Information Group Inc, 110 N Royal Street, Suite 200, Alexandria 22314, USA
Tel: (+1 703) 683 37 00   Fax: (+1 703) 836 02 97   Telex: 6819193
Tel: (+1 800) 824 07 68   Fax: (+1 800) 836 02 97
e-mail: info.us@janes.com

Send Asia enquiries to: *David Fisher – Group Business Manager*
Jane's Information Group Asia, 78 Shenton Way, #10-02, Singapore 079120
Tel: +65 6325 0866   Fax: +65 6226 1185
e-mail: asiapacific@janes.com

Send Australia/New Zealand enquiries to: *Pauline Roberts – Business Manager*
Jane's Information Group, PO Box 3502, Rozelle Delivery Centre, NSW 2039, Australia
Tel: +61 (0)2 8587 7900   Fax: +61 (0)2 8587 7901
e-mail: oceania@janes.com

Send Middle East enquiries to: *Ali Abdellatif Siali – Regional Sales Manager*
Jane's Information Group, PO Box 502138, Dubai, United Arab Emirates
Tel: +971 4 390 2335   Tel: +971 4 390 2336   Fax: +971 4 390 8848
e-mail: mideast@janes.com

Send Japan enquiries to: *Norihisa Fukuyama – Information Consultant*
Jane's Information Group, Palaceside Building, 5F, 1-1-1, Hitotsubashi, Chiyoda-ku, Tokyo 100-0003, Japan
Tel: +81 (0)3 5218 7682   Fax: +81 (0)3 5222 1280
e-mail: norihisa.fukuyama@janes.jp

Send India enquiries to: *T.C. Martin – Information Consultant*
Jane's Information Group, PO Box 3806, New Delhi 110049, India
Tel/Fax: +91 (0) 11 26 51 61 05
e-mail: india@janes.com

Send China enquiries to: *Jia Mao – Business Consultant for People's Republic of China*
Jane's Information Group, 78 Shenton Way, #10-02, Singapore 079120
Tel/Fax: +86 (0) 10 64 16 91 73
e-mail: china@janes.com

Send Canada enquiries to: *Geoff Mizen – Canadian Sales*
Jane's Information Group, 1400, 100 Queen Street, Ottawa, Ontario, Canada K1P 1J9
Tel: (+1 613) 566 3642   Fax: (+1 613) 566 3640
e-mail: geoff.mizen@janes.com

ADVERTISEMENT SALES OFFICES

Head Office
Jane's Information Group
Sentinel House, 163 Brighton Road, Coulsdon, Surrey CR5 2YH
Tel: (+44 20) 87 00 37 00   Fax: (+44 20) 87 00 38 59/37 44
e-mail: transadsales@janes.com

Tracy Attwooll, Advertisement Sales Manager, Transport
Tel: (+44 20) 87 00 37 41   Fax: (+44 20) 87 00 38 59/37 44
e-mail: tracy.attwooll@janes.com

USA/Canada
Jane's Information Group
110 N Royal St. Suite 200, Alexandria, Virginia 22314, USA
Tel: (+1 703) 683 37 00   Fax: (+1 703) 836 55 37
e-mail: transadsales@janes.com

Katie Taplett, US Advertising Sales Director
Tel: (+1 703) 683 37 00   Fax (+1 703) 836 55 37
e-mail: katie.taplett@janes.com

Sean Fitzgerald, Account Executive
Tel: (+1 703) 683 37 00   Fax: (+1 703) 836 55 37
e-mail: sean.fitzgerald@janes.com

Northern USA and Eastern Canada
Harry Carter, Advertising Sales Manager
Tel: (+1 703) 683 37 00   Fax: (+1 703) 836 55 37
e-mail: harry.carter@janes.com

South Eastern USA
Kristin D Schulze, Advertising Sales Manager
PO Box 270190, Tampa, Florida 33688-0190
Tel: (+1 813) 961 81 32   Fax: (+1 813) 961 96 42
e-mail: kristin@intnet.net

Australia: *Richard West (UK Head Office)*

Benelux: *Tracy Attwooll (UK Head Office)*

China and Hong Kong: *Tracy Attwooll (UK Head Office)*

CIS: *Tracy Attwooll (UK Head Office)*

France: *Tracy Attwooll (UK Head Office)*

Germany and Austria: *MCW (Media and Consulting Wehrstedt)*
Tel: (+49 34) 74 36 20 90   Fax: (+49 34) 74 36 20 91
e-mail: info@wehrsteddt.org

Israel: *Oreet – International Media*
Tel: (+972 3) 570 65 27   Fax: (+972 3) 570 65 26
e-mail: liat_h@oreet-marcom.com

Italy and Switzerland: *Ediconsult Internazionale Srl*
Tel: (+39 010) 58 36 84   Fax: (+39 010) 56 65 78
e-mail: genova@ediconsult.com

Japan: *Skynet Media, Inc*
Tel: (+81 3) 54 74 78 35   Fax: (+81 3) 54 74 78 37

Middle East: Tracy Attwooll (see UK Head Office)

Russia: Vladimir N Usov
Tel/Fax: (+7 3435) 32 96 23
e-mail: uvn125@uraltelecom.ru

Scandinavia: *The Falsten Partnership*
Tel: (+44 20) 88 06 23 01   Fax: (+44 20) 88 06 81 37
e-mail: sales@falsten.com

Singapore: Tracy Attwooll (see UK Head Office)

South Africa: Richard West *(see UK Head Office)*

South Korea: *JES Media Inc*
Contact: Young-Seoh Chinn
Tel: (+82 2) 481 34 11/3   Fax: (+82 2) 481 34 14
e-mail: jesmedia@unitel.co.kr

Spain: *Via Exclusivas, SL*
Contact: Julio de Andres
C/Viriato no 69SC, E-28010 Madrid
Tel: (+34 91) 448 76 22   Fax: (+34 91) 446 02 14
e-mail: j.a.deandres@viaexclusivas.com

For all other areas, contact Tracy Attwooll (UK Head Office)

ADVERTISING COPY
Linda Letori (Jane's UK Head Office address)
Tel: (+44 20) 87 00 38 56   Fax: (+44 20) 87 00 38 59; 37 44
e-mail: linda.letori@janes.com

For North America, South America and Caribbean only:
Lia Johns (Jane's US address)
Tel: (+1 703) 683 37 00   Fax: (+1 703) 836 55 37
e-mail: lia.johns@janes.com

Jane's Electronic Solutions

Access over 200 sources of Jane's news, analysis and reference covering defence, risk, security, aerospace, transport and law enforcement information with electronic solutions designed to meet the demands of your information needs.

Jane's electronic solutions offer you the choice of how you want to receive the data. Whether you want to integrate data into your organisation's network, access the data online or on CD-ROM, Jane's can provide you with critical information, quickly and easily in a format that suits your organisation.

www.janes.com

Accessible by IP address, for networking within organisations or by unique username and password, janes.com enables you to access Jane's information wherever you are in the world.

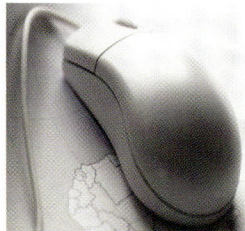

Why subscribe to janes.com:

- Search function
  Allows you to explore within the contents of all Jane's datasets by keyword and/or fielded search terms.
- Image Searching
  Allows you to search captions within Jane's data with results in the form of a thumbnail and caption.
- Latest News Extra
  Search for the latest news globally across all areas of defence, transport, aerospace, security and business.

- Active Interlinking
  Allows you to navigate via hyperlinks in records to other related product entries throughout Jane's content to which you subscribe, reducing your research time significantly.
- Browse function
  Allows you to view the available contents of Jane's datasets by Country, Image, Market, News/Analysis, Operational Guide, Organisation, and Systems & Equipment.

Jane's Libraries

Jane's Libraries offer you and your organisation comprehensive datasets in the areas of defence, risk, security, aerospace, transport and law enforcement. Each Library groups relevant information and graphics that can be cross-searched to ensure you find every reference you are looking for.

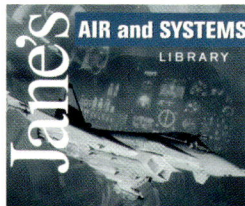

With Jane's Libraries you can:

- Pinpoint the information you need using a variety of basic and advanced searches.
- Export text easily into ASCII, dbase or comma-delimited formats for use in your own reports and presentations.
- Download JPEG photographs and technical line drawings for your own internal use.
- Print all text and graphics straight from the Libraries.
- Network the Libraries within your organisation to ensure easy access by all.

Libraries available:

Jane's Sea and Systems Library
Jane's Air and Systems Library
Jane's Land and Systems Library
Jane's Police and Security Library
Jane's Market Intelligence Library
Jane's Geopolitical Library
Jane's Transport Library
Jane's Defence Equipment Library
Jane's Defence Magazines Library

For information on how a Jane's Library could help your organisation please contact your local Jane's sales office.

Jane's Data Service

Jane's information on your intranet or controlled military network

Jane's Data Services brings together more than 200 sources of near-realtime and technical reference information serving defence, intelligence, space, transportation and law enforcement professionals.

Jane's Data Service provides you with:

- Full integration of data into your own secure environment
- Flexibility of choice with your selection of data
- Frequent updates via CD-ROM or FTP

- Re-usable data for internal presentations and reports
- Full integration with other data sources for cross-databank searching
- High-quality JPEG images

http://consultancy.janes.com

As a decision maker, you shoulder a heavy burden. Yes, you can access more information faster and more easily than ever before, but more information does not always help you to make better choices. So, how do you maintain your advantage in such a world?

You could build a sophisticated global network and collect, weigh and analyse the information it gathers. Or, you could simply turn to Jane's Consultancy.

What we can do for you:

- Market Intelligence
- Systems Analysis

- Military Assessments
- Security Risks and Red Teaming
- Expert Testimony

RADAR FREQUENCY/WAVELENGTH TABLE

Historic Radar Bands			NATO/EW Bands		
Band Designation	Frequency GHz	Wavelength cm	Band Designation	Frequency GHz	Wavelength cm
VHF	0.03-0.3	1000-100	A	0.03-0.25	1000-120
UHF	0.3-1	100-30	B	0.25-0.5	120-60
L	1-2	30-15	C	0.5-1	60-30
S	2-4	15-7.5	D	1-2	30-15
C	4-8	7.5-3.75	E	2-3	15-10
X	8-12	3.75-2.5	F	3-4	10-7.5
Ku	12-18	2.5-1.6	G	4-6	7.5-5
K	18-27	1.6-1.1	H	6-8	5-3.75
Ka	27-40	1.1-0.75	I	8-10	3.75-3
MM	40-100	0.75-0.3	J	10-20	3-1.5
			K	20-40	1.5-0.75
			L	40-60	0.75-0.5
			M	60-100	0.5-0.3

"AN" NOMENCLATURE MATRIX

MIL-STD-196E (Feb 1998), "Joint Electronics Type Designation System"
EQUIPMENT INDICATOR LETTERS

1st Letter (installation)
A - Piloted aircraft
B - Underwater mobile, submarine
C - Cryptographic
D - Pilotless carrier
F - Fixed ground
G - General ground use
K - Amphibious
M - Ground, mobile
P - Portable
S - Water, surface craft
T - Ground, transportable
U - General utility includes two or more general installation classes, airborne, shipboard, and ground).
V - Vehicular, ground (installed in vehicle designed for functions other than carrying electronic equipment)
W - Water, surface and underwater combined
Z - Piloted/pilotless airborne vehicle combined

2nd Letter (type)
A - Invisible light, heat radiation
B - COMSEC
C - Carrier -electronic wave/signal)
D - Radiac
E - Laser
F - Fibre optics
G - Telegraph or teletype (wire)
I - Interphone and public address
J - Electromechanical or inertial wire covered
K - Telemetering
L - Countermeasures
M - Meteorological
N - Sound in air
P - Radar
Q - Sonar and underwater sound
R - Radio
S - Special types, magnetic or combination of types
T - Telephone (wire)
V - Visual and visible light
W - Armament
X - Facsimile or television
Y - Data processing
Z - Communications

3rd Letter (purpose)
A - Auxiliary assemblies
B - Bombing
C - Communications (receive and transmit)
D - Direction finding and/or reconnaissance and surveillance
G - Fire control or searchlight directing
H - Recording/reproducing
K - Computing
M - Maintenance and test assemblies
N - Navigational aids
Q - Special or combination of types
R - Receiving, passive detecting
S - Detecting range and/or bearing, search
T - Transmitting
W - Automatic flight or remote control
X - Identification and recognition
Y - Surveillance and control
Z - Secure

Example: The SINCGARS AN/PRC119A is a portable (manpackable) radio communications device, model number 119A.

Executive Overview

The major event of the past 12 months (prior to February 2004) has been the operation in Iraq, known by various names by different members of the Coalition. There has already been much written about events in Iraq and it is not the intention to cover old ground, nor to indulge in a long discussion about what worked and what did not, so this overview contains some observations on some aspects of C4I which have emerged from that operation. I am conscious that this makes for a particularly Anglo/US-centric piece but the fact remains that the most rapid progress on the broadest fronts in the C4I world is being made in the US and this progress and these developments are based on operational experience.

This time last year we were waiting to see whether there would be an invasion of Iraq and, from the C4I point of view, how the digitisation process would work in a major campaign with a coalition force. It appeared that this would be the first real test of warfare in the network-centric era. Operations in Afghanistan had already demonstrated what had been achieved in bringing the sensor and the shooter together, in empowering both the man on the ground with control over weaponry and equally providing the commander many miles away with the ability to see for himself what was going on. But this was likely to be the first real test of warfighting in the digital age, in a joint and combined environment.

A major theme, which is reflected in the programmes of a number of nations, is the importance that situational awareness systems (SAS) are assuming. In the US this is principally demonstrated by the development of FBCB2 (Force XXI Battle Command Brigade and Below) from a battle management system relying on Enhanced Position Location Reporting System (EPLRS) to a major battlefield-wide SAS. Developments introduced in Kosovo and expanded in Afghanistan to introduce satellite communications into FBCB2, thus dramatically increasing its range and coverage, noted in this overview last year, were of immense value in Iraq, overcoming the speed of advance and distance problems that affected a whole range of systems. It has clearly been valued most highly for its SAS capability - hence its 'Blue Force Tracker' designation; at the Association of the United States Army (AUSA) annual meeting in Washington in October 2003 a succession of operational commanders were singing its praises after their experiences in Iraq, some of whom were happy to admit to initial scepticism. FBCB2 was provided both to the United States Marine Corps (USMC) and to UK forces to overcome problems of interoperability and it has been fitted in a variety of helicopter platforms, both in the US and now in the UK. The capability is being brought to the dismounted soldier with Northrop-Grumman's PDA version, displayed at AUSA. There remain some security problems and the need for a bridge between the tactical Internet and satellite-based elements of the network, but it is clear that this is one aspect of digitising the battlefield that has been a success. Having said that, it is worth noting that a senior UK field commander observed that whenever USMC officers visited his HQ in Iraq, they were keen to gather round an old-fashioned paper map to

The interior of the M4 command and control vehicle displayed at AUSA 2003 showing three of the workstations of the integrated ABCS (Patrick Allen/Jane's) NEW/1029568

discuss progress and plans, which is perhaps an indication either that the electronic map display has its limitations, or that there are still cultural barriers to its use to surmount.

We have heard less about the other elements in the US Army's suite of digital systems, the Army Battle Command System (ABCS), although it would appear that the Advanced Field Artillery Tactical Data System (AFATDS), which is well established, performed well. However, the pace of introduction of integrated platforms for ABCS has quickened. The M4 Command Vehicle was a dead programme, with United Defense attempting to sell the limited number of vehicles from the early production run around the world, the programme having been cancelled. However, the need for a CV which could keep up with the armoured formations and could be fitted with the ABCS suite rapidly became apparent to those formations preparing for operations in Iraq. The army reclaimed the vehicles and they were deployed to Iraq, where they provided tactical operations centres (TOCs) at formation level, fitted with an integrated ABCS suite. A similar programme in which demonstration and preproduction models were deployed to Iraq is the Army Airborne Command and Control System (A2C2S), which contributed to some notable long-range heliborne operations, particularly by 101AB Div. The development of TOCs with an integrated ABCS suite based on other platforms, such as Bradley and HMMVW, is now in train. Examples were on show at AUSA 2003.

Continuing with lessons identified from Iraq experiences, the speed of movement and distances covered during the Coalition advance exposed two particular weaknesses which are having an impact on the C4I front, one on communications and one on logistics. The communications ranges demanded were often too great for the tactical nets and networks provided by Combat Net Radio (CNR) and the EPLRS, there was insufficient capacity in beyond-line-of-sight systems, and the advance proceeded too quickly to establish a Mobile Subscriber Equipment (MSE) network. One solution is being sought through 'satcom-on-the-move'; another is the Expeditionary Tactical Communications System (ETCS), a USMC initiative which is to be tested in the Sea Viking 04 Experiment (SV 04). ETCS uses the Iridium satellite system, now leased by the US DoD, to provide push-to-talk voice and data netted communications that will communicate across beams and across satellites. This will be linked to a variety of USMC systems and with the addition of GPS capability will provide an inherent Position Location Information or 'Blue Force Tracking' capability.

The speed of advance also exposed weaknesses in logistics command and control. In order for combat supplies to be provided at the right place and time in a fast moving and fluid operation, logistic units needed the same tactical picture as the forward units, an accurate picture of the logistic situation and the ability to predict requirements up to 72 hours in advance. It would appear that the logistic element of ABCS, the Combat Service Support Command System (CSSCS), was not able to provide this and its development

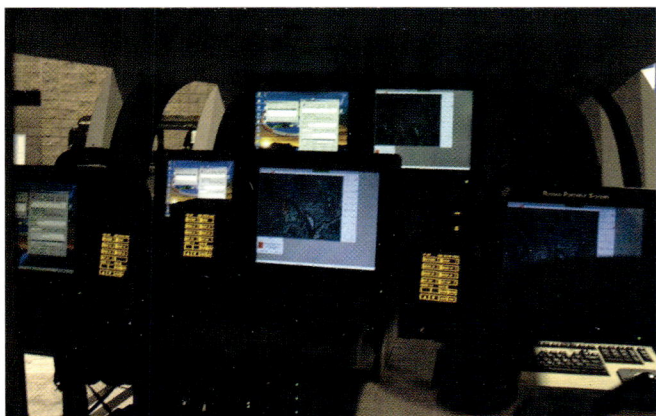

The interior of the Army Airborne Command and Control System, showing the workstations of the integrated ABCS (Patrick Allen/Jane's) NEW/1029543

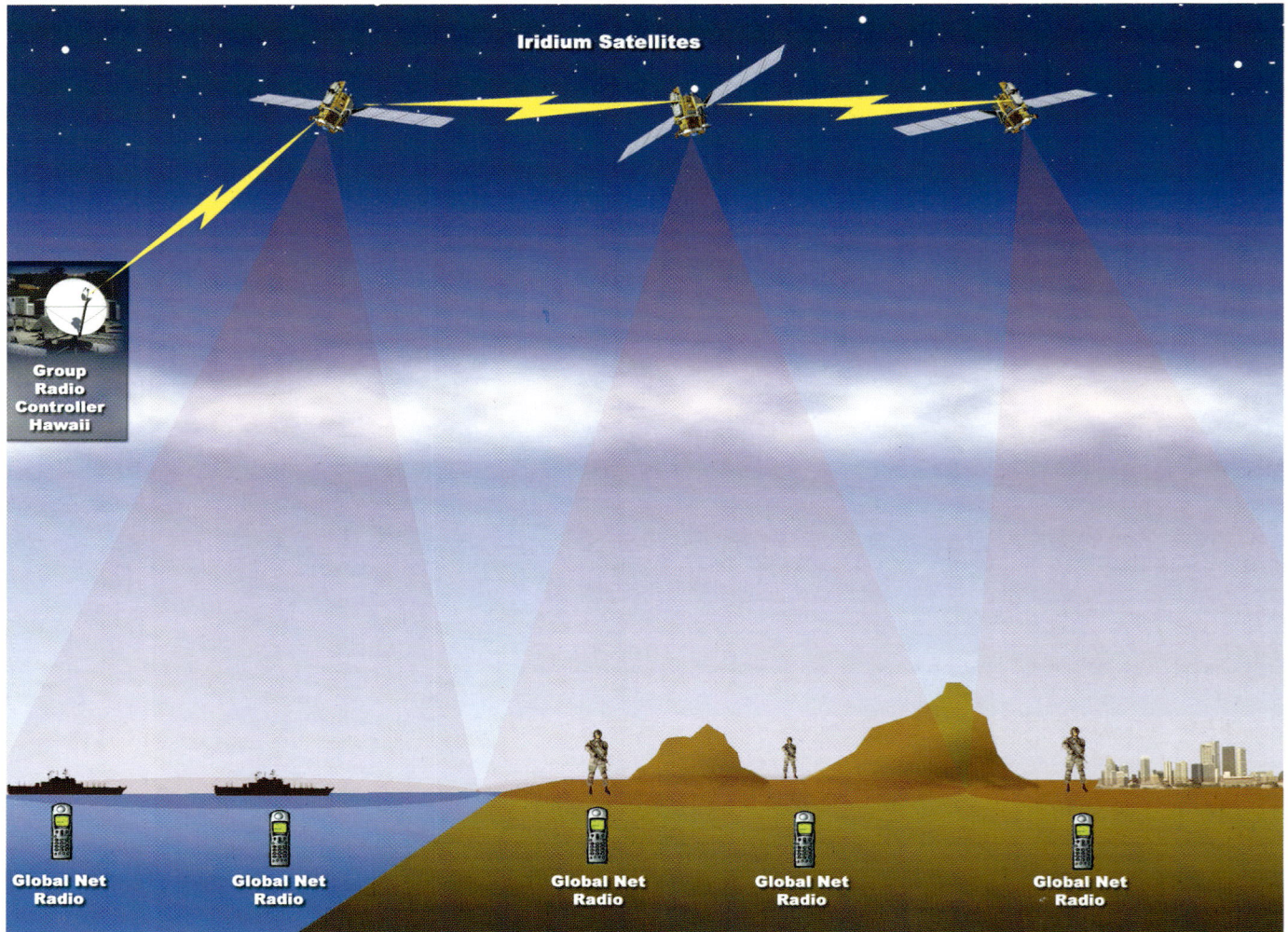

The USMC hopes the Expeditionary Tactical Communications System, which uses netted Iridium satellite phones, will help overcome the difficulties of Beyond-Line-of-Sight tactical communications. It will be tested during the USMC Sea Viking 04 Experiment (USMC Quantico)

NEW/0564079

has been halted. An alternative system, the Battle Command Sustainment Support System (BCS3), is being developed, drawing on experience from the use of comercial-off-the-shelf (COTS) software packages in theatre. The USMC is also pursuing a similar goal, which is particularly aimed at assisting the conduct of sea-based logistics. A Common Logistics C2 System (CLC2S) will be trialled on SV 04. This includes a rapid request system, a database of all supplies and their locations and a logistic planning tool which allows rapid analysis of logistic demands, allowing 'what if' planning and providing logistic go/no go decisions. CLC2S connects with and draws information from legacy systems and injects planning data into Command and Control for the PC (C2PC), the USMC tactical C2 application, making it available to commanders.

Another aspect of logistics command and control, thrown into sharp focus in post-operation debates in the UK, was the need for in-transit visibility (ITV) and asset tracking of stocks, both en route to and within theatre. The US Movement Tracking System from Comtech which provides this uses some of the same technology used in the Blue Force Tracker and, in fact, preceded it.

An increasing feature of military operations is the use of commercial equipment to provide communications architecture. Examples of this can be seen in the Iraq operations. The UK established a strategic communications architecture with what was named OSCAR, the Operational Strategic Communications Architecture. Following the reasoning that many of the locations to be served were fixed and secure - such as the JFHQ at Qatar - the mobility and robustness provided by military equipment was not required, but bandwidth was. The system was therefore heavily based on COTS equipment, with VSAT terminals and Proxima multimedia switches. This selective use of COTS released military equipment to be used further forward, providing greater connectivity at the tactical level. OSCAR, with a hub-and-spoke configuration rather than a true network, ultimately covered 8 countries, with 44 nodes, 30 satellite heads, 6 security domains, and provided access to 8 secure voice networks, with a 54 Mbyte information flow.

As another example, while this overview was being written, it was announced that a contract had been awarded for satellite communications equipment for a coalition military network being installed in Iraq. The network will use commercial satellites to provide secure voice and data services, plus voice-over-IP to integrate telephone and data on a common network, connecting coalition forces to each other and to the coalition backbone. The order is for a hub and 20 remote terminals and it will provide not only increased bandwidth but also will release tactical communications assets. This follows a pattern which has also been seen in Sierra Leone and the former Yugoslavia, with COTS systems installed as soon as the situation permits.

In a coalition, a common command and control system is fairly fundamental to the effective integration of the efforts of coalition members. The backbone of the US command and control network is the Secret Internet Protocol Router Network (SIPRNET), which provides web-based access to operation orders, situation reports, intelligence reports and so on. Access to SIPRNET is fairly critical to effective coalition operations with US forces, but it is a 'NOFORN' system - foreigners are not allowed access. An interesting development which enabled the Iraq Coalition to operate more effectively was the coming of age of the Coalition Wide Area Network (COWAN) just in time, as it overcame this difficulty and in one senior UK commander's view, Rear Admiral David Snelson, UK Maritime Component Commander (MCC) and Deputy Coalition MCC, was the key to Coalition success in the maritime area. The USA developed software which replicates SIPRNET web pages on COWAN provided the security clearances are appropriate. Fitted in US, UK and Australian ships, the USN used COWAN in preference to SIPRNET when there was a coalition formation. Equally the e-mail system on COWAN became the main e-mail system, and 'same time' chat was used as the main C2 net.

Northrop Grumman's hand-held FBCB2/Blue Force Tracking Satcom R-PDA device, showing the dual GPS/satellite antenna (right) and battery container (left) (Patrick Allen/Jane's) **NEW**/1029521

In order to overcome the problem of access to SIPRNET for land force headquarters, the UK rapidly developed, had approved and introduced 'X-net', which provided limited interoperability with SIPRNET to British headquarters, enabling them to exchange information with certain addressees and limited access to information. In at least one UK brigade access to SIPRNET was also available through US liaison officers.

On the intelligence collection front *Jane's Defence Weekly* reported on a USMC after-action report which re-emphasised the perennial problem of the conflict between the need to control and task scarce collection assets at the highest level and the speed of reaction required for their product to be of any use to the man on the ground. In a fast-moving battle there is no time for long-winded tasking and requesting procedures, but assets need to be allocated where the priorities are greatest. This conflict has always been there, both for intelligence collection and weapon systems. More development will be required to improve tasking and production cycles to try and square this circle, as well as meeting the Marines' desire for smaller, less sophisticated collection assets that can be pushed well forward with delegated control. Whether this was a complaint in the US Army as well or more a reflection of difficulties stemming from army assets supporting USMC formations and consequent lack of understanding is uncertain. However, the two services are inevitably going to undertake joint operations in the future, so this must lend impetus to the development of common or combined display systems for different surveillance assets which provide greater flexibility. The increasing use of reach back to control surveillance assets and interpret the product which is becoming possible as bandwidth availability increases may also draw the surveillance capability away from the tactical commander.

Moving away from Iraq, noticeable by their presence at recent shows have been 'C2 in a suitcase' systems, providing security, satellite communications, video capability, network hubs and basic IT support. Although these have been around for some time, increasing capability means that more and more frequently they are being marketed as the answer to the needs of those who must travel light but who require this type of capability - Special Forces and early entry headquarters elements seem to be the principal users.

Other miscellaneous events or programmes worthy of note are:
- Australia is embarking on a new communications programme, JP 2072, to provide a Battlespace Communications System (Land). Essentially this will upgrade and replace its current land communications systems to enable Australian forces to operate in the network-centred battlespace of the future.
- Taiwan is extending its C4I capability through the "Po Sheng" programme, particularly through developments in its Link 16 capability .
- The UK established the Network Integrated Test and Experimentation (NITEworks), a 3-year programme to investigate system integration and interoperability issues, in support of the drive to Network Enabled Capability.
- The first EU military C2 facility has been established at Northwood in the UK, providing a networked system to support an EU operation .
- The re-organisation of NATO's command structure and the establishment of the Allied Command for Transformation (ACT) alongside its US counterpart USJFCOM at Norfolk, Virginia, means that the Alliance will now benefit from the experimental and developmental work taking place there.

Acknowledgements

Welcome to the 2004-05 *Jane's C4I Systems* Yearbook, the sixteenth edition. Many readers, especially of this publication, are likely to have access to the electronic version of this product. This is now the priority medium for Jane's publications; those who are reading this in hard copy should note that it is only a snapshot at one particular moment during the year and that updating takes place on a continuous basis in the online product.

The editor is but one of a number of people who are involved with the production of this work. Although it might be invidious to single out individuals, I must acknowledge with considerable gratitude the help of all those at Jane's who have been involved this year, notably Belinda Dodman and Jacqui Beard of the collection staff, Peter Partridge, Managing Editor Industry and Government, and Neil Grace and the team in the production department, particularly Liz Glendinning who has had the onerous task of proof reading. My fellow editors John Williamson (*Jane's Military Communications*), Chris Foss (*Jane's Armour and Artillery*) and Martin Streetly (*Jane's Radar and Electronic Warfare*) together with Rupert Pengelley, the Group Technical Editor, have been a great source of advice,

information and mutual support. Patrick Allen has been a valuable provider of images and a cheerful companion at Exhibitions. As always, particular thanks are due to Mike Bryant, my Managing Editor, who has continued to give me much encouragement.

Comments, suggestions and further information are welcome and should be sent to:

The Editor
Jane's C4I Systems
Sentinel House
163 Brighton Road
Coulsdon
Surrey CR5 2YH
United Kingdom

e-mail: janesc4i@btinternet.com

Giles Ebbutt
February 2004

Lt Col Giles Ebbutt MA RM (ret'd)

Giles Ebbutt was educated at Magdalen College School and Merton College, Oxford. He joined the Royal Marines in 1971 and his early service was in 45 Commando RM, including Arctic Warfare training in Norway.

Having specialised as a Signals Officer in 1978, he went on to hold a variety of communications appointments during his military career, including that of communications staff officer of 3 Commando Brigade and command of both 3 Commando Brigade Headquarters and Signal Squadron and the Royal Marines Signal Training Wing, as well as an exchange posting with the Army at the Royal School of Signals, Blandford. He attended the Army Staff College at Camberley in 1985-86, and subsequent appointments, outside communications, included Fleet Amphibious Intelligence officer on the staff of Commander-in-Chief Fleet, and within the Directorate of Naval Operations in the Ministry of Defence. In late 1996 he commanded the multinational training team in the Baltic States as part of the Baltic Battalion project.

Giles Ebbutt retired from the Royal Marines in 1999 and very soon after was appointed Editor of *Jane's C4I Systems*. He is married with two daughters and currently lives outside Plymouth.

Glossary

This glossary is particularly comprehensive because of the broad nature of C4I systems and their involvement in a considerable range of military endeavours at sea, on land and in the air. In addition, the specialist knowledge of readers is unlikely to include all the technologies and disciplines mentioned in this book and so more terms are included rather than less. The target readership is international and terms very familiar to some readers will not be to others; no apology is made for including them, with explanations where necessary. The editor invites corrections or useful additions at any time.

A2C2	Army Airborne Command and Control
A2C2S	Army Airborne Command and Control System
A	Ampère
AA	anti-aircraft
AAA	Anti-Aircraft Artillery
AABNCP	Advanced Airborne National Command Post (USA)
AAC	Army Air Corps. The organisation which provides Aviation support to the British Army.
AAD	Army Air Defense
AADC	Area Air Defense Commander
AADCOM	Army Air Defense Command
AADCP	Army Air Defense Command Post
AAM	Air-to-Air Missile
AAW	Anti-Air Warfare
AAWS	Advanced Anti-tank Weapon System
ABCCC	Airborne Battlefield Command Control Centre (USA)
ABCS	Army Battle Command System
ABM	anti-ballistic missile
ABNCC	Airborne Command and Control
ABNCP	Airborne Command Post
AC	Alternating current, or air combat
ACCS	Army Command and Control System (USA) or Air Command and Control System (NATO)
ACDS	Advanced Combat Direction System (USA)
ACE	Allied Command Europe (NATO)
ACE ACCIS	Allied Command Europe (ACE) Automated Command and Control Information System
ACLOS	Automatic command to line of sight. Type of missile guidance, see under CLOS
ACM	air combat manoeuvre
ACMS	Automatic communication management system
ACO	Air Combat Order
ACWAR	Agile continuous wave acquisition radar
AD	1. air defence
	2. Assistant Director
ADat-P3	Automatic Data Processing (Standard) 3 (NATO)
ADAWS	Action Data Automation Weapons System (UK)
ADCIS	Air Defence Command Information System (UK)
ADDS	Army Data Distribution System (USA)
ADF	Automatic Direction-Finding
ADGE	air defence ground environment
ADI	1. Air Defense Initiative (USA)
	2. Australian Defence Industries
ADIMP	ADAWS Improvement Programme (UK)
ADIVS	air defence interoperability validation system
ADLIPS	Automatic DataLink Plotting System (USA)
ADOC	air defence operations centre
ADP	Automatic data processing
ADR	Air Defence Region (UK)
ADSI	Air Defense Systems Integrator
AEW	Airborne early warning
AF	Audio frequency
AFATDS	Advanced Field Artillery Tactical Data System
AFB	Air Force Base
AFC	Automatic frequency control
AFS	Automatic frequency selection
AFSATCOM	Air Force Satellite Communications (USA)
AFV	Armoured fighting vehicle
AGC	Automatic gain control
AGE	Advanced Graphics Engine, or Air/Ground Environment
AGES/AD	Air-to-Ground Engagement/Air Defence

AI	1. artificial intelligence
	2. Air intercept (for instance AI radar)
AIFV	Armoured infantry fighting vehicle
AIM	air intercept missile
AIO	Action information organisation
ALE	Automatic link establishment
ALGOL	Algorithmic language
Aliasing, Anti-aliasing	Aliasing is a term used in scanning technology to describe a number of undesirable effects produced by translating an original image into discrete pixels. These effects include jagged and/or crawling edges, gaps in thin polygons and a tendency for small polygons to blink on and off. As an example of one of these effects, when a straight edge displayed in raster format is rotated to an angle diagonal to the raster lines, a jagged edge (sometimes known as 'jaggies') will be seen instead of a straight line; a technique known as 'anti-aliasing' is employed to soften the ragged edge effect and to preserve the presented image as a straight line. Anti-aliasing techniques embrace measures to reduce all aliasing effects and make calculations at sub-pixel level for colour, texture, shading, fading and translucency in order to smooth the edges concerned by averaging the characteristics of the pixels actually displayed in accordance with the sub-pixel calculations
ALQA	Automatic link quality analysis
AM	Amplitude modulation
AMB	Air-Mobile Brigade
AMCM	Airborne mine countermeasures
AMDPCS	Air and Missile Defense Planning and Control System
AMETS	Artillery Meteorological System (UK)
AMS	Automatic marking system
AMTD	Adaptive moving target detection
Ångstrom	Name of Scandinavian scientist who gave his name to a unit of length. There are 10,000 Ångstroms in a μ and 1 million μ in a metre. Human visual acuity is from about 0.4 (violet) to 0.7 (red) μ, or 4,000 to 7,000 Ångstroms. Today, the Ångstrom is less popular and the nanometre is more often used instead, the visual region in this case being 400 to 700 nm
Anti-aliasing	See Aliasing
AO	Area of Operations
AOI	Area of Interest
AOR	Area of Responsibility
AOSS	Automatic ordnance scoring system
AOV	Artillery observation vehicle
AP	Armour-piercing
APB	Analysing printer base
APC	Armoured personnel carrier
APDS	Armour-piercing discarding sabot
APFSDS	Armour-piercing fin-stabilised discarding sabot
API	Application programming interface
APS	Adaptive processing sonar
APSE	Ada programming support environment
AQAP	Allied Quality Assurance Publication (NATO)
AR&M	Availability, reliability and maintainability
arc/arcmin/arc minute	Implies angle measurement, in the case of minutes the letters 'arc' are sometimes added to clearly distinguish between minutes of time (1/60 hour) and minutes of angle (1/60 of a degree). Used in measurement of resolution of the human eye and of visual display systems, for instance 'arcminutes per pixel' or 'arcminutes per Optical Line Pair'. See under Resolution
ARE	Admiralty Research Establishment (UK)
ARM	anti-radiation missile
Arty	Artillery
ASAS	All Source Analysis System (USA)

ASCS	AirSpace Control System (Chile)
ASDC	Alternative space defence centre
ASIC	Application specific integrated circuit
ASM	air-to-surface missile, anti-ship missile
ASMI	Airfield surface movement indicator
ASOC	air support operations centre
ASR	Airfield surveillance radar
ASR	Air Staff Requirement (UK)
ASW	anti-submarine warfare
ATACO	air tactical control operator
ATC	air traffic control
ATCC	air traffic control centre
ATCO	air traffic control officer
ATCRU	air traffic control radar unit
ATCCS	Army Tactical Command and Control System (US)
ATDL	Army Tactical DataLink
ATDS	Airborne Tactical Data System (USA)
ATE	Automatic test equipment
ATG	Automatic test guide
ATGW	anti-tank guided weapon. Now normally referred to as an ATM since current terminology uses the term 'Missile' for 'Guided Weapon', particularly in the USA
ATM	anti-tank missile or asynchronous transfer mode
ATO	Air Tasking Order
ATR	air transportable rack
ATS	Agile target system, or air traffic system, or aircrew training system
AUTODIN	Automatic Digital Network (USA)
AUTOVON	Automatic Voice Network (USA)
AWACS	Airborne Warning and Control System (USA)
AWDATS	Automatic Weapon Data Transmission System (UK)
BAI	Battlefield air interdiction. Not Close Air Support (CAS), not long-range interdiction, but interdiction of supply routes and other forms of communications and support routes and assets which are immediately concerned with the local battlefield
BATES	Battlefield Artillery Target Engagement System (UK)
BER	bit error rate
BISA	Battlefield Information System Application (UK)
BIT	Built-in test
Bit	Bit is an abbreviation of Binary digit, that is, one unit of magnetic memory used in computing, the small spot of material which is in one state or the other (in magnetic systems, either magnetised or not)
BITE	Built-in test equipment
BITS	Battlefield Information Transmission System
BMD	Ballistic Missile Defence, such as by employing ABMs
BMDO	Ballistic Missile Defence Organization
BMEWS	Ballistic Missile Early Warning System (USA)
BPSK	Binary phase shift keying
Buffer	RAM devices used for temporary storage of information such as before transfer to and from disk, to a printer, and so on. See Z Buffer
BVR	Beyond visual range. Applied to targets, particularly with respect to air defence fighter aircraft. In gunnery the equivalent term is indirect fire
Byte	A group of eight memory bits (qv) which can be used to designate one character. This provides two to the power eight (256) combinations which can be used to specify letters, numbers, symbols and so on, in coding systems such as ASCII, ANSI and so forth
C2	Command and control
C2PC	Command and Control for the PC
C3	Command, control and communications
C3I	Command, control, communications and intelligence
C4I	Command, control, communications, computers and intelligence
C4ISR	command, control, communications, computers, intelligence, surveillance and reconnaissance
C4IFTW	C4I for the Warrior
CAA	Civil Aviation Authority. The name of the Civil Aviation Regulatory Authority in a number of countries including the UK
CAAIS	Computer-Aided Action Information System
CACS	Computer assisted command system
CAS	1. close air support 2.Chief of the Air Staff (UK)
CATIES	Combined Arms Training Integrated Evaluation System
CATRIN	*Sistema Campale Trasmissioni Informazioni* (Italy)
CC	Close combat
CCD	Charge coupled device
CCIS	Command and control information system
CCTV	Closed-circuit television
CD	Compact disc
CDMA	Code division multiple access
CD-ROM	Compact disc - read-only memory
CEC	Co-operative Engagement Capability
CEP	Circular Error Probable, normally to a 50% probability level unless stated otherwise (50% CEP), a radius within which 50% of rounds, shells, bombs or other weapons should fall if sufficient are fired or dropped in the defined conditions for which the CEP is quoted to be statistically significant. For CEP figures to be meaningful, the weapon impact pattern should be essentially circular. If not, other statistical terms should be used such as for elliptical impact distributions, 50% RE (Range Error Probable) and 50% DE (Deflection Error Probable). The RE and DE figures essentially constitute the major axes of an ellipse in which 50% of impacts should occur if enough weapons are fired or dropped
CET	Combat engineering tractor
CEV	Combat engineer vehicle
CFAR	Constant false alarm rate
CGI	Computer-generated image/imagery. Application of computer graphics techniques to generate images, generally for transmission of real-time imagery to a related but separate display system
CHS	Common Hardware/Software (US)

CIC	Command (or combat) information centre
CIS	Command (or combat) information system
CIWS	Close-in weapon system
CLOS	Command to line of sight. Guidance principle used in guided missile systems where the missile follows the direct sightline to the target without any predictions of target movement such as lead angles or the use of proportional navigation equations. The missile can be manually guided to the line of sight by direct operator commands (MCLOS) where the operator tracks both missile and target and guides one to the other. Guidance can be fully automatic (ACLOS) such as from radar returns which track both missile and target and generate guidance signals independent of the operator, or semi-automatically guided (SACLOS) where the operator tracks the target and signals are automatically generated to bring the missile on to the operator's tracking sightline
CMOS	Complementary metal oxide semi-conductor. Used, amongst other applications, in computer systems, particularly for ROMs, PROMs and EPROMs
CNRI	Combat net radio interface
COC	Command operations centre
COE	Common Operating Environment
COIN	Counter-insurgency
COMCEN	Communications centre
COMINT	Communications intelligence
COMSEC	Communications security
CONUS	Continental United States
COTS	Commercial-off-the-shelf
CP	Command post
CPA	Closest point of approach
CPACS	Coded pulse anti-clutter system
CPOP	Command post/observation post
CPS	Cardinal points specification
cps	Characters per second
CPU	Central processing unit, for instance of a computer system
CRC	Command (or control) and reporting centre
CRP	Command (or control) and reporting post
CRT	Cathode-ray tube
CSCI	Computer software configuration item
CSI	Computer synthesised image/imagery. A term describing a technique where a real-world background image is synthesised by scanning and then digitising photographic information, compared to conventional CGI where the image is created completely from computerised information. The CSI image is often overlaid with targets which can either be projected on the scene by laser or other target projectors, or targets created by normal CGI techniques can be integrated into the scene from the start
CSRDF	Crew station research and development facility
CTAPS	Contingency theatre automated planning system
CVSD	Continuously variable slope delta (modulation)
CW	Continuous wave or chemical warfare
CZMCS	Combat Zone Mobile Communication System (Turkey)
DA	Design authority
DAMA	Demand Assigned Multiple Access
DARPA	Defense Advanced Research Projects Agency (USA)
dB	Decibel
dBm	Decibel relative to 1 milliwatt
DBMS	Database management system
DBWS	Database workstation
DC	Direct current
DCSA	Defence Communication Services Agency (UK)
DCA	Defense Communications Agency (USA)
DCS	Defense Communication System (USA)
DCTN	Defense Commercial Telecommunication Network (USA)
DDN	Defense Data Network (USA)
DE	Deflection error probable. The deflection (lateral) element of a weapon impact distribution, normally to a 50% probability level unless stated otherwise. See RE and CEP
DEW	Distant Early Warning (USA)
DF or D/F	1. direction-finding 2. Direct fire
DII	Defence Information Infrastructure (UK)
DII COE	Defense Information Infrastructure Common Operating Environment (US)
DISA	Defense Information Systems Agency
DISCON	Distributed Integrated Secure Communications Network (Australia)
DLMS	Digital Land Mass Simulation. The NATO system for digitising terrain (DTED) and feature (DFAD) information for mapping, targeting, simulation and other purposes. Although NATO-based, it is a series of digitising protocols applicable and used worldwide. So-called 'DLMS Level 1' data is about the equivalent content of a 1:250,000 topographical map and 'Level 2' to data contained on a 1:50,000 map. Superseded by the US DoD MIL-STD protocols but DLMS formats are still in common use
DLP	DataLink Processor
DLTS	DataLink Test Set
DMA	Defense Mapping Agency (USA)
DME	Distance measuring equipment
DML	Decision and modelling language
DMS	Defense Message Service (USA)
DMSK	Differential minimum shift keying
DMSO	Defense Modeling and Simulation Office (USA)
DMTI	Digital moving target indication
DNVT	Digital non-secure voice terminal
DOA	Direction of arrival
DoD	Department of Defense (generally refers to the USA but some other countries such as Australia use the title also)
DODIIS	DOD Intelligence Information System (USA)
DOTE	Director, Operational Test and Evaluation (USA)

DPCM	Differential pulse code modulation
DPSK	Digital phase shift keying
DRA	Defence Research Agency (UK)
DRES	Defense Research Establishment Suffield (Canada)
DSB	Double sideband
DSCS	Defense Satellite Communication System (USA)
DSN	Defense Switched Network (USA)
DSP	Defense Support Program (USA)
DSVT	Digital secure voice terminal
DT&E	Development test and evaluation
DTED	Digital Terrain Elevation Data (NATO)
DTIC	Defense Technology Information Center (USA)
DTMF	dual tone multifrequency. DTMF is the generic communications term for Touch-Tone
DTSS	Digital Topographic Support System (US)
DVL	Data/voice logger
E3	end-to-end encryption
ECM	Electronic countermeasures
ECCM	Electronic counter-countermeasures
EHF	Extra high frequency (30-300 GHz)
EIRP	Effective isotropic radiated power. This term describes the strength of the signal leaving a satellite antenna or the transmitting earth station antenna and is used in determining the C/N and S/N. The transmit power value in units of dBW is expressed by the product of the transponder output power and the gain of the satellite transmit antenna
EL	Electroluminescent
ELF	Extremely low frequency (0-3 kHz)
ELINT	Electronic intelligence
EM	Electromagnetic
EMC	Electromagnetic compatibility
EMI	Electromagnetic interference
EMP	Electromagnetic pulse
ENSCE	Enemy Situation Correlation Element (USA)
E-O	Electro-optical
EOW	Engineering order wire
EPLRS	Enhanced Position Location Reporting System (USA)
EPROM	Erasable programmable read-only memory
EROM	Erasable read-only memory
ESM	Electronic support measures
ETHERNET	A local area network that connects devices like computers, printers, terminals. Ethernet operates over twisted-pair or coaxial cable at speeds at 10 Mbps
EUROCOM	European Communications
EW	Electronic warfare or early warning
EWO	Electronic warfare officer
FAA	Federal Aviation Administration. The civil aviation regulatory body of the USA
FAAD	Forward Area Air Defense (USA)
FAC	Forward Air Controller
FACE	Field Artillery Computer Equipment (UK)
fax	Facsimile
FBCB2	Force XXI Battle Command - Brigade and Below (USA)
FCC	fire-control centre (or computer)
FCS	fire-control system
FD	full development (equipment project phase before production)
FDC	fire distribution centre
FDDI	Fibre distributed data interface
FDM	Frequency division multiplex
FDMA	Frequency division multiplex access
FEBA	Forward edge of the battle area
FEC	Forward Error Correction
FEP	Front End Processor
FET	field effect transistor
FFT	fast Fourier transform
FLIR	Forward-looking infra-red
Flop/s, kFlop/s, GFlops	Floating point operations per second, k = thousand, M = million, G = 1,000 million
FLOT	Forward Line of Own Troops
FLTSATCOM	Fleet Satellite Communications (USA)
FM	Frequency modulation

FMS	Foreign military sales (USA)
FO	Forward observer
FOC	Full Operational Capability, see also IOC and ISD
FOGM	Fibre optic-guided missile
FORTRAN	Formula translating system
FOV	field of view
FPS	fast packet switch
FRAME RELAY	A form packet switching, but using smaller packets and less error checking than traditional forms of packet switching (such as X25). Now a new international standard for efficiently handling high-speed, burst data over wide area networks
FS	Feasibility study
FSD	full-scale development (same as FD)
FSK	Frequency shift keying
FTP	File Transfer Protocol
FU	Fire Unit
FY	Fiscal/financial year
G	giga (one thousand million)
g	Gramme
g	Gravity (approximately 9.81 m/s/s or 32.2 ft/s/s)
GaAs	Gallium arsenide
GBCS	Ground-Based Common Sensors
GBS	Ground-Based Sensor
GCA	Ground-controlled approach
GCCS	Global Command and Control System (USA)
GCI	Ground-controlled intercept
GDU	gun display unit
GEADGE	German Air Defence Ground Environment
GEHOC	German HAWK Operations Centre
GEOREF	Geographic Reference
GFE	Government Furnished Equipment
giga	10 to the power 9, in other words 1,000 million. For instance GHz (1 GHz = 1,000 MHz), Gbyte
GIS	Geographic Information System
GKS	Graphic kernel system
GLONASS	Global Orbital Navigation Satellite System, the Russian GNSS system similar to the US GPS
GNSS	Global navigation satellite system a generic term for all systems such as the Russian GLONASS and the US GPS
GOLD	General online diagnostic
GOTS	Government-off-the-shelf
GP	General purpose
GPIB	General purpose interface board
GPS	Global Positioning System, the GNSS system controlled by the US DoD and Dept of Transportation but available for receivers worldwide
GPTE	General purpose test equipment
GRCS	Guardrail Common Sensor
GRDS	Generic radar display system
GRP	Glass-reinforced plastic
GSE	Ground Support Equipment
G/T	A figure of merit of an antenna and low noise amplifier combination expressed in dB. 'G' is the net gain of the system and 'T' is the noise temperature of the system. The higher the number, the better the system
GUI	Graphical user interface. Computer interface displays based on graphics as opposed to alphanumerics.
GW	Guided weapon
GWEN	Ground Wave Emergency Network (USA)
h	Hour(s)
HCI	Human/computer interface, in other words MMI in the computer field
HDTV	High-Definition Television. New generation commercial TV with 1,200 or more raster scan lines at between 50 and 60 Hz
HE	high explosive
HEAT	high-explosive anti-tank, a type of anti-armour round with a shaped HE charge, which, on detonation, uses the Monroe effect to focus the explosive force and to generate an intensely hot liquid metal slug which proceeds at high velocity and is able to penetrate armour
HEP	high explosive plastic
HEROS	*das Heeres Führungsinformationsystem zur Rechnergestützten Operationsführung in Stäben* (Germany)

HESH	high-explosive squash-head, a type of anti-armour round which is designed to explode in contact with (the 'squash' in the title) armour such as the outer shell of an AFV. Transfer of energy across the sheet of armour then causes metal to detach on the inner side ('spalling') which is the damage mechanism inside the target
HF	high frequency (2 - 30 MHz)
HFDF	High-Frequency Direction-Finding
HLL	high-level language (used in computer systems, for instance Ada, FORTRAN, COBOL, C and so forth)
HMMWV	High-Mobility Multipurpose Wheeled Vehicle (USA)
hp	Horse power
HPA	high-power amplifier
HQ	Headquarters
HT	high tension
HUMINT	Human intelligence
Hz	Hertz, the scientific unit for cycles per second
iaw	in accordance with
IC	Integrated circuit
ICBM	Intercontinental ballistic missile
ICS	Integrated Communications System (UK)
ICV	Infantry combat vehicle
IDE	Integrated Drive Electronics
IDF	Israel Defence Force
IES	Imagery Exploration System
IEW	Intelligence and Electronic Warfare
IF	Intermediate frequency
IFF	Identification, friend or foe
IFF/SIF	IFF/Selective Identification Feature
IFM	Instantaneous frequency measurement
IFV	infantry fighting vehicle
IGE	Instrumentation graphics environment
IJMS	Interim JTIDS Message System
IKAT	interactive keyboard and terminal
ILS	integrated logistic support or instrument landing system (VHF band aircraft landing aid)
IMETS	Integrated Meteorological System
IMINT	imagery intelligence
INS	inertial navigation system, a navigational system using a set of (normally three) accurately calibrated mechanical or laser gyros
I/O	input/output
IOC	initial operational capability
IOT&E	initial operational test and evaluation
IP	intellectual property, or initial production
IPDS	Imagery Processing and Dissemination System
IPR	intellectual property rights
ips	Instructions per second
IR	infra-red (wavelengths of between about 700 nm (the near IR) up to about 10,000 nm (10 µ or 10æ)
IRST	Infra-red Search and Track
ISB	Independent sideband
ISD	in-service date
ISDN	Integrated services digital network
ISLS	Interrogator sidelobe suppression
ISO	International Standards Organisation
ISTAR	Intelligence, Surveillance, Target Acquisition and Reconnaissance
IT	Information technology
ITP	intent to purchase
ITT	invitation to tender
IU	Interface Unit
IVD	interactive video disk
IVDN	integrated voice and data network
IVDU	intelligent visual display unit
IVIS	Inter Vehicle Information System
IVSN	Initial Voice Switched Network (NATO)
JCEWS	Joint Combat Electronic Warfare System
JFACC	Joint Force Air Component Commander
JFC	Joint Force Command(er)
JFLCC	Joint Force Land Component Commander
JFMCC	Joint Force Maritime Component Commander
JIC	Joint Intelligence Centre
JIES	Joint Interoperability Evaluation System (USA)
JINTACCS	Joint Interoperability of Tactical Command and Control Systems (USA)
JMCIS	Joint Maritime Command Information System
JORN	Jindalee OTH Radar Network (Australia)
JOPES	Joint Operations Planning and Execution System
JPITL	Joint Prioritised Integrated Target List
JSTARS	Joint Surveillance and Target Attack Radar System (USA)
JSOW	Joint Stand Off Weapon
JTAGS	Joint Tactical Ground Station
JTF	1. Joint Task Force 2. Joint Tactical Fusion 3. Joint Test Facility
JTFP	Joint Tactical Fusion Program (USA)
JTIDS	Joint Tactical Information Distribution System (USA)
JWIS	Joint Worldwide Military Command and Control Information System Program
k	kilo (one thousand)
KADS	knowledge acquisition data system
KB	Kilobyte
KB/S	Kilobytes per second
KBU	keyboard unit
KHz	Kilohertz (thousand cycles per second)
Km	Kilometre
KPS	knowledge processing system
KVDT	keyboard visual display terminal

LAADS	low-altitude air defence system
LAN	local area network
LAV	light armoured vehicle
LCC	local command centre
LCC	life cycle costs, the same as through life costs (TLC). The total costs of an equipment or system from inception through production, operational service to disposal
LCD	liquid crystal display
LCU	Lightweight Computer Unit
LED	light emitting diode
LEO	Low Earth Orbit
LEO SAT	Low Earth Orbiting Satellite
LF	low frequency (30,300 kHz)
LIU	LAN Interface Unit
LLAD	low level air defence
LLTV	low-light television, normally operates in the near-IR band using reflected energy between 0.7 and 1 µ (700-1,000 nm)
LO	local oscillator
LOB	line of bearing
LOC	local operations console
LOCE	Limited Operational Capability, Europe
LOMEZ	Low-Altitude Missile Engagement Zone
LOS	line of sight
LP	low power
LPD	Low Probability of Detection
LPI	low probability of interception
LPU	Line Processing Unit; Logical Program Unit
LRI, LRU	line replaceable item/unit
LRIP	Low Rate Initial Production
LSA	logistic support analysis
LSB	lower sideband
LSD	Large Screen Display
LSI	large scale integration
LTACFIRE	Light Tactical Fire Direction System
LTC	long-term costing
LUT	Limited User Test
LVA	large vertical aperture (radar)
M	mega (one million)
m	milli (one thousandth)
MAD	magnetic anomaly detector/detection
MANPADS	manportable air defence system
MART	mean active repair time
MASER	microwave amplification by stimulated emission of radiation
MASINT	measuring and signature intelligence
MATCALS	Marine Air Traffic Control and Landing System (USA)
MAWS	Missile Attack Warning System
max	Maximum
MBT	main battle tank
MCC	management, command and control, or manual control and counter, or mission control centre
MCCIS	Maritime Command and Control Information System
MCE	Modular Control Equipment (USA)
MCM	mine countermeasures
MCMV	mine countermeasures vessel
MCRC	mobile control and reporting centre
MCS	Maneuver Control System (USA)
MCW	modulated continuous wave
MEECN	Minimum Essential Emergency Communications Network (USA)
MEROD	message entry and readout device
MESAR	Multifunctional Electronic Scanned Adaptive Radar
MET	Meteorological
METEOSAT	Meteorological Satellite
MF	medium frequency (300 kHz to 3 MHz)
MFCS	Mortar Fire Control System
MHS	Message Handling System
MIDS	Multifunction Information Distribution System (NATO)
Mil Spec	military specification
MIL STD	military standard
MILNET	Military Network
MILSATCOM	Military Satellite Communications
MILSTAR	Military Strategic/Tactical Relay
MIPS	Million Instructions Per Second
MIRV	multiple independently targetable re-entry vehicle
MIS	management information system
MLI	mid-life improvement (same as MLU)
MLRS	Multiple Launch Rocket System (USA)
MLU	mid-life update (same as MLI)
MMI	man/machine interface
MMIC	monolithic microwave integrated circuit
MMW	millimetric wave
MoD	Ministry of Defence (term used in countries such as India, Israel and the UK. The US equivalent is DoD)
MODEM	Modulator Demodulator
MOOTW	Military Operations Other Than War
MOSFET	metal oxide silicon field effect transistor
MOTR	Multiple Object Tracking Radar
MoU	memorandum of understanding
MPA	maritime patrol aircraft
MRAAM	medium range air-to-air missile
MRAD	milliradian (1/1,000 of a radian or approximately 6/100 of a degree, or 3.44 seconds of arc)
MRCS	mobile reporting and control system
MRSR	Multi Role Survivable Radar
MRPV	mini remotely piloted vehicle
MSE	Mobile Subscriber Equipment (USA)
MSIP	Multi Spectral Imagery Processor
MSK	minimum shift keying
MTBF	mean time between failures
MTBCF	Mean Time Between Critical Failure

MTBMA	Mean Time Between Maintenance Actions
MTD	moving target detection
MTDS	Marine Tactical Data System (USA)
MTI	moving target indicator
MTTR	mean time to repair
MUX	Multiplexer
MWHQ	mobile war headquarters
n	nano (one thousand millionth)
NC3A	NATO Consultation, Command and Control Agency
NADGE	NATO Air Defence Ground Environment
NAEW	NATO Airborne Early Warning (System)
NAU	Network Access Unit
NALLADS	Norwegian Army Low Level Air Defence System
NAS	naval air station
NASA	National Aeronautics and Space Administration (USA)
NASAMS	Norwegian Advanced Surface to Air Missile System
NATO	North Atlantic Treaty Organisation
NAVSTAR	Navigational System Tracking and Range
NBC	Nuclear, Biological and Chemical (warfare)
NCA	National Command Authorities (USA)
NCD	Net Control Device
NCW	Network-Centric Warfare
NDI	non developmental item
NEACP	National Emergency Airborne Command Post (USA)
NGCS	NATO General Purpose Communications System
NiCad	nickel/cadmium (battery, as opposed to lead/acid)
NICS	NATO Integrated Communications System (NATO)
NIPRNET	Non-Secure Internet Protocol Router Network
NIS	NATO Identification System
NIU	network interface unit
nm	nanometre, nautical mile
NMC	Network Monitoring and Control
NMCI	Navy and Marine Corps Intranet (US)
NMS	NATO Messaging Service
NOAH	Norwegian Adapted HAWK
NOC	Network Operations Centre
NORAD	North American Air Defense Command (USA)
NRI	Net Radio Interface
NSR	naval staff requirement (UK)
NSVN	NATO Secure Voice Network
NTB	national testbed
NTC	national training center (US)
NTSC	National Television Standards Committee (US)
NTDS	Naval Tactical Distribution System (USA)
NWS	North Warning System (USA)
OA	operational analysis
OCA	Offensive Counter Air
OEM	Original Equipment Manufacturer
OOA	1 Object Oriented Analysis
	2. Out of Area
OODBMS	Object Oriented Database Management Systems
OOP	Object Oriented Programming
OOTW	Operations Other Than War
OP	observation post/point/position, for instance, for artillery observation by an AOV, manual observation, or an instrumentation/tracking
OR	operational requirement. Sometimes OR stands for Operational Research, but because of possible confusion with the term Operational Requirement, the term OA (see above) is generally used for this subject
ORD	Operational Requirements Document
OS	Operating System
OSF	open systems foundation
OSI	open systems interconnection
OTH	over the horizon
OTHB	over the horizon, backscatter
OTHR	over the horizon radar
OTHSW	over the horizon, surface wave
OTS	off-the-shelf
P3I	preplanned product improvements
PABX	private automatic branch exchange
PACCS	Post Attack Command and Control System (USA)
PAL	Phase Alternate Lines
PANDA	Personnel and Administration
PATRIOT	Phased Array Tracking to Intercept Of Target
PAWS	Portable ASAS Work Station, Portable Analyst Workstation
PAX	Passengers
PC	personal computer (normally implies IBM-compatible)
PCB	printed circuit board
PCM	pulse code modulation
PCMCIA	Personal Computer Memory Card International Association
PCU	portable control unit
PD	project definition. A phase in an equipment project after a feasibility study and before approval.
PEO	Program Executive Office(r)
PEP	peak envelope power
PERT	progress evaluation and review technique. Often summarised in the form of a chart or charts showing critical and other paths of various project stages and activities
PGM	Precision Guided Missile (or Munitions)
PHIGS	programmer's hierarchical interactive graphics system
PIU	Programmable Interface Unit
PJH	PLRS/JTIDS Hybrid (USA)
PLGR	Precision Lightweight GPS Receiver
PLRS	Position Locating Reporting System (USA)
PLSS	Precision Location Strike System (USA)
POI	probability of intercept
POSIX	Portable Operating System IX
POTS	production off-the-shelf

PPI	plan position indicator. Normally applied to a radar presentation in mapping orientation (in other words a vertical view of terrain)
ppm	parts per million
pps	pulses per second
PRF	pulse repetition frequency, of radars and other pulsed devices such as lasers used for range-finding or coding
PRI	pulse repetition interval
PROM	programmable read only memory
PSDN	packet switched digital network
PSI	pounds per square inch
PSSSU	power supply and system selector unit
PSTN	Public Switched Telephone Network
PSU	power supply unit
PTT	Posts, Telegraph and Telecommunications
PVC	polyvinyl chloride. A plastic material
PWHQ	primary war headquarters
QA	quality assurance
QCPSK	quadrature coherent phase shift keying
QPSK	quadrature phase shift keying
R&D	research and development
R&M	reliability and maintainability
RAM	random access memory; radar absorbent material
Raster	1.A scanning pattern where a line is drawn, generally on a phosphor-coated screen by an electron gun, followed by a rapid flick-back to draw the next line down. The speed at which the beam(s) can be modulated to change the colour and intensity, determines the number of pixels on each raster line. Many displays use an interlaced technique where only every other line is drawn during the first field scan, a second scan being needed to complete all of the information.
	2. A means of storing and displaying maps in digital format, produced either directly from the vector map or by scanning the original paper documents.
RATT	Radio Teletype
RAU	radio access unit
RCC	Rescue Control Centre (for SAR), or regional control centre
RCMDS	remote-control mine disposal system
RCS	range control station, radar cross section (equivalent radar reflective area in the frequency band specified)
RD&A	research, development and acquisition
RDBMS	relational database management system
RDP	radar data processor
RDT&E	research, development, test and evaluation
RE	range error probable
(I) REMBASS	(Improved) Remotely Monitored Battlefield Sensor System (USA)
RF	radio frequency
RFC	Request For Change
RFI	request for information; radio frequency interference; radio frequency interferometer
RFP	request for proposals (to meet an Operational Requirement (OR))
RFQ	request for quotations (to meet an Operational Requirement (OR), a more definitive stage than RFP since firm cost quotations are required). Similar to ITT
RGB	red, green and blue. The three primary colours which are used to make up a colour display. Implies full colour when applied to image generation or display systems
RHWR	radar homing and warning receiver
RIO	radar intercept officer
RIPP	radar information processing post
RISC	reduced instruction set computer
RITA	*Réseau Integre de Transmission Automatique* (France)
RIU	Radio Interface Unit
RM&T	reliability, maintainability and testability
RMS	root mean square
ROE	Rules Of Engagement
ROM	read only memory
ROTHR	Relocatable Over The Horizon Radar (USA)
ROV	remotely operated vehicle
RP	reporting post
RPC	Remote Procedure Calls
rpm	revolutions per minute
RPV	remotely piloted vehicle, a type of UAV
RRA	Radar Reflective Area, as RCS
RSDU	radar/sonar display unit
RSLS	receiver sidelobe suppression
RSRE	Royal Signals and Radar Establishment (UK)
R/T	receiver/transmitter, or radiotelephone, or RadioTelephony (in other words use of radios to transmit and receive)
RTTY	Radio Teletype
RWR	radar warning receiver
RX	Receive
s	Second
SACLOS	semi-automatic command to line of sight
SADA	Semi-automatic Air Defence system (Spain)
SAM	surface-to-air missile
SAR	synthetic aperture radar, or search and rescue
SATCOM	Satellite Communications
SAW	surface acoustic wave
SAWHQ	SHAPE Alternative War Headquarters (NATO)
SCAMP	Single Channel, Anti Jam, Manportable
SCC	sector command centre
SCDL	Surveillance Control DataLink
SCOTT	Single Channel Objective Tactical Terminal
SCRA	single channel radio access
SCSI	small computer system interface
SCU	Satellite Communications Unit
SDC	space defence centre
SDME	Software Development and Maintenance Environment

SHAPE	Supreme Headquarters Allied Powers, Europe (NATO)		**TBMCS**	Theatre Battle Management Core System
SHF	super high frequency (330 GHz)		**TCIM**	Tactical Communications Interface Module
SHINPADS	Shipboard Integrated Processing and Display System (Canada)		**TCP/IP**	Transmission Control Protocol/Internet Protocol. A set of protocols that link dissimilar computers across networks
SHORAD	short-range air defence			
SHTU	Simplified Handheld Terminal Unit		**TCT**	Tactical Computer Terminal (USA)
SICF	Système d' Information pour le Commandement des Forces		**TDM**	time division multiplex
SICPS	Standardised Integrated Command Post System		**TDMA**	time division multiple access
SIF	selective identification feature		**TDS**	tactical data system
SIGINT	signals intelligence		**TEMPEST**	Transient ElectroMagnetic Pulse Emanation Standard
SINCGARS	Single Channel Ground and Airborne Radio System		**TENCAP**	Tactical exploitation of national capabilities (USA)
SINCGARS SIP	SINCGARS System Improvement Programme		**TFT**	thin film transistor
SINS	ships inertial navigation system		**TGIF**	Transportable Ground Interface Facility
SIPRNET	Secure Internet Protocol Router Network		**THAAD**	Theatre High Altitude Area Defence
SIPS	software intensive projects		**TIBS**	Tactical Information Broadcast System
SIR	Système d' Information Régimentaire		**TLC**	through life costs, the same as life cycle costs (LCC)
SLBM	submarine-launched ballistic missile		**TMD**	tactical modular display
SLCM	submarine-launched cruise missile		**TOA**	time of arrival
SMARTT	Secure Mobile Anti Jam Reliable Tactical Terminal		**TOC**	tactical operations centre
SMTP	Standard Mail Transfer Protocol		**TOW**	Tube-launched, optically-tracked, wire-guided
SNEPC	Saudi Naval Expansion Programme, Communications		**TQM**	total quality management
SNMP	Simple Network Management Protocol. A protocol governing network management and monitoring of network devices and their functions		**TRAP**	Tactical Related Applications
			TRE	Tactical Receive Equipment
SOC	sector operations centre		**TREE**	transient radiation effect on electronics
SOP	standard operating procedure		**TRITAC**	Tri-Service Tactical Communications programme (USA)
SOW	Statement Of Work		**TRIXS**	Tactical Reconnaissance Intelligence Exchange System
SP	1. single phase		**TWS**	track-while-scan
	2. self-propelled		**TWT**	travelling-wave-tube
SPADATS	Space Detection and Tracking System (USA)			
SPADOC	Space Defense Operations Center (USA)		**UAV**	unmanned aerial vehicle
SPAWAR	Space and Naval Systems Warfare Command (USA)		**UHF**	ultra-high frequency (300 MHz - 3 GHz)
SPI	Software Process Improvement		**UKADGE**	United Kingdom Air Defence Ground Environment
SQL	structured query language		**UPS**	Uninterruptible Power Supply
SRAM	Static Random Access Memory		**USB**	(1) upper sideband
SRAAM	short-range air-to-air missile			(2) Universal Serial Bus
SSB	single sideband		**UTM**	universal transverse Mercator
SSBN	nuclear powered ballistic submarine			
SSK	conventional (diesel electric) submarine		**V**	volt(s)
SSM	surface-to-surface missile		**VCR**	video cassette recorder
SSN	nuclear powered attack submarine		**VDU**	video (or visual) display unit
SSR	secondary surveillance radar		**VF**	voice frequency
STA	surveillance and target acquisition		**VFMED**	variable format message entry device
STANAG	Standardisation Agreement. Specifically, NATO STANAG, normally given a unique number so that it can be quoted as a customer requirement in the same way as US Mil Spec or MIL-STD		**VFO**	variable frequency oscillator
			VGA	Video Graphics Array
			VHF	very high frequency (30 - 300 MHz)
STARS	surveillance target attack radar system		**VHSIC**	very high speed integrated circuit
START	Super High Frequency Tri-Band Advanced Range Extension Terminal		**VLF**	very low frequency (330 kHz)
STRIDA	*Système de Traitement et de Représentation des Informations de Défense Aérienne* (France)		**VLSI**	very large scale integration
			VME	virtual memory environment
STU	Secure Telephone Unit		**VOCODER**	voice encoder/decoder
SVGA	Super Video Graphics Array		**VSAT**	Very Small Aperture Terminal
SWHQ	static war headquarters		**VSWR**	voltage standing wave ratio
SXGA	Super Extended Graphics Array		**VTR**	video tape recorder
TACAMO	Take Charge and Move Out (USA)		**W**	watt
TACAN	tactical air navigation. An aircraft navigational aid giving range (DME) and bearing from a beacon		**WAN**	wide area network
			WIS	WWMCCS Information System (USA)
TACC	Tactical Air Control Centre		**WORM**	write once, read multiple
TACCIMS	Theater Automated Command and Control Information Management System (USA)		**WSO**	weapon system officer
			W/T	wireless telegraphy
TACOMS	Tactical Area Communications System		**WUXGA**	wide ultra-extended graphics array
TACS	Tactical Air Control System (USA)		**WWABNCP**	Worldwide Airborne Command Post (USA)
TADIL	Tactical Digital Information Link (USA)		**WWMCCS**	Worldwide Military Command and Control System (USA)
TADIX	Tactical Data Information Exchange System			
TADIXB	TADIX Broadcast		**YIG**	yttrium indium garnet
TADOC	tactical air defence operations centre			
T&E	trial and evaluation		**ZODIAC**	Zone Digital Automatic Communications network (Netherlands)
TAFIM	Technical Architecture Framework for Information Management			
TAOC	Tactical Air Operations Center (USA)		μ	micro (one millionth) or micrometer
TAOM	Tactical Air Operations Module		Ω	ohm, unit of electrical resistance or impedance
TAPS	Target Analysis and Planning System, Terminal Applications Processor System			
TARE	Telegraph Automatic Relay Equipment			
TBM	1. Theatre Ballistic Missile			
	2. Tactical Ballistic Missile			

Alphabetical list of advertisers

E

EADS, (Dornier GmbH)
D-88039 Friedrichshafen, Germany .... [2]

Users' Charter

T his publication is brought to you by Jane's Information Group, a global company with more than 100 years of innovation and unrivalled reputation for impartiality, accuracy and authority.

Our collection and output of information and images is not dictated by any political or commercial affiliation. Our reportage is undertaken without fear of, or favour from, any government, alliance, state or corporation.

We publish information that is collected overtly from unclassified sources, although much could be regarded as extremely sensitive or not publicly accessible.

Our validation and analysis aims to eradicate misinformation or disinformation as well as factual errors; our objective is always to produce the most accurate and authoritative data.

In the event of any significant inaccuracies, we undertake to draw these to the readers' attention to preserve the highly valued relationship of trust and credibility with our customers worldwide.

If you believe that these policies have been breached by this title, you are invited to contact the editor.

A copy of Jane's Information Group's Code of Conduct for its editorial teams is available from the publisher.

Quality Policy

Jane's Information Group is the world's leading unclassified information integrator for military, government and commercial organisations worldwide. To maintain this position, the Company will strive to meet and exceed customers' expectations in the design, production and fulfilment of goods and services.

Information published by Jane's is renowned for its accuracy, authority and impartiality, and the Company is committed to seeking ongoing improvement in both products and processes.

Jane's will at all times endeavour to respond directly to market demands and will also ensure that customer satisfaction is measured and employees are encouraged to question and suggest improvements to working practices.

Jane's will continue to invest in its people through training and development, to meet the Investor in People standards and changing customer requirements.

Jane's

FREE ENTRY/CONTENT IN THIS PUBLICATION

Having your products and services represented in our titles means that they are being seen by the professionals who matter – both by those involved in procurement and by those working for the companies that are likely to affect your business. We therefore feel that it is very much in the interests of your organisation, as well as Jane's, to ensure your data is current and accurate.

- **Don't forget** – You may be missing out on business if your entry in a Jane's book, CD-ROM or Online product is incorrect because you have not supplied the latest information to us.

- **Ask yourself** – Can you afford not to be represented in Jane's printed and electronic products? And if you are listed, can you afford for your information to be out of date?

- **And most importantly** – The best part of all is that your entries in Jane's products are TOTALLY FREE OF CHARGE.

Please provide (using a photocopy of this form) the information on the following categories where appropriate:

1. Organisation name: _____

2. Division name: _____

3. Location address: _____

   _____

4. Mailing address if different: _____

   _____

5. Telephone (please include switchboard and main department contact numbers, for example Public Relations, Sales, and so on):

   _____

6. Facsimile: _____

7. e-mail: _____

8. Web sites: _____

9. Contact name and job title: _____

   _____

10. A brief description of your organisation's activities, products and services: _____

   _____

   _____

11. Jane's publications in which you would like to be included: _____

   _____

Please send this information to:
Jacqui Beard, Information Collection, Jane's Information Group,
Sentinel House, 163 Brighton Road, Coulsdon, Surrey, CR5 2YH, UK
Tel: (+44 20) 87 00 38 08
Fax: (+44 20) 87 00 39 59
e-mail: yearbook@janes.com

Copyright enquiries:
Contact: Keith Faulkner
Tel/Fax: (+44 1342) 30 50 32
e-mail: keith.faulkner@janes.com

Please tick this box if you do not wish your organisation's staff to be included in Jane's mailing lists ☐

JC4I

COMMAND INFORMATION SYSTEMS

Joint
Land
Maritime
Air
C2 facilities

JOINT

Australia

Joint Command and Control System (JCCS)

ADI's generic Joint Command and Control System (JCCS) provides operational and tactical level functionality in a platform-independent system that can run in UNIX, SOLARIS and Windows environments. The software package is called Llama/Cheetah (Llama is the client and Cheetah is the server), was first developed in 1996 and uses a modular and open architecture, Java programming language and Internet technology, thus ensuring platform independence and rapid dissemination of information across networks. The package provides the following:

(a) Common Operating Picture
(b) Decision and planning aids, including closest point of approach, course and time to intercept, dead reckoning, sun/moon almanac, tide calculations, navigation plans, track groupings and OPNOTEs
(c) Linkage to encyclopaedic and related data
(d) Overlay and map imagery management; all standard shapes including arc, line, polygon, sector, polar box, text and symbology; simplified colouring, line width, shading and transparency; static overlays and dynamic overlays associated with a track; raster and vector images; georeferencing for accurate map projection; intuitive map layering techniques
(e) Briefing and presentation tools
(f) Interoperability with allied forces
(g) The ability to operate over a variety of bandwidths.

Specifications for Cheetah
(a) Architecture:
Windows-PC based
Client server architecture
Automatic synchronisation of databases
Supports up to 200 client workstations using a single server
Supports mobile operations.
(b) Communications:
Operationally effective over bandwidths from 75 bps to 1 Gbps
Protocol support including TCP socket connection and SMTP
Message formats support including ADFORM, OS-OTG, ADGE, USMTF and TRAP.
(c) Cheetah processing speed: Benchmarked at over 2000 contact reports per second
(d) Intuitive user interface:
Windows GUI
Quick action buttons for common user functions
Autopopulation of calculations.
(e) External application integration:
Windows NT error logging
Common messaging systems including Lotus Notes, Microsoft Outlook and Netscape mail
Internet/Intranet browsers
Distributed collaborative planning tools

Status
JCCS forms the basis of the ADF's JCSS, the JTFHQ (Afloat), HQIADS and JMDSS (see separate entries).

Manufacturer
ADI Ltd, Sydney, New South Wales.

UPDATED

Joint Command Support System (JCSS)

JCSS provides an integrated command, management and communications environment to support the command and control of Australian Defence Force operations, working at the secret level. It is the standard command support system used at the strategic and operational levels of command within the ADF, and at the tactical level within the Air Command Support System, which is a modification and customisation of the JCSS. JCSS is also installed and operational on HMAS *Manoora* and HMAS *Kanimbla* as part of the JTFHQ (Afloat).

In order to enhance situational awareness the system has been designed to handle large quantities of information from a variety of different sources and bandwidths. This is achieved by using open military message standards automatically to update databases and subsequent geographic display, analysis and dissemination, providing a common picture to all users. A secure communications network connects LANs within each command centre to one another and to other specialist headquarters. These include Headquarters Australian Theatre (HQAST) and Headquarters Strategic Command Division (HSCD).

The latest version of the system is based on ADI's generic Joint Command and Control System (see separate entry).

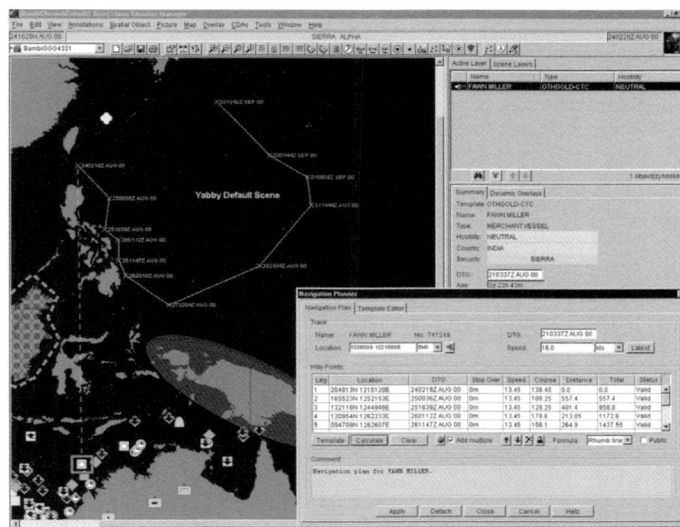

JCSS screenshot (ADI Ltd)   0116690

Status
The project has been running since 1994 over seven phases. It is currently installed on over 200 servers and 3,000 desktops in more than 40 installations across Australia.

Contractor
ADI Ltd, Garden Island, New South Wales.

UPDATED

Special Operations Command Support System (SOCSS)

In 1997, ADI was awarded the contract to develop a planning and mission support system, now known as the Special Operations Command Support System (SOCSS), for special operations and emergency services to Australia's Special Air Service Regiment (SASR). The development took place in three phases, with the final phase delivered in October 2001. SOCSS is based on ADI's Informavue software which is designed specifically for special operations. It has a client server architecture utilising Remote Method Invocation (RMI) and Jini, is fully deployable with a flexible configuration and can operate either networked or stand-alone. The system uses PCs/laptops running Windows™ NT or UNIX.

The system can be connected via both Local Area (LANs) and Wide Area Networks (WANs) and also to other secure networks using a wide variety of communications systems including PSTN, ISDN, HF, VHF, microwave links and satellite. It has the following features:

(a) A common tactical picture is provided using efficient data replication and messaging features over high- and low- bandwidth environments
(b) Intuitive drawing/planning tools with a user-friendly task-based interface and image import capability
(c) Able to develop floor plans, allocate areas of responsibility and plan routes
(d) Able to view automatically positional reports concerning assets and areas of interest and intelligence reports against a geographical map, including terrain features and Digital Terrain Elevation Data (DTED)
(e) Ability to import images, text, video and CAD models
(f) A configurable symbol set enables users to develop graphical orders rapidly
(g) Specialist tools including Line of Sight and Visibility Fans
(h) Collaborative 2-D shared plans and drawings, which can be transformed into 3-D walk-throughs
(i) Lotus Notes groupware templates
(j) Videoconferencing
(k) Recognised picture
(l) Event and intelligence logs.

Status
In operational use.

Contractor
ADI Ltd, Sydney, New South Wales.

UPDATED

Belgium

Belgian Military Operations CIS (BEMILOPSCIS)

BEMILOPSCIS is the strategic level CIS recently acquired by the Belgian Joint Headquarters for use in their crisis management centre. Based on EADS' JOCOP software (see separate entry) and compliant with the ATCCIS data model it is a web-based system that provides cartography, an Oracle database, a number of specialist applications, XML messaging and office automation using both MS and Lotus tools. Additional functions added by Prodata are planning, workflow, security and an electronic document management tool called Meridia.

Status

The system was delivered in November 2001 and achieved full operational capability in early 2002. Consideration is now being given to extending it to the single-service headquarters and to Belgian liaison detachments at SHAPE and to providing a deployable capability.

In March 2003 requests for tender were issued for the purchase, delivery, installation and maintenance of this extended capability, as Phase 2 of the project.

Contractor

Prodata (prime contractor for Belgian procurement reasons), Zaventem, Belgium.
EADS Systems & Defence Electronics, Vélizy-Villacoublay, France.

UPDATED

France

Java Open Common Operational Picture (JOCOP)

JOCOP is a software package produced by EADS as a private venture which will form the basis of a strategic or operational level CIS. It is at the core of the Belgian strategic CIS, BEMILOPSCIS (see separate entry), and in mid-2002 was under evaluation by the Bundeswehr and by Finland. Written in JAVA, including the cartographic viewer, it is not dependent on specific hardware or software platforms, so can be used with a variety of operating systems or database formats. Based on the ATCCIS GH4+ data model, which is implemented in a database, the system will provide a continuously updated operational picture, together with planning and office automation tools. The database is linked to a cartographic viewer which is able to read commercial and military GIS without a requirement to transform data. Other features include:

(a) Exchange of information using IP and XML format, including automatic XML data import into the data base using standard e-mail messages.
(b) Display uses APP6A symbology. A catalogue of additional icons is provided which can be amended by the user.
(c) Automatic alert when data is updated

Contractor

EADS Systems & Defence Electronics, Vélizy-Villacoublay, France.

VERIFIED

Germany

Jo-CCIS

The Jo-CCIS is an open (commercial-off-the-shelf) COTS-based C4I system. It has been developed with the aim of being suitable for all operations and tasks, for all levels of command and control, with a modular and scalable system architecture which can be easily customised to the required functionality and which provides scope for future enhancements. This generic system is designed for mobile and static, single and multiheadquarter command and control levels and consists to a large extent of COTS information and communication technology components. The open client-server architecture allows integration of new technology and additional functions. The system architecture provides a centralised access tier to the operational database using native input/output sequences, with inter-process communication between C4I applications using COM and OLE standards. Additional applications can access the access tier. Both single- and multiserver configurations are available, the latter using database replication mechanisms.

The system includes a Java programmed tool to support order of battle functions. Connected to the operational C4I database it provides structure viewing, editing and generating database objects for military facilities. The structure is displayed in a tree view and can be exported to other clients. Logistic information related to units, material or personnel is displayed in charts or tabular forms. The command hierarchy for own and estimated enemy or neutral forces can be manipulated by the operator using the Windows GUI and a planning facility is provided.

Military message handling is based on the functionality of Lotus Notes. A predefined set of database supported report and message templates is provided for the operator. The multilingual design and content of report forms can be

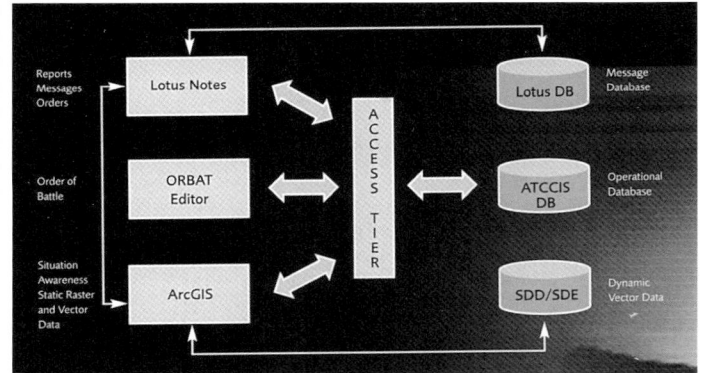

Jo-CCIS system architecture (EADS)   0143780

adapted to customer needs. Using communication protocols in strategic and tactical networks, formatted messages can be transferred between different command levels. The Lotus Notes front end enables the operator to use COTS functionality. Other features include:
(a) Document management including document replication
(b) War diary journal
(c) Sorted message views
(d) OLE link capabilities to other C4I applications.

The situation display provides interactive visualisation of own, enemy and neutral forces on digitised maps using standardised tactical symbols for units, boundaries, movements, areas and other user defined graphical objects. The information displayed on raster and vector maps is based on the content of the tactical database. The situation display provides direct information exchange with other C4I applications. As a result, the graphical content of formatted messages stored in the database can be automatically converted to the map display and vice versa. Using the ESRI® ArcGIS technology with the open programming capability ArcObjects, it is possible to adapt the GIS functionality to the specific requirements of different command and control tasks and levels. The GIS supports integrated data access and allows multilingual data entry and retrieval. Computing height information and displaying elevation data in 3-dimensional views is possible using extensions of the ESRI® ArcGIS package. Other capabilities include:
(a) APP-6 or APP-6a (2525E) symbol set. User defined symbols
(b) Adaptable legend control
(c) All common and various military raster and vector data formats
(d) Situation awareness
(e) Integrated multi-user data access
(f) Automatic client update
(g) Spatial data analysis.

Key System Features

(a) Windows-based system
(b) Multilingual user interface (currently available in English and Arabic)
(c) ATCCIS data model using ORACLE DE
(d) APP-6 or APP-6a symbolisation
(e) Handling of different user modes as life, simulation and exercise
(f) ADatP- 3 conforming data model
(g) Synchronous data replication of operational data for multiserver systems
(h) Database supported Situation Display
(i) Display features such as line of sight profiling, 3-D analyses
(j) Handling of formatted messages and operational orders using Lotus Notes
(k) Automated generation of formatted messages such as Situation Reports
(l) Messages and data flow easily adaptable to customer needs
(m) Order of Battle application including data entry
(n) Interoperability with Airspace Control Systems
(o) Interoperability with communication links including SatCom and HF
(p) Connection of mobile message clients
(q) Integrated display of the Recognised Air Picture.

Status

Jo-CCIS forms the basis of the CIS in the military headquarters of the UAE and of the 'OFEQ' system installed in Oman.

Contractor

EADS Dornier GmbH, Friedrichshafen, Germany.

VERIFIED

Israel

Tadiran Theatre Missile Defence Systems and Programmes

Tadiran Electronic Systems' Israeli Theatre Missile Defence Test Bed was developed by Tadiran as part of Israel's participation in the US Strategic Defence Initiative. The Test Bed has been extensively used and enhanced in a series of experiments and software improvements over seven years. The ITB is a unique Theatre Missile Defence facility for the Middle East arena, which can effectively and affordably be adapted to different customers' environments, doctrines, systems and concepts.

The ITB simulates TBM attacks against multiple targets, and the defence action taken by the weapon systems controlled by one or several Battle Management, Command and Control (BM/C2) centres. The system features extensive Human in the Loop (HIL) capabilities, creating the precise environment and conditions that will be faced in a real life scenario.

These features have made the Tadiran TMD simulator a proven, powerful, low risk tool for assessing the effectiveness of TMD defence systems, BM/C2 concepts, and for defining the human role in the TMD battle. The system supports the study of interoperability issues relevant to every stage of the TMD battle.

The Tadiran TMD simulator has made a significant contribution to the Arrow Weapon System's development being the primary prototyping and validation tool for the Arrow Weapon System BM/C2 algorithms. The ITB is also being used as the design and validation environment for the Israeli Boost Phase Intercept System BM/C3. The inherent flexibility of the ITB makes it easily adaptable to a variety of theatres, systems, threat scenarios and defence missions.

The TMD Test Bed provides support for the following activities:
(a) Definition of TMD goals
(b) Development of TMD doctrine
(c) Definition, evaluation and selection of TMD systems components
(d) Evaluation of candidate architectures
(e) Development of BM/C2 concepts, algorithms and Human-Computer Interface
(f) Development of human control requirements
(g) Study of Interoperability issues and tools among national and multinational TMD systems

Contractor

Tadiran Electronic Systems Ltd, Holon.

UPDATED

Korea, South

Theatre Automated Command and Control Information Management System (TACCIMS)

TACCIMS was built for the Combined Forces Command of the USA and South Korea. By March 1989 contracts worth some US$25.6 million had been let. TACCIMS is installed at various command posts and other locations in South Korea. It was designed as an integrated system composed of subsystems for information processing, communications, display, message handling and the translation of information in both English and Korean. Instead of simply providing equivalent English and Korean words on a word-for-word basis, the system translates messages and information according to the syntax of each language.

The complete information processing system comprises 400 computer workstations and graphics terminals as well as fibre optic Local Area Networks (LANs) in two main command centres. A communications network allows data transfer between all TACCIMS sites. The new system includes two US/Korean headquarters installations, at Command Centre Seoul, Command Post TANGO, 13 remote sites and seven transportable field units.

Status

In September 1998, TACCIMS integration into GCCS-A (see separate entry), was tested during Exercise Ulchi Focus Lens. GCCS-A was subsequently accepted into Korean service on 1 November 1999. Secure walls filter releasable data from the US GCCS-A into the combined system.

GCCS-A functionality has now subsumed the original TACCIMS.

Contractor

Lockheed Martin Mission Systems (prime contractor for GCCS-A), Gaithersburg, Maryland.

UPDATED

Malaysia

National C4I system

Malaysia's new strategic and operational level C4I system was shown for the first time in public at DSA 2002. This is a home-grown system developed by a local company, Systems Consultancy Services Bhd, to replace the previous system installed by the then GEC Marconi in 1996. Although little is known of this original system, it included both the command and control applications and the communications infrastructure. The former were found to be non-Y2K compliant, necessitating both hardware and software replacement.

The new system is PC-based, has Windows NT as an operating system and uses COTS hardware. Based on distributed databases and using web technology, it provides C2 facilities for all three services, together with an additional facility for the Intelligence branch, and is installed down to platform/battalion/air station level.

Each service manages its own operational picture at the service headquarters and passes it to the Joint Headquarters, which maintains an overall joint picture. Any subordinate HQ is able to access those aspects of the joint picture for which they have permission. Information is transmitted using a data format unique to the system, but this can be translated at single-service headquarters into other formats to assist interoperability with other nations. Positional information is displayed against spot imagery or digital mapping. Information on individual units, both background and immediate, is displayed on a web page and updated by the unit.

Communications are provided by the original Marconi infrastructure, which has been enhanced and, with a reported network capacity of 2 Mbytes, has ample bandwidth to allow all forms of communication including video conferencing.

SCS claims it to be the first home-produced C4I system in the region. It is believed that the Malaysian Ministry of Defence may now be considering an interoperable battlefield system.

Status

The system is in operational use in Malaysia. Following extensive tests and trials it received final operational approval in late 2001.

Contractor

System Consultancy Services, Kuala Lumpur.

VERIFIED

NATO

Allied Deployment and Movement System (ADAMS)

ADAMS is the NATO system for transport and movement planning and management. The changes in the European security environment in the early 1990s opened the way for a new NATO strategy based on flexibility, mobility and multinationality. ADAMS supports that strategy by facilitating the planning and management of the movement and transportation of the forces and supplies from NATO nations and their partners, either for defence or for peace support operations.

The main features of ADAMS are:
(a) Data communications network for the rapid exchange of mobility data and plans.
(b) Software for the deployment planning, mobility simulation and force tracking.
(c) Data on the forces, the transportation assets and the geographical infrastructure.

Network. ADAMS can be used in a stand-alone mode, but the key to multinational co-ordination in a crisis is the ability rapidly to exchange both background mobility data and plans, particularly the detailed deployment plans. The deconfliction of the national detailed deployment plans is an iterative process requiring extensive data communication between the national and NATO movement staffs. The end result is the consolidated, multinational detailed deployment plans. At present (mid-2002) the ADAMS Network connects all NATO nations and most NATO Headquarters.

Software. The software modules provide the ADAMS user with the tools to plan and manage the deployment operations. The main functions are database management, force planning, movement and transportation planning, sustainment planning, mobility simulation and force tracking. The software also includes conversion modules for interfacing between ADAMS and national mobility management systems.

Data. The ADAMS data is organised in a relational database with a separate Database Management Module. The three main categories of data are the Forces, the Transportation Assets and the Transportation Infrastructure. The force data include unit organisational structures, equipment holdings, descriptions of the equipment and supply items. Each nation maintains the data on its own units, modifies it according to the particular planning situation and ensures that the national detailed deployment plan is based on the right unit description. The transport data include all modes of transport: sea, air, road and inland waterways, including vehicles as well as infrastructure, that is ships, planes, trucks, rail cars and barges as well as ports and airports.

Force Planning. The result of the Force Planning process as it relates to ADAMS is the list of units, their destinations, priorities and the arrival times required to meet the operational objectives. At the start of the planning process the list is typically expressed in terms of generic units. These are subsequently replaced by the real units committed by the nations. This also includes initial recommendations as to Air and Sea Ports of Debarkation which become the starting point for national planning.

Movement, Sustainment and Transportation Planning. The Deployment Planning Module uses the force list as the starting point and extracts the data for the relevant units from the database. The planner splits and schedules the troops, unit inventory and accompanying supplies into movement components (advance parties and so on) tailored to the relevant mode of transport, and then assigns transport assets to the movements. The output at any stage of the process is a partial or complete detailed deployment plan.

Deployment Analysis. Although extensive assessment functionality has been built into the software ADAMS includes two special purpose analytical tools. The deployment display module provides the facility, on a geographical background, to examine the operational and logistic implications of the overall plan. It is specifically designed to support the multinational deconfliction process at the allied headquarters. The General Deployment Model is a detailed critical event simulation model. It takes as input either an incomplete or fully developed deployment plan and simulates the assignment of transportation assets and the scheduling of each movement.

Force Tracking. In the deliberate peacetime planning for potential NATO Contingency Operations the deconflicted multinational deployment plan is the end product of the ADAMS work. However, in the implementation of actual NATO deployments the process continues with the monitoring and management of the operation, tracking the forces and the transport assets. For 'near-realtime' tracking the deployment planning module allows the updating of deployment plans with the confirmation of departures and arrivals of component movements.

Contractor

There is no single contractor. The system has been developed through a project managed by:
NATO C3 Agency, The Hague, Netherlands.

VERIFIED

CRONOS (Crisis Response Operations in NATO Open Systems)

CRONOS is a system consisting of interconnected IS nodes and a wide-area network, evolved from an initial capability developed for deployment with IFOR in FRY. Currently the system, which operates 'System High' at NATO Secret level, consists of nearly 100 nodes NATO-wide with several thousand user workstations and over 1,000 mailboxes. Secure connectivity is provided between CRONOS and several national and coalition systems. Using Windows NT4, the system provides users with specific military information services including the ground, maritime and air pictures as well as office automation services.

Contractor

The project is managed by:
NATO C3 Agency, The Hague.

VERIFIED

Linked Operations-Intelligence Centres Europe (LOCE)

LOCE is a network supporting intelligence operations throughout NATO and is the declared intelligence backbone for operations in FRY. US developed and managed, it consists of a centralised set of servers containing databases and imagery information and a network supporting more than six hundred web-enabled PC workstations distributed among US and multinational users at all levels of command. Operating at the US Secret Releasable to NATO and NATO Secret classification levels, the system provides multimedia e-mail, bulletin board, TACELINT, secondary imagery, order of battle databases, network services, and a secure voice capability, and gives each user access to near-real-time (NRT), all-source, correlated air, ground and naval intelligence analysis and products. It supports I&W, current intelligence, collection management, and most aspects of the targeting cycle including nominations, air tasking orders, and battle damage assessments. It also provides the TBM data architecture supporting shared early warning among NATO and theatre components; this application provides early launch detection warnings of missile activity for disaster relief operations, while providing the military lead time for discretionary intervention. Specifically, it provides corresponding launch and impact data with illustrative areas. The LOCE Correlation Centre, located at the Joint Analysis Centre, RAF Molesworth, UK, functions as the US gateway for exchange of operational intelligence with NATO. USEUCOM has the responsibility for maintaining LOCE.

Originally designed as a NATO-only information dissemination system, LOCE has greatly aided the dissemination of US intelligence to coalition partners in the Balkans. LOCE has been used in support of operations in both Bosnia and Kosovo. It is also an element of the Battlefield Information Collection and Exploitation System (BICES).

Capabilities

(a) Fully Web Enabled
(b) Near-realtime sensor report correlation via the Intelligence Report Database (IRDB)
(c) Combined Order of Battle (COB) — Air, Missile, Electronic, Defensive Missile, Naval, and the NATO target data inventory
(d) Multi-national digitised secondary imagery exchange and manipulation
(e) Provides a common battlefield picture
(f) Integrated secure voice communications
(g) Theatre Missile Defense Warning
(h) DIA filtered 30 country Military Intelligence Integrated Database (MIIDB)
(i) DIA Counter-Proliferation Database

(j) Autonomous, stand-alone, "Fly-Away" capability to contingency areas through rapid deployment of the LOCE Mobile Correlation Center (LMCC)
(k) Satellite communications to include five X-band 2.4 metre remote ground stations and Ku-band commercial V-SATs
(l) Electronic mail functions, including a bulletin board text product server.

Status
In operational use across NATO.

Contractor
NATO C3 Agency, The Hague.
General Dynamics Network Systems (network management), Needham, Massachusetts.
Anteon Corporation (software and engineering support), Fairfax, Virginia.

UPDATED

Strategic C3 Systems

Background
NATO's strategic CIS have developed separately on either side of the Atlantic. Currently the two strategic systems are MCCIS in ACLANT and the developing ACCIS in ACE. The Bi-Strategic Commanders Automated Information System (Bi-SC AIS) aims to integrate these two systems to provide a single core capability. This common system will be managed within a single Bi-SC AIS management structure which is currently being established. The Baseline 0 of the Bi-SC AIS Core Capability will be established by mid-2004 and it will include the Baseline 1.0 of the ACE ACCIS Increment 1 (see below), which by then will be completed. Further Bi-SC AIS enhancement will be undertaken on an incremental basis to meet the requirements of the two strategic commands. Through alignment with US standards and operating environment it is intended that the Bi-SC AIS will be interoperable with the GCCS (see separate entry).

ACCIS
The Allied Command Europe (ACE) Automated Command and Control Information System (ACCIS) is an information system which will provide automated support of command and control by commanders throughout ACE, using common hardware and common software. Its overall objective is to provide the means to execute consultation, command and control, including rapid exchange of information, decision support and decision execution, as defined by the command and control cycle, and allow assessment and exchange of a combined air, land, maritime situation picture throughout NATO in peace and in crisis. The system is subject to continued evolution including the enhancement of capabilities, upgrades and the expansion of functional support. The ACE ACCIS Implementation Plan, which was approved in Nov 1998, lays down a programme of three increments.

Increment 1 is the first step towards the overall architectural goal by consolidating the existing system facilities into a more stable baseline of core capability services referred to as Baseline 1.0, onto which applications to provide Functional Area Services (FASs) may be integrated. Baseline 1.0 involves the establishment of robust wide and local area connectivity with built in security and system management services. These will be complemented by 'interoperability services' supporting both push and pull information exchange mechanisms and cartographic workshop facilities across all ACE ACCIS nodes. Increment 1 (Baseline 1.0) will be completed in May-June 2004.

Increment 2 will implement enhancements to the core capability services with distributed document management and FAS specific geographic services. These services are independent of each other and may be implemented in parallel. These enhancements together with continuing FAS integration will result in Baseline 2.0. However, Increment 2 will not be implemented as an ACE capability but as a Bi-SC AIS capability and the plans will be adjusted accordingly, as funds have only been authorised for Bi-SC AIS Core Capability enhancements.

Increment 3 will further evolve the FASs by building on the homogeneous application platform constructed in Increments 1 and 2 in a progressive, convergent way in conformance with ACCIS Target Architecture. These include Command Group and crisis management, Intelligence, Logistic and Operations (comprising joint, maritime, land and air) FASs. Although the full integration of identified FASs is planned for Increment 3, which will implement Baseline 3.0, the progressive integration of ongoing FAS developments may well occur during Increments 1 and 2.

The core services of Baseline 1.0 are:
(a) Windows 2000 OS services
(b) MS Office
(c) Information access/sharing
(d) Informal messaging (e-mail)
(e) Web services
(f) Collaborative software tools
(g) Document management and workflow systems
(h) Security services
(i) Enterprise management system
(j) GIS
These are being implemented in ACE command headquarters locations at Stavanger, Viborg, Casteau (2), Brunssum, Ramstein, Heidelberg (2), Naples (2), Verona, Madrid, Larissa and Izmir. By mid-2002 about 75 per cent were complete,

with the remainder expected to be completed by the end of the year.

The minimum hardware specifications are 266 MHz Pentium II CPU, 128 Mbytes RAM, 5 Gbytes hard disk.

MCCIS

MCCIS is a near-realtime C2 system focused on strategic and operational level use. In the early 1980s, the Supreme Allied Commander Atlantic (SACLANT) and the Regional Headquarters South Atlantic (RHQ SOUTHLANT) initiated projects to produce their own CCIS capability. The two projects merged in 1989 under the name Alpha CCIS. The US Navy Operational Support System (OSS) was selected as the foundation for Alpha CCIS after a detailed investigation of available CCIS solutions and products. An agreement was signed in 1992 between SACLANT on behalf of NATO and the Space and Warfare Command (SPAWAR) on behalf of the US Navy, to provide for the co-operative development of CCIS capabilities. As SACLANT took over maintenance responsibility for the product, there was a transition from government-off-the-shelf (GOTS) to commercial-off-the-shelf (COTS) and NATO-off-the-shelf (NOTS) capabilities. In 1997, the name was officially changed to MCCIS to better reflect the NATO wide usage of the system.

An open architecture system, its ability to operate over a wide range of command levels and proven interoperability with an expansive set of NATO, national and commercial formats and interfaces has led to its being the chosen platform for the management of NATO's Initial Common Operational Picture (COP).

In general, MCCIS provides the following functional capabilities:
(a) Use of predefined geographic areas
(b) Access to multiprojection, multiresolution maps
(c) Graphic display of positional data
(d) Automatic display update of positional data
(e) Access to positional information via several retrieval methods
(f) Automatic and user-controlled database update
(g) Access to database via forms and queries
(h) Automatic generation of formatted reports
(i) Interface with TARE, Link-11, Link-14 and Link-16
(j) Interface to other NATO C2 systems like ICC and JOIIS
(k) Interface to national systems like GCCS, NORCISSII, RDNCCIS, ACOM
(l) Ability to draft, review, edit, validate and release messages
(m) Ability to save, display and print screens
(n) Briefing support with multiprojector control
(o) Electronic mail for internal communications between users
(p) Integrated word processing, graphics, and spreadsheets
(q) Water space Management
(r) Hydrographic, Oceanographic and Environmental tools
(s) Web portal

MCCIS is installed and in operational use in 61 sites in the following nations: Canada, Denmark, France, Germany, Greece, Iceland, Italy, the Netherlands, Norway, Portugal, Spain, Turkey, United Kingdom and the United States. There are over 250 users.

NATO Message Service

The NATO Messaging Service (NMS) will replace the formal messaging service currently provided by TARE (Telegraph Automatic Relay Equipment), which is a self contained and obsolescent system. The NMS will be provided on a user workstation, with the server operating system being either Windows 2000 or Unix, or mixed Win2K (in local NMS) and Unix (in backbone and boundary); this remains to be finalised. There are two increments to NMS. Increment 1, which consists of X.400 informal e-mail, has already been implemented and was part of Baseline 1.0 for ACE ACCIS. Increment 2, which covers formal message traffic and therefore a variety of issues of security, accountability and interface to national systems, will be implemented in 2 phases:
(a) Phase 1. An Initial Operating Capability at NATO HQ, SACLANT (Norfolk, Va), SHAPE (Mons, Belgium), Northwood (UK), NATO CIS Operating and Support Agency (NACOSA) plus 20 remote users. This is expected to be complete by early 2004.
(b) Phase 2. The NATO-wide roll-out, which on the current (June 2002) programme is due for completion by late 2005.

NATO General Purpose Communications System (NGCS)

The NGCS is the communications system being developed by NATO to link the national defence communications networks of member nations for use in conditions up to and including conventional war. A separate Special Purpose System exists for the nuclear environment. NGCS is made up of three components:
(a) Packet transport component. This is Step 1 of the NGCS programme expected (in mid 2002) to be completed by the end of 2002, in line with the requirements of ACCIS Increment 1.
(b) Circuit switch component. At mid-2002 this was in the process of implementation.
(c) Real-time semi-permanent on demand component. (Intelligent Bandwidth Manager (IBWM))

Contractor

A variety of manufacturers provide hardware and software. Projects are managed by:

NATO C3 Agency, The Hague.

UPDATED

Netherlands

METIS and METIS 2000 meteorological information system

The Sigmex (now Almos) METIS has been implemented as a computer-aided system to support the work of meteorologists in the provision of reliable, short duration, weather forecasts to aid in all aspects of mission flight planning. It currently consists of over 20 years of special computer software. The hardware is based on DEC MicroVAX computers and Sigmex high-performance graphics equipment. However, the design is such that it can be adapted and implemented on a wide range of computer equipment.

The concept of METIS is to provide computer-assisted procedures in support of the meteorologist. The computer handles routine tasks, with data collated, analysed and presented as soon as it is available, thus enabling the meteorologist to produce faster and more accurate forecasts. The system is able to operate 24 hours a day, seven days a week, carrying out assigned tasks either on an automatic basis or manually under the control of an operating meteorologist. All of the relevant incoming weather data is collected in specially designed databases for inspection and analysis.

Any information held within the METIS system can be presented in a number of forms, that is, typically as a graphic alphanumeric screen display, or various types and sizes of colour or monochrome hard copy. The METIS system has three main data presentation formats; geographic, textual and miscellaneous graphical presentations. The geographic model allows the meteorologist to view data on a defined map area at a given level of detail, while the textual model allows database retrievals of meteorological and other stored data such as details of reporting stations. Lastly, special purpose user-defined formats, such as vertical analysis of the atmosphere, are available.

A feature of the system is the use of high performance intelligent graphics generators closely coupled to the local database central processor. By combining multiple screen output (graphics and text), with single tablet and keyboard input, the meteorologist has been provided with an optimised man/machine interface suited to the analytical nature of weather forecasting. The system is command driven with a choice of input methods and levels for users of different experience. Online help and prompting is available with immediate response to operator requests.

The application software is totally based on the international Graphical Kernel System (GKS). World Meteorological Organisation standard for message formats, weather station identifiers and display symbology are all incorporated in the design. The geographic database was derived from commercially available map data.

METIS 2000 is the latest version of METIS and runs on a PCI based 64Bit RISC server with an Oracle database. Visualisation tools are run on desktop PCs running Windows NT. This configuration allows users to integrate METIS with other office tools, such as Word, Excel and PowerPoint.

PC-METIS is a scaleable Windows based briefing terminal that can support most meteorological data streams including MDD, SADIS, ISCS and the NATO PES.

Data from all the major weather and communication satellites can be integrated into the METIS database. Image data in high and low resolution formats from GOES, NOAA, METEOSAT and GMS are supported. Alphanumeric data from METEOSAT (MDD and RDCP) and the ICAO SADIS and WAFS transmissions are also supported. User annotation can be added to the data for the generation of significant weather and other forecast charts.

Status

METIS, METIS 2000 (13) and PC-METIS (54) have been supplied to the Royal Netherlands Air Force for various locations. At least two METIS systems in ruggedised containers have been supplied to the Royal Netherlands Army. Both METIS 2000 and PC-METIS have been supplied to the Belgian Air Force and AFCENT.

Contractor

Almos Systems BV, Culemborg.

METIS displays

UPDATED

Norway

NORCCIS II

NORCCIS II supports the planning and execution of Joint (Land and Naval) operations at the strategic and operational level, including deployed HQs. It is CHOD (Chief of Defence) Norway's primary C2 system for the support of joint operations at the strategic and operational levels and is the most widely fielded C2 system in the Norwegian forces. It provides a Common Operational Picture (COP), including the Recognised Air Picture (RAP) via a Link 1 feed; the Recognised Land Picture (RLP) utilising information from own and external sources; the Recognised Maritime Picture (RMP) utilising information from maritime sources ranging from shore-based commands to units at sea. To display and exchange COP information, NORCCIS II is compliant with NATO standard symbology and recognises the message formats AdatP-3 and OTHT-Gold. It is capable of the management of the Rules of Engagement (ROE) and NATO Precautionary System (NPS) as part of Crisis Management, and can display information from Air Tasking Orders, Air Co-ordination Orders and TBMEW. The operating system is Windows 2000.

NORCCIS II consists of a set of integrated modules based on ACE ACCIS platform compliant customised COTS software packages:
(a) Naval
(b) Land
(c) Crisis Management
(d) Order of Battle Manager
(e) Situation Display/Map system
(f) MMHS and Directory System
(g) Message Processing
(h) Planning Module
(i) WEB Publish
(j) WEB Portal
(k) e-mail and MS Office

The Situation Display/Map System module is provided by Teleplan's MARIA Military Mapping Application modified by the NORCCIS II project. MARIA presents vector data, in the form of map 'templates'. Each template shows a grouping of map data such as trees or coastline best suited for display at a specified map scale. The following types of data can be displayed on the chosen map background:
(a) Database information stored via textview or message input.
(b) Text and drawings entered directly onto the map, including tactical graphics in acordance with APP-6A standard
(c) Height Profiles, Radar/Radio/Line of Sight Coverage presentations.

The Order of Battle Manager module, also based on MARIA, supports the users with a graphical representation of all units registered in the NORCCIS II database and their respective organisational structures. OBM is closely related to the Land and Naval Modules and the Situation Display (SD), and information can be exchanged between the modules. It is possible to display information about selected units in different ways; a tactical, geographical position in map, command relationship in ORBAT and detailed information about mission, operational status, personnel and equipments detailed information in textviews. In OBM the users are able to establish and update an ORBAT by using 'drag and drop' functions to organise sub-units and to select their command relationship.

The MMHS, Directory System and Message Processing is provided by Systematic's IRIS and HEKATE messaging and integration products (see separate entry).

The NORCCIS II WEB Portal provides support to C2 data consumers, some of whom may be users on connected systems. The following applications are available:
(a) COP
(b) Bulletin Board
(c) WEB publishing
(d) NewsFlash
(e) File browsing
(f) Outlook
(g) Jane's Online

The NORCCIS II Intranet is based on an unclassified IP network (NDDN and commercial assets) with the Thales TCE 621IP-Crypto for NATO SECRET communications. The NORCCIS II Intranet is connected to NIDTS/Cronos through a firewall and a mailguard, providing WEB and e-mail access.

NORCCIS II can be customised for various sizes and types of HQs due to its modularity. These range from Norwegian SF using NORCCIS II on a single laptop to major HQs comprising several hundred users.

Status
NORCCIS II has been in operational use since 1992. In addition there are installations in KFOR and at SHAPE, AFNORTH and NATO HQ.

Manufacturers
Systematic Software Engineering , Aabyhøj, Denmark.
Thales Communications AS, Oslo.
Teleplan AS, Oslo.

UPDATED

Spain

Amphibious Operations Command and Control Information System (AOCCIS (SICOA))

AOCCIS (SICOA is the Spanish acronym) is a system to support the Amphibious Task Force (ie naval) and Landing Force (LF) Commanders in conducting amphibious operations. It supports operational planning and command through the doctrinal phases of amphibious operations of embarkation, movement to the objective area, assault, subsequent operations and re-embarkation. The system is installed on board the command ship, but subsystems can be deployed ashore with the landing force. It receives maritime information from other onboard systems such as the ship combat system (tactical situation through Link 11 and own sensors (air, surface submarine and EW tracks)) and the Message Handling System (MHS) which exchanges information (structured and formatted messages) with other headquarters. The system is designed to accept interfaces with onboard communications equipment, the naval command system and NATO MCCIS. When the LF HQ is disembarked both subsystems will be linked via datalink and/or through a satellite, to enable a common picture to be maintained and planning to continue in parallel.

The COTS HW architecture of the onboard and disembarkable subsystems is built around a fast Ethernet Local Area Network (LAN); two servers, to improve the system's availability; the operator consoles; a projector with a large screen display; printers and scanners.

The AOCCIS provides graphics management functions (maps, overlays, distance/visibility calculations), database management (NATO LC2IEDM interoperable data model with amphibious operations extensions), operational order/plan editor, stowage planning utilities, tactical display (NTDS symbology and APP 6(A)), recording/playback and system HW/SW administration.

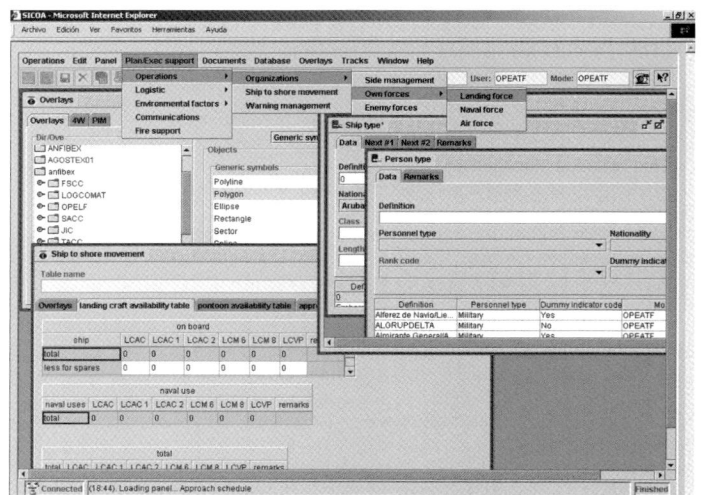

Screenshot from Indra's AOCCIS showing a number of planning functions including ship-shore movement (Indra)   0525544

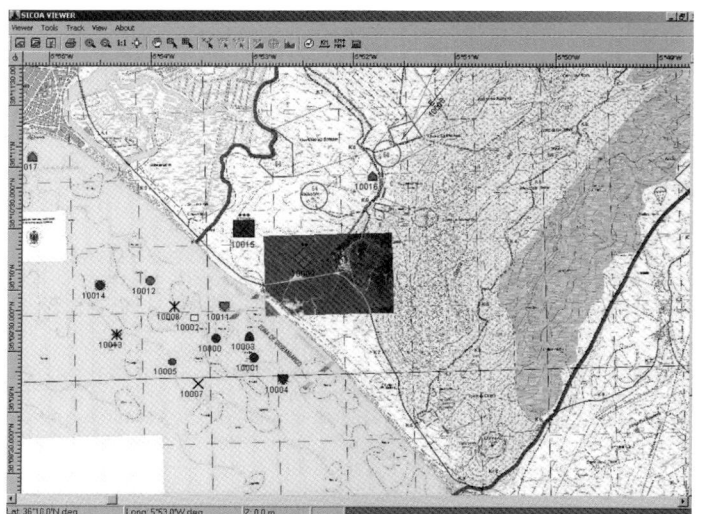

Screenshot from Indra's AOCCIS showing an integrated maritime and land picture and the use of overlaid air photographs (Indra)   0525545

Status
The AOCCIS has been operational on board the Spanish Navy assault ship LPD Castilla since September 2001 and the deployable is available to the Landing Force headquarters.

Contractor
Indra, Madrid.

UPDATED

Thailand

Joint Military Decision Support System

In 1993 the Thai Government decided to centralise control of the country's Army, Navy and Air Forces within the Royal Thai Supreme Command (RTSC), creating a Joint Command. Following extensive studies by consultants Booz Allen and Hamilton (BAH) it was decided to build a fully integrated command-and-control management information system to support Thai military decision-making and deployment. Central to the plan was the design and installation of local-area networks and mobile command posts all of which would be linked to a new joint operations centre based in Bangkok. In 1996, BAH awarded ADI a contract to develop the software for the Joint Military Decision Support System (JMDSS). The JMDSS is a key component of the RTSC C³I system. The JMDSS provides planning, reporting and monitoring functions to the Supreme Commander and his staff. It supports the three services - Army, Navy and Air Force - of the RTSC individually, jointly and in coalition. The system provides full interoperability with US Forces.

The capabilities of the system include:
(a) Display of current operational status in a graphical multitasking environment
(b) Near real-time situation awareness
(c) Military formatted messaging, compatible with US formats
(d) Multilingual support (English and Thai modes).

Development
The JMDSS was implemented in three evolutionary phases over a three year project development life cycle:
Phase One
(a) Rapid prototyping and implementation of isolated Local Area Networks (LANs)
(b) E-mail services to local users using US military formatted message formats
(c) Implementation of common office automation applications to support daily administrative activities
(d) User training on deployed application sets and network management.
Phase Two
(a) Internetworking of LANs, providing a seamless interface and data transfer between users
(b) Establishment of gateways to legacy mainframes at each service component and the RTSC
(c) Initial development of mobile command posts.
Phase Three
(a) System integration
(b) Final maturation of the LAN/Wide Area Network (WAN) infrastructure
(c) System testing.

Status
In operational use since 1996.

Contractor
ADI Ltd, Sydney, New South Wales.

VERIFIED

United Kingdom

Air Defence Systems Integrator (ADSI)

ADSI is a real-time tactical command and control system consisting of a set of software modules which integrate feeds from a wide variety of inputs and presents them on a single, correlated and fused display, thus integrating the recognised air picture. It will maintain up to 16 tactical datalinks, using joint forwarding standards, including Link 1, Link 11, Link 11B and Link 16. It accepts a wide variety of radar inputs, providing automatic track initiation and simultaneously combining tracks from multiple radars (up to 24) into a cohesive tactical picture. It will receive, correlate and fuse data from intelligence networks (up to 8), including TIBS, TRAP/TDDS, IBS Transmit, OTCIXS (OTH-Gold Format, Over-the-air, and GCCS) and others. It will correlate new data with previously received data, reducing duplicate information, and then correlate intelligence data against data received from tactical datalinks and radar inputs, fusing all for enhanced identification. It will automatically correlate the Air Tasking order to the real-time tactical air picture, adding ATO information such as mission number and weapons load to air tracks.

The ADSI tactical situation display provides surveillance, weapon, planning, simulation and system management facilities. A user-defined soft-switch area is provided, enabling immediate action switches to be available at a single keystroke. The situation display can use map data from raster and vector formats as a backdrop to the dynamic picture, and can also display Airspace Co-ordination Orders and other overlay information.

The system is hosted on COTS hardware components and is supplied in rugged, field-deployable containers.

ADSI has been developed by Advanced Programming Concepts, Austin, Texas, which is an Ultra Electronics Company.

ADSI Operational architecture (Ultra Electronics)   0109977

ADSI Tactical display (Ultra Electronics)   0109978

ADSI in a command console (Ultra Electronics)   0533836

Status

ADSI is in operational use in the US Navy, Air Force, Army and Marine Corps, with the UK and has been sold to other unspecified countries. Over 300 systems are currently deployed.

Contractor

Ultra Electronics Command and Control Systems, High Wycombe.

UPDATED

ASH

ASH is a multiservice Explosive Ordnance Disposal (EOD) information system for the collation and dissemination of data on explosive items and their disposal procedures. The fully integrated system allows EOD organisations to maintain an up to date record of devices and to provide operators in the field with information to assist in their reporting, identification, rendering safe and disposal. Hosted on a rugged laptop such as the Panasonic Toughbook, operators are able to access extensive textual and image information, which is updated centrally. The integrated database, management and analysis system offers a database of all known ordnance, capable of holding the details of more than 200,000 devices. The characteristics of ordnance can be compared on site with the database and up-to-date Render Safe Procedure options are provided.

The system is designed to be intuitive and logical in order to ease use under stress, and a common interface is provided across all levels.

A Windows 2000 version is to be released in 2004.

Status

In service with UK forces. Has been used operationally in Afghanistan and Bosnia. It is likely to be the EOD Battlefield Information System Application (BISA) in the UK's Digitisation Phase 2.

The system has been trialled by the German armed forces.

Contractor

EDS Defence Ltd, Hook, Hampshire.

UPDATED

Autonomous Link Eleven System (ALES)

ALES is a fully ruggedised, stand-alone, receive only Link 11 system which includes a Windows™-based tactical display and which allows units without organic sensors to receive a complete tactical picture. The ALES system includes HF and UHF radios, Link 11 Data Terminal Set, Crypto, GPS receiver, IBM-compatible PC and colour monitor. The ALES HCI includes:

(a) full colour tactical display
(b) user selectable display centre and range
(c) user selectable maps, grid lines and range rings
(d) display filters
(e) intuitive graphical user interface (point and click, pull down menus and so on)
(f) tote and mini-tote windows.

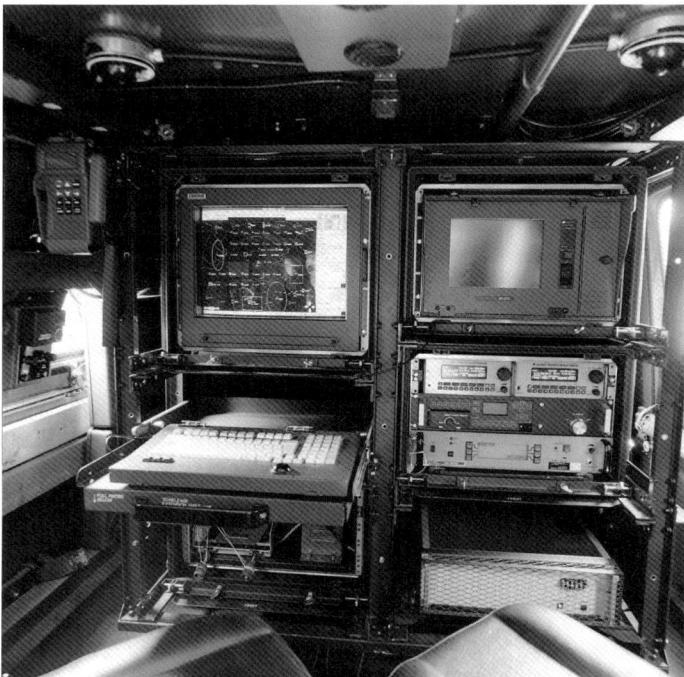

ALES installed in Hägglunds BV206D   0001002

ALES Link 11 software is based on IBM's Processor System (DLPS) software (see separate entry) and written in Ada. Display software is written in C++. ALES runs on an IBM-compatible PC under Microsoft Windows™. For benign environments, ALES can be supplied on standard COTS hardware.

Status

ALES was originally developed by Data Sciences Ltd, which was acquired by IBM in 1996. It is fitted in Hägglunds BV206D vehicles to provide support to AD units of 3 Commando Brigade Royal Marines.

Contractor

IBM Defence, Farnborough, Hampshire.

UPDATED

Corporate Headquarters Office Technology System (CHOTS)

CHOTS is a multimillion pound secure, standards-based, office information system. It was developed by the TOPIX consortium, led by ICL (now Fujitsu), to support the central administrative operations of the UK MoD. Other members of TOPIX are BICC, Coopers & Lybrand Deloitte, Data Logic and Hewlett-Packard.

The benefits and practicality of the secure office were proved in a CHOTS prototype developed and implemented for MoD by TOPIX in the early 90s. The principles established in the prototype are now applied to the whole of MoD headquarters. The secure office system was built as far as possible from commercially available software products mounted on UNIX hardware platforms from two different suppliers.

CHOTS provides office services in a secure environment to 20,000 users in over 50 sites, the implementation being phased over five years. A wide range of services is available to the user via a single terminal on the desk. Personal productivity tools such as word processing, spreadsheet, personal databases, diaries and business graphics, are provided by the ICL office system OFFICEPOWER.

Additionally, the users may use departmental applications that serve their immediate work group or department, for example, a database of information that is centrally available to any number of users. The database may be developed using the OFFICEPOWER User Defined Applications facility or, for more complex applications, relational database management systems such as Oracle, Ingres and Informix, are available.

As well as providing the basis for personal and departmental applications, CHOTS caters for the integration of corporate applications. TOPIX has implemented in CHOTS a secure registry system, an interface to MoD's signals systems and an electronic version of a corporate telephone and functional

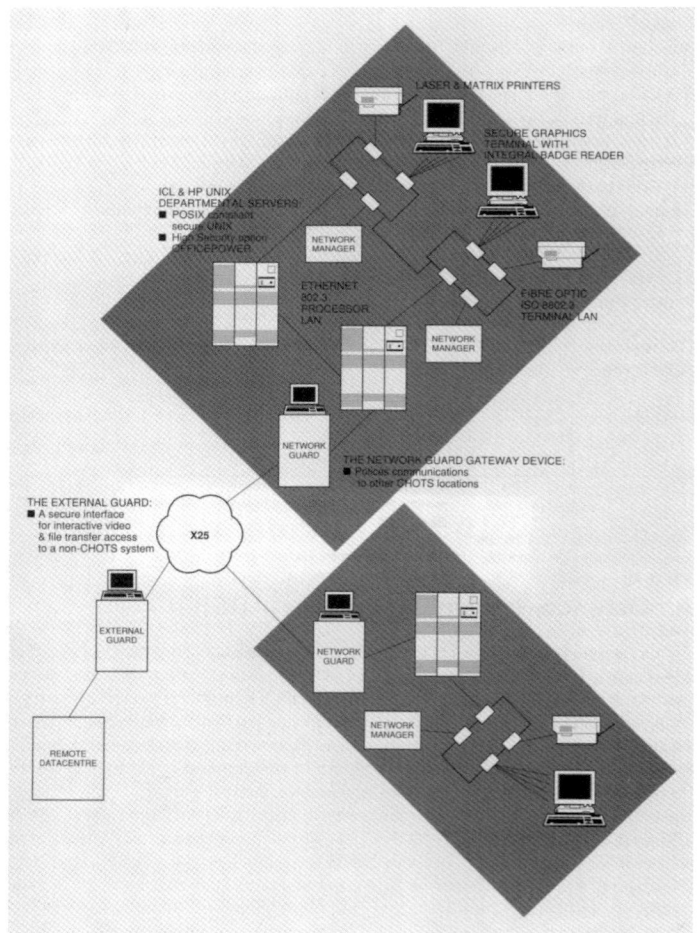

The CHOTS network

directory. All CHOTS users are linked by secure X400 e-mail across the MoD's Defence Packet Switched Network. Users with the correct clearances and authorisations are able to gain electronic access to remote data centres.

The security system includes secure UNIX and OFFICEPOWER with security enhancements. It has been independently evaluated to check that it meets the security requirements defined by MoD's security policy. The main CHOTS components are designed to achieve and maintain a confidence level approximating to UK Level 3, with some critical communicating components at UK Level 5.

Components conforming to international standards have been used in the development of CHOTS in order to maximise opportunities for upgrading its facilities. These include transport standards such as 802.3 and FDDI for local area networks and X25 for the wide area networks, as well as application standards such as X400 mail and X500 directory services. As new industry technologies become available, for example, processors with greater capabilities, they can be readily incorporated to provide improved price and performance.

Status
After its introduction in 1994 the development of the system has been somewhat chequered and was not entirely popular with users. Subsequently the system migrated to Windows NT and CHOTS Version 8 is now in operation. Gateways have been developed with many other MOD CIS, including JOCS (see separate entry), RAFCCIS (see separate entry) and the Dstl Analysis (ex-DERA) network.

Still in operation in 2003. As the UK Defence Information Infrastucture (DII) develops it will probably be replaced.

Contractor
Fujitsu Services, Basingstoke.

UPDATED

EU Operational HQ CIS

An IT system to support an EU Military Headquarters engaged on operations has been developed and installed at Northwood in the UK at a cost of £1 million, the money reportedly coming from funding ring-fenced in the UK defence budget for EU military developments. The design and implementation of the system was completed in ten weeks. As at June 2003 this was the first operational-level EU military headquarters CIS to be established; other countries in the process of creating similar facilities, to be known as Operational HQs (OHQs), are France, Germany, Italy and Greece.

Originally contracted for 107, subsequently increased by 98, the system consists of 205 PC workstations installed in flexibly configured space in one of the office buildings at Northwood and is entirely independent of all the other CIS on the site. The design is very resilient with cluster servers from Compaq, plus a webserver.. The security level is believed to be "EU Secret". There has also been a heavy investment in 3M wall displays and large plasma panels, together with a multimedia switch. A further 50-position system using Dell laptops has also been provided for a forward headquarters (FHQ) that could be deployed in theatre. This deployable system has identical packaging to the UK's deployable Joint Operational Command System (JOCS) (see separate entry). In addition to this deployable system, there has been provision for an element of the system to be deployed to Brussels in time of crisis for use by the EU Military Staff.

Both the OHQ and FHQ systems are Local Area Networks, the former with a gigabit Ethernet backbone. When linked by communications, which can be by any means providing it has a minimum 64 kbytes capacity, they form a WAN. If the communications are lost, the two will continue to operate as independent LANs and servers are automatically updated once connectivity is restored. If an element was deployed to Brussels, communications would probably be provided via ISDN lines, but this element would not be part of the WAN.

No military applications nor GIS were supplied with the system, which carries Microsoft Windows and Office 2000, and Microsoft Exchange Server. The employment of widely used COTS applications has been deliberate in order to facilitate the use of the system by staff officers required to operate it from a standing start, as they are most likely to be familiar with these applications. The UK has elected to use wholly Microsoft applications, but other countries may use different applications providing they meet the standards agreed by the EU Military Staff CIS Committee; it is suggested that the French are more likely to select Lotus applications.

The intention has been to provide a system separate to those already installed at Northwood Joint Headquarters site for use by EU staff if an EU operation is mounted, allowing the site to host a multinational staff without impinging on the national and NATO facilities already in situ.

Once there are a number of compatible systems established, the concept of use is that any FHQ can then work to any OHQ, depending on who is fulfilling those particular roles. Thus, a UK FHQ could work to a German OHQ, or vice versa; or elements of FHQs from two countries could combine and work to the OHQ of a third.

The system at Northwood has some additional potential value as a Command Post Exercise facility for training and for business continuity.

Status
Operational.

Contractor
EDS Ltd, Hook.

NEW ENTRY

Joint Operations Command System (JOCS)

The Joint Operations Command System (JOCS) provides a federated Command and Control system in British Strategic and Operational Headquarters to support all Joint and Combined Operations. Its hub is at the Permanent Joint Headquarters (PJHQ) at Northwood, North London.

JOCS is a multiterminal, multisite system based on commercially available software packages and hardware. Sites and servers are either static or placed in transit cases for transportation and deployment worldwide to support a single client or a major deployment of 50 to 60 clients. JOCS provides connections throughout the command structure including deployed Joint Force Headquarters (JFHQ) and Component Commanders Headquarters (CCHQs). HMS *Ark Royal, Illustrious, Ocean* and the designated alternate command ships (Type 22 Batch 3 frigates) are also fitted with JOCS installations. The system is also implemented in the Ministry of Defence, London, other government departments (OGD) and other major military headquarters overseas and in UK.

The system is designed to create and manage the Joint Operational Picture (JOP). This provides a common Situation Awareness picture encompassing Intelligence and Operational data common to all users. Users have the ability to associate information within the 'picture' and drill down to view it when required. Each level of command will have its own unique tailored view of the JOP that can be viewed throughout and over the network. This view of the picture can also be viewed via secure web technology. A high proportion of this functionality is provided by COTS products:

(a) Mapping and tracking – managed by ICS, C2PC, which uses digital maps and overlays including Shapefiles. These are provided by UK DMilSurvey, US NIMA, and other Military Geo specialists
(b) Office Automation – Microsoft Office 2000
(c) Secure Messaging – Nexor X400/500 with gateways for ACP 127 and SMTP, Systematic IRIS for AdatP3 formatted messages
(d) Databases – Oracle RDBMS with *Jane's* and substantial access to classified defence databases using Sybase
(e) System Management – CA Unicenter TNG, Sunrise
(f) Interfaces – OED, CHOTS, (giving links to SDAWN and CASH), ISIS, RAFCCIS, RNCSS, FOSCLE, USGCCS, CRONOS, ATacCS, NSTN MHS and other intelligence systems. (See separate entries)
(g) Hardware – Enterprise E4500 with SAN, Compaq DL360, SUN Ultra 60 & 80s and NT workstations throughout
(h) Imagery – GTE Remote View
(i) Conferencing including Audio/Visual MS NetMeeting
(j) Decision and Planning Aids. EDS Structure Tool, Orbat & Task Organisation creator
(k) Secure Intranet – Netscape servers, Browsers and Newsgroups. Verity search engine. With DEFO for Outlook providing secure e-mail utilising MS Exchange Server
(l) Security – 2 domains, web-based data publishing and labelling.

Status
Deliveries of JOCS Stage 1 started in 1996 to replace the EDS supplied Pilot JOCS (PJOCS) system which had supported PJHQ since its inauguration. Stage 3 was due for introduction in mid-2002 and a continuing programme of updates will follow. The system has supported operations in Afghanistan, the Balkans, East Timor, Middle East, Rwanda, Sierra Leone and most recently in Iraq.

Contractor
EDS Defence Ltd, Hook, Hampshire.

UPDATED

Olympus distributed collaborative planning system

The Olympus system is designed to support the planning for complex operations. This is achieved by use of high-resolution displays for the presentation of maps, software with an intuitive gesture-based user interface and applications optimised for collaborative working which allow individual staff cells to contribute to the development of a military plan.

The Olympus software is based on an industry standard geographic information system that can display all common types of map data. Against this backdrop, the user can enter lines, shapes and military symbols with a light pen. The touch-sensitive screen translates simple gestures into the desired symbology; voice commands can also be used to control many Olympus features. Olympus supports multiple layers on which information can be drawn. These are typically used for different operators to build up their own specialist views of the plan, which can then be called up at will, individually or in any combination. The final plans may be platted or electronically distributed ensuring timely, accurate and consistent information ready for action. The live Common Operational Picture can also be displayed.

Olympus can be used with any type of display, but use of a touch-sensitive screen allows the full capabilities to be exploited. Ultra have developed two displays specifically for Olympus. The horizontal Electronic Birdtable is a large area, high-resolution, horizontal computer display that has a built-in pen-operated digitiser. It displays maps at high resolution and has been developed to support larger co-operative group working such as formation command in static land headquarters and afloat. The display measures 1.5 m diagonally (approximately A0 map size) and has a resolution of 2048 × 1536 pixels; its overall dimensions are

UltraScribe in an AFV 436 command vehicle (Ultra Electronics)   0533833

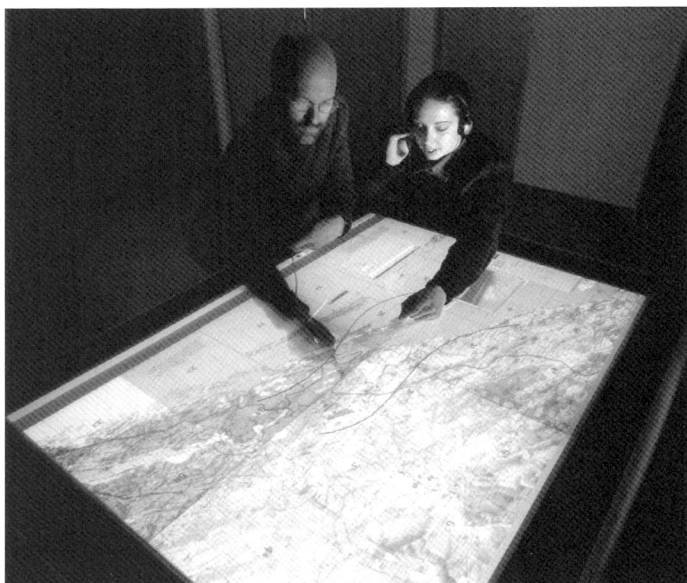

Olympus Electronic Birdtable (Ultra Electronics)   0533834

1.43 × 1.13 × 0.9 m. The smaller UltraScribe is a high resolution LCD with a large viewable area but requiring low power and weight. This workstation enables a single user to prepare data direct to different overlays by drawing directly on to a map and is designed for teams of up to three, such as in staff cells and at the tactical level, including in command vehicles. Overlays can then be shared on the network or called up on the Electronic Birdtable. The display measures 21 in diagonally (equates to a 25 in CRT) and has a resolution of 1600 × 1200 pixels; its overall dimensions are 545 × 460 × 54 mm and it weighs 13 kg.

Status

The Olympus system has been supplied to UK and overseas armed forces. It was used on Exercise Saif Sareea II in 2001 and demonstrated during Joint Warrior Interoperability Demonstration (JWID) 2002 and 2003.

Contractor

Ultra Electronics Command and Control Systems, High Wycombe.

UPDATED

WAH-64 Apache Ground Support System (GSS)

The Apache GSS provides facilities to enable the UK Apache fleet to be both operated and maintained in peace and war. The system consists of the Mission Planning Station (MPS) and the Maintenance Data Station (MDS), together with the Mission Data Preparation Facility (MDPF). Numbers of MDS and MPS can be varied as required; only one MDPF will support a group of deployed MPS as it meets the requirement for map data preparation for all the MPSs. Both MPS and MDS can exchange data electronically with other MPS and MDS either using physical connections or media transfer.

Maintenance Data Station (MDS)

The MDS enables all aircraft maintenance processes and requirements to be electronically tracked, forecasted and recorded, providing a computerised management capability for both 'at-aircraft' maintenance work and 'off-aircraft' support. A complete database is held of each airframe's maintenance history and

Apache Ground Support System (Aerosystems International)   0116326

current maintenance state, together with a record of the necessary maintenance processes to keep the aircraft available for operations and a forecast of actions required. The system can also host the Interactive Electronic Technical Publications View package, enabling a mechanic to draw down video of the maintenance process required and view it before undertaking the work. It is envisaged that in due course this latter capability will be available on palmtop hardware, allowing it to be used at the aircraft.

Mission Planning Station (MPS)

The MPS is designed to enable those planning WAH-64 missions to prepare and transfer data to the aircraft, with a conscious effort to reduce crew workload. The station is capable of supporting 32 aircraft, planning patrols of up to 8 aircraft each. Using maps generated by the MDPF, which can be supplemented by overhead satellite imagery, aircraft crews can plan their mission. The system provides all necessary information on battlefield data, threat warnings, intervisibility, engagement zones, communications details, transponder information and IFF settings. For each mission, the crew can select weapon and fuel loads and conduct 'what-if' planning to achieve optimum weapon load, fuel load and route. The system is based on the full performance model of the WAH-64, but as individual airframe data is collected it will be sensitive to the capabilities and limitations of individual airframes. A limited 3-D capability is provided in the form of a visualisation of the selected route, and the manufacturer is currently developing a more sophisticated fly-through capability. Once planning is complete, mission data is downloaded onto a Data Transfer Cartridge (DTC) for transfer to the aircraft. Each DTC, which has a capacity of 1 Mb, can carry data for 2 missions, but more than one can be carried in the aircraft and they can be changed in mid-flight. DTC will record mission data for post-flight analysis.

The GSS consists of COTS hardware using a Pentium processor hosting Windows NT 4.0, removable hard disc and CD ROM. DRS Technologies have provided much of the hardware, similar to their Genesis Commander workstation (see separate entry), FPR 16 remote colour flat panel display, and some of the ruggedised storage system. Some of the peripherals such as the printer are conventional COTS office equipment contained in ruggedised boxes. A PCU allows both DC and AC power to be used, and batteries permit stand-alone use.

MPS and MDS can be linked within the same location by a LAN. For wider communication, the GSS can communicate via a variety of communications carriers including Ptarmigan. For the future it is intended that communication will be achieved over the BOWMAN communications system, and that GSS will form a significant part of the Aviation Battlefield Information Software Application (BISA) for the UK's Digitisation programme. By integrating an Improved Data Modem with the MPS and using current radio communications, the manufacturer has also recently demonstrated near-realtime transmission of the output from the aircraft's Longbow radar overlain on the map display, together with the exchange of text messages. As originally reported in *Jane's International Defence Review* in June 1999, this capability is likely to improve target allocation and selection, the timely passage of tactical information, and in-flight retasking. Also currently under examination is the development of the system to make it interoperable with the Royal Navy's CSS, to allow full integration of the Apache in the amphibious environment.

In September 2002 the MPS with integrated IDM was involved in the Jenisys 1 trial, which successfully demonstrated links over VHF; beyond line of sight over HF; use of the Extendor payload as a relay; and the transfer of data from MPS over Link 16.

Status

80-90 systems are now in service with the UK Apache programme. 22 MPS are to be delivered for the UK Merlin Mk 3 programme in April 2003, incorporating a new GIS.

Contractor

Aerosystems International Ltd, Yeovil, Somerset.

UPDATED

WaveHawk/BattleHawk

WaveHawk and BattleHawk are tactical information management and display systems which have evolved from the naval Semi-Automatic General Operations Plot (SAGOP) first developed in 1990. Their open architecture and flexible integration capability make them particularly suitable for the upgrade of legacy platforms.

Description

The WaveHawk/BattleHawk systems use a compact ruggedised processor unit with an integral shock and anti-vibration system. The processor is a sealed unit, providing protection against the ingress of moisture, dust and other contaminants. BattleHawk is designed to operate under the severe environmental conditions normal to the fighting vehicle domain. The standard processor uses a 700 MHz CPU with solid-state memory, and 256 Mbytes RAM. The processor design is based on a stack with through-pin connectors suspended within a mechanically isolated support system. The standard stack comprises a power distribution/UPS unit, processor, solid-state memory, interface expansion, and video capture boards.

The systems are designed to be able to interface to multiple sensors and subsystems, automatically capturing as much tactical data as possible to reduce operator workload. The processor in standard format can accommodate eight external interfaces which can be expanded to meet further requirements if necessary. It can also be supplied as CAN-Bus compliant.

WaveHawk uses a number of different display formats ranging from 215 mm through to 1066 mm screen sizes, and 4:3 and 16:9 aspect ratios. The Operator's Display can be provided with touchscreens, and normally have a resolution of 1280 × 1024. BattleHawk display formats range from 215 mm for reconnaissance platforms through to 520 mm screen size for Command Vehicles, and all are provided with touchscreen capability, are daylight readable, and are provided with day/night settings.

The BattleHawk command terminal showing keyboard and trackball (WA Systems)
0528211

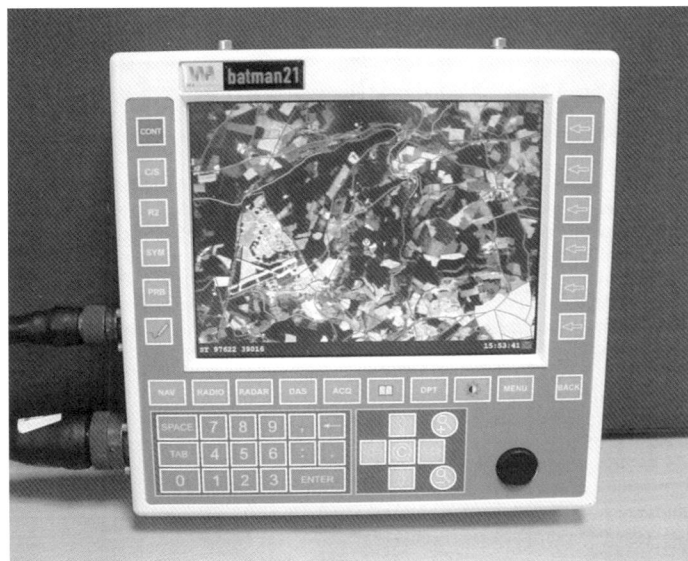

The BattleHawk UDT displaying aerial imagery (WA Systems)   0528201

WaveHawk is interfaced to the navigation subsystem as standard, and can additionally be interfaced to radar, sonar, datalinks, command and control and other ship's systems. If required, the system can be linked to other naval and land platforms and can transfer data via the message manager. Battlehawk is interfaced to the vehicle navigation subsystem as standard (INS and/or GPS), and can additionally be interfaced to EW, Sights, FCS, and other platform systems. It has a dedicated tactical communications module to aid report generation and communication management.

WaveHawk is provided with custom keypads, a tracker ball and keyboard to suit specific installation requirements. BattleHawk is provided with a custom human-computer interface, specific to each installation to aid intuitive operation, reduce training requirements and operator workload.

Development

SAGOP was originally developed to fill a perceived gap between information management systems then in service in 1990 and the advent of CSS (see separate entry). Emerging from support provided by WA Systems to the Fleet Operational Analysis Staff and the Maritime Warfare Centre, it provided a method of displaying the tactical picture, with tracks input either by datalink or by manual update. Developed specifically for the RN, the software was carried on COTS hardware. Twenty three systems were produced, which were principally installed in those platforms which were surface command capable, and it was particularly welcome in options with older equipment. Systems were moved between platforms as required and some were networked. Some still remain in service.

From 1995 onwards a commercial variant was developed with redesigned software and hardware to produce a fully ruggedised naval system which was marketed as SAGOP 2000. By 1999 this had evolved into a product which was offered with single or multiple displays of varying sizes and in a variety of hardware combinations. The HCI was designed to be very flexible, providing touchscreen facilities as well as trackball and keyboard.

WaveHawk

In 2000/2001 it was decided to incorporate the Electronic Chart Display Information System (ECDIS) in order to use marine charts as a backdrop to the

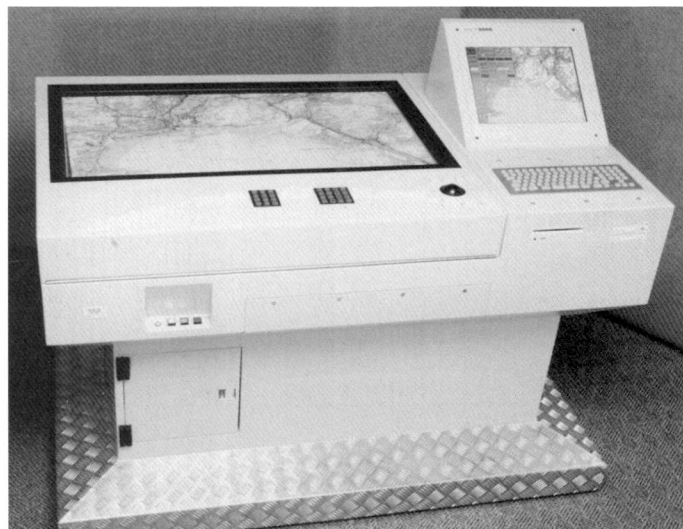

WaveHawk in the 'table' configuration, displaying a land map. (WA Systems)
0528204

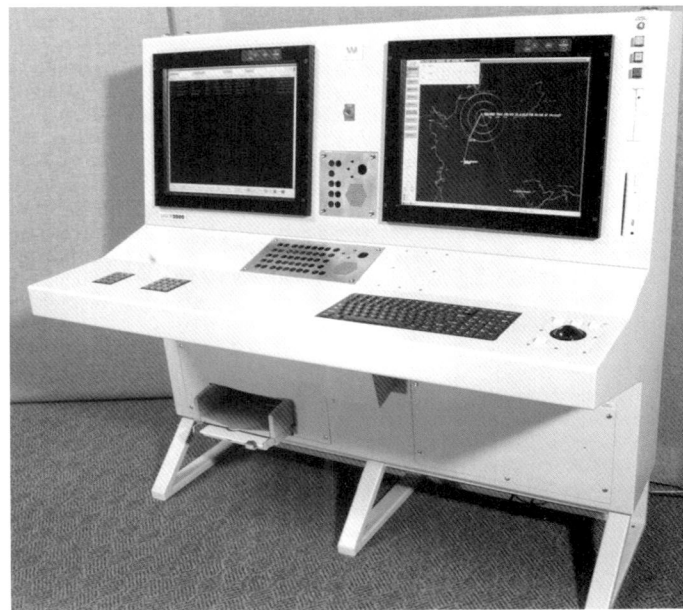

A dual display WaveHawk configuration (WA Systems)   0528207

tactical display. The system supports both raster (including UK Hydrographic Office ARCS) and vector (including S57 data and Additional Military Layers(AML)) images. The AML facility enables additional marine information to be displayed, such as salinity layers or beach profiles, which can be turned on and off as required. For littoral operations satellite imagery and land map images can also be incorporated. The ECDIS facility meets the necessary requirements for navigational use, and its integrity is maintained by using a separate processor for the navigational information. The system can display tactical information without this background, operating as a simple plotting table, or the addition of the tactical information to ECDIS provides a Warship ECDIS (WECDIS), offering a command decision and planning aid. The system has a number of serial interface ports that can be configured to accept track data from the platform subsystems, together with that provided by external datalink.

BattleHawk

WA Systems provided an early technology demonstrator on the Sultan and Scimitar CVR(T) platforms in 1999 of a Battle Management System (BMS) drawing on the technologies developed for their naval systems. This linked the organic vehicle sensors to the system and demonstrated data transfer between vehicles. Based on this trial, User Data Terminals (UDT) were designed specifically for use in CVs, concentrating on the needs of the operator to ensure that under battlefield conditions the user would be able to make the maximum use of the system's capabilities. Considerable emphasis was placed on touchscreen and hot key facilities to reduce operator load. Subsequently the BMS software was redesigned. The system was originally marketed as Batman 21, and subsequently renamed BattleHawk to demonstrate its commonality with WaveHawk. The core information handling techniques are identical, enabling the two versions to work together.

The system is intended for use at the tactical level. Each vehicle obtains Platform Derived Information (PDI) from the navigation, turret and targeting subsystems to produce a local tactical picture, generating reports and returns semi-automatically for transmission. Communication can be over radio, satellite and landline. Interfaces are available to both GPS and inertial navigation systems, and navigation data can be further processed by the system. Vehicles are linked together to provide a common tactical picture, with information flowing both up from individual vehicles and down from the command level.

Status

SAGOP remains in use in some RN platforms. SAGOP 2000 has been sold to unspecified Asian navies. A WECDIS system similar to WaveHawk is being fitted to the 'Wielingen' class frigates of the Belgian Navy as part of their update programme. WaveHawk is being strongly marketed into new NATO entrant countries.

In December 2001 *International Defense Review* reported that BattleHawk was being considered for adoption by the Polish Army. It too is being strongly marketed into new NATO entrant countries.

Contractor

WA Systems, Exeter.

NEW ENTRY

United States

Attack and Launch Early Reporting to Theater (ALERT)

The ALERT system was developed by Aerojet (now part of Northrop Grumman) under the direction of the US Air Force Space and Missile Systems Center as Talon Shield. ALERT is a ground-based processing station using satellite data to provide improved detection, identification and tracking of ballistic missiles in support of theatre missile defence. In addition to providing early warning to areas under attack, ALERT provides more accurate locations for targeting launch systems or cueing radars associated with interceptors.

The heart of the ALERT system is the Central Tactical Processing Element (CTPE), located at the National Test facility. CTPE is composed of COTS telemetry hardware and ONYX computers processing data from all the Defense Support Program (DSP) satellites simultaneously and performing multiple sensor tracking of tactical ballistic missiles. DSP has been has been a segment of NORAD's Tactical Warning and Attack Assessment program since 1970. The satellites use infra-red detectors to sense heat from missile plumes against the earth's background – see separate entry.

Status

ALERT is operated by the 11th Space Warning Squadron at Falcon Air Force Base in Colorado on a 24-hour basis, and is supported by Aerojet's Azusa and Colorado Springs facilities. Currently using Defense Support Program (DSP) satellites, ALERT will perform the same real-time tactical applications for future space-based sensor systems. Since 1998 the ALERT system has provided a back-up capability to USSPACECOM for launch detection messages for strategic missile warning.

Contractor

Northrop Grumman Electronic Systems, Azusa, California.

VERIFIED

Automated Deep Operations Coordination System (ADOCS)

The Automated Deep Operations Coordination System (ADOCS) is a joint mission management software application which originated as a Defense Research Projects Agency (DARPA) programme. It provides a suite of tools and interfaces for horizontal and vertical integration across battlespace functional areas. The maritime variant of ADOCS, the Land Attack Warfare System (LAWS), is the baseline for the Naval Fires Control System (NFCS). ADOCS is also a major segment of the intelligence application package for Theatre Battle Management Core System (see separate entry) functionality at wing and squadron level (TBMCS-NT).

The key integration functions within ADOCS are:
(a) The Counterfire Common Operational Picture (CF-COP) provides a near-realtime picture of the artillery battle. It allocates tube and rocket counter-battery resources for more efficient counterfire operations through digital integration from Joint/Combined level down to tactical firing units. CF-COP also includes munitions allocation and status.
(b) Joint Battlespace Management provides the capability to assess the impact of surface fires on airspace activity, thus enabling improved co-ordination between air and ground component commanders. Kill box management tools enable the operator to integrate ISR data and task offensive resources, thus improving the timing of strikes to coincide with enemy movements into and out of named areas of attack. Airspace deconfliction capabilities provide co-ordination at Joint and Combined levels to reduce the threat to air missions from friendly fire.
(c) The Coalition Coordination and Integration function assists the integration of coalition artillery for both the counterfire battle and other surface fires missions.
(d) Air Interdiction (AI) Planning and Execution assists in the effective employment of AI assets through the timely and improved information flow for the identification, assignment and nomination of AI targets. Critical air resources can be allocated in a more efficient manner through early assessment of potential and planned missions. This facility provides the ability to monitor the Integrated Tasking Order/Air Tasking Order (ITO/ATO) execution through all phases and provides immediate visibility into AI nominations throughout the targeting process, including 8- and 4-hour updates, thus enabling AI missions to be tuned and fires maximised.
(e) The Fire Support Coordination Measures Analysis function provides a means for assessing changes and movements of the Fire Support Co-ordination Line (FSCL) on current and planned missions in the ITO/ATO. It provides immediate visibility of targets exposed or covered by movements in the FSCL. It offers users the opportunity to assess the consequences of FSCL movement prior to commitment.
(f) The Battlespace Visualisation function uses tools that provide visualisation of co-ordination measures, ingress and egress routes, and air defence threats. It also enables the user to visualise friendly fire in 3-D space over any area. Battlespace geometries can also be overlaid with imagery and terrain data to improve situation awareness and planning.

Status

ADOCS was developed as part of the Joint Precision Strike Demonstration programme and emerged from the Theatre Precision Strike Operations Advanced Concept Technology Demonstration in 1998 as a 6-year programme funded with US$91 million. ADOCS is in widespread use in all US unified commands and in different guises in all four US services. It was used in support of operations in both Afghanistan and Iraq. The manufacturer claims it has more than 5,000 users across the US services.

Manufacturer

General Dynamics C4 Systems, Arlington, Virginia.

NEW ENTRY

Command Data Network System (CDNS)

The Command Data Network System (CDNS) BMS (BMS) is the core of a turnkey command and control system developed by General Dynamics. It provides nearly real-time situation awareness and command and control functions to support commanders, staffs and support personnel. CDNS BMS can provide a homogeneous, common command and control system that extends across organisational boundaries from division to the individual soldier and combat platform level. The software requires no legacy system modification to exchange selected information with other military services and agencies. It can operate on a wide variety of hardware and platforms including PDAs, laptops and various combat vehicles and is carried on a wide range of operating systems including Windows®, Unix® and Linux®. The look and feel of commercial applications is maintained throughout. In addition, CDNS BMS automatically networks over heterogeneous communication equipment and systems. A Micro-Inter-Networking Controller (MINC) provides the necessary interface to 'virtually all' fielded Combat Net Radio (CNR) systems to allow proper messaging. The CDNS BMS, combined with the appropriate modem, provides communication between non-homogeneous communication networks including:
(a) Low-bandwidth Combat Net Radios Networks (VHF, UHF, HF)
(b) Local Area Networks (LAN) — Wired/ Wireless
(c) Serial Connections

(d) Satellite Communications (SATCOM)
(e) Public Service Telephone Networks (PSTN)
(f) Field Telephone (trunk) Systems.

CDNS BMS is designed to support interfacing to existing (legacy) systems such as ASAS, AFATDS and MCS from the US Army Battle Command system (see separate entries). Once the physical connection for the interface is defined, a small interface software package that 'speaks the same language' as the legacy system is added to the CDNS BMS suite. Once the interface is started, the legacy system thinks it is communicating with one of its '"peers' and no modifications are therefore necessary to the existing legacy system.

Messaging

The CDNS BMS messaging function provides a suite of tools for composing, storing and sending a variety of message types. It supports the sending and receiving of both text and graphics over the available tactical communication media. Messaging support includes user ID, creating/editing, transmitting/addressing and receiving /storing messages. CDNS BMS does not need a prior knowledge of IP addresses for tactical units but uses a 'network discovery' process to establish connectivity. In addition, the software can support US MTF, ADatP3 and JVMF message formats. User ID support in CDNS BMS uses operational tactical unit symbology and 'natural naming' for operational users. It also provides message templates grouped by several operational categories including:

(a) Reports
(b) Requests
(c) Alerts
(d) Plans
(e) Orders
(f) COA

Each template supports attaching a file or object of either text or graphics and uses automatic selector fields as appropriate to minimise keyboard entry by the user. Frequently used templates can also be assigned to a workstation function key for rapid access. In addition, an XML toolset will support the generation of user-defined templates.

The system enables the user also to create all required text and graphics on the workstation and then transmit the entire plan or order (text and graphic overlays) to all addressees with a single "transmit" command. Several aids are included to speed up the report composition process: invalid entry protection is provided for report fields to reduce data entry errors; report location fields can be filled by pasting grid co-ordinates from the tactical situation display to a message field; many reports provide an automatic roll-up capability. Additionally, the CDNS BMS provides a message forwarding capability that enables users to choose the types of reports or requests to be automatically forwarded as they are received.

All CDNS BMS message formats provide an addressing feature that includes a list of net members to which the user's workstation is connected. The user can address the message to any of the addresses identified on the list. Choosing multiple destinations will cause CDNS BMS to multicast the object to the destinations selected. The address book feature provides a compiled list of all net members that have ever been in communications allowing the user to create distribution lists. Users can transmit a message by either selecting Send on a message form or by pasting an object onto the tactical icon of an available net member or distribution list. Each received message increments the 'unread message' counter at the receiving station and is visible to the user in the messages queues. An audible alarm can be set to notify the user that a message has been received as well as a visual alarm display for Flash precedence messages. Messages are time stamped according to the receiving station's clock and are stored in the program's database. Additionally, users can archive specific user-defined folders to the computer's hard drive for later review and retrieval, or to various storage devices for transport.

Situational awareness and display

The CDNS BMS tactical situation display (TSD) is a graphical display with an intuitive interface which provides a selection of graphic and tactical display aids that allows the user to tailor the display in real time. The display itself is organised so that all key functions are available at all times via the keyboard or pointing device.

All menus, submenus, and templates have a common 'look and feel'. CDNS BMS also has a full set of tools for the creation and modification of operational graphics and provides a library of preformatted message templates. Some message formats can automatically add symbology, such as enemy contacts or alerts, and overlays to the TSD. Map tools allow users to configure the map display to their needs with features such as scaling, dragging, centring, 3-D rotation, and heading orientation. CDNS BMS provides the facility for loading map data from a CD ROM in ADRG, CADRG, ASRP, or DTED formats. An optional ESRI GIS tool interface is available that enables use of many vector formats. CDNS BMS supports the use of APP 6A military symbols.

The CDNS BMS receives and stores position reports and automatically displays friendly locations. It also supports the situation awareness requirements for:

(a) Receiving, consolidating, and distributing position reports
(b) Converting and displaying position locations
(c) Maintaining and displaying tracks and track history
(d) Using the PLGR as a navigational aid.

The CDNS BMS, based on initialisation criteria entered by users, automatically collects position location information from an internal GPS or external PLGR. The software then converts this information into an Own Station Position Report (OSPR) and transmits it as a broadcast message to all users on its network. An OSPR can be generated based on time, distance travelled, degrees turned, proximity to a particular location/unit, or with every transmitted message from the reporting user. A Consolidated Position Report (CPR) provides a summary of all reported OSPRs from subordinate Task Org /ORBAT units. A CPR can be generated either manually or based on a time criteria established by the user at the echelon generating the CPR. The CDNS BMS converts the information contained in an OSPR or CPR to the appropriate military symbol for the unit reporting and displays that symbol at the proper location on the Friendly Unit overlay. In addition to the symbol, the date and time of the latest report may also be displayed. Individual unit locations or a centre of mass can be selected for display. A tabular list, the LOCSTATBOARD, of the most recent position reports of all units on the TSD is also maintained. The CDNS BMS also provides tools for the user to employ the information available from GPS for navigational purposes. In addition to being able to obtain an accurate current location, users can establish a route to a particular destination by entering up to 30 way points into a preformatted template. Distance and bearing information from a user's current location to a selected way point or another user's location can be displayed on the TSD. Users may also use the CDNS BMS to query the PLGR to provide a rate of progress from a given start point toward a selected way point measured over time.

Interoperability and Application Program Interfaces (API)

The purpose of the Application Program Interfaces (APIs) is to provide 'additional application software' access to the services provided by the system. CDNS BMS supports program-to-program interfacing via the mechanisms of:

(a) Component Object Model (COM)
(b) Distributed Component Object Model (DCOM)
(c) Open Data Base Connectivity (ODBC)

In addition, the software provides an application with the ability to post to the tactical map and to display the map centred around user-specified locations. User applications can also provide a CD ROM containing the map data over which the object is displayed. User applications can use CDNS BMS as a 'bearer service' to pass data messages for delivery. An additional software application can request the location file held by the LOCSTAT facility. The ODBC mechanism will be defined in the API to accomplish this. Location can be requested in any of the Grid Reference systems supported, which currently include DMS geographic, Decimal geographic, MGRS and UTM, and the system has supported the loading of theatre-specific, country-unique co-ordinate systems for particular countries/regions. CDNS BMS will also provide, to a requesting application, the current date and time and the facility for setting UDT calendar/clock at all workstations on a net.

Networking support

CDNS BMS allows users to establish and maintain a network composed of a variety of communications media. The software provides the networking between command posts, within CP, and to subordinate units of the force using existing field communications equipment. Networks are established through self-discovery. To leave a network (de-affiliate) in an orderly way, the exiting workstation tells the other net members its intention by 'saying goodbye'. Affiliation and de-affiliation of net members is achieved by automatic monitoring of the network traffic and modifying the net members list in response.

Several methods of sending data to single destinations or multiple destinations are supported. Any object in the system can be sent to other net members. The Comm Check feature allows the user to repeat the network discovery process to re-establish or confirm data communications. Additionally, the software provides a network topology chart that allows users to see the net members and their connectivity by net name. From this chart, the user can monitor path statistics.

The Comm Sit shows network connectivity on a georeferenced display. All communications within CDNS BMS are acknowledged by a short message that is returned to the sender. Intra-networking allows a net member to act as a relay point if communications are broken due to conditions or terrain as well as providing the capability of transferring messages from one network to a second network.

CDNS BMS architecture

The CDNS BMS architecture is layered and modular. The architecture provides loosely coupled layers that isolate the command level application layer from the networking and communications infrastructure which helps the system to be readily customised. The Command Overlay (CO) portion provides CDNS BMS user services. The Network System Suite (NSS) portion provides the CDNS BMS APIs necessary to implement the C2 application level objects defined in the CO. All major functionality is accessible to external applications via an industry standard COM interface. All databases in the system are accessible via an ODBC interface. CDNS BMS is written in C++.

Specifications

(a) Software

Internationalisation (UNICODE support, right-to-left languages like Arabic and Hebrew) multiple language support: English, French, Spanish, Chinese (Mandarin), Arabic
Instant messaging (chat)
Object-oriented technology
Standard Windows® features
Direct object manipulation
Runs on UNIX®, Windows®, and Linux®
Engine-based architecture
MODL object development language tool set
Tailorable user interface
Automatic message format translation
Fixed and function keys and touch panel
Message and journal search tools
Role-based access controls
HTTP interface (application becomes a web server)
Axion Spatial GIS interface (intervisibility, high-speed map rotation)

(b) Networking features
Network routing by discovery
No network planning
Any organisational configuration
Transparent multimedia networking
Natural name addressing
Dynamic load balancing
Distributed access to communications
Overhearing for net use minimisation
Internetting, Intranetting, Multinetting
Automatic voice-data contention (voice priority and net sharing)
Automatic voice and data on KY-57 networks
Automatic expansion, contraction

(c) Communications equipment supported
PRC-37A
PRC-77 VHF
RT-524 VHF
Clansman series: UK-RT 353, UK-RT 352 and UK-RT 351
PRC-840 VHF
RT-F200 (Raven) VHF
RT-F500 Wagtail (BAE)
RT-F700 Pintail (Racal)
NGR TADIRAN VHF
SINCGARS VHF R/T family
MBITR
RF-5000 Falcon (SM) series
RF-5800 Falcon (SM) II series
VRC-12 Series RJS (VHF)
PRC-117 D/F (Harris)
VRC-94F
SATCOM: Iridium, LBAND - MTS, LST-5, MST 20, PSC-5(B) (Spitfire) and WSC-3
Encryption devices: KY-99A Narrow-band HF R/T RT-F100 (Raven) and SATCOM, KG-84C / 5710
F modem / HF R/T, KIV-7 / HF modem HF R/T and KY-57 / HF R/T
ANDVT / KW5 HF R/T (PRC-104)
PRC-138 embedded modem HF R/T (RT-5020 or 5022)
MDM-3001 HF modem HF R/T
DB-independent query tools (decoupled from SQL)

(d) Platforms supported
Dismounted soldier – Rugged PDA Hand-held Pocket PC (Windows CE)
Combat vehicle,– ruggedised Laptop UNIX®, Windows®, and Linux®
TOC – desktops, laptops, ruggedised workstations

(e) Mapping system
Native Electronic Maps (EMAP)
Images: TIFF, BPM, JPEG, ADRG, CADRG
Spatial: DTED
OpenGL 3-D visualisation of DTED data, with emaps draped over terrain
Mercator, Universal Transverse Mercator (UTM), Universal Polar Stereographic (UPS), Modified Mercator, Equirectangular and Albers
ESRI MapObjects 2 Component
ASRP, ADRG, CADRG, CIB, CRP, GeoTIFF, GIF, JFIF (JPEG), MrSID, NITF, SVF, USRP, TIIFF, SUN, ERDAS, IMPELL, BIL, BIP, BSQ
Vector format: CAD, VPF, Arc Shapefile, ArcInfo Coverages, SDE layers, Grid Data, Attribute Tables
2-D and 3-D image mapping

Status
CDNS is used in a wide variety of BMS, including battleWEB (Canada); BCSS (Australia); the New Zealand Battle Laboratory; an unspecified SE Asian nation (understood to be Singapore); possibly in the Po-Sheng programme in Taiwan; a prototype SOF system in Jordan; in a high-level system in Venezuela. It is the BMS in the Commander's Digital Assistant, an interim spiral development in the US Land Warrior programme: mounted in a ruggedised Ipaq this was issued in two Brigades of 82 (AB) Div down to squad leader level for operations in Iraq.

A customised version provides the Common Battlefield Applications Toolset (ComBAT) and the Bowman Situation Awareness Module (BSAM) which provide the core battlefield information-system application (BISA) into which other functional BISAs will be integrated to form a unified battle command system for the UK Bowman project(see separate entry).

Manufacturer
General Dynamics C4 Systems, Fort Wayne, Indiana.

UPDATED

Defense Information Infrastructure (DII)

In 1992, the US DoD recognised that advances in technology and the associated change in the military environment required a new approach to information systems. The DII is not a single programme, but a capability resulting from the integration of individual information management programs across the DoD intended to:
(a) revolutionise information exchange defence-wide
(b) apply computing, communications, and information management capabilities effectively to the accomplishment of DoD's mission

(c) significantly reduce the information technology burdens on operational and functional staffs
(d) enable the operational and functional staffs to access, share, and exchange information worldwide with minimal knowledge of communication and computing technologies.
The DII is the web of communications networks, computers, software, databases, applications, weapon system interfaces, data, security services, and other services that meet the information processing and transport needs of DoD users, across the range of military operations. It encompasses:
(a) sustaining base, tactical, DoD-wide information systems and C4I interfaces to weapons systems
(b) the physical facilities used to collect, distribute, store, process, and display voice, data, and imagery
(c) the applications and data engineering tools, methods, and processes to build and maintain the software for Command and Control (C2), Intelligence, Surveillance, Reconnaissance and Mission Support users to access and manipulate, organise, and digest proliferating quantities of information
(d) the standards and protocols that facilitate interconnection and interoperation among networks
(e) the people and assets which provide the integrating design, management and operation of the DII, develop the applications and services, construct the facilities, and train others in DII capabilities and use.
The DII includes the information infrastructure of the Office of the Secretary of Defense (OSD), the military departments, the Chairman of the Joint Chiefs of Staff (CJCS), the defense agencies and the combatant commands. The DII includes information infrastructure regardless of its role or location, whether it is part of the enterprise infrastructure, the sustaining base, deployed, or afloat. The information interfaces to industry, government, academia and allies are also within the scope of the DII as are weapons systems interfaces to the DII. The DII is the responsibility of the US Defense Information Systems Agency (DISA).

Communications and computer infrastructure
The communications and computer infrastructure of the DII provides information processing and transport services. It includes the Defense Information System Network (DISN) for information transport; the Defense Megacenters for information system processing; the DII Control Centers that manage the DII network and systems; and Base and Deployed/Afloat Communications and Computer assets. Together, these elements form DoD's end-to-end capability for information distribution, processing, storage, and display. The DII Control Centers, operated co-operatively by DISA, the military services and defense agencies provide global, regional, and local control centers to manage the communications and computer infrastructure

Common applications
Common applications provide cross-functional, cross-organisation capabilities for personal and organisational messaging through the Defence Message System (DMS), and support electronic commerce through Electronic Commerce/ Electronic Data Interchange (EC/EDI). The DII Common Operating Environment (COE) (see separate entry) provides for integrated common support services, a corresponding software development environment for functional applications, and enables execution and integration of joint and military service mission applications. The Shared Data Environment (SHADE) supports interoperability of functional area applications at the data level among military services and functional areas as needed to conduct DoD's mission.

Defense Information System Network (DISN)
The DISN is the backbone worldwide communications network that will provide a full range of government controlled and secure information transfer services. It will extend to all areas of the globe and will exchange voice, data and imagery. The DISN infrastructure consists of a Continental US (CONUS) segment, including sustaining bases, European and Pacific theatre segments, a space segment and a deployable capability. DISN is the primary carrier for all DoD services, including the Defense Message System (DMS), GCSS and electronic data exchange/ commerce. (See separate entry.)

Defense Message System (DMS)
The DMS (see separate entry) is replacing the existing US AUTODIN with a worldwide, secure messaging system. It utilises international X400 messaging and X500 directory service standards using DISN as the carrier. In addition, DMS makes full use of COTS software for the front end.

UPDATED

Defense Information Infrastructure Common Operating Environment (DII COE)

The Common Operating Environment (COE) provides the foundation for all Defense Information Infrastructure (DII) (see separate entry) system architectures to enable operational realisation of the Command, Control, Communications, Computers and Intelligence for the Warfighter (C4IFTW) vision. The COE is an integration approach that enables rapid application integration, a point and click installation, and fast turnaround. It is also a product that provides 'pluggable', reusable, architecturally consistent, integrated components called segments. It is a process for distributing engineering across the US Department of Defense (DoD) supporting the construction of systems from components developed by comparable organisations.

Interoperability is the ability to share the same data in a consistent fashion, applying identical business rules to achieve coherent knowledge in a distributed environment. Achieving system interoperability is inherently difficult and costly, and the COE approach reduces both. However, achieving the benefits of the COE requires the willingness to impose, and accept, constraints for the sake of integration and interoperability. The COE approach to interoperability provides a single infrastructure for Joint and Service-specific systems, constraining system development efforts according to the level of integration required. The COE achieves interoperability primarily by using common software, predominantly commercial-off-the-shelf (COTS) products but also government-off-the-shelf (GOTS) software, together with common system engineering practices and some degree of data standardisation. The COE programme office manages a repository for publicly accessible data that is defined via the extensible Markup Language (XML). The repository is not intended to provide central control of the definitions, but to ensure that definitions are unique within a namespace and visible to all programmes that might benefit from their use.

The COE is a three-tiered architecture. The first layer, the Kernel, contains the operating system, operating system extensions, a common desktop, software installation tools and security extensions. The second layer, Infrastructure Services, includes the relational database, worldwide web, network and system management, communications and print services components. The third layer, Common Support Applications, includes mapping, correlation/fusion, collaboration, Common Operational Picture (COP) and enterprise resource planning components.

Shared Data Engineering (SHADE) is also included in the COE approach for interoperability through:
(a) a common modular data server architecture for COE segments and tools
(b) a common representation with common battlespace objects and the COE Data Emporium
(c) data interchange via XML
(d) data access services using data sharing techniques.
    Major systems developed within the COE include (see separate entries):
(a) Global Command and Control System (GCCS) and Global Combat Support System (GCSS)
(b) Army Battle Command System (ABCS) and Global Command and Control System - Army (GCCS-A)
(c) Global Command and Control System - Maritime (GCCS-M)
(d) Theatre Battle Management Core Systems (TBMCS)

More than 100 other systems and programmes across the US DoD, the Coast Guard, and the US Customs Service are also currently fielded or planned for use. Every major Service C4I system currently under development uses the same set of application programming interfaces (APIs). Future directions for the COE include real-time extensions, command and control personal computer (C2PC) functionality; a web-enabled visualisation environment; and 'componentised' COP Server internals for track management, synchronisation and correlation.

VERIFIED

Department Of Defense Intelligence Information System (DODIIS) Automated Message Handling System (AMHS)

The DODIIS AMHS provides improved automated message handling capabilities to the US intelligence community, and replaces the Modular Architecture for the eXchange of Intelligence (MAXI). It provides connectivity to and interoperability with other government agencies, tactical users, allies, defence contractors and other approved users.

AMHS (Automatic Message Handling System) processes AUTODIN message traffic and wire services, providing soft copy message generation, co-ordination, and release; content indexing, online message storage; and retrospective search capabilities. Survivability is provided by a distributed workstation architecture and an industry standard LAN, interconnecting the system nodes to provide flexibility and redundancy. The currently fielded version also supports remote users in a low bandwidth environment. Maximum use is made of menu-driven, user friendly software for both the generation and distribution of messages. The AMHS directly supports four external communication links: Communications Support Processor (CSP), the AGT Gateguard, wire services (AP, UPI, Reuters), and Foreign Broadcast Information System (FBIS). The AMHS also receives SMTP messages directly from the site LAN. The AMHS receives wire service traffic from three wire services: Associated Press (AP), United Press International (UPI), and Reuters. The wire service communication links are read only and are received via an asynchronous RS-232 line using American Newspaper Publishers Association (ANPA) message coding protocol. The AMHS also receives FBIS wire service messages via a read only link. DODIIS AMHS is fully certified, and accredited for System High at TS/SCI Level at 28 operational sites, providing discretionary access control, special processing for restricted access messages, audit trails, privilege/ login control and account locks.

The latest version of the AMHS may have been renamed the Multimedia Message Manager (M3).

Contractor
Boeing, St Louis, Missouri.

UPDATED

Dynamic Airspace Management System (DAMS)

The Dynamic Airspace Management System (DAMS) is designed for the management of airspace in dense environment where all services may be deploying aircraft, artillery, missiles, and air defence.It has been specifically designed to address the issues of deconfliction and fusion of missions. By providing a true three-dimensional (3-D)/ time varying model of the airspace, DAMS allows users to visualise and utilise airspace more effectively.

Automated viewing tools, such as 3-D view controls, elevation, and time filters allow users to interpret more easily densely populated airspace. Automatic conflict detection ensures that routes, orbits, and other Airspace Control Measures will not be assigned in a manner that could cause potential conflicts.

DAMS provides additional tools both to produce and automatically import United States Message Text Formatted Airspace Control Orders (USMTF ACO) for display and deconfliction,. The display of near-realtime track data, continuously rendered 3-D displays and interfaces to other C3I systems allow DAMS to be used as a tool for both planning and execution.

System features
(a) 3-D Viewing of Airspace Control Measures, near-realtime air tracks, and maps - zoom, pan, rotate
(b) automatic conflict detection
(c) near-realtime track display
(d) USMTF Airspace Control Order - automatic generation, automatic parsing utilities
(e) filters - altitude, class, object, time
(f) maps - world vector shoreline, worldwide database II, ARC digitised raster graphics, digital terrain elevation data
(g) online user documentation.

COTS standards
(a) Solaris, SunOS
(b) X-Windows/Motif
(c) Sun Common Desktop Environment (CDE)
(d) OpenGL
(e) C, C++
(f) National Imagery and Mapping Agency (NIMA) map products.

Status
DAMS has been sold commercially since 1992. It was deployed for use in both Bosnia and Haiti, and continues to be used for Joint and Combined exercises. Users include the US Air Force, US Army, and NATO. DAMS is currently an integral subsystem within the US Army as part of the Tactical Airspace Integration System (TAIS) (see separate entry), which is claimed by its manufacturer, General Dynamics, as the most advanced battlefield airspace management system in the US DoD.

Contractor
Raytheon Systems Company, Lexington, Massachusetts.

VERIFIED

Global Command and Control System (GCCS)

The Defense Information Systems Agency's (DISA) Global Command and Control System (GCCS) provides a single strategic joint command and control (C2) system for US Forces. GCCS assists Combatant Commanders and Joint Task Force (JTF) commanders in the maintenance of dominant battlefield awareness through a fused, integrated, near-realtime picture of the battlespace. GCCS provides information processing support in the areas of planning, mobility and sustainment to combatant commanders, the Services, and Defense agencies. It has replaced the World-Wide Military Command and Control System (WWMCCS). As a C4I system, GCCS includes multiple workstations co-operating in a distributed LAN/ WAN environment. Key features include 'push/pull' data exchange, data processing, sensor fusion, dynamic situation display, analysis and briefing support, and maintenance of a common tactical picture among distributed GCCS sites. GCCS is already fielded at a number of operational CINCs.

GCCS is composed of several mission applications built to a single common operating environment networked to support sharing, displaying and passing of information and databases. The GCCS infrastructure consists of a client server environment incorporating UNIX-based servers and client terminals as well as personal computer (PC) X-terminal workstations, operating on a standardised Local Area Network (LAN). The GCCS infrastructure supports a communications capability providing data transfer facilities among workstations and servers. The Secret Internet Protocol Router NETwork (SIPRNET), the secret layer of the Defense Information Systems Network (DISN) provides connectivity between GCCS sites. Remote user access is also supported via dial-in communications servers, or via telnet from remote SIPRNET nodes.

The baseline GCCS architecture consists of a suite of relational database and application servers. At most GCCS sites, the relational database server acts as a typical file server by hosting user accounts, user specific data, and site specific files not part of GCCS. The application servers host the automated message handling system, applications not loaded on the database server and other databases.

At each GCCS site, one application server is configured as the Executive Manager (EM) providing LAN desktop services. It also hosts applications not loaded on the database server. The EM server acts as the user interface providing access to GCCS applications through user identification and discrete passwords. GCCS software applications are categorised into two groups: Common Operating Environment (COE), and Mission applications.

GCCS mission applications

Joint Operation Planning and Execution System (JOPES)

The Joint Operation Planning and Execution System (JOPES) is the integrated command and control system used to plan and execute joint military operations. It is a combination of joint policies, procedures, personnel, training and a reporting structure supported by automated data processing on GCCS. The capabilities of the JOPES mission applications support translation of the National Command Authority's policy decisions into planning and execution of joint military operations. JOPES applications include:

(a) Requirements Development and Analysis (RDA) creates, analyses and edits Time Phased Force and Deployment Data (TPFDD)

(b) Scheduling and Movement (S&M) handles command and control information on deployment activity and status. It functions as a vehicle for the scheduling and tracking movement of TPFDD requirements

(c) Logistics Sustainment Analysis and Feasibility Estimator (LOGSAFE) assists logistics planners in determining sustained movement requirements during deliberate and crisis action planning

(d) Joint Flow and Analysis System for Transportation (JFAST) is an analysis tool that provides users with the ability to determine transportation feasibility of an Operation Plan (OPLAN) or Course of Action (COA)

(e) Joint Engineer Planning and Execution System (JEPES) provides planners with a method to determine requirements and/or adequacy of engineering support provided in OPLANs or COAs

(f) Medical Planning and Execution System (MEPES) provides contingency medical support information for allocating medical resources

(g) Non Unit Personnel Generator (NPG) functions are to assist in determining quantities of replacement and filler personnel

(h) Ad Hoc Query (AHQ) provides users with a means to develop, save, and print tailored queries extracting data from the JOPES core database

(i) Systems Support functions as the JOPES core database management subsystem for functional managers

(j) Airfields is an information retrieval application providing the user with the capability to access, extract, and print information from the Automated Air Facilities Information File database.

Global Reconnaissance Information System (GRIS)

GRIS supports the planning and scheduling of monthly Sensitive Reconnaissance Operations (SRO) theatre requests. The Joint Staff staffs these requests through the office of the Secretary of Defense, Central Intelligence Agency, and State Department for National Security Council approval. Incoming RECON 1/2/3/4 formatted messages are received by an automated message handling system, validated, and passed to the GRIS application for automated processing and database update. GRIS generates all RECON messages and also monitors the monthly execution of theatre reconnaissance missions approved in the previous month. GRIS is used by the Joint Staff and theatre commands exercising operational control (OPCON) over airborne reconnaissance assets.

Evacuation System (EVAC)

EVAC collects and displays information about US citizens located outside the United States as collected by US State Department embassies and consulates. It accesses the database server via TELNET operation from a GCCS-compatible client.

Fuel Resources Analysis System (FRAS)

FRAS provides fuel planners an automated capability for determining supportability of a deliberate or crisis action plan and for generating the time-phased bulk petroleum, oil and lubricants required to support an OPLAN. FRAS facilitates review of the fuel requirements of a proposed, new, or revised OPLAN and assesses adequacy of available resources to support crisis action planning. Requirements can be generated and analysis performed for the overall OPLAN, regions within the OPLAN, by Service and within Service by regions. Two or more OPLANs can be combined into a single OPLAN for analysis. The requirements generated can be varied through the use of intensity tables and consumption data extracted from the Logistics Factors File (LFF) or with Service-provided data system.

Global Combat Support System (GCSS). See separate entry.

GCCS and GCSS both need the Defense Information System Network (DISN) and the Defense Message System (DMS) (see separate entries). GCSS will rely on all components of the DISN for information transport services including voice, text, and imagery. DMS provides a secure, reliable, and accountable writer-to-reader messaging infrastructure at reduced cost.

UPDATED

....

Grenadier Beyond line of sight Reporting And Tracking (BRAT)

Grenadier Beyond line of sight Reporting And Tracking (BRAT) is a lightweight blue force tracking device employing the Global Positioning System (GPS), a low probability of intercept/detection waveform, and other national capabilities to provide a continuous, near-realtime, beyond line of sight tracking capability for critical assets.

The primary components of the Grenadier BRAT system include the transponder, a Hand-Held Terminal (HHT), a small (approximately 3.5 in) UHF transmit antenna and a GPS receive antenna. A planning computer is also used to create, manage and load the Grenadier BRAT system files prior to a mission. The transponder is small and light enough to be used in either man-packed, military vehicle or aircraft configurations. The man-packed option gives dismounted patrols the ability to report their location up the command chain. In this configuration the system uses a separate rechargeable battery pack; but when used in vehicles or aircraft the Grenadier BRAT can operate from vehicle or aircraft power.

The system works by first calculating positional information through the signal that it receives from GPS satellites. After location has been determined, Grenadier BRAT transmits its GPS position, along with unit identification and a brevity code via a special waveform. The waveform has very low probability of intercept and low probability of detection, and is also encrypted. It is generally indistinguishable from radio background noise, so it provides the user with security and reduced risk of exposure. To increase security further the GB transmission uses spread spectrum transmission.

Grenadier BRAT uses existing collection and dissemination architectures to provide its data to commanders, which can be combined with red-force data to provide a near-realtime view of both friendly and enemy forces on the battlefield. GB data can be displayed as part of a unit's Common Operational Picture on a variety of platforms, including GCCS-A, MCS and ASAS-RWS (see separate entries).

Status

The US Army procured an initial 400 Grenadier BRAT devices in 2001 as part of a Warfighter Rapid Acquisition Program (WRAP), providing some of them to US Special Forces involved in operations in Afghanistan, and exercised an option to buy 400 more in 2002 in response to further needs identified for these operations. From late 2001 the US Army Space Program Office worked with the US Army Europe(USAREUR) and the Aviation Engineering Directorate in Huntsville, Alabama, on an Air Worthiness Release (AWR) for the Grenadier BRAT system on the AH-64D Apache Longbow aircraft. By mid-2002 the AWR had been completed, certifying the simultaneous use of Grenadier BRAT and the Tactical Engagement Simulation System (TESS) onboard the AH-64D, which will allow pilots to conduct AH-64 training (via TESS) while tracking their aircraft on real world C3I systems. The initial fielding of Grenadier BRAT to USAREUR's 6th Squadron, 6th Cavalry Regiment (Apache Longbow) began in July 2002.

Contractor

US Army Space Program Office, Alexandria, Virginia.

NEW ENTRY

....

Joint Defensive Planner (JDP)

The Joint Defensive Planner (JDP) is a Theatre Battle Management Core Systems (TBMCS) mission application for preparing and evaluating Theatre Air and Missile Defence (TAMD) plans. Defence planners can import current friendly and enemy orders of battle from external data sources and review them using a map-based Graphical User Interface (GUI). Its web centric design allows deliberate and crisis action planning to be conducted in a collaborative environment. It provides the Area Air Defence Commander (AADC) (whether ashore or afloat) and supporting units with an extensive array of automated decision aids for counter-air, counter-missile, and anti-air warfare planning. These aids include: prioritisation of critical asset protection; analysis of defensive capabilities against differing threats; quantifying risks; generation of XML-based air defence plans and orders. Planners can use JDP's analytic weapon and sensor coverage utilities or a force on force simulation to evaluate alternative air and missile defence plans.

Status

Operational testing of JDP V.2 was completed in February 2002, and it was accepted for incorporation as an application in TBMCS. It is likely to be fielded in TBMCS v1.1 and is being tested for fielding in GCCS (see separate entry). It was used in the NATO Exercises Joint Project OPTIC WINDMILL 01 and 02, and ROVING SANDS 01.

Contractor

Northrop Grumman Integrated Systems, Arlington, Virginia.

NEW ENTRY

....

Joint Mission Planning System (JMPS)

The Joint Mission Planning System (JMPS) is the replacement for all US military unit mission-planning systems, providing an integrated planning capability for aircraft, weapon and sensor missions and intended for both fixed- and rotary-wing aircraft and UAVs, and ground vehicles in the future. JMPS comprises COTS, GOTS, and newly developed software. It uses software components as basic building blocks for the system, the JMPS framework providing the entire component management and infrastructure support needed to build such a component-based system. Mission application components can plug into the framework to create the functionality required for planning specific missions. The system interfaces with the Defense Information Infrastructure Common Operating Environment (DII COE), and JMPS will be certified and accredited to handle system-high operations at the Top Secret level and above. A server walkaway facility enables JMPS to be used on a laptop that can be disconnected and later reconnected to a network of JMPS planning stations. Upon reconnection, all changes will update the network data sources selected by the user.

Capabilities
(a) Task View. This provides the mission planner with a way to parse the air tasking order (ATO).
(b) Route Collaboration. The Route Collaboration tool allows the sharing of mission planning data and functions with other JMPS users in real time.
(c) Administrative Functions. These functions make the system uniform for groups of users. The system can default to the look and feel needed by each branch of the military. Preferences can also be set up for individual users.
(d) MIDB Interface from GCCS-M. The system interfaces with GCCS-M (see separate entry) for threat information from the Modernised Intelligence Database. This information feeds into the JMPS Threat Server to provide with near-realtime threat information.
(e) PTW. For US Navy aircraft planning, JMPS has an interface to the Precision Targeting Workstation. Through this interface, target co-ordinates and imagery can be included in the JMPS mission planning activities. JMPS also interfaces with the Strike Planning Folder system, which lets the strike planning organisation on an aircraft carrier access, view, and use information associated with the strike planning process.
(f) CAPS. The Combat Airdrop Planning System operates within JMPS to provide mission planning services.
(g) TOLD. Take-off and Landing Data provides standard departures and arrivals.
(h) Moving Map. JMPS will function as a moving map display during mission execution by plugging in a GPS receiver. While in moving map mode, the system records the flight path, which can be saved for later display during mission debriefs.
(i) Maps, Charts, and Imagery. The system can display any data published by the National Imagery & Mapping Agency (NIMA).
The JMPS provides a number of features for the operator:
(a) Multiple Windows. As many planning windows as there is memory available can be used to superimpose mission data on maps and imagery. The View Selector can set these windows to different views such as three-dimensional (3-D), 2-D, 2-D animated, tabular flight plans, or as briefing charts.
(b) Explorer. The Explorer provides click-and-drag access to all mission data and information which can be dragged onto any (or all) of the mission planning windows.
(c) Route Editor. The Route Editor creates a route or a series of points along a path and allows specific events to be inserted, such as observing a point or launching a weapon.
(d) Route Calc. The Route Calc tool automates fuel flows and flight times. The route calculations are based on one or more flight performance models (FPMs), which are provided for each type of vehicle.

Minimum system specifications
256 Mbytes of RAM
Pentium III CPU at 400 MHz or better
Microsoft Windows 2000 with Office 2000
The standard JMPS application runs under DII COE version 4.4

Status
An initial US$54 million investment split 50/50 USN/USAF will provide JMPS initial operational capability by March 2004. This will be V1, giving a general mission planning capability. A further US$20 million USN investment will provide additional combat capability including PGM planning and enables migration of TAMPS capabilities and other single platform MPS, such as those for EA-6B and AV-8B. Full operational capability is expected by 2006. Jane's sources indicate that JMPS will be trialled by the UK in 2003 for use in a future rotorcraft, possibly SABR. JMPS is also likely to be the MPS for JSF.

Contractor
Northrop Grumman Information Technology , San Pedro, California.

NEW ENTRY

....

Joint Services Workstation (JSWS)

The Joint Services Workstation (JSWS) is a single operator, transportable, reduced footprint, dismounted workstation variant of the Common Ground Station (CGS) (see separate entry), using the same hardware and software and providing the same functionality as the CGS. The system provides an Army/Air Force sensor and attack control capability to locate, track, classify and assist in attacking moving and stationary targets beyond the Forward Line of Troops (FLOT). It is a real-time, multisensor command, control, communications, computers, and intelligence(C4I) system which provides capabilities that support real-time surveillance; reconnaissance; situation awareness; target development; theatre missile defence; and battlefield visualisation.

The JSWS acquires, processes, displays and disseminates data from multiple real-time sensors including Moving Target Indicator (MTI)/Synthetic Aperture Radars (SARs); Unmanned Aerial Vehicles (UAVs); Imagery Intelligence (IMINT) platforms; Signal Intelligence (SIGINT); and Electronic Intelligence (ELINT). Specific sensor platforms include: JSTARS, E-8; Guardrail; Rivet Joint; ASAS; AFATDS; simulation; advanced imagery; Predator UAV; Outrider UAV; Hunter UAV; ARL; U2-R (ETRAC); other ground stations; TBMCS; ADSI; and AODA. The system includes a suite of communications equipment for secure radio, satellite, secure phone/fax, SIPRNET and LAN/WAN networks.

Data processing and management capabilities include real-time sensor processing; real-time geo-reference of data, maps, images and symbols; multi-image processing and exploitation; database management; and image product library. Fixed-site monitoring and surveillance are accomplished by drawing Areas of Interest and setting configuration options in a pop-up panel. Video and frame imagery windows are displayed in the requested area.

The JSWS has a scalable, open system architecture which provides a core foundation for the JSWS and which can be applied to a variety of systems requiring real-time interfaces, data processing, data management, geocritical information processing, and HMI. The architecture is scalable to provide a migration path for system capability upgrades; growth in processing power, storage, and interfaces; and simplified support and maintenance. Hardware is commercial computer servers, workstations, networking and industry standard interfaces.

The JSWS software is based on a real-time open-system architecture derived from the Common Ground Station using modules, or engines, that provide an object oriented design approach for applications development. The system software provides: a set of core components for C4I systems (processing engines for interfaces, data, graphics database, HMI); a set of Application Programming Interfaces (APIs) to simplify and accelerate software development; and portability across major Unix platforms.

Status
In August 2000 the US Army CECOM awarded a US $49.7 million JSWS production contract. The JSWS is deployed by all US services worldwide. According to the supporting data for the FY2002 US Army Budget Submission to Congress "CGS/JSWS has repeatedly provided high-value targeting and intelligence data to Field Commanders during contingencies (for example Operation Joint Endeavor), as well as during standard mission operations of fielded units."

Originally offered by Motorola's Integrated Information Systems Group. This Motorola unit was acquired by General Dynamics in October 2001.

Contractor
General Dynamics Decision Systems, Scottsdale, Arizona.

UPDATED

....

Joint Tactical Ground Station (JTAGS)

JTAGS provides timely and tailored reporting on tactical ballistic missile launches via in-theatre stereo ground processing of infra-red data from Defense Support Program (DSP) sensors. In doing so, the system supports the theatre missile defence architecture. In addition, space-based warning data and other tactical parameters are provided to end-users via existing communications networks. The ability of JTAGS to process and disseminate data within the theatre significantly reduces the risk of single-point failures, provides improved reporting timelines and minimises the loading on high-priority communications links between Continental USA (CONUS) and the theatre user. Dissemination of data is usually over UHF satcom links.

JTAGS is housed in a standard military 8 × 8 × 20 ft (2.44 × 2.44 × 6.01 m) shelter equipped with a standard mobiliser. It requires minimal military personnel to operate, is transportable by C-141 aircraft and can be operational within hours of deployment. Each unit has three portable 8 ft dish antennas. For redundancy during contingency situations, the system will deploy in pairs. During crisis situations, the system will conduct joint operations.

In order to reduce cost and accelerate fielding, JTAGS uses Commercial Off-The-Shelf (COTS) hardware with minor modifications to enhance transportability and deployment options, together with government off-the-shelf equipment.

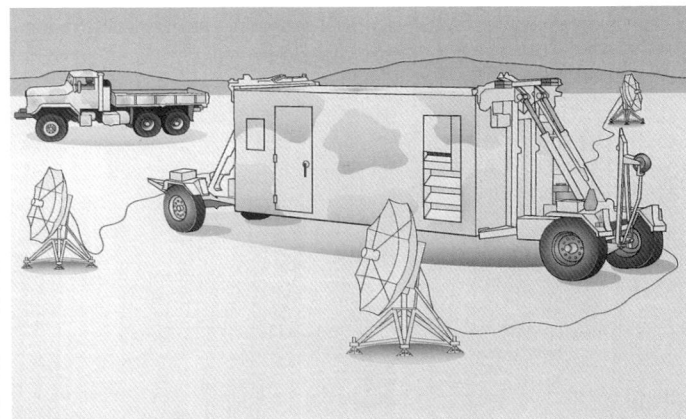

Artist's impression of JTAGS in theatre

The Joint Tactical Ground Station (Northrop Grumman)   0111711

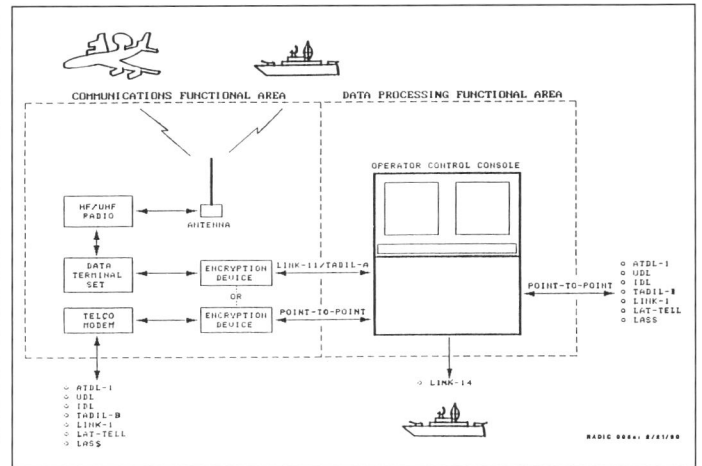
Operational block diagram of RADIC

The technical feasibility of JTAGS was validated by the US Army's Tactical Surveillance Demonstration (TSD) proof-of-principle prototype developed by Aerojet. The fixed-site system was successfully tested at White Sands missile range.

Status
In October 1994, the US Army awarded Aerojet an Engineering, Manufacturing and Development (EMID) contract to field the operational JTAGS. In 1997 five production units were fielded to the European Command (Germany), the Pacific Command (Korea) and the Army Space Command Continental United States. In the same year Phase I of the preplanned upgrades was commenced, providing Joint Tactical Information Distribution System (JTIDS) integration and data fusion with other sensors. This was completed in 2001. Work on Phase II upgrades to permit operation with SBIRS and DSP satellites and to provide coverage for both army theatre and air force strategic missions commenced in 1999 and will continue, with completion expected in 2005. The system will then be known as the Mobile MultiMission Processor (M3P).

Contractor
Northrop Grumman, Electronic Systems Division, Azusa, California.

UPDATED

Datalink protocols supported by the RADIC system include TADIL-A/-B/-C ATDL-1, Link-I/-4/-11/-14, MDBL, JSS, UDL, 412L, International DataLink (IDL) and SAGE. Standard interfaces available include MIL-STD-1397, MIL-STD-1553B and MIL-STD-188C.

Status
A version of the system is being used with the NATO ship-shore-ship buffer programme. The RADIC performs as a buffer and translator link between ground-based NATO Air Defence Ground Environment (NADGE) control centres and naval shipborne command and control systems to provide an extended common air picture for the NATO countries.

An associated development, the RADIL (ROCC/ AWACS Digital Information Link), was used by the US Air Force to buffer and translate tactical datalink information between E-3A aircraft using TADIL-A networks and joint surveillance system ground-based radar tactical information. RADILs have been provided for the Icelandic Command and Control Enhancement and the Saudi Arabian Peace Shield programmes.

Manufacturer
EDO Communications and Countermeasures Systems, Simi Valley, California.

VERIFIED

Rapidly Deployable Integrated Command and control (RADIC) system

The RADIC system interfaces to a variety of radar, sensor, communications and weapon systems to provide regional tactical situation monitoring and control.

The system has the capability to interface with adjacent and higher echelon C3 systems. RADIC systems can be installed in small vehicle-mounted shelters or employed in a fixed-site configuration. They can operate in various modes ranging from completely passive to a totally active role. The former is for higher level command, retaining complete operational control of air defence assets and group/ battalion level systems and providing identification, status and reference data. In the active role, RADIC provides group/low-level operational control of a designated sector receiving identification and reference data.

A typical system can include one or more display modules for operational functions, a data processing module, a peripheral equipment module, a communications or radio module and other C3 and/or simulation modules.

RADIC operator control console

Tactical Control System (TCS)

The Tactical Control System (TCS) is the software, software-related hardware and extra ground support hardware necessary for the control of the Advanced Concept Technology Demonstration (ACTD), Outrider Tactical Unmanned Aerial Vehicle (TUAV), the RQ-1A Predator Medium Altitude Endurance (MAE), Unmanned Aerial Vehicle (UAV) and future tactical UAVs. A planned Engineering Change Proposal (ECP) will provide TCS interoperability with the RQ-2A Pioneer tactical UAV. The objective is that TCS will have the capability of receiving Global Hawk and Dark Star High Altitude Endurance (HAE) UAV payload information. In addition, TCS incorporates the technical interfaces necessary for the dissemination of UAV imagery and data to 24 selected joint and Service C4I systems. TCS software is compliant with Joint Tactical Architecture (JTA), Common Imagery Ground/ Surface System (CIGSS), and Defense Information Infrastructure Common Operating Environment (DII/COE).

TCS software operates on current US service hardware: Sun/SPARC (Air Force), CHS-II/SPARC-20 (Army/Marine Corps), and TAC-N (Navy). The Air Force incorporates TCS software and selected components into existing RQ-1A GCSs. For the Army, TCS is an integral part of the High-Mobility MultiWheeled Vehicle (HMMWV)-based Outrider TUAV GCSs and Tactical Operations Centres (TOCs) at various echelons of command. For the Marine Corps, TCS is an integral part of the HMMWV-based TUAV GCSs and TOCs at various echelons of command. For the Navy, TCS is the control system for UAV operations from ships, submarines, and temporary sites ashore. Standard TCS interfaces to joint and Service C4I systems ensures UAV compatibility with fielded combat systems and supports connectivity to lower command echelons.

The TCS consists of six subsystems:
(a) line of sight antenna assembly
(b) integrated data terminal
(c) datalink control module
(d) computer
(e) synthetic aperture radar subsystem
(f) workstation.

The TCS is designed to provide a scaleable and modular capability to operate UAVs on existing computer systems and future C4I processing systems. It provides the flexibility to increase or decrease the systems operational capability by adding or removing electronic cards. This allows the TCS to be configured to meet the user's deployment or operational limitations. TCS scaleability permits the system to function at five discrete levels of TCS-to-UAV interaction ranging from receipt and transmission of secondary imagery and data, to full control and operation of a

UAV including take-off and landing. (Each level of TCS capability incorporates the functionality of all lower levels). The five levels of interaction include:

(a) level 1 receipt and transmission of secondary imagery or data
(b) level 2 receipt of imagery or data directly from the UAV
(c) level 3 control of the UAV payload
(d) level 4 control of the UAV, less takeoff and landing
(e) level 5 full function and control of the UAV to include take-off and landing

The TCS system operates primarily by internally and externally interfacing specific hardware and software to achieve the functionality of a specific configuration. Internal interfaces support interaction between the various hardware and software components within the TCS core and its associated subsystems. External interfaces involve inputs and outputs between the TCS system and supporting equipment. This functionality is available in four configurations:

(a) land-based (LB) HMMWV-shelter TCS production version
(b) ship-based (SB) TCS production version
(c) RQ-1A (Predator) system ground control station (GCS) version
(d) RQ-2A (Pioneer) shelter version.

Status

In November 1998, the team led by Raytheon Systems Company (RSC) was awarded the Tactical Control System Integration contract. Other members of the RSC team are EG&G and Aerodyne. In April 2001 level 5 interaction was achieved at Fort Huachuca, Arizona. In August 2001 Raytheon received a contract for US$11.2 million for further development of the software. A demonstration system was tested with a US Army Shadow 600 Tactical UAV in November 2001.

Contractors

Raytheon Systems Company (Prime), Falls Church, Virginia.

UPDATED

LAND

Australia

Battlefield Command Support System (BCSS)

Saab Systems Land Systems development started with AUSTACSS - a highly integrated suite of software and hardware designed to support Australian Army operations in wartime. The system integrated battlemap information, military messaging, office automation, a database, graphics and specific military applications into a single system. This evolved into the Battlefield Command Support System (BCSS) which through an evolutionary acquisition migrated to PC based hardware, Windows NT operating system and many commercial components integrated with the core system.

BCSS is a second-generation Land command support system that incorporates the functionality of the AUSTACSS system but also provides a solid platform for the integration of additional capabilities. It consists of rugged and portable PC workstations and servers, various peripherals, devices and bridges arranged into local and wide area networks. The topology is based on workgroups that roughly equate to command posts or elements within a command post. The operating system is Microsoft Windows. The system integrates battle map information, military messaging, office automation (using MS Office), a database, graphics and specific military applications into a single system. The environment incorporates commercially available products where possible and integrates them. Applications are developed within this integrated environment-making use of the features of the commercial packages, resulting in, from a user perspective, a seamless browser style interface. The Intelligence and GIS packages were developed by ADI Ltd, in the latter case using ESRI's ArcView software (see separate entry).

The system has been optimised for the needs of Brigade HQs, Battalion HQs and lower command levels and is now addressing the individual soldiers with Hand-Held Terminals.

To aid operational flexibility, the system can be dynamically constructed and reconfigured at short notice. Workstations can operate independently or as part of a networked workgroup or collection of workgroups. Communication on a LAN basis is done via Ethernet and fibre optic (FDDI) connection and WANs. Combat Net Radio is used for mobile units.

The primary BCSS functions are:
(a) Situational Awareness including GPS based positioning
(b) Terrain Analysis functions, both 2- and 3- dimensional
(c) Military Communications, with structured message formats over a variety of in service high and low capacity links
(d) Engineering Support functions
(e) Operational Planning tools for the generation of plans, orders and briefing materials
(f) Logistic Support and Capability Status tools to calculate logistics requirements and record and highlight the status of all units under command
(g) Intelligence databases and tools for analysis, recording and reporting.
Standards used include:
(a) NA TO APP6A symbology
(b) IRIS mission file
(c) USMTF, ASMTF and VMF Message formats
(d) ESRI compatible vector and raster map data
(e) SQL compliance
(f) Ethemet, MIL-STD-188-114 (CNR), serial (EIA-232), NMEA & PLGR GPS.

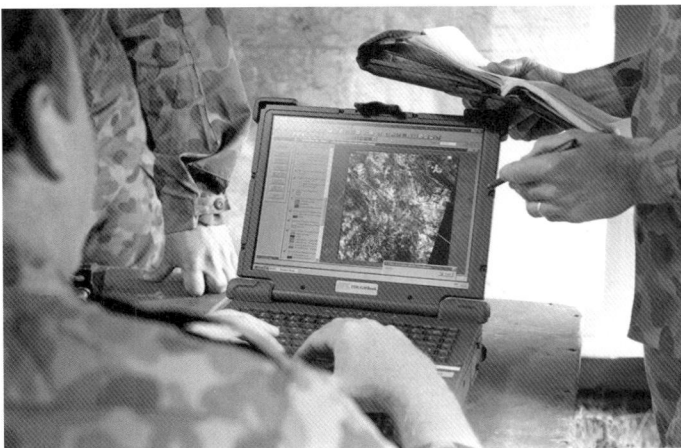

SAAB BCSS (SAAB Systems)   0116329

Status

BCSS was first issued to the Australian Army as a brigade-and-below battle management system in 1998, and is now in service to section level with armoured and mechanised elements of the 1st and 3rd Bdes, located at Darwin and Townsville respectively. It was used in support of operations in East Timor. In early 2001 the system was reported as being trialled by the New Zealand Army as a contender for the land component within its future JCCCS.

In mid 2003 it was reported that the latest software release (v6.1) was deployed from Divisional to Company level.

Contractor

SAAB Systems Pty, Mawson Lakes.

UPDATED

....

Very low-level air defence weapon Alerting and Cueing System (VACS)

VACS is an air defence alerting and weapons cueing solution. It contains sensor hardware which is integrated with system, display and communication software for cueing, alerting and messaging.

The system consists of a radar sensor, command post system and up to 12 weapons systems linked by Combat Net Radios (CNRs). A simulator is also provided for training purposes.

The VACS system:
(a) Detects and tracks with a high probability of detection of all short-range air-to-ground defence threats
(b) Distinguishes between fixed-wing and rotary-wing threats and identifies the specific type of rotary-wing threat (for example, Apache, Puma, UH-1) from its rotor blade signature
(c) Distributes and transmits target data to command post centres and weapons detachments
(d) Operates in varied climatic and geographic conditions.
(e) Has quick response times.
(f) Operates in all battlefield conditions and environments including night operations, jamming and threats from NBC exposure.
(g) Has comparable tactical mobility to the supported force
(h) Supports sustained, long-term operations.
(i) Has system components (for example radars, C3 systems) that can operate in an autonomous mode of operation or as part of an integrated network of systems.
(j) Has an integral IFF that can detect, confirm, classify and attain IFF status on every target in the battle space or as part of an integrated network of systems.
(k) Has reliable communications under high bit error rates and jamming conditions.

The field surveillance sensor used in VACS is the Lockheed-Martin PSTAR Surveillance Radar. This radar unit has been extensively tested by the US Army, Singapore and an unnamed European country. The US Army tests over a 15 month period included operational, technical, environmental and logistic support tests. Following these extensive tests, PSTAR was certified for integration into the US Army forces inventory and received the official nomenclature of AN/PPQ-2.

The Command Post System (CPS) can be located up to 500 m from the radar. It has an integral GPS and displays the tactical picture to the weapons commander and allows him to:
(a) Control the radar and set inhibit sectors
(b) Set up and display the locations of the weapons batteries and the vital point, minimum risk routes, safe lanes, unmask ranges and engagement boundaries
(c) Enter the weapon serviceability status
(d) Automatically allocate targets to the weapon batteries with user selectable threat assessment weightings
(e) Create, edit, store, receive and transmit text messages and standard pro formas
(f) Print the set up data, messages and unmask boundaries
(g) Run self-test diagnostics
(h) Record the operations for future playback and analysis.

The Target Data System (TDS) displays the tactical picture centred on the weapon location. It interfaces to the weapon stand and provides cueing tones to the operator to enable rapid acquisition of the allocated target. One radar together with a CPS can service up to 12 TDSs. The TDS allows the operator to:
(a) Set and display the locations of the target sensor, command post, weapons batteries, vital point, minimum risk routes and safe lanes

---

For details of the latest updates to *Jane's C4I Systems* online and to discover the additional information available exclusively to online subscribers please visit
jc4i.janes.com

(b) Create, edit, store, receive and transmit text messages and standard pro formas
(c) Manually select targets
(d) Run self-test diagnostics
(e) Rapidly send a fall to launch message to the command post.

The simulator allows the VACS functions to be simulated indoors and in close training areas without the requirement for the radar and aircraft. An instructor can define the number, type, IFF status, flight path, speed at each waypoint and detection time of the aircraft, as well as the direction and duration of jamming. Once set up, exercise scenarios can be stored for later use. It is also possible to train on real data recorded by the CPS.

Status

VACS was trialed in 1999, and was delivered to 16 Air Defence Regiment in 2001.

Contractor

BAE Systems Australia, Sydney, New South Wales.

VERIFIED

Canada

Athene Tactical System

The Athene Tactical System (ATS) adds to Iris a scalable, integrated network of computers with specific software applications designed to shorten Commanders' decision cycles. The decision cycle consists of the functional elements required to plan, monitor and direct troops during operations in the field. The ATS includes an upgrade to the Iris vehicular Local Area Network (LAN), the computers and software to assist Headquarters Staff to accomplish these high-level, generic functions faster, more accurately and with greater efficiency. Physically, Athene can be a single, stand-alone ATS Workstation or an integrated network of ATS Workstations, which provide an automated system to support all aspects of conducting Land Force operations in the field. The ATS makes use of Ethernet and supports client-server applications. It provides interoperability through both ADat P-3 messages and a QIP/MIP gateway. It is a vertically integrated solution comprising:

(a) The C2-specific software applications (a Canadianised adaptation of Thales France's Système pour le Commandement des Forces (SICF - see separate entry) product used by the French Army) that are used to store, retrieve, process and display tactical information
(b) A series of standard commercial-off-the-shelf office automation tools (Microsoft Office and Microsoft Project) that are used for document preparation, spreadsheet preparation and manipulation.
(c) A data communications software package based on, and interfaced with, the Iris Communication System's Tactical Message Handling System (TMHS) (a NEXOR-based X.400 electronic mail service with military extensions) (see separate entry)
(d) Militarised computer workstations composed of a Pentium III-based rugged computer unit (running the Microsoft WINDOWS NT4 operating system) a 20″ flat panel display and a full-sized keyboard with integrated pointing device
(e) An Athene Ethernet Unit (AEU) for each ATS-equipped vehicle that serves as both a three-port Fast Ethernet hub for the workstations and peripheral devices in the vehicle (for example, each ATS-equipped vehicle has up to three ATS

workstations or peripheral devices connected as a Fast Ethernet LAN within the vehicle), and a bridge between the ATS Fast Ethernet LAN within the vehicle and the Iris Communication System network for communication between vehicles
(f) Modifications to the Iris Communication System networking software to enable connection of the ATS components to the Iris network and ATS use of Iris as its bearer communication service.

Status

357 ATS workstations will be installed in 242 vehicles to provide automated command support tools for brigade and unit level commanders and their staffs. Scheduled to be delivered in early 2002, it was reported as having been trialled successfully, mounted in a BISON APC Command Post, in a Multilateral Interoperability Programme in late 2001 in Germany, and in use in 2 (CA) Mech Bde in March 2002.

Contractor

General Dynamics Canada , Calgary, Alberta.

UPDATED

Denmark

Danish Army Command and Control Information System (DACCIS)

The Danish Army Command and Control Information System (DACCIS) complies with the standards in the NATO Land Command and Control Information Exchange Data Model (LC2IEDM - formerly known as the ATCCIS data model). This model has been developed by several NATO countries as a Common Army Tactical Command and Control Information System and includes its own replication mechanism (ARM). The LC2IEDM is integrated into the NATO Multi-lateral Interoperability Programme (MIP). Using COTS products and real-time information sharing DACCIS provides a common land picture integrated with functions covering most aspects of combat, combat support and combat service support, and is intended to support all major C2 functions from corps to battalion level. It

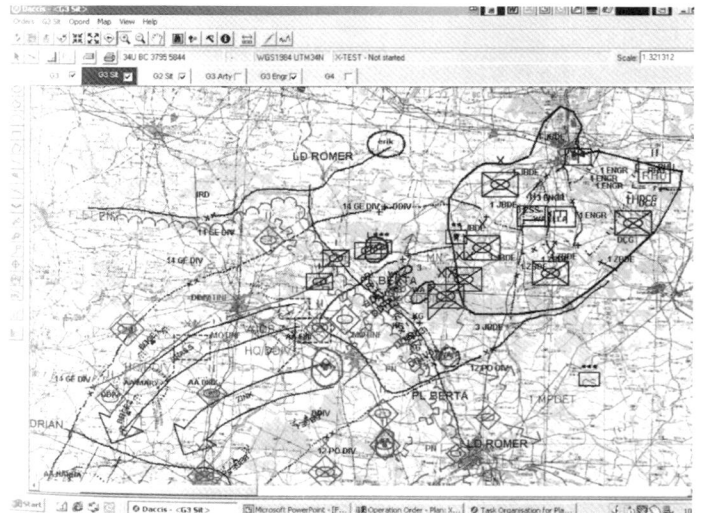

DACCIS Screenshot (Maersk Data Defence)   0533830

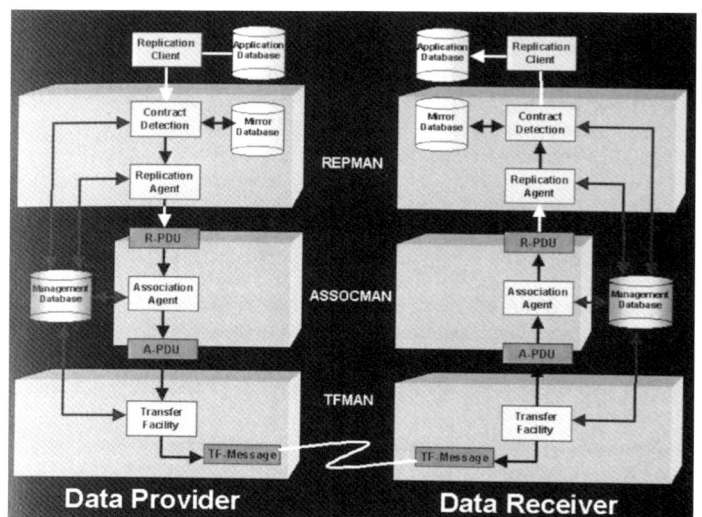

DACCIS Replication Architecture (Maersk Data Defence)   0533829

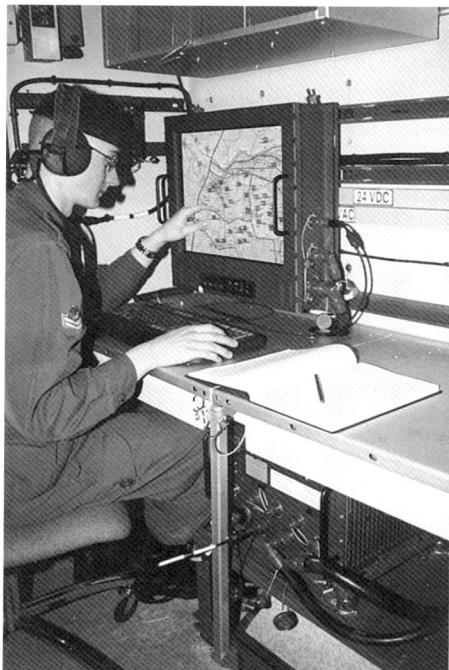

Athene Tactical System (General Dynamics Canada)
0120230

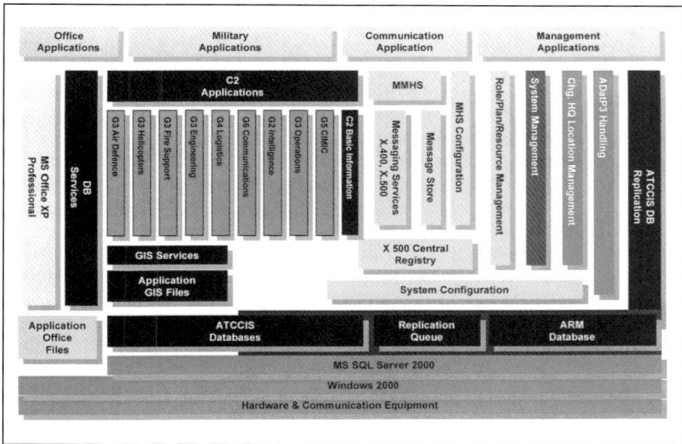

DACCIS System Architecture (Maersk Data Defence)   0533831

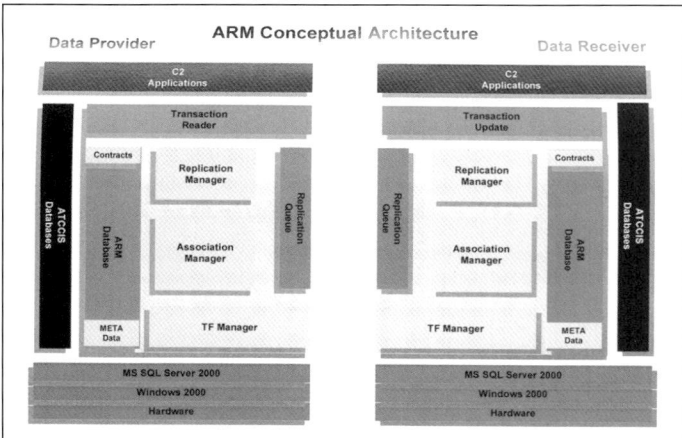

The DACCIS conceptual ARM architecture, showing the 3 software 'managers' (Maersk Data Defence)   0533832

does not include a communications system of its own, and will be carried over the current Marconi-provided Danish system which is reported as being upgraded to provide additional bandwidth. The standard software includes Microsoft Windows 2000 Server/Workstation, Microsoft SQL Server 2000 and Microsoft Office XP, with GIS based on ESRI's ArcGIS. DACCIS also includes message formats such as ADatP-3 and ADatP-32 in order to allow data distribution across systems and national boundaries.

Functionalities within the system include:
(a) Decision making and planning support
(b) Current operations
(c) Intelligence reports and assessments
(d) Fire support
(e) Air defence
(f) Engineers
(g) Aviation
(h) Air space management
(i) Logistics
(j) Communications
(k) CIMIC
(l) Integration to other systems. Version 2 is intended to be integrated with NBC-Analysis and ADAMS (see separate entries), and with an unspecified UAV system.

Real-time information sharing is provided through the ATCCIS Replication Mechanism (ARM). This provides for the replication of databases with staff officers able to access local copies of data while communications are disrupted and automatically ensures that databases are updated following a period of lost communications, such as when HQs are moving. This mechanism will also ensure interoperability with other nations who are using systems based on the LC2IEDM. ARM can manage replication of data to different recipients simultaneously and independently. Information exchange contracts are established between users to establish the flow of information through and across the command structure, by defining the type of information to be delivered and the intervals at which this is to take place. This enables information exchange to be limited to the minimum required to meet the operational need and thus reduces the possibility of information overload. The flow can take place both through the normal, hierarchical military command structure and also across non-hierarchical boundaries between adjacent units. Replication is controlled by three software 'managers' (see ARM graphic) which ensure that 'contracts' are conformed to, and that all data has been successfully transferred.

Both hardware and software is COTS-based. The system is based on MS Windows 2000 platform, with the database running on MS SQL Server 2000, although the latter can be migrated to other database systems such as Oracle or Informix. The Military Message Handling System is based on a product developed by NEXOR, but other MMHS such as IRIS can be integrated. Office functions are based on MS Office XP. Communications routers and switches are from CISCO.

The GIS is provided by ESRI, whose ArcGIS, Spatial Analyst and Image Analyser products (see separate entries), among others, have been integrated into the system. In the current version the GIS functionality is primarily used to show and edit the various digital overlays developed in the DACCIS application, but in version 2 the GIS will be used for analysis and support for decision making.

Status
In 2002 a development version of DACCIS was in use in a response cell of the Dansk Division with a full-scale issue to 1 Jutland Brigade. By the summer of 2003 it is expected that the Divisional HQ and four brigades, together with some divisional troops, will be fully equipped with the production version. The remaining divisional troops will be equipped subsequently, depending on when financial resources are available. Jane's sources suggest that ultimately a total of 200 server sets (each set being three connected servers) and more than 1,000 workstations will be procured.

In September 2003 Jane's sources indicated that the programme is running on schedule. In 2004 it was expected that DACCIS would be issued down to battalion HQ level. It was also indicated that Maersk were redesigning the architecture to make it more modular, and were investigating the possibility of developing the joint capabilities of the system. DACCIS forms the basis of a Maersk bid for a CIS contract for the Swiss Army, the result of which would be known by the end of 2003.

Contractor
Maersk Data Defence, Sonderborg, Denmark.

UPDATED

France

ATLAS automated field artillery fire support system

ATLAS is the latest generation of artillery C3I system to be procured for the French Army and is part of the French battlefield digitisation programme. It performs real-time firing sequence management and provides the applications needed to provide command and control of artillery regiments in the field, as well as collecting target acquisition and intelligence data. ATLAS is designed to provide full interoperability with the other systems deployed in the battlespace and with allied artillery systems (Germany, Italy, United Kingdom, United States). It is the successor to the ATILA system, developed by Thomson-CSF (now Thales), which has been in service with the French armed forces since the 1980s and has been exported to about a dozen countries.

The system has the following features:
(a) The system manages all artillery data and activity from gun to brigade or divisional level, including supporting or associated resources (target acquisition assets, meteorological stations, counterbattery radar).
(b) Message exchange formats:
Allied - ASCA
French Army - SICAT
Specific - MLRS, COBRA and so on
(c) High level of automated functions, including orders and reports, situation awareness (real-time display of unit status), safety control, target correlation, fire mission optimisation and fire unit allocation.
(d) Increased survivability through dynamic system reconfiguration to adapt to the operational situation and data backup between CPUs and vehicles.
(e) Modular ballistic kernel suitable for various weapons.
(f) A variety of computer configurations is available:
Portable touchscreen PC, dismountable from the vehicle (typically for use by a forward observer)
One TFT display PC
Two TFT display workstation
Ethernet networks of up to four workstations in two command vehicles.

Atlas automated field artillery fire support system   0052014

The interior of an ATLAS CP showing a double twin-screen workstation configuration (Giles Ebbutt)   0525548

(g) Connectivity is provided by fibre-optic links from guns to CP, by PR4G up to regimental level and over the RITA network (see separate entry) above that. Each CP can be used as a message relay using software routing.

(h) Interface with SICF (see separate entry) is provided and interface with SIR (see separate entry) is an aspiration.

(i) The modular software running on Windows NT4 supports all artillery activities. Functions can be adapted to user profiles and mission types.

Status

In November 2000 Thomson-CSF, now Thales, was awarded a €187 million production contract for eight artillery regiments, two Cobra counterbattery radar link units, and associated instruction and training resources. It also included capability sustainment services such as new munitions and interoperability solutions. As at mid-2002 the 40th Artillery Regt had been equipped with Atlas and sources suggested that further issues would be at the rate of two regiments per year. No information was available concerning a parallel HQ equipment programme.

In September 2003 the first full regimental system was ceremonially handed over, following the completion of technical and operational testing by the end of 2002. All seven French artillery regiments are to be equipped.

Contractor

Thales Communications, Gennevilliers.

UPDATED

FINDERS

FINDERS® (Fast Information, Navigation, Decision and Reporting System) is a tactical battle management system aimed at sub-unit level and below, developed as an off-the-shelf project by GIAT. It provides situational awareness to vehicle crews as well as command and control facilities to commanders. The system provides: graphic tactical display; a graphic and preformatted text messaging system; automatic update of situation with a continuous display of all platforms; assistance with contact reports and fire control; automatic maintenance of logistic status through interrogation of individual platforms, retrieving and collating ammunition, fuel, NBC and vehicle states.

There is a high level of commonality and extensive use of COTS in the component. The latest software packages are written in Windows NT, but are compatible with earlier versions. FINDERS can be used with any graphic visual

FINDERS® used with TacMaster® Tactical Terminal (AMSI)   0116931

display unit, although the French Army have selected the TacMaster® Tactical terminal from Matra Systèmes & Information (now EADS), which has an A4 size high-resolution touchscreen.

Communications can be over most VHF or HF digital radios, and are compatible with data rates of 4,800 bps down to 600 bps (the standard rate on PR4G, the French VHF radio, is 2,400 bps). Data is transmitted alongside voice traffic, and the system automatically routes traffic between different networks. At sub-unit headquarters level a router function provides an interface to the battalion-level C3 system (SIR in the French case), using Adat-P3 message format or a variant depending on the requirements of the customer.

Specific special-to-role software modules for mechanised infantry, armour, engineers, logistics and recce have also been developed and are added to the basic functions which are shared by all platforms. FINDERS can also be linked to the artillery ATLAS system (see separate entry). The recce module has improved mapping, route display, and navigation aid functions; intervisibility computation; management of sensors; and rapid reporting capability. It has been successfully integrated with optronics from EL OP, SAGEM and Thales. A further development interfaces with all the sensors mounted in the platforms in which it is fitted, using the MELISSA interface module used in the SIR system.

Status

First developed in 1995 for the UAE in its Leclerc MBTs, where it is known as LBMS (Leclerc Battle Management System); according to *Jane's International Defence Review* it has also been installed in Russian-supplied BMP-3s, and other UAE armoured vehicles are expected to be retrofitted. It has been selected by the French Army as its tactical BMS, SIT (Système d'Information Terminal). The SIT V1 programme covers AMX-10, VBL (contract signed May 2000) and LeClerc MBT (contract signed Nov 2000), with the VBCI AIFV expected to be included in the future. In July 2003 the AMX-10, VBL and VB2L versions were accepted; technical trials are due to take place in September 2003. Operational trials at squadron level are programmed for mid-2004.

FINDERS has also been proposed as a solution to BMS requirements for Swiss (Leopard MBT) and Belgian (reconnaissance vehicle) programmes.

Contractor

Giat Industries, Versailles.

UPDATED

Land Forces Logistic Information System (SILCENT, POSTE NOMADE)

The French Land Forces Logistic Information System (French acronym SILAT) provides the Commandement des Forces Logistiques Terrestres (CFLT) (Land Logistic Forces Command) at Monthery with the ability to control and monitor all logistic support for the French Army and consists of two principal components. SILCENT (Système d'Information Logistique Centrale), the central army logistics information system provides item control from main storage depots in metropolitan France and logistic flow control (Conduite des Flux Logistique (CFL)). It enables logisticians to organise and control supply, react to requirements and provide information on the status of supply. POSTE NOMADE or 'roaming station' is the information system provided to deployed forces or other locations in France. It is designed to work autonomously but it can communicate with SILCENT and other deployed forces via a messaging system to transmit or receive information. The automation of the system minimises the entry of data for deployed forces, but the architecture allows deployed logistics staff to continue to operate locally without constant communications with SILCENT. All items are bar coded, either by warehouse management systems in the base organisation or by deployed forces using the POSTE NOMADE, and are tracked by reference to the bar code.

SILCENT provides the following functions:
(a) Organisation of dispatch, including the selection of means of transport
(b) Flow conduct and control, providing item visibility and load content
(c) Reaction to an incident or emergency change allowing items to be redirected
(d) Interface with external systems for the reception or dispatch of requirements and reports.

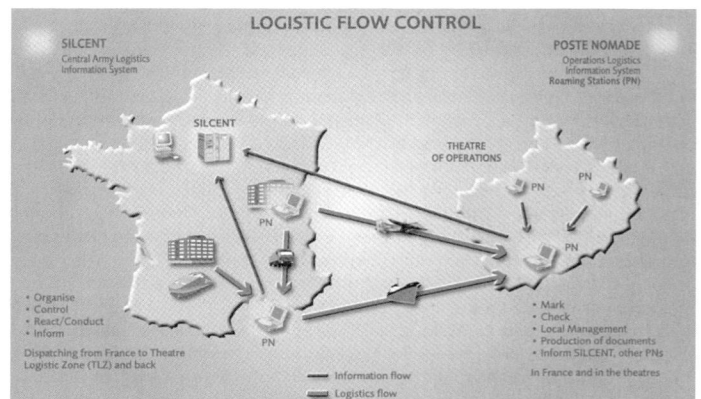

The French Land Forces Logistic Information System architecture showing the SILCENT and POSTE NOMADE relationship (EADS)   0525554

POSTE NOMADE provides the following functions:
(a) Marking of items for dispatch
(b) Checking of items during loading and unloading
(c) Local logistics management - control of storage areas, inventories and so on
(d) Provision of automated reports to SILCENT or other roaming stations
(e) Provision of information to local logistic staff
(f) Office facilities.

A Unix HP server using Oracle 8 provides the central SILCENT database. Workstations are PC-type using Windows NT, with the roaming stations holding a local database. In theatre, any roaming station can act as a server to others. Communication can be by a variety of means including floppy disk or e-mail and can be carried over a LAN or by satcom.

Status

Development of the SILCENT part of the project was begun in 1997 by Matra SI, now EADS, and was operational by May 1999 with the complete system established at approximately 12 sites. POSTE NOMADE was developed over the period 2000-2002 and was operational in January 2002. It has been used to support operations in Kosovo, Thessalonika and Afghanistan. It is now being developed to support the French Air Force and is also used at the joint level.

Contractor

EADS Systems & Defence Electronics, Velizy-Villacoublay.

VERIFIED

SICF Force Command Information System

SICF (Système d' Informaton pour le Commandement des Forces) is the French Army command information system from army corps to brigade level, which is designed to increase the efficiency of command posts, improving the reliability and speed of information transfer by eliminating intermediaries and allowing the automatic creation and diffusion of messages. It replaces the previous SIG system. The system consists of servers and workstations, connected by an Ethernet LAN. PC-compatible computers can also be connected to the network, together with a variety of peripherals. Communications are provided by the tactical communications network, in particular RITA (see separate entry), and modems allow direct connection to both civil and military infrastructures such as the RITTER military network, and to satellite communications (SYRACUSE, STARSAT, TANIT, INMARSAT).

The system provides two basic functions:
(a) a 'transport' function, managing received and circulated information
(b) a 'situation' function, maintaining an automatically updated database both for the general tactical situation and for separate staff branches.

Each workstation has applications to support common command functions and co-operative working, and specialist applications are provided for specific staff branches (intelligence, fire support and so on).

The UNIX-based software architecture is COTS-based, including ILOG VIEWS and IFC to manage alphanumeric and graphic dialogues; VERSANT object database; ORBIX object bus; and MTA EXCHANGE. It was developed by Thomson, now Thales, with X11 graphic interface, CLIO DBMS, X400 88 messaging, office e-mail (MS-mail, Exchange) and Internet (ESMTP). Workstations include the IFPS 7588 from IRTS together with their Rider Station ruggedised packaging (see separate entries).

The SICF programme began in 1985 and a number of prototypes were developed and evaluated. SICF V1 definition was achieved in 1994, with

An SICF workstation using the IRTS 7588 with tilting screen (French Army)
0083171

development beginning in 1996. Full operational capability was achieved in October 2000. The SICF V2 is currently in development and is expected to be in production by 2004. SICF V3 is currently at the definition stage. SICF is a command information system designed to meet the requirements of units in the field as well as in the office. It combines the efficiency of dedicated tactical applications and the flexibility of commercial firmware. The system can be deployed in shelters, in fixed or mobile infrastructures and includes industrial, ruggedised high-performance data processing equipment.

Status

SICF V1 is in operational service with the French Army. 40 HQs of differing sizes have been equipped and over 2,000 workstations have been fielded. SICF has been used in Bosnia, notably in the French-commanded Multinational Division SE. SICF new-generation is currently being fielded by the French Army. SICF has been chosen by the Canadian and Belgian armies.

A version of the software was selected for the Canadian Athene tactical system (see separate entry).

Contractor

Thales Communications, Colombes.

UPDATED

Système d' Information Régimentaire (SIR) (Regimental Information System)

The Système d' Information Régimentaire (SIR) is the battle management system for the French Army which is for use at battalion and brigade level, sitting above SIT V1 and below SICF (see separate entries) in the system hierarchy and interoperable with specialist systems such as Atlas and Martha (see separate entries). It makes extensive use of COTS hardware and software and has a modular and open architecture. Like similar systems it provides core functions which are common to all users:
(a) Tactical situation updates, including friendly and enemy pictures and NBC, with interfaces to SIT V1 and SICF.
(b) Preparation and dissemination of orders and reports including graphics.
(c) Map and terrain analysis using a variety of cartographic tools.
(d) ORBAT management.

SIR is a flexible system split into layers. The lower layer implements the common core software while the external layer implements specific functions. Dedicated applications ensure the specific requirements of the Army branches and major operational functions, including armour, artillery, air defence, engineers, infantry, aviation, logistics, intelligence. Communications are provided either via the Thales PR4G or RITA 2000 (see separate entries).

SIR will equip the Command Posts (CP) of battalions, regiments and brigades. Some functions will be useable while CP vehicles are on the move. There are two CP vehicle variants, each of which will be equipped with GPS. An air-conditioned shelter mounted on a TRM 2000 or GBC truck will accommodate two connected workstations. There is also a portable workstation for use in lighter vehicles. CP vehicles will be connected by optical fibre, creating a LAN.

The system has a UNIX operating system, XWindows/Motif and CORBA object-oriented database, and can also be carried on a Windows-NT platform.

Status

The first deployment of an operational system, SIR V2, took place at the end of 2002 with fielding continuing until mid 2005. SIR V3 is under development and the programme currently envisages fielding from 2005 to 2007, with the aim of equipping over 500 CV overall. Further programme phases may follow. EADS is also developing a helicopter-borne version in response to a French requirement for a SIR-equipped heli-borne CP, equipped with 2 or 3 workstations, possibly with two only able to be used in flight and the third when on the ground.

Contractor

EADS Systems & Defence Electronics, Vélizy Villacoublay.

UPDATED

TACTIS C2 system

TACTIS is a smart command and control system, developed to meet the mission planning, tactical information management and message handling requirements of modern armies. It provides a significant improvement in battlefield situation awareness and assistance in mission management. TACTIS is the latest of a series of systems already fielded in the French armed forces.

Key technological resources implemented in the TACTIS system include:
(a) advanced digital mapping and data processing
(b) advanced communication and messaging protocols
(c) observation (target acquisition and imaging)
(d) navigation and guidance (including commander and driver requirements)
The TACTIS system design was conducted with a particular emphasis on:
(a) tactical situation synthesis and display optimisation
(b) tactical terrain data analysis

Onboard TACTIS terminals   0001009

Battalion level TACTIS workstation   0001010

(c) mission planning information editing, in compatible form for in-vehicle use
(d) real-time mobility command and control, at battalion, squadron and platoon levels, of combat or logistics vehicles fitted with different grades of navigation systems (inertial, hybrid and GPS positioning systems)
(e) tactical messaging at battalion level and under, including encryption compatibility with existing radio communication networks
(f) interoperability, on an evolutionary basis, with existing C³1 systems
(g) simplicity of use
(h) flexible tactical database updating
(i) security and confidentiality.

The TACTIS system man/machine interface and operation are based on a high-performance digital mapping editor: tactical information, including the position of vehicles and operational situation data, is geographically referenced and overlaid on a digital map background. The map background and georeferenced tactical information are displayed at various scales, with the required degree of detail, as selected by the operator. Additional information relative to any georeferenced item (military site, area, vehicle or entity) may be displayed in appropriate symbology (APP6 NATO standard).

TACTIS system operation accommodates most types of digital mapping information, including:
(a) digitised map data
(b) vectorised map data
(c) digital feature analysis data
(d) digital terrain elevation data
(e) georeferenced satellite and aerial images
(f) standard databases.

In order to produce the required database ahead of operation of the TACTIS system, the geographical information from various sources may be synthesised and merged in GIPSY 2001, a geographical information production system developed by SAGEM(see separate entry). SENATER, a very high production throughput system of a similar type, was developed by SAGEM under contract from the French Army and entered operational service in mid-1994.

TACTIS implements secure data multiplexing protocols to interface with existing radio communication equipment and networks and so provide periodical position information updates and messaging capabilities for every fielded vehicle. At battalion level, TACTIS is compatible with the MESREG regimental message handling system (see separate entry). MESREG can be connected to all communication media available in the field. The messaging function is performed via encryption terminals at both ends, together with in-vehicle tactical display terminals.

TACTIS can be implemented in different versions, ranging from the mobile sheltered version with several workstations on line to the man-portable version, capable of in-vehicle operation. The design allows easy installation.

Status
CRECERELLE, the French Army battlefield observation drone implementing elements of the TACTIS family (ground station) is now operational. The INOS tactical situation management system (a member of the TACTIS family) is included in the SPERWER observation UAV system recently chosen by the Royal Netherlands Army.

Contractor
SAGEM SA, Defence and Security Division, Paris.

VERIFIED

Germany

ADLER artillery computer network

The ADLER (Artillerie-Daten-Lage und Einsatz-Rechnerverbund) artillery computer network enables the rapid passage of information for artillery command and control and reduces response times for fire missions of all kinds. The availability of fire support weapons systems is improved and the effect of fires on the target is enhanced by short-time high-intensity fire concentrations. The accuracy of fires is improved by using current target and fire unit status data, while the radio circuit load is reduced considerably by shorter transmission times. The reduction of routine tasks allows the available personnel to concentrate on command and fire-control tasks more effectively and the effect of enemy counter-battery fire is reduced through increased speed of deployment and redeployment.

The ADLER system and applications programs support:
(a) Co-ordination of target acquisition systems and weapon systems
(b) Calculation of the optimum use of weapons and ammunition for proposals of target engagements
(c) Timely availability of all data for fire direction and fire control and the conduct of fire missions
(d) Processing and distribution of commands and reports as well as target and status data
(e) Planning and deployment of target acquisition systems through determination of terrain profiles (coverage diagrams) using the digital terrain model
(f) Graphical display of maps, situation and other information.

ADLER hardware and software equipment allows two operators, each at one of two PC workstations within one shelter/vehicle, to work independently from each other and simultaneously on different tasks. The equipment includes PCs with graphical capabilities, printer and combat net radio for both voice and data communications. All this is mounted on installation kits for standard shelters and tracked vehicles of the German armed forces. Due to largely identical fittings for all ADLER shelters and vehicles in terms of computer and communications equipment as well as application software, the following can be achieved:
(a) Any ADLER function can be assigned to any shelter/vehicle
(b) Any artillery function can be performed independently at any workstation within a command post
(c) Any number of shelters can be linked to form a command post

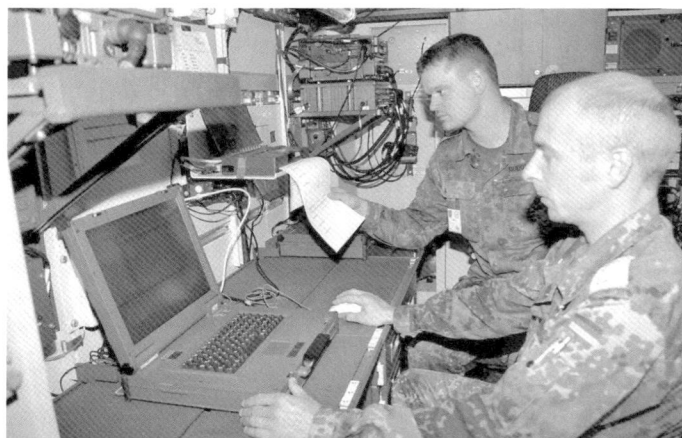

ADLER laptop workstations in a shelter (ESG)   1030078

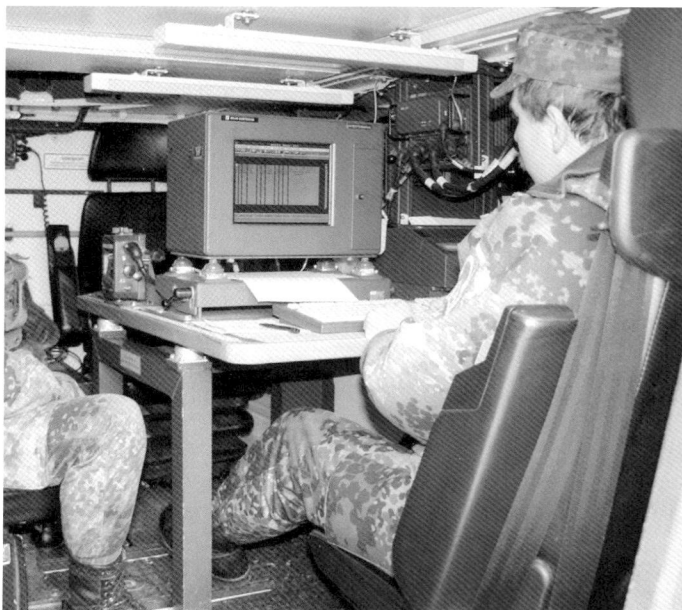

ADLER data terminal (ESG)   0528087

ADLER AFV (ESG) 0528088

(d) The change of command from one command post to another can be achieved without loss of capability

(e) An equivalent substitute can quickly be made available in the event of damage or equipment failure

(f) The logistic input for maintenance of the system is minimised.

Within a command post, all shelters and vehicles are interconnected via an Ethernet LAN. Beyond the command post LAN, the ADLER communications system provides automatic, secure real-time transmission and distribution of information by means of fixed format messages over VHF combat net radio or field wire links. Up to three radio circuits and one command post field wire network can be operated from every ADLER shelter. The transmission protocol supports mobile and flexible deployment on the battlefield and ensures:

(a) Transmission of a large number of mainly short messages in a hierarchically structured network with a large number of subscribers

(b) Access to all subscribers by means of function-oriented addressing

(c) Priority-controlled and collision-tolerant access to transmission media by means of a contention mode protocol.

Automated conversion functions on technical interfaces guarantee the rapid exchange of information, without operator intervention, between the artillery systems (for example downward compatibility with other national artillery systems) and with the army command and control system.

ADLER is involved in the international ASCA (Artillery Systems Co-operation Activities) Interoperability Programme. ASCA was formed by France, Germany, Italy, UK and the US to establish a technical and operational interface between their national artillery command and control systems. Via the ASCA interface, ADLER allows for the automated exchange of artillery fire support information for:

(a) Fire missions in current operations

(b) Fire plans

(c) Target acquisition data

(d) Fire support co-ordination

(e) Command and control

(f) Meteorological support

(g) NBC warning and reporting.

ADLER can apply this interface to support artillery formations to provide mutual fire support across a common boundary or when under the operational command/control of another nation.

Status

The development work by ESG took account of the results of investigations in the late 1970s on the efficiency of computerised support of artillery command, control and fire-direction systems. Based on an MOU, the US and Germany started an evaluation and test programme on the interoperability between an experimental ADLER system and the US TACFIRE artillery system (see separate entry). Interoperability was demonstrated successfully in 1985 at live firing trials. With the delivery of an ADLER prototype system for user trials, an intense trials and evaluation phase began, which included numerous system interfaces and fielding permission was finally granted in 1992. The trials and evaluation phase was then

followed by the preparation for series production, extensions of the functionality and adaptation of the equipment to the changed mission conditions of the German armed forces in multinational units and crisis reaction forces. The first production systems were fielded in May 1995; fielding is complete. In early 2000 ESG started an upgrading programme for both software and hardware. In March 2003 the upgraded hardware version (ADLER II) was integrated into ten shelters for the German artillery and the software was successfully applied during an operational evaluation of the ASCA interface in the first half of that year. The upgrade of all shelters and vehicles in the German artillery is planned to be completed by 2004.

Manufacturer

ESG, Munich.

NEW ENTRY

ARES rocket artillery employment system

Dornier, part of DaimlerChrysler Aerospace AG, now EADS Systems and Defence Electronics, developed a tactical artillery employment system especially for use with the Multiple Launch Rocket System (MLRS). It forms part of the Adler integrated artillery system (see separate entry) in the German Army.

ARES (*Artillerie-Raketeneinsatzsystem*) is a battery level system for tactical fire control of rocket artillery which provides real-time processing and data transmission of all fire-control information. The modularity and flexibility of the hardware and software also enables ARES to be used for command and control and fire-control tasks at higher levels of command (battalion/regiment), when connected to a higher level CCIS.

Description

The main system components of ARES are:

(a) battery Fire Direction Centre (FDC)

(b) platoon leader's vehicle

(c) communication components on board MLRS.

The main operational tasks of the ARES FDC are:

(a) assignment of launchers to firing positions

(b) determination of zones of fire

(c) selection of launchers for the execution of fire missions

(d) computation of numbers and locations of mean points of impact

(e) determination of amount and type of ammunition to be employed (optimisation of effect on target)

Battery FDC hardware installation in a standard shelter

ADLER soft-skinned shelter (ESG) 0528089

Battery FDC and platoon leader's vehicle

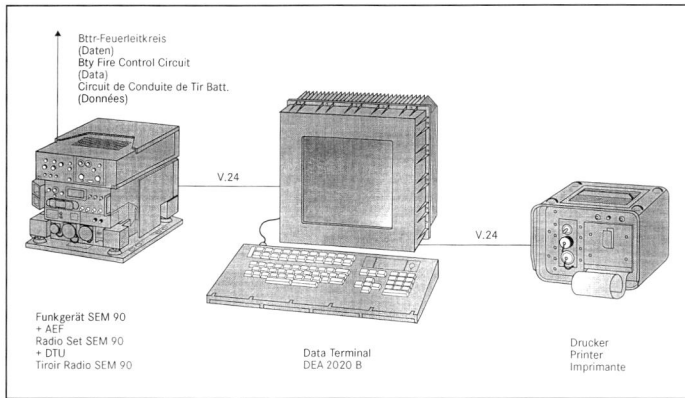

Platoon leader's vehicle hardware configuration

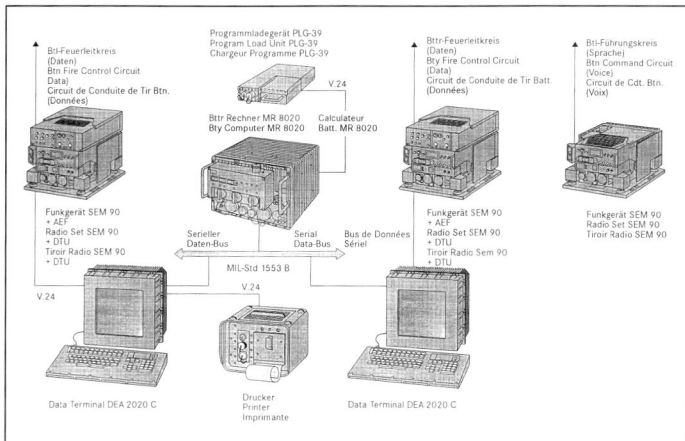

ARES hardware configuration for a battery fire direction centre

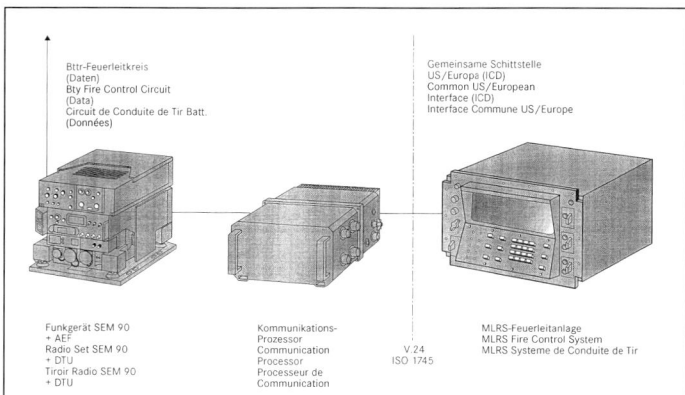

MARS/MLRS hardware configuration

(f) management and storage of all mission-relevant data
(g) issue of mission orders and all kinds of reports
(h) management of summaries (for example, status, firing positions, ammunition)
(i) communication with the other system components via data transmission.

The ARES communication components on board the MLRS launcher handle all the data messages to and from the MLRS and provide the interface between the onboard fire-control system and the national-specific data transmission procedures and protocols. Within the ARES system all mission orders, messages and information for command, fire control and ammunition supply are usually exchanged by data transmission using formatted data messages (HDLC protocol).

The battery FDC is housed in a standard shelter, mounted on a 2 tonne truck and contains a battery computer (MR 8020), program load unit (PLG-39), two data terminals (DEA 2020 C with stand-alone capability), printer, radio sets (SEM 90) with Data Transmission Units (DTUs) and a MIL-STD-1553B databus. Power is supplied by a generator on a two wheel trailer attached to the FDC.

A 0.9 tonne truck houses the platoon leader's data terminal (DEA 2020 B with stand-alone capability), printer and SEM 90 radio fitted with a DTU.

On board the MLRS is a KMP communication processor and SEM 90 radio and DTU.

Status
The ARES system has been in use with German Army rocket artillery units since 1991. A total of 39 battery fire direction centres, 38 platoon leader's vehicle configurations and 150 communication processors aboard the MLRS were delivered.

Contractor
EADS Systems and Defence Electronics, Munich.

UPDATED

FaKoM/Army Command and Control Equipment (ACE)
STN ATLAS ELEKTRONIK has developed a universal vehicle command and communication equipment FaKoM for combat vehicles. It is a modular and flexible system intended particularly for integration in existing vehicles and is available as a compact unit or in the form of separate modules to allow integration even if space is at a premium. Universal interface assemblies allow the adaptation of vehicle subsystems and radio equipment of different types and from various manufacturers. The fundamental hardware elements of the system are the tactical terminal, command computer, mission data terminal and driver's indication unit and the software is based on the STN ATLAS MCCIS (see separate entry). Tactical situations, commands, messages and alarms are transmitted quickly and without errors, digitised terrain maps are displayed on high-resolution colour screens. The export variant Army Command and Control Equipment (ACE) is available in different system variants.

Contractor
STN ATLAS, Bremen.

UPDATED

FAUST tactical CCIS
FAUST is the Bundeswehr battlefield command and control system for brigade level and below. It provides current information across all command levels and supports staff work. The system supports situational awareness with an electronic map showing the current situation of all subordinate military units, using conventional tactical symbology. All staff and material data are stored with their current status in a database; updating is carried out by AdatP-3 based messages. Secure communication links ensure the interconnection of all system components via user-selectable VHF, HF, SatCom, ISDN and GSM.

FAUST has simple and flexible configuration options for the different command levels (team to brigade) and user roles (e.g. S1- S4, S6, G1-G4, G6, J1-J4, J6). Because of the modular structure and the scalability of the software, the system can be used both in individual vehicles and complex command posts with several workstations. An integrated touch screen is provided for the use in combat vehicles and a multi-lingual user interface has been developed.

FAUST provides the following functions to staff at all command levels:
(a) Situational awareness across all command levels is provided by automatic position messages (adapted to the low bandwidth of available communications) and their display on the current situation map.
(b) Support of operation planning by terrain assessment using digital map information.
(c) Reliable and fast assessment of own forces by easy-to-use staff and materiel tables.
(d) Graphic display of tactical grouping and task organisation, with drag and drop facility to amend these and implement transfer of authority.
(e) Processing of non-military incidents in OOTW by means of formalised messages (accident, EOD, attacks)

The system provides interfaces for subordinate battlefield management systems of all branches of the Bundeswehr, such as SAFES for the medical service, RAFES for NBC defence and HERGIS for army aviation. For the import of mission-specific basic data it provides interfaces to the respective Bundeswehr technical information centres.

FAUST uses software technologies, such as Java, XML, CORBA as middleware and an object-oriented database management system. The software is optimised to provide high availability and reliable and fast operation across a variety of different military hardware platforms, ranging from the ruggedised Notebook up to the touch-screen computer.

FAUST integrated into a DINGO vehicle, shown at AUSA 2002. The processing and communications equipment is between the front seats, with the screen mounted on the front passenger dashboard (Patrick Allen/Jane's)   0526717

The FAUST screen mounted on the front passenger dashboard of the DINGO vehicle (Patrick Allen/Jane's) 0526716

Specifications

(a) Operating system: MS Windows NT 4.0 Workstation or MS Windows 2000 Professional

(b) COTS products: Borland Visibroker (CORBA); Versant OODBMS (object database); MilGeo PCMAP (map and situation display) (see separate entry).

(c) Options: MS Office; Adobe Acrobat Reader; software keyboard IMG Corp; MyTSoft; WinZip

(d) Communication:

Communication Servers from SEL and ATM

Comm-Module from ATM

SATCOM, GSM, LAN, WAN

(e) Vehicle interfaces have been developed for the following vehicles: Wolf, DINGO, Leopard 2 A5, Marder, Luchs, Fuchs, M113.

Status

FAUST has been developed in close co-operation with the Bundeswehr. Following a series of user field and development trials from 2000-2002, the system successfully completed a final trial in November 2002, conducted on the basis of an extensive peace-keeping mission scenario covering information exchange across five command levels and the employment of different communication links (VHF/SatCom/ISDN/GSM), and on completion of the trial was certified suitable for operational use. The Bundeswehr contingents at SFOR and KFOR (FRY) are expected to be equipped with FAUST from March 2003, and those in ISAF (Afghanistan) are expected to follow by mid-2003. It is expected that the system will be further upgraded as mission-specific functions become apparent.

Contractor

EADS Systems and Defence Electronics, Ulm.

NEW ENTRY

GEHOC HAWK battalion operation centre

GEHOC is an operations centre for the German HAWK surface-to-air missile system. Its main functions are datalink communication, air situation compilation, identification, threat analysis, weapon assignment and task surveillance.

The equipment is installed in a standard cabin and is land-, sea- and air-transportable. GEHOC has two identical workstations with high-resolution raster scan displays. The data processing system is a multiprocessor system and workstations are connected via a local area network.

German Hawk Operations Centre (GEHOC)

GEHOC software is written in Ada and is developed as an open system. It is composed of generic software modules, which are easily portable to different processors and operating systems and reusable in other systems such as SAMOC (see separate entry).

Status

In service since 1994.

Contractor

EADS Deutschland GmbH, Ulm.

VERIFIED

HEROS-2/1 mobile C3I system

The mobile C3I system Heeresführungsinformationssystem zur Rechnergestützten Operationsführung in Stäben -2/1 (HEROS-2/1) supports headquarters from corps down to brigade level during all phases of the command and control cycle. Its functionality provides the necessary effectiveness to achieve optimised employment of forces during both tactical and peacetime missions. The system is being deployed in the German Army and also forms the core of the EUROKORPS C3I system.

In addition to the various system functions such as message handling, situation display, drafting of orders or preparation of briefings, the MS Office software package (Word, Excel, Access, Project and PowerPoint) is integrated for routine headquarters work. The e-mail service has been implemented with Lotus Notes/ Domino software. Among other things, it supports the creation and addressing of e-mail messages for the distribution of documents, reports, situation graphics, commands and intelligence. The organisational structure is stored in the name and address book, which is used for tasks such as the identification of the addressees for the e-mail service and the assignment of access privileges. All functions can be selected from a standard operator interface and can be linked by means of common object-oriented data management.

The HEROS-2/1 database contains graphical data, tables and message data. The data processing system has a client/server architecture that has standard interfaces and is therefore impartial to products and manufacturers. This means that any component can be readily updated or replaced with products from other manufacturers, should the need arise. The operating system is Windows NT. The system's ability to process both formatted and non-formatted messages enables the system to adapt to changes in organisation, working practices and information requirements. It automatically converts graphical situation information to and from the NATO ADatP-3 message format. The geographical reference of the tactical situation is achieved through the use of vector and raster maps.

GeoGrid is used to display and process maps and situations. It works on the basis of the PCMAP map format, which is the standard format for maps used by the German Federal Armed Forces Geographic Office. Map data for almost every part of the world can be supplied on CD-ROM and directly processed by the system without any conversion. When electronic map data is unavailable, maps from non-military sources can be scanned or satellite maps can be entered into the system. GeoGrid's object linking and embedding (OLE) capability allows the exchange of information and data with MS Office products. Geogrid also provides the situation display, together with an open system interface for connection to the database and third-party systems that support the reception and transmission of situation information.

HEROS CCIS

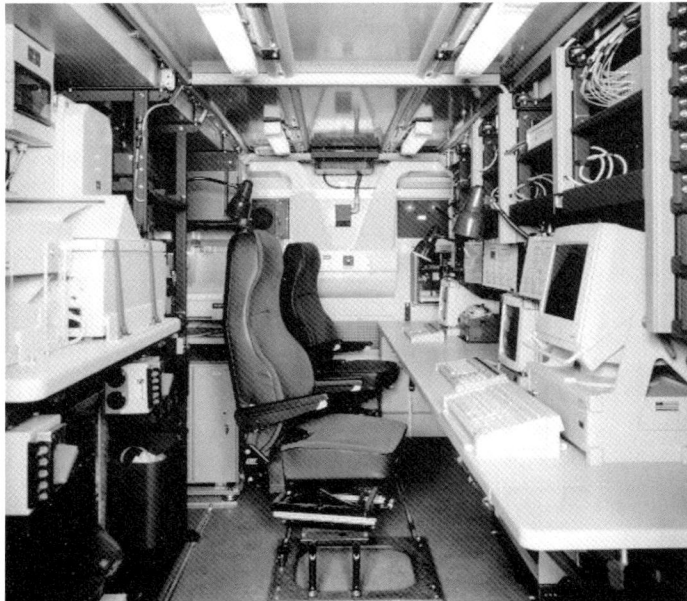

The interior of a HEROS shelter

Two database servers located in shelters form the central elements of the system and are synchronised and replicated to ensure adequate data consistency and dissemination. Transportable command post workstations are connected to these shelters via a tactical fibre optic local area network.

The external communication architecture of HEROS-2/1 is based on a concept which allows the system to communicate simultaneously with a multitude of analogue, digital and radio networks and via military and public satellites. Based on the NATO ADatP-3 messaging conventions and a technical interface developed as part of the Quadrilateral Interoperability Programme (QIP), messages can be exchanged automatically with the systems of allied partners. Message handling is based on the information that is already available in the system. The operator is provided with input templates designed for the different tactical and operational requirements.

The system can be configured at different levels, depending on customer requirements. The functionality can range from a simple information system based on office communication products and the integration of additional COTS products – such as for map, situation and reporting – to the full HEROS-2/1 Batch 2 system.

Status

HEROS-2/1 is being deployed in the German Army in three lots. Lot 1, which is Unix-based, was developed by Siemens and is in service. Lot 2, which has been developed by ESG in association with EADS, is based on Windows NT and received formal approval for service in November 2000. The first Lot 2 brigade set was delivered in 2001 and the first division was planned to be fully equipped by 2003, but according to *Jane's International Defense Review* this is likely to slip to 2004-5 for budgetary reasons. In Oct 2003 an order was announced for HEROS Lot 2 equipment for the HQs of the EuroCorps in Strasbourg and the GE/NL Corps in Munster; the functionality will be adapted for the specific requirements of the two HQs.

Austria and Switzerland have procured test configurations to gather experience for a potential procurement of HEROS-2/1, modified to their national requirements, although in late 2003 indications suggest that Switzerland will not be procuring the system.

According to *Jane's World Armies,* a deal signed in 2002 transferring to Poland Leopard 2 MBTs and associated equipment included a HEROS node and associated communications. This was destined for the Polish 10th Bde, which is assigned to 7(Ge) Armd Div as part of the ARRC. This is likely to be operational by early 2004.

Contractor

EADS Defence & Communications Systems (Prime), Ulm.
ESG, Munich.

UPDATED

leKomDEG

The leKomDEG (light communication data terminal) was introduced under a short-term procurement programme for the German IFOR/SFOR contingent in Bosnia to provide a mobile C2 capability. Mobile versions of the command and mission support component were integrated in the Luchs reconnaissance vehicle and versions for operational command and control were installed in the command posts. The system is based on the STN ATLAS ACE (see separate entry) and draws on the same firm's MCCIS software (see separate entry). It provides:
(a) Automatic report processing of free text messages and AdatP-3 formatted messages with automatic conversion of alphanumeric information into graphic information and vice versa.

(b) Automatic graphic tactical display.
(c) Position determination and automatic own position reporting.
(d) Processing and storing of operational and logistic data.
    The hardware of the leKomDEG is based on a ruggedised notebook with integrated interface and communication server as well as a GPS unit. It provides the interface to the sensor and communication systems required for a command system. The leKomDEG is particularly well suited for mobile operations and can be operated independently without connection to the external power supply for several hours.

Status

Still in operational use in Bosnia. Some installations also in service in Germany.

Contractor

STN ATLAS Elektronic, Bremen.

UPDATED

STN ATLAS Mobile Command Control and Information System (MCCIS)

STN ATLAS has developed a software system which underpins a number of low-level CIS, including ACE, TCCS, LINCE and leKomDEG (see respective separate entries). Interoperability with other command and control systems is assured by the use of internationally standardised protocols and reports and has been proven in various projects such as the Battlefield Interoperability Programme (BIP). MCCIS contains the following functions:

Map functions
(a) grid maps on any scale
(b) UTM grid indication
(c) online zoom
(d) automatic/manual scroll
(e) selective display of map information.

Tactical display and processing
(a) 5 main overlays (present situation, initial situation, firing plan, barrier plans, notes)
(b) tactical symbol library (dot symbols, lines, spaces, text for map)
(c) position-accurate display of own symbol and communication partner as tactical symbol
(d) staff functions (free text, personnel, material, medical and logistic situation).

Communication
(a) radio communication with dynamic network configuration
(b) INMARSAT communication
(c) automatic symbol-oriented graphic report collection in formatted reports
(d) individual, group and broadcast addressing
(e) status indication of report transmissions
(f) filing of reports.

Printing functions
(a) colour prints of maps and/or tactical data
(b) any scale of map
(c) printout of reports.

Administration tools
(a) map processing, administration and distribution
(b) symbol editor
(c) configuration tool for command support software MCCIS-C1.

Optional office functions
(a) MS Office
(b) other software.

Contractor

STN ATLAS, Bremen.

UPDATED

Tank Command and Control System (TCCS)

STN ATLAS ELEKTRONIK has developed a command and control system for the Leopard 2 MBT and other combat vehicles under contract to the Swedish Army. 130 systems have been delivered to Sweden. Based on the STN ATLAS MCCIS software (see separate entry), the TCCS is also connected to all relevant subsystems of the tank (fire-control system, navigation system) via a bus system. Furthermore, chassis and engine data are provided. Other sensors and systems can be connected via the flexible hardware and software structure. The interoperability with other C3I systems is achieved by using internationally standardised protocols and STANAG messages. The Built-In Test Equipment (BITE) allows quick and accurate fault localisation in the case of hardware and software faults. It also enables efficient and cost-effective maintenance and repair work.

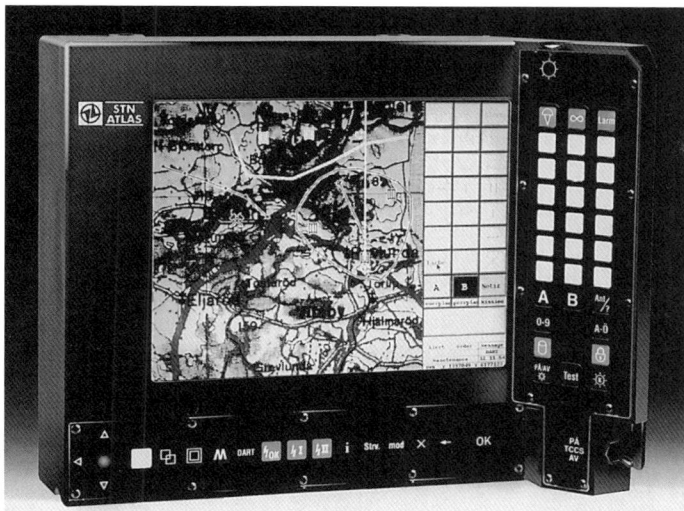

The TCCS display in the Swedish Leopard 2 programme

Status
In production.

Contractor
STN ATLAS Elektronik, Bremen.

UPDATED

International

Leopard Information Control Equipment (LINCE)

Based on the STN ATLAS ACE vehicle C2 system (see separate entry) and produced in partnership with Amper Programas, LINCE is a command and control system for armoured vehicles developed for the Spanish army. Intended for all command levels, designed particularly to interface with SIMACET (see separate entry) and using the ATCCIS (LC2IEDM) data model to maximise interoperability the system has the following principal characteristics:
(a) Automatic tactical data and messaging transmission and reception according to ATCCIS replication (ARM)
(b) High priority data transmission and reception (location and alarms) over VHF radio in TDMA mode
(c) Navigation system integration
(d) Cartographic management (digital, raster and vector)
(e) Operating system based on Windows.

LINCE provides the commander with a means of generating and transmitting reports whose content is provided by inputs from the platform systems, allowing minimal user input and a launch-and-forget message transmission system. This enables transmission of tactical data moments after it is acquired, with a significant reduction in the risk of operator-induced errors. The multi-function tactile display provided by LINCE gives the operator information in a number of formats. In the normal display mode, the operator is presented with a digital map view showing own position and tactical symbology (standardised symbols according to APP6) representing friendly and hostile forces (as well as unknowns), task information, and navigation information. Alternative display modes enable the display as a viewing port for sensors, or as a text viewer for Reports and Returns. All of this data can be updated in near real-time when the LINCE system is used in conjunction with voice/data radio systems. The system can also be configured to provide flexible tactical data permitting the efficient transfer of tactical information within a local tactical group, onward transmission to higher echelons, and sideways transmission to flanking units.

Status
In Spain, 219 LEOPARD 2E MBTs and 16 ARVs will be equipped with this system, due to be in service by 2003. AMPER are also marketing the system under the name Tauro.

Contractors
AMPER Programas , Madrid.
STN ATLAS Elektronic, Bremen.

UPDATED

Medium Extended Air Defence System (MEADS)

The Medium Extended Air Defence System (MEADS) will be a highly mobile, low to medium air defence system designed to replace the HAWK and PATRIOT PAC-3 air defence system. As part of the army air and missile defence architecture, the system will be compatible and interoperable with other US Army air defence

systems and will interface with joint and allied sensors and battle management command, control, communications, computers and intelligence (BM/C4I) networks.

The MEADS battalion will consist of three firing batteries and a headquarters battery. Each battery will have nine launchers controlled by a battery tactical operations centre and each launcher will be equipped with twelve hit-to-kill missiles. Two X-band Multifunctional Fire-Control Radars (MFCRs) and one UHF surveillance radar are intrinsic to the MEADS standard fire unit, with both providing 360° coverage. MEADS will introduce the concept of plug-and-fight, through which MEADS takes control of external sensors associated with other air defence systems to provide alerting and cueing information to any tactical operations centre in the battalion. As a national option, MEADS may also take control of other air defence interceptors. Netted and distributed battle management command, control, communications, computers and intelligence (BM/C4I) will permit battle elements to join in or break off to protect forces on the move.

Status
NAMEADSMA, a chartered organisation of NATO, awarded dual contracts for the Programme Definition and Validation phase in 1996. NAMEADSMA selected MEADS International to develop the system in 1999 after rejecting a protest by Raytheon. Headquartered in Orlando, Florida, MEADS International's participating companies include MBDA Italia, EADS/LFK (Lenkflugkörpersysteme, a subsidiary) in Germany and Lockheed Martin in the United States. The US, Germany and Italy finance the programme in shares of 55, 28 and 17 per cent.

In 2001, based on use of the PAC-3 missile, MEADS International began a US$216 million Risk Reduction Effort programme to reduce technical and cost risk and ready the system to enter a Design and Development (D&D) Phase. In 2003, MEADS International submitted a solicited proposal for the D&D Phase. The D&D contract is planned to begin in 2004 and would extend the programme, currently completing the Risk Reduction Effort phase, for seven years. In a contract milestone demonstration in 2003, the system demonstrated its ability to acquire, classify, track and destroy simulated aircraft and missile targets in a successful System Level Interface Demonstration. The Risk Reduction Effort is on schedule for completion in 2004.

Contractors
MEADS International, Orlando, Florida.

NEW ENTRY

Ireland

S-TBMS (SINCGARS Tactical Battlefield Management System)

S-TBMS is a map based real-time position tracking and command messaging system. Developed for the Irish Army to take advantage of the procurement of SINCGARS (see separate entry) and working on a COTS PC (standard or ruggedised) running MS Windows NT or 2000, the system is a battlefield management system for use at the tactical level. Using ARC-GIS (see separate entry) and embedded GPS, individuals or groups can be tracked in near-realtime against a digital map background using full military symbology. The information is distributed using the integrated networking capabilities of SINGARS, providing a common operating picture at all command levels, with information filtering as required. A variety of overlays (situation, event, obstacles) can be generated using C-TBMS (see below), and movements can be recorded and played back for analysis. The system also provides a tactical messaging capability, using standard military message formats as well as free text. Alerts are triggered automatically, and message handling and distribution can be configured by the user.

In conjunction with S-TBMS, C-TBMS (Command Tactical Battlefield Management System) provides the map based tools for battlefield planning and analysis. In addition to overlay creation, spatial and key terrain analysis tools are

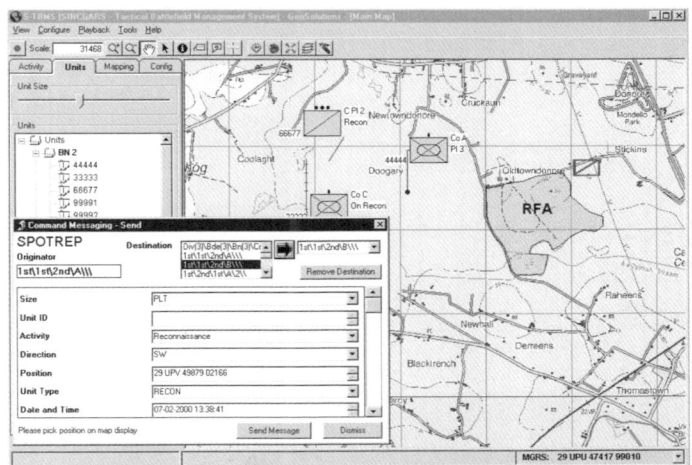

S-TBMS Screenshot (GeoSolutions)   0121454

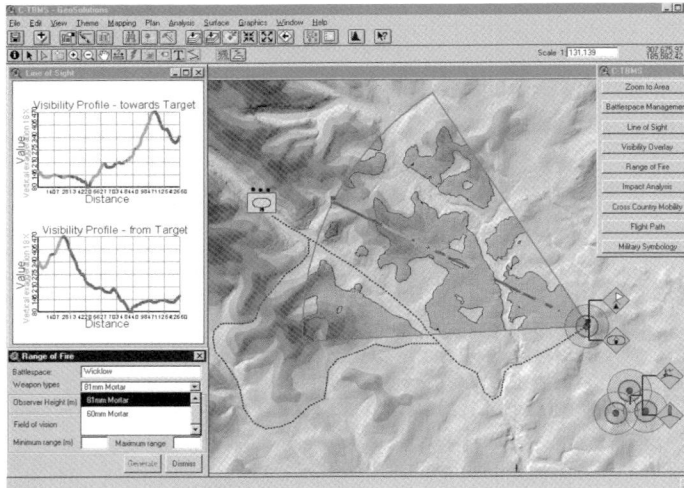

C-TBMS Screenshot (GeoSolutions)   0121455

provided, enabling line-of-sight, intervisibility, range, impact assessment and visualisation analysis to be carried out.

Status

In service with the Irish Army. Offered by ITT as a package with SINCGARS.

Contractor

GeoSolutions, Dublin.

UPDATED

Israel

Citron Tree

Citron Tree, the Battle Management/Command, Control, Communication and Intelligence (BM/C3I) system centre of the Arrow Weapon System, processes sensors' and external sources' data and optimises interceptions in real time. It is a fully automated system with human intervention capability available at every stage. Citron Tree is capable of interoperability with other TMD weapon systems and Israel Defence Force's C3I systems. Citron Tree has full simulation capability for training, as well as recording and playback for comprehensive post mission debriefing. The system was proven during the successful test of the Arrow in September 1998.

Ten Citron tree operator battle stations are housed in a 12.1 m shelter. The Arrow battery command team sits at the central bank of three workstations, with the battery commander (who reports direct to the IDF Air Defence Commander) in the centre flanked by an intelligence officer and a systems officer. The other seven workstations are typically used by intercept operators, the number of which depends on the threat level. Each intercept operator concentrates on a specific threat area.

Status

In operational use.

Contractor

Tadiran Electronic Systems Ltd, Holon.

NEW ENTRY

Tadiran's Citron Tree BM/C3I for the Arrow Weapon System   0085289

COMBAT artillery command, control and communication system

COMBAT is a command, control, communication and information system for field artillery, now in its third generation. It automates all artillery and communication procedures and substantially shortens artillery unit reaction time, increasing lethality and survivability.

The system has a modular design and possible configurations include a comprehensive divisional artillery C3 system; a battalion level system (up to 6 gun batteries); a battery level system (up to 8 guns); and a single artillery computer at battery level. It uses the ETC 2000 Tactical Computer (see separate entry) with a communications controller for up to four networks and the Hand-Held Computer (HHC) (see separate entry) for forward observers and at the gun. Both these are Elbit's own terminals. It includes links to meteorological stations, electronic goniometers and laser range-finders, muzzle velocity radars, target acquisition radars, navigation and laying systems. All types of artillery platforms are supported, including self-propelled guns, towed guns, mortars and rocket launchers (including MLRS). The system supports the concept of the autonomous gun, providing onboard ballistic computation and mission management.

The system provides the following functions;

(a) Fire support planning and fire plan distribution
(b) Monitoring of fire units status
(c) Target analysis and selection of fire unit and engagement to task
(d) Target data management
(e) Firing zones, safety areas and co-ordination procedures
(f) Storage, management and distribution of ballistic data
(g) Firing orders processing and distribution
(h) Survey computations
(i) Tactical graphical display, using raster, vector or DTM maps

Status

In service with the Israeli armed forces and claimed by the manufacturer to be in service with 14 other armies worldwide.

Manufacturer

Elbit Systems Ltd, Haifa.

UPDATED

··

DACCS–BOMBARD Divisional Artillery Command-and-Control Systems and Artillery Battalion Fire-Control System

DACCS and BOMBARD, either separately or combined, integrate Division Artillery C4I with Battalion C4I and transform fire control into an integral part of the C4I distribution network. The system enables rapid processing of Fire Support Requests (FSR) and Fire Support Allocation (FSA) at battalion, division and regimental levels.

Artillery commanders at all divisional levels are provided with a true, updated picture of enemy forces and their counter-battery activities. Firing units, fire-support elements and target-acquisition assets can be used more effectively, as commanders can access operational and logistics status in real time. DACCS-BOMBARD provides digital, secure communications which can be connected to existing equipment. The major benefits of the system are:

(a) accomplishment of more firing systems in the allotted time
(b) more efficient use of ammunition
(c) fast and accurate response to urgent fire requests
(d) rapid amassing of fire capability
(e) effective counter-battery fire
(f) presentation of vital graphic and alphanumeric data to each station in real time
(g) real-time dissemination of battle orders, overlays, target analysis, logistic status and available resources

DACCS – forward observer sending new mission   0052017

DACCS – battalion fire direction centre, working inside the APC   0052016

(h) assistance to all command levels by recommending the most efficient use of existing resources

(i) highly reliable and secure communications

(j) firing data calculations down to individual weapon

(k) connectivity to auxiliary systems such as TPQ 36/7, meteorological systems and UAVs.

Divisional Artillery Command and Control System (DACCS)

A modular distributed artillery and Battle Management System for use at all command levels up to the divisional level. DACCS supports all procedures, calculations and data communication within the artillery loop, from forward and firing echelons up to divisional level. The system serves artillery commanders in the pre-planning phase and supports all artillery engagements during battle, assisting manoeuvring and firing echelons in co-ordination with and in response to changing battle demands in the shortest possible time.

BOMBARD–Artillery Battalion Fire-Control System

An artillery force multiplier for all levels of command and control systems up to battalion, BOMBARD ensures optimum utilisation of resources at every engagement. It provides rapid fire planning for 'on-target' first-strike capability whatever the condition. BOMBARD integrates multifaceted ballistic computations with multinetwork communications.

DACCS-BOMBARD building blocks
Tactical Communication Controller (TCC 2200)

A sophisticated device capable of transferring messages simultaneously on several different channels, allowing a computerised node to be connected to different operational networks. It has been designed to be compatible with various customers' existing radio equipment. It supports four channels as a stand-alone server or two channels via a PCMCIA interface card (PC - COM 2200).

Laptop computer

A full MIL-SPEC PC laptop based on state-of-the-art technologies providing highest processing speeds all in an ergonomically designed package adaptable to tank, APC or shelter environments. The laptop can be optionally equipped with a two-channel communication controller and a GPS receiver, which transforms the computer into a high-power stand-alone C4I communications centre.

Hand-held Pentop computer

An ergonomically designed compact PC computer providing an ideal solution for low-level echelons or for the soldier on the move. Packed into its small size are powerful, sophisticated computing capabilities. It is also available with a two-channel communication controller and GPS receiver.

Compact, fully militarised printer

Software packages to assist commanders in critical pre-combat and post-combat tasks:

(a) calculating ballistic data for individual weapon

(b) supporting commanders in decision-making process, presenting recommendations during planning and conducting battle engagements

(c) controlling communication flow.

Status

The system is currently operational with a number of unidentified armed forces. In 1998 Tadiran was awarded a US$120 million contract when the system was selected by the Swiss Army for its artillery division.

Manufacturer

Tadiran Electronic Systems Ltd, Holon.

UPDATED

EL/S-8825 Generic Command and Control Station (GCCS)

The GCCS is a ruggedised and portable Control Station which can be used to control both UAVs and their payloads as well as other battlefield sensors. It can be vehicle- or ship-mounted. It is designed to receive real-time video and data from a variety of sensors and payloads. The GCCS is a portable, ruggedised dual display system based on two Pentium PC workstations, which are mounted on a common backplane, using commercial-off-the-shelf (COTS) circuit boards.

Features

(a) Operating systems: Windows 95/NT

(b) LCD-TFT colour displays, 12 in and 15 in

(c) Two 4.0 Gbytes hard disk drives

(d) Digital recording system

(e) CD-ROM

(f) GPS

(g) Internal UPS

(h) Operating panel includes keyboard and pointing device

(i) Power: 28 V DC or 115/230 V

(j) Physical: Dimensions 670 × 360 × 260 mm (W × H × D); Weight 30 kg max

Major Functional Characteristics

(a) Receive, process and display real-time video and data

(b) Output the received image to a recorder

(c) Capability to record real-time video on internal RAID system

(d) Image freezing and processing

(e) Store and retrieve images

(f) Display a digital map together with symbols and data overlays

Contractor

ELTA Electronics Industries Ltd(a subsidiary of Israel Aircraft Industries Ltd), Ashdod.

VERIFIED

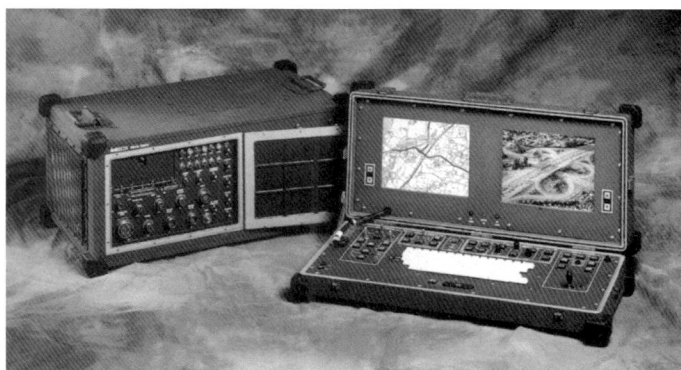

EL/S-8825 Generic Command and Control Station (GCCS)   0009934

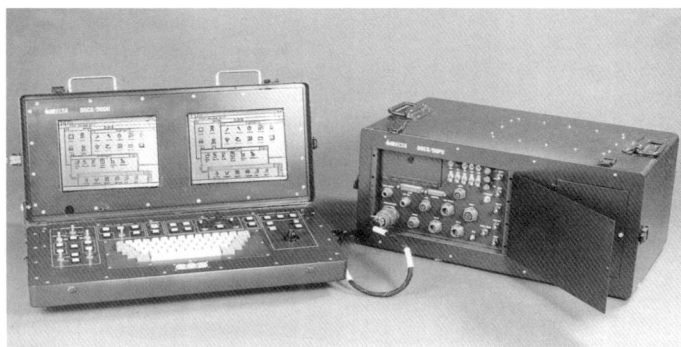

EL/S-8825 Generic Command and Control Station   0052018

TACDIS Divisional C4I Battle Management System

The TACDIS Divisional C4I battle management system covers a broad range of applications including G3 (operations), G2 (intelligence) and logistics. The system presents own-force and enemy-situation picture and co-ordinates EW and divisional artillery operations. TACDIS is designed with built-in modularity and flexibility incorporating Commercial Off-The-Shelf (COTS) units. Primary system services are:

(a) Electronic messages (alphanumeric and graphic) to disseminate information, data and orders among the various users throughout the deployed echelon: preformatted/free text messages (for example, location, unit status, intelligence reports and fire orders); message acknowledgement status (machine/man); message retrieval and message database queries

(b) map display and graphics overlays to present tactical information, allowing users to converse while looking at an identical situation picture or battle plan: 2-D maps in multiple scales; 3-D displays; battle plan and unit overlays

(c) unit position reporting to maintain an accurate friendly-forces situation picture (GPS-generated or manual update)

(d) unit database management

(e) G3 services (friendly-forces situation picture) and manoeuvre control to assist in the creation, display and update of the situation picture, including unit position, location, function, status and inventory, plans development, order issuing, activity monitoring and information analysis.

(f) G2 services (enemy-situation picture) to support G2 staff in formulating the enemy-situation picture, including deployment and identification, evaluation of new information in light of previous data, deployment history, alternate possibilities, modification of existing battle plans, efficient intelligence reports handling and intelligence-resource management

(g) liaison services to allow commanders to maintain close contact with other functional units operating in the divisional theatre

(h) information communication infrastructure: to maintain connectivity between all users via tactical radio networks, Local-Area Networks (LANs), radio telephone links and Wide-Area Networks (WANs); to provide sophisticated self-learning capabilities of network connections; to afford a full solution for communication on the move; to connect with a customer's existing communication equipment; to offer Wide-Area Communication System (WACS) for systems dispersed over large areas through the use of HF networks.

TACDIS common building blocks (see separate entries)
Tactical communication controller
A device capable of transferring messages simultaneously on several different channels, allowing a computerised node to be connected to different operational networks. The communication controller is designed to be compatible with the customer's existing radio equipment. It supports four channels as a stand-alone server or two channels via a PCMCIA interface card.

Laptop computer
A full MIL-SPEC PC laptop in an ergonomically designed package adaptable to tank, APC or shelter environments. The laptop can be optionally equipped with a two-channel communication controller and a GPS receiver.

Hand-held Pentop computer
Equipped with optional GPS, in an ergonomically designed package, providing a solution for low-level echelons or for the soldier on the move. It is also available with a two-channel communication controller and a GPS receiver.

Contractor
Tadiran Electronic Systems Ltd, Holon.

VERIFIED

Italy

Artillery Tactical Automatic Fire System (ATAFS)

The Artillery Tactical Automatic Fire System (ATAFS) is a C3 system for battery and battalion covering all the main and auxiliary functions required for artillery. It allows the core of the artillery system, for example the artillery battalion Fire Direction Centre (FDC), to interface with all other units. On one side, the surveillance equipment components gather data from the outside world - the enemy forces and dispositions, and the area of operation. On the other side, humidity, wind speed and direction, air temperature and other relevant information about atmospheric conditions are collected and processed by the meteorological station. A link between the artillery battalion FDC and the divisional HQ allows the exchange of intelligence messages. Functional interfaces are automatically defined and adjusted at start-up during software initialisation.

ATAFS allows suitable communication management by sending and receiving messages among the different units and sub-units of the whole system. Message broadcast can be both automatic and controlled by the operator depending on the configuration of the system. Standard messages, such as orders, requests or reports can be handled simultaneously by the operator.

Depending on the configuration, defined during initialisation of the system, ATAFS automates the command and control functions performed by artillery battalion units, according to NATO procedures.

The standard configuration has three basic levels: FDC level (battalion and battery), gun level and forward observer level.

At the FDC level, ATAFS consists of:
(a) NATO 3 equipped shelter
(b) Equipment for M113 and M577 armoured vehicles
(c) IMAGE computer unit
(d) Tactical display computer unit
(e) Thermal printer, hard disk and video terminal
(f) Power supply unit
(g) Radio communication unit
(h) Local area network
    At gun level the system comprises:
(a) Equipment compatible with weapons system including M106, M114, FH70, M109G, M109L, M109A1, M11 OA1, MLRS and others

(b) Gun display unit
(c) Data repeater unit for second operator
(d) Muzzle velocity radar
(e) Local area network
(f) Radio communication unit
(g) Intercom system.
    A forward observer is provided with:
(a) Vehicle installation
(b) Digital message generator
(c) Radio communication unit
(d) Local area network
(e) Laser rangefinder

In the FDC Battalion configuration, ATAFS interfaces with the Bty FDC, battery commander, forward observer, meteorology station, divisional headquarters, artillery battalion and surveillance component. In the FDC Bty configuration, ATAFS is interfaced with the guns, battery commander and forward observer.

The main command and control functions are target analysis, assistance in fire planning in real time, operational status reports, ammunition reports, muzzle velocity updates and assistance in preparing orders, including optimal fire commands. Computing functions provide automatic fire data verification between battalion and battery FDCs, full survey functions, local meteorology messages and data correlation and aggregation.

Status
Possibly in service with the Argentinian Army.

Contractor
AMS, Chelmsford.

UPDATED

....

CATRIN army field C3I system

CATRIN - *Sistema Campale di Trasmissione ed Informazione* (Battlefield Surveillance, Data and Voice Transmission and Intelligence) has been designed and developed for the Italian Army. The system has three parts: SOATCC - *SOttosistema Avvistamento Tattico Comando e Controllo* (Tactical Air Target Acquisition Command and Control Subsystem), SORAO - *Sottosistema Ricerca e Acquisizione Obiettivi* (Battlefield Surveillance and Target Acquisition Subsystem), which are described below, and SOTRIN, the area communications network to provide information bearers.

SORAO
The SORAO Battlefield Surveillance and Target Acquisition comprises a suite of active and passive sensors capable of covering the whole area of combat. Data acquired by the sensors is sent to data correlation centres - CCD - where aggregation and distribution of data is performed.

The SORAO functions can be summarised as:
(a) battlefield surveillance
(b) target acquisition
(c) data acquisition, correlation and distribution
(d) artillery fire support.

Typical sensors include long and medium range drones, mini RPVs, counter-fire radar, heliborne radar, laser rangefinder and thermal infra-red rangefinder. The unmanned drones perform aerial surveillance and intelligence data acquisition and mini RPVs allow the continuous monitoring of the battlefield situation. These are mainly devoted to target detection and location in order to present the operational situation on displays for information management at the different headquarters levels.

One of the main airborne sensors is the CRESO. Together with short- and medium-range UAVs, the CRESO system is designed to accomplish two different but complementary battlefield objectives. These are real-time detection of mobile ground targets and the surveillance and measurement of radar emissions (EMS).

CRESO stand-off heliborne radar provides a ground surveillance capability in supporting ground operations at the tactical level. Its primary tasks are to detect and locate moving targets (mainly columns of vehicles and tanks) in a zone 60 km across and least 50 km beyond the FEBA and to detect hostile radar transmissions. CRESO is a tactical system conceived for medium-range ground surveillance in order to minimise helicopter exposure time proportional to helicopter height and hence to the square of the distance to be covered. The system takes advantages of orographic masking effects by use of suitable prefixed pop-up points and protected low-flying paths.

The system consists of four main elements: the helicopter-borne platform, the sensors (MTI and ESM systems), the self-defence suite and the data analysis and distribution station.

New ADI, HSI, datalink controls and two data display have been accommodated on the helicopter platform. The centrally located head down display unit displays the infra-red images collected by the navigation FLIR, which has been installed on the roof of the helicopter cabin. A second, larger display is mounted at the co-pilot's position and presents information collected by the ESM system, including radar warning functions to improve the crew's spatial awareness and self-defence capabilities.

All the information gathered by the different sensors is processed by the Data Correlation and Aggregation Centres - *Centro di Correlazione Dati* (CCD), which provide the higher level formations with force location, potential and threat. The

CATRIN operations shelter

CCD provides data processing in accordance with the NATO ADATP-3 formatted messages, data correlation and aggregation and the display of the tactical situation using digital topographic maps.

The CCD is based on a standard NATO UE02 shelter, which is ground- , helicopter- and air-transportable. The shelter meets NBC, EMC, EMP, TEMPEST requirements and FINABEL and MIL-STD environment requirements. It includes three main consoles to process the data, video and geographical images based on the IMproved Architecture Graphical Engine - IMAGE computer family.

SAOTCC

The SOATCC is an integrated and sheltered subsystem of CATRIN to exercise:
(a) command and control of Anti-aircraft Artillery (AA) units;
(b) command and control of Army Light Aviation (ALA) units;
(c) alert of VSHORAD weapon systems and other command posts (AA excluded);
(d) management and control of airspace assigned to the Army Corps (close co-ordination with Air Force).

SOATCC elements are separated into three functional areas:
(a) surveillance: it includes an integrated net of sensors;
(b) air defence: it includes Command and Control Centres for AA units;
(c) army light aviation: it includes Command and Control Centres for ALA units.

For surveillance the radar sensor network associated with the SOATCC uses a number of two-dimensional mobile radars (RAT-30C) and one or more three-dimensional mobile radars (RAT-31S). These provide continuous airspace surveillance from low and very low altitude up to the lower band of high altitude within the airspace over the Army Corps. Sensors are connected to a Reporting Centre (RC) where a Track Data Fusion is performed using multi radar tracking techniques to generate the Recognised Air Picture (RAP). The RAP is used either to exercise tactical control of AA units or to guide ALA aircraft. In the Air Defence and ALA areas, Command and Control is conducted by a number of mobile sheltered centres located at three different echelons of the Army Corps.

AA Command and Control Centres comprise:
(a) An Air Defence Cell located into the Corps MAIN to implement the Army Corps C-in-C directives for air defence;
(b) AA Command Operation Centre for command activities at the Corps Artillery Command Post level;
(c) AA Tactical Control Operation Room for tactical control activities at the Corps Artillery Command Post level;
(d) AA Battalion Command and Control Centre containing three elements:
  (i) AA Battalion Command Operation Centre for command activities;
  (ii) AA Battalion Tactical Control Room for tactical control activities;
  (iii) AA Battalion Surveillance Operation Room for fusion of track data from the connected Battery Command Posts.

ALA Command and Control Centres are:
(a) ALA Cell. Directed by the ALA assigned officer, it implements the Army Corp C-in-C directives for ALA force utilisation including mission planning and resource management. The ALA Cell constitutes the interface element, from the technical-operational point of view, between MAIN, where it is based and the ALA Command Operation Centre at the ALA Regiment Command Post level.
(b) ALA Command Operation Centre. Managed by operations, training and logistic assigned officers; it provides airspace management in close co-ordination with

Air Force, ALA unit utilisation including mission tasking and logistic command activities using data from the ALA Logistic Centre (CILO) where logistic orders are carried out. It is assigned to ALA Regiment Command Post.
(c) ALA Airspace Control Operation Room. Managed by the air traffic supervising assigned officer; it provides direction and monitoring of airspace organisation and air traffic control in emergency or overloaded situations; it is assigned to ALA Regiment Command Post.
(d) ALA Control Centre. Exercises air traffic control on assigned area and ALA mission tactical control when tasked; it receives directives and tasks from ALA Regiment Command Post.
(e) ALA Squadrons Group Command Centre. Managed by operations, training and logistic assigned officers; it provides preparation of missions tasked by higher echelon, planning of own group missions, control of assigned air units and logistic command activities using data from a local Logistic Centre (CILO) where logistic orders are carried out. It is assigned to ALA Squadrons Group Command Post.
(f) ALA Airfield Traffic Control Room. Exercises air traffic control on the airfield assigned zone; it is assigned to ALA Squadrons Group Command Post.
(g) CILO complex. Logistic centres where logistic orders are carried out to support command posts; they are located at ALA Regiment and ALA Squadrons Group Command Post level.
(h) COPA complex. Based on the PLRS (TDMA) radio network, it calculates the position of airborne friendly units on a real time basis and provides robust and effective radio links for secure communications; it receives directives and tasks from the ALA Regiment Command Post.
(i) COMIX airborne post. Provides control and monitoring of helicopter status during the mission using the COPA capabilities, displays the mission scenario to the Mission Commander and permits the control and co-ordination of the activity in progress in the mission zone until completion; it is assigned to ALA Squadrons Group Command Post.

Due to its performance and capabilities, the CRESO system is a candidate for integration into the NATO Allied Ground Surveillance (AGS) system.

Contractor
CATRIN is produced by a consortium consisting of the Italian firms: Agusta, Italtel, Telettra, AMS.

VERIFIED

SIACCON Automated Command and Control System

SIACCON (Sistema Automatizzato di Commando e Controllo) is a COTS-based distributed C2 system that provides battle management support across a wide range of functions in the Italian Army. It is based on the generic Marconi Automated CCIS (MACCIS), versions which are in use in several countries, and uses a range of military and civilian standards.

The system is scalable both in hardware and software terms, enabling its use at all levels from company to corps. The core system remains the same at all levels, but higher headquarters have a more comprehensive range of operational functions. A typical Command Post (CP) configuration consists of a fusion centre and one or more CP cells. The fusion centre contains a database server or servers to store all the operational data managed by SIACCON; a server or servers to provide generic services such as Mail, Web, DHCP and DNS services; and telecommunication and encryption devices. C2 workstations are located in the CP cells. Each CP has a LAN, and these are connected via tactical communications links. The system uses the LC2IEDM common data model for database replication, enabling interoperability with other systems. It accepts NATO formatted messages.

SIACCON provides a common operational picture to all users, with live locational information provided via GPS. The system provides functional support across the full range of staff branches and also includes specific functionality to support Operations Other Than War (OOTW).

General purpose capabilities include:
(a) Digital mapping
Vector and raster mapping
Navigational functions
Terrain Analysis
3-D view
(b) Formal message handling
(c) Inter/intranet
(d) Multimedia management
(e) Co-operative working
(f) Office automation
(g) e-mail

The following COTS products are integrated in the system:
(a) MS Windows NT
(b) Oracle Relational Database Management System
(c) MapObjects GIS (see separate entry)
(d) IRIS formal messaging system (see separate entry)
(e) MS office

Status
SIACCON V1, which was UNIX-based, was introduced in the early 1990s.The current Windows NT version of MACCIS, which is SIACCON V1AW, was introduced into the Italian Army in 2000 and is in use down to company level. An

upgrade to Windows 2000 is likely to take place in 2004.

The Bulgarian Army has one division equipped with a MACCIS system similar to SIACCON, with different functions, limited to air defence, NBC and artillery. This limitation in deployment is possibly for financial reasons. The Brazilian Army has one division equipped with the UNIX-based version of MACCIS, but in late 2003 this was understood to be in the process of being upgraded to V1AW. Chile has three divisions equipped with MACCIS with SIACCON functions; a fourth division may be similarly equipped in future.

Manufacturer

Marconi Selenia Communications, Genoa.

NEW ENTRY

Netherlands

Command and Control Workstation (C2WS) Framework

The Command and Control Workstation (C2WS) will be the platform on which all new RNLA C2 systems will be implemented. Whenever feasible, existing systems will migrate to the C2WS Framework. The C2WS Framework will deliver

(a) An application framework: the C2 Console. This application framework provides the seamless integration of applications to the user. The ISIS 3.0 (see separate entry) application is currently the main application, but applications for fire support, intelligence and air defence are under development. The C2 Console also hosts generic functionalities such as office automation, intranet and internet. It provides a single entry point for all applications and facilitates the integration of information from different sources.

(b) A generic component for data exchange. This will provide a solution for the requirement of 'zero dependency' of systems. It provides a loose coupling of data producers and data consumers. The producers publish their information and the consumers are able to subscribe to data published. A COTS product is used to implement the publish/subscribe mechanism.

(c) Mechanisms and tools for data storage and handling. A 'Common Operational Picture (COP)-Catalogue' has been developed to provide an overview of available information. Through the COP-Catalogue the user can create his own COP by subscribing to the relevant contexts. He contributes to the COP of other users by publishing contexts in the COP-Catalogue.

(d) A Synchronisation Service. The C2WS Framework provides a Synchronisation Service to all applications. This service prevents the loss of data when connections are temporarily lost. It also allows a user to continue working when not connected to the network. The Synchronisation Service will synchronise the relevant data after reconnecting to the network. Applications running on the C2WS framework are independent of entities like servers and central databases.

NEW ENTRY

Flycatcher Mark 2

Flycatcher Mk 2 is Thales's new Hybrid Weapons Control Centre for (V)SHORAD. Its main characteristics are:

(a) Integrated search and fire control sensors
(b) Air Defence and Command Centre modes
(c) SHORAD and (V)SHORAD control capability
(d) High mobility and rapid deployment.

Description

Flycatcher Mk 2 can control up to three guns and one missile system simultaneously. In addition, the system can exchange tactical information with higher command and control levels, contributing to a wider air defence network. It can be used in two different operating modes. In the air defence mode, surveillance is optimised for a very short system reaction time, while the system can control up to three medium calibre guns and a missile system. In the Command Centre mode it will typically operate in a network with other Flycatcher Mk 2 systems, providing the operator with a longer range (up to 50 km) surveillance coverage, thus supporting the coordination of surface to air missiles up to missile ranges of 18 km. The extended surveillance coverage also contributes to the identification of targets and can provide an unlimited number of weapons with warning data.

The system, which is housed in two shelters, is transportable by air in, for example, a C130; by truck; or by train.

Status

Three sold to Venezuela in 1999.

Contractor

Thales Nederland, Hengelo.

VERIFIED

Integrated Staff Information System (ISIS)

The Integrated Staff Information System (ISIS) is the Royal Netherlands Army (RNLA)command information system for use at brigade level and above. It is based on commercial off-the-shelf hardware and software, runs on the Windows NT operating system and uses Microsoft Office for office software, linked to a Tactical Messaging System (TMS) providing e-mail. Also standard are the Visual C++ and Visual Basic software, Oracle database management system, and Tensing GIS. ISIS is based on a number of synchronised databases to provide a common tactical picture for all users.

The ISIS tactical data communications network in the field will connect seamlessly to the strategic network using the X.400-based Microsoft Exchange e-mail system as another standard. ISIS Ethernet-based tactical LANs use special data communications access point boxes, named cluster boxes, which contain the active network components. Sixteen vehicles or subnetworks can connect to one box using fibre optic cable or a wireless network. The cluster boxes use 16 to 64 kbits/s synchronous datalinks to connect command post LANs via tactical communications systems to the ISIS wide area network. The tactical data communications network is independent of the type of communications bearer used, so not only the RNLA's ZODIAC tactical mobile digital radio relay system can be used but also others such as RITA 2000 or AUTOKO, as well as PTT, SATCOM or digital combat net radio.

Status

In operational use in the RNLA. Used in the trial of the NATO Immediate Reaction Task Force Land (IRTF(L)) concept in late 2001 and by HQ AMF(L) on Exercise 'Strong Resolve' in March 2002.

ISIS version 3.0 is currently under development by the C2 Support Centre, the 'software house' of the RNLA, with the aim of being operational by the first quarter of 2004. This will be built on Command and Control Workstation (C2WS) Framework (see separate entry). This version was used as the Netherlands contribution to the MIP trials in September 2003.

UPDATED

New Zealand

Morfire

Morfire is a computer based, networked C2 system for mortar platoons. The system provides and manages real-time data transfer between mortar fire controllers, command posts and mortar positions in digital format over combat net radio using burst transmission, while retaining the option of virtually simultaneous voice communications. The system supports fire planning, fire support co-ordination measures and management of target records; STANAG formats are retained; and free text messaging is provided. Target engagement times were reduced in trials in 1995 by up to 5 minutes, and using the target record facility it is claimed to yield a 400 per cent increase in platoon neutralisation capability. The system consists of the following elements:

Command Post Computer (CPC). This is a hand-held lightweight ruggedised battery powered IBM compatible computer running MS DOS in ROM. It is configured by proprietary modular software to perform ballistic calculations, handle data transfer between system elements and facilitate command and control. It has a daylight readable 230 × 64 pixel display with adjustable backlighting and has hardware and software interfaces to combat net radio and the mortar display unit. It provides survey and orientation data for individual base plates and multiple mortar lines and computes firing data with meteorological corrections. It enables the simultaneous engagement of four targets by individual base plates and/or multiple mortar lines and the convergence or distribution of fire. It can manage up to two fire plans with individual countdowns to H and L hour, providing audio visual prompts at base plate, and allow the automatic application of safety zones. Power is provided by 3 × AA batteries which will provide 30 hours of continuous use or 80 hours intermittent use.
Dimensions: 236 mm × 128 mm × 43mm. Weight: 750 g including batteries.

Forward Observer Data Entry Terminal (FODC). With the same characteristics as the Command Post Computer, this is configured to support the generation of Fire Missions, Fire Plans and Target Records and to handle data transfer. It has hardware and software interface to combat net radio and optionally to GPS and a target acquisition system, typically a tripod-mounted laser range-finder. It enables a single Mortar Fire Controller to handle multiple target engagements and conduct fire planning for two fire plans with a proforma display time management and target record store, provides a constant update on the status and availability of all mortars and ammunition in the battalion and will manage safe fire zones. All friendly force locations can be held in the memory and the system provides the automatic input of own location with target data. Screen layouts are in Fire Orders 'Aide Memoire' format.

Mortar Display Unit (MDU). This is a battery powered data terminal providing high-speed data transfer between command post and individual barrels, with supporting voice communications, using either radio or line, over a maximum range of 2 km in the latter case. It can be automatically configured as a CP or

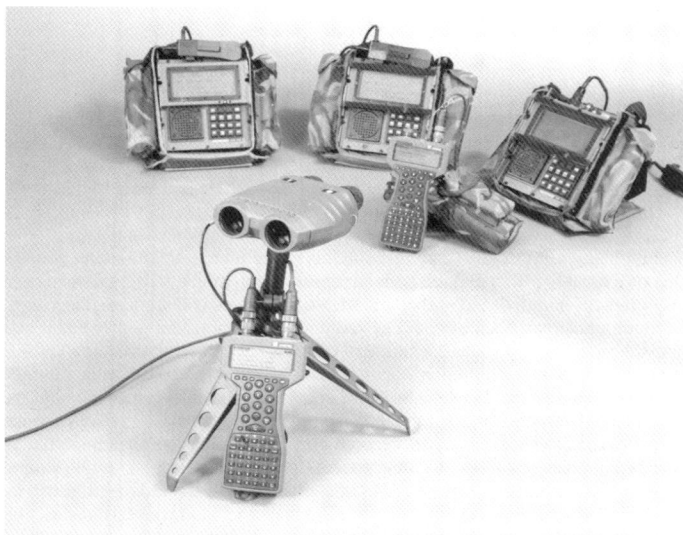

The MORFIRE System. Foreground: FODC with Leica Vector 4 laser rangefinder. Background: Three MDUs and a CPC (Marine Air Systems) 0095950

mortar display when connected and powered. It provides a display of complete fire orders and can also keep ammunition records including the management of mixed lots. It has a simple keypad for data entry, with a daylight-readable large character display (150 × 65 mm viewing size) with adjustable backlighting, which is invisible from 10 m at night. When fitted with a GPS interface it can input and broadcast barrel location and orientation. Power is provided by 3 × D Alkaline batteries which will provide 30 hours of continuous use or 60 hours intermittent use.
Dimensions: 230 mm × 260 mm × 100 mm. Weight: 3.5 kg.

Burst Radio Modem. Developed by the manufacturer and built into all data terminals, this uses broadcast to user programmed address groups or individual callsigns and a network management and message handling system software package is present in all terminals.

Specifications
(a) Display/Screen
  Morfire: 240 × 64 pixel full graphics LCD screen with integral backlight and EMI shield 8 lines × 40 characters
  MDU: 4 lines × 20 characters LCD with integral backlight and EMC shield
(b) Power source
  Morfire: 3.6 V Ni-MH battery
  MDU: 3 × D cell batteries
(c) Operating system
  Morfire: MS DOS 6.22in ROM
(d) Microprocessor
  Morfire: 32 bit Intel 386 EX, clock speed 25MHz
  MDU: MC68302C
(e) Communications
  J1 Burst Radio Modem
  J2 RS-232
(f) Memory
  RAM: 2 Mbytes of low power D RAM, battery supplied
  Disk: Non volatile compact flash disk, 16 Mbytes

Status
In production. A number of unspecified Asian, European and Middle Eastern countries have ordered the system.

Manufacturer
Marine Air Systems (NZ) Ltd, Wellington, New Zealand.

UPDATED

Norway

ComBatt

ComBatt is a mobile CCIS designed to support HQs from divisional to platoon level. Operating in conjunction with the EriTac communications system (see separate entry) and using a variety of COTS products, COMBATT provides a common operational picture and battle management tools, including:
(a) planning tools, including the use of a special XML tool
(b) directives/order distribution
(c) routine staff functions
(d) tactical e-mail
(e) a battle map overlay which is common for all
(f) route and deployment planning
(g) war diary
(h) information handling

(i) integrated communications and C2IS management system
(j) distributed databases
(k) The system provides a GPS-based situational display and a fighting state and logistic situation display through the use of Status Flags' or 'pop-up' information. Updating can be either automatic or manual
(l) A fire-support application
(m) A basic intelligence application.

The Software Architecture for ComBatt is based on the Defence Information Infrastructure (DII) Common Operating Environment (COE) architecture. Essential NATO standards are supported: ATCCIS Generic Hub 3 database, AdatP-3 Baseline 11, STANAG 562 Message Formats, MIL-STD-2525A 'Common Warfighting Symbology', App9 Plans and Orders, Defence Message System (DMS) and ACP 123. The operating system is Windows NT with a number of COTS applications including MS Office, CORBA/IDL, DBTools.h++, MapObjects (see separate entry), and Visual C++.

Status
In use with the Norwegian Army, where it is known as NACCIS. It is used down to battalion level, except in the artillery where it goes down to OP, FDC and the guns. As at mid-2002 the Norwegian Army was considering whether to expand the system down to platform level, particularly for its Rapid Reaction Forces, to increase functionality to include Logistics, Engineer applications, and increase the Intelligence functionality. Such a programme would be likely to extend to 2005/06. In mid 2003 there was no information to indicate when this expansion decision would be taken.

Contractor
Kongsberg Defence Communications, Asker, Norway.

VERIFIED

Ground Based Air Defence Operations Centre (GBADOC)

GBADOC is a C2 post for co-ordinating ground-based air defence assets. It will control MSAM and SHORAD/VSHORAD systems and can be fully integrated with the former through standard NATO links. The use of Links 1, 11, 11B and GBDL has been proven in a variety of exercises, and the system is also prepared for Links 16 and 22. The software runs on a suite of COTS UNIX and Windows NT platforms, interconnected by a LAN or WAN. The universal GBADOC cell is software configured at log-on to support various engagement or force operations functions. Its basic configuration is as a tactical control element, but it can also be software configured as a current operations cell, a future operations cell or as a tactical communications cell. The open architecture and modular structure will allow the system to be upgraded to a complete AOC. The standard cell exists in various container sizes, including HMMWV, 13 ft and 20 ft shelters, enabling it to be mounted on a variety of vehicles.

Functions
Engagement
(a) Local RAP
(b) Manual/automated ID handling
(c) Terrain and battlespace visualisation
(d) ACO integration and ACM visualisation
(e) GBAD status information
(f) Threat evaluation and ordering matched against assets
(g) Weapon selection and co-ordinated assignment.

Force operations
(a) Web-based GBADOC intranet
(b) Land situation picture
(c) Tools for plans/orders/directives
(d) Intelligence and logistics planning and support tools.

Interior of GBADOC Tactical Control Element (Kongsberg) 0143790

Communications

(a) Voice and video to all GBADOC cells
(b) Voice services to all subordinate and lateral units, and superior headquarters
(c) Data link configuration and management.

Status

In 2001, Kongsberg signed a US$2.7 million contract with the Norwegian Air Force to produce a preproduction unit. Further orders may follow once trials are complete. The GBADOC TCE is being marketed internationally in conjunction with Raytheon.

Contractor

Kongsberg Defence & Aerospace AS, Kongsberg, Norway.

VERIFIED

Poland

TOPAZ

TOPAZ is a field artillery command and control system which uses a number of WB Electronics products linked by communications. It can be used up to Divisional level. It provides the following features:

(a) Automation of all functions from observation posts through command posts to guns, including fire planning and execution
(b) Digital communications
(c) Automatic ballistic calculations
(d) Tactical displays
(e) Logistic status display and calculations.

The system consists of the following elements:

(a) Forward observer - the PCJ-9650 set (see separate entry for PC-9600)
(b) Guns - the LIOD message display unit. This unit can display up to 24 alphanumeric characters in 3 rows of eight, controlled by a remote computer linked through a serial RS-232 port via the FONET intercom. Data flow is reduced by only sending ASCII codes to the display. The remote computer can also control the nature - brightness, character attributes etc - of the display
(c) Gun commander - PC 9600 or DD-9620 (see separate entry)
(d) Command Post - The BFC202 battlefield computer (see separate entry)
(e) Communications, provided either by radio or line. Radio sets used include the Thales PR4G (see separate entry) in command posts, the Radmor R3501 (as in the PCJ-9650), or other Radmor radios such as the RRC 9500. In vehicle installations the components are integrated via the FONET digital intercom.

Status

In production and in use in the Polish Army.

Contractor

WB Electronics, Warsaw.

VERIFIED

Singapore

FIRECON

The FIRECON Automatic Gun Control System is an artillery fire-control system which interfaces with sensors and hardware at gun detachment level.
FIRECON enhances artillery operations in :

Planning and Control of Firing Mission
Ammunition Handling System
Monitoring and Control
Tracking and Monitoring of Sub-system
Muzzle Velocity Tracking
Ballistic Computation and Correction
Alert Management
Ammunition Resource Management
Navigation and Positioning
Meteorological Correction
Digital Battlefield Information

FIRECON is powered by in-house software architecture known as 'Common Application Platform' (or CAP) from Singapore Engineering Software, a subsidiary of Singapore Technologies Electronics. CAP facilitates complete scalability and flexibility for easy customisation to suit any user's needs.

Status

In production.

Manufacturer

Singapore Technologies Electronics, Singapore.

VERIFIED

Spain

COAAAS - Anti Aircraft Artillery Semiautomatic Operations Centre

COAAAS has been designed to integrate the Spanish Army's entire anti-aircraft weapon inventory. Two levels have been developed.

COAAAS-L is for the control of low and very low altitude weapons systems, including guns, MANPAD and SHORAD systems and is intended for the Air Defence Artillery (ADA) platoon command post. The complete system will detect and identify aircraft flying within its coverage area, using its own resources, conduct track management, threat evaluation, weapon assignment and the management of firing units. Each system consists of two very low altitude RAVEN radar systems with low output power (in the order of 20 W) and a detection range of 20 km, an Engagement Control Unit (UCE) mounted in a vehicle, and a terminal for each assigned weapon system hosted on a ruggedised laptop. The UCE consists of a single console for the operator, which provides remote control of the radar as well as communications links to the radars, the weapons display terminals and other operations centres. Detection information passed by the radars is merged by the UCE, which conducts threat assessments, and alerts and allocates the weapons systems. Positive direction of fire units can then be conducted.

COAAAS-M. Alternatively, the system can be installed at the ADA battalion Fire Direction Centre (FDC) to link a number of UCEs together to ensure the effective use of procedural fire control, and to control medium-range weapons systems. In this configuration it is integrated to the upper level of the Air Defence System through Link 11B and can also be integrated with local 3-D radar. It also provides the commander with resource and deployment planning support.

Status

The Spanish Army has four operational COAAAS-L units, and a second production run of five further units is in progress. The first COAAAS-M unit has been completed and is operational.

Manufacturer

Indra, Madrid.

UPDATED

The elements of
COAAASL (Indra)
0142035

SIMACET

SIMACET is the Spanish Army command and control system which will link all levels of command down to battalion headquarters. It has been designed as a single system, with a common core and a number of sub-systems. At the core is the Land Command and Control Information Exchange Data Model (LC2IEDM) with a Spanish national extension, together with the ATCCIS replication mechanism (ARM), which enables the same information to be replicated at all command posts. The system provides a common operational picture at a level of detail commensurate with the level of command of the user, and a variety of staff planning and communications tools. Information is exchanged either via satellite (Hispasat), the RBA area communications system (see separate entry), or over combat net radio (largely PR4G). Interoperability with allies is achieved through the use of LC2IEDM and ARM protocols and Adat-P3 messaging. Extensive use is made of COTS software including Windows 2000, Oracle, Lotus Notes and Lotus Domino Server.

Tools

(a) ANTARES Cartographic tool. A GIS manager which provides geographical information support in a variety of formats including raster and vector, hipsometric mapping, digital terrain data modelling and an object data base
(b) ANTARES Tactical tool. This provides all aspects of the common operational picture, including orbats and situational awareness, and a wide range of planning data. It also provides planning tools and specific functionality across all staff branches, including fire support and NBC
(c) PLEYADES. An LC2IEDM orbat management tool
(d) HERACLES. A cartographic tool to enable import and export from/to cartographic formats and the management of graphical objects
(e) CASTOR. An LC2IEDM system management tool
(f) ALTAIR. Management of the LAN
(g) POLUX. ORACLE database management
(h) HELIOS. An LC2IEDM automatic document management tool
(i) ORION. An LC2IEDM reports and graphics tool.

Status

In July 2000 a prototype entered service with the Spanish Army. By December 2001 an updated version (V2) was in service at one major headquarters at Valencia and with nodes at 2 × corps, 3 × divisions, 6 × brigades and 10 × battalions. IOC was achieved in May 2002. V3 was due to enter service during 2002 and it was planned that by November 2002 a further 8 × brigade and 21 battalion headquarters would be equipped, plus the Spanish Marines. In late 2003 this was not confirmed.

The German Army is evaluating the system as a possible replacement for HEROS II. The Swiss Army is also evaluating the system.

Amper is also marketing the system as the Zodiaco C2IS.

SIMACET is in use in the Spanish Reaction Corps HQ, one of the NATO High Readiness Force HQ (Land).

Contractor

Amper Programas, Madrid.

UPDATED

Sweden

SKER Ballistic Computer

The SKER Ballistic Computer provides a C2 system for artillery units, allowing the passage of information and ballistic calculation between forward observers, command posts and guns. It can be utilised at battery level, in a battalion or as part of an artillery division system. Typically, SKER is mounted in an off-road vehicle at the battery command post.

The system consists of the SKER Fire Control Computer and the GDU (Gun Display Unit). The Fire Control Computer is the key component. This 32-bit computer, programmed in Ada, makes ballistic calculations, computes fire direction and fire control. It also handles voice and/or data communications with the command centre. The target information is processed in the computer and the individually calculated fire parameters are transmitted to the GDU at the gun site. SKER is claimed to be unique in that its computer incorporates all the necessary command and communication facilities required to interface with any upper-echelon command and control system and thus provides a variable medium for the transmission of orders and information. Communications with SKER can be by voice or data using combat-net radio, including frequency-hopping systems, or field cable.

The SKER fire-control computer can supply up to eight guns with individual firing parameters. Azimuth, elevation, fuse settings and charge data are all computed on an individual basis along with trajectory integration for each gun. Any combination of guns, howitzers and mortars with associated ammunition may be included in the battery. The computer also handles smart ammunition.

Ballistic data for a variety of weapon types and ammunition can be stored permanently in the memory. The ballistic database can be updated by simply reprogramming a printed circuit board and slotting it in. To facilitate future upgrading, the basic version comes with an extensive spare memory capacity which can be extended when required.

The GDU is used to present calculated firing parameters to guns, howitzers and mortars and can accommodate all types of ammunition. To make the unit suitable for use with mortars, where it must be battery powered, it has been designed to consume very little energy. Target seeker data for smart ammunition is transferred to the GDU and fed into the ammunition without the need for extra equipment and can also be calculated on the spot within the GDU. The GDU communicates with the SKER Fire Control Computer via field cable or radio in the form of either voice or data, and it can be used to receive orders and transmit reports within the battery. The GDU can also be interfaced with the GPS navigation system as well as various types of target acquisition equipment.

Status

The Swedish Army has ordered a large number of SKER battery fire-control systems as part of a major artillery and mortar modernisation programme.

Contractor

SaabTech Systems AB, Järfälla.

VERIFIED

Vehicle Command and Control System (VCCS)

The VCCS is a computerised command and control system that is primarily designed to assist individual MBT/AFV commanders to make decisions based on all relevant data from Combat Net Radio, Fire-Control System and, where available, integrated warning sensors. However, the individual vehicle systems can be linked together to provide a basic tactical level C2 system.

The VCCS provides a single display unit both for tactical information and images from available sensors presented as overlays to a background digital map. An integrated command and control environment is created comprising major subsystems of the vehicle such as:
(a) Navigation system
(b) Fire-control system including turret position and laser rangefinder
(c) Combat Net Radio
(d) Video from thermal imagers and TV-cameras
(e) Drivers display
(f) Vehicle Support System (BIT, fuel, ammo and so on)
(g) Sensor- and Countermeasure System

The VCCS is provided with a set of functional tools such as navigation handling; map and symbol handling; overlay handling; message handling; and target handling, which allow the user to perform operational tasks such as mission planning, fire planning, combat supervision, orders preparation and distribution, and reporting. Orders, positions and other information is transmitted automatically or sent by the user to other vehicles and shown on their VCCS displays, providing low-level C2.

The VCCS can be hosted on a combined computer and presentation unit, TMAP. The TMAP provides a ruggedised Intel-based computer with an integrated touch panel colour display.

Specifications

(a) IBM compatible PC with Intel Pentium 166 MHz CPU and 32 Mbytes DRAM.
(b) 80/300 Mbytes Solid State Flash memory (IDE/ATA compatible).
(c) Integrated 12.1in LCD/TFT display, SVGA 800 × 600 pixel and 3 × 6 bit colour.
(d) 17 integrated PC-compatible function keys.
(e) Video input, PAL&NTSC (for display in window or full screen).
(f) Touch panel.
(g) Integrated ethernet connection.
(h) Two independent RS-232 and four independent RS-422 interfaces for connection to external devices.
(i) Connections to external PC keyboard and PC mouse.
(j) Software drivers for Built-In Test (Windows and NT).
(k) Software drivers for integrated function keys and pointing device (Windows and NT).

Options

(a) VGA input (to use as display for external PC).
(b) CAN-bus.
(c) MIL-1553 interface.
(d) ARINC interface.
(e) NVG adaption.
Other workstations are available, ranging from a 6 in display for the driver's station to a 16 in situation display with separate keyboard and pointing device.

Status

A limited number of systems have been purchased for the Swedish Army.

Contractor

SaabTech Systems AB, Järfälla.

VERIFIED

Switzerland

Fieldguard artillery fire direction system

Fieldguard is an all-weather fire direction system for cannon and rocket artillery. It detects the influences of weather and weapon conditions on the trajectory by radar tracking rounds or rockets in flight. The best firing data is then immediately determined by the system computer and transmitted automatically to the weapons. Fieldguard is compact and modularly designed, providing the flexibility for mounting on any wheeled or tracked vehicle.

The system is self-contained. It consists of a radar tracking unit, a digital computer, an operator console for man/machine dialogue, digital data transmission with weapon display units and a power supply.

Fieldguard communicates with various kinds of sensors such as drones, battlefield radars and forward observers. The latter, for example, are equipped with portable target location equipments and connected to Fieldguard by radio or field cable. The target co-ordinates will be located by the laser-goniometer and transmitted to Fieldguard by the digital message device.

Fieldguard receives a firing order with target co-ordinates. The computer calculates the firing data for the basic gun or the ground launcher, based on the stored ballistics, and transmits it automatically to that weapon display unit. Up to three pilot rounds are fired, tracked by the radar and destroyed in flight in order not to alert the target. Fieldguard extrapolates the remainder of the trajectory and then computes the firing data of the adjusted trajectory individually for each weapon. Fire for effect then takes place.

Fieldguard system deployed operationally

Fieldguard on a medium 4 × 4 truck (Oerlikon Contraves)   0123403

Without previous pilot firing, Fieldguard can check and update best firing data by tracking one round within a salvo or one rocket within a ripple of fire for effect.

Status
In service with at least six countries including Germany which deploys it with the 110 mm Light Artillery Rocket System. The system is known to be used with the Avibras ASTROS II artillery rocket system.

Manufacturer
Oerlikon Contraves AG, Zürich.

VERIFIED

Skyguard anti-aircraft fire-control system

Skyguard is the fire-control system that Oerlikon Contraves has developed as the successor to the Super Fledermaus system that they have marketed so successfully in various versions for many years.

Major differences between Skyguard and its predecessors, apart from a general modernisation of design techniques, are the provision of two radars (one for search and one for tracking) in place of one and the use of a digital instead of an analogue computer. Another important difference is the use of pulse-Doppler type radar in place of pulse radar.

Skyguard (left) being used to control two twin 35 mm towed anti-aircraft guns with two four-round Sparrow launchers

Skyguard fire-control system   0001020

Main system components are an Ericsson UAR 1021 search and tracking pulse-Doppler radar, a TV tracking system, a digital computer, a control console and a power supply system. The whole system is contained in a fire-resistant reinforced glass fibre polyester cabin that can be mounted on a trailer or a wheeled or tracked prime mover and is air-transportable. The power supply system is built into the main equipment but can be removed for external operation and is automatically refuelled directly from cans.

A closed-circuit TV system provides for optical tracking and automatic TV tracking. An optical sight on a rotating chair is provided for visual target acquisition with provision for target indication by means of a flashing strobe on the PPI.

As part of the programme to augment the combat value of the Skyguard, Oerlikon Contraves has designed various technical innovations and developed accessory equipment. These primarily ensure optimal acquisition, tracking and combat of low-flying aircraft under extreme topographical conditions. Besides X-band radar and TV-tracking, Skyguard is now available with fully coherent Ka-band radar. This ensures improved target resolution in terrain where the use of radar is very difficult. At the same time it increases the redundancy of the automatic tracking possibilities. With the additional Ka-band radar Skyguard remains easy to operate. The Ka-data also appears on the Combat Display. The Ka-band radar can be easily and quickly retrofitted to all existing Skyguard fire-control units.

The Oerlikon Contraves computer operates in real time to perform threat evaluation functions, calculate ballistic data for guns and command signals for missiles, aid the target tracking operation and monitor and test the entire system.

In addition to the PPI and R-trace displays already mentioned, the control console incorporates a tactical display with numerical readout, the TV tracking monitor, rolling-ball control for PPI markers, joystick control for manual tracking and a matrix panel for data input and output.

A Search Radar Data Extractor (SRDE) by Oerlikon Contraves AG in Zürich is available and the system can be integrated into Skyguard's fire-control unit. It enhances the degree of automation and the time required for target acquisition and tracking is shortened and threat evaluation automated.

The SRDE system is used for the following main tasks:
(a) Automatic first-target alert coupled with automatic target acquisition.
As soon as the first-target alert is acquired by the search radar, operators are automatically alerted. Thereupon, the target is automatically passed on to the tracking radar which keeps on tracking it. Despite very high clutter suppression of the Skyguard's pulse-Doppler search radar, alerts will also be unavoidably triggered by false targets in the shape of moving objects (motor cars, ventilating fans and so on). Such false targets are specifically marked and can at any time be wholly or partially erased at the operator's discretion.
(b) Automatic threat evaluation regarding targets in the combat area or those already acquired by the search radar. With the track-while-scan system it is possible to track several targets simultaneously as well as to evaluate automatically the respective degree of threat. Skyguard's digital system

Close-up of the new man/machine interface units for the Oerlikon Contraves Skyguard III fire-control unit before installation (C F Foss)   0127331

computes, among other things, the threat degree for each individual target, based on the angle of the approaching target, velocity, range and so on. Targets are shown simultaneously on the PPI in the order of priority corresponding to threat involved.

As a matter of course, the data supplied by Skyguard's integrated IFF unit are also included in the threat evaluation.

Skyguard is able to control medium-calibre guns and/or guided missiles. The Oerlikon Contraves designed missile launcher uses the proven undercarriage of the Oerlikon Contraves 35 mm gun and carries four Sparrow or Aspide missiles in containers, as well as the target tracking and illuminator radar antenna. The launcher is directed either remotely from the Skyguard system or locally by the launcher operator.

For crew training, there are two different simulators available: one is integrated in the system, while the larger one is an additional unit connected to it called Training Simulator 2 (TS2).

Skyguard III

In mid-2002 an upgrade, designated Skyguard III, was launched. This upgrade involves stripping down the existing Skyguard system and replacing a number of older, obsolete subsystems with new ones. These include: a new dual-mode surveillance radar, new identification 'friend or foe' system, new processors, a new man/machine interface, new electrical drives and new and more compact electronics. Various sensors can be mounted with the tracking radar, including TV/infra-red cameras and a laser rangefinder. Two crew members are provided with new flat touchscreen displays that can be rapidly removed from the trailer-mounted unit and operated from a safe distance of up to 500 m. The operator uses one and the TV operator, who carries out the target engagement, uses the other.

Oerlikon has designed Skyguard III to detect and track not only aircraft and helicopters but also much smaller targets such as air-launched weapons and unmanned air vehicles. One Skyguard III FCU typically controls two GDF twin 35 mm anti-aircraft guns although it can be used to control up to four gun or missile units, or a mix of guns and missiles.

Status

Skyguard was developed by Oerlikon Contraves as a private venture project and, after evaluation, it has been adopted by the armed forces of more than 20 countries including Austria, Germany, Spain and Switzerland. More than 500 systems have so far been ordered.

Most customers use the system with the Oerlikon Contraves GDF twin 35 mm gun but it is also used with the Bofors 40 mm L/70. The Italian Army selected Skyguard for use in the Spada point defence missile system.

Manufacturer

Oerlikon Contraves AG, Zürich.

VERIFIED

....

Skyshield35 Ahead Air Defence System Fire-Control Unit (FCU)

Skyshield35 Ahead Air Defence System has been developed and built as a private venture by Oerlikon Contraves to meet the requirements of present and future low-level air defence. To counter attacking air targets at ranges of up to 10 km, Oerlikon Contraves provides a gun missile weapon mix system, integrating the missile launcher into the Skyshield35-Ahead Fire-Control Unit (FCU). This gun/missile integrated system is a powerful combination layered air defence system, based on a single FCU for managing both guns and surface-to-air missiles.

The Skyshield35 Ahead FCU offers maximum crew protection and is of small, compact, lightweight and modular design. It consists of an unmanned sensor unit and a detached command post housing two operators (three operators if

The Skyshield35 FCU showing the sensor unit with the multiple beam antenna search radar on top of the multisensor tracking unit (radar plus TV/Laser/FLIR EO Module), with the command post shelter behind (Oerlikon Contraves)   0123402

integrated with a missile launcher). It is suitable for controlling the 35 mm twin field air defence guns, types GDF-003 and GDF-005. If the GDF-003/005 guns are upgraded to fire the Ahead ammunition, the user has available the full Ahead capability. Such Ahead upgraded 35 mm GDF guns are still able to fire the conventional types of 35 mm ammunition.

Description

The Skyshield35 Fire-Control Unit consists of two main units, a remote-controlled sensor unit and a command post.

The sensor unit consists of a multiple beam search radar, a tracking radar and signal processing equipment, together with an electro-optical group, tracker and drive electronics, a data processing group and a power supply unit. Optional items include an IFF unit, a high-resolution TV camera instead of standard TV camera, an FLIR camera instead of TV camera, a laser range-finder and an optical sight.

The command post consists of a command console, a tracking console, a data processing group, an intercom group, a power supply unit and a container assembly. The subunits of the command post are housed in a container which can be displaced from the sensor unit for operation. To provide additional flexibility they can also be installed separately, for example in covered bunkers or emplacements for permanent installation, or in special customer supplied containers for temporary or permanent installation.

A fibre optic link is provided between the sensor unit and command post for transmission of data, voice, Doppler tone and synchronisation signals: up to 500 m (optionally 1,500 m). A twisted pair of field wires connect the command post and the guns to transmit gun angles (relative to horizon), the fire order and Ahead fuze setting time to the guns as well as for transmission of mean muzzle velocity and status signals from the guns to the command post: up to 500 m (optionally 1,500 m). Also provided is an intercom system that connects the command post and the weapon and other stations as required.

The built-in Training Simulator (TS) supports the operational training of the crew. This consists of a software package running on the command post computer and does not need any additional hardware. A fixed number of preprogrammed training exercises are provided. Additional exercises may be defined by the operators by modifying scenario parameters.

Status

Development is completed and the Skyshield35 Fire-Control Unit (FCU) is ready for production. It is being offered as an upgrade to the Skyguard system (see separate entry).

Manufacturer

Oerlikon Contraves AG, Zurich.

VERIFIED

The complete Skyshield35 system showing both missiles and guns (Oerlikon Contraves)   0123400

Turkey

BAIKS - Field Artillery Battery Fire Direction system

ASELSAN have produced two artillery fire direction systems in the BAIKS series, the original 7400 and the more recent BAIKS-2000.

7400

The 7400 Field Artillery Battery Fire Direction System is designed to provide the functions of fast and accurate ballistic computations for a wide range of artillery weapons, fire support co-ordination, message transfer in digital format and ammunition accounting at battery level.

The system consists of a battery fire control computer, a communications control unit, forward observer and fire support officer's digital message devices and gun display units. The system units are linked by digital communications using tactical radios or field cable. The system can also be linked to a tactical fire control system. Messages are digitally encrypted and automatic acknowledgement, error detection and correction capabilities are provided to reduce errors and improve reliability.

The battery fire control computer receives target information and calls for fire in digital format either from the forward observers or from the battalion fire direction centre computer. The battery computer computes the firing data for each gun (up to eight guns) and transfers the firing commands to the gun display units.

The battery fire control computer and the communication control unit can be mounted in a vehicle or dismounted in a stationary command post. The digital message devices and the gun display units are lightweight handheld units, which differ only in their custom keyboards and software, leading to savings in logistics and maintenance. The computer and the communications control unit operate from any 24 V DC source including the vehicle batteries. The hand-held message units are powered by internal rechargeable batteries.

System Units
These include:
(a) AC-7401 Battery Fire-Control Computer
(b) CU-7401 Communication Control Unit
(c) DT-7401/FO Forward Observer Message Unit
(d) DT-7401/DU Gun Display Unit
(e) DT-7401/FS Fire Support Officer Message Unit
(f) DT-7401 BS Battery Commander Message Unit

Functional Capabilities
Fire-control computer
(a) Conduct two simultaneous fire missions
(b) Target location and weapon displacement in grid, polar or laser polar co-ordinates and shift-from-known point techniques
(c) Storage of targets, known points and forward observer locations
(d) Compute firing data for high and low angle trajectories
(e) Compute firing data for distribution of fire using a broad range of ammunition types
(f) Execute fire plans
(g) Storage of no-fire areas
(h) Conduct time-on-target mission
(i) Receive, store and process met message.

Communication control unit
(a) Functions as the system communications processor
(b) Supports digital communications via 3 channels (HF, VHF, UHF radios or wired lines).

Digital message devices
(a) Entry, transmission and reception of digital messages
(b) Message encryption
(c) Communication via radio/wired line and RS-232 interfaces.

BAIKS-2000
BAIKS-2000 is a further development of the 7400 system. It automates the fire direction processes in both towed and self-propelled artillery. The system also provides the Company Fire Support Team Headquarters and Forward Observers with digital capabilities for target acquisition, fire support planning, co-ordination and execution, as well as data communications via tactical radios and field cables. The fire direction software uses the NATO Armament Ballistic Kernel (NABK). The system can easily be configured to support different unit organisations (for example battery-based with 4, 6 or 8 guns, platoon-based battery with 2 platoons, each having 2, 3 or 4 guns) as required by the tactical situation. The system digitally integrates the field artillery batteries and the fire support command elements with the other fire support and/or other functional area C3I systems. BAIKS-2000 units can digitally communicate with tactical fire direction systems located in various fire support and manoeuvre unit command posts, artillery meteorology systems, target acquisition radars and other fire support systems. It supports data communication protocols in compliance with MIL-STD-188-220B and MIL-STD-2045-47001

The system consists of the following components:
(a) The Fire Direction Computer (FDC), which together with a number of peripherals is used at the Battery/Platoon Fire Direction Centres
(b) The Platoon Leader's Digital Message Unit (PLDMU) is used by platoon leaders in platoon-based batteries and by Battery Executive Officers in battery based operations
(c) The Gun Commander's Digital Message Unit (GDMU) which is either installed in the weapon platform or used as a hand-held unit by the gun commander
(d) The Company Fire Support Officer's Computer (FSOC) which is integrated into the Fire Support Team Headquarters vehicle
(e) The Forward Observer Digital Message Unit (FODMU) is a hand-held terminal used by Forward Observers
(f) Tactical Data/Internet Communications Unit (VIA). The VIA supports the data communications protocol defined in MIL-STD-188-220B and provides data communications via tactical radios and field cable. Within BAIKS-2000, VIA is used at Fire Direction Centres and Company Fire Support Headquarters. The HT-7243 terminal (see below) has built-in data communications modules compatible with VIA
(g) Network Connection Unit (NCU). The NCU enables the FDC and the FSOC to be connected to the VIA using Ethernet.

Gun Commander's Digital Message Unit (GCDMU)
The GCDMU is hosted on the ASELSAN HT-7243 Hand-Held Terminal (see separate entry) with specifically developed application software. Gun Commanders can use this device either as a hand-held unit or it can be installed on the weapon, which is the usual practice for SP howitzers. It will perform the following basic functions:
(a) Enables the gun commander to communicate digitally with the platoon leader and the battery or platoon fire direction centre
(b) Receives and displays firing data, commands and other formatted and free text messages sent by the fire direction centre or the platoon leader
(c) Sends gun status, fire mission reports and other formatted and free text messages to the fire direction centre and the platoon leaders
(d) When integrated digitally to a muzzle velocity radar, it receives muzzle velocity data from the radar and sends it to the FDC
(e) Sends firing data (azimuth, elevation and fuse setting) digitally to automatic gun laying, automatic loader and automatic fuse setting systems where in use
(f) Receives gun position and pointing data directly from on-board navigation/positioning systems and sends it to the fire direction centre.

Platoon Leader's Digital Message Unit (PLDMU)
The PLDMU is also based on the ASELSAN HT-7243 Hand-Held Terminal with specifically developed application software. It can be used by platoon leaders in platoon based batteries or by Battery Executive Officers in battery based operations. It can be used either as a hand-held unit or it can be installed in a vehicle, and it performs the following basic functions:
(a) Displays all fire mission related data
(b) Enables the user to communicate digitally with the battery fire direction centre and gun commanders
(c) Enables the platoon leader to prepare and send reports and commands as formatted or free text messages
(d) Receives and displays formatted or free text messages forwarded by the FDC and GCDMU
(e) Displays mission related information over the digital map of the battlefield; eg battery centre, gun positions, fire support co-ordination measures, target locations, observation posts, known points, command posts, friendly and hostile units, terrain data.

Company Fire Support Officer's Computer (FSOC)
The FSOC provides the Company Fire Support Team Headquarters with the capability for fire support planning, co-ordination and execution in a digitally automated environment, as well as data communications with the other fire support elements. The FSOC is a ruggedised computer installed in the Company Fire Support Team Headquarters vehicle and consists of a high-speed processor unit and a high-resolution colour monitor. It performs the following basic functions
(a) Sends/relays local or forward observer-originated target data and fire requests to tactical fire direction command posts, the battery fire direction centre or a company/battalion mortar fire direction centre
(b) Receives/sends formatted and free text messages
(c) Perform fire support planning and co-ordination tasks
(d) Displays fire support co-ordination measures, targets, observer locations, known points, registration points, command post locations and other tactical information on a digital map
(e) Enables the Company Fire Support Team Headquarters to communicate in a digital format with the Forward Observers, artillery fire direction centres (battalion, battery and platoon) mortar fire direction centres (battalion/company mortars) and manoeuvre battalion Fire Support Co-ordination Centre.

Forward Observer's Digital Message Unit (FODMU)
The FODMU is also based on the ASELSAN HT-7243 Hand-Held Terminal with specifically developed application software for forward observer tasks. The FODMU can be used either hand-held - which is the usual practice- or powered with an integrated battery or it can be fixed to a vehicle and powered with external (vehicle) power source. The FODMU performs the following basic functions:
(a) Transmits 'Calls for Fire' and subsequent fire mission messages.
(b) Receives commands/orders and transmits target/fire mission related data via digital communication
(c) Enables a forward observer to communicate with Company Fire Support Team Headquarters, battery/platoon Fire Direction Centre, manoeuvre battalion Fire Support Co-ordination Centre, and battalion/company mortar Fire Direction Centre via formatted and free text messages
(d) Receives target position/ direction data generated by electro-optical target detection equipment used by forward observers and incorporates this data into the appropriate messages sent to the command posts
(e) Displays data such as mission area, fire support co-ordination measures, targets, observer locations, known points, registration points, command posts, friendly and hostile units, critical terrain data on a digital map.

Status
The 7400 entered production in 1990 and it is in use in the Turkish armed forces. Trials of BAIKS-2000 have been completed, the system is in production and the first deliveries to Turkish army units were made in 2003.

Manufacturer
ASELSAN Military Electronic Industries Inc, Ankara.

UPDATED

Skywatcher

Aselsan's Skywatcher is a tactical air defence command and control system that takes air threat information from a variety of radars, produces a recognised air picture in real time and assigns the available air defence weapons to selected targets. All-weather capable, it consists of AD command posts at formation level, interface units for long-, medium- and short-range AD radars, and interfaces for AD weapons systems. It has open system architecture, giving scope for sensor and weapon development, and modular hardware and software on a distributed architecture. The principal features and capabilities of the system are:

(a) Production of a real-time air picture by combining track information from a variety of sensors
(b) Distribution of the air picture to weapons systems
(c) Track identification using IFF mechanisms
(d) Real-time monitoring of the battlespace using digital maps
(e) A variety of analytical tools overlaid on digital maps
(f) Display and distribution of system unit status including location, operational readiness and equipment status
(g) Manual, semi-automatic or automatic target-weapon matching
(h) Secure communications provided by the TASMUS area communications system (see separate entry)
(i) Simulation and replay for training and analysis.

Status

Believed to be in operational use in the Turkish Army, although no recent information has been received.

Contractor

ASELSAN Military Electronic Industries Inc, Ankara, Turkey.

UPDATED

United Kingdom

Air Defence Command Information System (ADCIS)

ADCIS is a tactical C3 system which provides a fully integrated command information system for air defence units. It operates over the British Army's Ptarmigan tactical area communications system and the Clansman VHF combat

ADCIS manpack terminal (AMS)
0525553

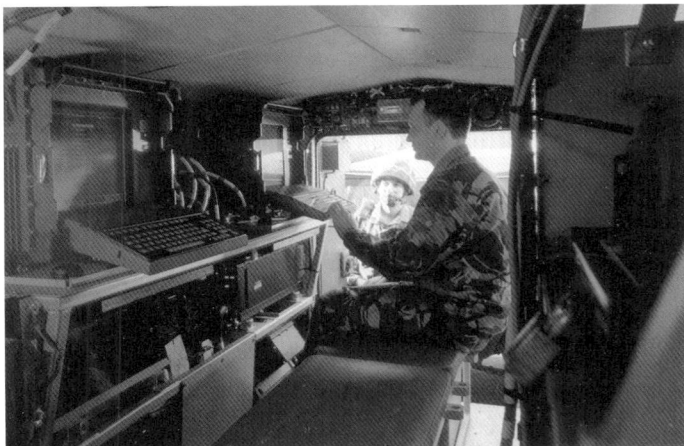

ADCIS in Land Rover (AMS)   0525552

ADCIS vehicle terminal (AMS)   0525550

ADCIS manpack terminal linked to Clansman PRC 351 VHF radio (AMS)
0525551

net radios. Its main function is to provide effective command and control of army air defence weapons which, for the British Army, include Rapier, Starstreak and Javelin. ADCIS will also allow full and safe utilisation of airspace by friendly aircraft including helicopters. This is accomplished by the timely passage of control orders to ground-based air defence weapons. ADCIS is installed in Land Rovers and wheeled and tracked armoured vehicles, and has a manpack version.

Supporting the message-passing functions of ADCIS is a series of software modules designed to assist staff by the automation of a number of procedures carried out at headquarters. They include the ability to plan future air corridors and air defence weapon sites, accounting for air defence ammunition stocks, warning troops of hostile air raids, NBC contamination and timely dissemination of information to allow the efficient employment of the weapon systems.

The system uses military hardware developed from the commercial environment to reduce the development risk and is centred on 58 military specification VAX Model 860 computers (supplied by Raytheon) running software based on Ada.

Status

In May 1988, GEC-Marconi (now Alenia Marconi Systems (AMS)) announced that it had won a £90 million fixed-price contract for the development of ADCIS following competitive feasibility and engineering studies. For the main development contract, a number of major subcontractors (Raytheon, CDC, Cossor, Siemens Plessey, DEC, H-P, SEMA and Logica) were used.

The system has been fully delivered to the British Army and is in service.

In September 1999 it was announced that AMS had been awarded a contract worth over £4 million to enhance ADCIS, providing for changes in flexible operations, point-to-point communications, control of preplanned airspace and general airspace control procedural enhancements. ADCIS will be replaced in the short term by the GBAD-BRIC (see separate entry) and in the long term by the GBAD BISA.

Contractor

AMS, Camberley, Surrey.

UPDATED

Air Defence Siting Computer

The Air Defence Siting Computer (ADSC) is a planning tool to assist in the optimal siting of Rapier and other SHORAD systems. ADSC offers the user a simple way of storing and analysing data on many potential air defence weapon sites, Vulnerable Points, Routes and Defended Areas. It can produce optimised deployment selections for any number of weapon sites against any user-defined threat. The speed of calculation allows the user to try out a variety of different threat options to identify optimal deployment. The simulation of attacks against the chosen deployment option then allows the robustness of the defence to be tested. The system will generate reports for detachment commanders and staff and full integration or data exchange with other AD C3I systems is also possible.

The ADSC uses Windows NT/CE OS and can be integrated with other MS Office applications such as Word and PowerPoint. It can use a variety of mapping data, including ASRP, CRP, ADRG, CADRG, VMap, DFAD, DFAF and CIB. When available, site data can be updated and scanned images, diagrams and digital photographs added to the database to support briefings. Survey data can also be collected through the use of integrated GPS and laser rangefinders.

Status

ADSC has developed from a computer-based siting planning aid originally introduced into the Royal Air Force in 1986 for use with Rapier. This was updated in 1994, entering service as ADSC in 1996. It was subsequently updated in 1999. In 2000 it was also purchased for use by the Royal Artillery and Royal Marines, for use with Rapier, Javelin and Starstreak. ADSC is included as the site planning software in the MBDA Jernas AD system, a contract for which was signed with Malaysia in 2002.

Manufacturer

Cunning Running Software Ltd, Romsey, Hants.

NEW ENTRY

ATacCS Enhanced Liaison Computer (ELC) (EDS)   0132530

Army Command Support Application Suite (ACSAS)

ACSAS - the Army Command Support Application Suite - is a suite of integrated software tools designed to provide information systems support to army staffs at the four star down to one star levels of command, both in barracks and in the field where elements will also be deployed to unit level. It provides tools to support the G1 (Personnel), G2 (Intelligence),G3 (Operations and Plans) and G4 (Logistics) staffs, and succeeds the previously non-integrated Projects AP3 (G1), GP3 (G2/G3) and QP24 (G4). ACSAS makes use of a client-server architecture and integrates COTS products in accordance with the UK MoD's Common Operating Environment which sets out a number of system procurement requirements and provides a flexible basis for all future CCIS, including the ability to support Joint/Allied operations. Information is presented to the staff on a geographic information system with NATO military symbology, from which further data is accessible from the relational database, allowing the monitoring of status direct from the electronic map display. Among the toolsets provided are those to support: information management; planning - including manning, movement, sustainability, Intelligence Preparation of the Battlefield (IPB), creating ORBATS and TASKORGS, and analysing the feasibility of options; creating and distributing orders and directives; and intelligence analysis, correlation, asset management, requirements management and target analysis. A synchronisation matrix is provided and troops can be tracked into/out of theatre, and through the casualty train.

Status

ACSAS entered service in July 2000. It is being upgraded with annual releases.

Contractor

BAE Systems, Christchurch, Dorset.

UPDATED

Army Tactical Computing System (ATacCS)

First designed as the Interim Ace Rapid Reaction Corps Information System (IARRCIS) for NATO's ARRC, the British Army Tactical Computing System (ATacCS) is a highly integrated modular package supporting Divisions, Brigades and Supporting Arms, including Special Forces, Signals, Logistics, NBC, Helicopter and Engineer units. It combines commercial-off-the-shelf (COTS) information technology with purpose-designed specialised elements to meet military requirements. Based on MOTS rugged hardware and COTS software, it is claimed that ATacCS offers more rapid development, reduced cost and reduced procurement risk, while improving overall effectiveness and interoperability. It provides an integrated system that is flexible, yet secure and resilient, that can be rapidly adapted and deployed in response to dynamic situations. By building up network components, ATacCS can be configured for a wide range of possibilities, from systems comprising a single LAN, located, for example, in a forward HQ, to numerous LANs across deployed HQs, barracks and liaison teams, with fibre optic interconnections and trunk bulk transfers.

ATacCS provides secure fibre optic battlefield Local Area Networks (LANs), which interconnect secure, rugged PCs and provide users with a variety of software

tools. Dispersed HQ locations based on LANs communicate over a Wide Area Network (WAN) using X25 services (including Ptarmigan). ATacCS uses a commercial X400 message handling system to deliver messages across LAN and WAN, providing a cohesive corps infrastructure. Support for standard military message text formats enables interoperability between ATacCS and other systems and databases. The system is designed to ensure effective message handling in a frequently changing, secure network. It may be used in barracks with commercial equipment, thus providing a common operational and non-operational environment. Common software and communications interfaces link peacetime locations and the field, providing users with software packages and tools.

In the field an HQ may typically be split over several locations, for example Main, Rear, and Forward, each using local group oriented PCs and peripherals. The ATacCS components are used to form small LANs which correspond to cells in the HQ, providing staff officers manning those cells with information and software support to co-ordinate various functions such as operations, planning and intelligence, artillery, logistics and communications. Liaison officers remote from the main HQ are equipped with portable rugged computers with built-in Ptarmigan capabilities to give direct file transfer with the main ATacCS system. Combining a flexible message handling system with easy to use graphical management software, the system can be rapidly and easily configured to support traditionally difficult operational activities such as the management of fragmented headquarters, role shadowing between alternative headquarters, change of command and general changes due to movement.

Using interchangeable network components, the ATacCS system is designed to be extremely flexible and configurable for the widest variety of possible deployment scenarios and user requirements. It is based on secure LANs of rugged PCs, which can be interlinked across a WAN. This provides users with the building blocks to develop many different configurations, ranging from simple LANs to corps-wide integration.

The ATacCS fibre optic networks can support up to 1,922 users per LAN. These battlefield LANs communicate over the WAN using the X25 Ptarmigan Packet Switched Network and handle bulk transfers over the Ptarmigan circuit switched network, SATCOM or public switched telephone network. The system integrates commercially available and easy to use X400 message handling, X500 directory and network management software with personal productivity tools that support standard military message text formats (ADatP-3) to facilitate interoperability. Resilience is added by the use of duplicate message stores and dynamic alternative routeing.

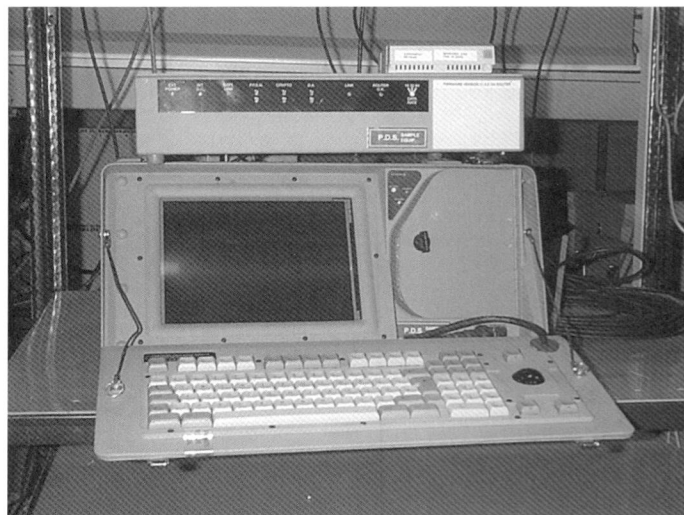

ATacCS Network Computer and Router (EDS)   0132531

ATacCS ApplicationServer (EDS)   0132532

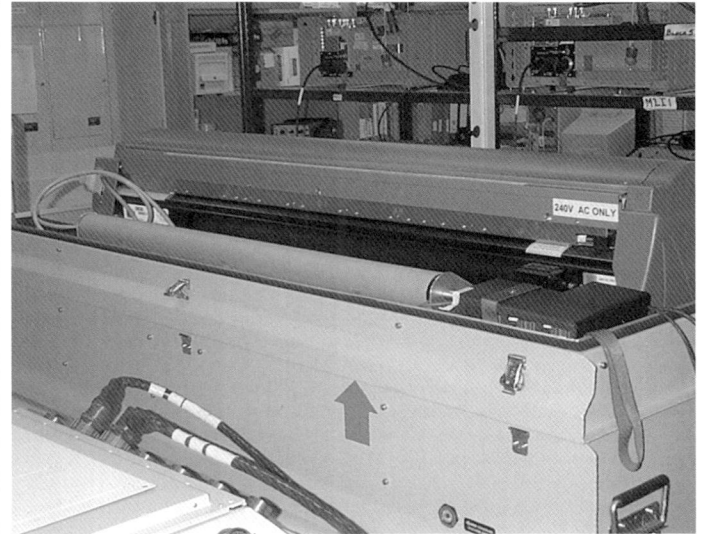

ATacCS A0 Plotter (EDS)   0132533

Originally, IARRCIS provided Windows, DOS and UNIX environments with integrated Microsoft Office productivity tools, together with an Oracle client/server environment for strategic user applications. ATacCS, fielded in early 1999, replaced the UNIX Server, Windows client with an all Windows NT4 solution, providing the productivity of Windows-compatible products. The use of widely available packages provides software which is easily maintained and upgraded in line with commercial developments and changing user requirements.

The basic ATacCS hardware building blocks are a desktop computer, laptop computer, fibre optic hub, portable printer and a tape back-up unit. The network computer is a rugged IBM-compatible desktop personal computer comprising a Pentium P133 processor with integrated Ethernet and SCSI controllers, two removable hard disk drive bays and an integral 800 × 600 TFT flat screen and keyboard with integral trackball. Full/half-length expansion slots are available. LAN and WAN connections are available directly to the computer. WAN connection is facilitated by direct connection to Ptarmigan through packet-switch and circuit-switch access. Power can be drawn from a wide range of primary power sources and 1 hour continuous use from integral batteries is possible. Expansion slots allow the computer to be adapted for a wide range of specific customer applications utilising COTS cards.

The laptop is the Enhanced Liaison Computer (ELC), introduced under a separate contract in 1999. The ELC has a Pentium III 700 MHz processor with integrated Ethernet and SCSI controllers, a removable hard disk drive, CD-ROM and LS120 floppy disk, keyboard with integral trackerball and mouse, and 15 in colour screen. It has 512Mb of RAM and a 1,024 × 768 display. One PCMCIA and full length ISA expansion slot are available for additional card support. Like the desktop, power can be drawn from a wide range of primary power sources and at least 1 hour continuous use from integral batteries is possible.

The fibre optic hub is designed specifically for battlefield LAN connectivity and is supplied with integral stand-by battery support and brackets for rack, bulkhead or worktop mounting. It provides eight fibre optic Ethernet ports with hermaphroditic rugged fibre optic connectors. An additional connection is available for cascading to another hub. Up to four hubs may be cascaded while still forming a single Ethernet repeater. Management facilities are available through an SNMP compatible agent, accessed through any of the Ethernet ports. The hub will operate from AC and DC supplies, requiring only 15 W and is provided with surge suppression to protect the unit from fluctuations resulting from unstable supplies.

The rugged tape back-up uses the universally available, proven and rugged DC6000 style, 0.25 in cartridges with capacity to 1 Gbyte. The unit has a high-speed SCSI interface and is directly compatible with commercial tape drives. Built-in power-on self-test is an added feature. This unit may also be configured for use with CD-ROM or removable hard disk drives. The back-up unit will operate from AC and DC supplies, requiring only 15 W and is provided with surge suppression, to protect the unit from fluctuations resulting from unstable supplies.

The portable printer is designed specifically for battlefield printing where compactness and low weight are important and is based on ink-jet technology with the advantages of quiet operation, low power consumption and single or multiple cut-sheet feed. The user has the option of monochrome or colour. The interface is an industry standard Centronics parallel. The printer will operate from wide ranging AC and DC supplies, requiring only 20 W and is provided with surge compression to protect the unit from fluctuations resulting from unstable supplies.

Specifications
Desktop computer
(a) Processor features: P133 Pentium in socket 7, options for AMD K6-2 to 300 MHz
(b) System: up to 512 Mbyte DRAM
(c) Serial ports: 2 × RS-232
(d) Parallel port: 1
(e) Drive bays: 2 × Ultra SCSI available
(f) CD-ROM: optional or tape streamer
(g) Disk options: up to 9 Gbyte
(h) Floppy disk drive: 35 in slimline 1.44 Mbyte
(i) Display:

Full colour SVGA screen, 800 × 600, 10.4 in TFT display option up to 1,024 × 768
Option for 12.1 in TFT flat panel
IP65 keyboard with full travel keys and integral trackball
(j) Power supply: autoranging, integral, 100-264 V AC 47 to 63 Hz, 24-32 V DC
(k) Battery operation:
1 h (configuration dependent)
Four full AT and three half AT expansion slots
Passive backplane
(l) Other facilities include:
Optional PCMCIA
External SCSI port (SCSI-2)
External video port to 1,280 × 1,024 × 65 K colours
Rugged fibre optic Ethernet interface
External mouse and keyboard ports
Customer-defined connectivity options
Direct Ptarmigan Packet and Circuit Switch interface.

Enhanced Liaison Computer
(a) Processor: 700 MHz FC/PGA. Two SODIMM memory sockets; integral SCSI (external) and IDE controllers; LVDS flat panel interface; 4 Mb graphics adaptor; soundblaster compatible audio
(b) Storage: Integral slimline CD-ROM, 3.5 in high capacity disk drive (LS120), 2.5 in RHDD (to 12 GB or greater)
(c) Display: Integral 15 in TFT full colour (1,024 × 768 resolution)
(d) Interfaces: PCMCIA slot (type III) and full size AT expansion slot for extra card support; external colour video; audio output; IEEE parallel (printer); 2 serial ports; SCSI port
(e) Power supply: Integral 1 h (minimum) battery backup using smart Lithium Ion batteries
(f) Optional interfaces: Built in Ptarmigan interface for direct connection to UK trunk system and SCRA; 5 port USB hub and 2 extra serial ports; video grab/ camera output; audio output
(g) Proven application support: MS NT4 and Office, Oracle, UK selected military applications
(h) Other supported facilities: external keyboard/trackerball
(i) Weight: 12 kg (25 lb)
(j) Dimensions: 420 × 290 × 110 mm.

Fibre optic hub
(a) IEEE 8802.3 (FOIRL) compliance
(b) 8 fibre optic Ethernet ports
(c) Expansion (stacking) port
(d) Ethernet link status and activity indication
(e) BootP, SNMP management
(f) Battery-backed operation
(g) Power supply autoranging, integral, 100-264 V, 47-63 Hz, 24-32 V DC
(h) Weight: 9.5 kg
(i) Dimensions: 43 5 × 210 × 88 mm.

Rugged peripherals unit
(a) 5.25 in drive bay capacity (RHDD, CD-ROM, QIC)
(b) SCSI-2 interface
(c) Power-on self test
(d) Front panel power and status indicators
(e) Weight: 8.5 kg
(f) Dimensions: 330 × 230 × 180 mm.

Portable printer
(a) Centronics parallel connection port
(b) Monochrome or colour
(c) Single or multiple sheet feed
(d) 600 × 300 dpi resolution
(e) PCL3 printer command language

(f) Power supply autoranging, integral, 100-264 V, 47-63 Hz, 24-32 V DC
(g) Weight: 13 kg
(h) Dimensions: 480 × 215 × 290 mm.

Large screen display
(a) 1,280 × 1,024 × 65 K colour multisync
(b) 16 in viewable area
(c) Battery-backed operation
(d) Free standing or wall mounted
(e) Weight: 15 kg
(f) Dimensions: 440 × 330 × 100 mm.

Status
ATacCS was first fielded in 1995. In 1996-97 UK urgent operational requirements led to further contracts, and the enhanced version was then developed in the period 1997-99. In 1999 the ELC was introduced. Currently, the majority of Army HQs and the Royal Marines are equipped with ATacCS, together with HQ ARRC. ATacCS provides the basic structure for the UK's Land Digitisation Stage 1, and elements will probably exist well into Stage 2. A Remote User Terminal capability is available for use at battle group level in conjunction with SCRA.

Contractor
EDS Defence Ltd, Hook, Hampshire.

VERIFIED

Battlefield Artillery Target Engagement System (BATES)

BATES (Battlefield Artillery Target Engagement System) is a semi-automatic data processing and fire-control system developed for the British Army. It features distributed data processing at each command level, having computing power and functionality appropriate to the level involved. Information is conveyed around the system using digital communications.

Development
Development of the system began in 1976 with a feasibility study with full project definition beginning in 1980. At this time Marconi was responsible for the hardware and Scicon was responsible for the software. Spending restrictions and software difficulties meant the system, which was scheduled to enter service in about 1985, slipped two years. In October 1985, Marconi Command and Control Systems took over responsibility for both hardware and software when it was awarded a £100 million contract for Phase 1 of the programme. Hunting Hivolt was responsible for vehicle installations while Marconi Secure Radio Systems was responsible for encryption.

The development of the system for conventional weapons was completed in December 1992 when the UK MoD accepted the system from GEC-Marconi (now Alenia Marconi Systems (AMS)). Delivery of approximately one third of the total UK hardware requirement (around 200 systems) was included with the system development task. The integration of other sensing systems and new weapon platforms into BATES, such as Phoenix, COBRA (see separate entry), MLRS, BMETS and AS90 was subsequently completed.

Description
The hardware includes hardened digital processors, plasma visual display units, data entry units, gun display units and communications interfaces with the Clansman combat net radio and the Ptarmigan tactical network.

The function of the system is to accept inputs from observers and artillery staff using sensors such as radar, sound ranging systems and reconnaissance devices. Data is then stored, analysed and displayed to assist the artillery commander in

The interior of a BATES vehicle

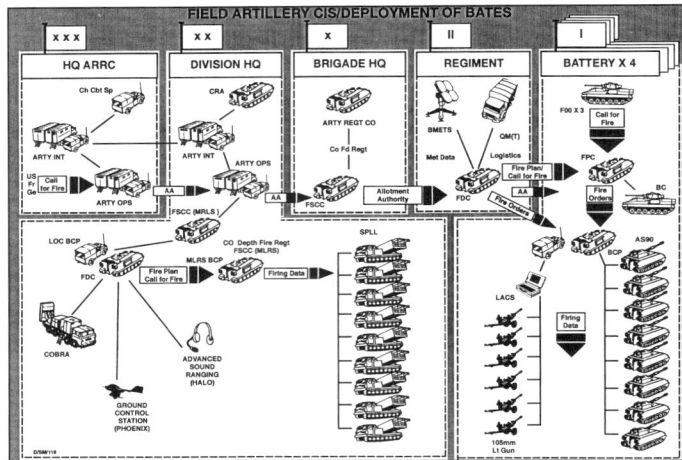
BATES schematic diagram

Dismounted BATES (AMS) 0525549

decision making. Net and trunk communications systems link target acquisition with artillery headquarters and the fire units. The automatic data processor allows forward observers to feed target data to the command posts and speed up the engagement process by reducing the human links in the chain.

Status
The development of the system for use with the specified conventional was followed by a series of user trials. The UK Ministry of Defence accepted 'Conventional BATES' in December 1992.

In 1993, GEC-Marconi (now AMS) was awarded a further production contract for the manufacture of additional BATES hardware to complete the current UK requirement. Deliveries were completed in 1996. Also in 1996, successful integration with Counter Battery Radar (COBRA) was achieved, and in 1998 integration to Phoenix and BMETS completed the suite of interfaces to sensing and weapon platforms offered by the system.

In 2001 BATES software V2.6 was introduced to the UK Field Army, which provides a claimed 'step improvement' to core functionality and general operation. Improvements to both hardware and software continue to be made under UK MoD Post Design Support tasking.

International interoperability development also continues which will ultimately enable BATES to interface with the equivalent French, German, Italian and US artillery command and control systems.

Contractor
AMS, Camberley, Surrey.

UPDATED

ComBAT and P BISA

ComBAT and P BISA are the two main components in the basic battle management system which is being procured as part of the UK's Bowman project, although it will contribute to the wider Digitisation of the Battlespace (Land) programme. The system will consist of the Common Battlefield Application Toolset (ComBAT), Infrastructure (I) and armoured Platform Battlefield Information System Application (P BISA), commonly known as "CIP". CIP is three interrelated projects procured as a single entity.

ComBAT

ComBAT is built on the infrastructure of the Bowman Situational Awareness Module (BSAM) (see below) originally designed for the Bowman programme by General Dynamics C4 Systems (GDC4S). Both BSAM and ComBAT utilise the Command Data Network System (CDNS)™ (see separate entry) Battle Management Software as the core and have been customised to meet the specific needs of the British Army. ComBAT provides improved situation awareness, messaging, planning, logistics, GIS, intelligence, and office automation tools and has been developed for use in armoured fighting vehicles, individual soldier systems and staff configurations. It is designed to operate on individual terminals either stand-alone or as part of a local area network and all levels from combat platform to corps. The software will reside on terminals within all cells of each headquarters at the formation, unit, and sub-unit level, as well as on platform terminal devices such as the Commander Crew Station (CCS) and Battle Group Thermal Imager (BGTI). Additionally, ComBAT will reside on commander's terminals at each of these levels wherever they are provided.

ComBAT is a fully integrated battle management software suite that combines an enhanced BSAM module with existing 3rd party software tools. The BSAM enhancements focus primarily in the overall improvement of the situational awareness, messaging, and planning functions within BSAM including the Commander's War Diary and intelligence, G1 and G4 planning. Software tools will provide additional functionality for intelligence, logistic planning and operationally critical assets/resource management and usage monitoring.

The Commander's War Diary

ComBAT will extend the functionality of the Commander's War Diary to include the storage of entries into the activity log as well as archiving the log to be searchable by the user at a later date. Log entries will be searchable by Date/Time Group (DTG) band, formation, unit, sub-unit and general text. The Commander's War Diary contains default elements that will automatically be entered into storage including the activity log, all orders issued and received, and emergency burial details. In addition, other user-specified items can also be selected for automatic storage.

G1 Planning

Requiring a variety of computational tools to allow for proper planning as well as effective measurement, ComBAT is enhanced with a number of features to enable more accurate G1 planning including the ability to predict:
(a) Number of casualties
(b) Needed medical support
(c) Enemy prisoners
The enhanced personnel functionality will allow for additional details to be included for each member of the organisation including emergency burial details, KINFORM and NOTICAS (the UK procedures for passing details on casualties).

G4 Planning

ComBAT provides enhanced logistics planning with the integration of force tracking capabilities including tracking operationally critical assets and resources and monitoring their usage against predictions.

Enemy Force Movement Prediction

Utilising engineering intelligence data, ComBAT will calculate enemy and own force movement speeds to assess the impact of terrain on the channelling of the flow, particularly with respect to the creation of choke points. ComBAT can be used to analyse movement through the terrain to support the IPB process by taking into account restricted and severely restricted areas of going, man-made obstacles, slope and height of terrain and availability of roads to support movement.

Route Management

ComBAT will provide a tool for displaying route information in a graphical format. Operating on route networks and engineering intelligence information, the software can predict the movement profile against time of a given convoy. In addition, the route management software will import national movement characteristics and incorporate these characteristics into its model.

P BISA

The Bowman Situation Awareness Module (BSAM), has evolved from the Command Data Network System (CDNS)™ Battle Management Software (see separate entry)and is known as the Platform Battlefield Information System Application (P BISA). BSAM was originally developed to meet the Bowman provisions for networking, messaging and information display, and provides an automated command and control system that can be employed from platform to corps at any location on the battlefield. It provides immediate capabilities for the UK's Digitisation of the Battlespace (Land) programme. Although it is designed for use with the Bowman communications equipment, BSAM is also capable of operating with the UK's legacy communications systems without the need for equipment upgrades. In the case of the legacy Clansman system, the only additional requirement for interoperability is the use of the GDC4S Micro Inter-networking Controller (MINC), which provides modem connectivity.

The BSAM provides automated support for commanders and staff officers to represent textually and graphically the information necessary to appreciate the current situation and to develop future operations. It provides the capability to create all required text and graphics on user workstations and then transmit the entire plan or order (text and graphics) to all addressees.

Messaging support includes the creation, editing, transmission, addressing, receipt and storage of messages. The BSAM supports both predefined (using ADatP3 syntax) and user-defined message templates.

The BSAM fully supports the Bowman APLNR service requirements for receiving, consolidating and distributing position reports; converting and displaying position locations; maintaining and displaying tracks and track history; and using the Bowman APLNR as a navigational aid.

Status

Training on the Bowman system for the initial battalion (1st Battalion, the Royal Anglian Regiment) began in July 2003. The first Brigade to be converted will be 12 Mechanised Brigade. The first brigade-level trial, which will be conducted using two mechanised battle groups and a brigade headquarters, is expected in March 2004. Bowman is expected to be formally declared operational later that month. A second (armoured) brigade trial is planned for late 2004. A rolling programme of conversion will take place thereafter, together with the integration of further BISAs as part of the Digitisation programme.

Contractor

General Dynamics UK Ltd (Prime), Oakdale, South Wales.

NEW ENTRY

....

GBAD Bridging Capability

The Ground-based Air Defence Bridging Capability (GBAD-BriC) is one of the first Battlefield Information System Applications (BISAs) to be introduced into the British forces as part of the digitisation process and, as the name implies, is a bridging capability until the introduction of the GBAD BISA in 2007. It will support the siting, deployment and control of GBAD assets. The front end is based on the in-service ADSC (see separate entry). It will allow procedural (24 hour planning) and dynamic procedural (15 minute planning) control of airspace. Airspace Control Measures (ACM) are imported as part of the Air Control Order (ACO) or entered as requests and passed as required, restricted firing arcs. Weapon engagement zones are calculated at each fire unit and weapon status is available and displayed as required throughout the system.

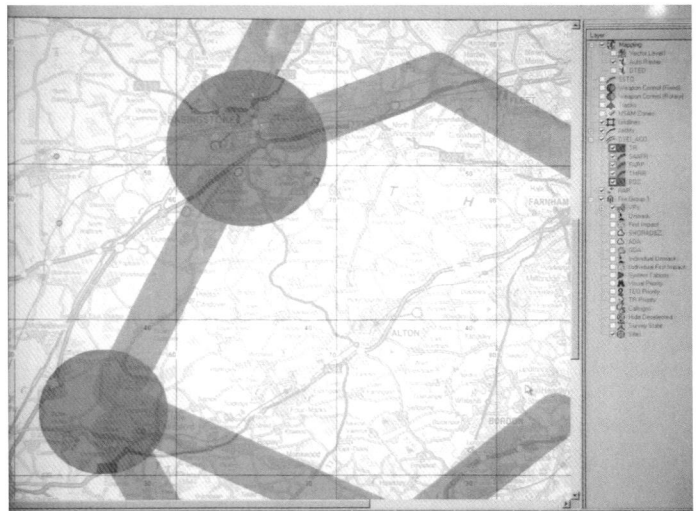

Screenshot from GBAD BriC workstation showing air corridors
(Patrick Allen/Jane's)   1027031

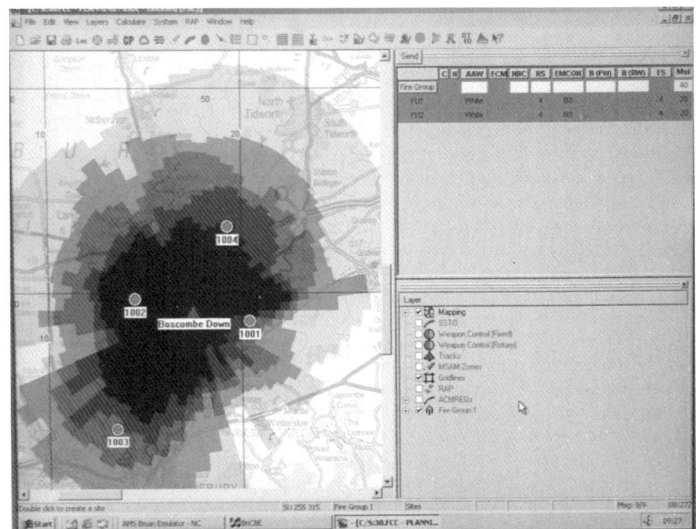

Screenshot from GBAD BriC workstation showing threat levels and defensive coverage (Patrick Allen/Jane's)   1027029

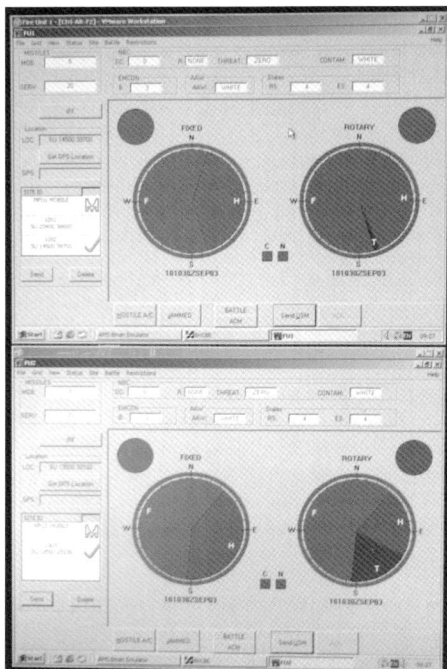

Screenshot from GBAD BriC workstation showing status of two fire units
(Patrick Allen/Jane's)
1027027

At formation headquarters the system will provide a control mechanism for the GBAD community; import and distribute airspace control orders; distribute IFF codes; approve airspace control requests; and provide air raid and GBAD information to non AD units. At sub-unit level the system can provide planning tools to produce the optimal weapons sites layout; provide control of multiple fire groups and fire units; provide positive control over fire units; provide immediate access to weapon status; and enable the fast issue of orders. At weapon-platform level it provides an immediate indication of weapon control status and access to orders; fast status reporting; access to textual and graphical deployment orders; and can be given an optional specific display for airfield defence.

The system is compatible with a range of communications systems, including analogue and digital radios, GSM, mobile and fixed telephone systems. It runs on COTS hardware using Windows OS.

Status
AMS, as prime contractor, was awarded a contract worth approximately £2 million to develop the capability in October 2002. The first Factory Acceptance Test (FAT) was successfully completed in June 2003. The second FAT is due in late October 2003, with fielding due in 2004.

The system is being marketed under the name Meerkat by AMS.

Contractor
AMS (Prime), Frimley, Surrey.
Cunning Running Software Ltd, Romsey, Hants.

NEW ENTRY

GUNZEN Mk 3

Based on the Zengrange ZHC2000 Hand-Held Computer, the GUNZEN Mk 3 provides firing data and support functions for artillery and ground to ground rocket systems. The Mk 3 has an improved user interface and faster computation than existing GUNZEN computers. The system is designed to meet the requirements of DEFSTAN 00-35 Pt2 and Mil-Std 810E for dismounted troops in the ground mobile role. The backlit display and membrane keyboard allow operation in all conditions day and night.

The computer may be programmed for any weapon system and ballistics can be changed or upgraded in under one minute. Full ballistic computation is carried out for up to 8 weapons including the application of meteorological data and the checking of safety areas and crest clearance. Engagement may be carried out for two simultaneous missions.

The key functions provided by the GUNZEN Mk 3 are:
(a) Engagement of missions by grid polar or recorded target
(b) Calculation of dispersed fire (parallel, converged, linear, circular, rectangular or range and lateral spread)
(c) Target storage for 200 targets, five defensive fire lists and two fire plans
(d) 50 observers locations and 50 own troops locations
(e) Meteorological data storage for two meteor messages and up to 70 registration corrections
(f) Comprehensive safety and crest calculation and storage
(g) Ammunition stock storage and calculations
(h) Comprehensive survey calculations and storage.

Specifications
(a) Size: 182 mm × 260 mm × 50 mm
(b) Weight: 1.9 kg
(c) Battery life: 18 h continuous use

The Zengrange GUNZEN Mk 3
0122988

Status
In service with Brazil, Japan, Malaysia, Mexico and being actively marketed elsewhere. The UK MoD bought the previous version, GUNZEN Mk 2, and is considering GUNZEN Mk 3. An expected contract with the Royal Netherlands Marine Corps is currently (September 03) in abeyance due to financial restrictions. In partnership with BAE Systems GUNZEN Mk 3 is being supplied to an unspecified country in Asia and, in partnership with RDM Technology from the Netherlands, the system is being supplied for a Philippine army modernisation programme.

Manufacturer
Zengrange Defence Systems Ltd, Leeds.

UPDATED

Lightweight Artillery Computer System (LACS®)

The Lightweight Artillery Computer System (LACS®) is a simple, portable artillery fire-control system which provides C2 facilities between observers, guns and command centres and enables the rapid passage of digital fire-control information. It is a different system to the similarly named one from AMS (see separate entry). It will provide ballistic firing data calculated to the nearest Mil for both azimuth/bearing and elevation. Communications can be over VHF radio systems including the Jaguar, PR4g, SINCGARS and Panther systems among others. The system consists of the following components:

Forward Observers Data Computer (FODAC). This provides artillery forward observers with the means to call for fire from brigade, battalion or battery assets.
Facilities:
(a) Initiation and control of all types of fire mission
(b) Creation and implementation of quick-fire plans

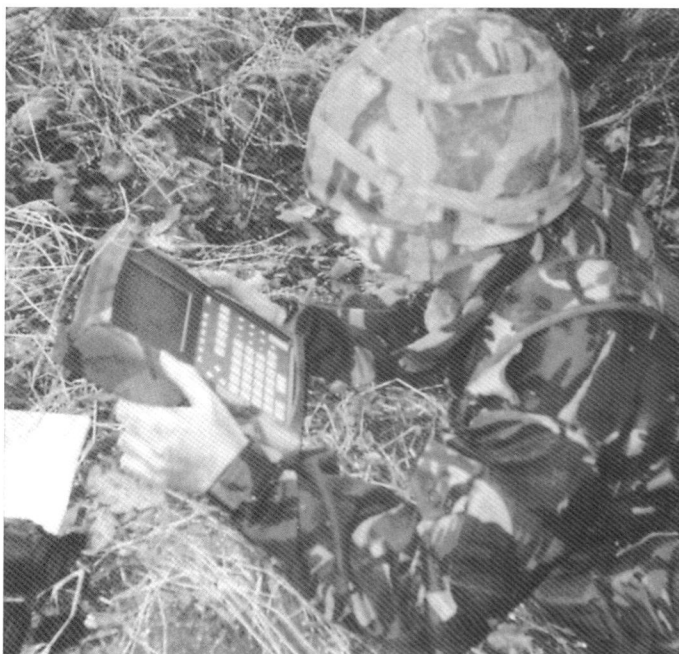

LACS® forward observer's data computer (Zengrange)
0116927

(c) Storage of up to 200 target records, 5 defensive fire lists, 2 fire plans, 50 observers' locations, 50 own troops' locations
(d) Tactical message management
(e) Easily integrated with electronic target acquisition systems, such as the Leica TAS 10, TAS 2000, Vector 1500/4500 laser binoculars and GPS receivers including Rockwell Collins' PLGR/SPGR
(f) Supports observer safety routines.
Weight 1.9 kg, dimensions 18 cm × 26 cm × 4.5 cm. AC or DC power supply, or lithium battery with up to 15 hours life (continuous use).

Brigade/Regimental Computer Unit (RCU)/Battery Computer Unit (BCU). Located at unit or battery headquarters level, this allocates resources to fire missions and tasks.
RCU facilities:
(a) Status display of all unit assets
(b) Processes all fire missions
(c) Allocates resources to specific tasks
(d) Controls unit fire plans
(e) Collation and management of tactical and logistic reports.

LACS® Regimental/Battery computer unit (Zengrange)   0122986

LACS® firing data display unit (Zengrange)   0122987

BCU facilities:
(a) Processes incoming calls for fire
(b) Automatic management of fire plans
(c) Controls progress of fire missions
(d) Manages battlefield tactical and logistic reports.
Weight 6.8 kg, dimensions 40 cm × 26.7 cm × 10.2 cm. AC or DC power supply, or lithium battery with up to 6 hours life (continuous use). A similar computer can be carried by the battery or unit commander. Both can be connected to a ruggedised printer.
Data is passed via the Firing Data Control Unit (FDCU) to the Firing Data Display Unit (FDDU). Located at the guns, this displays firing data provided by the RCU or BCU. It can be fitted internally for SP weapons, or mounted on a tripod beside towed weapons and can be used independently as a gun platform loudspeaker for 2-way communications by line or radio to and from the command post. Facilities:
(a) Provides line or radio communications to the BCU
(b) Internal memory for constant Final Protective Fire (FPF) data storage
(c) Manages and transmits status messages to BCU.
Weight 8 kg, dimensions 28 cm × 2.8 cm × 20 cm.

Status

In service in an unspecified Middle East country and being actively marketed elsewhere. In partnership with RDM Technology from the Netherlands, the system is being supplied for a Philippine army modernisation programme.

Manufacturer

Zengrange Defence Systems Ltd, Leeds.

UPDATED

Lightweight Artillery Computing System (LACS)

A rugged laptop-based artillery fire-control system used by light forces in forward areas to provide data to battery command posts. Uses BATES (see separate entry) software. Should not be confused with the very similarly named system from Zengrange (see separate entry).

Status

Entered service with British Army 105 mm Light Gun regiments in 1999.
AMS, Camberley, Surrey.

NEW ENTRY

MiniMOST

The MiniMost military satellite communications terminal is designed to operate with Skynet, NATO, DSCS and other SHF constellations. The system comprises a 1.7 m petalised antenna mounted on a motorised tripod mount assembly. The associated electronics are housed in two robust transit cases. Designed to operate up to 128 kbps, the terminal provides reliable voice and data communications under challenging conditions and has recently exceeded its design aims by providing 384 kbps of voice and data over the Skynet 4 satellites.

Status

Originally delivered by Matra Marconi Space (now Astrium) in 1999, two systems are in service with the UK MOD. Especially designed for use by Special Forces, the system has been used extensively in Europe and other parts of the world.

Contractor

Astrium, Stevenage.

NEW ENTRY

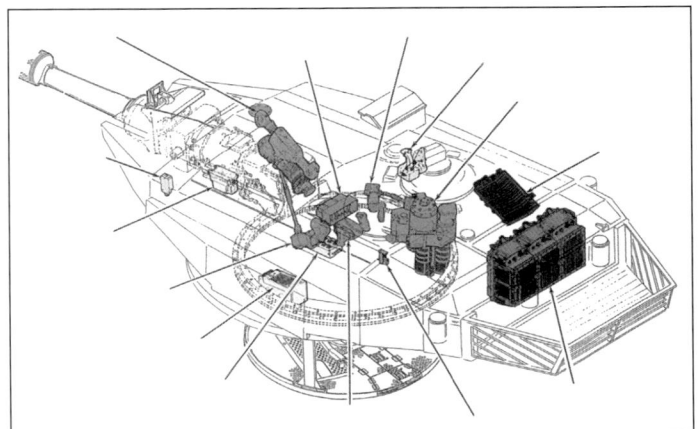

MiniMOST (Astrium)   0101592

RYAN

RYAN is an application designed to support personnel management in an operational theatre, especially in Operations Other Than War (OOTW). It originated with an Urgent Operational Requirement to support the British Army in its deployment into Kosovo. The system is designed to run on most desktop or laptop PCs, with support for small networks, which allows varying numbers of RYAN installations to be deployed as necessary to meet the operational need. The application is easily extensible and tailorable, and can be readily adapted to meet the specific needs of a deployment should the standard facilities need enhancement.

Facilities

RYAN stores information about people such as own or allied forces servicemen and women, civilians working with the military force, visitors to an operational theatre, casualties or Prisoners of War (PWs). As it has few mandatory fields, RYAN can provide support from a basic service requiring little data entry to a data-rich personnel management facility.   Typically, a military force has available encyclopaedic information about its service personnel. RYAN can hold such data and use this as the basis for a theatre record for each person. Apart from basic information such as name, rank, number and date of birth, this could include a history of deployment to the operational theatre, including location within theatre, casualty history and a photograph. RYAN also stores basic information about units and medical facilities in order to support the personnel function.

For casualties, RYAN stores the following information:
(a) The medical unit currently treating the casualty
(b) The ward and bed to which the casualty is assigned
(c) A brief medical diagnosis (not full medical records or any medical-in-confidence information)
(d) Categories for type of casualty, including: nuclear biological, chemical, disease, battle injury, accident, severity of injury/illness
(e) A record of official Notification of Casualty signals and the status of NOK notification
(f) A record of casualty evacuation, emergency burial or return to unit
(g) A full history of changes to the casualty record
Among the items of information RYAN stores for PWs are:
(a) Unique identification number
(b) Name
(c) Gender
(d) Date of birth
(e) Service
(f) Rank or title
(g) Status (officer, other rank, medical personnel, religious personnel, civilian)
(h) Address
(i) Photographs
(j) Power served
(k) Service number
(l) Unit
(m) Engagement
(n) Family and NOK details
(o) Other information as required by the Geneva Conventions

Formation HQ Support

Staff responsible for personnel matters at a Formation HQ can obtain lists of people whose details meet certain criteria (for example, all females in theatre, all persons with a given set of skills, all casualties with battle injuries). They can also obtain summaries of the locations of people at a given time, or summaries of the locations of all people in a given unit. Other facilities available include summaries of medical unit occupancy, planning reinforcements at individual level and details about PWs held.

Operational Location Support

An Operational Location (OpLoc) is a place where the movement of service personnel and other people of interest is monitored. These include Theatre Reception Centres, Places of Embarkation (for theatre), Divisional and Brigade rendezvous points and Unit HQs. At OpLocs, RYAN provides quick methods of displaying the known details of a person and for creating records for people previously unknown. Where a person's unique identity string is held on a magnetic stripe card, RYAN supports a card reading device connected to the serial port of the computer. RYAN already understands the layout of UK service cards, and can read not only the service number, but also other information such as name, rank and unit from the card. It can be taught to recognise unique identities on other types of card, from company ID cards to credit or banking cards. RYAN can be set up to process people rapidly through an OpLoc, recording their reason for transit, the unit (if any) they are accompanying, the date of transit and the location of the OpLoc.

Medical Unit Support

At a medical unit, RYAN can be used to:
(a) Track the location of each casualty in the medical chain
(b) Record medical unit, department name and bed number
(c) Record a history of the casualty's progress
(d) Record the final destination of the casualty
(e) Record brief diagnoses on patients
(f) Obtain summary and filtered lists of casualties by type or location
(g) Record details of incidents that led to the injuries or illnesses
(h) Raise formal signals to be sent to the home country to record the casualty and to instigate the notification of the NOK

Prisoner of War holding area support

At a PW holding area, RYAN stores the information about PWs, and provides the ability to produce reports in formats that meet the requirements of the various Geneva Conventions on the treatment and handling of PWs. In addition, detailed records can be filtered and sorted to support statistics-gathering activities.

Management of the application

RYAN provides comprehensive management facilities, including:
(a) Replication of information between RYAN installations
(b) Exchange of information with other systems
(c) Archiving of old information
(d) Housekeeping
(e) User management
(f) Maintenance of pick lists

System details

RYAN has a client/server architecture. The back-end server database holds most of the database tables. The front-end client application supplies the screen forms, the reports and the query definitions used to support them. By installing the RYAN server database on one computer in a network and the client application on all the computers (including the one with the server database), up to six users can share the same server database simultaneously. RYAN uses record level locking to ensure that only one user can update any one item in the database at any one time.

The system will run on Windows 98, Windows NT4 or Windows 2000, and is designed to run on any PC - desktop or laptop - that meets the minimum specification:
(a) 100 MHz Pentium processor (300 MHz or faster recommended)
(b) 64 Mbytes RAM (128 Mbytes recommended)
(c) 20 Mbytes free disk space for client application
(d) 500 Mbytes free disk space for server database (sufficient for details about 150,000 people)
(e) 1 free serial port for Magnetic Card Reader (optional)
(f) 1 free serial port for PW chip reader/writer (optional)
RYAN has been written using Microsoft Access. The installation CD comes with versions to run with Office 97 (RYAN 97) or Office 2000 (RYAN 2000), with Office XP support due imminently. Should no version of Office be installed on the target computer, there is an option to install the runtime version of Access 97, allowing RYAN 97 to run.

Status

In service with the British Army, where it is designated as AP3, as a project in Digitisation Stage 1. It has been deployed operationally to Bosnia, Kosovo, Afghanistan and the Gulf, and has also been used in support of Op FRESCO, the provision of fire-fighter cover in UK.

Contractors

The project is jointly attributed to:
BAE Systems, Christchurch, Dorset.
AMS (a BAE Systems/Finmeccanica joint venture), Camberley, Surrey.

NEW ENTRY

....

TACISYS

TACISYS is a geographic tactical information system which provides geographical support to staff in formation headquarters. It assists in the Intelligence Preparation of the Battlefield (IPB) process, and enables it to be completed in less than the time previously required, with a higher degree of accuracy and providing new categories of analysis such as route analysis and real-time 3-D mission rehearsal. It also provides the ability to support electronic C4I systems. Two standard ISO containers versions are currently available, either 4.27 m or 6 m in length, and other configurations can be provided to meet customer needs.

Data Input

TACISYS can import or directly read and use all major geographic data formats, commercial and classified imagery sources and the capability can be extended easily where required. All the major data formats are supported, and where digital data is not available TACISYS can scan and geo-reference existing paper maps to create its own data. Supported data formats include:
(a) DIGEST – including the ASRP, CRP and VMAP
(b) Non-DIGEST – such as ADRG, CADRG, NITF
(c) Commercial – including TIFF, SPOT, LANDSAT, AVHRR, CGM and others
(d) Proprietary formats – including ESRI, ERDAS, Marconi Integrated System, AUTO-CAD, Intergraph

Data Preparation

TACISYS has extensive data manipulation and integration capability including the ability to:
(a) change the geographical reference of the data by co-ordinate or data transformation and re-projection
(b) create new data by feature extraction or digitisation
(c) directly edit existing data
(d) produce orthophotos and mosaics
(e) extract data from source data including feature and digital elevation models

The interior of the TACISYS container (Ultra Electronics) 0543467

(f) carry out other corrections on the data such as sensor and atmospheric corrections

(g) geo-rectify a wide range of geographically related data

Data Analysis

The integration of the platform raster and vector GIS software gives TACISYS considerable data analysis capability. The analysis tools are specifically designed to be flexible in application, which enables TACISYS to support operations in both peace and hostilities. The system enables the operator to analyse and extract information from raster and vector data in 2-D and 3-D, and data analysis tasks include terrain analysis, time/goings and cross-country movement.

Imagery Intelligence

TACISYS has a fully integrated set of tools which enable it to analyse, manipulate and extract information from most battlefield imagery sources. The results of the analysis allow extraction and annotation of data for other processes within TACISYS or by other users and systems. Uses of imagery within TACISYS include:

(a) provision of current information to update base vector and raster mapping

(b) provision of a source of data for substitute mapping

(c) 3-D stereo extraction of features and information

Visualisation

The visualisation tools produce a much more realistic representation of the terrain and of the results of analysis, and make the following capabilities possible:

(a) accurate position planning of key assets such as air defence weapons or target designators

(b) rehearsal of routes and approaches by means of drive-through or fly-through of 3-dimensional representations of the terrain. This picture can include modelling of the influence of sensors such as radar

Map Production

TACISYS has the capability to produce standard mapping products in both electronic and hardcopy formats, which can be rapidly produced substitute map products, updates to existing maps or new maps. Substitute mapping and data editing tools allow the production of image maps and addition of new detail to existing maps. The system can also merge data from a variety of sources to produce new mapping. These are complemented by standard mapping templates and layouts that ensure that the end result of any TACISYS task is properly presented to the user.

Output

TACISYS has extensive data publishing capability in both softcopy and hardcopy formats. Softcopy formats include direct support for a wide range of commercial

A TACISYS workstation (Ultra Electronics) 0543468

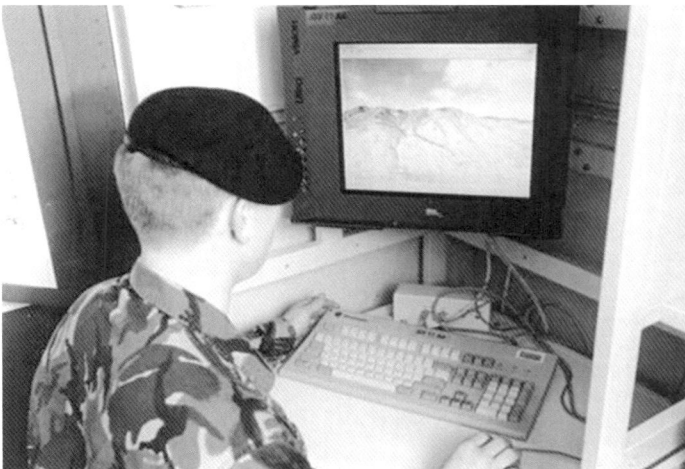

A 3-D map with tactical overlay and grid, produced by TACISYS (Ultra Electronics) 0543469

standard products, including ESRI, ERDAS, MapInfo and others. The results of 3D-visualisation analysis can be saved to videotape, or to computer-readable formats including Apple Quick time or Microsoft AVI. Hardcopy output includes paper and film in colour and black and white from A4 to AO.

Hardware

The system can either be run on two high specification PC workstations running the Microsoft Windows NT® Operating System or a combination of Silicon Graphics and Sun RISC workstations running UNIX operating systems. Two Ultrathinvision™ colour LCD flat panel displays are fitted as standard and additional CRT monitors are included for dual-headed stereo imagery analysis. For disk storage the system is supplied with a 90 Gbyte RAID array per workstation as standard, with options to extend and increase the storage available to 1 Tbyte.

Software

The software suite is based around COTS software from ESRI, ERDAS and Marconi Integrated System, including the following core products:

(a) ESRI ArcView

(b) ESRI ArcInfo

(c) ERDAS Image

(d) Marconi Integrated System Socet Set

These applications can be customised or specialist GIS applications produced using ESRI MapObjects. (See separate entries for ESRI products). The suite also includes Adobe and MS Office packages.

Status

The first prototypes were produced to support the initial NATO deployment of IFOR to Bosnia in late 1995. The operational experience gained from the deployment of the prototype in Bosnia enabled a final procurement specification to be developed for the production TACISYS. The UK is believed to have procured a total of 11 operational systems and delivery was completed in the summer of 1997. These systems provide the integral specialist support to UK Divisional HQs; a national Geographic Support Group for expeditionary operations; integrated specialist support to HQ ARRC; and a NATO Theatre Geographic Support Group. TACISYS has been deployed operationally to FRY, Afghanistan and the Gulf.

Contractor

Ultra Electronics Command and Control Systems, High Wycombe.

NEW ENTRY

United States

Advanced Field Artillery Tactical Data System (AFATDS)

AFATDS is a multiservice automated command and control system of mobile, multifunctional nodes providing automated planning and execution capabilities to fire support Operational FACilities (OPFACs), and Independent User Centers (IUC). IUC's are remote terminals, which allow commanders and selected fire support personnel to monitor fire support operations and issue guidance and directions from widely dispersed battlefield locations. AFATDS will operate at the Fire Support Element (FSE) and Fire Support Co-ordination Centres (FSCC) of the supported manoeuvre force, and Field Artillery Command Posts (FACP), Fire Direction Centres (FDC) and selected Field Artillery (FA) elements throughout the command structure. It can process over 400 fire missions per hour. It is one of the software applications within the ATCCS (Army Tactical Command and Control System (see separate entry). It interfaces with JSTARS (Joint Surveillance Target Attack Radar System) and the artillery fire-control systems of Germany, UK and France (ADLER, ATLAS, BATES).

AFATDS provides the commander with:

(a) integrated, responsive and reliable fire support

(b) vastly improved flexibility in providing inputs for items such as commander's criteria and priority of fire information

(c) a distributed database for all OPFAC systems which will insure that they are all operating with the same information

(d) a user friendly set of screens which can utilise a help program. The system also has a set of training software available

(e) the ability to attack the right target, using the right weapon system, with the right munitions, at the right time.

In addition to vastly improved communication capability and interface with most other systems, AFATDS supports the following field artillery functional areas:

Fire support planning

(a) develop fire support planning guidance
(b) develop fire support plans
(c) determine the Commander's concept of operations
(d) develop and monitor the FA logistic plan
(e) determine target acquisition capabilities
(f) co-ordinate Meteorological operations
(g) co-ordinate survey support
(h) develop the FA support plan.

Fire support execution

(a) process targets
(b) report FA status
(c) analyse FS attack systems
(d) analyse and perform Target Damage Assessments (TDA)
(e) develop order to fire
(f) prepare order to fire
(g) conduct FA sensor operations.

Movement control

(a) control fire support movement
(b) control FA movement
(c) prepare FA movement requests.

Field Artillery Mission Support

(a) control FA supplies
(b) control FA maintenance
(c) control FA personnel.

Field artillery fire direction operations

(a) determine firing unit capabilities
(b) process fire missions
(c) report fire mission status.

AFATDS hardware

(a) Reduced Instruction Set Computer Transportable Computer Unit (RISC TCU): This is the processing unit, keyboard and trackball. The unit contains a 1 Gbyte or larger hard disk drive and contains from 144 to 208 Mbyte of RAM. It contains an internal and external LAN card, and a Magneto Optical Disk Drive (MOD) which contains the optical disk, a CD-ROM drive, and a 1.4 Mbyte floppy drive. The MoD has a 650 Mbyte capacity, the CD-ROM has a 600 Mbyte capacity

(b) Super High Resolution Display (SHRD): Is the monitor for the system and has a 16 in screen. (1,280 × 1,024 colour)

(c) Uninterruptable Power Supply (UPS): Provides filtered power to the TCU, MOD and monitor. It also provides up to 30 minutes battery back-up power in case of external power loss

(d) Printer: Depending on the configuration of the system there will be from one to three printers

(e) External Tactical Communications Interface Module (TCIM): Each system comes with two TCIM's. Each TCIM allows for two digital nets. The nets can be a combination of radio and two or four wire

(f) AC/DC Converter/Charger: AFATDS can operate from either 110 V AC, 220 V AC commercial/generator power or 28 V DC vehicle power. Depending on system configuration there can be from one to three Converter/Chargers issued

(g) Lightweight Computer Unit AN/GYK-37: The AFATDS software is also capable of being run from a Pentium LCU.

AFATDS software

The software was developed by Magnavox and operates on common hardware at all Field Artillery (FA) elements throughout the battlefield. HP Unix, SCO Unix or Solaris operating systems can be used. The software is stored on the hard drive inside the TCU and performs the following functons:

(a) Initialisation: The system allows a system administrator to create and maintain different user accounts. A user account is defined as a set of privileges assigned to a particular system or group of systems. Additionally, it allows for a password for each system or group. It also allows the set up for any other devices used by the system. The communications parameters can be set here for current operations as well as future plans

(b) Database: Databases can be built, stored, edited and deleted by the operator. Information in the database is very important and should be built at the highest level. The system uses top down planning. The database is used to access firing unit information, ammunition and other logistical data

(c) Fire Support: In order to process fire missions AFATDS uses unit information. The system will initiate fire missions, subsequent corrections (SUBS), shot, splash, rounds complete, End Of Mission (EOM) and Mission Fired Reports (MFR). It will also look at which system is the best to accomplish the mission, which ammunition will complete the mission with the fewest rounds and which unit is in range to fire. It will also automatically update ammunition counts when the mission is fired and a MFR is received. The system will also create, maintain

and send fire plans. These plans can be built at any OPFAC and sent to whoever needs them for storage, refinement or execution

(d) Guidance: Guidance in AFATDS is in six main categories, Target, FS Attack, Unit and Sensor, FA Attack, C³, and Miscellaneous. Some of these are record keeping tools, others allow the information to be placed in plans and others are used in processing fire missions. In fire mission processing, AFATDS uses target categories which in turn use target types and filters to determine which unit can fire and whether the target should be engaged. The guidance part of AFATDS is the commander's ability to influence how he wants the battle to be fought and how to control his fire support assets

(e) Movement control: Manages and co-ordinates the movements of FA units including fire units, radar and MET. This activity identifies FA movement requirements, prepares movement requests in the form of movement tables and co-ordinates those requests with the appropriate force level headquarters

(f) Mission support: Field Artillery Mission Support includes the functions to logistically support the Field Artillery system through the creation and maintenance of supply inventory files, supply requests and reports. All of these are used by the commander to manage supply activities for the FA. The output of this process will provide a current inventory of selected classes of supply (classes III, V and VIII). Supply requests can then be generated as necessary. The system has a program for limited maintenance which can be performed by operators using a maintenance disk in the optical drive.

The most recent version, AFATDS 97/98 will receive the Air Tasking Order (ATO) from CTAPS, parse it and provide a platform for CAS requests, ATO, input, and deconfliction. AFATDS is composed of a common suite of hardware and software employed in varying configurations at different operational facilities (or nodes) interconnected by tactical communications. Both hardware and software can be being tailored to perform the fire support command, control and co-ordination requirements at any level of command. Its modular software architecture, use of Common Operating Environment (COE) and decentralised processing approach make AFATDS effective, flexible, survivable, mobile and easy to maintain.

Status

AFATDS went into full-rate production for the US Army in 1996. A total of 3,266 systems is on order. AFATDS completed LUT in October 1997, and has subsequently been selected for use by the USMC. The tested hardware and software did not support USMC mobility requirements. However, new functionality to address these problems will be provided.

In June 2001 Raytheon announced a US$1.8 million contract to supply Portugal with AFATDS to equip two army battalions.

Contractor

Raytheon Systems Company, Falls Church, Virginia.

UPDATED

··

Air and Missile Defence Workstation (AMDWS)

The Air and Missile Defence Workstation (AMDWS) provides a common air and missile defence planning, staff planning, and situational awareness tool to air defence and army units at all echelons of command. It is the air/missile defence component of the Army Battle Command System (ABCS). It provides a common air/missile defence operational planning tool for air defence commanders at all echelons of command (battery to theatre) for all air/missile defence weapon systems (Stinger Based SHORAD, PATRIOT, THAAD and so on). Although a component of the Air/Missile Defence Planning and Control System (AMDPCS), it is fielded as a component of AMDPCS to ADA Brigades and with each air/missile defence system. Through digital linkages with the various air defence weapon systems and the joint air surveillance net, the AMDWS provides the ABCS with the air component of the Common Tactical Picture at the Division and Corps echelons of command. The AMDWS provides interoperability between all components of the air/missile defence force and the ABCS. In addition, the AMDWS provides interoperability with the air planning components of the US Air Force/US Navy Theatre Battle Management Core Systems (TBMCS).

Status

A prototype system has been fielded and enhanced software versions are under development. Version 2 is expected to be fielded in late 2002 as part of ABCS version 6.2. V2 improves the functionality on the loader, allowing maps to be loaded more easily. It also provides easier-to-use menus.

Contractor

Northrop Grumman Mission Systems, Reston, Virginia.

UPDATED

··

AN/TSQ-179 Common Ground Station

The AN/TSQ-179(V)1 JSTARS Common Ground Station (CGS) is a product improvement of the LGSM (see JSTARS entry). It includes all the functionality of the LGSM plus extensive technological improvements. It incorporates additional mission functionality into a fully mobile targeting, battlefield management and surveillance system, providing:

(a) real-time surveillance

(b) reconnaissance
(c) situation awareness
(d) target development
(e) theatre missile defence

The CGS receives Imagery-Intelligence (IMINT), Electronics-Intelligence (ELINT) and Communications-Intelligence (COMINT) information of enemy forces across the forward line of own troops via multiple sensors and Intelligence Broadcast Networks (IBNs). IMINT sensor interfaces of the CGS include JSTARS, secondary image dissemination and Unmanned Aerial Vehicle (UAV) (both video and telemetry). ELINT and COMINT data from IBNs is accessed through the Commanders' Tactical Terminal (CTT)/Joint Tactical Terminal (JTT). Data sources include the JSTARS aircraft, Guardrail, U2, Rivet Joint, UAV Ground Control Station (UAV GCS), Apache Longbow and Airborne Reconnaissance Low (ARL). Global Positioning System (GPS) provides the CGS with current time and location, which is the basis for correlation of sensor and IBN data.

CGS output, in Tactical Fire Direction System/Advanced Field Artillery Tactical Data System (TACFIRE/AFATADS) or All Source Analysis System (ASAS) (see separate entries) message format, is provided to command, control, communications and intelligence nodes via secure or non-secure wire or radio. This enables the development and execution of plans for integrated battle management, surveillance, targeting and interdiction.

A standard CGS system consists of a mission vehicle, lightweight multipurpose shelter containing mission equipment, support vehicle and two trailer-mounted generators. The mission and support vehicles, which tow the trailers, are heavy variant HMMWVs. The CGS can deploy from movement to operation in 15 minutes, using only the six crew members. The equipment (including communications) can be dismounted from the HMMWV for semi-permanent site installation, and its components can also be split into transportable loads for rapid deployment independent of vehicles. The transportable suite is contained in up to four transit cases which, when opened, become a configured workstation with table and chair.

Each CGS contains a Remote Workstation (RWS) in the truck cab and connections for interfacing up to four other RWS at remote locations. The RWS has the same functional capabilities as the internal CGS operator workstation, in addition to providing CGS functions for in-unit training (for example CGS sensor and message traffic stimulation and other training functions). The RWS modes of operation are remote mode, training mode, and CGS workstation mode. In remote mode, all the capabilities of internal workstations are available, except transmitting intelligence, targeting, and tasking messages. Training mode is operation of the RWS as a lesson control workstation for in-unit training. The CGS workstation mode provides all the capabilities of an internal CGS workstation, except audio. The RWS also provides an interface for additional displays.

CGSs can function independently or may be interconnected to other CGSs over a fibre optic LAN allowing their multiple databases to be integrated. The CGS hardware and software architectures also facilitate Pre-Planned Product Improvements (P3I), such as additional sensor interfaces, additional command and control interfaces, enhanced processing and display capabilities and growth to other platforms via technology insertion. Standards for the CGS automated data processing architecture comply with US Army C4I Technical Architecture standards.

Status
In operational service with the US Army.

Manufacturer
General Dynamics Decision Systems, Scottsdale, Arizona.

UPDATED

AN/TSQ-221 Tactical Airspace Integration System (TAIS)

The AN/TSQ-221 Tactical Airspace Integration System (TAIS) is the principal airspace management component of the US Army Battle Command System (ABCS) (see separate entry). It provides the airspace management, planning and execution tool at Division, Corps and Echelons above Corps. It also provides interoperability with the airspace planning components of the Theatre Battle Management Core System (TBMCS - see separate entry), with USAF and USN systems, with allied forces and with civilian air traffic control.

The TAIS is housed in rigid wall shelters on two HMMWVs and is fully self-contained. Typically, TAIS deploys one shelter with a division or Corps TOC while the other shelter can be remotely located to perform air traffic services. Each TAIS shelter is identical in form, fit and function and receives the recognised air picture through several communication links providing near-realtime situational awareness for both friendly and enemy air activity. It provides the following facilities:
(a) Near-realtime integrated air and ground picture
(b) Air track and battlespace geometric overlays
(c) Generates alerts for aircraft flight movements in and out of areas with Air Control Measures (ACM)
(d) Air Co-ordination Order (ACO) display in 2 or 3 dimensions
(e) ACO/Air Tasking Order (ATO) management
(f) Near-realtime flight following
(g) View of 2-D/3-D picture from any angle.
The following interfaces and links are contained in TAIS:
(a) Army Battle Command System
(b) Air Defence System Integrator
(c) AWACS TADIL-A Link

TAIS on HMMWV with trailer (General Dynamics) 0143064

(d) TBMCS
(e) CTAPS
(f) FAAD C2I
(g) TADIL-A Link
(h) TADIL-B Link
(i) TADIL-J Link
(j) International Civil Aviation
(k) FAA
The following communications systems are integrated in TAIS:
(a) AccessNet Intercom System
(b) Enhanced Position Location Reporting System ((EPLRS)
(c) AN/PSN-11 Precision Lightweight GPS Receiver (PLGR)
(d) AN/PSC-5 Spitfire Tactical Satellite Radio
(e) Rockwell Collins 95S HF/UHF Receiver
(f) KY—68 Digital Secure Voice Terminal
(g) AN/VRC-90 VHF Radio
(h) AN/VRC-92 VHF Radio
(i) AN/VRC-83 VHF/UHF Radio
(j) AN/URC-200 Multiband Radio
(k) AN/ARC-220 HF Tactical Radio
(l) Quick Erect Antenna Mast (QEAM)
(m) Mobile Subscriber Equipment (MSE)
(n) Secure Telephone Equipment (STE).

Status
In service in the US in III and XVIII Corps, in Europe and South Korea. Being considered by the USMC.

Contractor
General Dynamics Decision Systems, Huntsville, Alabama.

UPDATED

Army Battle Command System (ABCS)

The Army Battle Command System (ABCS) is the all-embracing name for a number of US army command and control systems covering the strategic, operational and tactical levels of command, the development of which is fundamental to the migration of the US Army to its ultimate goal of a fully digitised force. ABCS consists of three principal subdivisions and other supporting systems, all of which are the subject of one or more separate entries:

Global Command and Control System - Army (GCCS-A)
This is the army component of the Joint GCCS, providing information and understanding at the theatre level, including the joint environment.

Army Tactical Command and Control System (ATCCS)
This is the integration of five primary functional area control systems providing situational information and decision support to the battlefield operating systems (BOS) from corps to battalion echelons. ATCCS includes AFATDS, MCS, AMDPCS, ASAS and CSSCS. ATCCS and its components are discussed in a separate entry.

Force XXI Battle Command Brigade and Below (FBCB2)
FBCB2 enables the combat, combat support and combat service support forces at and below brigade level to share a common picture of the battlefield, while passing digitised combat information, orders and graphics vertically and horizontally. The Tactical Internet (TI) is the network of radios and routers that provide the links to connect the FBCB2 platforms. It consists of the Enhanced Position Location Reporting System (EPLRS), the Single-Channel Ground and Airborne Radio System (SINCGARS) and the Internet Controller Router. (EPLRS and SINCGARS are covered in separate entries.)

Digital Topographic Support System (DTSS)
DTSS is an automated battlefield system that provides geospatial data in digital format for use on ABCS systems.

Integrated Meteorological System (IMETS)

The IMETS provides high-resolution current and prognostic meteorological data and weather effects.

Tactical Airspace Integration System (TAIS)

The TAIS is the principal airspace management component of the Army Battle Command System (ABCS). It provides the airspace management planning tool for commanders at Division, Corps, and Echelons Above Corps (EAC) echelons of command. Through digital linkages with the various users of airspace and other command and control systems, the TAIS provides the ABCS with the airspace component of the Common Tactical Picture at the Division, Corps, and EAC echelons of command. The TAIS also provides interoperability with the airspace planning components of the USAF/USN Theatre Battle Management Core Systems (TBMCS - see separate entry) and with the airspace planning and air traffic control systems of allied forces.

VERIFIED

Army Tactical Command and Control System (ATCCS)

The Army Tactical Command and Control System (ATCCS) is one of the components of the US Army Battle Command System (ABCS) (see separate entry). The component parts of the ATCCS synthesize incoming information and distribute it from battalion to corps, linking with FBCB2 below and GCCS-A above.

The Manoeuvre Control System (MCS)

This is the conduit through which all the other automated systems collate and send their information at division, brigade and battalion levels. MCS consists of a network of computer workstations that integrate information from subordinate units with that from other ATCCS battlefield functional areas to create a battlefield common tactical picture (see separate entry).

The All Source Analysis System (ASAS)

This is the intelligence system for division, brigade and battalion levels. It organises and fuses combat information from a multitude of sources, from individual soldiers to reconnaissance satellites, providing a constant update on enemy status (see separate entry).

The Advanced Field Artillery Tactical Data System (AFATDS)

The AFATDS automates the control of all tactical indirect fires, including air force strikes and naval gunfire. It helps commanders determine the optimum firing platforms and munitions mix to defeat enemy targets. It automates artillery fire planning and co-ordination and provides the manoeuvre commander with artillery information (see separate entry).

Air-Missile Defence Planning and Control System (AMDPCS)

This system is the air defence component of ATCCS and is used to provide third dimension situational understanding. It consists of two subordinate systems, the Forward Area Air Defence Command, Control, Communications and Intelligence system (FAADC3I) and the Air Missile Defence Work Station (AMDWS). AMDPCS integrates air defence fire units, sensors and C2 centres into a coherent system capable of defeating/denying the low altitude aerial threat. The FAAD C3I integrates air defence, air defence fire units and sensors to defeat and deny enemy low-altitude air threats and cruise missiles. It also automates air defence planning and equipment status reporting. AMDWS is the ADA tool that provides the air and missile defence planning and air situational understanding and is used from the ADA battery to theatre echelons. It is also the air missile defense planning and control link to joint/allied C2 systems. It provides direct connectivity to and interoperability with the Joint Defense Planner, a theatre level air and missile defence planning tool (see separate entries).

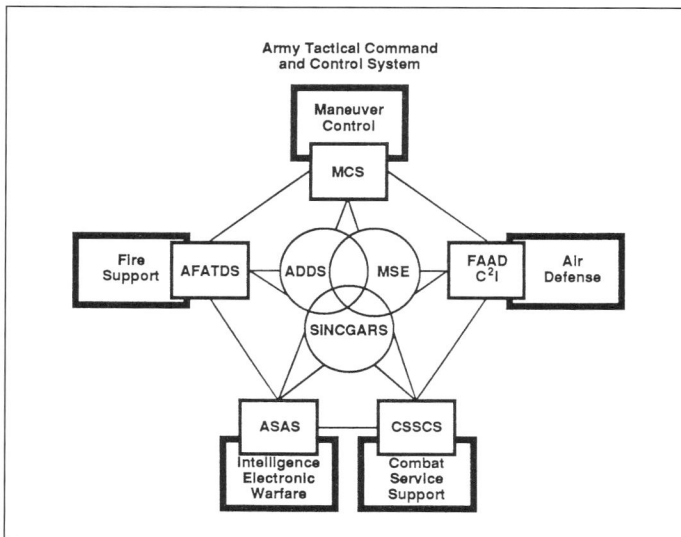

The Combat Service Support Control System (CSSCS)

This is the logistics arm of ATCCS. It generates current supply, maintenance, transportation, medical, and personnel status information, as well as planning projections for future operations. The CSSCS provides desired command and control information to the CSS and Force Level Commanders and their staff. It is based on data received from the CSS Standard Army Management Information Systems (STAMIS) and subordinate CSS command and logistical, personnel, medical and financial staff elements. In addition, CSSCS exchanges logistical and tactical information with the other four Army Tactical Command and Control Systems (ATCCS) and Battlefield Functional Area (BFA) Command and Control Systems which are consistent with the Force Level Control System (FLCS) software design concept. STAMIS and BFA information is posted to the CSSCS database, supporting the generation of reports, projections and administrative/logistics orders, to aid decision making and planning (see separate entry).

VERIFIED

Battery Computer System (BCS)

The Battery Computer System (BCS), an advanced fire-control system developed for the US Army, replaced the current artillery digital automatic computer (FADAC) and extended the capability of TACFIRE, the tactical fire direction system. It provides computerised control of artillery fire at the battery level. A real-time command, control and communications system, it is capable of providing simultaneous computations for up to 12 howitzers or guns and handling three concurrent fire missions.

Development

Norden Systems (now part of Northrop Grumman Electronic Systems) of Norwalk, Connecticut, USA, was awarded a multimillion dollar contract for development of the BCS in September 1976. GEC-Marconi (subsequently Alenia Marconi Systems, now AMS) of the UK was Norden's principal subcontractor and incorporated much of the experience gained in the design and development of the British FACE field artillery computer equipment. A production contract was awarded to Norden in April 1980.

Description

BCS is a portable, single-unit system containing 256 kbytes (4.7 Mbits) of memory capacity. By providing direct digital access for forward observers and to battalion TACFIRE, it extends that system's capability to the firing battery level. It also operates in an autonomous mode and can be used for fire planning and the engagement of moving targets.

The BCS is composed of the Battery Computer Unit (BCU) located at the battery or platoon Fire Direction Centre (FDC) and one Gun Display Unit (GDU) for each weapon in the battery. The BCU is the electronic centre for fire control. It maintains a digital communication link with fire direction officers, fire support officers, forward observers, weapon section chiefs, target acquisition systems (for example, counter battery radars) and meteorological data systems. It performs all required computations for first-round accuracy, with results appearing on a 1,728-character plasma display. The BCU contains a data management system, which maintains target, fire unit, ammunition, meteorological, map, forward observers and other system information to simplify the FDC tasks. The central processor within the BCU contains 256 K × 32 position × 18 bit words of memory, with error detection and correction circuitry for fail-safe reliability. The time taken to compute a fire mission with individual data for all 12 weapons is 2 per cent of the time of flight to the target.

The BCU houses a magnetic tape cartridge for software program and data storage, the former permitting central processor programming in under 1 minute. It interfaces with existing US Army COMSEC equipment and printers. Data is transmitted to the guns in 1 second. The BCU communicates on the battlefield with WD-1 wire and AM/FM radio.

The communications control device is driven by a microprocessor controller and offers high-speed direct access to the memory module. The BCU contains the main

Army Tactical Command and Control System

```
                    Army Tactical Command
                    and Control System

                         Maneuver
                          Control

                           MCS

   Fire                                          Air
  Support   AFATDS   ADDS    MSE    FAAD        Defense
                                     C²I

                         SINCGARS

              ASAS              CSSCS

          Intelligence        Combat
           Electronic         Service
            Warfare           Support
```

ATCCS architecture and battlefield functional areas (US Army)

The battery computer unit, one of two elements of the BCS, acting on information from the forward observer, automatically computes the firing data and displays it on the unit's screen. Also displayed is the status of the battery's howitzers. The soldier will select a gun for the mission and transmit the firing data to the gun display unit located at the howitzer

Mission assignments for the BCS are received from forward observers who digitally transmit target information to a battalion, TACFIRE or directly to the battery computer unit located at the field artillery level

Combat Service Support Control System screenshot, showing ORBAT on the left and unit logistic status by category on the right (Patrick Allen/Jane's) 0526726

Combat Service Support Control System (Patrick Allen/Jane's) 0526725

power switches and self-test status panel, so that faults in individual elements can be easily isolated and corrected.

The GDU is the final link in the BICS communications chain. Its place is with each weapon in the firing battery. The GDU consists of three assemblies, a Section Chief Assembly (SCA) and two Gun Assemblies (GA), plus a signal and power distribution unit. All elements are of solid-state sealed construction.

The SCA is a personal tool for the section chief (weapon commander), the man who commands the weapon crew. Slightly larger than a commercial handheld calculator, the SCA is connected by cable to the signal and power distribution unit, giving instant access to all gun-related data and commands. The device contains its own memory bank for sequential or direct access to stored fire mission information. It is equipped with audible alarms and visual cueing to alert the section chief to fire missions and check fires. An eight digit display shows positive indication of gun commands under all ambient conditions. A single button sequences the display of gun orders.

The two identical GAs on the weapon show the gunner and assistant gunner separate elevation and deflection displays. They are connected by wire to the signal and power distribution unit and receive the commands directly.

Status

Deliveries began in November 1981. The multiyear production contract is valued at about US$97 million and in June 1980 Norden awarded GEC-Marconi Ltd (now AMS) a US$20 million contract for the supply of BCS GDUs. These units were based on the MC1800 microprocessor which was also employed in the latest version of the FACE field artillery computer equipment. In March 1986, GEC-Marconi (AMS) was awarded a further contract worth over £5 million for GDUs for use with the BCS. This order, for over 1,000 GDUs, followed six previous contracts and brought total value of GDU contracts up to over £23.5 million. Deliveries for the latest contract started in September 1990. In addition to being deployed by the US Army, BCS is also deployed by the US Marine Corps and foreign armies. BCS is also fielded to Multiple-Launch Rocket System/Army Tactical Missile System (ML-RS/ATACMS) and Lance batteries and battalions. Over 1,400 BCS have been fielded. In 1999 the AFATDS (see separate entry) programme was concentrating on integrating the BCS with AFATDS.

Contractors

Norden Systems Inc (part of Northrop Grumman Electronic Systems) (prime contractor), Norwalk, Connecticut.
AMS, Frimley, Surrey.

UPDATED

Combat Service Support Control System (CSSCS)

CSSCS, the CSS component of the Army Tactical Command and Control System (see separate entry), is designed to provide a picture of unit CSS requirements and capabilities, by collecting, processing and displaying information on the status of units, supplies, logistics assets, logistics tasks and battlefield situational awareness. The system provides situational awareness of a variety of supply classes, and maintenance, transport and medical functionality are also provided. It particularly tracks key items of supply, services and personnel through comparison of a variety of databases with those items identified as critical by the commander. Maintenance, transportation and medical reporting are also provided.

Data is collected from multiple sources, both through automatic interfaces to other ATCCS elements and through manual entry, and is distributed to other CSSCS nodes, mainly by using Mobile Subscriber Equipment (MSE) as the primary means of communication. The system processes the information based on standard planning factors, the task organisation and established support relationships, thus enabling large amounts of data to be distilled and presented in a usable form. Selected information is also forwarded to other systems within ATCCS to provide commanders with a common logistics picture. An interface to the Movement Tracking System (MTS) provides near-realtime updates of CSS vehicle (for example trucks, ambulances) locations. Convoy movements, for example, can

be received from the MTS equipped vehicles through an MTS control station, processed by CSSCS, displayed on the ABCS Common Tactical Picture (map) and relayed to the FBCB2 for map displays . CSSCS exchanges near-realtime data with FBCB2 (see separate entry): when unit-level data is entered at a brigade CSSCS node it is automatically transmitted and aggregated up the command chain, making data re-entry unnecessary. When Task Organisations changes are entered into the system, CSSCS automatically reconfigures the reporting relationships so that company level data is aggregated accordingly.

Information is graphically portrayed using a colour code to indicate status, which can be projected out to four days, using a combination of planning factor and manually generated estimates. Staffs can compare the supportability of up to three possible courses of action at the same time, each extending for four days, with variables including task organisation, operational tempo and geographical constraints. Users have considerable freedom to modify and tailor output reports to suit local requirements.

CSSCS software has been re-engineered to make it capable of running on a web-technology/PC Windows platform. CSSCS is also capable of running in a client/server mode with the PC client connected to the CSSCS UNIX-based server. The CSSCS Light system is highly platform independent and is derived from the evolving client/server designs. This client approach allows, for example, any Battlefield Automated System (BAS) to connect to a CSSCS server and execute CSSCS client software directly on the BAS device. CSSCS has also been redesigned to provide a web browser based interface.

The CSSCS incremental development supports the DII COE evolutionary migration to a client-server architecture. CSSCS has installed Joint Mapping Tool Kit (JMTK) to provide the Common Tactical Picture (CTP)/Common Operating Picture (COP). CSSCS also interoperates with the Joint Common Data Base (JCDB), and employs a Common Message Processor to enable receipt and transmission of US Message Text Format (USMTF) and Joint Variable Message Format (JVMF) messages.

Status

In service with the US Army, providing the enabler for the logistic support of digitised formations. First fielded with 4th Infantry Division in 1995. In operational use in support of US operations in Bosnia.

Contractor

Northrop Grumman Mission Systems , Carson, California.

UPDATED

Enhanced Position Location Reporting System (EPLRS)

EPLRS acts as the digital backbone of the Tactical Internet (TI) and is an integrated C3 system that provides near-realtime data communications, position/location, navigation, identification and reporting information on the battlefield. EPLRS provides a means for data distribution and position/navigation both vertically and horizontally. The system supports the five functional areas (Manoeuvre Control, Air Defence, Fire Support, Intelligence/ Electronic Warfare (IEW) and Combat Service Support) of the Army Tactical Command and Control System (ATCCS). The system utilises error correction coding and is capable of supporting multiple communications channel operations and has an automatic relay capability. It is a fundamental element of FBCB2 (see separate entry).

The Enhanced Position Location Reporting System (EPLRS) Very High Speed Integrated Circuit (VHSIC) Radio Set (RS) (AN/VSQ-2) is a state-of-the-art, line of sight, data-only digital radio system operating in the 420-450 MHz UHF frequency band. It serves as a position location, navigation, identification and communications system. Its primary components are the Network Control Station (NCS) and the RSs. The NCS is the centralised control element that is used for system initialisation and monitoring and control of the EPLRS VHSIC network. The RS is the radio receiver transmitter provided to the EPLRS VHSIC users. The Appliqué System architecture utilises EPLRS VHSIC to provide Wide Area Network (WAN) connectivity across the Single Channel Ground and Airborne Radio System (SINCGARS) System Improvement Program (SINCGARS SIP) networks from platoon level to brigade level. Operational units are equipped with RSs to provide a tactical WAN backbone for the TI. This radio provides secure, jam-resistant digital communications and accurate position location capabilities for the user. Its use of frequency hopping (512/s), spread spectrum technology (eight frequencies between 420 and 450 MHz) embedded communications security (COMSEC) module (KIV-14) and adjustable power output provides secure communications with a low probability of intercept and detection. It has Built-In-Test (BIT) functions that are activated when power is turned on.

The antenna used with the EPLRS VHSIC RS is an omnidirectional dipole. The planning range is 3 to 10 km between radios depending on power output settings and terrain. The RS provides transmission relay functions that are transparent to the user. The maximum distance the RS can cover is based on 3 to 10 km distance between each radio and the maximum number of radios in the needline link. Duplex and group addressed needlines will automatically use up to four (group addressed) or five duplex RS relays (hops) to establish a needline link from endpoint to endpoint. Local area coverage for these needline types is limited to two hops (1 relay) and will typically cover a battalion sized area. Extended area coverage will use either four for carrier sense multiple access (CSMA) or six for multisource group (MSG) and point-to-point (P-P) hops to cover up to a division sized area. Each RS has variable power output from 0.4 to 100 W.

Three major functions of the EPLRS VHSIC include:

(a) Position Location: The RS operator is able to enter requests and receive position location updates in the form of a 10 character alphanumeric code related to the Military Grid Reference System (MGRS). Such data is constantly updated and stored at the NCS and is provided to each RS upon request. In addition, the RS operator may request the MGRS position of or range/bearing to another RS in the net. Expected practical position accuracy is a result of varied terrain, reflections, multipath diffraction and scatter

(b) Navigation: Battlefield lanes, corridors, or zones (such as minefields or contaminated areas) may be specified at the NCS, or by any RS user. Automatic alerts will be transmitted to any RS approaching, crossing into, or exiting the boundaries of these areas designated as lanes, corridors, or lanes. Also, advisories or range and bearing information can be provided to RS operators to assist them in navigating to predetermined checkpoints, landmarks, or other units in the network

(c) Identification: Included in the data stored at the NCS is the military identification (MILID) of all RSs in the net. An RS operator may request the identification of an unknown RS by providing the MGRS co-ordinates of that unknown RS or the range/bearing to that RS, to the NCS. At the NCS, positive identification of all RSs in the net assists the NCS operator in co-ordinating the operation of all RSs.

Specifications

Characteristics
Frequency band: UHF, 420-450 MHz
Frequency hop: 512/s over 8 frequencies
System architecture: synchronous TDMA, frequency and code division multiplexing
Typical system size: 300-900 terminals in a division with up to 4 NCS's
ECM: spread spectrum, frequency hopping error and correction
Security: embedded crypto, transmission and dual level communication security
Data rates: 2 types of low data rate, (LDR) needlines with rates to 14,400 bps. Three types of high data rate, (HDR) needlines with rates to 100+ kbps.
Navigational aids/services: More than 20, including navigation, zone alerts, lane guidance, friendly ID, time, and so on.
Dimensions: 14 × 10 × 5 in
Prime power: 28 V DC, 16 W
Weight: 17 or 26 lb with batteries
Output power: selectable 100, 20, 3, or 0.4 W

EPLRS Micro-lite

In 2002 Raytheon introduced EPLRS Micro-lite, which provides the same functionality as EPLRS as well as Voice over Internet Protocol (VoIP). It is a repackaged EPLRS designed to maintain waveform network security and software compatibility with the full-size EPLRS. Power is supplied via an external (remoted) battery pack or via an external power supply from fixed, vehicle or air platform power sources, supplied via the interface connector pins.

Characteristics
Frequency range: 420-450 MHz (optional extensions)
Data rates: Variable from 57.6 kbps-525 kbps
Interfaces: RS-422, RS-232, 10/1000 base-T Ethernet, VoIP
Weight: Approx 12-14 oz without power supply
Power output: 5 W
Dimensions (H × W × D): 5 × 2.75 × 1 in

Status

EPLRS originated as a programme started in 1975 by the US Marine Corps called Position Location and Reporting System (PLRS). Subsequently, the army took over management of the programme and it has gradually been improved. By adding Very High Speed Integrated Circuit (VHSIC) and commercial-off-the-shelf technologies EPLRS performance has tripled and its cost halved. The army bought 1,816 radio sets of the VHSIC configuration under a Low Rate Initial Production (LRIP) contract. A more capable version called the System Improvement Program (SIP) has more memory and more processing power and has significantly increased reliability.

Recent improvements have been through a Value Engineering Change Proposal (VECP) of the EPLRS based on the use of 1990's technology and continued system improvements to meet digitisation requirements. RSC was awarded a multiyear contract in December 1997 to produce an additional 4,431 EPLRS radios. This is a joint service procurement with the army acting as the acquisition executive. The award consists of 2,880 units for the army, 758 for the ANG, 764 for the USMC and 29 for the USN. The current version has increased data rates and provides for the elimination of a centralised net control station.

In October 2000 authorisation was granted to increase the US Army procurement ceiling by a further 4,739 equipments to 12,896.

EPLRS Micro-lite is due to be evaluated by the US Army in early 2003.

Contractor

Raytheon Systems Company, Marlborough, Massachusetts.

UPDATED

Force XXI Battle Command Brigade-and-Below (FBCB2)

FBCB2 is the lowest level of the US Army Battle Command System. It is a tactical system which permits near-real time exchange of locational and C2 information through the use of Joint Variable Message Format messages and consists of application software and mounted Appliqué+ V4 computers. Users are digitally

BDI Qualcomm OmniTRACS satellite aerial mounted on roof of HMMWV (Patrick Allen/Jane's)   0538781

A close-up of Balkan Digitisation Initiative (or Blue Force Tracking) display hardware mounted in a HMMWV, shown at AUSA 02 (Patrick Allen/Jane's)   0526724

Balkan Digitisation Initiative hardware in use with 212 Military Police Company, US Army in Kosovo. The display and keyboard are from Kontron Fieldworks; the PLGR +96 GPS is in the bracket on the left (Patrick Allen/Jane's)   0538779

BDI-equipped HMMWVs on patrol in Kosovo. The Qualcomm OmniTRACS aerials can be seen with camouflage covers on the right rear of the roof of each vehicle (Patrick Allen/Jane's)   0538727

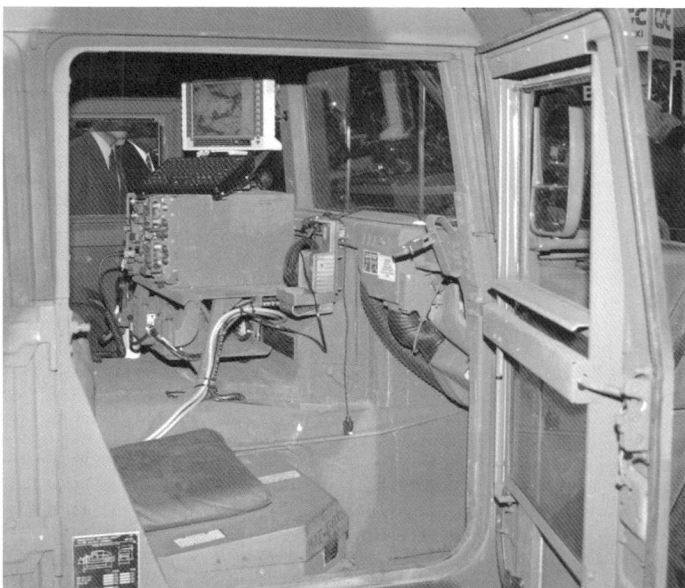

The Balkan Digitisation Initiative (or Blue Force Tracking) hardware mounted in a HMMWV, shown at AUSA 02 (Patrick Allen/Jane's)   0526701

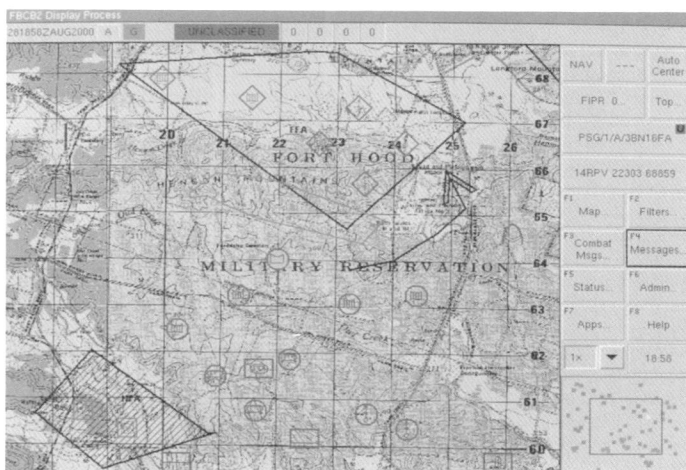

FBCB2 in HMMWV (TRW/Mark Arnold)   0116688

FBCB2 screen (TRW)   0116687

FBCB2 dismounted (TRW)   0116689

connected via a 'tactical internet' wireless communications architecture. The system is a fundamental part of the US Army's drive towards digitisation, bringing the benefits to the tactical level.

Each FBCB2 derives its own location via GPS (embedded within the communications equipment) and automatically updates and broadcasts its current location to all other FBCB2-equipped platforms. The radios also transmit and receive C2 messages such as orders, overlays, and reports. The composite picture, which shows near-realtime locational data, both friendly and enemy, is presented on a colour map background. A fully integrated package of battle management applications cover planning, manoeuvre, intelligence, logistics and fire support functions, at the appropriate command level. The package also includes a communications management tool with which the Tactical Internet can be configured and the use of radio bandwidth optimised; this tool can also be used for network management of Tactical Operations Centre LANs. The 'tactical internet' uses the same protocols and standards as the Internet, and in the development of the programme SINCGARS, EPLRS and Harris PRC-138B have been used for communications.

In late 2002 a 3-D graphics system for the system was revealed, providing a 'fly-through' capability.

A system based on FBCB2 has been developed for use in the Balkans to provide own force locational data. This is the Enhanced Information System (EIS), part of the Balkan Digitisation Initiative (BDI), also known as Blue Force Tracking. This was developed following the capture of a US patrol which had strayed into Kosovo in the period before the NATO ground operation. The EIS integrates the Rockwell-Collins PLGR +96 handheld GPS and FBCB2 display software with the commercial Qualcomm OmniTRACS Satellite Mobile Communications System, using OmniTRACS transceivers.

Status

Elements of both III Corps (including 4 Inf Div and 1 Cav Div) and the IBCT have been equipped with FBCB2, although it is yet to receive final approval from O&TE for full-rate production. In late 2002 it was expected that V7 of the software would be released shortly. Once full production is authorised, a total of 59,522 systems are expected to be installed in the US Army in the period up to 2011.

3 US Inf Div were issued with the system for operations in Iraq, together with supporting units. The USMC has also bought a limited number for units engaged in the same operations. *Jane's Defence Weekly* reported that about 50 systems were to be supplied for use by forces of a coalition partner, almost certainly the UK.

Contractor

Northrop Grumman Mission Systems, Carson, California.

UPDATED

Forward Area Air Defense Command, Control and Intelligence (FAAD C2I)

FAAD C2I provides near-realtime targeting and C2 information, accurate and timely identification of targets, alerting of SHORAD and force elements, cueing of SHORAD weapons and interoperability with allied and joint AD C2 systems. It is a network that connects command posts, weapons and sensors within AD units, is one of the five components that comprise the Army Tactical Command and Control System (ATCCS) and in essence is an automated system for providing command, control, targeting and other information to the air defenders on the battlefield. The software can also be hosted on the Air and Missile Defence Work Station (AMDWS) (see separate entry).

Its linked Ground-Based Sensor (GBS) provides air surveillance, target acquisition and target tracking information to the weapons in the AD unit, and additional data is provided from external sources such as USAF E-3 and USN E-2. The system provides the means for collecting, storing, processing, displaying and disseminating information, providing an automated, timely picture of the air defence battle throughout the divisional area. Threat tracks cause alerts, automatically cueing fire units to targets. Computer displays allow commanders access to databases for the air picture, situation reports, enemy assessments, and friendly forces, the amount of data base access varying at each SHORAD echelon, and the system will provide several air defense overlays, each automatically and dynamically updated from a source system, displayed over military maps, overhead imagery, or false-coloured terrain (2-D and 3-D).

The FAAD C2I system consists of common hardware, software and communications equipment, all mounted in a HMMWV variant. The software performs the air track and battle management processing functions. The communications equipment consists of the Single Channel Ground and Airborne Radio System (SINCGARS), the Joint Tactical Information Distribution System

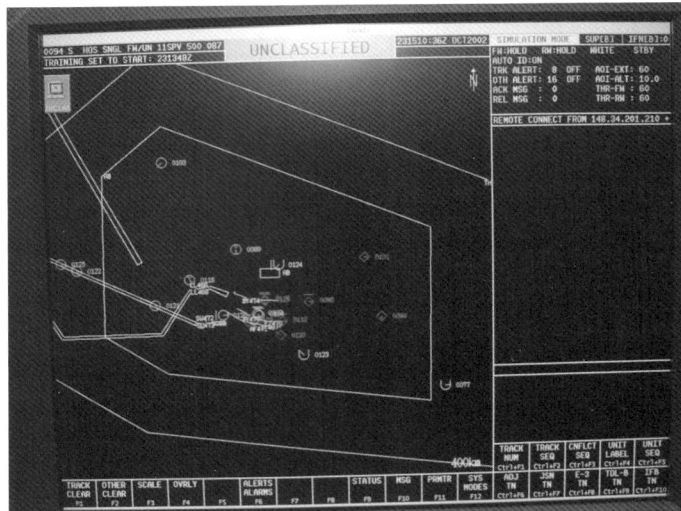

FAADC2I display (Patrick Allen/Jane's)   0526728

FAADC2I (Patrick Allen/Jane's)   0526727

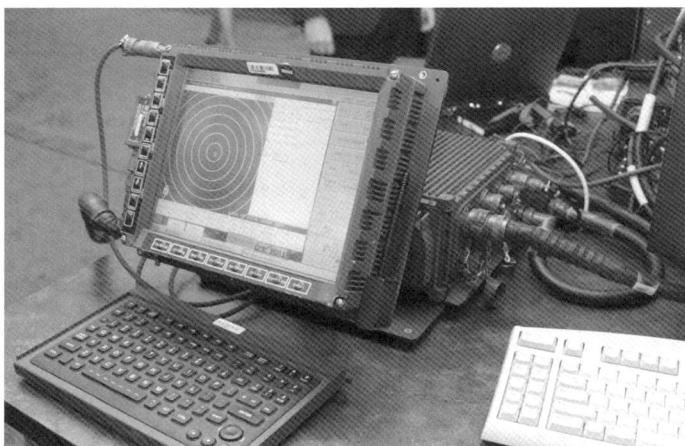

FAADC2I screenshot showing tracks and safe corridors (Patrick Allen/Jane's)   0526729

(JTIDS) and the Enhanced Position Location Reporting System (EPLRS) (See separate entries).

To enable individual weapons systems to see the air and ground pictures on a single display, FAAD Co-host has been developed. This is essentially FBCB2 with a minimal version of FAADC2I software hosted on the same display equipment.

Status

The first operational test of the FAAD C2I system was a Limited User Test (LUT) conducted in early 1993. Following this the US Army made a low rate initial production decision to procure and field the FAAD C2I system to one light division. A new version of the software was tested in 1997 (version 4R), and a further version, (5.2) was tested in early 2000; no information is available on the results of that test.

In 2002 FAADC2I was in series production for the US Army and had been fielded in digitised formations and the Interim Brigade Combat Teams.

Contractors

Northrop Grumman Mission Systems, Reston, Virginia.

UPDATED

Global Combat Support System - Army/Tactical (GCSS-A/T)

The Global Combat Support System - Army/Tactical (GCSS-A/T) is being developed to provide accurate and consistent supply chain information from the depots to the front line. The new browser-based system will replace the US Army's Standard Army Management Information Systems (STAMIS), the current suite of logistics systems. A COTS hardware solution using Windows NT, it will establish interfaces with other CSS automated systems to reduce the amount of data entry required.

GCSS will create a technical environment and process to integrate economically existing computer-based systems software and hardware using the Common Operating Environment (COE) and shared data environment (SHADE). It will also expand the GCCS Common Operational Picture (COP) to accommodate combat support information, and will also provide 'split base-reachback' capabilities from the foxhole to the sustaining base to allow the combatants to be 'deployed/sustained' by electronic means. One of the components of GCSS is the Electronic Commerce (EC)/Electronic Data Interchange (EDI) infrastructure initiative. This initiative enables electronic access to goods and services in a timely and efficient manner via the Electronic Commerce Infrastructure (ECI).

GCSS provides online connectivity to NIPRNET/SIPRNET web, applications, and data. Its development is expected to have heavy user participation, and through incremental improvement will evolve to a position where it is independent of specific hardware and is interoperable. It will provide single log-on (Public Key Infrastructure (PKI) capability to selected combat support data sources. To provide integrated access to combat support information, GCSS will engineer data access services to identify, locate and access combat support data in legacy sources; translate, integrate, and fuse that data; and provide the resulting information to the requestor. These services will hide the details of data access from applications like the COP and the Joint Logistics Advanced Concept Technology Demonstration (ACTD) tools, while ensuring that the information they need is available in the appropriate format at the right time and place.

GCSS goals are to:
(a) Provide all echelons with ready access to combat support capabilities and personnel
(b) Provide a combat support infrastructure that is responsive to the Joint Task Force Commander's needs
(c) Provide a flexible and adaptive open computing environment
(d) Enable interoperability and integration across combat support areas and from combat support to the combat environments

(e) Integrate and implement an information infrastructure that provides end-to-end information connectivity and access.
  The system is being developed in three stages:
(a) Initial Operating Capability. The IOC will be developed through integration and modernisation of existing systems to produce a single enterprise database consisting of six modules:
  - Supply/Property
  - Maintenance
  - Ammunition Supply
  - Supply support activity
  - Integrated materiel management
  - Management (integration of modules and data exchange with other systems)
(b) Enhanced Operational Capability (Wholesale and Retail integration).
(c) Full Operational Capability. FOC will implement all required interfaces with automated systems in the joint community, national sustaining base and allied systems.

Status

Prototypes have been fielded at Fort Hood, Texas.

Contractor

Northrop Grumman Mission Systems, Reston, Virginia.

UPDATED

Global Command and Control System - Army (GCCS-A)

The Global Command and Control System-Army (GCCS-A) is the US Army component of the Joint GCCS (see separate entry) and is its strategic and theatre level C2 system. For Theatre commanders, GCCS-A provides a Common Operational Picture (COP) and associated friendly and enemy status information, force employment planning and execution tools (receipt of forces, staging, intra-theater planning, readiness, force tracking, onward movement and execution status), and overall interoperability with Joint and Coalition systems and the divisional-level systems incorporated in the Army Tactical Command and Control System (see separate entry). Developed from application programs in the Army WWMCCS and other legacy systems, GCCS-A consists of ADP hardware, software and communications system components that are tailored and integrated to support strategic, operational and theatre functional requirements. The overall software architecture is based on the four-layer DOD Joint Technical Architecture (JTA) and a client-server architecture based on the DII COE. The commonality in architecture facilitates integration of the GCCS-A and the GCCS. GCCS-A uses the Common Hardware Software II hardware and consists of a mix of UNIX database servers, Windows-NT PC user workstations and laptops configured to specific site requirements. The system architecture links users via Local Area Networks (LANs) in client/server configurations with an interface to the Secret Internet Protocol Router Network (SIPRNET) for worldwide communication.

Status

In operational use in all major US Army headquarters and formations.

Contractor

Lockheed Martin Mission Systems(Prime), Gaithersburg, Maryland.

VERIFIED

Inter-Vehicular Information System (IVIS)

The Inter-Vehicular Information System (IVIS) is a digital command information system embedded on the M1A2 Abrams MBT. It is designed to exploit the latest developments in digital electronics and software technology enabling battlefield situations to be quickly and accurately assessed. IVIS can be interfaced through gateways to other higher echelon C^2 systems such as MCS, AFATDS and CSSCS. Its main application areas are in: continuous, accurate, friendly force location; accurate target location; enhanced navigation; and the enhanced ability to mass friendly forces.

Analysis has quantified the benefits of IVIS. Crews are 98 per cent more accurate in reporting their own locations and IVIS equipped platoons are 59 per cent more accurate in reporting target locations. There is also evidence of improvement in mission effectiveness with: over 50 per cent reduction in planning time at crew, platoon and company levels; a 25 per cent increase in platoon offensive missions completed; 34 per cent in defensive missions; 42 per cent improvement in mission execution time at platoon level; and a 33 per cent reduction in mission execution time at company level.

The IVIS commander's tactical display is a thin film EL unit which presents: current menu layer; warnings and cautions (from the system's built-in test and self-test subsystem); date and time group; vehicle heading; eight-digit grid co-ordinate; grid map surrounding current location; highlight area for incoming messages; display of IVIS messages; and legends for menu/option select buttons.

Interfaces to onboard tank systems provide Far Target Designation to include accurate target positions; and co-ordination between driver and commander with 'steer-to' information.

Plans, reports and overlays on the system include: Log-on (Initial, Response, Complete); MEDEVAC report; contact report; call for fire report; spot report (ground and air); own operations overlay 1; own operations overlay 2; operations overlay 1 update; operations overlay 2 update; higher echelons operations overlay 2; fire support overlay; fire support overlay update; enemy overlay; enemy overlay update; obstacle overlay; obstacle overlay update; ID report; situation report; and position report.

A gateway interface is also available to the Tactical Fire Direction system (TACFIRE) and Airborne Target Hand-off System (ATHS). Included are fire rescue grid, end of mission, position update request, message to observer, forward observer command, observer's location, spot report and situation report.

Status

Fielded in M1A2 tanks and demonstrated in applique versions at the US National Training Center, Twenty-nine Palms, California. Also available in M1A2 for the Kingdom of Saudi Arabia and State of Kuwait.

Contractor

General Dynamics, Land Systems Division, Sterling Heights, Missouri.

VERIFIED

Manoeuvre Control System (MCS)

The MCS is the central command and control system for the manoeuvre elements in battalion through corps echelons. It consists of a network of computer workstations that integrate information from subordinate units with that from the ATCCS (see separate entry) battlefield functional areas. This creates a force level information database referred to as the Battlefield Common Tactical Picture.

The MCS's databases maintain and display, in text and graphic formats, critical current information on friendly and enemy forces, drawing on information held in a wide range of supporting combat and combat support systems in addition to the primary ones of the ATCCS. By using common decision graphics, which include map overlays and battle resources by unit, the commander and staff can identify possible courses of action. Using these tools, the commander can determine the appropriate course of action. The staff then uses the MCS to prepare and send warning orders, operation orders and related annexes.

From battalion through corps, the rapid exchange of information through the MCS gives all CPs the same picture of the battlefield. This, along with the ability to query both local and remote databases, helps commanders to synchronise the battle.

Status

In 1980, the US Army fielded the first MCS system with limited command, control, and communications capabilities to VII Corps in Europe. In 1982, the Army awarded a five-year contract to continue MCS development, and by 1986 MCS software had evolved to Version 9, also fielded in Europe. Despite unsatisfactory test results the Army awarded a second five-year contract that resulted in Version 10, which was fielded in October 1988.

In 1988, the Army awarded a contract for the development of Block III software Version 11, but in 1993 stopped the development because of programme slips, design flaws, and concerns with cost growth. The programme was reorganised in April 1993 to develop software Version 12.01 using software segments salvaged from Version 11.

In September 1996, the Army awarded a contract to initiate development of the next version of MCS. This effort, the Block IV MCS, is being developed by Lockheed Martin and involves substantially different software, including the required Defense Information Infrastructure Common Operating Environment. The Army postponed Initial Operational Testing & Evaluation (IOT&E) of Block III in November 1996 due to software deficiencies but in lieu of this a Limited User Test was conducted from October to November 1996 to establish a Block III baseline and identify software problems requiring correction prior to IOT&E. This also supported the Army's procurement of MCS for the training base prior to successful completion of IOT&E.

The Army conducted MCS Block III IOT&E in June 1998 during a Division Command Post Exercise at Fort Hood, Texas. This test included live Tactical Operations Centers (TOCs) at division, brigade, and battalion echelons equipped with 47 MCS workstations. These tests were only of limited success for a variety of reasons, and employing MCS with the realistic tactical dispersion and displacement of a dynamic battlefield is expected to further degrade operational performance.

As a result, it was decided not to field the Block III version. In 1999 limited development and procurement of Block IV was authorised and the programme was rebaselined on this Block. Further problems in development caused the planned IOT&E in 2001 to be postponed, and this may now take place in 2003. The system is now evolving as an integrator of the information held on the other components of the Army Battle Command System (ABCS) (see separate entry) to provide the Common Tactical Picture. Further development Blocks will increase the interoperability of the MCS with other systems.

Contractor

Lockheed Martin Mission Systems (Prime), Gaithersburg, Maryland.

UPDATED

Movement Tracking System (MTS)

The (MTS) provides a tracking and messaging system for vehicles, and has been developed principally to improve the C2 of Combat Service Support vehicles. There are two main components to the system: a mobile unit which is mounted on the vehicle and a control station which monitors the vehicle locations. Both components use the same basic communications software and hardware, although the control station equipment has a larger display and faster processor. Communications between the two is provided by secure packet data over commercial satellites, automatically transmitting vehicle GPS positions. The system incorporates digital maps in the vehicles and allows two-way messaging. The mobile terminal equipment is Paravant's RVS-250 (see separate entry).

The MT-2011 satellite transceiver can be mounted either on vehicles or on other platforms (such as containers). It has the following characteristics:

(a) **Dimensions:** 8 × 8 × 4 in
(b) **Weight:** 3 lbs
(c) **Output:** <5W
(d) **Frequency:** L-Band
(e) **Interface port:** RS-422

Status

In 1999 Comtech was awarded a contract worth up to a then US$418 million to supply the MTS. The initial US Army requirement is for approximately 33,000 mobile units and 7,000 control stations. The first formations to be equipped (in 2002-2003) are those of III Corps, together with some National Guard units which have a responsibility to support III Corps.

Contractor

Comtech Mobile Datacom Corporation (Prime), Germantown, Maryland.

NEW ENTRY

....

Strategic Readiness System (SRS)

The US Army is developing a new integrated strategic management and readiness reporting system which combines both the definition of unit goals and objectives with the collection of data. Achievement of each objective is measured and evaluated against a red/amber/green scale. Ultimately this will provide a simple consolidated readiness status report for commanders.

The US Army has broken down unit missions into objectives and measures that are incorporated into a Balanced Scorecard. Each organisation within the Army will build its own mission map and Balanced Scorecard. Each Scorecard identifies specific elements and objectives that define readiness at the relevant level which, when completed, provides a one-page snapshot of organisational performance against strategic priorities: people, readiness, transformation and sound business practices. The Scorecard uses lagging and leading indicators to measure progress toward meeting each unit or formation's strategic goals and objectives. Achievement of each strategic objective is measured with two to three metrics and evaluated using the red/amber/green methodology.

There is no duplication of reports or data entry. Once data is entered into the source database the SRS is updated through automated data linkages. The system sits behind the Army Knowledge Online network.

Status

The SRS is planned to be in operation down to divisional level by late 2003, and to battalion and independent company level by mid-2004. The UK Ministry of Defence has shown interest in the system.

NEW ENTRY

....

Team Care Automation System (TCAS)

Team Care Automation System (TCAS) is an automated medical command support system which has been developed from the widely used Command Data Network System (CDNS) (see separate entry). By making use of the situational awareness, tracking, report generating, messaging and interface tools available in CDNS, TCAS provides medical staffs with a means of collection and dissemination of medical information and of control of the medical support system.

The system interfaces with new and existing medical support hardware allowing the system to track patient condition and update medical records. Typical configurations and interfaces may include a Personal Identification Card (PIC) reader designed to store patient information and medical history, a vital signs monitor and a life signs monitor/printer. Users can use TCAS and the PIC to update a patient's current condition and treatment status in the field to be stored on the card or distributed to other users.

Manufacturer

General Dynamics C4 Systems, Fort Wayne, Indiana.

NEW ENTRY

THAAD Battle Management Command, Control, Communications and Intelligence (BMC3I) system

The US Army's Theater High-Altitude Area Defense (THAAD) weapon system will provide defence against theatre/tactical ballistic missile attack. The THAAD weapon system includes the THAAD Missile, the THAAD Launcher, the THAAD Radar and the THAAD Battle Management/Command, Control, Communications and Intelligence (BMC3I) system. Litton Data Systems (now Northrop Grumman) is responsible for the development of the THAAD BMC3I, which integrates and controls the THAAD weapon system in execution of its battlefield mission.

The THAAD BMC3I provides an integrated set of engagement and force operations capabilities in a Netted, Distributed and Replicated (NDR) architecture that ensures the system will perform well under a variety of stressing battlefield conditions. This NDR architecture includes a data distribution system that supports both local and remote THAAD launchers and radars and is managed in real time by the BMC3I to achieve the mission objectives specified when planning the defence. This NDR architecture calls for two shelter configurations – a Tactical Operations Station (TOS) and a Launch Control Station (Information Systems). When coupled together, these are known as a Tactical Shelter Group (TSG). These shelter configurations support three BMC3I element roles on the battlefield: (1) the Tactical Operations Centre (TOC) comprises either one or two TSGs and is the manned portion of the system where defence planning and engagement decisions are made (2) the Sensor System Interface is made up of one TSG, where BMC3I processing supports remote launcher and radar management and the THAAD distributed network (3) the Communications Relay is made up of one Launch Control Station (Information Systems) that supports remote launcher management and communications relaying.

The THAAD BMC3I provides battle management of weapon system resources, assigning those resources in an optimised fashion to achieve specified battlefield mission objectives, including decision aids that assist in the development of defence plans to satisfy mission objectives in the TOC. The THAAD communications subsystem includes the use of JTIDS for THAAD net and Joint Net communications, a Mobile Subscriber Equipment interface for TADIL-B interoperability and voice capabilities, SINCGARS for voice communications and

A modified SICPS shelter mounted on an HMMWV houses either a Tactical Operations Station (TOS) or a Launch Control Station (LCS)   0009936

A Tactical Operations Station (TOS) contains three high-performance HP-735 processors and two operator stations

the Joint Tactical Terminal for intelligence data. The TOS contains two manned workstation computers and a central computer. The LCS contains an additional computer, a communications interface unit and the communications equipment. All of this equipment is contained in a SICPS shelter mounted on a HMMWV. The computer in each case is the HP735, ruggedised as the Super Bobcat in the US Army's CHS programme. This computer is being replaced by the CHS-2 Sun Ultra Enterprise 4000 to provide symmetric multiprocessing for improvements in response time and capacity performance.

Status

The Demonstration/Validation (Dem/Val) software (over 400,000 lines of Ada applications code) and hardware (a total of 13 stations) have been delivered to the THAAD prime contractor, Lockheed Martin. The THAAD BMC3I has met or exceeded requirements in all system tests to date, including seven THAAD flight tests at White Sands Missile Range. THAAD is currently in the Program Definition and Risk Reduction (PD&RR) phase, as part of which the contractors are conducting the flight test program. Initial production is expected to begin in 2006.

Contractor

Northrop Grumman Electronic Systems, Agoura Hills, California.

VERIFIED

For details of the latest updates to *Jane's C4I Systems* online and to discover the additional information available exclusively to online subscribers please visit
jc4i.janes.com

Australia

Integrated Submarine Combat System (ISCS)

Boeing (Australia), formerly Rockwell, is prime contractor in a multinational team developing an Integrated Submarine Combat System (ISCS) for the 'Collins' class (471) submarines being built for the Royal Australian Navy by the Australian Submarine Corporation.

Description

The ISCS design is based on the integration of a high-performance databus-oriented architecture and proven underlying technologies established through individual subsystems performance. Rockwell's fibre optic databus, known as the Expanded Service-Shipboard Data Multiplex System (ES-SDMS), forms the backbone of the system. Updates to the various technologies have been made in hardware and software since 1987, for use on the RAN's new 'Collins' class submarine project.

The ES-SDMS databus interconnects all high-performance equipment including seven identical multifunction consoles, sonar signal processors, a command plot, dual-weapons data converters and dual-system supervisory units containing global bus assets. The ESM suite, search and attack periscopes, navigation resources including global positioning and inertial navigation systems, navigation radar and other sensor systems are all fully integrated into the system.

The system is as much an information management and processing system as a weapon control system. The ISCS is functionally organised to match the natural progression of a tactical situation – surveillance, track prosecution and support, motion analysis and classification – and guidance for search optimisation and counter-detection avoidance activities. It manages the launching and guidance of tube-launched weapons, including Mk 48 torpedoes and harpoon missiles with interfaces to the weapon discharge system.

Major components and their manufacturers include:
(a) C3 hardware and software – Boeing (Australia)
(b) integrated sonar – Thomson-ASM (now Thales) (France)
(c) weapons control, consoles and command plot –Loral Librascope Corporation (USA)
(d) tactical software – CSC Australia Pty Ltd
(e) logistics support – Scientific Management Associates Pty Ltd (Australia)
(f) electronic support measures – Watkins-Johnson Co and ARGOSystems Inc (USA)
(g) periscopes – Barr & Stroud Ltd (UK)
(h) inertial navigator – Litton Systems (now Northrop Grumman) (USA)
(i) navigation radar – GEC-Marconi (now BAE Systems) (Australia)

Status

The ISCS completed its development phase and began land-based integration testing in early 1992. All six boats have been delivered. However, there was considerable dissatisfaction with the performance of the combat system and the Raytheon CCS Mk2 has been selected to replace it.

Contractor

Boeing Australia Ltd, Brisbane, Queensland.

UPDATED

MINTACS Mine Warfare Tactical Command Software

MINTACS is a software suite which enables a Mine Warfare Force commander and staff to plan missions, monitor tasking and conduct the detailed assessment of mission progress. It provides a number of planning tools and a tactical display which provides the MW situation to the operator.

Planning tools include:
(a) Minehunting and minesweeping planning and assessment. These use the geometry of the segment and environmental data from the tactical display. Using probabilities and times for operations, the time for a specific level of clearance to be achieved or the level of clearance that can be achieved in a specific time can be calculated.
(b) Route survey database. MINTACS can integrate with route survey capture assets, such as side scan sonar and manage and display the resultant data.
(c) Clearance Diving. The Clearance Diving Planning and Assessment tool includes the more complicated coverage rates of a diving team, and calculates the time required to achieve desired levels of clearance at different rates. A planning tool is also available for managing dive team recompression and dive schedules.
(d) Force Clearance tool. MINTACS calculates the combined clearance for a total MCM Force. It takes results from the Minehunting, Minesweeping and Clearance Diving tools and calculates the combined clearance on a Q-Route or area. This is displayed as a colour-coded (as per NATO doctrine) clearance percentage along the Q route on the Tactical Display. The Force Clearance

Module also calculates the threat to transiting ships before and after clearance has taken place. These results are shown graphically down the route and can be viewed on the Tactical Display.
(e) The Force Schedule tool manages the scheduling and tasking of MCM ships and other assets. It can pull in the task set created in the Force Clearance module enabling the scheduling of tasks to suit the availability of assets and the operational cycle.
(f) The Operations Log keeps a record of operational events automatically from the Tactical Display and allows for operator-entered activities.

The Tactical Display presents the Mine Warfare situation to the operator and includes the following features:
(a) Environmental:
Raster and vector chart display
Standard GIS pan, zoom and scroll
Selectable display layers
Distance measure
Cursor latitude, longitude
Support for most civilian and military GIS formats including S57 using the S52 display standard.
(b) Tactical:
Q-Routes and segments
Clearance diver search areas
Minefields and threat mine densities
Mine contacts
MCM vessels and other assets
Environmental conditions
Clearance and threat display on segments
Any other point, line or polygon feature.

MINTACS provides the capability to send and receive signals in a variety of standard formats either automatically or by user request. All signals are stored for future reference and recall. The system has a suite of post-mission analysis tools including the ability to replay the mission tapes from the MCM assets.

It has been developed on Microsoft Windows NT and is available as a stand-alone application or as a network application suite. The stand-alone variant is suitable for laptop or individual workstation computers. The network variant can be installed on large classified networks. MINTACS is currently configured to use Microsoft SQL Server database, but can use any major relational database.

Status

In service with the Royal Australian Navy, by whom it has been used operationally in East Timor and during MCM operations in the clearance of Umm Qasar in 2003. It has been selected by the RN as its replacement Minewarfare Tactical Support System.

Manufacturer

Solutions from Silicon Pty Ltd, Chatswood, New South Wales.

NEW ENTRY

Canada

Maritime Command Operational Information System III (MCOIN III)

MCOIN III is a shore-based command and control information system that replaces the outmoded MCOIN II at the Maritime Forces Atlantic headquarters (MARLANT) in Halifax and upgrades the automated operational capacity to the Canadian

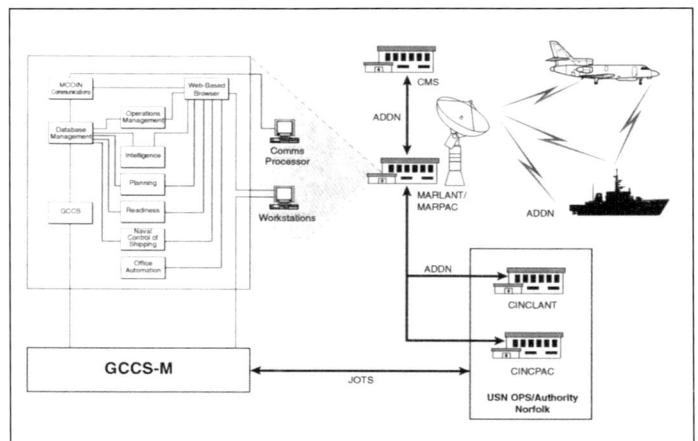

MCOIN III - External Connectivity   0109965

MCOIN III - Internal Connectivity   0109966

Maritime Forces Pacific command (MARPAC) and the Chief of Maritime Staff in Ottawa. The system processes military messages and a variety of sensor data, enabling the production of a Recognised Maritime Picture (RMP) for Canada. This information is then provided not only to the MARLANT command, but also to MARPAC and Ottawa. Classified and unclassified information can also be disseminated to other military and non-military agencies and organisations.

MCOIN III is an open-architecture client-server system using OTS components including web-based technologies for the user interface. Among the OTS components are GCCS-M(JMCIS) (see separate entry) software developed for the USN; IRIS, a military messaging COTS product (see separate entry); Oracle RDMS product suites; Netscape and MS Internet Explorer; and MS Office 97.

The system is designed to interface with: the Canadian Automated Data Defence Network (ADDN); the DND Joint Command Control and Information System (J2CIS); the Canadian Land Forces Command and Information System (LFCIS); the Canadian Air Force Command Control Information System (AFCCIS); Allied Command and Information Systems, particularly those of the US and NATO; and other government departments.

Status
In operational use from March 2001.

Contractor
Macdonald Dettwiler, Ottawa, Ontario.

VERIFIED

SHipboard INtegrated Processing And Display System (SHINPADS)

Designed principally for the Canadian Patrol Frigate (CPF), SHINPADS comprises 29 computers and a serial databus. The computers used are AN/UYK-505s, AN/UYK-502s and 16 bit architecture machines.

Four cables running the length of the ship, one on each side and two in the middle, connect the ship's weapon systems and the computers in a local area network configuration. The system can continue to function at 100 per cent after the loss of any two of the four cables.

Five AN/UYK-505 computers act as mainframes and they and the other computers compose the system databus. All subsystems have access to the ship's information-gathering activities to provide redundancy through versatility.

Serial data transmission is used to allow rapid data processing and the bus is rated at 10 MHz. Some parallel connections exist between individual weapons systems to serve as back-ups. During normal operation, with all the sensors and weapons working simultaneously, the bus operates at approximately 10 per cent of its capacity. Computer architecture and software are designed to take advantage of the bus capacity, distributing traffic and functions to use the system effectively or to reconfigure in the event of a loss.

The software package consists of three million instructions in the naval computer language CMS-2. Two hard disks are used for data storage and the software can be reconfigured to continue running a mission despite the loss of a key subsystem unit. The software also allows easier upgrading of subsystem hardware.

Standard digital equipment and peripherals, as well as software, are used in the CPF electronics system and a standard display used at each workstation enhances the system's reconfigurability. The system uses 15 AN/UYQ-501 workstations.

The CPF is equipped with Link 11 which allows the ship to integrate its air defence with that of other NATO vessels. Digital communications and tactical displays can be exchanged between ships and targets can be switched from one fire-control system to another. With each vessel having a complete database of everything acquired by other ships, the fleet becomes a large-scale version of SHINPADS, able to transfer target responsibility if a unit is disabled or removed from the network.

Status
SHINPADS is in operational service aboard the 'Halifax' and 'Tribal' class ships.

Contractor
Lockheed Martin Canada, Montreal, Quebec.

VERIFIED

Chile

SP100 command and control system

The SP100 command and control system is a distributed and modular system, built around high-performance microprocessors. It is characterised by its advanced man/machine interface which can be connected to almost any sensor and weapon on board through specialised interfaces. The system provides a means of maintaining a common tactical picture in each unit of a force.

Description
The system is based on a set of multifunctional consoles (CONTAC) (rather than a central processor) which work as the basic processing nodes of the system. The processing is distributed among the consoles enabling each configuration to perform a specific tactical role. All the consoles are furnished with the necessary modules to perform any tactical role and each console has the capability to assume the functions of any other. These and other redundancies, such as the duplication of the Local Area Network (LAN), allow the system to achieve a high availability rate.

The system architecture is modular and information is distributed among the specialised computers making up the system's master components (preprocessors, multitrackers, graphic processors and so on). Modular software has been developed in accordance with DOD-STD 2167A. The combination of methodology and development tools, plus the use of standard parts and components, allows modular development and easy maintenance. The system can be adapted to any platform, including submarines and aircraft. New capabilities and the integration of new sensors, weapons or computers can be achieved without redesign of the system, requiring only the addition of new consoles or workstations for additional operators, thus keeping expansion costs low. The manufacturer claims that the system modularity allows for an almost unlimited expansion capacity.

The Tactical Consoles (CONTAC) are the main processing nodes that provide the man/machine interface for the system. Their purpose is both the compilation as well as the appraisal of the tactical picture. Each CONTAC will:
(a) process tactical information with specialised processors (80486DX2/66 and TMS34020), enabling the system to interact with sensors and weapons and simultaneously manage up to 512 tracks, 64 reference points, 100 bearing lines, eight tactical gridlines, 64 acquisition sectors, two tote windows, a readout window, an information window and radar video of 1,024 × 1,024 pixels (128 levels) and sectors, areas, zones, axes, signals, messages, history of tracks, history of bearings, and other data
(b) present tactical information which is displayed on high-resolution colour screen (1,600 × 1,200 pixels). The tactical elements are displayed with relative or absolute movement, centred or off-centred, with multiple windows, zoom and other features
(c) accept operator's commands through an efficient man/machine interface composed of a high-resolution screen, touchscreen, a qwerty alphanumeric keyboard and a trackball.

On the tactical picture, the operator can apply filters, create geometrical elements, define tactical gridlines and create synthetic maps. A number of other utilities specially designed for a rapid and accurate tactical situation assessment are also included. Tactical operations are co-ordinated via a datalink, enabling tracks to be assigned to a certain unit or designated to a weapons system. Inter-unit tactical data interchange and positive control of a specific unit can also be achieved. Some decision-making aids are also provided such as blind sectors, mutual interference sectors and exclusion zones, and the CONTAC can also record and replay exercises for further analysis.

SP100 architecture   0001004

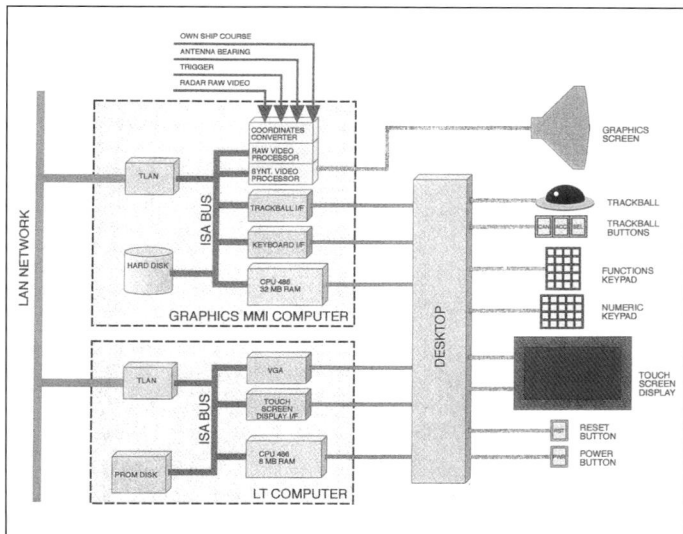

CONTAC architecture   0001005

The Navigation Computer's (CONAV) mission is to provide continuously updated 'own ship's' data (position, course, speed, wind direction and speed, roll, pitch and temperature) to the rest of the SP100 system. This data can be provided either by onboard sensors, estimated by the CONAV or can be manually entered by the operator. Each data source is selected remotely by a CONTAC operator.

The Tactical Preprocessor (PPTs) is the interface between any tactical sensor and the SP100 system. The PPT takes information from any source sensor and converts it to SP100 format for storage in the tactical database. This allows the SP100 to be used with a variety of different sensors. Weapon Preprocessors (PPAs) provide the interface between the weapons system and the SP100, converting data into the format required by the specific weapons system. The manufacturer claims that this has enabled SP100 to be integrated with a variety of weapons system including Aerospatiale's MM38 and MM40 Exocet and IAI-Elta's Barak.

The service computer provides the means to monitor the performance of important auxiliary functions such as:

(a) storage of computer's configuration to print and plot the tactical panorama
(b) record and replay of naval exercises
(c) permanent storage of tactical information, such as synthetic maps, merchant ships' positions and other information
(d) storage of signals and messages
(e) keeping a log of the SP100 computer's status, to help the system's recuperation in the event of a power supply fault
(f) maintaining the Tactical Data Base (BADET), to restore the tactical information after a power supply fault.

Status
In service with the navies of Chile and Ecuador.

Contractor
SISDEF Ltda, Quintero.

UPDATED

Denmark

Standard Flex 300 C3I System

The SF300 C3I System is an electronic data system for ships. It uses standard consoles, computers and interface units to integrate subsystems such as weapons, sensors and communications. It allows all peripheral systems connected to the databus to be operated and controlled from a central position by standard consoles. The software is programmed in Ada. The system consists of four main components.

The Standard Processor is used to supply sufficient application processing power in the entire C3 system. Each system comprises several standard computers physically dispersed in order to have functions available where needed and performed where it is most convenient in terms of system design and databus load. Computers are constructed from one or more of only five different microprocessor printed circuit boards utilising the industry standard for multiprocessor 32-bit architectures, the VME-bus.

The Standard Interface Unit (SIU) has two main purposes. Its primary function is to perform the necessary conversion and adaptation of data formats and transmission protocols between subsystems and the Local Area Network (LAN). Secondly, the SIU performs local processing (for example, validation and preprocessing) of data before it enters or leaves the network. TERMA claims these features imply that all known and future subsystems can connect to the C3I at reasonable cost, thus making the system fully open ended.

The LAN is a Carrier Sense Multiple Access/Collision Detection (CSMA/CD) type of network and is standardised as the IEEE 802.3 LAN, compliant with the ISO

Open System Interconnection Architecture. The system may have several cables for redundancy.

The Standard Console is a general operator's workstation, from which normal operation of all connected weapons and electronics systems, including the C3I Bus system itself, is performed. The console is equipped with two high-resolution 508 mm (20 in) daylight colour raster displays, each capable of being divided into four separate areas (windows) for display of information from various sources. It is operated through several input devices: two software-controlled flat-panel displays with a set of menu and function keys which change with the current task, two trackerballs for operation of the displays and for fire control, a qwerty keyboard with numeric keypad and a number of fixed-function hard keys. The duplicated input device allows dual operation of the console. Two multiprocessor computers are contained in the Standard Console; one takes care of the presentation system while the other performs local C3 functions and network interface.

The system is now being upgraded. In 2000 a £17M contract for 6 units was signed with Terma as the prime contractor together with Systematic and Infocom. In June 2002 it was announced that LynxWorks would be supplying its LynxOS real-time operating system for the upgrade. The principle reasons for the upgrade are to overcome hardware obsolescence and the associated support difficulties. The technology was considered to be outdated and minor operational changes required expensive testing. This also coincides with the introduction of the Royal Danish Navy's new STANFLEX 3500 class. The upgraded system will be known as C-Flex, and will be based on a generic system developed by Systematic, which is also being used by the Danish Army to upgrade its Air Defence system .

In the future system, the SF300 Intranet will have a number of servers and databases. Each containerised weapon system will have its own Sub-System Interface Units. Radar and video information will be digitised and transported on a parallel, separate TCP/IP network. New Sub-System Interface Units (SSIU) will replace the standard interface units currently employed with weapons and peripheral sensors. This fulfils an operational requirement for a 'graceful degradation' in the system, allowing sub-systems to function independently in the event of action damage or malfunction.

The Local Area Networks will be upgraded to an intranet - using TCP/IP - with fibre-optic cables, routers and switches. It is a redundant system and each sub-system will have its own interface unit to the net. Routing is carried out by the system itself if a degradation of the system occurs. Other disks or servers can take over. The ships will be fitted with fibre-optic cables throughout.

The Standard consoles will be replaced by multifunction, flat screen MS Windows workstations, with additional RGB features for sonar applications, from Infocom. They will have the same height and width as the old consoles but half the depth. The C3 and data base computers will be powerful Solaris servers in a cluster figuration, which will provide scalability and redundancy, and where possible radar video will be digitised at the sensor.

The first version will provide:

(a) Picture compilation with unlimited tracks (up to 1000 tracks with no performance reduction)
(b) Correlation of tracks with all sensor and data link sources
(c) Weapons C2
(d) AAW Threat Evaluation and Weapon Assignment (TEWA) functionality - hard and soft kill
(e) Link 11 and DanLink.

The second version will have extended functionality:

(a) Management of CIWS
(b) MCCIS integration
(c) Graphic TEWA presentation
(d) Large screen displays
(e) Mine laying functionality

Standard Flex 300 C3I standard console

(f) Helicopter control through a COTS helo planning tool
(g) Inclusion of Link 16 and 22.

Status

The Standard Flex system was sold to the Royal Danish Navy for the SF300 *Flyvefisken* class, in two series of seven ships, and for the four SF3000 *Thetis* class.

The first version of C-Flex is expected to undergo sea trials in late 2003, with the remaining 5 platforms of the first stage of the contract fitted in 2004-2005. Thereafter the remaining SF300 and the SF 3000 platforms will be refitted. The more capable version of C-Flex is to be fitted to the SF3500 Flexible Support ships when built, the contract for which is worth more than DKr100 million. This was signed in June 2003.

Contractor

Terma A/S, Naval and Ground Systems Division, Tåstrup.

UPDATED

Finland

Finnish Integrated Coastal Defence C4I System

The Finnish Integrated Coastal Defence System is designed to enhance defensive firepower against sea invasion. It is capable of engaging all surface targets and even helicopter air assault formations in the speed range of 0 to 200 m/s, using both dedicated fixed coastal artillery batteries, mobile and SP artillery units and anti-ship missile batteries. The primary means for target surveillance, acquisition and tracking are the Sea and Air Surveillance (SAS) systems and the Coast Artillery Forward Observation Posts (CAFOP) are dedicated for fire control. These CAFOPs are especially advantageous when repelling violations against territorial integrity by warning fire. The system can also be used for maritime traffic control and navigational piloting.

The system is flexible and easily adjustable according to the local situation. It consists of four principal elements:

(a) the Sensor System for target surveillance, acquisition and tracking is based on sea and air surveillance posts and stations and the CAFOPs. These are equipped with manned or unmanned optical and FLIR observation instruments, laser range-finders and, for example, track-while-scan radar systems, locally or remotely operated. The system provides the needed intelligence data on the nature, activities, location and movements of enemy targets and friendly forces for both operational, tactical and fire direction/control purposes

(b) command posts, for district and local sea surveillance, area responsibility, firing units, and so forth, are equipped with customised, computer-assisted display, control and planning desks with synthetic, alphanumeric and graphic displays with or without map underlay, joysticks, keyboards, control keys, printers and so on. For fast-changing situations and threats in coastal defence, Coastal Artillery Fire Direction Centres are provided with sophisticated, ground mobile means able to concentrate and disperse the defensive fire in an optimal way according to the available firing units and the nature of the targets

(c) the Coast Artillery Battery Computer System situated in the CABCPs. This consists of the ground mobile CA ballistic computer with separate Gun Computers (GC), easily read Gun Displays and auxiliary GD units on the fixed, mobile or SP gun sites. The BC provides continuously, that is, with 0.5 or 1 second intervals, individually calculated firing data for each freely dispersed, deployed and aimed gun. The number of guns can be any reasonable figure (for example, six, with another six backed up). The BC and GC compute the following firing parameters for the free firing areas: azimuth, elevation, type of ammunition, shell, fuze and its setting, time of flight, mode of fire, number of

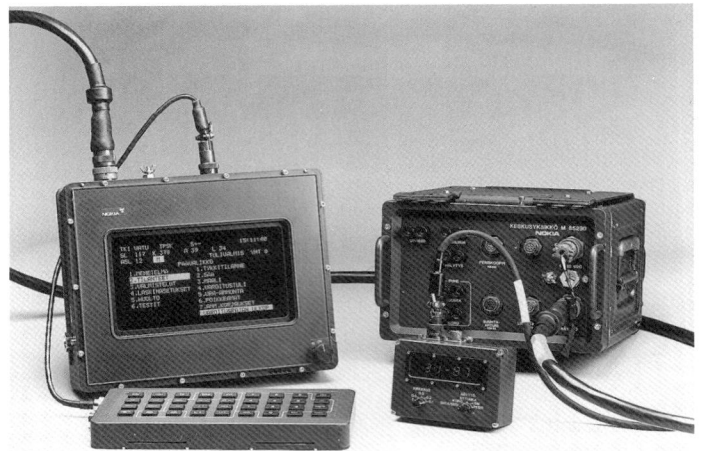

CA Gun Computer M85250 and Auxiliary Gun Display Unit M85180 (foreground, right) ensure accurate and fast handling of the individual gun to maximise fire effect against any fixed or moving coastal target

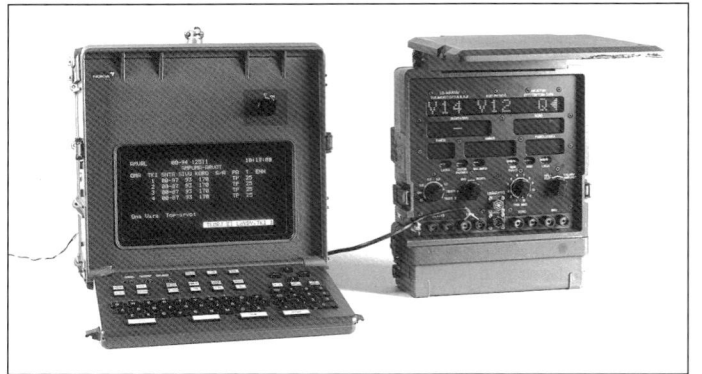

Battery Computer M85270 and Gun Display Unit M85320 for coastal artillery

rounds etc using AMETS data, muzzle velocity-radar and fire adjustment inputs. The algorithms used include all the contemporary ballistic knowledge to ensure maximum accuracy of each gun

(d) the Nokia (now Patria Finavitec) Message Switching Network Communication System (see separate entry) provides a secure burst data transmission network with individual and group addressing possibilities and automatic encryption/decryption. It integrates the different subsystems of the overall coast defence and fire-control system for different flexible fire control and firing configurations. The network uses message switches and general and special purpose message terminals.

The Coastal Defence System interacts with AS missile and field artillery units, higher echelon tactical centres, naval units and with AD units. The system's flexibility makes it very adaptable according to each individual customer's requirements.

Status

This was Nokia's third-generation coastal defence system. The first, RADAL, was developed in the 1960s and saw service from 1968 to 1987. The second-generation, RAVAL, was developed in the 1970s and has been in service since 1980. This latest system's prototype was fielded in 1993 and serial production began in 1994. The Nokia Special projects division was sold to Patria in 1999. As at March 2003 no new information on this system had been received.

Contractor

Patria New Technologies Oy, Tampere.

UPDATED

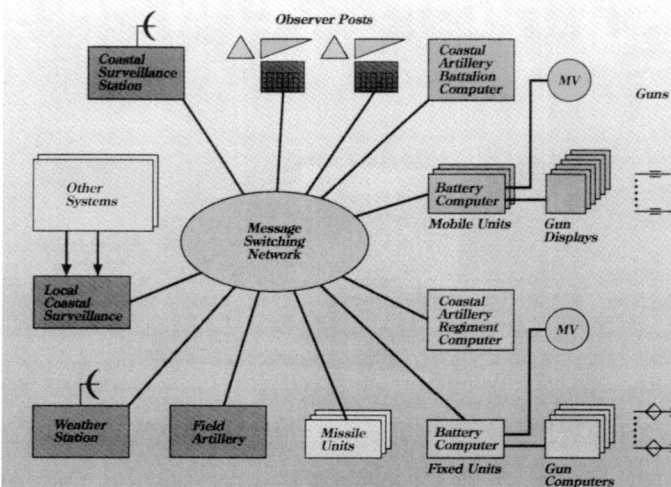

Finnish coastal defence and surveillance system

France

Aide au Commandement de la Marine (ACOM)

Aide au Commandement de la Marine (ACOM) is the global Command and Information system for the French Navy. The system has 5 configurations depending on command organisation and operational use, at sea as well as ashore: SYCOM for Navy Headquarters, Shore Commanders and Naval Operations Centres, AIDCOMER for Naval Staffs at sea at CTF or CTG level, OPSMER for ship commanding officers, TRELO for small ships without a command system (TRELO includes a read-only Link 11 subsystem), and CENTAC for the French Navy Atlantique 2 MPA. It is installed in 50 ships and submarines and 15 shore command facilities, plus training facilities and technical support centres. It provides national and multinational interoperability including an intranet and e-mail. All the configurations use the same core software for common applications and the same relational database. They are linked together through the existing French Navy communication systems: shore to shore through the navy data net, ship to shore through military or civilian satcom systems or standard HF systems for small ships.

Based on COTS workstations and the UNIX operating system, ACOM is a modular system, including configurations of one workstation for the smallest ships, up to fifteen workstations for naval operations centres and twenty five workstations on board aircraft carriers. When installed on board as a command support system it is connected to the combat system, and fuses tactical information coming from Link 11, 14 or 16 with intelligence from the RMP.

ACOM provides a Wide Area Picture (WAP) which complements the real-time situation generated by ship sensors or tactical data links. This is drawn from shore-based surveillance radars, humint and elint sources, allied sources, reports from naval forces at sea, reports from MPAs and reports from air and land components. The system supports the exchange of information with all shore command centres and naval units at sea with automatic data transfer from database to database when satcom facilities are available, and with other systems and subsystems requiring the maritime situation, for example mission planning systems for aircraft. It also includes planning and decision aids and naval logistic planning tools. It provides immediate recording of data and events, allowing immediate reports and modification of orders, and can also provide simulation and replay facilities.

ACOM uses the Oracle 8 relational database management system on board ships and in all shore SYCOM and CENTAC sites. This database contains two types of information. The first is encyclopaedic data, such as ship, weapons and equipment characteristics and photographs, human and geographical permanent information such as airport or harbour descriptions, EEZ and territorial waters limits, all drawn from both open and national intelligence sources. It includes descriptions of military sites ashore, tactical information such as ROE, classification criteria, ship duties and task organisation, ephemeral data and astronomical tables. The second is information related to operational formatted messages, tracks, graphics, and operational areas such as tactical screens and subnotes. The system can handle three messages received per minute with an average size of 6,000 characters per message; and has an online storage capacity of three months of intelligence situation with 10,000 tracks, 200,000 positions, 10,000 operational areas, 3,000 messages, and one month of tactical situation. Offline storage capacity is practically unlimited. The permanent database is managed by a national technical centre and regularly updated by dissemination of the national reference database.

The system is fully interoperable with equivalent NATO and US systems and can read both ADaT P-3 and OTH-GOLD. Specific national formats are also provided, particularly for SSBN operations.

Specifications

(a) SUN Microsystems graphic workstations (SPARC Architecture)
(b) SOLARIS Operating System
(c) Ethernet TCP/IP Local Area Network
(d) X25 Wide Area Network with X 400 Message Handling System
(e) ORACLE 8 Relational Database Management System
(f) AppliX Office Automation Tools
(g) VERITY Search Engine for Documentation
(h) X11, Motif and ILOG VIEWS-based Human Machine Interfaces
(i) Electronic Charts Visualisation Server
(j) Object oriented design with C++ code development.

Status

Version 3 of the ACOM software is operational in the French Navy. Version 4 has been developed, was trialled in the aircraft carrier *Charles de Gaulle* and is now being installed in all other platforms. This is a completely new version with a new software architecture with three main layers, connected through two standardised Application Programming Interfaces (API). The first one connects the Common Operating Environment including Oracle 8 DBMS, HTML, X11/Motif, TCP/IP and X400/X25 technologies to the 'ethical core software' including a new cartographic server, an object general management C++ software and the ACOM Web Browser. This layer is connected through an API to the Mission Application Software layer, which contains: a new geographical server including all the standardised geographical formats, 3-D satellite imagery and the new IHO ECDIS ENC 57 format; a new man-machine interface; new air operations applications such as the air mission preparation application; a land situation management application; and a textual documentation application which allows the attachment of textual information to all objects managed in the ACOM system the creation of specific document folders.

Contractor

EADS Systems & Defence Electronics, Vélizy Villacoublay.

UPDATED

SENIT combat management system

Versions of SENIT CMS have been installed in most French naval vessels and in vessels of a number of foreign navies. The latest are the SENIT 8 family, a variant of which is installed on board the *Charles de Gaulle* nuclear-powered aircraft carrier, and SENIT 2000 which has been derived from SENIT 8 by DCN for use in FPBs. In the latter the operating modes are designed specifically for coastal warfare, with special emphasis on anti-surface weapons, passive detection, tactical datalinks and fast response to 'pop-up' air threats. Both versions comply with all relevant

SENIT 2000 combat management system 0009937

NATO interoperability requirements, including tactical data Links 11, 16 and 22, and provide 3-D fusion of data from co-operating sensors which complies with the USN Co-operative Engagement Concept (CEC). The system has an open architecture based on a dual-redundant tactical network interfacing with all weapons and sensors, plus navigation and communications equipment, using protocols including ATM. A high level of functional integration is achieved as CIC and bridge operators have access to combat system, navigation, communication and platform management functions. The system also supports other tasks including mission planning and management, debriefing and onboard simulation. The layered software allows all application modules to be independent of CMS hardware and combat system configuration.

SENIT 8 and 2000 make extensive use of commercial-off-the-shelf (COTS) hardware and software components including: flat panel multifunction consoles (allowing relocation to any combination of operator duties), Hewlett-Packard PA-RISC processors, standard operating systems (UNIX and POSIX), X-Windows and MOTIF graphical interfaces and ADA (tactical processes) and C ++ (operator interfaces) programming languages. DCN claim that SENIT 2000 represents the next generation of combat management systems in terms of data processing capability and that this will be the first time a processing capability similar to that of a frigate is implemented on such compact vessels.

Status

SENIT 2 (2/3 Univac 1212 computers): French *Suffren* class. Also in CV *Foch*, now Brazilian *Sao Paulo* (possibly updated to SENIT 8).
SENIT 3 (2 Univac 1230 computers): French 'Tourville' class
SENIT 4 (1 Iris 55 computer and 7 consoles): French 'George Leygues' class
SENIT 7 (also known as STI) (2 MLX 32 computers and 5 colour consoles: Saudi Arabia 'Al Riyadh' class
SENIT 8/01: Three of the 'Georges Leygues' class
SENIT 8/05: *Charles de Gaulle*.
SENIT 8/: French and Italian 'Horizon' class frigates currently under construction (2 in each country) and the 2 French 'Mistral' class LHD/command ships currently under construction.
SENIT 2000 was selected by the Royal Norwegian Navy in 1997 as the integrated weapon system for modernisation of 14 existing 'Hauk' class FPBs. The first 3 platforms were handed over in Nov 2001. It was also fitted in the first of the new 'Skjold' class FPB with an option of outfitting a further six.

Contractors

DCN International, Paris.

UPDATED

Submarine Tactical Integrated Combat System (SUBTICS®)

SUBTICS® is based on advanced weapons and sensors and is adaptable to any type of submarine, whether a new build or refit. In addition, its modern design and modular build enable the system to be readily updated and adapted. SUBTICS® has the following key features:

(a) signal processing from a variety of sensors including: acoustic (sonar) and non-acoustic (optical, optronic, ESM, radar and communications) for detection, tracking, location and identification of vessels
(b) tactical situation display (track association, fusion, synthesis, trajectography and management)
(c) support for tactical analysis, decision making and action management encompassing both the geographical and tactical environment
(d) target engagement and weapons command and control (including weapon choice, launching and guidance).

SUBTICS® has a high level of operational capability and a long-range capacity, both in detection and weapon launching due to the use of external sensors

SUBTICS® on board (UDS International/Gabriel Martinez)   0130056

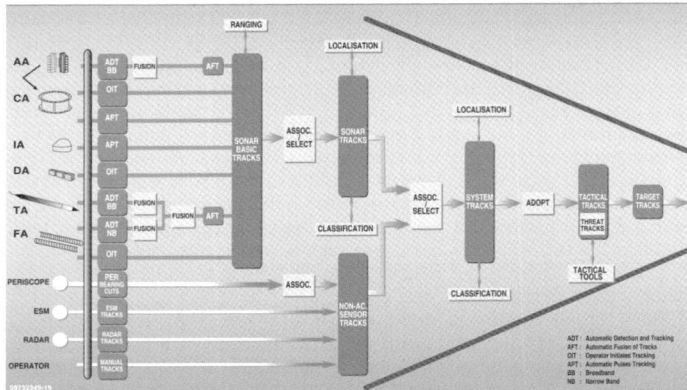

SUBTICS® track management data flow (UDS International)   0122931

(including flank array and towed array). It also provides an anti-surface ship missile capability. Advanced automatic and interactive processing, together with a modern man/machine interface provide a clear and intuitive display of the tactical situation. A high degree of operational availability is achieved by the use of functional and hardware redundancy, reconfiguration flexibility, online management of system status and performance and the use of advanced standard technology. The use of common hardware and software and interface standards, together with COTS technology provide for easy system maintenance and reduced life-cycle and training costs. These features enable the system to be both adaptable and have the flexibility for growth and upgrade.

Specifications
Hardware and Software
Signal processing: TMS 320 C 30 processors on a speed ring network
Data processing: PowerPC VME boards
Multifunction consoles: 2 high-definition 19 in colour monitors with ruggedised PowerPC workstations.
Software standards: POSIX/UNIX, TCP/IP, X WINDOWS/MOTIF, ANSI C/C++ languageThe system complies with DOD STD 2167 for software development and utilises an Ethernet redundant system network.

Status
SUBTICS® was selected by Pakistan for fitting on board three *Agosta* 90B submarines: the first was commissioned in 1999, the other two are to be commissioned in 2003. The first combat system has been successfully tested at

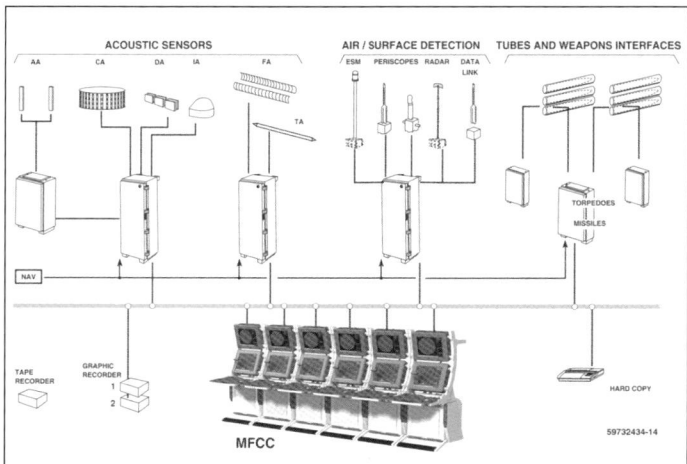

Typical SUBTICS® block schematic diagram (UDS International)   0122930

The latest generation of SUBTICS® system (UDS International/Gabriel Martinez)
NEW/1022639

sea. SUBTICS® is to be fitted on two Chilean Navy *Scorpene* submarines, and in June 2002 it was announced that the system would be fitted on two *Scorpene* submarines for the Royal Malaysian Navy.

Contractor
Underwater Defence Systems International (UDSI), Sophia Antipolis, France.

UPDATED

TACTICOS combat management system

In December 1990, soon after the acquisition of Hollandse Signaalapparaten by Thomson-CSF, now Thales, a corporate decision was taken to merge the best elements of Signaal's STACOS and Thomson's TAVITAC 2000. The new system was designated TACTICOS (TACTical Information and COmmand System) and has been marketed since 1991.

Description
TACTICOS is a combat management system comprising command and control, command support and fire-control facilities for anti-air, surface, anti-submarine and electronic warfare as well as naval gunfire support. It allows the command team to assess and monitor the tactical situation, to plan and co-ordinate naval operations and to control the sensor and weapon systems.
TACTICOS tasks include:
(a) short and long-term planning of maritime operations
(b) control of online surveillance sensors, datalinks and communication equipment
(c) presentation of raw or digitised sensor information
(d) compilation and presentation of the tactical situation in the air, surface and subsurface environments, including the EW situation

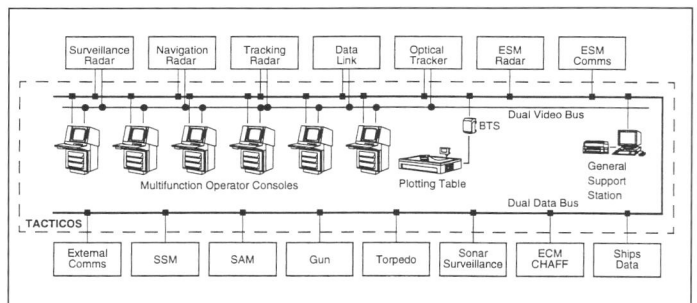

TACTICOS within a combat system

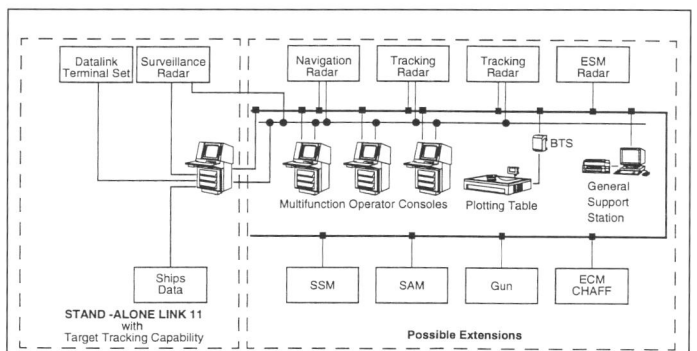

TACTICOS-based stand-alone Link 11 system including target tracking capability and an example of possible extensions

Multifunction Operator Consoles (MOC) on board a Hellenic Navy Meko 200 frigate

(e) decision support for the command team
(f) automatic threat evaluation and threat ranking
(g) automatic target designation to weapon systems and tracking/illumination sensors
(h) assistance in operations, including navigation and the direction of aircraft and other assets under control
(i) Embedded training in a fully synthetic or mixed (synthetic and real tracks) environment.

It contains scenario and simulation facilities for wargaming, planning and training. TACTICOS can be used in a variety of warships, ranging from destroyers and frigates down to fast attack craft size, as well as a retrofit on existing ships.

Being an open system, TACTICOS is able to host a wide variety of open market software in the areas of message handling, damage control, meteorological data processing, map handling, relational databases, planning systems, TMCCIS-like software and so on.

TACTICOS is based on a distributed computer architecture, applying a multinode, multiprocessor concept in a battle damage-resistant configuration. The nodes are hosted in the Multifunction Operator Consoles (MOCs) and Bus Terminal Servers (BTSs) and contain software, written in Ada and C++ and includes the SPLICE distributed database and data communication software that operates alongside a real-time version of UNIX. The console hardware is based on unmodified commercial off-the-shelf workstations, and hardware is linked through a redundant Local Area Network (LAN) formed by Ethernet, Fast Ethernet or ATM. A redundant video distribution bus collects the radar, television and/or infra-red (TV/IR) videos and distributes these to the various operator consoles, where a selection can be made out of the available videos for presentation on the console.

Some subsystems, such as Thales' latest sensor and tactical datalink subsystems may be linked directly to the LAN and video bus. Those that cannot be so linked use the BTS to interface with the LAN and the video bus. The BTS contains spare interface card positions that will allow NTDS, MIL-STD-1553B, RS-432 or any other subsystem interface, to link into TACTICOS.

The standard MOC Mk 2 features one or two 19 in colour raster scan displays, each 1,280 × 1,024 pixels. On the monitors a real-time layered presentation is possible of maps, radar video, tactical tracks, tactical figures and TV/IR video, as well as windows to control the online sensor, weapon and datalink subsystems. In addition to the Sigma node, the MOC contains a RISC graphics host processor with a Winchester disk drive and operator input devices consisting of a panel with programmable keys, a rollerball (with a joystick as an optional extra), a qwerty keyboard and five protected keys. The MMI provides a graphic user interface conforming to the X-Window and OSF/ Motif standards. The MOC can also be delivered as a three-operator conference console and can be equipped with a large screen display (up to 80 in).

Other hardware includes an online plotting table for automatic and manual plotting on various types of charts and a General Support Station (GSS). The latter is a standard commercial workstation with a 16 in colour raster scan display and a laser printer and is used for system monitoring, database support and training.

The system is also offered for retrofit on existing ships. The MOC can be equipped with local subsystem interfaces as well as air and surface target tracking capabilities. This allows low-cost solutions with the growth potential towards a complete TACTICOS. This concept is used on the Turkish 'Knox' class frigates and the Portuguese 'Joao Belo' class frigates.

Status
TACTICOS is claimed to be "currently installed on board more than 70 ships in 11 countries" in platforms ranging from small patrol boats to frigates. Selected in 2001 by the Turkish Navy for installation in the four FPBs of the Kilic-II programme.

Contractor
Thales Nederland, Hengelo.

UPDATED

TAVITAC NT naval Combat Management System

TAVITAC NT (New Technology) is the latest naval combat management system developed by Thales Naval France. The system benefits from the operational experience gained with the French Navy SENIT systems, programmed in a large part by Thales engineers and from the export market with TAVITAC systems for frigates and fast patrol boats. TAVITAC NT is a technological evolution of the TAVITAC 2000 system being installed in the French Navy *La Fayette* class frigates. Operational functions include:

(a) compilation and display of the tactical picture and the Recognised Maritime Picture (RMP) based on data from ship sensors or from offboard sensors via datalinks.
(b) threat evaluation and weapon allocation and engagement
(c) co-ordination and direct control of the fully automatic deployment of anti-air weapons for self-defence
(d) datalinks and communications management
(e) command support system functions
(f) navigation support
(g) simulation and training facilities and system monitoring.

TAVITAC NT features an open architecture that allows customers to adjust the combat system to their requirements. It is a fully redundant system organised around a duplicated FDDI (or Ethernet) local area network, which endows the system with large growth potential in terms of processing capacity, performance and adaptability to a variety of sensors and to hard- or soft-kill weapons and new functions. It can be used both for tactical combat management and command support.

Tactical/Real-Time. The system uses multifunction consoles with large high-resolution colour flat screens. Each console houses a tactical computer which provides redundancy when the system has two consoles. Redundancy increases with the number of consoles leading to a highly reliable system. The data provided by the sensors and weapons are managed in a database fully distributed on each tactical computer. TAVITAC NT also uses System Interface Units (SIU) to interface with subsystems. All the hardware units are based on the same commercial-off-the-shelf (COTS) products, thus facilitating the replacement of computers by a more powerful future generation.

The software is fully programmed in Ada in compliance with the methodology defined by DOD-STD-2167A. The systematic use of COTS software standards such as X11, Motif, POSIX 1003.1-1003.4 and 1003.4a/UNIX real-time and VMEbus ensures portability of applications and consequently the growth potential of the system without modification of its architecture. The software includes a kernel, which is used for all applications of TAVITAC NT and specific functions which are adapted to the missions of the ship. The system has the capacity to handle an average of 500 tactical tracks corresponding to an average of 4,000 tracks generated by 1,000 real-time objects.

Command Support System (CSS)/Non Real-Time. The TAVITAC NT architecture provides the connections to CSS. Since a CSS handles non real-time data and makes wide use of Relational Databases, the CSS is based on standards such as Java and Corba. This makes possible the exchange of formatted data (OTH-T Gold, Adat-P3) as well as text, pictures and video and secure e-mail.

For further details see *Jane's Naval Weapons Systems*.

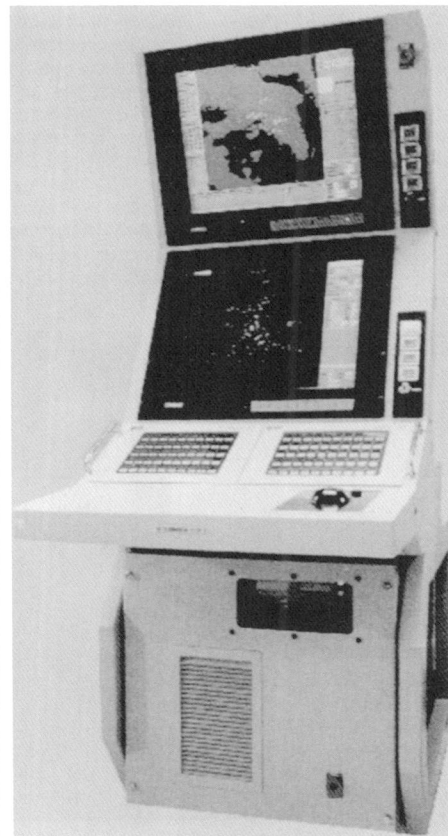

Cyber multifunction console
0098806

Status

In service with the Kuwait Navy 'Um Almaradim' (Combattante I) class. In production for an additional export customer.

Contractor

Thales Naval France, Bagneux.

UPDATED

TAVITAC/TAVITAC 2000/STI tactical data handling systems

In the early 1970s, development began of a new tactical data handling system, SENIT 5 (*Système d'Exploitation Navale des Informations Tactiques*) (see separate entry), for the French Navy, which eventually decided not to purchase it. However, Thales Naval France (ex-Thomson-CSF NCS France) decided to continue development of the system 'commercially'. Originally the system was known as *Vega Nouvelle Generation* or Vega 3C but subsequently it was named TAVITAC, an acronym for *Traitement Automatique et VIsualisation TACtique*.

Development was completed in 1978 and the system was subsequently sold abroad, the last major sale being in 1986 to China. However, in the early 1980s the manufacturers recognised the necessity of exploiting the latest electronic concepts and began a substantial redesign of the system. In its place they have developed a modular one which was first revealed in 1988 as TAVITAC 2000. The TAVITAC 2000 system forms the basis of the STI (*Système de Traitement d'Informations*). In 1990, Thomson-CSF and CPM (*Centre de Programmation de la Marine*) were awarded a contract to develop TAVITAC 2000 into STI for the *La Fayette* class light frigates on which it is now operational.

TAVITAC is a tactical picture compilation system which uses data from ship sensors and can also integrate data from offboard sensors through a Link 11 datalink system. It also designates targets to weapon systems.

The TAVITAC system has a federated architecture based originally upon the CIMSA-SINTRA 15M05 computer and later upon the 15M/125X as used in SENIT 6. The original computer had a memory of 64 to 128 kbyte 16-bit words. The number of computers varies with the operational requirement but in frigate-size vessels, such as the F 2000S purchased by Saudi Arabia, there are two computers. A ruggedised disk storage system is used to maintain the database. The software, written in LTR 2, can be adapted to associated sensors and operational requirements. It can be produced upon commercial computers which are compatible with the Thales processors used in TAVITAC.

For display purposes TAVITAC uses the modular GVM system, which has vertical and horizontal consoles. The vertical display systems may also incorporate a fire-control console while the horizontal ones can have two or three operator positions. The primary displays may be supplemented by television or plasma displays for alphanumeric data. The consoles can be driven by a dedicated computer or they may include integral processing.

Each console includes a combined radar-synthetic image generator with a renewal memory of the associated computer and a video compression device. The console also includes a 40.64 cm (16 in) or 55.88 cm (22 in) rectangular or circular monochrome or raster scan PPI display, while the MMI includes order keyboards, video control panel and rollerball control. The screens can be monochrome or colour, circular or rectangular and the display capabilities are adjustable to customer requirements with up to 200 tracks and 100 different symbols. The Type F 2000S display systems have square, monochrome PPI displays.

Standard functions in the GVM consoles include automatic tracking with manual or automatic initialisation, rate-aided manual tracking as well as manual entry of tracks and special points. The consoles can also display optical and ESM bearings, identify tracks, evaluate threats, designate targets and aid tactical navigation. Some six extensions are available to the system's capabilities including automatic engagement of air targets, tracking and direction of ASW targets and weapons, tactical situation recording and replaying, storage of coastline details for display on PPIs and training facilities in association with a video simulator.

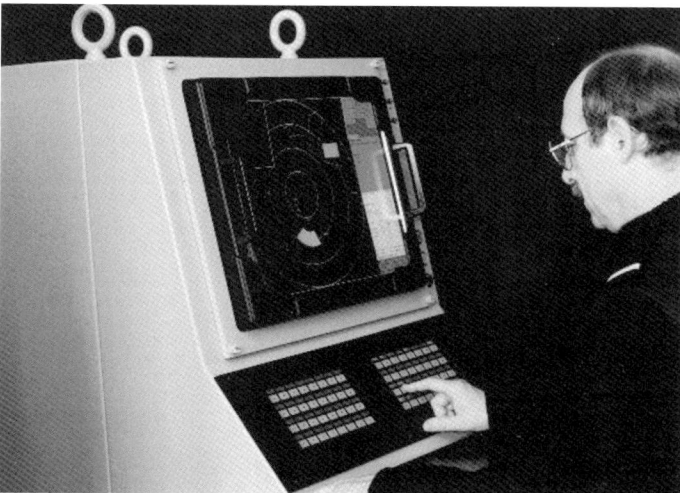

TAVITAC 2000 high-definition raster scan display console

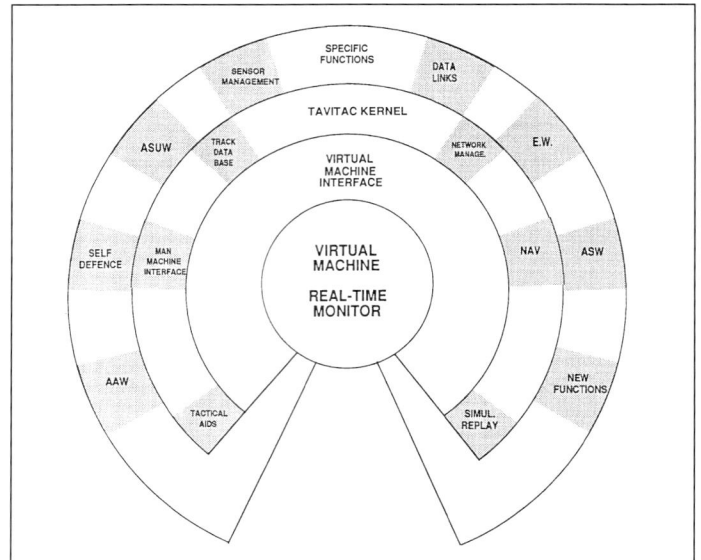

The software comprises a kernel which is valid for all implementations of TAVITAC 2000

For a small system using a single console there is automatic tracking of 16 targets. For a medium-sized system for a frigate or a destroyer, between two and six consoles will be used and each will automatically track between 16 and 32 targets. For the largest warships, up to 10 consoles will be required and will be able to track a total of up to 200 targets. In the Type F 2000S there are five consoles and an E7000 automatic plotting table in the CIC. A sixth console on the bridge presents tactical data to the ship's commander.

The TAVITAC 2000 system has a distributed architecture using the Thales MLX-32 computer which is built around a Motorola 68030 microprocessor. The system uses Ada software language, the UNIX System V operating system and features a duplicated VMEbus 5 Mbytes/s Ethernet-standard local area network. A ruggedised disk storage system offers a database management capability for map displays, ship resources and management.

The system includes the Vista raster scan software-driven display consoles which incorporate the Motorola 68020 microprocessor. The consoles, with high-resolution colour displays and a powerful radar scan converter, allow the radar video data to be superimposed upon the graphic picture. The displays use a 48.26 cm (19 in) colour 1,024 × 1,280 pixel screen which is increased to 2,000 × 2,000 pixels at the graphic controller. The Vista console is capable of processing up to 800 tracks and displaying 300 of them together with 30 circles, 30 ellipses, 500 vectors and 2,000 shoreline vectors. Options include a second 19 in television monitor. The MMI facilities include two programmable keyboards and a trackerball, with four associated buttons. Features which may be added include a second radar scan converter for window displays of further raw radar video and further MMI facilities.

Up to 15 display consoles and network interface units can be run from two computers. In the *La Fayette* class there are two MLX-32 computers, one of which provides redundancy and another two to interface with sensor and weapon systems as well as Link 11. There are six Vista consoles and a Precilec E8000 tactical table linked by an Ethernet dual-redundancy LAN.

There are plans to supplement TAVITAC with MAIDTAC, an expert system for evaluating the tactical situation using artificial intelligence techniques. MAIDTAC seeks to detect the enemy tactical plan and proposes solutions to defeat such a plan.

For further details see *Jane's Naval Weapons Systems*.

Status

TAVITAC has been purchased by China, Colombia, Saudi Arabia and Tunisia. All these systems are integrated with a Vega fire-control system to create a Thomsea system. TAVITAC 2000 is operational on French Navy 'La Fayette' frigates and has been ordered for the Saudi Arabian F3000, the South African MEKO A-200 and the Singapore and Taiwan 'La Fayette' classes.

Manufacturer

Thales Naval France, Bagneux.

UPDATED

TSM 2061

The TSM 2061 is a tactical C3 system for minehunting sonar. It permits the control and display of operational data from a variety of individual or combinations of sonar sensors for minehunting, surveillance and minesweeping, providing flexibility in the selection of sensor systems.

The system consists of a ruggedised multifunction console including one or two 19 in high-resolution flat colour monitors (1,024 × 1,290 pixels); an operator desk with alphanumeric keyboard, numeric keypad, rollerball, function keys and an X-windows graphic controller; a Helmsman flat colour display; CD recorder; and a colour printer. Other peripherals such as a plotter are also available.

The system performs the following functions:
(a) mission planning and task management and operational control
(b) positional location of vessel
(c) positioning of detected objects or underwater devices
(d) data display and recording
(e) chart and object database management
(f) electronic chart display

Status
In service with the navies of France ('Eridan' class), Pakistan ('Munsif' class), Singapore ('Landsort' class), South Korea ('Yang Yang' class), Denmark ('Flyvefisken' class).

Contractor
Thales Underwater Systems, Sophia Antipolis.

UPDATED

Germany

Integrated Tactical Management/Control System (TMS/TCS)

The Tactical Management/Control System is an integrated display and control system for mission systems (sensors and actuators) in surveillance aircraft. It is primarily but not exclusively designed for use in Maritime Patrol Aircraft (MPA). The system is fully modular and software configurable thus easily adaptable to customer needs without major system redevelopment. The primary requirements which govern system configuration are:
(a) number of identical multifunction workstations
(b) number and type of mission systems to be controlled.

The system offers integrated processing and control for a variety of modern mission systems. It has a fault tolerant system architecture design with distributed processing and display functions. Maximum use is made of (ruggedised) Commercial-Off-The-Shelf (COTS) equipment.

Flexible crew task allocation is possible via the identical multifunction operator workstations. The interface design is based on the use of colour windowing techniques and a keyboard/trackball combination for all display and control functions. This design is based on the experience of Dornier (now EADS) from a comprehensive number of ground-based and airborne applications which are in operational service with the German and other armed forces.

Dornier has been involved for many years in national and international programmes for the integration of complex mission equipment sets into aircraft. The list of programmes includes:
(a) the German BR1 150 Atlantic MPA (production, maintenance, several equipment updates)
(b) the NATO AWACS aircraft (equipment installation, aircraft maintenance)
(c) the German Open Skies aircraft (equipment production and integration).

The experience gained from these programmes, together with comprehensive conceptional and design study work under government contract, has resulted in a modern system which fulfils most recent NATO customer requirements while having sufficient growth potential for the future.

Status
Under development. Technical demonstrator in use for customer verification of concept. A possible contender for use in the German-Italian MPA 2000 project.

Contractor
EADS Deutschland, Friedrichshafen.

VERIFIED

Israel

Coastal Surveillance System (CSS)

Elisra's proposed Coastal Surveillance System (CSS) is designed to detect, track and analyse sea traffic along the coast and to distinguish general sea traffic from and identify, unusual or suspect vessels or floating objects, either moving or stationary. The system will acquire data about the movements and behaviour of these elements to enable coastguard vessels to perform interception and identification with controlled and optimal manoeuvring.
CSS is based on several surveillance stations located near the shore and a C2 centre. The system is designed to enable the integration of surveillance stations in order to cover the entire coast. It is also designed to be able to incorporate growth potential for additional equipment.

Each surveillance station performs a continuous radar and optronic lookout over a predetermined region of the sea. The surveillance stations are fitted with radar operating in the E/F- and I-bands, an IFF interrogator, optronic equipment for day and night vision, a processor and display for station management, a means for recording the various data and appropriate communications equipment.

The functions of the shore surveillance stations, under C2 centre control and guidance, are to generate a picture of the local maritime situation (within the ranges

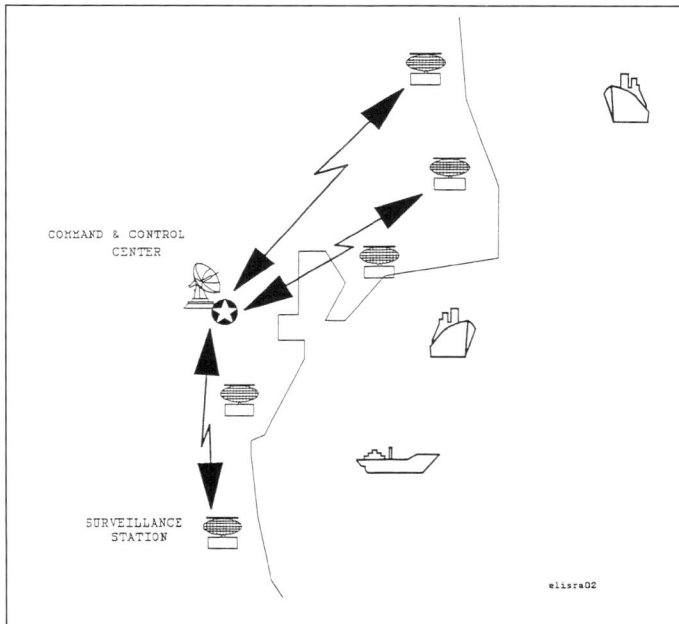
Coastal Surveillance System configuration

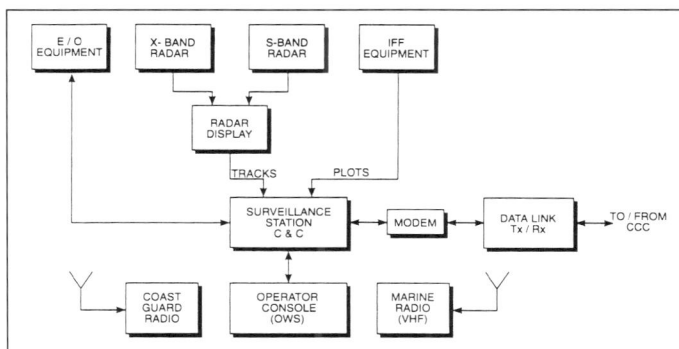
Surveillance station system configuration

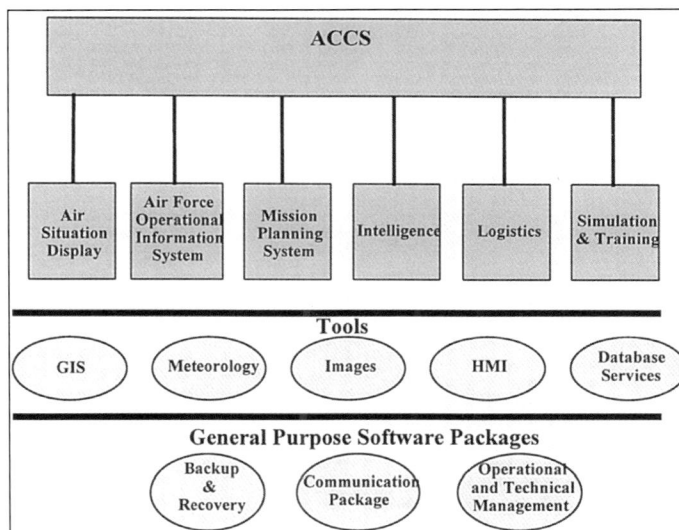
ACCS main functions   0052053

of its radar and optronic sensors) and to direct the interception of the coastguard vessels in its region of responsibility.

Management and control of the CSS is performed by the C2 centre. Three main elements are included in the centre:
(a) communication with the surveillance stations and commanding headquarters, coastguard, port authorities, the Ministry of Transport, shipping companies, fishing authorities and so on.
(b) control and management of the CSS
(c) a database of environmental conditions and their effects, vessels registered in the country and the behaviour of sea traffic in its various forms near the shores. The database can be used for regular analysis of the various operations near each shore segment for accumulating operational and deployment experience for the CSS.

Contractor
Elisra Electronic Systems Ltd, Bene Beraq, Israel.

VERIFIED

ENTCS 2000 Combat Management System

The ENTCS 2000 is a fully distributed Naval Combat Management System designed to assure 'knowledge superiority' over potential enemies, support common Tactical Picture (CTP), shorten decision cycles and execute rapid and accurate weapon engagement in the task force.

Based on open architecture, easily upgradable COTS building blocks and Windows NT software, its distributed architecture results in enhanced redundancy, with no single point of failure. The system's modular design makes it suitable for a broad range of naval platforms from small patrol boats to frigates. It is also easily installed in Maritime Patrol Aircraft (MPA), helicopters and land-based installations.

Using multifunction consoles radar, EW, guns, missiles, sonar and chaff can all be integrated, together with navigation and communication capabilities. Over 1,000 tracks can be included in the system's tactical picture.

More detailed information can be found in *Jane's Naval Weapon Systems*.

Status

In service in the Venezuelan Navy (modified 'Lupo' class). Earlier versions are in service in the navies of Israel ('Eilat' class) and Singapore ('Victory') class.

Contractor

Elbit Systems Ltd, Haifa.

UPDATED

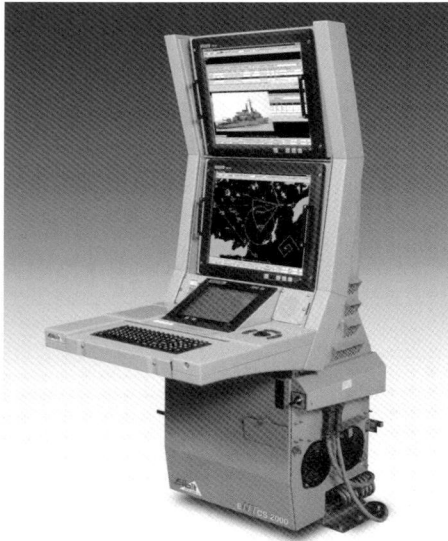

ENTCS 2000 Console
(Elbit Systems)
0130754

Reshet naval command and control system

Reshet is a shipborne tactical command and control system for small and medium size vessels. It provides the command with an updated and comprehensive picture of the overall tactical situation, based on data from 'own ship's' sensors and information derived from consorts and/or shore-based facilities via datalinks. The system is also integrated with weapon systems to support the ship's offensive and defensive operations.

Description

Several versions have been designed but all are composed of one or more Operating Consoles. Older versions included a central processing unit and a data management terminal, while the latest versions are based on distributed architectures. Operational requirements met by a typical Reshet system include: display and application of all available data from various ship's sensors; common tactical situation picture for CIC operators and for the command; tactical assessment; co-ordination of task force actions; AA, surface and ASW support; over-the-horizon targeting.

A typical Reshet multifunctional system provides the following facilities:

(a) automatic tracking of multiple targets from search and fire-control radar video data

(b) exchange of tactical information and messages within data-link network stations

(c) execution of triangulation and correlation of data from various sensors

(d) classification and threat evaluation

(e) navigational calculations and other computations for operational purposes

(f) high resolution tactical display with maps, grid, targets and bearing lines, datum points, danger zones, future positions, track histories and other tactical data, super-imposed on raw radar video

(g) comprehensive display, editing and data manipulation facilities such as zoom, off-centering, true/relative motion display, declutter and so on

(h) anti-submarine warfare calculations and display

(i) designation of tracked target to weapon systems and other units

The typical system consists of processing modules, a number of consoles used as: the target acquisition and designation position, tactical editor console and commander's console, and datalink equipment. This configuration is sufficient to provide the display of data from the ship's own radar systems, automatic tracking, synthetic symbols, datalink management and display and target designation capability. A modular and open architecture allows the use of different type of sensors (surveillance and fire control radars, electro-optical directors) while supporting various types of weapons simultaneously. Integration with other onboard systems such as the EW system is possible. The Reshet system's modularity allows simplified versions to be configured for smaller ships with a reduced number of consoles, and the Human Machine Interface can be tailored to specific needs.

The full version (titled Reshet+), which is a complete command and weapon management system, is under development as an extension of the Reshet system. It is a COTS based combat management system that provides weapon fire control as well as command and control and tactical display features.

Status

Probably no longer in production, but may remain in service with Israeli and other naval forces.

Contractor

Israel Aircraft Industries Ltd (IAI), Ben Gurion International Airport.

UPDATED

Italy

IPN command and control systems

IPN is a series of command and control systems for naval operations. The systems are able to gather, correlate and filter the information coming from ship sensors, communications and data networks and to display clearly the tactical situation. In addition to this, the systems, which can be configured according to the operational requirements of the user, provide assistance to the command primarily in the following areas:

(a) situation evaluation

(b) threat evaluation and weapons control

(c) management of electronic countermeasures

(d) aircraft and helicopter control

(e) conduct of ASW operations

(f) datalinks management

(g) database management.

The IPN-10 and IPN-20 systems are configured in a series of central units, engineered in small cabinets and a number of display consoles, according to the function required. The early IPN systems are based on the NDC-160 and CDG-3032 computers and on display systems consisting of the three-operator horizontal console and the single operator vertical console. The IPN-S is the latest system in the series and its performance has been considerably enhanced by the introduction of the new computers of the MARA family and MAGICS display consoles (see separate entry).

The MARA family of computers is characterised by advanced hardware and software technology, very high modularity which enables the use of a small range of hardware module types to construct systems with a wide range of processing power, memory and interface capacity, strong support for fault detection, fault tolerance and maintainability. The MARA computers use the same software that runs on a Digital VAX commercial computer and use the Ada language.

The MAGICS display consoles are characterised by high modularity enabling the composition of differing configurations to fulfil the requirements of the combat system and platform control, use of the raster scan technique and bit map concept associated with high-resolution colour monitors and the use of the MARA family of computers as display processors.

IPN-S mono-monitors
installed in a corvette
operations room (AMS)
0132915

Multiple operator horizontal console for IPN-10 system   0009780

Status

IPN systems are in production and are in service in a variety of platforms in the Italian Navy; in the Spanish 'Santa Maria' class (FFG); in the Ecuador 'Esmeraldas' class (FSG); in the Taiwan 'Lung Chiang' class (PCFG); and in the Venezuelan modified 'Lupo' class (FFG). A Bharat-modified variant is in the Indian 'Delhi' class (DDG). In April 2002 a US$15 million contract was awarded to AMS to upgrade the command systems on two Malaysian 'Laksamana' class (FSGM) (corvettes) with IPN-S.

Contractor

AMS, Rome.

UPDATED

MM/SSN-714 minesweeper data processing system (MACTIS)

Under the designation MM/SSN-714, DATAMAT developed a digital navigation and plotting system for the Italian Navy's 'Lerici' and 'Gaeta' class of mine countermeasures ships. The main functions of the SSN-714 MACTIS (Minehunting ACTion Information Subsystem) are:
(a) automatic computation and presentation of the ship's current position
(b) display of the tactical situation
(c) analysis and presentation of target characteristics
(d) location of surface targets
(e) event recording
(f) operations planning
(g) guidance of surface and underwater craft.

The system is based on a computer of the same type or similar to that employed in the SACTIS submarine AIO system, namely a Rolm MSE 14 machine. In the SSN-714 system, this is interfaced with recording units, display units, controls, printer and ship's sensors. The principal items in the last of these categories are radar, sonar, compass, log and various navigation aids. The operator's display has a vertical screen CRT and a keyboard for communications with the system and there are supplementary data readout display units and associated input controls on the bridge and in the operations room.

MACTIS is the central subsystem of the Integrated MInehunting Combat System (IMICS).

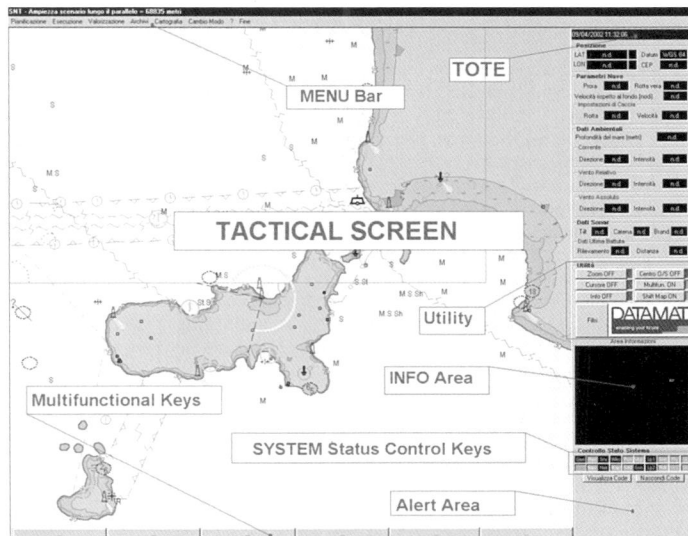

MM/SSN-714(V)3 system architecture (Datamat)   *NEW*/0593726

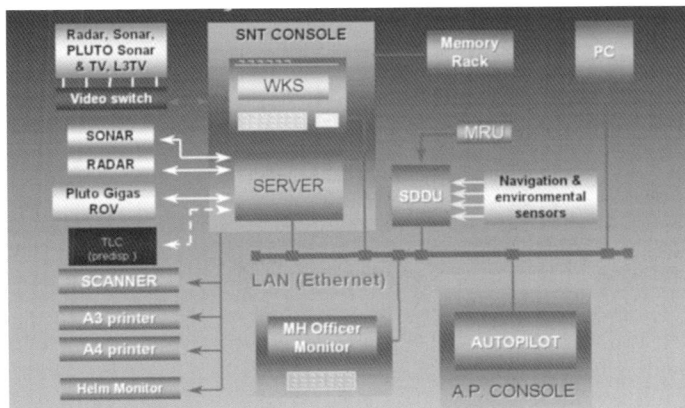

Graphic display for MM/SSN-714(V)3 system (Datamat)   *NEW*/0593727

Version 3 of the system has been developed to upgrade the 'Lerici' class, in place of the MM/SSN-714(V)1 system. The aim has been to adapt to a new Combat System Configuration; to prepare for planned future upgrades; to adapt operational software to new CMM NATO standards; to solve problems associated with the previous combat system configuration, including high maintenance costs, small recording capability, an outdated MMI and the lack of electronic cartography; to increase the level of integration with the Autopilot which may provide scope for manpower reductions; and to improve navigation sensor availability and precision in computation of ship position.

The main functions of the MM/SSN-714(V)3 are:
(a) operator control of all functions from the main console
(b) operator support during all phases (planning, execution, results analysis and exploitation) of a Mine Countermeasures Mission
(c) automatic ship plotting
(d) navigation control (through integrated autopilot)
(e) data and video interfaces with platform devices including radar, sonar and ROV
(f) common use function always available:
  - information display
  - multimedia Data Base
  - self-monitoring (alarm generation)

The MM/SSN-714(V)3 system is based on COTS hardware and software (including Ethernet LAN, Pentium CPUs, Windows NT, C++, Visual Studio) for increased performance, user friendliness and reduced maintenance costs. Its open architecture is based on the use of a main console with high-resolution colour display, standard MMI devices (QWERTY keyboard and Track-ball), a secondary console with high-resolution colour display and a set of fixed action keys, a bridge display, a dedicated unit (SDDU, Smart Data Distribution Unit) for interfacing navigation and environmental sensors, an Autopilot card to control automatically ship motion, an Ethernet BUS for a complete integration of consoles, Autopilot and SDDU to increase system availability and reduce manning requirements, two colour printers and an A4 scanner.

Status

In service with Italian minehunters. 'Lerici' has been updated to V3, and the remaining three platforms of the class which currently have V1 will be updated to V3 "shortly". V2 with 'Gaeta' class.

Contractor

DATAMAT, Rome.
Calzoni, Milan.
Fiat Avio, Turin.

UPDATED

Netherlands

DAISY and STACOS systems

DAISY/STACOS is a modular digital action information system designed to be used in ship combat information centres. DAISY forms the heart of the SEWACO sensor, weapon control and command system (see separate entry).

DAISY, which is a designation of the Royal Netherlands Navy, uses core command and control software developed by the RNethN. STACOS is the Thales Nederland designation for the tactical command system which includes Thales Nederland-developed operational software. STACOS ultimately became the basis of TACTICOS (see separate entry).

The DAISY/STACOS tasks include:
(a) presentation of raw or digitised information from primary and secondary radar
(b) compilation and presentation of tactical air, surface and subsurface pictures
(c) datalink operation

(d) designation of targets to weapon systems
(e) automatic threat ranking and evaluation and weapon assignment
(f) automatic changeover from trackers in case of target crossings
(g) assistance in operations, including ASW helicopter direction and tactical navigation.

DAISY/STACOS systems have been delivered in a variety of configurations and with a number of upgrades in architecture and applied technology. Earlier DAISY/STACOS systems are configured from a data handling cabinet and a number of display consoles appropriate to the operational functions of the system. The data handling cabinet contains the data handling computer, a sensor data distribution unit and one or more video extractors. In these systems, the data handling computer is a Thales Nederland SMR family machine and operational programs are written in the RTL-2 language. These systems use 40 cm diameter main monochromatic Labelled Plan Displays (LPD) and additional 7 and 15 in raster scan alphanumeric displays. These elements are configured in two types of console: Vertical Display Console (VDC) and Horizontal Display Console (HDC) or 'conference' type.

More recent DAISY/STACOS systems are composed from one or more command and control computer complexes; full-colour, high-resolution raster scan consoles; and a redundant databus. The architecture of these systems represents a major increase in the level of integration, of automation and in operator-friendly man/machine interfacing.

In this architecture, sets of system tasks are defined as 'system functions' (logical subsystems). Every system function can be independently accessed from any workstation in the system. Data communications between system functions and workstations is handled by the data communication network, including the high-speed databusses and the communications software package, COMPASS. The computer complex in the system consists of Thales Nederland's SMR family computers and, optionally, additional commercial-off-the-shelf processors. The complex provides fail-safe and fail-soft facilities through duplications of critical parts. Installation of the complex is decentralised to achieve optimum damage control.

Operator Stations

To optimise operator flexibility, the display subsystem consists of SIgnaal General-purpose High-resolution Tactical (SIGHT) workstations. These workstations are of the colour raster scan type. Simultaneously different types of data sources for display can be handled: graphic data (characters, vectors and so on), radar data, image data (maps, sonar images).

Storage for maps, manuals, procedures and so on, can be provided using laser optical disc devices. The workstation enables multiple system access, facilitating operator system access through modern windowing techniques. Each workstation can be deployed for any of the system functions within the security limits of any given operator. General-purpose Operator Stations (GOS) workstations are no longer supplied, having been replaced by Multifunction Operator Consoles (MOC) since 1989. These provide stronger graphics facilities and more configuration options.

The databus system provides the data highway between the major components of the system. It comprises a multiple databus based on Ethernet. The busses (up to four) may be applied in parallel (capacity up to 40 Mbyte) or in a redundant fail-safe configuration.

Ship's INtegrated COmmunication System (SINCOS)

Any SIGHT workstation can also control the integrated external and internal communication system. A new communication system, SINCOS 1200, has been developed by Thales Nederland. This system is specifically suitable for the use of GPW, since it adapts itself automatically to the functions required by a given GPW operator. The external part covers all military frequency bands from VLF up to UHF. The HF and VHF/UHF bands have full ECCM capability based on frequency hopping.

The internal part of SINCOS uses the digital FOCON-A fibre optical bus for the distribution of voice and equipment control signals. SINCOS has built-in automatic reconfiguration facilities for fail-safe/fail-soft operation.

Status

DAISY and STACOS systems have been delivered for the navies of Argentina, Belgium, Canada, Greece, India, Indonesia, Malaysia, Netherlands, Nigeria, Peru, Portugal, Thailand and Turkey. It is in production for the Hellenic Navy. SINCOS 1200 is in production for the Royal Netherlands Navy.

Contractor

Thales Nederland, Hengelo.

UPDATED

....

SEWACO sensor, weapon control and command system

Under the SEWACO designation, Thales Nederland produces integrated sensor, weapon and command systems of varying configurations intended for corvettes, frigates and higher level combat ships. Full advantage is taken of the integration of the various subsystems. System functions can be carried out using distributed data processing, depending on the operational status of the combat system components. The heart of the system is the DAISY/STACOS C2 system (see separate entry).

The first integrated SEWACO system was developed for the Royal Netherlands Navy frigates of the *Tromp* class. This system has been followed by new generations of SEWACO systems in which new technologies have been incorporated, for example for the sensors, computers and displays. In the latest generation, SEWACO FD, the processing power for the command and control software is fully distributed over the system components.

A SEWACO system in general comprises four major parts:
(a) sensors consisting of primary 2-/3-D radar systems for long-, medium- and short-range air and surface warning, secondary radar sensors such as IFF and helicopter transponders, hull-mounted sonar systems and passive sensors such as ESM and IR sensors. Automatic initiation and tracking of air and surface targets is released by built-in track capability or additional primary and secondary video extractors
(b) a combat management system evolved from DAISY via Foresee and STACOS to TACTICOS (see separate entries), comprising operator consoles, plotting table, conference display and the command and control software package
(c) weapon control systems comprising active and/or passive target tracking and/or illumination systems, weapon interface systems for interfacing with hard-kill and soft-kill weapons and close in weapon control systems
(d) a communications system with internal and external communication equipment and automatic datalink systems.
Functions of the SEWACO system include:
(a) air, surface and subsurface warning
(b) compilation and display of tactical air, surface and subsurface situations
(c) threat evaluation and target engagement in AAW, EW, ASUW, ASW and NGS
(d) navigation including ASW helicopter direction
(e) communication including tactical datalink operation
(f) onboard scenario-driven training
(g) organisation and system management.

SEWACO I

The principal sensors of the SEWACO I system in the *Tromp* class guided missile frigates (now sold by the RNLN) are the Thales Nederland 3-D multitarget tracking search and target designation radar and the combined search and tracking radars of the WM25 fire-control system, which is integrated into SEWACO I and the Thales Nederland ESM system. The 3-D radar also incorporates a slotted wave-guide antenna for IFF/SIF facilities. Among the weapons associated with SEWACO I are Tartar, Standard SM2 MR, the Harpoon SSM, the NATO Sea Sparrow point defence system and a twin 119 mm (4.7 in) gun turret.

SEWACO II

This system has been delivered to 10 Royal Netherlands Navy's standard frigates and for two Hellenic Navy 'S' class frigates. Sensors include the Thales Nederland LW-08 and ZW-06 radars, the combined search and tracking radar of the WM25 weapon control system and ESM system, two optical target designation sights and an automatic datalink. Non-Thales sensors include IFF, a helicopter transponder system and the AN/SQS-505 sonar set. The SEWACO II system consists of Thales Nederland's Digital Action Information SYstem (DAISY), WM25 weapon control system, including a Separate Tracking and Illumination Radar (STIR) on which a TV camera is mounted.

SEWACO IV

This is the designation of the SEWACO system supplied for the *Westhinder* class of escorts for the Belgian Navy. No official details have been revealed but the following data derived from published sources corresponds with what is known of these new vessels and their weapons and equipment fits.

SEWACO VII   0009940

Sensors comprise a DA-05 air and surface warning and target indication radar, WM25 combined search and tracking radars, a commercial Raytheon surface search/ navigation/ helicopter control radar, IFF/SIF, helicopter transponder, a French-made ESM system, datalink, optical director and sonar. Weapons controlled by the SEWACO system include NATO Sea Sparrow surface-to-air missiles, Exocet surface-to surface missiles, dual-purpose gun, anti-submarine rocket launcher, ASW torpedoes and ECM systems. The weapon control and combat information equipment consists of the WM25 FCS for missiles, gun, and A/S rockets, three horizontal displays for combat information functions, each with its own separate alphanumeric tote display. A sonar display console, radar plotting table with true-motion computer and one video extractor for automatic air and surface tracking complete the system.

SEWACO V

This designation is given to the SEWACO system for the mid-life conversion of the six *Van Speyk* class frigates, sold to Indonesia. Sensors include an LW-02 air warning radar, a DA-05 air and surface warning and target indication radar, a Decca navigation radar with two displays, IFF and helicopter transponder system, two optical target designation sights, two sonar sets, an ESM system and an automatic datalink. The action information system is of the DAISY type and includes four horizontal tactical display consoles with alphanumeric displays and a plotting table. The weapons fit comprises an OTO Melara 76/62 dual-purpose gun, controlled by an M45 weapon control system, Seacat launchers controlled by an M44 system, a Harpoon SSM system, ASW torpedoes and ECM equipment. The ship will be equipped with an ASW helicopter.

SEWACO VI

The first of two Royal Netherlands Navy's *Jacob van Heemskerck* class frigates was commissioned on 16 January 1986. The SEWACO system for these vessels comprises: Thales Nederland LW-08 extended range and elevation cover version, a stabilised DA-05, ZW-06 and Goalkeeper search radar sensors; multiple video extractor systems; a Sphinx ESM system; Ramses ECM system; two optical target designation sights; the GMCS guided missile control system, consisting of three STIRs (one 1.8 m and two 2.4 m scanners) with high-power coherent transmitter chains. CW illumination for Standard SM1 MR and for Sea Sparrow missiles is provided by each STIR. The system provides a high level of autonomy by means of automatic threat evaluation and STIR and weapon selection for automatic target engagement. The DAISY system includes six vertical consoles, two horizontal consoles and a sonar display console and primary and secondary video extractors. In addition to the normal DAISY functions it has extensive air defence software facilities. IFF, VESTA and datalink are also integrated in the SEWACO system.

The weapons are the SM1 MR to be launched by Mk 13/4 launcher, Sea Sparrow from eight-cell launcher, Harpoon surface-to-surface missiles, RBOC countermeasures, Mk 46 torpedoes and the EX83 30 mm gun portion of the Goalkeeper CIWS.

SEWACO VII

This version has been nominated for the Royal Netherlands Navy's eight *Karel Doorman* class frigate (M-frigate), the first of which was commissioned in January 1992. SEWACO VII is the first of a new generation of SEWACO systems with an increased level of integration and automation. The system includes Thales Nederland LW-08 long-range D-band radar with lightweight pedestal, SMART-S 3-D F-band radar and Goalkeeper search radar sensors; multiple video extra for systems; VESTA helicopter transponder system and the hull-mounted sonar PHS36. Non-Thales Nederland sensors include IFF, navigation radar, ESM/ECM and the towed-array sonar, Anaconda.

The action information system comprises duplicated central computer complexes for C2 and database management, colour workstations, VDUs and high-speed database. The weapon control is performed by the Multiple Weapon Control System (MWCS), a system very similar to SEWACO VII GMCS, for control of NSSM, launched from a 16 cell Mk 48 VLS and of OMCG 76 mm. The CIWS Goalkeeper is fitted for short-range air defence. The communications suite is fully integrated into SEWACO VII and comprises the integrated internal and external communication system SINCOS 1200.

SEWACO VIII

SEWACO VII was designed for the *Walrus II* class submarines of the Royal Netherlands Navy and is based on the Gipsy C2 system. It consists of seven identical Display and Computer Consoles (DaCCs). The built-in computer is of the SMR-MU type; the 406 mm (16 in) Plan View Display (PVD) provides a high-load synthetic picture together with compressed radar video or sonar video and the control panel is a multipurpose unit. The use of identical DaCCs offers maximum flexibility.

A Central Control Unit (CCU) is used to regulate the mutual data transfer between the DaCCs and the sensors and weapons. For this function two SMR-MU computers are provided, one active and the other a hot standby machine. The large amount of data from the sonars is handled by two extra SMR-MUs, also housed in the CCU. Three types of sonar are fitted (towed array, flank array and circular array) and other sensors and data sources include a noise analyser, ESM facilities, radar, periscopes and position finding equipment.

The weapon control system can control a mixed load of weapons for subsurface and surface engagements and the ship's launching system consists of two mutually independent sections, each of which is controlled from a Launching System Control Panel (LSCP). The interface with the weapons (modern subsurface missile) is formed by two identical distribution cabinets. All hardware necessary for the integration of these weapons is included in the distribution cabinet, avoiding the necessity for additional equipment.

SEWACO FD

Further development has led to SEWACO FD (Fully Distributed). In each SEWACO FD system the distributed processing power replaces the centralised computer complex(es). The processing power is formed by a Sigma node built into every multifunction operator console (MOC Mk III) (see separate entry) and the Thales Nederland-made subsystems. The C2 software package, TACTICOS (see separate entry), is allocated to the available nodes. The SPLICE (Subscription Paradigm for the Logical Interconnection of Concurrent Engines) distributed database and data communication infrastructure enables, during run time, dynamic reallocation of functions over the nodes online. SEWACO FD includes SEWACO IX and SEWACO XI.

Status

SEWACO systems are in operational use on board the Royal Netherlands Navy's S-class, L-class and M-class escorts; the Belgian Navy's 'Wielingen' class of escorts; the Argentinean 'Almirante Brown' class DDG and 'Espora' class FFG; and the Hellenic Navy's S-class frigates. Similar SEWACO systems, with slightly varying sensor and weapons fits, are in operational use on board the Turkish Navy Track 1 Meko 200 type of frigates, the Portuguese Navy 'Vasco da Gama' Meko type of frigates, and the Hellenic Navy Meko 200 type of frigates.

SEWACO IX is installed in the RNLN 'Alkmaar' MCMV and SEWACO XI in the 'De Zeven Provinciën' DDG. Other SEWACO FD systems are operational on the Turkish Navy's 'Barbaros' class frigates and 'Yildiz' class FACs; the FACs of the Oman Navy's 'Muheet' programme; the Qatar 'Vita' class FACs; the German F124 'Sachsen' class FFG; the Turkish 'Kilic' FAC and the Indonesian Navy 'Todak' class FAC.

Contractor

Thales Nederland, Hengelo.

UPDATED

Norway

MSI-80S fire-control system

The MSI-80S is a weapon control system for small ships and was initially specified and developed for the Royal Norwegian Navy's *Hawk* class fast patrol boats.

The system provides a multitarget capability using low-cost conventional navigational radars, stabilised electro-optical sensors for night/passive operations and target filtering/tracking techniques. Homing surface-to-surface missiles, wire-guided torpedoes and an AA gun can be controlled individually or simultaneously.

Tracking and guidance problems are solved either automatically or semi-manually by man/machine integration. Data from all available sensors can be used at any time and in any order.

A three-man operating console is provided, this being crewed by:
(a) tactical operator, who initiates and controls target data calculations and the overall tactical situation plot;
(b) weapon control operator, who selects, controls and designates the guidance mode(s) of the relevant weapons;
(c) passive sensor operator, who operates the passive sensors and feeds relevant data to the target tracking system.

A fourth position is provided at the console for the commanding officer.

A horizontal 584 mm (23 in) CRT is used for the presentation of raw radar and computer-generated data and a 305 mm (12 in) CRT is provided for the display of alphanumeric data. The latter can also serve as a back-up for the main tactical display (23 in).

Status

No longer in production. In service in the Norwegian and Greek navies. Being replaced in the former by SENIT 2000 (see separate entry).

Manufacturer

Kongsberg Defence & Aerospace AS, Kongsberg, Norway.

VERIFIED

MSI-90U submarine basic command and weapons control system

MSI-90U has been designed to comply with the requirements of the Royal Norwegian Navy and the German Navy for the Norwegian 'Ula' class and the German 'U212' class submarines.

The basic design philosophy has been to achieve high, all-round capabilities, system flexibility and availability, with limited support costs, through a structured design of the functional parts of the system. This design highlights the importance of redundancy and independent operation of the different subsystems.

MSI-90U is a software-based command and weapons control system which uses distributed processing, a high-capacity serial data transmission system (Local Area Network – LAN), and multifunction operator consoles to achieve a high degree of capability, flexibility and availability. The main features of the system are:
(a) distributed data processing using the 32 bit KS-900 general purpose computer designed for real-time processing, based on the commonly used Motorola

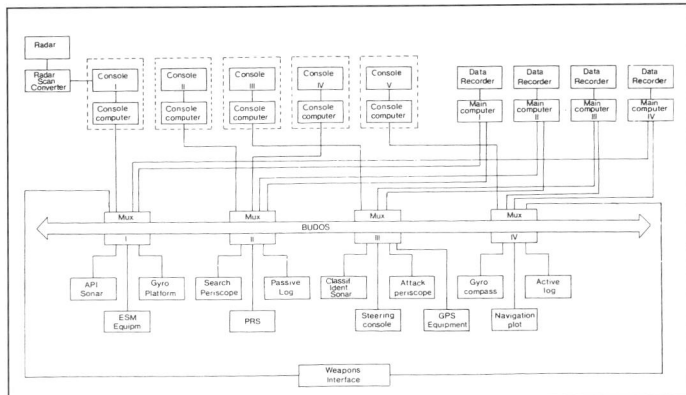

Block diagram of a typical MSI-90U configuration

68020/68040 range of microprocessors. The modUlar KS-900 is designed for programming in Ada, C and Pascal;

(b) a high-capacity LAN is used for data communication between subsystems of MSI-90U, between MSI-90U and sensors and sensor-to-sensor communication;

(c) multifunction operator consoles allowing every operator to have access to all information in the system as well as permitting the number of operators to be adjusted to suit the current tactical situation. That means that one operator can operate the complete system with one console in, for example, a patrol situation. In the standard MSI-90U configuration four identical multifunction operator consoles are used, which can be configured to different console (work) modes. The number of consoles can be adapted to customer requirements;

(d) built-in redundancy allowing subsystems to operate separately in a graceful degradation fallback mode.

The LAN, named BUDOS, is based on Ethernet and can interface sensors/ subsystems via Ethernet, RS-422 or NATO STANAG 4156. Via BUDOS any subsystem can communicate with any other connected subsystem. Connection of new subsystems can easily be made by means of spare interfaces on the existing BUDOS multiplexers, or by adding more multiplexers.

Four multifunction operator consoles (each with its own console computer, KS-900), three main computers (KS-900) and two or three KS-900 weapon computers (depending on weapon type and customer requirements) ensures adequate redundancy.

Each console houses its own computer and information is presented on two high-resolution colour raster scan displays. The Man/Machine Interaction (MMI) is via a programmable entry panel (a plasma display with touch-sensitive overlay matrix), a trackerball with associated control buttons and a standard alphanumeric keyboard.

The MMI concept of the system, including layout of the display pictures, has been subject to extensive MMI studies and trials in co-operation with the Norwegian and German navies.

The system is designed to carry out sensor integration covering target motion analysis, classification and identification and weapons assignment and control. A number of supplementary facilities are also provided, including:

- tactical evaluation and navigation
- threat evaluation
- engagement analysis
- preprogrammed movements
- sound trajectory calculations and presentations
- predicted sonar ranges for 'own ship' and hostile ships
- presentation of geographical fixed points and areas
- data recording and simulation for training purposes.

MSI-90U multifunction consoles

Status

The MSI-90U is operational on board all six Norwegian 'Ula' class submarines with original delivery taking place between 1988 and 1992. During the period from mid-1994 to April 1997, the MSI-90U underwent a joint German/Norwegian upgrade programme. This programme was concluded with a successful Demonstration of Operability (DoO) for representatives from the Norwegian MoD, Norwegian Navy, German MoD, Bundesamt für Wehrtechnik und Beschaffung (BWB) and German Navy in the factory based test stand (in which real MSI-90U equipment was connected to sensor, torpedo and scenario simulators) on 9 April 1997. The upgrading programme was concluded within the contractual time scale.

Series deliveries of the MSI-90U for the German 'U212' class, as well as upgrading of the systems on board the Norwegian 'Ula' class submarines, are scheduled to take place between 1997 and 2005. The upgrade (MSI 90U Mk 2) is based on COTS technology and includes the MFC-2000M compact multifunction console, which has UltraSPARC processors and high-resolution colour flat panel displays. Its software is also based on commercial components and the asynchronous transfer mode (ATM) network protocol. The MSI 90U Mk 2 is capable of simultaneous automatic and operator-interactive Target Motion Analysis (TMA) computation for up to 25 targets and can conduct simultaneous firing preparation and guidance of up to eight torpedoes together with simultaneous preparation and control of up to four missiles.

Kongsberg was also awarded a contract with the Italian FINCANTIERI Naval Shipbuilding Division for the delivery of three MSI-90U systems for the planned Italian Navy's U212 A submarine, two of which will be installed on the submarines presently under construction, while the third system is destined for Italy's Naval Training Centre. The contract carries an option for two more MSI-90U systems. Deliveries to FINCANTIERI are scheduled to take place between 2001 and 2003.

Contractor

Kongsberg Defence & Aerospace AS, Kongsberg, Norway.

VERIFIED

MSI-3100 multiple integration system

MSI-3100 is a fully integrated and compact system originally designed for the combat information centre in Norwegian frigates.

The principal functions performed by MSI-3100 are surveillance, target detection and tracking, threat evaluation and allocation of track sensors and weapons. Auxiliary functions include Electronic Support Measures (ESM), manoeuvring aid and tactical co-ordination and support. Functions are carried out automatically taking into account manual override or support at any stage. The system conducts action information and enables the ship to counter air, surface and subsurface threats concurrently. This is achieved by a distributed control of all significant sensors and weapons assigned to AAW, AWW and ASW warfare areas. The decentralised arrangement also permits independent management of each warfare area. The full integration of sensors and weapons alleviates decision making and reduces reaction times. From the instant of target detection and confirmation, 'gun settled and ready to fire' is accomplished in less than 10 seconds.

The heart of MSI-3100 comprises two double function-oriented consoles, manned by a total of four operators. Tactical pictures are shown in colour on high-resolution raster scan displays. The radar video is digitised, thus permitting 'picture freeze' when desired for splash-spotting and entry of firing corrections. Operator inputs are entered via a joystick, a programmable touchpanel or a standard keyboard.

Status

No longer in production. In operation in all the 'Oslo' class frigates.

Manufacturer

Kongsberg Defence & Aerospace AS, Kongsberg, Norway.

VERIFIED

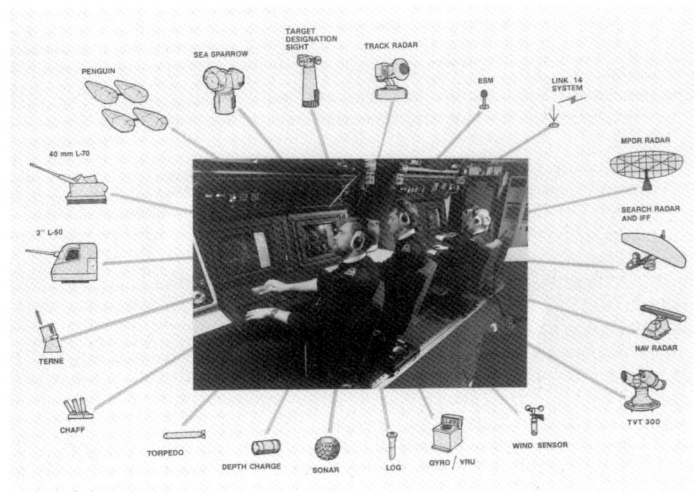

MSI-3100 multiple integration system

Russian Federation

IKASU Integrated shipboard weapon control system

The IKASU automated weapon control system is designed to integrate the control of anti-ship and antisubmarine armaments. Developed from separate ASM and ASW systems with considerable use made of common hardware and software, one control system now incorporates the functions of both. The control of both systems can be achieved from a single workstation, and full target data and armament status can be displayed on the commander's standby panel if necessary. The system is designed to control a variety of anti-ship missiles and various torpedoes.

Status
In service with the Russian and other foreign navies.

Contractor
Granit Central Research Institute, St Petersburg, Russia.

VERIFIED

IKASU Commander's Workstation
0110492

Sweden

9 LV 200 Mk 2 weapon control system

The 9 LV 200 Mk 2 combat information and weapon control system is the digital successor to the analogue 9 LV 200 and is suitable for use in all types of naval and offshore patrol vessels of 100 tons and upwards. It is optimised for data gathering, compilation and tactical evaluation for the control of one dual-purpose gun (57, 76 mm or larger), one air defence gun (single- or twin-mounted 20 to 40 mm) and surface-to-surface missiles. There are also options for torpedo control.

Description
A typical 9 LV 200 Mk 2 installation consists of:
(a) an X-band search and surveillance radar with frequency agility or MTI, with or without on-mounted IFF
(b) a Ku-band radar and director for gun fire control. IR, TV and laser sensors are options
(c) tactical and missile control console for overall viewing, threat evaluation and analysis, datalink control and target designation to guns, missiles or torpedoes

KD Handalan of the Royal Malaysian Navy, with Bofors 57 mm and 40 mm guns and Exocet MM 38, all controlled by a version of 9 LV 200 Mk 2

(d) a control console, mainly for air surveillance and target designation to the director and fall of shot observation; this console can also be used for fire control against surface targets
(e) a surface gun console primarily for gun fire control against surface targets tracked by the surveillance radar; this may also be used as an air surveillance back-up console
(f) two 'bridge pointer' target designators for use by lookouts for direct optical target designation or acquisition, typically against surface-to-surface missiles
(g) wind sensor.

Status
In service since 1978 with a variety of navies including Malaysia. It has been superseded by the 9LV Mk 3 (see separate entry).

Contractor
SaabTech Systems AB, Järfälla.

VERIFIED

···

9 LV Mk 3 Combat Management System (CMS)

Following the 9LV Mk2, the 9 LV Mk 3 series has been developed for a variety of shipborne (surface and submarine) CMS applications. It has the capacity to correlate data from various surveillance sensors, including a new-generation search radar and an advanced ESM surveillance system. The design also allows operation in a multiple threat environment by integrated air defence functions employing SAMs, guns and ECM. The modular design permits enhancement as well as simple breakdown into smaller systems to fit any size of ship. This is achieved by local area network internal communication, a distributed processor concept and the modular hardware and software block design.

CeCots Operator Console (SaabTech Systems)   0130053

Royal Swedish Navy 'Göteborg' class corvettes are equipped with 9 LV Mk 3

A typical 9 LV Mk 3 system comprises:

(a) a 2- or 3-D surveillance for automatic plotting of air and surface targets

(b) a CMS system with functions for tactical analysis, target management, situation picture compilation and presentation

(c) a duplicated data storage system where all the subsystems have access to the latest data

(d) a gun fire-control system with the 9 LV Mk 3 TWT tracking radar and a choice of electro-optical sensors as options.

The system can be extended by adding one or more of the following options:

(a) additional gun fire-control channels as above

(b) EW systems comprising sensors and countermeasures, which can be integrated with the gun fire-control system to achieve optimised counteraction in a multiple threat situation

(c) interfaces to a range of weapon systems including SSM, SAM and torpedo systems

(d) datalink interfaces including Link 11 and Link Y.

Data processing capacity is distributed over the system. The system will be fitted with workstations corresponding to the number of subsystems involved. The workstations are multifunction consoles with software-controlled man/machine interfaces to enable reallocation of functions between different workstations.

The latest version, the 9LV Mk3E incorporates COTS components and includes an automatic air defence function (Air Defence Co-ordination), claimed to be unique in naval C3, which presents advanced threat evaluation and weapon assignment/control giving high kill probability. Other applications include AAW, ASuW, ASW, MW, MCM and a wide range of command functions. EW and communications systems can be integrated as needed.

Associated with the 9LV Mk3E is the CeCots operator console, which provides operation of all interfaced subsystems where functionality can be fully integrated. When it forms part of the 9LV Mk3E CMS the software is based on MS Windows 2000. Specifications are as follows:

(a) Two 20 in full-colour high-res Main Tactical Displays

(b) One 20 in full-colour high-res Touch Input Display

(c) Two trackballs or trackball and joystick

(d) Alphanumeric keyboard

(e) Weapon Safety switches

(f) Weapon firing pushbuttons, interface for firing pedal

(g) Audio interface - headset or speakers

(h) Space for optional control panels

Status

Over 50 systems have been contracted in: the ANZAC frigates for the Royal Australian and Royal New Zealand navies; Danish 'Flyvefisken' (Stanflex-300) class multirole ships, 'Thetis' class ocean escort and 'Niels Juel' FFG; Finnish 'Rauma' and 'Hamina' fast attack craft; the Omani 'Al Bushra' FAC; the UAE 'Ban Yas' class; the Pakistani 'Tariq' frigate upgrade; the Swedish 'Göteborg' and 'Visby' class corvettes and 'Götland' class submarines.

Contractor

SaabTech Systems AB, Järfälla.

VERIFIED

MARIL shipboard command and control information system

MARIL is a family of shipboard CCISs designed to meet the operational needs of different types of ship, ranging from offshore patrol vessels and fast patrol boats up to ships of destroyer size.

Description

The MARIL is a multidisplay, computer-aided system for collecting, processing, evaluating and transmitting tactical data, commands and messages. It can be subdivided into computer and display processors; input/output communication interface; operator input and display facilities and video extractors. The CCIS architecture is based on units incorporated into the Censor 900V series computer system and the DS 86 graphic display system. The computer used in the MARIL operates as a combined display and operational calculation computer. A number of displays can be connected to a single computer and auxiliary computers and databases can also be connected to the system. A number of fallback modes of operation are included.

Memory capacity of each computer is 512 kbyte, 32-bit words and a high-speed disk storage unit as well as peripherals used for program production. Computer-controlled data cartridge recorders are used to enter operational programs and store geographical reference information which can be called up for display online. Selected tactical data can also be recorded for subsequent evaluation and analysis.

Information in various formats (such as synchro, analogue and digital data) is converted by the input/output subsystem to and from formats suitable for the computers. Dedicated front-end processors are used for preprocessing, formatting, transmission control and so on. Track data and messages sent to and from co-operating units are transmitted as serial data messages via onboard UHF/HF radio equipment. The datalink terminals are integrated into the system. A typical transmission rate is 1.2 kbytes/s for UHF, with other rates selectable. Enciphering and deciphering equipment can be connected online.

A typical operations suite can include vertical and horizontal Operations Consoles (OC) and vertical General Operations Plot Consoles (GOPC). The horizontal OC is used primarily for picture compilation (detection, tracking, identification) and target allocation to weapon systems, plus weapon co-ordination and control. The console can have up to three operator positions, each provided with a functional keyboard, trackerball or joystick control and controls for internal/ external voice communication. One or two alphanumeric data displays and a radar selector panel are provided on the console. The PPI (16 in diameter) presents true motion-stabilised raw and extracted radar pictures.

The GOPC display is intended for command and control positions which, in addition to a 23 in graphic display, incorporate an alphanumeric data display, a functional keyboard and a trackerball. It presents a computer-generated situation picture that includes tracks with history marks, bearing and vector lines, map and chart information and geographical reference grids. Display information (graphic and alphanumeric) is stored in the computer memory. The display processors scan the display file and control display generation. Pictures that incorporate raw radar information are generated on a time-sharing basis.

A flexible keyboard design permits identical keyboard layouts to be used throughout the system.

Status

MARIL is used on the Swedish Navy 'Norrköping' class missile boats, on the coastal corvettes of the *Stockholm* class and on the training ship 'Carlskrona'. The system has also been exported to a NATO country for use in corvettes and frigates.

Contractor

SaabTech Systems AB, Järfälla.

VERIFIED

MARIL aboard a 'Stockholm' class coastal corvette

STRIKA coastal surveillance and weapons control system

The STRIKA coastal surveillance and weapons control system has been developed for use by the Swedish coastal artillery. It operates at a number of levels including brigade, battalion and battery. STRIKA incorporates the following functions:

(a) information collection from radar, radio, visual reporting posts, seabed magnetic sensors and so on

(b) information processing including: target tracking, correlation/multisensor tracking, identification, display functions, recording and playback, generation of tactical alarms and database/tote functions

(c) command and control including: tactical analysis, firing option analysis, weapons allocation and electronic warfare management

Operations position at a fixed STRIKA site

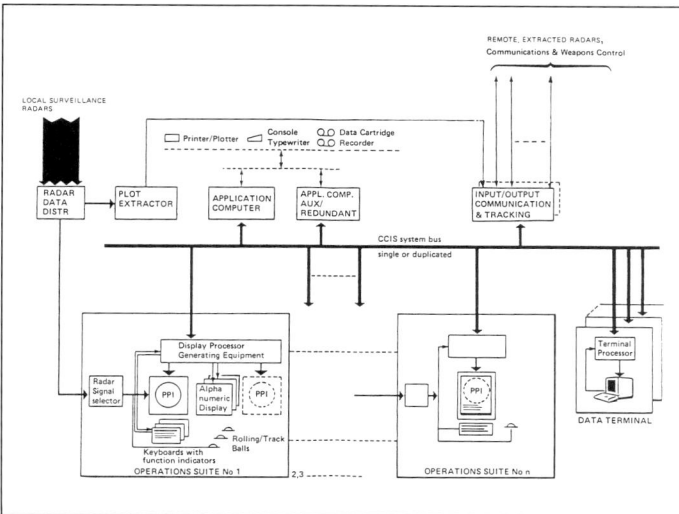

STRIKA system block diagram

(d) distribution/reporting of target data (including situation plots), tote data and optional text (operator's notepad) via ground-to-ground communication network or radio datalinks

(e) simulation and training.

An extended and enhanced version of STRIKA is used as a C2 system on naval base level (Naval Command Centre). It features expert functions for operator decision support.

Description

The STRIKA technical philosophy emphasises the same applications software on all levels and similar operational equipment on all levels (both stationary and transportable), so as to simplify technical and operational training, documentation, spares holdings and maintenance.

Information in various formats is converted by the input/output subsystem and dedicated front end processors are used for preprocessing, formatting, transmission control and so on. Track data and messages sent to and from co-operating units are transmitted as serial data messages via the ground-to-ground communications network and UHF/VHF/HF radio equipment. The radio datalink terminals are integrated into the system.

Status

Operational in Sweden. No longer in production.

Contractor

SaabTech Systems AB, Järfälla.

UPDATED

United Kingdom

ADAWS 4, 10 and 12 Action Data Automation Weapon Systems

ADAWS is a family of action data automation and fire-control systems developed for the Royal Navy (RN). The original ADAWS 1 system used Poseidon computers and was fitted in Batch 2 DLGs. All later ADAWS variants are based on ADAWS 2, which was fitted on HMS *Bristol*. ADAWS 2 was developed by Ferranti Naval Systems, now part of AMS, with the latter providing the display equipment. The system uses FM1600 series digital computers and the Mk 8 display system.

Development of ADAWS began in 1965. The development team, now part of AMS, worked as contractors to the UK Ministry of Defence (Navy) in respect of the RN Type 42 destroyers. In total, 12 ships of this class were originally planned (two for the Argentine Navy and 10 for the RN).

Subsequent changes in the RN fleet, due to the loss of HMS *Sheffield and* HMS *Coventry* in 1982 and the addition of Batch 3, comprising four ships launched between 1980 and 1982, led to a UK complement of 12 ships. Due to the elapse of some 10 years between the laying down of the first of class and launch of the latest and because of design changes affecting size, armament and sensor fit, individual ADAWS fitted in Type 42 ships differed appreciably in detail. All Batch 1 Type 42 now have their ADAWS System at a common standard and Batch II and III Type 42 now have the ADAWS IMprovement Programme (ADIMP) Command System (see separate entry). All Royal Navy ADAWS fitted ships now have a common software called ADAWS 12. This uses a ship-specific configuration file to accommodate the differing weapon and sensor fits.

The ADAWS family of systems now in service consists of ADAWS 4 (in the Type 42 destroyers) and ADAWS 10 (in the RN carriers). (ADAWS 3 was designed for the CV01 carrier which was cancelled, ADAWS 5 was fitted to eight 'Leander' class frigates equipped with the Ikara ASW missile, ADAWS 6 was fitted in the RN carriers before they were upgraded to ADAWS 10, and ADAWS 7 and 8 in Batch II and Batch III Type 42 before they were upgraded to ADIMP). Specific benefits of the modular nature of these systems are reductions in the development effort required and in the time necessary for system proving.

The ADAWS 4 was produced in two slightly different versions for the Royal Navy HMS *Birmingham* Type 42 class and the 'Hercules' class supplied to the Argentine Navy. Inputs to the system include the search radars 1022, 996 and 1007, two Type 909 Sea Dart tracker radars, EW passive direction-finding equipment, datalinks 10 and 11, Type 2016 or Type 2050 sonar and various manual input facilities.

The central data processing complex includes two FM1 600 computers, with an Alenia Marconi Systems Mk 8 digital display system for the presentation of tactical and other information. The computers provide up to a total of 2 × 256 k words of core store. They gather data from all the ship's sensors, assemble and correlate this information from other vessels obtained via datalink and present it to the command teams on appropriate display consoles.

All information from 'own ship' and consorts in datalink contact will be accessible at any console and on a selective basis, for example, either air situation or underwater situation. The computers also include programs to assist in the evaluation of the relative threat posed by different targets, thus assisting the command in the selection of targets and assigning weapons, particularly the twin-headed Sea Dart missile system. When action is joined, the computers also provide control of ship's weapons, in addition to providing all the aiming and fire-control computations for the Type 42 Sea Dart missiles. ADAWS also includes fire control for the 114 mm (4.5 in) automatic gun mounting and the calculation of intercept instructions to be passed by radio to fighter aircraft and anti-submarine helicopters. Other weapons interfaced include jammer, chaff launcher and the STWS ship torpedo weapon system.

Status

In service with the navies of Argentine and Chile, the RNZN and the RN. A major update programme for the RN was started in 1989 called ADAWS Improvements or ADIMP (see separate entry) and a further update programme, ADAWS 2000 (see separate entry), was initiated in the late 1990s.

Contractor

AMS, Camberley, Surrey.

UPDATED

ADAWS 2000 ADIMP (colour) command system

ADAWS 2000 is a development of the ADAWS Mod 1 (ADIMP) system (see separate entry). It uses the ADAWS Mod 1 software and either the F2420 processor (for commonality with existing ADAWS systems) or a commercial-off-the-shelf (COTS) RISC processor (for new builds). The displays use COTS technology, with a workstation class processor interfacing to a radar scan converter. An open systems architecture is used.

ADAWS 2000 offers a 1,000 track database with over 100 track updates per second. It interfaces to a maximum of seven radars, IFF, sonar, missile and gun systems, datalinks and EW equipment. In addition to weapon control and Threat Evaluation and Weapon Assignment (TEWA), it employs data fusion techniques to present a clear tactical picture from multiplatform inputs over large operational areas.

COTS hardware and software is used with modular software (Ada, C and Assembler) for each function and common infrastructure software. The display infrastructure is based round a UNIX operating system, with X-Windows and Motif for display. Radar scan conversion is used for the multiple radar inputs and a powerful graphics engine drives single or multiple-screen colour consoles, which feature CCTV input. The system uses a keyboard with additional function keys and rolling ball, with soft keys and pull-down menus displayed on the colour monitor. The latest version uses PC-based hardware and 46 cm flat-panel colour displays, and is designed to enable existing CRT display consoles to be updated in situ. Other input devices such as electroluminescent touchpanels are also supported,

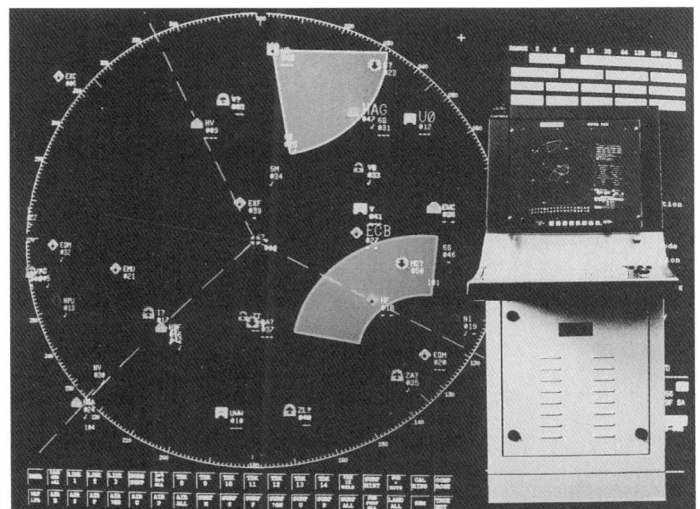

ADAWS display

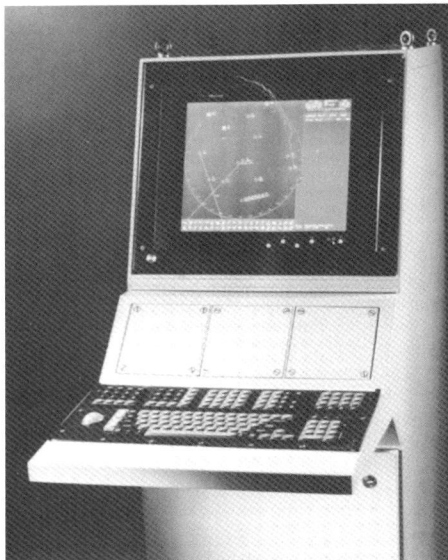

ADAWS 2000
0110488

as are other displays such as tote only and large screen displays. The system has a single or dual-redundant Ethernet local area network which connects to the dual redundant Combat System Highway.

Status

ADAWS 2000 technology can be used for either full or partial upgrades to existing ADAWS and CAAIS systems. Royal Navy Type 42 Batch II and III and 'Invincible' class carriers have a number of ADAWS 2000 displays in order to support the introduction of Link 16. HMS *Invincible* is being converted to ADIMP (colour) standard. The existing displays have been converted by removing the CRT and associated electronics and fitting an electronics rack and 51 cm flat-panel colour display. Replacement three-operator position tactical plots are also fitted. ADAWS 2000 (ADAWS Mod 1 Colour variant) is operational on HMS *Ocean*, L12, the new RN Landing Platform Helicopter (LPH), and on a number of other RN ADAWS fitted ships. It is also fitted in the replacement Landing Platform Docks (LPD), HM Ships *Albion* and *Bulwark*.

Contractor

AMS, Camberley, Surrey.

UPDATED

ADAWS IMProvements (ADIMP)

A major upgrade programme to the ADAWS naval command system (see separate entry) was started in 1989 called ADAWS IMProvements or ADIMP. This programme upgraded the combat system in the Batch 2 and Batch 3 Type 42 destroyers and the Royal Navy aircraft carriers. The system is known as ADAWS Mod 1 or ADAWS 20. The basic hardware and software are common across all the ships, with different numbers of displays and the software configured for the appropriate weapon and sensor fit.

Development of the system was conducted by AMS (then GEC-Marconi S31). THORN-EMI Electronics (subsequently Racal, now Thales) was a subcontractor. All development activities were completed in 1995. Enhancements to the system are planned to support further combat system changes, including the addition of Link 16.

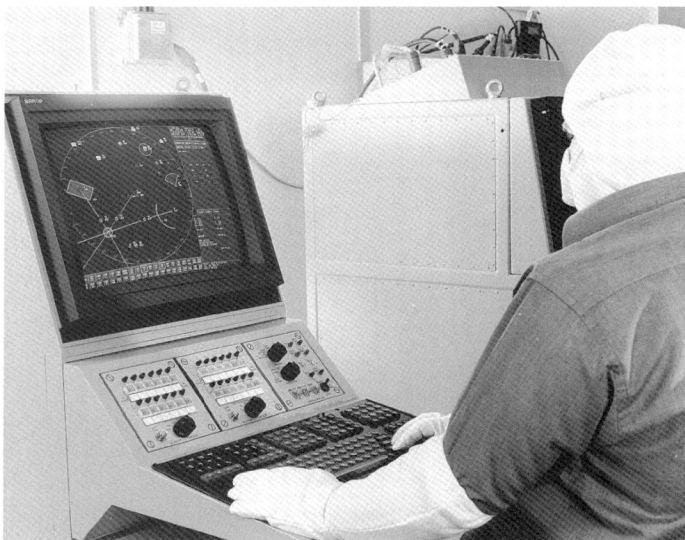

ADAWS display console

The ADIMP programme involves replacement of the FM I 600 computer by the far more powerful F2420 processors. The display system is enhanced to enable a much larger number of synthetic track markers to be displayed. The system track handling capacity is multiplied by four and the processing power available is over 10 times greater. A standard single operator display console is introduced. This has a labelled plan display, tote display, keyboard, rolling ball and space for two versatile control system modules. Each ship also has two tactical displays for use by the command team.

As part of the overall improvements to the ship's weapon systems, ADAWS is interfaced to a new range of weapons and sensors via a Combat System Highway (CSH). This is a dual-redundant bus, which uses a polling protocol to ensure timely transfer of combat system data. The radars now interface to track extractors which output data onto the CSH. AMS provides Outfit LFA, which interfaces to the Type 996 air search radar. Thales provides two types of automatic radar track extractors for the programme: Outfit LFC interfaces to the Type 1006/1007 navigation radar (two fitted on carriers) and Outfit LFB interfaces to the Type 1022 long-range surveillance radar. In Outfit LFC, radar and helicopter transporter contacts relating to the same contact are correlated as a combined track. In both outfits, tracks and status messages are reported onto the CSH. The company also supplies the radar track combiner, Outfit LFD. This analyses the track extractor output from each radar by reading the track messages from the CSH. It correlates all the information, identifies those messages from different radars which relate to each real world target and establishes accurate track information. The combined data is retransmitted onto the CSH as the ship's recognised air and surface picture to be read by the ADAWS command system and the datalink processing system. Other equipments interfaced to the CSH include the ESM sensor UAF or UAT, Sonar 2050 and the 909 tracking radars. The ADIMP System is now being upgraded by the introduction of a number of ADAWS 2000 (see separate entry) displays, software upgrades and enhancements to the datalink processing system as part of the introduction of Link 16.

Status

In service with the Royal Navy.

Contractors

AMS, Camberley, Surrey.
Thales Defence Information Systems, Wells, Somerset.

UPDATED

AMS tactical datalink systems

Datalinks have traditionally been supplied as an integral part of a computerised command/mission system. However, units not so fitted are unable to play their full part in the overall effectiveness of the force. AMS have developed three datalink systems that enable all units to participate in a force datalink net, whatever type of command/mission system may be installed, including units with a manual system or none at all.

AMS datalinks use a voice-frequency bandwidth for the automatic exchange of tactical data between participating units, thus ensuring every unit in a force has access to the latest information available to any individual unit.

The AMS DataLink Processing System (DLPS) is capable of operation as a completely independent, stand-alone system with its own display terminal, or it can be coupled to, or integrated with the unit's command/mission system. The equipment is based on a military Argus computer and is housed in a fully ruggedised enclosure. A display and keyboard can be provided and the equipment may be interfaced directly with a command/mission system. The equipment also connects to an HF, UHF or VHF radio transmitter/receiver capable of voice channel operation. Data received over the links is stored in the DLPS database where it may be designated for transmission over the datalinks. The DLPS database has a capacity of up to 120 tracks, including tracks received over the links, 'own ship' tracks, reference points, bearing lines and so on. Any of the tracks may be selected for display.

The picture shown on the colour display consists of computer-generated symbols showing positional track data and track labels in a Labelled Plan Display (LPD) format. Extensive use of colour is used for track identification. The equipment is capable of processing and displaying data up to 2,000 data miles from own unit position. The displayed picture may be centred anywhere within the operational area. Readout areas are provided for the display of command messages, alerts and operator control.

In the standard DLPS, the operator communicates with the system by means of a purpose-designed, alphanumeric keyboard containing special purpose keys and an electronic cursor for indicating positions on the display. The system provides prompt menus to the operator.

If the DLPS is working in association with a command system then the datalink may be set up and controlled from this system and link-derived data will be displayed in its own establishment format and symbology.

The AMS MultiLink Processing System (MLPS) is a receive-only Link 11 system. MLPS is a modular system that can be configured in a number of different ways to meet the requirements of individual platforms. It can be configured to receive any particular subset of Link 11 messages. The Link 11 messages currently handled are in accordance with STANAG 5511, Edition 2, Amendment 3.

The naval Data Terminal Set (DTS) enables a ship to participate in a Link 11 network as either net control or a picket station. It can operate over UHF or HF radio, including Independent SideBand (ISB). Once configured and initialised by the operator, the DTS operates in accordance with the net transmission protocol. It modulates and demodulates the signal to and from the radio, including diversity with ISB and controls transmit/receive switching. It tests for, acquires and tracks signal presence and carries out Doppler correction and synchronisation. It also generates tests for and responds to control codes and applies error detection/correction. Message data is transferred to and from the host computer through an NTDS interface. ATDS is available as an alternative.

The operator interface for control, monitor and test purposes can be both local and remote to the data terminal. It comprises a keypad and display with suitable selection and prompt sequences. In a net control station this panel would also be used to insert the list of roll-call addresses.

The terminal incorporates online facilities to check itself and the correct operation of the link. It also incorporates offline test and monitor facilities, which enable a fault to be located speedily and rectified by replacement of plug-in cards.

Most of the functions of the DTS are implemented in microprocessors, programmed in read-only memory during manufacture. Future enhancements in modulation, demodulation, coding and protocol can be accommodated by changing the stored program so that the effective life of the basic equipment may be extended.

Status

To date over 80 naval datalink systems including Link 11, Link 10, Link Y, Link 14 and Link Z have been delivered. AMS has also supplied datalinks for air defence purposes, including NATO Link 1 and airborne platforms.

Contractor

AMS, Camberley, Surrey.

VERIFIED

....

CACS 1 and 5

The Computer Assisted Command System (CACS) was the first of a family of new command systems for the Royal Navy, incorporating modern computers and a new operator interface. It provided a substantial increase in system capability which can be directly employed to enhance the command functions necessary in a modern warship. Both the hardware and software are constructed in modular fashion so that further system development and expansion can be implemented after installation on board and experience at sea, enabling the system performance to keep pace with changing and increasing requirements.

The origin of CACS can be traced back to 1977, when Ferranti (subsequently GEC-Marconi and now AMS) was responsible for a study on behalf of Director Surface Weapons Projects (Naval) of proposed improvements to the AIO system of the Type 21 frigates at half-life refit. It had been appreciated by the Naval Staff that the operational environment and the projected weapon and sensor fits of those ships had outgrown the capabilities of an AIO system that had been established in the late 1960s.

Towards the end of this study, MoD(N) extended the scope of the work to encompass the Type 22 frigates, as it had become clear that these new frigates would also need an enhanced AIO capability. A number of design concepts were presented to DSWP(N) who decided that, despite the extra development and production costs involved, the design that involved the greatest changes to system hardware architecture was the one that would best meet their requirements. The resulting development contract was placed with Ferranti in 1979 and detailed work on the design of CACS 1 was initiated, with Ferranti nominated as design authority for the first time for equipment of this type. The development work was carried out in close co-operation with the MoD at ARE, Portsdown.

The CACS system architecture is based on a facility consisting of dual F2420 computers. These are connected by a distributed highway to several display positions, each of which contains its own Argus M700/20 microprocessor system to optimise operator effectiveness and minimise angle point failures.

The consoles are based on the Series 9 display system which, with greatly increased quality and capacity, supersedes the Mk 8 display technology, used in ADAWS. There is extensive use of video retiming and scan compression techniques. The principal input device is the light pen, which replaces most of the rolling ball and keyboard functions.

CACS employs an all-vertical display concept which, besides offering the operator a more comfortable and efficient working position, overcomes severe ship fitting constraints and the problems associated with conventional horizontal dual-operator displays.

The most significant advance from the user's viewpoint was the implementation of the 'prompt' interface for operator communication with the system. For CACS 1 the interface was further developed to include the use of light pens and two TV monitors, one for tote data and one for handling operator inputs. With this interface the tote displays information relevant to the current task in hand and indicates 'prompts' or choices, which, when used in conjunction with the light pen and/or keyboard, allows the operator to input data and control the system in a logical

General view of the operations room of HMS Boxer

sequence. The need to remember complicated manual injection sequences was eliminated, with a consequential reduction of operator training requirements.

A further version of CACS was developed for the four Batch 3 Type 22 frigates. CACS 5 was designed to interface, via the Command System Highway (CSH), with the new range of weapons incorporated on these ships including the Harpoon surface-to-surface missile system, GSA8 gun system and Goalkeeper CIWS.

Status

CACS 1 is operational in the Type 22 Batch 2 frigates (only one remains in RN service), and CACS 5 is operational in the Batch 3 Type 22s. Both these systems are now in a Responsive Adaptation (RA) programme. Alenia Marconi Systems (AMS) is prime contractor and design authority.

Contractor

AMS, Camberley, Surrey.

VERIFIED

....

Combat Advice System (CAS)

The CAS has been developed to aid ship command decision making under stress. It is a computer system that stores naval tactics and applies these to current real-time scenarios. The tactical picture is received from onboard and offboard sensors via the command system, or direct from a tactical highway and the embedded knowledge is applied to the threat situation. Ship weapon system status is received by the CAS either by means of the ship highway or by manual input. Any conflicts, such as blind weapon arcs, are then evaluated and an optimum course is offered by the system to resolve these problems. The time available to implement the course change is also provided, along with an explanation of how the decision was derived. The command team is thus presented with both tactical information and solutions. The system can be modified on board without specialist help when either tactics or weapon systems change.

The software design allows immediate access to the knowledge held in the CAS database. This knowledge is compared with the ship's position, course and weapon system status. The software modules manage the information to ensure that a course recommendation minimises radar cross-section, whilst maximising the effectiveness of the ship weapon systems. The recommended course is then routed to both the chaff system and to either the surface-to-air missile system or anti-air guns, depending on the ship fit, to optimise the planned response. In making its recommendation, CAS ensures that SAM tracking radars are not in a blind arc. Other problems or conflicts, such as friendly aircraft within the weapon engagement zone, are highlighted as a summary of actions to be considered.

A rugged processor assembly using industry-standard Motorola 68040 processors and Inmos transputers provides the platform on which to run the software. The configuration of an electroluminescent panel and a trackerball provides a simple-to-use man/machine interface. All screen formats can be accessed by a single touch of the panel, which can also be configured as a qwerty keyboard for updating information held within the system.

CAS's full-colour tactical situation display is configured with labels and display formats compatible with those of the ship command system to ensure that no misinterpretation is possible. Tote information provides the course recommendation and conflict summary. Other information, such as windspeed, ship speed and direction and specific track data, is also displayed. The CAS will not allow a solution to be considered that involves a particular weapon system if that weapon system is not available. The status of all ship systems availability can be displayed on the electroluminescent panel and combat air patrol and ASW status can also be shown.

For details of the latest updates to *Jane's C4I Systems* online and to discover the additional information available exclusively to online subscribers please visit

jc4i.janes.com

The acceptance of a computer recommendation demands understanding of the logic used in forming that recommendation. To give the command confidence, the course recommendation display builds a picture of 'no go' areas. The best course is clearly shown from the resultant clear lanes. These 'no go' areas are built from known blind arcs, radar cross-sections and ship manoeuvring capabilities. A ship manoeuvre preview facility is also provided. This projects the tactical situation, based on the current course and speed of known tracks. The resulting relative positions are then displayed. Conflicts, alerts, course recommendations and explanations can also be displayed at the operator's request.

CAS is configured for each ship class and existing console designs can be adapted to install the CAS displays and processors. The man/machine interface is designed to be compatible with existing command system display formats. This ensures that no confusion arises in the command assessment of the tactical picture. All CAS builds contain identical foundation software. The ship fit information for a specific vessel is subsequently loaded into the CAS and includes data concerning 3-D blind arcs and ship performance as well as default tactical rules. The CAS is also available as a software package that can be included in the ship's command system, with its displays available as windows on the Command System consoles.

Status
A variant of CAS was selected by the Royal Navy for installation on CVSG and Type 42 ships. No longer in production.

Contractor
AMS, Camberley, Surrey.

UPDATED

··

Command Support System (CSS)

The Royal Navy Command Support System (CSS) is a fully automated information system designed to achieve a radical advance in the effectiveness of maritime operations. It supports the planning and conduct of maritime and amphibious operations, from initial planning through command and control of the operation to post-operation analysis. The system provides access to and use of the latest available information, allied with a suite of automated processing and display tools, comprehensive planning and decision aids and all of the enabling hardware, software and support. A communications network management system enables effective ship-shore and ship-ship exchange.

CSS supports all operational and administrative planning tasks for force and unit activities ashore, afloat and in the air. It processes, manages and displays up-to-date positional intelligence and other tactical information and is capable of storing, retrieving, manipulating and distributing large amounts of tactical and encyclopaedic data. It operates in near-real-time, with extensive facilities to produce and manage the wide-area picture. This is achieved through communication links to the ship's command system and to many contributing systems ashore and afloat. Message handling, Secure Web and office automation tools help speed up decision-making, planning and dissemination of information, allowing users afloat to share tactical and strategic data with those ashore. The many planning and decision aids include overlays of military symbology, briefing facilities, waterspace management, emission control planning, satellite vulnerability and logistics planning. Dedicated applications provide for control and support of amphibious operations. The system is designed to be extremely flexible while providing high performance, ease of use and inherent security.

CSS comprises a network of workstations and servers. Designed around multiple sustainable hubs which provide high availability and growth potential while retaining low-cost network interfaces, it provides the flexibility of a client-server design with delivery of appropriate performance. The architecture provides resilience and scope for future expansion.

The majority of installations use standard commercially available hardware. Particularly demanding installations are provided with enhanced protection. The majority of displays are 20 in, but a full range is available to support differing needs; large screen displays and projectors can be used to support briefing and presentation of shared information. The latter are particularly useful in tactical shore-based headquarters, where large screens are less practical. Where space is limited, laptops are provided. Each workstation can support all user facilities and provides a common look and feel, enabling flexibility in manning and organisation. Access control enables users to access CSS from any workstation, providing true multifunction capability. Use of open standards ensures maximum adaptability and helps contain through-life costs.

CSS operates in a managed information network. Centralised network management optimises the extensive communications functionality through ship-based and land-based channels, including improved effectiveness of satellite communications. The remotely managed wide-area and local-area networks also offer productivity tools, such as e-mail for transfer of briefing and planning information between ships and commands. CSS also supports the integration of coalition networks, using web-based replication software such as Lotus Domino.

CSS employs an evolutionary open-systems architecture using industry standards based on Solaris, HP and Windows NT. Over 70 per cent of the system comprises commercial off-the-shelf products, tailored and integrated to meet the naval operational requirements. A key component of this is Northrop Grumman's ICS and C2PC track management software (see separate entries), which manage geographic and location data. ICS is derived from an equivalent US Navy program, JMCIS, where it provides the unified C4I software baseline. It offers display facilities, automated aids and connectivity. Communications software is based on

Nexor and ICL product suites together with Compucat CMX message handling and a wide range of e-mail, web and tactical interfaces. The messaging capability is augmented by Systematic's IRIS suite (see separate entry) to assist with AdatP3 formatted messages.

Each system is scoped to the operational requirements of a particular unit. All systems have the same basic software suite available, and all are capable of supporting the full system functionality. Further, CSS is designed to be scalable and is easily tailored to suit particular needs. A typical frigate fit employs five workstations with associated services and communications; submarines have one or two each; the system for the replacement assault ships [LPD(R)] has 72 workstations for naval task-group and amphibious landing-force command. This flexibility is underpinned by a modular design, able to accept new developments as they become available and to cope with changes in the environment and operational requirements. This has been evident in the production of fully integrated and deployable CSS systems for use by mobile commands and HQs. These deployable systems expand the fixed ship CSS systems by an additional 45 seats in CVS platforms and an additional 10 in destroyer-size platforms.

Functionality
Recognised Picture - Maritime and Land Pictures
The display and management of tracks covering all environments (space, air, maritime surface, maritime sub-surface and land). Tools provide track management and track kinematic calculations - speed/time/distance, relative velocity, cross-fixing and so on, plus a number of simple tools and utilities. CSS uses a combination of ICS and C2PC applications to provide the maritime and land pictures and uses APP6a symbology.

CSS will work with the following map formats:
(a) DNC
(b) DTED Level I
(c) ADRG & CADRG
(d) Vmap Levels 0, 1, 2 & Uvmap
(e) WVS
(f) GEBCO
(g) CIB
(h) ASRP
(i) HCRF
(j) S57 (by conversion to VPF)
(k) VPF 96

Decision and Planning Tools
General Tools
(a) Office Automation. MS Office and MS Project. Numerous military and naval templates are provided in Word.
(b) NT Explorer.
(c) JOP. For Junior Officer Pilot functions
(d) Web Browsing.
(e) Plan Manager. Similar to a standard File Manager, Plan Manager is central to the planning process. It assists in the configuration control of the planning process within CSS, giving the user control over the ownership, access, distribution and authorisation of all CSS plans.
(f) Threat List/Friend List. To assist the user in creating snapshots of information for operations and exercises. The Threat List and the Friend List are the basic building blocks around which much of the staff planning will be constructed.
(g) Ship Commanding Officer, Ship's Company and Personnel Record. The Ship Commanding Officer form lists all COs in seniority order for particular exercises or operations. The Ship's Company form lists numbers in individual ships, identifying particular specialists.
(h) Rules of Engagement (ROE). Based on pre-defined plans, this application produces the ROE Stateboard. It allows the planning of ROE for the force.
(i) Communications. This provides communication planning tools for amphibious and maritime operations as spreadsheets. It includes resource bookings, information-exchange requirements, frequency allocation, and distribution of circuits, equipment and aerials. In addition it is integrated to a facility for creation of the OPTASK COMMS. Connectivity diagrams for communications planning are also provided under this application, together with path profile calculations.
(j) Callsign. This displays relevant maritime and land callsigns, with the information being controlled centrally. Callsign changeover is also enforced centrally.
(k) Stock Levels. This feature allow the user to maintain an overview of critical stocks and displays graphical representations for briefing purposes.
(l) 3-D Graphics. The 3-D Graphics application provides a sophisticated overlay facility integrated with ADatP3 messaging. It integrates with the Wide-Area Picture (WAP). It also provides terrain cross-section, site selection and intervisibility, and weapon and sensor coverage.
(m) Geographic Mutual Interference. This implements 2-D+ Height Band mutual interference calculations on 2-D objects in 3-D Graphics.
(n) Routes. Facilities to cover sea and air routes and entry/departure plans
(o) NBC. This tool provides NBC forecasting.
(p) Navigation. This provides navigation calculations and a tidal calculator.
(q) Station Planning. This provides 2 W and 4 W dispositions and sector screen from ATP 1(C).
(r) Harbour Defence. Harbour and ship search planning is provided, making use of the Resource Scheduler (see below). Deck search planning is provided through user annotation of desk plans. In addition a tote capability is provided to store lists of available staff, current security status and search progress.
(s) Search Planning. A search analysis tool is provided.
(t) Collaborative Planning.

Maritime Applications
(a) Maritime Task Organisation. This application assists the user in constructing a hierarchical structure, breaking the force into components in accordance with the rules laid down in ATP 1(C).
(b) Duties and Responsibilities. This application plans the allocation of duties and responsibilities to the components of a Task Organisation.
(c) Programme Planning. This application provides facilities to generate a ship's programme, including assignment of resources. Constraints warn the operator when resources are double-booked or when ships cannot get from one area to another within the time allowed.
(d) Resource Scheduler. The Resource Scheduler provides scheduling for a flying programme, ASuW, AAW and ASW. The application assists the user in generating plans. Resources are based on data from the ADatP3 standard. The resulting schedule information is used to populate the relevant signal messages.
(e) Preplanned Response Planning.
(f) Weapon Effort Planning.
(g) EW. This application provides maritime EW planning facilities to provide a frequency-to-jammer allocation tool and also includes EMCON planning, EW tasking, RADFREQ plan, satellite countermeasures and mutual interference avoidance planning.

Amphibious Applications
(a) Land Operations. A Land Task Org/ ORBAT tool. The Land Task Org tool provides filtering and display of units on the Wide-Area Picture (WAP) and a link to unit status data. In addition, tools are provided for relative-strengths calculations, combat effectiveness, Formal Estimate and resource scheduling.
(b) Onload/Offload. This tool provides onload and offload scheduling for amphibious operations. It assists the operator in identifying how to load amphibious shipping. It will produce the Helicopter Employment Assault Land Table and the Surface Assault Schedule.
(c) Fire Support. Provision of planning for naval gunfire support and field artillery, in particular the preparation of reports and orders integral with an amphibious operation. It includes extensions to 3-D Graphics plan object types to allow classification of objects such as roads and minefields to support the field engineering aspects of amphibious warfare.
(d) Amphibious EW. This provides an amphibious EW resource siting and allocation tool, as well as land and comms EMCON planning, including sensor coverage, frequency deconfliction, analysis of enemy emissions and EW planning.
(e) Amphibious Intelligence. Planning for intelligence gathering. It includes an Enemy ORBATs database and other intelligence, allocation of resources to tasks/targets, and target data.
(f) Engineering Planning. This application will provide field engineering calculation spreadsheets with resource scheduling. Field Engineering Plans and Reserved Demolition Plans are covered. Calculation spreadsheets are independent of geography.
(g) Area Bookings. This provides tools to allow examination of space bookings, showing the details of missions, making use of areas, and allowing the user to de-conflict usage. It also allows the user to identify features such as avalanche danger areas.

Environmental Applications
(a) Forecaster's Workbench. The Forecaster's Workbench provides a comprehensive suite of applications for visualising, analysing and modifying meteorological and oceanographic data. It accepts gridded binary data from numerical weather-prediction models, coded observational data and satellite images. The workbench produces a range of products including the surface analysis chart and overlays of Metoc data for display on the WAP.
(b) Acoustic Propagation Forecasting. The Acoustic Propagation Workbench provides the functionality to analyse and fuse modelled, observational and climatological oceanographic profile data in order to specify the environment for propagation modelling. The workbench incorporates propagation loss models covering the full range of operational frequencies and water depths. It produces a variety of products including probability of detection cross-sections.
(c) Radar/Radio Propagation and Met Utilities. The Radar Propagation Workbench provides a common front end to create radar, VHF and UHF communications, ESM and Line of Sight scenarios using modelled, observational or climatological data. The workbench includes three RF propagation models, each of which may be populated with the created scenarios. It produces a variety of products including probability of detection cross-sections.
(d) Meteorological Utilities. A variety of stand-alone utilities are provided. The utilities will calculate a variety of parameters, including sun and moon rising/setting; morning civil twilight/evening civil twilight; evaporation duct height; heat stress index; wind chill factor; QNH and QFE; and infra-red visibility.

Messaging
(a) Message Handling Application. This provides ACP 127 messages direct to the user. The ADatP3 message-formatting package is integrated with the Message Handling package.
(b) E-mail (X 400). Both interorganisational and interpersonal e-mail.
(c) Text Comparison Tool. In a number of instances, CSS will receive a particular message from a sending unit at intervals, with each report giving the latest status. The Text Comparison Tool will allow the operator to identify the information that has changed since the last report. Any message can be compared.

(d) ADatP-3 Messages. A variety of ADatP3 messages are automatically processed by CSS. Much of the information contained in these messages is used to update the database so that the users can view the information within the appropriate application.

Status
CSS first entered service in 1997. It is now fitted to the majority of RN ships including submarines, to Royal Fleet Auxiliaries, in operational shore HQs and in training establishments. As noted above, fixed ship systems are expanded when a Force HQ is embarked in a CVS or destroyer. It is also in service with the Royal Marines, including HQ 3 Commando Brigade RM. Version 7, which completed the amphibious capability, is currently in service. V7.1 is due in autumn 2003 and it includes upgrades to ADatP3 functionality and further enhancements from user feedback. Further upgrades are planned.

Contractor
EDS Defence Ltd, Hook, Hampshire.

UPDATED

DCB submarine tactical data-handling and integrated fire-control system

Development
To reduce the manpower requirements of attack teams, the fitting of digital computer systems in the Royal Navy's nuclear submarine fleet began in the early 1970s with the combat information System DCA. System DCB was developed from System DCA in the late 1970s when a second digital computer, for fire control was integrated into the system. Development of DCB has continued with regular software updates to incorporate new sensors, weapons and tactics. System DCB was subsequently upgraded and entitled DCB (REHOST).

Description
System DCB is a well-proven tactical data handling system with integrated fire-control facilities. The system comprises a two-position combat information console and a two-position fire-control console installed in the control room; weapon interface and battery monitoring equipment's installed in the weapon compartment; and the combat information and fire-control computers (two F2420 computers) and associated electronics, installed in the computer room. All console positions are provided with two CRT displays, one pictorial and the other for tabular tote information. The main operator interaction with the system is through light pen selections on the displays, supplemented by dedicated switches. The submarine's sonars, periscopes, ESM and navigation equipments are linked to the combat information computer by S^3 (Serial Signalling System) interfaces and synchro and resolver inputs, which are converted in computer peripheral units and input as digital data. The computer processes the received sensor data into tracks. Control of track association can be exercised by the operator. Additional information such as classification and intercept data can also be input by the operator in coded digital form from the keyboards. Target motion analysis, using bearing-only techniques, is carried out for all contacts. The tracking solutions and the sensor data received are compiled into a tactical picture that is shown in plan, time/bearing or time/frequency form on the combat information console pictorial displays. To assist in the evaluation of the tactical picture a variety of tactical calculations can be carried out at operator request. The results of these calculations, together with amplifying data on contacts, can be presented on the tote displays. Contact data are also output to the Automatic Contact Evaluation Plotter (ACEP). Digital data on selected targets are passed to the fire-control computer where fire-control solutions for either wire-guided torpedoes or anti-ship air flight missiles are calculated. The solutions are shown in plan form on the pictorial displays at the fire-control console, while the weapon settings are shown in tabular form on the tote displays. Weapon settings, tube orders and weapon orders are passed to the weapon interface equipment as digital words on duplicated serial highways. The weapon interface equipment decodes the digital data and applies power, switch signals, discharge gear drive pulses and guidance commands to the weapons and tubes as necessary.

After firing, data feedback received from the wire-guided torpedoes is passed to the fire-control computer on duplicated serial highways. The fire-control computer decodes the feedback data and generates steer commands to the weapon if required, as well as updating the fire-control information on the displays. Duplication of the serial highways for weapon data and the cross-linking between the weapon compartment units provides a choice of fallback modes to maintain a weapon firing capability.

DCB REHOST
The DCB REHOST system maintains the functionality of the DCB system while removing the main obsolescence problems. Based around commercial-of-the-shelf (COTS) hardware and a Proprietary Software Emulation package, it achieves this by utilising the following: the F2420 processors are replaced by Sun Micro Electronic SPARC processors running the existing DCB Software using the F2420 emulation software. These processors are connected to new operator workstations, located in the DCB consoles, via a Local Area Network. The existing monochrome CRT displays, for each operator position on both the fire-control and action information consoles (one pictorial and one for tabular information with light pen interface) are replaced by a single 20 in colour flat screen display with a

touchscreen overlay. This configuration mimics the operation of the light pen and the CRT and therefore gives the benefit of minimal retraining for the operator. The magnetic tape decks used for data recording will be replaced by Digital Audio Tape (DAT) and CD-ROM will be used for program loading. The existing input/output devices are maintained enabling the DCB REHOST system to interface to all the sensors and weapons handled by the DCB system. The DCB REHOST system uses an open architecture based on SUN SPARC processors and the UNIX operating system. The evolution of the DCB system onto this architecture builds on the well proven functionality and reliability of the DCB software, while enabling additional applications, produced using the latest software development methods and languages, to be easily added.

Status
In service with Royal Navy 'Trafalgar' and 'Swiftsure' class submarines or replaced by SMCS under mid-life refit programme.

Contractor
AMS, Camberley, Surrey.

UPDATED

DCC submarine tactical data-handling system

DCC is the tactical data-handling and fire-control system developed for the Royal Navy's 'Upholder' class of SSK, now sold to Canada as the 'Victoria' class. The system was derived from the System DCB (see separate entry) in service with the UK's nuclear submarines.

DCC comprises three console positions in the control room, supported by computer equipment in the computer room and weapon interface equipment in the forward part of the submarine. The main computer is fully redundant; a second computer is provided to run the system should a fault occur in the primary computer, thus ensuring that the facilities of DCC are maintained.

Each operator position provides two cursive display units, one for pictorial displays the other for tabular tote data, a light pen and a fire-control order and status unit keyboard. The main human computer interface with the system is provided by the light pens, but selection and switching facilities for fire control are provided on the console keyboards. Each operator position provides full combat information and fire-control facilities allowing increased flexibility in the manning of the system, such that a single operator can control the complete system.

There is close integration between the sensors and the DCC system, with sensor data being passed to DCC via a digital databus. The sensors passing tactical and environmental data to DCC include: long-range passive sonar, medium-range passive/active sonar, sonar intercept, passive ranging sonar, search and attack periscopes and ESM equipment. The sensor data is collated, processed and presented in plan, time/bearing or time/frequency form on the pictorial displays of the console with supplementary or amplifying data being shown on the tote displays. The processing includes target motion analysis and tactical situation and the decision making processes.

The weapons fit for the 'Upholder' included dual-purpose wire-guided torpedoes (anti-submarine/anti-ship), anti-ship air flight missiles and mines. The fire control for these weapons is integrated into the system and includes target designation, weapon and tube preparation, weapon launch and, where appropriate, post-launch guidance.

Weapon settings, tube orders and weapon orders are passed to the weapon interface equipments which apply power and switch signals to the tubes and weapons during the preparation and launch phases. After firing, tell-backs from the wire-guided torpedoes are passed to the central computer which generates steer commands as required and also updates the fire-control information on the displays. Duplication of the highways for weapon data between the computer and weapon interface equipment and the cross-linking between the units of the weapon interface equipment provides a choice of fallback modes.

Status
DCC was installed in the four 'Upholder', now 'Victoria', class SSKs built for the UK and sold to Canada, and is being brought to operational readiness before the boats leave the UK. However, the system is due to be replaced when the vessels arrive in Canada.

Contractors
AMS, Camberley, Surrey.
Ultra Electronics Ltd, High Wycombe, Buckinghamshire.

VERIFIED

DCG Rehost (DCG(R)) submarine tactical data handling system

DCG Rehost (DCG(R)) is a tactical data processing system which has been designed, developed and built to support the command system in selected UK submarines. The system exploits commercially available computing units to achieve performance, portability and reliability within a low lifetime cost and short procurement lead times. It is based on the AMS Maritime Acoustic and Navigational Tactical Aid (MANTA) (see separate entry).

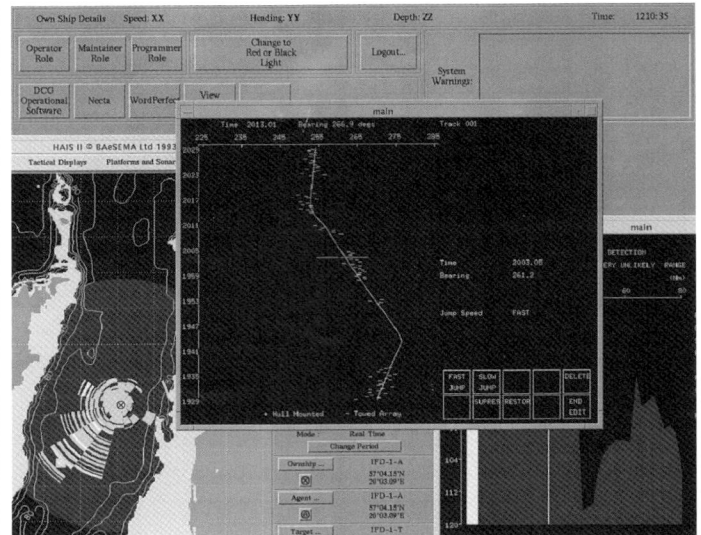

DCG(R) tactical display

System DCG(R) replicates the full functionality of its precursor, System DCG, including its interface to the DCB AIO computer. It also includes substantial improvements in operability, performance, overall capacity and upgrade potential over the previous system. It can operate independently, although it is linked to the main submarine command (action information) computer from which it derives data describing the submarine's own movement and the perceived environment. It subjects these data to rigorous and intensive analysis to produce a more accurate picture of the tactical situation.

While the system was developed originally for UK submarines, connections to any automated command system may readily be achieved using standard interfaces. Alternatively, the manual data entry capability enables system DCG(R) to be used independently.

The system is based entirely on commercial-off-the-shelf (COTS) hardware repackaged to meet the environmental and electromagnetic compatibility requirements, allied with substantial re-use of existing software. It provides a distributed open systems processing environment, based on networked Sun SPARC workstations running UNIX and, the network is easily reconfigurable to deliver high-system availability and reliability. Each workstation has a high-resolution flat panel colour display. Each user is able to select from various applications within DCG(R)'s open systems software environment, including DCG itself. The initial configuration of the system includes a number of new utilities, in addition to the full DCG functionality. The design also includes substantial expansion capability, permitting future inclusion of new control room facilities either as integrated functions with access to tactical picture data, or as stand-alone functions running within the overall environment. It has commonality with the next generation Submarine Command System (SMCS Releases 6 and 7), which offers a cost-effective migration path for software ultimately destined for inclusion in SMCS.

The software design includes a substantial proportion of the original DCG software, converted to run under UNIX and X-Windows and incorporating a high-performance object oriented track database. The large amount of software re-use capitalises on the reliability, user familiarity and proven algorithms resulting from over 10 years of DCG software support and development. This is combined with the use of Windows, point and click user interaction, improved screen resolution and colour displays to achieve considerable operability enhancements over system DCG.

Status
A total of 10 systems is in service with the Royal Navy.

Contractor
AMS, New Malden, Surrey.

UPDATED

Fleet Operational Command System - Life Extension (FOCSLE)

FOCSLE is designed to support all the UK Maritime Headquarters and provide the Commander-in-Chief Fleet and his subordinate commanders with enhanced command and control facilities, including the display and manipulation of an integrated dynamic recognised maritime picture with up to four concurrent maps displayed simultaneously, waterspace and mutual interference management, full message handling facilities, secure management and distribution of classified data, automated database updates, decision and planning aids, briefing facilities with large screen displays, exercise support facilities and comprehensive office automation facilities. FOCSLE takes account of the current practices of operational staff in the execution of their command and control tasks, and the prime objectives of the system are to ensure easy and rapid assimilation of information by the users, to release the users from time-consuming repetitive tasks, to furnish consistent unobtrusive interfaces with other maritime systems and to reduce time to learn and training costs. The system provides the Fleet staff with the means to acquire, store, manipulate and display data in support of surface, sub-surface and maritime air operations and exercises, and to interoperate with other nations and allies.

The system integrates all maritime CCIS facilities, including direct links to the Naval Signal Telegraph Network Message Handling System for automatic receipt, despatch and signal message handling, and automatic database updates from messages. It also exchanges data and e-mail via CHOTS (see separate entry) and is interoperable with RAF CCIS (see separate entry) and JOCS (see separate entry). FOCSLE has an operational database with user and predefined totes for frequently used information and incorporates encyclopaedic databases, such as *Jane's*. Automatic preparation/generation of messages from database information is possible.

FOCSLE uses a Compartmented Mode Workstation (CMW) UNIX operating system and provides full system and network management, including audit and analysis facilities. The underlying features of the system provide compartmentalised multilevel security, and the system security facilities include both mandatory and discretionary access control. User identification and authentication by password is required before logging on to the system and users are given clearance based on staff title which reflects their security clearance and defined role. Similar mechanisms oversee data processing to prevent unauthorised access. The system handles data security and the control and integrity of the database. In particular it provides correct security labelling for data transferred to and from other systems. All data in the system is labelled and printouts are given the correct security label. FOCSLE is accredited to UK security confidence level 3.

The FOCSLE client server architecture is intended to provide a balance between performance, data consistency and cost. The system makes maximum use of COTS hardware and software and the open system environment is capable of being expanded in both capability and capacity. The Common Operating Environment (COE) used reflects most of the UK MoD major system procurement requirements, including vendor independence, reliance on standards, flexibility to respond to changing requirements, future-proofing against developing technologies and the need to improve interoperability.

Status
The US$30 million FOCSLE contract was awarded in 1995 and interim operational capability was achieved in 1996, followed by full operational capability in 1998. This provides interoperability with additional external systems and full system functionality within an accredited secure environment. The principal areas of enhancement have been an encrypted wideband area network to connect remote sites, remote site installation, and the full development of the operational database.

Contractor
AMS, Camberley, Surrey.

UPDATED

....

KAFS submarine AIO and FCS

KAFS was a commercial development of Outfit DCC (see separate entry). The system provides a wide range of operational facilities including sensor data handling, target motion analysis, picture compilation, fire control, tactical calculations and data recording, so that the commanding officer may carry out the wide variety of tasks for which a modern submarine can be employed. A comprehensive onboard training facility, enabling training in both AIO and fire-control procedures to be carried out in harbour or at sea, is also included. The system hardware and software has been specifically designed in a compact, modular manner to provide a wide range of operator facilities which can be adapted or expanded as necessary to meet individual system requirements.

The equipment configuration is an operator console in the control room linked by a MIL-STD-1553 databus to the weapon control equipment in the torpedo room.

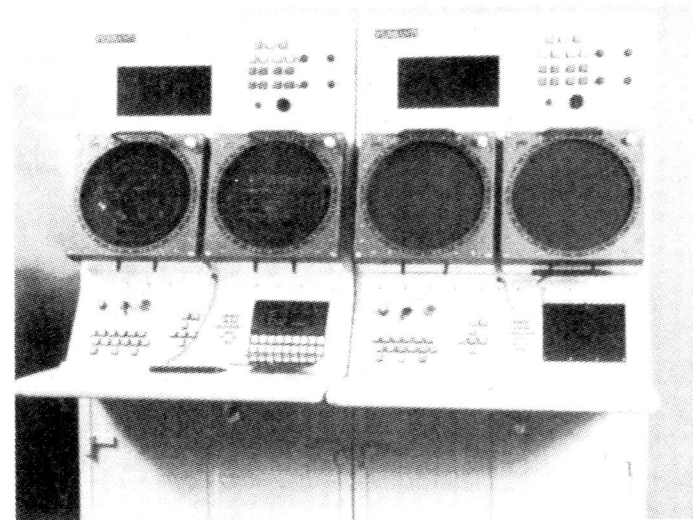

Double KAFS console

The torpedo room equipment comprises a local control panel that provides common services, together with two weapon interface units and two tube switching units. The weapon interface and tube switching are provided on a sided basis, port and starboard, so that in the event of a failure in the equipment of one side the other remains operational. Data from the sonar, search and attack periscopes, ESM, radar and navigation sensors is also passed to the system via the databus.

The control room console provides two identical operator positions mounted above two electrical sections that house the central computer and all the associated electronics. Full AIO and fire-control facilities are provided at each console operator position and either operator can control the whole system, thereby allowing one position to be shut down in patrol state. Each console position has two CRT displays, two command data panels and a weapon control and system status panel. The two display areas provide the operator with a label plan, time/bearing or vertical section display and a tote display providing pages of data and selections to enable him to interact with the system using a light pen. Track data, intercept alarms and tube and weapon status are displayed on the AIO and fire-control command data panels. Tube and weapon preparation orders are passed to the torpedo room using the weapon control and system status panel; this also has facilities for manual guidance post launch, system control as well as status indications.

The weapon control equipment in the torpedo room, which is microprocessor-based, provides the interfaces to the tubes and weapons. Tube and weapon status is displayed on the local control panel which also provides weapon discharge and control facilities in fallback mode, independent of the central processor. The local control panel also provides facilities for weapon battery heating, together with monitoring to detect hazardous conditions. The weapon interface and tube switching units apply switch settings and power supplies to the tube and weapons during the preparation and launch phases. After-launch guidance commands, if appropriate, are also routed to the weapons via these units.

Status
In service with the Brazilian Navy in HDW Type 209 submarines.

Contractor
AMS, Camberley, Surrey.

VERIFIED

....

Maritime Asset Planning System (MAPS)

The Maritime Asset Planning System (MAPS) is a system that applies artificial intelligence to the problems encountered in the management of forces at sea. These advanced programming techniques make it possible to implement the user's current tactical doctrine. The modular design of the software enables the MAPS system to be tailored to any level of force command, any combination of problems and to any type of fleet or area of operations.

MAPS enables command at sea to be exercised more effectively by providing assistance with: command appreciation, resource planning, formation and screen design, replenishment and logistics, geographical setting, operational readiness, signal message production, rules of engagement, route planning, force disposition design, flying programmes, electronic warfare tasking, environmental conditions, effects of changes in resources and plans, water space management and weapon effort planning.

The system is embedded within ADAWS (see separate entry).

Status
In service, as part of ADAWS, with the Royal Navy, and with at least one other navy.

Contractor
AMS, Camberley, Surrey.

VERIFIED

....

NAUTIS command and weapon control systems

NAUTIS is a versatile command and weapon control system. A single NAUTIS console will fulfil the requirements of smaller warships. In larger systems, multiple NAUTIS consoles are networked to provide fully distributed processing with database replication at each console.

NAUTIS applications include air, surface and underwater warfare, amphibious warfare and maritime policing/EEZ patrol. NAUTIS has many capabilities, which include command display, radar surveillance, sonar surveillance and analysis, threat evaluation and weapon assignment, ESM/ECM management, gunfire control, missile control, tactical datalinks, real-time operational planning, helicopter and fighter control and command advice.

NAUTIS is self-contained, compact and simple to install. All the maintenance is carried out from the front. Its key features are that it is air-cooled and operates from standard ship electrical supplies with a low magnetic signature. It has integral EMC shielding and satisfies all relevant environmental standards. No special tools or test equipment are required for onboard maintenance and it has a comprehensive built-in test.

The systems architecture comprises a dual-redundant Command System Highway (CSH) and distributed database to eliminate single node failures. The compact design provides easy installation and upgrade with online diagnostic monitoring equipment. The system is modular and includes a portable software infrastructure running on commercial-off-the-shelf (COTS) hardware and software. There is a library of reusable software applications, which technology evolution and easy adaptation to a customer's specific requirements. The Human Computer Interface (HCI) includes: high-resolution, windowed colour display CRT and flat panel display options, combined display of synthetic and radar video, alphanumeric tote windows, CCTV and Infra-Red (IR) video windows, menu-driven control using touch sensitive panel, extensive system database, simulation and onboard training and multimission operations.

The system is suitable for MCMVs and Surface combatants ranging from Patrol Vessels to highly capable Frigates.

Status

The general variant of NAUTIS has been purchased by Malaysia, Brunei (both NAUTIS 2), New Zealand and Thailand. The minehunter variant has been purchased by Australia, Japan, Saudi Arabia, Spain and the USA, and by the UK for

NAUTIS-M fitted in HMS Sandown, the first of 12 Royal Navy single-role minehunters

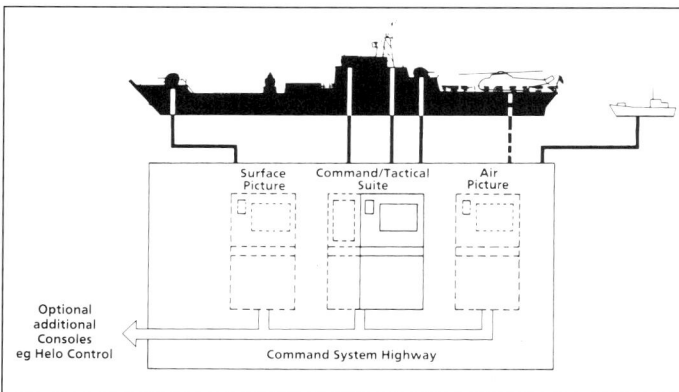

A typical NAUTIS layout for amphibious warfare and logistic supply ships

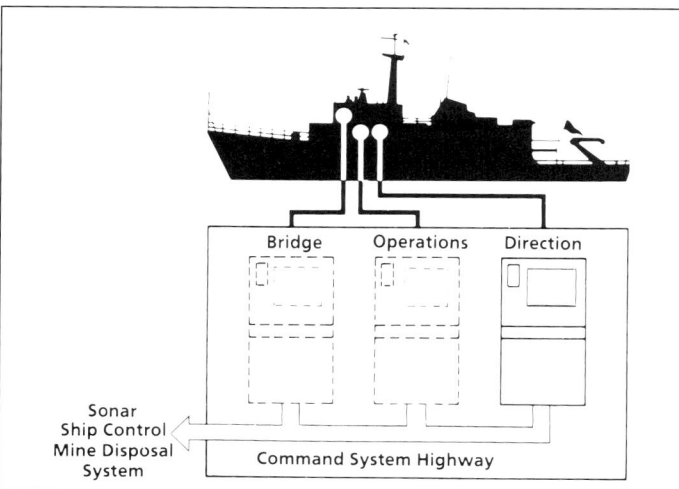

NAUTIS-M for minehunters and minesweepers

the RN 'Sandown' class minehunters. NAUTIS 3 has been selected by the UK for the upgrade of the RN 'Hunt' class MCMV and by Turkey for their new MCMVs.

In January 2003 *Jane's Defence Weekly* reported that the CACS command system on board 2 ex-RN Type 22 Batch 2 frigates to be sold to Romania would be upgraded with NAUTIS 3.

In September 2003 it was reported that NAUTIS 3 had successfully completed factory acceptance tests and remained on schedule for harbour and sea acceptance trials, following ship fitting, later in the same year.

Contractor

AMS, Camberley, Surrey.

UPDATED

Submarine Command Systems (SMCS)

The Submarine Command System (SMCS) features an integrated suite of tactical picture, weapon management and fire-control functions supported by oceanography, onboard training and navigation facilities.

SMCS is capable of handling large inputs of data from sensors and weapons, providing the combat team with the means to operate the submarine to maximum effectiveness. Its processing and large database ensure that SMCS presents a clear, accurate and all-round real-time representation of the tactical picture at all times. The system allows new technology to be incorporated without disturbing the complex application software and now uses commercial-off-the-shelf (COTS) hardware and software. The system is flexible enough to be fitted in any submarine type, diesel or nuclear.

SMCS is a high-capacity open architecture distributed processing command system based on a dual-redundant fibre optic Local Area Network (LAN). It interfaces to the submarine's sensors and tactical weapons, providing the control of the weapon discharge systems and functions to support the full range of combat system management tasks. The functions include: Track Motion Analysis (TMA); tactical picture compilation; oceanographic data analysis; weapon management; weapon command tactical aids; onboard training; data recording and replay.

The capacity of SMCS for handling and storing data is claimed to be considerably greater than that of any current system. Many simultaneous TMA solutions can be handled. An important consideration in the functional design has been to automate as much of the routine of track management as possible, leaving operators free to concentrate on the more critical aspects.

Operator interface is by high-resolution colour graphics displays at a number of MultiFunction Consoles (MFCs), with interactive plasma panels and pucks (similar to a mouse). Secondary workstations are also available where space is particularly limited.

SMCS technology is based on COTS components, ruggedised to suit the submarine environment. The use of commercial standards ensures widespread and long-term availability and support for the key components, as well as ease of upgrade. The main processor is the Intel 80386/486, with the possibility of upgrading as this family of processors is enhanced, while fast processing is performed by Inmos transputers, which can also be upgraded. The system architecture allows this substantial processing power to be easily increased, if necessary, to accommodate future requirements and there is sufficient spare physical capacity to do so.

SMCS software is written in Ada, while Occam is used for the transputers. This approach yields a highly modular system, which facilitates enhancements.

Submarine Command System (SMCS)

Fundamental to the SMCS architecture is the high-bandwidth, dual-redundant, fibre optic LAN based on the IEEE 802.5 token passing protocol, with built-in self-healing. To the LAN are connected the processing nodes and the MFCs, which also contain processing capability.

An important feature of SMCS is its inherent resilience to failure. All processing nodes are duplicated, with one in use and the other on standby, and the MFCs are identical. The standby nodes are automatically switched into use in the event of a failure being detected in the other.

A number of key features of SMCS include:
(a) a high degree of flexibility and expandability
(b) a substantial increase in usable processing power
(c) improved operability, to ease the growing burden on the crew in the modern submarine environment
(d) improved reliability to provide a 'non-stop' system
(e) the capability to match envisaged and likely improvements in weapons and sensor technology.

The flexibility of the basic system architecture has been demonstrated, not only by the different submarine fits being produced, but also by the adoption of successor-based systems for the Royal Navy's surface ships, where the detailed operational requirements are very different.

Versions of SMCS can be readily produced to suit almost any type of submarine, to form the key central element of the combat system.

Status
The system is fully operational in the Royal Navy's 'Vanguard', 'Trafalgar' and 'Swiftsure' classes of submarine. The latest variant, termed SMCS NG (Next Generation), is due for installation in HMS *Torbay* early in 2004 as a replacement for the existing SMCS equipment.

Contractor
AMS, New Malden, Surrey.

UPDATED

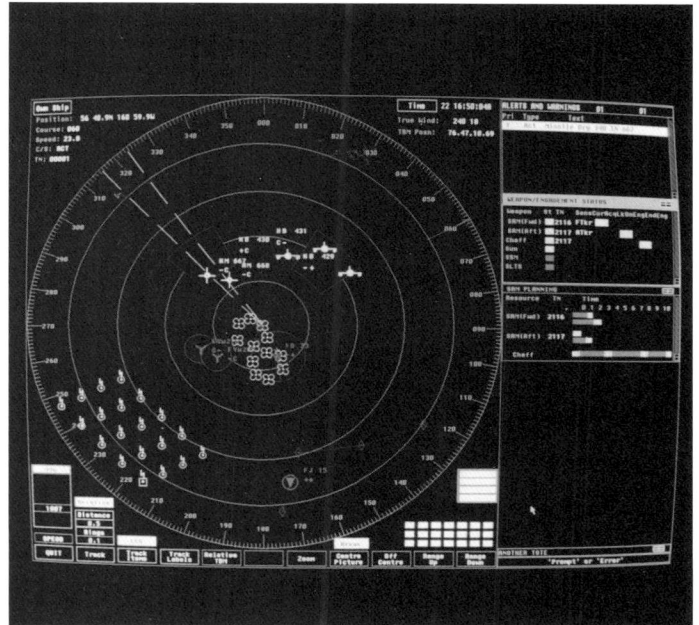
SSCS display screen

Surface Ship Command System (SSCS)

The Surface Ship Command System (SSCS) or DNA(I) is a powerful and responsive command system which provides standard functions to support the command team, together with facilities for the prosecution of specific maritime warfare functions in the ASW, ASuW and AAW environments.

The following functions are a selection of those provided as standard in SSCS:
(a) navigation support – Blind Pilotage (BP), including preparation of BP plans, collision avoidance, intercept solutions, 'own ship' positioning, selection and conversion of chart projections, spheroids and datums, presentation of 'own ship' data
(b) tactical picture compilation – sensor management and co-ordination of 'own ship' and force surface and subsurface sensors, sensor data management, datalink functions, track management with a very large track database, air, surface and subsurface picture management
(c) situation assessment and planning – replay analysis, quick look analysis of weapons system operations, target motion analysis, threat assessment, command planning, decision aids, support to tactical doctrine and rules of engagement
(d) platform and weapon system direction and control – weapon system status monitoring, weapon assignment, target indication and engagement, control of hard and soft kill weapons, 'own ship' manoeuvring and evasive steering, helicopter direction and control
(e) training – the ability to create, run, record and modify complex scenario exercises for individual operators, command subteams or the whole command team with combinations of real, prerecorded and transient exercise data
(f) a human/computer interface which provides maximum ease of use for operators, assisting the command team to manage the combat system effectively.

The principal components of SSCS comprise: MultiFunction Consoles (MFCs), Local Area Network (LAN), Input/Output Nodes (IONs), Signal Switching Unit (SSU), Radar Video Distribution Unit (RVDU) and peripheral devices.

MFCs provide each member of the command team with operator facilities which include a high-resolution colour display. Every MFC contains a complete consistent

SSCS on board HMS Westminster

set of processed data from which the selected tactical picture is presented. An operator can locally select a role and can carry out any command system task. A fibre optic LAN connects all MFCs to two dual-redundant IONs and the RVDU. The RVDU provides interfaces to radar videos and distributes them in such a way that any operator at any MFC can independently select any radar video (or CCTV) for display. Each ION provides general processing and interfaces to the Combat System Highway (CSH) for data exchange with most combat system equipments. The IONs link to non-CSH combat system equipments via the SSU. The SSU is adaptable to offer a variety of interface protocols and thus link a wide variety of combat system equipments into the dual-redundant SSCS system structure. Peripheral devices include a maintainer's facility, a bridge display, a printer, and a contact evaluation plotter

The design of SSCS ensures that the system will recover automatically in the event of a single-point equipment failure, including temporary loss of the system main power supply. During recovery, system data is preserved and continuity of service of command system functions is assured.

SSCS-21
A variant of SSCS, SSCS-21, has been developed as a Private Venture (PV). Three principal PV outputs have been fed into SSCS-21: Common Infrastructure on UNIX (CI/UX), which enables more cost-effective implementation of existing applications and infrastructure software in a UNIX-operating environment; the use of a PC-based display-drive using the Windows NT operating system; and the development of human computer interface software for fire-control applications. The standard SSCS uses over 200 32-bit processors, including 80386,80486 and Pentium cards and T800 transputers. SSCS-21 uses fewer engine-processor boards for applications execution. As a result, the number of processor card sets is rationalised. Whereas the current SSCS server nodes need 12 cards of nine separate types, the equivalent SSCS-21 server needs four cards of two types. Only one engine/processor card is now needed for applications processing, compared to 10 cards of eight separate types.

Similar rationalisation is applied to display processing: Motorola 68040 and bespoke Racal cards are replaced by an off-the-shelf PrimaGraphics card set, reducing the number of display processors from 10 to 5. Three console types for SSCS-21 are envisaged: a double-headed operator station with two flat-screen displays and foldaway keyboard; a twin-screen side-by-side console; and a portable notebook configuration for 'plug-in' applications. All are baselined with the NEC 48 cm flat-screen colour display.

The existing software architecture has been re-engineered to allow transition to an open environment based on UNIX and Windows NT. The CI/UX infrastructure sits above UNIX: a bridge layer allows third-party applications running on UNIX to come across to CI/UX.

Status
SSCS is operational in all Royal Navy 'Type 23' frigates and a derivant is fitted to the Republic of Korea Navy's 'Okpo' (KDX-I) and KDX-II class destroyers. In October 2002 it was announced that AMS had been awarded a contract in excess of £35 million for three systems for KDX-II Batch 2, the first hardware to be delivered in March 2004 with deliveries spread over 40 months.

SSCS now provides the baseline architecture and functionality for the Type 45 command system. Area defence functionality will be ported from the existing ADAWS system (see separate entry) to provide an overall evolutionary low risk approach for the new vessels.

Contractor
AMS, New Malden, Surrey.

UPDATED

WSA-420 and CAAIS 450 series fire-control and action information systems

These systems are based on and incorporate many of the proven features of WSA-4, WSA-400 series and CAAIS now in service with the Royal Navy and other navies.

The system variants are:
(a) WSA-421 – fire-control system (receiving target indication from an external source)
(b) WSA-422 – fire-control system with limited AIO facilities
(c) WSA-423 – integrated fire-control and AIO system
(d) CAAIS 450 Series – AIO systems.

Display and processor upgrades for these systems are available using technology developed for ADAWS 2000 (see separate entry). The CRT displays can be replaced by flat panel colour displays, with an associated electronics rack. A commercial-off-the-shelf (COTS) processor retaining the proven CAAIS software and providing room for expansion can replace the FM1600E processor. All systems can include onboard training aids, such as radar echo generators and simulated weapon tell-backs.

WSA-421 fire-control system
This is designed primarily for frigates and corvettes fitted with an independent, but linked, AIO/C1C system such as the CAAIS 450. Two tracking heads provide the WSA-421 with two independent channels of fire. The fire-control prediction is performed by a Ferranti FM1600E computer. The system accepts target indication from the ship's AIO system.

WSA-422 fire-control system with limited AIO
Designed primarily for vessels of patrol craft size to provide weapon control and surveillance tracking for deployment of 35 to 127 mm calibre guns and/or surface-to-air missiles and optionally, surface-to-surface missiles, the system uses the FM1600E computer.

The main sensor is the ship's surveillance and/or navigation radar. A tracker radar and OFD which can mount a laser range-finder in addition to the EO or CCTV is also fitted, thus providing two independent channels of fire and enabling two gun mountings to be controlled. Although the tracker radar will normally be paired with the main gun, the flexibility of the system enables any sensor-target pairing to be achieved. The composite weapon system console is manned normally by a single operator, with provision for up to three operators in active conditions and houses all displays, indicators and controls for surveillance tracking, target designation, control of weapon sensors and of the armament. Manual rate-aided tracking, with computer generated symbology, is available on up to 60 targets. The power of the FM1600E computer is such that additional facilities maybe provided if required, such as a remote-controlled, stabilised video tracker, comprising CCTV and laser range-finder, to assist tracking in poor ECM environments.

WSA-423 integrated fire-control and AIO system
This system is designed for vessels of patrol craft and corvette size to provide weapon control and AIO/CIC facilities for the deployment of guns (35 to 130 mm), surface-to-air and surface-to-surface missiles.

WSA-423 is an extension of the WSA-422 system with added AIO facilities to provide an integrated system.

CAAIS 450 Series AIO systems
This series is designed to provide AIO facilities for a range of vessels of FPB size or larger. Provision is made for passing target indication data to associated fire-control systems, which could be a subset of WSA-422.

CAAIS 450 can accept tactical data from two surveillance radars, from sonar and ESM systems. Auto-tracking of targets is provided on both radars. Although the FM1600E computer's capability includes driving up to eight, 305 or 406 mm (12 or 16 in) LRDs, the exact display fit will depend upon the particular application. A basic configuration would have three displays: two for picture compilation and one for command appreciation. Additional display requirements can be envisaged for

Ferranti combat system WSA-422 system console

such functions as ASW, helicopter control, ESM/ECM management or datalink management. The software provided is generally derived from the CAAIS programs. However, due to the versatility of this configuration, which results from the use of a powerful computer, specific applications may require that some special software is needed so that the full capability of associated weapons and sensors can be realised.

Status
WSA-420, CAAIS 450 and WSA-423 series systems are in service in the navies of Brazil, Ecuador, Egypt, Greece, India, Kenya, S Korea and UK. No longer in production. Some CAAIS systems are currently being updated with commercial-off-the-shelf (COTS) processors and colour flat-panel displays.

Contractor
AMS, Camberley, Surrey.

UPDATED

United States

AN/USQ-123(V) Common Data Link-Shipboard Terminal (CDL-N)

The Common Data Link-Shipboard Terminal (CDL-N), AN/USQ-123 (V), is the multifunction shipboard datalink terminal installed aboard aircraft carriers and other amphibious vessels to support reconnaissance/surveillance missions. CDL-N is an automated communications node that acquires datalinked signals from airborne reconnaissance vehicles (manned or unmanned) and distributes the received signals to the appropriate shipboard intelligence user for processing/exploitation. Once signals are acquired, CDL-N automatically tracks the aircraft-to-ship signal through two 1 m dish antennas located fore and aft.

From its location in the ship's detection and track room, CDL-N receives data over the Airborne Common Data Link from sensors providing digital imagery, SIGnal INTelligence (SIGINT), Infra-Red (IR) and radar. It supports the Advanced Tactical Airborne Reconnaissance System (ATARS) and Battle Group Passive Horizon Extension System (BGPHES) used by a Carrier Battle Group (CVBG). These missions provide a CVBG with complete surveillance reconnaissance coverage within a 300 n mile radius.

ATARS is a multisensor reconnaissance package carried by the F/A-18 aircraft. The sensor package consists of medium- and wide-field electro-optical sensors and an IR sensor. The Airborne Common DataLink transmits imagery from the F/A-18 to the carrier through the CDL-N to the Joint Services Image Processing System-Navy (JSIPS-N) at the rate of 137 Mbs.

The BGPHES is a passive Electronic Surveillance Measures (ESM) system carried by an ES-3A aircraft. The airborne Common DataLink transmits ESM sensor data through the CDL-N to the BGPHES surface terminal at the rate of 10.71 Mbyte/s.

As CDL-N operates at X- and Ku-bands and at multiple data rates (10.71, 137, and 274 Mbyte/s), it is capable of supporting missions and platforms equipped with compatible airborne datalinks. Aerial platforms in the conceptual or development stages that will be compatible with CDL-N include the Maritime UAV, the Endurance UAV systems, and the SH-60R LAMPS BK II helicopter. These have imagery, SIGINT and radar sensor packages and transmit data at 10.71, 137, to 274 Mbyte/s to CDL-N. CDL-N handles one full-duplex ATARS or BGPHES datalink at a time, but can be expanded to two independent full-duplex links supporting simultaneous missions.

The CDL-N is equipped with the necessary command channels that provide all executive functions, voice communication (CVSD-Continuous Variable Slope Delta modulation), and 10 user channels for prime mission equipment. The CDL-N provides up to 25 return link channels with individual rates of 50 kbytes/s to 42 Mbyte/s.

In CDL-N, the operator establishes link configuration, performs mission monitoring and does maintenance and diagnostics via software control.

Specifications
(a) Frequency: X- and Ku-bands
(b) Uplink rate: 200 kbytes/s
(c) Downlink rates:
    10.71 Mbytes/s
    137 Mbytes/s
    274 Mbytes/s
(d) Modulation:
    uplink BPSK
    downlink O-QPSK
(e) Bit error rate: 1×10^{-6} without encryption
(f) Forward error correction coding
(g) Variable-depth data interleaving
(h) Jam resistance: direct sequence spread spectrum command link at 200 Kb/s
(i) Control, status and data interfaces for encryption devices
(j) Command channels (with encryption)
(k) Executive functions:
    voice - CVSD
    10 prime mission channels

(l) Return channels:
low rate (10.71 Mbytes/s)
high rate (274/137 Mbytes/s)
voice (CVSD)
up to 25 user mission channels
(m) Online fault indication:
single drawer 90 per cent
two drawer 95 per cent
three drawer 97 per cent
(n) Expandable to four independent and simultaneous full-duplex links.

Status

In operational use by the US Navy

Manufacturer

L-3 Communications Systems-West, Salt Lake City, Utah.

UPDATED

Area Air Defense Commander Capability system

The AN/UYQ-89 Area Air Defense Commander (AADC) Capability system was developed as a proof of concept by the Johns Hopkins University's Applied Physics Laboratory. General Dynamics Advanced Information Systems (GD AIS) is responsible for its production and further evolution. It is a network-based C2 system that provides a three-dimensional display to enhance situational awareness. It allows a joint staff to respond quickly to changes in an enemy's order of battle (EOB) and course of action, as well as the friendly EOB and the prioritised defended-asset list. The AADC Capability system can generate a new air-defence plan within minutes, using fewer personnel (typically 20 to 30) than at present.

The Area Air Defence Commander (that is the individual fulfilling that role) is responsible for planning, executing, and co-ordinating air and missile defence operations in an integrated air-defence environment theatrewide. The AADC Capability system is a battlespace management system designed to improve the AAD Command's battlefield readiness by analysing the capabilities and intentions of enemy ballistic missile and air forces, comparing them with allied assets in the theatre, and creating 'what-if' scenarios to assist in the rapid development of air defence plans.

The system is designed to display a three-dimensional, graphically rich battlespace, with clearly discernible friendly air defence assets and enemy ballistic missiles, land attack weapons, and air fighters. Information regarding the enemy's assets, the constraints of his weaponry, and likely courses of action are continuously analysed and paired against the capabilities of friendly forces. This enables the rapid development of air defence plans, and allows timely response to new and changing enemy courses of action. The objective is to improve radically battlefield response times and automate a laborious process that was previously carried out manually, allowing plans to be created in minutes and then distributed once approved.

The system allows commanders to monitor the action in real time, using wide-screen, high-definition displays that show the battlespace three-dimensionally. Personnel can quickly engage in real-time fly-throughs over and around battlefield assets to examine the space from any angle. Both the planning and the current operations displays can be viewed simultaneously, giving commanders the ability to alter a friendly force air defence laydown and then quickly assess the operational effect of that change.

The AADC system is supported by a wide range of SGI computing and visualisation products. Each installation includes a 32-processor SGI Origin® 3400 server, four SGI Onyx® 3200 visualisation systems (including one for system redundancy), eight Silicon Graphics Octane2™ visualisation workstations, and one Silicon Graphics 02+™ graphics workstation. In addition to the display technology, videoconferencing and e-mail capabilities are built-in, enabling other commanders to be easily included in the planning process. The SGI AADC technology displays objects in the theatre as they really are. The use of visual representations enhances situational awareness, particularly useful when participants are under stress and time constraints. Realistic colour-coded icons are universally recognisable, allowing for rapid grasp of the operational situation. On the displays and large 'reality screens' aircraft look like aircraft, and the friendly can easily be distinguished from the enemy. The AADC allows the user to see the engagements occurring, and the system will present him with a schedule of engagements that he can execute or not in a complex battlespace.

Status

The AADC technology developed by JHU Applied Physics Lab was initially tested in 1998 during Fleet Battle Experiment Charlie and in the Theatre Missile Defence Initiative. The system was then used as part of a full-scale, multinational fleet exercise during RIMPAC 2000. Conducted off the Hawaiian coast, the AADC Capability was installed in USS *Shiloh*, which was acting as the anti-air warfare commander for the USS *Abraham Lincoln* carrier battle group. In July 2000, the US Navy awarded GD AIS a contract to produce an AADC Capability Engineering Development Model. However, positive input from users in the fleet and the immediate requirement to address threats to US forces and homeland defence led the navy to decide to field more rapidly the AADC prototype system and in October 2001 the service redirected its programme to permit rapid fielding during FY02 and to implement a preplanned product-improvement programme. As a result, General

Dynamics had to make a transition rapidly from a prototype to a production system and deploy it on an accelerated basis.

AADC has been installed and fielded in the command ships USS *Mount Whitney* and *Blue Ridge* as well as in USS *Shiloh*. In the latter platform the system was first deployed operationally with the *Abraham Lincoln* Battlegroup in July 2002. At least 17 other systems are to be procured by 2007; recipients probably include other 'Ticonderoga' class vessels as part of the Cruiser Conversion Programme, the Fifth Fleet command centre in Bahrain, and other land-based sites. In December 2002 GD AIS was awarded an initial US$21 million contract for low-rate initial production and engineering support.

Contractor

General Dynamics Advanced Information Systems (Prime), Arlington, Virginia.
SGI (Silicon Graphics International), Mountain View, California.

NEW ENTRY

Combat Control System Mk 2

The Mk 2 Combat Control System provides comprehensive processing and evaluation of all combat data and co-ordinated operational control of all sensors, weapons and combat manoeuvres. The system's architecture has resulted in lower costs, higher reliability and easier operation and maintenance than its predecessors.

The system provides for surveillance of a large number of contacts and the simultaneous engagement of multiple threats. The surveillance process is highly automated due to the large amount of navigation, contact tracking, communication link and weapon telemetry surveillance data that must be processed and evaluated. The results of the automatic processing are presented to system operators in user-friendly pictorial and natural language forms. The highly automated approach frees the operators from perfunctory tasks and allows them to manage by exception. This results in a reduction in the number of operators required for surveillance and allows the assignment of additional operators to the prosecution of contacts which are a threat to the ship's mission. The contact prosecution process, while operator intensive, includes the display of tactical options and recommendations to aid operators in attacking threats using torpedoes and missiles. In addition, the system provides support for performance monitoring and fault localisation, combat data recording and reconstruction and operation and maintenance training.

Central to the system is a set of distributed computers which are embedded in each of the hardware elements. The computer design is modular and therefore expandable by inserting additional processing and memory cards. The system comprises the following hardware elements: a sensor data converter/processor with two computers, multiple combat control consoles with one computer each and a weapon data converter/processor with two computers, all interconnected by a dual-combat databus. The system hardware is completely redundant and can therefore sustain single failures of any type without any loss of tactical capability. The equipment is very compact and will fit down a 650 mm hatch.

The sensor data converter/processor provides:
(a) interfaces with the navigation sensors, the tracking sensors (including passive/intercept/active sonar, radar, electronic support measures, TV and periscopes) and the communication link
(b) automatic surveillance and support processing
(c) combat data recording.

Combat control consoles provide a multifunction display of sensor video and processor generated graphics and alphanumerics. The main display surface is a 45 Hz non-interlaced raster scan colour CRT with 1,280 × 1,024 picture element resolution. Each console independently provides for the simultaneous display of video from any two sensor sources plus the overlay of graphic and alphanumeric annotation. Radar, navigation charts and annotation may be combined for use in combat navigation. The console contains a 430 × 220 mm (17 × 8.5 in) touch-interactive plasma display which is used as a tactical summary display and for interaction with the weapon launch compartment personnel during preparation for a launch.

The weapon data converter/processor provides:
(a) automatic contact prosecution and support processing
(b) interfaces with the launch tubes and weapons
(c) emergency weapon setting, launching and control.

A plasma display like the one in the combat control console is included and is used for interaction with the console operators during preparation for a normal launch and for local control during emergency operations.

The combat databus provides communication between the other elements of the system at a 1 Mbit/s serial rate over each half of a dual bus in the MIL-STD-1553B format. Data is broadcast on the bus from each computer to all other computers simultaneously, so that all computers receive the data at the same time. Each half of the bus has the capacity to provide all of the required system communication. The bus is available in both copper and fibre optic forms.

The sensor and weapon data converter/processor performance monitoring tasks combine to provide a double check on system failures and in conjunction initiate automatic switching to redundant hardware/software in the event of a single point of failure. Switching processors is simplified as program loading and database updates are not required. Also, switching halves of the databus is simplified by switching all communication from one half to the other in the event of any failure on the half that is providing system communication.

Status

The latest installation of CCS Mk 2 is a three-phase programme for transforming various existing legacy submarine combat systems (BSY-1, CCS Mk 1, CCS Mk 2 and DWS-118) to a common, more capable and flexible COTS/Open System Architecture (OSA). The use of COTS/OSA technologies and systems will enable rapid periodic updates to be made to both software and hardware. COTS-based processors will allow computer power growth at a rate commensurate with the commercial industry.

Phase I (CCS Mk 2 Block 1C - FY 2000) introduced automated Strike engagement planning capability (ATWCS) and 'Virginia' class data distribution and services. Phase II (CCS Mk 2 Block 1C ECP-4 - FY 2002) introduces advanced weapons improvements and processing with the installation of 'Virginia' class equivalent COTS processor, replacing the existing UYK-43 computer with COTS hardware and supporting the introduction of co-ordinated strike warfare (Tactical Tomahawk missile and weapon control system), improved anti-diesel littoral torpedo (ADCAP CBASS), and improved mining (ISLMM) capabilities. Phase III (CCS Mk 2 Block 1C ECP-5 - FY 2007) installs 'Virginia' class weapons-launch improvements and provides an at-sea, end-to-end launcher testing capability. The first phase of Mk 2 Block 1C is complete. The first installation of Phase 2 was completed in FY 2002.

In July 2003 it was announced that the Royal Australian Navy had awarded a US$32.4 million contract for 5 CCS Mk2 as replacement command systems for the 'Collins' class submarines.

Contractor

Raytheon Integrated Defense Systems, Portsmouth, Rhode Island.

UPDATED

Global Command and Control System - Maritime (GCCS-M)

GCCS-M has evolved from the Joint Maritime CIS as part of the development of the C4I for the Warrior concept within the Department of Defense Common Operating Environment. It is the US Navy's primary strategic Command and Control System. GCCS-M enhances combat capability and aids in the decision-making process by the retrieval, processing, management and display of data on the status of neutral, friendly and hostile forces, in order to execute the full range of Navy missions in near-realtime. Data is exchanged via external communication channels, Local Area Networks (LANs) and direct interfaces with other systems.

The GCCS-M system is comprised of four main variants, Ashore, Afloat, Tactical/Mobile and Multi-Level Security (MLS) that together provide command and control information to combatants in all naval environments.

Ashore variant

The Ashore variant provides a single, integrated C4I capability to land-based forces in support of the combat requirements of the Chief of Naval Operations (CNO), Fleet Commanders in Chief (FLTCINCs) and Navy supported Unified CINCs (CINCUSPACOM and CINCUSAJFCOM). The Ashore variant provides near-realtime weapons targeting data to submarines, supports real-time tasking of Maritime Patrol Aircraft (MPA) assets, and supports the force scheduling requirements of the Navy (from CNO to the squadron level).

Ashore software has been made available to Joint Task Force (JTF) command centres, aircraft carriers and other fleet flagships, and selected Army, Air Force and Marine Corps Component Commander command centres. In addition, it has been provided to NATO Maritime command centres (as the core of the North Atlantic Command and Control Information System (NACCIS)), US Coast Guard, and Military Sealift Command (MSC) command centres.

The Ashore variant's hardware independence and scalability allows it to be used as the primary C4I support system, from the largest command centre to a single workstation within a mobile command centre. The Ashore variant operates in compliance with the DII COE, which ensures it is fully interoperable with all modern US Joint and Naval C4I systems. Partnership in the NACCIS project ensures continued interoperability with current and future NATO and Allied C4I systems.

Afloat variant

The Afloat variant provides a single C4I capability to sea-based forces. It supports the C2I mission requirements of:
(a) Commander Joint Task Force (CJTF)
(b) Joint Force Navy Component Commander
(c) Joint Force Air Component Commander (JFACC)
(d) Numbered Fleet Commanders
(e) Officer-in-Tactical Command/Composite Warfare Commander (OTC/CWC)
(f) Commander Amphibious Task Force (CATF)
(g) Commander Landing Force (CLF)
(h) Ship's Commanding Officer/Tactical Action Officer (CO/TAO).

The Afloat programme is a phased, evolutionary acquisition programme incorporating the functionality of many systems within the COE. It functions in a networked, client/server architecture featuring standard commercial hardware components and software applications. Software components are comprised of core service modules, linked with mission applications through Application Program Interfaces (APIs).

Tactical / Mobile Variants (TMV)

GCCS-M Tactical/Mobile Systems include both fixed sites – Tactical Support Centres (TSCs) and Tactical Mobile Variants (TMVs) – Mobile Operation Command

and Control centres (MOCCs), Mobile Ashore Support Terminals (MASTs) and Mobile Integrated Command Facilities (MICFACs). These sites provide the capability to plan, direct and control the tactical operations of Joint and Naval Expeditionary Forces (NEFs) and other assigned units.

TSCs are fixed-site C4I systems that will evolve to the GCCS-M architecture based upon NT PCs. Evolution will be in compliance with the DII COE, air-ground, satellite and point-to-point communications systems, Wide Area Network (WAN) capabilities, sensor analysis capabilities, avionics and weapons system interfaces, and facilities equipment.

MOCC is a rapidly deployable, self-contained, C4I system that can be transported in two fleet-configured P-3 aircraft for contingency operations. MAST and MICFAC are miniaturised mobile facilities designed to support a theatre commander or naval liaison element ashore. MAST provides a basic C4I capability for rapid deployment to remote locations. The MICFAC is a robust C4I system that is deployable and can support a commander's staff ashore. MASTs and MICFACs were expected to be upgraded to a common architecture and redesignated as Joint Mobile Ashore Support Terminals (JMASTs) beginning in FY02.

TSCs, MOCCs, MASTs and MICFACs are interoperable with other GCCS-M platforms, joint, NATO and allied.

MultiLevel Security (MLS)/ OED / RADIANT MERCURY

GCCS-M MultiLevel Security (MLS) enables operators in a joint/coalition environment to access, retrieve, process, and disseminate all necessary information for maintenance of a consistent Common Operating Picture (COP). MLS will use existing communication networks, integrated with current (OED and RADIANT MERCURY) and future technologies into the GCCS-M SCI architecture.

Capabilities

GCCS-M provides the following capabilities:

Core

(a) provides a single integrated Command, Control, Communications, Computers and Intelligence (C4I) system that receives, processes, displays, maintains and assesses the unit characteristics, employment scheduling, material condition, combat readiness, war fighting capabilities, positional information and disposition of own and allied forces
(b) provides afloat tactical commanders with a timely, authoritative, fused and common tactical picture with integrated intelligence services and databases
(c) plans, directs, and controls the tactical operations of forces under the operational commander's control.

Briefing support/Office automation

(a) provides a UNIX- and NT-based, multiscreen and multinode briefing display and control application; enables integration of office automation
(b) provides Unix- and NT-based Applications for building Maritime Patrol Craft (MPA) briefs for US Navy and Allies in STANAG 4283 format
(c) prepares, displays and prints briefing text.

Database

(a) maintains an authoritative history of information in a relational database format
(b) provides database maintenance capabilities for TSC segments. Provides the capability to extract data from received United States Message Text Format (USMTF) messages and populate various databases.

LAN/WAN

(a) displays GCCS-M segments that are presently loaded on each workstation; provides a user-oriented network monitor
(b) provides tools for anchor desk capabilities through a WAN.

Mission operations

(a) provides contact location data and precise Over-the-Horizon Targeting (OTH-T) data to submarines equipped with TOMAHAWK Missile variants
(b) provides combat capabilities for surface, air, and subsurface platforms. Provides those forces with and integrates the waterspace picture for timely asset management
(c) detects and displays threat information for warfare commanders embarked on GCCS-M-equipped platforms
(d) correlates, maintains and analyses tracks
(e) analyse tactical platform sensor data for dissemination to other fleet units
(f) correlates single link attributes or single emitter ELINT tracks
(g) provides a graphical post-ASW mission replay capability.

Mission planning

(a) provides Navy Command and Control Systems (NCCS) Ashore units as well as units afloat with the ability to acquire, analyse, control, and disseminate pertinent ASW mission planning data
(b) provide safety of flight planning for MPA to/from operational areas and for co-ordinating MPA turnover on station.

Imagery

(a) supports the full range of imagery requirements including viewing, mensuration, transmission/receipt and output
(b) supports near-realtime receipts and transmission of tactical imagery data to/from Anti-surface warfare Improvement programme (AIP) aircraft.

Intelligence

Provides comprehensive military intelligence data and message applications.

Logistics
Provides capability to assess short-term operational sustainability requirements for naval forces afloat.

Meteorological/oceanographic
Provides applications and tools to process environmental data received from METOC production or regional centres.

Communications
Provides continuous C4I/SR services to assigned US and Allied Maritime Patrol Aircraft (MPA), special mission aircraft and other ASW forces operating independently or as part of a Battle Group.

JMCIS was initially implemented on high-performance UNIX workstations, but the increase in processing capability of the Intel PC processor family and the maturity of the Windows NT and JAVA/Web multi-user operating systems allows considerable cost savings and technical advantages. Of the approximately 250 JMCIS segments, about 160 have migrated to new platforms in GCCS-M development.

GCCS-M uses common communications media. The Secure Internet Protocol Router NETwork (SIPRNET), Non-secure Internet Protocol Router NETwork (NIPRNET) and the Joint Worldwide Intelligence Communication System (JWICS) provide the necessary WAN connectivity. JMCOMS will provide the WAN connectivity for the Afloat and Tactical/Mobile GCCS-M systems.

GCCS-M expanded functionality
The evolution of the maritime command information system will continue beyond the migration of JMCIS to the PC environment. While GCCS-M will provide the same functionality as JMCIS 2.2, it will also support new features:
(a) integrated profiler capability
(b) Combat Direction System (CDS) Combat Track Association (CTA) with MIDB 2.0 (JULIET) Order Of Battle (OOB).
(c) JAVA Image and Video Exploitation (JIVE)
(d) Joint Message Handling System (JMHS) PC features, including flat file UNIX/NT interface.
(e) faster and improved security features
(f) improved track correlation
(g) extension of system to NT/PC environment allowing user to operate tactical and non-tactical standard applications
(h) enhanced message processing capability
(i) web-based interface to Naval Status of Forces (NSOF) data and Chief of Naval Operations' Consolidated History File (CHF).

Status
GCCS-M Afloat is installed on more than 300 ships and submarines throughout the US Navy. GCCS-M Ashore has been installed at 80 sites including the Office of the Chief of Naval Operations; five Fleet CinC headquarters; Keflavik, Iceland; two Unified CinCs (USAJFCOM and USPACOM); four Fleet High Level Terminal (FHLT) sites; and four Submarine Tactical Terminal (STT) sites. Conversion to GCCS-M continues.

Four GCCS-M Tactical/Mobile, 14 TSCs and eight Mobile Operational Command Control Centers (MOCCs) are operational worldwide. Four MAST Units are operational in support of Mobile Inshore Undersea Warfare (MIUW) units and four MICFACs are operational in support of Fleet CinCs and numbered fleet commanders. The process of upgrading and redesignation to JMAST is in progress.

The governments of Australia, Canada, France, Germany, Italy, Japan, South Korea, Netherlands, New Zealand, Saudi Arabia, Spain and the UK have entered into agreements with the US government to purchase GCCS-M variants. GCCS-M also forms the basis for the MCCIS (see NATO strategic systems entry) and is therefore available to all NATO countries.

UPDATED

Litton Naval Tactical Data System (LNTDS)

The Litton Naval Tactical Data System (LNTDS) is designed to improve naval and coastal defence through real-time exchange of tactical information among various land-based and shipboard sensors, co-ordinated at a shore-based command centre, to bring the most effective weapons to bear on enemy forces. The system collects, analyses, evaluates and displays tactical information from radar sites, combatant ships, patrol aircraft and the national Air Defence Ground Environment (ADGE) system to assist the commander in making decisions. The LNTDS also exchanges evaluated track information with participating naval units and provides a medium for ordering or commanding naval forces and tactical air and weapon systems in accomplishing the mission. The LNTDS will respond to any seaward attack prosecuted by aircraft, surface ships, or submarines, or a combination of all three.

In South Korea, LNTDS uses Link 11 to connect three *Ulsan* class frigates (each having one console) to a shore headquarters, the Fleet Command Centre (FCC), that includes seven consoles. The frigates lead groups of smaller craft in carrying out missions to detect and intercept ships trying to penetrate South Korean coastal waters, particularly enemy infiltrators and intelligence gatherers. Five shore radars feed the FCC over two links – Link 11 and the Inter-Site DataLink (ISDL). The FCC is also connected to the main air control centre.

Status
LNTDS has been operational in South Korea since 1995.

Contractor
Northrop Grumman Electronic Systems (Navigation Systems Division), Woodland Hills, California.

UPDATED

Model RWS-901 high-shock-resistant ruggedised VME colour workstation

The Model RWS-901 rack-based HP743i/744i colour VME workstation has been supplied to the US Navy for use in command and control centres. The compact, general purpose, rack-mounted VME RWS-901 workstation is qualified for MIL-STD-901D, Grade A, Class 1 shock. It is an NDI/COTS product designed for the demanding requirements of shock, vibration and EMI/EMC and offers a solution for command and control applications deployed in severe environmental conditions.

The VME card chassis is designed to mount on slides within a standard 19 in equipment rack. Within the chassis is a shock and vibration isolated subchassis holding a 14 slot VME (6U × 166 mm) card cage, four half-height peripheral slots and two 350 W power supplies. The power supply outputs may be paralleled to power a single backplane, or may be connected to separately power two independent backplanes. With the supplies connected separately, one or both can provide UPS capability. The VME card cage is accessible from the top of the chassis and the four slot half-height peripheral bay is accessible via the hinged front door facilitating maintainability. The I/O panel is removable and is easily punched and marked for specific applications. The VME card chassis accepts any industry standard processor and card set and it can be provided with any industry standard half-height peripheral including hard drives from 2-9 Gbytes, mounted in rugged removable canisters.

The CRT Monitor, R9010R/20, provides high-resolution colour display capabilities, uses industry standard video inputs and features automatic synchronisation from VGA to 1600 × 1280. Its microprocessor stores up to 40 programmable formats to automatically size and centre the presentation. Aydin is also under contract to supply its new ruggedised 51 cm (20 in) colour Active Matrix Liquid Crystal Display (AMLCD) as the display in its RWS-901 for a US Navy shipboard application.

The KeyBoard/Trackball Unit (KBTU) is a standard qwerty 121-key keyboard with integral three button trackball. The keyboard is fully ruggedised and all electronics, switches and connectors are sealed against dust, liquids and general contamination. The KBTU can be supplied as a Sun Type 5 compatible or as an IBM PC AT PS/2 compatible.

Status
In use in the US Navy in both surface ships and submarines.

Contractor
Aydin Corporation, Birdsboro, Pennsylvania.

UPDATED

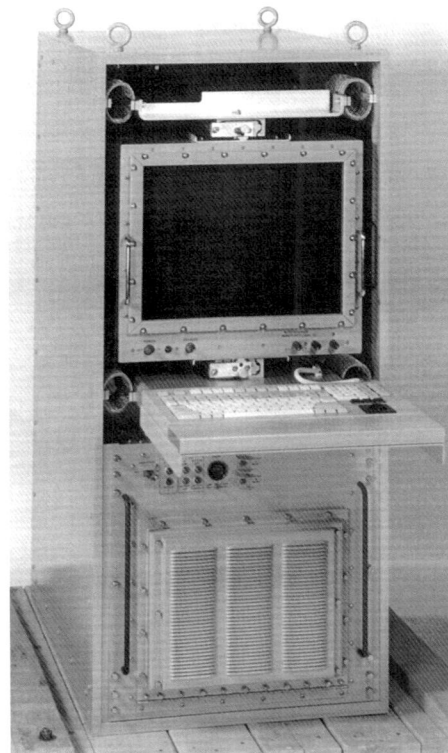

Model RWS-901 workstation
0009942

Tactical Command Management System (TCOMS)

The Tactical Command Management System (TCOMS) is a multi-operator, data link buffer, translation and command management system. By design, TCOMS can be part of the information grid for Network-Centric Warfare. TCOMS functionality is similar to that provided by the Advanced Tactical Data Link System (ATDLS) and the Command and Control Processor (C2P). It provides all branches of the military with a real-time air surveillance data exchange capability in order to optimise tactical control and overall force co-ordination. TCOMS also provides mutual early warning of air, surface and subsurface activities.

TCOMS is capable of transmitting, receiving, cross-telling and forwarding data between Link 16, Link 11, multiple Link 11B channels, Link 1 and Link 14. TCOMS is expandable to Link 22 and other data links (for example, VMF, ATDL-1, OTH-T Rainform Gold). TCOMS is a scalable, modular, and networked architecture of state-of-the-technology commercial products that is engineered for growth, and can be operated from one or more consoles. The Commercial-Off-The-Shelf (COTS) hardware (with an option for TEMPEST upgrades) provides an overall TCOMS operational availability of greater than 99.7 per cent and an overall system MTBF above 2,000 hours with minimal on-site spare parts.

Through EDO's Tactical Local Area Network (TacLAN) interface (or other existing interfaces), TCOMS is capable of interfacing to a range of legacy and forward-deployed, fixed or mobile command and control systems, providing backward compatibility. Through EDO's Communications Local Area Network (ComLAN), TCOMS can remotely control all equipment in a communications subsystem, including all voice and data HF and UHF radios, antennas, switches, cryptos, Data Terminal Sets and other modems, Have-Quick, telephones and so on.

TCOMS includes the following capabilities: data archive and replay; data reduction; operator-action archive and replay (VCR-like); simulation of onboard and offboard tracks and management messages; graphical creation of simulation scenarios for training; data filtering for transmit, receive and forwarding; manual track entry; automatic correlation and gridlock; and communication control. The software functionality is distributed across each operator console. All operator functions can be run at any console. Each console maintains duplicate databases for real-time display updates and data manipulation.

TCOMS has been validated by the US Naval Centre for Tactical Systems Interoperability (NCTSI) and by NATO.

Specifications

(a) Message processing: (USA) OS 411.3; (NATO) STANAG 5501; STANAG 5511 Edition 3; STANAG 5601
      Electrical: (USA) MIL-STD-1397A

TCOMS installation   0024505

(b) (USA/NATO) MIL-STD-188-203-1A; MIL-STD-188-203-2 (Ethernet, IEEE 488) RS-232, IEEE 802.3
(c) Software development: (USA) DOD-STD-2167A.

Status

TCOMS has been deployed to customers throughout the world under various designations, including the Tactical Data Processing System, Link and Radar Display System, Tadil Processor and Display System, Mobile Universal Link Translator System. In 2001, TCOMS began deployment within Command and Reporting Centres in NATO as the Ship-Shore-Ship Buffer.

Planning is underway to add the following capabilities to TCOMS: integrated radar video and auto tracker, integrated TV/IR video, integrated IFF, tactical air control (air traffic control), Int and EW, weapons management, mission planning, and 3-D displays.

No new information received, but believed to be still in use.

Contractor

EDO Corporation, Combat Systems Division, Chesapeake, Virginia.

UPDATED

AIR

Australia

Joint Mission Management System (JMMS)

The Joint Mission Management System (JMMS) is a ground-based support system providing software coverage for the entire aircraft mission cycle. The feature set covers mission planning and preview in both 2-D and 3-D environments, real-time update of mission and control data between base and deployed platforms, post mission replay, analysis and reporting. JMMS is constructed using a component-based architecture and can be extended to address the requirements of various aircraft types including fixed-wing, rotary-wing and UAVs. In addition JMMS acts as an interface to related systems such as those for logistics and maintenance support. The system either accepts tasking orders and intelligence data from the Defence communications infrastructure to establish the basis for the mission plan, or an operator can build a mission plan entirely from keyboard entry. The mission plan is then developed taking into account aircraft performance, weapon and sensor performance and the tactical situation. JMMS can suggest flight routes taking these elements into consideration, or flight routes can be created manually by the operator. Once complete the flight routes can be previewed in both the 2-D and 3-D environments. JMMS then allows data loads to be prepared for transfer to the aircraft via data-load devices, managed using JMMS. While the aircraft is in flight, data can be updated via aircraft to aircraft datalink, aircraft to ground datalink, or voice communications. Following the mission, recorded data is downloaded for analysis in JMMS.

System Functions

JMMS provides support to the entire mission cycle including:
(a) Mission tasking
(b) Premission planning
(c) Mission briefing and preview
(d) Preparing aircraft data loads
(e) Real-time mission control
(f) Post-mission analysis, replay and input to future briefs
(g) Post-mission report generation and data distribution.

Operations Planning and Mission Tasking

JMMS supports the operational planning process, tracking aircraft and crew status on a scale of readiness from annual to daily. It is configured to accept and process a specified set of tasking orders, loading the data automatically into a draft mission plan and automatically generating an initial Mission Brief based on this data. The system will store, for operator review, messages that cannot be processed automatically. Relevant data from these messages can be manually entered into a mission plan. The planning task continues to develop a full Mission Brief. Part of this is the latest tactical picture, which can be viewed in the 2-D or 3-D environment and includes:
(a) Map data
(b) Overlay data
(c) Disposition of own forces
(d) Disposition of other forces
(e) Imagery data (such as scanned-in maps, satellite data or links to video data).

Where sensor, weapon or operational ranges are known for an object they can be displayed in both 2-D and 3-D environments; the operator can view the mission plan in a 3-D environment and can conduct a fly-through. The coverage volumes are modelled using basic equations to represent the known radiation pattern, or range of a given sensor or weapon system. The represented coverage will be modified to reflect the impact of surrounding terrain; weather conditions that affect sensors can also be specified (clear, light rain, heavy rain and other effects). Dynamic overlays representing the coverage of sensors and weapons for tracks of interest, including own aircraft, move with the tracks throughout the preview.

Threat Plans

JMMS will automatically generate threat plans based on the developed flight plans (including transit routes) and the situation awareness picture of the area of operations. Threat plans can be dynamically updated as threat data and disposition change during the planning process.

Tote System

A flexible tote (or form) system enables creation of a wide range of operator-defined totes. Totes interact with the Situation Awareness Picture and a resource database. The totes can be created and amended online without interfering with mainstream system operations. A range of analysis tools is provided allowing instant and operational activity charts to be produced.

Platform Configuration and Performance Plans

An aircraft performance database stores performance and configuration data such as performance envelope, rates of climb, weight and balance, weapons, sensor performance, fuel loads and fuel use rates. The data is used to develop the aircraft mission configuration, to validate mission flight plans and to run mission preview fly-throughs. A flight model for the specific aircraft, based on fuel and weapon load data, is used in the 2-D and 3-D environments for Mission Preview. The operator will be informed whether or not the current flight plan is possible given the aircraft performance parameters. Having set the weapon load and the way-points, JMMS will then use the airframe performance model to perform fuel calculations. JMMS also uses weapon and sensor range data for the aircraft to preview the weapons and sensor coverage during the mission. The performance information will be presented in table form, as an overlay in the 2-D environment and as a coloured volume in the 3-D environment. This will enable the operator to review and analyse the predicted performance of the platform for a given mission prior to take-off.

Flight Plans

Flight plans, including minimum lethality routes and attack profiles, can be developed using on-screen planning tools offering a combination of point and click, pick lists and direct entry features. The flight planning tools can correlate route planning with platform configuration and performance data from the aircraft parameters' database and the situation awareness picture to validate the flight plans and feed into threat plans, communications plans and the overall Mission Brief. When complete, the flight plan can be previewed in both the 2-D and 3-D environments.

Decision Aids

JMMS includes a range of decision aids including threat assessment, intercept calculations, bearing and range, alerts, closest point of approach and radar coverage calculations. The decision aids are configured to include aircraft performance tables and can be developed to meet specific operational requirements. The user interface to the decision aids is prompt-driven. The output from these tools can be textural, or where appropriate the result of a decision tool calculation will be displayed in the 2-D or 3-D environment.

Communications Plans

Communications plans are developed on JMMS and uploaded into the aircraft. The communications plan includes frequencies for HF, VHF and UHF (and schedules if applicable to the mission), data to manage secure communications and IFF codes. Where required, the communications plan is used to create configuration data for both the airborne communications equipment and ground-based radios within the system.

Geographic Database Management

The database within JMMS is designed to be compatible with a wide range of commercial mapping data standards. Where unique onboard data storage and handling requirements exist, a set of data converters is created to allow success transfer of data between the JMMS and the aircraft. Typically the following standards are applied:
(a) The Raster map is compliant with the DIGEST Format.
(b) The Terrain Elevation Database is compliant with the DLMS/D⁻ED Format. This allows the terrain elevation data to be displayed as near 3-D scenes.
(c) The Vector Database is compliant with the VMAP Format.
Operators can reference different maps online and select them for display. Battlespace awareness data (tracks, threats, overlays) are then displayed over the selected base map.

Mission Control Functions

The mission can be controlled based on observations made by the crew during the mission, tactical options and information provided by other systems. Real-time transfer of information is effected via a datalink. The data is transferred to the onboard displays and shown as overlays. The system is able to transfer 'snapshots' captured by onboard sensors back to the ground station or another aircraft to aid in situational awareness or target hand-over. The initial data load on the aircraft can be updated during flight from the ground via the datalink. Mission control can also be effected via voice communications and in this case, the update information from JMMS can be manually updated onboard by the crew.

Post Mission Analysis and Reporting

The mission can be reviewed at various speeds and with various forms of filtering and ageing applied to the data. Mission Replay can be viewed in either a 2-D or 3-D environment. Mission data can be analysed using the analysis tools and aids provided by JMMS. An effective graphical analysis capability is provided as part of JMMS. The downloaded mission data can be used to:
(a) update EW threat databases and libraries
(b) update the recognised air and surface picture
(c) analyse mission performance
(d) compare planned mission profile with actual
(e) brief crews for future missions to the same area of operation
(f) feed data into the plans for subsequent missions
(g) generate formatted post mission reports.

System Structure (Physical and Functional)

JMMS can be implemented as a three-level system offering aircraft operations support at:
(a) Bases (Fixed Site JMMS)
(b) Deployed sites (Deployable JMMS)
(c) The point of aircraft tactical operations (Minimal Configuration JMMS).

The first level provides a network for fixed sites comprising PC workstations and peripherals, data load devices, LAN, and network and radio communications. The second level provides the same functionality installed on laptop PCs for deployed sites, with the option of network and radio communications where available. The third layer provides a single ruggedised laptop PC, data insertion device interface, mission data transfer interface and communications port for forward deployment of a minimal number of aircraft. Step-up configurations, allowing for the planned advance of command post operations, can also be supplied.

System Architecture
The JMMS Architecture focuses on three main components:
(a) Application software for mission planning, briefing and analysis functions
(b) Computing environment including servers, desktop PCs, laptops, printers, scanners and large screen projection facilities; the equipment is further divided into fixed-site and deployable assets
(c) Radio communications system that provides complementary functionality to that of the aircraft communications systems (including cryptographic equipment); the communications system hosts both voice and datalink communications between JMMS and the mission aircraft.
While the JMMS solution provides both fixed site and deployable configurations, the software applications remain common, regardless of physical implementation.

Hardware Architecture
JMMS runs on standard commercial hardware. Fixed sites are based round a system server, with additional mass storage devices for data archiving, connected via a LAN to desktop computers. A typical installation will use two 19 in racks to hold:
(a) system servers
(b) UPS
(c) communications hub
(d) radio interface
(e) mass storage devices
(f) cryptographic equipment
   A typical desktop computer will have a local hard drive, substantial RAM, a 20 in screen and a high-performance graphic card, supported by a variety of peripherals. The deployable systems are contained in transit cases, with built-in 19 in racks where necessary and use laptops plus a similar range of peripherals. TEMPEST protection is an option.

Software Architecture
The software architecture of JMMS provides Applications Programming Interfaces (APIs) between the Applications, Support and System layers, providing flexibility at each functional layer. The extension of JMMS, to accommodate different aircraft types, is enabled via a core of common functions supplemented by platform-unique data and functions for each aircraft type.

Status
In operational use by the RAAF. In 1996, ADI was awarded the contract to supply the mission planning and support system for the Royal Australian Air Force's P-3C Orion maritime patrol aircraft and JMMS has developed from this initial system.

Contractor
ADI Ltd, Garden Island, New South Wales.

UPDATED

Chile

AirSpace Control System (ASCS)

The AirSpace Control System (ASCS) takes track information generated by a Track Acquisition System in a Radar Site and displays it to operators at the Air Defence Operational Centre (ADOC). The System consists of two main subsystems:
(a) Track Acquisition Subsystem located at the Radar Site
(b) Display Subsystem located at the ADOC.
Both subsystems can communicate by a variety of alternative methods, including fibre-optic cable and microwave links. The Display Subsystem consists of a number of consoles and two servers connected to a Local Area Network (LAN). Each console has a high-resolution monitor and can be configured for specific operator displays. Each console is capable of graphically representing the airspace status of those tracks generated by the Track Acquisition Subsystem. The console displays the estimated picture as well as that provided by IFF.

Track Acquisition Subsystem
The Track Acquisition Subsystem capabilities are:
(a) Acquisition of primary plots
(b) Acquisition of secondary plots
(c) Extraction of track information from plot data and sending of track information to the ADOC
The main components of this subsystem are:
(a) Radar Interface. This unit is used to adapt all the radar signal, video, trigger and bearing to the system requirements. It is also capable of transmitting the adapted radar signals to the plot extractor. This unit needs just one in-site calibration focused to error minimisation.
(b) Primary Plot Extractor. This unit extracts the primary plot directly from the video radar. It is fully automatic and does not need any special adjustments or calibration. All the primary plot information is sent to the multitracker unit through the LAN.

(c) Secondary Plot Extractor. This unit extracts secondary plot (IFF codes) directly from the video radar. It is fully automatic and does not need any special adjustments or calibration. All the secondary plot information is sent to the multitracker unit through the LAN.
(d) Multitracker. This unit receives primary plot information and IFF codes from the network and processes these data in order to extract new tracks. At the same time this unit tracks the old contacts in order to refresh the picture data. All the track information is sent to the media interface through the network.
(e) Media Interface. This unit serves as the interface between the LAN and the communication media.

The Display Subsystem
The display subsystem is installed at the Air Defence Operational Centre (ADOC). The main components of this subsystem are:
(a) Media Interface. This provides the interface between the communication media and the LAN.
(b) Database Servers. These use Windows NT and Commercial Off-The-Shelf (COTS) hardware and contain the tactical database. They work in a redundant configuration where each unit serves as back-up to the other.
(c) Communication Servers. These enable data exchange among consoles and servers.
(d) Display Consoles. These consoles are the Human—Machine Interface (HMI) between the operator and the system. The display console is based on COTS hardware. Each console includes a 19 in high-resolution display, a keyboard, a mouse and a network interface.
(e) Projection System. The projection system allows the supervisor of the system to select one of the available consoles to be displayed.

Console Capabilities
Basic Operations:
(a) Representation of geographical situation
(b) Environment control through login/logout access and role identification
(c) Display control by marking preferences and special conditions
(d) Communication among consoles for information exchange
(e) Administration of alarms and signs generated in the HMI
(f) Detailed track-information window.
Display of graphical elements:
(a) Tracks
(b) Fixed points
(c) Flight charts
(d) Areas.
Absolute and relative measurements with cursor and graphic elements:
(a) Measurements in geographical co-ordinates
(b) Measurements in polar/Cartesian co-ordinates
(c) Measurements of interceptions
Display of auxiliary elements that are used as visual reference by the operator:
(a) Line and bearing/range readout
(b) Range marks and rings display
(c) Interception geometry
Display of maps and grids:
(a) Geographical maps with different information layers
(b) Predefined grids with different layers.

Status
Available.

Manufacturer
SISDEF, Quintero.

VERIFIED

France

Air Base Operations System (ABOS)

Air Base Operations System (ABOS) is an information and communication system for air base operations, providing operational information for the management of air base resources. Its main functions are:
(a) Interface with other CIS such as:
   Local mission planning and debriefing systems
   Air Command and Control Systems (ACCS)
   Other applications using the XML exchange format
(b) Management of air activity and distribution of operational information:
   Aircraft/pilot availability
   Airspace co-ordination
   Flight orders
Information provided by the system includes:
(a) Weather reports
(b) Airbase operational conditions
(c) Tactical picture
(d) Planned and current missions
Data originating at squadron level can be passed up through the various levels of command, while operational information and directives can be passed down. This system enables staff to manage their resources more effectively.

Within an air base the system will consist of a number of user sets, each operational as a LAN within a building, with these connected to make a base network.

Status

In March 2001 it was announced that SAGEM had been awarded a contract to install the system at French Air Force bases. This programme is known as Système d'Information et de Communication des Opérations des Bases Aériennes (SICOPS - bases). Deployment is in progress.

Contractor

SAGEM SA(Defence and Aerospace Division), Paris.

UPDATED

AMASCOS Maritime Patrol System

The AMASCOS (Airborne Maritime Situation Control System) suite is an airborne multisensor maritime patrol system designed for real-time tactical situation build-up and update, as well as a decision aid to operators. Its modular design approach makes it possible to integrate the system on board most fixed- or rotary-wing aircraft.

There are three versions of AMASCOS (designated as AMASCOS 100, 200 and 300) that correspond to three broad categories of mission requirements ranging from constabulary tasks to multithreat war fighting. The typical AMASCOS configuration integrates Thales equipment (Radar, Chlio FLIR, DR3000 ESM, TMS 2000 acoustics, FLASH dipping sonar, Magnetic Anomaly Detector (MAD)), Link Y, NATO (Links 11,16,22) or national communications, TOTEM GPS/INS, and TOPDECK cockpit, but its modular architecture enables each system to be tailored to specific requirements.

AMASCOS 100

AMASCOS 100 is designed for constabulary missions and those at the low end of the conflict spectrum such as:
(a) Exclusive Economic Zone (EEZ) control, fisheries control, maritime traffic surveillance, anti-smuggling, anti-drug operations, anti-piracy, immigration control, pollution survey
(b) Search and Rescue (SAR)
(c) Sovereignty.
Light aircraft are generally used with one or two operators working in collaboration with the flight crew, but the system which weighs a maximum of 250 kg can also be installed in land- or ship-based helicopters.

AMASCOS 200

AMASCOS 200 is a more capable multimission system for:
(a) Anti Surface Ship Warfare (ASSW)
(b) Electronic Intelligence (ELINT)
(c) Communications Intelligence (COMINT)
(d) Joint littoral warfare
(e) Joint surveillance (land and sea)
(f) Maritime surveillance (EEZ control, SAR).
AMASCOS 200 adds an ESM or RWR system and the capability to carry out anti-ship missile attacks either independently or in conjunction with other units (OTHT). The system can be operated by two or three personnel and will best fit fixed- or rotary-wing aircraft of the 8 tons class.

AMASCOS 300

AMASCOS 300 is the most capable system and is designed for:
(a) Anti-Submarine Warfare (ASW)
(b) Anti Surface Ship Warfare (ASSW)

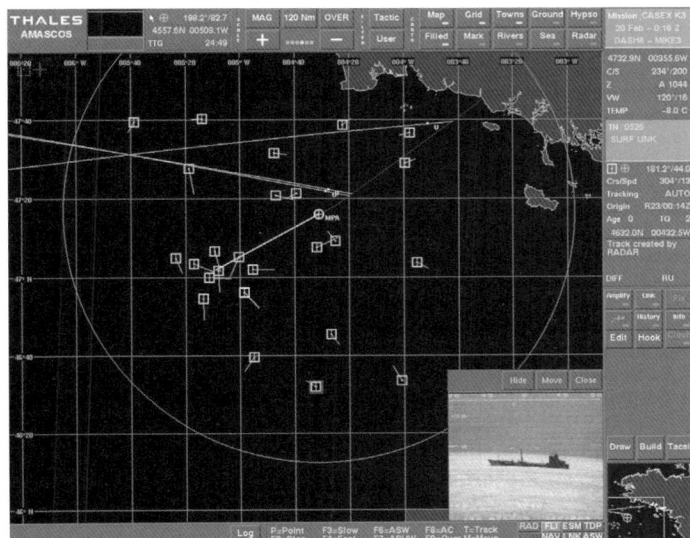

AMASCOS display   0102014

(c) Electronic Intelligence (ELINT)
(d) Communications Intelligence (COMINT)
(e) Joint littoral warfare
(f) Joint surveillance (land and sea)
(g) Maritime surveillance (EEZ control, SAR).
AMASCOS 300 adds acoustic equipment and the capability to carry out attacks with torpedoes and depth charges. Installed in fixed- or rotary-wing aircraft of the 10 tons class, the system can be operated by three or four personnel. It can also be expanded and fitted on board multimission long-range Maritime Patrol Aircraft, with seven or more operators.

Status

9 AMASCOS 100 systems are in use in Indonesia aboard IPTN NC-212 Aviocar MPA and NBO 105 helicopters. Installed in Pakistan Dassault Breguet MPA. Selected in 1998 for installation in UAE BOMBARDIER DASH-8 patrol aircraft. In June 2001 it was announced that a €50 million contract had been awarded for the supply of 3 CN235-220 aircraft equipped with an unspecified AMASCOS system for the Indonesian Air Force. In September 2002 the MELTEM contract with Turkey, worth in the region of US$400 million, was announced for the supply of AMASCOS-based systems for the Turkish Navy and Coast Guard. The contract is to supply nine systems into existing modified CN 235 aircraft and an additional 10 systems to be integrated into new platforms.

Contractor

Thales Airborne Systems, Elancourt.

VERIFIED

DOREMI

DOREMI is a target data management system. It consists of three main functions.
(a) The creation and management of target folders - creating and managing target folders on map backgrounds of different scales. These folders include imagery, target graphics, source document, elements of mission, Desired Main Points of Impact (DMPI) and any additional data.
(b) The targeting function: positioning targets on the map background, target system analysis (selecting one or several targets within a target system), creating new targets and creating or updating the geographical theatre.
(c) The battle damage assessment function: analysing strike damage, updating target folders, evaluating collateral damage and evaluating the efficiency of an air campaign.
Data can be imported and exported according to a number of different formats, including to and from formatted messages (such as MISREP or ATO). Folders can be supplied in paper format or HTML files which are exploitable by any standard computer system. Computer Aided Design (CAD) tools allow further information to be added to images, according to the user's needs.

Specifications

SUN workstation
Optical fibre Ethernet local area network
Colour laser printer
DLT disk driver
Server station with 3 external disks, one each for the application, targeting data and geographical data
Customer station with 1 external disk.

Status

In service with the French Air Force.

Contractor

SAGEM SA(Aerospace & Defence Division), Paris.

VERIFIED

MARTHA air defence command and control network

MARTHA (MAillage des Radars Tactiques et des systèmes d'armes pour la lutte contre les Helicoptères et les Aeronels à voilure fire) is being developed by Thales Air Defence and will be the next-generation air defence solution for the French Army, deployed at the Corps level and below. Its overall concept is based on a three-level architecture designed to provide efficient real-time airspace and air defence management on the battlefield. MARTHA will be responsible for the command, alerting and fire co-ordination of all anti-aircraft weapon systems, will co-ordinate and manage the deployment and siting of air defence assets and will provide commanders with continually updated status reports of air defence forces. As an airspace management tool, the primary task of MARTHA is to ensure the safety of friendly aircraft as well as providing regular updated information on the tactical air situation.

Each of MARTHA's three levels of co-ordination (section/battery, division/regiment and army corps) will have its own specific type of command and control centre. These are designated NC1, NC2, and NC3 (for Niveau de Co-ordination, or Co-ordination level 1, 2 and 3 respectively). These centres are designed to meet all

Interior of a MARTHA centre showing the operators' consoles on the left
NEW/0509723

operational co-ordination requirements at their particular command level. These include the capability to carry out functions such as:
(a) overall control of all air defence operations
(b) information collection, dissemination and distribution
(c) real-time adaptation of preplanned air situation
(d) control and co-ordination of weapon systems.
MARTHA centres are integrated inside vehicle-borne shelters that provide full NBC and electromagnetic protection and are air transportable.

Each MARTHA centre is connected via secure communication links to other centres at its own level as well as with an assigned centre at the next higher co-ordination level. A subset of centres NC1, NC2 and NC3 can be deployed in response to a specific operational mission.

The NC1 co-ordination centre is designed for section-, battery- or platoon-level operations primarily with a wide range of short- and very short-range anti-aircraft guns and SAMs. It will be responsible for alerting and air defence against low-flying aircraft, helicopters, and UAV. In French Army service it will co-ordinate the fire of man-portable MBDA Mistral missile systems. Each NC1 is capable of co-ordinating the fire of a platoon of up to eight anti-aircraft weapons which may be identical or of various types. In case an NC1 is damaged or destroyed its weapons can immediately be reassigned to other operational NC1s in the area, providing considerable operational flexibility. A single display provides both the air situation and the status of air defence weapons. The NC1 is effective only within a small area defined by the range of its sensors and weapons, typically some 20 km, but receives a wider picture from the NC2 co-ordination centre.

The NC2, normally deployed at division/regiment level, is intended to cover a geographical region or zone and the command and control centre may either be attached to the divisional HQ or to the air defence regiment. The NC2 is responsible for the exchange of information between other army units, the army air corps and the air force reporting networks. It will provide co-ordination of the full range of air defence weapons, and tracks and continuously updates the overall air situation, as well as monitoring in real time the status of its associated air defence assets on the battlefield. Two separate display consoles, one for the air situation and one for the tactical battlefield situation, are provided. A major function of the NC2 is radar data fusion, dissemination and distribution.

MARTHA's higher-level co-ordination centres, the NC3s, will be located at army corps headquarters. The NC3s will serve as data collection, correlation and planning tools. They will permit manoeuvre co-ordination between the army in the field and other supporting tactical air forces. They will be linked to each other and to the divisional level, as well as to the army C3 network (SICF).

The short-range communications links at the NC1 level are provided by secure Thales PR4G radios. A more complex communications network, the Multifunctional Information Distribution System (MIDS) will be used at the NC2 and NC3 levels, which is still under development but will almost certainly include Link 16.

Due to the modular design of the network, Thales Defence Systems has adopted a 'bottom up' building block approach in its development of MARTHA. This is claimed to be more pragmatic than a 'top down' design philosophy and allows earlier effective deployment of first-level systems.

Status
In early 2000, it was announced that the French DGA (Delegation Generale pour l'Armement) had awarded the then Thomson CSF Airsys a contract for the supply 49 NC1 MARTHA air defence command and control systems for delivery to the French Army over a four year period. Of the 49 systems, 18 are in the NC1 Roland configuration to support the Euromissile Roland short-range Surface-to-Air Missile (SAM) systems and 31 in the NC1 Mistral configuration to support the MBDA Mistral very short-range SAM systems. As well as covering the supply of the truck mounted shelter type NC1 systems, the contract also covers the supply of logistics support. Typically, the NC1 Roland will co-ordinate four Roland SAM systems while the NC1 Mistral will co-ordinate up to nine Mistral SAM systems.

In 2001, a second contract was issued to define the other co-ordination centres and command posts in the MARTHA network. Key objectives include optimising all of the French Army's air defence systems, light aviation, field artillery, UAVs and 3-D radar systems and ensuring interoperability with other French Army command information systems and C3 systems deployed by French and other NATO forces

such as the German HFLaFuSys, the Italian CATRIN, the UK ADCIS (Air Defence Command Information System) and the US FAADC3I (see separate entries). It will integrate the new Eurosam SAMP/T (Sol AirMoyenne Portée/Terre) SAM which will replace the current Improved HAWKs used by the French Army, and will provide connection with the Air Force command and control system SCCOA (Système de Commandement et de Conduite des Operations Aeriennes) (see separate entry).

Contractor
Thales Air Defence, Paris.

UPDATED

SCCOA (Système de Commandement et de Conduite des Operations Aériennes)

SCCOA (Système de Commandement et de Conduite des Operations Aériennes) is the new French Air Command and Control System. It embraces a number of aspects of C2 including surveillance, force generation and management, air campaign planning, preparation and control of missions, national air space control and is part of the French component of the NATO ACCS programme. The system consists of sensors, operations centres and a communications network, has been designed with an open architecture and covers at least 50 sites. The software may include an updated version of STRIDA (see separate entry). It will be interfaced with MARTHA (see separate entry).

Status
The decision to embark on the programme was taken in 1989. It has proceeded in three stages. Stage 1 (completed 1999) and Stage 2 (started 1997) were essentially conceptual design and development phases and were undertaken by EADS. In 2001 the prime contract for Stage 3, which covers the provision or upgrade of C2 facilities, sensors and communications, including a deployable capability, was awarded to a joint venture of Thales and EADS called MOSS. Initial capability is due by 2005 and Stage 3 should be complete by 2006. Stage 4 (to start the same year) will cover further integration with NATO ACCS.

Contractor
MOSS (A Thales/EADS joint venture) (Prime).

VERIFIED

STRIDA air defence system

STRIDA (*Système de Traitement et de Représentation des Informations de Défense Aérienne*) is the French air defence data handling system. It consists of a network of stations covering French territory with the following main functions:
(a) detection and identification of aircraft moving in French airspace by long-range, low-level and air-base radars plus AWACS
(b) threat evaluation and dissemination of early warnings. The air situation is centralised and synthesised in the Air Defence Operations Centre (ADOC)
(c) updating of active means (aircraft and missiles) status in every Sector Operations Centre (SOC)
(d) weapons selection, engagement and automatic intercept guidance
(e) aircraft recovery to airbases
(f) control of military operational and training flights
(g) co-ordination with the air traffic control system to ensure identification and spacing of operational military flights with respect to general air traffic.
The research and development programme for the STRIDA programme began in 1956 under the responsibility of Service Technique des Télécommunications et des Equipements aéronautiques (STTE). The first stations were fitted with specialised IBM/CAPAC real-time computers and Sintra VISU II display subsystem. They became operational in 1963. Alcatel is responsible for system integration, automatic data processing and software.

STRIDA consists of different types of operations centres (CAOC, SOC/CRC, CRC, ARP) which exchange digital messages by a special telecommunication network, the RA70 and RESEDA, associated with electronic switching stations.

At ARP and CRC levels, signals coming from 2- and 3-D long-range radars (Palmier) and electronic scan height-finding radars (SATRAPE with phased-array

A STRIDA air defence reporting centre and communications node collocated with a 3-D radar in the Alps

VISU IV consoles produced by the SINTRA subsidiary of Thomson-CSF for the STRIDA air defence system

antenna) are used for providing the labelled air situations needed at SOC/CRC and CAOC levels. All information and orders to be exchanged between these different centres for command and control purposes are transmitted through the RA70, RESEDA network.

The STRIDA network is connected to the NADGE, UKADGE, GEADGE and SADA systems to provide complete coverage of Western Europe. AWACS provides low-altitude detection and over-the-horizon early warning. STRIDA is also connected to STRAPP (STRIDA-APPROCHE), which integrates low-altitude radars; the French ATC CAUTRA for co-ordination of military and civil air traffic control; the Tactical Forward Air Post; the HAWK SAM control centre and the French Navy NTDS (SENIT).

The data handling equipment of a typical latest generation CRC station mainly consists of:
(a) an EMIR radar data extractor using a programmed extraction concept
(b) a high-power processing system using IBM main frame computer and UNIX workstation
(c) a display subsystem including from 20 to 30 operational positions.

Each operator position or console designed for one controller and his assistant is composed of:
(a) a plan view display for presenting raw video (local radar raw video) and data generated by the processing system (synthetic view of the air picture – mainly tracks). A 40 cm diameter screen is used
(b) one or several monochrome or colour screen(s) (diagonal from 13 to 35 cm and capacity up to 4,000 characters) for presenting detailed information on certain subjects (tracks, intercepts) or data received in alphanumeric form (operational status, flight schedules)
(c) several keyboards and a rolling ball (or a joystick) for selecting the data presented and entering functions and data
(d) control keyboards and panels for communications (radio, telephone, interphone) and secondary radar.

Status
Since 1963, there has been a continuous programme of improvements for software and hardware during the implementation of the whole military programme, every new centre having to be interoperable with the others. The last ones are equipped with IBM mainframe computers, UNIX workstation cluster for the processing part and SINTRA VISU IV and UNIX workstation display subsystems. Medium- and high-altitude coverage is achieved. The low-level coverage with STRAPP is now near completion. AWACS is now completely integrated. The next step entails the extension of the low-level coverage of the system by the integration of tactical operations radars and SAM activities.

An air defence operations centre (ADOC) of the STRIDA system

In 1980, SINTRA (subsequently Thomson, now Thales) was chosen to supply the STAC (*Système Traitement Automatique au Cabine*) and to design the AL73 (*Adapteur de Liaisons 73*) for tactical radar processing and HAWK activities management at each mobile tactical and SAM control centre. Data to and from STRIDA main centres are automatically processed and transmitted by the STAC and AL73 system.

STRIDA software is now mostly written in Ada (Software Release C.0) and can also be run in a tactical centre, the Alcatel DIAMS (Deployable Interoperable Air Command and Control Modular System) providing additional CAOC functions such as air planning and tasking (ATO/ACO) and intelligence.

Upgrades in progress in 1996 involved the use of COTS computers, low-altitude radar integration and assignment of datalinks to co-operating platforms.

STRIDA is understood to be fully operational.

Contractors
Thales Air Defence, Bagneux.

VERIFIED

Germany

EIFEL command, control information system

The preliminary design work on EIFEL, an electronic command and control system for operational readiness of the German Air Force, began in the second half of the 1960s. A system was first set up to support selected functional areas of the Air Force by introducing data processing into the tactical command loop. Siemens was awarded the contract to develop and supply this system which was handed over to the Luftwaffe in 1973.

At that time EIFEL consisted of three computer centres, each of which was equipped with a Siemens 4004 processor plus associated peripherals such as disk storage and printers. The operating system was BS1000 and software specially oriented to EIFEL, such as a databank and communication system, was developed for this application. The system trials and the initial user trials proved successful and, in 1978, the Luftwaffe introduced the system for operational service under the designation EIFEL 1.

Following the reorganisation of the NATO forces in the Central Europe sector, EIFEL 1 was chosen as the command, control and information system to support all new Allied command posts. EIFEL 1 was converted to the modern OS BS2000 in 1994 and is now known just as EIFEL. It runs on Siemens Nixdorf H90 mainframes. Currently there are four computer centres, 600 terminals, mostly PC-type with a modern Graphical User Interface (GUI) and the system is implemented at over 300 sites in central Europe.

Contractor
EADS Deutschland, Munich.

VERIFIED

EIFEL position

German Air Defence Ground Environment (GEADGE)

GEADGE is the acronymic title of the air defence network installed in the southern part of Germany. The system, for which Hughes Aircraft Company (now Raytheon) was nominated prime contractor in 1979 under a contract worth more than US$150 million, replaced the obsolescent 412L radar network operated by the former West German Air Force.

The system integrates new and existing long-range surveillance radars into a single network based on four centralised command centres and embraces

A German captain monitors the air traffic over the southern portion of Germany at a GEADGE site

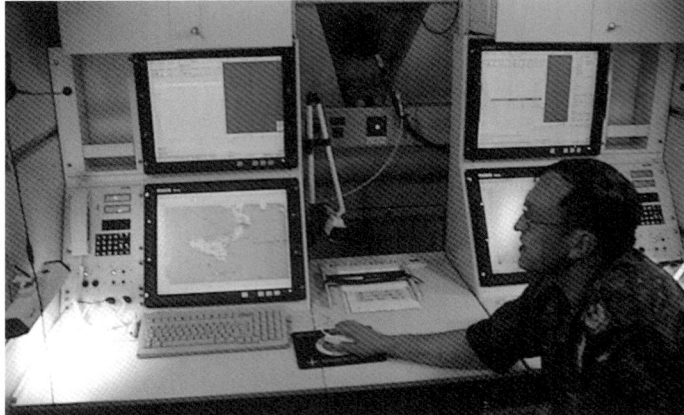

The interior of the SAMOC Operations Support Cell (JJL/Jane's)   0111472

manned and unmanned fixed and transportable radar systems. GEADGE also receives radar data directly from E-3A AWACS early warning aircraft. The southern portion of Germany was not included in the original NATO Air Defence Ground Environment (NADGE) system but GEADGE fills the gap left in that system and connects directly with it.

In addition to fixed and transportable gap-filler radars, the system utilises two of four new permanently located radars known by the manufacturer's name of Hughes Air Defence Radar (HADR). These are advanced 3-D, multirole radars that will automatically detect, classify and report on targets intruding into their coverage area.

Hughes (now Raytheon) has also supplied HMP-1116 mini-computers, H-5118M central computers and HMD-22 display and control consoles, as well as being responsible for software, installation and integration.

Status

The first two sites became operational in 1983 and 1984 and the final two were operational in 1985. Final system acceptance took place early in 1986. Hughes (now Raytheon) completed an upgrade to sites to receive AWACS target data in 1988 under the NATO AEGIS programme. GEADGE will be updated over the next three to four years as part of the NATO ACCS programme.

Contractor

Raytheon Company, El Segundo, California, USA.

UPDATED

SAM Operations Centre (SAMOC)

SAMOC is a ground-based air defence management system which is designed for operations both within NATO and within CJTF operations. A standard operational SAMOC system will consist of six cells housed in transportable shelters (mounted on 5 tonne trucks): four Fire Direction Centre/Operations Planning Cells (FDC/OPCs) with three workstations each and two Operations Support Cells (OSCs) with two workstations each. The minimum configuration is one OSC and one FDC/OPC. All operator stations have identical capabilities regardless of their location and all functions and roles can be operated at multiple operator stations simultaneously. The system can be divided while remaining operational to permit tactical moves, providing there are two OSCs, and this capability also provides redundancy in the event of malfunction or loss.

The main functions are:
(a) Force Operations, including situation analysis, theatre missile defence and cluster defence planning, communications planning, deployment monitoring and control, and staff functions for which there are several tools such as radar coverage diagrams, terrain databases, route proposals and timelines.
(b) Engagement Operations, including air surveillance (with air picture compilation and identification), mission control (with target analysis and target allocation) and engagement evaluation.
(c) Training & Simulation, including scenario generation, simulation control, operator training support, defence design evaluation support, and a Distributed Interactive Simulation (DIS) interface to tie in with the US-supplied Co-operative Air & Missile Defence Exercise Network (CAMDEN).
(d) General Services, including system management (human machine interface and local resource management) and communications (standard NATO datalinks, non-NATO datalinks, command and control information system connectivity, voice communications).

SAMOC uses primarily commercial-off-the-shelf (COTS) hardware and software, including DEC workstations, BarcoView 20 in rugged flat panel displays (see separate entry), two fibre optic digital data interface token ring local area networks (for data and for voice), and CORBA, GenaMap, Oracle, Unix and IRIS-MFS

software. Programming languages used are Ada and C++, while the contractor states that the entire system has a 'Windows look-and-feel'. The system's multilink processor ensures that each SAMOC can interface with the following protocols:
(a) Link 16 to link with Patriot, CRCs, AWACS, fighters, naval systems and future SAM systems such as MEADS
(b) Link 11B to link with Patriot, CRCs and the EADS-supplied German HAWK Operations Center (GEHOC)
(c) Army Tactical Data Link-1 (ATDL-1) to connect with GEHOC, HAWK fire units and the EADS-built Roland fire distribution centre
(d) Low Level Air Picture Interface (LLAPI) to connect with army air-defence surveillance and command and control systems
(e) The open architecture enables integration with non-standard datalinks.

The total number of subordinated fire units depends on the SAMOC configuration and the type of systems connected. Under the Luftwaffe concept of operations, a SAMOC will be able to plan and co-ordinate the operations of up to 12 Patriot fire units, eight HAWK PIP-II fire units and 20 Roland fire units.

Status

The preproduction system was ordered in 1997 and was delivered in 2000. Following extensive trials it was successfully tested in 2001.The development consists of 2 phases. Phase 1 is the basic system which has the same capabilities as the pre-production version. It is expected that series production for this will commence in 2002; there is a known requirement for a total of five systems for the Luftwaffe, one to be based at the GEAF training facility in the USA and one each for the four Luftwaffe SAM Wings at Husum, Bad Sülze, Burbach and Manching. Phase 2, which will be the final version, is to be developed in parallel by mid-2003, adding additional functionalities such as enhanced communications (Link 16 capability and ADAT P-3 message format), planning tools and TBMD capabilities, with the intention of ultimately delivering the more comprehensive capability. In December 2002 a €28 million contract was awarded by the German MoD for Phase 2.

Until a contract has been confirmed with the GEAF, the equipment will not be marketed abroad. However, as at July 2002 interest had been shown by UK, Czech Republic, Estonia, Hungary, Japan and Poland.

In 2003 *Jane's Sentinel* reported that a system had been ordered by Lithuania in 2002 for delivery in late 2004. In July 2003 the UK MoD selected the EADS/MBDA consortium as one of the two competitors for the UK Ground Based Air Defence (GBAD) programme; the CARACAL system which EADS is proposing as part of the solution is largely based on SAMOC.

Contractor

EADS Systems & Defence Electronics, Ulm, Germany.

UPDATED

International

Headquarters Integrated Area Defence System

In 1996, ADI was awarded a contract to upgrade the Air Defence Operations Centre and the Tactical Air Command at the Five Power Defence Arrangements (FPDA) Headquarters Integrated Air (redesignated Area in 2000) Defence System (HQ IADS) located at the Royal Malaysian Air Force base at Butterworth in Malaysia. The existing air environment system (A-FOCSS), also produced by ADI, was modified to include specialist air defence functions, and the new HQ IADS system was delivered within eight months of contract award.

The upgrade uses commercial software and hardware to provide a complete system for monitoring, planning and coordinating air defence operations and tasking air assets. It includes simulation, recording and fast replay for training and post exercise analysis. ADI claims that the system was the first deployed command support system capable of providing cost-effective 3-D visualisation of the air picture on a personal computer.

The specialist air defence functions of the HQ IADS solution include:
(a) Live data feeds from a number of air defence radar systems

(b) Attack planning tools such as route planning, 'fly the plan', deconfliction and automatic air tasking orders

(c) Totes displaying air raid warnings, ground and naval radar status, mission details, fighter status and alert status

(d) Threat assessment including alerts on zone infringements, resource levels and base weather

(e) Alarms on selected categories and track strengths.

Status

In operational use.

Contractor

ADI Ltd, Sydney, New South Wales, Australia.

VERIFIED

Israel

Mission & Debriefing Ground Terminal (MDGT) mission planning system

The MDGT is a Unix-based mission planning system which enables pilots and mission planners to view intelligence and target data and imagery, calculate aircraft configurations and loads, plan a mission, rehearse using three-dimensional terrain views and download the mission data to a portable data cartridge (DTC) for loading into the aircraft avionics. On completion of the mission data can be downloaded from the DTC for post-mission debriefing. The system has the following features:

(a) Friendly Motif driven Graphic User Interface

(b) Object graphics operations and editing

(c) Map tool and advanced map features

(d) Various map data base support (raster, vector, satellite, images, DTED, DMA maps)

(e) Separate graphic overlay logic

(f) Intelligence and tactical operations

(g) Graphic tactical tool

(h) Overlay manipulation

(i) Mission administration

(j) SMS data administration

(k) CNI data administration

(l) Graphic mission editing and planning

(m) Mission settings and squadron defaults

(n) Simultaneous display of several routes

(o) Automatic fuel consumption and time calculations by aircraft model

(p) Images scanning and manipulation

(q) Targets image bank

(r) Display file generation (avionic graphics interface)

(s) 2-D and 3-D dynamic mission rehearsal

(t) Operational and maintenance data debriefing

(u) Aircraft configuration

(v) Mission and map colour printing

(w) Text reports and pilot pocket pages

(x) DTC loading and reading.

Specifications

(a) MDGT Hardware:
HP 715/75 UNIX workstation, or can use other UNIX providers
CRX-24 graphic board (12 × 2 + 8 graphic overlays
8 Gbyte disk
128 Mbyte memory
19 in colour display
4 Gbyte DAT back-up tape
CD-ROM
Printer and scanner
DTC loader

(b) MDGT Software:
Tailored to DOD-STD-2167A
Object Oriented Design
UNIX (HP-UX) operating system
X Windows and Motif Graphic User interface
PEX 3-D Application
ANSI-C programming language
Informix© relational database
Advanced developing environment: HP Soft bench©, UIM/X© GUI builder

Status

In use with Israel Defence Force in both fixed- and rotary-wing platforms.

Contractor

Elbit Systems Ltd, Haifa.

VERIFIED

Italy

Mobile Command and Control System - C2M

C2M is a mobile, ground based Air Operation and Control System based on the NATO Initial CAOC Capability (ICC) Software. The system consists of four shelters with two primary modules, an Operational Control and Tactical Command Module (CCOA) and Tactical and Air surveillance Module (CCTA). C2M provides the following capabilities:

(a) air mission planning

(b) air space management

(c) command and control resource management

(d) defensive, offensive and support missions

(e) surveillance

(f) RAP generation and distribution

(g) Threat evaluation; target and weapon assignment

(h) SAM control

(i) mission control

(j) recording and data reduction

(k) simulation

Technical Characteristics:

(a) COTS open ADP architecture

(b) autonomous or integrated modes of operation

(c) large communications capability

Status

In service with the Italian Air Force.

Contractor

AMS, Rome.

UPDATED

Multirole Air defence Display Subsystem (MADS)

The Multirole Air defence Display Subsystem (MADS) is an intelligent multipurpose workstation based on Commercial Off-The-Shelf (COTS) components and a COTS operating system of non-proprietary standard design. It can be configured as a single operational position or as a control station integrated into the local area

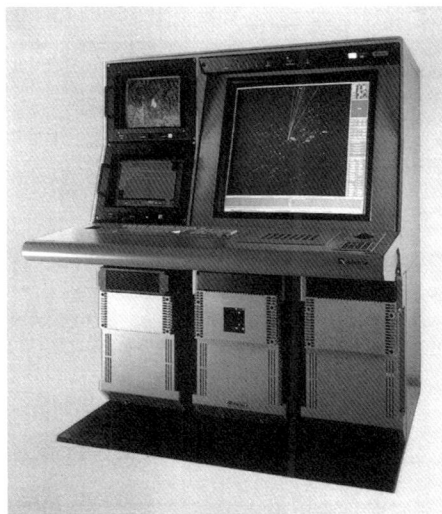

Multirole Air defence Display Subsystem (MADS)

MADS in use (AMS)   0132918

network of a control centre. It is able to carry out a wide range of operational functions required for radar management, surveillance and command-and-control applications.

In the surveillance and command-and-control fields, its operational multirole capability complies with the main operational requirements of synthetic radar data presentation and map generation and presentation.

Main features include:

(a) Multipurpose console for radar management, surveillance and command and control
(b) Open architecture design
(c) COTS hardware
(d) High-performance graphics engine
(e) Radar scan converter
(f) Modular software architecture supported by COTS operating system
(g) Standard I/O interfaces
(h) Fully compliant with NATO console standard.

Contractor
AMS, Rome.

VERIFIED

Japan

BADGE air defence system

Base Air Defence Ground Environment (BADGE) is a computerised air defence system designed to provide the information gathering data processing and display functions required for umbrella protection against aerial attack on Japan.

BADGE sites extend from the northernmost tip of Hokkaido to the southern extremity of Okinawa Island and there are believed to be at least 28 surveillance radar locations. Japanese airspace is believed to be divided into four air defence sectors with a direction centre for each: Western Air Defence Force Control Centre (ADFCC) Kasuga; Central ADFCC, Iruma; Northern ADFCC, Misawa; and Okinawa ADFCC.

The US$56 million system was largely built in Japan for the Japan Self-Defence Force and became operational in 1969. The prime contractor was the Hughes Aircraft Company (now Raytheon) which supplied much of the equipment for the first installation as well as being responsible for system and equipment design. Most of the subsequent manufacture, however, was carried out by Japanese industry.

The system comprises radars that will automatically detect, track and identify airborne targets over Japan and a large area of the surrounding ocean, computers to process the radar data and evaluate threats and other computers to process and furnish data on weapon availability, intercept geometry and related measures. All of this is displayed, together with the processed radar data, on complex displays to the appropriate interceptor or missile controllers. The original computers belonged to the Hughes H330 series.

BADGE is similar in many respects to the Florida system in Switzerland (see separate entry).

In 1984, Japan initiated an extension and upgrading programme called BADGE-X which increased the coverage and provided for the integration of the Japanese HAWK surface-to-air missile batteries with the existing defence systems.

VERIFIED

NATO

Air Command and Control System (ACCS)

The NATO Air Command and Control System (ACCS) was conceived almost 20 years ago to respond to the challenges of the Cold War. Since then, the system concept has undergone significant modification to adapt to the changing geo-strategic situation and advances in technology. The ACCS is designed to be a fully integrated and interoperable command and control system to support the planning, tasking, execution and air surveillance of all tactical air operations throughout NATO Europe from Eastern Turkey to Northern Norway. The system will comprise an in-place backbone sized to meet day-to-day peacetime requirements and those of the early stages of crisis. The additional capacity to deal with an intensified crisis in any region will be created by augmenting the backbone with the Deployable ACCS Component (DAC). The DAC will also be used for out-of-area peace support operations and to support Combined Joint Task Force (CJTF) operations, as authorised by the North Atlantic Council. All NATO nations, including the three new members, participate in the ACCS programme, with the exception of Iceland and Luxembourg who nevertheless are involved in the funding process. The programme is intended to combine, and automate, at the tactical level the planning, tasking and task execution of all air operations.

Two of the principal features of the ACCS will be its open architecture and the emphasis which is being placed on off the shelf products. Both are intended to permit evolution of the system without the need for major developmental effort.

Additional requirements in the areas of Theatre Missile Defence and Airborne Ground Surveillance are already being addressed.

The implementation of ACCS will be evolutionary, with the initial phase intended to satisfy a specific Level of Operational Capability (LOC). Subsequent phases will be determined within the overall long-term programme, taking into account operational requirements, technical risk and the financial situation. Initial LOC1 implementation will occur in 16 operational locations, including deployable elements. Implementation will begin with a software acquisition and validation phase at 4 sites in Belgium, France, Germany and Italy, followed by a replication phase during which installation will occur at the remaining 12 operational sites. A number of existing NATO Air Defence Ground Environment sites will be closed with ACCS implementation, resulting in reduced operational and maintenance costs. The LOC1 implementation was approved by the North Atlantic Council in May 1994. The subsequent release to industry in December 1996 of the Invitation For Bid was followed by bid reception and bid evaluation which was completed in December 1998.

Status

The first contract, worth some US$500 million was signed in July 1999 with Air Command Systems International (ACSI), a French-registered company formed by two shareholders, Raytheon of USA and Thomson-CSF (now Thales) of France. It provides for the development and testing of the ACCS system core software over a 69-month implementation schedule. Installation will initially occur in validation sites in Belgium, France, Germany and Italy. Thereafter, replication sites will be implemented in the remaining nations in the programme. Work associated with the French and Italian contracts includes the development of a Combined Air Operations Centre (AOC), and Air Control Centre (ACC), a Recognised Air Picture Production Centre (RPC) and Sensor Fusion Post (SFP). Under the Belgium contract, ACSI will develop an ACC, a RPC and a SFP. Under the German contract, ACSI will develop an AOC.

In early 2001 it was announced that EADS had been awarded a contract worth DM35 million to develop some of the software applications. This will be based on the Dornier mission planning system DIPLAS, which is in service with all wings of the German Air Force. The system is to be expanded to allow the planning, command and control of large-scale multinational air efforts in co-operation with the other services. Among the tasks involved are the determination of flight routes which take exclusion zones and minimum flight altitudes into account, co-ordinate the sequence of takeoffs and landings and also ensure the separation from civil air traffic, plus the advance planning of available aircraft, repair times and fuel quantities.

In June 2001, Thales Communications of Norway and CS Communication & Systèmes of France won a contract to supply Voice Communication Systems (VCS) to the four host nation sites for validation. The system incorporates all types of narrow and broadband traffic into the same network. At that time the full scale implementation of ACCS VCS was expected to take place from 2004 to 2010.

In August 2003 *Jane's Defence Weekly* reported that the critical design review had been completed; a further review later in the year will focus on Link 16 and multilink interfaces.

Contractor
Air Command Systems International (ACSI), owned by ThalesRaytheon, Paris.

UPDATED

Iceland Air Defence System (IADS)

In 1990, Hughes Aircraft (now Raytheon) was awarded a contract by the US Air Force Electronics Systems Division (ESD) to build a new NATO air defence system for Iceland. This provided command, control and communication systems to improve the overall air defence capabilities of Iceland and the surrounding North Atlantic region monitored and protected by NATO forces.

Description

IADS consists of four radar stations located on Miðnesheiði, Bolafjall, Stokksnes and Gunnólfsvíkurfjall; a Control and Reporting Centre (CRC) located at the Keflavik Naval Air Station; an alternative CRC co-located with the Iceland software support facility; and a voice and digital system providing on-island and off-island communications.

Adjacent air defence systems in Norway, the UK, Canada and the USA are networked into the system, thus linking the existing North American system and the NATO Air Defence Ground Environment (NADGE) together.

Hughes (now Raytheon) supplied operator workstations based on its AMD-44 product line. The AMD-44 is a flexible system using a full-colour, high-resolution (four million pixels on a 20 × 20 in screen) display.

Status

IADS is fully operational.

In 1999 a programme to integrate Link 16 into IADS was initiated, after initial funding approval from NATO. The improvements will provide a secure, ECM Resistant Communications System (ERCS) datalink and voice capability between the Control and Report Centre (CRC) at Naval Air Station Keflavík and NATO aircraft in the Military Air Defence Identification Zone (MADIZ). Use of Link 16 in the IADS will also provide the CRC with air and weapon control, surveillance and information management capabilities. The Project will include additional software and hardware. Thales Raytheon Systems (TRS), in conjunction with the Icelandic company, Kögun Ltd, submitted the only bid, which was approved in December 2002. It is expected that the building and installation of the updates will be completed by mid-2005.

Contractor
Raytheon Company, Tewksbury, Maryland, USA.

UPDATED

NATO Air Defence Ground Environment (NADGE)

NADGE is the name given to the system that links together the national air defence systems of 10 European countries. They are Belgium, Denmark, Germany, Greece, Italy, Netherlands, Norway, Portugal, Spain and Turkey. French participation is via STRIDA (see separate entry) and is limited to the use of, and contribution to, the reporting and control functions.

Status
The NADGE system is obsolescent and is to be replaced by NATO ACCS (see separate entry).

UPDATED

Radar Integration System (RIS)

The Radar Integration System (RIS) is a system that integrates all elements of a NATO Air Defence Ground Environment (NADGE) and a Control and Reporting Centre (CRC). It uses an expandable open systems architecture for hardware and software featuring redundant Ethernet Local Area Networks (LANs). RIS integrates the air defence display subsystem, the H5118 Central Computer (CC) and most NATO radars, such as RSRP (Radars for the Southern Region and Portugal), 2-, 3-D and coastal radar.

The RIS controls and collects radar sensor data from local and remote radar heads of various types and forwards this data to the main processing element (H5118) within the NADGE Control and Reporting Centre (CRC) to support the multiple radar tracking function. The radar plot and mask data is also processed and displayed on the existing air defence consoles.

RIS provides an open system architecture that will accommodate the migration of all currently allocated air defence functions from the CRC to the RIS.

Features of the RIS include:
(a) interfaces with any radar and to a wide variety of air defence systems
(b) monitors, diagnoses and controls all resources
(c) redundant system
(d) high MTBF, low MTTR
(e) error tolerant software
(f) composed of commercial-off-the-shelf (COTS) hardware and COTS software operating system
(g) distributed intelligence and a capability to accommodate distributed switching
(h) applications software in Ada
(i) growth capability to accommodate external site communications (Links)
(j) built-in test capability.

The RIS design provides flexibility to accommodate interfaces to any radar and a wide variety of air defence systems. Currently, RIS is installed in Denmark, Greece, Turkey and at the NATO Programming Centre in Glons, Belgium. The radar interfaces include 2-D radars with video extractors, coastal and 3-D radars and

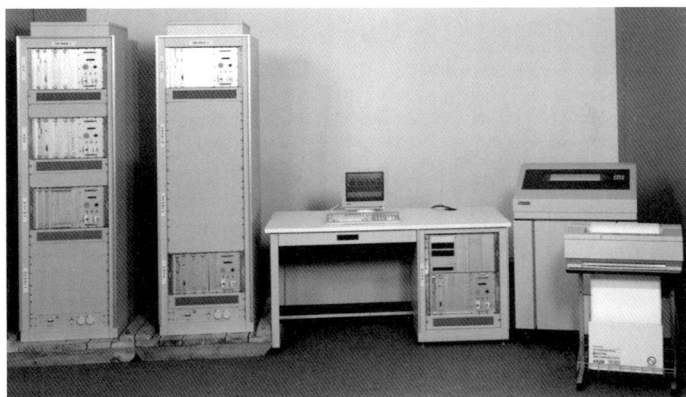

Radar Integration System (RIS)   0009943

medium power radars. The system provides distributed intelligence and a capability to accommodate distributed switching. The LAN architecture provides TCP/IP services that allow interfacing with other networks, including gateways with protocol conversion facilities without restructuring the system. The RIS controls, monitors and diagnoses all system resources connected to it.

RIS is furnished in a redundant LAN configuration providing a high MTBF. Modular hardware configuration provides easy access for corrective maintenance and a low MTTR. COTS hardware and operating system software is used and application software is in Ada. The software has been developed in accordance with MIL-STD-2167, MIL-STD-2168 and AQAP 13 using CASE tools and will support ISO 7498 standard and standard terminal interfaces. The RIS uses internationally recognised standard design that allows the application of the International Standard Organisation-Open System Interconnection (ISO-OSI).

RIS is a NATO procurement structured to allow all NATO nations and agencies to procure any RIS element, in any quantity, as an option to the current contract. Complete sites, individual elements, installation, on-site support, spares, training and documentation are listed as fixed price options in the current contract. This permits users to tailor a custom system to meet specific needs, at the best possible price, without the administrative delay and cost associated with further procurement through ICB.

The system is designed to accommodate additional existing radars as well as other new radars. Extra radars can be accommodated by the incorporation of additional User Node Interface Modules (UNIMs) to support the radar required. New radars are easily accommodated by a standard UNIM. Radars with proprietary hardware interfaces can be accommodated by a UNIM specifically designed for the required hardware and software interface requirements.

While RIS interfaces to the existing NADGE site data processing equipment, other existing processors can be accommodated with hardware and/or software changes. Additionally, a new UNIM can be provided to completely replace the functionality of the air defence processor.

RIS will interface to the existing NADGE site displays and off-the-shelf plasma and colour displays can be provided that either replace or augment the NADGE displays. Large-screen displays can also be provided for existing NADGE displays and as new displays.

While the system is currently designed to be installed in fixed NADGE sites, the RIS elements can be installed into shelterised configurations to accommodate mobile or transportable requirements, such as the interim Deployable ACCS Component.

Status
The RIS project undertaken by AYESAS for NATO is a major programme managed by the NATO Maintenance and Supply Agency (NAMSA) on behalf of the nations of Denmark, Greece and Turkey. The system is also being installed at the NATO Programming Centre in Glons, Belgium. The contract, worth "over US$20 million" was signed in 1991.

The system was successfully fielded at 14 sites in Denmark, Greece and Turkey, plus the Belgian site, and is believed to have been completed in 1998.

Contractor
AYESAS (Aydin Yazilim ve Elektronik Danayi A.S), Ankara.

UPDATED

Oman

Oman air defence system

To make the most effective use of Oman's defence force, a master command and control network has been provided to give the necessary early warning and tactical control and command link to the defence force. The then Marconi supplied the bulk of the radar, communications and command and control equipment needed.

Description
The communications system has two main centres, connected by strategically sited terminal and repeater stations. These link the air defence operations centre with the two sector operations centres, each of which has its own surveillance radar station.

Defence centres in the vicinity of Muscat in the north and, along the border with Yemen in the south are linked by a tropospheric scatter system running the length of the country and fed by a short-haul line of sight radio links. This network is used to convey processed data and high-priority communications from radar sites and sector operations centres to the main operations centre.

Status
A £38 million contract to extend and update the Oman Integrated Air Defence System was awarded to Marconi in 1985 and the system became operational in

For details of the latest updates to *Jane's C4I Systems* online and to discover the additional information available exclusively to online subscribers please visit
jc4i.janes.com

The interior of an air defence shelter

1989. The major items of new equipment were two Martello S713 long-range 3-D radars, their associated MACE display and data handling systems. Part of the contract involved updating and expansion of existing Sector Operations Centres (SOCs) and Control and Reporting Centres (CRCs), as well as the provision of a new CRC. In 1999 a contract was let to Alenia Marconi Systems (now AMS) for a new and updated air defence system. This included radars, communications and CRCs. In addition, AMS was to provide a new simulator system for an air defence college.

Contractor
AMS, Camberley, Surrey.

UPDATED

Saudi Arabia

Peace Shield

Peace Shield is the C3I system developed for the Royal Saudi Air Force (RSAF). The system consists of a Command Operations Centre (COC); five Sector Command Centres (SCCs) collocated with Sector Operations Centres (SOCs) at the main operating bases; two base operations centres, one collocated with an SCC/SOC. In addition, it integrates 17 combined long-range radar and remote-controlled air/ground radio communications sites; and an associated telecommunications network. Central command is executed from Riyadh, while sectors are controlled from Dhahran, Raif, Tabuk, Khamis Mushait and Al Kharj. The total system includes 164 sites and more than 1,600 communications circuits. The communications network includes HF radios, a mobile telephone network, microwave line-of-sight radio systems, a store-and-forward message-switched network and a kingdom-wide fibre optic system. The Peace Shield system also links the networks of the Royal Saudi Land Forces (RSLF) and the Royal Saudi Naval Forces (RSNF).

Data from the RSAF's 17 General Electric AN/ FPS-117 long-range 3-D radars and six Northrop Grumman AN/TPS-43 tactical radars feed the system, together with data from 10 AWACS ground entry stations. The RSLF's AN/TPS-43 radars, the Raytheon Improved HAWK air defence missile system and the radars of the RSNF are also integrated into the system.

The system integrates much existing radar and communications equipment with Hughes AMD-44 workstations, Hughes HDP-6200 large-screen displays, a modern data processing architecture and advanced software. Much of the technology is similar to that developed for IUKADGE.

Status
The procurement of the Peace Shield system was conducted on a government-to-government basis with the Electronic Systems Division (ESD) of the US Air Force managing the contract. Hughes Aircraft Company (now Raytheon) was awarded a contract worth US$837 million by ESD for the Peace Shield programme. (Total programme cost was US$5.6 billion.) Peace Shield became operational in late 1996 and is being continually upgraded under a maintenance and development programme.

Contractor
Raytheon Company, Garland, Texas, USA.

UPDATED

Spain

Fully Integrated Tactical System (FITS)

The Fully Integrated Tactical System (FITS) is an airborne tactical system for Maritime Patrol applications. FITS integrates mission sensors, aircraft navigation and communications, and weapon control systems when applicable. It is designed to be installed over a wide variety of platforms and can be used for a range of maritime applications.

Air Segment
FITS has an open architecture design which is adaptable to different sensor configurations depending on a customer's requirements, including search radar, infra-red/electro-optic (IR/EO) turret, ESM/ELINT, acoustics, MAD, Automatic Identification System (AIS), IFF interrogator, SATCOM and line of sight communication/datalink. Sensor integration can include a variety of sensor models and manufacturers. Sensors already integrated with FITS include the Sea Vue (Raytheon) and the EL/M-2022 (Elta) series of surveillance radars, and the Star SAFIRE I/II (Flir Systems) IR/EO turrets. The main functions of FITS are:
(a) Provide an intuitive Human-Machine Interface (HMI) based on standard X-Windows Graphical User Interface (GUI).
(b) Provide control of mission sensors, from any of the operator consoles.
(c) Display of sensor video and data
(d) Create and maintain a tactical situation picture over digital map/chart symbology (Tactical Situation Window)
(e) Create and maintain a tactical database, with automatic mission data recording
(f) Provide information and control of aircraft navigation to perform effective search planning
(g) Provide multisensor tracking capability (Data Fusion)
(h) Provide decision aids to the operator (Tactical Computations)
(i) Management of communications and tactical data exchange (radios, datalink Link-11 and Link-16) with the Mission Support Centre or any co-operative unit.
(j) ASW-specific functions (sonobuoy and weapon inventory management, launch patterns)

Tactical operators can exercise sensor control and other functionalities by means of multifunctional consoles (MFC), which are connected to central processors through a high-speed local area network. Typical system sizes are 6+ consoles for P-3 Orion, 4 consoles for C-295 ASW, 2 for CN-235 and 1 for C-212. Each MFC is fitted with its own processor, a high-resolution colour display, two multifunction touch panels (for sensor control and tactical functions), a keyboard and a trackball. The MFCs are physically and functionally identical, although they can be reconfigured according to an operator's assigned role. Information is provided to tactical operators in a multiwindows display, which includes a tactical situation window, video windows (live IR/EO video or recorded video, both IR/EO and radar), sensor windows (radar, ESM, acoustics, MAD) and other special windows (such as navigation data, datalink and communications control, display management). Windows can be resized and moved on the screen.

There are two redundant central processors: the Tactical Management Processor (TMP) which acts as system server and main data processor and the Tactical Interface Computer (TIC) which manages real-time information exchange between sensors and the aircraft navigation and communication equipment.

FITS also incorporates a data transfer unit for data loading into the system (pre-flight mission data) and downloading (for post-flight analysis). The system includes video recorders to allow recording of radar and IR/EO video signals and the replay of recorded images. A cockpit display is also included to provide the flight crew with the tactical situation, navigation information and sensor video feeds. An associated control unit allows the pilot video source selection and tactical symbology management. Additionally, a colour display is provided in the ordnance area, when appropriate to the aircraft type and configuration, for sonobuoy and armament management.

Hardware and Software
FITS uses ruggedised commercial-off-the-shelf (COTS) hardware, integrated through a high-speed Local Area Network (LAN). It has been designed following an Open Systems Architecture (OSA) concept, using standard interfaces, providing flexibility for the addition of new sensors and equipment. The FITS application

A four-console FITS installation (EADS CASA)   *NEW*/1029868

One of the multifunctional consoles in a four-console FITS installation, showing touchscreen and keyboard (EADS CASA)   ***NEW**/1029867*

software has been developed in C++ language, with a modular approach to maximise modifiability and reusability. A variety of commercial software has been used in the development: Unix and VxWorks operating systems (although software portability allows migration to other operating systems, such as Linux), the geographical information system (GIS) and the relational database management system.

Ground Segment

The Mission Support Centre (MSC) consists of a local network of COTS hardware, with a set of commercial and specific software applications. The MSC is tailored for the specific customer requirements, providing a range of capabilities from a compact system for a Fisheries Surveillance operator, to a complete facility for a military service responsible for Maritime Patrol and ASW/ASuW missions. The MSC provides the following functionalities:
(a) Mission planning and preparation, including the generation of the pre-flight data insertion package (PDIP) for data transfer into the FITS on board.
(b) Briefing (tactical situation, inventories, sensor-specific briefings, weather reports, NOTAMs and checklists) and debriefing (reports, mission replay)
(c) Tactical information exchange with the aircraft (voice, data and image) for an automated update of the tactical situation picture (mission follow-up).
(d) Mission analysis for each sensor (ESM/ELINT, acoustics, Radar/FLIR video analysis) and using multisensor data fusion techniques
In addition to the integrated mission system (airborne and ground segments), EADS CASA offers a Tactical Trainer which allows tactical crew to operate FITS and mission sensors in predefined scenarios through an identical HMI to the one on board the aircraft. The system simulates the operation of the sensors and the aircraft navigation.

Status

An integrated mission system, TDMS, the predecessor of the current FITS, has been in operation in the Maritime/Fisheries Surveillance role on board two CASA CN-235 of the Irish Air Corps (IAC) since 1994.

CASA (now EADS CASA) continued the definition and development of Integrated Mission Systems, mainly through a succession of company-funded technology demonstration programmes. In early 2001 FITS was demonstrated in flight in a C-295 demonstration aircraft. This same model, the C-295 equipped with FITS in a four-console ASW/ASuW configuration and a customised sensor suite, was selected by the United Arab Emirates. In December 2001 a contract was awarded by Mexico to upgrade 8 C-212 aircraft including the installation of FITS.

FITS in the MPA/ASW role has been retrofitted to five P-3Bs of the Spanish Air Force. This modernisation programme includes the upgrade with FITS and a complete set of new sensors and equipment. It is to be fitted to nine of a batch of 12 P-3A aircraft acquired by Brazil from the US. FITS may also be under consideration for other P-3 upgrade programmes, such as New Zealand.

The mission system for the recently contracted US Coast Guard's CN-235 MPA as part of the Deepwater programme will also include FITS technology associated with the integration of mission sensors and navigation/communication systems for the aircraft.

Manufacturer

EADS CASA, Madrid.

UPDATED

I-ARS

The I-ARS ('previous to ACCS' NATO systems) system forms a part of the Spanish air defence system and provides data processing and presentation for a new generation of Air Command and Control Centres, replacing SADA 2000-CPS/SADAC (see separate entry). The system integrates Link 11 and uses commercial hardware for processing, communications and displays (workstations). It also includes COTS software (HMI, NATO validated compilers, databases) for all supporting functionality.

Operational Functions
(a) Interceptor Control
(b) Radar Data Acceptance and Process
(c) Radar Data quality and supervision control
(d) Remote Radar Management and Control
(e) Active Tracking, fusing Short-, Medium- and Long-Distance 2-D and 3-D Radar data
(f) Threat Evaluation
(g) Air Space Management
(h) Interoperability with Surface-Air Missile Control (SAM)
(i) Track Management
(j) Identification and Flight Plan Processing
(k) Track Exchange with other adjacent centres
(l) Track Exchange with Tactical Naval Network (NTDS).

Support Functions
(a) Offline Graphical Editor for Exercise preparation and storage
(b) Real-Time Simulation of air routes including radar clutter and noise effects
(c) Simulation at the message level of all radar and external sources data (Link 11, Crosstell, SAMs/SHORADs)
(d) HW/SW Monitoring and Diagnosis for failure detection, isolation, performance monitoring and reconfiguration
(e) Configurable Recording of operational and input data
(f) Playback: Replay of real air situation synchronised with voice recording
(g) Analysis and report formatting of operational data
(h) Automatic and manual processing chain switch between main servers
(i) Integrated Database with security management.

Human/Machine Interface
(a) Multirole console
(b) Access to secure e-mail
(c) 'On-line' integrated help (CBT)
(d) Local map generator
(e) Personalised workstation configuration
(f) Standard symbology.

Command Extensions and Tools
Force Deployment Planning
(a) Aircraft, radar and SAM/SHORAD deployment display on GIS cartography
(b) Edition of radar, SAM/SHORAD and aircraft characteristics and deployment
(c) Multiradar surveillance coverage computation at selectable altitude levels
(d) SAM/SHORAD Surveillance and Fire coverage computation at selectable altitude levels
(e) Action radii of deployed aircraft at different profiles (BBB, BAB, ABA, and so on).

Operations Planning Decision Aids
(a) EDIPO
Air Operations Plan management and edition in structured standard sections. ATO generation for all NATO air missions (SEAD, AI, SAR).
(b) EDIACO
Georeferenced Graphical edition of all NATO standard ACMs. Spatio-temporal operative validation of ACMs and ACOs.
(c) PLAN VIEWER
Time and logistic validation of air operations plans. Weapons Effectiveness Analysis for optimal target assignment.

Route/Air Mission Planning
(a) Detailed air route planning on GIS georeferenced cartography
(b) Aircraft independent and reconfigurable
(c) Aircraft manoeuvres and envelopes characterisation editor
(d) Plant and Profile graphical display and edition of air routes
(e) Multisegment (take-off to landing) A/A and A/G missions
(f) Computation of Time/Distance/Fuel and non-specified altitude/speeds
(g) Automatic insertion of missing (required) segments within the air route
(h) Previsualisation and printing of preconfigurable Pilot Book.

Specifications (Typical medium-size centre configuration)
(a) General
Situational Display (SIT): 1,200 n miles × 1,200 n miles
Active + Passive Tracks: 500+100

An I-ARS control room (Indra)   0142036

Radar Plots/10 s: 5,000
Air Missions: 50
Flight Plans: 500
Radar: 10
Radar interfaces: DDE/DTE/HDLC/ASTERIX
ATC Link: 1
Link 1: 4
Link 11B: 8
Link 11: 2 (no SSSB)
Operational + Data Entry + Planning Working Positions: 20+10+5
(b) Software Development
NATO validated Compilers: ADA, C/C++, JAVA
HMI: X-WINDOWS, MOTIF, GIS
(c) Hardware Platform
Operating Systems: UNIX, WNT
64 bits technology: SUN, DIGITAL, PC
(d) Network
Protocol: TCP-UDP/IP
Support: ETHERNET, FDDI and ATM
Monitoring: Network traffic supervision
(e) Database management
Relational and Distributed: ORACLE + MASTER-MASTER replication
Interface with ADA: PRO-C
Local and remote access to system data: SQL-NET, INTRANET
Tools for easy generation of forms and reports: J-BUILDER, J-DEVELOPER, J-ODBC

Status
The first I-ARS system has been operational since March 2001, at Zaragoza Air Base. A second system is currently being installed at the Canary Islands, which is expected to be fully operational from January 2004. Similar systems to I-ARS are installed in several South American countries.

Contractor
Indra, Madrid.

UPDATED

....

Spanish air defence system (SADA 2000 - CPS)

The SADA 2000-CPS system forms part of the Spanish air defence system, covering mainland Spain and the Balearic Islands.

Prime contractor for the SADA system, originally called Combat Grande, was COMCO Electronics Co, a company jointly and equally owned by Hughes Aircraft (now Raytheon) and CESELSA (now called Indra). The Combat Grande programme was designed to automate Spain's manual air defence system by developing a Combat Operations Centre (COC) and Sector Operations Centres (SOCs) and by the modernisation of a number of long-range radar and communications sites throughout the Spanish peninsula.

SADA has been considerably upgraded over the years and notable among the enhancements made by Indra are the integration of surface-to-air missile groups, the integration of F/A-17 interceptor aircraft and the recent upgrading of the SADA radar sensors and display systems (SADA 2000).

The CPS system, designed by Indra, is a new element which has been added to the operation of the previous SADA. It uses new COTS computers and software, parallel operator array display systems and a digital communication control system (VCCS). Software for the system is written in Ada, both data processing chains feature two 32-bit, general purpose computers and the display consoles are controlled by 32 bit microprocessors.

Status
SADA has been operational since 1976. The CPS has been in operation since 1992.

Contractor
Indra, Madrid.

UPDATED

One of the eight long-range radar sites in the Spanish air defence system

Sweden

STRIL/STRIC air defence system

STRIL is a fully automatic air surveillance and operations control system operated by the Swedish Air Force. It is based on air defence sectors, each having a Sector Operations Centre (SOC) which receives radar data from static and mobile Control and Reporting Centres (CRC).

Inputs to the system come mainly from high- and low-level air surveillance radars as well as AEW (FSR890), but a back-up visual reporting service is included to supplement the radar data and to replace it if the radar input is blocked. Information from all these sources is fed into a central data store whence it is extracted for selective presentation to controllers having specific territorial assignments. Although primarily an air defence system, STRIL is linked to coastal and anti-aircraft artillery and missiles systems. Currently, the forces controlled by the system are the interceptor aircraft, AA guns and Improved HAWK surface-to-air missiles.

The STRIL centre is also linked up to the civil defence organisation and can alert them and warn industry or the civilian population.

If the threat cannot be countered by the forces under the control of one centre but can be countered by those of another centre, the system provides both voice and transmission links for distribution of information to other centres. Both narrow and broadband microwave links are employed for the exchange of data between centres.

CelsiusTech has developed an export version which is known as Airguard.

Development
In October 1990, a contract was signed between CelsiusTech Systems AB (now SaabTech Systems AB)and the Swedish Defence Matériel Administration (FMV) regarding the design, production and delivery of a new generation of air defence system for the Swedish Air Force. The contract represents the largest individual project in the computer system business ever carried out in Sweden.

The project, known as STRIC (or STRIL/C-90), covers hardware and software for a number of air defence centres located all over Sweden. In addition, the project includes extensive services such as installation, documentation, training and maintenance preparation. Approximately half the project costs are for software design.

STRIC will be part of the future system for air defence and surveillance which will gradually replace the present STRIL 60. The primary use of STRIC will be in the command and control of military aircraft, the co-ordination of air defence efforts and in air traffic control of transport aviation, and it will also be used to alert the civilian population to imminent air raids by ordering the activation of the air raid alarm. In peacetime STRIC will be used for the surveillance of Sweden's air space. From the middle of the 1990s and well into the 2000s, STRIC will be the heart of the air defence command and control systems of the Swedish Air Force.

The present system for air defence command and control, which dates back to the 1960s, has been updated gradually, but a large part of the equipment is becoming outdated and, consequently, expensive in operation and maintenance. This is, in itself, a strong enough reason for equipment renewal, but other factors, too, have been important in the decision to procure STRIC. Examples of such factors are:
(a) introduction of new sensors (ground, airborne radar)
(b) command and control of new aircraft (Gripen)
(c) higher demands on performance, capacity and flexibility
(d) integration with new weapon systems.

Among the fundamental tasks to be solved in an air defence centre are surveillance of the air space and elimination of threats. For this purpose, the personnel in the air defence centre must be able to produce a reliable situation picture of all aviation activities in the airspace and, based on this picture, analyse potential threats and incidents, and counteract against intruders or enemy aircraft. Doing this efficiently requires access to modem sensor, communications and computer technology.

STRIC gets its information from a number of sources, mainly consisting of radar stations. The information is transferred to the centre via modern data communications networks and is then processed in the STRIC computer system.

Airguard is the export version of STRIL/STRIC   0009945

This provides support functions for operators and decision makers in their surveillance and command tasks. A key function in this is multisensor tracking, which makes it possible to have refined automatic detection and tracking of targets by weighing together information from several sensors.

Description

The technical solution of STRIC is based on CelsiusTech's system Concept 2000 and draws on the company's experience of developing command systems for the Danish, Finnish and Swedish navies where the software in each system comprised approximately one million lines of code written in Ada. While designing these command systems, the requirements of systems such as STRIC were taken into account. Thus, much of the software is already developed and tested and can be used in STRIC.

The technical solution is largely based on commercial standards and products. The backbone of the STRIC system is a Local Area Network (LAN) and a telephone switch. The LAN, a standard Ethernet, links together the computer nodes in the system. The computer system is composed of commercially available products from the IBM RS 6000 POWER stations range. A number of operator workstations are connected to the LAN and an important feature of the workstation concept is that any operator position may be used for any operator role. The system gives the user access to the specific operation interface (input and presentation) required for the selected operator role.

All centres in STRIC will be identical as regards the technical system and its software. This offers great advantages in, for example, training, maintenance and operation. In addition, the technical architecture with standardised interfaces makes it easy to expand or upgrade the STRIC system.

During different stages in war, crisis and peace, the needs of operators may change as circumstances differ. Within the limits of the total number of operator workstations the distribution between different sites may, therefore, alter. The operator workstations are movable and will, in peacetime, to a certain extent be used for training and incident readiness.

In order to realise a project of the size of complexity that STRIC represents, the use of advanced methods and tools for development and project realisation are necessary to facilitate high project reliability and SaabTech Systems AB is employing a new and powerful development environment called APEX generating software in Ada.

Status

The STRIC system has been operational since 1998 and is organised in the form of two airborne radar groups with the associated ground organisation. In 2001 two modification contracts worth SKr41 million and SKr28 million were announced, both to improve integration with the Gripen aircraft and Erieye AEW platform (see separate entry). In 2003 a further modification contract worth SKr73 million was announced for unspecified improvements.

Contractor

SaabTech Systems AB, Järfälla.

UPDATED

Track Reporting System

SaabTech Systems' Track Reporting System (TRS) is a high performance, cost effective system for air surveillance, air defence and air traffic control. The TRS is highly configurable in a number of centres and in a number of workstations. The TRS is a hierarchical system built up from a number of interconnected Superior Air Defence Centres (SADC), Air Defence Centres (ADC) and Track Reporting Posts (TRP), all of which use the same basic hardware and software and is configured in accordance with the specific task of the Centre. Each centre assembles, processes and presents an individual Recognised Air situation Picture (RAP). A uniform RAP is compiled in the highest level using a special developed correlation function. It is presently adapted to use commercial-off-the-shelf (COTS) hardware and software.

Each TRS may either be configured as a Track Reporting Post (TRP) or an Air Defence Centre (ADC) or a superior ADC (SADC). The TRP is the lowest level in the hierarchy and is used to compile an air situation picture for the superior ADC. The ADC's have the same functionality as a TRP with the addition of Weapon Control functions. The ADC's report the air situation picture to the SADC.

The overall operational tasks of a TRS include:
(a) Production of a clear, current and complete air situation picture for the airspace covered by the system;
(b) Fusion of overlapping sensors;
(c) Identification of targets;
(d) Efficient management of air defence resources
(e) Command and Control of air defence operations;
(f) Identification of air movements to own air defence batteries;
(g) Interchange of the air situation picture with associated centres;
(h) Weapon assignment;
(i) Fully automatic, semi-automatic or manual operation.
In order to carry out these tasks, the TRS includes the following technical functions:
(a) Threat evaluation in defended areas
(b) Trial intercept calculations for Ground Controlled Intercept (GCI) and Firing Units
(c) Ground Controlled Intercept calculations with Stern Cut-off and Pursuit attack calculations;
(d) MultiSensor Tracker (MST);

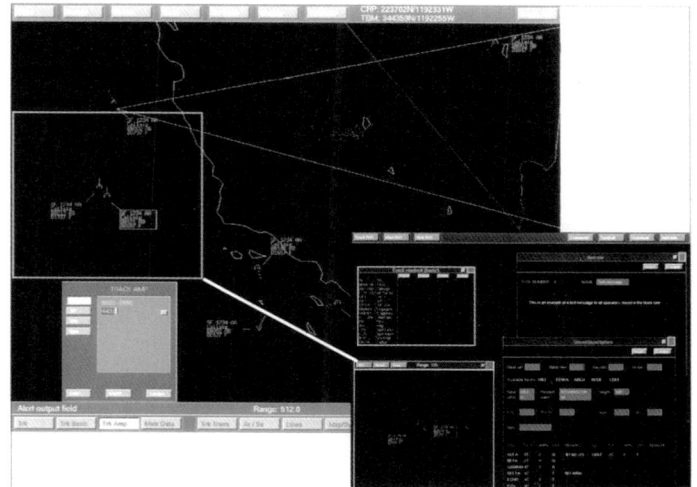

SaabTech Systems' TRS utilises a two-screen workstation. Information in selected areas can be exchanged and displayed in windows 0084491

(e) Firing Unit (FU) close control support with the ability to send target data and commands to the FU and receive status data;
(f) Passive Detection Unit (PDU) close control support with the ability to send command and track data to the PDU and receive strobe data;
(g) Strobe tracking and triangulation with the ability to maintain tracks at the intersections of strobe crossings received from subordinate TFS and from PD Units and those created locally. The triangulation calculations are handled separately from the MST;
(h) Raw radar presentation with the ability to display one radar with raw video and at the same time display plot information from all connected radars.

Status

TRS was developed for unspecified customers from South East Asia and the USA but is no longer in production.

Contractor

SaabTech Systems AB, Järfälla.

UPDATED

Switzerland

Florako air defence command and control system

FLORAKO will replace FLORIDA (see separate entry), which was supplied during the 1960s by the former Hughes Aircraft Company, now part of Raytheon Systems. Under the eight-year contract, a Raytheon/Thales joint consortium will provide new radar sensors, new communications equipment throughout Switzerland and will completely renovate the existing command and control centres. Phase 1 of the contract, for risk reduction and prototyping, was awarded in 1997. Phases II /III, which included the core system design, new primary and secondary surveillance radars, air defence hardware equipment and software, and new voice and data communications was awarded to the consortium in November, 1998. The Phase IV/V contract was awarded a year later and includes new multifunction military radars, new monopulse secondary surveillance radars, modifications to military radar sites and integration of the new radars into the FLORAKO system. Raytheon is responsible for the global system architecture and system integration and is providing new hardware and software for the Swiss command and control centres, data and voice communication systems and the Swiss Airspace Management Cell. Raytheon has subcontracted the data and voice communication system to Siemens Switzerland. Thales is providing new primary and secondary surveillance radars and some new software for the command and control centres. Oerlikon-Contraves, as a subcontractor of Thales, will modify the Swiss military radar sites and assist in the installation and integration of the new radars.

In late 2000 it was announced that BarcoView would supply PVS 5600M PCI graphics controllers, which would be integrated with a HPC360 computer driving a Sony 20 × 20 in high resolution CRT display as part of the operational display console.

Status

The system is still under construction. The first radar stations were installed in summer 2001. The system is expected largely to have replaced the Florida system by mid-2003 and the project is expected to be complete by 2005.

Contractor

Raytheon, Marlborough, Massachusetts.
Thales Air Defence, Bagneux.

UPDATED

Florida air defence command and control system

Florida is the name given to a computerised air defence command and control system accepted and declared operational by the Swiss military authorities in April 1970. General contractor for the system was the Hughes Aircraft Company (now Raytheon Systems).

Description

The system consists of several military radar stations with 3-D radars and air defence direction centres, including computers, display consoles and associated equipment. Information from the radar station is fed into conversion equipment in underground air defence direction centres and processed in turn by a high-speed general purpose computer.

This computer automatically establishes speed, heading and altitude of an unidentified intruder. Display consoles present a constantly updated picture of the aircraft's flight track as well as information on the various weapons available, their launch ranges, velocities, armament, restrictions and time-to-intercept.

Should the target be identified as an immediate threat, the air defence commander can electronically request interceptor aircraft or surface-to-air missiles to intercept and can also alert the civil defence organisation.

The radar used is a long-range three-dimensional radar with a planar array antenna using the Hughes elevation frequency scanning technique. This provides simultaneous range, bearing and altitude data. An IFF system is incorporated. Track acquisition and updating is automatic.

Processed information is fed to air defence direction centres where it is accepted by a computer. This takes in information from the missile sites, airfields and other military installations and stores it for use in selecting defensive measures.

It can also simulate air battles for training and instruction and can be used as a general purpose data processing centre.

The computer also stores information on weapons available, their launch ranges, velocities, armament, restrictions and time-to-kill. The computer, the Hughes H-3324, is of the same type as those associated with the radars but has a much larger storage capacity.

In the direction centre, the display consoles are arranged to display only that information which is pertinent to the tasks of the operator concerned. The principal functions are:

identification officer – responsible for designating radar tracks as friendly or hostile aircraft

interceptor director – deals with controlled interceptor and the assigned hostile aircraft. Specific data about this intercept is presented, including attack geometry, altitude, speed, time to intercept and so on

air traffic co-ordinator – concerned with military aircraft requesting clearance to cross civilian airways

missile officer (surface-to-air) – in close contact with the missile batteries and assigns targets to the appropriate site

chief of air defence – concentrates on deployment and the threat. Total composite air situation and threat boundaries are important to him.

Status

The Florida system became operational officially in April 1970. In the 1979-80 period the Swiss government studied methods of updating the system, particularly in the sphere of providing air surveillance and ground control of interceptor aircraft in the lower airspace. Field tests and evaluations were held for competing systems, among them the Hughes VSTAR and the Alenia Difesa MRCS-403 mobile air defence system with its 3-D radar RAT-31S. In early 1982 it was announced that under a Swiss air defence programme called Taflir, a version of the Northrop Grumman AN/TPS-70(V)2 radar, known as Vigilant, had been ordered and was delivered in 1984 for evaluation. These trials were understood to have been successful and it was reported that an order was placed for systems to be delivered, commencing in the late 1980s. However, no official confirmation of this has been made.

The system is now being replaced by the Florako system (see separate entry).

Contractor

Raytheon Systems, Fullerton, California, USA.

VERIFIED

Turkey

Turkish Air Force Mobile Radar Complex C3 element

AYESAS is currently under contract to supply the command, control and communications element of the Turkish Mobile Radar Complex. The air defence C3 system provides extensive capability using computer, display and communications technologies, packaged in a sheterised, highly mobile configuration.

The system provides for multiple radar input of both analogue and digital plot data including IFF. It provides automatic datalink communications for Link 1, Link 11, and ATDL-1 with a growth capability to Link 16. The user-friendly displays provide a complete implementation of all air defence functions including a comprehensive weapons allocation capability and an automatic intercept control (guidance) capability. Manual and automatic identification is provided.

Extensive communications facilities are incorporated: HF, VHF and UHF radios with Have Quick II and direction finder; a secure microwave subsystem for radar

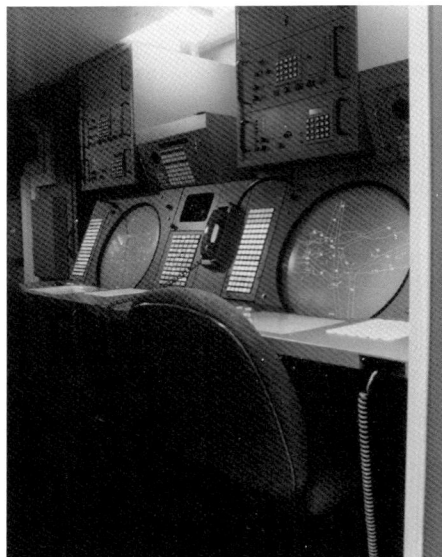

Turkish Mobile Radar Complex C3 system shelter

and radio remoting; a secure microwave subsystem for voice and data connection to national telephone and telegraph services; as well as voice and data cryptographic equipment, facsimile and teletype equipment. A communications panel is provided at each of the air defence consoles for radio, landline, telephone and intercom voice communications. A modern, digital communications switch supports intercom, interphone and telephone services.

Tracking features include: automatic track initiation; automatic tracking using multiple radar inputs; automatic jam strobe triangulation and track initiation; and command tracking based on interceptor guidance calculations.

Identification features include: complete IFF/SIF Mk XII implementation (Modes 1, 2, 3A, 4 and C); automatic ID area and safe corridor identification; automatic flight plan correlation; and complete identification taxonomy.

Weapons features include: automatic threat evaluation; automatic weapon assignment (SAM and interceptor); and automatic interceptor guidance.

There are three subsystems. The Radar Subsystem interfaces Air Defence and Air Traffic Control radars and accommodates multiple radar inputs. The ruggedised, VME-based Computer Subsystem employs RISC computers with expandable memory and interfaces. It uses distributed architecture and an industry standard Ethernet LAN for internal communications. The Display Subsystem provides monochrome PPI with multiple modes. Colour raster, large-screen displays and plasma panel displays are also available.

Status

The programme contract, which is worth approximately US$220 million, was awarded in 1990.

Contractor

AYESAS (AydinYazilim ve Elektronik Sanayi A.S.), Ankara.

UPDATED

United Kingdom

Advanced Mission Planning Aid (AMPA)

The Advanced Mission Planning Aid supports mission planning, targeting and tactical air drop for both fixed- and rotary-wing aircraft and provides a method of receiving the Recognised Intelligence Picture (RIP), Air Tasking Order (ATO) and Airspace Control Order (ACO) as well as informal communications. It can operate either stand-alone or networked. It can support a variety of aircraft types, as over 80 per cent of the mission planning software is common.

Unix-based, the typical configuration consists of a number of workstations on a LAN, providing planning and briefing facilities. Data is held in a RAID (Redundant Array of Independent Disk Drives). When deployed, data for the operational area is held in external disks and is updated by reachback using web browsers. A squadron will typically have a separate LAN, and this will be joined to other squadrons and to the Wing HQ by either a LAN or WAN, with replicated databases. In the event of lost communications or independent operations, AMPA will function on its own until reconnected, when the databases are updated automatically.

Strategic connectivity is provided through interfaces to operational CIS such as RN CSS and RAF CCIS and through them to the UK strategic system, JOCS. Connectivity with CSS was proven in 2001 from an RN CVS supporting Harrier GR7. Connectivity with RAFCCIS was proven during JWID 2001. The RIP, air tasking and e-mail communication all pass via these interfaces.

The system tools allow detailed mission planning with all possible weapon loads, including Alarm, Brimstone, Paveway I, II and III and Storm Shadow, as well as supporting a number of reconnaissance pods. Target identification, threat avoidance, adherence to ROE and deconfliction are supported and a 'flythrough' capability allows visualisation of the mission. The system enables rapid replanning

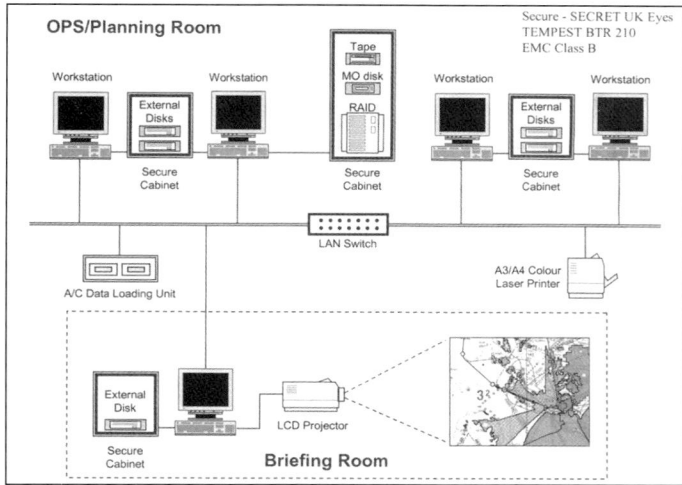

A typical configuration for AMPA in the non-deployed environment (EDS) 0132528

to take place to meet changing circumstances and has increased the operational flexibility and tactical freedom of pilots, through the greater confidence in deconfliction that it provides.

Status
The AMPA entered UK operational service with Harrier squadrons in 1996 and with Tornado squadrons in late 1999. It has been used operationally supporting both these aircraft types in Iraq and Kosovo. Limited functionality was provided for C130J in 1998 and this was upgraded in 2003. Functionality to support rotary-wing platforms is available and has been used by Sea King (W) Mk 7 and aircraft from the Joint Helicopter Command during operations in Iraq.

Manufacturer
EDS Defence Ltd, Hook, Hampshire.

UPDATED

JTIDS Air Platform Network Management System (JAPNMS)

The JTIDS Air Platform Network Management System (JAPNMS) is the first UK-fielded equipment capable of performing real time, dynamic management of a JTIDS Link 16 (TADIL-J) network. The system, which entered service at Royal Air Force Air Defence Ground Environment (ADGE) sites in July 1998, comprises a number of Network Manager Stations (NMS), providing the operator display and communications facilities and Ground Buffers containing the Link 16 terminal, amplifier and antenna.

Each NMS provides the necessary functionality to execute the responsibilities of the Link 16 Network Manager. This enables the operator to view, select and modify predefined network plans to suit the operational scenario and communications needs and providing assistance to promulgate that plan to all likely network participants. Once the network is active, the network manager is presented with details of time-slot usage and requests and a clear indication of the status and relative connectivity of each participant in the network. The operator is subsequently able to reassign time-slot blocks, relay capacity and cryptographic variables in real-time to maintain the robustness and coherence of the required communications and operational requirements. For peacetime operations, the

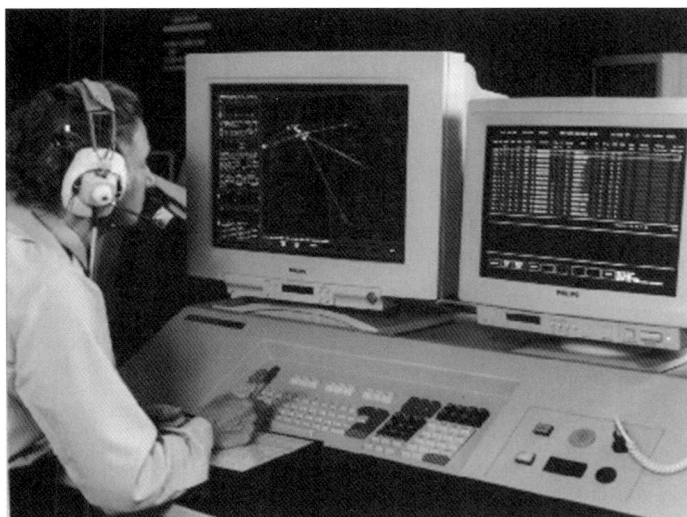

JAPNMS Workstation 0052051

NMS automatically alerts the operator to any violation of the JTIDS frequency clearance constraints.

A Transportable JTIDS Facility (TJF) based on JAPNMS has also been developed. TJF offers a range of automated tools for planning a Link 16 network, monitoring network operations, policing frequency clearance criteria and dynamically adjusting network capacity in real time. It will eventually provide an extensive implementation of the tactical Link 16 C2 and surveillance messages to provide OPNET managers a view of the result of their actions and to provide weapons controllers with major interaction on the tactical information nets. Two independent operator positions, each having a twin-headed multifunctional workstation, are housed with associated communications and terminal equipment in an environmentally protected, tempest and NBC-proofed, 20 ft ISO shelter. The system also includes the facility to enable additional remote operator workstations, such as a deployed HQ entity, to access TJF functionality.

Status
JAPNMS entered service with the Royal Air Force in July 1998. TJF achieved Initial Operational Capability in September 2001.

Contractor
Thales Defence Information Systems, Wells, Somerset.

UPDATED

Lightweight Mission Support System (LMSS)

The Lightweight Mission Support System (LMSS) has evolved from Thales's Mission Support System (see separate entry). The workstation incorporates a Digital Alpha processor and is housed in a rugged case, sealed to resist severe environmental conditions, whilst remaining lightweight and fully portable. LMSS provides stand-alone briefing and mission support facilities with the capacity to add additional equipment as required. Additionally, it generates appropriate data-loading media for equipment on aircraft and ships.

Secure communications and data/digital image transfer links may be established with controlling authorities via satellite, military or public service communication links. Deployed units are thus regularly updated with intelligence, meteorological, force composition and movement data. The high-resolution flat screen colour VDU displays intelligence databases, briefing material, sensor prediction and target profiles. Tasking messages in signal format are automatically processed and displayed as briefing documents. The Accessories Pack includes a printer, Digital TLZ09 8 Gb DAT drive and Wide Area Network (WAN) Unit to provide printing, data back-up and networking facilities.
LMSS has the following functions:

Information management
A comprehensive set of databases is updated either automatically by processing messages received via a communications network, or manually following inputs from external agencies. Information currently includes:
(a) Friendly force and threat plots
(b) Air Order of Battle and Mission Plans
(c) Environmental and meteorological data
(d) Intelligence and Force Movement
(e) Updates to ships via satellite
(f) Threat radar parametric data
(g) Aircraft performance data
(h) Jane's publications and photographic images.
(i) Operational support and intelligence data transfer to deployed units.

Mission planning
LMSS provides an extensive set of data for mission planning. Applications include:
(a) Automatic tasking signal processing

The Lightweight Mission Support System (LMSS) 0001036

(b) Route and operating area planning

(c) Deconfliction

(d) Fuel planning

(e) Sensor performance prediction

(f) Preflight data insertion for land-based and shipborne aircraft and helicopters.

Briefing

A briefing folder comprising a collation of pre-defined forms can be generated automatically from the mission plan. When deployed, a portable briefing option is available.

Mission reporting and debriefing

An initial analysis of a mission, formulated as a structured signal, can be transmitted to controlling and other interested agencies via an operational command and control network. Where appropriate, facilities can be provided to analyse mission recorded data. Both of these rapid-action processes can be used to assist in the planning of subsequent sorties.

Data communications

A key function of LMSS is to act as a data communications node within a secure operational network. Data, graphic files and photographic images can be distributed to individual users within a network. Broadcast and telephone facilities are also available.

Specifications

Technical

(a) 500 MHz DEC Alpha processor

(b) 256 kbytes of RAM

(c) 21 RS-232 serial ports

(d) One parallel port

(e) Twisted pair Ethernet

(f) Twin removable 3.4 in 43 Gbytes hard disk drives

(g) 2.3 in 1 44 Mbytes floppy disk drive

(h) 3.8 in 1,024 × 768 TFT LCD display (1,280 × 1,024 optional)

(i) Integrated autoranging power supply unit

(j) 85-260 V AC 47 - 440 Hz

(k) 21-32 V DC

(l) UPS support for up to two minutes at full load

(m) Detachable IP65 and EMC sealed keyboard

(n) EMC qualified to OEF Stan 59-41 Land class A.

Physical

(a) Weight less than 18 kg

(b) Dimensions 500 × 350 × 200 mm (when stored in padded transit case).

Environmental

Operating:

(a) Temperature: 0 to + 50°C

(b) Humidity: 30% to 75%

(c) Altitude: −300 to 40,000ft AMSL

Non-Operating:

(a) Temperature −30 to +60°C

(b) Humidity. 8% to 95%

(c) Altitude −300 to 4,000 ft AMSL

Status

In operational use with the UK Royal Air Force and Royal Navy.

Contractor

Thales Airborne Systems UK, Crawley, Sussex.

UPDATED

***

Lychgate intelligence processing and dissemination system

The Lychgate intelligence processing and dissemination system connects intelligence staff at Headquarters Royal Air Force Strike Command at High Wycombe with its counterparts in the UK Ministry of Defence and other services, as well as to front-line squadrons of the Royal Air Force deployed in the UK and overseas. The system provides an advanced and rapid means to receive, collate, analyse and disseminate multimedia intelligence data from a wide variety of sources. Lychgate's deployable elements replicate the configuration at remote sites and are used to provide up-to-date intelligence data to deployed commanders.

A client-server architecture supports locations mainly within the UK from one main server site located at HQ Strike Command. This is connected to remote sites via encrypted WAN links, which ensures a comparable level of service to all users. The system will automatically interface with several military messaging networks. Applications allow users to profile, process, separate and distribute a variety of message types. Full e-mail and 'chatter' facilities are provided.

A comprehensive set of OA tools is supplied, based on the Applixware standard. The icon interface and the use of multiple windows will allow the 'cut and paste' of information from various databases, as well as the production of briefing and presentation materials. Users can interact with Lychgate's Oracle database

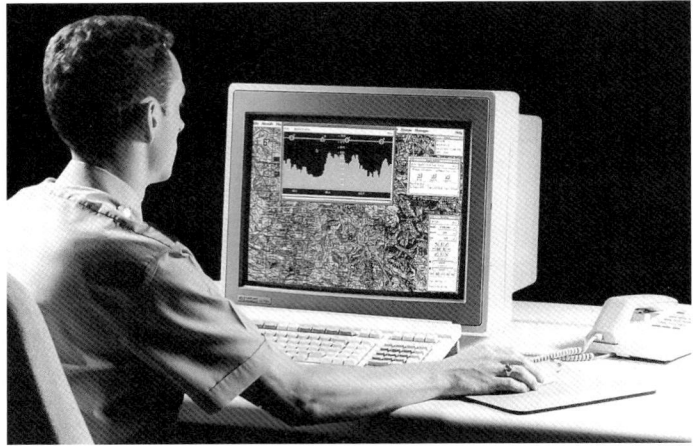

Lychgate intelligence processing and dissemination system   0001035

management system via a range of commercial-off-the-shelf (COTS) query and search tools. Information, including that from government and commercial databases can be displayed graphically using a series of analysis applications.

Lychgate supports mapping from both DMA and Mil Survey. The GIS is easily integrated with applications and supports the required map formats. The system uses a generic COTS mission support tool, the Air Force Mission Support System (AFMSS) (see separate entry), which is in service with many armed services and special operations forces worldwide. It supports georeferenced imagery through COTS software called Digital Imagery Exploitation and Production System (DIEPS), which was selected to provide commonality with the image production community.

Status

In operational use.

Contractor

AMS, Camberley, Surrey.

UPDATED

***

RAF CCIS

RAF CCIS is the single service command and control system which has replaced two systems previously used for air command and control within the RAF, the UKAIR Command and Control Information System (UKAIR CCIS) and the Air Staff Management Aid (ASMA). A parallel development, the Station and Deployed Forces Operational Management Project (SDOMP), was intended to replace the Station Operational Management Aid (SOMA), significantly extending its functionality and coverage, but since the command and control function relies heavily on status reporting from units, it was decided that a common database would support these two projects. It was also realised that the command and control functions could be satisfied along with station and Out Of Area (OOA) functionality through a single software application. The resulting project was named RASDA, which is hosted on RAF CCIS.

RAF CCIS is a Windows NT and Windows 2000 network delivering e-mail, military messaging, office automation, intranet, and operational database applications. It also acts as a platform for other applications, some of which are currently under development. It is installed at over 60 sites in the UK and overseas. The size of the installation at each site varies from a few workstations to hundreds of workstations with many servers. The users are drawn from the RAF and air elements of the joint operational community at squadron, station and headquarters levels, and the system is managed from a single operations centre at HQ Strike Command at High Wycombe.

There are four 'core' RAF CCIS sites which form an e-mail backbone and connect to the Automated Message Routing And Distribution (AMRAD) military messaging service. There are approximately 20 other major RAF CCIS sites that, together with the core sites, have servers for e-mail, file/print, messaging, directory, intranet and operational database. Server clustering is used within the fixed network to provide increased resilience between the Exchange and File/Print servers. OOA sites all have Exchange servers so that e-mail can continue to be exchanged locally when wide-area communications are lost.

RAF CCIS sites in the UK generally use the RAF's Local Data Communications Network (SECRET) (LDCN(S)) for local area networking. All sites are interconnected via the Defence Fixed Telecommunications System (DFTS) Secret LAN Interconnect (SLI) service or via bridged links to other nearby sites. On deployment, RAF CCIS uses the RAF's Deployable Local Area Network (SECRET) (DLAN(S)) for local networking with satellite links back to the UK. There is also a dial-in solution allowing users access to RAF CCIS applications via a portable satellite terminal.

The type of RAF CCIS installation at each site varies according to the number of workstations required. NT Workstations (NTWs) can be subdivided into 3 types: desktop (NTD), portable (NTP) and TEMPEST-approved (NTT). Any site with 3-6 NTWs is normally considered a small site. A small site will have a local server, which acts as both a Backup Domain Controller (BDC) and File/Print server. An Exchange Server at an adjacent medium or large site will provide the e-mail service. The remote site will be accessed across the SLI or via a LAN bridge. A site with 7-23

NTWs will normally be classed as a medium site, and a site with more than 24 NTWs a large site. Both types of site will have a full server farm consisting of a File/Print Server, an Exchange Server, and a Backup Domain Controller. The File/Print and Exchange servers are clustered together. Large sites will have extra disks, and may have more than one RASDA Application Server.   A deployed site will have a combined File/Print and BDC server, together with an Exchange server.

RAF CCIS operates with Windows NT on all servers and user workstations but the system is planned to migrate to Windows 2000/XP (or a later version). Some server functions are already provided on Windows 2000. The browser used is MS Internet Explorer.

RAF CCIS was designed from the outset to be deployable. However, none of the hardware, all of which is COTS, is ruggedised. Standard PCs and notebooks are used as deployable workstations. The deployed servers are housed in air-transportable containers that act as server racks in the deployed location. Larger deployments will include servers, workstations, printers and network infrastructure, but smaller deployments will have a scaled-down infrastructure.

Hardware

The minimum standards for COTS workstations are:

NTD/NTTs:
(a) Capable of running Windows NT and Microsoft Office Professional
(b) Minimum RAM 128 Mbytes
(c) Minimum Hard disk size 3 Gbytes
(d) Fitted with floppy and CD drives (ordinarily disabled in software)
(e) Capable of storing SECRET data (RAF CCIS uses removable hard drives to achieve this)
(f) LDCN SECRET compatible fibre NIC
(g) Minimum monitor size 17 in
(h) Minimum resolution 1,024 × 768.

NTP:
(a) Capable of running Windows NT and Microsoft Office Professional
(b) Minimum RAM 64 Mbytes
(c) Minimum Hard disk size 3 Gbytes
(d) Fitted with floppy and CD drives (ordinarily disabled in software)
(e) Minimum resolution 1,024 × 768.

RASDA

RASDA is a globally distributed database providing services to local RAF CCIS users. These services are provided to the users via their local instance of the RASDA Application Server. Essentially the RASDA Application Server takes the raw information from the global RASDA database and presents this to the users in a form suitable to meet their expected needs. In the current version of RASDA all the Application Servers offer identical services to the local users. It is envisaged that in the future, specialised local services may be developed for certain locations. (Possibly for external system access). The NT Workstation provides the actual means of accessing the RASDA Application, through a standard Microsoft Internet Explorer browser.

Military Message Handling Service

The RAF CCIS Military Messaging Handling System (MMHS) architecture is hierarchical in nature. At the root of the hierarchy is AMRAD. The MMHS comprises four RAF CCIS hub nodes, which in turn connect to multiple Military Messaging servers: these hub servers also act as Military Messaging servers for the sites at which they are located. Each Military Messaging server is co-located with an MS Exchange server which provides the interface into both the formal military messaging service and the informal e-mail service. Users access both services using MS Outlook on their workstations, with IRIS for Outlook providing the user interface for military messaging. When viewing or creating military messages, the user sees a messaging form similar to the e-mail form on MS Outlook but with all the additional header fields required for formal messaging. In addition, users have access to the IRIS Body Editor that provides facilities for viewing and creating standard formatted military messages using templates. On RAF CCIS, an interface between IRIS and RASDA allows users to exchange data directly between formatted messages and the operational database.

The Military Messaging Servers provide an X.500 service, which enables users to draft and release military messages using Military Plain Language Addresses (PLAs). At Out Of Area sites, RAF CCIS Baseline Exchange Servers provide the X.500 service.

External Interfaces

RAF CCIS interfaces to various external systems such as CHOTS and JOCS (see separate entries) via dedicated gateways (e-mail, web and application), Windows 2000 terminal services and Trust Relationships. The type of mechanism employed depends upon the trust in the external system and the facilities required from the external system. The terminal server provides external system users with access to the full capabilities of RAF CCIS. These external interfaces significantly enhance the information reach for RAF CCIS users, and enable external systems to access RAF CCIS services.

Status

RAFCCIS is being implemented in four increments: Baseline Architecture (in service); RASDA (in service); Military Messaging (initial capability in service); and a further stage, currently under development.

Contractor

Fujitsu Services (Prime), Basingstoke.

UPDATED

S362 display system

The use of commercial-off-the-shelf (COTS) hardware and software, open system architecture and state-of-the-art application packages enables the S362 display system to be utilised for a wide variety of radar-based tasks in a multitude of configurations.

Applications

Air defence workstations (CRP-SOC-ADOC). In an Air Defence role, typical facilities are: Radar Presentation (labelled plots and tracks, jamming strobes and sectors, maps, range rings, range and bearing line, single- or multi- radar tracking, interception vectors); Totes (met information, aircraft readiness, A/f status, radar position and status, flt Plans, oporders, threat assessment).
Air traffic control workstations (Airport – ATC)
Radar management workstations - naval/maritime situation displays (VTMS – EEZ monitoring)

Configurations

Single or dual display workstations
CRT/flat screen displays
Stand-alone or networked systems

Platforms

DEC Alpha/Unix
PC Windows NT

Contractor

AMS, Chelmsford, Essex.

UPDATED

UK Air Defence, Ground Environment (UKADGE)

UKADGE is the UK's national air defence command and control network which produces the Recognised Air Picture (RAP) for the UK Air Defence Region (UKADR) within NATO. The main element is the Integrated Command and Control System (ICCS) which accepts inputs from fixed and mobile land-based radars, British and NATO Airborne Early Warning (AEW) aircraft and British and allied warships. It can also exchange data with the NATO Air Defence, Ground Environment (NADGE) and the French STRIDA II (see separate entries).

ICCS structure

The primary command centre, the Combined Air Operations Centre (CAOC) is at the hardened facility at High Wycombe. This is backed up by a standby facility at Bentley Priory. Overall control is provided by the primary Control and Reporting Centre (CRC) at Neatishead in Norfolk and at Buchan in Scotland, with standby facilities at Boulmer in Northumberland. The CRC is served by a number of Control and Reporting Posts (CRPs) which receive radar and ESM data for the compilation of the Recognised Air and Surface Picture (RASP). All units hold a local and full air picture and have the capability to exercise tactical control of air assets. In addition, the flexibility of the system allows the control of assets using the air picture and communications of a remote site. The ICCS elements and other major Royal Air Force locations are interlinked by UNITER, a countrywide highly resilient and secure ground-based data and voice communications network (see separate entry). Satellite links are also provided from remote sites.

ICCS hardware/software

Hughes HDM 4000 four-colour large-screen displays are provided at the units, together with the key ICCS data handling system whose processors are based on commercial DEC VAX (LSD) minicomputers using software written in Fortran.

Each unit has two VAX 8650 computers, which are used for input/output processing, tracking and weapon control. In addition, a full simulation facility is provided. The data handling system has two databases, one for tactical data compilation and the other a resource data catalogue, which are linked to every UKADGE site.

A Digital Data Network (DDN), supplied by Siemens Plessey Systems (BAE Systems), incorporates high-speed RISC processors for the front end and uses packet-switching techniques to distribute data between the sites over the UNITER network. These are X6883 links to the radar and X25 links to the ICCS sites, airfields and the UK host air traffic control computer at West Drayton. Software for the DDN is written in Coral 66 or RTL2.

The principal workstations at the ICCS sites are the Marconi Universal Consoles. These consoles can be readily reconfigured for one- or two-person operation and the role of each console is determined by the Position Description Language software which is written in Coral 66. Local processing power at each console is provided by a Marconi Locus 16 distributive processor incorporating 15 microprocessors.

Each console is equipped with three displays: a 56 cm (22 in) four-colour graphics display, a 38 cm (15 in) interactive tabular display and a 38 cm (15 in) tote tabular display. In addition, an Enhanced Data Display for tote information and handling has been developed by the RAF, which provides additional information management capabilities.

The Universal Consoles are each linked to the ICCS data handling system by a shared simplex high-speed serial receiving channel and a private duplex low-speed transmitting channel, time multiplexed with as many as three other consoles.

Comprehensive voice communications are also provided by two dedicated modules resident in the processor unit.

The UKADGE system has been enhanced by the addition of full Link 11 capability and the JTIDS Air Platform Network Management System (JAPNMS) provided by Thales. JAPNMS performs real-time, dynamic management of a JTIDS Link 16 (TADIL-J) network. (See separate entry).

UCMP

UKADGE is due to be updated by the UKADGE Capability Maintenance Programme (UCMP), and ultimately by UK/NATO ACCS. IBM has been selected as the prime contractor for UCMP, with introduction starting in 2004. The new system makes extensive use of commercial-off-the-shelf (COTS) components. The new system is designed to provide displays and controls for 160 operators with functions that include:
(a) Radar interfacing
(b) Air track generation and management
(c) Graphical map and data display and control
(d) Voice and data communications
(e) Database replication and management
(f) System management

Status

Fully operational. RAF Buchan will relinquish its role as a CRC by Nov 2004, although the facility will continue to provide data.

Contractors

Prime contractor for UKADGE was UKADGE Systems Ltd, London, a company owned equally by the major subcontractors:
Raytheon Systems Company (previously Hughes), Fullerton, California, USA.
AMS, Chelmsford, Essex.
Prime contractor for UCMP is
IBM, Farnborough.

UPDATED

United States

Air Force Mission Support System (AFMSS)

The AFMSS provides automated mission planning support for fixed- and rotary-wing aircraft and guided munitions. AFMSS will operate as either a stand-alone system, or can be linked with other command information systems to enable planners to select an optimal route through hostile territory.

AFMSS comprises software tools that support aircraft and weapon mission planning. A key feature of AFMSS is the unique software architecture that allows new aircraft or new missions to be easily added to the system without changing the basic planning tools. These basic tools include the maps (uses MCG&I) and route planning, threat analysis, terrain and target analysis and mission preview, all of which work with different types of aircraft. Intelligence, weather and operational data may be automatically fed into the AFMSS databases for use by the system. The core software is combined with tailored, platform unique Aircraft/Weapon/Electronic (A/W/E) modules and weapon system specific Flight Performance Modules (FPM) to provide a Mission Planning Environment for each aircraft type. A Common Low Observable Autorouter (CLOAR) module is available for low observable aircraft to plan routes to minimise threat exposure. The hardware architecture is also modular and is based upon commercial hardware, which allows for easy upgrades.

For aircraft with electronic data transfer capability, aircraft-unique hardware peripherals are used to prepare data transfer devices for uploading mission information into aircraft computers The crew member can either print out his entire mission or input the data into a cartridge for the onboard computer.

The primary outputs of the Mission Planning subsystem are the Combat Mission Folder and Data Transfer Device. The Data Transfer Device is used to upload mission information to aircraft onboard computers and weapons systems. The Combat Mission Folder tools provide the functionality to produce any combination of maps, images, and predictions for inclusion in the Combat Mission Folder. These include:
(a) strip charts
(b) overview chart or form
(c) user-defined series of pages
(d) multiple copies of pages in mission
(e) take-off/landing data cards
(f) radar predictions
(g) perspective views of ADRI imagery and target imagery
(h) warning messages
(i) plan views of ADRI imagery and target imagery.

AFMSS is available in several hardware configurations, all comprising commercially available off-the-shelf hardware components. The open architecture provides:
(a) scalability and extendibility
(b) maximum use of NDI software
(c) industry standard commercial interfaces
(d) single or multiseat configurations.

There are currently two versions of AFMSS, one based for UNIX hardware platforms and one for PC platforms:

Mission Planning System (MPS)

This is a UNIX-based planning system that is used primarily by the F-15E, F-117, B-1, B-2, B-52, Global Hawk and F-16s using PGM. The hardware is COTS based and comes in at least 3 configurations: MPS II, MPS V and MPS VI. The core software is developed and maintained by BAE Systems North America and the hardware by Rugged Portable Systems.

Portable Flight Planning Software (PFPS)

This is a PC based planning system that is used by A-10, F15C, F-16, C-130, C-141, C-5, C-17, E-3, E-8, RC-135. PFPS uses primarily laptop computers with the aircraft specific Data Transfer Devices. The core software includes FalconView, developed by Georgia Tech Research Institute (see separate entry), and other packages developed by Tybrin Corporation.

Status

AFMSS is operationally certified by the US Air Force and is currently deployed around the world supporting the US Department of Defense. Additional systems have been developed and delivered for the governments of Israel, Italy, Saudi Arabia and Taiwan. It is planned that all US Air Force and US Navy mission planning systems will be migrated to a Joint Mission Planning System (JMPS) (see separate entry), although the timescale for this would appear to have slipped from 2002 to at least 2004. The AFMSS software continues to be developed in parallel to work on JMPS.

Contractor

BAE Systems North America (Information & Electronic Warfare Systems), Nashua, New Hampshire.

UPDATED

Air Mobility Command C2 Information Processing System (AMC C2 IPS)

The Air Mobility Command Command and Control Information Processing System (AMC C2 IPS) is an integrated, command-wide system that supports AMC airlift missions around the globe with a centralised control/decentralised execution philosophy.

IPS provides direct command and control decision support to various echelons of the US Air Force, including the Tanker Airlift Control Element (TALCE), the Air Mobility Element (AME), the Air Mobility Unit (AMU) and the Wing Operations Center/Tanker Task Force (WOC/TTF). By accommodating the wide range of activities and volume of work performed at each US Air Force level, IPS facilitates the distribution of information across the network. Communication interfaces to the Global Decision Support System (GDSS) allow the exchange of schedule and status information with the Tanker Airlift Control Center (TACC). Other interfaces offer integration with the Computer-Aided Aircrew Scheduling System (CAASS), G081 and, in the future, with the Aerial Port Command and Control System (APACCS) and Contingency Tactical Automated Planning System (CTAPS) In addition. IPS configurations are easily transportable for rapid deployment anywhere in the world. The supported echelons are connected over available military communications systems.

IPS provides a co-ordinated, consistent database for scheduled, inbound, resident and outbound missions. Operators process database updates and message transmission so that person-to-person communication is reduced and surge capabilities are improved. When a problem is detected in the Local Area Network (LAN), each workstation can store operator inputs until the network is restored. Real-time data updates help reduce ground times and improve management of transportation, maintenance, aircrews, logistics and combat communications. AMC C2 IPS complies with the US Message Text Format (USMTF) and communication processing ensures end-to-end message delivery. IPS availability is better than 98 per cent in the field.

IPS can be tailored to site type-specific requirements. At fixed sites, such as the en route node, IPS supports airflow and is used to manage resources and report status. At the fixed AMU, IPS is used to plan, schedule and execute tasked missions and to manage unit resources, capabilities, and limitations. In deployable operations, such as at the Air Mobility Element (AME), IPS is used to interface with the Air Operations Centre (AOC), thereby providing a focal point to plan, schedule, task, and execute strategic and theatre air mobility operations. IPS is used in the deployable TALCE to support deployed AMC onload, en route and offload operations where support is insufficient or non-existent.

The AMC C2 IPS consists of commercial-off-the-shelf (COTS), non-TEMPEST hardware; identical components are used for both fixed and deployable configurations. Each IPS node has a Communications Processor (CP), a file server, workstations and a nodal data network. The CP handles all messaging between the file server and external communications media. It logs all received messages; queues the messages according to precedence; and handles message pre-emption, format conversion and routeing. The file server controls and co-ordinates mission data and maintains consistency among all workstations. The workstations provide full support for mission monitoring, planning and scheduling functions, including message generation and review, display generation, data entry validation and interactive graphic aids. A user-selectable sorting and filtering capability allows individual screens to be tailored to a user's needs. The distinctive field colourings help operators to manage by exception; warning or alarm events are set off in yellow and red (for example, non-mission-capable aircraft are shown in red).

The CP, file server and workstations are interconnected via a fibre optic cable LAN in a nodal data network. Individual base buildings are bridged by either a backbone fibre ring or the infrastructure of the airbase 10BaseT LAN.

Status

The IPS has been converted to open systems standards. IPS, originally programmed in Ada, will now comply with POSIX and MOTIF standards. An X400 communications interface and FIPS-compliant Structured Query Language (SQL) standards were also being implemented. The system platform is the DEC Alpha Processor. Future software releases will allow planners to create and flow multiple requirements to complete mission schedules and apply and monitor AMC resources. New displays and databases will support air refuelling missions. Computer-based training will keep operators up to date on system upgrades.

Over 150 sites are equipped with AMC C2 IPS.

No new information received, but believed to be still in use in late 2003.

Contractor

Computer Sciences Corporation, El Segundo, California.

UPDATED

Ground Theater Air Control System (GTACS)

GTACS provides the ground-based C2 elements of the Theater Air Control System supporting air operations performed by the Combat Air Forces. It provides the JFACC with the means to plan, direct, and control air operations and co-ordinate these air operations with ground, naval and coalition forces.

Major GTACS functions include:
(a) Command planning and direction
(b) Aircraft control and warning
(c) Close air support co-ord and control
(d) Airspace management
(e) Airborne airstrike co-ord and control
(f) Ground target sensor surveillance
(g) Tactical airlift

GTACS battle management elements include:
(a) Air Operations Center (AOC)
(b) Control and Reporting Center (CRC)
(c) Control and Reporting Element (CRE)
(d) Air Support Operations Center (ASOC)
(e) Tactical Air Control Party (TACP)

The major components of the GTACS are the AN/TPS-75 Radar System and the AN/TYQ-23 Modular Control Equipment (MCE). The AN/TPS-75 Radar provides a real-time radar airspace picture and data in support of the battle commander and the Ground Theater Air Control System (GTACS) via radio, telephone, microwave relay or satellite communications link. The AN/TPS-74 radar is a AN/TPS-43E radar whose performance has been greatly improved through the installation of two major modification kits: an ultra low sidelobe antenna and a shelter modification kit which includes a completely new signal processor.

The keystone of the Ground Theater Air Control System (GTACS) is the AN/TYQ-23 Modular Control Equipment (MCE) (see separate entry), an automated computer-based information system operating in a proprietary environment that provides a variety of automated information functions such as aircraft surveillance, flight follow, control and communication functions within the GTACS. The MCE comprises a number of Operations Modules (OM) that contain data processing, tactical communications and display equipment. The MCE is undergoing a number of upgrades to provide a JTIDS Class II terminal, SATCOM, Joint Tactical Air Orders (JTAO) and Air Tasking Order functionality. The JTIDS Modules (JM) are modified S-711 communication shelters (US Army Regency Net) designed to interface with the MCE OM. JM shelters contain various communications equipment and interface with MCE via Tri-Service Tactical (Tri-Tac) link. The JTIDS Module upgrade will provide 5 JTIDS Modules, which will be interfaced with the MCE OM to establish a Link-16 capability. Northrop-Grumman's Expert Missile Tracker (EMT) will provide integrated capability to detect/track TMDs via AN/

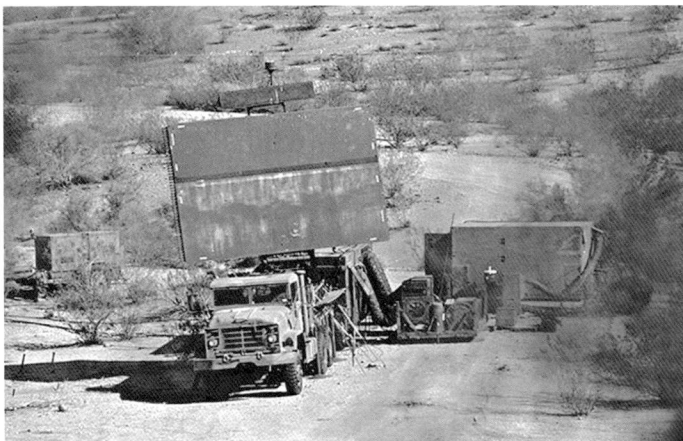

The AN/TPS-75 radar is the primary sensor for the GTACS system   0085275

TPS-75 Radars, and interface to Tactical Air Operations Modules at CRCs/CREs. Additional equipment includes the Automatic Radar Evaluation System and Jammer Test Set (ARADES IV) by Northrop-Grumman, which is a portable PC based test set that performs calibration, performance evaluation (including ECCM) and maintenance diagnostics for the AN/TPS-75 Tactical Radar.

Status

In operational service with the USAF. The Theater Air Control System is in the process of being modernised and in its new incarnation will be called the Battle Control System.

UPDATED

Information Warfare Planning Capability (IWPC)

The US Air Force is developing an integrated suite of software tools to provide a full-spectrum, offensive and defensive, Information Warfare planning capability. IWPC operators will develop IW courses of action for the Joint Force Air Component Commander (JFACC) and nominate IW 'targets' for inclusion into the Master Air Attack Plan and the Joint Integrated Prioritised List (JIPL). These tools are:

InfoWorkSpace™

This is the primary collaboration tool utilised by IWPC. It provides a virtual workspace with 'buildings', 'floors' and 'offices'. Each 'office' is a virtual meeting space supporting real-time communication between users by voice or text chat and complete with virtual whiteboard, bulletin board and filing cabinet. IWS allows multiple servers to be amalgamated into a federation of associated users.

Analyst Collaborative Environment (ACE)

ACE provides users with a simple, collaborative interface to a variety of intelligence data sources and analytical tools to support analysts in finding and collating related intelligence. This interface allows multiple IW analysts to share information and knowledge through the use of analyst teams, shared target folders and query profiles. ACE can automatically populate folders with any media, text or source information as it becomes available.

Collaborative Workflow Tool (CWT)

CWT provides a common online checklist to track and record a tailorable sequence of tasks to guide a collaborating team through a project. Tasks can be allocated to a person or organisation and functionality can be limited by user roles.

Collaborative Planning Tool (CPT)

CPT is a task-to-action tool that allows an IW planner to import or create an IW plan, associate targets to planned actions, and export the plan for use by other planning nodes. Targets can be imported from the Modernised Integrated Database (MIDB) or manually added. Plans are maintained in XML format to support sharing with other planners.

Information Warfare Visualisation (IW Viz)

IW Viz provides the ability to visualise collaboratively the results of IWPC applications. IW Viz currently uses OpenMap and Jmap mapping packages, producing a map with multiple layers of detail. The IW Viz layer adds IW targets to the map and will evolve to using the Common Operational Picture (COP) applications.

NAIC/AFRL Tel-Scope Network Model

This model provides a platform to interact dynamically with links and nodes data to evaluate the effects of IW actions on a communications network. Users define the network by importing a links and nodes file with network topology and functions, then perform 'what if' drills to identify potential IW targets or alternate communication routeing for use by IW planners. Targets identified by Tel-Scope can be exported into CPT and IW Viz.

Course of Action Tool (Coat)

COAT is a prototype application that provides a preliminary view of concepts for managing and selecting Courses of Action. COAT allows the user to create, modify and delete COAs. Once a collection of COAs are defined, one of the key issues is identifying the correct COA (or collections of COAs) needed to address a specific situation. COAT provides the capability to compare and contrast COAs based on criteria specific to the current situation.

Target and Guidance Interface Facility (TGIF)

TGIF is the primary interface tool for interacting with TBMCS (see separate entry). It provides tools for importing, generating and exporting target lists. These can be generated either by querying the MIDB or manually entering targets. Using TGIF, the IW planner can compare and highlight differences between target lists.

Status

Currently in development, the integration of the tools suite is programmed to be complete by early 2004.

Contractor

General Dynamics Advanced Information Systems (Prime), Arlington, Virginia.

NEW ENTRY

Marine Air Traffic Control and Landing System (MATCALS)

MATCALS is a fully automated, all-weather, expeditionary terminal Air Traffic Control (ATC) system used by MACS to rapidly establish communications, take-off, landing, and other ATC services required for Visual Flight Rules (VFR) and Instrument Flight Rules control of aircraft at remote area landing sites. MATCALS integrates with other Marine Air Command and Control Systems (MACCS) and federal agencies, such as the Federal Aviation Agency. It provides the ability to expeditiously move combat aircraft throughout the Amphibious Objective Area without regard to the effects of weather. ATC and landing automation reduce air traffic handling and management time, allowing more time for mission response and task accomplishment. Thus, it supports an increase to aircraft sortie rates and directly contributes to extending an aircraft's time-on-target. The system provides for integration of the ATC and landing systems into the total MACCS interfacing by means of automated transfer.

Description

MATCALS has three primary subsystems:

(1) the AN/TSQ-131(V) Control and Communications Subsystem (CCS) with Communications Control Group, radios, computer software, multimode displays, and peripherals

(2) the Air Traffic Control Subsystem (ATCS) consisting of AN/TPS-73 Airport Surveillance Radar (ASR) and various peripheral equipment

(3) the All-Weather Landing Subsystem consisting of the AN/TPN-22 PAR, the AN/UYK-44 Computer, and peripheral equipment.

Other related systems include the AN/TSQ-120A/B ATCC Towers, the AN/TSQ-216 RLST, the AN/TPN-30A MRAALS, and other related support items that contribute to the safe and expeditious flow of air traffic at expeditionary airfields and remote landing sites.

AN/TSQ-131(V) Control and Communications Subsystem

The AN/TSQ-131(V) CCS is a transportable ATC radar facility. It houses two sets of four radar operator positions for the AN/UYQ-34(V)2 Processor Display Set (PDS), a supervisor position, and the required equipment for data processing, voice communications, and data communications. The AN/TSQ-131(V) CCS integrates data received from the AN/TPS-73 ASR, the AN/TPN-22 PAR, other Marine ATC (MATC) systems, including AN/TSQ-120A/B ATCC Towers, AN/TRN-44 Tactical Control and Navigation (TACAN) Sets, AN/TPN-30A MRAALS, and external ATC agencies into a unified ATC System. Voice communications provide coverage of the following nets: Ultra High Frequency (UHF), Very High Frequency (VHF), and High Frequency (HF), and Amplitude Modulation (AM) and Frequency Modulation (FM) radio bands. Components of the AN/TSQ-131(V) CCS include:

(a) AN/UYQ-34(V)2 processor display set

(b) TD-1089/UYQ-4 modem

(c) AN/UYQ-41 digitizer-switching set

(d) AN/UYQ-42 control-distribution set

(e) HD-1099/TSQ environmental control unit

(f) RO-572/TSQ-131(V)1 line printer data processing

(g) AN/GMQ-31 wind measuring set

(h) AN/GRC-171(V)1/2 UHF radio sets *

(i) AN/GRC-211 VHF radio set *

(j) AN/URC-94(V)2HF (AM)/VHF (FM) radio set

(k) MEP-006A/MEP-806A generator sets

(l) AN/GSH-60 recorder-reproducer set

(m) AN/USH-26(V) signal data recorder-reproducer set

(n) AN/USQ-94 bus access set.

* AN/GRC-171(V) and AN/GRC-211 radio sets in the AN/TSQ-131 CCS Radar Facility and the AN/TSQ-120B ATCC Tower will be replaced by the AN/ARC-210(V) EP radio system.

Air Traffic Control Subsystem

The ATCS is a transportable, tactical ASR subsystem for MATCALS. It provides AN/TPS-73 ASR, Secondary Surveillance Radar (SSR), and Autotracker functions to the Radar Controllers assigned to the ATC Detachment of the MACS. During

Interior view of MATCALS

operations, the ATCS is unmanned. It is controlled from the AN/TSQ-131(V) CCS through the ATCS remote-control panel. The ATCS design includes numerous redundant functions to ensure continued independent operation in case of failure of one system. The ATCS features end-to-end online performance monitoring, self-alignment capability during operation, fully integrated Built-In Test and Built-In Test Equipment (BIT/BITE), and online repair capability of the fail-soft ASR transmitter.

AN/TPS-73 Airport Surveillance Radar

The AN/TPS-73 ASR is an S-band non-linear frequency modulated system. It contains a solid-state transmitter that generates a 10.3 and a 100 µs pulse. The transmitted pulses can be any of 20 different frequencies in the 2,705 to 2,895 MHz range in 10 MHz steps. It uses a digital receiver to decode and interpret radar returns. The AN/TPS-73 ASR can detect 1 m^2 targets at ranges from 0.5 to 60 n miles and altitudes to 60,000 ft above ground level. The AN/TPS-73 ASR receives beacon plot and video data from the SSR, performs radar-to-beacon correlation and synchronisation, and forwards this data to the Autotracker.

Secondary Surveillance Radar

The SSR is an L-band monopulse beacon with a Mode 4 capability for Identification Friend or Foe. The SSR can detect targets at ranges up to 120 n miles. Each of the two SSRs contains two solid-state transmitters that independently power the sum and omni-beams of the monopulse antenna. Three logarithmic receivers provide signals that are processed and sent to the AN/TPS-73 ASR for synchronisation with the AN/UYQ-34(V)2 PDS.

Autotracker

The Autotracker accepts AN/TPS-73 ASR and SSR synchronised video data from the AN/TPS-73 ASR. It detects and tracks up to 600 air targets, correlates AN/TPS-73 ASR and SSR targets and develops digital track data. The serial databus of the MATCALS is used to transmit this data to the AN/TSQ-131(V) CCS for use by controllers. Additional components of the ATCS include:

(a) AN/UYQ-34(V)2 processor display set

(b) C-11515/UYQ-41 operator control unit

(c) HD-1099/TSQ environmental control unit

(d) MEP-006A generator set

(e) AN/USQ-94 bus access set.

AN/TPN-22 Precision Approach Radar

The AN/TPN-22 PAR is a transportable, computerised, pencil-beam, three dimension (3-D), track-while-search, precision approach radar system used to execute multimode, automatic, precision approach and landing of tactical aircraft. The AN/TPN-22 PAR uses phase and frequency scanning techniques in an electronically steered beam antenna array to provide data at a high rate for detection and automatic tracking of up to six aircraft simultaneously in the approach and landing airspace. The frequency range is 9,000 to 9,200 MHz. The AN/TPN-22 PAR has 46° coverage in azimuth, 8 ° (−1 to +7) angular coverage in elevation, and 750 ft to 10 n miles coverage in range. The AN/TPN-22 PAR operates as an integral data acquisition and processing computer subsystem in concert with the AN/TSQ-131(V) CCS and AN/TPS-73 ASR for simultaneous manual, semi-automatic, and automatic aircraft approach and landing operations. The AN/TPN-22 PAR has been upgraded with a solid-state modulator. Components include:

(a) AN/UYQ-34(V)2 processor display set

(b) C-11515/UYQ-41 operator control unit

(c) AN/UYK-20X(V) general purpose data processor

(d) AN/USH-26(V) signal data recorder-reproducer set

(e) MEP-006A generator set

(f) HD-1099/TSQ environmental control unit

(g) AN/USQ-94 bus access set.

AN/TSQ-120A Air Traffic Control Central

The AN/TSQ-120A ATCC is a transportable ATC tower facility, which provides 360° of visual observation of aircraft within a designated control zone, both on the ground and in the air, and visual control over ground vehicles in the vicinity of the runway(s). Control is accomplished through use of radio communications and visual aids. Aircraft operations are co-ordinated with remote facilities and agencies by use of telephone and intercommunication control systems. The AN/TSQ-120A ATCC Tower consists of:

OK-312/TSQ-120A Operations Central Group

The OK-312/TSQ-120A Operations Central Group (OCG) is a tower cab that provides 360° of visibility for controller observation. There are three operator positions for control of radio transmitting and receiving operations, and for telephone communications. Other controls and indicating equipment, such as overhead speakers, crash alarm, fire detector, wind direction velocity indicators, environmental control, and intercom are readily available to all three operating positions. There are provisions for installing and using the C-10363/URN control indicator, the C-8534/TRA-45 TACAN remote-control indicator, and the C-10194/TPN-30 control indicator or C-10195/TPN-30 remote control as required by the MACS. Radio equipment in the OK-312/TSQ-120A OCG Tower Cab includes:

(a) C-10618/TSQ-120A receiver-transmitter

(b) C-7999/GRC-171(V) UHF radios - 2 each

(c) C-8314/GRC-211 VHF radio - 1 each.

OW-81A/TSQ-120A Terminal Group

The OW-81A/TSQ-120A Terminal Group (TG) contains radios, telephone equipment, recorders, intercom, and signal and power distribution systems that provide required communications information to the operating positions in the tower cab. In addition, a maintenance console and workbench are provided for maintenance personnel to simultaneously select voice communications on any or all of the system radios, select voice communications on any one of the ten system telephone lines, and monitor audio input/output signals to the radios.

Radio equipment
Radio equipment in the OW-81A/TSQ-120A TG includes:
(a) SA-2257/TSQ-120 switching matrix
(b) J-3638/TSQ-120 interface unit
(c) AN/GRC-171(V) UHF radio sets - 5 each
(d) AN/GRC-211 VHF radio sets - 3 each
(e) AN/URC-94(V)2 HF (AM)/VHF (FM) radios - 2 each
(f) AN/VRC-82(V)2 radio set
(g) audio patch panel.

OA-7621(V)/FSA-52(V) Landline Selector Group
The OA-7621(V)/FSA-52(V) Landline Selector Group is a communication control system used to operate multichannel landline communications in conjunction with radio communications. It provides six ring-down signaling lines, two voice call-up signaling lines, and two selective signaling lines.

Recording equipment
(a) AN/GSH-60 recorder reproducer
(b) CDD-1000 digital deck Automatic Terminal Information Service (ATIS) recorder.

Antennas
(a) TACO D-2118 UHF-VHF antenna - 3 each
(b) AS-1729/VRC VHF antenna - 2 each
(c) AT-1011/U HF antenna - 2 each
(d) crash net antenna.

AB-1236/TSQ-120 Tower
The AB-1236/TSQ-120 Tower is a portable, field-erected structure that supports the OK-312/TSQ-120A OCG Tower Cab at an elevation of 8, 16, or 24 ft above the ground. An inclined stairway with handrails, rising around the tower perimeter and leading to a platform at the top level, is provided for personnel access to the OK-312/TSQ-120A OCG Tower Cab. The design of the tower allows it to surround the OK-312/TSQ-120A OCG Tower Cab at ground level, providing clear access for raising and lowering the OK-312/TSQ-120A OCG Tower Cab.

OA-8883/TSQ-120 Storage-Transport Group
The OA-8883/TSQ-120 Storage Transport Group (STG) provides storage and transport for those components external to the OK-312/TSQ-120A OCG Tower Cab and OW-81A/TSQ-120A TG that do not pack-out in either shelter. The OA-8883/TSQ-120 STG consists of a transport pallet and a tower container box for the AB-1236/TSQ-120 Tower.

AN/TSQ-120B Air Traffic Control Central
The AN/TSQ-120B ATCC Tower provides the same essential services as the AN/TSQ-120A ATCC Tower. However, some operational, embarkation, and reliability enhancements have been incorporated. This includes secure voice capability incorporated with the AN/UYQ-41 digitiser switching set and racks for TSEC/KY-58 and TSEC/KY-75 speech security equipment. The OW-81/TSQ-120A TG shelter is replaced with a standard-sized shelter (8 ft wide, 8 ft deep, and 10 ft high). The equipment storage pallet (part of the OA-8883/TSQ-120 STG) is deleted. Component equipment previously stored on the pallet is packed-out in one of the two shelters, and in the tower scaffolding box for embarkation.

AN/TRC-195 Control Central
The AN/TRC-195 CC provides a limited tower capability for remote-site operations. It contains four 20 Hz telephone lines, two UHF, one VHF-AM, one HF/VHF-FM, and one crash net radio, and a wind measuring set powered by a single MEP-003 generator or equivalent power source. AN/TRC-195 CC Tower full operating power requirement is 3.0 kW, the emergency operating power requirement is 1.5 kW. The AN/TRC-195 CC Tower communication systems are capable of encrypted transmitting and receiving. The unit is transportable by forklift (no mobiliser), and is usually loaded into and employed from the back of a Highly Mobile Multi-purpose Wheeled Vehicle (HMMWV). However, the AN/TRC-195 CC Tower can be loaded on a variety of vehicles. The radios (except for the AN/VRC-82 radio), speech security equipment, and wind measuring set are provided from other MATC systems to enable full operational capability. The 28 V power supply is provided with each system. The assembled unit has a nylon top and four mast assemblies. The system includes the following MATCALS equipment:
(a) AN/GRC-171(V) UHF radio sets - 2 each
(b) AN/GRC-211 VHF radio set
(c) AN/URC-94 HF (AM)/VHF (FM) radio set
(d) AN/VRC-82 radio set
(e) AN/GMQ-31 wind measuring set
(f) TSEC/KY-58 Crypto, for UHF, VHF and VHF/FM bands 3 each
(g) TSEC/KY-75 Crypto, for HF band
(h) Trio Labs 28 V power supply.
* The AN/TRC-195 CC Tower is being replaced by the AN/TSQ-216 RLST.

AN/TSQ-216 Remote Landing Site Tower
The AN/TSQ-216 RLST will be introduced through new production, replacing the AN/TRC-195 CC Tower. The AN/TSQ-216 RLST has been developed as an interim tower, to be used when the AN/TSQ-120A/B ATCC Towers are unavailable. It

USMC's MATCALS with control tower

provides for a rapid emplacement and expeditious establishment and withdrawal of communications and related capabilities required for VFR services. Its communications systems provide coverage of the following nets: UHF, VHF, and AM radio bands; tactical command, combat information and detection, air defense alert or communication co-ordination HF nets; and VHF-FM base defense and crash nets. The AN/TSQ-216 RLST is transported on a heavy HMMWV and consists of a mounted shelter and trailer.

AN/TRN-44 Tactical Control and Navigation Set
The AN/TRN-44 TACAN is a transportable, dual-channel navigational aid which provides TACAN-equipped aircraft with range, bearing and station identification information effectively within a 200 n miles radius. It is used for both en route navigation guidance and as an instrument approach aid. It has 126 operating channels in X mode and 126 operating channels in Y mode; transmitting and receiving in the frequency range of 962 to 1,213 MHz. It can provide distance information for as many as 100 aircraft and provides an infinite number of aircraft with azimuth information and station identification. The AN/TRN-44 TACAN can be remotely controlled and monitored, and incorporates an external 1° monitor. The shelter is air conditioned and heated for environmental control. The AN/TRN-44 TACAN requires primary power of 120/208 V, 60 Hz, 3 phase, 4 wire. Power consumption is 18.7 kW.

AN/TPN-30A Marine Remote Area Approach and Landing Set
The AN/TPN-30A MRAALS is a two-person transportable, all-weather landing system which transmits azimuth, elevation angle, and range data to specially equipped aircraft. The airborne system translates the data and provides glideslope, localiser, range, and range rate information to the pilot's indicators. The AN/TPN-30A MRAALS transmits azimuth, distance, and elevation data in the K-band frequency range, 15.412 to 15.688 GHz, and distance measuring equipment and station identification data in the L-band frequency range of 962 to 1,213 MHz, as well as 15 Hz TACAN bearing data to provide 360° of bearing information. The AN/TPN-30A MRAALS can be set up in one of two configurations, co-located or split site. The co-located configuration is employed at landing zones and uses one AN/TPN-30A MRAALS to provide azimuth, elevation, distance, and station identification data. The split site configuration is employed at airfields and airports and uses two AN/TPN-30A MRAALS; one at the end of the runway (aligned with the runway centreline) to provide azimuth data, and one parallel to the runway (parallel to the designated touchdown point) to provide elevation and range data. In the co-located configuration, the AN/TPN-30A MRAALS can be remotely controlled (up to 1,000 ft) using field wire by the C-10195/TPN-30 remote control. The C-10194/TPN-30 control indicator may be operated remotely by cable and provides status information. In the split site configuration, the C-10194/TPN-30 remotely controls and provides status of the two AN/TPN-30A MRAALS, which are synchronised with field wire.

AN/TSM-170 Maintenance Repair Group
The AN/TSM-170 Maintenance Repair Group (MRG) consists of four shelters, which contain workbenches, test equipment, cabinets, tools, and other equipment necessary for section maintenance of MATC equipment. All shelters allow some degree of flexibility to accommodate changed maintenance demands based on mission and equipment configuration.

OA-9141/TSM-170 Auxiliary Equipment Repair Group
The OA-914/TSM-170 Auxiliary Equipment Repair Group (AERG) provides the work space necessary for the maintenance of the Environmental Control Units (ECU), diesel generator sets and other designated support equipment.

OA-9142/TSM-170 Communications Equipment Repair Group
The OA-9142/TSM-170 Communications Equipment Repair Group (CERG) provides workspace and parts storage for the maintenance of all MATC communications equipment.

OA-9143/TSM-170 Radar Equipment Repair Group
The OA-9143/TSM-170 Radar Equipment Repair Group (RERG) provides workspace and parts storage for the repair of the AN/UYQ-34(V)2 PDS, its associated hardware, and other radar component equipment.

OA-9144/TSM-170 Electronic Module Repair Group

The OA-9144/ TSM-170 Electronic Module Repair Group (EMRG) provides micro-miniature repair capabilities for the maintenance of printed circuit boards. It also contains space for maintenance management functions including the maintenance data system computer.

Status

MATCALS is in operational service with the USMC. Developmental Test (DT) and Operational Test (OT) for MATCALS were completed in FY85. DT for the AN/TSQ-216 RLST was completed in February 1998 and OT was completed in June 1998. The AN/ARC-210(V) EP radio system is replacing the AN/GRC-171(V) and AN/GRC-211 radio sets in the AN/TSQ-131 CCS radar facility and the AN/TSQ-120B ATCC Tower, beginning in FY00, with an estimated completion date of FY06. The AN/TSQ-216 RLST is replacing the AN/TRC-95 CC Tower. Future plans call for the replacement of MATCALS with new production ASPARCS. This lightweight, highly mobile, ATC system with advanced aircraft technologies will replace MATCALS as it reaches its service life limits. Two AN/TPN-30A Marine Remote Area Approach and Landing Systems (MRAALS) have been procured by the government of Japan.

Contractor

BAE Systems North America (Prime), Rockville, Maryland.

UPDATED

SAIC mission planning systems

A full range of aircraft Mission Planning Systems (MPS) is available from Science Applications International Corp (SAIC). All are based on SAIC's "unique" map database technology and are configured to meet the specific needs of each customer. Systems are customised for each aircraft and user interfaces are claimed to operate the way pilots normally plan missions. Operator training can be embedded directly into the system.

The MPS allow air crews to plan missions using digital maps, imagery and elevation data. All information can then be loaded on data transfer devices for input to the aircraft avionics suite. No additional downloading is necessary in the cockpit, thus enhancing rapid response and eliminating possibility of error. Planning outputs are also printed on strip maps and kneeboard cards. The MPS seamless map database contains raster maps, satellite imagery, reconnaissance photos, DTED and vector data. Flight plans can be filed electronically with appropriate authorities. The systems can be used either at the aircraft's home base or at remote locations.

The systems calculate all the necessary parameters of the flight, including effects of weapons load on weight and drag. Performance is recalculated after each weapon release, climb, descent or air-to-air refuelling. Safety calculations such as bingo fuel, point of safe return, and point of safe divert are automatically generated. Multiple aircraft can be planned in a single mission. Each MPS also includes built-in error detection algorithms to provide warnings when calculated values exceed operational limits. In addition, the weapons load rules of the specific military user are incorporated to prevent unauthorised configurations.

The systems have aircraft-specific modules so all flight manual limits, aircraft configuration data and performance data are included. Aircrews use the system to plan a 'minimum risk route' with a computerised map, complete with icons representing threats, targets, landmarks and so on.

The digital map database may be generated from a variety of sources, including ADRG (ARC Digital Raster Graphics), other international standards such as Spot Image and Intergraph, or standard paper maps. Map scales run from 1:100 million to 1:8,000. Using Digital Terrain Elevation Data (DTED), the systems can perform radar terrain masking calculations and calculate and display three-dimensional perspective views of the mission terrain.

All systems are fully network capable. This allows the MPS to share common data and to send mission plans from one unit to another.

The systems can be run on a range of hardware suites, ranging from standard workstations to notebook computers. Installations available include Windows/Windows NT and DOS, as well as X-Windows/Motif for UNIX systems.

Status

The company has already delivered more than 2,500 mission planning and support systems for military aircraft around the world. Aircraft already using these systems include F/A-18, A-4AR, AV-8B, F-111, MH-60, MH-53E, AH-1, KC-13OR, KC-135 and V-22 in Argentina, Australia, Finland, Italy, Malaysia, Spain, Switzerland and in the USA.

Manufacturer

Science Applications International Corp (SAIC), McClean, Virginia.

VERIFIED

Theatre Battle Management Core Systems (TBMCS)

The TBMCS combines three 'legacy' systems—CTAPS: Contingency Theater Air Planning System, CIS: Combat Intelligence System, and WCCS: Wing Command and Control System—into one. It provides the means to plan, direct and control all theatre air operations and to co-ordinate with ground and maritime elements, and can be tailored to large- or small-scale operations.

TBMCS functionality includes intelligence processing; air campaign planning, execution and monitoring; aircraft scheduling; unit-level maintenance operations; unit- and force-level logistics planning; and weather monitoring and analysis. Threat evaluation tools allow users to perform modelling actions of potential threats to help determine their lethality and assess the probability of detection and engagement. Target selection tools give mission planners automated capabilities for selecting targets and developing weapons solutions to destroy them, matching aircraft and munitions to the mission. The system provides a capability to receive and process imagery data. It is a significant aid in the production of the Air Tasking Order.

At the force level, TBMCS supports the JFACC through the Air Operations Center (AOC) and Air Support Operations Center (ASOC). At the unit level, it supports the Wing Commander through the Wing Operations Center (WOC), Maintenance Operations Center (MOC), and Squadron Operations Center (SOC).

Status

Fielding of TBMCS began in October 2000 and was planned to replace CTAPS in all theatres by March 2001. Also procured by the USMC. Introduced into service with the Royal Australian Air Force in 2002.

Contractor

Lockheed Martin Mission Systems, Gaithersburg, Maryland.

UPDATED

C2 FACILITIES

Australia

Joint Task Force HQ (Afloat)

ADI was awarded a contract in September 2000 to provide a design for the integration of seven C4I systems that comprise the Australian Joint Task Force Headquarters (JTFHQ) (Afloat). This capability has been developed at short notice to fulfil the Australian need for a headquarters afloat for the Commander Australian Theatre with both a tactical and logistic command capability. The deployable headquarters provides broadband communications between the deployed headquarters, the mainland ADF headquarters and forces under command. Central to this capability is the Joint Command Support System (JCSS) (see separate entry). The system provides situational awareness to the deployed headquarters through the Theatre Broadcast System (TBS) that allows JCSS database downloads and transmission of video data. Also a Parakeet switch (see separate entry) provides secure voice communications and a data communications interface to a TS LAN, JCSS LAN and the ship's administrative LAN. This facility will enable a joint task force commander to direct operations from the ship at both the tactical and operational level.

Status
The system was installed in HMAS MANOORA between January and May 2001 and was handed over in June 2001. An identical capability was installed on the sister ship HMAS KANIMBLA and was completed in June 2002.

Contractor
ADI Ltd, Garden Island, New South Wales.

UPDATED

HMAS *Manoora* (ADI Ltd)   0137497

The JTFHQ (Afloat) in HMAS Manoora (ADI Ltd)   *NEW*/0561763

Italy

Control and Reporting Post (CRP)

The AMS Control and Reporting Post (CRP) is a mobile tactical air defence system designed for deployment in the dynamic and hostile battlefield environment. It provides all the air defence functions to cope with friendly and unfriendly aircraft. It comprises an RAT-31 SL 3-D S-band NATO Class 1 radar and an operations section based on two workstations equipped with two colour 29 in displays. It has an ADP open architecture system using Local Area Network (LAN) and Commercial Off-The-Shelf (COTS) ruggedised components.

The CRP is able to perform operational functions of tracking, using state-of-the-art MultiHypothesis Tracking (MHT) techniques, identification and air mission control, and the system can be integrated with higher-level air defence centres.

Its operational flexibility enables the CRP to perform all the activities necessary to support both military and civilian air traffic control.

Status
In production.

Manufacturer
AMS, Rome.

VERIFIED

MRCS-403 Mobile Reporting and Control System

The MRCS-403 is a tactical air defence system designed for deployment in the dynamic and hostile field environment encountered in tactical warfare. It provides all the air defence functions necessary to cope with friendly and hostile aircraft.

Modular and flexible, the MRCS-403 can operate as an indepencent air defence operations command and control centre or as an integral part of an extensive air defence network.

The MRCS-403 functions include:
(a) Detection, identification, three-dimensional location and tracking of aircraft
(b) Target designation and control of interception
(c) Target designation to surface-to-air missiles and guns
(d) Reporting of air situation to air defence headquarters
(e) Navigational assistance
(f) Data collection and data recording
(g) Simulation.

The main components of the MRCS-403 are an operations shelter, a 3-D radar and a prime power source. In the basic configuration, the AMS RAT-31S radar is provided, but the radar interface unit has been designed to be easily adapted to meet the requirement of any 3-D radar.

The operations centre consists of a dual computer complex, three modular display consoles, a computer peripheral set, a communications control unit, ground-to-air communications equipment and a datalink interface for plot reception.

Specifications
(a) Track capacity: up to 100
(b) Controlled airfields: up to 4
(c) Display synthetic video maps: 2
(d) Simultaneous intercept/recoveries: up to 4
(e) Interceptor aircraft types: 2
(f) Operator positions: 5

Status
Available. In service in Italy and Austria.

Manufacturer
AMS, Rome.

UPDATED

MRCS 403 system deployed   0052054

NATO

High-Readiness Forces Land Headquarters (HRF(L) HQ)

The German and Dutch binational corps-level headquarters is being transformed to a High-Readiness Forces Land Headquarters, or HRF(L) HQ, as part of the European efforts to provide NATO with at least three HRF(L) HQ plus two lower-readiness follow-on HQ certified for sustainment missions. As a HRF(L) HQ the German/Netherlands (GE/NL) Corps HQ is designed to act either as land component command in a major combined joint task force operation or as corps HQ, and is intended to be capable of conducting "effective command and control in a first-entry operation throughout the full mission spectrum, inside or outside NATO territory, within 20 to 30 days of receiving the order to deploy".

Its capabilities include:

(a) aerial deployment within five days of two large recce teams to the port of debarkation and to the operational area

(b) aerial deployment within 10 days of an initial C2 element (ICE) and the nucleus of the rear support command (RSC)

(c) deployment within 20 days of the full HQ and supporting units.

The equipment for the ICE is based in Mercedes-Benz 4 × 4 Geländewagen, 4 × 4 Unimog and Volkswagen 4 × 4 Transporter vehicles. It can be fully operational 24 hours after arrival on-site, typically 10 days after the deployment order, and would require just four to five sorties by C-130 or C-160 aircraft to be air-transported to the area of operations. Manned by 55 people, the ICE includes a mission-tailored tactical command post (CP - including intelligence, operations, communications and command support cells); communication and information systems (CIS) support, providing long-distance communication and information systems connectivity; a field kitchen; diesel generators for power supply; a medical facility; and accommodation tents. The total airlift effort for the two recce teams, the ICE and the nucleus of the RSC is reported to be 30 C-130 or C-160 sorties.

These elements would sustain operations until they were reinforced by the full corps HQ which would deploy from its peacetime location in Münster, Germany (probably by both sea and air) within 20 days after an activation order. The complete HQ includes a main CP and a forward CP, both being fully mobile and the latter serving to sustain C2 operations while the main CP is being moved to a new location (which could typically take place every 48 hours in case of frequent relocations during a high-intensity conflict), but they can also be combined to create a more spacious working environment, such as in a static peace support operation).

The heart of the CP is the joint operations centre (JOC), supported by three separate clusters (intelligence, combat service support and C2 support) located in close proximity. The JOC is housed in the newly developed modular insulated tent system (MITS) supplied by Actum Projects and NEDAM Roermond. The 12.5 × 9.4 m MITS includes standard interfaces to attach eight trailer- or truck-mounted office shelters. The latter contain specialised cells such as the air operations co-ordination centre (AOCC), deep operations, fire support, air manoeuvre, special forces, information operations, current operations and plans. The MITS is carried in four 20 ft containers carried on two DROPS (dismountable rack off-loading and pick-up system) trucks with trailer. One contains the complete tent, one the furniture and computers, one the power generators and one the heating system. It is reported that a trained team of 11 people can build up the facility within 6 hours, while disassembly takes a claimed 5 hours. The MITS provides an optimal C2 working environment, a flexible and modular layout, and a self-sustaining capability in terms of power and CIS. Facilities include extensive fibre optic LAN/WAN connections and phone hook-up points throughout the CP; synchronised heating and air conditioning to enable operations in any climate; and full office support facilities such as printers and copiers. The central MITS work-floor includes the JOC operations room with space for 15 staff officer workstations plus a three-workstation command bridge. It faces a situational awareness wall featuring a large screen display for electronic presentations plus maps and status boards to the left and right, showing enemy and friendly forces' orders of battle and combat effectiveness assessments. The HQ's information system is the German Army's HEROS (see separate entry), but an alternative option is to adopt the RNLA's new ISIS system (see separate entry).

View of the Joint Operations Centre of the GE/NL HRF(L) HQ (Timo Beylemans)
0111823

Aerial picture of the GE/NL HRF(L) HQ deployed non-tactically. Visible are the JOC (centre, under urban camouflage, briefing tent behind), the intelligence cluster (below right) and the command support cluster (right). (Timo Beylemanns) 0111822

Under an agreement dating back to the formation years of the HQ (1993-95), Germany is responsible for providing information systems while the Netherlands has the lead for communication systems. CIS equipment is maintained and operated by a dedicated 487-strong bi-national CIS battalion based in Eibergen and Garderen, the Netherlands, and there is also a dedicated bi-national staff support battalion based at Münster, Germany, to provide security, logistic, medical and engineering support. Signals equipment is available to support a span of control that could include:

(a) Up to four divisional-size major subordinated units

(b) Eight (when fully mobile) smaller subordinated units (the latter could increase in number during a static operation such as KFOR)

(c) The HQ's two organic support battalions.

The CIS battalion will have nine rapid CIS elements (RACEs) assigned that would be detached to the subordinate division(s) and other CPs in order to provide connectivity.

The HQ is connected to a series of NATO-wide information sources, such as the Joint Operations Intelligence and Information System (JOIIS), Linked Operations Centres Europe (LOCE) and the Battlefield Information Collection and Exploitation System (BICES). For internal communications, the HQ is creating a master domain WAN linking together smaller LANs that serve the main CP, forward CP, staff support battalion, CIS battalion, the RSC and the peacetime HQ. The WAN will be accessible for subordinate units, and will be connected through firewall to other domains, such as the Bundeswehr Intranet.

External connectivity is expected to be supplied principally by the new Netherlands-supplied TITAAN (Theatre-Independent Tactical Army/Air Force Network) (see separate entry), which will include COTS fibre optic networks, server vehicles and LAN boxes and will support 250 users on laptops in the main CP. TITAAN should support video teleconferencing, secure and non-secure Voice Over Internet Protocol (VOIP) and a range of data communication functions. These include a mission secret network, a mission-unclassified network, office automation, an ATCCIS-compliant land information system, and a military message handling system. There will also be a cellular phone system and HF radio as back-up.

A satellite communications capability (of up to 512 kbytes/s) is expected to be in place by the end of 2002. Initially, this will be based on an interim system for which the contract for 12 trailer-mounted satcom terminals (Ku-band and C-band) was signed in March 2002 with Xantic, an Australian-Netherlands company formed by a merger between Station 12, SpecTec and Telstra Global Satellite.

By 2004, these should be replaced by 17 mobile satcom terminals providing up to 2 Mbytes/s, to be procured under the Netherlands Milsatcom programme. At the end of February 2002, the Netherlands Defence Ministry announced that a consortium led by ND SatCom (Friedrichshafen, Germany) had been selected over a team led by US company Harris and Siemens Nederland to supply these terminals as part of a wider €50 million (US$44 million) programme. The ND SatCom consortium includes L3 Communications, Stork Aerospace (Fokker Defense) and Thales Nederland.

Additional external connectivity can be provided by:

(a) Line of sight (up to 30 km) 3 × 1 Mbyte/s microwave links (taken from the Thales Zodiac system but modified to support VOIP)

(b) Existing ISDN land lines

(c) Up to three dedicated 30 × 30 km cellular phone networks fielded by the CIS Battalion

(d) German Army-supplied data-capable HF-A and HF-C radios as a back-up.

Status

All six HQs have successfully achieved full operational capability (FOC). The UK-led Allied Command Europe Rapid Reaction Corps (ARRC) was successfully evaluated in February-March 2002. The German/Netherlands (GE/NL) Corps qualified for interim operational capability status in March 2002, was operationally ready in June 2002 and achieved FOC in November 2002. The five-nation Eurocorps, 3 (Turkish) Corps, the Italian Reaction Corps and the Spanish Reaction Corps achieved FOC between September and December 2002.

The Danish/German/Polish Multinational Corps Northeast and a Greek HQ are preparing to become follow-on HQ, aiming to achieve FOC by 2005.

UPDATED

Poland

Automated mobile command and control post WD-94

The WD-94 is an automated mobile command and control post which can be used in different army formations and public services, with the software and equipment varying according to role, although it seems principally to have been designed for AD command and control. When used in this role it is designed for Wing Commanders and Antiaircraft Battery Commanders. Mounted in a container body on either a Honker 2324 or Mercedes cross-country vehicle, it has three crew (driver, commander and operator), and the manufacturer claims it can be deployed in less than 12 minutes from the move.

The WD-94 system provides:

(a) Automatic reception and display of the tactical air picture
(b) Automatic target allocation
(c) Voice and data communication in tactical radio networks and digital area communication systems
(d) Automatic identification of the vehicle position.

The post contains the following equipment:

(a) Three ground-to-ground VHF radios
(b) Integrated system for land navigation (inertial navigation and GPS)
(c) Board power generator and battery 27 V DC
(d) Heating, ventilation and fire detection systems
(e) Two computerised automatic workstations
(f) Antenna system (including UHF wide-band antenna with 6.5 m mast)
(g) Real-time application software using QNX operating system.

Manufacturer
Przemyslowy Instytut Telekomunikacji, Warsaw.

VERIFIED

Dunaj - Polish Air Force AD C2 System

In July 2002 *Jane's International Defense Review* reported that the Polish Air Force and Air Defence (WLOP) had declared operational its first new control and reporting centre (CRC) equipped with the Dunaj sector air-defence command-and-control (C2) system. The centre, known as 21st ODN, is located in Pyry outside Warsaw. Its Dunaj system was developed by the Przemyslowy Instytut Telekomunikacji (PIT) (Telecommunications Research Institute), and most of the software by the Filbico company. Operational consultancy and integration was supervised by the WLOP's automation centre.

The Dunaj system is claimed to be fully compliant with NATO Consultation, Command and Control Agency (NC3A) requirements for the future NATO Air Command and Control System (ACCS). It consists of LAN-connected workstations that form:

(a) The CRR-20 subsystem, which produces the recognised air picture (RAP), taking in data from a variety of sensors, including at least three fixed air-defence radars, several mobile stations and other assets.
(b) The CSD-20 sector command team subsystem, which is responsible for supervision of fighters, ground-based air-defence and electronic warfare systems.

Within its area of responsibility, a Dunaj system works with forward-located C2 assets including PIT's DL-15 mobile ground-controlled intercept (GCI) post, two of which are in WLOP service. It is also linked with ground-based air-defence units, wing operations centres and with the new air traffic co-ordination system installed at the Minsk-Mazowiecki and Poznan-Krzesiny air bases. In the future, it will be linked to a Filbico-developed air operations co-ordination system for corps-level army headquarters, known as CKOP. It will also be linked with the Polish Navy's Leba-2 C2 system, while it will be possible to exchange data with neighbouring

Inside the 21st ODN in Pyry, a view of the CRR-20 subsystem's positions (Grzegorz Holdanowicz) 0140035

NATO assets, in particular CRCs located in Denmark and Germany and the combined air operations centre (CAOC) in Kalkar, Germany.

Status
The 21st ODN is the first of four NATO-standard CRCs that will replace nine old-style sector command posts by 2004. The 22nd ODN in Bydgoszcz will be ready by the end of 2002, 31st ODN in Poznan in 2003 and 32nd ODN in Krakow in 2004. In January 2003 *Jane's Defence Weekly* reported that the 22nd ODN had been "recently handed over". The four networked CRCs are to be supervised by the WLOP's air operations centre (COP), which on 1 January 2002 replaced the commander-in-chief's main command post. The COP is co-located with the 21st ODN and the air sovereignty operations centre (ASOC), which operates as an integral part of the COP. In the future it is to be upgraded to a nationally operated CAOC. The US Air Force Electronic Systems Centre and the Polish Ministry of Defence are investigating the option of extending the ASOC system with additional combat capabilities.

After 2006, Poland is planning to retain only two ODNs (those in Bydgoszcz and Krakow). Krakow would then be reconfigured into an ARS (Air operations centre, Recognised air picture production centre, Sensor fusion post), and Bydgoszcz into a CARS (CAOC plus ARS) element of the ACCS. The 21st ODN would be kept as a training facility, while the fourth centre in Poznan could potentially be reformed into a mobile back-up system.

Contractor
Przemyslowy Instytut Telekomunikacji, Warsaw, Poland.

UPDATED

Russian Federation

Mainstay Airborne Early Warning and Control (AEW&C) system

The Mainstay is an AEW&C system based on the Ilyushin Il-76 aircraft. Previously known as A-50, the aircraft carries colour CRT displays, EW systems, IFF transponder/decoders and other equipment from the Ministry of Radio Equipment Industry. One version has been modified by Israeli Aircraft Industries.

See *Jane's Aircraft Upgrades* - Beriev A-50 - for further details.

UPDATED

Singapore

Armed Forces Command Post (AFCP)

The AFCP is designed to enhance the command structure of the Singapore Armed Forces (SAF). Track data from ground-based surveillance assets are linked to the AFCP by fibre optic cables and landlines, while encrypted data from air platforms are down-linked to the nearby Air Force Systems Command (AFSC) for landline/fibre optic link to the AFCP. This allows the SAF to collate and integrate data in real time. Co-ordination between the Ministry of Defence (MINDEF), the AFCP and the AFSC is enhanced by a network of underground tunnels connecting the facilities.

The AFCP programme commenced in the late 1980s to coincide with the relocation of MINDEF in 1989 to a purpose-built facility on Bukit Gombak hill. This involved the design and construction of a hardened complex extending underground, with elaborate earthworks and revetments protecting the structure's surface.

Defence of the complex is believed to be assigned to the 5th Battalion Singapore Infantry Regiment.

VERIFIED

Turkey

Armoured Command and Control Centre

The ASELSAN Armoured Command and Control Centre is a high-mobility tactical command and control system, developed to serve as a peacetime and wartime command post for deployed tactical units. This system is mounted on a 4 × 4 Light Armoured Vehicle and is very similar in capability to ASELSAN's soft-skinned Mobile Command and Control Centre. The modular structure allows the system to be adapted for different requirements.

The Armoured Command and Control Centre can be a part of a command and control system or stand alone. The system collects and evaluates operations and intelligence data and tracks the situation in the area of concern at the level it is deployed. The data is collected using various communications subsystems and is

Aselsan ACCC
0092759

Aselsan Satcom Terminal for the Mobile Command and Control Centre 0092763

*Aselsan Mobile
Command and Control
Centre*
0092762

evaluated using command and control software which has geographical information system, database management, image enhancement and battlefield information collection functions. Up to four consoles can be configured to perform specific applications such as situation assessment, communications control and monitoring, image analysis and sensor management. The system evaluates various sensor data that is transmitted from ground surveillance radars, surveillance thermal cameras, buried and camouflaged sensors. Power supply for the centre is provided by a portable generator and air conditioning is also available.

Specifications
(a) Command and Control software:
   GIS
   DBMS
   Image enhancement
   Battlefield information collection
(b) Communications:
   ASELSAN 9600 VHF Frequency Hopping Radio
   ASELSAN SK-4000 Digital Encrypted Radio
   SatCom Terminal
   Leased line
   GPS Receiver
   ASELSAN data terminal
   ASELSAN field telephone
(c) Communication monitoring and audio recording system:
   HF/VHF/UHF Monitor Receiver (20-1,200 MHz)
   Audio Tape Recorder

Status
The Armoured Command and Control Centre is in production and operational service.

Manufacturer
ASELSAN Military Electronic Industries Inc, Ankara.

VERIFIED

Aselsan Mobile Command and Control Centre

The ASELSAN Mobile Command and Control Centre is identical in function and capability to their armoured command and control vehicle (see separate entry) except that this system is mounted on a 5 tonne truck and is designed for use at battalion level and above. The prime components are: shelter, shelter lift device and a trailed generator providing power for the system and air conditioning. The modular structure allows the system to be adapted for different requirements and it can be a part of a wider command control system or operate stand-alone.

The system collects and evaluates operations and intelligence data and tracks the situation in the area of concern at the level it is deployed. The data is collected using various communications subsystems and is evaluated using command and control software which has geographical information system, database management, image enhancement and battlefield information collection functions. The computer infrastructure includes a command and control console, communication monitoring console and specific application consoles which can be configured to perform specific applications such as situation assessment, communications monitoring, image analysis and sensor management. The system evaluates various sensor data that is transmitted from ground surveillance radars, surveillance thermal cameras, buried and camouflaged sensors. Power supply for the centre is provided by a portable generator and air conditioning is also available.

Specifications
(a) Command and Control software:
   GIS
   DBMS
   Image enhancement
   Battlefield information collection
(b) Communications:
   ASELSAN 9600 VHF frequency hopping radio
   ASELSAN SK-4000 digital encrypted radio
   SatCom Terminal
   Leased line
   GPS receiver
   ASELSAN data terminal
   ASELSAN field telephone
(c) Communication monitoring and audio recording system:
   HF/VHF/UHF monitor receiver (20-1,200 MHz)
   Audio tape recorder

Status
The Mobile Command Control Centre is in production and believed to be in operational use in the Turkish armed forces.

Manufacturer
ASELSAN Military Electronic Industries Inc, Ankara.

VERIFIED

United Kingdom

MultiRole Operations Cabin (MROC)

The MultiRole Operations Cabin (MROC) provides a complete air defence command, control and communications capability within a deployable, single ISO cabin.

Employing a flexible, open systems architecture, a single MROC can be configured to satisfy a wide range of operational requirements, while the modular

An interior view of the AMS MultiRole Operations Cabin (AMS)   0525547

design enables a number of MROCs to be combined to meet the demands of the more complex configurations. Typical applications include reporting posts, control and reporting posts, control and reporting centres, sector operations centres and air defence operations centres.

The system supports a comprehensive range of air defence functions including picture compilation, identification/recognition, threat evaluation, resource management, weapons assignment and air mission control.

The system's design allows considerable flexibility by offering a wide range of internal layouts based on an extensive range of commercial-off-the-shelf (COTS) platforms. The HCI is Windows-based.

Specifications

(a) Picture compilation:
  primary and secondary radar plot and strobe processing
  manual track facilities
  automatic track initiation, maintenance and cancellation
  flight plan storage and maintenance
  safe corridor storage and maintenance
  defended area management
(b) Identification/recognition:
  manual facilities
  automatic correlation of tracks with flight plans and safe corridors
(c) Threat evaluation:
  automatic evaluation against defended areas
  threat track alerting
(d) Resource management:
  resource management of aircraft at airfields and on CAP
  resource management of SAM and AAA assets
(e) Weapons assignment:
  automatic weapons suggestions
  manual assignment
(f) Air mission control:
  broadcast control
  fighter target pairings
  loose control
(g) Additional facilities:
  airspace management
  RASP dissemination
  tactical datalink integration
  simulation and training facilities
  recording and replay facilities
(h) MROC configuration:
  20 ft air conditioned ISO cabin on a trailer or fitted with mobilisers
  trailer-mounted main and standby generators
  cross-site power and signal cables
  up to 5 multifunction COTS raster scan workstations
  data entry terminal
  fully integrated communications system
  VHF radios with mast, antennas and feeders
  UHF radios with mast, antennas and feeders
  ground-air-ground radio control equipment
  PTT compatible telephone switch
  HF/SSB radio with whip and V-antennas
  SATCOM connections for remote data transfer
(i) Typical capacities for basic MROC:
  reportable tracks 250
  radar sensors 4
  flight plans 400
  simulated tracks 100
  steerable simulation targets 40
  controlled missions 24
  maps and grids 25
  airfields 6
  SAM batteries 6

(j) Expansion capability:
  multiple MROCs can be combined to provide larger systems giving a greater level of interoperability
  supplementary displays can be added into dedicated facilities
  all the capacities listed above are adaptable to meet particular customer requirements

Status

The MROC concept was based upon equipment delivered to NATO to fulfil command and control needs in Portugal under the Portuguese Air Command and Control System (PoACCS) programme. The MROC cabins provide the Stand-Alone Control Facility (SACF) at Montejunto.

MROC is also in production for non-NATO customers in a control and reporting configuration.

Contractor

AMS, Camberley, Surrey.

VERIFIED

Tactical Air Control Centre (TACC)

The Tactical Air Control Centre (TACC) is highly mobile, flexible and designed to provide sustainable tactical direction to Rapid Reaction Forces engaged in global expeditionary operations. The equipment is produced by Thales in co-operation with Northrop Grumman who provide the real-time air applications software originating from their combat proven USAF and USMC mobile air control systems.

Description

The TACC requirement was formed against a background of geopolitical change and the associated military restructuring to cope with an increasing global commitment of forces to a wide range of combat, support or diplomatic tasks. The TACC will play a vital part in supporting the Rapid Reaction Air elements dispatched to such operations by orchestrating the minute-by-minute conduct of an air campaign, providing real-time tactical direction of assigned weapons systems, and conducting, when necessary, the management of an air defence battle. To accomplish these responsibilities, the TACC performs the following operational functions:
(a) creation and dissemination of a Recognised Air and Surface Picture (RASP)
(b) tactical direction to offensive / support aircraft and real-time re-tasking
(c) air defence battle management
(d) airspace management
  TACC features several crucial design characteristics:
(a) rapid and cost effective deployment
(b) flexible configurations
(c) optimum layout of facilities for the operational teams
(d) high-grade data fusion
(e) maximum use of COTS products
(f) low-risk approach to software development.

The current TACC design comprises 19 multifunction operator workstations, along with intelligence and higher formation CIS terminals, over 30 voice and data communications radios and interfaces for a variety of national landline circuits. All of the equipment is packaged into 10 standard ISO-style containers. Two of these contain the operator display facilities and two house the supporting communications and computer equipment. A single, expandable ISO shelter, known as the Battle Management Module is also available to provide the space necessary for the TACC commander and the supporting planning staff. Three further shelters are used to package sets of the power generation system and two others are used for spares and technical workshops.

With the modular design of these principal elements, it is possible to create different TACC equipment configurations to match the situation. For example an immediate capability can be provided using just three containers, offering six operator workstations, a mix of radios for voice and infra-red communications, interfaces for sensors and CIS inputs and power. Along with its local sensors, this will provide the ability to build the RASP and provide sufficient control capacity for the build-up phase of a campaign. As the situation develops, additional containers can be added to increase the control or planning capability until all the nine containers comprising the complete system are deployed.

The complete system can be packed into 7 × C130 loads; however, the immediate response can be airlifted to the operating area by two C130 Mk III aircraft. Once in theatre, the TACC can be rapidly redeployed either by road, rail, sea or by helicopter lift. With a variety of mobility features, such as fibre optic cabling and integral mobilisers, the TACC can be rapidly readied for deployment and re-erected at a new site.

A very sophisticated display and data handling system is at the heart of the TACC system and this provides the functionality necessary for the operators to execute their respective picture compilation or weapons control tasks

It features an ORBIX-based, client server architecture employing the latest Sun Servers feeding, in the ACC and BMM configuration, 19 × Sun Ultra operator workstations. The real-time applications software is supplied by Litton Data Systems in Ada and C++. This updated, re-engineered software provides the multisensor tracker, a track data manager, recognition tools and multidata link processing to create the picture, along with threat warning, asset management and weapons control packages to aid the weapons teams. The real-time segment also includes a simulation facility. With off-line scenario generation, dynamic track control and representation of all data connections, this faithfully represents all the TACC functionality for effective operator training. The information derived from the Display and Data handling software is presented to the operators on newly developed graphic and tote displays.

The TACC is equipped with the communications necessary to interface with other sensors and intelligence platforms, other maritime and land tactical control formations, weapons platforms and higher formations. To that end, it currently employs NATO Links 1, 11A, 11B (TADIL A and B) and Link 14 and can interface with up to 15 different types of active/passive sensors. Moreover, there is a mix of secure, insecure and ECM-resistant UHF / VHF radios and up to 10 HF radios. Operator access to communications is via a programmable, 10.4 in, colour LCD touchscreen Communication Control Panel (CCP). This is positioned below the main display of each TACC workstation, with connections for headsets and handsets and a loudspeaker. The voice system is complete with two digital voice recorders each recording ground-to-ground and ground-to-air exchanges for each Operator's CCP.

The architecture and existing capacity of the TACC is designed to allow growth and customisation for other Services' or Nations' applications. For example, a SAM/SHORAD weapons control package is immediately available for inclusion, and the integration of a Theatre Ballistic Missile Defence software package is being explored that will allow asset planning, the display of hostile missile launch and impact points, and the time lines for respective weapon engagement possibilities. Moreover, plans are underway to install a campaign planning tool to provide an Air HQ capability, and software for flight strip handling and en-route deconfliction is being evaluated to build-in the potential for civil or military Air Traffic Control applications.

Status

Scheduled to enter service with the Royal Air Force in March 1999, this programme was beset by delays. In January 2003 it was announced that the equipment had been delivered to RAF Boulmer for use by No 1 Air Control Centre.

Contractor

Thales Defence Information Systems (prime), Wells, Somerset.

UPDATED

United States

ABCCC III (Airborne Battlefield Command Control Center III)

ABCCC III is a containerised command and control centre designed to be carried by US Air Force EC-130E aircraft. Its principal role is to co-ordinate strike aircraft in air-to-ground missions. However, the system is not limited to this function and can be used to complement virtually any command and control agency on the battlefield. As part of the Tactical Air Control System, it improves communications between widely dispersed combat units and maintains contact with the air support operations centres usually co-located with each army corps. It can also communicate with E-3A AWACS aircraft.

Development

The ABCCC III system replaces the ABCCC II which was an upgraded version of a system developed in 1964. The ABCCC III squadron is operated by the 7th Airborne Command and Control Squadron based at Keesler Air Force Base, Biloxi, Mississippi, which is part of the 28th Air Division and reports to the Tactical Air Command, Langley, Virginia.

It was announced in April 1988, after a 15 month demonstration, that Unisys (Paramax Division) had won (over TRW) the contract to produce the first of eight ABCCC IIIs. Initial year value of the contract was US$13.5 million.

The interior of an ABCCC III container

Self-contained command and control module being loaded on an EC-130 ABCCC III aircraft

Description

Twelve battle staff consoles, each with a high-resolution (1,024 × 1,280) CRT multicolour display, are provided in the ABCCC III container. ABCCC III has a large database of current information necessary for battle management, including map data and location/status information on own and enemy forces.

Controllers keep track of air and ground forces using situation awareness displays. Map displays are provided as software from a ground-based map support station which maintains a digitised map database supplied by the Defense Mapping Agency. With worldwide coverage, the digitised maps are based on the 1:2,000,000 scale jet navigation charts. The map support station is able to produce digitised maps between 1:500,000 and 1:50,000 scale for areas of interest. The database for any mission covers 2,048 × 2,048 n miles, with enhanced detail in the centre 1,024 × 1,024 n miles.

The map databases are stored on one of four 200 Mbyte 5.25 in optical disks which also store the operating software, real-time mission history and online playback. System initialisation and mission planning data, including maps, communications plans and the Air Tasking Order, are prepared by portable ruggedised mission planners for export via optical disk to the container. For redundancy, the system uses two AN/UYK-44(V) (designated CP-2025/USC-48) processors which are connected to a serial databus. The 12 operator displays are driven by a Unisys-enhanced display generator (CP-2025/USC-48) that communicates with the processors.

Consoles for two communications operators are provided in the container. The communications suite consists of UHF, VHF AM, VHF FM and HF radios, the numbers of which vary with the configuration. Satellite communications terminals are also provided, as is a JTIDS terminal.

The airborne maintenance technician also has a console which houses a centralised built-in test system from which he is able to monitor the entire system.

Automated switching for all 15 console positions is provided by a 64 kbit CVSD Communications Distribution Group (OA-9400/USC-48).

Status

Integration and operational testing was completed in December 1990. Paramax has delivered four production units some of which were deployed with Operation Desert Storm. Four more units were scheduled for delivery in 1991.

In 1993, a JTIDS capability was added to ABCCC III.

There are seven modules which are carried in aircraft operated by 42nd Airborne Command and Control Squadron out of Davis-Monthan Air Base, Arizona.

In 2002 it was announced that the functions of the ABCCC were to be transferred, principally to E-8 JSTARS and E-3 AWACS (see separate entries), and the squadron was deactivated in September 2002.

Contractors

Paramax Systems Corporation (a Unisys company)(Prime Contractor), McLean, Virginia.

Astronautics Corporation of America(displays), Milwaukee, Wisconsin.

NORDAM (container), Tulsa, Oklahoma.

UPDATED

Airborne Warning And Control System (AWACS) E-3 Sentry

In the early 1960s the Electronic Systems Center of Air Force Material Command at Hanscom Air Force Base, Massachusetts set in motion plans for an airborne radar and command and control system that would be able to look down on low-flying strike aircraft. The result was the E-3A Sentry, a modified Boeing 707-320B aircraft re-engined with Pratt & Whitney TF-33 turbofans and carrying the Westinghouse (now Northrop Grumman) AN/APY-1 radar, an IBM 4PI-CC-1 data processing system and nine multipurpose consoles. The first AWACS aircraft were delivered to the US Air Force in 1977 and by 1981, 24 aircraft had been handed over. The 552nd AWACS Wing (now the 28th Air Division) at Tinker Air Force, Oklahoma became operational in 1978.

Two of the multipurpose consoles aboard the US Air Force AWACS aircraft

AWACS requirements for the Continental US (CONUS) depend on the air defence alert condition. In peacetime, a total of five AWACS aircraft would be deployed on a rotational basis, one at each of the US region operations control centres of the Joint Surveillance System (JSS). Of these, four are in the CONUS region and the fifth is in Alaska

In December 1978, NATO's Defence Planning Council gave approval for the procurement of E-3A Sentry aircraft. The version ordered by NATO contained a number of improvements, notably a new radar with a maritime surveillance capability (Northrop Grumman AN/APY-2), an improved computer (IBM 4PI-CC-2) and a Joint Tactical Information Distribution System (JTIDS) terminal. A total of 18 of these aircraft were delivered to NATO, the first in January 1982 and the last in April 1985. One was subsequently lost in a crash in Greece in July 1996. In addition, 10 US/NATO-type aircraft (nine operational and one test version) were delivered to the US Air Force at Tinker AFB by June 1984.

In 1983, Boeing was awarded a major contract to upgrade the original 24 US E-3A aircraft. The improvements included the installation of ECM-resistant Have Quick secure voice communications equipment, a JTIDS terminal, five additional multipurpose consoles, five additional UHF terminals, some maritime surveillance capability, improved computer, new software and self-defence missile hardpoints. These aircraft are now referred to as E-3Bs.

The 10 aircraft built in the US/NATO configuration have also been modified with the addition of five extra UHF radios, improved maritime surveillance capability and improved software and are now designated E-3Cs.

All of the E-3 aircraft have a distinctive mushroom-shaped radome protruding from the top of the fuselage. This houses the AN/APY-2 radar antenna mounted back-to-back with a complementary IFF/SSR antenna. Radar range is in excess of 320 km (200 miles). The radome normally rotates at 6 rpm giving a scan rate of once every 10 seconds. Operating frequency of the AN/APY-2 is in the NATO E/F-band (10 cm) and there are seven operating modes. On any azimuth scan, the surveillance volume can be divided into as many as 32 subsectors, each with its own set of operating modes and conditions. These modes can be accommodated on subsequent scans or rearranged to vary the type of coverage for any given area of interest or to accommodate changes in operating conditions. The seven modes are:
(a) Pulse Doppler Non-Elevation Scan (PDNES); this provides surveillance of aircraft down to the surface using pulse Doppler, with narrow Doppler filters and a sharp beam, to eliminate ground clutter. Target elevation is not measured
(b) Pulse Doppler Elevation Scan (PDES); radar operation in this mode is similar to PDNES, but target elevation is derived by electronic scanning of the beam in the vertical plane
(c) Beyond-The-Horizon (BTH); the BTH mode uses pulse radar, without Doppler, for extended range surveillance where ground clutter is in the horizon shadow
(d) passive; the radar transmitter can be shut down in selected subsectors while the receivers continue to process ECM data. A single strobe line passing through the position of each jamming source is generated on the display console
(e) maritime; this involves use of a very short pulse to decrease the size of the sea clutter patch to enhance the detection of moving or stationary surface ships. An adaptive digital processor adjusts automatically to variations in sea clutter and blanks land returns by means of computer-stored maps of land areas
(f) test/maintenance; control is delegated to the radar technician for maintenance purposes
(g) standby; radar kept in a warmed operational condition ready for immediate use. Receivers are shut down.

PDES and BTH can be used simultaneously or alone; either or both can be active or passive, as required. PDNES may be used with the maritime mode. Blanking commands in BTH and pulse-Doppler modes can be used in each of the 32 subsectors, thus enabling the maximum potential of the system to be concentrated in those subsectors of greatest interest.

A key element in the overall AWACS system is the Boeing-developed interface adaptor. This provides a common time reference for use throughout the system and interconnects the radar, IFF/SSR, avionics, data processing, communications, navigation and guidance instruments and display and control subsystems.

Status
The first E-3 entered US Air Force service in March 1977, preceded by more than 10 years of competitive fly-offs, prototype design and development. The last of 34 US AWACS aircraft was delivered in June 1984. NATO deliveries began in early 1982 and the eighteenth and last NATO E-3 was delivered in 1985.

A new, long-term multistage improvement programme for the E-3 began with a US Air Force award of the ICON (Integration Contract) to Boeing in May 1987. ICON will equip both US and NATO E-3s with an electronic support measures (ESM) passive surveillance capability, and block enhancements to the US E-3 fleet.

The latter improvements for the US fleet include upgrading the Joint Tactical Information Distribution System (JTIDS) to TADIL-J (Tactical Digital Information Link-J), increasing the computer and adding the ability to use the Global Positioning Satellite (GPS) system to pinpoint AWACS' location anywhere in the world. Development of these enhancements was completed n 1994 and production of modification kits has been authorised by the US Air Force.

In March 1989, Boeing was authorised to begin production of Have Quick A-NETS, an improved communication system that provides secure, anti-jam radio contact with other AWACS, friendly aircraft and ground stations to a degree not previously available. Retrofit of the US fleet was completed in late 1994.

In mid-1989, Westinghouse (now Northrop Grumman) received a US$223 million contract from the USAF for the full-scale development phase of the RSIP upgrade for the USAF E-3 AWACS fleet. This contract includes the design, development and flight test of the improvements to the AN/APY-1 and -2 radars to maintain operational capability against the growing threat from smaller radar cross-section targets, cruise missiles and ECM. In addition, significant improvements in the man/machine interface, and reliability and maintainability are included.

In 1993, NATO's AWACS modernisation programme intensified. Throughout the year, Boeing delivered production-quality, upgraded computer systems to NATO as part of the Memory Upgrade Programme. Under this effort, Boeing upgraded the existing IBM CC2 computer to the CC2E model. The new system uses advanced technologies to increase the system's memory by nearly 400 per cent.

In 1993, NATO also awarded three major upgrade contracts to Boeing. In January 1993, the company received a US$294.6 million procurement and production contract as part of the Mod Block 1 phase of NATO's modernisation programme. A US$35.5 million follow-on contract to install and test Mod Block 1 hardware was received in May.

Mod Block 1 covers three important mission system-related enhancements:
(a) colour displays, which will improve the form and usability of incoming situational information
(b) Have Quick radios, which will enhance UHF communications by adding antijam features
(c) a version of the Joint Tactical Information Distribution System (JTIDS), called Link 16, which will increase the amount of information that can be collected and distributed among other AWACS, allied aircraft and ground stations.

In 1994, Boeing received contracts from NATO and the US Air Force, worth US$16.8 million and US$127 million, respectively, to produce ESM kits for their E-3 aircraft. The most significant upgrade yet developed for the E-3, ESM was developed by Boeing under joint funding from the US Air Force and NATO.

ESM enables the AWACS to detect, identify and track electronic transmissions from ground, airborne and maritime sources. Using the ESM system, mission operators can determine radar and weapon system type.

In October 1995, the first US E-3 aircraft was equipped with ESM. US Air Force personnel at Tinker Air Force Base, Oklahoma, will install the remaining ESM kits on US aircraft during scheduled depot maintenance; the NATO retrofits will be done by DaimlerChrysler Aerospace.

Also in October, the NATO N-1 aircraft travelled to the alliance's main operating base in Gellenkirchen, Germany, for inspection and Initial Operational Test &

E-3 Sentry (AWACS) take-off showing rotating radome of AN/APY-2 early warning radar

Evaluation (IOT&E) testing. This marked the completion of RSIP installation on N-1. N-1 is an E-3 assigned by NATO to support enhancement programmes.

The Peace Sentinel programme for Saudi Arabia began in 1981. It included five AWACS aircraft and six E-3 derivative (KE-3) in-flight refuelling tanker aircraft, along with spare parts, trainers and support equipment. In 1984, the Saudi government exercised an option to increase the tanker order to eight.

The first Saudi E-3 was delivered in June 1986, with deliveries of the remaining E-3s and tankers completed by September 1987. In addition to building the aircraft, Boeing assists in operating and maintaining the AWACS and tankers in Saudi Arabia. Boeing has issued subcontracts to three Saudi companies to assist in this support work. In 2001 it was reported that the aircraft's mission computer and its software were being upgraded to the same level as that of the USAF fleet. Two aircraft were to be retrofitted in 2001 and the remaining three in 2002.

In March 1986, Boeing was invited to respond to a proposal by the UK seeking replacement candidates for the Nimrod airborne early warning system. Boeing proposed the existing AWACS standard configuration, with system improvements such as CFM-56 engines, and other enhancements to respond to particular Royal Air Force specifications.

In September 1986, the UK Ministry of Defence narrowed the field of seven AEW candidates to the E-3 and Nimrod. In December, the MoD announced its decision to purchase a minimum of six E-3s. A final contract was signed in February 1987.

France joined with the UK in its evaluation of competing systems, continuing French interest in the capabilities of the E-3 that dates back to the late 1970s. In December 1986, Boeing submitted a proposal for an E-3 purchase to the French Ministry of Defence. In February 1987, the French government announced an order for a minimum of three E-3s to be configured similarly to those planned for the UK. Both sales were made directly by Boeing to the respective governments. The two countries also announced they would jointly manage their AWACS procurement and established the Joint Anglo-French Management Office.

By the third quarter of 1987, both countries also had exercised contract options for one additional E-3 each, bringing to 11 the total of AWACS on order for delivery beginning in the first quarter of 1991. The first UK AWACS was delivered in March 1991, the first French AWACS in May 1991. France received its final E-3 in February 1992, and the UK deliveries were complete in May 1992. Along with the ability to operate with the worldwide AWACS fleet, the UK and France gained acquisition cost savings associated with purchases in the same period of time. Enhancements added to meet each country's mission requirements include a probe system to augment the existing boom receptacle for in-flight refuelling, a digital recorder for mission audio transmissions and improved radio equipment.

Additionally, the UK system includes ESM and the radar is upgraded with Maritime Scan-to-Scan (MSSC) Capability. The UK has also joined the USA in initiating the RSIP radar upgrade. France has adopted the US (Boeing) ESM system as its first major E-3 upgrade. Plans are under way to define the production and installation programmes. The cost of the UK's fleet, including support equipment, was approximately US$1.3 billion and the contract price for the French fleet was about US$550 million.

In US and NATO operations, the aircraft have significantly surpassed the standard for mission readiness, demonstrating an availability level of 95 per cent.

The AWACS fleet has been in operational service in Asia, Europe and the Middle East. During the 1991 Gulf War, 11 US Air Force AWACS aircraft were deployed to Saudi Arabia, supplemented by three more from airbases in Turkey, two on standby in the UK, one on alert at Tinker Air Force Base, Oklahoma and five AWACS owned by the Royal Saudi Air Force. When the allied air strikes began against Iraq, the AWACS role shifted from defence to a variety of tasks including surveillance, directing air strikes, interdiction of Iraqi aircraft, co-ordination of air-to-air refuelling flights and protection of high-value aircraft conducting intelligence and ground surveillance. Some 845 AWACS sorties were flown for a total of 10,500 hours. The AWACS fleet monitored 120,000 coalition sorties - 2,000 to 3,000 per day.

NATO E-3s have been a major factor in the UNs' ability to monitor and enforce the 'no-fly' zone over Bosnia-Herzegovina. From July 1992 to September 1995, E-3 aircraft, including UK and French AWACS assets, flew nearly 5,000 missions totalling more than 39,000 hours in support of UN objectives in this region.

Production of the 707 airframe ended in May 1991. Following extensive studies of the most suitable follow-on aircraft for the AWACS mission, Boeing announced in December 1991 that it would offer a modified 767 commercial jetliner as the new platform for the system. Wind tunnel tests in 1992 corroborated engineering

The RAF AWACS differs in appearance from the US, NATO and Saudi Arabian fleets in having ESM wing pods and the refuelling probe over the cockpit

estimates on stability, control and performance after the addition of the AWACS rotodome and confirmed rotodome location.

The 767 provides several advantages over the 707. Because of its wide-body configuration, the 767 offers 50 per cent more floor space and nearly twice the volume of the 707. The two-man flight crew and high-reliability twin engines provide economic advantages as well. More than 400 Model 767 aircraft are in service with 48 of the world's airlines, so there is a wide base of suppliers, spare parts and support equipment. The aircraft continues in production at a high rate.

Although Boeing is offering a new platform, the basic AWACS mission equipment will take advantage of the proven avionics currently employed on board operational AWACS aircraft. The 767 has mission capability comparable to the 707 and is interoperable with it.

The 767 AWACS is produced at the company's plant in Everett, Washington, with modifications occurring in Wichita, Kansas. Installation of mission equipment, painting and testing is done in Seattle. Japan has now joined the AWACS community with orders for four 767 AWACS, which were delivered to the JASDF in 1999.

In August 1995, Boeing announced a contract from the US Air Force to develop and demonstrate an infra-red sensor suite for AWACS that will enable E-3 crews to detect and track Theatre Ballistic Missiles (TBMs).

Priced at about US$43.5 million, the contract was awarded by the US Air Force Electronic Systems Center at Hanscom Air Force Base, Massachusetts and is an acquisition programme funded by the US Department of Defense Ballistic Missile Defense Organisation.

As prime contractor for the effort, called the Extended Airborne Global Launch Evaluator (EAGLE) programme, Boeing will work closely with its major subcontractor, Texas Instruments, to design and procure a prototype sensor system that is based on non-developmental subsystems, for eventual integration into the overall EAGLE sensor suite. The team will test the new suite in laboratory, field and airborne environments, and install the equipment on board Test System 3 (TS-3), an Air Force E-3 aircraft that is based at Boeing Field in Seattle and used to support the testing of AWACS enhancements.

The company will provide the US Air Force with the information needed to transition the sensor into a production configuration that is fully integrated with existing E-3 mission systems.

The EAGLE sensor suite comprises five elements: two passive infra-red sensors to detect and track TBMs; a laser ranger to determine the distance to launched missiles; an inertial navigation system to ensure a missile's location is known in standard earth references; and a computer system to compute launch points, trajectories and impact points. The computer system also will format missile information for use on JTIDS.

In use, EAGLE's infra-red sensors will detect missiles early, during the ascent phase of their trajectories. They then cue the eye-safe laser ranger to generate a highly accurate trajectory that can be relayed to terminal and point defences, as well as to other elements capable of destroying missiles and launchers. Single-beam cueing of ground-based assets or airborne interceptors eliminates their need to search for threats, increasing the number of targets that can be evaluated, and maximising the time available for defenders to investigate and respond to each situation. The end result will be a more flexible, responsive network of co-operative assets, and more effective battle management.

The latest modification to the UK E3 is the Radar System Improvement Programme (RSIP), which is a joint US, UK and NATO radar hardware and software upgrade. This programme is designed to improve the pulse-Doppler radar sensitivity and resistance to ECM, as well as increase reliability and maintainability. The RSIP modification will increase the radar's sensitivity, including the development of new waveform and processing algorithms to restore the target tracking and stand-off ranges delivered in 1997, but were decreased by the reduction in RCS of fighters and cruise missiles. This modification was completed on the UK fleet in 2000, together with an upgrade to the Global Positioning/Inertial Navigation System (GINS). In 2001 an upgrade by Rockwell Collins to the HF radio system on the same aircraft was reported as completed.

In March 2001, *Jane's International Defence Review* reported that Boeing had been awarded a US$25.5 million contract to install a Global Positioning Inertial Navigation System (GINS) capability into the mission system and flight deck of the four French E-3Fs, and the aircraft's altitude measurement system is also to be upgraded. In February 2002 it was announced that Northrop Grumman had been awarded a US$61 million contract to provide RSIP kits for the French aircraft.

As at March 2001, NATO's fleet is undergoing a mid-term modernisation programme which includes:
(a) fourteen new workstation consoles with flat-panel displays offering a Windows-like feel
(b) a mission computing system with an open-system architecture allowing future upgrades to the hardware and software. Multisensor integration will improve the reliability and accuracy of the tracking process and target identification

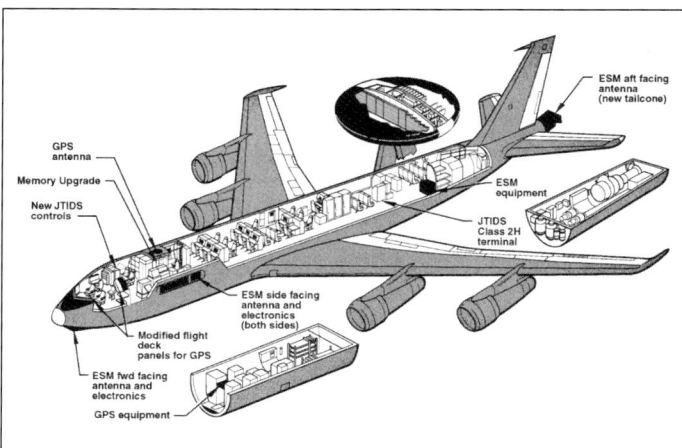

AWACS Block 30/35 system diagram

(c) digital communications systems to improve crew access to available radio links and provide automatic record and replay of communications and display data. Satellite communications will be integrated into the mission system offering a wider range of improved OTH communications via satellite links

(d) GINS

(e) broad spectrum VHF radios to support improved interoperability with Eastern European nations' air and ground forces

(f) an improved IFF transponder and an IFF Mode S interrogator.

In December 2002 it was announced that Boeing had received a contract modernisation to complete the NATO mid-term modernisation programme, increasing the value of the contract to 'more than US$1.3 billion'. The initial operational test programme on the first modified aircraft is due to be completed in late 2003. If successful, modification of the remainder of the fleet will begin in 2004.

In August 2003 *Jane's International Defense Review* reported that the AWACS Programme Office anticipated a successful Milestone B decision on the Block 40/45 upgrade, allowing it to enter the system development and demonstration phase. The upgrade will include the introduction of COTS computers and networking technologies, implemented in an open-systems architecture. The new processors will replace the proprietary mainframe computers in the E-3's mission suite, which will provide an intuitive, object-oriented point-and-click system to replace the panels of hard-push buttons and toggle switches. The aim is to reduce the workload, preventing operators from becoming task-saturated, and allow the air-battle managers to evolve to a broader command-and-control role, including greater support for the rapid prosecution of time-critical targets.

The modular Block 40/45 software includes three key components:

(a) The Multi Source Integrator (MSI), performing multisensor data fusion and combat identification. MSI accepts real-time data provided by the E-3's radar and IFF system, which it correlates with inputs from offboard sources received via tactical datalinks. It then generates a unique fused track with associated hostile or friendly identification for each target in the surveillance volume, permitting the construction of a Single Integrated Air Picture. This improves operator situational awareness and reduces the likelihood of fratricide.

(b) The Primary AWACS Display (PAD), a graphical user interface that presents the comprehensive air picture to the operator. PAD is written in Java, and can run on numerous operating systems (including UNIX, Windows 98, Windows NT and Windows 2000). The new system will be able to display much higher track loads, overlaid on full-colour maps, with additional data such as ground-unit locations and geographical features added to the picture. For example, the ability to show mountain ranges can indicate why a track may have been dropped because of terrain masking.

(c) Data Link Infrastructure (DLI), which provides prioritised target-report scheduling over current and future tactical datalinks. DLI ensures that data relating to the highest-priority threats are transmitted early, fast and often.

The Block 40/45 programme exploits the Tactical Display Framework (TDF) developed by Solipsys (now part of Raytheon). The TDF has already been implemented as part of the NORAD (North American Aerospace Defense Command) Contingency Suite.

Contractors

Boeing Defense & Space Group, Seattle, Washington.

Northrop Grumman Corporation, Electronic Systems, Baltimore, Maryland.

UPDATED

....

AN/MSN-7 Tower Restoral Vehicle (TRV)

The AN/MSN-7 Tower Restoral Vehicle (TRV) is a mobile Air Traffic Control (ATC) tower that is used to provide critical ATC tower services at temporary and bare bases, alternative off-base landing areas, or fixed air bases after loss of permanent facilities. Transportable by C-130 aircraft, the TRV is sturdy enough to be driven over rough terrain. The TRV can be set up and operational in 10 minutes and fully deployed in less than 90 minutes.

The TRV consists of a primary vehicle mounted with an operations shelter and a support vehicle that carries the generator. Both are M-1113 High-Mobility Multipurpose Wheeled Vehicles (HMMWV). An optional configuration would use a single HMMWV with a trailer to carry the generator. The standard Air Force configuration uses a 30 kW generator, but the TRV can operate on a generator as small as 10 kW. The TRV is configured to operate on standard worldwide utility power sources, but in emergency cases when external power fails, it can operate on the HMMWVs power for a limited time. An auto-sensing and auto-switching capability triggers the power subsystem. In the transport mode, the TRV roof is retracted and can be fully extended in 55 seconds. The shelter itself provides three fully compatible operator-controller positions.

The TRV was developed for the US Air Force by Tracor Aerospace Inc (now BAE Systems North America)and is designed specifically to meet the needs of US Air Force deployment operations. The TRV can be deployed to provide critical ATC tower services at temporary and bare bases or alternate off-base landing areas for extended or initial use. It also may be used to rapidly restore limited ATC tower services at tactical air bases after loss of more permanent facilities. Operations of the mobile tower may extend up to 30 days without preventative maintenance.

The Telegenix PROCOM-1540 system at the heart of the TRV supports three Communication Access Units (CAU), seven ground-to-air UHF/VHF radios and two land mobile radios (VHF FM) (optional SINCGARS radios). In addition, six connections for external radios, 12 landline/telephone connections, multicoupler (UHF/VHF) dual deck voice recorder (31 channels), maintenance monitor and

AN/MSN-7 Tower Restoral Vehicle (TRV)   0001006

configuration station, and crash phone are supported, together with radio-telephone patching, unlimited conferencing and intercom. The flexibility of the multi-Radio Interface Adaptor cards (RIA) and multi-Telephone Interface Adaptor cards (TIA) allows deployment in varying field configurations with confidence.

The Procom 1500 Series Mobile Communications Systems were cesigned from the ground up for transportable applications, where it is essential to have maximum flexibility in the ability to field configure to many different types of radio, telephone, crypto and other devices. PROCOM-1500 Series components satisfy the requirements of MIL-STD-810E, MIL-STD-461C, are powered by MIL-STD-1275 sources, and are at home hard-mounted on vehicles with Type IV (all terrain) mobility. As well as the AN/MSN-7 TRV, these systems are currently f elded with the USMC AN/TSQ-216 Remote Landing Site Tower (RLST) and USACOM Joint Forces AN/MSQ-126 Forward Deployed Communication Center (FDCC).

Features

(a) high MTBF and low MTTR

(b) no single point of failure

(c) system NVRAM remembers the last configuration for ultra-fast start-up

(d) operator position adjustable multicolour indicators and backlighting for day or night operations

(e) built-in test, allows isolation of fault down to LRU, reported at the operator position and to system maintenance workstation PC

(f) modular, distributed, fault tolerant design allows for easy expansion and repair

(g) hot Swappable LRUs allow on the fly expansion or repair

(h) MIL-STD-810, MIL-STD-461 tested compliance

(i) front accessible circuit card indications, alignment controls and jacks

(j) operating temp −28 to +65°C

(k) non operating temp −57 to +85°C

(l) Multi-Interface Telephone Interface Card including 2W PABX/CO, 4W voice, auto ringdown and many others

(m) military telephone precedence / override tones

(n) Multi-Interface Radio Interface Card including simplex/separate wire key, ground, voltage and/or contact closure keying, radio/telephone patching

(o) any mix of operator position, radio, and telephone ports up to 60 ports

(p) integrated crash phone system

(q) radio activity detect and display.

PROCOM-1500 Series components:
Communications Access Unit (CAU)

This provides operator access and control of the communications assets in the system. An integrated speaker and port for an external jackbox supporting dual headset/handset operation with instructor/student override or dual function. Jackboxes supporting a variety of handset, headset and footswitch PTT connectors are available.

Digital Interface Adaptor Card (DIA)

This provides the system interface with the CAU utilising two, two-wire digital ISDN link remotable up to 1 km over WD1 field wire. The ISDN link provides two 64 kbit/s bidirectional audio channels and two 16 kbit/s bidirectional data channels, one used for system internal signalling and the other is available for remoting/switching RS-232 data.

Multi-Telephone Interface Card (TIA)

This provides SS1-SS4 selective signalling circuits with 1, 2 or 3 digit addressing to operator position, manual/auto privacy modes, PABX/CO, auto-ringdown, loop in / ring-out. DTMF in / ring out, 2 W/4 W voice in/out, voice out / ring in, ring in / voice out, E&M Types I and V. TRI-TAC interface via TA-838A emulation, including precedence functions. onboard recorder interface configurable for TX, RX, or both.

Multi-Radio Interface Adaptor Card (RIA)

This provides voltage/ground/dry, simplex and direct wire keying, optically coupled inputs to sense voltage/ground/dry, simplex and direct wire squelch break, PTT confirmation, crypto clear/encrypted modes. Onboard recorder interface configurable for TX, RX, or both. Integrated multicoupler control.

Status

The TRV entered production in 1997 and there are 17 systems fielded. It has been used to support operations in Afghanistan during Op ENDURING FREEDOM and was used to support helicopter control during the Winter Olympics at Salt Lake City in February 2002.

Contractor

BAE Systems North America, Austin, Texas.

UPDATED

AN/TYQ-23(V) Tactical Air Operations Module (TAOM) Modular Control Equipment (MCE)

The TAOM/MCE is a transportable automated air C2 system for the coordination and control of aircraft and air defence weapons. The basic system element is the AN/TYQ-23 Tactical Air Operations Module (TAOM). A single TAOM, housed in a standard ANSI ISO shelter, contains all the air command and control equipment needed to perform the air defence function, including a full range of tactical digital datalinks. System sensors and prime power equipment are external to the MCE shelter. Four operator consoles are located in each TAOM and each is fitted with a multicolour display. A single shelter that weighs about 6,800 kg with all the TAOM equipment installed provides full system functional capability.

The ANTYQ-23 system provides facilities for:
(a) accepting inputs from search radar and IFF systems
(b) performing automatic track correlation, acquisition, identification, classification, tracking, threat evaluation and weapon selection and assignment
(c) receiving and processing track information, orders, command and status data received via digital data links from other command and control systems with digital infra-red capabilities — these datalinks are via TADIL-A, TADIL-B, one- and two-way TADIL-C, ATDL-1, NATO Link 1, MTACCS and JTIDS
(d) processing inputs from operator consoles for the entry, deletion, or modification of stored information and for the initiation of appropriate actions both within the TAOM and for external transmission
(e) displaying on operator consoles, the real-time tactical air situation based on all system inputs, both manual and automatic.

The present system design permits the interconnection of up to five MCEs through the use of fibre optic cables, although it is envisaged that four will be used as a Control and Reporting Centre (CRC) and two as a Control and Reporting Element (CRE). Up to four TAOMs will be interconnected to form a US Marine Corps Tactical Air Operations Centre (TAOC). Cables in 500 m lengths allow the dispersion of TAOMs for tactical considerations or because of terrain constraints. Interfacing radars can be located up to 2 km from the TAOM when connected by fibre optic cables. Radar/TAOM separation of up to 50 km can be achieved using equipment installed in the TAOM narrowband secure radio link.

Modular Control Equipment shelter

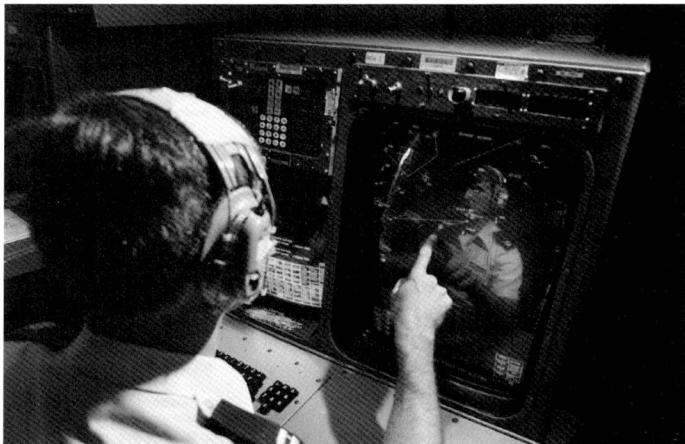

The MCE operator console units provide the man/machine interface system operator with real-time situation and auxiliary display information   0001007

A Pre-Planned Product Improvement (P3I) contract has provided the following:
(a) SATCOM interfaces
(b) JTIDS and TADIL-J interfaces
(c) SINCGARS radio interface
(d) Various force management system interfaces
(e) Replacement of the operator console unit's firmware with UNIX-based DII COE Ada code.

The US Marine Corps has also undertaken a TAOM upgrade programme that includes the following:
(a) new-open-architecture, COTS-based workstations
(b) new laser printer
(c) replacement of the voice control access unit hardware with a windowed software version
(d) addition of an electro-optic local area network
(e) implementation of the TADIL-J J3.6 message set (Link 16)
(f) replacement of the mass memory unit
(g) replacement of the operator console unit firmware with UNIX-based DII COE Ada code.

Upon completion of this modification, these TAOMs will become the AN/TYQ-23 Version 4 (V4).

Prototype AN/TYQ-82

To meet the US Marine Corps' requirement for a stand-alone TADIL-J capability that will support the TAOM and other MACCS elements, Litton (now Northrop Grumman) developed the AN/TYQ-82 Tactical Data Communications Processor (TDCP). Based on a highly mobile, reduced-footprint platform that is scalable and configurable, the TDSCP is an open-architecture, DII COE-compliant system with a full TADIL-J (Link 16) capability and the full command and control operational functionality from the AN/TYQ-23. The TDCP system comprises the following:
(a) COTS (Sun Ultra 2) operator workstation
(b) JTIDS Class 2H terminal
(c) multichannel interface unit
(d) DII COE-based software hosted on the operator workstation.

Housed in a Gichner 1497A shelter mounted on an M1097 HMMWV, the TDCP can be directly interfaced to external radars providing an electronic warfare and IFF/SIF surveillance capability and can display and transmit TBM data when interfaced to TMD-modified radars. To provide additional external operator positions, the system will support multiple workstations via a 100 MHz Ethernet electro-optic converter for interfacing to a Local Area Network (LAN).

The US Marine Corps successfully completed operational testing of two AN/TYQ-82 prototypes early in 1998.

Status

In this joint development programme, the US Marine Corps procured a total of 42 TAOM units. The US Air Force purchased 95 sheltered MCEs, plus additional unsheltered systems for use at operator and maintenance training organisations and at the USAF software support centre. All deliveries were completed by the second quarter of 1995. Systems are currently operational with US forces in Korea, Japan, Germany, Italy and Kuwait. The TYQ-23 was deployed operationally in the Balkans.

In 1998 two new units were delivered to an unspecified NATO nation. In 2002 a US$12 million contract was awarded for USAF operator console unit upgrades. The work was to be completed by September 2004.

Contractor

Northrop Grumman Electronic Systems, Navigation Systems Division, Agoura Hills, California.

VERIFIED

Army Airborne Command and Control System (A2C2S)

The Army Airborne Command and Control System (A2C2S) is a helicopter-based (UH-60) C2 system. Development started during the 1990s with the aim of replacing AN/ASC-15B/C consoles which had previously been fitted in command variants of the UH-60. The system provides five reconfigurable/removable user stations and two large common displays and enables commanders to continue to exercise command and control while on the move. The A2C2S can also be used in a static mode in which quick-erect ground antennas are used. In the future other host platforms may be used including C-130, and a very similar system is being developed for use in HMMWV and AFVs.

The system can be divided into three elements: the A kit, which provides the necessary modifications to the host platform; the B kit which provides the processing and display equipment; and the communications suite.

A Kit

The A kit consists of the following:
(a) Platform modifications
    Modified cargo floor
    Additional floor structure
    Internal interface panels
    Electrical cables
    Antenna cables
(b) Additional A2C2S antennas
(c) External interface panel

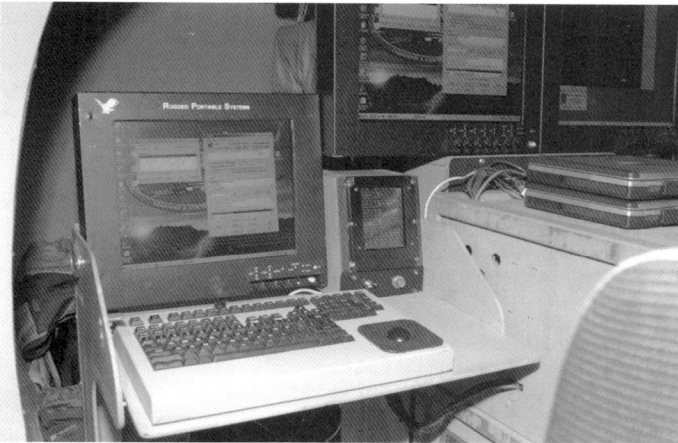

The forward port side position, normally the Intelligence position, in an A2C2S mockup at the AUSA 2003 Exhibition, showing the workstation and, above and to the right, one of the 20 in common displays (Patrick Allen/Jane's)   **NEW**/1029542

The view from the rear of an A2C2S mockup at the AUSA 2003 Exhibition, showing (L–R) the Operations, Command and Fire Support workstations, with beyond these the Intelligence workstation (L) and the two common displays. The smaller screens with yellow displays are the communications control boxes, giving the users access to communications circuits (Patrick Allen/Jane's)   **NEW**/1029543

Connections ground mast
Ground power connections
TFOCA (fibre cable) to TOC
(d) A2C2S Interface Panel (AIP) (located in pilot compartment)
Intercom interface
High frequency radio
Power interfaces

B Kit
The B kit consists of the following:
(a) Displays
18 in operator display (× 5). Displays any of 11 processors through KVM
20 in common display (× 2). Hosts A2C2S application software
Embedded Processor: Pentium II / 266 MHz
440 Mbyte solid-state hard drive
Ethernet access (10/100BaseT)
1280 × 1024 pixel resolution
(b) Keyboard/Video/ Mouse(KVM) switch. Allows operators to use any computer processor from any workstation
(c) GPS (× 2). Provide location and time information to ABCS and radios
(d) Co-Site Mitigation Assembly. Filters various frequencies to reduce interference and splits GPS signal
(e) Rugged notebook computer (Dolch Notepak). Host BFT-A (Blue Force Tracking), or A2C2S software.
400 MHz Pentium II rugged notebook computer
Windows NT operating software
Ethernet connection
13.3 in colour display
XGA 1024 × 768 pixel resolution
Supports auto sensing AC power from 47 Hz to 63 Hz and direct DC power of 10 to 20 volts
Battery can operate up to approximately three hours on a full charge
Standard I/O interface ports include parallel printer, two serial ports, external video, keyboard, mouse and infra-red ports
Keyboard: Rubber-coated with backlit keys
(f) Router. Provides LAN/WAN connectivity and network security
(g) Crew Access Unit. Operator's access to intercom and radios
(h) MPU × 2. 3 SPARC and 1 Pentium per MPU hosts ABCS software

Communications Suite
The communications suite consists of the following (see separate entries in most cases and *Jane's Military Communications*):
(a) 4 × SINCGARS-ASIP (VRC92F) (voice/data). Provides primary tactical voice and tactical internet
(b) 2 × HAVEQUICK II (2 × ARC-231). (LOS UHF/VHF voice/data)
(c) 1 × SATCOM DAMA (1 × ARC-231). (NLOS UHF voice/data)
(d) 1 × ARC-220 HF Interface
(e) 1 × NTDR. Primary upper Tactical Internet
(f) 1 × EPLRS. Lower Tactical Internet

Software
A2C2S is capable of running a range of tactical software applications. Apart from the basic five components of the Army Battle Command System (MCS, AFATDS, ASAS, AMDWS, FBCB2 - see separate entries) it can also host others, including CSSCS, C2PC, Falcon View and GCCS-A (see separate entries).

Operators at each of the workstations can access the various software packages loaded onto the MPU. Each workstation is capable of accessing any of the BAS software packages and multiple workstations may view the same BAS simultaneously. This is made possible by the KVM. The laptop and each of the processors in the MPU are connected as an input to the KVM. The A2C2S workstations are connected as outputs and the KVM provides the switching capability that allows the user to select which input he controls on the workstation. An application hosted on the laptop allows remote control of the KVM to control the common displays' output.

Status
Two proof-of-concept systems were built and delivered to 4th Infantry and 101st Airborne Divisions in July 2001/January 2002 respectively. Raytheon was subsequently awarded a US$110 million contract for Low Rate Initial Production. The first three production models were due to be delivered to 4th Infantry Division in January 2003.

In the event the demonstration systems deployed to Iraq in early 2003 with 4 Inf and 101 AB Divs and were used successfully in operations during the advance on Baghdad, including the co-ordination of the air assault on Najaf launched from Kuwait, a distance of 248 miles. Subsequently (April 2003) three preproduction models were also deployed, two to 4th Inf Div and one to 1st Cav Div, and at the time of writing (November 2003) all systems remain deployed in Iraq.

The current version hosts only the ABCS software applications. Integration of others will be the subject of further development packages.

Contractor
Raytheon Company, Huntsville, Alabama.

UPDATED

E-6A/B airborne command control and communications platform

The E-6A/B, derived from the Boeing 707, is a command, control and communications (C3) platform with two roles. The E-6A's Take Charge and Move Out (TACAMO) mission provides multiple C3 links for Emergency Action Message (EAM) relay from the National Command Authority to strategic and non-strategic operating forces. Designed to support a robust and flexible nuclear deterrence posture into the 21st century, the E-6B incorporates Airborne National Command Post (ABNCP) equipment from retiring US Air Force EC-135Cs E-6As became fully operational in 1992 and E-6B initial operational capability began in 1998.

The ABNCP task of the E-6B derives from the original Strategic Air Command (SAC) Post Attack Command and Control System (PACCS). This was an airborne command post capable of assuming command of the US bomber and missile force and executing emergency war orders at the direction of the US president. These EC-135 aircraft, a modified KC-135 tanker airframe, operated under the name of 'Looking Glass', and flew with a battle staff of C4I specialists and experienced Minuteman crew and launch officers under the direction of a SAC general officer. These aircraft were continually airborne from 1961-90, and the final mission was flown in 1998 when the task was handed over to the E-6B.

E-6A/Bs transmit and receive secure and non-secure voice and data at very low, low, and high frequencies, and also via UHF line of sight and satellite communications systems. The E-6A can deploy a 28,000 ft trailing-wire antenna and a 5,000 ft short trailing-wire antenna for Very Low Frequency (VLF) communications with submerged ballistic missile submarines. With in-flight refuelling, the E-6B ABNCP is capable of providing up to 72 airborne hours of decision-level conferencing, force management, situation monitoring, and communications support.

Background
Early in the development of the SLBM force it was recognised that some new form of survivable communication method would be required, and also that some form of expedient, interim solution was necessary for the near term. Extremely low-frequency (ELF) communications technology appeared to offer the needed solution if a capable and survivable transmitting facility could be designed and built. The design concept for this facility was based on the use of a very large antenna grid that was driven by many transmitters located in hardened capsules. This design concept was known as SANGUINE and was, for some time, considered the long-term solution to communicating with the SLBM force. TACAMO was

E-6B TACAMO   0055008

conceived as an expedient, interim solution for survivable communications to the SLBM force. This interim TACAMO concept exploited existing very low-frequency (VLF) communications technology by placing a VLF transmitter in an aircraft.

The TACAMO concept utilised a technique for deploying a substantial vertical VLF trailing wire antenna from a large aircraft. This involved flying the aircraft in a tight circular orbit, which stalls the long wire antenna, causing it to hang in a near vertical configuration. High wire verticality is essential for VLF communications to submerged submarines because only the vertically polarised component of the VLF signal provides sea penetration to reach submerged VLF receive antennas. A Lockheed C-130 aircraft was used as the airborne platform because it had the payload capacity to carry the large and heavy VLF transmitting equipment and perform the critical orbiting manoeuvre. This demonstration system became known as TACAMO I and served as the prototype for an improved version known as TACAMO II. This latter version was operational from 1964-68 and in turn became the basis for the third-generation TACAMO III.

By the late 1970s it had become clear that hardened and dispersed concepts like SANGUINE could not provide a long-term solution even if political problems and environmental concerns could be overcome. The development and deployment of MIRVs had greatly increased the vulnerability of all fixed site types of military facilities and TACAMO was therefore officially designated as the primary survivable submarine communication system, as opposed to its previous interim status. In early 1982, the navy issued the ECX RFQ for a replacement TACAMO aircraft, and Boeing's proposal was a derivative of the E-3A (AWACS) airframe equipped with CFM56 engines. This combination exploited many of the useful military features of the E-3A such as the high-quality electrical power and cooling subsystems. It also exploited data from the 707/CFM56 prototype and the USAF KC-135R re-engining program. Mission avionics would come from the EC-130's but would be reconfigured and upgraded to enhance overall system capabilities, maintainability, and reliability.

E-6A description
The aircraft is painted white for thermal protection but has the radar and infra-red signatures typical of commercial transports. Its flight characteristics also make it nearly indistinguishable from commercial aircraft.

The forward portion of the cabin is devoted to living and rest accommodation for extended missions and self-sufficient operations. Communication central, where the mission crews perform their functions, is located over the wing in a low-noise environment, isolated compartment. Major equipment and mission avionics components are installed in the rear of the aircraft where they are isolated from communication central, yet readily accessible for maintenance. Two lower lobes are used for additional equipment installations and storage of deployment spares and other provisions. In-flight access is provided to both lower lobes and personnel ground entry is provided by hatches and self-contained ladders through the forward lower lobe. A position and viewing port to monitor trailing wire deployment is provided in the aft lower lobe.

The flight deck avionics, including the modern ring laser gyro navigation system, flight management computer and weather radar were specifically selected to give the E-6A greater operational flexibility in a stressed environment. Extensive EMP hardening is incorporated in the form of console, rack and wiring shielding, as well as filters and protective devices. SIMOP (simultaneous operation) capability was similarly given special attention by the addition of special filters and improved antenna isolation.

In 1977, an Avionics Block Upgrade (ABU) Program and Orbit Improvement System (OIS) Program were initiated to improve the performance of E-6A Aircraft and its communications equipment. Major avionics equipment to be added to the E-6A as part of this were a High Power Transmitter Set (HPTS), Global Positioning System (GPS), and Extremely High Frequency (EHF) Milstar. The HPTS is a VLF transmitter that contains a 200 kW Solid State Power Supply (SSPS) and Dual Trailing Wire Antenna (DTWA). The Orbit Improvement System incorporates modification of the E-6 Flight Management System and the installation of an autothrottle to improve the position of the HPTS antenna wires for optimum transmission.

E-6B description
The E-6B programme was established to upgrade TACAMO operational capabilities and cross-deck a subset of the Strategic Command's (STRATCOM) EC-135 Airborne Command Post (ABNCP) equipment to the E-6A aircraft. The E-6B Command Post Modification enables STRATCOM to perform current and projected TACAMO and ABNCP operational tasking. The conversion involved the integration and installation of the following systems into the E-6 aircraft:

(a) Airborne Launch Control System (ALCS) which operates through Ultra High Frequency (UHF) C³ radios, enabling the E-6B to function as an Airborne Launch Control Centre. The ALCS system allows determination of missile status in silos, launch, or change in missile assignments

(b) UHF C3 Radio Subsystem adds three UHF transceivers that support 1,000 W full-duplex transmissions using amplitude modulation (AM) or frequency modulation (FM). It provides: UHF frequency division multiplex (FDM) (three full-duplex groups of 15 channels each), ALCS, conventional UHF AM line of sight (three half-duplex channels), and/or Fleet satellite communication (SATCOM) phase shift keying (one receive-only channel)

(c) Digital Airborne Intercommunications Switching System (DAISS) provides automated audio distribution and equipment control/configuration among the communications equipment supporting the ABNCP mission and access to the TACAMO equipment

(d) Military Strategic Tactical And Relay (MILSTAR) Airborne Terminal System provides Extremely High Frequency/Super High Frequency/UHF connectivity through the survivable MILSTAR satellite system

(e) Mission Computer System enhances message handling and processing by providing user-friendly operations for message receipt, edit, storage, and transmit; identifying emergency action messages; and routing data among peripherals (printers, keyboards, and so on)

(f) UHF SATCOM Receive System upgrade replaced the OE-242 antenna controller with a more reliable and supportable unit

(g) Time/Frequency Standards Distribution System provides retrieval and distribution of the accurate universal co-ordinated time from the global positioning system. Time of day, 1 pulse/s and precision 5 Mhz reference signals are distributed to Very Low Frequency (VLF) and UHF communications equipment to provide accurate reference timing

(h) High Power Transmit Set replaces the existing 200 kW VLF high-power amplifier and dual Trailing Wire assembly, providing increased capabilities (including low-frequency transmission spectrum) with significant reliability and operability improvements

(i) Three dual-redundant MIL-STD-1553B databusses accommodate future modifications to the E-6B weapon system.

E-6C Enhancement Programme
The E-6C Theatre Enhancement Programme will enable the E-6C to have global mission capabilities. It will add equipment to the E-6 to meet FAA/ICAO evolving 'commercial-size' aircraft operational requirements. Some of the equipment being considered for inclusion into the E-6 airframe to make this aircraft a theatre command, control and communication aircraft are: a Traffic Collision Avoidance System (TCAS), Differential GPS, an altitude alert system, and the Dual Satcom (Inmarsat) system. New mission supporting communication systems and upgrades are also being considered for future use on the E-6.

Status
A total of 16 E-6B systems are being procured. The first E-6B was delivered in 1997 and assumed its dual-operational mission in October 1998. The E-6 fleet will be completely modified to the E-6B configuration by 2003.
For further details, see *Jane's Electronic Mission Aircraft*.

Contractors
Boeing Aircraft Company.
Raytheon Company.

VERIFIED

Joint Tactical Terminal (JTT)

The JTT is a family of software-programmable intelligence radios that can operate at security levels above 'Secret'. They are designed to provide a common communications and display system for intelligence, providing the means to receive a variety of intelligence broadcast networks such as NRTD, TDDS, TADIXS-B, TIBS and TRIXS within the Integrated Broadcast Service (IBS), as well as data from Tactical Data Links and the Secondary Imagery Dissemination System (SIDS).

The JTT full service version ('Senior') operates in the 225 to 400 MHz UHF band and provides half- and full-duplex, SATCOM and LOS operation on 5 and 25 kHz channels. A single JTT terminal is configurable from 4 receive/0 transmit channels to 12 receive/4 transmit channels. The JTT receiver/exciter is packaged in a single

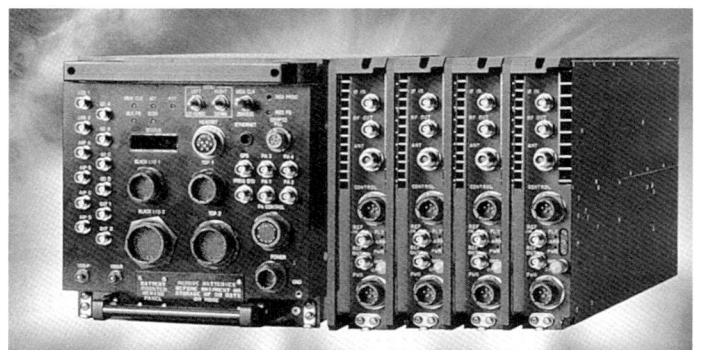

The Joint Tactical Terminal full service version (Raytheon)   0143778

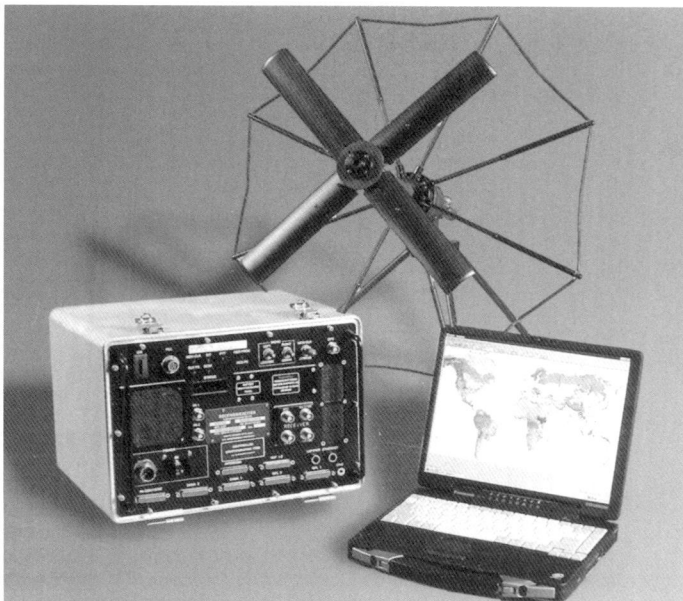

The Joint Tactical Terminal (Briefcase), showing terminal and laptop (Raytheon)
0143777

full-ATR and each JTT power amplifier in a ¼ATR. Modulations supported include: 4-ary FSK at 32 kbits/s, Vinson compatible FSK at 32 kbits/s, BPSK and SBPSK at 1.2, 2.4, 4.8, 9.6 and 19.2 kbits/s, DEQPSK, OQPSK, and SOQPSK at 1.2, 1.6, 2.4, 4.8, 6.0, 9.6,19.2 and 32 kbits/s. JTT provides anti-jam Have-Quick II frequency hopping and adaptive null steering in the TRIXS network. It also includes maximal-ratio quad-diversity antenna combining in all SATCOM modes of operation to support shipboard applications and operates with a variety of tactical data processors, including existing MATT and CTT units. Datafiltering, formatting and correlation are all internal and the terminal includes a built-in GPS receiver for establishing network time and supporting Geographical Data Filtering.

The Common IBS Modules for COMSEC and IBS Format Processing will be made available to the migration terminals (CTT, MATT) as well as some legacy terminals (SUCCESS, TRE) to provide continued operability with the changing IBS networks.

A receive-only briefcase system, the JTT (Briefcase) (JTT(B)) has also been developed. This uses a ruggedised laptop connected to the terminal by an Ethernet LAN, which allows multiple computers to deceive data in parallel. The computer serves as the Tactical Data Processor both to process and display intelligence data and control the terminal.

Future intended developments include migration to full JTRS compliance and the development of a hand-held variant.

A comprehensive radio interface to the JTT has been developed, called the JTT Control Client. It is a software CIBS-Module that works in both Windows NT™ and Solaris ™ environments.

Status
On behalf of the US Joint community, the U.S. Army's Communication and Electronics Command awarded a JTT system design and development contract to Raytheon Company in September 1997. In late 2003, low-rate initial production of 551 units was continuing steadily and will be completed by September 2004. As of July 2003, 226 JTT 'Senior' units had been delivered. Raytheon's contract also includes the JTT(B) and 58 of 59 systems had been delivered by July 2003. The portable system reportedly performed well in Afghanistan.

Both JTT Senior and JTT(B) were deployed in support of Operation Iraqi Freedom.

Contractor
Raytheon Company, Network Centric Systems, St Petersburg, Florida.

UPDATED

M4 Command and Control Vehicle (C2V)

The C2V is a tracked, armoured vehicle designed to provide an automated tactical command post for mobile armoured operations. At the corps and division level, C2V operates as a Tactical Command Post (TAC) and at the brigade and battalion level as the Tactical Operations Centre (TOC).

The C2V is mounted on a modified M993 MLRS chassis. It is designed to be survivable against nuclear, biological and chemical threats and electromagnetic environmental effects. The C2V houses a crew of two and has workspace for four staff officers. The C2V has an inter/intra communications capability, allowing both internal and external communications via a wireless LAN. The LAN will function while the vehicle is on the move and is capable of sending and receiving data with the M1A2 tank, M2A3 Bradley as well as linking to the Army Tactical Command and Control Systems (ATCCS). The vehicle has an onboard power system (43 kW), environmental control unit and an overpressurised NBC system. The C2V Mission

Command and Control Vehicle (C2V)   0055006

Module is built by Logicon and integrates four computer workstations with army standard communications equipment. The C2V Mission Module uses the army's common hardware and software standards. Design features include:
(a) shock and vibration-isolated equipment mounts, capable of withstanding rail transportation
(b) power and signal distribution systems protected against EMI/EMP
(c) maximum consideration of human factors in workstation design.

The work stations are equipped with combinations of the ABCS subsystems, appliqués, voice radio and intercom features. For example, the S-2 station may have an All Source Analysis System (ASAS) terminal, the S-3 stat on a Manoeuvre Control System (MCS) terminal and the fire support station an Advanced Field Artillery Tactical Data System (AFATDS) terminal. The operators can relay data to each other from their stations and maintain voice communications within the vehicle and over the Vehicle Net Radio (VNR). In addition, there will be at least one appliqué system component for the Tactical Internet (TI), plus VNR and position locating equipment available as well. An onboard Local Area Network (LAN) links the different systems with each other and the vehicle communications system.

To provide additional communications, there are also the Vehicular Intra/Inter Communications System (VIICS), VNR via Single Channel Ground and Airborne Radio System (SINCGARS) and Mobile Subscriber Equipment (MSE) for inter- and intra-vehicle voice communications. MSE with facsimile (FAX) is available for access to the Army Common User System (ACUS). The vehicle is equipped to operate in a wireless LAN whether functioning as a part of the TAC or main CP. The vehicle can also use wire and cable links to other C2Vs, other C2 platforms, for access to LANs and Wide Area Network (WANs) as required.

The C2V has a variety of communications antennas. In addition to whip antennas, there is an erectable tall mast antenna for long-range communications while stationary. The mast antenna cannot be used on the move. For stationary operations, a Standard Integrated Command Post System (SICPS) shelter can be placed at the rear of the vehicle to extend the CP enclosure.

Status
The major procurement programme of over 400 vehicles was cancelled in late 1999. Around 25 vehicles were produced for the US Army and remained available. Jane's sources understood that the Norwegian Army was interested in them.

In April 2003 *Jane's Defence Weekly* reported that in mid-2002 United Defense, which was storing the vehicles, was asked to provide as many as possible for US operations in Iraq. 15 were provided to: US V Corps (3); 3rd US Inf Div (3); 1st US Cav Div (4); 3rd Armd Cav Regt (2); 1st US Armd Div (3). Most subsequently were deployed on operations. The remaining 10 were retained for spare parts.

Contractors
United Defense LP (Prime), Ground Systems Division, York, Pennsylvania.
Logicon Inc, Tacoma, Washington.

UPDATED

National Airborne Operations Centre (NAOC)

The National Emergency Airborne Command Post (NEACP) programme was established in 1962 as a method of providing a secure airborne C³ platform for the National Command Authority (NCA) during a nuclear crisis. It augments the National Military Command Centre in the Pentagon and the Alternate National Military Command Centre, located at Site R in Pennsylvania. The original NEACP aircraft were EC-135Js, heavily modified KC-135s, but were replaced by the E4A in 1974. The E-4B (a militarized version of the Boeing 747-200) evolved from the E-4A, the first of which was delivered to the USAF in January 1980 and by 1985 converted all aircraft to B models. All E-4B aircraft are assigned to the 55th Wing, Offutt Air Force Base, Nebraska.

The E-4B serves as the National Airborne Operations Centre for the NCA. In case of national emergency or destruction of ground command control centres, the

E4B National Airborne Operations Centre   0055011

aircraft provides a modern, highly survivable, command, control and communications centre to direct US forces, execute emergency war orders and co-ordinate actions by civil authorities. In August 1994, the E-4B assumed the additional role of supporting the Federal Emergency Management Agency's request for assistance when a natural disaster occurs. The E-4B would be tasked to fly the FEMA Emergency Response to the disaster site and become the FEMA command and control centre until the emergency team's own equipment and facilities can be set up. With E-4B support the emergency team's response is a matter of hours as opposed to days. With this change of role, the NEACP was renamed the NAOC.

Air Combat Command (ACC) is the Air Force single-resource manager for the E-4B and provides aircrew, maintenance, security and communications support. The joint chiefs of staff actually control E-4B operations and provide personnel for the airborne operations centre.

The main deck is divided into six functional areas: a NCAs' work area, conference room, briefing room, an operations teamwork area and communications and rest areas. An E-4B crew may include up to 114 people, including a joint service operations team, an ACC flight crew, a maintenance and security component, a communications team and selected augmentees.

The E-4B has electromagnetic pulse protection, an electrical system designed to support advanced electronics and a wide variety of new communications equipment. Other improvements include nuclear and thermal effects shielding, acoustic control, an improved technical control facility and an upgraded air conditioning system for cooling electrical components. An advanced satellite communications system improves worldwide communications among strategic and tactical satellite systems and the airborne operations centre. In addition, the E4B has VLF, LF, MF, HF, VHF, UHF, L Band, SHF and EHF communications. The equipment fit is being continually upgraded and a new three-antenna radome was fitted in early 1996.

To provide direct support to the NCA, at least one E-4B is always on alert at one of many selected bases throughout the world.

In October 2002 it was announced that DRS Technologies had been awarded a contract (maximum value US$ 1million) to provide Enhanced Command Consoles (ECC) for one of the platforms. These will serve as the primary workstations controlling the audio communications subsystem. The ECC is a ruggedised Sun® UltraSPARCIii™ workstation with a 21 in AMLCD flat panel display, a remote media access bay and a keyboard/trackball. System deliveries are expected to be completed in 2003.

UPDATED

··

Reconfigurable Command & Control Platform (RC2P)

Raytheon's Reconfigurable Command & Control Platform (RC2P) is an air deployable and rapidly erected command post that has drawn on the design, development and lessons learned from the Army Airborne Command and Control System (A2C2S) (see separate entry).

The prototype RC2P was jointly funded by Raytheon and the Depth and Simultaneous Attack Battle Lab (D&SBL) at Fort Sill for evaluation as a Light Field Artillery Tactical Operations Centre (FA TOC) Concept Experimentation

The removable laptop workstation at the head of the RC2P table, with the CCB next to it (Patrick Allen/Jane's)   *NEW*/0526720

The RC2P viewed from the end of the table looking towards the back of the vehicle. The larger screens mounted on the vehicle can be seen together with the four main workstations and the laptop at the head of the table. Beside each workstation is a communications control box. The whole is enclosed by the tent attached to the vehicle (Patrick Allen/Jane's)   *NEW*/0526718

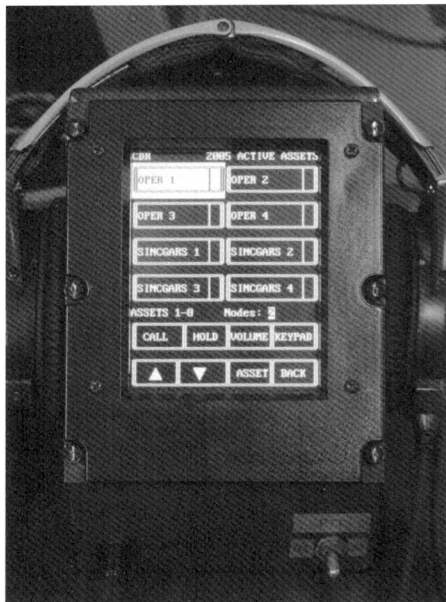

The touchscreen communications control box (CCB) provided for each RC2P staff member. The eight available circuits can be selected from the screen (Patrick Allen/Jane's)
NEW/0526719

Programme (CEP). The Ft. Sill CEP validated the use of the A2C2S architecture in a ground system and employment of the RC2P as a Light FA TOC. The prototype RC2P includes a Multiple Processor Unit (MPU) and a Keyboard Video Mouse (KVM) switch. These components, which are the same as A2C2S components, along with the RC2P's modular applications and communications suite, suggest that the RC2P can be used in support of a variety of roles including: Homeland Security, Light Air Assault TOC, FA TOC, Light Tactical Command Post (LTAC), Corps TAC - Forward (CTAC-F), Light Tactical Operations Centre (LTOC) and Light Digitised Operations Centre (LDOC).

The prototype design consists of a primary vehicle containing communications equipment and five workstations to include one in the cab for operations when on the move. The workstations are software reconfigurable to support a wide variety of digital command and control systems to include the Army Battle Command System (ABCS)(see separate entry), the USMC equivalent to ABCS or civilian systems. A removable laptop is also included at the head of the command table. Each workstation has a communications control box (CCB) providing access to up to 8 circuits. A VIPER™ 5 Kilowatt under-the-hood generator in conjunction with a 200A alternator provides power for self-sustained operations. A secondary or support vehicle may be added to the configuration to provide environmental conditioning, additional transportation for the crew, an alternate source of power or additional communications/operational equipment as desired. The support vehicle may also be equipped with the VIPER™ to provide power for extended periods of time and to reduce noise and vibration in the primary vehicle. Raytheon claim average set up and tear down times of under ten minutes by day and under twelve minutes at night. The RC2P can be carried in a C130 and underslung by a UH60L Blackhawk.

The RC2P makes maximum use of currently available US DoD hardware and software. The prime mover is the standard cargo HMMWV (target is M998, prototype is on an M1097), and the system may be integrated into other vehicles. Communications equipment includes standard military radios and antennas. Possible radios include, but are not limited to; SINCGARS, EPLRS, NTDR PSC-5 SATCOM, PLGR, AN/PRC-150, and AN/PRC-117F.

The ABCS software runs on a ruggedised MPU and uses the KVM switch to support reconfiguration based on mission requirements. For example, if used by an artillery unit, it could have multiple workstations running the Advanced Field Artillery Tactical Data System (AFATDS) software, while if used by a manoeuvre unit, it might have multiple workstations running the Manoeuvre Control System (MCS) software.

Status

Still at the prototype stage. Much of the concept seems to be mirrored in the Light Digital Operations Centre (LDOC) developed in-house by the US Army and trialled, apparently successfully, by 101(AB) Div on Ex MILLENIUM CHALLENGE 02.

Contractor

Raytheon Systems Company (C3I Systems), Huntsville, Alabama.

NEW ENTRY

SPAce Defense Operations Center (SPADOC)

The Space Defense Operations Centre (SPADOC) is located in Cheyenne Mountain and serves as a fusion centre for the space control mission. SPADOC is responsible for protecting DoD, US civilian, and allied nation space systems. SPADOC fulfils its mission responsibilities primarily through monitoring space and space-related activities, informing members of the space community of unique space-related events, and planning possible defensive countermeasures. To achieve its objectives, SPADOC specifically monitors and reports abnormal or unusual space activity, and recommends the necessary follow-on steps to specific organisations. SPADOC also analyses possible threat attack information, determines the time and location of the attack, and identifies both the space system under attack as well as the method and type of attack taking place. Finally, SPADOC advises specific organisations of which US space systems are vulnerable to attack or are likely to be targeted for attack.

SPADOC communicates with organisations owning or operating space-based systems through various secure and insecure communications means. SPADOC, a key centre of operations under USSPACECOM, routinely communicates with other USSPACECOM operations centres and component commands for routine status information. In the event of a space threat, SPADOC will communicate directly with specific satellite system owners/operators to preclude delay in transmission of critical warning messages.

The primary method of secure connectivity between SPADOC and all space system owners/operators is the Space Defense Command and Control System (SPADCCS). SPADCCS is a communications network using hard copy messages to and from SPADOC and space system owners/operators.

Status

The SPADOC system is fully operational. In 2001 it was reported that block 4 (SPADOC 4) was operational. The addition of SPADOC 4 increases the capability for database management and database size. New computer hardware will allow for cataloging of 30,000 on-orbit objects - this is about three times the prior capability. In addition to enhanced database capability, the system provides enhanced sensor tasking and orbit propagation capabilities.

Contractor

Lockheed Martin Command & Control Systems, Colorado Springs, Colorado.

VERIFIED

COMMUNICATION SYSTEMS

Joint
Land
Maritime
Air
Satellite

Australia

Joint project 2043 High Frequency Modernisation Project (HFMOD)

This is an Australian Department of Defence project designed to upgrade long-range HF radio communications, replacing the three individual service legacy systems, some aspects of which have been in use since the 1960s. In 1993 it was decided to combine the three systems into one that covers all of continental Australia as well as 2,000 km out to sea. Communication with Australia's deployed forces will be achieved by improving and automating the radios on ships, planes and vehicles used by the navy, army and air force to provide facilities similar to a public telephone system. Boeing (Australia) is the prime contractor.

The Modernised High Frequency Communications Wide Area Network provided by the HFMOD Project will link command centres and deployed aircraft, ships and land forces. Redundancy will be built into the system allowing it to continue to provide the required level of service with less than all the stations in operation. The new system will consist of four transmit and receive sites (called nodes) at North West Cape (Western Australia), Darwin (New Territories), Townsville (Queensland) and in the Riverina Region (New South Wales). These will be connected to two specially built control centres in Canberra. The system will use Automatic Link Establishment (ALE) capable transceivers, and will provide voice, data, facsimile, e-mail and imagery over HF services. The HF communications suites in ships, aircraft and land units will be upgraded and the maximum use will be made of COTS products.

Part of the project has been subcontracted to Braintree Communications, whose solution was a customised synchronous terminal server that offered an integrated multifunction solution for serial and parallel networking applications. The solution provides an integrated Ethernet network access to multiple serial devices. It allows connection of up to eight synchronous ports to a Local Area Network (LAN) and is designed for transaction and/or data processing with Ethernet attached devices. The solution is remotely managed and diagnosed via Simple Network Management Protocol (SNMP) and can be configured, managed and upgraded either locally or remotely over the network. These features allow the network manager to examine and adjust the terminal servers for optimum performance. The solution's remote management features allow the network manager to install and configure the network according to a specific environment using Application Programming Interface (API) capabilities, claimed by Braintree to be unique.

Status

Two network definition contracts were let in August 1995, one to Rockwell Systems Australia Pty Ltd (now Boeing Australia) and the other to Telstra teamed with the then GEC-Marconi Ltd. These were completed in 1995 and were followed by project definition studies, which were completed in 1996. The contract for implementation was awarded to Boeing (Australia) in 1997. Initial operating capability is due in 2004, with full implementation by the end of 2005, when the fixed network will be complete and some mobiles in operation. The project will cost in total more than A$200 million.

Contractor

Boeing (Australia) Ltd (prime), Brisbane.

UPDATED

Parakeet tactical satellite and trunk communications system

Project Parakeet provides the Australian Army and Air Force with a mobile, integrated, secure, tactical trunk communication system to meet the land environment's requirement for a transportable high-capacity system to be used within and between Joint Task Forces. It provides users with voice, telegraph, data and facsimile services. The Parakeet system interfaces with Combat Net Radio (VHF and HF (Raven and Wagtail))and the strategic secure voice, facsimile and data networks. It also provides facilities for access by data users to the Australian Army Battlefield Command Support System (BCSS) (see separate entry) and interfaces to allied and civilian communication networks.

The heart of the Parakeet system is the Circuit Switch Assemblage, which incorporates an integrated eight-trunk port circuit and packet switch, cryptographic equipment and a range of bearer equipment. Access to the network is provided by a robust and versatile set of man-portable switch and multiplexer frames. These are deployed to areas of concentrations of users, and connected to the Circuit Switch Assemblage over optical fibre. This concept is designed to reduce significantly the amount of twisted pair cable used in the headquarters.

The switches in the network are interconnected by a variety of wideband transmission bearers, including satellite communications, radio relay and optical fibre. While satellite communications links are currently limited to 512 kbits/s, radio relay and optical fibre capacities are 2 and 4 Mbits/s respectively. Network connectivity is designed to provide a reasonable level of wideband redundancy, or alternate routeing. In addition, a limited number of HF radio and commercial

Parakeet SATCOM Terminal Assemblage (STA) (BAE SYSTEMS)   **NEW**/0594891

telecommunications bearers are available. The headquarters is to be equipped with a message centre for common user message transfer services. The Parakeet system is designed to meet EUROCOM standards. Individual circuits operate at 16 kbits/s. Interfaces to STANAG gateways are provided.

The Radio Relay Assemblage comprises three UHF multichannel radios which are connected to an associated CSA to provide trunk connections to other nodes. These radios can be used as repeaters should an extended path be required.

The SATCOM Terminal Assemblage (STA) allows operation with X-band military satellites and C-band and Ku-band commercial satellites. The STA provides two Group Traffic channels and three orderwire channels. Each Group Traffic channel has a maximum data rate of 2 Mbits/s. The data interfaces for the two Group Traffic channels can be selected from either two optical fibre EUROCOM interfaces (to connect to the Parakeet Network) or two RS-422 electrical interfaces for generic data input including Asynchronous Transfer Mode (ATM) data. The STA is connected to a CSA via fibre optic cable. It can be connected directly to another STA or to a strategic site.

The Communications Control Assemblage provides system control and management functions for a deployed Parakeet system.

Status

Phases 1, 2 and 3 of the project consisted of concept development, system definition and equipment definition studies and are complete. Phase 4 has been delivered, equipping an operational level headquarters, three Joint Task Forces, deployed logistic elements and two alternate airfields. This phase is subdivided into two. Phase 4.1/4.2 covers the procurement of circuit and packet switch subsystems, a transmission subsystem, including radio relay and optical fibre, and a maintenance subsystem. The contract for this phase was signed in March 1994. Introduction into service of this phase is complete. Phase 4.3 covered the provision of a satellite communications subsystem consisting of 18 STAs and one fixed satellite control station. Delivery to Defence units was completed in May 1996. Refurbishment of the two prototypes to production standard was completed in September 1997 and delivery of this equipment was completed in December 1997. Phase 6 addressed the procurement of additional equipment including a further12 STAs for issue to units not equipped under Phase 4, equipment enhancement (to the satellite communications subsystem) and two new capabilities - Wireless Local Area Networks (mobile cellular telephone communications) and a Deployable ATM Hub assembly (DAHA). A sole source contract for this phase was signed with BAE Systems (Australia) on 31 May 1999. All elements of Phase 6 were delivered by the contracted schedule date.

The system was successfully deployed in support of peacekeeping operations in East Timor in 2000, providing communications between deployed elements and to the Australian mainland. The system has been deployed in support of Australian forces in Afghanistan, Iraq and the Solomon Islands.

Manufacturer

BAE Systems Australia, Edinburgh Parks, South Australia.

UPDATED

Brazil

Military integrated communications

Astrium has provided a system under contract for the Brazilian Armed Forces using the X-band transponders on the latest Brasilsat satellite.

Astrium has also supplied shipborne communications terminals and a number of land-based tactical mobile terminals for use by the Brazilian Navy, Army and Air Force, together with a central hub station and network management facilities. The complete system provides voice, data and fax communications to deployed units throughout Brazil, including remote areas and coastal regions.

As prime contractor, Astrium worked closely with a Brazilian company, Promon Eletrônica to implement the system.

Status

The system was completed in late 1998.

Contractor

Astrium UK, Stevenage, UK.

VERIFIED

Denmark

VCS 2000 Voice Communication System

Building on their successful ICS 2000 system (see separate entry) INFOCOM (now Maersk Data Defence)has developed a voice communications system using the DCS 2000 switch, designed for ground-air-ground and ground-ground communications. The system can support over 100 operator positions and can be used for the integration of geographically distributed centres. A flexible and adaptable system, it provides a number of features:

(a) Multiple communication circuit arrangements
(b) Operator role allocation
(c) Concurrent use of up to 14 direct access radio or ground assets from the operator audio unit (OAU)
(d) Parallel connectivity of OAUs
(e) Control of local or remote radio operating parameters
(f) Direct access keys
(g) Direct access conference facility
(h) Telephone access via submenus
(i) Direct access intercom
(j) Integral instant/short-term voice recording

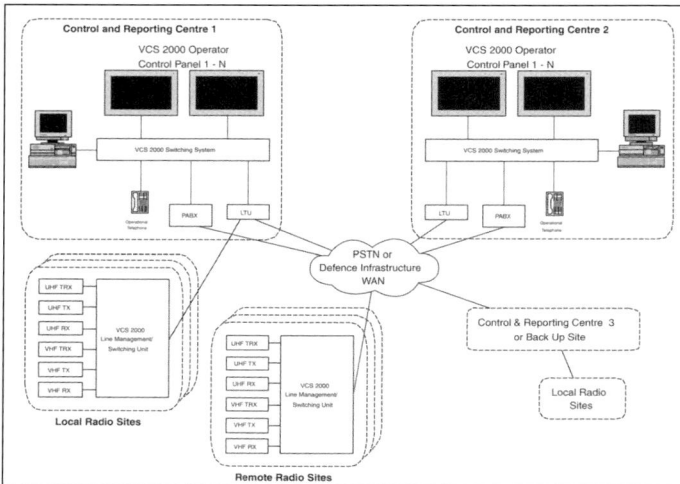

VCS 2000 - typical system architecture (INFOCOM)　　　0109976

VCS 2000 with 3 OCPs on desk and 2 OAUs above (INFOCOM)　　　0109975

(k) Analogue and digital PDSTN connection capability
(l) Interfacing with ISDN lines with primary or basic rate access
(m) Frequency management

Operator positions are either provided with a touch screen control panel with an active matrix LCD 12 in display module, known as the Operator Control Panel, with up to 60 keys which can be configured and customised to any user requirement, or an audio unit based on the TSS 2000 for headset connection. The latter has a built-in speaker and an LCD which displays the frequencies or channel numbers of the last five received radio calls. Operator positions are connected, for audio, in a star configuration to the DCS 2000 via fibre optic cables. The system assets are controlled via a LAN, which can be an industrial standard Ethernet. The network uses COTS media attachment units and hub units. System management services are provided by a Windows NT software program based on the INFOCOM CP2000.

Status

Selected by the Royal Danish Air Force to upgrade the NATO Air Defence Ground Environment (NADGE) infrastructure in Denmark. Development started in 1998 and was completed towards the end of 2000. In January 2002 the prototype of a similar system was delivered to the Finnish Air Force for use in requirements definition.

Contractor

Maersk Data Defence, Sønderborg.

VERIFIED

France

Ramses strategic nuclear communication network

The Ramses (Réseau Amont Maille Stratégique et de Survie) network is designed to convey communications between the French President, Prime Minister, Ministry of Defence and the French nuclear forces. The digital network, used by all three services, carries telephone, telegraph and digital data transmissions. Highly strengthened, it can withstand various forms of attack including nuclear electromagnetic pulse.

Ramses is the result of co-operation between the French Defence Staffs, the Délégation Générale pour l'Armement (DGA) and French industry. It is based on the THOMPAC multiservice high-speed asynchronous packet switching system. The main features claimed for THOMPAC are the optimum use of the transmission bearers; simplified network architecture; interconnection with all types of existing links; connection with all current and future terminals; and reliable and secure transmissions. A major segment of the Ramses network is the Astarte airborne communications relay aircraft.

Ramses uses a meshed network of THOMPAC switches linked by line of sight microwave or tropospheric scatter radio relays and fibre optic cables, providing redundant paths for reliability. An interface to satellite communications is also provided. The system carries telephone, facsimile and data traffic, at rates from 256 to 2,048 bit/s, for fixed and mobile subscribers. Encryption is provided for classified data.

Status

In service since the beginning of 1988. Believed to be still operational.

Contractor

Thales Communications, Colombes.

UPDATED

Ramses network　　　0055041

SOCRATE tri-service network

The SOCRATE (Système opérationnel constitué des réseaux des armées pour les télécommunications) network is designed to be the ground-to-ground backbone of the French armed forces' communication networks. This fully digital network will:
(a) link all sites used by the French armed forces
(b) allow interconnection with other French and allied infrastructure networks (NICS-NTTS) and French airbase communications facilities
(c) integrate with the existing network.

Thomson-CSF Communications (now Thales) was chosen by the French Military Procurement Agency (Délégation Générale pour l'Armement (DGA-STEI) to handle the project management and to design, develop and supply the main components of the system and, in particular, the multiple-service ATM switches and the network management and security system. SOCRATE has a sophisticated network security control and supervision subsystem and uses a meshed network of THOMPAC 2G switches linked by line of sight microwave links and fibre optic cables. Connection to existing and future networks and users is provided through its ISDN interfaces. Data rates range from 64 kbit/s up to 34 Mbit/s.

Status

In May 1998, Thomson-CSF Communications delivered and put into operation an initial system comprising about 30 switching sites. Deployment continued until 2002, for a planned 120 switching sites serving 250 access points.

Contractor

Thales Communications, Colombes.

UPDATED

THOMPAC 2G ATM switching equipment

THOMPAC 2G has been designed to set up integrated services strategic networks. It manages voice, data and video transmission and utilises for example, ISDN PBX or LAN interconnection, as well as ATM-UNI, for ATM LAN connection.

The equipment employs CCITT ATM techniques and specific defence mechanisms for maximum network reliability. Dynamic, non-hierarchical and decentralised routing algorithms allow a traffic load distribution which accommodates priority and system damage.

ATM techniques allow single-mode transmission of information, independent of type. On entering the ATM network, information is broken down into fixed size bit packets (or cells). Each cell contains a header allowing identification of the associated communication. The header facilitates routing by a hardware controller as soon as a cell reaches a network node. The controller deduces an output direction from the header and pieces the cell in the corresponding transmission queue. Transmission is then effected as soon as all the cells ahead in the queue have been sent on the right link. At the network output, the information is restored to its original form. Within the network, the cells are dynamically multiplexed, both on the internode links and in the internal node queues. This is designed to give optimum use at all resources, natural handling of traffic bursts and dynamic allocation of transmission capacities according to the instantaneous requirements of the users.

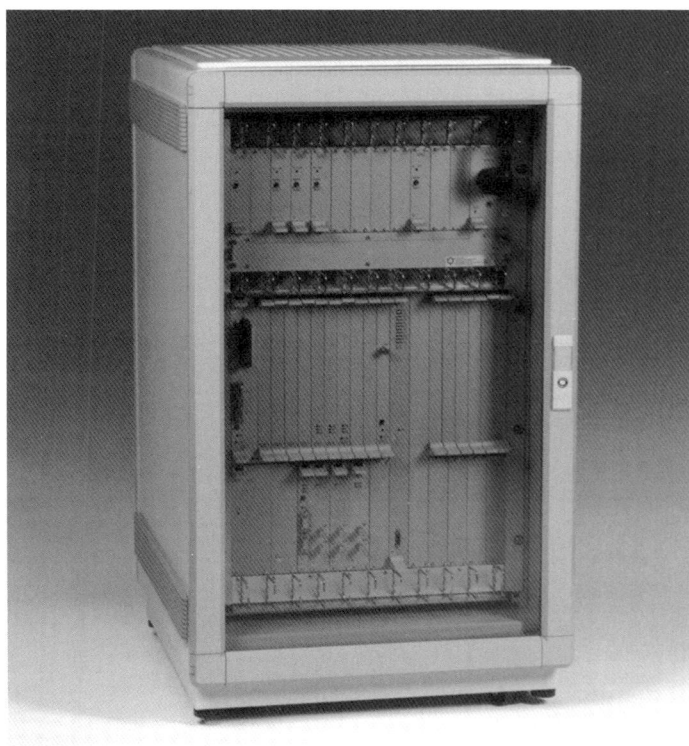

THOMPAC 2G equipment 0055051

The THOMPAC 2G network comprises switching nodes interlinked by high-rate links (digital microwave or fibre optic, for example). Currently operating at 34 or 155 Mbits/s, these links will eventually use higher capacities.

A THOMPAC 2G network includes management centres offering a complete set of control facilities (configuration, status and alarm reporting) and security features (access control, COMSEC and TRANSEC management). Internetworking within existing networks and PBXs is possible with adapted signalling, numbering plans and routing.

Status

Field trials were completed in 1998, by which time about 300 switches were in use. About 400 access and transit switches have been installed in SOCRATE, a multisource ATM transit network for the three French armed forces (see separate entry).

Manufacturer

Thales Communications, Colombes.

VERIFIED

Germany

Multiple Adaptive HF Radio System (MAHRS)

The MAHRS HF radio system provides fully automatic and fast short-wave radio data and voice services in the 1.5 to 30 MHz frequency bands. Packet radio type data transmission - including FEC and ARQ error protection - combined with a high speed modem, ALE, automatic radio channel selection and the capability to change link parameters during transmission (ALM), are key characteristics of the MAHRS capability to function without the attention of a human operator.

The software package forming the system's communication computer supports the handling of a spectrum of services including file transfer, e-mail, fax, ACP-127 messages and other formats. Automatic information exchange, with external message processing systems and local message preparation, storage and retrieval, is supported. The multilink concept behind MAHRS provides automatic routeing of messages between different communication services and enables the integration of MAHRS into C²I systems, for example in a client-server structure.

The system has an online cryptographic capability and TEMPEST design. MAHRS is available in mobile, fixed and airborne versions. Transmitting power ranges from 100 W to 1 kW, depending on the transmitter used.

MAHRS naval version 0120332

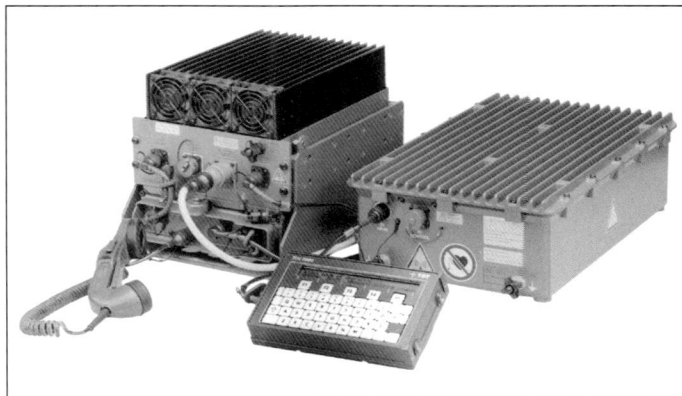

MAHRS HR 7400/M vehicle version 0120328

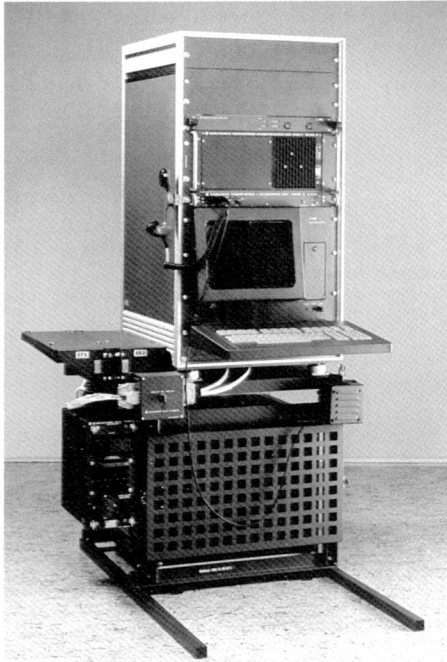

MAHRS FTA army version
0120333

Status
Mobile and fixed stations have been in operation in the German armed forces since 1995.

Contractors
EADS Radio Communications Systems , Ulm.

UPDATED

SECCOM® Military message handling system

The SECCOM Military Message Handling System provides seamless e-mail functionality in combination with military protocols and procedures, following the NATO standard STANAG 4406. The system is based on an open client/server architecture using commercial-off-the-shelf (COTS) hardware and software, where necessary ruggedised, tempest hardened and complemented by specific military/governmental components. Existing communications infrastructures are used. Gateways to services such as fax are provided as well as links to commercial e-mail systems such as MS Exchange or Lotus Notes. A specific ACP 127 gateway connects the modern e-mail service to legacy communication services and networks. Tactical message handling is supported by gateways to HF, VHF and UHF radios. Transparent interfaces to all major networks and network services (analogue, TCP/IP, ISDN, X.25, ATM, SATCOM, radio links) are provided.

The SECCOM Message Router software module automatically selects the appropriate network, taking availability, cost and security aspects into account and automatically reroutes messages where network degradation has occurred. It also provides logging and tracing functions for all messages routed through the system. Addressing is provided by a standard X.500 Directory Service, including schemes such as shadowing. Management functionality allows for Web-based management of local and remote servers. Security services can be integrated via line encryption, network (IP) encryption or Public Key end-to-end services using nationally approved encryption methods and devices. SECCOM Secure Connect, a military security gateway, can be installed as a black box between 'red' and 'black' security domains thus bridging the air gap between classified and unclassified subsystems. This gateway has been evaluated by an independent German government authority according to ITSEC E3 criteria.

SECCOM Modules
Military Client
The messaging client is a software package which runs on the user's workstation. The client allows the creation, transmission and receiving of messages and interacts with the Message Transfer System described below. The messaging client is a Windows application. Although similar to traditional mail clients used in a Windows or Internet environment, the military version includes a number of additional elements such as SIC, Special Handling Instructions or Security Policy which are defined in NATO STANAG 4406. In addition, military security labelling is provided.

Message Transfer System
The Message Transfer System forms the (software) basis of the messaging system. It consists of Message Transfer Agents (MTA) which route the messages from one sender to one or more addressees. Intelligent routing strategies can be configured which can automatically select an alternative channel, such as via another communication network, in the case of communication failure.

Directory Server
A Directory Server (DSA) based on ITU-standard X.500 Directory Service stores and distributes addresses and other general information. The Global Directory Server used implements completely the X.500 (93) standard and is accessed via LDAP protocol. The directory information is automatically replicated to all servers in the system.

SECCOM Message Router
The SECCOM Message Router (SMR) forms the core of the MMHS. All messages passing through a server are analysed by the SMR and treated individually, according to predefined rules. The functions of the Message Router include:
(a) Routing. Comprehensive routing options can be defined, including message routing to different gateways and networks, routing depending on message priority, alternative routing in case of network failure, rule-based message delivery (alternative recipient)
(b) Mail Lists. Messages can be replicated automatically according to predefined expansion lists, avoiding manual duplication
(c) Security (Firewall). Security functions within the Message Router control messages with respect to networks, users and services which prevent, for example, sensitive information being transmitted over insecure services such as fax. Violations are recorded
(d) Journaling/Logging. Comprehensive logging permits message tracing throughout the system. Message journals, system error logs and system status logs are automatically generated
(e) Closed User Groups. Management of closed user groups is supported to ensure confidential communication
(f) Network-specific Functions. Additional functions of the SMR are provided for legacy networks and services not offering the standard features of a Messaging System. An example is the ASCII check for services limited to text handling, such as teletype or narrow band services via radio links. Message attachments containing coding other than ASCII will be rejected in this case.

System Management
Management functions are controlled via graphic user dialogues which ensure efficient system installation and fast reconfiguration, as well as permanent monitoring of message exchange. The SECCOM Management module allows control of the entire system, including all processes, error supervision, message logging, and maintenance of system and network files. SECCOM Management is implemented as a Web-based application and allows the supervision of remote servers.

Boundary Protection Device
In addition to the SECCOM Message Router, an optional security gateway (SECCOM Secure Connect) can be integrated as an application fire-wall for controlling the interface between classified and unclassified domains. SECCOM Secure Connect is a separate black box and has been evaluated according to ITSEC E3 criteria by an independent German government authority.

ACP 127 Gateway
The ACP 127 Gateway ensures interoperability between modern message handling and military legacy teletype services until they can be replaced. The gateway converts messages between X.400 and ACP 127 formats and offers all the functions required for automatic message exchange. A Web-based management module is provided for operation, supervision, and manual generation of ACP 127 messages.

Fax Gateway
The Fax Gateway is used for addressees equipped with fax devices only. It automatically converts MMHS messages to fax messages and vice versa and generates MMHS reports.

Integration of Office Products
SECCOM offers automatic gateways to major commercial group ware and e-mail solutions including Microsoft Exchange Mail, Lotus Notes Mail and Internet (SMTP) Mail.

Network Connectivity
All major communication networks can be used for message exchange such as:
(a) analogue switched networks (PTT)
(b) ISDN (PTT, MIL)
(c) X.25 (PTT, MIL)
(d) TCP/IP networks
(e) military wide area networks
(f) military tactical networks (HF, VHF, UHF)
(g) SATCOM.

Hardware
Hardware can be selected from a range of commercial products, based on Intel/MS and UNIX platforms and includes:
(a) servers
(b) standard client workstations
(c) peripheral equipment such as printers and scanners
(d) standard LAN equipment
(e) standard IP routers
(f) crypto devices (link encryptors, IP encryptors, Public Key encryption using smart cards)
(g) SECCOM Secure Connect gateway for interfacing restricted and non-restricted domains.

Status

In operational service with the German armed services. The German Navy operates a system covering all Navy sites. German Army C3I systems use SECCOM messaging components. German Air Force Control and Reporting Centres use SECCOM as their secure message handling system.

The submarines of an unspecified NATO country are being equipped with SECCOM components.

Contractor

EADS Systems & Defence Electronics, Friedrichshafen.

UPDATED

International

Secure Automated Military Messaging System (SAMMS)

Secure Automated Military Messaging System (SAMMS) is an automated message handling system based on COTS hardware and software that is a secure version of the Mercury messaging system (see separate entry). It provides the following features:

(a) message switching and store and forward facilities
(b) message preparation, dispatch and delivery facilities
(c) physical and logical connection to a range of network types, including Land and Naval radio networks and strategic Wide Area Networks
(d) support for the military messaging protocols: ACP126, ACP127 and ACP128
(e) message preparation and reception
(f) client / server LAN based messaging support
(g) graphical user interface with online context-sensitive help
(h) integration with the Microsoft Windows NT environment
(i) designed for unskilled operation
(j) multilingual support. The baseline version of SAMMS is English, but it can be provided in other languages including Arabic.

Configurations

SAMMS Standalone. The SAMMS Standalone is the base software application for Message Preparation and Message Switching operations. It can be configured, at run time, to operate as a Message Preparation terminal or, by adding the appropriate communications interfaces and message routeing information, as a store and forward Message Switch.

SAMMS Network.The SAMMS Network extends the SAMMS message preparation facilities into the Local Area Network environment. It uses client/server architecture to provide message preparation and store and forward facilities to clients positioned on the Local Area Network. There are three components to the SAMMS Network configuration: SAMMS Server, SAMMS Client and SAMMS Gateway. The SAMMS Server is the central application that manages the flow of messages within the SAMMS system, monitors message timeouts and central SAMMS Administration facilities. The SAMMS Client provides message preparation/reception positions on the Local Area Network. The SAMMS Gateway provides the connectivity of the SAMMS Network environment to the outside world, which in many cases is to a SAMMS Message Switch.

Features

(a) Store and Forward Message Processing
    All messages to and from a SAMMS terminal are saved to disk prior to the message being placed into a transmission queue for delivery.
(b) Graphical User Interface
    Using Microsoft Windows, the SAMMS primary user interface consists of two main windows: the Control Centre and the Audit Trail. Other data dictionary windows are also presented to the user from the main window. The Control Centre provides the user with the central point for managing the messaging and communications interfaces that are defined within the system, while the Audit Trail provides the central point for recording and viewing activities that have occurred on the SAMMS system.
(c) Message Switching
    A standard feature of all SAMMS terminals is the ability to switch messages between channels that have been defined on the system. The user can configure the terminal's routeing rules, using a variety of categories to indicate how the SAMMS terminal should handle/switch messages being processed by the system. By extending this Message Switching functionality, SAMMS can be configured to provide interfaces to a number of communications interfaces simultaneously. This is particularly useful for connection to a number of crypto, modem or radio interfaces, such as on board a naval vessel.
(d) Communications Interfaces:
    Asynchronous RS232. The SAMMS ASYNC channel provides an RS232 interface for connecting external communications systems to the SAMMS terminal, typically done at the SAMMS Standalone or the SAMMS Gateway in the Network configuration. From this RS232 interface radios, modems and encryption devices can be connected directly to SAMMS.
    TCP/IP Windows Socket. The SAMMS Socket (Windows TCP/IP interface) provides a mechanism for connecting SAMMS terminals, Standalone Message Switches, Standalone Message Preparation and SAMMS Network Gateways together using the TCP/IP networking protocol. This can be used over either a Local or Wide Area Network.

(e) Message Preparation
    Considerable preparation and formatting assistance is provided, and data can be imported from other Windows applications such as Microsoft Word and Excel. ACP127/8, Miltope Tiger formats and the Systematic IRIS formatted message generation product for incorporating ADatP3 message content into the text of a SAMMS message are all supported. Data files can be attached.

System Requirements

The SAMMS software will run on any standard IBM-compatible PC running Microsoft Windows™ XP, 2000, NT or 95/98. It has the following system requirements:

(a) IBM PC or 100 per cent compatible 400 MHz or higher Pentium
(b) 128 Mbyte RAM (minimum/recommended)
(c) 4 Gbyte hard disk
(d) SVGA Monitor and card capable of running 800 × 600 256 colours under Windows NT
(e) Mouse
(f) 3.5 in floppy drive
(g) CD-ROM

Status

SAMMS has been sold to the Norwegian, Romanian, Turkish and UAE navies.

Contractor

Aeromaritime Systembau GmbH, Munich.
BAE Systems Australia, Canberra.

UPDATED

Israel

ADS-21 Audio and Data Distribution System

The ADS-21 is an advanced digital switching and control system for fixed or mobile headquarters. It provides high-quality audio and data distribution utilising all-digital switching and DSP. The system has flexible control and interface capabilities providing positive connectivity to internal, external and auxiliary sources. The decentralised system has high survivability and is suitable for shipboard audio and data distribution, command and control systems and ATC.

Characteristics

The ADS-21 provides the following connectivity:
(a) internal (intercom) networks;
(b) external (radio) networks;
(c) telephone lines (2W/4W, CB/LB);
(d) recording systems;
(e) PA systems;
(f) data connectivity.

The system interfaces to encryption equipment and can maintain Black/Red separation. In addition, it provides shared access to radio and landline on a priority basis. Analogue and digital interfaces enable integration to a wide variety of existing systems.

Status

The ADS-21 is in service with the Israel Navy, Israel Air Force and in several other armies.

Contractor

Tadiran Electronic Systems Ltd, Holon.

Tadiran ADS-21
0055013

VERIFIED

EL/K-1850 Integrated Data Link Network

The EL/K-1850 Integrated Data Link Network is a wideband integrated microwave communication network designed for a variety of ground-to-ground, ground-to-air, ship-to-air and air-to-air applications. The IDL Network can be tailored from a variety of Data Terminals, including Ground Data Terminals (GDT), Air Data Terminals (ADT) and Video Receiving Assemblies (VRA), enabling the user at each terminal to receive and/or transmit analogue data, video and/or digital data. Typical applications:

(a) Command and control for UAVs
(b) Transfer of data from Remote Imaging Sensor, ESM Sensor or CSM Sensor
(c) Back-up to communications links
(d) Beyond horizon communication by relay
(e) Data Tx/Rx for Special Mission Aircraft
(f) Command and Control communication network.

Features

(a) Long range: up to 360 km LOS
(b) Extended range beyond LOS via Relay
(c) RF Frequency: C-Band, X-Band
(d) Wide bandwidth: Span up to 1.5 GHz
(e) Antennas omnidirectional or directional, parabolic dish or planar array
(f) Full-duplex communication
(g) Carry one or a mixture of the following:
   Command or Telemetry Data
   Video, black and white or colour
   Data: from 64 kbps up to 280 Mbps
(h) Back-up navigation - range and azimuth, a coherent loop between the received command and the transmitted telemetry data enables range measurement at the opposite data terminal
(i) ECCM/LPI mode (SpSp) for protected communication
(j) Remote-control capability via standard databus
(k) Tracking capability
(l) BIT.

Network Subsystems

The EL/K-1850 network consists of the following subsystems:

(a) Ground Data Terminal, mobile or fixed installation, for ground or shipborne applications in several configurations:
   EL/K-1861 - Mobile or Fixed Ground Data Terminal
   EL/K-1862 - Compact Ground Data Terminal
   EL/K-1871 - Downsized Ground Data Terminal
(b) Air Data Terminal - EL/K-1865 Air Data Terminal for airborne installations
(c) Video Receiving Assembly - EL/K-1863 Video Receiving Assembly (receive-only)
(d) Relay
   Combining two terminals into a relay station for extending communication range
   The Air Data Relay (ADR) combines two Air Terminals and the Ground Data Relay (GDR) combines two Ground Terminals.
   Data Terminal Configuration
   The standard basic configuration of each data terminal includes:
(a) Antenna (Omni or Directional)
(b) Transceiver (with synthesized LO) and Power Supply
(c) Front-End (Rx Filter/Preamplifier and Tx Power Amplifier).

Optional additional features

(a) Spread Spectrum modem for protected communications
(b) Video compression/decompression unit for transmitting and receiving video
(c) Digital data modem for transceiving digital data
(d) Data encryption module for data security.

Characteristic Specifications

(a) RF Frequency: C-Band or X-Band
(b) Antennas: Omnidirectional or Directional
(c) Transmit Power: 2 W, 10 W, 25 W or 100 W
(d) Receive Mode: Command Data Mode
(e) Transmit Mode: Video with Telemetry (TX/TV) or Telemetry only (TM), Medium rate or High capability Data Mode (Option).

Status

Available. May be in service with the Israel Defence Force.

Contractor

ELTA Electronics Industries Ltd(a subsidiary of Israel Aircraft Industries Ltd), Ashdod.

UPDATED

....

Tactical Intranet Geographic Dissemination in Real-time (TIGER)

Tactical Intranet Geographic dissEmination in Real-time (TIGER) is an interest and location-based dissemination system. Its dissemination protocol, based on the "Publish and Subscribe Broker Concept", enables the dissemination of geographic and interest-based information in real time. TIGER connects C4I applications from the theatre command level down to the single vehicle level and combines management and control capabilities with information and system

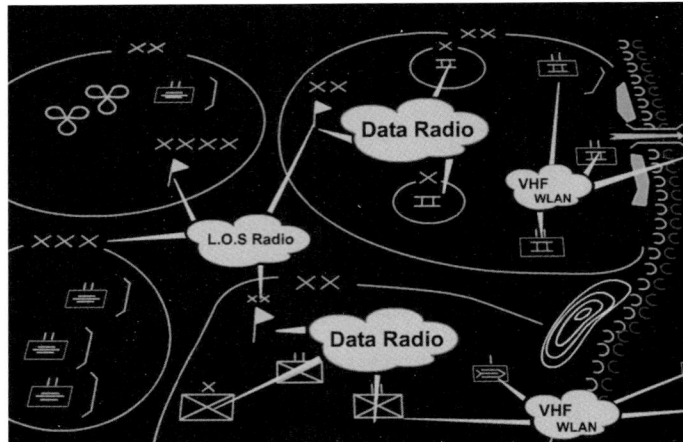

A graphical representation of a TIGER network (Elbit)   **NEW**/0569676

security, resulting in a dissemination system which can support thousands of independent workstations.

Based on a modular and scalable architecture concept its architecture enables both standard and dedicated applications to co-exist over the same infrastructure, which reduces development time and cost. The component-based architecture allows every layer and/or component to be upgraded or replaced without affecting other layers or components. TIGER incorporates a unique proprietary network layer, optimised for military tactical environments which allows it to integrate various communication media - LAN, WAN, cellular, data radio, tactical radios, WLAN, satellite - characterised by large scale military hierarchical topology, into a single unified and fully distributed network.

Characteristics

(a) Subscribers receive all, but only, relevant information, according to defined topics of interest (including subject-related and geographic-related interest)
(b) Can support a large scale (thousands of units) distributed tactical network in which each unit functions as a router
(c) Self-forming and self-healing - automatic and adaptive learning of highly dynamic network topology
(d) Multilayer security architecture supports all existing military-oriented communication channels
(e) Automatic, dynamic, adaptive, and optimised routeing of messages
(f) Simulation and test bed - theoretical simulations of protocols and algorithms as well as comprehensive system test simulation facilities.

System Architecture Components and Capabilities

(a) Automatic, Geographical Publish and Subscribe (AGPS) Module
   A distributed network of brokers transfers information to geographically distributed units, according to their specifically defined topics of interest, current location and other parameters. The principle behind this module is that information producers and consumers do not have to be aware of each other. The AGPS consists of Stations, Brokers and Gateways. In order to reduce the tactical network's information flow, brokers serve as intermediate servers. They gather subscribers' topics and areas of interest and disseminate this information. Each broker is responsible for a group of stations and serves as the mediator to the rest of the network. The gateway operates as an 'enhanced' broker, connecting the various networks. This divides the network into clusters and reduces message flow.
(b) Tactical Message Oriented Middleware (TMOM)
   TMOM enables the seamless transfer of messages between military forces, over different communication channels to consumers and C4I applications. As soon as the TMOM receives a message, it transfers it to its destination in the most efficient and secured manner while making optimum use of available communication channels. It provides a claimed end-to-end, 100 per cent guaranteed delivery using 'store and forward' message transfer techniques; adaptive acknowledgement and retransmission mechanism; flow control; multiple priority queuing; obsolete message replacement and expired message deletion; survivability (the ability to work in an autonomic regime in any subnet which is temporarily disconnected from the Tactical Intranet); recovery mechanism, such as after reconnection to the network, after hardware failure and/or node destruction
(c) Network Infrastructure
   The Network Infrastructure combines the large scale (thousands of units) distributed tactical sites and the multiple heterogeneous communication media into one network. The network layer ensures that all routeing information is up-to-date even in the most dynamic connectivity environments but its 'technical' overhead (control messages) is low even for VHF tactical networks.
(d) Media Adaptors
   The media adaptors perform the adjustments for the optimal operation of the system across each communication medium. They isolate the higher communication layers from the communication media. This plug-in concept makes it possible to add or change communication media without affecting the software of the higher layers.
(e) Communication Controller
   The communications controller includes a variety of customisable MAC protocols that enable optimal use of tactical radio networks and provides advanced Forward Error Correcting Coding (FECC).

Status

According to *Jane's International Defense Review* in August 2002, TIGER will provide the network communications layer for the battle management system being developed as part of the IDF Sidre Berashit ('new beginning') digitisation project.

Manufacturer

Elbit Systems Ltd, Haifa, Israel.

UPDATED

Italy

SICRAL system

SICRAL (Satellite Italiano per Comunicazione Riservate) is an integrated satellite communications system which will provide secure telecommunications for the Italian Ministry of Defence and civil security forces for both domestic needs and remote operations. The programme consists of a space segment with one operating and one back-up multifrequency satellite in geostationary orbit, and the control and ground segments comprising satellite control equipment and diverse types of terminals: fixed, mobile, transportable, man-packs, shipborne, airborne. SICRAL carries 9 SHF, UHF and EHF transponders. It is positioned at 16.2° east. The interconnectivity of terminals is enhanced by the satellite's cross-strappable communication bands which allow terminals operating on different bands to communicate with each other. The satellite weighs a total of 2.5 tonnes at launch and can carry a payload of 330 kg. SICRAL is interoperable with the NATO IV, Fltsatcom, DSCS and Skynet systems and most of the channels of the Syracuse and Hispasat systems.

As part of the SITAB consortium, Alenia Spazio is responsible for the system's general architecture, and also developed and constructed the satellite and fixed and mobile receiver terminals. BPD Difesa é Spazio managed the launch contract, and is also responsible for the satellite's propulsion system. Nuova Telespazio has constructed the control centre.

Status

Launched on 7 February 2000 on the same Ariane vehicle as the UK Skynet 4F, and in service.

Contractor

Alenia Aerospazio, Rome.

VERIFIED

SIDAS - Strategic Integrated Digital Automatic System

The Strategic Integrated Digital Automatic System (SIDAS) is a multichannel infrastructure system intended as a backbone network for national defence and security organisations. It can be implemented in a number of solutions, according to communication needs and geographical, environmental and installation constraints.

The system is based on functional elements such as a nodal subsystem providing communication and trunk switching services, a fixed access subsystem providing local access services, a mobile access system handling mobile users and interconnected through radio base stations and a management subsystem at national, regional and area levels. The nodes are connected to one another by means of high-capacity links (up to 34 Mbit/s), while the access nodes gain access to the transit ones through medium- and low- (8 and 2 Mbit/s) capacity links.

The SIDAS is provided with features that have been recognised as fundamental for such systems:

(a) it can be arranged in full-grid topology networks and all the radios can be supplied in hot standby configuration, thus allowing for the necessary degree of robustness and reliability
(b) any link can be correctly sized so as to minimise the necessary bandwidth
(c) it is a fully secure system with COMSEC components such as bulk or even single-channel encryption
(d) it can be easily planned, managed and controlled by a unique, easy to operate network management system
(e) it can be connected to the PSTN and satellite systems through suitable single or multichannel interfaces
(f) it is able to interface and integrate the MIDAS systems (see separate entry), working either as a simple transmission medium for their mutual interconnections or interoperating with them at functional and operational level
(g) other than the voice and data services, it allows for transmission and switching facilities expressly designed for C3I systems and wideband video applications.

The telecommunication equipment includes CD140 series circuit/packet switching exchanges and associated MT330/MT400 series combiners/multiplexers; CF100/CM100 series digital bulk and end-to-end encryption equipments and relevant portable loading devices from the FG100 series; MTH48 SHF high-capacity radio relay; MH200 SHF troposcatter radios; MH400 series UHF medium capacity radios; MF15 SHF and MH900 millimetre wave radios for short-range, medium-capacity links; MSH100/400 series VHF radios for very low-capacity links; MT300 series optical and cable line terminating units; a complete family of auxiliary elements that, other than those already mentioned for MIDAS, use the AS108 reference clock, the MD321 high-order multiplexer and the PN123 digital cross-connect.

The control equipment includes data terminals, switchboard operator consoles (PGE100 series) and supervisory computers all based on rugged industrial or commercial platforms.

Marconi supplies a number of application software packages suitable for managing and monitoring the SIDAS system and integrates them with protocols such as TCP-IP, with relational databases and graphics packages; the electronic key distribution system can be integrated as well.

Status

Most equipment is in service with the Italian and other armies.

Contractor

Marconi Selenia Communications, Genoa.

VERIFIED

NATO

NATO Integrated Communications System (NICS)

The NATO Integrated Communications System (NICS), much of the first phase of which was completed by the end of 1985, was the largest commonly funded infrastructure project ever undertaken by the Alliance, emphasising the importance placed by NATO members on the need for effective and full consultation, command and control.

The first two decades of the existence of NATO saw its communication facilities evolve primarily to support a defence policy and strategy based on the concept of a massive nuclear retaliation to any Soviet or Warsaw Pact aggression against the Alliance - the so-called 'tripwire' approach. Communication was based on a series of point-to-point links to enable nuclear release orders to be transmitted to operational commanders responsible for launching the nuclear retaliatory forces. Communication for most other aspects of command and control was extremely primitive. There was a telegraph link on the High Command (HICOM) network through SHAPE which connected the Standing Group representatives at NATO headquarters in Paris with the Standing Group itself in Washington, but the Secretary General's only means of communicating with any NATO head of government was via the communication facilities of the various national delegations. Following France's decision in 1965 to leave the integrated command structure of the Alliance, and largely at the instigation of the UK and US delegations to NATO, a special committee of defence ministers was established with a number of working groups of senior national experts set up to examine ways of improving the effectiveness of the Alliance. An important subgroup of one of the many main working groups was given the task of looking at the whole area of information exchange, including the communication equipment and systems which would be needed to permit the flow of this information.

One of the main recommendations of the subgroup, known as the MacNamara Committee, was for the establishment of a NATO-wide system linking its headquarters (being moved from Paris to Brussels) with the capitals of all the member nations and with each of the major NATO commanders.

At about the same time, the wider concept of a complete NATO Integrated Communications System was being created at the Communications Division of the SHAPE Technical Centre in The Hague. The Technical Centre had been working on the idea of a common user meshed grid network for Allied Command Europe (known as ACENET) which would link all important military and civilian users throughout the Alliance to provide message traffic as well as secure and

NATO satellite ground terminal which serves SHAPE near Mons, and NATO Headquarters near Brussels

Control and switching centre of NATO satellite terminal

One of NATO's satellite ground terminals contained within a radome

unclassified telephone calls and the transmission of data in various forms. ACENET was revised, expanded and presented as the basis for a proposed new NATO Integrated Communications System (NICS), a programme which was approved in 1971 by the North Atlantic Council and established under a new management agency known as NICSMA, currently renamed as NACISA (NATO Communications and Information Systems Agency). From July 1996, NACISA was integrated with the SHAPE Technical Centre in one agency under the name NATO Consultation Command and Control Agency (NC3A).

The NICS concept envisaged the creation of a meshed grid-type common user network for voice, telegraph and data traffic. A meshed grid system was chosen to provide a high degree of alternative routeing at times of high message traffic levels or at times of damage to part of the network. It also provided increased survivability through dispersion and redundancy, and increased performance capability through a high degree of network automation. Where possible it was planned to exploit existing equipment and systems in use within NATO, such as HF and microwave radio links, satellite systems and national military and civil communication systems. Where necessary, new links and facilities were to be incorporated to replace obsolete equipment and improve performance. NICS was designed as a highly sophisticated, circuit-switched system with a ground and space segment, and with operational automatic message switching centres with switching nodes and gateways. The aim was to link all military, political and civil emergency organisations within the Alliance.

Such was the magnitude and complexity of the project that it was decided to deal with the most important and critical elements of the programme in an initial phase, leaving the remaining part of the system for following stages dependent on the availability of funds, implementation progress and other factors.

The implementation of the first phase of the NICS, at an estimated cost of about US$500 million, was completed in the early 1980s, and included the following major elements: the Initial Voice Switched Network (IVSN); the Telegraph Automatic Relay Equipment (TARE); the NATO III satellite subsystem; a narrowband secure voice network; terrestrial transmission systems and the system integration project. However, because of a number of major implementation delays, the first phase was not completed until late 1985.

The IVSN is a dedicated telephone switching system. It comprises 24 operational access switches together with two additional switches for training and software development. The complete network was designed to provide facilities similar to those of the US AUTOVON system. The contractor for the IVSN was ITT North Electric. All switches were installed and taken over by NATO. The training switch is located at the NATO Communications Training Centre near Rome and the software maintenance switch at the NATO Headquarters at Evere.

The network provides automatic telephone facilities to about 12,000 subscribers throughout the NATO area, although a fully expanded network could easily service 600,000 subscribers. The IVSN is configured as a meshed analogue grid network. The transition to a fully digital network was expected to extend into the 1990s.

The TARE is designed to provide a network consisting of 18 operational TARE message switches, plus two additional equipments for training and software development. The network is a message store-and-forward processing system and provides telegraph services similar to the US AUTODIN system.

The system, the prime contractor for which was Litton Data Systems, is based on the use of analogue circuits with the capability of carrying digital traffic. It is a two-computer system using a communications processor and a message processor, each duplicated for improved reliability.

A contract for 20 TAREs was placed in 1976 with the first equipment due for delivery in September 1978 and the remainder being installed at roughly three-monthly intervals, with completion expected during 1985. However, Litton had problems in delivering equipment meeting the contractual performance and reliability standards. Despite the difficulties, all 17 operational TAREs were activated in the NATO area by mid-1987. An eighteenth TARE was scheduled to be added to the system in 1991. The TARE network was subsequently re-engineered to meet current reduced requirements with five sites to be closed leaving 13 TARE sites operational.

The NATO III SATCOM system of the NICS consisted of four operational orbiting satellites and 21 static ground terminals, including 12 ground terminals provided under the NATO II programme. In addition, a transportable satellite terminal has been provided. It is a digital system, the first all-digital network within the total NICS. The system's satellites, for which Ford Aerospace was the main contractor, were launched between 1976 and 1984. Work on upgrading the NATO II ground terminals and completion of the new stations, for which Ford Aerospace was again the main contractor and Standard Elektrik Lorenz the system contractor, was originally scheduled for completion by 1983. The first new ground terminal was accepted in October 1982 and the final station in October 1985.

The NATO III satellites were replaced by a constellation of two NATO IV spacecraft. The first, NATO IVA, was launched in January 1991 and the second, NATO IVB in December 1993. The launch timetable was selected to allow a 10-year lifetime for the constellation. The NATO IV satellites are of the UK Skynet type and were produced by British Aerospace and Matra Marconi. The launch vehicle was the McDonnell Douglas Delta II rocket. The NATO SATCOM ground terminals are being modified to exploit fully the new satellites' communication capabilities.

The manual secure voice network, designed to provide a secure voice capability for about 1,600 nominated subscribers, has been automated and become part of the IVSN. Together with four-wire telephone switchboards which have now been located at major user locations, it is a complete, dedicated, manually switched network. Main contractors were Siemens AG and AEG-Telefunken of Germany with Page Europa in Italy producing the switchboards.

Later (in the mid-80s), the secure voice system, complemented by the US-made STU IIA and (made in the Netherlands) SPENDEX 40 voice crypto equipment, was loaned to NATO to provide end-to-end secure voice communications through the NICS. Main manufacturers of this equipment are Siemens, Motorola and Philips. In a third stage, NATO is implementing the Narrow Band Secure Voice II programme with the modern crypto equipment ELCROVOX 1-40, STU IIB and SPENDEX-40M procured from the same manufacturers respectively. The terrestrial transmission system largely comprises existing NATO-owned subsystems which are being upgraded and extended. For example, seven subprojects have been completed which expand the ACE HIGH network in Europe. The ACE HIGH system was closed by the end of 1996. It is being replaced by the National Defence Systems. Where the latter are not available, leased PTO circuitry will be used in the interim.

Many more projects have been initiated to provide the necessary transmission support for the NICS. Included is the CIP-67 microwave radio network which provides links in the Central Region where a large number of NATO subscribers are located. CIP-67, the acronym for a NATO Communications Improvement Programme which had its conceptual birth in 1967, is now completed. It was implemented by a multinational consortium of ITT's Federal Electric Corporation, Siemens AG and Marconi Italiana. It is a network of 53 stations configured to provide alternative routeing, many of which are housed in bunker-like hardened buildings.

The NICS also makes extensive use of civil Post and Telecommunications Authority networks which, it is estimated, carry about a third of NATO message traffic.

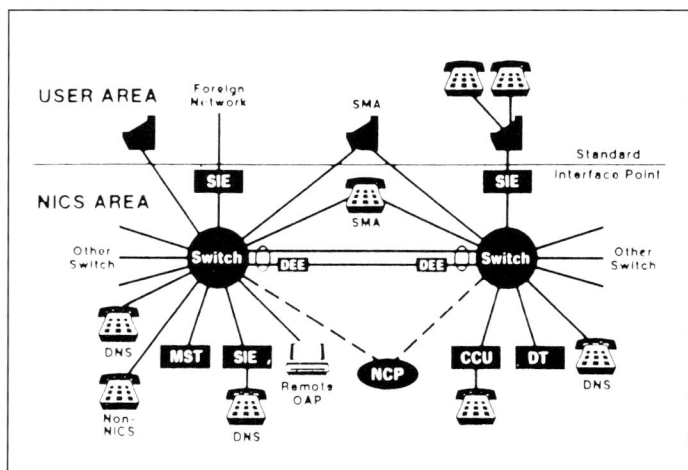

IVSN system configuration

The SubSystems Integration Project (SSIP) is the most complex of all and provides the ancillary facilities needed at each site to permit total interaction within the NICS. In all, there are over 300 principal and secondary sites within the NICS, each of which involves different configurations and different local interfacing into national systems.

The NICS IVSN and TARE expansion programme improved the quantity and quality of service to subscribers, provided enhanced survivability through the improvement of the switches, increased the interoperability with national systems through the use of common standards and through agreed interface equipment and/or procedures, and increased security with new cryptographics and cryptophonic equipment.

By early 2001 the number of operational TARE switches had been reduced to nine. IVSN is to be replaced by the NATO General-purpose segment Communication System (NGCS).

VERIFIED

Netherlands

Netherlands Armed Forces Integrated Network (NAFIN)

With the Netherlands Armed Forces Integrated Network (NAFIN), the entire Dutch armed services will have its own telecommunications network for speech, data and fixed connections. In 1995, the Ministry of Defence entered into a contract with Nortel Networks for the turnkey implementation of NAFIN phase 1B: the Transmission Network.

NAFIN has been designed to combine in a single network all existing military connections, now still using beam transmitters and leased lines, together with a host of new connections - such as, for example, data traffic. In view of the fact that present day long-distance telecommunications almost exclusively use optical techniques, the Ministry of Defence has laid approximately 2,600 km of its own fibre-optic cable.

Halfway through 1995, a start was made on implementing the NAFIN transmission network, applying the latest technology: Synchronous Digital Hierarchy (SDH). The chief characteristic of an SDH network is that speech and data signals can be transported at very high speed. In the event of calamities such as a break in a cable, automatic protection mechanisms ensure that traffic can be dealt with via another route.

The NAFIN network will connect about 150 locations spread over the whole of the Netherlands. With a view to optimum manageability and efficient use of network capacity, a layered topology of rings was chosen. The 2.5 Gbits/s main ring for transporting inter-regional traffic runs centrally through the Netherlands. To this main ring 14 geographically dispersed 622 Mbits/s access rings are connected. All military locations are either directly or indirectly connected to these access rings. Through the application of the most modern SDH technology to a layered ring topology, the Dutch armed forces will have one of the world's most flexible and reliable networks at its disposal.

NAFIN consists of an almost 3,000 km (1860 mi) fiber SDH (SONET) backbone covering the Netherlands. NAFIN connects about 350 military objects with cross-border connections to NATO partners. In the future, NAFIN will also provide mobile and satellite connections for global coverage.

The MoD implemented NAFIN with transmission hardware supplied by Northern Telecom. For the interim LAN Interconnect facility, NAFNET, the MoD implemented a Frame Relay-based router network with seven 2220 Nways BroadBand Switches in the backbone. The MoD chose routers for LAN and backbone access.

NAFIN started NAFNET with IP and SNA over Frame Relay. Their existing 15000-node SNA network will be phased out but must co-exist with the IP traffic during the transition period. SNA traffic flows using separate Data-Link Control Interfaces (DLCIs) over Frame Relay, allowing users to connect to the hosts and eliminating conflicts between SNA and IP traffic. The maximum speed at the SDH level is set to 2.5 Gbps until higher speeds are needed. Access will be supported for speeds starting with 64 Kbps. All PABXs will be connected to NAFIN on the SDH layer initially with connections to ATM in the future. The final solution to the LAN Interconnect Service will be based on ATM and on MultiProtocol Over ATM (MPOA) standards. The MoD started a market survey for the ATM solution in 1997.

Status

NATO is providing around one fifth of the programme cost. The project is being managed by the Royal Netherlands Air Force.

In mid-1995 Northern Telecom BV was awarded a US$40 million contract to supply Synchronous Digital Hierarchy (SDH) transmission equipment for use in the NAFIN. The Nortel SDH equipment operates at up to 2.5 Gbit/s.

It is understood that the system became fully operational in 1999.

VERIFIED

Theatre-Independent Tactical Army and Air Force Network (TITAAN)

At the end of 2001 the Netherlands Ministry of Defence (MoD) announced that it would not pursue the planned US$100 million mid-life update of the ZODIAC communications network system (see separate entry). Instead, the MoD launched the procurement of the tri-service US$110 million Theatre-Independent Tactical Army and Air Force Network (TITAAN). According to the Dutch State Secretary for Defence Procurement, the ZODIAC system was insufficient for data communication and was not capable of supporting operations over long distances and in mountainous areas. TITAAN will be a modular system capable of handling large volumes of voice, data and video over long distances and in any terrain. It is to be based on civil and NATO standards and proven technology.

The basic TITAAN module is reported to be a LAN, (cable-connected or wireless) including all user terminals, peripherals and system equipment such as servers and routers. A total of 98 of these modules will be acquired to equip 91 command posts (battalion-level and above) in the Royal Netherlands Army (RNLA) and seven command posts (squadron-level and above) within the Royal Netherlands Air Force Tactical Helicopter Group. To build a WAN, 61 mobile mast-mounted transceiver radio stations will be procured. The requirement for satellite communications terminals will be met under a separate project.

TITAAN will interface with the RNLA's Battlefield Management System which is being acquired for battalion-level and below, and with the Royal Netherlands Marine Corps' planned NIMCIS equipment. The RNLA Command and Control Support Centre will be tasked with the integration of the various subsystems to be procured from a variety of suppliers.

TITAAN - Headquarters I (GE/NL) Corps

As an interim measure, communication capabilities within HQ 1 (GE/NL) Corps (see entry for High-Readiness Forces Land Headquarters (HRF(L) HQ)) are already being upgraded by the introduction of satellite communication, mobile telephone communication, a mobile LAN capability and video teleconferencing, worth approximately US$13 million. In January 2003 the TITAAN assemblage in the HQ was reported by *Jane's International Defense Review* to include the following key elements:

(a) A WAN/LAN configuration
(b) Bearer systems: primarily satcom but also 1,024 kbit/s UHF secure radio relay and land lines
(c) Voice service (based on a total of 700 Cisco Systems IP Phone Model 7960 VOIP telephones)
(d) Video teleconferencing (based on the PictureTel system by US company Polycom, featuring Sony displays and a voice-recognition capability which automatically pans the camera to the person who is speaking. At the moment there are six units plus the associated master control unit, but this will grow to 12)
(e) Up to three dedicated 30 × 30 km cellular phone networks
(f) Secure HF back-up (based on 12-year old German Army Rohde & Schwarz HF-C radios available in 100 W and 1 kW versions).

The heart of the WAN/LAN configuration is the Mobile CIS Control Center (MCCC). This acts as a help desk for all problems and includes a Compaq server with two Xeon processors of 1 Gbyte each, running Aprisma's Spectrum network management tool. This tool (including the SpectroSERVER and SpectroGRAPH subtools) provides instant visibility of how the network is set up, which links are active and where problems are occurring. All satcom-network management, currently a separate tool, will ultimately be integrated into this system. The WAN can connect the Main CP, Forward CP, the peacetime HQ, the combat support battalion, the CIS battalion, the Tactical CP, and up to four major subordinate commands. All telephone links are via the LAN (using Voice-over-Internet-Protocol (VOIP)).

Two Xantic satellite dish units deployed with HQ 1 (GE/NL) Corps (Timo Beylemans)   *NEW*/0532479

For details of the latest updates to *Jane's C4I Systems* online and to discover the additional information available exclusively to online subscribers please visit
jc4i.janes.com

The interior of the Xantic satellite terminal indoor unit (Jane's/IDR)   **NEW**/0532481

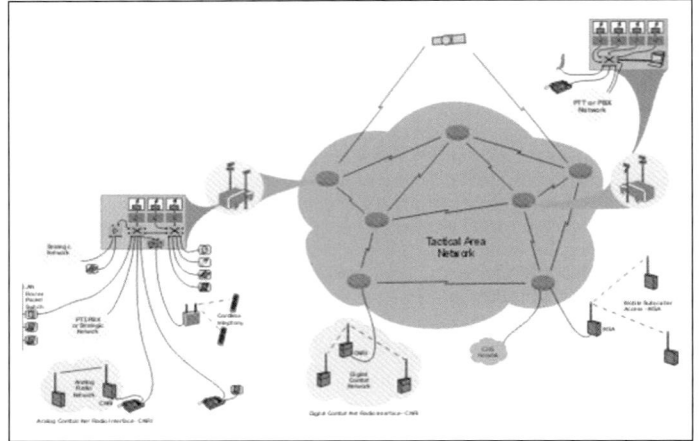

EriTac system   0079950

Command-and-control information, planning and support systems such as LOCE, ADAMS (see separate entries), BICES or JOIIS can be run on any workstation, and digital maps and aerial photos, if required overlaid on each other, can be displayed using the two 3M model MP8755 projector systems mounted overhead in the Joint Operations Centre (JOC).

Satcom communications are provided by Xantic, a Netherlands-Australian satcom provider formed from a merger between Station 12, Telstra Global Satellite, SpecTec and KPN Broadcast. Six terminals were operational by November 2002 and a total of 12 were expected to be in place by February 2003. The system will be in service for at least four years until replaced by a new NL armed forces military satcom system.

The Xantic system is based on Xantic's Broadband WAN (BB WAN) product, which is described as a dedicated network service via satellite. According to Xantic, a BB WAN network is based on VSAT (very small aperture terminal) technology and is suitable for multimedia applications (voice, fax, data, e-mail, internet, file transfer, video teleconferencing). The network configuration would enable any site in the network to communicate directly with any other site with a capacity of up to 512 kbit/s. The system procured for TITAAN is based on mobile trailers (outdoor units) and inside military shelters supplied by Fokker Special Products (indoor units) and includes VSAT satellite dishes and dish controllers from Vertex Satcom Systems. It is capable of operating on C-band or on Ku-band; switching is achieved by changing the feedhorn and installing a Ku-band subreflector. The outside is equipped with an automatic de-icing system.

The indoor system is based on the SkyWAN IDU 3000 system from German-based ND SatCom. Fully secure communications are achieved with a TCE 600 series crypto device from Thales Communications AS and a Datacryptor 2000 hardware encryptor from Thales e-Security Ltd. The system includes a Model 2398 9 kHz-2.7 GHz spectrum analyser from US company IFR Systems for satellite link management, together with a satellite frequency distribution system from DEV Systemtechnik, a Compaq server and uninterrupted power supply.

RACEs (Rapid CIS Elements)

The RACE elements are platoon-sized, fully-mobile units that include facility control, cable/generator group, radio relay group, HF radio group, satcom terminal group, and a video teleconferencing capability. The unit supporting HQ GE/NL Corps includes two RACE elements for the HQ, one RACE for the Rear Support Command, and four RACE elements to connect subordinated divisions or brigades. The latter provides the command post of a subordinate formation division with satcom connectivity and a TITAAN-based LAN, including formations from a nation with no NATO C2. This is partly because TITAAN and the applications run on it are based on Microsoft Windows 2000, significantly reducing the training and familiarisation requirement.

The Main RACE, configured to support the Main CP, includes two TITAAN server vehicles (one active, one in hot standby) equipped with Compaq servers. These vehicles incorporate a Compaq uninterrupted power supply (able to bridge a power cut of 20 min), a Cisco Systems-modified Compaq server which manages the VOIP telephone traffic, a Cisco Systems switch, a Compaq file and print server (which stores all the user data), a storage rack with six hard disks (with growth space for six more), and a Cisco Systems gatekeeper (for connecting the VOIP calls to the outer world). There is also a facility-control shelter and a help-desk shelter.

UPDATED

Norway

EriTac communications network

The multipurpose EriTac system provides automatically switched voice and data communications. Typical applications are tactical area communications networks, air defence and tactical air direction networks, and command post communications networks.

The heart of the system is the CPX200 tactical switch. The CPX200 has five EUROCOM D1A ports with trunk rates of 256, 512, 1,024 or 2,048 kbit/s. The ports support a free mix of channels with bit rates ranging from 16 to 256 kbit/s. The digital interfaces provide reliable error protected data communication with channel bit rates up to 76.8 kbit/s using standard EUROCOM class 4 FEC techniques for local and network connections. Without using error correction, channels of 256 kbit/s user data are supported. Switches may be configured and monitored either locally from the front panel or from a PC-based MS2000 network management station. This displays all the transmission links in the network and enables operators to supervise and authorise changes and to check and change subscriber data in any of the switches. The management station can be connected anywhere in the network as a subscriber. The system can be divided into a number of sub-nets which automatically have full functionality. Interfaces with both civil and other military networks can be achieved either through cable or radio link. The free numbering system enables subscribers to reconnect anywhere in the network using the same number. CNR can also be added to the network, either using the Digital Access Point for analogue radios or the CNR Interface for digital equipment. In the latter case the contractor offers a MultiRole Radio (MRR) in handheld (under development), manpack or vehicle borne form. The MultiLine Terminal (MLT) provides direct single-button user access to radio nets, conferences and intercom circuits. The MLT additionally supports dialled access to all users in the network. Staff cell users communicate via the MLT using a split headset with microphone. The noise-cancelling headset enables simultaneous monitoring of and conferencing with, several radio nets and/or conferences. Access to radio sets or other user circuits is available via dedicated or general dial pad keys on the MLT front panel.

The system has a variety of Electronic Protection Measures including delta modulation, flood search and automatic alternative routeing, automatic output power control combined with directional antennas, and frequency hopping.

The data communications features allow exchange of sensor and weapon data, which is particularly appropriate for air defence, enabling integration of sensors, weapons and communications at fire-control and weapon sites. Tactical air direction is also supported through the access provided to different radio nets for positive control of the variety of agencies involved.

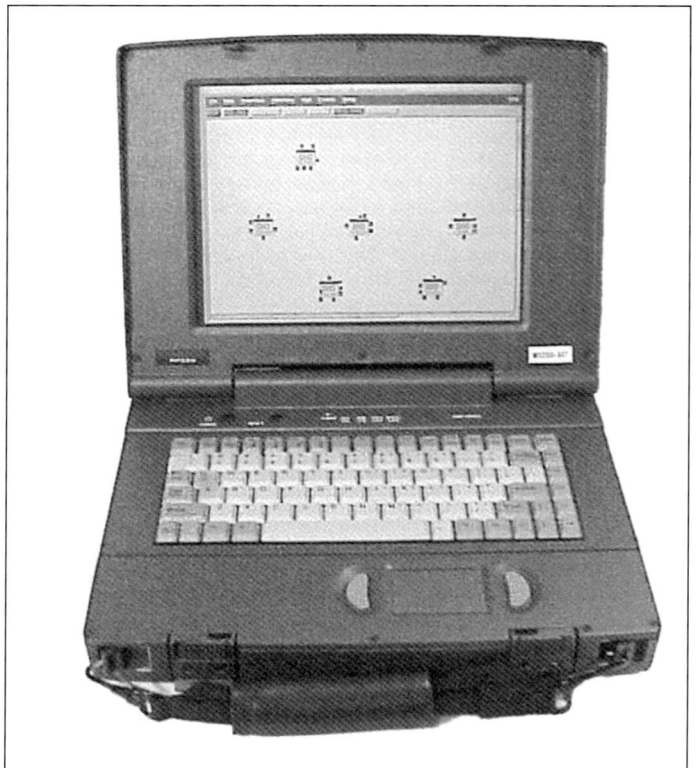

MS200 network management station   0079958

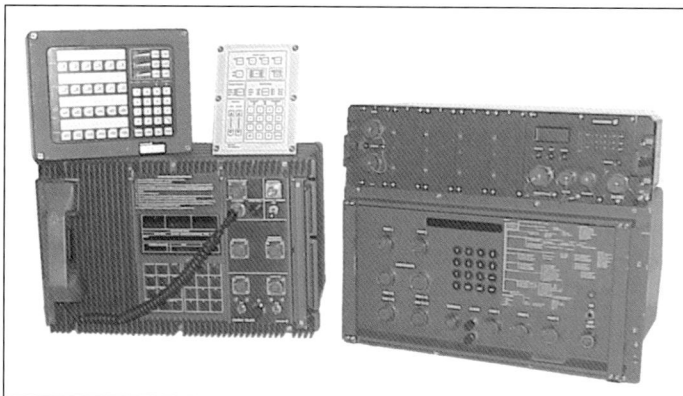

EriTac equipment   0079949

MRR
0079956

Dispersed command post cells can also be supported by connecting by cable, fibre optic or radio, with the system's flexibility of configuration and access allowing staff to access any part of the network. Any cell can therefore be used as any type of command post element, permitting tactical flexibility and redundancy.

Status

In use in the Norwegian adapted Hawk system (NOAH), Norwegian ground launched AMRAAM (Norwegian Advanced SAM System - NASAMS), Norwegian SHORAD (Norwegian Army LLAD System - NALLADS), and in conjunction with Patriot in several countries. Used by the Norwegian Army as a command area communications network. Claimed to have been sold widely in the Middle East and Africa, including Egypt, and in Eastern Europe. Also in use in Finland.

In December 2002 and January 2003 it was announced that contracts worth respectively NKr95 million and NKr400 million had been awarded by Oman and Kuwait. The former was for an EriTac system and the latter for an EriTac system integrated with Kongsberg's MultiRole Radio, to be known as the Al-Ameen tactical area network. The Kuwait programme will be fielded over a 3-year period.

Contractor

Kongsberg Defence Communications, Billingstad.

VERIFIED

United Kingdom

ATRACKS

ATRACKS is a PC-based Automatic Tasking, Relay and Acknowledgement System for secure in-flight retasking of helicopters operating in support of land forces. It enables ground and sea-based agencies to prepare helicopter tasking messages including target positions, landing sites and load details, send them to airborne helicopters and obtain an automatic message acknowledgement. It also enables details of current tasking to be provided to helicopters rejoining a network. GPS

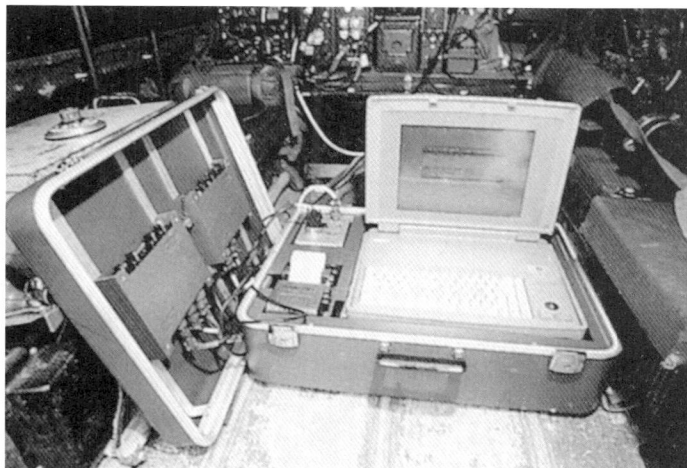

ATRACKS Vehicle Fit (MASS Consultants Ltd)   0109969

tracking of aircraft is passed automatically and locations can be displayed either on a tote or as a map overlay. The system uses existing VHF and HF radios to provide a secure data transmission carrier ground-to-ground, ground-to-air and air-to-air, and can also act as a bridging unit whereby two networks can be linked at one ATRACKS unit in order to pass messages seamlessly from one network to the other. Preformatted standard tasking messages are used and there is a free text capability. Messages can be stored in the PC for subsequent recall.

ATRACKS can be fitted in ships, aircraft and both static and mobile ground units. There is also a lightweight battery-operated manpack version.

Each ATRACKS node comprises:
(a) A rugged lightweight, low-power encryption device containing the Pritchell 2 encryption chip identical to that used in the Bowman VHF and HF radio family. The MASS encryption device is fully CESG accredited and is claimed to be the first Pritchell device to have been cleared for operational use.
(b) A rugged lightweight modem designed specifically for very low power consumption in the ATRACKS manpack configuration. Baud rates up to 2400 bps may be selected and the modem will support many standard protocols.
(c) A rugged lightweight O'Neill printer with re-chargeable battery.
(d) A rugged data terminal:
    For aircraft and manpack variants, the MBM Technologies LT450 'Termite' terminal is used (see separate entry). This small lightweight clamshell laptop is cleared to Land Class A EMC standard and has a NVG compatible 6.2 in TFT screen and full qwerty keyboard. It is cleared for flight operation and is capable of battery or external power operation.
    Vehicle and HQ ATRACKS nodes use the 'Scorpion' Bowman Management Data Terminal made by DRS (formerly Paravant Corp) with 13.1 in colour screen, removable 2Gbyte HDD, removable CD-ROM and replaceable battery pack. (See separate entry.)
(e) Aircraft nodes also comprise two other small boxes for radio switching and separation of red and black data.
(f) Vehicle and HQ nodes have a MASS designed Power Distribution Unit allowing operation of all ATRACKS components from either a 24-28V DC vehicle or ship supply or a 240/110V AC mains supply.

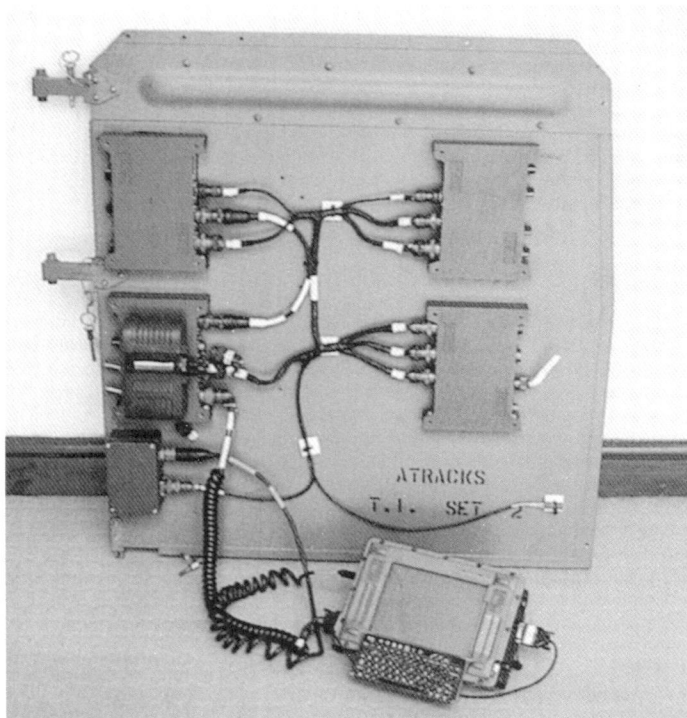

ATRACKS Helicopter Equipment (MASS Consultants Ltd)   0109968

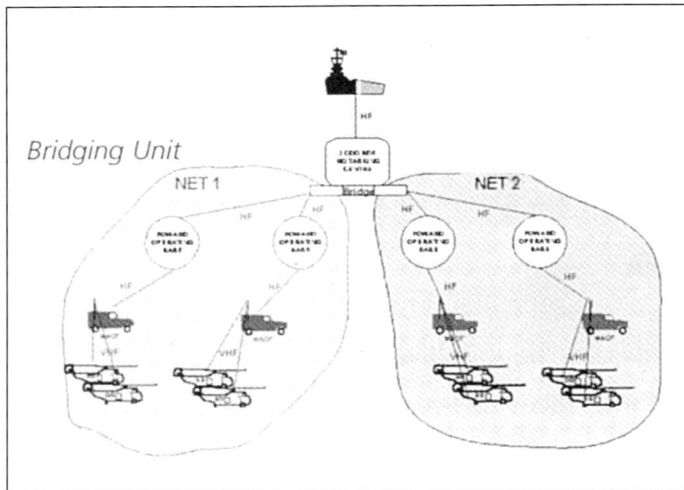

Bridging Unit operation (MASS Consultants Ltd)   0109967

The ATRACKS manpack version at DSEi 2003 (Patrick Allen/Jane's) *NEW*/1026996

(g) In addition, MASS has developed a battery maintained Fill Gun to load the Pritchell encryption codes.

Status

Early exploratory and development trials were conducted between 1997 and 2001, in a stop-start programme utilising underspend funding. In August 2002 qualification testing was funded but in December MASS was asked to consider delivering a production version as an Urgent Operational Requirement for the use in operations in the Gulf. Following receipt of a contract in mid January 2003, 10 ATRACKS sets (6 aircraft and 4 ground) were manufactured and delivered by late February. Thereafter, production was continued culminating in the delivery of 84 full ATRACKS systems, incorporating a totally revised software suite by late June. As at early July 2003 14 air systems were being fitted to Sea King Mk4 helicopters, including five in Iraq and 16 ground systems will be fitted to vehicles. The system may be considered as an interim fit on other battlefield helicopters until a policy on Bowman Air Radio data is decided and implemented.

It is also likely to be fitted in the UK LPH HMS *Ocean* and in the UK LPD(R), although funding issues still remain to be clarified. The Republic of Ireland Navy trialled the modem in 2002.

Contractor

Mass Consultants Ltd, St Neots, Cambridgeshire.

UPDATED

....

Bowman communications management system

BAE Systems has developed the BOWMAN Communications Management System (BCMS), which plans and manages the BOWMAN tactical radio communications system. The BCMS has been developed to run on Windows 2000 platforms with a minimal footprint, but can also be hosted on Windows NT, Windows 98 and Windows XP. It will also run on a Windows Emulator running on Lynux. The software has been developed in Microsoft Visual C++, making use of COTS GIS products while providing open interfaces for the export and import of data in ASCII, CSV, XML and BAE Systems' own data format. The BCMS suite comprises:
(a) The Communication Management Information System (CMIS), which provides tools for system management
(b) Local Area Subsystem Management Information System (LAS MIS), which provides tools to support LAS management
(c) Initialisation Manager, used to initialise terminals and radios.

CMIS
Force definition and GIS
CMIS enables a manager to produce communications plans using a Windows™ Explorer-style Graphical User Interface (GUI) to manage organisations, combat radio nets, platforms and radios. Full support for cut and paste and drag and drop editing is provided, with right-click menus available. Extensive use is made of the tree view to present a familiar hierarchical view of the planned ORBAT. Mapping functionality allows the manager to visualise geographical locations of communications assets using APP-6A military symbols and features zoom and pan functionality. Additional information can be displayed, including shading for Frequency Restricted Areas and Net Service Areas.

Network planning and management
CMIS allows the communications manager to plan and initialise the Tactical Internet by identifying platform routers, planning terminal and end-user locations, and generating routeing tables for the planned deployment. Net connectivity analysis based on Digital Terrain & Evaluation Data (DTED) permits further optimisation of the internet facility. An SNMP interface facilitates real-time feedback from networked assets, alerting the manager about congestion conditions and link failures.

Spectrum management
The frequency assignment engine allows the manager to perform automatic assignment to combat net radios operating in fixed frequency, frequency hopping, free channel search and clear hail modes. The engine performs a number of checks, including:
(a) Co-site and far-site interference
(b) Adjacent channel separation
(c) Interoperability with legacy radios
(d) Frequency restricted areas and barred bands
(e) Exclusive channels
(f) Effects of groundwave and skywave propagation modes
(g) Spurious emissions and responses
(h) Quietened channels
(i) Inter-modal checks
(j) N-signal intermodulation products
Frequency allotment creation, edit and import/export facilities are also provided.

Crypto planning and management
CMIS facilitates the planning of key material in conjunction with a key variable management subsystem. Compromise analysis capabilities are included to allow the manager to determine the potential impact of a compromise and support planning for the recovery activities.

Fill generation and distribution
Once radio configuration has been planned, the fill data is generated electronically for later use with the radio fill device. Additional support for the fill distribution process is provided in terms of assistance with fill logistic planning.

Analysis tools
The Probability of Successful Operation (PSO) map allows the communications manage to visualise the potential coverage for an RF transmitter. Statistical methods are applied to determine the PSO at each zone on the map as defined by a radial and angular step entered by the user.

Grey-scale representation is supplemented by cursor position data supplied in the status bar. DTED terrain data is used in conjunction with path-loss prediction algorithms.

Rapid Path Profile Analysis (RPPA) allows the manager to visualise terrain data and assess its impact on communications provision. When used in conjunction with the PSO analysis tools, CMIS provides the user with functionality to automatically sight the transmitter used in the PSO analysis and locate receivers for this purpose for analysis and trouble shooting of communications provision.

Detailed analysis of the frequencies assigned to nets can highlight any potential frequency deconfliction problem to the CMIS user. These analysis tools can be run with current and historical data, and can be used for planned 'what if' scenarios during the communications planning stage.

Checkers
A variety of checking tools are provided, allowing the user to validate data entry and highlight inconsistencies. Tools include automatic checking of communications plans, frequency allotments, equipment files and frequency assignments.

LAS MIS
The LAS MIS receives the initial LAS plans, identifying communications assets and WAN configuration, from the CMIS. It provides additional tools to plan:
(a) User control devices
(b) LAS topology
(c) Staff intercom
(d) Permitted nets
(e) LAS emissions
The LAS MIS allows the LAS manager to perform co site analyses in order to determine potential electromagnetic interference problems for radios within a LAS. It generates initialisation data from the LAS plan. This data can be subsequently downloaded to the BOWMAN Network Access Units (BNAU) and BOWMAN Gateway Equipment.
The LAS MIS provides the means to monitor the LAS. Monitoring capabilities include:
(a) Presence of platforms
(b) Physical connectivity of platforms (BNAUs) that comprise the LAS
(c) The status of UCDs (users affiliated with appropriate subscriber profile)
(d) Membership of call groups and staff intercom groups
(e) Status of radios within the LAS

(f) External nets available within the LAS
(g) Status of routers within the BNAU
LAS managers can modify some of these functions using on-line control functionality provided by the LAS MIS and can also remotely control any radios within the LAS.

Initialisation Manager
The Initialisation Manager accepts initialisation files generated by the CMIS and LAS MIS. These files may then be transferred to removable media or downloaded directly to communications equipment such as terminals, VHF radios, network access units or gateway equipment.

Status
In June 2002 BAE Systems was contracted by General Dynamics UK, the Bowman prime contractor, to supply BCMS. It was to be delivered as a series of phased releases, culminating in the final delivery of the full system in December 2003.

Contractor
BAE Systems, Bristol.

NEW ENTRY

....

DPABX deployable communications system

The DPABX system is a deployable voice communication solution based on the Nortel Networks Meridian PABX product. Cogent has used the Mini Meridian package to design a PABX for the tactical communicator.

DPABX is a highly modular system with a 'first in and build up' capability, 'plug and play' configuration, optimised telephone wiring system and the possibility of further enhancement for wireless voice and data communications if required. In two primary packages it provides over 100 digital/analogue/ISDN and field telephone interfaces with the flexibility of up to 10 analogue trunk links and 9 Primary/Basic rate digital trunks. The remaining modular system building blocks provide a power distribution system with DC and multinational AC power supply input, an Uninterruptible Power Supply (UPS), and a stand-alone air conditioning unit (ACU) for environmental system management.

The DPABX can be enhanced with the addition of Nortel Networks' smallest Passport multimedia switch. This changes the military voice switch into a communications system capable of supporting a small to medium size stand-alone military headquarters, without any increase in the physical size of the system. The Passport 4460 provides the core of a data switching system, providing interfaces to satellite, radio or troposcatter systems and the provision of intelligent bandwidth management by voice compression. The Passport 4460 can be equipped with up to 60 rear link voice channels, 2 segmented LAN interfaces, primary and basic rate ISDN trunk and serial interfaces. This is contained in a small lightweight package fully integrated into the existing DPABX package.

Specifications
DPABX (potential configurations)
(a) Up to 48 Digital Subscriber Interfaces
(b) Up to 64 Analogue Subscriber Interfaces
(c) 2 Field Telephone Interfaces
(d) Up to 10 ISDN subscriber interfaces
(e) 1 Operator console Interface
(f) Up to 3 Primary Rate Trunk Interfaces (EI/TI)
(g) Up to 6 Basic Rate Digital Trunk Interfaces
(h) Up to 10 Analogue Trunk Interfaces

Passport 4460 System Enhancement
(a) 6 - 60 Digital channels of compressed voice-over rear link
(b) 2 - 8 Analogue channels of compressed voice-over rear link
(c) 1 - 2 10/100-baseT segmented LAN interfaces
(d) 1 - 5 V11/V28/V34 Serial Data Interfaces
(e) 1 - 2 EI Trunk Interfaces
(f) 1 - 2 Basic Trunk Interfaces

System Ancillaries
(a) Digital Telephone Handsets
(b) Analogue Telephone Handsets
(c) ISDN Telephone Handsets
(d) Digital/ Analogue/ISDN Breakout Boxes
(e) Power Distribution System
(f) Uninterruptible Power Supply
(g) Air Conditioning Unit

Status
In use in the British Army, the Royal Marines and the Tactical Communications Wing (TCW) of the RAF.

Contractor
EADS Telecom UK (Cogent Defence and Security Networks), Newport, South Wales.

NEW ENTRY

HF Frequency Management Tool (HF FMT)

The Frequency Management Tool (FMT) is designed to provide a complete frequency planning and allocation service for all types of HF assets within a region of interest. It allows the automatic and integrated planning of HF nets and provides HF frequency sets to cover both tactical and strategic requirements for Automatic Radio Control Systems (ARCS), fixed-frequency and frequency-hopping nets. It can cater for radio assets from different manufacturers and fixed and mobile HF platforms with multiple equipment fits. An X-Windows graphical user interface incorporates a map of the area displaying networks, stations and terrain features.

The user can define assets in terms of location, equipment type, sensitivity and RF power output, antenna characteristics, co-site restrictions and barred frequencies. The user defines which radios are required to communicate with each other. The FMT provides a frequency plan for the area which is designed to minimise interference and maximise communications. The plan accommodates both point-to-point and net communications in fixed-frequency or hopping modes for up to 1,650 links or nets. The output is a complete listing of frequencies or hopsets for each link or net.

The system's planning algorithms make use of groundwave propagation prediction using the Bremmer method and skywave propagation prediction and HF noise prediction using ITU-R recommendations. A frequency assignment algorithm, which gives a figure of merit to each link, is improved to a combined optimum through an interactive process.

The FMT system is run on a Sun Sparc Station with twin 150 MHz processors, under the Solaris 2.5 Unix operating system, Sybase DBMS and Motif Window environment. The user interface is a mouse-driven, motif-style environment, allowing the user to run assignments, input, edit and output data and view tables, maps and help screens. The FMT allows the user to create realistic propagation scenarios based upon a database of actual equipment, antenna and station parameters.

To help users plan their networks, the FMT can display a map of the area displaying network paths, stations and areas of operation, and can display predicted signal-to-noise ratios for each network, for each timeblock and for each frequency block.

Status
In service with the Turkish Armed Forces in 1997.

Manufacturer
Marconi Selenia Communications Ltd (formed in 2003 following the acquisition of the Marconi plc defence business by Finmeccanica), Chelmsford, UK.

NEW ENTRY

....

JTIDS Portable Capability

The JTIDS Portable Capability (JPC) is a fielded system developed to meet the operational needs of monitoring and actively managing JTIDS/MIDS Link 16 networks. It provides a comprehensive network monitoring and management suite. It also hosts a detailed situational awareness display, data reduction and analysis tool, alerts package and complete mission recording and replay facility.

Description
In the basic configuration JPC consists of a terminal and a network management unit. These units can be connected by either fibre optic or any bearer capable of supporting TCP/IP. A number of optional modules exist which provide a multi TDL capability along with organic voice communications (V/UHF and HF). JPC integrates with a number of JTIDS terminal types including Class II UKADGE, MIDS LVT and AN/URC 138 and allows the TDL manager to monitor the TDL architecture as well as the data being exchanged over it. It utilises the Link 16 J-series message set to rectify dynamically any faults identified or to cater for effects of changing mission profiles on the information exchange requirements. JPC also provides a comprehensive Situational Awareness (SA) suite which allows operators or battlespace commanders to view the multidomain tactical picture.

JPC hardware is designed to be modular, utilising either a single rack housing a UK Low Volume JTIDS Terminal (UKLVT) or a double rack housing a Class 2 Ground Terminal. The Link 16 Terminal Module is controlled by the Remote Processing and Display Module which can be located up to 3 km away over a fibre optic cable or may be located anywhere worldwide if encrypted modem communications is employed. Secure JTIDS voice may also be utilised from the Remote Processing and Display Module over these distances. An optional, separate Communications Module houses a UHF radio and an HF radio. This is also controlled by the Remote Processing and Display Module and may be located up to 100 m away. Growth potential is provided to accommodate a Link 11 module, which is co-located with the Link 16 Terminal Module, and has the same remoting capabilities. Each type of module is packaged as a stand-alone, rugged unit and includes air cooling where required. All modules are man-luggable. Front and back panels allow the contained equipment to be fully protected during transit. The equipment can be installed in the existing UK MoD-owned Mobile JTIDS Cabin or an all-terrain vehicle such as the Pinzgauer, as well as in ships and aircraft (see below).

The Processing and Display module hosts Ael's Link 16 network monitoring and management software – ATLAS. This provides all the tools required for JTIDS Network Managers dynamically to assess and manipulate the network to improve its performance, to detect any network violations and finally to review and record how well the network performed.

ATLAS features:
(a) Real-time Situation Awareness Display of air, surface, subsurface and ground picture with:
Complete world map
Topographical data
TACAN/DME points
JU and surveillance picture
Track Label/Tote information summary
(b) Network Connectivity Monitoring with:
Actual Connectivity Display
Theoretical Connectivity Display
Pseudo Connectivity Display allowing 'what if' changes to be made
Automated generation of J-Series messages to change connectivity
(c) Timeslot Management with:
Graphical manipulation of network time slot structure
Graphical display of timeslot allocations
Complete network violation checks for created timeslots
(d) Network Monitoring including real-time monitoring of:
Network performance/usage
Timeslot utilisation
Network/CAA violations (for example detection of illegal transmissions/TSDFs)
Man-readable datalink message display
(e) Post operation analysis including:
Complete record and replay capability
Network performance summary
Violation summary
Network summary data provided as Microsoft Excel spreadsheets.

Tanker fit

An early version of JPC was temporarily fitted to RAF TriStar tankers in 1999. This capability provides:
(a) Situational Awareness for the aircrew, enabling them to position the aircraft in support of current activity
(b) Receiving aircraft can 'see' where the tankers are
(c) Tankers can see friendly and hostile track data
(d) Fuel off-load status from each Tanker is transmitted on the network
(e) Dynamic planning is possible according to current situation requirements.

In 2001 in response to an Urgent Operational Requirement to support operations over Afghanistan, a JPC system was provided using AN/URC-138V, a GOTS Link 16 terminal, a Barco MPRD126 High Bright/NVG display, and a small keyboard (discovered in use in a US police department's vehicles) and was installed in both TriStar and VC10 tankers.

The TriStar fit has the display, hard-mounted in a bespoke structure and keyboard sited for operation by the Co-pilot and/or Flight Engineer, with the keyboard stowage at Flight Engineer station. The control switches and indicators are mounted below display. The VC10 fit has the display and keyboard sited for operation by the Navigator, with the display mounted in bespoke structure stowable for take-off/landing, the keyboard stowed in a soft pouch and the JTIDS control switches all at the Navigator's station. The crate containing the terminals is mounted in the forward freight bay.

Ship fit

Also in response to a UK Urgent Operational Requirement, a JPC system is currently providing a non integrated Link 16 System for the RN's CVS and some Batch II and III Type 42 destroyers. This followed a successful trial on the Type 23 frigate HMS *Richmond* in early 2002. This solution utilises the Network Control and Initialisation Data Preparation Facility (NCIDPF) and is an interim fit until Data Link Processor System (DLPS) functionality is available. A stand-alone solution using terminals no longer required for Sea Harrier is also being fitted to Batch I Type 42 destroyers. For the former, displays are provided in the Operations Room and FlyCo and, in the latter, three displays are provided in the Operations Room for the AWO/Command, the PWO (see illustration) and TPS.

Status

In service with the RAF, one JPC is operated by staff from the UK Data Links Operations Centre (DLOC) which is the specialist datalink element supporting the UK Joint Force Air Component Headquarters (UK JFACHQ) based at RAF High Wycombe. In October 2001 the JPC was deployed to Oman during Exercise Saif Sarea II to carry out theatre level TDL management and supply a situational awareness display to the JFAC HQ. A second JPC is operated by Number 1 Air Control Centre (No 1 ACC) at RAF Bulmer to support deployed operations.

The RAF tanker fit is currently in operational use.

The ship interim fit is installed in the CVSs HMS *Illustrious* and *Ark Royal* and in the Type 42 Batch 2 and 3 destroyers HMS *Edinburgh, Manchester, Southampton, York* and *Nottingham*. The stand-alone solution is installed in HMS *Cardiff*.

JPC had also been sold to NATO (SHAPE) for delivery by the end of 2002, to be operated from a mobile datalink van during exercises, and to Norway for delivery in 2003, where it will operate as a ground station for the testing and monitoring of the JTIDS network for F-16s.

The network management system is to be incorporated into the future Type 45 frigate and in submarine upgrades for Greece, Italy, Turkey and Kuwait.

Contractor

Aerosystems International Ltd, Yeovil, Somerset.

UPDATED

Link Inter Operability Network (LION)

The Link Inter Operability Network is a testing facility developed in conjunction with the UK MOD. It simulates a Link network and enables interoperability tests to be conducted between geographically separated sites using encrypted ISDN lines, using the NATO agreed Standard Interface for Multiple Platform Link Evaluation (SIMPLE) (STANAG 5602) format for data transmission. It therefore avoids the use of live radio transmissions and saves air trials while still using real hardware, which is COTS. A central scenario facility provides scenario data to individual rig interface units, controlling the sensor interfaces and the datalink interfaces at each individual rig. It simulates Link 11 and Link 16, and has an expansion capability for Link 22 and JRE. Current platforms contracted include E3D, Tornado F3, Sea King AEW and T42/CVS; future platforms include ASTOR, Typhoon and the new UK Tanker.

The system can also communicate through an international ISDN line to allow interoperability testing between Link equipped forces in other countries. The facility is also provided as a mobile package in a Pinzgauer vehicle, which is air portable in a C130.

Status

In current use, run under joint Industry/ UK MoD management as a Public Private Partnership. In March 2003 it was announced that LION had been ordered by Norway.

Contractor

Aerosystems International Ltd, Yeovil, Somerset.

UPDATED

The LION Pinzgauer (Aerosystems International)   0116327

An internal shot of the LION Pinzgauer (Aerosystems International)   0116328

Marconi Managed HF System

The Marconi Managed HF System comprises traffic, frequency, link and asset management functions linked together by a network control system, eliminating the need for skilled radio operators. The system provides transparent subscriber-to-subscriber connectivity and supports multiple radio station nodes, which can be accessed by remote subscribers via telephone lines or LANs. This enables radio assets to be shared by many subscribers or configured as dedicated circuits.

The network control system is partitioned into a number of key areas:

Frequency Management
Each node uses a set of frequency management tools and deterministic techniques to produce pools of frequencies for the stations. At each node predictions are also produced of the LQA for each frequency, and for each link, from that node to all other nodes in the net. The stations scan these pools of

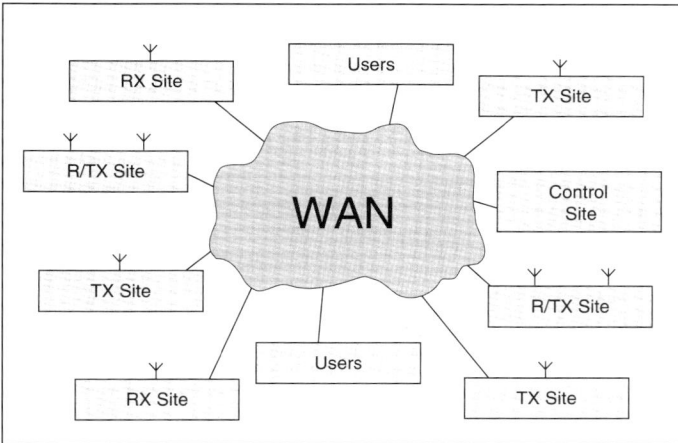

Simplified network architecture

frequencies looking for traffic. As they scan, they evaluate each frequency for the quality of overheard traffic and also for noise. In addition, whenever a station is involved in traffic, the channel performance is evaluated. Each station in the net thus builds up a picture of the HF environment. To protect the net from jamming and interception, the pools of frequencies are changed regularly. In addition, a station will avoid channels which have high noise or interference.

Link Management

The link establishment uses a synchronous ALE scheme, which allows for rapid linking and reduced on-air traffic. Three types of traffic transfer are provided:
(a) Confirmed point-to-point
(b) Non-confirmed point-to-point
(c) Broadcast:

The frequency for the message transfer is selected using the link quality information maintained by the Frequency Management facility. Forward error correction is included on all transmissions and a set of adaptation techniques, including those listed below, are used to maintain the link:
(d) Dynamic block size adjustment
(e) Interleaving depth adjustment
(f) ARQ scheme for confirmed link
(g) Auto repeat for non-confirmed and broadcast messages.
(h) Change of frequency for confirmed links.

Traffic Management

The system supports a variety of traffic types from simple text formats through to IP Datagrams. An analogue Open Channel is also available to allow any voiceband signal to be passed. Later additions will include a range of standard text format interfaces, e-mail input direct from PC and digital voice.

Asset Management

To ensure the system makes best use of its assets, arbitration functions operate within the stations at a node to ensure traffic loading is spread amongst the stations. Automatic recovery from failure conditions is also provided. Event and traffic logging allow for detailed off-line analysis of system performance.

Time Management

The system is based around a synchronous ALE scheme, and distribution and management of time is vital. A time standard is not required for system operation and one station in the net is designated as the master. The other stations are synchronised with the master, and remain synchronised by extracting the embedded time signals from messages. There is no need for regular time signals from the master station and the stations in a net can remain in synchronisation for several days without traffic.

An objective of this system is to use internationally recognised standards, in order to provide the customer with procurement flexibility, interoperability and an operating system that has an identifiable future upgrade path. Some of the standards used within the network are as follows:
(a) MIL-STD-188-141A/B
(b) FED-STAN-1045 and 1052
(c) CCITT V.24
(d) CCITT X.25
(e) CCITT X.400
(f) CCITT T4, Group 3 fax
(g) SNMP - Network Management
(h) TCP/IP - OSI links
(i) RS-232/422/485.

Status

A version of the system, known as KV90, is in service with the Swedish Army and Navy following trials during 1998-2000 on a prototype system numbering more than 20 stations, including fixed site, transportable shelter and ship-mounted installations. The system has been used in Kosovo by SWERAP, the Swedish Rapid Reaction Unit, both for operational and welfare communications back to Sweden.

Manufacturer

Marconi Selenia Communications Ltd, Chelmsford, Essex.

UPDATED

MATELO HF communications

The Maritime Air TELecommunications Organisation (MATELO) HF communications system provides SSB voice, Morse and secure telegraph traffic to support NATO communications between ground tasking agencies and mobile users within the Eastern Atlantic (EASTLANT) area. It forms a combined system with Strike Command Integrated Communications System (STCICS). Together, they provide three HF transmit/receive/technical-control complexes within the UK, providing survivability and spatial diversity. MATELO traffic originates from two control centres based in London and Scotland. STCICS traffic, which is mainly voice, originates from control authorities at numerous airfields and headquarters sites.

MATELO minimises operator involvement when establishing circuits between ground and mobile users and by providing remote control of transmitter and receiver sites from the control centre. This allows the remote transmitter and receiver sites to be unmanned and permits the equipment to be pooled and allocated to circuits on demand, rather than dedicated. The MATELO system interfaces directly to the Royal Air Force secure fixed telecommunications system to provide on-demand connections between the MATELO/STCICS control sites to reduce costs and allow more flexible connections.

Flexible use of receivers and transmitters is assisted by the inclusion of a Receive Antenna Distribution Equipment (RADE) so that any receive antenna may be connected to any one or more receivers. Similarly dual transmit antenna exchanges are provided to route traffic from any transmitter to any antenna or RF test load with 'daisy-chained' connections between the two exchanges. The radio control and switching system routes the traffic from the user to and from the radio frequency systems and provides monitoring and control of all hardware at the remote sites.

Status

Operational. Now being subsumed into the Defence High Frequency Communication Service, a Public/Private Partnership with VT Merlin.

Contractor

VT Merlin (for DHFCS).

UPDATED

MATELO 0010442

MPS2000 message handling system

The MPS2000 is a family of products which can be used as building blocks in a countrywide communication network. It consists of a stand-alone or networked Message/Management Terminal (MMT), a MiniSwitch fulfilling smaller-sized network requirements and a Network Message Switch capable of driving large multinode client-server systems. The Open System Integration (OSI) architecture on which the MPS2000 is based, gives the capability to integrate applications, information and systems from different sources into a cohesive, productive environment.

MPS2000, which is portable across a wide variety of hardware platforms, automates the preparation, control, transmission and reception of messages at fixed and mobile sites. It offers fast, accurate, accountable and efficient message delivery across all communications bearers. Its modular design, which includes the communication interface modules, allows the system the flexibility of connecting existing equipment and thereby preserving the investment. The modules can easily be interchanged, providing upgrade ability to fulfil future expanding systems requirements. The system uses industry standard Graphical User Interfaces (GUI) thereby reducing training costs as no specialist knowledge is required to operate the system.

MPS2000 provides multilevel access giving flexible control of system security. The system design has segregated the message handling software from the database, operating environment and data storage. This segregation allows third-

MPS2000, which is portable across a wide variety of hardware platforms, automates the preparation, control, transmission and reception of messages at fixed and mobile sites

party software and data storage to be individually certified which makes the system an affordable and secure message handling system.

The MMT is a message preparation terminal operating in MS-Windows, Windows NT or UNIX environments. It is designed to operate as a self-contained message handling system or as part of an integrated network.

The MiniSwitch offers the same GUI message preparation facility as the MMT with the additional ability to relay and distribute messages automatically. Message relaying can act as a gateway between separate networks operating different message formats with automatic format conversion.

The Network Switch has the capability of driving a large number of high-volume traffic ports within a client/server environment to fulfil the requirements of major communications networks. This Windows NT or UNIX-based system interconnects with any ACP-based message handling system including both the MiniSwitch and MMT. The latter operates as the client terminal to the Network Switch.

Status
MPS2000 has been selected for use by the Royal School of Signals at Blandford and by the Royal Navy for HMS *Ocean, Albion* and *Bulwark*. It has been selected as the ACP-127 messaging interface for the UK's recently awarded Naval Afloat Message Coherency (NAMC) programme.

Contractor
BAE Systems, Plymouth, Devon.

UPDATED

Pilot Direct Broadcast System (PDBS)

The Pilot Direct Broadcast System (PDBS) is a programme to explore and develop a full Direct Broadcast System capability for British forces. The aim of the system is to deliver large volumes of military information to multiple, geographically dispersed, end users in theatre, tailored to their specific requirements, on a worldwide basis over a one-way, secure, satellite broadcast system. Typical examples of uses include the broadcast of video briefings, dissemination of knowledge from defence information systems and transmission of data from civil and military geographical applications. Users will be able to specify their information requirements, while information providers will supply relevant information and field commanders will authorise and control user accesses.

PDBS equipment deployed on Ex Saif Sareea II (Jeremy Flack)   ***NEW****/0111543*

Status
PDBS was first trialled during Exercise *Saif Sareea II* in Oman in October 2001. In July 2002 *Jane's International Defense Review* reported that under Phase 1 (Initial Operating Capability), the PDBS management system had been established at the Permanent Joint Headquarters (PJHQ) in Northwood in the UK, and was complemented by two primary and one secondary receive suites, variously equipped with Vertex 2.4 m and 1.2 m dishes. The latter was expected to become the standard fit for the receive suites, the larger size being used for the In-theatre Transmission System (ITS) which was to be added later in 2002 under Phase 2 (Mid Operating Capability) of the development programme, together with a link into the Joint Operations Command System (JOCS) (see separate entry) at Northwood. In 2003, under Phase 3 (Full Operating Capability), a PDBS ship fit is due to be tested, comprising a server, operator screen, disk storage and ship-system's interface. However, during JWID 2002 in May 2002, PDBS equipment was installed aboard the CVS HMS *Ark Royal* to create an Operational Database Rapid Reaction Facility (ODBRRF), enabling data to be injected directly into its C2 and electronic warfare systems via the ship's SCOT satellite terminal (see separate entry) and the Skynet 4F satellite (see separate entry).

PDBS was used in support of the 3 Commando Brigade RM's deployment to Afghanistan in mid-2002. Product information (such as mapping and imagery data) was delivered in the form of digital video disks to the PDBS management system located at Northwood. It was then broadcast to a primary receive suite at the deployed headquarters, where the data was burned onto CDs for distribution.

Contractor
BAE Systems (Prime), Christchurch, Dorset.

NEW ENTRY

Project CORMORANT

Project CORMORANT will provide the UK's Joint Rapid Reaction Force (JRRF) with a comprehensive modern communications capability, filling a notable capability gap. Designed to link all components of a Joint Force, the system will enable the JRRF to deploy and operate its Wide Area Network (WAN) communications system in either peacekeeping roles or in a fully operational military deployment. The system is fully containerised and can be operated in either vehicle mounted or dismounted mode. Each small HQ is designed to scale up in line with the requirements of a particular operation. Additional CORMORANT vehicles can be added so that larger communities of users can be accommodated if required. As further HQs are established the system scales up, and the network management system is used to plan and implement the evolving network topology. In a large operation CORMORANT will support full and seamless networking across all environments. Within the network the system will route traffic based on least cost routing paths, and in the event of network failures or battle damage the network switching system will select alternative routing paths. The overall management of the network can be segmented or centralised at the commander's discretion.

The system consists of a circuit switch, a network management system, an ATM switch, link hardening and encryption, radio and fibre bearers, an uninterruptible power supply, and a tactical gateway. Services include fully networked secure and restricted voice and data facilities. The extensively COTS-based switching architecture provides CORMORANT with a resilient, self-healing communications networking capability and a set of software programmable gateways allows the system to communicate effectively with a wide range of peer systems. In addition to the primary terrestrial radio bearer (AN/GRC-245(V) radio from Ultra Electronics Tactical Communications Systems (previously CMC Electronics) other short-range radio bearers and fibre optic capabilities are provided and a troposcatter system (the TDS-502V produced by Comtech systems) provides 'over the horizon' communications for extended links. Strategic communications can be achieved back to UK via a variety of options including SATCOM and commercial bearers. The Management Information System provided for CORMORANT is a suite of COTS products integrated within a single tactical enclosure. These management applications are all web-browsable, thereby ensuring a common user interface throughout. The CMIS is designed to support the full operational cycle from planning and Network lay-down to operation, maintenance and repair activities and, ultimately, recovery and post-operation procedures.

The vehicle selected is the Bucher (now MOWAG) Duro, which has been selected by the company to maximise airportability capacity. Although when selected the vehicle had no commonality with other UK military vehicle fleets, following operations in Iraq it is now in service with UK forces. Each vehicle has an integrated cabin installation, provided by Marshall Specialist Vehicles, but all of the equipment is designed to be dismounted from its vehicle so that an HQ can be established in either tented accommodation or a building. Autonomous system operation is achieved by a fully redundant generator set. The generator set provides reliable uninterruptible power while also facilitating the use of 'scavenged' power from a variety of different power source standards.

A CORMORANT network can consist of the following vehicle-mounted (or dismounted) installations:
(a) Local Area Support module
(b) Core Element
(c) CLASp. A hybrid of CE and LASp
(d) Bearer Module
(e) Long-Range Bearer Module (Tropo)
(f) CORMORANT Management Information Systems

(g) Interoperable Gateways
(h) Tactical Fibre Optic cabling
(i) Short Range Radio.

Status

Originally, Initial Operational Capability (IOC) was expected in July 2002 with Full Operational Capability to be delivered in April 2004. IOC is now expected in May 2004, with final deliveries by the end of 2005. In April 2002 the first vehicles were reported to have been delivered for fitting.

Contractor

Cogent Defence Systems (Prime) (Part of EADS Telecom), Newport, Wales.

UPDATED

United States

AN/TSC-122 Communications Central

The AN/TSC-122 Communications Central is a highly flexible, self-contained HF communications system that offers a cost-effective solution for worldwide tactical mission scenarios. User-friendly, computer-controlled AUTODIN message processing and system control/monitor make the AN/TSC-122 available for traffic within minutes. Housed in an EMP-hardened S-250 shelter, the AN/TSC-122 can be deployed via commercial unit cargo vehicle or equivalent prime mover, rail, helicopter, or cargo aircraft such as a C-130. It can operate as a 1 kW fixed station or a reduced-power mobile unit and is designed to convert easily from one configuration to the other.

The AN/TSC-122 capabilities include:
(a) interoperability with existing deployed systems
(b) backward compatibility with Defense Communication System (CS) entry interim operating standards
(c) new MIL-STD-188-110CN/2 single-tone modem capabilities
(d) new MIL-STD-188-141A Automatic Link Establishment (ALE) and Automatic Link Quality Analysis (ALQA) capabilities up to 10 stations and up to 20 frequencies.

Description

The AN/TSC-122 Communications Central is an all-solid-state, 1 kW HF communications system featuring Rockwell's Spectrum 2000 radio hardware and MDM-2001 multimode modems. The message input/output and control/monitor functions are automated using separate IBM PC-AT compatible computers. Simple, intuitive menu screens and preset configuration tables allow control of all radio, modem and link parameter establishment tasks by using a single computer terminal. A second interchangeable unit provides record traffic connectivity with AUTODIN or interforce interfaces. Full radio computer control can be accomplished from the vehicle cab or from a remote location up to 16 km (10 miles) away.

The antennas for the system include a 4.88 m (16 ft) whip and three sloping-V antennas - two for diversity receive and a single transmit that can be configured to provide NVIS to multihop skywave connectivity.

The Spectrum 2000 hardware includes two four-channel receivers with preselectors capable of operating in space diversity and one four-channel exciter with post-selector, all implemented in a 178 mm (7 in) rack unit. A 1 kW power amplifier and power supply and a silent, fast tune 1 kW antenna matching unit provide the necessary RF amplification to communicate in a stressed environment. Radio control and monitoring are provided using a multifunction audio panel capable of sideband selection, microphone interface, speaker or headset selection for audio monitoring and an independent TEMPEST-approved computer.

Radiotelephone and modem interfaces are provided by two signal terminal units and three MDM-2001 multimode modems. The telephone interfaces include two- or four-wire conversion, VOX operation and are compatible with the TA-312 and TA-838 field telephones and a variety of tactical switchboards. The MDM-2001 is interoperable with most high-speed multitone, single-tone, eight-channel voice

The AN/TSC-122 is a 1 kW HF communications system housed in an EMP-hardened S-250 shelter

frequency carrier telegraph and various-speed single-channel radio teletypewriter modes.

The AN/TSC-122 uses three modems to satisfy the interoperability requirements for 10 system modes. Each modem, under computer control, provides up to eight different modes of operation. Excluding the 110 mode, each modem is also capable of handling two modes simultaneously. Through the use of LSI technologies (one microprocessor chip per modem), each modem can be programmed for a maximum of 17 possible configurations.

The AN/TSC-122 can be operated in the clear and secure voice and data modes. The system uses the KY-65A and KG-84A or KG-84C cryptographic units. Message/data entry storage and printing is accomplished using a TEMPEST-approved printer and laptop computer.

Specifications

Operational characteristics
(a) Transmitter power: 1 kW PEP or average RF output adjustable in 256 steps
(b) Frequency range: 2-29.9999 MHz
(c) Tuning increments: 100 Hz
(d) Tuning mode: fully automatic
(e) Frequency stability: 1 part in 10^9/day over specified temperature range
(f) Frequency tune time: transmitter - 1 ms (exciter nominal); receiver - 1 ms (nominal)
(g) Modes of operation: USB, LSB, AME, CW/FM and 4-channel ISB
(h) ISB information types: voice, data, teletypewriter
(i) Remote control: full radio and ALE control from a 16 km (10 mile) location
(j) Telephone: local TA-838; DTMF signalling, dialling and ringing; 2- or 4-wire conversion; VOX key capability
(k) Modem: 8-channel VFCT (1,290 equivalent)
(l) Modes: single-channel FSK (MD-522 equivalent); time-diversity (MD-1142 equivalent); 16-tone programmable in accordance with MIL-STD-188; high-speed programmable (MD-1061 equivalent); wire-line and high performance, 39-tone/single-tone in accordance with MIL-STD-188-110, Change 2
(m) Data interface: channels; programmable FSK, 16 channels; MIL-STD-188-114
(n) Internal message I/O: AUTODIN modes I, II, V (Category I, II, III certified); 75-2,400 bit/s HF; AN/UGC-74/129 interoperability; 75-1,200 bit/s asynchronous; KG-84A/C for secure traffic
(o) Antennas: 4.88 m (16 ft) whip, short/long-range sloping-V with receive space-diversity capability
(p) Power requirements: 120/208 V AC ±10%, 47-63 Hz commercial and/or military power and 28 V DC for mobile voice operation

Environmental characteristics
(a) Temperature: −45.6 to +49°C, operating; −51 to +68.3°C, non-operating
(b) Operating humidity: tropical (to 95% RH)

Physical characteristics
(a) Shelter: S-250
(b) Weight: <1,633 kg (3,600 lb) with equipment and personnel
(c) Dimensions: Length: 2.21 m (87 in); Width: 2.01 m (79 in); Height: 1.78 m (70 in)

Status

First produced in 1990 for the US Army and Marine Corps. No longer in production.

Contractor

Rockwell Collins, Cedar Rapids, Iowa.

VERIFIED

AN/USC-42 UHF SATCOM and LOS communication set

The AN/USC-42 Miniaturised Demand Assigned Multiple Access (DAMA) set is a downsized member of the TD-1271 terminal family. It achieves interoperability with the US Navy's TD-1271B/U multiplexer and AN/WSC-3 and the AFSATCOM system. The Mini-DAMA will function in nine operational modes. Among them is 25 kHz UHF SATCOM, where the system will support navy TDMA-1 network operations and non-TDMA communications. On 25 kHz LOS channels, it will support short-range tactical communications. On 5 kHz UHF SATCOM channels, it

AN/USC-42(V)1 set

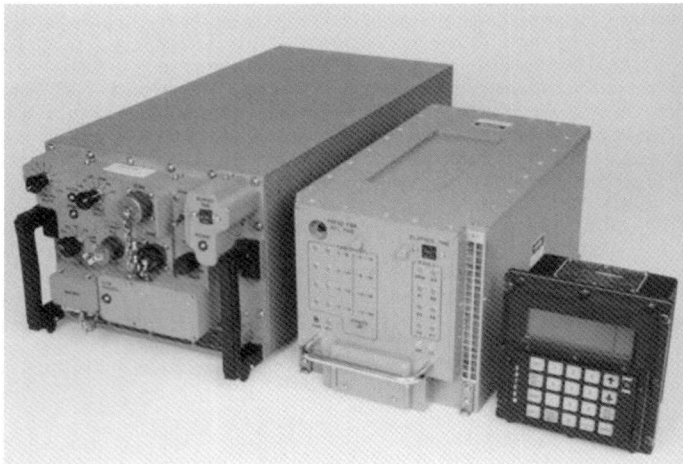

AN/USC-42(V)3 set

will interoperate with US Navy non-TDMA communications, US Air Force DAMA network operations, US Air Force non-TDMA communications and AFSATCOM I network operations.

The US Navy's Fleet Broadcast (FLTBDCST), CUDIXS/NAVMACS, SSIXS, OTCIXS, secure voice, and TACINTEL subsystems use Mini-DAMA for data exchange. Mini-DAMA modem/receiver/transmitters come in two configurations: 483 mm rack for ship and shore installations ((V)2) and as a 1 ATR long package for aircraft ((V)3). Principal components of the Mini-DAMA system include an integrated Modem/Receiver-Transmitter(R/T), a Power Amplifier (PA) and (in the aviation version) an external Display Entry Panel (DEP). The Mini-DAMA provides two, simultaneous, full-duplex, satcom or LOS channels. Data rates are up to 9.6 kbps over 5 kHz channels and 56 kbps over 25 kHz channels and it provides 8 full-duplex or 16 half-duplex baseband i/o ports per channel. The terminal is controlled through one of four remote ports using commercial computers, a DEP, or VT-100 across serial interfaces, ethernet LAN, or 1553 bus.

Status

Development contracts awarded in 1989 for terminals for submarine and airborne applications. Engineering models in mid-1992. A US$10 million production contract was awarded in March 1994. A follow-on contract for US$2.3 million for Mini-DAMA terminals for the US Navy was awarded in 1999.

Manufacturer

Titan Systems Corporation, San Diego, California.

UPDATED

AN/USQ-125(V) Link 11 data terminal set

The AN/USQ-125(V) is a programmable Link 11 (TADIL-A) data terminal set which provides all required modem and network control functions in a TADIL-A/NATO Link 11 system using either HF, UHF or SATLINK radio equipment. The terminal can also be utilised as the Signal Processor Controller (SPC) in a Link 22 tactical data link system by loading appropriate Link 22 software into the Link Processor. The AN/USQ-125(V) meets the data terminal set requirements of MIL-STD-188-203-1A. The AN/USQ-125(V) also provides the new Link 11 Single Tone Waveform (SLEW). With the SLEW, Link 11 can operate effectively in HF single-sideband channels in the presence of skywave multipath interference. The AN/USQ-125(V) may be operated in either waveform as a picket or net control station in a TADIL-A net. As a net control station, the AN/USQ-125(V) accepts addresses from the tactical data computer or from remote-control facilities.

The AN/USQ-125(V) provides all the modes of the Link 11 (TADIL-A) system including net control or picket, high and low data rate, net test, net sync, short broadcast, long broadcast and full-duplex (for single-station system tests). Doppler correction circuits, which operate independently on both sidebands, are operator selectable in conventional waveform operation.

The AN/USQ-125(V) is available either as a self-contained package in a 19 in rack-mountable unit or as a set of two 6U-VME cards. The rack-mounted unit includes a power supply, BIT indicator and connector set on a read panel. The VME card set can be hosted in any VME operating environment. The AN/USQ-125(V) link control processor card can operate as a VME slave or controller. The parallel (NTDS) and serial (ATDS) computer interface cards always operate as slaves to the processor card.

All system control, address selection, link monitoring and self-test functions are externally controlled from a computer or terminal over a MIL-STD-188-114/RS-232C compatible asynchronous remote-control interface. Operational remote-control software programs, which run under DOS, UNIX or WINDOWS, are available. The AN/USQ-125 (V) remote-control facility is a set of menus displayed on the remote terminal. All the menus are stored in AN/USQ-125 (V) memory and the only function of the remote terminal is to tell the AN/USQ-125(V) which menu item is selected. Selection on the remote terminal is by touchpanel, mouse, trackball, cursor and so on. To put the AN/USQ-125 (V) into operation with a minimum of operator action, a single key is provided which selects a preset operating mode.

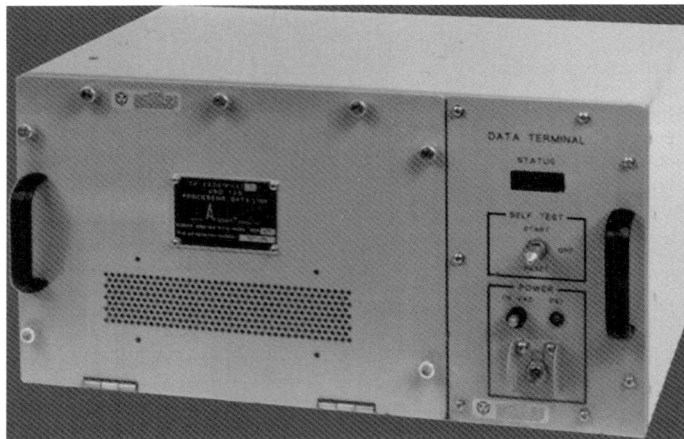

AN/USQ-125 Link 11 data terminal set　　　　0055016

The building blocks of the AN/USQ-125 (V) are a link control processor card, an NTDS data converter card and an ATDS data converter card. The rack-mounted unit may be supplied with either or both of the NTDS and ATDS cards. Built-in test provision includes loopback functions that verify operation of the system and can be used to isolate faults in other components of the Link 11 system. An 'echo memory' allows the tactical computer, either serial or parallel, to perform a single-station POFA test with data encryption. The AN/USQ-125-(V), whether NCS or picket, maintains a log of all activities for each PU in the net and displays a summary screen on the remote-control facility indicating signal quality and interrogation statistics. As an option, additional link quality information and the power spectrum can be displayed for either received sideband for each PU in the net. This option is called the Extended Link Quality Analysis (ELQA). A Maximum Useable Frequency (MUF) prediction function for HF links is also provided. This function is useful since the Single-Tone Waveform performs well on skywave paths.

An optional 2,400 bps, full-duplex, RS-232C compatible wireline/satellite interface transmits and receives compatible TADIL-A data in digital form. TADIL-A data may be sent over satellite, wireline or other tactical circuits. A software option provides operation in either the 'digital' mode, the 'conventional' mode, or in a 'mixed' mode, where some pickets operate in the digital transmission mode and some in the conventional HF or UHF mode in the same net.

Another wireline interface option is the AN/USQ-125(V) 'Split DTS' configuration. Data terminal sets at an operations centre and at a remote ground entry station are interconnected by a wireline, satellite, or terrestrial 2,400 bps link. The 'Split DTS' appears to the Link 11 net as a single participating unit meeting the specified system timeouts. The 'Split DTS' concept eliminates the need for expensive, conditioned audio lines between remote ground entry stations and operations centres and operates irrespective of the time delays encountered between the local and remote sites.

Status

In service with the US Navy.

Contractor

DRS Communications Company, LLC, Wyndmoor, Pennsylvania.

UPDATED

DCS Mediterranean Improvement Program (DMIP)

DMIP satellite upgrades are designed to replace the ageing terrestrial Defense Communications System Network which consisted of analogue microwave and tropospheric scatter communications facilities from the 1960s. The eight Turkey-based DMIP earth stations located at major US and Turkish General Staff (TGS) facilities are designed to provide combined high-speed voice and data transmissions services over a broad range of digital bandwidth options.

Inside the Golbasi facility

Within Turkey, the DMIP network interconnects the Turkish PTT and US and TGS facilities with voice and data circuits relayed via an Intelsat satellite. Turkish locations include Sinop, Cakmakli, Izmir, Marmaris, Incirlik, Erzurum and Pirinclik, each linked to a 'hub' facility in Golbasi.

The Turkish PTT monitors and controls the DMIP network from a Network Control Centre at Golbasi. This 'hub' facility also contains the engineering service channel order-wire which will maintain voice contact with all TGS, PTT and US technical control facilities in the network.

The initial DMIP satellite connection was cut over in May 1990. It links the Golbasi satellite communications centre, located about 40 km south of Ankara, with the Incirlik military airbase at Adana.

Status

Comsat Systems Division was selected prime contractor by the Turkish PTT for systems engineering and integration of the new network in March 1989.

As prime contractor and systems integrator, Comsat is responsible to the Turkish government for systems and technical requirements. This includes network design, equipment selection, procurement and implementation. Working with Comsat are three Turkish firms, Endem, Toker and Yelpro, which are supplying installation and facilities construction, in-country logistical and administrative services. Satellite Transmission Systems Inc is providing radio frequency terminals and monitoring and control portions of the system.

Contractor

Comsat General Corporation (part of Lockheed Martin), Bethesda, Maryland.

UPDATED

Defense Information System Network (DISN)

The DISN is the backbone worldwide communications network for the US DoD for voice, data, and imagery. The DISN infrastructure consists of a CONtinental United States (CONUS) segment including sustaining bases, European and Pacific theatre segments, a space segment, and a deployable capability. DISN is the primary carrier for all DoD value-added services such as the Defense Message System, the Global Command and Control System, and Electronic Commerce/ Electronic Data Interchange capabilities. Whenever possible, telecommunications services are used from the Federal Telecommunications System (FTS2000) contracts; however, the many military unique features of DISN require that DISN be a DoD acquired and controlled network.

The DISN uses a three-layer model to define the different areas of network management (NM) responsibility. The top management centre is referred to as the Global Control Center (GCC) which is operated by the DISA C4I Network Systems Management Division (D31). The GCC provides management oversight for the deployed networks of the Defense Information Infrastructure (DII) for which DISA has NM responsibility. The second layer comprises the Regional Control Centres (RCCs). The RCCs are responsible for the day-to-day operations of the networks under their immediate control. They are geographically oriented with several centres dispersed across the US, a centre located at the DISA European facilities to cover Europe, and another located at the DISA Pacific facilities to cover the Pacific assets. The RCCs are responsible for the DISA assets within their areas and operate as peers to each other. The RCCs and the GCC are responsible for DISA assets only. The third layer of the hierarchy model is the Local Control Centres (LCCs) which belong to the individual subscriber communities. These management centres control or monitor the assets owned by the individual service/agencies connected to the WANs.

The Defense Information System Network provides a wide range of information services to DoD users, including voice telephony, formal messaging, data networking and video.

(a) Data. The DISN provides interoperable, secure Internet Protocol (IP) and Asynchronous Transfer Mode (ATM) data communications services. The Unclassified but Sensitive IP Router Network (NIPRNet) provides seamless interoperability for unclassified combat support applications, as well as controlled access to the Internet. The Secret IP Router Network (SIPRNet) is the US DoD's largest interoperable command and control data network, supporting the Global Command and Control System (GCCS), the Defense Message System (DMS), (see separate entries) collaborative planning and numerous other applications. Direct connection data rates range from 56 kbps to 155 Mbps for the NIPRNet, and up to 45 Mbps for the SIPRNet. Remote dial-up services are also available, ranging from 19.2 kbps on SIPRNet to 56 kbps on NIPRNet. The DISN ATM Services (DATMS) provide unclassified ATM services and the DISN ATM Services-Classified (DATMS-C) provide secret ATM services to support unique customer requirements at data rates from 1 Mbps to 155 Mbps.

(b) Secure Voice. Secure voice services are provided using the Joint Staff Defense Red Switch Network (DRSN). This global, secure voice service provides the President, Secretary of Defense, Joint Chiefs of Staff, combatant commanders and selected agencies with command and control secure voice and voice-conferencing capabilities up to the Top Secret SCI level.

(c) Transport Services. The DISN provides a variety of voice, video and data transport services for classified and unclassified users in the continental United States (CONUS) and overseas (OCONUS). It supports customer requirements from 2.4 kbps to 155 Mbps (OCONUS) and 2.5 Gbps (CONUS). Its best-value network solutions include inherent joint interoperability, assured security, redundancy, high reliability/availability, 24/7 in-band and out-of-band network management, engineering support and customer service.

(d) Video Services (VS). The DISN provides interoperable dial-up and dedicated subscriber services for point-to-point and multipoint video teleconferencing. In addition to connecting unclassified Video Teleconferencing Facilities (VTF), DISN VS can support up to, and including, Top Secret bridging requirements.

(e) Voice and Dial-up Services. The DISN provides global voice services through the Defense Switched Network (DSN), a worldwide private-line telephone network. Multilevel Precedence and Pre-emption (MLPP) capabilities on the DSN utilised by command and control users ensure that the highest-priority calls achieve connection quickly, especially during a crisis. The DSN also provides global data and video services using dial-up switched 56 kbps or 64 kbps Integrated Services Digital Network (ISDN) services. Secure voice services are provided by the Secure Telephone Unit, Third-Generation/ Secure Terminal Equipment (STU-III/STE) family of equipment that provides end-to-end encryption over non-secure DSN circuits. Interfaces are provided between strategic and tactical forces, allied military networks and Enhanced Mobile Satellite Services (EMSS).

Status

DISN is being procured at a total cost of US$5.6 billion, and went into full rate production in 1997. Procurement and development continue.

UPDATED

Defense Message System (DMS)

The Defense Message System (DMS) is the designated messaging system created by the US Defense Information Systems Agency (DISA) for the Department of Defense (DoD) and supporting agencies. It is a commercial-off-the-shelf (COTS)-based application providing multimedia messaging and directory services using the underlying Defense Information Infrastructure (DII) network and security services. It has replaced AUTODIN for general services traffic. The system will continue to develop through a combination of the refreshment of commercial technology and the development of unique capability to meet DoD requirements.

Installed and operational at 270 military installations worldwide, DMS provides message service to all DoD users (to include deployed tactical users), and interfaces to other US government agencies, allied forces and Defense contractors. DMS provides two grades of enabled service: high and medium. High Grade Service provides organisational messaging/record traffic and replaces incompatible, unsecured e-mail systems. Medium Grade Service, a protected messaging capability for individuals, leverages the installed base of COTS e-mail products that are administered as standard network applications across DoD. It was to have replaced the legacy secure messaging system, AUTODIN, by September 2003 as well as replace more than 45 disparate e-mail systems in use within the DoD.

DMS Transition Hubs (DTH) provide a continuing capability to continue to satisfy legacy messaging requirements, allied and tactical interoperability and emergency action message dissemination. They also support Top Secret/Collateral messaging.

The programme was established in response to US Joint Staff validated requirements for an integrated common-user, writer-to-reader organisational and individual messaging service. The twelve requirements are: connectivity/ interoperability, message delivery, timely delivery, confidentiality/security, sender authentication, integrity, availability/reliability, training, identification of recipients, message preparation support, storage and retrieval support, and distribution determination and delivery.

DMS and Multilevel Information System Security Initiative (MISSI) Components

E-mail Client (User Agent, UA): The client is the user's interface to the DMS. The client allows users to compose, digitally sign, encrypt (if required), and transmit messages. The client is the means for a user to receive, decrypt, and read messages. The client is interoperable with the DMS directory service providing users with the ability to address messages to anyone in the DoD directory. This directory access module is frequently referred to as an Integrated Directory User Agent (IDUA). Other DMS components that access the DMS directory have IDUAs.

E-mail Server or Groupware Server: The groupware server is the component that serves e-mail clients. These servers hold the mailboxes for e-mail clients, store local directory information for use by its clients, and can be configured to supply the groupware capabilities such as calendars, workflow and groupware applications available in today's commercial electronic mail product suites. The groupware server is the primary provider of interoperability services for SMTP based users (that is defense contractors, academia).

DMS Message Switch: A Message Transfer Agent (MTA), or message switch, stores and forwards messages across a fully interconnected switch fabric called the Message Transfer System (MTS). The switches prioritise message transfer based on each message's grade of delivery (derived from the message precedence) and provide error notification, event logging, message transfer, message delivery, delivery and non-delivery notifications, and audit trail services.

DMS Directory Service: The DMS Directory Service provides DoD wide directory capability. It consists of a distributed repository of directory information in Directory Services Agents (DSAs) and an administration tool. This information, known as the Directory Information Base (DIB), represents in aggregate, a global DoD directory.

Directory information, including user information and their associated security certificates, are replicated among DSAs located in the DMS backbone infrastructure or within the local enclave. Replication increases performance by storing pertinent directory information as close to the user as possible. Thus the directory system supports global access to directory information from all services, DoD agencies and selected non-DoD agencies (those with formal agreements with the department) with distributed administration. DSA administration is accomplished through an Administrative Directory User Agent (ADUA) which allows authorised users to add, modify, and delete entries as well as to perform maintenance of the knowledge information shared between DSAs to control replication and chaining.

Certification Authority Workstation (CAW): The CAW is used for the generation, management and distribution of keying material, including the programming of FORTEZZA cards. The CAW generates and posts user certificates to the X.500 Directory System. These certificates contain information that binds the public key of a user with the user's identity. The CAW also generates Certificate Revocation Lists (CRL) that identify certificates that have been revoked for administrative reasons and Compromised Key Lists (CKL) that identify certificates that represent a potential security risk (emergency revocations).

FORTEZZA Card: The FORTEZZA card, a Level 2 compliant cryptographic module, is the DoD Public Key Infrastructure's (PKI) current high-assurance mechanism. This PCMCIA card provides the high-assurance cryptographic services to the DMS applications. The card holds a user's private certificate and provides integrity services via digital signature (using the Digital Signature Algorithm), confidentiality services via encryption/decryption (using the SKIPJACK Algorithm, and the Key Exchange Algorithm), and authentication services via possession of the card and a Personal Identification Number (PIN) for card access.

Mail List Agent (MLA): The MLA provides group (distribution list) addressing capability for DMS. This capability is similar to an Address Indicator Group (AIG) used in the AUTODIN system and similar to the use of a distribution list or group address list found in commercial e-mail products. This function allows the originator to have a message delivered to a group of recipients by addressing a single name (the mail list). The MLA expands the mail list, adds the designated recipients to the message routing list, and provides a cryptographic mechanism for the recipient to open the message (a token).

Profiling User Agent (PUA): The PUA provides an organisation with a profiling service supporting customisable onward delivery of incoming messages. The PUA may act as the recipient for all incoming Organisational messages. When a message originator sends a message to Organisation X, the addressing and security information retrieved from the DMS directory is that of Organisation X's PUA. Once the PUA has received a message, it will automatically redistribute the message to the appropriate action officers and other parties based on the contents of the message and rule sets (profiles) established by the enclave's PUA system administrator. Starting in Release 2.1, the PUA also serves as an output device capable of releasing a message on behalf of an organisation in support of Domain FORTEZZA concepts.

Management Workstation (MWS): The MWS provides for remote monitoring and configuration of DMS products, fault and performance analysis, message tracking, and customer service.

Multi-Function Interpreter (MFI): The MFI is the primary means of providing interoperability with AUTODIN users that have not migrated to DMS. In addition, it provides interoperability with allies, tactical users, and other government agencies using legacy formats. The MFI provides translation services (availability defined by release) for approved legacy message formats such as JANAP 128.

High Assurance Guard (HAG): The Guard (sometimes referred to as the 'DII Guard') provides secure guard services between security domains (Secret and Unclassified). The Guard ensures messages released from a higher classification domain have been properly labeled and protected enabling the message to be switched to the lower classification domain. For example, a Guard separating the Secret domain and Unclassified domain will ensure that a message from the Secret Domain is labeled Unclassified before transmitting it to the Unclassified Domain. Messages labeled Secret will be prevented from leaving the Secret domain. The Guard can enforce local policy (via site/enclave selectable Guard configuration options such as Access Control Lists and filter settings) on messages entering the higher classification domain from a lower domain. The HAG does not perform virus checking for technical, performance and DII defense-in-depth architecture reasons (every message would have to be decrypted to do the virus scan; DoD enterprise licenses are already available for servers and clients). Like AUTODIN, the HAG cannot ensure a message that has been labelled Unclassified by the originator does not, in fact, include classified information.

Registrar: Registrar is a Windows NT-based software tool that facilitates remote registration. It will support high- and medium-assurance PKI registration and maintenance. The Registrar permits authorised users to register new users and add, modify or delete permissions for existing users. It cannot cut new FORTEZZA cards; a CAW (on a secure operating system) is required for that function. Working in conjunction with a Certificate Management User Agent (CMUA, a card update utility) the FORTEZZA card can be updated at the end user desktop.

Status
Ground work for the DMS began in 1988 and the US Air Force has executive agency responsibility for DMS procurement. The design and architecture for DMS were developed by DISA, with security technologies developed and tested by the NSA at Fort Meade, Maryland. DMS became operational in September 2000.

The full range of DMS operational capabilities are achieved through co-ordinated product releases. Each release is focused on a critical aspect of DMS and builds new capabilities and updates established products as part of an integrated system. Release 3.0, which began fielding in July 2002 replacing Version 2.2, addresses essential Intelligence Community requirements and provides automated access controls for compartments, codes words and caveats. Following the full fielding of Release 3.0, the DTHs are scheduled to be closed in FY03.

Contractor
Lockheed Martin Mission Systems (Prime), Owego, New York.

UPDATED

··

Defense Switched Network (DSN)

The DSN is the US worldwide interbase telecommunications system that provides end-to-end common user and dedicated telephone service, voice-band data, and dial-up VTC for the Department of Defense. The DSN is the switched circuit telecommunications system of the DISN. It provides switched dial-up secure and nonsecure voice, voice-band data, and video services to authorised users throughout the Department of Defense. It replaced the Autovon system.

Secure voice service is provided by the Defense Red Switch Network, a separate secure switched network that is considered part of the DSN, and the STU-III/STE family of equipment that provides end-to-end encryption over nonsecure DSN circuits.

VERIFIED

··

Improved Data Modem (IDM)

The IDM is a high-speed digital datalink modem that can pass near-realtime targeting data between ground-based observers, military aircraft (fixed and rotary wing), attack teams and artillery fire direction centres. It will also act as a gateway between networks, enabling ground troops (such as FACs) equipped with data radios to pass data and images to aircraft. The IDM supports the following missions:
(a) Suppression of Enemy Air Defences (SEAD)
(b) Close Air Support
(c) Forward Air Control (FAC)
(d) Air combat Joint Air Attack Team (JAAT)
(e) Intraflight Data Link (IDL)
(f) Situational Awareness Data Link (SADL)
(g) Command and Control.
The IDM provides four half-duplex radio channels that can transmit data at rates of 75 to 16,000 bps. Transmission or reception is possible on any of the four radio channels simultaneously. Each channel can be configured into one of three communications ports: analogue, digital and secure digital.

Analogue port
The IDM analogue port provides nonsecure analogue data transmission and reception via Continuous Phase Frequency Shift Keying (CPFSK) or duobinary Frequency Shift Keying (FSK) modulation and demodulation.

Digital port
The IDM digital port provides non-secure digital data transmission and reception via Amplitude Shift Keying (ASK) modulation and demodulation.

Secure digital port
The secure digital port provides secure digital data transmission and reception via a KY-58 COMSEC device. It is capable of NRZ-L digital data modulation.

IDM hardware
The IDM hardware consists of four Standard Electronic Module Format E (SEM-E) modules mounted in the IDM chassis, the Digital Signal Processor (DSP) (two per IDM), the Generic Interface Processor (GIP), and the Power Converter. Three additional SEM-E slots are reserved for future capabilities.

Digital signal processor
The DSP card uses a TMS320C30 digital signal processor to modulate and demodulate messages. Each DSP card can modulate or demodulate two half-duplex channels simultaneously. The DSP module is able to interface with a variety of radios, including the following:
(a) ARC-164 (HAVEQUICK II)
(b) ARC-182
(c) ARC-186
(d) ARC-201 (SINCGARS)
(e) ARC-210
(f) ARC-222 (IDM software provides DRA function)
(g) SINCGARS SIP
(h) KY-58.

Improved Data Modem 0055029

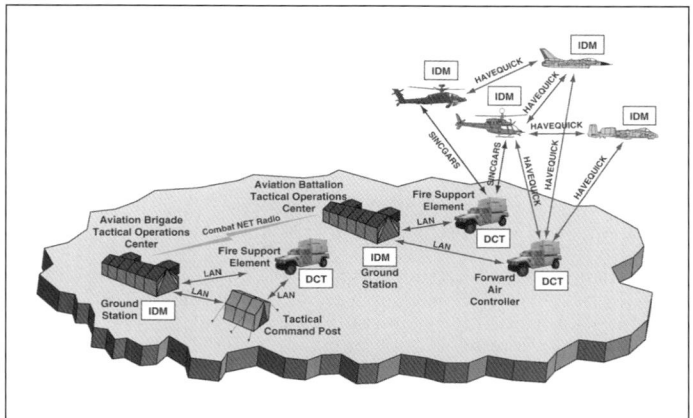

Improved Data Modem operations 0055030

Generic Interface Processor (GIP)

The GIP module uses an Intel 80960MC RISC microprocessor capable of simultaneously processing four half-duplex channels of AFAPD, MTS, TACFIRE, or VMF messages. The GIP receives and transmits messages and configuration commands, processes data and processes configuration commands.

Power converter

The power converter accepts and filters 28 V DC from a variety of host platforms. It provides +/− 5 V DC and +/− 15 V DC power to the DSP and GIP.

IDM software

The IDM includes three computer software configuration items (CSCIs):

The user interface/link protocol CSCI performs link-level and protocol-level routing and processing for the IDM and is downloaded to the GIP during power-up. Messages transmitted and received will conform to one of three selected communications protocols: TACFIRE, AFAPD and MTS. The user interface/link protocol CSCI provides user interface functions, IDM configuration and status information, Standard Memory Load Verification (SMLV), wraparound, message translation from MIL-STD-1553A or B bus and Built-In Testing (BIT).

The GIP and DSP Boot PROM CSCI, which perform initial boot-up operation and power-up testing of the GIP and DSPs.

The Modem CSCI performs modulation, demodulation and physical-level radio interface functions for the IDM. Modem software resides on the DSP and communicates with the user interface/link protocol CSCI on the GIP module. CPFSK and ASK data modulation and demodulation are supported.

Specifications

Analogue port (CPSK or duobinary FSK modulation)
(a) Data rate: 75, 150, 300, 600, 1,200 or 2,400 bps
(b) Data rate deviation: 0.1%
(c) Tone pairs: 1,300/1,700, 1,300/2,100, or 1,200/2,400
(d) Tone frequency deviation: 0.1%
(e) Input signal level: 0.1 to 6 V (RMS)
(f) Input impedance: 150 or 600 Ω (balanced, high impedance unbalanced)
(g) Output signal level: 10 mV to 1.5 V (RMS) (programmable)
(h) Output impedance: 150 Ω, balanced load

Digital port (ASK)
(a) Signal type: modulation
(b) Data rate: 75, 150, 300, 600, 1,200, 2,400, 4,800, 8,000, 9,600, or 16,000 bps
(c) Data rate deviation: 0.1%
(d) Input signal level: 0.1 to 25 V (peak-to-peak)
(e) Input impedance: >2,000 Ω
(f) Bandwidth characteristics: 0 - 20 kHz linear channel
(g) Output signal level: programmable 1-12 V (peak-to-peak)
(h) Impedance: <50 Ω

Secure port (MIL-STD-188-114, single ended)
(a) Signal type: NRZ-L
(b) Signal level: ± 5V
(c) Data rate: Up to 16,000 bps

Physical
(a) Dimensions (D × H × W): 9 × 7.4 × 5.3 in
(b) Weight: 14 lbs

Photo Reconnaissance Intelligence Strike Module (PRISM) IDM

A fifth circuit card has been developed to provide the ability to transmit and receive images via the IDM. In this configuration the equipment is known as PRISM. The main features include:

(a) The ability to capture, compress/decompress, and receive/transmit imagery data for target handover/identification and near-realtime BDA and reconnaissance
(b) Imagery capture rates of four images per second
(c) Capture and transmit images from tape playback, so the pilot can record the event in a hostile environment, then playback and transmit when safe
(d) Accessible FLASH memory for ease of upload of pre-mission images or maps for viewing en route, and post-mission download for analysis
(e) Command controlled over the 1553 bus or by RS-232
(f) The ability to transmit colour or monochrome images gives the flexibility to transmit colour radar, FLIR, day/night television, or any other video source
(g) Display of thumbnail image directories for quick review of images to reduce selection and transmission times
(h) Programmable image compression ratios for both Wavelet and JPEG that allows the proper compression ratio as determined by the mission
(i) A programmable ability to predetermine image capture rate and time
(j) Single image shot mode
(k) Abort of image transmission and reception to return radio to voice mode for high-priority voice communications.

IDM-501/IDM Junior™

Originally designed for UAV applications, the IDM-501 provides the same functionality as the IDM including the PRISM option but is half the size and weight, and generates half the heat. This has been achieved by combining two boards onto one, reducing the basic requirement for slots to two. Four slots are available, leaving two unoccupied for PRISM or other options.

This version weighs 7.5 lbs without and 9 lbs with PRISM, and measures 8.9 × 7.4 × 3.5 in (D × H × W).

Status

The IDM was originally developed by the US Naval Research Laboratory for the F-16. It is now in full rate production for a projected total of over 4,000 units. The IDM is currently in operational service with the US Air Force on F-16 Block 40 and Block 50 aircraft. IDMs are also installed on a variety of international and US platforms, including other F-16 aircraft, the Longbow AH-64D and WAH-64, EA-6B, OH-58D, JSTARS, A2C2S, UH-60Q, E-2C, Jaguar GR1/3 (UK) and several UAV platforms.

GKN Westland has procured 70 for incorporation in the UK MoD's WAH-64 Longbow Apache programme, where they have been used to integrate with Aerosystems International's Mission Planning System.

Contractor

Symetrics Industries, Melbourne Florida.

UPDATED

....

Joint Tactical Information Distribution System (JTIDS)

The Joint Tactical Information Distribution System (JTIDS) is a US joint service command and control support system providing secure jam-resistant communications and embedded navigation and identification for land, sea and air platforms. Using Time Division Multiple Access (TDMA) technology, it provides high-capacity networking among diverse airborne and surface users. It allows all stations to share an integrated awareness of the combat situation for friendly forces as well as detected threats in real time, using the tri-service multinational message catalogue TADIL-J (mirrored in STANAG 5516). It can thus provide a language-independent method to ensure co-ordinated operations on a multinational battlefield. The data received can be displayed in both symbology and language of the host nation's platform. Thus, data received by a UK system would use English as the language for a given message; an Italian system could display the same message in Italian.

JTIDS Class 1 Terminal in use at an AEGIS (NADGE) site in Norway

JTIDS operates on 51 frequencies in the 960 to 1,215 MHz frequency band, sharing with Tacan but strictly avoiding the IFF transponder frequencies which are also in the band. The JTIDS TDMA scheme breaks up time into 7.8125 ms time slots and allocates these slots to users, based on projected traffic demand. Users employ their time slots to transmit while listening during all other times, thus enabling relays of opportunity, to maximise the probability of message reception. Participants routinely inject information into the network through their regular broadcast slots without necessarily knowing who needs the information, using a broadcast-oriented architecture. Each transmission consists of a pulse stream consisting of a synchronisation preamble, a header to define the type of message and the message itself. Both predefined messages and free format messages are permissible in the system. All users can continuously monitor the data available and retain information that they require. Typical JTIDS messages from data sources provide identity, location, altitude, speed and heading. Status messages provide such information as target acquisition, weapon or stores availability, fuel remaining and equipment status.

The TDMA architecture of JTIDS provides for a highly accurate time synchronisation of the terminals which is used to support the position location function. JTIDS terminals receive messages within time slots and utilise the message to measure the Time Of Arrival (TOA). This information is then combined with the terminal's estimate of position to form the pseudo-range observation for the terminal's Kalman filter. The filter processing of the sequence of pseudo-range observations from sources with superior position accuracy permits each terminal in the community to maximise navigation performance. The distribution of precise geodetic position data establishes a JTIDS geodetic community that is correlated to the best available geodetic position reference. This periodic reporting of secure position and identification also serves to identify JTIDS users to all other JTIDS elements, even elements which operate in a receive-only mode. Identification is not limited to JTIDS users. The identification and position of other units (friendly or enemy) can be fed into the JTIDS network by radar sites, intelligence posts and other participants.

JTIDS is designed to survive the highest levels of enemy radio electronic combat. The links are encrypted with the latest approved crypto devices. Jam resistance is achieved by multiple techniques through fast hopping over all frequencies, direct sequence spread spectrum of the waveform and Reed-Solomon forward error correction. The result of the signal processing gains from this combination of techniques means that the JTIDS omnidirectional radiation pattern can offer as much signal improvement as if it were being transmitted by a highly directional antenna (such as parabolic dish), without the difficulties of beam pointing or the limitations of single path links.

The initial emphasis in the development of JTIDS hardware was on Class 1 terminals developed from 1974. Developed by Hughes Aircraft, first deliveries took place in 1977 and resulted in terminals on board the Boeing E-3A AWACS aircraft, in ground terminals in the UK Royal Air Force IUKADGE system and in transportable shelters designated as Adaptive Surface Interface Terminals (ASIT) used by the US Army and US Air Force ground control facilities. The terminal was also used in the NATO Air Defence Ground Environment. The Class 1 terminal comprises rack-mounted components occupying almost 0.26 m, weighing

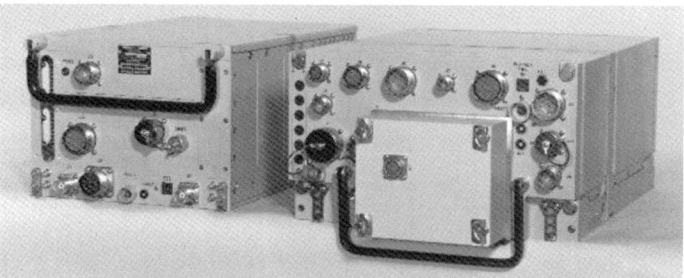

F-15 Class 2 JTIDS Terminal. The assembly on the right is the Data Processor Group which includes an interface unit and a digital data processor. The unit on the left is the receiver/transmitter

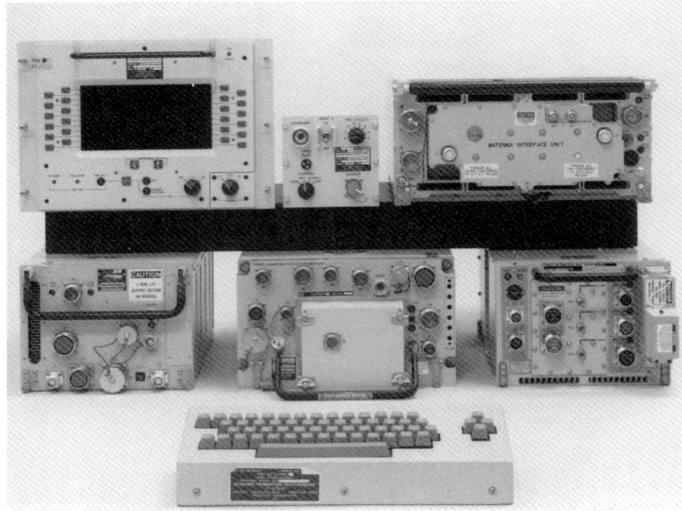

Class 2H Terminals as used aboard the US, UK and French E-3 AWACS

approximately 192 kg and designated AN/ARC-181. An improved model of this equipment was completed in 1978. The AN/ARC-181 Terminal comprised three electronic units, together occupying approximately five ATR rack spaces and a controller. Hughes later developed a privately funded JTIDS Class 1 Terminal with the aim of size and weight reduction. This resulted in the Hughes Improved Terminal (HIT) with a reduction of some 40 per cent in volume and a corresponding weight decrease. These HIT terminals were standard equipment for US Air Force and NATO E-3A AWACS aircraft.

In 1974 the US Air Force was designated as lead service and the development responsibility for JTIDS was assigned to a Joint Program Office (JPO) at Hanscom Air Force Base, Massachusetts. The JPO took over responsibility for the then current Air Force Class 1 programme and was assigned the responsibility for developing all future terminals for the US Air Force, Army and Navy. The first JPO development programme was for the Class 2 TDMA Terminal which was designated for the F-15 and F-16 aircraft and for US Army applications. After a competitive procurement process, the contract for development of this terminal was awarded to a leader-follower team consisting of GEC-Marconi, serving as prime and systems integrator and Rockwell Collins as follower.

Editor's note: Before discussing the Class 2 Terminal development it may be helpful to clarify the names of the contractors. GEC-Marconi Hazeltine Corporation was originally, for the purposes of the JTIDS development, the Singer Company's Kearfott Division. It subsequently became Plessey Electronic Systems, then GEC-Marconi Electronic Systems Corporation until, in 1996, it purchased the Hazeltine Corporation. Rockwell International's Collins Government Avionics Division is now that company's Collins Avionics & Communications Division. For simplicity the short form of their current names, that is GEC-Marconi and Rockwell Collins, are used. GEC-Marconi is now BAE Systems North America.

Class 2 Terminal full-scale development was started in January 1981. Initially, 46 terminals were procured, 18 for F-15 aircraft, one for an F-16 aircraft and the remaining 27 for army land-based use. Under subcontract, Rockwell Collins supplied transmitter/receiver subsystems to GEC-Marconi, which developed the processor and input-output units.

Subsequent contracts were let to GEC-Marconi for the development of a high-power version called the Class 2H and a smaller version called the Class 2M. The Class 2H Terminal was intended for new command and control applications and to replace the HIT terminal in the US E-3A aircraft. The Class 2M Terminal is designed for US Army ground users. The leader-follower arrangement between GEC-Marconi and Rockwell Collins has continued for these subsequent versions of the Class 2 Terminal.

In addition to the Class 2 Terminal development programme being run by the JPO, the US Navy, under the cognisance of the JPO, was developing a variant of JTIDS utilising a technology called DTDMA. A development contract for a family of terminals was let to a joint venture of Hughes and ITT in 1981. However, in October 1985, Secretary of State John Lehman cancelled this development to lower costs and increase commonality between the US services. The US Navy at that time joined the Air Force and Army in the development of the Class 2 and Class 2H TDMA Terminals. The US Navy is installing the Class 2 and Class 2H Terminals in its F-14D and E-2C aircraft, major combatant ships and the US Marine Corps Tactical Air Operations Module (TAOM) and air defence control shelters. The programme achieved a significant milestone in February 1994 when the US DoD approved the Class 2 Terminal production.

The UK also participated in the development of JTIDS and let development contracts for the design of equipment for its Air Defence Variant of the Tornado. In 1983, the UK Ministry of Defence let a contract to GEC-Marconi for the development of an interface unit for the Tornado and the IUKADGE ground air defence system. The UK and France also co-operated in the purchase of the E-3-D version of the AWACS Sentry aircraft for which these countries purchased Class 2H Terminals. All production terminals for the E-3 and Tornado ADV aircraft have been delivered.

Class 2 Terminal AN/URC-107(V)

The AN/URC-107(V) Class 2 Terminal was originally developed for the US Air Force McDonnell Douglas F-15 Eagle. Subsequent versions of the terminal have been developed for the F-14 and UK Tornado ADV.

The Class 2H Terminal incorporates a more extensive Built-In Test (BIT) capability than the Class 2 to simplify maintenance and overall logistic support. BIT will detect 98 per cent of the out of specification faults, isolate 95 per cent of the detected faults to the faulty LRU, isolate 98 per cent of them to a group of three Shop-Replaceable Units (SRUs) and 95 per cent of these to the failed SRU, all without requiring external test equipment.

Class 2M Terminal AN/GSQ-240

The Class 2M Terminal includes a number of US Army-needed features not available in the Class 2, including: a different host interface standard (the international X25 local area network interface, instead of MIL-STD-1553B), operation from a 28V DC vehicle generator and integral blower for installation without forced air conditioning. The Class 2M does not incorporate integral voice channels or TACAN functionality. Since it was designed five years after the Class 2 Terminal, technology advances allowed for integrated cryptographic keying control and for more efficient repromulgation relay and more extensive over-the-air initialisation and rekey functions. The Class 2M also has the capability for slot-by-slot switching from the normal 200 W transmit power to a 42 W transmit level. It also has the ability to switch from frequency hopping to single-frequency mode, on a slot-by-slot basis, if necessary.

MIDS

In 1987, NATO started definition of a terminal to meet its requirements. The NATO terminology for the programme is Multifunctional Information Distribution System (MIDS). The definition of MIDS is governed by two STANAGs, 4175 and 5516, which define the waveform and message requirements. The specifications essentially parallel the US requirements and JTIDS. Development of this equipment has been managed by an international programme office under the cognisance of an international steering committee. Five nations - France, Germany, Italy, Spain and the USA - are co-operating in this development, with the USA acting as the designated host (lead) country.

The MIDS-LVT programme was established to design, develop and deliver Link 16 tactical information system terminals that are smaller and lighter than, and fully compatible with, JTIDS Class 2 terminals. The participants have worked together to develop, test and produce a tactical information system terminal with reduced size, cost and weight. This is intended to make the MIDS-LVT more readily available for use in a wider variety of airborne platforms as well as maritime and ground applications. As an integral part of the programme's acquisition strategy, the MIDS-LVT system implements an Open Systems Architecture and uses performance specifications to allow for continuous competition between MIDS vendors throughout the production phase. This strategy supports the implementation of transAtlantic competition after the US and European contractors have completed qualification efforts and established full-rate production capabilities.

There are three variants of the MIDS terminal. MIDS-LVT(1) will be used by US Navy, Marine Corps and Air Force aircraft as well as by those of European nations. MIDS-LVT(2) will be used by army combat systems such as Patriot for both the US Army and military forces of France. MIDS-LVT(3),also known as the Fighter Data Link, is already operational and has been deployed with US Air Force F-15 fighter and strike aircraft.

According to the US Space and Naval Warfare Systems Command (SPANAW) as of May 2002 the US Defense Department planned to purchase 2,705 MIDS terminals up to the 2011 fiscal year. This number excludes spares and is subject to revision during each budget cycle depending on available funding. At that time the planned purchase by type was: 1,880 MIDS-LVT(1); 97 MIDS-LVT(2); and 728 MIDS-LVT(3) terminals.

SPANAW is the US Defense Department's contracting agency for the MIDS-LVT programme. The acquisition strategy includes having two US vendors and one European vendor. The US vendors are Data Link Solutions (a limited liability company comprising BAE Advanced Systems and Rockwell Collins) and ViaSat. The European vendor is EuroMIDS, a consortium comprising four companies - one from each of the European MIDS participating nations. The companies comprising EuroMIDS are Thales(France), Marconi Selenia Communications (Italy), Indra (Spain), and EADS (Germany).

At May 2002 the United States MIDS-LVT requirements were competed for annually between the US vendors with awards made using best value source selection procedures. European partner nation requirements for MIDS-LVT

US Navy shipboard terminal. The Terminal Electronic Cabinet Group incorporates the major elements of the Class 2H type terminal in a single cabinet

The Class 2 Terminal is a two-box system consisting of a Data Processing Group (DPG) and a Receiver/Transmitter (R/T), and has a TACAN capability equivalent to the ARN-118. Since JTIDS operates in the TACAN and IFF frequency bands, it is equipped with extensive interference protection features which effectively monitor radio frequency emission and prevent interference in the event of an 'out of specification' transmission.

The DPG consists of four elements: Digital Data Processor (DDP), Interface Unit (IU), Secure Data Unit (SDU) and battery. The DDP is common to all Class 2 Terminals and performs message processing, scheduling and network protocols required for the terminal operation. The IU is unique to each weapon system and provides the interface between the terminal and host platform. Any processing which is peculiar to a given user is performed in the IU. The SDU provides the cryptographic codes used to encrypt and decrypt data. The battery powers the terminal memory to maintain initialisation data during standby mode or to the SDU when loading cryptographic variables. The Class 2 Terminal incorporates extensive Built-In Test capability to simplify maintenance. It is designed to achieve 98 per cent fault detection and 95 per cent isolation to a faulty line-replaceable unit.

Class 2H Terminal AN/URC-107(V)4

The Class 2H Terminal is a high-transmit power version of the Class 2 Terminal. It was originally developed for US E-3 to replace the Class 1 equipment and has subsequently been provided for the UK and French E-3. It is currently being integrated into the US Marine Corps Tactical Air Operations Module (TAOM), US Navy E-2 Hawkeye air surveillance systems and US Navy major combatant ships.

It consists of a DPG, R/T, High Power Amplifier Group (HPAG) and, when required by a particular version, a Cockpit Display Unit (CDU). The DPG and R/T are common with the Class 2 equipment. The HPAG provides a 1 kW output and incorporates an antenna IU which can be customised to the application. The CDU, as used in the E-3 AWACS, incorporates a plasma graphic display and provides operator control and status information of the terminal.

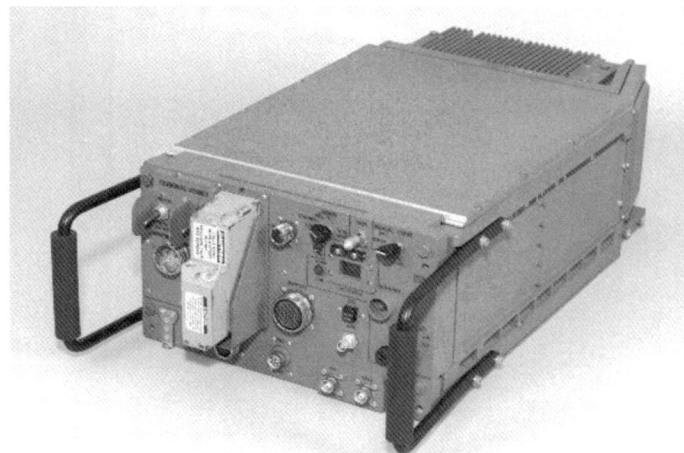

JTIDS Class 2M Terminal for US Army ground users

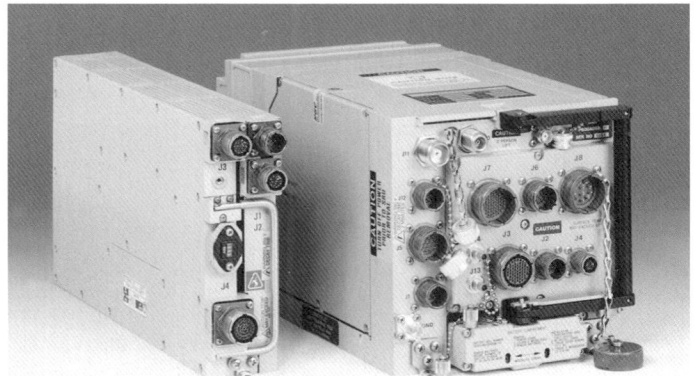

The MIDS-LVT terminal will incorporate the same functional capabilities as the JTIDS Class 2 terminal, with significant decreases in size, weight, power and cost

NEW/0044796

requirements are contracted to EuroMIDS on a directed sole source basis asset fourth in the Program Memorandum of Understanding between the five MIDS nations. SPANAW says competition between the three MIDS vendors for combined US and European requirements could begin as early as the FY 2005(US).

Others

A variety of transportable, portable and other JTIDS facilities have also been developed. (See separate entries).

Status

The initial (DT&E/IOT&E) US Air Force testing, along with US Army initial operational assessment, was completed at Eglin Air Force Base in March 1987, with JTIDS utility receiving high marks. However, concerns about hardware reliability led to additional follow-on IOT&E testing, which was completed in June 1989. As a result of this extensive test programme the F-15 JTIDS Terminal was approved by the US Department of Defense for Low-Rate Initial Production (LRIP) in September 1989. As part of the Reliability Growth Test (RGT) programme mandated by the US Defense Acquisition Board, GEC-Marconi was required to meet a 400 hour Mean Time Between Failures (MTBF) requirement before additional LRIP terminals could be procured. The company exceeded that threshold in November 1990, two months ahead of schedule.

The first production Class 2 Terminal for the Tornado F-3 aircraft was delivered in January 1993. This was the first of a total of 53 terminals ordered under the present contract. Of that total, six are ground terminals for the JTIDS Air Platform Network Management System (JAPNMS) and the remaining 47 will be installed in active duty Panavia Tornado F-3s as part of an in-service update programme. The Tornado JTIDS flight trials programme, which began in 1986, has demonstrated relative navigation, secure voice, integral TACAN and data exchange with real and simulated C2 units. On an earlier contract, GEC-Marconi had delivered the Class 2H Terminals used on the UK E-3 AWACS aircraft, which are now in service at RAF Waddington.

JTIDS terminals were mounted in the back of JSTARS aircraft during Operation Desert Storm to allow crews to monitor the air battle situation. The system tracked friendly as well as hostile aircraft and on three occasions prevented possible mid-air collisions between Coalition aircraft. JTIDS terminals logged more than 350 hours of flying time, becoming a piece of 'mission-essential' JSTARS equipment, whose operation had to be verified before the pilots would take off.

The Class 2M Terminals, the Net Management Station and Dedicated JTIDS Relay Unit (the latter two systems mounted in US Army Standard Integrated Command Post shelters) completed technical testing at Fort Huachuca, Arizona, in September 1992.

In March 1995, Rockwell was awarded the 60 per cent majority share of the JTIDS full-rate terminal production by the DoD. At that point the company had been contracted to supply 172 Class 2/2H terminals with spares. The following year the company also won a 60 per cent share of the JTIDS full-rate terminal production business awarded by the DoD.

In mid-1997, GEC-Marconi Hazeltine (now part of BAE Systems) announced a US$24.7 million order from Electronic Systems Command, Hanscom AFB, for 34 Class 2M terminals and associated spares and services. This contract was scheduled to be completed by 2001.

In early 2000, the US Air Force extended a contract with Rockwell for Class 2/2H terminals. At that time, the company had provided 142 terminals to the USAF and 388 terminals worldwide.

As at May 2002 the platforms for US MIDS-LVT(1) terminals were USN F/A-18 ships, EA-6B, and USAF F-16, B-2, and Airborne Laser/OSD JITC. 257 units had been ordered, including spares. The potential market included USN P-3, H-60, MH-60R, MH-60S, and USAF B-1, B-52, and F-22. Data Link Solutions had been contracted in the Lot 1 buy for 45 units, the Lot 2 buy for 134 units and the Lot 3 buy was then pending. ViaSat had been contracted in the Lot 1 buy for 45 units, in the Lot 2 buy for 33 units and the Lot 3 buy was pending.

The platforms for US MIDS-LVT(2) terminals were US Army Patriot, THAAD and AASROM. 43 units had been ordered, including spares. The potential market included US MEADS. ViaSat had been contracted in the Lot 1 buy for 13 units, in the Lot 2 buy for 30 units, and the Lot 3 buy was pending. The platforms for US MIDS-LVT(3)/Fighter Data Link were USAFF-15A/B/C/D/E. 786 units had been ordered, including spares. The F-15 aircraft was a potential user and the contractor was Data Link Solutions. The European MIDS-LVT platforms were EuroFighter 2000 (EF-2000) and Rafale. 344 units had been ordered. The potential market was Tornado, ships, Typhoon and ACCS and the contractor was EuroMIDS.

MIDS-LVT US Foreign Military Sales (FMS) included Australian F/A-18, Belgian F-16, Danish F-16, Norwegian F-16, and Swiss F/A-18. A total of 17 units had been

MIDS full-scale development terminal

ordered. The potential market was for 14,235 units. The contractor for units ordered to May 2002 was Data Link Solutions. MIDS-LVT Direct Commercial Sales had been made to the UK EF-2000, Spanish EF-2000, German EF-2000 and Dutch F-16. 107 units had been ordered. The potential market is included in FMS potential market. DataLink Solutions had supplied 104 units (12 Spain EF-2000, 16 Germany EF-2000 and 76 UK EF-2000). ViaSat had supplied 3 units (Netherlands F-16).

Contractors

Rockwell Collins Government Systems, Cedar Rapids, Iowa.
BAE Systems, North America, Wayne, New Jersey.
EURO-MIDS.

UPDATED

LST-5D (DAMA) secure TACSAT transceiver

The LST-5D (DAMA) is a transceiver/LST-5 upgrade with the Titan Linkabit Demand Assignment Multiple Access (DAMA) modem embedded. The LST-5D (DAMA) provides wideband (25 kHz) and narrowband (5 kHz) DAMA satellite communications in the UHF range and is interoperable with the LST-5B/C/E, URC-101, -110 and -200, HST-4, MST-20, WSC-3 and PSC-3. Depending on radio and modem, encryption is ANDVT, KG-84, or KY-57/58.

All existing LST-5B/C/E transceivers, of which more than 4,500 units had been fielded as of mid-1994, can be upgraded to the LST-5D (DAMA) configuration. Power is via BA-5590/U, BB-590/U or LSAD-100 AC/DC. Accessories include the AM-7175/URC 200 W PA and the PTPE-100/101 preamp.

Specifications

(a) Frequency range: 225-399.975 MHz
(b) Channel spacing: 5 and 25 kHz
(c) Frequency accuracy: ±1 ppm
(d) Weight: 11.7 kg approx

Status

Procured by USAF, USN, US Coast Guard, US Customs.
No longer in production.

Contractor

General Dynamics Decision Systems (originally produced by Motorola), Scottsdale, Arizona.

VERIFIED

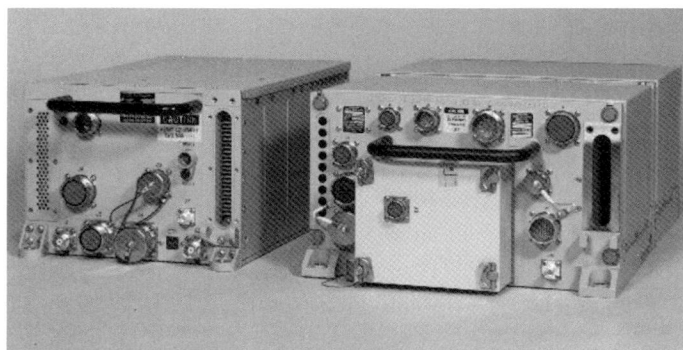

Class 2 Terminal developed for the Tornado F-3 ADV

LST-5D (DAMA)

LST-5E lightweight secure SATELLITE/LOS terminal

LST-5E (AN/PSC-10) is a small UHF tactical SATCOM/LOS transceiver with embedded COMSEC. It provides the user with a single unit for high-grade half-duplex secure voice and data over both wideband (25 kHz) AM/FM and narrowband (5 kHz) 1,200 bit/s BPSK and 2,400 bit/s BPSK. The terminal has all-weather capability and is compatible with other AM or FM radios that operate in the 225 to 400 MHz frequency band.

The embedded COMSEC provides interoperability with most existing COMSEC devices. The LST-5E meets the JCS standards for narrowband secure voice (LPC-10E) and data at 2,400 bit/s compatible with the ANDVT/KYV-5.

LST-5E performs advanced key management with receive over-the-air-rekey (SARK) from both a KY-57/58 and an ANDVT/KYV-5. Both the COMSEC and the radio transceiver can be operated independently of one another to provide the user with the flexibility to secure other VHF/UHF radios or use other external COMSEC devices via the X-Mode interface. The terminal provides the ability to preset up to six complete operating modes, including all radio frequencies, offsets, COMSEC operating modes, key variables, data rates and so on.

Current applications for the LST-5E include manpack, vehicular, airborne, shipborne, remote and fixed stations.

The LST-5E utilises existing accessories for the LST-5B/C equipment including mounting hardware and shock trays. A modification kit is also available to upgrade existing LST-5B or LST-5C radios to the configuration of the LST-5E.

Specifications

(a) Modes: AM and FM, voice, cipher, data and beacon; 1200 BPSK, 2400 BPSK, 2400 SBPSK data, non-differential or differentially encoded data
(b) Frequency range: 225-399.995 MHz
(c) Channel spacing: 5 and 25 kHz
(d) Power output: 2-18 W adjustable in 2 W steps for FM, PM; 2 or 5 W AM
(e) Power source: BA-5590/U; BB-590U; LSAD-100 AC/DC supply
(f) Operating temperature: −30 to +50°C
(g) Dimensions: 109.2 × 152.4 × 276.9 mm (H × W × D)
(h) Weight: 5 kg

LST-5E terminal

LST-5E in use

Status

The LST-5E was operational in Bosnia, NATO having procured 106 units for IFOR. No longer in production.

Contractor

General Dynamics Decision Systems (originally manufactured by Motorola), Scottsdale, Arizona.

VERIFIED

LYNXX transportable Inmarsat-B earth terminal

The LYNXX transportable terminal is understood to have been the world's first commercial Inmarsat-B mobile earth station. It was introduced by ViaSat Technology Corporation in 1992 and received Inmarsat-B Type Approval in 1993. A LYNXX unit provides global access to the latest Inmarsat all-digital technology, with on-demand, dial-up service and ISDN connectivity. The LYNXX features as standard 56/64 kbit/s voice, high-quality 16 kbit/s voice, support for 9.6 kbit/s Group III fax, medium-speed data to 16 kbit/s, Group IV fax, compressed video and encryption for secure communications.

A remote antenna kit allows the antenna to be located at up to 15.24 m from the rest of the equipment.

Status

Supplied to and used by the US Special Operations Command. Current operational status is uncertain.

Contractor

L-3 Communications, Satellite Networks, Hauppauge, New York.

UPDATED

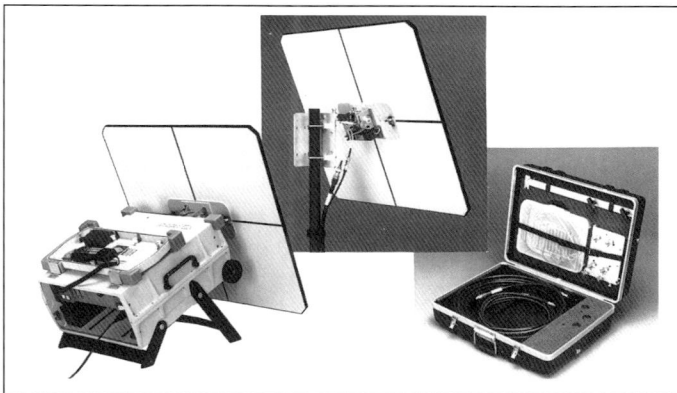

LYNXX remote antenna kit (right), LYNXX antenna mounted on pole (centre) and LYNXX antenna (left)

Mercury Wideband Network Radio

The Mercury Wideband Network Radio is an all-embracing term covering the projects that have resulted in the development of the Near Term Data Radio (NTDR) for the US Army and the High Capacity Data Radio (HCDR) for the UK Bowman project, using common technology known by the manufacturer as 'Technos'. By using networking protocols that do not require fixed base stations, the network can be formed and manage itself automatically, and allow for the movement of stations throughout the network without loss of connectivity. The WNR transports up to 288 kbps of user information for each cluster of users, backbone channel or point-to-point connection. Each unit operates at a maximum power of 20 W, providing typical operational range of greater than 20 km, and because each unit is capable of acting as an information relay point, the effective system range can be up to several hundred kilometres. The equipment has been designed using an open architecture, commercially available plug-and-play hardware and rapidly programmable software for future modifications.

WNR is designed to fit into the same space currently occupied by the EPLRS radio (see separate entry) and to use the existing EPLRS mounting tray. The radio uses two antennas, one for the normal UHF operating band and one for the embedded GPS receiver. Three user interfaces are provided: an RS-423 for commercial computer equipment; an RS-422 for router or host user equipment; and Ethernet for single user, vehicle LAN or command post applications.

Specifications

(a) Size (H × W × D): 138.4 × 235 × 348 mm
(b) Weight (including IK): 9.5 kg
(c) Adaptable Tx power: 0.25 MW to 20 W
(d) User data rate: 288 kbits/s
(e) Receiver sensitivity: −100 dBm WB-97 dBm NB
(f) Frequency band: 225-400 MHz
(g) Tuning increments: 0.625 MHz WB.125 MHz NB
(h) Speed of service (3 Hops): 0.75 s

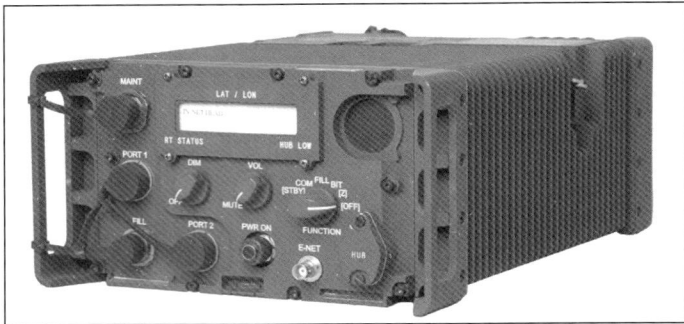

Mercury Wideband Network Radio (ITT)   **NEW**/0594413

Status

NTDR was extensively trialled by the US Army during 1997-99 as part of the digitisation programme and was subsequently selected as the data radio for the US 4th Infantry Division and 1st Cavalry Division, the first and second digitised divisions. The HCDR has been selected as the data radio for the UK Bowman programme. The technology has also been incorporated into modified, country-specific systems for the Canadian, Croatian, Dutch, German, Italian and Swedish armed forces.

Manufacturer

ITT Industries, Aerospace/Communications Division, Fort Wayne, Indiana.

VERIFIED

MX512PA Airborne data terminal set

The MX512PA, is a programmable Link 11 (TADIL-A) data terminal set which provides all required modem and network control functions in a TADIL-A/NATO Link 11 system using HF or UHF radio equipment. The MX-512PA meets the data terminal set requirements of MIL-STD-188-203-1A. The MX-512PA may be operated as a picket or net control station in a TADIL-A net. As a net control station, the MX-512PA accepts addresses from the tactical data computer or from its own front panel. The MX-512PA provides all of the modes of the Link 11 (TADIL-A) system, including net controller or picket, high and low data rates, net test, net sync, short broadcast, long broadcast and full-duplex (for single-station tests). Doppler correction circuits, which operate independently on both sidebands, are operator selectable.

The MX-512PA also operates in the improved Link-11 Waveform (SLEW) mode. The data terminal set may be controlled externally by a computer over a MIL-STD-188-114 or RS-232C compatible asynchronous control interface. The MX-512PA may also be controlled from a separate remote-control panel using menus standard to the MX-512PA DTS family. It also provides, as an option, a MIL-STD-1553B data bus interface which can provide control data and accept status and link quality data. The MX-512PA is a programmable DTS. All modem, network control and link monitoring functions are performed digitally in microprocessors using a modular multiprocessor architecture. Selection of the conventional Link-11 or Improved Link-11 Waveform is made over the remote-control interface. The single-tone waveform for Link-11 provides improved performance in HF Link-11 networks on a single sideband HF channel. Single-tone Link-11 uses an 8-phase modulated 1,800 Hz tone. Adaptive equalisation is used to demodulate the signal under the severe multipath conditions typical of HF propagation paths. Robust error detection and correction codes are used to provide enhanced message throughput.

The MX-512PA provides, as an option, a 2,400 bps, full-duplex, RS-232C satellite/wireline interface which transmits and receives compatible Link-11 data in digital form. Link-11 data may be sent over satellite, wireline, or other tactical circuits. The MX-512PA can be operated in either the 'digital' mode, the 'conventional' mode, or in a 'mixed' mode (a gateway), where some pickets operate in the digital transmission mode and some in the conventional HF or UHF mode. The DRS MX-512PA provides, as an option, link quality analysis indicators which include multipath spread, fading bandwidth, net cycle time since last reply, and tone power spectrum for each participating unit in the network. Using these indicators, an operator can troubleshoot equipment failures and configuration set-up problems in the net and determine when HF propagation problems require a change of radio frequency. Built-In Test provisions in the MX-512PA include loopback functions which verify operation of the system.

Specifications

(a) Data rates: 1,800 bps
(b) Tone library
conventional waveform: 15 data tones plus Doppler tone (TADIL-A)
single tone waveform: 1,800 Hz tone, 8-phase PSK
(c) Computer data interface: serial tactical computer interface
(d) Computer control interface: MIL-STD-188-114 at 4,800 bps
(e) Remote-control interface: RS-232C compatible at 19.2 kbps or 2,400 bps
(f) Audio interfaces: 600 Ω balanced (Rx and Tx).

Contractor

DRS Communications LLC, Wyndmoor, Pennsylvania.

UPDATED

SDS-1 command and control switching system

The SDS-1 is a field-proven, digital command and control crypto gateway switch. Designed for military command and control facilities requiring highly reliable secure communications, it is a multilevel precedence digital switching system. It provides full-feature service to a community of users in a secure (physically protected) communications environment. All trunk circuits, whether analogue or digital, are encrypted prior to transmission outside the dedicated secure area.

The SDS-1 is a modular, expandable, high-isolation, redundant, non-blocking switch. Design emphasis is on reliability and interface flexibility. The switch provides intra/intercommunications among subscriber lines, attendant consoles, trunks and data ports. It features: extremely high cross-talk isolation; unlimited digital conferencing; MLPP call processing (to subscriber level); all-digital TEMPEST command consoles and telephones; integrated networking, including Automatic Number Identification (ANI); multilevel security operation using Security Access Level (SAL); and full EPABX service capabilities.

Among the secure interfaces supported are: AUTOSEVOCOM (KY-3); STU-II (KY-71); STU-III; Tri-Tac (KY-68/78); Vinson trunk terminal; SATCOM (KY-57/58); cellular STU-III; HF/SSB (KY-65/75); LMR (KY-57/58); 56/64 kbit/s trunk (KG-84); and T1/LBT1 (KG-81/94). The non-secure counterparts are AUTOVON/DSN, PSTN, base telephone, JCSAN, COPAN, UHF LOS and SATCOM, cellular phone, HF/SSB, LMR, 56/64 kbit/s trunk and commercial T1.

NACSIM 5100A peripheral equipment supported includes: the IST 40-line red/black digital telephone; the ST-40 40-line, MLP-2 48-line and SLP-2 six-line digital telephones; the DSA-1 digital speaker assembly; the RBC red/black controller; the IST 40-line and LCC 160-line command consoles; the DPM-1 16-channel digital telephone multiplexer; and the DPA/DTA remote subscriber interface controller.

Status

Used in the Command and Control Switching System (CCSS) depot support programme.

Manufacturer

Raytheon Systems Company, Richardson, Texas.

UPDATED

SDS-1 switching system

SINCGARS (export) Communications Management System (CMS)

The SINCGARS CMS is a communication key, frequency, network and distribution management system for SINCGARS (export) ATCS users. The CMS software enables users of SINCGARS (export) radios to plan and manage communication security keys, frequencies, networks and distribution assets from corps to squad level. The software, which is hosted on a PC running Windows® XP, integrates these separate communication-planning functions.

CMS provides essential communication planning activities for secure communication needs, including compromised recovery. By using CMS key management functions, the user is able to generate keys, obtain a record of their use for accountability purposes, make sure that correct key types are assigned and verify that planned COMSEC equipment is interoperable. Key management security doctrine is implemented by performing these communication planning activities.

The frequency management functions of CMS allow the user to make better use of the RF spectrum, thereby reducing frequency interface and conflicts. The system permits the user to reconfigure frequency plans based on battlefield conditions and to manage frequency allocations, allotments, restrictions and assignments. As part of its planning function, CMS allows the user to match frequency assignments with current and future deployment scenarios.

The network management functions of CMS assist in the planning of communication asset allocation, the connectivity of users, and the interoperability of communications equipment allowing the user to plan initial, active and contingency networks.

Features of the menu-driven Windows and colour graphics system include:
(a) password security
(b) modem capability
(c) purge database capability
(d) support for several printer and plotter interfaces
(e) data transfer to other ITT CMS systems
(f) Options include a communications-operations document.

Manufacturer

ITT Industries, Aerospace/Communications Division, Fort Wayne, Indiana.

UPDATED

TIGDL stowed for movement (L-3 Communications)   0130751

Tactical Interoperable Ground Data Link (TIGDL)

The Tactical Ground Data Link (TIGDL) is a Common Data Link (CDL) interoperable surface terminal that provides the capability to receive sensor data from an airborne platform. The TIGDL is divided into two functional groupings, the Tracking Antenna Group (TAG) and the Control Processing Group (CPG). The TIGDL is the current generation of ground data link terminals and is designed to be more cost effective and performance scalable than previous generation ground terminals. The TAG includes a trailer-mounted 6 ft diameter parabolic antenna with tracker, antenna pedestal, power conditioning and RF/IF electronics. The trailer is capable of being towed by a High Mobility Multi-Wheeled Vehicle (HMMWV). The CPG consists of a notebook computer and a VME chassis of digital electronics. The CPG is mounted in a standard 19 in rack that may reside in a shelter mounted on a HMMWV. The TAG and CPG are connected via two fibre optic cables that provide for easy connection and remoting of the antenna as desired.

TIGDL cost effectiveness is achieved through the utilisation of commercial-off-the-shelf (COTS) components and standard architectures such as VME/VXI and Asynchronous Transfer Mode (ATM) where applicable. The TIGDL consists of more than 50 per cent COTS components and provides flexibility and scaleability through its modular design to add and/or delete functions depending on use requirements and constraints. The use of standard COTS architecture also provides the advantage of continuing development without costly investment.

The TIGDL satisfies the CDL Class 1 specification for the ground terminal supporting the full waveform for the CDL command link (surface to aircraft) and all three CDL return link (aircraft to surface) data rates. The TIGDL operates in the X-band frequency range and provides for the addition of Ku-band to accomplish dual-band operation. Upgraded data rates are also available.

The TIGDL interfaces to data link users via COTS standard ATM fibre optic interface compliant with the Common Image Ground/Surface Station (CIG/SS). A CDL 'legacy' interface is available in the TIGDL. This alternative can be provided either instead of, or in addition to the ATM interface.

The TIGDL can also operate as the Flexible Information Dissemination System (FINDS) ground station to provide secure transmission of data such as imagery to an aircraft relay for retransmission to the portable/mobile FINDS ground receiver.

Key features

High Bandwidth Digital Data Link
CDL compatible
Low cost COTS
Upgradable to higher data rates
Rapid communication of high volume

Specifications

Physical characteristics

	TAG	CPG
Size	165 (L) × 85 (W) × (142 (H) in Deployed) (102 (H) in Transport Mode)	24 (L) × 19 (W) × 21 (H) in
Weight	3,440 lb	100 lb
Power	1,200 W Three Phase 120/280 V 50/60 Hz	400 W single phase 120 V 50/60 Hz

Performance characteristics

	TIGDL	Enhanced
Command link		
Data rates	200 kbit/s, 2 Mbit/s	200 kbit/s, 2 Mbit/s
Modulation	Spread spectrum BPSK	Spread spectrum BPSK
Power amp	5 W SSPA	5 W SSPA
Bit Error Rate (BER) -6	1×10^{-6}	1×10^{-6}
Return link		
Data rates	10.71, 137, 274 Mbit/s	10.71, 137, 274 Mbit/s
Modulation	Offset QPSK	Offset QPSK
BER	1×10^{-6}	1×10^{-6}
COMSEC	Yes (10.71 Mbit/s)	Yes (10.71 Mbit/s)
Antenna		
Size	6 ft parabolic reflector	6 ft parabolic reflector
Frequency	X-band	Dual-band (X and Ku)
Azimuth capability	Continuous	Continuous
Slew rate	20°/s	30°/s
Operation wind	40 mph	70 mph
Remoting	500 m	10 m
Other		
2 Band operation	Optional	Yes
Frequency tuning	CDL 5 MHz steps	CDL 5 MHz steps
BIT	To module level	To module level
Link audio	Yes	Yes
UHF	No	Yes
ACA (Audio control assembly)	No	Yes
Aircraft position display/ maps	No	Yes
Ancillary equipment	Remote Spectrum Analyser only	Yes
Mission Record	No	Yes
ATM User I/F	Yes	Yes
CDL User I/F	Optional	Yes

Status

In operational service.

Contractor

L-3 Communications-West, Salt Lake City.

TIGDL Components (L-3 Communications)   0130750

VERIFIED

US Joint Interoperability Test Command (JITC) C4I testing systems

The US Joint Interoperability Test Command (JITC) employs a suite of capabilities to include the Joint Interoperability Modular Evaluation System (JIMES) for active Link 11/11B/16 testing and the Joint C4I Operational Assessment Team (JOCAT) for assessing interoperability at Joint and Service/Agency (S/A) tests and exercises.

JIMES

JIMES is a computer system designed to support testing of US DOD tactical C4I systems' JIMES Link 11/11B/16 standards compliance and interoperability in joint (US) and combined (US/Allied) operations. JITC uses operationally configured system hardware and software for data link testing. JIMES is equipped with a sensor stimulator and is capable of providing participating systems sensor tracking simulation. JIES records all the exchanged messages and produces a formatted report to facilitate post-test analysis of interface operations and system performance. For Link 11/11B/16, the JIMES interfaces with participating systems through standard communications and cryptographic equipment. For Link 11/11B/16 testing, connectivity between the remote C4I systems and their JTIDS terminals is achieved through a gateway system produced by Space and Naval Warfare Systems (SPAWAR) in San Diego. Systems are connected to JIES through leased T-1 grade telephone lines or commercial dial-up lines.

JOCAT

The JOCAT is a modular, deployable, equipment and personnel team that provides JITC the capability to conduct real-time and near-realtime (NRT) and analysis of C4I systems, to include national, strategic, and tactical data transfer for the Joint Data Network (JDN), Joint Planning Network (JPN), and the Joint Composite Tracking Network (JCTN). Additionally, JOCAT will support assessment of the Family of Systems (FoS) for Joint Theatre Air and Missile Defence (JTAMD) and system interoperability assessment and certification during CINC field exercises. The JOCAT will provide the ability to conduct NRT analysis of collected data and critical intelligence feeds essential to the evaluation and assessment of the ability of C4I systems to provide decision makers an accurate and timely single integrated air picture (SIAP). During real time, the JOCAT provides the capability to view a set of displays (graphic and text based) and the information portrayed on the displays and voice nets to determine items of interest for analysis in near-realtime. During near-realtime, the analysis is based on discrete events and anomalies noted under the direction of the real-time operator. The near-realtime analysis is also based on specific analysis to support the exercise requirements (such as JTAMD critical operational issues or objectives based on the exercise support plan).

UPDATED

LAND

Australia

Project RAVEN

Project RAVEN provided the Australian Army with a fully integrated tactical radio network of HF and VHF radios incorporating ECCM capabilities. It was designed to replace a variety of ageing high frequency (HF) and very high frequency (VHF) equipment in service with the Australian Defence Force (ADF). The system has been fielded. The RAVEN system interfaces with equipment from Projects WAGTAIL, PINTAIL, PARAKEET and Battlefield Command Support System (see separate entries) to form part of an integrated command, control and communications system to support the ADF.

The RAVEN radio system is divided into the communications subsystem (VHF, HF and common ancillaries), the frequency management facility subsystem and the maintenance subsystem.

The communications subsystem is based on versions of the RAVEN radio system which is described in a separate entry.

The frequency management subsystem provides a total HF and VHF frequency allocation and distribution system throughout the battlefield area. It provides field commanders with accurate and reliable net radio information to set up and maintain effective communications. The subsystem consists of a frequency management processor and an electronic distribution system. The frequency management processor assigns HF and VHF frequencies to all nets in a deployment and provides other radio net information allowing operators to efficiently manage communications with their radios. The initial shelter-mounted component of the FMF sub-system was replaced in late 1999 by the Frequency Management Application Software Version 2 (FMAS 2). This is a modernised software band system.

The maintenance sub-system provides a repair capability at both field and base repair levels, making extensive use of automated fault diagnosis and testing. It has the capacity to support both RAVEN and Project WAGTAIL (see separate entry) combat net radio equipment. The Base Repair Facility is located at a Government owned contractor operated facility at the Defence National Storage and Distribution Centre at Moorebank in Sydney, and this was accepted in 2001.

Status
Initial production of the HF radios commenced in May 1987 and has been completed. VHF radio production commenced in the latter half of 1993. The VHF radios were introduced into service in 1994. The final VHF transceiver was delivered to the Australian Army in July 1997.

In the late 1990s, a further tranche of RAVEN HF radios was acquired as Phase 1 of Project WAGTAIL, and further VHF radios and ancillaries were acquired in Phase 2 with Project RAVEN funding.

Contractor
BAE Systems Australia, Sydney, New South Wales.

VERIFIED

....

Project WAGTAIL

Project WAGTAIL provided combat net radios for the Australian Defence Forces that supplement and are interoperable with radios procured under Project RAVEN (see separate entry). This project represents the final phase in the ADF's upgrade of its combat net radio system. It also provides the intercommunication harnesses for combat vehicle fleets. The project was divided into two phases:

Phase 1
The acquisition of additional high frequency (HF) RAVEN radios.

Phase 2
This phase consisted of the acquisition of both radios and vehicle harnesses. The contract was awarded in December 1996 to Siemens Plessey Electronic Systems Pty Ltd, now BAE Systems Australia. BAE Systems has provided its Australian Advanced VHF radio (AAVR) which is a second generation development of the RAVEN combat net radio system and has significant commonality with the RAVEN VHF radios. Key features of the equipment include: KY-57/KG-84C compatible COMSEC with full over-the-air re-key capability; internal error correction and interleaving data rate adaptor for operation at standard rates of 150 to 9,600 bits/s; asynchronous, synchronous and isochronous data formats; an MTBF of >5,000

hours; enhanced power management for improved battery life; dynamic display allowing full radio configuration and operational status to be shown; channel scanning; COMSEC key mapping with selective zeroise; secure remote channel selector; and built-in loudspeaker amplifier. Phase 2 also acquired additional RAVEN VHF vehicle radios and associated ancillaries; these radios were sourced under the Project RAVEN contract. Also part of Phase 2 of the project was the acquisition of Royal Ordnance Vehicle Intercommunication System (ROVIS) digital intercommunication harnesses for M113s. These were supplied in 1998 by BAE Systems under a separate contract which was signed in June 1996. (In Jan 2003 it was announced that the element of BAE Systems producing ROVIS had been acquired by the Chelton Gp and renamed Chelton Defence Communications Ltd.)

Status
Production commenced in May 1997 with deliveries from May 1998 to September 2000. The equipment entered service during 1998-99.

Contractor
BAE Systems Australia, Sydney, New South Wales.

UPDATED

Austria

IFMIN area communications system

IFMIN (Integrierte Fernmeldeinfrastruktur) is a digital communications network for the automatic switching and transmission of different communications services (telephone, fax, teleprinter and data). Stationary and mobile elements form a meshed net, which can be adapted depending upon military need and situation. IFMIN was accepted by the Austrian Army in 1993. ARGE IFMIN of Austria was the prime contractor with Siemens Plessey Systems (now BAE Systems) of the UK supplying the key communications equipment based on the company's MultiRole System, MRS 2000 (see separate entry). The equipment includes circuit and package switches, network management facilities, PABX interfaces and other ancillaries. The system combines static, remotely controlled strategic switching nodes with mobile, tactical switch detachments. The system is also linked to Austrian Air Force command centres.

Status
Operational in the Austrian Army. In 2001 it was announced that BAE Systems had been awarded a contract for an unspecified amount to provide system care and the possibility of further enhancements.

Contractor
BAE Systems, Christchurch, Dorset.

UPDATED

Belgium

BAMS combat net radio system

The BAMS CNR system is designed for tactical VHF applications and has built-in protection against EW.

The radios can be operated in three different configurations: as a 5 W manpack; or as a 5 or 50 W vehicular or helicopter station. A modular construction has been chosen for the 5 and 50 W versions.

The CNR system can include a transmitter/receiver unit with removable control display, a 50 W RF power amplifier unit and the vehicular unit.

These building blocks can be used in combination with a radio system management set up with key gun, battery packs, mounting fixtures, interfaces to vehicle intercoms, audio ancillaries, remote-control systems and a variety of antennas and battery chargers.

The main BAMS radios cover the frequency range from 30 to 108 MHz, and the manufacturers say that it is the first military radio to comply with the new ARFA (Allied Radio Frequency Agency) recommendations. The radio can be tuned in steps of 25 kHz and uses a channel bandwidth of 25 kHz.

For details of the latest updates to *Jane's C4I Systems* online and to discover the additional information available exclusively to online subscribers please visit
jc4i.janes.com

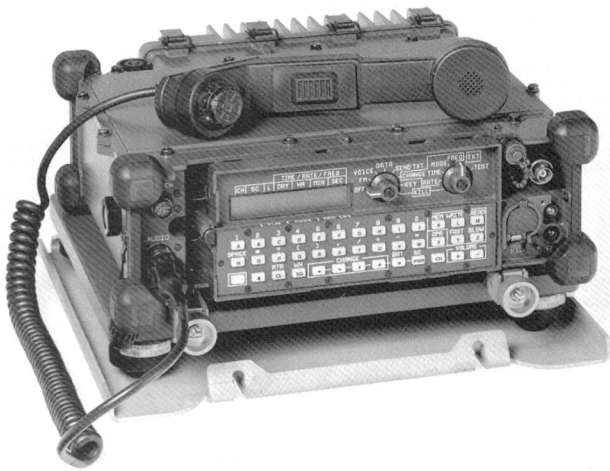

BAMS 5 W vehicular configuration

Transmission capabilities are Frequency Hopping (FH) at over 200 hop/s and fixed frequency. There are from 1 to 3,120 hop set frequencies. In fixed frequency both analogue (FM-F3E) and digital (F) frequency modulation can be used. FH and F, which both use a digital modulation technique, can also be encrypted (FHS or FS). FM has been implemented for compatibility reasons and provides for voice communication between VHF radios of the previous generation and those of the new.

As well as voice, BAMS has digital data capability to link up with computer systems and for the transmission of printed messages through teletype and facsimile. The user can connect the digital system with bit rates of 75, 300, 600, 1,200, 2,400 and 16,000 bit/s. An error correction technique is used for bit rates of 75 to 2,400 bit/s.

In addition, a text communication capability allows the transmission of short messages of a maximum of 300 characters, which can be prepared on the control display of the BAMS radio. For links suffering from severe interference, an ECCM 'slow' function is used. If the on-the-air time has to be reduced, a burst mode of transmission can be selected with the 'fast' anti-ESM function. Both methods employ adapted error correction techniques optimised as a function of the FH parameters. In severe jamming conditions when voice is disrupted, it is still possible to send a 300-character message, or switch over to Free Channel Selection (FCS) mode when it is inconvenient to input messages on the keyboard. In FCS mode the transceiver is continuously scanning the hop set frequencies and at each transmission the least disturbed frequency is selected and used. An automatic printout is available when a printer is connected to the system's RS-232 port.

Parallel to the development of the radio system, a System for the Management of Frequencies and key codes (SMF) was developed. The radios of different nets are loaded by means of key guns.

The digital modulation technique applied in the BAMS radio is Generalised Tamed Frequency Modulation (GTFM), allowing a greater concentration of the transmitted power into the relevant channel. According to the manufacturer, this gives excellent adjacent channel power characteristics.

The BAMS concept of the removable control display panel allows remote-control operation and makes it possible to change the control display panel without altering the basic transmitter/receiver. For example, a simplified control display panel without an alphanumeric keyboard can also be used. The BAMS radio can be integrated automatically into RITA or other tactical networks via an interface unit (CII).

The behaviour of the radio sets in collocation is such that no external filters or sophisticated frequency management systems are required. Optional proximity filters are available to be embedded in the 50 W power amplifier.

BAMS meets STANAG 4292, MIL-STD-461C and MIL-STD-1275 and FINABEL 2C10. SECAN approval is in progress for the NATO crypto module, for which the BAMS radio has been made TEMPEST-proof. The manpack set with battery cover measures 113 × 307 × 323 mm and weighs 7 kg with battery cover and antenna.

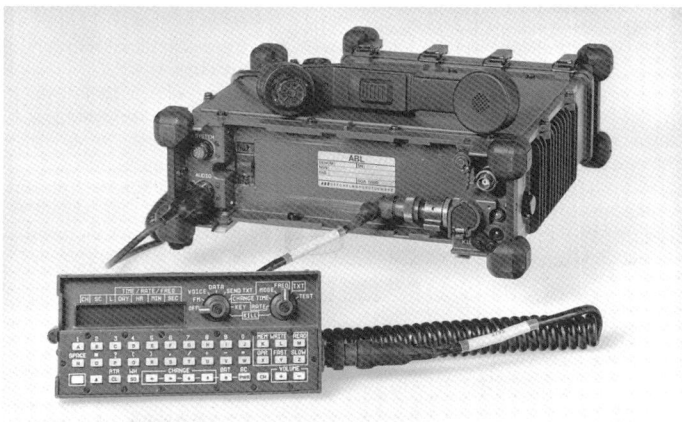

BAMS transceiver with detachable control unit/digital message device

BAMS manpack

The low-power vehicular version measures 160 × 318 × 330 mm and weighs 10 kg. Its high-power counterpart measures 331 × 350 × 320 mm and weighs 30 kg.

Specifications
(a) Modes: analogue and digital voice, digital data, text (4 receive and/or transmit, 10 receive only messages)
(b) Frequency range: 30-108 MHz
(c) Channel spacing: 25 kHz
(d) Preset channels: 7 + 1 guard for FH; 7 + 1 guard for F and hailing
(e) Clock stability: 1×10^{-6}
(f) EW modes: FH (optional plug-in module): encryption (optional plug-in module)
(g) DMD: slow rate as ECCM; high rate (burst) as anti-ESM
(h) FCS: automatic

BAMS Hand-held
Since the original range was produced, the BAMS hand-held has been added. This is a portable VHF radio with a power output of 1 W, a range of 2 to 5 km and a frequency range of 45 to 79.97 MHz . It weighs 1.135 kg and has a 90 cm whip antenna. Power is provided by NiCad battery (10 h) or alkaline cells (15 h).

Status
Development, funded by the Belgian government, started in 1983. The Belgian Army placed a substantial order at the end of 1989 for first deliveries in 1992. Export sales of the equipment were being pursued. The total Belgian requirement was around 7,000 sets. The equipment was fielded by the Belgian Army beginning in 1993 and was officially adopted at the end of 1993.

The BAMS Association originally comprised NV Alcatel Bell Telephone, Antwerp; Alcatel Bell-SDT SA, Charleroi; Thomson-CSF Electronics Belgium SA (now Thales Communications), Tubize; SAIT Systems SA, Brussels.

Contractor
BAMS Association, Brussels.

UPDATED

BAMS 50 W vehicular installation with local remote control

Canada

Iris communications system

The Iris System is the Canadian Army's new digital communications system, comprising a complete revision to current voice and data distribution using radio, telephone, computers, custom software, fibre-optic and wireless networks, trunk, satellite, and encryption and management applications. Iris supports land tactical and strategic communications capabilities. It is made up of more than 200 types of equipment, including 15,000 radios, 1,500 data terminals and three major software applications. Components of the system are being installed in approximately 6,400 armoured and soft-skinned vehicles. Each of the systems or subsystems can operate independently; however, the full operational capability of the Iris System is met when they operate as a whole.

An Iris System deployment enables commanders at all levels to have access to a fully integrated, secure communications system, providing voice and data facilities. Typically, at the unit level, this could be provided by a combination of Combat Net Radio (CNR) and the Information Distribution System (IDS) and may be further linked to the Iris Trunk System (ITS) and the Long Range Communications System (LRCS).

Information Distribution System (IDS)

The IDS forms the heart of the Iris System. This system integrates all the components into a unified tactical command, control and communications system at the HQ site. It serves users at all levels in a variety of vehicle configurations. A Tactical Command Post (Tac CP), formation HQ, designated armoured vehicles, and other selected command vehicles use IDS to access secure and non-secure voice and data communications resources.

Combat Net Radio (CNR)

The Iris System CNR consists of a full range of tactical radio systems. These include radio net, point-to-point, ship-to-shore, air-ground-air, long range, and voice and data communications covering the HF, VHF and UHF bands. The equipment can be employed in vehicles, aircraft, ship or manpack configurations. The CNR subsystems provide users with the means to transmit and receive voice and data messages utilising the Tactical Message Handling System (TMHS), and the employment of Radio Nodes (RN).

The Combat Net Radio Primary (CNR(P)) is a VHF FM-radio which can be operated in fixed-frequency or frequency-hopping mode. These radios have built-in voice and data encryption and are deployed as manpacks, and low-power and high-power vehicle-mounted units. CNR(P) units are also installed in ships and helicopters for joint operations. The Lightweight Assault Radio (LAR) is a VHF FM radio with voice and data encryption. It can communicate with the CNR(P) in fixed frequency mode. This is the smallest, hand-held, front-line radio available to the army. The Combat Net Radio High Frequency (CNR(HF)) is used for guard nets, special tasks, and long-range communications. It uses an external crypto device and can be deployed as a manpack or mounted in a vehicle. Ancillary equipment includes a headset with active noise reduction and a battery charger capable of charging different sizes of battery simultaneously.

Iris Trunk System (ITS)

The ITS allows users to access other users of the trunk system through the various nodes and the IDS. It operates over fibre optic cables and UHF and SHF Line-of-Sight Radio Relays (LOS-RR). The switching capability of the ITS extends the range of tactical communications links across tactical boundaries and through the long-range capability of the Iris System to strategic, allied or commercial networks.

The ITS consists of a grid of Trunk Nodes, which are switching centres for voice and data traffic. Each Trunk Node comprises two line of sight Radio Relay Vehicles which are controlled by a Trunk Switch Vehicle. Each Radio Relay Vehicle can establish up to five line-of-sight radio links to headquarters and to other trunk nodes. The ITS permits signals resources to always retain a link of two trunk resources in communication with each other, while another is able to relocate. The Trunk Switch Vehicle also acts as the interface with trunk nodes of Allied Forces and the Canadian Forces Strategic System.

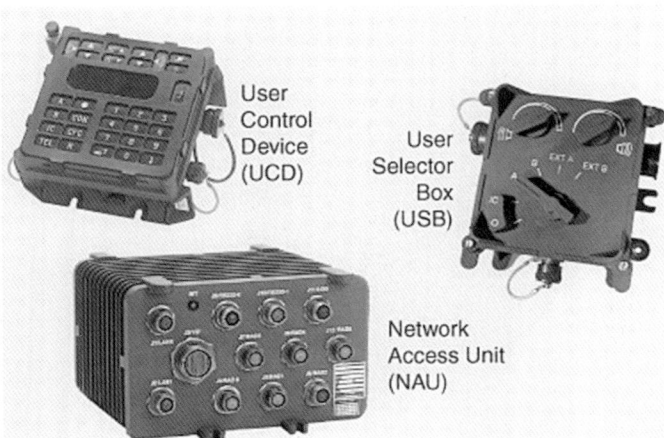

User Control Device (UCD)

User Selector Box (USB)

Network Access Unit (NAU)

Information Distribution System (Computing Devices Canada)   0120232

Long Range Communications System (LRCS)

The LRCS provides extended range communications to commanders in the field by the use of both tactical and strategic resources. It is made up of both satellite and HF facilities which can support operations independently or together as a system. The LRCS is divided into two main groups, the Gateways and the Field Detachments.

Gateways

The Gateway is the hub of a network of communication links. It allows each detachment to communicate with others, and provides access to strategic communication networks and long-range HF links. There are two Gateway sites consisting of the control consoles, antennas, encryption, signals and computer equipment required to engineer, integrate, and monitor channels.   The Gateways can be divided into two parts: the Tactical Interface Equipment (TIE), which is used to interface the strategic networks to an Iris deployment; and the Strategic Interface Equipment (SIE), which is used to provide end users with connectivity to strategic networks. The systems support secure voice and data communications over extended ranges not covered by other Iris components. The two Gateways are located on the east and west coasts of Canada and each one is able to route all traffic between strategic communications networks. The Eastern Gateway contains three SATCOM Systems, one each for C-band, Ku-band, and X-band; and four 1 kW High Power HF Systems. The Western Gateway contains two SATCOM Systems, one each for C-band and Ku-band, but does not have X-band capability like the Eastern Gateway; and four 1 kW High Power HF Systems.

An independent facility within each Gateway handles the Tactical Interface Equipment (TIE) and Gateway Control Facility (GCF). The TIE is the interface point between deployed Iris System elements and the strategic networks. Secure traffic associated with Iris System deployments is processed at the TIE. The TIE serves as a data switch to transfer data and voice traffic, as required. The GCF is integrated to control and monitor the Gateway subscribers, communications, equipment, facilities and the connections to strategic networks. The GCF includes multiple workstations with full control over the non-TIE assets. The GCF can also communicate with operators and equipment for non-secure voice co-ordination of communications links.

Field Detachments

The Field Detachments consist of the following:
(a) TLRCT Detachment
The TLRCT Detachment provides a mobile satellite vehicle with a Tactical Long Range Communication Terminal (TLRCT). This provides the facilities to transmit and receive modulated RF carriers between ground stations and transponder satellites. The TLRCT equipment supports voice and data communication between isolated or tactically deployed units and distant strategic voice and data networks. It integrates a variety of voice and data signals received from local landline or fibre optic sources for onward transmission to other deployed TLRCTs, or a Gateway. A field interface to strategic networks can be connected directly to a TLRCT and provide an interface between a small headquarters and the TIE through satellite.
(b) Medium Communications Terminal Detachment
The MCT Detachment provides extended range HF secure voice and high-speed data communications with one of the Gateway sites, and other MCT Dets, and is also compatible with the CNR(HF).

Additional Capabilities

(a) SATCOM Networking. The LRCS provides a virtual full mesh SATCOM connectivity of up to twelve deployed units through each Gateway and connectivity between Gateways.
(b) High Frequency Networking. HF networks may consist of half-duplex point-to-point radio links or radio nets with a maximum of eight stations sharing the assigned HF.
(c) Mixed SATCOM and High Frequency Networking. The Gateways provide the capability to patch circuits between the HF and SATCOM links. The SATCOM and HF systems are designed to operate together without degradation caused by mutual interference.

Iris System Management (ISM)

The Iris System provides the signals organisations of the Canadian Army with the ability to perform communication management functions completely transparent to the elements that they support. This ensures a continuous and reliable communication system despite dispersion of resources, hostile activity, frequent tactical moves and planned connectivity changes.

ISM is made up of three subsystem management tools and the Tactical Message Handling System (TMHS). The TMHS provides e-mail service for the user and a transport layer for the system management tools to function. The subsystem management tools are, the Communication Management System (CMS), the Cryptographic Material Management System (CMMS), and IDS Network Management (INM). Additional support for ISM comes from the IDS Network Services (INS) and IDS Operating To System Extension (IOSX).

Tactical Message Handling System (TMHS)

The Iris System TMHS provides users with secure, tactical messaging capability through the data terminals assigned to them. It provides electronic messaging and a store and forward capability for mobile users. The segment interfaces with CF strategic and allied messaging systems. Users are able to transmit and receive messages utilising either CNR, TCS, or ITS links. Incorporated into the TMHS are the preformatted common reports required for field operations as well as the ability to write free-text messages. TMHS is divided into the Alpha Domain, consisting of the full Iris System excluding CNR, and the Beta Domain consisting of the CNR.

Combat Net Radio (Computing Devices Canada)   0120231

Communication Management System (CMS)

The CMS is an integrated and automated Iris radio system software designed to facilitate communications management functions, providing a single level of operational capability to serve division, brigade and unit levels. CMS functionality is divided into three roles: System Executive Planning (SEP), Operational System Control (OSC), and Facility Controller (FC). The SEP normally exists only at the highest deployed level, although it may have lower level detachments for co-ordination and its primary role is the overall planning of communications resources to support operational requirements. The SEP also interacts with the system management processes of allies, host nation telecommunications systems and the strategic communications systems. The SEP issues communication instructions and orders, and then supervises those instructions through the OSC. The OSC is the signals management cell which has technical and tactical control over communications resources within its defined area of operations. It exists at all HQs above the unit level. The FC exists at Unit HQ level and above, Technical Control Vehicles (TCV) and Trunk Switch Vehicles (TSV) and is responsible for the detailed real-time technical control of system equipment in the vehicle or complex.

The CMS roles can communicate between themselves (SEP to OSC, OSC to FC, and FC to FC). In addition, other Information Distribution System (IDS) software can access the CMS database and request plans. The CMS will also accept data, solicited and unsolicited, from other IDS sources. The CMS provides automated fixed-frequency and frequency-hopping analysis and assignment and the Communication Planning and Control (CPC) application of the CMS enables the planning and managing of communications resources and provides the ability to monitor their status. The Cryptographic Material Management System (CMMS) works closely with CMS to provide the management and planning of system wide cryptographic materials.

Iris Situation Awareness System (SAS)

Situation awareness is the process of monitoring tactical circumstances by recording and tracking the positions, movement and operational status of friendly and enemy units, organisations and actions. The Iris Situation Awareness System uses global positioning systems, digital communications and specialised software programs to gather and distribute data.
Iris SAS consists of three basic elements:

Positioning Data

Iris SAS uses the Navigation Set Satellite Signal AN/PSN-11(V)1, or Precision Lightweight GPS Receiver + 96 Iris (PLGR+96 Iris) to receive NAVSTAR Global Positioning System (GPS) signals which it uses to calculate, store and distribute position data.

Communications

Iris SAS uses Iris Communications System network resources, radios and data terminals to distribute, store and display situation awareness data.

Situational Awareness Software

Iris SAS can use PLGR+96 Iris internal software or data terminals equipped with the Situational Awareness Module (SAM) to establish and maintain a network and compile, display and distribute situation awareness data. The basic features of SAM are:
(a) A Tactical Situation Display (TSD) consisting of electronic topographic maps overlaid with graphic displays of real-time situation data
(b) Menus and tools for the creation and management of TSD graphics
(c) Message services and communications resource management for the distribution of data
(d) Communications and GPS equipment interfacing and software configuration tools to customise module to user's preferences
(e) Tools to manage the distribution of position data within the Iris SAS network
(f) A library of predefined electronic forms for creation of orders, plans, reports and requests.

Iris SAS network equipment configurations are based on task and command responsibilities. The following are two examples of possible Iris SAS configurations:
(a) Tactical Vehicle. Equipped with PLGR+96 Iris and an Iris vehicle LAN including Control Indicator (CI), Network Access Unit (NAU) and radio resources. This configuration would automatically transmit the vehicle's position over radio to be used as part of situation awareness data
(b) Command Vehicle. Equipped with PLGR+96 Iris, a data terminal operating the SAM and an Iris vehicle LAN and radios. This configuration would gather, display, distribute and store situation awareness data for use by command elements.
Using the above configurations, tactical elements could transmit positioning data calculated by the PLGR+96 through radio net links to SAM equipped vehicles, and the SAM plots and displays position data on an electronic map using military symbology. Networked command terminals simultaneously share the same situational picture; SAM command and control features can be used to create and distribute electronic reports, requests and orders.

Status

Iris began entering service in late 1999 and fielding was due to be complete by 2002. Parts of the system are in service with the USMC, and it will form part of the UK Bowman programme.

Contractor

General Dynamics Canada, Calgary, Alberta.

UPDATED

MESHnet™

MESHnet™ is an integrated voice and data, vehicle-based Command Post communications Network (CPCN) and Tactical Intercom System. It allows radio accessibility regardless of location in a command post/tactical operation centre together with increased survivability due to its multiple fibre optic links between vehicles. Multiple intercom circuits are provided and the whole system is managed from a single workstation running facility control management software. The system consists of four main components - a Network Access Unit (NAU), a User Control Device (UCD), an Audio Interface Unit (AIU), and a MESHnet™ Electro-Optic Module Media Converter. One cable connecting the NAU and UCD provides for integrated voice and data services.

A MESHnet™ vehicle, aviation or deployed node system is composed of one NAU and up to nine UCDs. The UCDs are connected to the NAU using the MESHnet™ Vehicle Internal Data System (VIDS). The VIDS provides an internal Voice Data Network (VDN) and a 10/100 MBPS Ethernet Local Area Network (LAN). The MESHnet™ AIU connects to one audio connector of the UCD providing additional audio services including operation over legacy slip-rings connecting vehicle commander to driver, access to vehicle audio alarms, a Two-Wire Intercom (TWIN) for inter-vehicles communication using field telephone wire. The UCD and AIU support the connection of a variety of passive and noise cancelling headsets, handsets and loudspeakers. Two or more MESHnet™ systems are connected to each other via fibre optic cable between the NAUs, using a MESHnet™ Vehicle External Data System (VEDS) to form a command post or headquarter communication network. The connectivity between the NAU and the fibre optic ports is accomplished using the MESHnet Electro-Optic Module and expanded beam, hermaphroditic connectors or TFOCA II Connectors. Meshed connections between NAUs provide a fault tolerant, distributed network centric system in either a string, ring, star, or meshed arrangement, so physical connectivity of the system is arbitrary; the system self-organises to provide user-to-user connectivity. Network repair is transparent and immediate, as the network automatically reroutes data and voice services over redundant links. The network can withstand multiple link and node losses with 'no' degradation of voice or data services.

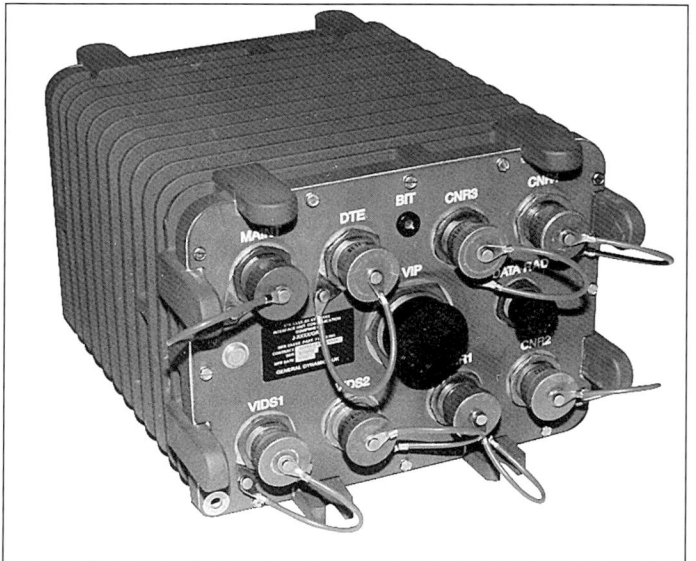

MESHnet™ Network Access Unit (General Dynamics Canada)   0137495

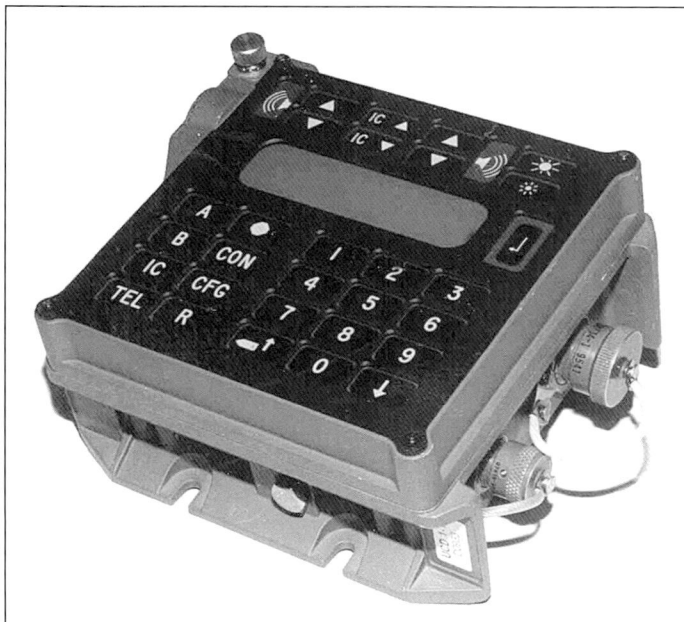

MESHnet™ User Control Device (General Dynamics Canada) 0137496

Facilities

(a) Radio Communications. MESHnet™ provides distributed access and remote control for multiple users to combat net radios. Radios can be remoted up to 1.5 km via fibre optic cable, providing increased survivability and reduced vulnerability. Users can map individual function keys to either specific vehicle-mounted radio nets or combat net radios located anywhere within the network.

(b) External Telephone Interfaces. MESHnet™ telephony provides external connectivity to wide area tactical systems such as MSE and TRI-TAC, NATO systems such as EUROCOM D/1, STANAG 4206 Multi-channel Tactical Digital Gateway and Tactical ISDN STANAG 4578, and commercial telephone system through secure communications equipment and H.323 Voice Over Internet Protocol (VoIP). MESHnet™ uses flexible voice encoding schemes such as 16/32 KBPS CVSD and 64/128 KBPS PCM to provide end-to-end digital voice connectivity through external system gateways.

(c) Telephone Services. MESHnet™'s telephone system is self-contained and fully distributed and does not rely on any external equipment to operate. As soon as connectivity to another MESHnet™ system is established, telephone services are immediately available to the operator, using the MESHnet™ VEDS. This reduces the load on trunk systems such as MSE. All the features found on the standard office digital telephone are embedded in the MESHnet™ system, including caller ID, call forwarding, conference calling, calls hold, and call transfer. If a link should fail, MESHnet™ establishes an alternative path without operator intervention. This is claimed to be normally achieved without any discernible loss of communications during the switchover.

(d) Enhanced Intercom Capabilities. MESHnet™ has four intercom operating modes which operators can select from their User Control Device (UCD). A Hot Mike feature allows for hands-free operation of the Intercom:

Standard 'all informed' intercom service is typically used by crews and staff members within a vehicle or node.

Staff intercom provides multiple internal subgroups distributed between vehicles or nodes.

Override mode provides a broadcast capability for use in emergencies within a vehicle or node. Override temporarily suspends all other communications for the announcement. Once the PTT is released, the intercom defaults back to the operator's original communications setting. Override works across both operating modes.

HQ Override mode provides a broadcast capability that extends throughout the VEDS Command Centre.

(e) Tactical Internet. Through the Network Access Unit (NAU), MESHnet™ provides voice and data communications linking internal vehicle communications with combat radio, wireless wide area data radios and mobile area tactical communication systems including line, microwave and satellite communications.

(f) MESHnet™ Facilities Management System. The MESHnet™ Facility Management System is used to configure the system, monitor system performance, and provide status and alert information using SNMP Internet Protocols. The System allows a Facility Controller to assign radios to specific nets irrespective of location or frequency and can configure parameters for radios with a remote-control capability. The facility controller can restrict access to radio nets on a user-by-user basis. Status and alert information includes the arrival and departure of vehicles as well as changes in network topology. The Facility Management System operates on a connected workstation and a single workstation can be used to manage all network requirements.

User Control Device

The User Control Device provides the principal user interface with access to intercom, telephone, and radio services for voice. It has a vacuum fluorescent display, a 29-key backlit keypad, two independent audio ports, a data port and two connection ports for the system's local area network. A binaural headset permits each crew member to access and mix two independent audio channels from intercom, telephone and radio. The device uses a digital signal processor that can digitise voice at 16, 32, 64 and 128 kbits/s. The control unit also provides an interface to both serial and Ethernet digital data devices. Digital data devices connect to the UCD general purpose serial data port, which supports several interfaces and protocols or the MEU Ethernet Port. It provides user access to telephone radio intercom service including override, MSE and TRI-TAC, together with 15 m remote access to a vehicle.

Network Access Unit (NAU)

The NAU is a high-speed data packet switch. It provides radio access, vehicle and headquarters intercom, embedded telephone services and connectivity to US Legacy and NATO C4I systems. Each NAU can be connected to four other NAUs creating redundant links, increased data throughput and network survivability. The NAU provides both the internal dual ring Local Area Network (LAN) and the external mesh connected Local Data Network (LDN). The LAN supports the connection of up to a total of nine User Control Devices (UCD).

Network Access Unit-Trunk (NAU-T)

The NAU-T provides digital voice and data switching for the Local Area Network (within vehicles), Local Distribution Network (between command post vehicles) and the Trunk Distribution Network (between headquarters). It is designed to be installed in a variety of tactical communications vehicles and provides a multichannel trunk capability using the MESHnet™ tactical router. NAU-Ts are linked together using the Trunk Distribution Network (TDN), which can be carried on a variety of media, including tactical fibre optic cable, Line of Sight Radio Relay (LOS-RR) or SATCOM links. Each NAU-T provides 3TDN ports. Multiple independent NAU-Ts may be clustered together inside a vehicle to provide up to 6, 9 or 12 trunk ports. Clustering provides a highly redundant distributed architecture that ensures high availability.

Audio Interface Unit (AIU)

The Audio Interface Unit provides the Local Area Network with a connection to the Two-Wire Intercom Network (TWIN) and interfaces with the LAN via the UCD.

Electro-Optic Module Media Converter (MEOM)

The MEOM converts signals from the NAU into optical signals for transmission on optical fibres in the LDN or LAN. It can be used to extend the LDN for up to 1.5 km.

Status

MESHnet is currently in service as part of the Canadian Army's C$1.4 billion Iris Tactical C3 System in at least 6,000 vehicle installations of various types.

In 2001, General Dynamics Canada competitively won the UK MoD Bowman (see separate entry) contract to modify, deliver and field 8,400 MESHnet™ systems as the Bowman Local Area System. The system will also be installed in a variety of Royal Navy warships. The Bowman MESHnet™ systems will be built under licence by General Dynamics United Kingdom Ltd.

In mid 2003, 32 systems had been delivered to the US Army for a variety of tactical operations centres ranging from Brigade to Corps main HQ. At the same date 24 had been delivered to the United States Marine Corps (USMC) for use in Direct Air Support Centres (DASC), Air Defence Communications Platforms (ADCP) and Tactical Air Command Centres (TACC).

Contractor

General Dynamics Canada, Communications, Command, Control & Integrated Sensor Systems, Calgary, Alberta.

UPDATED

The MESHnet™ Audio Interface Unit provides the Local Area Network with a connection to external audio sources (General Dynamics Canada) **NEW**/1030912

Denmark

Terma Air Defence C3 system

Terma A/S has developed a C3 system as part of the Danish Enhanced Hawk project. Housed in a box-body it consists of three identical workstations and is connected to sensors and weapon systems in a wide area network through radio access points. All sensors deliver track information to feed a real-time air picture. Weapon systems are positively controlled by operators. A battle management system is also provided to support planning, logistics and other combat support activities. The system can be adapted to a variety of weapons.

Status

In operational use with the Danish Army and Air Force since 1994 and 2001 respectively, interfaced with Hawk and Stinger. In operational use since 1997 with the Austrian Army and Air Force, interfaced with Mistral.

Contractor

Terma A/S, Lystrup.

UPDATED

Operator positions - TERMA AD C3 system (Giles Ebbutt)   0109964

Finland

Patria tactical message switching network

Designed for military headquarters and command posts, this highly mobile and compact tactical message switch can be used to build up packet switching type mobile message communication networks. It facilitates the use of message terminals more efficiently and with higher security compared to their use in a hierarchical point-to-point network.

The switch consists of the Message Switch Unit (MSU) and the Data Terminal Unit (DTU). MSU takes care of the automatic routeing of messages in the network. DTU is used to enter the network parameters. After that it can be used like normal message terminals but with enhanced functional operater/machine communication and interfacing capabilities.

MSU has the following interfaces:
(a) DTU-MSU interface RS-422
(b) eight separate message terminal channels
(c) one 41-pin MS connector with all eight message terminal channels
(d) two serial interfaces (RS-422 or RS-232)
(e) external AC power interface
(f) external DC power interface.

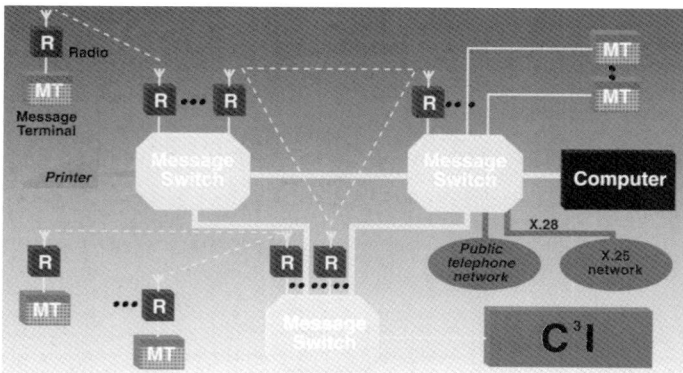

Fig 1 Message switching network

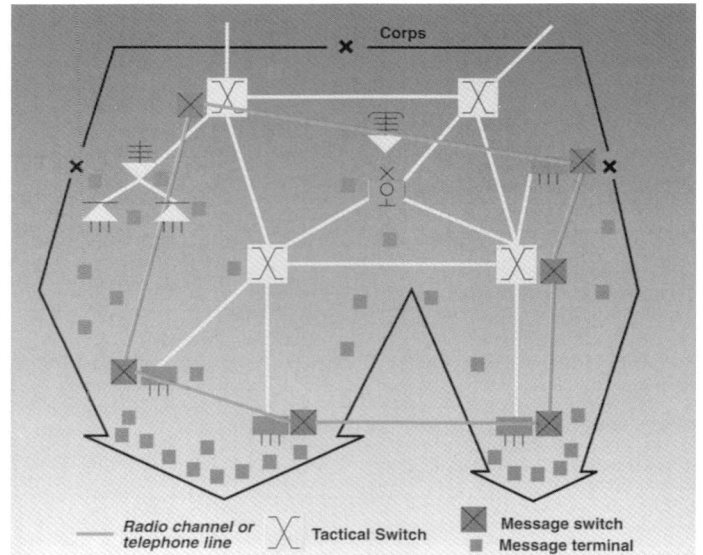

Fig 2 Typical brigade communication network using message terminals and switches

An extension MSU adds eight message terminals and two more serial ports. One message terminal can accommodate up to 64 individual addresses.

The serial ports have their own individual and group addresses and can be used to connect a printer or an auto-dial modem to the MSU. In addition, X.28 interface protocol allows communication with X.25 packet.

The DTU has a 12 × 72 or 20 × 57 character electroluminescent display with a touchscreen, a conventional keyboard and the following interfaces:
(a) computer interface RS-422 with two separate ports (110 to 9,600 Baud); two switches can be interconnected via this interface
(b) printer interface RS-232 (110 to 9,600 Baud for ASCII characters and 50, 100, 150 and 200 Baud for telex characters)
(c) RS-422 DTU-MSU interface (9,600 Baud)
(d) external DC power interface.

The short burst message terminal family consists of palmtop, laptop and tabletop versions for tactical data communication. All the terminals are microprocessor-based and are intended for transmission, reception and editing of messages in hostile environments. The transmission is via radio or wire-connected links and is always encrypted.

An example of a packet switching message terminal network is shown in figure 1. Message switches receive encrypted messages simultaneously from different message terminals, other switches, field computers, sensors or public telephone networks, X.25 networks, and so on. Using the flood search capability they direct the messages automatically first to the switch nearest the receiving station and from there to the receiving station. The difference between this and a traditional hierarchical communication network is that the transmission path between the sending and receiving stations can contain 'hops' using radio channels or telephone lines. In radio communication, the frequencies in the different sections of the path can vary. This makes it possible to build message communication networks which are transparent to the enemy.

The network is configurable online. Message terminals, switches and entire command posts can be removed from, and added to, the network without disturbing its operation. During manoeuvres the switches can be connected to the

Message switch and data terminal unit

The field computer family (rear row), the message terminal family (front row) and the message switch (middle, right)

network via radio channels and during static periods wire-connected channels can be established: a station needs only to communicate with the nearest switch to become part of the network. Compared with hierarchical point-to-point networks, this network allows shorter communication ranges and consequently lower power outputs, leading to better jamming resistance and making the location establishment more difficult.

A typical brigade level tactical communication network is shown in figure 2. Message terminals are distributed to the forces at the forward battle area. Message switches are allocated in battalion and brigade level command posts in star, tree or loop-like nets and connected to the overall C3I system through the tactical switches of the area communication network.

Any message terminal station can send messages to any other station in the C3I system. Using the group addressing or broadcast capability, orders, important messages and sensor information can be distributed practically at the same time to any station at any level of hierarchy. Provided the receiver has a message terminal available, this is achieved regardless of whether a radio (on any frequency) or telephone channel is being used.

Status
The current status of this equipment is uncertain.

Contractor
Patria New Technologies Oy, Tampere.

UPDATED

France

MESREG message handling system

MESREG is a battalion level message handling system designed to aid command and control by automating message handling within deployed forces. It is able to use all available operational communications links including local net (Ethernet), HF or VHF (digital or analogue radionet), satellite links (Inmarsat, Syracuse), switched telephone networks and mobile phone networks. Short- and long-range combined links provide added flexibility. A PC card ciphering unit provides security and an integrated digital map coupled with a GPS receiver is available to locate the station in real time. The battalion main communications centre integrates the radio subnetwork stations and these provide a link to the other battalions and up the chain of command. Messages are in ADatP-3 format or informal (text, drawings, fax, data files). Message routeing is entirely automatic and is performed by broadcast or station to station, with or without acknowledgement. Stations can be used both as a message switchboard and terminal.

MESREG Message Handling System   0055035

Specifications
(a) Operating system: Windows NT
(b) Software: Office Pro, Windows Draw Applications: ADatP-3, Mail, X400 UA, facsimile, administration
(c) Communications:
  Ethernet, TCP/IP, NFS
  (radio) protocol - FED STD 1052. ECC - Reed Solomon. max. BER better than 0.2, Digital synchronous port:16 kbit/s, BPSK
  (Modem) 1,800 kbit/s
  (Rates) 1,800 kbit/s VHF analogue or 75, 150, 225, 300 kbit/s HF
  Telephone network, group III facsimile with ECM, 9,600 kbit/s with fall back to 2,400 kbit/s, mobile phone network,4,800 kbit/s and 9,600 kbit/s
  RS-232 ports up to 19,200 kbit/s
(d) Digital maps: scales from 50,000 to 5,000,000
(e) Power supply:
  (AC) 230 V −50 Hz
  (DC) 24 V vehicle
(f) Environment: 0-40°C (climatic), 25 g, 11 ms (shocks), EN 98020 (EMC)
(g) Dimensions: (W × D × H) 380 × 450 × 190 mm
(h) Weight: 12.5 kg

Status
Over 800 MESREG systems are in operation with the French Army.

Contractor
SAGEM SA (Aerospace & Defence Division), Paris.

VERIFIED

NMS 2000 Network Management Software

The NMS 2000 network management software provides a real-time, automatic command and control capability for a communications network and follows NATO TACOMS 2000 recommendations. The modular architecture enables the system to be readily adapted for use by ground forces and for other operational requirements including peacekeeping operations, humanitarian missions, multiple out of area deployments, a major single-army engagement or as part of a coalition. NMS 2000 enables the preparation and deployment of a typical network, which can include up

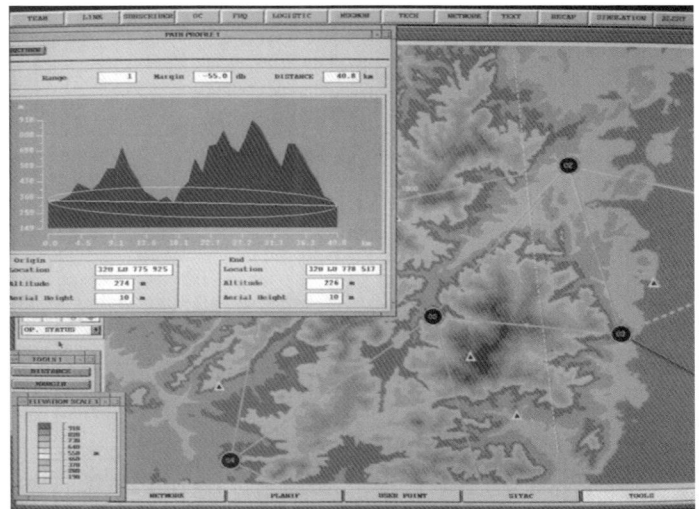

NMS 2000 screen shot   0055039

NMS 2000   0055037

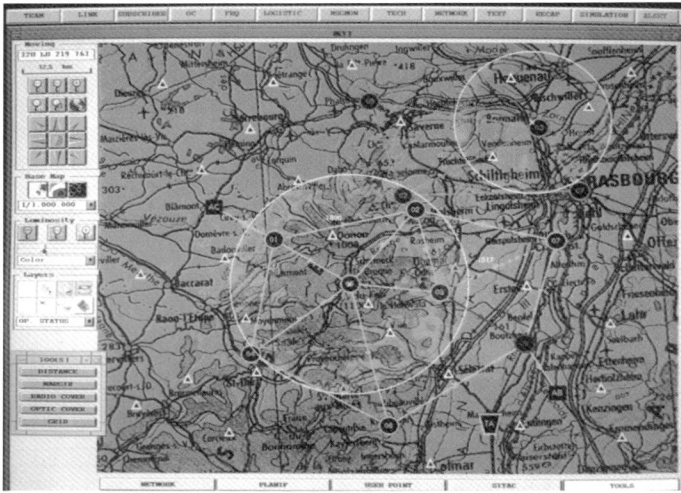

NMS 2000 screen shot illustrating network configuration   0055038

to 200 nodes, 600 links and 5,000 subscribers. The supported network can be deployed in multiple areas, covering up to 40,000 sq km each, anywhere in the world. The use of a user-friendly Graphical User Interface (GUI) considerably enhances the software's useability and flexibility.

The NMS 2000's architecture is fully compatible with the three-level System Executive and Plans (SEP) EUROCOM model, Operational System Control (OSC) and Facilities Control (FC). For greater operational flexibility, the NMS 2000's functions are split into two units:

(a) NMC 2000: the network management centre at the heart of the system. NMC performs all functions required at SEP and OSC levels

(b) FCT 2000: the facilities control terminal, which provides control and supervision of all equipment comprising the network elements.

Status

In use with the French Army in the RITA 2000 area communications system (see separate entry).

Manufacturer

Thales Communications, Colombes.

VERIFIED

Poste Radio de 4ème Génération (PR4G) VHF system

The PR4G is the French Army's tactical radio communications system designed for a range of applications stretching from use by ground troops to deployment in weapon systems. The PR4G system comprises transceivers, frequency and key management components and operational peripherals. It is protected against listening-in, spoofing, direction-finding, or jamming and is hardened against the electromagnetic effects of a nuclear explosion. It is claimed to be the only system with frequency hopping, free channel search, integrated digital encryption and data transmission functions built into each transceiver. The four different transceivers at the heart of the system - manpack, vehicular, hand-held and airborne - all offer alert transmission, break-in facility, authentication of correspondent, link-test, hailing and data transmission protocols.

Key management and frequency attribution are automatically ensured by the frequency and key management system. Frequencies and keys are generated at divisional level by a Frequency and Key Management Unit (FKMU). The FKMU manages all networks for a division according to operational data and is designed to optimise spectrum utilisation for frequency allocation. The data generated by the FKMU are then loaded in a Fill Device (FD) and passed down at regimental level; at this level, a Frequency and Key Copy Unit (FKCU) copies the contents of one FD into eight other FDs. The FDs are then used to enter automatically key and frequency data into each transceiver, by briefly connecting them to the front panel.

The vehicular radio (TRC9500) can be fitted to any type of carrier, from light and armoured vehicles to tanks - the Leclerc in particular - or weapon systems. The manpack and vehicular versions entered service with the French Army in 1992. The PR4G radio sets will act as the main support for data transmission for most of the forward-area French Army weapon systems and command networks. The airborne version is primarily designed for the helicopters of the French Army Light Air Wing (ALAT) such as the Gazelle, Puma and Tiger. The hand-held version is designed for combat troops and for operational situations where personnel cannot carry heavy equipment.

TRC9100 hand-held radio

The 2 or 4 W TRC9100 has five operational modes - FH, free channel search, mixed mode, digital FF and analogue FF. It also has a built-in microphone/loudspeaker and TRANSEC and COMSEC modules. Designed to combat interception, jamming, listening-in and DF, the 1 kg unit's FH capability extends over the 30 to 88 MHz band. Features include digital message transmission, hailing, selective call, alarms transmission, correspondent authentication, remote control and BITE.

Autonomy is 8 to 30 hours depending on battery type. It is compatible with the other PR4G radios.

TRC9200 manpack radio

The 0.4 or 4 W TRC9200 has the same operational modes, FH frequency range and family compatibility as the TRC9100 (see above). It has up to 2,320 channels, built-in TRANSEC and COMSEC modules and can memorise seven networks. It also has a built-in 50 to 4,800 bit/s data adaptor with error detection and correcting codes. The 6 kg unit has a 12 to 24 hour autonomy depending on the battery type and is powered from 11 to 30 V DC. Features include hailing, SELCALL, over the air rekeying, alarms transmission, break-in, correspondent authentication, multimode rebroadcast, remote control and BITE.

TRC9300 A/B/C modular vehicle station

TRC9300 A/B/C dual-fit configuration for the PR4G family offers the following combination in the same volume as the AN/VRC-12/46: the TRC9300 A, a single high-power modular 50 W station; the TRC9300 B shared 50 W amplifier allowing a 5 + 50 W configuration which is useful for rebroadcasting applications; and the TRC9300 C dual-fit 50 W station.

TRC9500 vehicle transceiver

The TRC9500 has the same operational modes, FH frequency range, family compatibility, data capability and features as the TRC9200 (see above). The 0.5, or 40 W unit has a fast/medium hopping rate and up to 2,320 channels. It is powered from 18 to 30 V DC.

TRC9600 airborne radio

The 0.5, 5 or 10 W TRC9600 has the same operational modes, FH frequency range, family compatibility and data capability as the TRC9500. It also has similar operational features, with the addition of a built-in homing function and Vinson KY-58 compatibility. The medium/fast hopper weighs around 8 kg and is powered from 28 V DC.

F@stnet

In 2001 Thales introduced the latest version of the PR4G, F@stnet. It is physically smaller than its predecessor (close to half the volume), but uses a new 64 kbits/s waveform which provides substantially higher data throughput, and has embedded GPS. Simultaneous voice and data transmission is possible, and all necessary functions are provided to enable a tactical internet to be implemented, including an IP router and an Ethernet interface. The radio is completely backward compatible with previous PR4G versions and is available in both manpack and vehicular versions.

Status

In the early 1980s, the Délégation Générale pour l'Armement (DGA) commissioned a feasibility study for the PR4G system which, at the time, was known as the PRF-Future Radio Set. The development phase started in 1986, when the French government telecommunications research department (SEFT) of the DGA took over programme management. Thomson-CSF (now Thales) was appointed as prime contractor for the overall system and for the development of the manpack and vehicular versions of the transceiver. TRT was contracted to develop the airborne version. Following in-depth analysis by programme management, representatives from the French Army and the prime contractor, system specifications were revised in 1988 achieving a 10 per cent reduction in the cost of the main system elements.

On the strength of this reduction in costs and the findings of a new international market study, production start up was announced in April 1989. In the meantime, Secre, Sextant, CEIS and Elno had been selected to develop the system peripherals. Following the results of technical trials and experiments conducted on prototypes of the manpack and vehicular versions, the DGA awarded the first contract for series production to Thomson-CSF in June 1990.

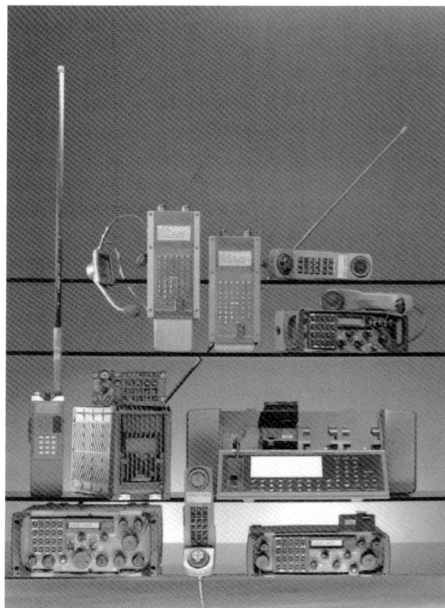

PR4G system versions
with peripherals

The PR4G F@stnet manpack radio, seen alongside the original model (Jane's/IDR)
***NEW**/0111885*

The contract provided for the delivery of 500 manpack and vehicular PR4G transceivers starting in the second half of 1991. In that year, the PR4G was scheduled to start replacing the TRPP13 and TRPP11 VHF radio sets that had been in service with the French Army since the end of the 1960s.

The orders planned by the DGA provide for the delivery of tens of thousands of these two versions of the PR4G over a 10 year period. By mid-1990, the French Army had committed itself to 46,600 transceivers, with 4,500 destined for other users and 94,000 peripherals were on order. The hand-held and airborne versions of the PR4G and the different frequency and key management peripherals were in the development phase in mid-1990. The overall programme cost was of the order of FFr10 billion.

In late 1991, the French Ministry of Defence awarded Thomson-CSF (now Thales) a US$268 million supply contract.

Thomson won a FFr1 billion contract in early 1992 to supply more than 8,000 PR4G transceivers to the Royal Netherlands Army. Around 70 per cent of the hardware, including the SOTAS system, will be supplied by the Thomson subsidiary Signaal (now Thales Nederland). The PR4G reportedly overcame SINCGARS and BAMS for this contract. In March 1993, several hundred PR4G transceivers were contracted to be supplied to the Spanish Army through AMPER and 1,477 were bought in mid-1995. Shortlisted by Swiss Groupement de l'Armement for Switzerland's tactical VHF radio replacement programme. In November 1993, a US$100 million deal was announced to supply a system based on the PR4G to the United Arab Emirates. Part of this order will equip the 388 Leclerc tanks being supplied by Giat. The following year the Finnish MoD awarded a contract for the supply of the system for use by the Finnish Army. The latter requires several thousand radios over the next few years.

The PR4G was selected in mid-1995 by the Swiss Groupement de L'Armement for Switzerland's tactical radio programme, with a major contract awarded in late 1996. In November 1996 the Polish Ministries of Defence and Industry selected the PR4G to replace the VHF tactical equipment of the Polish armed forces. In October 1998 the PR4G was selected by Greece and Egypt. Also in use by the Sultanate of Oman.

The PR4G has now been selected by 24 countries including 12 European and 8 NATO countries accounting for over 70,000 radios.

In April 2002 it was reported in *Jane's International Defense Review* that a US$159 million contract for over 5,000 F@stnet versions had been awarded by the French procurement authorities, for delivery in 2004.

Contractor

Thales Communications, Colombes.

UPDATED

....

RITA

RITA is an automatic integrated transmission network developed on an evolutionary basis to meet the needs of tactical C³I up to the end of the century. It has been adopted as standard by the French and Belgian armed forces and has been in field service since 1982. In 1985, RITA was chosen by the US Army to meet its MSE requirements when Thomson-CSF was awarded a contract, co-ordinated by GTE, to equip 26 US Army divisions between 1986 and 1992.

Development

The original concept and baseline for RITA commenced in 1974 under the sponsorship and control of the Section d'Etudes et de Fabrication des Télécommunications (SEFT) and the Direction Technique des Armements Terrestres (DTAT).

Description

In the design of RITA, emphasis was placed on a number of key service requirements for which simple and efficient solutions have been provided. These parameters include: mobility of forces; total area coverage; mobility of subscribers (from one node to another); permanent and automatic directory number on the total network; high traffic capacity - speech, data, graphics (including telephoto), signalling, teletype; subscriber-to-subscriber direct dialling; speech and data encryption up to and including top secret; automatic and simple operation; provision for new services at a later date and the use of fully militarised components.

To meet these requirements a number of fundamental solutions were adopted, including a fully meshed network architecture, call-diffusion routeing, the digitisation and full ciphering of the system and the provision of an automatic military radiotelephone service.

The nodal stations include stored program electronic switches, usually located at geographically high points, which are normally interconnected by line of sight digital microwave links. Military satellite links can be used instead if required.

The nodal centres are usually deployed in support of a command post. The centre equipment is entirely self-contained and is supplied installed in standard shelters. Normal grid spacing between nodes is 40 km. Although local subscribers in the command post area are connected to the main switching centre by cable or field wire, a remote concentrator is provided, where needed, to provide service to groups of subscribers at adjacent units. Either cable or radio link can be used to connect the concentrator to the main switching centre. A mobile radio interface unit is collocated with the main switching centre to provide service to mobile subscribers.

Call-diffusion routeing, which has been widely tested in the field, makes it possible to access any subscriber in the network with a fixed directory number regardless of the switching centre to which he may actually be connected. Network switching and transmission are arranged on a 6-bit PCM basis, providing 64 quantisation levels for speech with a nominal single-channel bit rate of 48 kbit/s and a transmission highway bit rate of 1,152 kbit/s.

RITA is a quasi-synchronous network with high-stability local clocks. In an integrated network of this kind, synchronisation, signalling and information transfer are processed by time division multiplexed PCM on the various internodal highways. Digital data transmission ensures high-quality communication and suitability for encryption.

Analogue subscribers (except those in a main switch local area) are connected to the network through concentrators which also perform the analogue channel digitising and multiplexing functions. Each concentrator serves up to 54 analogue subscribers. The concentration ratio is close to 2:1.

The main switching centres interface directly with 1,152 kbit/s digital highways. Each switching centre will accept a maximum of 12 × 1,152 kbit/s PCM ports, with an extra one reserved for local analogue subscribers. The 12 ports can be assigned to six interswitch highways and six concentrators, or to any combination of the two. The switching centre uses a closed numbering plan with a fixed seven-digit subscriber telephone number. Less than 1 second is required to establish an end-to-end connection.

A four-level pre-emption plan is provided with four levels of priority: P2 for subscribers with precedence and pre-emption; P1 for subscribers with precedence but without pre-emption; N for normal subscribers without precedence or pre-emption; and R for restricted subscribers without precedence or pre-emption and restricted to a fixed geographical area.

These levels and other subscriber facilities are software assignable. A mobile subscriber calling in for the first time will have his directory number automatically added to the switch memory. When he moves to another exchange area and calls in, his directory number is automatically transferred to the new switching centre. Automatic call transfer is provided for absent subscribers, as well as calling subscriber identification, subscriber localisation and alert broadcast.

Radio access point

Data processing plays an essential part in the fully digital RITA system. Microprogrammed computers (15 M/125) control the circuit switches, the store and forward message switching centres and the Cecore. The computer is a 16-bit processor having a RAM storage capacity of 512 kbytes, dual floppy disks and a magnetic tape mass storage. Data Transmission is achieved using VF modems connected to concentrator analogue ports: up to 2,400 bits/s. Direct digital subscriber: up to 48 kbits/s.

Digital Microwave Radio (DMR)

DMR is used for the backbone internode transmission facility, as well as for 'down-the-hill' links. Modular design provides the flexibility to meet different applications, as well as a wide choice of frequency ranges and reduced maintenance. The DMR can be operated by personnel with little training. Each radio terminal comprises a common equipment package for each of the frequency ranges. Frequency generation is by synthesiser locked to a crystal standard.

Ranges

225-400 MHz (1,392 RF channels, 15 W output)
400-960 MHz (4,472 RF channels, 10/5 W output)
1.35-2.7 GHz (10,792 RF channels, 2.5/1.5 W output)

Channel capacity

digital: 24 PCM channels, including 1 channel for signalling and 1 for synchronisation
analogue: 12 to 120 FDM channels
Order-wire channel: 300 Hz-2.4 kHz

Upgrade

Thales have now developed the TRC 4000 high data rate radio which provides communications up to 50 km LOS between RITA nodes with a frequency range of 4.4-5 GHz, and up to 35 km 'down the hill' using a frequency range of 14.62-15.229 GHz.

Interfacing with other networks

Manual Switchboard. (CMM)
    This provides an interface for subscribers with two-wire Local Battery (LB) telephone sets, LB networks and Common Battery (CB) networks.
NATO-RITA and RITTER-RITA Interface
    This provides automatic access to the RITA system for subscribers connected to NATO networks or to the French Army RITTER network and allows transit communications in the RITA system between two allied networks (NATO Standard AC-225).
Mobile Radio Interface.
    The omnidirectional mobile radio terminal is interconnected with the switching centre in such a way as to provide mobile subscribers with the same automatic dial calling privileges as fixed subscribers. When leaving one coverage area, subscribers are automatically handed over to a contiguous radio terminal station. The mobile radio terminal employs continuously variable slope delta modulation with syllabic compression and provides for connection of up to 25 subscribers to each node. Mobile radio frequencies are pooled and assigned on a demand basis to optimise traffic capacity. RF power output and receiver threshold are automatically optimised. The mobile subscriber telephone is operated in the same manner as the fixed subscriber telephone set.
Mode: FM with 50 Hz adjacent channel spacing. Frequency range: 40 full-duplex pairs around 90 and 80 MHz are available but not all used due to interference from adjacent friendly mobile radio base stations, interference, jamming, and so on. Connection to network: 9 grouped lines. Max number of connectable subscribers per mobile fixed station: 25
Modular Transmission Terminating Equipment
    This includes a switching centre/cable interface equipped for a maximum of six digital highways, a concentrator/cable interface, digital microwave baseband/ cable interface and a regenerative PCM repeater (battery from PCM terminal).

Network command

Installed in transmission command posts, the network command centre (Cecore) provides real-time display of traffic flows in each of the network branches, as well as an immediate indication of faults and congestion or other degraded conditions. An additional function performed by Cecore provides for the automation of RF frequency planning. A network command centre consists of four specialised cells. Cell 1 supervises the data processing system and manages the exchanges between the network and the operation cells, and contains a data bank (15 M/125 computer, desk, teleprinter). Cell 2 supervises the command network (two alphanumeric consoles, one graphic console, one teleprinter). Cell 3 manages the frequency plan (alphanumeric console, teleprinter). Cell 4 manages the nodal centres (movements, potential) (alphanumeric console). The Cecore interface to RITA consists of two full-duplex high-speed data channels.

Status

RITA is fully operational and in series production. In 1983 the first of three French Army Corps was fully equipped. The system has been supplied to Belgium and it has been selected by the USA for use in its MSE. In 1989, Thomson (now Thales) was contracted to enhance the data transmission capacity of the system. An advanced version was being developed for use in the 1990s in the French HADES nuclear missile programme. In 1991, HADES was scheduled to be equipped with a fast frequency-hopping radiotelephone. In mid-1995, the system's mid-life update was in the engineering phase. By 1996, an upgrade to ISDN compatibility was in progress. Alcatel was also involved in a project to add mobile digital data capability to RITA.
    A new version, RITA 2000, is now replacing RITA (see separate entry).

Contractors

Thales Communications.
Bell Telephone Manufacturing Company.
Lignes Télégraphiques et Téléphoniques (LTT).
SAGEM SA.
Société Anonyme des Télécommunications (SAT).

UPDATED

RITA 2000 tactical communications network

The RITA 2000 trunk system is designed to provide vital links between battle networks, national and allied tactical headquarters and the high-command echelons of Metropolitan France. Its applications include major area cover type deployments and lighter types of deployment, such as force projection or rapid intervention.
    RITA 2000's key elements are its lightness, modularity and installation flexibility. RITA 2000 conforms to most civilian standards and protocols and is ruggedised to withstand all military environments, even the most hostile conditions. RITA 2000 enables the battlefield use of Commercial-Off-The-Shelf (COTS) civilian equipment, such as terminals, faxes, multimedia PCs, videoconferencing systems and Private Mobile Radio (PMR). All communication stations are equipped with the ATM Tactical Switch (ATS 2000). The CHANG compact-ciphering unit provides trunk encryption. The 2000 Network Management System (NMS) provides commanders with network planning, implementation, command, inspection and management tools. RITA 2000 offers either automatic radio integration for mobile subscribers, equipped with PR4G radio sets or manual radio integration.
    Characteristics:
(a) first tactical ATM multimedia network in the world supporting armed forces communications on the digital battlefield
(b) multimedia communications (voice, data, internet, video teleconferencing, fax and e-mail services)
(c) offers a wide variety of interfaces (EUROCOM, ISDN, ATM, NATO standards) for full compatibility with existing and future networks
(d) multibearer secure links (satellites, microwave, fibre optic)
(e) MMR concept (SCRA, CNR) and integration of COTS ISDN terminals (DECT, GSM, TETRA).
    For further details see *Jane's Military Communications*.

Status

In operational service with the French and Belgian armies. In 2001 the French defence procurement agency placed a €78 million order for the development of new capabilities (Internet Protocol, high data rate) and the supply of about 100 new stations for the French Army by 2003.

Contractor

Thales Communications, Colombes.

VERIFIED

RITA 2000 in operation   0055042

Rubis network

    The Rubis digital radio communication network provides secure communications using Tetrapol technology for the French Gendarmerie throughout France. The system has a Department-based structure integrated into a national system, managed at Departmental level and co-ordinated at the National Management Centre near Paris. Each Department has a main switch and a number of secondary switches, linked by trunk lines. Radio base stations provide access for mobile terminals, and PABX provide access to other networks. The system has a full range of terminals, including vehicle, motorcycle, portable, handheld and fixed, the latter with either radio or wired links. There is a common radio unit consisting of a digital radio transceiver unit, a vocoder, a remote control device, and for vehicles and base stations connection to a computer terminal. Encryption is by means of keys generated on a nationwide basis. All usual facilities are provided for voice communications. Data features include access to the Gendarmerie national database, national electronic mail (X400), and electronic file transfer. The total system consists of 600 infrastructure sites, 740 microwave links, 6400 fixed stations, 17,900 mobile terminals, 1,350 portable terminals, 11,400 handheld terminals and 11,700 data terminals.

Status

The original contract for the system was placed in 1987 with Matra Communications and, over the course of its development and deployment, it has cost FFr3 billion (US$408.4 million). Following the merger of France's Aerospatiale Matra and Germany's Daimler Chrysler Aerospace to form EADS in 2000, the programme has been run by EADS Defence and Security Systems (EDSN) (now EADS Telecom, part of EADS Defence & Security Systems). The French Ministry of Defence marked the completion of the Rubis network in March 2001.

EDSN has also sold its Tetrapol secure networks technology to a wide variety of police and military users.

Contractor

EADS Telecom (part of EADS Defence & Security Systems), Bois d'Arcy.

UPDATED

Germany

AUTOKO 90

AUTOKO 90 will form the backbone of the German Army's future digital communication network for transmitting speech, facsimile and various types of data and is the replacement for the analogue AUTOKO I/II. Since the entire battlefield will be covered, the efficiency of C3I systems will be significantly enhanced as well. AUTOKO transmission links are encrypted at trunk level and are safeguarded against monitoring, intrusion and manipulation. Compared with its predecessors, the system offers significantly better mobility and flexibility and, via digital gateways, improved interoperability with other nations' systems, together with a 40% reduction in operators.

The heart of the network is the MKS 200 multicommunications system, a digital node and branch switch that also permits digital transition from EUROCOM networks into ISDN and provides SATCOM gateways. Since MKS 200 is deployed as a common unit in networks for mobile air defence as well as in the mobile headquarters of higher NATO commands, EADS (then Siemens) took over the respective activities for the product from Philips Kommunikations Industrie AG (PKI) at the beginning of 1995 in order to safeguard this national core competence. The development of the system included in the region of 280 MKS 200 switches, 440 CTE131/134 line terminating units and 91 power generation/air conditioning systems.

Status

The first delivery of AUTOKO 90 equipment was made in October 1995. Based on a DM400 million procurement contract awarded in July 1996, series deliveries

Inside view of an AUTOKO 90 trunk node (EADS)　　　　　　**NEW**/0511436

started in September 1997, and full operational capability was achieved by early 2002. The equipment has been in operational use in FRY (KFOR/SFOR) since April 2001.

Contractor

EADS Deutschland, Munich.

UPDATED

Bigstaf command post communication system

Bigstaf (marketed commercially as CPnet) is a jam-resistant, wireless and/or fibre optic based command post voice and data communication system for use by the German Army at Brigade, Division and Corps levels. It is designed to connect command centres with AUTOKO and provide interoperability with AUTOKO and HEROS and the US MSE and MCS. Bigstaf users will either operate 51 and 60 GHz millimetric radio links, or fibre optic LAN links, a combination intended to provide maximum operational flexibility.

The system core consists of internet-worked Network Access Units (NAUs). Each NAU is a combined EUROCOM switch and Ethernet hub. This network access unit will provide analogue and digital subscribers with EUROCOM D/1 interfaces and Ethernet fibre interfaces. The system server is designed to provide call establishment and signalling, conference control and 16 or 32 kbit/s delta-modulation for analogue lines. Reported basic capabilities include: sole user circuit; switched hot line; three precedence levels; barred trunk access; conference calls; and broadcast.

The TDMA radio element of the system provides 1 W spread spectrum capability. There are also up to four gateways at each command post for integrating analogue or digital, tactical and wide area networks, including satellite access.

The system is self-configuring and self-healing. If part of the network is lost through battle damage, if part of the command post moves or if additional NAUs are inserted, the system will automatically reconfigure itself.

Status

SEL Defense Systems (now Thales Communications) delivered the first systems to the German Army at the end of 1998.　Thales reported contract activity in 2001. In October 2002 a €20 million contract for the supply of a second batch of BIGSTAF fibre optic network equipment was announced by Thales. This was to complete the equipment of the Crisis Reaction Force of the German Army.

Contractor

Thales Communications, Stuttgart.

UPDATED

HRS 7000 tactical HF communications system

The HRS 7000 tactical HF communications system provides data and conventional speech capabilities and is claimed to be suitable for the fast, secure and reliable burst transmission of data. It is of particular interest to long-range reconnaissance and other special mission forces.

The system covers the 2 to 30 MHz range. It has integrated cyphering, automatic frequency, time, address and crypto management and ECCM and uses Echotel and FEC techniques.

The HRS 7000 system comprises the HRC 7000 fixed or semi-mobile base station and the HRM 7000 mobile and portable mobile stations.

The modular HRC 7000 consists of receiving and transmitting units and a central control unit with a high degree of automation. It contains modules from the mobile station to simplify logistics.

The modular HRM 7000 can be configured as a manpack or vehicular station, or operated remotely.

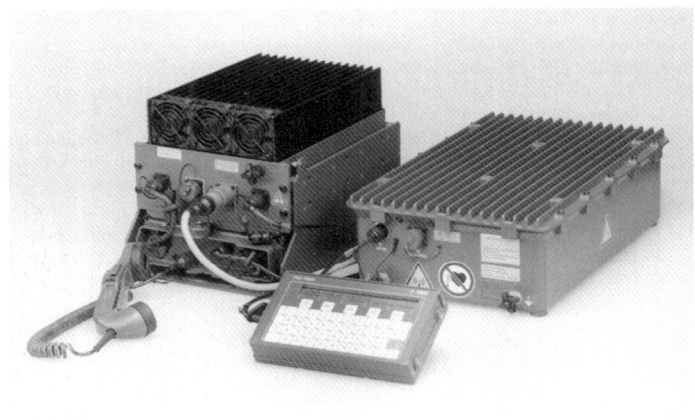

HRM 7000 vehicle station　　　　　　0055027

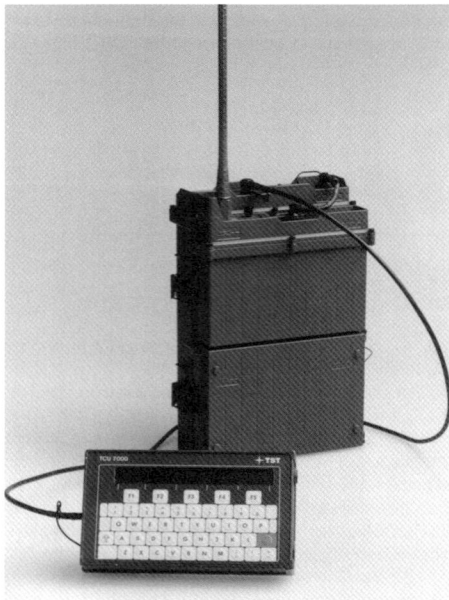

HRM 7000 HF Manpack
0055028

Operating sequences are automated by modern high-performance processors and the management of frequency, time and cypher contributes to reducing the operator's workload. Operation is menu-controlled via the terminal or a PC.

Specifications
(a) Modes: data, voice
(b) Frequency range: 2 - 30 MHz
(c) Power output: 30 W and 400 W
(d) Manpacks: Volume 6 litres approx; Weight 8 kg approx.

Status
In service since 1997.

Contractor
EADS Radio Communications Systems, Ulm.

UPDATED

Greece

Hermes 2 tactical communication system

HERMES is the area tactical communication system in use with the Hellenic Army. Following a year's successful operation, HERMES was deployed and became operational throughout Greece in late 1994, providing significant enhancements to national and NATO communications facilities.

Mobile LOS radios from the Siemens CTM family provide the backbone of the system. Siemens Plessey Systems (now BAE Systems) of the UK is responsible for the switching and network management systems, the latter employing the company's MultiRole Switch (MRS) technology.

To support the HERMES network, Siemens supplied a 30-month maintenance package, together with training, documentation and logistic support to complete the Hellenic Army's familiarisation with the operation and maintenance of the system.

In September 1999 it was reported that BAE Systems was to play a major role in a programme to implement the second phase of HERMES. The HERMES 2 project, to be managed by Hellenic Aerospace Industry SA (HAI), extends the existing HERMES 1 area communications system to meet the communications needs of the Hellenic First Army. It is designed to provide high levels of improved communications functionality, including automatic message switching, packet switching and highly flexible mobile subscriber access. HERMES 2 both replaces the existing APOLLO system and provides military communications links to Greek air defence systems.

BAE Systems' contribution to HERMES 2 is again based on the manufacture and supply of a range of MRS 2000 area communications switches, complete with its integral management system and the incorporation of this technology in a Single Channel Radio Access (SCRA) supplied by DASA. The communication services being provided include voice, telegraph, facsimile, high capacity data transfer, message switching and message handling. HERMES 2 also features gateways to enable the system to interconnect with other national and NATO systems and networks.

Contractor
BAE Systems, Christchurch, Dorset.

UPDATED

India

Army Radio ENgineered radio (AREN)

AREN is a tactical area radio communications system which provides Indian ground forces with a secure, computerised area grid communication network. It first entered service in the early 1990s and includes HF and VHF radios at various command levels, microprocessor-controlled radio relay systems, mobile analogue and digital microwave tropospheric scatter systems.

A significant element of AREN is the Bharat Electronics Ltd (BEL) truck-mounted, shelterised trunk exchange known as the Automatic Electronic Switch. This can handle 192 digitised voice, 256 teleprinter and 32 digital data channels. The equipment's facilities include multilevel priority pre-emption, automatic disconnection of defective or inactive teleprinters and data terminals and short-test good quality route selection.

Contractor
Bharat Electronics Ltd, Bangalore.

UPDATED

Army Static Switched Communication Network (ASCON)

ASCON was developed to integrate the static military telecommunication infrastructure with the mobile tactical communication networks. It is a digital, fully automated, secure, reliable and survivable static communication system based on microwave radio, optical fibre cable, satellite and millimetric wave communication equipments. Fax, Telex, data transfer and video are also available. The third-generation system is currently under development.

ASCON, which can be interfaced to the tactical Army Radio ENgineered radio network (AREN) (see separate entry), is believed to link all army commands and to include at least 52 locations. ASCON's existing microwave links are being replaced with fibre optic cable.

NEW ENTRY

Israel

CNR-9000 radio system

Tadiran's involvement in tactical VHF goes back to the mid 1960s when the company began production under licence of the PRC-77 and VRC-12 radios. From 1970-1975, Tadiran developed and released upgrades to these products (25 kHz channel spacing, silicon transistors replacing tubes and germanium transistors). Encouraged by this success, at the beginning of the 1980s Tadiran developed its VHF-88 family, a new line of frequency-hopping and secure combat net radio sets. At the time, the VHF-88 was the first deployed, fully operational VHF frequency-hopping radio system. In 1989, Tadiran introduced the CNR-900, its second-generation frequency-hopping radio system, providing five times higher frequency-hopping rates, as well as enhanced data transmission speed and capabilities with the same basic radio performance. In 1999 the CNR-9000 was introduced.

The CNR-9000 is a third-generation frequency-hopping VHF radio system which is produced in six manpack and vehicle-borne configurations, plus an additional airborne version. These are all based around the common transmitter/receiver, the RT-9001. There has been considerable effort to make operating easy, with a simple menu-driven display, to reduce operator workload in combat.

The system has evolved from Tadiran's CNR-900 system. New features include an automatic synchronisation technique which eliminates the need for a master station, and orthogonal nets with internet synchronisation. Mutual interference is greatly reduced. Options include an extended frequency band to 108 MHz, internal integrated GPS and an external GPS antenna connection. GPS information is displayed on the radio display, giving the operator a constant statement of his location and the GPS functionality also allows the operator both to transmit his position automatically and to display the location of other stations on his net.

Also offered as an option is an integrated communications controller, the CC-9000, which provides data services either to a single user with a PC or to a group of users on a LAN, who can run applications such as e-mail or file transfer. This will extend the reach of the LAN to the Combat Area Network (CAN) served by the CC-9000 and, by connecting CC-9000s back-to-back routing functionality, can be achieved between different CANs. These functions provide a tactical internet.

The variants are:
PRC-930 Manpack radio set
PRC-930HP High-power manpack radio set
VRC-906 Low-power vehicular radio set
VRC-920 High-power airborne radio set
VRC-950 High-power vehicular radio set
VRC-980 Dual-fit, high-power vehicular radio set

VRC-990 Dual-fit, high-power vehicular radio set with additional external antenna. Further details can be found in *Jane's Military Communications*.

Specifications

(a) General
Frequency: 30.000-87.975 MHz (optional: 30.000 to 107.975 MHz)
No of channels: 2,320 (3,120 optional) at 25 kHz spacing
Modulation: F3 simplex, voice, analogue and digital data
Modes of operation: Fixed frequency: Clear and COMSEC. Frequency hopping: ECCM/COMSEC
Preset channels: 100
Display: Menu driven LCD, large graphic display, that presents a wide variety of radio status

(b) Power Source
Manpack: 12 V DC nominal
Primary battery: TLS-020 Lithium battery, TLS-0151 A Lithium battery, BA-3791 Alkaline battery
Rechargeable battery: TNH-2012 NiMH battery, TNC-2188 NiCad battery, TLI-9380 Lilon battery
Vehicular: 24 V DC nominal per MiL-STD-1275
Airborne: 28 V DC nominal per MiL-STD-704A

(c) Environmental
Operating temperature: −40 to +65°C
Immersion: up to 1 m of water for 2 h

(d) Physical
Manpack dimensions (W × D × H) (including battery): 226 × 245 × 86 mm
Vehicular dimensions (W × D × H):
VRC-906/VRC-950: 245 × 300 × 195 mm
VRC-920: 245 × 300 × 195 mm + 146 × 1 00 × 115 mm(C-6200)
VRC-980: 345 × 280 × 245 mm
VRC-990: 345 × 280 × 245 mm + 147 × 268 × 135 mm (AM-9050)
RT Weight: 3 kg

Status

In service with the Israel Defence Force, at least one unspecified Latin American country and possibly one or more Asian countries.

Contractor

Tadiran Communications Ltd, Holon, Israel.

UPDATED

....

GRC-408

In the late 1970s, Tadiran introduced the TAD-200, its first generation of line of sight multichannel radios - operating in Band II (610 to 960 MHz) and featuring FDM (frequency division multiplexing) and TDM (time division multiplexing) capabilities. The GRC-400 family followed the TAD-200. Comprising the GRC-404 (Band I: 225-400 MHz), the GRC-406 (Band II) and the GRC-408 (Band III: 1,350-1,850 MHz), it provided new features and higher data rate transmission capability. Over a thousand GRC-408 radios were integrated in the USMC DWTS (Digital Wide-Area Transmission System) project. The GRC-408A, a Band III radio, further enhanced the capabilities and features of the GRC-408. It provided higher data rates (up to 2 Mbytes/s), higher transmission power and built-in forward error correction. This model was integrated into the US/CECOM IMSE and the Spanish RBA programmes. The most recent model, the GRC-408E, evolved from the GRC-408A and the GRC-408HC+, providing 8 Mbytes/s data capabilities and spectral efficiency over the extended frequency Band III (1,350-2,690 MHz), while maintaining full backward compatibility with earlier models.

Status

In production. In use by the Israel Defence Force, US and Spanish forces.

Contractor

Tadiran Communications Ltd, Holon.

VERIFIED

....

GRC-2000C

The GRC-2000 is currently the fourth-generation radio relay of the Tadiran legacy of line of sight multichannel radios (MCRs). It is a multi-channel ECCM radio relay designed to provide fast frequency hopping (FFH), spectrally-efficient line of sight radio relay links for use in hostile, jamming environments. The equipment is combat-proven and is claimed to be suitable for use in enhanced ECCM digital backbone applications in wide area networks. The GRC-2000 operates in NATO Band IV, meeting EUROCOM interfaces and traffic rates (beyond 2 Mbits/s). It has automatic antenna positioning for rapid tactical deployment of links under field conditions.

Status

The GRC-2000 has been successfully deployed by the Israel Defence Forces operating under severe, jamming, field combat conditions. Several military users,

including a European customer, have selected this ECCM/FFH equipment as their digital radio relay backbone for tactical wide area networks. The GRC-2000 has been used each year in the Partnership for Peace (PfP) programme.

Contractor

Tadiran Communications Ltd, Holon.

VERIFIED

....

HF-6000

Tadiran's involvement in tactical HF goes back to the mid 60s when the company started the production, under licence, of the PRC-74 (20 W manpack) and the GRC-106 (400 W mobile) radios. Substantial research and development carried out by Tadiran during the 70s resulted in a new HF series, the HF-700 radio system, introducing innovative features to improve the usability of HF in the battlespace. Among these features was 'Active Squelch', allowing the HF to be integrated in a vehicle intercommunication systems without degrading the overall communication quality. In the mid 80s, Tadiran introduced its third generation of HF radios, the HF-2000, which was equipped with a tactical automatic link establishment system, the AUTOCALL, and powerful full-band frequency hopping and voice and data security.

The HF-6000 radio system provides a family of HF radios built around the RT-6001 receiver/transmitter. The family consists of 6 sets.

PRC-6020. 20 W manpack set. Complete unit weighs only 3.9 kg
VRC-6020. 20 W vehicular set. The same set in a vehicle mount, with a vehicle antenna and powered from the vehicle's DC battery.
VRC-6100. 100 W vehicle/fixed station set. The 6020 with a 100 W power amplifier and vehicle adaptor.
VRC-6200. 100 W vehicle/fixed station set. Although based on all the same components as the rest of the range, this set is housed in a single box.
GRC-6400/6600. 400 W and 1,000 W fixed station sets built around the RT-6001.
Common capabilities include:

(a) Automatic Link Establishment. The optimal frequency is selected automatically from a continuously evaluated frequency set
(b) Digital squelch
(c) Selective calling
(d) The following are all optional capabilities:
Data communications. A built-in high-rate modem uses a variety of wave forms at data rates from 50 b/s to 4,800 b/s
(e) COMSEC features. Secure voice, data and burst transmissions
(f) ECCM features. Frequency hopping over the entire band at more than 15 hops/s. Automatic synchronisation (no master station)
(g) Vocoder
(h) Dual frequency. Reception and transmission on different frequencies
(i) Preprogrammed FLASH burst transmissions
(j) Adaptive high-rate modem. Matches data transmission rate to the quality of the link
(k) Adaptive power control. Matches power output to communications conditions.

Status

In March 2002, *Jane's Defence Weekly* reported that Tadiran had received a US$8 million contract for "several hundred" units for an undisclosed Asian country, which may have been the originator of a further US$25 million order for an "existing Asian customer" announced in mid 2003. Other unspecified Asian and Latin American countries are also likely to have ordered the equipment, according to announcements in 2002 and 2003.

Contractor

Tadiran Communications Ltd, Holon.

UPDATED

....

Tadiran TCC 2200 and PC COM 2200 Tactical Communication Controllers

The Tadiran Tactical Communications Controllers provide mobile, reliable connectivity for C4I applications. Tactical communications protocols are supported to provide connectivity under all conditions, together with automatic internet and intranet message relaying and routeing.

Description

The TCC 2200 is a PC-based, four (logical) channel front end communication controller for network data transmission and reception, designed for message routeing between local applications and tactical networks.

The PC COM 2200 is a PCMIA-packaged two-channel tactical communications controller which transforms any user workstation into a powerful wireless terminal capable of routeing between local and tactical networks.

Both the PC COM and the TCC 2200 share the following prime characteristics:
(a) end to end message transmission and reception
(b) end to end acknowledgement and retransmission

Tadiran TCC 2200 Tactical Communication Controller 0055049

PCMIA card
0105884

(c) packeting and de-packeting of messages
(d) adaptive routeing and relaying
(e) 3 priority levels of message transmission
(f) media access protocols (CSMA, MS-TDMA)
(g) connectivity and topology learning
(h) error detection and correction algorithms
(i) positive datalink acknowledgement
(j) media interfaces: HF, VHF, UHF, tactical radios, analogue or digital 2W/4W land line, microwave or radio telephone links.

Both controllers are optionally available with encryption capabilities, a built-in GPS receiving module and activation of Automatic Link Establishment (ALE) for HF channels.

Status
In use in the Tadiran artillery C4I systems DACCS and Bombard (see separate entries).

Contractor
Tadiran Electronic Systems Ltd, Holon.

UPDATED

Italy

Mobile Integrated Digital Automatic System (MIDAS)

The Mobile Integrated Digital Automatic System (MIDAS) is a tactical multichannel system developed to meet communication requirements and tactical needs of army forces at brigade, division or army corps level. The entire system can be installed on tactical vehicles, or fitted in shelters when a higher level of NBC/EMP protection is required.

Digital techniques are used throughout the system allowing full integration of services and the use of digital encryption methods, both at trunk and subscriber level. Telephony, telegraphy, data transmission and facsimile are supported.

Networks ranging from hierarchical to full grid-type can be configured with network performance, such as system capacity, subscriber facilities and system redundancy, tailored to meet specific requirements.

MIDAS is based on functional system elements, like communication trunk and access nodes, where the required functions of subscriber interfacing, multiplexing, encryption, transmission and switching take place. In addition to the multichannel

network, subsystems are provided performing functions such as combat net radio interfacing, long distance tropospheric scatter transmission, supervision and network management, network maintenance and power supply. The tasks of each functional system element are performed by one or more autonomous functional vehicles (or shelters).

The telecommunication equipment includes CD110 series circuit/packet switching exchanges and associated MT310 series multiplexers; CF100/CM100 series digital bulk and end-to-end encryption equipments and the relevant portable loading devices from the FG100 series; MH300 series VHF and UHF multichannel radio relays and the MFI5 SHF radio relay; MT300 series optical and cable line terminating units; a complete family of auxiliary elements such as the EOW terminal/concentrator (AS339), tactical digital cross-connect (AS111), STANAG interface (MT303), G703 interface (AS415), reference clock (AS110), and the radio access interfaces (RA110 series).

The control equipment includes the data terminals (AS101/107), the alarm collection unit (AT151), the supervisory computers of the GPC series and the key generation system (KS100) which can be integrated with the EKD electronic key distribution system.

Marconi also supplies a number of application software packages tailored to tactical applications and suitable for planning, deploying, monitoring and managing the MIDAS system in whatever version it is implemented. Commercial transport protocols for computer communication, relational databases and graphics packages are used to implement the various supervisory levels.

Status
Most equipment is in service with the Italian and other armies.

Contractor
Marconi Selenia Communications, Genoa.

VERIFIED

Korea, South

MSC-500K Tactical Communication System

The MSC-500K is an area communications system providing point-to-point trunk communications, mobile radio equipment and access to external systems including combat net radio and MSE. It is also known as the SPIDER Network. The system consists of a network of trunk nodes, access nodes and unit level switches, providing connectivity and access for a variety of communications types. Equipment is housed in air-conditioned shelters with EMI shielding. Power is provided either by towed 10 kW or man-portable 750 W generators. Key features of the system are:
(a) Digital switching providing transit/access functions
(b) Voice and data communications
(c) Flood searching and adaptive routing
(d) A fixed and deducible directory
(e) Fully integrated radiotelephone
(f) Packet switching capability
(g) Security facilities including traffic and signal encryption, checking of personal code on reaffiliation and preaffiliation of authorised subscribers.

System components
Switching Equipment
(a) Tactical Digital Switch (TDS) (TTC-95K)
    Provides flood searching and adaptive routing
    Serves up to12 digital trunk groups and 240 subscriber terminals
(b) Packet Switching Unit (PSU) (TTC-610K)
    One per trunk/access node
    One 1,024 kbytes/s digital trunk group per trunk/access node
    Four X.24 ports
    Four 256 kbytes/s digital trunk groups per RAP/RSU
(c) Remote Switching Unit (RSU) (GTC-620K)
    Serves up to 60 subscriber terminals
    Two 256 kbytes/s digital trunk groups to PSU/RSU

Transmission Equipment
(a) High-speed radio transceiver (HRT-21)
    Transmission capacity up to 4 Mbytes/s
    Frequency range: bands I, II, III
    Nominal range: 50 km
    Frequency hopping
(b) Tactical Microwave Radio (GRC-650K)
    Transmission capacity: 256, 512, 1,025 kbytes/s
    Frequency range: 1.71-1.84 GHz
    Nominal range: 15 km depending on terrain

Mobile Radio Equipment
(a) Radio Access Point (RAP) (TRC-660K)
    Eight digital (16 kbytes/s) radios
    Range: 8-20 km depending on terrain
    Two 256 kbytes/s digital trunk groups to PSU/RAP

(b) Mobile radio terminal (VRC-680K)
Digital (16 kbytes/s)
Automatic power control
Automatic reaffiliation
Frequency range: 230-290 MHz

Network Management Equipment

(a) System Control Computer (SCC) (TYQ-692K)
Network management, planning and configuration
Automatic dissemination of reports and orders to/from nodes
RAP frequency management, planning and distribution
LOS frequency management, automatic terrain analysis and path profiling
(b) Node Control Computer
Node management.
Node performance reports to SCC
Node connectivity and status display

Interface/Subscriber Terminals

(a) Combat Net Radio Interface (TTC-640K)
Provide access to analogue CNR, (HF, VHF, UHF) and digital frequency hopping CNR
Automatic and /or manual communication set up
(b) Digital Voice Terminal (TTC-630K)
Digital (16 kbytes/s)
2-wire with data port to interface data terminal
(c) Analogue Interface
Provides MSE analogue interface

Status

Development began in 1989 and the system is now in use by the South Korean land forces. It is believed that this will eventually provide all area communications down to unit level. The MSE interface will allow connectivity with US forces.

Contractor

Samsung Thales Co Ltd, Seoul, Korea.

UPDATED

Netherlands

ZODIAC mobile telecommunications system

ZODIAC (Zone Digital Automatic Communications network) has been fielded as the mobile tactical area communication system for the Netherlands First Army Corps.

Description

The technical parameters are in accordance with the EUROCOM D0 AND D1 requirements. This implies digital switching of 16 kbit/s channels.

Analogue signals are coded by means of digitally controlled delta modulation. The switch has a connecting capacity of 16 digital time division multiplexed groups. Each group can consist of 16, 32 or 64 channels. Several types of in-band signalling and common channel signalling can be served. Selection is made by the facility controller. The channel switching matrix is of a modular expansible design and is a non-blocking network.

Interior of Zodiac Communications Shelter

ZODIAC mobile telecommunications

The central control unit of the switch has a central processor which is also used for the front end processors that are incorporated in the I/O cards. The memory of the central control unit contains all the necessary programs and data read from a back-up memory. The switch has an operator position and a facility controller position, the functions of which can be interchanged or combined. Extensive BITE facilities are embedded in the switch.

Other pieces of equipment involved in the system are multiplexers, store and forward systems and digital secure terminals. The last allow voice communications but can also function as modems for telex, data or facsimile equipment. Using this terminal, all the types of traffic can be encrypted to realise end-to-end encryption.

ZODIAC can be connected to RITA, AUTOKO, Ptarmigan, Tri-Tac and the NATO systems. The ZODIAC was also supplied in a derived version to NATO.

Status

ZODIAC is currently in service with the Royal Netherlands Army and the joint Royal Netherlands/German Army Corps.

At the end of 2001 the Netherlands Defence Ministry (MOD NL) announced that it would not pursue the planned US$100 million mid-life update of the ZODIAC communications network system. Instead, MOD NL launched the procurement of a tri-service US$110 million Theatre-Independent Tactical Army and Air Force Network (TITAAN - see separate entry). According to the Dutch State Secretary for Defence Procurement, the ZODIAC system was insufficient for data communication and was not capable of supporting operations over long distances and in mountainous areas.

Contractors

THALES Communications BV, Huizen.
General Dynamics Communication Systems (design and production of central control unit, matrix and programs), Needham Heights, Massachusetts, USA.
Philips Crypto BV, (design and production of cryptographic equipment), Eindhoven.

UPDATED

Norway

TADKOM tactical area communications system

TADKOM is a tactical communications system for automatic switching and transmission of voice and data. The system is used as an area network at brigade, division and corps level as well as the communication part of command and control systems. It was developed by Alcatel Telecom Norway, now Thales Communications AS, and NFT Ericsson Communications under the terms of a contract placed by the Norwegian government for the Norwegian Army. Fielding started in 1987 and after being updated several times is still in operational use.

TADKOM is designed to use small autonomous and ruggedised units with high capacity. This gives flexibility in changing network topology according to operational requirements. All system equipment is designed for installation in combat vehicles without the need for expensive shelters or other environmental protection.

The automatic system is based on EUROCOM recommendations and is fully digital using delta modulation at 16 or 32 kbit/s. The system offers a variety of subscriber facilities including priority call, ring back, line groups, sole user lines with automatic re-routeing and so on. A free numbering scheme allows all subscribers to keep their telephone number independent of their physical location in the network.

Two elements in the system are the tactical switch TDS 300 and the multiplexer DMU 200 as well as the new generation TAS 350 switch from Thales. The TDS and TAS switches provide both circuit and packet switching. All trunk connections can be equipped with crypto modules built into the TDS 300 switch. The switch has a non-blocking capacity of 1,024 channels, in the form of eight TDM ports. These

TADKOM trunk/access node

ports can be connected freely to other TDS/TAS switches or to DMU 200 multiplexers.

Subscriber access to the network is by two- or four-wire connections to the DMU 200 or TAS 3500. This can interface a free mixture of digital and analogue subscriber terminals such as automatic telephones, manual telephones, PABX lines, PTT lines, facsimile and so on. In addition high-speed digital connections can be made to the TAS 350 for connection of IP routers and so on. One DMU 200 multiplexer has a capacity of 15 subscriber lines. Two multiplexers may be stacked to achieve a capacity of 30 subscriber lines via one TDM connection. One TAS 350 access switch has a capacity of up to 30 subscriber lines. Several access switches can be connected in a ring or daisy chain structure to form a command post network.

A TADKOM network is built up by switching nodes linked in a grid structure by use of transmission equipment. Each node consists of one or more TDS 300 switches. The transmission equipment may be line of sight radio relay, satellite, optical or electrical cables. The TDS 300 supports network services enabling the subscriber to move freely in the network. Each subscriber is identified by his telephone number only and not by physical location.

The system provides communication security by bulk encryption of radio links as well as end-to-end encryption. Resistance to EW is achieved by use of saturation routeing, error correction protocols and ECCM radio links.

The Data Access Point (DAP) terminal is the general terminal for voice and low speed data communication, with a DTE terminal for telegraphy and facsimile. The DAP has two wire subscriber loops of the same length (up to 8 km) as analogue telephones using conventional field cable. The DAP can be connected to EUROCOM "K" line interfaces at both 16 and 32 kbit/s. The DCE performance is 50 to 2,400 bit/s and 9.6 kbit/s. DTE interface is RS-232C.

The Single Channel Radio Access (SCRA) is an integrated part of TADKOM. Access to the TADKOM is through the Radio Access Point (RAP) which controls four separate radios simultaneously and has the necessary interface to the

Data access point terminal

Tactical network control unit

switched network. The RAP is connected to the TADKOM via the DMU 200 multiplexer. The SCRA system gives full subscriber facilities to mobile users. It can handle 16 kbit/s voice, 300 bit/s and 2.4 kbit/s asynchronous data and 2.4 and 16 kbit/s synchronous data.

The TADKOM system also includes a tactical message handling system. This system is based on the STANAG 4406 standard which defines additional military enhancements to the X.400 standards.

Status

The TADKOM system was in use in the Norwegian Army in the 1990s and as the communication part of the Norwegian Adapted HAWK Air Defence System. Developments to the system were also begun in the first part of the 1990s and the Norwegian Army is now implementing a number of improvements including the new MultiRole Radio (MRR) family that will act as both a CNR and an SCRA subsystem for TADKOM. The MRR system was fielded in 2001.

Contractors

THALES Communications AS, Oslo.
Kongsberg Defence Communications AS, Billingstad.

UPDATED

TDS 300 switch

Spain

RBA

The RBA (Red Básica de Área or Basic Network of Area) is an integrated automatic digital network communications system catering for static, semi-mobile and mobile users. The system is modular, has a high MTBF and is secure. The system consists of a number of nodes (*Rioja*) connected by multichannel radio links, providing coverage of the operational area of the supported formation. This provides communications access for headquarters irrespective of their location in relation to each other. Semi-mobile and mobile subscribers are connected to the network via access nodes, and the network can also be interfaced (via stations called *Asturias*) with permanent telephone systems or satellite links. The network is managed from dedicated management stations called *Murcia*.

The tactical area network system functions with circuit-switching channels at 16 and 32 kbit/s. Packet switching uses X.25. Subscribers can be semifixed or mobile for voice and data. This system provides trunk signaling in a separated channel (Eurocom) and loop signaling. Routing involves saturation, and network synchronisation is plesiochronous to a high-stability clock. External interfaces include NATO networks, strategic and civil networks, an integrated services digital network Primary Access Rate at 2,048 kbit/s, and X.25 networks with an X75 gateway. The automatic switching and management systems ensure that alternative communications paths are found in the event of the loss of part of the network.

Both Thales and Tadiran equipment is used, but all of the software, network controls and integration has been developed by Amper. Among the systems are Tadiran's digital subscriber terminal, the TA-359, a digital two- or four-line telephone terminal with voice codifications at 16 or 32 kbit/s, data Class 1 to 4 Eurocom D/1, and Class 5 as an add on. Data interface is with the synchronous or asynchronous RS-232, or RS-422 synchronous terminals. A multiple optical-line terminal unit, the MT-323, provides three channels from 256 to 2,048 kbit/s (Eurocom B), three channels Engineering Order Wire (EOW) and two data channels V10/V11. Tadiran's GRC-408A, provides a frequency range from 1,350 to 1,850 MHz. Data is at 256, 512, 1,024 and 2,948 kbit/s with an access method interface, and the service channel multiplex offers voice, remote control and automatic control power. A modular tactical switch is digital and Eurocom compatible, featuring flood circuit routing, extended range of services and interfaces.

Status

A contract for Pta28 million was awarded in 1995, reduced to Pta23.1 million in 1996, for the development of the system. The first equipment entered service in 1999 and the programme was due to be completed by 2002. The introduction of the system has reportedly resulted in an average of 40 per cent reduction in manpower and equipment over the predecessor 'Mt Olympus' system.

Contractor

Amper Programas, Madrid.

VERIFIED

Turkey

Combat Zone Mobile Communications System (CZMCS)

The CZMCS, one of the largest tactical mobile communications systems deployed in NATO countries, is designed to provide army corps and subordinate echelons of Turkey's Second and Third Armies with a reliable and quickly relocatable voice data and video communication system.

The system, configured in a grid topology, is based on a number of nodal points interconnected with digital tropospheric scatter links and digital line of sight

Nodal point

Tropo terminal deployed

LOS Terminal Deployed (Page Europa SpA)   0125362

microwave spur links. CZMCS is composed of more than 220 NATO II and NATO III shelters, including tropo terminal shelters, line of sight shelters, digital switchboards shelters, communication system control shelters, maintenance vans, plus tropo antenna trailers, military trucks and support vehicles.

The telephone system is fully automatic in accordance with EUROCOM standards, using the TDM switch technique; voice, telegraph, facsimile and data transmission can be automatically routed through the network. Security is provided at link and channel level.

All the equipment supplied has high reliability and requires minimum maintenance; extensive use of BITE and automated fault diagnosis reduce the downtime at extremely low levels. Redundant configuration (hot standby with automatic changeover) is used throughout the systems at all levels from RF to VF.

The mobility of the system is such as to allow the deployment and tear down of any part of it in1 hour with a crew of four, without the use of special test equipment.

Status

Page Europa, as prime contractor, designed, engineered and implemented on a turnkey basis the entire CZMCS project by integrating different equipments selected on the NATO market. Marconi's contribution to the programme included approximately 80 digital exchanges, 80 multiplexes and 300 radios. Contract value to Page Europa for CZMCS is over US$120 million. The system successfully passed Acceptance Test in November 1989 and is presently in operation.

Contractor

Page Europa SpA (Prime), Rome, Italy.

UPDATED

....

TASMUS

TASMUS (Taktik Saha Muhabere Sistemi) is a tactical area communications system which provides a common picture of the battlefield in near-real time and shares data among and between battlefield operating systems. It facilitates fusion and display of intelligence information to commanders at all levels and handles the rapid exchange of targeting data from sensor to weapon systems. TASMUS is designed to form a mobile, survivable, flexible and secure tactical network to support all the communication needs (voice, data and video) of tactical commanders. TASMUS communications services include voice (clear/encrypted); data (asynchronous, synchronous, X.25 packet data), graphics, video, images, file transfer; fax and video conferencing. TACOMS information services include: the provision of a near-real time picture of the battlefield; digital map and geographic

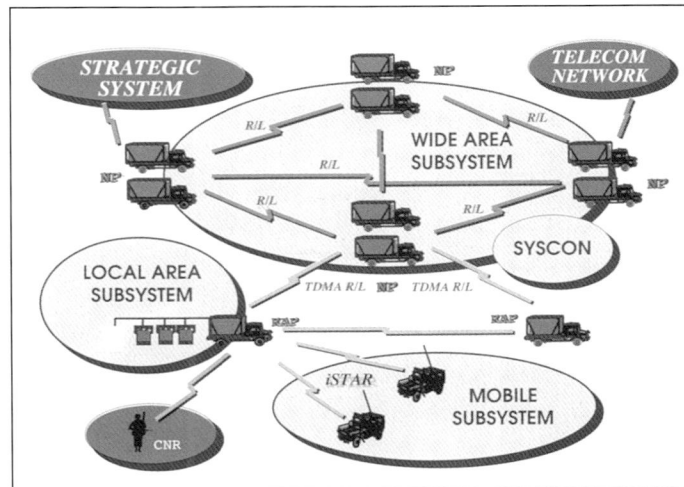

TASMUS system architecture (ASELSAN)   0137504

information; meteorological information; intelligence reports; and logistic information. Switching in the system is based on Asynchronous Transfer Mode and ISDN.

The system architecture consists of three interconnected subsystems, the Wide Area Subsystem (WAS), Local Area Subsystems (LAS) and the Mobile Subsystem. The WAS is provided by point-to-point communications with ATM and ISDN switches provided by Netas (Nortel Networks, Turkey). The Mobile Subsystem is provided by the iSTAR radio family (mobile and personal subscriber terminals) which offers radio networking, simultaneous voice and data, near-realtime packet switching, and GPS. The iSTAR family (gateway radio) also provides the access between WAS, LAS and Mobiles through nodal and radio access points respectively. The nodal points provide access to strategic communications and the PSTN. The radio access points provide access to CNR. A system control facility is also provided.

Shelters are provided by Page Europa. ASELSAN has integrated the system.

Status
Trialled in 2001 and now in production for the Turkish Army. Sources suggest that the Netherlands and Switzerland may be interested in some aspects of the system.

Contractor
ASELSAN Military Electronic Industries Inc (Prime), Ankara.

VERIFIED

United Kingdom

Bowman

The concept of Bowman dates from a 1989 General Staff Requirement (GSR) for a system to replace the ageing Clansman radio system, which was subsequently modified to accommodate post Cold War scenarios. Bowman will provide a tactical, secure voice and data communications system for joint operations. The procurement programme has had a long and chequered history, with a number of consortia involved in the development and bidding process. This process culminated in the failure in 2000 of the preferred bidder, Archer, a consortium of Racal, ITT and Siemens Plessey, to deliver the requirement within budget, and its resultant sacking by the UK MoD. The subsequent bidding process for the contract was won by a consortium led by Computer Devices Company (CDC) UK, now General Dynamics UK Ltd.

The complete contract involves more than 48,000 radios and more than 30,000 computers being installed in more than 30,000 platforms, together with the necessary training. The Bowman system is a fundamental part of the UK Digitisation process, as it will provide the carriers for the passage of data between the various software applications involved.

System Components
The following components will make up the Bowman system. The equipment is covered in greater detail in *Jane's Military Communications*.

Personal Role Radio (PRR)
This is a small UHF radio supplied by Marconi which will be issued down to the individual level in the infantry and in other situations where communications to the individual is required. Production is in progress and the radio is in operational use. It was reportedly a great success in Afghanistan and Iraq in 2002 and 2003. The USMC has purchased several thousand.

VHF Radio
These will be the Bowman Advanced Digital Radio (ADR) family, supplied by ITT. The ADR family shares common RF and baseband circuitry and software to support mixed secure voice/data RF nets, IP networking and GPS position reporting. The radios are equipped with embedded IP routers and GPS. The ADR family is fully software-reprogrammable. The hardware architecture is designed to enable

The Bowman High Capacity Data Radio (ITT)   *NEW*/1030918

The Bowman Advanced Digital Radio Airborne (ADRA) - UK/ARC-341 (ITT)   *NEW*/1030915

specific national crypto to be easily implemented. The family members are as follows:

Bowman Advanced Digital Radio Plus (ADR+) (Vehicular and Manpack Configuration)
ADR+ is the VHF vehicle and manpack radio member of the Advanced Digital Radio (ADR) family of VHF radios developed initially for the UK Bowman programme. ADR+ is fully compatible with the ADRP VHF Portable radio and the ADRA VHF Airborne radio. It uses a rechargeable Lithium-ion smart battery which, together with power-saving modes, provides a long operational time between battery recharges. It is designed for tracked, wheeled and shipborne environments.

Specifications
(a) Vehicle power: through radio mount
(b) Manpack power: Smart rechargeable lithium-ion battery
(c) Size: 185 × 88 × 234 mm
(d) Weight: 3.4 kg
(e) Temperature range: −40 to +71°C

Bowman Advanced Digital Radio (Portable) (ADRP)
ADRP is the lightweight VHF Portable member of the ADR family. It features a remote MMI for improved flexibility and is fully compatible with the ADR+ and the ADRA. A smart Lithium-ion battery provides extended operational deployments.

Specifications
(a) Power: Smart Rechargeable Lithium-ion battery
(b) Temperature range: −40 to +71°C
(c) Size: 44 × 94 × 194 mm
(d) Weight: 1.2 kg

Bowman Advanced Digital Radio Airborne (ADRA) - UK/ARC-341
The ADRA provides a Bowman VHF radio capability to helicopters and fixed-wing aircraft that is fully compatible with the ADR+ and the Bowman VHF Portable Transceiver (VPT). It provides secure anti-jam voice and data facilities and operates in a shared voice/data environment using Carrier Sense Multiple Access (CSMA) techniques. The ADR features MIL-STD-1553 remote control and serial RS-422 control for legacy platforms without MIL-STD-1553 capability. It is derived from, and has the same installation footprint as, the ARC-20m SINCGARS Airborne radio. A flexible software architecture will allow the ADRA to assume a variety of roles.

Specifications
(a) Primary power: 28 V DC per MIL-STD-704
(b) Temperature range: −55 to +90°C
(c) Hold-Up Power: Lithium-ion 3.6 V 1.6A

The Radio Control and Display Unit (RCDU)
This provides a remote-control and display facility to the ADRA. The RCDU supports control of up to two ADRA radios, the LCD display and keypad being shared between the two. Dedicated Function Control and Preset Selection switches are provided for each radio to allow rapid access to regularly used functions. Less common functions can be accessed by means of the keypad using a simple menu structure. The RCDU is designed for use in aircraft that are not fitted with MIL-STD-1553 bus control. Discrete control lines from the RCDU control power on/off switching for each radio and emergency crypto purge (E-Purge). The keypad layout and associated menu structures are very similar to those used on the ADR+. The LCD display, keypad and switch legends are provided with backlighting compatible with Night Vision Goggle (NVG) and are clearly readable in full daylight.

Specifications
(a) Power: Aircraft 28 V DC power supply
(b) Temperature range: −40 to +71°C
(c) Size: 86 × 146 × 85 mm
(d) Weight: < 1.5 kg

HF Radio
This will be the HF Harris Falcon II radio.

High Capacity Data Radio (HCDR)

The traffic loading resulting from Digitisation will exceed the capacity which radios can provide. As a result an additional layer is required to give the necessary capacity. This layer will be provided by a secure, data-only radio known as the HCDR, deployed within the operational area to form self-configuring networks. The data service will automatically route its messages along HCDR links. This requirement will be met by a variant of the ITT Mercury Wideband Network Radio (see separate entry). Its predecessor, the NTDR, is utilised by the US Army as the backbone for its first two digitised divisions and the HCDR itself will provide the data backbone for the UK's Bowman Digitisation Programme.

The HCDR implements networking protocols that do not require fixed-base stations. Without a network control station, these protocols enable the system to form and manage self-organising networks, or clusters, and allow for dynamic movement of data across the battlefield by the user without loss of connectivity or need for user intervention. The HCDR is designed using an open architecture, commercially available plug-and-play hardware with rapidly programmable software and can be connected to any battlefield system via standard interfaces using standard IP protocols.

Specifications

(a) Power: 28 V DC per MIL-STD-704
(b) Temperature range: −51 to +71°C
(c) Size: 194 × 41 × 297 mm
(d) Weight: 16 kg

Ancillaries

Thales will provide all the radio audio ancillaries from their Arrowhead range.

MESHnet

Provided by General Dynamics, this IP based system is the core of the information distribution system. (See separate entry).

Battle Management System

The basic Bowman battle management system will consist of the Common Battlefield Application Toolset (ComBAT), Infrastructure (I) and armoured Platform Battlefield Information System Application (P BISA), commonly known as "CIP". CIP is three interrelated projects procured as a single entity. (See separate entry entitled 'ComBAT and P BISA').

Data terminals

L-3 Communications, Ruggedised Command and Control Solutions, will provide Personal User Data Terminals (PUDT). In October 2002 it was announced that Bowman Management Data Terminals (BMDT), Vehicle User Data Terminals (VUDT) and Dismountable User Data Terminals (DUDT) will be supplied by Paravant (now DRS), based on its Scorpion ruggedised portable computer. (See separate entries)

GPS

Rockwell Collins will provide more than 37,000 embedded and hand-held GPS equipments. These will be a key element in providing situational awareness for commanders, providing data for automatically updated displays and thus reducing the considerable amount of locational traffic currently passed on military nets.

Network management

BAE SYSTEMS will provide the Communications Management Information System (CMIS) which will allow network managers to control the system (see separate entry) and the Bowman Logistic Information System (BLIS) which will provide essential information to enable the system to be sustained.

Apache connectivity

In September 2003 it was announced that Aerosystems International had been contracted to create the software solution known as the Bowman Apache Mission Planning System (BAMPS) which will enable airborne UK Apache helicopters to exchange data messages with the Bowman network using IDM and a modified Network Access Unit (NAU) (see MESHneT).

Status

In development. Preliminary climatic trials involving light equipment scales (manpacks only) took place in 2003 with hot weather trials in the US and tropical trials in Brunei. Arctic trials in Alaska are timetabled for early 2004.

Initial technical field trials with 40 Land Rover-borne installations began in March 2003. Training for the initial battalion (1st Battalion, the Royal Anglian Regiment) began in July 2003. The first Brigade to be converted will be 12 Mechanised Brigade. The first brigade-level trial, which will be conducted using two mechanised battle groups and a brigade headquarters, is expected in March 2004. Bowman is expected to be formally declared operational later that month. A second (armoured) brigade trial is planned for late 2004. A rolling programme of conversion will take place thereafter, together with the integration of BISAs as part of the Digitisation programme.

Contractor

General Dynamics UK Ltd(Prime), Oakdale, South Wales.
ITT Industries Defence Ltd , Basingstoke, Hants.

UPDATED

Frequency Assignment Management Equipment (FAME)

In mid-1989 SD (now EDS) was awarded a contract by the MoD to develop and supply Frequency Assignment Management Equipment (FAME) for the British Army.

The system is designed to meet the current and future frequency management requirements in the VHF and UHF bands of the Ptarmigan digital tactical area communications network system.

FAME uses off-the-shelf rugged microprocessors and is programmed in Ada. In addition to dynamic frequency assignment employing state-of-the-art algorithms, FAME offers access by remote users and a link planning facility including path profile analysis.

Status

FAME was accepted into service in October 1992 following extensive trials, including operational deployment during the 1990-91 Gulf War. In 2002 it was still in service as an integral part of the Ptarmigan system.

Contractor

EDS Defence Ltd, Hook, Hampshire.

VERIFIED

MultiRole System 2000 (MRS 2000)

The MultiRole System 2000, or MRS 2000, is a mobile and intelligent communication system that offers secure and survivable networks for armed forces. With powerful switches and extensive subscriber facilities, MRS 2000 can configure area networks of all sizes and complexity, ranging from small tactical links to large strategic networks. The required configuration can be built up by selecting from a small number of modules. Small in overall size and simple to operate, MRS 2000 equipment has the mobility needed for rapid deployment. The system offers many levels of service, and the basic MRS 2000 voice network can be expanded to handle communications protocols such as internet and Ethernet through multiplexing, packet switching and a distributed data system.

MRS 2000 circuit switch

The MRS 2000 provides a wide range of tactical and strategic circuit switches. The switches can be used in headquarter access or trunk roles in hierarchical chain of command or mesh-connected area networks. MRS 2000 networks can be mobile for tactical operation, static for strategic purposes or hybrid systems where mobile tactical elements move freely within a compatible static backbone network.

Switches can be installed in small, highly mobile, soft skin or armoured fighting vehicles. Alternatively, for strategic applications, switches can be rack-mounted in equipment rooms or bunkers.

A switch consists of a row of modules mounted in a rugged, man-portable case. Common host modules provide central processor, memory and timing functions while other modules can be selected to provide subscriber or trunk interfaces. Subscriber terminals can be digital or analogue and may be connected to the switch directly or through multiplexers.

The case also houses the rubidium frequency standard, when required and all the necessary power supplies, including power to voice terminals and ring tone for analogue telephones.

Switches are managed through Facility Control Terminals (FCTs). These can be simple data terminals or personal computers providing additional capability, such as the off-line preparation of deployment data and subsequent rapid downloading into the switch database.

MRS 2000 task oriented man/machine interface   0010447

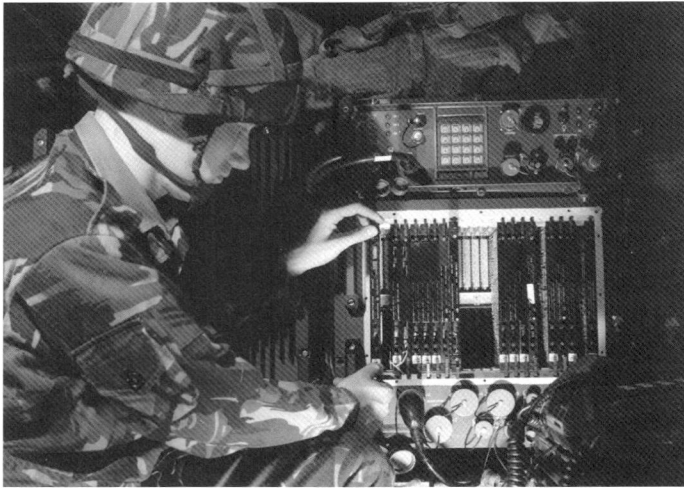

MRS 2000 high-speed distributed data system   0010448

Control may be exercised locally by the FCT or remotely from an FCT at another switch. MRS 2000 Network Management Processors (NMPs) are available to manage large or complex networks.

Circuit switches will operate with CCITT PTT systems, including mobile cellular radio and civil PABXs. Analogue telephones interoperability with other, similarly equipped, military systems is effected through STANAG 4206 multichannel digital gateways and STANAG 5040 single-channel analogue gateways. Switches can operate over most transmission media, including microwave, Satcom and fibre optic links.

MRS 2000 Combat Net Radio Interface (CNRI)

The CNRI connects between the area network circuit switch or multiplexer and a combat net radio and provides dial-up or operator-controlled connections. A CNRI connects directly to an MRS 2000 circuit switch or multiplexer as a subscriber termination. The interface to the combat radio net is via the audio input of the radio.

Combat Radio Diallers (CRDs) provide telephone type keypad functions which enable any combat net radio user to obtain automatic, dial-up access to and from circuit switch networks. In addition, the CRD also provides for selective calling of individuals or groups of other net users. A CRD connects into the audio input of a combat net radio and usually derives its power from the radio, although a battery pack can be provided.

In many operational scenarios, the ability to link forward combat nets with the trunk network is essential. Calls across the trunk network, between different nets, can be made by combat radio users connected to different CNRIs. MRS 2000 CNR interface units provide a cost-effective and flexible solution to these requirements. The interfaces are easily installed and will operate with in-service HF, VHF and UHF radio equipment.

Combat radio users, connected to the trunk network by CNRIs, are allocated unique directory numbers and can gain access to network features such as conference, precedence, pre-emption and internetwork gateways.

The use of CRDs provides increased functionality within existing combat radio nets. In addition selective calls between combat net users, dialled collective and all stations calls are also provided.

MRS 2000 advanced combat net radio interface

The MRS 2000 Advanced Combat Net Radio Interface (ACNRI) allows calls to be made between different combat nets and between these combat nets and the MRS 2000 network. Access is via ACNRI at a radio access point that can support several simultaneous calls. Combat net radio subscribers can move freely because they are not tied to a single ACNRI and no operator action is required.

A radio access point consists of an MRS 2000 circuit switch, a number of radios and an ACNRI to control each radio. The circuit switch and radios can be existing equipment. At least four radio channels can be supported, determined by the frequencies and co-siting properties of the radios and antennas. The radio channels are automatically allocated to traffic and signalling as required. The radio access point does not need a dedicated circuit switch and can be located at any circuit switch near to the required combat net. The combat net radios can use frequency hopping and encryption for security and resistance to ECM and should operate in the 30 to 88 MHz band for optimum range coverage.

Each ACNRI manages radio operations including parameters such as frequency sets. The ACNRI communicate with each other and with the circuit switch to set up calls and allocate them to traffic. An MRS 2000 circuit switch can support two radio access points, each with separate radios, and a number of trunk groups and local subscribers at the same time. A radio access point can be installed in any suitable container or vehicle such as a Land Rover. An ACNRI allows combat net subscribers to call directly any other mobile or static subscriber in the network and gives combat net subscribers comparable MRS 2000 network access and facilities. Special protocols are available to provide quasi-duplex operation on data.

Combat net subscribers are not tied to a single radio access point and can roam freely between areas covered by other radio access points without operator action. Each subscriber has a unique personal directory number which can be located anywhere in the network. Subscribers can affiliate as static or mobile subscribers and use the same subscriber profile. The network recognises that a subscriber is static or mobile from the parameters of the line and allows access to all available and permitted facilities.

To take full advantage of ACNRI, the combat net radios must have dialler facilities. The BAE Systems Combat Net Radio Dialler (CRD) provides a telephone type keypad so that any combat net radio subscriber can obtain automatic dial-up access to the MRS 2000 network and selective calling of other combat net subscribers. A CRD connects to the audio input of a combat net radio and derives power from the radio or from a separate battery pack.

MRS 2000 DECT link

The Digital European Cordless Telecommunications (DECT) link system provides wireless voice communication for headquarters staff and similar locations where users require to move around within a small area. Users can make and receive calls to other wireless terminals and can access all the features of the MRS network.

The system provides instant communications for important headquarters users without the need to lay cable to connect telephones to the switch. Full remote wide MRS subscriber features can be provided to the DECT link user within moments of enabling a trunk link to connect the switch into the network.

The DECT standard combines high capacity and frequency efficiency with a self-organising air interface. Frequency planning is unnecessary - an important feature when frequencies are likely to be restricted and movement frequent. Terminals monitor the entire frequency range and decide on the most suitable carrier and time slot, which are allocated by the base station. The system tolerates channel interference using automatic intracell handover. Speech is encoded at 32 kbit/s using an ADPCM encoding scheme.

The base station provides two lines of connection to an MRS circuit switch or multiplexer. The connection uses a standard analogue loop, enabling the base station to be located some distance from the switch. Each base station can support up to six mobile terminals. A number of base stations may be deployed in the same area to support more mobile terminals.

The mobile terminal consists of a compact, light handset together with a cradle which incorporates a battery charger, the batteries providing up to 40 hours use between charging. Subscribers can make local calls to others on the same base station, as well as make and receive calls to and from other subscribers in the MRS network. Once they have gained access to the MRS network through the base station, users can access the full range of MRS facilities. Pressed keys allow communication with half-duplex subscribers (such as combat net radios) and conference calls.

The base station can be programmed to direct incoming calls from MRS to one or more handsets. The person who answers the call can forward it to another terminal on the same base station.

MRS 2000 Digital Voice Terminal (DVT)

The MRS Digital Voice Terminal (DVT) provides voice and data access to all the facilities of the MRS circuit switch network and provides data access to a packet switch.

Tough, compact, lightweight and simple to operate, the DVT is designed for reliable performance under severe battlefield conditions. It is easy to maintain and offers high reliability. Adaptable siting arrangements allow it to be desk- or wall-mounted.

DVTs are connected to circuit switches or multiplexers by a four-wire cable up to 4 km long. A two-wire interface is available as an option. Power may be supplied from circuit switches and multiplexers via the cable phantom, or provided locally. MRS DVTs can be connected 'back-to-back' providing sole-user circuits between subscribers.

The full range of MRS features is available to DVT users, plus last number redial. Sounder volume and indicator brightness can be controlled from the keypad and a whisper mode is provided. A handset presser switch is provided for conferences, use in noisy environments and for half-duplex radio links using the Combat Net Radio Interface (CNRI).

Teleprinters, facsimiles, PCs and other data devices can be connected to the data port. Calls can be set up in voice or data mode, and the mode can be changed while calls are in progress.

The DVT digitises voice information using Continuous Variable Slope Delta (CVSD) modulation, which is user selectable at 16 kbit/s or 32 kbit/s rates. When required to operate at 16 kbit/s (for example, to make a call over a radio net), the DVT is automatically set to the correct rate by the circuit switch.

MRS 2000 small circuit switch and switching multiplexer   0010449

For terminal-to-terminal or host-to-terminal communications, data mode calls can be set up and cleared down either manually or automatically. Alternatively, a call may be set up in voice mode for positive identification and then switched to data mode for traffic. Data rate and mode are controlled from the DVT keyboard. Synchronous and asynchronous data modes of up to 19.2 kbit/s are provided, with error protection to ensure data integrity.

MRS 2000 high-speed distributed data system

The Distributed Data System (DDS) is a new, successfully demonstrated feature that greatly enhances the data throughput performance of an MRS 2000 network. DDS provides a very fast bearer overlay for data communications to meet the needs of advanced command and control systems without compromising the ability to work in a tactical environment. DDS offers a faster and higher capacity solution than that offered by earlier packet switches.

Any existing MRS system can be easily upgraded for DDS by the simple exchange of multichannel group cards in the circuit switches where high-capacity data transfer is required.

DDS can be applied throughout an entire MRS 2000 network or only to those circuit switches that carry large volumes of data. The DDS network is linked by synchronous Point-to-Point Protocol (PPP) trunks. These trunks have Forward Error Correction (FEC) applied on a link-to-link basis to correct random errors. The block error protection feature of PPP efficiently corrects burst errors.

The upgraded card design manages the available multichannel bandwidth and the allocation of channels to voice and data. The data channels are bit interleaved into a single high-speed channel.

The DDS subnetwork is survivable and allows large amounts of data traffic, between HQ and the theatre of operation,, to be carried successfully at data rates considerably faster than with conventional switches. Speeds greater than ISDN rates can be provided.

DDS uses embedded communications processors that combine high-density processing power with embedded serial communications controllers. The data switches use industry standard management information bases that can be interrogated locally or remotely using standard Simple Network Management Protocol (SNMP).

DDS is inherently self-managing and needs minimum modification to an existing system. The high-speed data can be managed by a separate terminal or integrated with the management system at workstations. Dial-up calls can be made via customer application programmes.

The DDS uses an enhanced range of embedded communications processors that combine high-density processing power with embedded serial communications controllers. Their design has already considered the possibility of Asynchronous Transfer Mode (ATM) in the operating system, hence the use of DDS has a confirmed future within the general industry migration to an ATM systems solution. The DDS is designed to work with the higher transmission rates that will be provided by the new generation of ATM ready trunk radios.

MRS 2000 multiplexer

The MRS 2000 multiplexer provides a method of extending full trunk network service to isolated headquarters and other small groups of subscribers and has a wide range of subscriber interfaces. It combines 30 channels into a single bit stream for transmission as a EUROCOM loop group over wire, fibre optic or radio carriers. Alternatively the 30 channels can be shared between two 15-channel multiplexers, interconnected via the extension port, which allows them to operate in separate locations interconnected by cable or a radio relay link. On the loop side a wide range of analogue and digital equipments and interfaces can be connected. All termination types can be mixed within a single unit and multiplexers can be interconnected on the group or loop sides to provide simple non-switched or partially switched digital networks.

MRS 2000 multiplexers can operate over radio relay, microwave, troposcatter, Satcom, conventional cable and fibre optic transmission systems and can interwork with civil/CCITT networks and provide interface points between mobile tactical systems and static infrastructure communications which are often based on civil equipment. The STANAG 5040 interface provides a gateway to other similarly equipped systems. MRS 2000 multiplexers can be integrated with existing civil or military communications to meet initial needs for security and more resilient communications. Subsequently, these systems can be built up by adding switching, mobile access, improved data and other facilities to the level required.

The multiplexers can be diagnosed to line card level from the keypad and the faulty line cards replaced by front access, without taking the multiplexer out of service.

MRS 2000 Network Management Processor

The Network Management Processor (NMP) is an automated support tool designed to enable the effective management of trunk communications networks. It provides communications managers with the comprehensive facilities necessary to plan and configure networks rapidly and easily to meet changing operational conditions.

Management of a trunk network is divided into three levels:
(a) System Executive and Plans (SEP). This is the top level of network management. The MRS 2000 NMP gives planners rapid access to all important

MRS 2000 packet switch 0010450

network data such as locations, connectivity, status of communications, frequency allocations, reserves and states of readiness. It provides planners with the facilities to generate and amend network plans and orders.
(b) Operational System Control (OSC). Second management level OSCs implement the orders and directives from SEP, monitoring and controlling some five to ten nodes in their allocated areas of responsibility. They have appropriate access to the NMP management database and control of their own part of it. OSC managers can manage remote switches through the NMP.
(c) Facilities Control (FC) is responsible for the direct technical control of nodal switches and, where appropriate, their connected equipment.
At each level, the NMP provides the requisite management tools, either directly for the SEP and OSC roles or via a remote connection to a circuit switch for facilities control.

The MRS 2000 NMP is a single 19 in unit with disk backing store. It houses a set of modules, several of which are common to other MRS 2000 units. Each NMP provides six data access ports serving either local or remote managers' terminals or printers. Depending on function and workload, managers may be equipped with simple data terminals, PCs or full workstations. NMPs automatically interrogate switch processors, collecting network data and traffic statistics, and replicating switch databases. This data transfer is carried in the signalling channel and does not use traffic channels. Access to and ownership of data can be strictly regulated under password access control.

In a typical manager's position, a PC running Microsoft Windows can display data from two NMP ports, with a third port terminating a printer. This enables the PC to show two simultaneous NMP windows, typically connectivity graphics and related text. Formats, terminology and language can be customised to meet specific needs. New combinations and formats can be created as required. The network database enables managers to extract and bring together related information from many plans, records and displays. Data changes are automatically carried through into related plans and records. For survivability, data is held in any or all NMPs as decided by network Planners.

MRS 2000 packet switch

The MRS 2000 packet switch is designed specifically for data communications between host computers, data terminals, servers and workstations in mobile tactical networks. It also provides low error rate communications for teleprinters, other character mode message terminals and message switches.

Unlike conventional virtual circuit networks, an MRS packet switched network has what the makers claim to be a unique 'connectionless' datagram routeing system. This enables packets to be independently routed, rather than following a single virtual circuit path which can be interrupted. All routes in the network are available and regardless of the routes taken packets will be delivered and sorted into the correct order. MRS 2000 packet switching also incorporates EUROCOM Class 4 data error protection.

MRS 2000 packet switches provide both trunk and access switching functions. A typical MRS 2000 packet switch consists of a single, rugged, portable unit suitable for operation free standing or installed in tracked or wheeled vehicles. The unit may be rack-mounted. It contains a range of modules, including a host set which is common to those in the other MRS switches and a range of line interface cards, all interconnected through a backplane. MRS 2000 packet switches are interconnected by packet trunks within the circuit switched network. This adds the enhanced packet switching service to the capacity, connectivity and communications security of the MRS 2000 circuit switched network. It also provides subscribers anywhere in the circuit switched network with dial-up access to the packet switching service over circuit switched channels.

MRS 2000 combat net radio interface units   0010452

Character mode subscribers, typically teleprinter and message switches, are primarily concerned with asynchronous message traffic. These access the packet switch through an inbuilt X.28 Packet Assembly/Disassembly Device (PAD). Packet subscribers, typically host computers, servers, workstations and remote data terminals access the switch via an X.25 interface. Both direct and circuit switched access allow for X.25 (packet) and X.28 (character mode) equipment to be connected. At the same location, subscribers can be directly connected to the packet switch or can obtain access by dialling for packet switch service through the integrated circuit switched network.

Integrated management of circuit and packet switching is by Facility Control Terminals (FCTs). These can be simple data terminals or PCs, control may be exercised locally or remotely. MRS Network Management Processors (NMPs) can provide remote FCT functions and replicated databases for connected packet switches.

The MRS 2000 packet switch incorporates an X.75 gateway. This provides an interface for packet data traffic between an MRS 2000 packet switched network and other, similarly equipped, civil and military packet switching systems.

MRS 2000 small circuit switch and switching multiplexer

The small switch/switching multiplexer is an integrated addition to the MRS 2000 range and provides cost-effective access for brigade-level or smaller headquarters. The case is the same size as the CTM 300 radio relay equipment and houses modules which are identical with the MRS 2000 circuit switch, giving benefits of operational and logistic commonality. A variety of configurations is possible using the 240 channel non-blocking capacity; ranging from eight 30-channel trunk groups up to 180 subscribers via multiplexer access. Its configuration as a switching multiplexer has two trunks and 36 directly connected subscribers.

The small circuit switch/switching multiplexer can operate without the need for management terminals using data prepared in barracks and stored on non-volatile memory. When online management is required, switches are managed through a Facility Control Terminal (FCT). These range from simple data terminals to personal computers, such as a laptop PC using a Windows-based man/machine interface.

In common with the MRS 2000 circuit switch, the small switch/switching multiplexer will operate with CCITT PTT systems, including mobile cellular radio, civil PABXs and analogue telephones. Interoperability with other, similarly equipped, military systems is effected through STANAG 4206 multichannel digital gateways and through STANAG 5040 single channel analogue gateways. Via the MRS CNRI, it can automatically access combat radio networks.

MRS 2000 Task Oriented Man/Machine Interface (TOMMI)

The MRS 2000 circuit switch is configured and managed by a process called facilities control. This process requires the operator to connect a terminal or personal computer to the circuit switch to access the built-in text based controls. The Task Oriented Man/Machine Interface (TOMMI) is a software package that will display these text-based controls in a 'user-friendly' format to ease the management of a circuit switch. The program controls built into a circuit switch are also adequate to manage a small network. TOMMI can manage a small network from any circuit switch in that network.

The management philosophy is to minimise the requirement for operator skill. The operator does not need to know and understand complex text commands, so dedicated staff with specific experience are not required. Set up and reconfiguration of a circuit switch or a small network is via menu-driven default selections which guide the operator through task-oriented displays.

The application platform is a laptop, notebook or palmtop computer built to commercial, industrial or military specification, thereby reducing procurement costs and easing maintenance. The computer can be connected permanently or connected only when activity is needed and can manage any circuit switch from any other in the network. The interface to an MRS 2000 circuit switch is via the standard personal computer serial controller. The TOMMI software is written in a high-level language and so is relatively simple to upgrade as specific requirements develop.

Specifications

MRS 2000 circuit switch
(a) Network and user facilities
    Multilevel precedence and pre-emption at any 3 from 10 levels multisite conference and broadcast (random, predetermined and meet-me) with a wide range of additional options.
(b) Group terminations
    Switching capacity: 480 non-blocking channels

Number of terminations: up to 16 30-channel trunk or multiplexer groups; up to 84 direct subscribers (96 in stand-alone switch); up to 420 subscribers via multiplexer access; wide range of alternative configurations available.
    Trunk or loop groups: 16, 32 or 64 channels at 16 kbit/s or 32 kbit/s
    Group interface (operator configurable): EUROCOM trunk or loop group, interface point A; STANAG 4206-4212 multichannel digital gateway
(c) Subscriber terminations
    Voice interfaces: delta voice terminal (digital telephone) with V24/V28 data port; combat net radio interface; loop disconnect, MF and magneto telephones; analogue PABX/PTT; wireless access for HQ staff
    Facsimile: analogue groups 1-3; STANAG 5000
    Data OSI level 1: V10/X.26 unbalanced circuits
(d) Physical and electrical
    V24 physical
    Interoperates with:
    V11/X.27 balanced circuits
    V28 electrical
    RS-232 physical and electrical
    X.20/X.20bis asynchronous
    X.21/X.21bis synchronous (suitable for DTE access to packet switch)
(e) Data rates
    Asynchronous: 50-4,800 baud; teleprinter or 75-4,800 bit/s data with RMVD (EUROCOM Class 2)
    Synchronous: 8/16/32 kbit/s, EUROCOM Class 1; up to 4.8 kbit/s with RMVD, EUROCOM Class 3
(f) System functions
    Multichannel trunk group signalling: common channel using EUROCOM IB4 or HDLC
    Multichannel loop group signalling: in-band, EUROCOM ID3
    Digital loop signalling: EUROCOM ID3 on digital loops
    Network synchronisation: plesiochronous; bit integrity period 24 h; clock stability better than 1 in 10^{-9}
    Network routeing: delegated with route preference table (automatically generated)
    Subscriber location: self-learning, frequently called list with flood search back up
    Switch operator facilities: local and remote facility control and fault diagnosis to module level; offline configuration/data preparation; cluster management for multishelf switches
(g) Physical characteristics
    Dimensions: 400 × 484 × 444 mm
    Weight: 42 kg
    Power: 28 V DC, typically 140 W; AC options available
    Environment: : −31 to +55°C (operating); up to 95% RH at 40°C; −40 to +85°C (storage)
    EMC: MIL-STD-461 C

CNRI
Switch interface: EUROCOM ID3, 16 kbit/s or 32 kbit/s
Radio interface: FSK tones to BELL 202, 150 baud asynchronous with error correction
Supervisory tones: EUROCOM ID3
Line termination
Electrical: EUROCOM ID4, 4-wire connection
Line rate: 16 kbit/s or 32 bit/s
Voice coding: EUROCOM IA8, 16 kbit/s or 32 kbit/s

Radio interface
Frequency response: 300-3,400 Hz
Outputs: 48 mV peak-to-peak max into 600 Ω balanced; 24 mV peak-to-peak max into 300 Ω unbalanced
Inputs: 2.8 mV to 2.8 V peak-to-peak into 600 Ω balanced; 1.8 mV to 1.8 V peak-to-peak into 1,000 Ω unbalanced

Physical characteristics
Dimensions: 120 × 276 × 250 mm
Weight: 4.8 kg
Power: 28 V DC, DEF STAN 61-5 Part 6, 3 W

Combat Radio Dialler (CRD) - Radio interface
Frequency response: 300-3,400 Hz
Outputs: 48 mV peak-to-peak max into 300 Ω balanced, 24 mV peak-to-peak max into 300 Ω unbalanced
Inputs: 1.8 mV to 1.8 V peak-to-peak into 1,000 Ω unbalanced
Signalling: FSK tones to BELL 202, 150 baud asynchronous with error correction
Supervisory tones: EUROCOM ID3
Power: from radio harness or 9 V battery pack
Dimensions: 48 X 162 X 82 mm (with battery box, length 243 mm)
Weight: 0.65 kg (1.15 kg with battery box, incl batteries)

General characteristics (CNRI & CRD)
Environment: : −31 to +55°C (operating); up to 95% RH at 40°C; −40 to +85°C (storage); immersion proof
EMC: MIL-STD-461 C

DECT link
Facilities. Local calls, calls to and from MRS network, call barring, call hold, call forward, abbreviated dialling (10 stored numbers), speech volume, ringer volume,

MRS 2000 circuit switch   0010453

affiliation and deaffiliation of hand sets to base station, range warning, PIN protection for system changes.
Network interface: 2-wire analogue loop MF or loop disconnect signalling

Wireless interface
(a) General
Standard: DECT
Frequency range: 1,880-1,900 MHz
Channel separation: 1,728 kHz
Modulation: GFSK
Encoding rate: 32 kbit/s
Range: up to 300 m in free air; up to 50 m in buildings
(b) Power
Base station and charger: 220/230 V AC, 50 Hz
Mobile terminal battery: up to 40 h (standby); up to 6.5 h (in traffic)
(c) Environment
Operating: 0 to +55°C
Storage: −10 to +60°C
Humidity: 20-70%
(d) Dimensions
Base station: 170 × 180 × 35 mm
Handset: 175 × 55 × 25 mm
Handset weight: <200 g

DVT
(a) Line interface
Electrical: EUROCOM ID4, 4-wire connection (2-wire available as option) with phantom power
Line rate: 16 kbit/s or 32 kbit/s
Signalling: 8 bit cyclically permutable code words, EUROCOM ID3
(b) Voice interface
Encoding: EUROCOM IA9 and STANAG 4209, 16 kbit/s or 32 kbit/s
Supervisory: EUROCOM ID3
(c) Data interface
Electrical interface: CCITT V10 and V11; EIA RS-449, MIL-STD-188-114A
Interface types: telegraph terminal; X.20bis; X.21; X.21bis; EUROCOM 'J' (V25bis); STANAG 5000 facsimile; Hayes signalling
Data processing: asynchronous, 50-4,800 bit/s with RMVD, (EUROCOM IA9 Class 2); isochronous, 2,400-4,800 bit/s with RMVD (EUROCOM IA9 Class 3); synchronous 1,200-19,200 bit/s with BCH forward error correction (EUROCOM IA9 Class 4); asynchronous 2,400-19,200 bit/s with BCH forward error correction (EUROCOM IA9 Class 4); synchronous 16/32 kbit/s (EUROCOM IA9 Class 1)
(d) Power input
Phantom supply: 20-56 V DC (48 V nominal)
Local supply: 5.5-28 V DC
Power consumption: 1.5 W off-hook; 0.35 W on-hook
(e) Physical characteristics
Dimensions: 241 × 171 × 99 mm
Weight: <2.8 kg
(f) Environment
Lightning protection provided
Operating: −30 to +55°C
Storage: −57 to +71°C
Humidity: 98% at 65°C
Rain: 125 mm/h
Immersion: 0.9 m for 2 h
Reliability
MTBF: 50,000 h at 40°C

High-speed distributed data system
Network and user facilities: direct access to the circuit switch; remote access via extension router/hub; dial-up calls via customer application programmes

Error correction: random errors; forward error correction by EUROCOM Class 4; burst errors; CRC detection within HDLC with retransmission of error blocks
Capacity: up to 12 DDS access points at each circuit switch
Hardware technology: VLSI/CMOS with FPGA and SMT
Traffic access rates: IEEE-802.3 LAN at 10 Mbit/s; synchronous PPP at up to 256 kbit/s; asynchronous PPP at up to 76.8 kbit/s
Network data channel: up to 512 kbit/s can be configured by combining 16 channels of a 1 Mbit/s or 2 Mbit/s multichannel group
Network capacity: each circuit switch with DDS supports a total switching capacity up to 1,500 datagram packets/s
Data interfaces: PPP V11; LAN to IEEE-803.2
Dimensions: the card assembly fits in an MRS circuit switch

Multiplexer
(a) Group terminations
Groups terminated: one
Channel bit rate: 16 or 32 kbit/s
Group structures: 16 × 16 kbit/s - 256 kbit/s; 16 × 32 kbit/s - 512 kbit/s; 32 × 16 kbit/s - 512 kbit/s; 32 × 32 kbit/s - 1,024 kbit/s
Group interface: EUROCOM A
(b) Subscriber terminations
Voice interfaces: delta voice terminal (digital telephone) with V24/V28 data port; CNRI; loop disconnect, MF and magneto telephones; analogue PABX/PTT STANAG 5040
Telegraph/data interfaces: V24 (V10); 50 to 4,800 bit/s asynchronous (EUROCOM Data Class 2); 300 to 4,800 bit/s synchronous (EUROCOM Data Class 3); 8/16/32 kbit/s synchronous (EUROCOM Data Class 1); X.20/X.20bis 50-4,800 bit/s asynchronous; X.21/X.21bis 300-4,800 bit/s synchronous; RMVD error protection up to 4,800 bits
Facsimile: analogue Groups 1-3; STANAG 5000
(c) Signalling
Groups: in-channel using EUROCOM ID3
Digital loops: EUROCOM ID3
Analogue loops: Loop disconnect, DTMF or magneto LB
Network synchronisation: slaved to switch or stand-alone
Hardware technology: VLSI/CMOS, semi-custom ASICs
(d) Physical characteristics
Dimensions: 178 × 484 × 454 mm
Weight: 21 kg max
Environment: −31 to +55°C, up to 95% RH at 40°C (operating); −40.C to +85°C EMC MIL-STD-461C (storage)
Power input: 28 V DC (DEF-STAN 61.5 part 6), typically 40 W 100/230 V, 50/60/400 Hz

Network Management Processor (NMP)
(a) Facilities
Network monitoring: switches are monitored to provide automatic detection of hardware failures, link failures and switches joining or leaving the network
Network statistics: traffic loading, grade of service, failures and so on, are automatically monitored and stored and can be viewed when required
Access control: password protection plus additional access controls at the data level provide security
Online database: powerful database with data replication between NMPs for accessibility and survivability. Switch databases are also replicated by NMPs
Screen formats: data driven displays. A range of layouts, terminology, language, special schema or free format displays can be supplied as required
Screen editing: user-friendly screen editor with automatic prompting for data entry
Choice of terminals: colour graphics PCs, workstations or standard VT100 terminals can be used
Multiple windows: available on PCs and workstations
Flexible displays: displays are data driven to simplify changes to headings, prompts, validation criteria and so on
Predeployment data preparation: switch commands can be entered in advance and stored in NMP command files, which can be rapidly downloaded to switches when they join the network
Planning facility: enables managers to create and amend plans for network deployments
Graphics displays: graphics of the current network highlight failed links and planned links which are missing. Plans for future deployments can be checked via graphics
(b) Connection:
NMP to switch. High integrity HDLC datalink. Does not use network traffic channels
workstation interfaces standard VT100/VT240/RS-232
backing store - standard disk drive or solid-state disk up to 600 Mbytes
(c) Physical characteristics
Dimensions: 399 × 482 × 390 mm
Weight: 26 kg
Power: 28 V DC or optional AC input, typically 65 W
Environment: −31 to +55°C, up to 95% RH at 40°C (operating); −40 to 85°C, EMC MIL-STD-461 C (storage)

Packet switch
(a) Network and user facilities:
direct access or via circuit switched network
packet or character mode access
dial up or hot line data virtual calls

MRS 2000 digital voice terminal   0010451

(b) Switching capacity

Data ports: up to 48 ports for use as trunk, access or gateway, maximum 16 trunk ports and 6 gateway ports

Virtual calls: up to 255 per port, total 512 per switch

Traffic: up to 64 kbit/s per port, total 512 kbit/s

Switching rates: in excess of 200 packets/s

Circuit switch interface: 2 × 32 channel EUROCOM trunk groups; single or dual-homed

(c) Data interfaces

Direct access: X.25 (packet mode); X.28 (character mode)

Internetwork: X.75 gateway

Circuit switch access: X.25 EUROCOM, data classes 1, 3 and 4 (packet mode); X.28 EUROCOM, data classes 2 and 4 (character mode)

Data rates: 2.4 to 64 kbit/s (direct access gateway and packet mode); up to 9.6 kbit/s (character mode)

Circuit switch access: 2.4 to 32 kbit/s (packet mode); up to 9.6 kbit/s (character mode)

Packet trunks: up to 32 kbit/s or 19.2 kbit/s with additional error protection (EUROCOM Data Class 4)

Network routeing: delegated routeing with automatically generated route preference table and congestion avoidance

Subscriber location: self-learning, frequently called list with flood search backup

Hardware technology: VLSI/CMOS, semi-custom ASICs, FPGAs and SMT

Switch operator facilities: local and remote facility control and fault diagnosis to module level; offline configuration/data preparation; cluster management for multishelf switches

(d) Physical characteristics

Dimensions: 400 × 484 × 444 mm

Weight: 42 kg

Power: 28 V DC, typically 90 W; AC options available

Environment: −31 to +55°C, up to 95% RH at 40°C (operating); −40 to +85°C (storage)

EMC: MIL-STD-461 C

Small switch/switching multiplexer

(a) Network and user facilities: multilevel precedence and pre-emption at any 3 from 10 levels, multisite conference and broadcast (random, predetermined and meet-me) with add-on, line grouping, group hunting, camp-on busy with automatic call back, call hold/second call for enquiry/transfer (forward), call diversion (transfer), abbreviated and compressed dialling, deducible directory, operator-initiated semi-permanent circuit, automatic restoration of user-selected circuits, switched hot line, closed user groups, subscriber assistance and call queuing, controlled affiliation, trunk barring (3-level), operator-configurable group interface.

(b) Group termination

Switching capacity: 240 non-blocking channels

Number of terminations: up to 8 30-channel trunk or multiplexer groups; up to 36 direct subscribers (48 in stand-alone switch); up to 180 subscribers via multiplexer access; wide range of alternative configurations available

Trunk or loop groups: 16, 32 or 64 channels at 16 kbit/s or 32 kbit/s

Group interface (operator configurable): EUROCOM trunk or loop group, interface point A; STANAG 4206-4212 multichannel digital gateway

(c) Subscriber terminations

Voice interfaces: delta voice terminal (digital telephone) with V24/V28 data port; CNRI; loop disconnect, MF and magneto telephones; analogue PABX/PTT

Facsimile: analogue groups 1-3; STANAG 5000

Data OSI level 1: V10/X.26 unbalanced circuits

Physical and electrical: V24 physical; interoperates with V11/X.27 balanced circuits, V28 electrical, RS-232 physical and electrical, X.20/X.20bis asynchronous, and X.21/X.21 bis synchronous (suitable for DTE access to packet switch)

(d) Data rates

Asynchronous: 50-4,800 baud teleprinter or 75-4,800 bit/s data with RMVD

Synchronous: 8/16/32 kbit/s, EUROCOM Class 1; up to 4.8 kbit/s with RMVD EUROCOM Class 3

(e) System functions

Multichannel trunk group signalling: common channel using EUROCOM; IB4 or HDLC

Multichannel loop group signalling: in-band, EUROCOM ID3

Digital loop signalling: EUROCOM ID3 on digital loops

Network synchronisation: plesiochronous; bit integrity period 24 h; clock stability better than 1 in 10^9

Network routeing: delegated with route preference table (automatically generated)

Subscriber location: self-learning, frequently called list with flood search back up

Switch operator facilities: local and remote facility control and fault diagnosis to module level; offline configuration/data preparation; cluster management for multishelf switches

(f) Physical characteristics

Dimensions (H × W × D): 290 × 445 × 420 mm

Weight: 25 kg

Power: 28 V DC, typically 75 W; AC options available

Environment: −31 to + 55°C, up to 95% RH at 40°C (operating); −40 to + 85°C (storage)

EMC: MIL-STD-461 C

Status

The MRS 2000 system in various forms is in service with the British Army and with national defence forces in Asia, Australasia, Europe and the Middle East, for strategic and tactical applications. In 1990, MRS 2000 was selected as the basis for the Swiss tactical area communications system. It is the core of the Austrian Army's major area communication system IFMIN and is an element of the Greek Army's Hermes II area communications system.

Contractor

BAE Systems, Christchurch, Dorset.

UPDATED

Ptarmigan Area Communications System

Ptarmigan is the UK's second-generation voice and data battlefield communication system designed to meet the needs of mobile warfare. Originally developed for British forces in Germany as the area communications system to replace BRUIN after the demise of Project Mallard in the 1970s. Its basis is a survivable network of processor-controlled automatic switches and radio centrals interlinked by multichannel radio links over distances up to 25 km. The trunk radio parts of the system carry 16 kbits/s channels in TDM streams of 256 or 512 kbits/s. Distributed management facilities together with fast set up and tear down times are intended to give high mobility to the system. The system's major access node consists of three vehicles and provides connection to the trunk network for a number of subscribers located typically at a major formation headquarters. For survivability it is connected to two separate trunk switches. The secondary access switch is installed in wheeled or armoured vehicles and serves groups of up to 30 subscribers such as brigade headquarters. The secondary access switch is capable of operating either connected to the network or isolated from it. Smaller, more mobile headquarters and isolated users can gain access to the full range of network facilities using SCRA single-channel radio links to radio centrals in the trunk network. These mobile subscribers can transfer from one SCRA radio central to another without any knowledge of the network configuration or of radio frequencies. The system is managed through a number of system control computers which contain network planning data and are connected to the network at trunk nodes. They are accessed by means of data terminals and provide a common network planning database available to all system managers and node commanders. Store and forward capabilities are located at several Ptarmigan nodes so that an automatic message

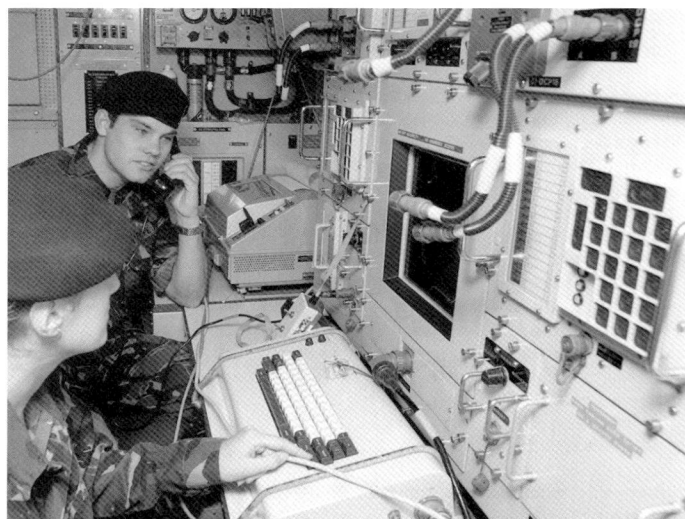

Interior of the Ptarmigan trunk switch   0084898

The Pinzgauer-mounted (SCRA(Switching Central)(Air-Portable), SCRA(SC)(AP)) consisting of the primary communications container vehicle (right) and the power generation and support vehicle (left) (BAE Systems)   **NEW**/0528083

handling facility is always available in spite of the dynamic nature of the network. If headquarters are disconnected, messages can be stored for several hours and delivered when links are re-established. Calls to and from allied formations are possible via tactical interfaces connected by radio relay links to trunk nodes. Embedded into the Ptarmigan network is a packet switched network. The primary packet switches are integrated in the trunk switches, while secondary packet switches are at access nodes.

The manufacturer says that this was the world's first tactical packet switched network to offer X.25 virtual circuit service with survivable datagram transnetwork routeing. Ptarmigan user facilities are available corps-wide for data as well as voice. They include four-digit abbreviated dialling, three-digit compressed dialling, call holding, conference and broadcast calls, call forwarding and transfer, precedence calls and hot line and sole user capability.

Status

Ptarmigan system design was the result of detailed studies and feasibility trials carried out jointly by the UK MoD, Royal Signals and Radar Establishment (RSRE), the British Army Royal School of Signals and industry. The Plessey Company (subsequently Siemens Plessey Systems and now BAE Systems) was appointed prime contractor and system design authority for Ptarmigan in 1973, with responsibility for engineering development of the complete system. The initial development programme was followed by a series of production contracts worth some £500 million. They covered the provision of the full range of Ptarmigan items from small individual equipments, such as subsets, to major vehicle-mounted installations such as switches and SCRA radio centrals. A major phased enhancement programme to provide high-integrity packet switched data, including mobile X.25 packet access, international interfacing and the development of equipment for use in armoured vehicles commenced in 1984 and was completed in 1992. BAE Systems is now the appointed design authority for supporting the system throughout its post-design phase. This covers the full range of support services from components and equipment up to network level. During the 1991 Operation Granby, Ptarmigan was deployed extensively throughout the operational area with extended satellite trunk links, and was heavily used by British and allied forces. The system gained further worldwide recognition when deployed in support of the International Peace Implementation Force (IFOR) in Bosnia. Total investment in Ptarmigan by mid-1992 was approaching £1 billion. In August 1993, a £22 million contract was awarded to modify the system to allow deployment over long distances with satellite links. This modification was fielded in early 1998.

In late 1994, the company announced a £17.5 million award for post-design services. In 1995, the company was also made prime contractor to a programme to supply Multipurpose Communications Equipment (MCE) enabling data terminal users to communicate over Ptarmigan. In early 1996, BAE Systems announced a contract to supply a Ptarmigan modelling facility that will be used to assess the current system and predict the effects of enhancements and new deployments. Integration testing of Ptarmigan with the Interim ACE Rapid Reaction Corps Information System (IARRCIS) was completed by April 1996. In May 1997, the company won a contract to upgrade Ptarmigan's vehicle-mounted power system and was also awarded a project definition to study the technical issues involved in providing a general purpose trunk access port for the system.

In early 1998, BAE Systems won a £25 million contract to enhance Ptarmigan with Mobile Access to Ptarmigan Packet Switching (MAPPS) capability. A combined circuit and packet switch, based on MRS 2000 (see separate entry) will enhance the data handling capacity of the SCRA Central function for 30 installations and will give improved access to mobile subscribers, allowing deployment independently of the main Ptarmigan network. Fifteen of these are existing Bedford vehicle container and pallet installations, redesignated as the SCRA (Switching Central)(Standard) and fifteen will be new, air-portable installations (SCRA(Switching Central)(Air-Portable), SCRA(SC)(AP)), installed in Pinzgauer vehicles. Each Pinzgauer-hosted MAPPS capability requires two vehicles. The first is the primary communications container vehicle while the second supplies power generation and support. The first three pairs were delivered in early 2001. A £17.9 million contract was also awarded to provide a further five years post design services. In January 2003 it was announced that delivery of thirty MAPPS installations was complete.

The MAPPS programme was extended in October 1998 to include the design and installation of General Purpose Trunk Access Port (GP-TAP) software enhancements. These provide software to support interoperability with friendly nations' communications systems in coalition operations and with other UK systems such as CORMORANT and FALCON. GP-TAP was successfully embodied in Germany and the UK in mid-2002.

Contractor
BAE Systems, Christchurch, Dorset.

UPDATED

PVS 5300 HF SSB radio system

The PVS5300 system provides a range of tactical HF manpack and vehicle radios based on the PTR5300 transceiver. The vehicle low-power station has had a 100 W amplifier and ATU added to the range to provide a full military specification 100 W output vehicle station. The radio is microprocessor-controlled and covers the 2 to 30 MHz range. Operating from 12 V power supplies in the manpack role with a transmitter output power of 10 W PEP, the transceiver unit weighs 3.8 kg. The PTR5300 can also be operated from all types of wheeled vehicle from either a 12 V power source or a converter to enable operation from 24 V DC. The PTR5300 utilises a speech processing module which, with ALC, provides whisper operation while maintaining high mean output power.

The PTR5300 offers 280,000 channels at 100 Hz spacing. Any of the channels can be selected by entering the frequency from the radio front panel keypad. The radio's memory can store 10 preselected channels. Operation of the internal antenna tuning unit is fully automatic.

Groundwave range of the radio in manpack role is typically 40 to 50 km, day and night in rolling terrain, with skywave ranges in excess of 2,000 km. In the 100 W vehicle role groundwave ranges are very much extended with, depending on antenna and terrain, ranges of in excess of 100 km recorded.

The PTR5300 is self-diagnostic to module level. The liquid crystal readout provides an indication of faults using preformatted codes. The radio has a range of ancillaries and can be used with a digital message terminal for burst transmission, and an encryption unit.

The radio measures 79 × 215 × 235 mm without battery and 79 × 215 × 305 mm with battery.

Specifications
(a) Modes: J3E USB/LSB, J2A CW, H3E AM
(b) Frequency range: 2-30 MHz
(c) Number of channels: 280,000 at 100 Hz intervals (10 stored)
(d) Power output: 10 or 1 W PEP (manpack, vehicle low-power station); with vehicle amp and ATU nominal 100 W
(e) Temperature range:
   operating: −20 to +70°C
   storage: −35 to +85°C
(f) Immersion: 1 m for 2 h
(g) Altitude
   operating: 2,500 m
   non-operating: 7,500 m
(h) Weight
   excl battery: 3.8 kg
   incl 4 Ah battery: 5.7 kg
   100 W amp: 8 kg

Status
In production in the 1990s and exported to countries in Africa, Asia , Europe and the Middle East. In 1995 IDR reported that a likely customer for a then £12.5 million contract was Zimbabwe.

Contractor
BAE Systems , Christchurch, Dorset.

UPDATED

PVS 5300 HF SSB radio system   0010458

RAVEN 2V combat net radio system

The RAVEN 2V combat net radio system provides multirole battlefield communications from a range of manpack, vehicle and base stations with co-siting multiple-radio capabilities. Voice, data, teleprinter and facsimile communications are protected by an extensive range of advanced Electronic Counter-Counter Measures (ECCM). These include frequency hopping, digital encryption, remote control, frequency offset and burst-data transmission. Most of the VHF range is based on the PTR 4411 transceiver but in addition there is a hand-held radio for use at squad or patrol level. The HF set is now marketed separately as the PTR 5300.

As a manpack the radio can be powered by primary or rechargeable secondary batteries, and can be clipped into a vehicle. As a permanent vehicle installation, the transceiver and associated equipment are mounted on a shock-absorbent tray to form a low- or high-powered mobile station. The station can be connected to a variety of vehicle harnesses such as ROVIS, AN-VIC or those of other US and European manufacture. A remote handset can be connected to the manpack transceiver, using a field cable up to 5 km long, providing voice and transmitter key control. By connecting a remote radio control system, full intercom and radio station operation is provided, again at up to 5 km. An extended front panel is also available.

Operating in the 30 to 88 MHz frequency band, the radio provides 25 kHz channel spacing and an additional 12.5 kHz offset facility. The programmable channel facility allows up to 16 fixed frequencies and hop nets (plus one operator's channel) to be preset before beginning a mission. The stations offer four power levels (50 W, 5 W, 0.5 W and 100 mW) which are selectable via the keypad. In data mode, RAVEN 2V allows exchange of both synchronous and asynchronous data from a range of rates up to 16 kbit/s, essential for modern weapon command and control and data systems.

The handheld features are slightly different. This is a pocket-sized transceiver weighing 750 g complete with battery. Like the others it operates in the 30 to 88 MHz frequency band and provides 2,320 channels at 25 kHz spacings. Ten channels can be preset and can be selected via the radio's keypad. Blade, compressed helical and whip antennas are available, with the antenna type used being dependent upon the required operating range. Both shoulder holster and waist mounting carrying pouches can be supplied.

The HF set has fully automatic tuning over the 2- 29.9999 MHz frequency range, with 100 Hz spacing. This allows 280,000 useable channels. Ten automatic channel settings are available. Transmitter output power is 10 W and the automatic antenna tuner is built in. The set can be used in manpack, vehicle or base station roles. The set can either be clipped into a vehicle or fully mounted using the vehicle power supply; a 100 W adaptor is available.

Specifications
VHF
Technical
(a) Frequency: 30-87.975 MHz
(b) Channel spacing: 25 kHz (50 kHz interoperability mode) with optional +12.5 kHz frequency offset

RAVEN 2V VHF compact 0010460

RAVEN 2V VHF manpack 0010461

(c) Modes of operation: FM voice (F3E); FM data 16 kbit/s (F1D); FM analogue data (F2D)
(d) Voice transmission: digital voice 16 kbit/s (EUROCOM D1 standard)
(e) Data transmission: data MIL-STD-188-144, V24 or V28 CCITT, synchronous or asynchronous (50-9,600 .bit/s with forward error correction)
(f) Deviation: ±5 kHz (±10 kHz interoperability mode)
(g) Squelch: analogue voice - 150 Hz tone; digital squelch
(h) Co-siting: within 3 dB of non co-siting sensitivity at frequency separation of 10%
(i) Power supply: 20-33 V DC (QSTAG 307 compliant)
(j) Transmitter Power output: 4 levels 50 W (with vehicle adaptor), 5 W, 0.5 W and 100 mW

Physical
(k) Weight: 13 kg (manpack configuration)
(l) Dimensions:
Vehicle station: 225 × 405 × 347 mm
Manpack: 86 × 252 × 229 mm

Environmental
(m) −40 to +70°C (operation); −40 to +85°C (storage)
(n) immersion: 1 m for 2 h
(o) freefall: 1.2 m.

ECCM data summary
(p) Full band hopping: up to 2,320 channels
(q) Barred bands: any combination of channels or bands can be barred
(r) Nets: 17 (8 store A, 8 store B and operator).
(s) COMSEC: provision for wide range of internal or external cryptos, 16 kbit/s access for external equipment (for example, Vinson KY57).

Hand-held
(a) Frequency range: 30-88 MHz with 25 kHz channel spacing
(b) Channels: 10 preset in 30-88 MHz range (basic operation); 2,320 channels after the introduction through the keyboard of a password followed by the Rx and Tx frequencies (extended operation)

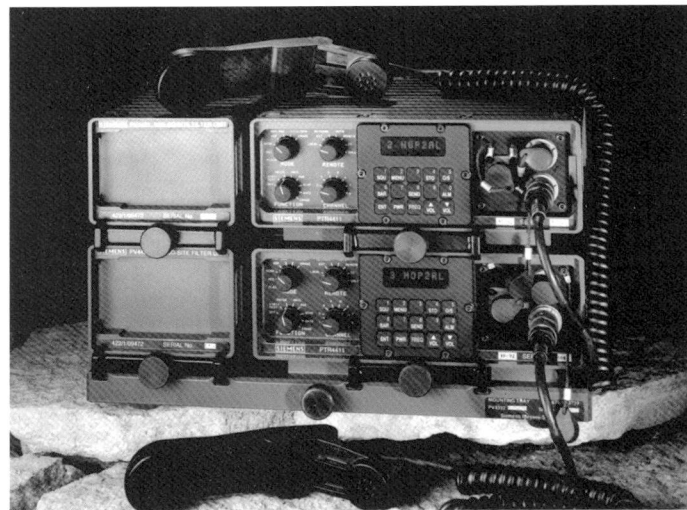

RAVEN 2V VHF vehicle and base station 0010462

RAVEN 2V hand-held in shoulder holster 0010459

(c) Operation mode: simplex and half-duplex (F3 or XMOD)
(d) Power input: 12 V DC
(e) Size: 218 × 65 × 32 mm
(f) Weight: 750 g with batteries
(g) Temperature range: −30 to +55°C
(h) Power output: 1 W ±1 dB at 50 Ω.

HF
(a) Frequency range: 2-29.9999 MHz
(b) Channel spacing: 100 Hz intervals (280,000 channels)
(c) Preset channels: 10
(d) Modes: J3E (USB and LSB), J2A, H3E
(e) Antenna Tuning: Automatic in 0.5 s
(f) Power output: 10 W (100 W vehicle adaptor)

Status
In production. A version was purchased by the Australian Defence Force as Project RAVEN (see separate entry). Sold in small numbers under licence to unspecified customers worldwide. Sold to the UN for use in Sierra Leone. Sold to NATO as part of a contingency pool. Used as the HALO (see separate entry) communications system.

Contractor
BAE Systems, Christchurch, Dorset.

VERIFIED

United States

AN/VRC-100 ground/vehicular HF communications system

The AN/VRC-100 is a ground or mobile high-frequency communication system which uses advanced Digital Signal Processor (DSP) technology with embedded Automatic Link Establishment (ALE), serial tone data modem and anti-jam (ECCM) functions. The AN/VRC-100 is a fully integrated 'plug and play' multimode voice or data communications system configured in a portable case. The AN/ARC 220 is the airborne version and it has identical capabilities.

Applications
The Rockwell AN/VRC-100 is an easy to operate, multifunctional, fully digital high-frequency communication system intended for use in tactical operations centre, air traffic control and vehicular applications such as the High Mobility MultiPurpose Wheeled Vehicle (HMMWV). It is built from Rockwell's successful AN/ARC-220 airborne communications LRU's, a power supply and an audio interface, all packaged in a metallic case.

The AN/VRC-100 provides secure and non-secure voice and data communications, with Automatic Link Establishment (ALE) and Electronic Counter-CounterMeasures (ECCM) capabilities, in the 2.0 to 29.9999 MHz frequency bands. Its primary use is for beyond line of site communication between aircraft, mobile communication centres and other ground and airborne assets. Both standard audio voice and Advanced Narrowband Digital Voice Terminal (ANDVT) are supported. Embedded Automatic Link Establishment (ALE) with linking protection, Electronic Counter-CounterMeasures (ECCM) and combined ALE-ECCM capability are provided in both voice and data modes of operation. A spare card slot is provided in the Receiver/Transmitter LRU for embedding platform specific applications in the AN/VRC-100.

The (V) 1 version includes all functionality, while the (V) 2 removes the ECCM and NSA link protection, for exportability. An audio data port is provided to support interface with the Automatic Target Hand-over System (ATHS). Digital data interfaces compatible with MIL-STD-188-114A allow transmission and reception of data from the Improved Data Modem (IDM) and data secured by either an AIRTERM (KY-100) or a TACTERM (USC-43). Upper SideBand (USB), Lower SideBand (LSB), Amplitude Modulation Equivalent (AME) and Continuous Wave (CW) emission modes are provided. The AN/VRC-100 can be controlled via either the control display unit (provided) or externally from a MIL-STD-1553 databus system controller.

System description
The AN/VRC-100 ground radio uses the AN/ARC-220 airborne Receiver/Transmitter (R/T), Power Amplifier/Coupler (PA/Coupler) and Control Display Unit (CDU) LRUs without modification. Its metal case, with removable top, provides easy access for removal of the LRUs. All controls and radio I/O are located on the front panel.

The AN/VRC-100 case is constructed of 0.09 in aluminium, with primary structural support provided by side rails within the case, which are also used as hold down points. Internal brackets and welded corners provide added rigidity. When hard mounted, the AN/VRC-100 meets vibration requirements for HMMWV ground mobile wheeled vehicles as well as shock requirements for paved roads and Parryman 3 cross-country terrain when tested in accordance with MIL-STD-810E, Method 514.4. Holes in the AN/VRC-100 are provided for drainage.

The AN/VRC-100 provides audio data port and a front-panel speaker for receive audio signal and it provides the capability for H-250 handset operation. The system is powered from 24 V DC vehicular power IAW MIL-STD-1275 or from a 115 or 220 V

AC 50/60 Hz power source. The system 24 V DC power is brought out to a front panel connector for powering accessory equipment. Three 15 ft three-wire power cords (one DC, one 115 V AC and one 220 V AC) are provided with each system. The AN/VRC-100 has syllabic squelch, which provides noise-free voice-activated reception. It provides an embedded modem, as well as embedded ALE (with linking protection) and ECCM. The modem functions are in accordance with the MIL-STD-188-110A and STANAG 4285 with Appendix E. Data rates can range from 75 to 1,200 bits/s in FSK and from 75 to 2,400 bits/s in PSK. The AN/VRC-100 provides ALE in accordance with MIL-STD-188-141A. Linking Protection (LP) is also provided in accordance with ASQB-OSE-TR-92-04 Level AL-1 and an NSA algorithm. The AN/VRC-100 provides ECCM in accordance with MIL-STD-188-148A and CR-CX-0218-001 (army enhanced). ALE for both forms of ECCM is provided. Both forms of ECCM utilise a KGV-10 to provide the randomised sequences required supporting frequency hopping. A single card in the R/T, containing all TRANSEC functions, is removable to support Foreign Military Sales (FMS). The AN/VRC-100 is ready for integration into tomorrow's digital platforms. MIL-STD-188-114A interfaces provide the means to integrate the AN/VRC-100 with the Army's Improved Data Modem (IDM) and Advanced Narrowband Digital Voice Terminals (ANDVT). Global Positioning Satellite (GPS) time and position interfaces allow for position reporting and quick, accurate ECCM initialisation. Operator generated messages up to 500 characters in length can be sent or received utilising the embedded MIL-STD-188-110A modem. Datafill may be provided by any one of the following: a DS-101D interface (supporting the AN/CYZ-10), an RS-232 interface (using PC software like Rockwell's CPS™ mission planning tool), or a MIL-STD-1553 interface. This allows mission data (frequencies, scanlists, hop sets, prestored messages) to be loaded into the radio with a minimum of effort, allowing the operator to quickly reinitialise the AN/VRC-100 for rapidly changing situations. Keyfill is performed using a separate DS-101D port.

The Coupler provides 175 W PEP/100 W average. Transmit power output levels of 175, 50 and 10 W are selectable by operator control. The AN/VRC-100 tunes selected narrowband and broadband antennas open and shorted loop antennas and 50 Ω resistive loads. All frequencies within the HF band are tuned nominally within 1 second (3 seconds maximum). A learned preset tuning feature allows virtually instantaneous tuning of frequencies previously tuned by the system. A receive-only antenna port allows the AN/VRC-100 to interface with a separate antenna to optimise receiver performance. The control display unit within the AN/VRC-100 provides an intuitive operator interface to the transmit and receive system. The passive liquid crystal display provides 4:1 improvement in sunlight readability over comparable CRTs while in darkness, an adjustable backlight maintains an 84 per cent uniformity across the entire screen to ensure night vision goggle compatibility.

Specifications
General
Frequency: 2.0 - 29.9999 MHz in 100 Hz steps
Channels
 20 user-programmable simplex or half-duplex
 20 programmable simplex or half-duplex
 20 programmable automatic link establishment (ALE) scan lists
 12 programmable ECCM hopsets (with ALE capability)
Modes: USB and LSB - voice and data, CW and AME
ALE: IAW MIL-STD-188-141A with linking protection
ECCM: IAW MIL-STD-188-148A (with ALE) IAW CR-CX-0218-001 (army enhanced with ALE)
Modem: IAW MIL-STD-188-110A, STANAG 4285
Power: 28 V DC (550 W transmit), 117 V AC (880 VA transmit) or 220 V AC (880 VA transmit)
Reliability: 1,000 h MTBF min
Mechanical: 22.25 × 22.875 × 8.75 in. (L × W × H); 88.0 lb

Receive
Characteristics: IAW MIL-STD-188-141A
Sensitivity: 0.5 mV (−113 dBm) max for 10 dB (S+N/N)
Noise reduction: IAW CCIR 445-1

Transmit
Characteristics: IAW MIL-STD-188-141A

AN/VRC-100 HF Communications System   0055017

Power output (3 Levels)
  175 W PEP (100 W average)
  50 W PEP and average
  10 W PEP and average
Tune time: 1 s nominal, 3 s max

Control
Display: monochrome passive LCD; 6 lines of 21 characters
Message size: 500 characters max (transmit or receive)
Character size: 0.2 in high
NVG compatible: IAW MIL-L-85762F
Alternate control: Via MIL-STD-1553B

Secure operation
With data sources: KY-100 (AIRTERM), KY-99 (MINTERM), USC-43 (TACTERM), KY-75, IDM, ATHS, GPS (time and position)

Status
The AN/VRC-100 is in service with the US Army, 100 units being delivered in 1999/2000. The AN/ARC 220 is in service with US Army aviation helicopters.

Contractor
Rockwell Collins, Cedar Rapids, Iowa.

VERIFIED

Dismounted Line Of Sight (DLOS) radio

Contingency Communications Package (CCP)

GTE (now General Dynamic C4 Systems) repackaged the Mobile Subscriber Equipment (MSE) system (see separate entry) for much smaller deployments than required by the original MSE contract in response to the needs of airborne and light forces. Designed for early entry, a typical Contingency Communications Package (CCP) or Light Contingency Communications Package (LCCP) can be deployed in two C141B sorties. The CCP which features all the system and subscriber features present in MSE, can be deployed as a fully operational communications centre within 30 minutes, enabling wire and mobile subscribers to communicate not only with each other but also via tactical satellite to anywhere in the world. Combat net radio and NATO interfaces further expand CCP's breadth of interoperability. A CCP can support a task force CP/airfield and manoeuvre brigade headquarters to include the brigade main and step-up CPs.

The CCP, because of its inherent attributes of size, mobility and functionality, is especially appropriate for contingency airfields where operations must be transported in and established rapidly. A tactical airbase contingency would have immediate interoperability with other services, its mission support base, and a means to handle data traffic without compromising voice communications.

CCP's mobility has been made possible by combining several of the standard MSE functions into fewer shelters. This has been aided by engineering redesigns of many components into significantly smaller packages having the same functionality as their predecessors.

The CCP system is made up of a number of building blocks: Force Entry Switch (FES), Line Of Sight radio (LOS(V)X), Dismounted Extension Switch (DES), Dismountable Line Of Sight radio (DLOS(V)1), Dismounted Line Termination Unit (DLTU), Remote Terminal (RT), and Dismounted Combat Net Radio Interface (DCNRI).

Force Entry Switch (FES)
In addition to its role as a voice and data communications switch, the FES also functions as the radio access point for mobile subscribers. This combining of functions into one vehicle-mounted shelter is made possible by several factors: the MSE routeing subsystem has been repackaged into a significantly reduced

Remote terminal

envelope (RSSD); magnetic tape drives, teletypewriter, video display and call service position have been replaced by a computer workstation; a single six-row switch matrix is utilised; and a dual loop key generator provides twice the secure loops within the same form.

The FES provides a management facility for the node centre, with means to remote the management function, should that be required. The FES is also where interfaces to other networks are established.

The principal features of the FES are:
(a) 10,001 lb vehicle, XM-1097
(b) S-250E shelter
(c) AN/UYK-86(V)3 workstation, electroluminescent display, keyboard, printer, floppy disk drive
(d) C/3 packet switch with connectivity for two LAN, six X25 general purpose hosts, four high-speed digital trunk groups (internodal trunks), one dial-in (16 kbit/s)
(e) group logic unit
(f) six-row nest with downsized switching controller group
(g) four RT-1539 radio sets
(h) KGX-93A/HGF-96/eight KG-112 dual density loop key generators
(i) up to eight KG-194A encrypted digital trunk groups, two uncrypted
(j) capable of having one supergroup
(k) orderwire
(l) processor, L-3212A
(m) downsized routeing subsystem
(n) up to 117 local digital subscribers: three LTUs with 35 each, J-box with 12
(o) interfaces: MSE, TTC/TYC-39, TACSAT, SB-3865, NATO, CNR, AUTOVON, commercial office
(p) remote operating capability (up to 9.15 m) via dismounted remote terminal
(q) 10 kW trailer-mounted diesel generator, PU-753M.

Line Of Sight radio (LOS(V)X)
The LOS used for CCP differs from its MSE counterpart in that there are three separate dismountable LOS radios and three separate dismountable Transmission Interface Modules (TIMs) housed in the shelter. These radios link the FES to other extension switches, creating a communications network.

Principle components of LOS(V)X are:
(a) 10,001 lb vehicle, XM-1097
(b) S-250 shelter
(c) three Transmission Interface Modules (TIMs)
(d) three AN/GRC-226(V) radios: one baseband unit, one Band I unit and one Band III RF unit
(e) three Band I and three Band III antennas
(f) three 15 m masts
(g) three transit cases
(h) radios, TIMs, and KY-57s dismountable into three DLOS(V)1 stacks
(i) 10 kW trailer-mounted diesel generator, PI-753/M.

Dismounted Extension Switch (DES)
Contingency missions often require the quick establishment of small, remote command posts. The DES can be either cabled to an FES (up to a mile (1.6 km) away) or remoted via an LOS radio to provide as many as 16 local subscribers voice and data network access and 10 internodal trunks. It also provides interfaces to both commercial telephones and combat net radio, should the need arise. The DES consists of an SB-4303 switch and a transit case for the TIM and encryption components. Subscribers can be connected by junction boxes and cabling or by direct connection to a terminal board on the back of the smallboard.

The DES has the following components and features:
(a) smallboard, SB-4303 (capable of stacking three SB-4303s)
(b) trunk encryption device, KG-194A
(c) Transmission Interface Module (TIM)
(d) wire access for 16 digital loops and 10 digital trunks (up to 52 digital loops with three stacked SB-4303s)

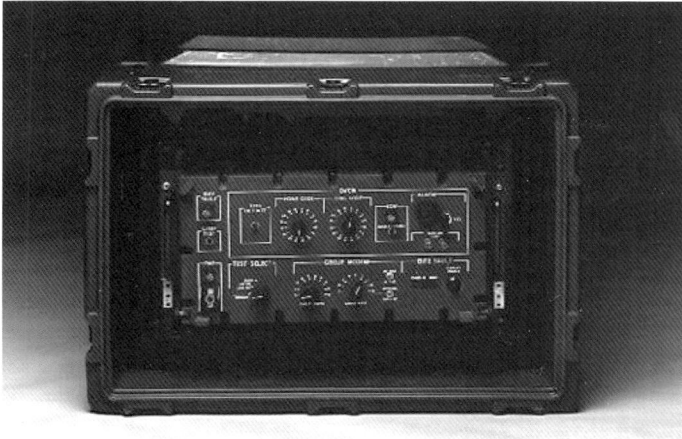

Dismounted Line Termination Unit (LTU)

(e) interfaces: MSE, TACSAT, CNR, NATO, AUTOVON, commercial office, RS-232 I/O port
(f) transit case
(g) orderwire capability
(h) TA-1035/U DNVT as operator's position.

Dismountable Line Of Sight radio (DLOS(V)1)

The DLOSs are the individual radio links that tie the CCP components into a network. Each mounted in two transit cases, they are easily and quickly transported for use with the FES and DES assemblages. With the addition of a directional antenna in place, the DLOS is functionally complete.

The DLOS(V)1 contains:

(a) AN/GRC-226(V) radio set: one baseband unit, one Band I unit and one Band III RF unit
(b) transmission interface module
(c) one Band I and one Band III antenna
(d) 15 m mast
(e) transit case.

Dismounted Line Termination Unit (DLTU)

Packaged as small, portable units, the DLTUs are the subscribers' interface to the FES and thus to the network. The dismounted digital LTU multiplexing device provides connectivity to the FES for up to 35 digital subscribers, while the analogue version supports AUTOVON, commercial office, TA-838, TA-341, 20 Hz ring down, DTMF and other analogue interfaces.

The DLTU comprises a line termination unit (CV-4180/T(V)2), a transit case and AC/DC power supply.

Remote Terminal (RT)

Designed for use with an FES, the RT is a workstation console which can be remoted from the FES. This allows node planning and management activities to be accomplished outside the shelter in, typically, a nearby command tent, giving the switch operator full access to the shelter interior. The RT also provides an operator position so that no personnel need be present inside the shelter.

Features of the RT are:

(a) electroluminescent display
(b) standard ASCII-compatible keyboard
(c) cable set
(d) volume control for CSP headset
(e) interfaces to AN/UYK-86(V)3
(f) configurable as either switch operator's console (CSP) or SCC-2 Network Control Terminal (NCT)
(g) transit case.

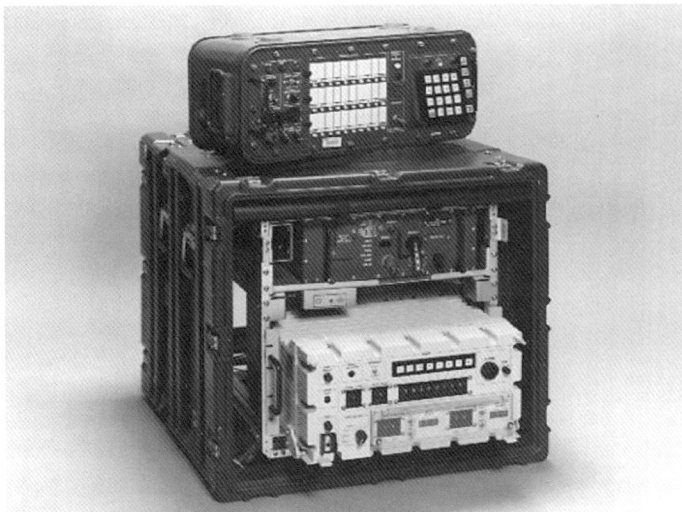

Dismounted Extension Switch

Force Entry Switch (FES)

Dismounted Combat Net Radio Interface (DCNRI)

The DCNRI can be used with either an FES or a DES. This means that the CCP provides seamless communications between the tactical and combat pipelines.

DCNRI comprises a secure digital net radio interface (KY-90), SAFK power supply, transit case and provision for mounting user-supplied AN/VRC-46 or -90 combat net radio on top of the transit case.

Status

Several CCP units were successfully battle tested with coalition forces during the 1990-91 Gulf War and fielding of CCP to the US Army was completed in 1993.

Contractor

General Dynamics C4 Systems, Taunton, Massachusetts.

VERIFIED

Digital Technical Control /Tactical Data Network

The Tactical Data Network (TDN) augments the existing USMC Marine Air Ground Task Force (MAGTF) communications infrastructure to provide the MAGTF commander an integrated data network, forming the communications backbone for MAGTF tactical data systems and Defense Message System (DMS). The TDN system consists of a network of gateways and servers interconnected with one another and their subscribers via a combination of common user long-haul transmission systems, local area networks, single channel radios, and the switched telephone system. This network provides its subscribers with basic data transfer and switching services; access to strategic, supporting establishment, joint, and other service component tactical data networks; network management capabilities; and other services such as message handling, directory services, file sharing and terminal emulation support.

The Digital Technical Control (DTC) provides a deployable digital technical control capability for communications support organisations organic to a MAGTF. It assists in the installation, operation, restoration and management of individual circuits and digital trunks and provides the primary interface between subscriber systems used to transport voice, message, data and imagery traffic. The DTC provides the interface to the Defense Information Systems Agency (DISA) Standardised Tactical Entry Point (STEP), to a JTF headquarters and other JTF component headquarters. The DTC also provides the MAGTF interface to its major subordinate commanders and interfaces from there to respective regiments and groups.

The TDN Gateway will be deployed at the Marine Expeditionary Force (MEF) and Major Subordinate Commands (MSC) level and will provide access to the Non-secure Internet Protocol Router Network (NIPRNET), Secret Internet Protocol Router Network (SIPRNET) and other service tactical packet switched networks. It will be in a Heavy-variant High Mobility Multi-Purpose Wheeled Vehicle (H-HMMWV) mounted shelter for mobility and will include a second H-HMMWV for support.

The TDN Server (also called the Data Distribution System) will be deployed to the MEF and MSCs and down to the battalion /squadron level. It will be in three transit cases and will be two-man portable. The TDN will enable USMC tactical users to transition from AUTODIN to the replacement system, the Defense Message System (DMS) (see separate entry). TDN will interface with and provide Battalion to Regiment subscriber access via the Enhanced Position Location Reporting System (EPLRS) (see separate entry). The TDN will use the currently fielded and available transmission media within the USMC inventory for interconnecting all

The interior of the Digital Technical Control container (General Dynamics). *Equipment key: 1. Power control panel; 2. GPS station clock; 3. Channel service unit/data service unit; 4. Ethernet hub; 5. Dial-up modems (2); 6. Commercial circuit switch call service position; 7. KY-68/TSEC digital subscriber voice terminal; 8. Tactical circuit switch call service position; 9. Oscilloscope; 10. Communications patch panel; 11. Short-haul modems (10); 12. COAX/T1 Patch panel; 13. Conditioned diphase interface modem CV-2048M; 14. Group patch panel; 15. Transmission resource controller; 16. KIV-19 trunk encryption devices; 17. Local workstation; 18. Uninterruptible power supply; 19. Printer; 20. Loop patch panels (30); 21. Multiple rate voice card assemblies; 22. Tactical circuit switch; 23. Analogue transmission test switch; 24. Commercial circuit switch shelves (8); 25. Administrative PC workstation processor; 26. Digital voice orderwire TSEC/KY-57; 27. Administrative PC display; 28. Low datarate multiplexer; 29. Line conditioning equipment; 30. Storage drawer; 31. Patch cord storage area; 32. Single row nest; 33. Transition unit nest assembly (2); 34. KIV-7HS data encryption devices (4); 35. BIT error rate tester; 36. Fibre optic modem alarm/control panel; 37. Local workstation monitor*
0143703

nodes as well as newer transmission media. The design is driven by interface requirements with existing tactical systems.

Tactical Data Network (TDN) Server AN/TSQ-228
Major component features:
(a) Cisco 7206 Router
   Wide range of port adaptors (provides flexibility in deployment).
   6-slot chassis for port adaptors exceeds requirements with room for expansion.
   Hot swappable port adaptors reduce down time.
   Redundant power supplies increase availability.
(b) HP KAYAK
   Pentium III architecture with 500 MHz processor, 384 Mbytes RAM and DAT drive.
   15 in LCO display, lightweight high performance.
   Removable 18 Gbyte hard drive supports separate drives for NIPRNET and SIPRNET.
   Thermal alarms and multiple cooling fans for operation in high heat environment.
(c) CISCO Catalyst 2924M
   24-port switch with spare slot expansion VLAN assignable provides discretionary access.
   SNMP management allows network control and monitoring.
   Web-based application is easy to use for configuration and monitoring.
(d) DNE CV-2048M NRZ to CDI converter
   Hot swap module provides maintenance without powering down.
   Single configuration port, a smart interface to configure up to four modules.
   Compact/low weight 2U package for easy transport.
(e) Capacities
   Up to 96 10/100BaseT users
   Four EPLRS (balanced CDI) up to four Encrypted Serial Network Connections (RS-530)
   Two Ethernet Network Connections (TFOCA).
(f) Technical characteristics
   Weight: 160 to 200 lbs per case
   Size: Network services and LAN access cases 27 × 22.5 × 36.5 in
   User access and UPS cases 21 × 29 × 36 in
   Power: Less than 1,500 W
   Operating temperature: 0 to 40° C

Tactical Data Network (TDN) Gateway AN/TSQ-222
Major component features:
(a) NET Promina 400
   Efficient bandwidth utilisation maximises throughput on tactical links.
   Flexible network interfaces provide stand-alone compatibility with DISA-STEP.
   Scaleable, modular platform eases system upgrades.
(b) CISCO 7206 Router
   As above, plus dual routers for security, separate networks for SIPRNET and NIPRNET.
(c) COMPAQ Proliant 1600R
   Pentium III architecture with dual 550 MHz processors, 1Gbyte RAM and 4 Mbyte V RAM.
   15 in LCD display.

Internal Level V Raid with 4 × 18.2 Gbyte hard drives for flexibility.
   Redundant power supplies provide increased availability.
(d) DNE CV-2048M NRZ to CDI Converter - see above.
(e) Technical characteristics
   Weight: 3,890 lbs
   Power: 3 Phase 208V <3.5 kW per phase
   Operating temperature: 0 to 40° C
   Data capacity (per enclave):
      4 EPLRS (balanced CDI)
      Encrypted serial links (RS-530)
      4 Ethernet network connections (TFOCA)
   Multiplexing:
      16 RS-530 ports
      2 SA trunks

Digital Technical Control (DTC) AN/TSQ-227
Major Component Features:
(a) Compact digital switch
   TRI-TAC digital transmission groups ensure interoperability with legacy systems.
   DTG-toT1 conversions enable transition from TRI-TAC to commercial systems.
   COMSEC parent switch provides requisite services to SB-3865
(b) REDCOM IGX-C
   Industry-standard interfaces provide ISDN capability to tactical forces.
   Modular architecture optimises sizing for each application.
   Multilevel precedence pre-emption allows full 5-level MLPP throughout the network.
(c) NET Promina 800
   Efficient bandwidth utilisation maximises throughput on tactical links.
   Flexible network interfaces ensure compatibility with DISA-STEP and USAFTDC.
   Scalable, modular platform eases system upgrades.
(d) DNE FCC-100 (V)9
   Flexible card set provides compatibility with fleet shipboard multiplexers.
   SNMP management allows network-wide control.
(e) Technical characteristics
   Weight: 9,150 lbs
   Power: 3 phase 208 V <3.5 kW per phase
   Operating temperature: 0 to 40° C
   Circuit switching capability
      Loop subscribers
         Up to 120 TRI- TAC
         Analogue/Digital
         Up to 304 2-wire
         Analogue (POTS)
         Up to 116 ISDN loops
      Trunks
         Up to 11 TRI-TAC DTG/T1-E1
         Up to 13T1/PRI
   Multiplexing capability
      Up to 416-port FCC-100
      Up to 42 port interfaces
      Up to 10 SA trunks

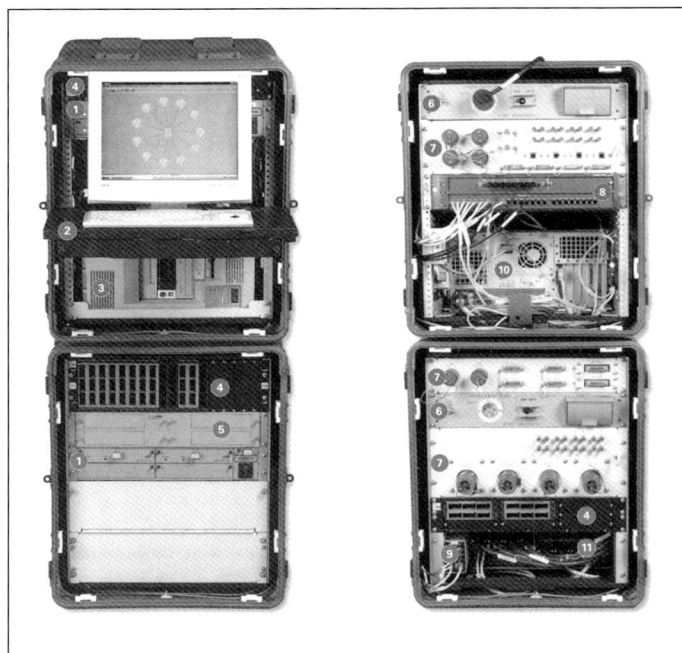

Tactical Data Network Server (General Dynamics). *Equipment key: 1. Conditioned diphase interface modem CV-2048M; 2. Slide-out worksurface and keyboard; 3. KAYAK XU processor; 4. Communication patch panel; 5. Data encryption device KIV-7HS; 6. Power entry panel; 7. Signal entry panel; 8. Ethernet switch CISCO Catalyst 2924M; 9. Media converters; 10. Rear of Kayak computer; 11. Router CISCO 7206*
0143702

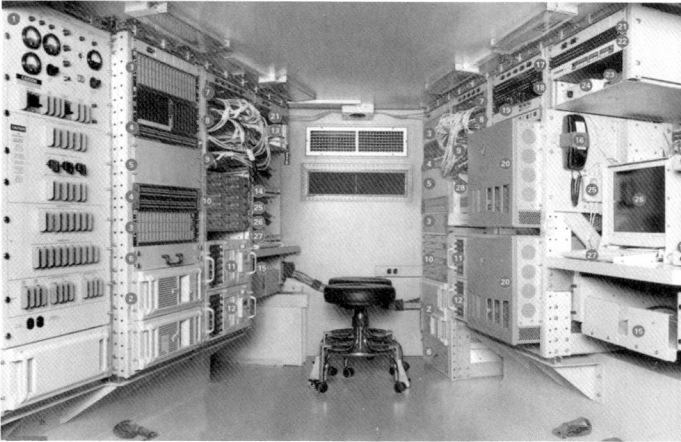

The interior of the Tactical Data Network Gateway container (General Dynamics). Equipment key: 1. Power control panel; 2. Uninterruptible power supply; 3. Communications patch panel; 4. Loop patch panels; 5. KIV-7HS data encryption devices; 6. Patch cord storage area; 7. Ethernet switch CISC Catalyst 2924M; 8. Media converter; 9. Router CISCO 7206; 10. Conditioned diphase interface modem CV-2048M; 11. Network Management System (NMS) processor; 12. Internet services provider (ISP); 13. KG-175 inline network encryptor; 14. Secure terminal equipment (STE); 15. Digital subscriber voice terminal (DSVT); 16. Telephone; 17. Group patch panel; 18. KIV-19 trunk encryption devices; 19. Fibre optic modem alarm control panel; 20. Transmission resource controller; 21. Coax patch panel; 22. Environmental integrator; 23. Dial-in modem; 24. Printer switch; 25. Keyboard, Video and Monitor switch (KVM); 26. Workstation monitor; 27. Printer

Status

The production contract was awarded on 12 January 1999 to GTE Government Systems, now General Dynamics C4 Systems. The production contract includes both the Digital Technical Control (DTC) System and the Tactical Data Network (TDN) System and is for 31 TDN Gateways and 561 TDN Servers. The development is taking place in 3 Blocks: Blocks I and II have been combined in the initial production and fielding is underway, starting with I MEF and ending with the USMCR in 2003, while the Block III enhancement, which is yet to be approved, is dependent on developing technology that will support multiple levels of security on the same networks. The projected time frame for the Block III enhancement is FY03-FY04.

In September 2002 it was announced that General Dynamics had been awarded a US$19.5 million modification contract to upgrade the TDN Server by installing RAID-5 memory capability to selected units, allowing recovery of data in the event of disc drive failures, and by adding server capacity to Major Subordinate Command systems. Options to outfit the remainder of the data distribution servers were included to be exercised as additional budget funds became available.

Contractor

General Dynamics, C4 Systems, Taunton, Massachusetts.

VERIFIED

....

Digital Wideband Transmission System (DWTS)

The US Marine Corps DWTS is the principal communications system used by operational formations to communicate ship-shore and between ships. It is made up of four main elements – ITT unit level circuit switches (the 150-line AN/TTC-42 and the 30-line SB-3865), a basic radio system with the Raytheon/Unisys AN/TRC-170 as its nucleus, the digital distribution system and the AN/MRC-142 LOS UHF radio relay.

The Marine Corps will receive 134 AN/TRC-170s under contracts awarded in January 1990. Deliveries of the AN/MRC-142 were started in July 1991. Loral TerraCom (now L-3 communications) was awarded a US$11.1 million contract for 34 AN/MRC-142s in January 1990 and provided up to 500 sets.

The MRC-142, designated SC190 by Loral, is a specially adapted version of the Tadiran GRC-400 mobile digital radio relay system and is normally installed in the back of a HMMWV. Although Loral has full design and production rights, it is acting as system integrator for the Marine Corps contract and has retained Tadiran as a subcontractor. Subsequently, the Canadian Marconi Company has provided the Band 3 radio subsystem portion of the DWTS. This consists of an upgraded version of the AN/GRC-226 tactical radio.

Status

Since June 1998, the full DWTS installation is being installed in all US amphibious ships. A report by the US General Accounting Office in 2001 indicated that there had been problems with the system. A Product Improvement Programme (not necessarily connected with these problems) consisting of a Shore Mount Accessory Kit (SMAK), an Uninterruptible Power Supply (UPS) and a robust multiplexer was to be fielded in 2002.

UPDATED

MSE Network Management Tool (NMT)

The US Army was completely fielded with Mobile Subscriber Equipment (MSE – see separate entry) in 1992, with the network management for the system being provided by the SCC-II (AN/TYQ-46). Although meeting all original MSE network management requirements, changes in both US Army doctrine and the radio spectrum environment in which MSE operates made it necessary for the SCC-II to be enhanced and this has led towards the Network Management Tool (NMT). Specific enhancements include:
(a) a downsized MSE network management capability in a single workstation, deployable in two transit cases, as opposed to two S-250E shelters
(b) a radio network planning and engineering capability to allow the establishment of MSE VHF, UHF and SHF radio networks, without conflicting with other radio emitters operating in the same area
(c) a network management system configuration to allow location of NMT workstations within signal brigade/battalion command posts, as opposed to isolated SCC-II S-250E shelters
(d) distribution and replication of network planning information among NMT systems, in addition to network status/configuration data as provided in MSE today
(e) NMT workstation hardware with substantial growth potential at low additional cost.

NMT allows deconfliction of SINCGARS and MSE MSRT radios by segregating VHF frequencies for SINCGARS and MSE MSRT use. MSE assemblage site selection is performed by an automatic tool that optimises access switch (for example, SENs and LENs) location based on battlefield location and the specific requirements of units using the MSE wire-line voice service. MSE Radio Access Unit (RAU) site selection is achieved by use of an optimisation routine which determines RAU location based upon SYSCON-identified MSRT areas of operation. Optimal network backbone switch locations are automatically established based on RAU and access switch locations. LOS radio links are engineered to connect backbone switch locations, as well as remote RAUs and user access switches, based upon predicted LOS link reliability.

This planning process is accomplished over a map background generated from standard Defense Mapping Agency (DMA) data. The NMT database contains information on 64 characteristics of each known potential assemblage site, with typically 28,500 sites supported.

NMT software has a structured code architecture. Various application code components are written in C, C+, Fortran and Ada. The software runs in a Commercial-Off-The-Shelf (COTS) environment, which includes the Solaris Operating System, MOTIF, Dataviews, Ingres RDBMS and Prior GKS.

The system uses a client/server automation architecture, each NMT system having a single data/communication server and up to eight operator positions per NMT. All the NMT workstations can be initialised either as a data/communication server or as an operator position, or as a special case of both modes simultaneously for operation as a single workstation NMT system.

The data distribution architecture is specifically engineered to address the unique characteristics of both tactical operations and the MSE network. NMT operational planning is distributed to and replicated at, each NMT system in a given MSE network. The data distribution mechanism is built from a combination of commercial code and code specifically written to meet the requirements of operation in a narrowband, high-noise, long-delay, high-mobility MSE network.

NMT interfaces with the MSE Packet switch Network (MPN) for interNMT communications, using TCP/IP protocol over IEEE 802.3. It also interfaces with the MSE circuit switch network for NMT to switch and NMT to RAU communications, using MSE's unique messaging over auto-dial/auto-answer logic. Non-essential mission-related communications are supported by electronic mail over the MPN, using DoD standard mail protocol (SMTP).

The hardware configuration of NMT is based on an HMMWV with an S-250 shelter and a number of workstations. These S-250 shelters house standard MSE equipment required for interface with MSE network elements. Workstations are used as either data/communication servers or network management operator positions. Operator positions are remoted from the shelter into US Army SYSCONs by IEEE 802.3 LAN and 4-wire diphase dial-up connections.

The number of workstations can be varied according to mission requirements, up to eight per NMT system. Nominal configurations are three workstations per MSE signal battalion unit, four workstations per MSE signal brigade unit, one workstation per MSE area battalion unit.

A single workstation configuration is used regardless of data/communication server or operator position application. The NMT workstation consists of four major components: 19 in high-resolution colour monitor, main computer unit, mass storage expansion unit and two associated transportation cases. The main computer unit is a Sun Microsystems Sparc 5 processor with 32 Mbyte RAM and a 535 Mbyte hard disk. The mass storage expansion unit includes a fixed 535 Mbyte hard disk, a removable 535 Mbyte hard disk, a CD-ROM drive and a tape drive.

Status

The NMT was delivered in 1997.

Contractor

General Dynamics C4 Systems, Taunton, Massachusetts.

VERIFIED

Single Channel Ground and Airborne Radio System VHF (SINCGARS-V)

SINCGARS-V is a family of manpack, vehicular and airborne VHF/FM radios that features high resistance to surveillance, intercept and jamming. The radios are able to interface with TACFIRE, Light Weight TACFIRE, Patriot, Chaparral, SHORAD, FAADS, MLRS and other primary weapon systems. They also interface with the two other primary components of the tactical internet; EPLRS and Appliqué. They are interoperable with, and are designed as a direct replacement for, the AN/PRC-25/77 and AN/VRC-12 family of radios. SINCGARS is also compatible with C-6709, GSA-7, GRA-7, GRA-39, VIC-1, VIC-2, IVIS and VIS intercom systems. The radio family is interoperable with communication security devices KY-57/TSEC, KY-58/TSEC, KY-99A MiniTerm and KY-100 Autovon. In its single-channel mode, SINCGARS is compatible with NATO systems.

The SINCGARS family includes:
AN/PRC-119A — manpack (replacing AN/PRC-25/77)
AN/VRC-87A — vehicular short-range (replacing AN/VRC-53/64)
AN/VRC-88A — vehicular short-range dismountable (replacing AN/GRC-125/160)
AN/VRC-89A — vehicular long-range/short-range (replacing AN/VRC-12/47)
AN/VRC-90A — vehicular long-range (replacing AN/VRC-43/46)
AN/VRC-91A — vehicular long-range/short-range dismountable (replacing AN/GRC-125/160 and AN/VRC-46)
AN/VRC-92A — vehicular dual long-range/retransmit (replacing AN/VRC-45/49)
AN/ARC-201A, B, C, D — airborne set in panel-mounted, dedicated remote and MIL-STD-1553B remote configurations (replacing AN/ARC-54/131, AN/ARC-114 and AN/ARC-186)
RT-1702C — SIP
RT-1523E — Advanced SIP
RT-1702E — Advanced Tactical Communications System (ATCS)
Spearhead — ATCS handheld transceiver.

At the heart of all the SINCGARS configurations is a basic 5 W Receiver/Transmitter (R/T). This provides frequency modulated transmission and reception on 2,320 channels over the 30 to 88 MHz range. There are eight single-channel and six frequency-hopping preset channels available. Within the basic R/T is the ECCM frequency-hopping capability which gives the radio a hop rate in excess of 100 frequency changes per second. Hop set data can be loaded via a front panel connector from a SINCGARS ECCM fill device, MX-18290, or over an RF link. The frequencies available for any given hop set are extremely flexible and are selectable from one frequency or any combination up to all 2,320 channels. In addition, any channel or group of channels can be excluded from a given hop set. An internal data module provides digital data transmission/reception from 600 bit/s to 16 kbit/s for digital message devices, facsimile or teletype. It also provides an FSK capability at 1,200/2,400 Hz for TACFIRE and similar systems. Upgrades include forward error correction using the Reed Solomon algorithm at data rates of 1,208, 2,400, 4,800 and 9,600 bit/s. This also includes a packet data interface in accordance with MIL-STD-188-22A.

All radios are currently in production for the US Army. The Systems Improvement Program (SIP) model and Lightweight Advanced SIP, provides secure voice and data capability that is backward compatible with radios using the KY-57 TSEC communications security device.

A vehicular amplifier adaptor with shockmounting unit with the same footprint as the AN/VRC-12's MT-1209, is used in nearly all vehicular configurations. The Amp Adaptor, AM-7239, will accept one or two R/Ts and one 50 W RF power amplifier AM-7238. The SIP and ASIP vehicular amplifier adaptor also includes an InterNet Controller (INC) which provides interfaces to tactical internet functions and backbone systems as well as the capability of adding an Ethernet interface. With these combinations, installations may be assembled ranging from a single 5 W short-range to a dual 50 W long-range configuration. A smaller Single Radio Mount (SRM) has been developed and fielded. In this configuration, the AN/VRC-87C and mounts, one 5 W R/T is used in vehicles where space is a consideration. This model may be installed as a 50 W radio with the addition of an external RF power amplifier. Either or both R/Ts in the vehicle installation can be dismantled for

AN/VRC-87E vehicular, short-range radio (ITT)   **NEW**/0594414

AN/VRC-89E vehicular, long-range/short-range radio (ITT)   **NEW**/0594415

conversion to a manpack system by the addition of a battery case, battery and 1 m whip antenna. Total weight of the manpack radio when equipped with ECCM and comsec capabilities is about 9.4 kg (20.7 lb). A 3 m antenna is available for extended manpack range.

For tracked vehicle installations, a new low-profile, end-fed, 1.8 m, broadband antenna, the AS-3916, has been developed. For wheeled vehicle installations, a High Voltage, Centre Fed (HVCF) antenna, the AS-3900, is used. Either antenna may be used on tracked vehicles.

Airborne SINCGARS

The SINCGARS airborne radio system, the AN/ARC-201A, B, C and D is currently in production for the US Army for installation in rotary- and fixed-wing aircraft. The AN/ARC-201A has an 80 per cent commonality with the ground system. The RT-1478/ARC-201A is a multiplex MIL-STD-1553B bus remotable unit.

SINCGARS airborne radios are available for foreign military sales in exportable versions that are compatible and interoperable with both types of FMS SINCGARS ground radio systems.

Airborne SINCGARS has been selected for use in the UK's Bowman tactical communications system.

SINCGARS Tactical Communications System (TCS)

In early 1992, ITT and the US Army initiated the SINCGARS System Improvement Program (SIP) that targets specific improvements to the SINCGARS. These improvements address the need for improved situational awareness, improved data throughput and speed of service on the battlefield.

The SIP radio was the basis for the 1995 SINCGARS production contract for the US Army and US Marine Corps. Awarded in 1995, the contract called for almost 36,000 SIP ground units to be built in 1996 and 1997. Specifically, the following features have been implemented:
(a) embedding Global Positioning System (GPS) position information and user information in all voice and data messages to provide reporting of friendly force positions for situational awareness
(b) implementation of advanced forward error correction techniques to significantly increase throughput over the existing radio, while extending communication range and improving the interference protection of the radio against co-site and jamming
(c) implementation of a new frequency-hopping packet data waveform with less on-air transmit time, which reduces transmission overhead and improves throughput
(d) implementation of an improved channel access algorithm that allows mixed voice and packet data operation on a common net with minimal impact on voice operation at high packet data throughput rates
(e) deployment of an InterNet Controller (INC) that provides for packet radio relay nodes across the battlefield for both horizontal and vertical integration of command and control elements. In addition, the INC provides a programmable interface for protocol conversion to support the seamless flow of data across the battlefield

Immediate and accurate position reporting, on the battlefield, is more important than ever to provide situation awareness and to reduce fratricide. The new radio has added features that provide an automatic and cost effective means to achieve this. A new feature of the improved radio is the reporting of GPS position embedded in voice and data messages. Through an interface to an external GPS receiver, or via an optional integrated GPS receiver, the new SINCGARS radio establishes the user's position and embeds this information into voice and data messages. The position information from the radio net members is then made available for display external to the radio.

The improved SINCGARS radio employs an interface that allows the radio to access GPS position information from an external GPS receiver. For the US Army, this interface is for the Precise Lightweight GPS Receiver (PLGR) but can easily be made to accommodate most available GPS receivers. The external GPS receiver attaches directly to the rear of the radio in the manpack configuration, or to the front

AN/VRC-90E international vehicular, long-range radio (ITT)   **NEW**/0594416

Airborne SIP (MIL-STD-1553B) RT-1478D/ARC-201D   0010470

of the Vehicular Adaptor for easy access in vehicular installations. Additionally, ITT has provided the capability to share the GPS information with at least four radios in a vehicular installation so that only one GPS receiver is needed per installation.

Under some operational scenarios, the added weight associated with a stand-alone GPS receiver is not desirable. ITT has addressed this by designing the new radio to be capable of accepting an integrated GPS receiver that plugs directly inside the radio. This provides the SINCGARS user with the flexibility of deploying with either an internal or external GPS receiver.

With either the external or internal GPS receiver, the radio transmits the position information with every voice and data message. In addition to the GPS position information, the radio embeds a user identification which is entered at the front panel of the radio. The radio can also be programmed via the front panel to automatically transmit this information at specified time intervals, or after significant movement.

In addition to position information, the radio uses the extremely accurate GPS time to maintain Frequency-Hopping (FH) synchronisation net time. Using GPS for FH sync time virtually eliminates the problems associated with operator entry or net time drift.

The SINCGARS on-air waveform has been improved to pass more data, more accurately, and under more severe environments. ITT has also added an asynchronous RS-232 interface to expand the long list of data terminals with which SINCGARS is compatible.

Enhancements to the SINCGARS waveform include advanced Forward Error Correction (FEC), a reduced COMSEC synchronisation overhead, and elimination of unnecessary interleaving overhead. ITT has implemented a Reed-Solomon error-control algorithm that has already completed preliminary testing at Fort Huachuca, Arizona. Test results have verified the capability of the error control algorithm to provide a significant increase in data capacity over the current SINCGARS/TACFIRE system. This provides an eight-fold increase in throughput while maintaining a communication range of 30 km and significantly increasing the tolerance to partial band interference.

The new error control algorithm is combined with reduced COMSEC synchronisation to reduce on-air time for each message. The reduced on-air time increases the overall data throughput and speed of service for data on the battlefield.

A new channel access algorithm has also been incorporated into the improved radio to allow operation in mixed voice and packet data mode on a single radio net. The algorithm is designed always to give the voice user access priority. Following

any given voice message, pending data messages are held off to allow the voice user time to respond in a conversation. When the channel has been sensed idle for a given period of time, the data terminal is allowed access to the channel. An algorithm is used for data access to minimise data packet collisions. This combination of improved channel access protocols allows voice users to communicate reliably, while allowing excess channel capacity to be used efficiently for data communications.

The final key aspect of the SINCGARS System Improvement Program (SIP) is the addition of an InterNet Controller (INC) that provides routeing of data packets both within (IntraNet) and between (InterNet) radio nets. The INC is a battlefield networking product that performs the functions of bridging, routeing and transporting of data. It will be used primarily in brigade and below applications to provide the base networking infrastructure. The INC functions in a manner similar to a packet switch, maintaining routeing and connectivity tables to relay data quickly and effectively through the SINCGARS forward area battlefield Internet.

The INC provides for reliable transport and routeing of data, both horizontally (across similar echelon levels) and vertically (to higher-level or lower-level elements) in the forward area battlefield through the bridging, routeing and gateway functions. Currently, information transfer between two commanders at the same echelon must go up the command chain by voice to a common commander and then back down the other side of the chain. This type of information transfer is inherently slow, prone to errors and makes inefficient use of the limited communication resources on the battlefield. The INC provides an effective bridge between like echelon commanders. It also provides an interface between the radio net and other backbone networks, such as the Mobile Subscriber Equipment (MSE), so that data traffic can seamlessly flow back to higher level element commanders.

A programmable interface enables the INC to access a wide variety of tactical command and control systems. The INC can be programmed to interface other systems via existing protocols and will support protocol conversion to allow these systems to benefit from the improved data relaying provided by the SINCGARS forward area battlefield internet.

Based on widely available commercial hardware, the INC can be incorporated into every SINCGARS Dual Vehicular Adaptor. Since the Dual Vehicular Adaptor is widely deployed across the forward area at major command posts and within commander vehicles down to the platoon level, incorporation of the INC provides high-density relay points for packet transmission among and between nets. The deployment of the INC, at virtually every node, affords the SINCGARS user a level of connectivity that is necessary to achieve horizontal and vertical integration of command and control links. The INC provides the potential for future expansion and interface flexibility through a daughter-board option which permits incorporation of special purpose interface types as needed.

Lightweight Advanced SINCGARS SIP (ASIP)

In 1997, the US Army ordered 35,000 new Lightweight Advanced SINCGARS SIP following a 'winner takes all' competition with General Dynamics Land Systems. In

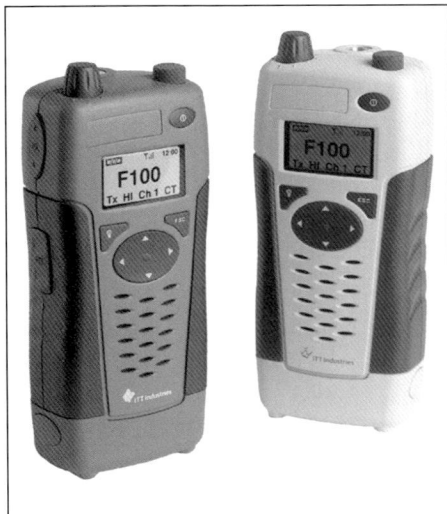

Spearhead hand-held radio (ITT)
NEW/0594418

AN/VRC-92E international dual long-range radio (ITT)   **NEW**/0594417

Receiver/transmitter RT-1523(C)/U   0010475

1998, ITT began production of an advanced version of the SIP. The cornerstone of this system is the ASIP RT which weighs approximately 8 lb with integrated BA-5590 lithium battery. This lightweight radio is suited for light tactical, extremely mobile forces anticipated for deployment on the 21st century battlefield. The upgraded design includes two TI TMS 320C56 digital signal processors with expandability for a third. Enhancements have been made in the RT's ability to operate in high RF noise environments and to synchronise to other transmitters. In conjunction with improvements to the CSMA algorithm, the reliability of voice and data communications has been significantly increased. The ASIP RT is highly integrated, and built using state-of-the-art surface mount techniques further increasing the reliability and thereby reducing life cycle costs. Power management techniques have been incorporated which significantly increase the battery life and further reduce the individual soldier's load. The ASIP RT is expandable to include an integrated GPS, integrated data router and advanced high capacity waveforms to further increase the data throughput and pass voice traffic more efficiently.

The ASIP Vehicle Adaptor Assembly (VAA) has also been redesigned to increase reliability and reduce costs. In addition, the Army's standard tactical internet data router, the InterNet Controller (INC), has been redesigned to include more memory and to increase the host data rate for improved performance. The ASIP VAA also includes an optional Ethernet interface to increase connectivity between SINCGARS nets for further horizontal and vertical integration in the forward battlefield. A new half-sized vehicle adaptor has been designed specifically for the ASIP radios.

Specifications
(a) Frequency band: VHF 30-87.975 MHz
(b) Frequency channels: 2,320
(c) Frequency Hop (FH): hops on 2-2,320 frequencies100 hops/s
(d) System architecture: FM voice or digital data
(e) ECCM: spread spectrum, FH, Forward Error Correction (FEC)
(f) Security: external COMSEC (KY-57, KY-58) for non-ICOM ground/airborne radios; embedded COMSEC in ICOM ground and SIP airborne radios
(g) Data rate: to 16 kbit/s (ICOM w/FEC 4.8kbit/s: SIP w/FEC 9.6 kbit/s), FSK
(h) Prime power: manpack +12 V DC (BA-5590); vehicular and aircraft 24 V DC
(i) Output power: manpack up to 4.5 W; Vehicular up to 50 W; Airborne 10 W
(j) Radio size: 10.7 wide, 3.4 high, 14.8 in deep
(k) Radio weight: 3.6 kg with battery

Spearhead Hand-held Radio
In 2002 ITT introduced the Spearhead hand-held radio, which is both SINCGARS and JTRS compatible. It weighs less than 500 g, including battery and antenna and can operate in any combination of clear, encrypted and single channel or frequency-hopping modes.

Export Versions
There are two versions of the SINCGARS radio family available for international customers. The Tactical Communications System (TCS) was introduced in 1996 and is built around the RT 1702E (V) with a built-in COMSEC module and internet

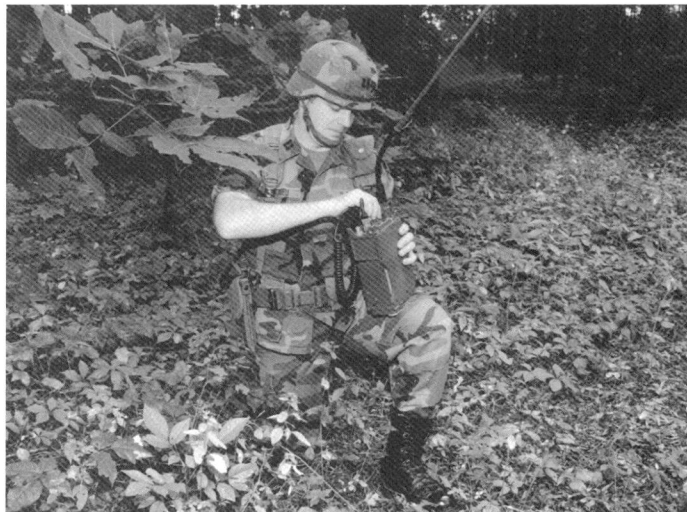

Two ASIP SINCGARS atop a SINCGARS SIP   0010471

controller in the VAA to route data across the battlefield. The latest system is the Advanced TCS which includes the following features:
(a) optional embedded Global Positioning System (GPS) information in all voice and data messages for reporting situational awareness
(b) advanced forward error correction to significantly increase throughput while extending range and improving protection from co-site and jamming interference
(c) improved frequency-hopping packet data waveform with less on-air transmit which reduces transmission overhead and improves throughput
(d) implementation of improved channel access algorithm and waveforms that allow mixed voice and packet data operation on a communications net with minimal impact on voice operations at high packet data throughput rates
(e) improved power management techniques which significantly increase battery life
(f) reduction in size and weight of 50 per cent
(g) free channel search.

Bowman
In 2001 the UK selected the SINCGARS-based Advanced Digital Radio Plus (ADR+) as the Bowman VHF radio solution, reflecting the involvement of ITT in the Bowman programme and the spin-off from Bowman requirements back into SINCGARS development. (See separate entry for Bowman).

Status
SINCGARS is in production for US Army, US Marine Corps and international customers.

ITT SINCGARS TCS was introduced to quantity production in 1996 (see above). These radios are presently in service with Italy, Morocco, Slovakia, Brazil, Spain, Estonia, Hungary, Bosnia, Ukraine, Georgia, Uzbekistan, Chile, Egypt, Japan, Jordan, the Saudi Arabian National Guard and the Marines of Bahrain, Kuwait and Thailand. In addition, the Republic of Ireland and Taiwan have placed significant orders. In February 2001 the New Zealand Army placed an order worth US$11.9 million (including supporting equipment and training).

SIP systems are in production and currently being delivered. ASIP production started in 1998. The US Army Acquisition Objective is 238,970 radios (230,348 ground radios) required for full fielding.

Manufacturer
ITT Industries, Aerospace/Communications Division, Fort Wayne, Indiana.

VERIFIED

ASIP SINCGARS   0010472

ASIP SINCGARS in operational use (ITT)
0129618

MARITIME

Denmark

Integrated Communications Systems (ICS) 1000/2000

ICS 1000

ICS 1000 is a maritime integrated communications system which handles all internal and external voice and data communications. Originally developed for the Danish Navy Standard Flex 300 class with the first system being ordered in 1988, the principal design features are:

(a) Digital system with fibre optic cables for connection of all user terminals and interfaces
(b) Radio remote control of all radios, allowing an unmanned radio room
(c) Full integration with the ship's Combat Management System

At the heart of the system is a high capacity Digital Communication Switch. It is non-blocking and can handle multiple large open duplex conference networks with several hundred simultaneous users. The Subscriber Station is a standard multipurpose user terminal; all are identical and are configured from the set-up terminal or via the combat management consoles. Each user is allocated the communication networks required for his role. An automatic message handling system is also provided, allowing the drafting, release, transmission and reception of formal message traffic, and providing full control of the available external communications resources and integration with the communications plan software.

ICS 2000

ICS 2000 is INFOCOM's (now Maersk Data Defence) development of the earlier system. A new switch, the DCS 2000, has been developed, which has a capacity of

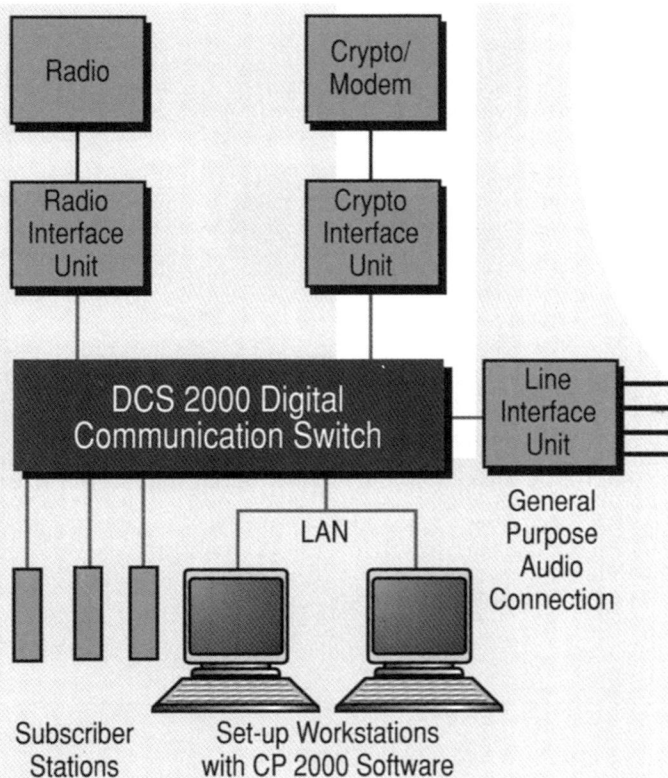

ICS 2000 - Star Configuration (INFOCOM)   0109973

TSS 2000 (INFOCOM)   0109972

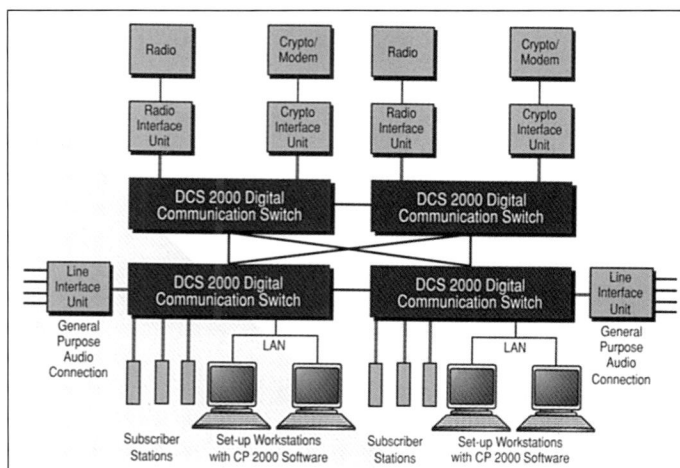

ICS 2000 - Ring Configuration (INFOCOM)   0109974

2048 channels each with 64kbps data rate, but it retains the conference net capability of its predecessor. The channels can be configured for any bandwidth required depending on the system design. Both data and voice channels can be handled by the same switch. The system is controlled and managed by the Communications Processor (CP) 2000 software package, which is designed to operate on a COTS PC Windows NT platform, connected to the DCS via a LAN. The CP software package is divided into four parts:

(a) System set-up, monitoring, access control, BIST
(b) Communications configuration
(c) Remote control of radio and crypto equipment and allocation of radio resources
(d) Automatic Message Handling System

Subscribers access the system via the Tactical Subscriber Station (TSS 2000). This can be configured to operate with a split headset and to monitor up to 14 networks simultaneously. The graphic display and function keys give the operator access to the required network with single keystrokes. Allocation of networks to keys is configured from the set-up workstations. The Combat Management System can be connected to ICS 2000 via the LAN.

The system can be configured either to operate in a star or ring configuration. The former is used for small and medium size installations, and as all the modules in the ICS 2000 which present a single source of system failure are duplicated, if a module fails the hot standby module automatically takes over. The MHS has a duplicated database to avoid loss of data from equipment failure or battle damage. Ringed or meshed networks are used for larger vessels where DCS configured as switching nodes distributed through the ship and connected by a high-speed ring or meshed dual network provide higher survivability; if part of the system is damaged the remainder will continue to operate. Separation of secure and insecure traffic is achieved by using two DCS 2000 connected only through crypto equipment.

Status

ICS 1000 has been installed in a variety of platforms, including 14 Standard Flex 300 vessels, 4 *Thetis* class frigates, 10 fast attack craft and fast patrol boats of the Danish Navy.

As at July 2001 ICS 2000 had been selected for the mid-life update of the *Nordkapp* class Coastguard vessel for the Royal Norwegian Navy and for one new vessel, with the first delivery in July 2001 and for two fast attack craft and the Air Cushion Vehicle protoype for the Finnish Navy, with the prospect of three further ACV fits in 2003/4. The Royal Swedish Navy has also selected the ICS 2000 for their new *Visby* class stealth corvette. For this solution the ICS 2000 is managed by the C3 LAN network on a common PC, with the TSS as an integrated part of the C3 system consoles.

Contractor

Maersk Data Defence, Sønderberg.

VERIFIED

France

THOMNET multiservice communication network

THOMNET is a secure, local multiservice network that supports ship internal communications (voice, data, video) and provides access to external transmission facilities. It provides subscribers with standard and specific communication services including telephony, intercom, conference, order and alarm broadcasting, data transmission and access to radio resources. A federative network, it has been

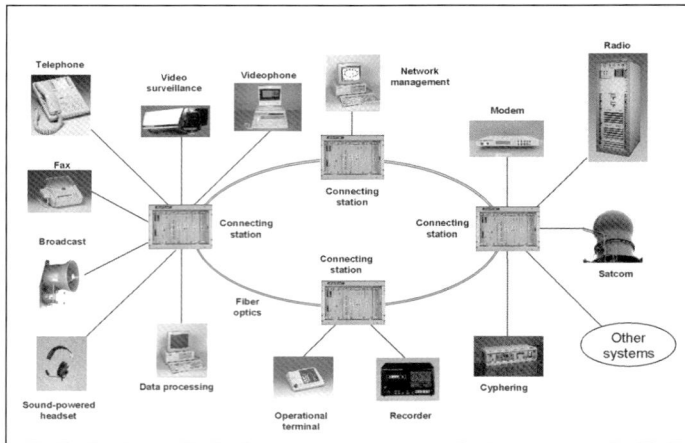

THOMNET system architecture   0055050

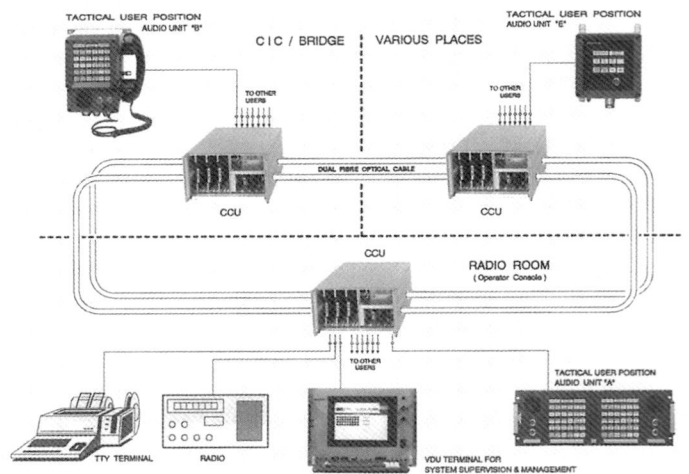

Schematic of ASYM 3000(A)

designed to meet the communication requirements of surface ships, submarines, naval bases and command centres. A THOMNET network can be included into a FICS system (see separate entry) to provide the backbone high-speed multiservice network required to support all internal communications and access to all external communications. The network exploits a distributed architecture, consisting of communication stations that are interconnected by 155 Mbits/s ATM ports. The recent upgrade to ATM technology makes it possible to support broadband services while maintaining those services that meet specific naval requirements and constraints. THOMNET architecture is based on the interconnection of a number of stations by one or more optical fibre loops and meshing links designed to provide a very high survivability in case of multiple failure or severe battle damage. An automatic reconfiguration mechanism switches to alternative paths offered by the meshed network topology. This multiservice communication network has a family of user terminals (voice/data and multimedia terminals including TOMA S0/S2, TMM and TMFP). THOMNET administration is provided by a PARTNER (see separate entry) management system. The THOMNET administration function incorporates both network and internal communication supervision. This centralised administration offers services such as configuration, security, surveillance and performance management. Standardised interfaces include ISDN, X.25 and Ethernet.

Specifications
(a) Circuit switching mode data rate: 32-128 Mbit/s depending on configuration
(b) Packet switching mode data rate: 32 Mbit/s; 200 packets/s switched per connecting station
(c) Remote supply of terminals in accordance with ISDN standard (40 V)
(d) Synchronous data interfaces: V11, VI8, X.25, X.400
(e) Asynchronous data interfaces: V11, V24, V28, X.3, X.28, X.29
(f) ISDN interfaces: I430, 1441, 1451, S0, S2
(g) Video surveillance: H261, S0, S2
(h) Coding G712 (PCM 64 kbit/s)
(i) Optical fibre: multimode fibre (monomode optional), 62.5/1 25 microns, 1,300 nm; maximum internodal distance 3,000 m (multimode)
(j) Network capabilities: 1-4 loops; up to 32 connecting stations per loop, 254 modules for a 4-loop network; up to 64 ports according to the ISDN type (S0, S2, X.25) and 12 Ethernet ports per connecting station; analogue access (interfacing of baseband signals with radio, CRYPTO, modem); digital access (interfacing of data, telegraphy and digital plain/secure voice signals)

Status
Although no longer marketed under the same name, THOMNET's first application was in the SGD large distribution support system on the French Navy's aircraft carrier *Charles de Gaulle*. It is also supplied to other navies, for both shipboard and ground applications. The main examples are the UK's 'Albion' class LPD, the 'Al Riyadh' class frigates for Saudi Arabia and the Belgian upgraded 'Wielingen' class frigate. It has also been selected for the United Arab Emirates naval communications upgrade and as part of the Fully Integrated Communications System (FICS) for the 'Horizon' destroyer, the UK T45 destroyer and the French Projection and Command Ship programmes.

Contractor
Thales Communications, Colombes.

UPDATED

Germany

ASYM 3000(A) communications control system

Based on the traditional communication systems technology of the 1970s, AEROMARITIME firstly developed the ASYM-2000 system and from this evolved the computer controlled ASYM-3000 Communications Control System to satisfy the tactical communication requirements of naval vessels. The system provided integrated tactical control and distribution both internal and external

communication functions throughout the ship, and has been installed on more than 60 naval vessels of varying size (fast patrol boats, corvettes and frigates).

The ASYM 3000(A) is a third-generation communications control system which provides additional features such as digital audio processing, distributed topology, high-speed 50 Mbit/s fibre optical cables, higher capacity, programmable operating functions, redundancy and survivability, and BITE. Its principal design objective is to be the link between previously independent communications equipment for internal/external communications and defined user positions, providing sophisticated switching, radio exchange and intercom facilities to satisfy the specific voice/data communications requirement of any user position, achieving a fully digital audio/data communications network. In addition the ASYM 3000(A) includes an overall control and status monitoring facility for system management.

The latest version of ASYM 3000(A) has been upgraded to ATM technology to form the Digital Broadband Network (DIBNET). The DIBNET subsystem provides external voice and data communications in plain or encrypted modes as well as tactical internal communications at all relevant positions throughout the vessel.

The Communications System Manager COSYMA is a software/hardware enhancement to the basic ASYM-3000A and provides the addition of reliable overall Communications System Management. The COSYMA subsystem includes a number of workstations which provide overall system and network management including full control of all major radio equipment. By a gateway provided to the Combat Management System (CMS) all functions of the workstations can be fully or partially allocated to Multi Function Consoles of the CMS.

ASYM 3000(A) - DIBNET with COSYMA can handle audio, video, data, message handling, crypto, radio control and frequency management.

Status
In use with military customers.
Recent contract awards for a complete Integrated Communications System using ASYM-3000 or ASYM-3000(A) include the navies of Indonesia, South Korea and Thailand, and the Royal Netherlands Navy, Turkish Navy and Royal Norwegian Navy.

Contractor
Aeromaritime Systembau GmbH, Munich.

UPDATED

Italy

Integrated radiocommunication system for frigates and aircraft carriers

Marconi's communication system for frigates and aircraft carriers includes a full complement of wideband and tunable antennas for transmission and reception over the LF to UHF frequency range. Facilities are provided for ship-to-ship, ship-to-shore and ship-to-air communications and for operation in the civil VHF maritime band. It is also compatible with civil telephone networks and has full interfaces with other civil radio networks.

Features include the use of HF/UHF multicouplers and antenna filters, automatic antennas/equipment switching, remote control, the use of fibre optics, time division mux-demux to user terminals, centralised system management and supervision, frequency management by computer, ECCM voice and data communication, public address and no-break emergency primary power supply.

HF/UHF operation is possible with NTDS, TADIL-A (Link 11) and other specialised networks and the system has been designed in compliance with EMI, HERO, RADHAZ and other NATO specifications.

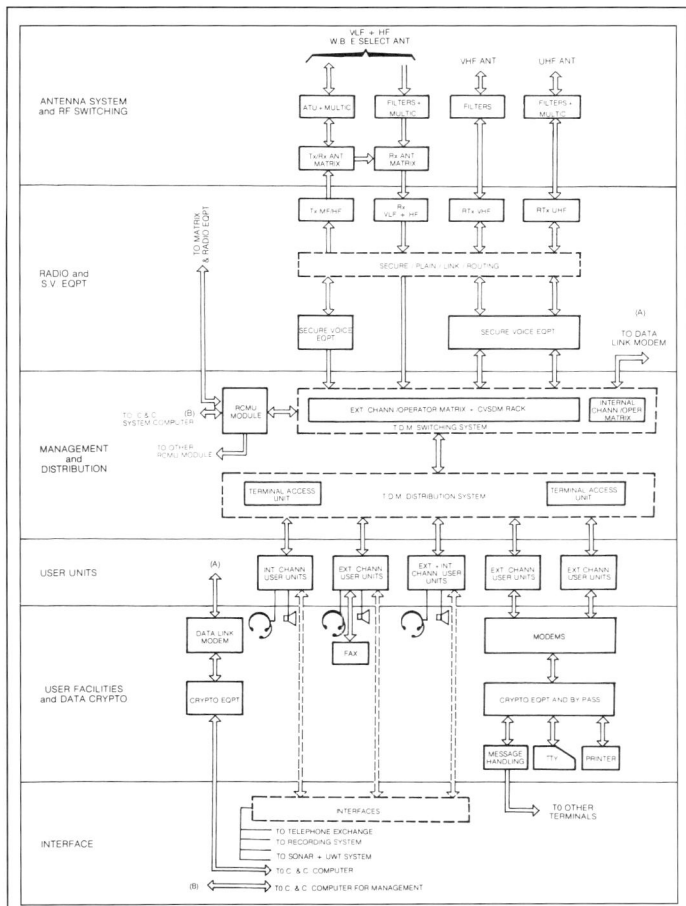

Typical integrated communications system for frigates

Specifications

(a) Frequency range: LF to UHF
(b) Power output: 20, 30, 100 or 1,000 W
(c) Types of service: analogue and digital voice, data, facsimile
(d) Antenna types: wire, whip, wideband, bent dipole, discone
(e) User capacity (internal and external): 256

Status

In service on board 'Garibaldi' class aircraft carriers and other units of the Italian Navy.

Manufacturer

Marconi Selenia Communications, Genoa.

UPDATED

Integrated radiocommunication systems for corvettes

Marconi's communication systems and subsystems for use aboard corvette-type ships provide for ship-to-shore, ship-to-ship and ship-to-air communications, with a full complement of wideband and tunable antennas for transmission and reception from LF to UHF frequency bands. The equipment is assembled in preconfigured racks and is available for different RF power levels and a wide selection of operating modes.

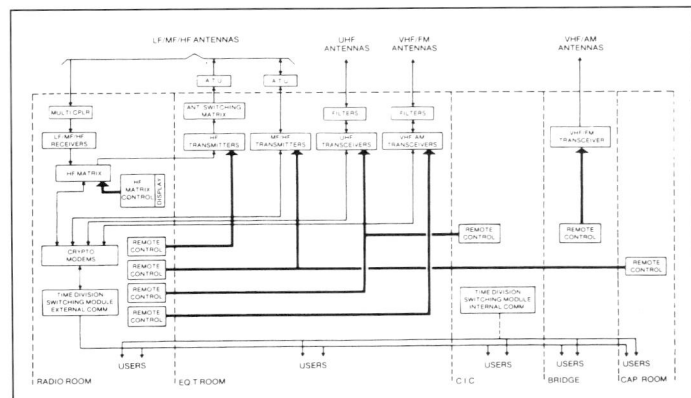

Diagram of typical integrated radiocommunications system for corvettes

1 kW transmitter racks

Other facilities include external channels/user and conference intercom, TDM switching, the use of serialised data transfer to minimise interconnection requirements, centralised management and supervision and a no-break emergency primary power supply.

Special services available include civil VHF maritime communication, VHF communication with landing forces and an internal telephone system compatible with public telephone networks. The systems comply with EMI, HERO, RADHAZ and other NATO specifications.

Specifications

(a) Frequency range: LF to UHF
(b) Power output: 20, 30, 100 or 1,000 W
(c) Antenna types: wire, whip, bent dipole, discone
(d) Service types: analogue and digital voice, data, facsimile
(e) No of external comms: up to 64
(f) No of users of internal comms: up to 32
(g) No of users: up to 128

Status

Produced in a number of variants, these systems have been adopted by the Italian Navy for a large number of minor naval units.

Manufacturer

Marconi Selenia Communications, Genoa.

UPDATED

United Kingdom

Advanced Ship-Shore Automatic Telegraphy System (ASSATS)

The ASSATS system is a first-generation automatic HF networking system. It is designed to allow up to 1,000 maritime mobile stations to pass their telegraph traffic into a shoreside network with minimum delay and minimum radiation.

It is a store and forward messaging system capable of 'file and forget' operation where, except for the most unlikely circumstances, once a message has been entered into the ship's system, intact delivery to the ultimate addressee is assured without further human intervention.

ASSATS is automatic in operation, adaptive in channel choice, is synchronous in its network control and is highly corruption-resistant. The main elements of the system are: ship station, shore receive site, shore transmit site, and network control site.

The operating sequence is as follows:

(a) an assessment receiver at the shore station monitors in sequence all allocated channels specified for use at the time by the tables held in store. It dwells on each channel for a predetermined length of time, recording the noise level and the presence or absence of interference on each of the 16 in-channel FEK frequencies
(b) a sounding transmission is broadcast by the shore station, informing all ships of the results of the channel assessment activity. Each ship receiving the sounding broadcast measures its signal strength and calculates the ship to shore path loss. This enables the ship to assess the probability of successful communication and the Effective Radiated Power (ERP) necessary
(c) dependent on the results of ship assessment, the ship transmits its traffic on the chosen frequency in a time-slot determined from stored tables, which indicate the times when the shore station will be listening out. Any necessary ARQs are obtained automatically and the transaction is completed by an

acknowledgement of receipt from the addressee contained in a subsequent sounding broadcast.

Operation is entirely automatic and does not rely on the presence of experienced HF operators at either end of the link for its successful operation.

The system is able to select the best operating frequency (from a pool of 256 channels) and the best in-channel tone-pair for the frequency shift keying modulation, taking account of propagation conditions, noise levels and interference.

Network operations are governed by stored tables of allocated times and frequencies for each station's use and controlled precisely by accurate clocks. This minimises abortive communication attempts, reducing the crucial time-to-establish-communication overhead. It also ensures that low-powered or distant members of the network get a fair share of network access time and are not blocked by the more powerful or more local members.

The basic timing of the network for access purposes is achieved by time-slicing each minute of the day into eight slots of 7.5 seconds for the sounding broadcasts. These are then further subdivided into three sub-slots of 2.5 seconds duration for the listening-out cycle. Ship stations listen out for sounding broadcasts at the 7.5 second points on frequencies found from their stored tables. Similarly, they know, from the same tables, in which 2.5 second slots the shore station receivers will be listening out for their traffic transmissions and their ARQ requests. The shore station can listen out on as many as six frequencies simultaneously in each 2.5 second sub-slot to enable the anticipated traffic level to be maintained and to achieve a rapid response from the shore station and so minimise ship transmissions.

ASSATS, in addition to using propagation prediction and sounding methods, uses a combination of methods to cope with path loss, multipath signals, receiver noise level, impulsive noise and co-channel interference, and fading:

(a) path loss is countered by adaptive selection of power based on ship's estimation of path loss by its receiver signal strength

(b) multipath phenomena are contained by the use of a relatively low keying rate which is normally 110 baud, but an alternative rate of 75 baud is available

(c) natural noise experienced on a clear, non-fading HF circuit is usually considered to give rise to random, single-character errors. To combat this effect the system employs Golay encoding (23, 12, 3), for forward error correction of up to three errors per block. The reduction in data rate caused by the coding inefficiency is considered to be an acceptable price to pay for the improvement achieved

(d) fading and fading impulsive interference are assumed to give rise to error bursts. For this purpose the system uses deep interleaving to convert the bursts to single character errors, correctable by the Golay process

(e) in addition, complete frames are subjected to cyclic redundancy checking. Uncorrectable errors give rise to ARQs, but only the corrupted frames are repeated.

A predetermined limit is set on the number of ARQs which can be requested and if ambiguity remains thereafter, the system will conduct a majority vote on the available versions of a message, choosing the most likely candidate version for onward transmission to the ultimate addressee. In the event that an acceptable version is not received, the system will make another attempt on a new frequency.

Status
In service with the Royal Navy.

Contractor
AMS, Chelmsford, Essex.

UPDATED

Automatic Message Handling Assistance (AMHA)

EDS has provided communications operators in the Royal Navy's Type 42 destroyers, Type 23 frigates, the Ocean Survey Vessel, HMS *Scott*, the Hydrographic Vessels and Auxiliary Oilers with Automatic Message Handling Assistance (AMHA). This is claimed to be the first project to involve an operational system in front-line warships based entirely on commercial-off-the-shelf products.

The AMHA solution supports the Royal Navy strategy for a common fleet-wide ACP 127 messaging capability and in the Type 23 operators benefit from fully distributed automatic message handling system functionality. AMHA gives real-time performance and high levels of reliability and resilience. It also offers the potential to alleviate the problems associated with dissimilar databases.

With the capability for fleet-wide expansion, EDS's solution for AMHA has been demonstrated exchanging data with the UK, NATO and US Message Handling System and the generic, open systems design of the NT-based Command Support System (CSS) (see separate entry).

Status
Delivery of the first AMHA began in April 1997 with the first installation on board HMS *Scott* achieved in July 1997. The installation programme for the 10 Type 42 destroyers began with HMS *Nottingham* in May 1998. Delivery of AMHA for the first Type 23 frigate occurred in January 1998 and a second delivery was made in early 1999. These were upgraded in November 2001. The NT-based version for the RN's Hydrographic Vessels was completed by EDS for installation by THALES in June 2002.

Contractor
EDS Defence Ltd, Hook, Hampshire.

UPDATED

DIMPS/SAMHADS

DIstributed Message Processing System (DIMPS) provides automation of the preparation and control of the message handling process utilising message formats ACP127, ACP126 and JANAP128 from a range of ship and submarine radio and satellite equipments. The following features are supplied with both DIMPS, the surface ship version and the similar Submarine Automated Message Handling And Distribution System (SAMHADS):

(a) audible and visual alarms for priority traffic

(b) filtering of unwanted messages

(c) data rates of 50 to 9.6 kbit/s

(d) automatic preparation and distribution facilities

(e) full message accountability

(f) message analysis on reception

(g) designed to fit standard equipment racks and environmental specifications

(h) provides major reduction in operator workload and manning levels.

Status
Supplied to the Royal Navy for fleetwide use on surface vessels. Also fitted to the Auxiliary Oil Replenishment (AOR) vessels RFA *Fort Victoria* and RFA *Fort George*. SAMHADS is fitted to Royal Navy submarines. DIMPS is to be progressively withdrawn from service, starting in 2004.

Contractor
BAE Systems, Plymouth, Devon.

UPDATED

DIMPS/SAMHADS provides automation of the preparation and control of the message handling process

ICS3/ICS4 (AN/URC-109) integrated communication system

ICS4 (AN/URC-109) is the latest version of the broadband naval communication system ICS3. The latter is now in service with more than 60 vessels worldwide. These systems provide tactical radio communication with other ships and aircraft and strategic radio communication with the command on shore.

This is a naval medium- and high-frequency radio system employing a unique but well proven architecture. It is a broadband system which allows virtually instant frequency changing with better than 2.5 per cent spacing between transmit and

ICS receivers

MF/HF exciter outfits

ICS active antenna

receive frequencies and virtually unlimited spacing between transmit frequencies. An advanced bus control makes it possible to reconfigure the system rapidly in the event of battle damage. The system handles all modes of signalling used in ship-to-ship, ship-to-air and ship-to-shore, as well as sonar communications to provide a comprehensive external communications facility.

The transmitting subsystem covers the MF and HF (240 kHz to 30 MHz) bands, employing a broadband architecture for the HF circuits, plus one or two MF/HF narrowband channels if required. A power bank technique permits simultaneous radiation of a number of frequencies from a single broadband antenna, with rapid change of frequencies and power levels. Narrowband channels are routed using a similar exciter and power amplifier to either an HF whip antenna tuner or an MF wire antenna tuner. The separation between adjacent HF channels can be reduced to as little as 50 kHz. Narrowband HF and MF channels can be incorporated to suit individual requirements.

The receiving subsystem covers the frequency spectra from VLF (10 kHz) to HF (30 MHz). Signals received by a small active antenna are routed via broadband active multicouplers to the receivers where they are converted to a suitable baseband for distribution to the many user positions. The antenna distribution unit can feed up to 36 receivers without passive splitters, tunable multicouplers, preselectors, notch filters or patch panels.

A MIL-STD-188-141A Automatic Link Establishment (ALE) facility has been added as a modular upgrade for the US Navy LHD. The Marconi embedded ALE sub-system is a plug-in module for the digital H1550 receiver and is controlled from a PC platform running Windows 95 or Windows NT.

With the addition of a COTS modem, a new STANAG 5066 data link protocol can be used to allow reliable data communications over HF. Standard PC-based communications applications (for example, e-mail) can use the HF medium in a seamless manner through the use of STANAG 5066.

Status

Over 40 Royal Navy ships are fitted with ICS3, including all new capital ships. ICS3 is also operational in Royal Netherlands Navy, Hellenic Navy and Nigerian Navy frigates. ICS4 was competitively selected for the US Navy's LHD combined assault ships under the AN/URC-109 nomenclature. By mid-1991, three systems had been supplied to the US Navy, with a fourth in final test.

Some ICS3/4 technology was offered for the US High-Frequency Anti-Jam (HFAJ) programme. In 1987 Marconi, teamed with Rockwell, won a US$450 million contract to supply a number - believed to be eight - of prototype HFAJ systems. HFAJ could have eventually been worth US$3 billion in the period to 1997. In early 1988, the Pentagon cancelled the HFAJ follow-on contracts. Later, ICS4 was sold to the French Navy for evaluation and integration with Thomson-CSF's (now Thales) SPIN naval frequency-hopping system. Other customers for ICS4 include the Royal Malaysian Navy for its 'Lekiu' class frigates, the Chinese Navy and the Royal New Zealand Navy for its 'Leander' class frigates.

Marconi Selenia Communications was formed in 2003 following the acquisition of the Marconi plc defence business by Finmeccanica.

Manufacturer

Marconi Selenia Communications Ltd, Chelmsford, Essex.

UPDATED

Infra-Com system

The Infra-Com System is a two-way, audio, infra-red communication system for use on board ship. It is used for key wireless communications for onboard locations including the bridge, machinery spaces and flyco areas but is equally suitable for other applications, where secure, interference-free communications are required. The system, which meets the strict environmental requirements of naval systems, allows communication between mobile operators and the ship's communication system and comprises three main units: fixed-base unit, infra-red antennas and mobile units.

The fixed-base unit employs several infra-red antennas, located throughout the area to be covered for communication. Only one fixed-base unit is usually required, although several can be used in the same area or can share antennas. The unit communicates with the infra-red antennas via two twisted pair, balanced line circuits, which comply with industry standard RS-422. It also interfaces to an analogue voice circuit.

Infra-red antennas are mounted at locations to cover the desired area for communication and transmit and receive infra-red signals for the operators' mobile units. The antennas are connected on a bus system, allowing large numbers of devices to be attached and facilitating simple extension to the area of coverage should this be required.

The mobile unit comprises a headset, transceivers and battery mobile set. To ensure optimum infra-red coverage, the small, lightweight mobile unit is integrated to the top of the headset headband. Several types of headset are available, ranging from full protection from noisy environments to single-sided lightweight for use on the bridge. On the flyco unit both internal and external communications can be made simultaneously. The battery pack that powers the mobile unit is attached to a belt, worn around the waist. Interconnections between the headset, battery and mobile unit are effected by a suitable plug and socket, allowing easy disconnection of a battery pack without the use of tools.

All controls are designed to allow operation by the user while wearing anti-flash or NBCD gloves.

Status

In production. In 2000 it was reported that Azdec Limited has been awarded a UK MoD contract to supply the system for the Royal Navy.

Contractor

Azdec Ltd, Sholing, Southampton.

VERIFIED

Marconi integrated naval communication systems

Marconi has developed a family of integrated naval communication systems that feature a combination of high technology and competitive cost to provide versatile, ECCM-resistant, HF transmitting and receiving systems for vessels of all sizes and operational roles.

The systems have been designed to match a modern navy's stringent financial and manpower budget, but much-needed operational features to enhance communication management in today's harsh electronic environment have been incorporated.

Of particular note is the computer-controlled bus which enables the entire ship's communications, including VHF, UHF and SHF, to be controlled from a central

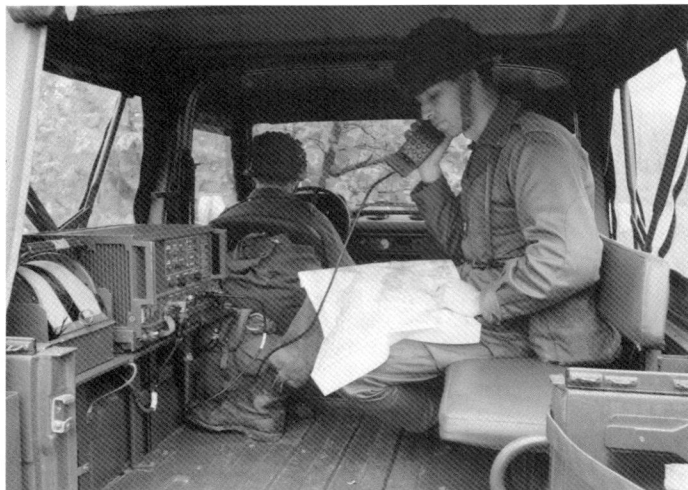

A typical drive unit

position. The bus allows full status monitoring, rapid reconfiguration of COMPLANs, fault reporting using Built-In Test Equipment (BITE), and comprehensive EMCON facilities.

A significant benefit of these systems is that they can also be installed in shore stations to provide a single range of common equipment throughout the fleet communications network. This reduces the need to support a wide range of different equipment and thus reduces the cost of spares and support services.

System features include:
(a) frequency management system allows low probability of intercept
(b) new solid-state amplifier
(c) new antenna matching unit
(d) digital drive and receiver
(e) flexible power management system
(f) modular construction
(g) simple to use
(h) optional broadband antenna system
(i) optional ECCM capability (frequency hopping).

Systems can be configured as narrow or broadband with many options, including a choice of 500 W or 1 kW solid-state amplifiers. Operational requirements and available space will dictate the systems best suited for any particular vessel.

For smaller vessels, a narrowband system feeds into whip-wire antennas. The equipment can be preprogrammed with up to 512 frequencies. The antenna matching unit and pre/post-selector unit are common to both the transmit and the receive chain. SELCAL and broadcast facilities are available and, for data transmission, ARQ, FEC and other powerful error detection and correction systems can be provided.

For larger vessels the transmit system can comprise three, five, seven or nine channels. The outputs of the amplifiers are coupled, via a hybrid, into a broadband antenna. The receive channel comprises a preselector and an SSB/ISB receiver. Any number of receivers may be connected to a single active antenna. The systems can be operated in either simplex or duplex modes.

An optional frequency management unit is available for use with either of the systems, which permits the automatic set up and acquisition of the best operational channel at the time requested. This fast-acting system enables the vessel to use optimum transmission levels, so significantly reducing the chance of interception and increasing channel integrity.

The operational role and size of vessel determines the architecture and facilities required in the control system and the following choices are available:
(a) a low-capacity, low-cost system using serial links from a central controller to each user position and controlled equipment
(b) a system with a higher capacity for a medium-sized system using a bus architecture from either a single or multiple controller. The bus can be duplicated for increased reliability
(c) for a major war vessel a network architecture with distributed controllers and multiple bus connections is available which ensures a high degree of integrity.

All these systems are easy to install and can use fibre optic links for increased data security.

A MIL-STD-188-141A Automatic Link Establishment (ALE) facility has been added as a modular upgrade for the US Navy LHD. The Marconi embedded ALE sub-system is a plug-in module for the digital H1550 receiver and is controlled from a PC platform running Windows 95 or Windows NT.

With the addition of a COTS modem, a new STANAG 5066 datalink protocol can be used to allow reliable data communications over HF. Standard PC-based communications applications (for example, e-mail) can utilise the HF medium in a seamless manner through the use of STANAG 5066.

Status
In service with several navies.

Manufacturer
Marconi Selenia Communications Ltd, Chelmsford, Essex.

UPDATED

Rationalised Internal Communications Equipment (RICE 10)

RICE 10 is an operational multi-user internal naval communications system which provides the required functionality while maintaining flexibility, speed and a capacity for survivability in a damage situation. Facilities available include shipwide broadcast, alarms, point-to-point communication and conference. These can be combined with the external communication components to form a fully integrated communications system. Developed by Redifon MEL (now Thales), RICE10 has the capacity to interconnect up to 512 users and is designed around a reconfigurable mesh-based architecture.

RICE10 provides interphone, intercom, broadcast, conference and alarm facilities. It accommodates up to 64 nodes with each node supporting up to 15 voice user units, known as outstations. Each node is capable of interfacing to three other nodes. Nodes are interconnected via 8 Mbit internodal links to form a mesh architecture. The use of a mesh architecture provides multiple paths between users and greatly enhances the survivability of the system in the event of node failure or mesh damage. The mesh network is managed through software running on the nodes in conjunction with one or more portable Maintainer Console Units (MCUs). An MCU is connected to a node in the system via an RS-422 link. Communications information is set up at a shore facility (PC-based) and downloaded via the MCU during refit. Changes to the communications information can be instigated on board at any time using the MCU.

BAE SYSTEMS is responsible under subcontract to Thales for the software elements of the node, MCU and the shore facility. The node software was developed in CORAL 66 under MASCOT; the MCU and shore facility software was developed using Ada and the Yourdon design methodology.

Status
RICE10 is installed in the UK Royal Navy's LPH HMS *Ocean*, the 'Swiftsure' and 'Trafalgar' class submarines and possibly other UK platforms as well.

Contractors
Thales Communications, Crawley, West Sussex.

UPDATED

....

Remote Control and Management System

The Marconi Selenia range of RCMSs has its roots in a number of systems supplied to the Royal Navy, other world navies, and a number of non-military customers throughout the last three decades. The systems have been designed to provide full remote control and management of communications assets, particularly radio equipment.

Control and management is available from either single or replicated control sites which are designed to automate the management process and reduce the required operational manpower. Multiple unmanned remote sites with the communications assets are controlled via either PSTN land lines, or a number of wide area network standards. Control information and management/status revertive information is passed to/from the remote sites using TCP/IP or other high-level Open System Interface standards. Automated system scheduling is a feature of the system, allowing automatic frequency changes on HF equipment, allocation of directional or omnidirectional antennas, linking of all assets in the traffic chain, and automatic reconfiguration of the system in the event of equipment malfunction.

Control can be affected either from a separate control site, or from one of the sites containing the controlled equipment. The control facilities can be run on a single mode computer system, or duplicated for reliability. Additional control terminals can be installed at each remote equipment site to provide local control, either as back-up, or as an additional facility.

Suitable interfaces are available to handle equipment from a range of suppliers already in service on radio or other communications sites. New equipment can be incorporated into the control suite for a particular customer. Control and revertive information is available to/from radio equipment (receivers and transmitters), antenna matrices, audio matrices, antennas, modems, multiplexers, site services such as diesel generators or other standby power, intruder or fire alarms, and any other equipment, provided it has a remote systems control interface.

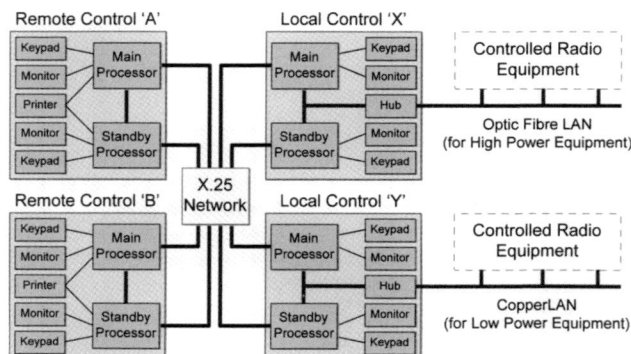

Schematic of typical RCMS

NEW/0006154

The system is able to interface to both serial and parallel remote-controlled equipment, protocol converters being used to provide the translation to a standard format used throughout the system.

Status

The equipment is in service with a number of users worldwide. The Royal Navy uses the system in a number of applications including the United Kingdom Maritime Coastal Communication System (UKMACCS) (see separate entry) and the Royal Navy HF Transmitting system (outfit KSX), with a number of enhancements. This system has dual-control sites, as does the Royal Netherlands Navy control system installed on a number of NATO HF transmitter sites.

Manufacturer

Marconi Selenia Communications Ltd (formed in 2003 following the acquisition of the Marconi plc defence business by Finmeccanica), Chelmsford, UK.

NEW ENTRY

Royal Navy JTIDS Ship System (RNJSS)

The prime contract for the full development and supply of the Royal Navy Joint Tactical Information Distribution System (JTIDS) Ship System (RNJSS) was awarded in 1995. The contract was extended towards the end of 1995 to include the provision of a Satellite Tactical Data Link (STDL) capability. The final acceptance trial took place in November 1999 with a two-platform trial between the Land Based Test Site (LBTS) at Portsdown and the CVS HMS *Illustrious*.

The RNJSS provides the means for RN ship platforms to become JTIDS subscribers with the capability to exchange real-time tactical data with airborne JTIDS units. The RNJSS provides secure, jam-resistant, high-capacity digital data and voice information distribution and accurate relative navigation. The major subsystems of the RNJSS are the JTIDS Class 2H Terminal, the Antenna Subsystem, and the Network Control and Initialisation Data Preparation Subsystem (NCIDPSS). The NCIDPSS provides a Human Computer Interface (HCI) which allows an operator to perform JTIDS network planning, initialisation, monitoring and dynamic network management.

Status

The RNJSS will be fitted to all CVS, Type 42 Batch 2 and 3 platforms, the LBTS and to training establishments.

Contractor

BAE Systems, Christchurch, Dorset.

UPDATED

SR(S)7392 HF broadcast multichannelling equipment

The SR(S)7392 HF broadcast multichannelling equipment programme is designed to increase broadcast traffic capacity for the Royal Navy.

The system is intended to provide high-capacity, shore-to-ship communications using the latest technological standards, increase broadcast speeds and maintain interoperability with allied forces. This has been achieved by the introduction of new, high-speed, serial modems (STANAG 4285) and time division multiplexers, supported by a Message Handling System (MHS) on intelligent messaging terminals which are adapted to replace the existing slow-speed teleprinters. BAE Systems, the prime contractor, has integrated a number of COTS products, modems, multiplexers, messaging software, computer hardware and printers to form the ship and shore elements of the programme. The latter include broadcast control stations, system control points and transmitter stations in the UK and overseas.

Status

The contract was awarded in 1998 and the equipment entered Royal Navy service following the achievement of Fleet Weapons Acceptance in July 2002. Ultimately, all RN seagoing platforms will be equipped.

Contractor

BAE Systems, Christchurch, Dorset.

NEW ENTRY

United Kingdom Maritime Automatic Coastal Communication System (UKMACCS)

UKMACCS is a control system developed for the Royal Navy, providing a full remote-control capability over unattended HF coastal communication stations. This control system became fully operational in mid-1987 and has provided an uninterrupted service since that date.

There are two separate control centres, a primary centre located in London and a secondary centre at a naval establishment in Yorkshire. The HF coastal communications comprise separate receiver and transmitter sites located in Scotland and Cornwall. The controlled equipment at each site includes a number of HF drive units, high-power amplifiers, RF combining equipment, air cooling systems and an antenna exchange. A number of HF receivers and an antenna exchange are located at each receiver site.

The control system provides full control of all equipment, at each remote site, from either of the two control centres. Indications from the station service equipment at each remote site, including fire and intruder alarms, are automatically monitored and a notification or any alarm condition is provided at the control centres. A purpose-built supervisory console is installed at each control centre allowing full control of all equipment by means of command inputs to menu-driven VDUs. The console also provides a control facility for an audio matrix unit, allowing the allocation of radio channels to various operators or services. Provision is also made for the monitoring and display of received signal strengths allowing the optimisation of reception on each channel by the selection of appropriate directional aerials.

A full dual-mode system was provided to meet the requirements for a high level of system security. At each control centre a pair of system control processors are installed, each capable of controlling the whole system. Pairs of remote peripheral interface units are fitted at the remote sites, the failure of one of a pair not affecting the overall operation of the system. Synchronous modem datalinks connect each remote site to both control centres, duplicate links ensuring that the high level of system security is maintained.

All the hardware and software development associated with this control system was undertaken by the Software Engineering Department of GEC-Marconi Communications (now AMS). Other control systems were also supplied under the same contract, these systems following a similar design philosophy. One system is located on the Isle of Portland, providing remote control of the Royal Naval communication stations covering the English Channel. A second system is centred on Faslane in Scotland and provides control of two transmitter stations and three receiver stations.

Status

The system was completed in 1987 and is still in operation.

Contractor

AMS, Chelmsford, Essex.

UPDATED

United States

AN/SSQ-33 Ship Automated Communication Control System (SACCS)

The AN/SSQ-33 provides an advanced ship radio communication network and circuit management tool. This tool is capable of supporting ship communication planning, implementation, monitoring and control. It also controls and manages virtual networks required by the Navy Communication Support System (CSS) architecture and Joint Global Grid networks. As the local radio network manager installed on each ship, the AN/SSQ-33 interfaces with the US Navy Automated Digital Network System (ADNS) and the Automated Integrated Communication System (AICS). The ADNS and AICS support a global, multilevel secure network and communications system for both Joint Force, Joint Service and US Navy ship and shore communication and control systems.

Development of the AN/SSQ-33 began in 1982. The system was then upgraded under a 1992 contract to use the TAC-3 hardware platform, Oracle and UNIX technologies in conjunction with an open architecture design and development approach, provide a reusable object capable of command and control of force radio and network communications. The AN/SSQ-33 system provides automated, reliable and robust circuit connectivity and HF, UHF VHF and SHF radio network monitoring and reconfiguration.

The AN/SSQ-33 features include:

(a) remote SNMP compliant legacy equipment control

For details of the latest updates to *Jane's C4I Systems* online and to discover the additional information available exclusively to online subscribers please visit

jc4i.janes.com

The AN/SSQ-33 SACCS uses the latest TAC-3, Oracle, and UNIX technologies

(b) seamless integration of network manager with tactical communication manager
(c) automated circuit build
(d) centralised operator control
(e) equipment/circuit status monitoring
(f) nomenclature equipment utilisation
(g) data driven, object oriented methodology for maximum system flexibility
(h) integrated bandwidth management, control and error resolution
(i) extendible device control interface bus
(j) legacy switches configured for unique platform requirements
(k) equipment database tailored to shipboard assets
(l) X-Window/Open Systems Foundation (OSF) Motif Graphical User Interface (GUI)
(m) point and click operations, trackerball input.

Status
In service. The system was rehosted to TAC-3 consoles on LHA, LHD, CVN, SSN and Aegis ships in 1992.

Contractor
Northrop Grumman Integrated Systems, Ocean Springs, Mississippi.

VERIFIED

Extremely Low-Frequency (ELF) communications programme

The concept of using extremely low-frequency radio signals to communicate with submerged submarines was first suggested over 30 years ago. However, the extremely large antenna size and environmental worries prevented its introduction until recently.

ELF signals can travel great distances with low loss and can penetrate seawater to considerable depths. In practice, an ELF consists of one or more shore-based transmitters, operating at around 40 to 80 Hz, connected to long horizontal wire antennas (either just above ground or buried for additional security) that are earthed at each end. Orthogonal antennas are used to provide omnidirectional radiation patterns. The transmitted signals are sensed by an antenna on the submarine and decoded by a sophisticated, computer-based receiver. As the bandwidth is low at ELF, the message transmission rate is necessarily very slow but, even by employing a simple three-letter code system, any of a great number of messages can be transmitted in reasonable time.

The most favourable site in the US for the installation of the system is in northern Michigan or north western Wisconsin where the low-conductivity bedrock formations in the Laurentian Shield greatly enhance propagation.

In 1969, the US Navy constructed an experimental transmitter in Wisconsin to study propagation and environmental effects of ELF. The antenna consisted of two 22.5 km long pole-mounted lines at right angles. In 1976, a message-handling capability was added and a small quantity of shipboard receivers were built and installed to prove conclusively that an ELF system would perform as anticipated.

Encouraged by this trial, the navy planned to construct an operational ELF system with a completely buried antenna and redundant transmitters. Codenamed Sanguine, this vast system would have been highly resistant to blast overpressure and could have absorbed a moderate number of direct nuclear hits. However, in 1975 a defence analysis group reached the conclusion that the increasing accuracy and number of Soviet nuclear warheads could neutralise the system. Sanguine was eventually cancelled.

The navy, undeterred by this setback, persisted and developed a more modest buried system with above-ground transmitters and a pole-mounted antenna 45 km long. This system, called Seafarer and installed at Clam Lake, Wisconsin, immediately ran into opposition from the residents of Wisconsin and the environmentalist lobby, who were concerned about the environmental impact of the system and health issues. Despite numerous navy-sponsored biological studies showing no adverse environmental or ecological effects from ELF and, a 1977 study carried out by the National Academy of Sciences giving ELF a clean bill of health, the project was cancelled in 1978.

In 1981, President Reagan ordered the Wisconsin transmitter to be reactivated and upgraded to operational status and, at the same time, ordered the Department of Defense to conduct a study of ELF requirements. That study resulted in the conclusion that ELF would enhance the US strategic C3 posture. It recommended that, in addition to the upgrade of the Wisconsin system, a supplementary 90 km system should be constructed at the nearby KI Sawyer Air Force Base on the Upper Peninsula of Michigan. Congress approved funds for the project in 1982 and research and development resumed. Construction commenced at both sites in 1983 but in January 1984 a court injunction halted further work pending the preparation of a supplemental Environmental Impact Statement (EIS) to evaluate health effect studies of ELF fields performed since the original EIS was filed in 1977. A study was carried out by the American Institute of Biological Sciences which reaffirmed the results of previous investigations and the EIS was filed in 1985. Construction restarted following the cancellation of the injunction by the US Court of Appeals. The Supreme Court subsequently upheld that decision.

By the end of 1986, both stations were completed. Prototype submarine receivers were delivered in April 1985. The first test in May 1985 aboard a submarine of the Pacific Fleet was successful. In subsequent tests, submarines in the Mediterranean, the western Pacific and on patrol under the North Polar ice cap have successfully received signals from the Wisconsin station.

There are four segments to the Wisconsin/Michigan ELF communications system: the Broadcast Control Segment (BCS), the Message Input Segment (MIS), the Transmitter Segment (TS) and the Receiver Segment (RS).

The main input port is the BCS, which is controlled by the Commander, Submarine Forces Atlantic (COMSUBLANT) in Norfolk, Virginia. The MIS, the secondary input port for ELF messages, is located at KI Sawyer AFB and can take over from the BCS should that become disabled. It can, if necessary, pre-empt the BCS.

The TS comprises the ELF stations at Wisconsin and Michigan which normally operate synchronously but can operate independently when required. Two frequency bands are used, 40-50 Hz and 70-80 Hz and each transmitter facility uses commercial prime power from its local utilities companies. In the event of power failure, each site has back-up diesel generators. Each transmitter facility has installed two spare back-up power amplifiers and an uninterruptible power system to ensure necessary reliability to continue transmitting in case mechanical or power failures occur. The TS is a soft, surface deployed subsystem with ECCM and electromagnetic pulse protection. However, it is not expected to withstand a hostile, physical attack.

The RS is located on SSBN and SSN submarines, although the BCS and MIS have receivers for monitoring functions. Signals are normally sent to the BCS, but the MIS can also input messages and then to the TS via dedicated communication lines. Both the BCS and MIS include a message entry element consisting of a data terminal set (teletype Model 40), a link encryption device (KG-84) and a datalink selector panel. This arrangement allows the operator to enter messages into the message queue at the master transmitter facility via telephone company lines. The BCS and MIS also include an order-wire for operator-to-operator communications between facilities. The master transmitter facility is in contact with the BCS and maintains the message queue and automatically updates the back-up queue at the slave facility.

At each transmitter facility a processor element converts the encrypted ELF message from the message processor into drive signals for the power amplifiers, monitors antenna current and controls the transmitter facility master/slave protocol.

The antenna arrays at each transmitter facility consist of antennas oriented north-south and east-west. Two pairs of power amplifiers exist at each facility, one pair for each antenna configuration. Only two power amplifiers are used at any one time, one for each antenna and each is rated at an output level of 660 kW.

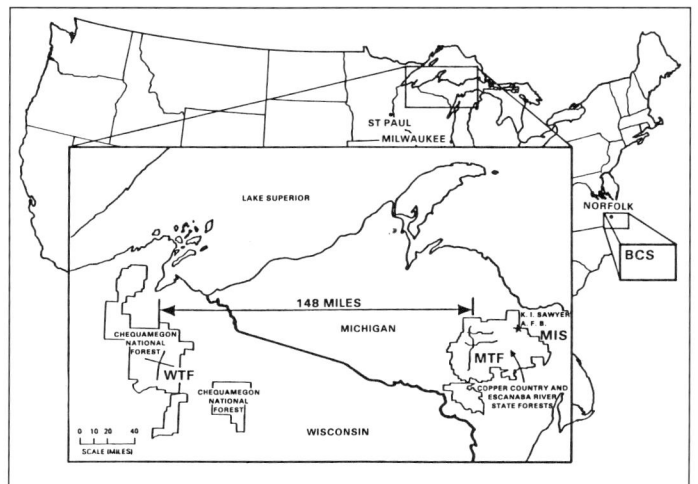

ELF shore sites

Signals are picked up via the OE-315 towed antenna which are then fed to the ELF receiver terminal group via an in-line amplifier. The receiver performs analogue and digital signal processing to detect any message that may be present. The receiver terminal group (OR-279(XN-1)/BRR) consists of five main units: the preamplifier, receiver timing and interface unit, the combined processor and key generator unit, time and frequency junction box and navigational interface junction box. To extract the ELF message, the receiver has first to filter out atmospheric and ocean noise and eliminate interference caused by the submarine's onboard power systems.

The ELF system is synchronous, requiring that the receiver knows accurate time information relative to the transmission and time compensation has to be allowed for the ELF signal propagation delay. A band-spreading key stream is generated and removed from the message, which is then decrypted. By using an embedded AN/UYK-44 militarised reconfigurable processor, many of these functions are performed digitally.

Status
In service with the US Navy. All US Navy submarines are fitted with ELF receivers as standard equipment.

Contractor
General Dynamics Network Systems (prime), Needham Heights, Massachusetts.

UPDATED

Integrated Digital Communications system

The shipboard integrated communications system is designed to provide integrated, secure digital communications. The functions of radio distribution, tactical intercom, administrative intercom and public address are provided with this single integrated system. Apart from functioning as a PBX, the system interfaces with the ship's external communications system, both secure and insecure; sound powered telephones; underwater telephones; shore telephone circuits; tape recording circuits; the public address system; and helicopter intercoms. The software control package provides the necessary call processing, conferencing, inter-computer communications, online diagnostics and maintenance features to support the hardware configuration.

The system is normally configured with two identical switches, physically separated within the ship. Where two switches are installed those terminals designated as operationally vital can be connected to both switches. Should the primary switch fail or suffer damage the secondary switch automatically takes over the load, transparent to the user. Three types of terminal with differing levels of facilities are provided: a command terminal that includes a dual-channel capability; a system terminal; and an operational terminal with limited facilities. There is a waterproof outdoor version of the latter. Terminals are connected by twisted pair copper cable; fibre optic connectivity is also possible.

Status
The system is currently supplied to the Canadian Navy as the Shipboard Integrated Communications System (SHINCOM). The AN/SSC-502(V) is a two-switch version, the AN/SSC-504(V) a single switch version, which is provided with manual back-up. In 2001 DRS was contracted to upgrade the systems in the 'Halifax', 'Iroquois' and 'Protecteur' classes of ships. The system has been installed on the aircraft carrier USS *George Washington*. It is also in use in the Venezuelan Navy and in several other countries. A compact version for smaller warships and submarines is also available.

Manufacturer
DRS Technologies, Parsippany, New Jersey.

VERIFIED

MarCom 2000 automated communication switching system

The MarCom 2000 product line is a modern digital time division matrix switching system used either in the integration or replacement of both legacy and modern systems for shipboard interior, shipboard/shore exterior, air traffic control, large conferencing calls, dispatch and command and control applications.

The MarCom 2000 consists of a 19 in rack-mountable VME chassis which houses the controller and interface cards. The controller is a COTS based computer card. For interfaces to radios and cryptographic equipment, only two programmable card types are used to handle the wide variety of legacy analogue and digital protocols, which reduce life cycle costs. The system can be controlled by a variety of end-user management software, from TELNET messaging to ase changes and not new interface cards or new software. The system is modular to handle configurations from submarines to aircraft carriers or major en route air traffic control centres.

For integrated voice applications, the MarCom 2000 can either replace or integrate with 'stovepipe' systems such as intercom, radiotelephone and sound power. The system includes a single integrated terminal which the operator can select and control access to either interior or exterior (radio) communication

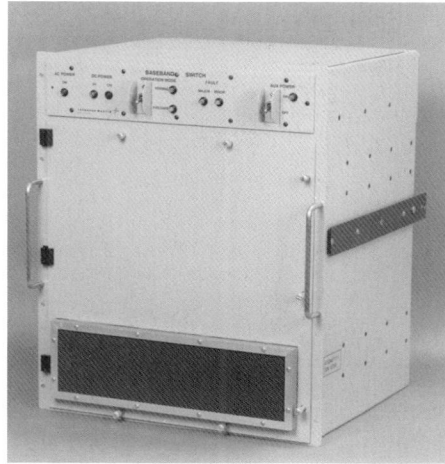

MarCom 2000 automated communication switching system
0010473

systems. The operator can access any number of interior stations, unlimited conferences or radio nets simultaneously, while maintaining the proper security required.

Specifications
Digital Time Division Multiplex (TDM) switch
(a) Electrical
Time slots: 2,048 digital 64 kbit/s (enables aggregate throughput >300 Mbit/s)
Interfaces: industry standard ISDN BRI and PRI; 8 bit u-law (converted to 12 bit for conferencing; analogue Simple Network Management Protocol. Adding or changing connected equipment can now be handled through datab interfaces vailable for sound power, radio and plain telephone circuits
Power supply: 24 V nominal (18-32 V DC), uninterruptible power (UPS) available
Cross talk: >92 dB
(b) Physical: modular 19 in rack- or bulkhead-mounted node sized for application specific interfaces ruggedised for harsh environments
(c) Environmental
Operating temperature: –10 to +55°C
Humidity: 95% at 40°C
EMI: MIL-STD-461B
EMP: MIL-STD-461B

User stations
(a) Electrical
Interface: industry standard ISDN 2B + D
Channels: two 64 kbit/s B; one 16 kbit/s D channel per user station
Power: supplied by TDM matrix
Audio power output: 2 W nominal, optional 12 W
Microphone input: 1.5 mV into 200 W
Monitoring: binaural headset interface monitors both B channels
(b) Environmental: ruggedised for shipboard application; shipboard shock MIL-S-901 D, Grade A, Class 1, Type A
(c) Operating temperature: –25 to +70°C
(d) Humidity: 95% at 40°C

Status
In production. Installed in USN 'Los Angeles' class submarines and 'Aegis' class destroyers, and with several unspecified navies. Selected in November 2001 for installation in the US first of class LPD 17 with an option for three further shipsets.

Manufacturer
L-3 Communications, Communication Systems-East, Camden, New Jersey.

VERIFIED

Programmable Integrated Communications Terminal (PICT)

The Programmable Integrated Communications Terminal (PICT), is a user voice terminal designed to operate with Integrated Services Digital Network (ISDN) switches to support both interior and radio shipboard communications.

The system is capable of interfacing with non-ISDN switches simply by replacing interface circuit(s) and/or firmware.

The PICT integrates four voice communications functions in one user terminal, replacing the Interior Communications (IC) dial telephones, the voice net terminals, the tactical intercoms and the Radio Communications System (RCS) voice terminals. PICT supports four shipboard ISDN interfaces, two with the IC switch and two with the RCS switch. Each ISDN interface can connect two voice circuits simultaneously with switch support. The system permits the user to monitor four circuits simultaneously in any combination of IC and RCS. It includes an embedded monitor-speaker and an embedded intercom speaker/microphone and provides two combination jacks for handset/stereo headset, including Push-To-Talk (PTT), one each for the user and supervisor and a separate jack for a footswitch PTT.

Northrop Grumman's Programmable Integrated Communications Terminal (PICT)
0085274

Access is provided by two USDN interfaces for up to three IC phone/net circuits simultaneously. The fourth circuit is reserved for incoming intercom calls. Interphone calls utilise standard dial telephone features with multidigit dial, one-button speed-dial, forward, override, conference and hold. One-button access is provided to voice nets, phones, announcing systems, and so on, via the IC switch, with manual answer for incoming phone calls. The audio is routed to the hand/headsets and to the monitor-speaker.

Intercom calls are handled through one-button access to the destination intercom terminal. Automatic answer is provided as receive only for incoming intercom calls, with the receive audio immediately routed to the intercom speaker. 'Handsfree' transmit capability is selected at the destination terminal by the operator and is controlled by the originating terminal operator PTT.

Access is provided by two USDN interfaces for up to four RCS circuits simultaneously. The user can request transmit secure/plain and is provided with receive cipher detect for each radio channel. The PICT maintains separation between the IC circuits and the RCS circuits, as well as between the transmit/receive for all circuits. For convenience, the IC/RCS receive audio can be combined at the headset earphone(s) and at the monitor-speaker.

Specifications
Environment
Shock: MIL-STD-901, Grade A
Vibration: MIL-STD-167, Type 1
Temperature: 0 to 50°C
Relative humidity: 5 to 95%
EMI: MIL-STD-461, CS02, CS06, RE01, RS01
Dimensions (W × H × D): 6 × 6 × 3 in (case); 7.5 × 7.5 × 1 in (front panel)
Weight: approximately 5 lb
Power: 8-11 W from the switch(es) (depends on interface configuration)
Connectors: rear-mounted on stand-alone PICT; side, top, bottom, or rear-mounted with optional enclosures

Contractor
Northrop Grumman Integrated Systems, Ocean Springs, Mississippi.

VERIFIED

TDLS-1000 series Tactical Data Link System

The TDLS-1000 system features a field-proven configuration of Link 11 equipment and software along with use of Commercial-Off-The-Shelf (COTS) elements. The baseline TDLS is designed for shipboard use and employs COTS hardware that is deployed in US and NATO fleets. The system software is UNIX-based and includes a human/machine interface implemented under the X Windows and OSF/Motif. The TDLS accepts GPS data using either the NMEA-0183 commercial interface standard, or the US Navy AN/SRN-25 interface. The TDLS can exchange tactical information with other systems using RS-232/422 or Ethernet. The Link 11 message set used with the TDLS has been approved for operational use by the US Naval Centre for Testing and Systems Interoperability (NCTSI). The TDLS-1000 console measures 1,289 × 610 × 1,162 mm and weighs 125 kg. System capabilities:
(a) Configuration options: receiver only, transceiver, UHF and HF
(b) Interfaces: radio, GPS Ethernet and DTS
(c) Operating environments: UNIX, X Windows and Motif
(d) TDLS-1000 datalink message set options: TADIL-A/Link 11; TADIL-J/Link 16; TADIL-B/Link 11B; Link 14; Link 1; TACLAN; Tlink; JMCIS; and custom datalinks
(e) Supported datalink and Ethernet messages: own unit; air; surface; sub-surface; ASW; special points; ESM; track management; voice call sign and voice control; IFF/SIF/DIF; area of probability; Data-Link Reference Point (DLRP); aircraft control handover; ASW summary; plain text; timing; weapon/engagement status; and command.

Status
Baseline TDLS and derived configurations have been fielded in shipboard, aircraft, mobile shore- based, and command centre environments.

Contractor
EDO Corporation, Combat Systems Division, Chesapeake, Virginia.

VERIFIED

AIR

Israel

GIGA-Links

Giga-Links is the name given to Elisra's family of digital communication products for the transmission of high data-rate data generated by wideband imaging sensors. Typical applications include airborne observation sensors such as high-resolution electro-optical cameras and thermal imaging scanners, as well as high-rate ELINT sensors.

The systems operate at microwave frequencies and use a modular concept to meet user needs: a single-channel implementation for rates of up to 150 Mbit/s and parallel channels for higher rates. Compression techniques are used to reduce the data rate when frequency bandwidth is constrained or when higher rate sensors are used.

The channelised airborne transmitter consists of frequency sources, QPSK modulators and solid-state power amplifiers. Error correcting codes are used to improve the link performance. A compact and lightweight design is used to meet the specific airborne or RPV requirements. High reliability is achieved using qualified parts, proven design and controlled manufacturing processes.

The ground-based station consists of a sensitive receiving chain with auto-tracking capability, coherent QPSK demodulators, digital processing and capability for interfacing with high-density tape recorders, laser beam recorders and other display devices.

For those applications requiring data compression, dedicated units are added at the transmitting and receiving ends, using a sophisticated adaptive compression algorithm implemented in VLSI. The sensor output data rate is reduced by up to a factor of four, while maintaining the excellent image quality required for reconnaissance applications.

An extensive monitoring and test capability is included in both the aircraft and the ground-based station for diagnostic purposes.

Specifications
(a) Communication data rate: up to 150 Mbit/s per channel (modular architecture)
(b) Compression: variable length DPCM with up to 4:1 data reduction
(c) Error control: Hamming block code or convolutional coding/threshold decoding
(d) Modulation: coherent QPSK
(e) Frequency band: per customer requirements
(f) Typical bit error rate: 1^{-7} to 10^{-9}
(g) Transmitted power: up to 10 W per channel (solid-state)

Contractor
Elisra Electronic Systems Ltd, Bene Beraq.

VERIFIED

*A multichannel airborne
image compression unit*

United Kingdom

MASTER

The MASTER airborne satellite communications terminal was developed in a collaborative venture between Astrium (then Matra Marconi Space) and the UK Ministry of Defence. The aim was to exploit satcom to provide long-range military aircraft with beyond line of sight communications superior in capacity and reliability to that available using HF radio.

Operating in the 7/8 GHz band, MASTER incorporates a high-performance antenna and modem designed to provide a robust anti-jam capability. The frequency-hopping spread spectrum modem is capable of multiplexing duplex data at up to 64 kbits/s. The system is fully compatible with MIL STD 1553B and MIL STD 704E.

Status
Flight trials were successfully completed by the Royal Aerospace Establishment in 1989. Further development has provided increased technical capability and an improved physical design, reducing system size and weight. The MASTER airborne terminal will be installed on the upgraded Nimrod maritime patrol aircraft fleet.

Contractor
Astrium UK, Stevenage, Hertfordshire.

VERIFIED

Project UNITER

Project UNITER is a secure survivable fixed network for use by the Royal Air Force. Stage 1 of the project, which had as its prime contractor Marconi Communications (formerly GPT Ltd), involved the establishment of a voice communications network between a number of sites within the UK Air Defence Ground Environment (UKADGE) air defence system. Stage 2 involved the implementation of a digital voice and data network between around 60 sites in the UK. This phase has five distinct elements - a circuit switch capability, a packet switch capability, message handling, local networking and network management.

In 1987, a number of contracts were let in the second phase of UNITER. With GPT again the prime contractor, British Telecom was selected to supply some 90 digital switches and 14,000 secure telephones. STC Defence Systems won a £20 million contract to supply optical transmission systems for the project. Honeywell Bull was selected to supply 40 DP56 mini-computers, 500 terminals and a range of application software to form the basis of the message handling subsystem known as CUDS, or Common User Data Services. GPT Video and Data Systems provided over 70 of its 4193 Series X25(84) packet switches for the Packet Network Subsystem, along with its Data Network Management Centres. The subcontract for the Network Management System was subsequently let to Hewlett Packard.

In 1990, GPT was commissioned to provide all the required UNITER Civil Works to ensure that sites reached operational capability in the required timescales. Following a number of competitive procurement activities on behalf of MoD, contracts were let at a number of stations covering ductwork, remote building work services and construction of a number of hardened and soft buildings to contain the communications infrastructure.

Status
The system is in operational use and now forms part of the Defence Fixed Telecommunications System (DFTS). Since 1998 Marconi has managed and maintained the UNITER network on behalf of the DFTS prime contractor, INCA.

Contractor
Marconi, Coventry, West Midlands.

UPDATED

Royal Air Force Tactical Trunk System (RTTS)

RTTS is designed to provide the UK Royal Air Force (RAF) with a highly secure and survivable system using commercial-off-the-shelf (COTS) equipment with specific military enhancements. This COTS equipment includes Nortel Networks' NW300 intelligent multiplexers and the Meridian family of PABXs. RTTS is a 10-node meshed network, carrying secure voice and data communication.

RTTS will enable the RAF to establish a command and control centre in any required location to interface fully with its tactical and strategic communication networks and with the local PTT facilities.

The RTTS network makes extensive use of fibre optic and radio connectivity, managed by an intelligent network management system with in-built redundancy.

The system has recently been upgraded to enable the extension of the RAF office IT environment into operational theatres, providing support for real-time operational IT systems. The upgrade included:
(a) An increased number of rear-link speech circuits with improved voice quality
(b) The capability to accommodate rear-link data rates up to 2 Mps
(c) Provision of a core LAN infrastructure (Deployable LAN(DLAN)) identical to that in UK to enable rapid migration of IT systems into theatre.

The rear link secure speech upgrade provided a new Secure Voice management enclosure. The equipment within the enclosure includes a Nortel Networks Passport 50, which has been closely integrated with the existing RTTS PABX, allowing maximum utilisation of satcom bandwidth. The DLAN solution includes Ethernet switching at each site to provide the bandwidth and control required to extend IT systems into theatre. WAN connectivity is provided using the existing RTTS Passport equipment, allowing the system to become fully multimedia enabled and to carry voice and data over the existing RTTS infrastructure.

Status

RTTS is in operational use.

Contractor

Cogent Defence and Security Networks Ltd, Newport, Gwent.

UPDATED

United States

CP-1516/ASQ and CP-2228/ASQ Automatic Target Hand-off System (ATHS)

The CP-1516/ASQ Automatic Target Hand-off System (ATHS), which was first demonstrated in 1985, is a battlefield mission management system used in conjunction with a control and display unit and up to four standard HF, VHF or UHF radios to provide a tactical C^{3}I network. The digital communication network can provide for stores management, target handovers and other similar functions to be passed to airborne, artillery and ground forces in short radio bursts which are difficult for the enemy to detect or jam. The ATHS has demonstrated interoperablity with the Improved Data Modem, Battlefield Computer System, Digital Messaging Device and AFATDS.

The CP-1516 features a recall capability for 12 previously received messages and allows the transmitting of preformatted messages or free-text messages using an alphanumeric keyboard. Non-volatile memory in the unit retains all critical information in the event of a power loss. In addition, the CP-1516 maintains the current status of up to 10 active airborne missions and two preplanned missions.

Various control/display unit options are available for data entry and display. The CP-1516 is fully compatible with the McDonnell Douglas AH-64 Apache data entry panel and TADS/PNVS display, the Bell OH-58 AHIP control/display and mast-mounted sight display and the JOH-58 light combat helicopter control/display unit.

The newest member of the TDM family, the CP-2228/ASQ (ATHS II) TDM 200, entered production in 1997. It is an upgraded and improved version of the CP-1516/ASQ. The TDM-200 was developed to provide additional capabilities to meet the more stringent environmental requirements of fighter and close air support aircraft, while also meeting the needs for future datalink applications.

The TDM-200 is capable of transmitting and receiving FSK from baud rates of 75 to 1,220 and digital data from 75 to 16,000 bit/s. The higher-frequency operation dramatically reduces transmission time, thus making it more difficult to detect and jam. It has four ports and up to four modems which can simultaneously transmit or receive messages.

Mission data and operational flight program data may be programmed via the MIL-STD-1553 databus or a digital data loader. It is fit and form backward compatible with earlier CP-1516/ASQ models.

Status

Over 750 ATHS are installed on various platforms which include the AH-64, OH-58D, JOH-58, ANG F-16, MH-60 A/K, MH-47 D/E, RAN SH-70B and Belgium Aeromobility A-109 aircraft. The TDM-200 is fully developed and has completed qualification testing for the US Marine Corps AV-8B aircraft in the close air support mission.

CP-1516/ASQ ATHS works with a variety of MIL-STD-1553B control/display units

Specifications

ATHS and ATHS II Features Comparison

Operational features	CP-1516/ASQ (ATHS)	CP-2228/ASQ (ATHS II)
Modulation format	FSK	Same plus digital baseband/diphase and DCT FSK tone sets
Transmission rates	75, 150, 300, 600, 1,200 b/s	Same plus 5, 8, 9.6 and 16 kbyte/s digital
Radio interfaces	4 ports, 1 modem	4 ports, 2 to 4 modems*
Simultaneous users	1 port active at a time	Same as modems
TEMPEST	Yes	Yes
Host vehicle interface	MIL-STD-1553B	Same plus MIL-STD-1553A
Programming language	PLM	C
Reprogramming method	Depot, via card edge	On-aircraft, via 1553 bus
Physical characteristics		
Power	35 W	17 W
Weight	4.54 kg (10 lb)	4.54 kg (10 lb) max
Size**	137.2 W × 167.6 H × 203.2 D (5.4 × 6.6 × 8.0)	Same
Mounting	Hard mount	Same
Cooling	Convection	Same
Spare card slots	None	Four (If 4 modems implemented)

Note(s): *ATHS II standard configuration contains two modems. Up to two additional modems can be added in prewired card slots.
**Millimetres (inches)

Contractor

Rockwell Collins Government Systems, Cedar Rapids, Iowa.

VERIFIED

SCOPE Command HF Global Communication System

SCOPE (System Capable of Planned Expansion) Command programme replaces all existing US Air Force high-power HF ground stations, including SCOPE Control, SCOPE Pattern and SCOPE Signal III radio systems, with a system that provides subscribers with a communication network having the operational simplicity, dependability and connectivity comparable to commercial telephone services. The system will consist of at least 14 worldwide HF stations interconnected through various transmission media and, will increase overall operational and mission capabilities while reducing costs. Its open architecture design and the Communication Control and Management System (CCMS) permit flexibility to meet changing mission and force requirements, 'lights-out' (unmanned) site operation and easy expansion of equipment or additions to the network. The HF Global Communications System is specifically intended to support four missions:

(a) United States Air Force (USAF) Global - Supports a wide range of users by providing air-ground-air, ship-to-shore, broadcast, and Automatic Link Establishment (ALE) capability to various DoD customers.

(b) Mystic Star - Provides HF communications for the President, Vice-President, cabinet members, and other senior government and military officials while aboard Special Air Mission aircraft.

(c) SITFAA - A Spanish/English/Portuguese language network supporting North, Central and South American Air Force users in 18 countries. Provides voice and data HF links.

(d) DCS HF entry - Provides HF communications services for tactical units in areas of the world where DCS connectivity is unavailable or insufficient.

SCOPE Command major equipments include operator consoles, circuit switching equipment, HF radios, RF matrixes and antennas. Each station consists of standard PC workstations using CCMS audio interfaces. CCMS provides fully automatic local as well as remote site, operation and maintenance and local- and wide-area network (LAN/WAN) control.

The switching subsystem is all digital, non-blocking and features unlimited conferencing, modular sizing and precedence function and includes expansion capability up to 2,016 lines.

The HF radio equipments include Rockwell's Spectrum DSP Receiver/Exciter, Model RT-2200. The radios have Automatic Link Establishment (ALE) and Link Quality Analysis (LQA) capability and are adaptable to future ECCM waveforms. Thirteen different HF modem waveforms ensure a high backward-and-forward capability and mission interoperability. The transmit subsystem includes 4 kW solid-state power amplifiers for HF coverage to all aircraft regardless of range capabilities, a high-power transmit matrix, a combination receive/multicoupler antenna matrix and omni and RLP antennas.

The CCMS control subsystem uses a modular, open system design to manage and control all system operations, including those at remote sites. CCMS maximises commercially available standards-based software and the IBM OS/2 multitasking operating system. It employs LAN software, servers, hubs and routers to support unlimited LAN/WAN networking.

System features include:

(a) ALE and LQA compatible

(b) open-ended system architecture and system control management

(c) user-friendly, automated resource management and control of local and remote sites

(d) fully redundant architecture with automatic cut-over

(e) centralised control from any location/site

(f) Collins' high-performance MIL-STD-188 radios and modems

(g) MIL-qualified performance

(h) Lincompex and other link enhancement techniques.

Unique features of the programme are the System Integration Laboratory (SIL) and test-bed that are used to confirm the system design approach and therefore result in a low-risk programme. This system, with radios, PAs, HF modems, antennas and other system equipment, forms a fully functional station to support SCOPE Command and other system applications. This facility provides the capability to verify the baseline design, test interface compatibility and perform functional verification tests.

Additionally, the SIL will provide online support for system and network operation and maintenance. This will enable Rockwell engineering and logistics support to help US Air Force personnel solve problems in real time.

The modular, open architecture system and CCMS design produce a system and network that are expandable for both equipment upgrades or additions, or adding another site to the network.

In March 2002 the Rockwell HF Messenger was successfully tested over the network. Messenger provides a secure e-mail capability at up to 9.6 kbit/s over the HF network. In November 2002 this capability was extended with a demonstration of worldwide HF e-mail into the US DOD SIPRNET and NIPRNET and the Internet.

Status

SCOPE Command was scheduled for full operational capability in 2002. The operating commands/services responsible for the 14 stations include the US Air Force's Air Mobility Command (AMC), Air Combat Command (ACC), Pacific Air Forces, (PACAF), United States Air Forces Europe (USAFE), Air Force Space Command (AFSPC), and the US Navy.

Contractor

Rockwell Collins, Cedar Rapids, Iowa.

UPDATED

Strategic Automated Command Control System (SACCS)

The Strategic Automated Command Control System (SACCS) network is the primary network for the transmission of Emergency Action Messages (EAMs) to the warfighting commanders in the field. It is the prime communications link between the CINC USSTRATCOM and their nuclear missile forces, as well as to other offensive and defensive forces worldwide. The system provides critical secure (TS) command control information, such as EAMs, FDMs, situation monitoring, current intelligence, force status, operations monitoring, warnings, strategic replanning and redirection, and damage/strike assessments. The system also provides SIOP (TS-ESI) messages to HQ ACC, AMC, AFRC and ANG for their deployment.

SACCS is located in the CINCSTRAT command post, strategic command centers, missile launch-control centers, launch-control facilities, and at strategic aircraft sites. The system was initially fielded in 1963 and was updated by SACDIN in 1988. The SACCS Data Transmission Subsystem provides primary command and control capability for receiving and transmitting secure Emergency Action Messages (EAMs) (highly structured, authenticated messages primarily used in the command and control of nuclear forces), Force Direction Messages and various informational-type messages from the NCA to and from the CINC United States Strategic Command, and to the strategic nuclear missile and bomber forces. SACCS also provides SIOP messages to Air Combat Command, Air Mobility Command, and the Air National Guard and Air Force Reserve for their effective deployment of strategic bombers, reconnaissance aircraft, mobilisation aircraft, and tanker support aircraft worldwide. SACCS is operationally capable of providing full-duplex, secure data communications over any level of projected message activity. The network currently has 93 active nodes located across the country with 20 additional Air National Guard sites slated to come online sometime in the next year. The network is composed of a hierarchical series of processors consisting of two Subnet Communications Processors (SCP), eight Base Communications Processors (BCP), numerous Hardened User Terminals (HUTE) and Collocated User Terminals (CUTE), and the associated lines linking the processors together. The SACCS network also serves six external interfaces: Data Processing Subsystem (DPS), Air Force Global Weather Center (AFGWC), Command Center Processing and Display System (CCPDS), AUTODIN, Air Force Satellite Communication (AFSATCOM), and 616A (a survivable Low-Frequency Communications System). In 1956 General Curtis LeMay, Commander-in-Chief of SAC, saw a need for improving SAC's command and control system. A co-ordinated effort was undertaken by government and industry to provide this system. The project was designated 465L, and was the predecessor to the current Strategic Automated Command Control System network. In the mid-1960s, SAC procured the 465L system, which was designed to be survivable and to provide rapid transmission, processing, and display of information to support command and control of SAC's geographically-separated forces. By the mid-1970s, SAC began a programme to replace the ageing hardware of the SACCS network and provide improved capabilities. The new system, the SAC Digital Network, was a subsystem of the Worldwide Military Command and Control System. It provided secure two-way communications between the National Command Authority, CINCSAC, SAC aircraft bases, and missile combat crews during peacetime and trans- and post-attack periods. Special emphasis was placed on SACDIN's ability to maintain an end-to-end message delivery time of not more than 15 seconds for

Emergency Action Messages. Total system cost for SACDIN, including a later modification, was nearly US\$1 billion. After formal acceptance, the system became known as SACCS.

SACCS is intended for use in a day-to-day and pre-attack environment. While it is a relatively high-speed and timely system, it is not survivable because of its reliance on 'soft' land lines for data transmission. In 1999 it was reported that desktop terminals known as the Strategic Automated Command and Control System Desktop Terminal, or SDT, had been introduced for use on the system, running COTS operating systems and applications.

This upgrade process is reported to have continued through to 2003, when the system was described as "providing the means for entry, transmission and distribution of emergency action messages, force direction messages and routine data messages between higher headquarters, unit command posts and missile alert facilities for 1,500 customers at 132 locations throughout the USA."

UPDATED

Theatre Deployable Communications (TDC)

Theatre Deployable Communications (TDC) is a ground-to-ground communications infrastructure designed to transmit and receive voice, data and video communications securely to or from wireless, satellite or hard-wired sources. It is designed to supply timely, secure and reliable data communications for deployed US Air Force units. The system is both mobile and modular, enabling the US Air Force to tailor the system to its specific needs and to transport the system anywhere in the world, with considerable reductions in airlift and manpower over previous systems. It has absorbed the legacy TActical Secure DAta Communications (TASDAC) system and interfaces with other area systems such as TRI-TAC and MSE, as well as with commercial networks.

Basic Access Module

The Basic Access Module is a flexible, scalable and configurable module that provides both voice and data functions for a deployable communications operation. In its basic configuration it has a voice switching chassis that implements a private branch exchange (PBX), which is configurable to meet customer mission scenarios. Local subscribers are provided telephone access via two-wire analogue circuits for Plain Old Telephone Service (POTS - WECO 2500) and compatible products such as faxes, modems and STU-IIIs. When the circuit switch is equipped with a Basic Rate Interface (BRI) card, access is provided to ISDN digital devices such as phones and video teleconferencing (VTC) units. For large concentrations of users, up to eight modules can be interconnected to form a large voice switch and segmented LANs. Some of the call features offered by the voice switch include call forward, call transfer, call park and un-park, speed dialing, last number redial and wake-up call. Trunk features include MLPP, caller ID and Autovon trunk class of service. Data switching /routing expansion enhancements are available as options.

Crypto Module

The Crypto Module uses KIV-19 Trunk Encryption Devices (TEDs) for long-haul bulk encryption of off-base trunks. The KIV-19 is smaller and lighter than its KG-194 equivalent and can handle data rates from 9.6 kbps to 13 Mbps. The Crypto Module accepts four red data streams and produces four black encrypted data streams in both NRZ (RS-422) and CDI (Conditioned Diphase, CX-11230) formats. The NRZ interface operates for data rates from 9.6 kbps to 13 Mbps. The CDI interface operates at discrete data rates from 32 to 64 kbps balanced and from 72 to 2,048 kbps. The Crypto Module also contains a Primary Reference Source (PRS), which provides a station-clock timing reference for the network. The PRS produces STRATUM 1 timing derived from GPS timing. The PRS has a GPS receiver and antenna with an option for redundancy. The included rubidium oscillator can provide STRATUM 2 timing in the event GPS timing is lost.

Microwave Module

The Microwave Module provides a 4 DS1 capacity wireless LOS transmission link between user nodes and hubs over a maximum distance of 5 km. Two of the DS1s connect directly to the ISDN PRI Switched Circuit Network (SCN) backbone. The two remaining DS1s connect to the Datagram Switched Network via an integrated IP router. The router has two integral T1 CSU/DSUs to route IP datagrams to/from two of the serial DS1 connections on the radio baseband assembly. The four DS1 signals are multiplexed within the radio's baseband assembly and then are interfaced with the RF assembly. The full duplex microwave link operates in the 14.4 to 15.35 GHz frequency band. Operators configure and monitor the radio and link through the radio admin port on the I/O distribution frame using a standard PC or laptop. The Microwave Module is capable of interfacing with the Basic Access Module, the Legacy (PTT) Module and any of ICAP's voice or data modules using T1 and 10BaseFL interfaces.

P-MUX Module

The P-MUX Module provides multiplexing and demultiplexing of voice, data and message traffic. This multiplexing function creates bandwidth efficient connectivity between the deployed base and off-base locations. The P-MUX Module will generally be located at the primary hub of a deployed base. The heart of the P-MUX Module is a digital multiplexer. The multiplexer accepts voice and data traffic and multiplexes it for transmission to off-base locations. Conversely, the multiplexer demultiplexes aggregate off-base traffic onto independent voice and data lines. Basic configuration includes a dual- port T1/PRI module, echo-cancellation and voice-compression features, high-speed synchronous serial, low-speed async/ sync serial and high-speed trunk modules.

Crypto Interface Module

The Crypto Interface Module provides data security for IP backbone and local network devices (such as hubs and switches) utilising the Internet Protocol (IP) to exchange data from nodes of a larger network and external military facilities. The electrical interfaces available are 10BaseFL, high-speed synchronous serial (EIA-530), EIA-232 serial and Conditioned Diphase. The Crypto Interface Module houses an Ethernet router that provides access to the IP backbone and serial trunk encryption devices allowing secure access for remote communications. The trunk encryption devices provide compatibility with KG-81 family (-81, -94, -194, and so on) and KG-84 family (-84A, -84C, KIV-7) encryption devices.

Legacy (PPT) Module

The Legacy (PPT) Module provides connection between Switched Circuit Network backbones and external switched circuit networks. This module is used by those customers requiring high-capacity voice circuits. The module has the capability, through the addition of optional boards, to expand up to 192 non-blocking or 256 line device ports. Circuit boards can also be added to provide analogue FXO (foreign exchange office) trunks that can be used for a simple interconnection to two-wire commercial telephone company (TELCO) subscriber lines anywhere in the world. The capability includes SF (single frequency) signalling, E&M signalling, Loop Start Ring Down (LSRD) signalling, ISDN BRI (U interface). KY-68/TRI-TAC Interfaces are also available as options. The module also has the capability to provide an SCN channel connection to a secure push-to-talk (PTT) radio that may be used for dial-up UHF SATCOM connectivity.

Red Hub Module

The Red Hub Module provides a secure interface for computer workstations and local network devices (hubs, switches and so on) using the Internet Protocol (IP) to exchange classified data between nodes of the network. The electrical interface to the local secure LAN user is 10BaseT. Encryption of classified IP data is performed by an Ethernet encryption system; the data are transmitted to other nodes using 10BaseFL interface. The Ethernet encryptor provides network encryption security designed to Secure Data Network System (SDNS) standards and is endorsed by NSA to handle classified (Type 1) data up to the Top Secret level. All usable electrical interfaces to the module components are brought to the I/O distribution frame and are individually protected from electrical surge and lightning. All components are fully SNMP manageable using any standard SNMP network management platform, such as HP OpenView.

Status

In service with the US Air Force.

Contractor

General Dynamics Decision Systems, Scottsdale, Arizona.

VERIFIED

SATELLITE

France

SPARTACUS Manpack Tactical Satellite Terminal

The Station Portable Autonome de Raccordements TACtiques Utilisant un Satellite (SPARTACUS) (Autonomous Manpack Terminal Using a Satellite for Tactical Connection) is a highly flexible manpack satellite terminal. Developed in co-operation with MMS (UK), it was the first manpack tactical X-Band terminal used by the French Army. Also identified by the nomenclature TRC 846.

As a military X-band terminal (7-8 GHz), the TRC 846 is compatible with all X-band geostationary satellites (TELECOM 2, SKYNET, NATO, HISPASAT, DSCS and so on). It is capable of point-to-point, duplex or simplex links at 2,400 bits/s or 4,800 bits/s. COMSEC and NETSEC crypto facilities are embedded. It can be rapidly deployed (<10 minutes) and is housed in a manpack unit (20 kg).

Specifications
(a) Size: 27 × 28 × 23 cm/41 × 33 × 23 cm
(b) Weight: 20 kg with battery
(c) EIRP: >39 dBW typ
(d) G/T: >7 dB/K at 10° elevation
(e) Operating time (25°C): (transmission) 1 h, (reception) 3 h

Status
After an international request for proposals launched in June 1991, the Délégation Générale pour l'Armement awarded Thomson-CSF (now Thales), teamed with the then Matra Marconi Space, a contract to supply 20 SPARTACUS stations in June 1995. In service with the French Army.

Contractor
Thales Communications, Colombes.

VERIFIED

SPARTACUS TRC 846 Tactical Satcom   0055055

Syracuse satellite communications system

Syracuse is the satellite-based communications system used by the French armed forces. Syracuse uses the French Telecom 1 and Telecom 2 multimission satellites.

Syracuse 1 was the first French government satellite network. Its implementation dates back to January 1980. It uses two repeaters on board each of France Telecom's Telecom 1 satellites and, in 1995, was providing round the clock communications between 26 earth stations (fixed stations within France, transportable stations and tactical stations). In January 1987, it was decided to initiate the Syracuse II programme. The Syracuse II system is designed to assure

Installed shipborne radome-enclosed satcom antennas

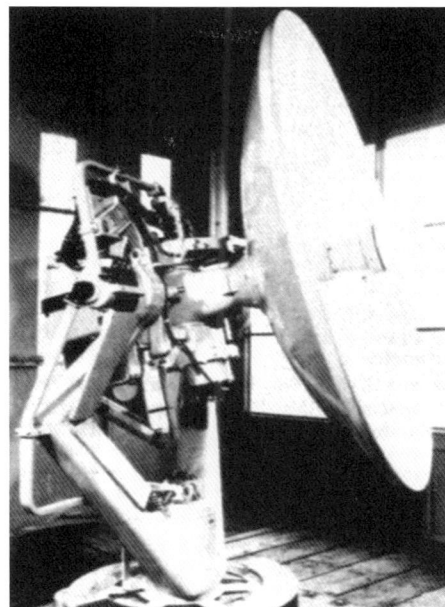

Shipboard triaxially stabilised antenna

the continuity of services provided by the existing Syracuse 1 system, while offering more than double the capacity available from its predecessor, a significant increase in the number of ground stations, greater protection against hostile action, the use of lighter stations and automated management of all system resources.

The Syracuse II military payload consists of five transparent military repeaters capable of operating simultaneously. The satellite offers a global coverage antenna, a fixed spot beam antenna covering central Europe (including France), a steerable spot beam antenna and a secure link for anti-jam purposes and operational messages.

The Syracuse II ground segment comprises three fixed metropolitan stations, updated Syracuse 1 mobile stations, additional stations of the type already in service and stations of a new type. The latter are lighter for enhanced mobility and can be carried aboard medium tonnage surface vessels or submarines. There are approximately 100 stations in total.

These stations communicate via the French Telecom 2 satellite and systems deployed by allied nations and NATO member states. The stations handle 75 bit/s telegraphy, 2,400 and 16,000 bit/s telephony and 75, 2,400 and 16,000 bit/s and N × 64 kbit/s, datalinks.

The central fixed stations serve as nodal points, exchanging information with the mobile stations designated TL, T, VL, N, NL, SM. In addition, they perform network synchronisation, receive telemetry signals from the satellite and send back necessary command signals. These stations feature 8 and 18 m antennas with periscope feeds. They additionally perform satellite angle-tracking. The TL light tactical station can be mounted on a four-wheel drive vehicle (ACMAT-VLRA) or transported by a Transall C-160 or Hercules C-130 type aircraft. The antenna is of the offset type with a diameter of 1.3 m. The T transportable station can be transported by truck (ACMAT TPK 650 SH) or airlifted in a Transall aircraft. The antenna is of the Cassegrain type with a diameter of 2.8 m. The VL light vehicle station can be relocated on lightweight commonalised or four-wheel drive vehicles. The dish antenna's reflector is 0.9 m in diameter. The NL lightweight naval station equips low-tonnage vessels (from 800 tonnes). It features two cabinets and a three-axis stabilised Cassegrain antenna with a diameter of 1 m. The S submarine station

Fixed satcom ground terminal electronic equipment

Fixed satcom terminal

Transportable satcom terminal

is distinguished by its two-axis stabilised radome-enclosed periscopic antenna. The N naval station equips medium- and large-tonnage ships. It features a shelter installed on the top deck and two antennas. Each antenna is a three-axis stabilised Cassegrain type with a diameter of 1.5 m. A new kind of station, dubbed the Light Vehicle 256, is derived from the VL station and provides a range of data rates from 9.6 to 256 kbit/s (with potential growth to 2,048 kbit/s). The dish antenna's reflector is 1.8 m in diameter.

In November 2000, the French Ministry of Defence chose Alcatel Space as prime contractor for the new generation Syracuse III. Alcatel Space is also specifically responsible for satellite construction, including in-orbit delivery, as well as the control centre, mission centre and the extension of ground stations in France. Alcatel Space will call on Thales for all specific ground segment developments, plus security-related ground and space equipment, as well as repair and maintenance of the Syracuse system. The contract is worth approximately €1.4 billion.

The Syracuse IIIa satellite, due to be launched in December 2003, will use a commercial Spacebus platform, hardened to resist nuclear attack. The communications payload will operate in two frequency bands:

SHF 4 spotbeams, 1 global beam, 1 metropolitan France beam; nine 40 MHz channels.

EHF:2 spotbeams, 1 global beam; six 40 MHz channels.

The jamming resistance and reconfigurability required for military missions are provided by an active antenna and latest-generation digital processor. The additional EHF capacity will initially be dedicated to satellite-France links to free the SHF payload for communications with forces deployed abroad. According to *Jane's International Defense Review* the German armed forces will employ part of the capacity under an accord signed in September 1999.

In early 2002, Thales and Alcatel were awarded a contract worth US$17.6 million for the supply of naval ground terminals for installation aboard two of the French

Tactical satcom station

Horizon destroyers currently under construction and for two French Navy transport ships. Further orders are expected. The terminals will be able to handle both military and commercial (X-band and C-band) bands automatically.

Status
Telecom 2A, carrying the first Syracuse II elements, was launched in late 1991 and Telecom 2B was launched in April 1992. Telecom 2C was launched in 1995 and the latest, Telecom 2D, in August 1996. Delivery of the Syracuse II system was complete by 1997.

Contractors
Alcatel Espace(prime), Toulouse.
Thales Communications, Colombes.

UPDATED

TANIT tactical earth station

TANIT was designed and developed to meet the need for rapid deployment of reliable satellite earth stations by tactical forces. TANIT stations carry medium- and high-speed trunks from 9.6 to 2,048 kbit/s, depending on individual user's requirements and satellite resource availability.

TANIT is compatible with all X-band satellite families and interconnects civil or military switches or multiplexers. TANIT carries all types of trunk traffic: voice, data, fax, images, scanned photos, secure transmission and so on. It can mesh network nodes and provides remote monitoring and control capabilities. The system is designed for ease of transportation and installation and can be erected in 30 minutes by two people.

Specifications
Antenna		
(reflector diameter)	1.6 m	2.4 m
Power amplifier	SSPA 40 W	SSPA 40 W
EIRP	55.5 dBW	59 dBW
Station G/T	15.5 dB/K	19 dB/K
Total weight	140 kg	250 kg
	(3 cases)	(4 cases)

TANIT can optionally be provided with a simultaneous multiple transponder capacity, remote monitoring and control, automatic tracking and a 24 V DC power supply.

Status
TANIT is in service with the French Army.

Manufacturer
Thales Communications, Colombes.

VERIFIED

TANIT Tactical Earth Station   0055048

Italy

Page Europa Rapid Acquisition Satellite Terminal (PE-RAST 4.8T)

Terminal
The PE-RAST 4.8T is an autonomous, self-powered and highly transportable X-band satellite communications terminal, designed to establish point-to-point or multipoint high-capacity links among forward mobile units, area backward units and static centres. Composed of an ACE-II shelter, a 4.8 m parabolic antenna on

PE-RAST 4.8T deployed 0001014

trailer and a skid-mounted twin generator set, the terminal can be transported by road (two 5 ton trucks) and air (helicopter, C-130 or similar aircraft) and can be deployed and put into service in 45 minutes from arrival on site by a crew of three.

The PE-RAST 4.8T can operate with any geostationary or quasi-geostationary military satellite in the X-band. The tracking system is either automatic (enhanced step-track mode) or performed by computer stored ephemeris (program-track) or it can be manual.

The station provides the simultaneous transmission of :

(a) up to four QPSK carriers (four × 2 Mbit/s) for a total of up to 480 voice circuits or any desired equivalent mix of voice, TTY and data circuits

(b) up to four spread spectrum CDMA (SSCDMA) carriers which can accept any combination of data, fax, TTY and voice circuits from 9.6 kbit/s to 256 kbit/s even under severe jamming conditions.

Traffic data rate and carrier bandwidth are flexible. Single link and multilink (multicarrier) connections can be established in any portion of the X-band. Type, number of circuits and traffic data rate of each carrier is adaptable to user needs. Flexibility is enhanced by a rapid reconfiguration of user circuits performed in the field by a personal computer or by a centralised station.

PE-RAST 4.8T is capable of connection and interface to a wide range of military and civil circuits and terminals through digital flexible multiplexers; it provides access and connection (voice, fax, data and TTY services) to groups of local subscribers, to subscribers of a PABX and to remote subscribers via standard CCITT digital data streams. Integrity and security of the links are assured even under severe jamming conditions by SSCDMA modems and bulk encryption (DISBEE). The station is designed to EMI/EMC specifications for military equipment.

Antenna

The PE-RASA 4.8T satellite antenna system has been designed for use with the PE-RAST terminal but is available as a stand-alone equipment for use in transportable, high-capacity satellite ground terminal applications over the frequency band 7.25 to 8.4 GHz.

Composed of a reflector assembly, a feed assembly and an antenna control unit, the PE-RASA system is mounted on a two-axle trailer equipped with adjustable ground-mounting fast pads with pneumatic devices for quick set up and stabilisation. The reflector assembly is a 4.8 m three-piece shaped aluminium reflector, whose surface roughness is within 0.4 mm RMS achieved by a unique and patented lathing technique. The reflector is equipped with a built-in hydraulic system for folding/unfolding the reflector wings during transportation/operation.

Using a Cassegrain feed assembly the antenna has electro-formed horns and filters specially designed to achieve low intermodulation products. The feed allows both single and multiple carrier communications. The antenna positioning system consists of azimuth/elevation actuators based on high-precision screw rods with AC motors/reductors, which allow azimuth panning of ±45° and elevation panning from 0 to 90°.

The antenna control unit, complete with attitude systems, allows the following operation modes:

(a) automatic (enhanced step-track mode)

(b) computer stored ephemeris (program track)

(c) manual.

Designed to EMI/EMC specifications for military equipment, the PE-RASA 4.8T can be towed by truck on paved road at speeds of up to 80 km/h or cross-country at up to 25 km/h. It can also be transported by fixed-wing aircraft (C-130 or similar), helicopter; sea and rail. It can be deployed and put into service in 20 minutes from arrival at a new site by a two-man crew in a 40 km/h wind.

Specifications

PE-RAST 4.8T

(a) Uplink EIRP: >83 dBW

(b) Downlink G/T: >26 dB/K

(c) Monitor And Control System (MACS): local control and monitor of all equipment and functions; remote control by ruggedised PC

(d) HPA: TWT wideband 1 + 1 in soft/fail configuration (4 kW)

(e) Up-down converters: ultra-low phase noise; 7.9-8.4 GHz, 1 kHz spacing (transmit); 7.25-7.75 GHz, 1 kHz spacing (receive)

(f) QPSK modem: flexible data rate up to 8 Mbit/s; FEC and service channel included

(g) Multiplexers: digital flexible speed from 75 bit/s to 2,048 kbit/s

(h) Frequency reference standard: dual rubidium oscillator with automatic changeover and ruled by GPS timing signals

(i) Shelter: ACE-II shelter with full MIL-SPEC EMI/EMC protection. All electronic equipment including the HPA are contained in the shelter and mounted with suitable shock absorbers

(j) Air conditioning: redundant split-system

(k) Power unit: transportable, fully redundant diesel genset, skid-mounted with local/remote-control and redundant automatic changeover (2 × 42 kVA; 3 × 380 V C 50 Hz)

PE-RASA 4.8T

(a) Size: 4.8 m (16 ft) diameter

(b) Feed: Cassegrain with electro-formed horn and filters

(c) Sky coverage: from 0 to 90° continuous elevation; from −45 to +45°continuous azimuth

(d) Positioner: elevation over azimuth positioner with AZ/EL actuators based on high-precision screw rods and AC motor/reductors

(e) Frequency: 7.9-8.4 GHz transmit; 7.25-7.75 GHz receive

(f) Polarisation: circular

(g) Intermodulation: <−135 dBm, two tones

(h) Deployment time: 2 persons in 20 min with up to 40 km/h wind

(i) Tracking: step-program-normal track

(j) Power handling: 5 kW CW

(k) Weight: 5,300 kg (trailer incl)

(l) Environmental:
Temperature: operating from −30 to +50°C plus sun radiation
Air humidity: up to 100% with condensation
Wind: up to 80 km/h, gusting to 100 km/h (operation); up to 150 km/h, gusting to 180 km/h (survival)

Status

The PE-RAST 4.8T and PE-RASA 4.8T have been designed and produced by Page Europa as part of a NATO contract awarded by NC3A for the supply of transportable SATCOM ground terminals to cover the NATO ACE SHF SATCOM needs. Six systems were delivered to NATO in 1996/97 and are presently in full operation in various countries.

Manufacturer

Page Europa SpA, Rome.

VERIFIED

....

Transportable Satellite Ground Terminal (TSGT)

The Transportable Satellite Ground Terminal (TSGT) is a compact, airliftable, road transportable and offroad-capable deployable SATCOM terminal. The system design includes one foldable antenna, mounted on a steerable trailer.

The purpose of the TSGT is to provide a satellite communications system that can be rapidly deployed. It enables the establishment of reliable and secure military communications in support of NATO command posts. The TSGT can be configured to operate in the X-, C- and Ku-bands and is able to communicate with Intelsat, Eutelsat and NATO satellites. The system can operate with four simultaneous QPSK carriers or two Electronic Protection Measures (EPM) carriers. The TSGT can be conveyed overland using a single 5 ton truck. It can also be moved by air using a suitable military transport aircraft without the aid of any special tools. Once the TSGT has arrived on site, it can be set up in one hour by a three-man crew including satellite acquisition. After the initial set-up the TSGT operates in an 'unmanned' mode. When a change of operating band is required, the reconfiguration can be achieved within one hour without any additional tools.

The TSGT consists of the following main components:

(a) Antenna trailer

(b) Communications equipment enclosure

(c) Twin set of diesel generators.

The station is capable of the simultaneous transmission of up to 4 QPSK carriers (4 × 2 Mbytes/s) and up to 2 EPM carriers, which can accept any combination of data, fax TTY and voice circuits from 9.6 kbytes/s to 256 kbytes/s even under severe jamming conditions.

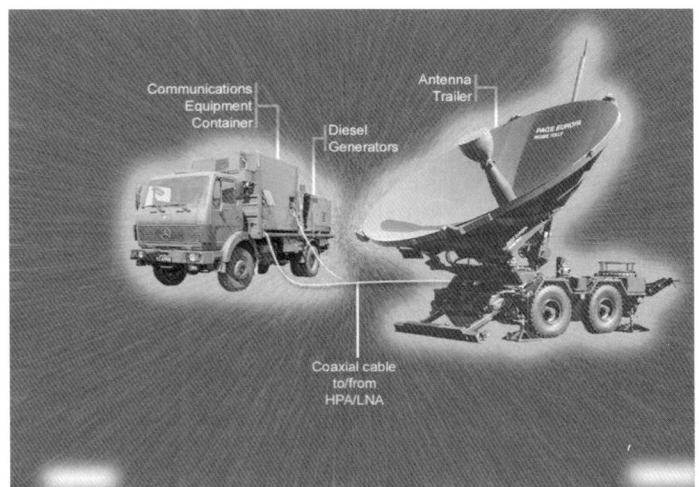

The TSGT in the operating position (Page Europa SpA) 0125365

The TSGT with the antenna stowed (Page Europa SpA)   0125366

The TSGT can interoperate with any geostationary or quasi-geostationary military satellite operating in the X-, C- and Ku-bands. The tracking system is either automatic (enhanced step-track mode), performed by computer stored ephemeris (program-track), or manual. A Digital cross Connect (DXC) Multiplexer provides up to four G703/G70 plus E1 signal inputs/outputs that are time-slots cross connected. The DXC provides a drop/insert function for two N × 64 kbits data circuits. The carrier bandwidth is optimised using the QPSK modem facilities.

The TSGT is able to connect and interface up to four G703/G70 plus E1 signals of military and civil circuits and terminals through DXC multiplexers. Signal access to the TSGT can be extended to a remote command post via a fibre optic connection. The user interfaces are available on a Fibre Optic Line Termination Unit (FOLTU).

The equipment container and/or shelter is provided according to the user specification. All electronic equipment excluding the HPA (RF section) is contained in the shelter and mounted with suitable shock absorbers.

Specifications
(a) Up link:
    C-band 71 dBW (min. EIRP)
    X-band 74 dBW(min. EIRP)
    Ku-band 74 dBW(min. EIRP)
(b) Down Link:
    C-band 23 dB/OK (min. G/T)
    X-band 28 dB/OK
    Ku-band 29 dB/OK
(c) Up to 4 QPSK carriers (4 × 2 Mbytes/s)
(d) Up to 2 EPM carriers
(e) Dual Rubidium oscillator with automatic change-over and ruled by GPS timing signals.
(f) Antenna: 4.8 m parabolic dish foldable during transportation. Cassegrain feed with high gain and low side-lobes.
(g) Tracking: Step track, programme track and manual. Full performance in wind speeds up to 80 km/h, reduced performance up to 120 km/h, survival in the stow position up to 150 km/h.
(h) Generator: Transportable, fully redundant 2 × 25 kVA diesel genset, skid mounted and provided with enclosure for outdoor operation and cold-start devices. Local/remote control with redundant automatic change over Mains/Genset and Genset 1/Genset 2 is provided. An uninterruptible power supply is available on request.
(i) Operational environmental conditions: temperature range from -30 to +50°C; humidity up to 100% under rain and snow conditions.
(j) Road transportation speed: Up to 80 km/h on paved roads, 25 km/h cross-country

Status
TSGTs X-band are in service with NATO.

Manufacturer
Page Europa SpA, Rome.

UPDATED

NATO

NATO IV satellites

NATO first became involved with satellite communications in the 1960s through the US Initial Defense Communications Satellite Program (IDCSP). The NATO I programme made use of two transportable X-band earth stations, known as MASCOT, with 27 IDCSP satellites. The NATO II programme consisted of NATO's first dedicated satellites with two medium capacity spacecraft similar to the UK's Skynet 1. One was launched in 1970 and the other in 1971. In 1973, a contract was awarded to Ford Aerospace (now Loral) to design and provide three NATO III satellites. These were launched consecutively in 1976, 1977 and 1978. A fourth satellite (NATO IIID), with doubled RF power, was launched in 1984.

In the mid-1980s, NATO decided to replace its third-generation system with a new series of satellites based on an existing operational satellite design - the UK-built Skynet 4. The UK Ministry of Defence was charged by NATO with the procurement of two satellites, designated NATO IVA and NATO IVB, and in February 1987, British Aerospace Space Systems (now part of Astrium) was

The NATO IV satellites are based on the Skynet 4 spacecraft

NATO satellite launch from Cape Canaveral

appointed prime contractor, with Matra Marconi Space (also now part of Astrium) as the communications payload contractor. It was the first time that NATO had selected, from international competition, a UK-led industrial team to provide its front line spaceborne communications system.

NATO IV spacecraft are directly based on the Skynet 4 military communications satellite. This employs a three-axis, stabilised platform design which was originally developed by a UK-led European consortium of companies for various European Space Agency (ESA) programmes.

The NATO IV satellite comprises a service module which provides the spacecraft with housekeeping and orbital control functions, and a communications module which carries the communications payload. Two rigid solar array wings, each comprising three solar array panels which unfold in space, rotate about their longitudinal axis to face and follow the sun, providing electrical power for the satellite.

The communications module is mounted under the service module and houses transponders and the antennas which point permanently towards the earth. The entire spacecraft has a mass at launch of 1,430 kg and a dry mass of 731 kg. The modular construction of the spacecraft permitted parallel assembly of the payload and the service module at different sites prior to full integration.

The communications payload comprises an SHF package of three transponders, each of 40 W power providing four channels at bandwidths of 60 to 135 MHz; and a UHF package of two 25 W transponders, each one serving a channel of 25 kHz bandwidth. SHF transmit and receive antennas provide a variety of footprints from spot beam to full cover of that part of the earth visible from the satellite. The UHF antenna is an earth-cover helix which is deployed before the spacecraft reaches its operational orbit.

An important feature of NATO IV satellites is their ability to survive electronic interference - both spacecraft employ signal processing and anti-jamming features.

Status
NATO IVA was launched on 7 January 1991 on the debut launch of a McDonnell Douglas commercial Delta II 7925 ELV; it is fully operational. NATO IVB was launched, again using a Delta II from Cape Canaveral, on 7 December 1993 and was declared fully operational in mid-1994.

Contractor
Astrium Ltd, Stevenage, Hertfordshire.

VERIFIED

Spain

HISPASAT

In September 1992 and July 1993 the first two HISPASAT spacecraft were launched into geostationary orbit. Built by Matra Marconi Space (now part of Astrium) the hybrid satellites (comprising both military and civil communications transponders) provide fixed and flexible coverage for the Spanish military. The X-band payload is based on equipment developed for the UK's Skynet 4 series of spacecraft.

In addition to the space segment, Astrium was responsible for the provision of ground terminals to the Spanish military including SCOT naval terminals, MANPACKs and tactical terminals.

Status
HISPASAT has been used to provide links between Spanish UN military operations and the Spanish mainland. HISPASAT 1c was launched in 1999, and HISPASAT 1d in September 2002. Both the latter satellites were built by Alcatel Space Industries.

Contractor
Astrium Ltd, Stevenage, Hertfordshire, UK.
Alcatel Space Industries, Paris.

UPDATED

United Kingdom

4.5 m Transportable Military Communications Satellite Terminal

The 4.5 m trailer-mounted parabolic reflector antenna has been designed to provide high-datarate communications. Designed for X-band operations it can also be modified for Ku-band operation and has transmit power of up to 1 kW. The electronics are housed in an EMP protected cabin mounted on the same trailer as the antenna. The system is specifically designed to provide secure long-haul voice and radar datalinks over the Skynet 4 satellite constellation. However it can also operate with any SHF satellite and in its Ku-band version meets the current satellite regulations.

The antenna is nominally 4.5 m in size but a 6 m version is also available. The antenna can be aligned within a claimed 15 minutes of arrival on site and the system readied for use within a claimed 2 hours by a two-man team. The system can be deployed by road or by air (CH-47 or C-130). It has been designed for multirole applications and has been used as a Transportable TT&C facility for satellite control.

4.5 m Transportable military communications satellite terminal (Astrium)
NEW/0528498

Status
This product has been in service with Qinetiq, Defford since 1996.

Contractor
Astrium, Stevenage, Hertfordshire.

UPDATED

Clansman integrated communication network

Clansman is an integrated communications network based on a family of radios. Each unit can also form the basis of an autonomous communication network.

The Clansman unit provides communication facilities over long and short distances for the infantryman, armoured vehicles, gun and missile batteries, beach landings, parachute drops and some ground-to-air links. Access to wider coverage area communication systems used for high-level command and control is made simple with Clansman radios.

Designed to replace some 24 separate equipments previously operated by the British Army, the Clansman family consists of seven basic sets: two HF, four VHF and one UHF. In both the HF and VHF groups there is an additional set which is a basic radio with a clip-on RF amplifier to provide higher power output. With these, the total number of sets in the range is nine; six manpack and three vehicle radios.

The HF sets (PRC 320, VRC 321, VRC 322) operate in the frequency range 1.5 to 30 MHz, the VHF sets (PRC 349, PRC 350, PRC 351, PRC 352, VRC 353) in the 30 to 75.975 MHz band and the UHF (PRC 344) in the 225 to 399.5 MHz range. VHF sets use 25 kHz channel spacing to double the number of tuning points in the band. However, the UK/VRC353 vehicle radio also operates at 50 kHz channel spacing to ensure interoperability with existing British and NATO equipment. With the exception of the UK/PRC349 and UK/PRC350 manpack radios, automatic rebroadcast is available on all Clansman VHF sets. This is achieved by the receiving set recognising a sub-audio tone from the transmitting Clansman set. Non-Clansman sets can communicate with Clansman in the normal way but will not operate the automatic rebroadcast system, giving Clansman VHF rebroadcasting nets a measure of protection from unwanted interference.

The HF sets normally operate in the single sideband voice and continuous wave modes but amplitude modulated double sideband is also provided to ensure interoperability with existing equipments. All three sets can be set up accurately by the operator on any frequency in the HF band at 100 Hz intervals. An adaptor is provided to allow the radios to use radioteletype where needed. A narrowband Morse facility, which has been found very useful for long-range patrol work, is also built into the HF sets.

The UHF set is a lightweight manpack for ground-to-air communicators. It is interoperable with existing airborne equipment in this band and uses 50 kHz channel spacing.

In the design of the Clansman range, particular attention has been paid to ElectroMagnetic Compatibility (EMC) with the suppression of broadband noise, harmonics and other spurious effects from both the transmitters and receivers. Thus all the vehicular sets can be used together in the same vehicle with the minimum amount of mutual interference.

Radios can be used either by themselves in vehicles or with the Clansman harness which gives additional facilities to both single and multiset installations. Particular attention has been given to the ability to cope with variations in vehicle supply voltage and transients. The control harness consists of a system of control and junction boxes which permits control of two or three radio sets from various positions in the vehicle. It also provides intercommunication between any two sets and remote-control facilities up to 3 km, using lightweight army signalling cable (DT10) and up to 5 km using two leads of CT10 cable.

The Clansman manpack radios can also be used as vehicle stations in a clip-in role. An installation kit is employed, together with the radio, to make use of the vehicle antenna and batteries and to provide suitable fixings.

A range of Clansman test gear, including a tape-sequential automatic tester, has been produced, so ensuring minimum maintenance load in the field. A field-repair test kit is available from the appropriate radio set manufacturer for diagnostic testing and alignment of sub-units and plug-in modules of the radio and to assist in

Clansman radios (left to right) UK/PRC352, UK/PRC349, UK/PRC350 and UK/PRC351

Triple set UK/VRC353 installation

fault-finding down to sub-unit level. A range of antennas and antenna tuning units is also available.

The VRC 353 can be made secure with the addition of the BID 250, and the PRC 352 with the BID 300, although the latter radio set requires a minor modification.

Status

Initial design studies began in the mid-1960s with user trials taking place in 1970-71. Fielding of the equipment began in the latter half of the 1970s and is now complete.

Total value of Clansman orders has not been disclosed but total sales by the four manufacturing companies are thought to have exceeded £650 million, about half of which were for export.

The Clansman range is no longer in production although it is still in operational use with UK forces, amongst others. Its increasing unreliability was highlighted during operations in Kosovo. It is due to be replaced in the UK by Bowman (see separate entry).

Contractors

BAE Systems (HF and UHF Manpack Radios), Christchurch, Dorset.
Racal Radio Ltd (now Thales) (VHF Manpack Radios), Bracknell.
MEL (now Thales) (HF Vehicle Radios), Crawley.
GEC-Marconi Defence Systems Ltd (now BAE Systems) (VHF Vehicle Radios), Secure Radio Division, Portsmouth.

VERIFIED

....

Containerised RElocatable Satellite Terminal (CREST)

The Containerised RElocatable Satellite Terminal (CREST) provides satellite communication facilities for users who need ground terminals which can be easily relocated from one operating site to another if requirements or priorities change. The terminal has a claimed set up and tear down time of two or three days and is transportable by C130.

CREST consists of an autotracking antenna, with antenna-mounted RF electronics and a standard ISO shipping container. Systems can be configured to match individual customer requirements and are available in C-, X- and Ku-band configurations. Typical communications capacity is 2,048 kbit/s. It provides high data rate communications and, in C- and Ku-band configurations, meets all the performance criteria for the major satellite operators (for example, EUTELSAT and INTELSAT). X-band configurations are fully compatible with military SHF systems, such as Skynet, NATO and DSCS.

The container, which may be located up to 100 m from the antenna, is partitioned into two areas. The equipment area is air conditioned and provides an office-type environment. Standard 19 in racks house the indoor electronics such as modems, baseband multiplexers, antenna control and test equipment. A storage

CREST

area provides facilities for securing the antenna and outdoor equipment during transport and for the storage of spares and tools.

The antenna can be installed on a pad foundation or on concrete piers or blocks. Assembly of the 4.5 m antenna can be carried out by two people, normally within two days, without the need for a crane or any special tools or jigs. The container needs only to be sited on a level surface.

CREST is equipped with the Siemens Integrated Monitor And Control System (SIMACS), a comprehensive monitoring and control facility based on proven COTS products. SIMACS monitors all major system parameters and alarms, and controls all the main system functions. A logging facility is also included, providing full traceability of alarms and events. SIMACS provides for remote supervision of the CREST equipment, either from elsewhere on the operational site or from the other end of the satellite link. The graphical user interface is 'intuitive'. SIMACS can be integrated with other Window-based packages, providing a comprehensive system covering a wide range of customer requirements.

Specifications

The following gives typical performance characteristics for systems using a 4.5 m antenna at C-, X- and Ku-bands. These represent a subset of the available CREST configurations and the system can be configured to meet most customer requirements.

C-Band
Transmit band: 5.925-6.425 GHz
Receive band: 3.7-4.2 GHz
EIRP: 64 dBW (50 W SSPA)
Figure of merit (G/T): 25 dB/K (45K LNC)

X-Band
Transmit band: 7.9-8.4 GHz
Receive band: 7.25-7.75 GHz
EIRP: 64 dBW (40 W SSPA)
Figure of merit (G/T): 25 dB/K (11 OK LNC)

Ku-Band
Transmit band: 14.0-14.5 GHz
Receive band: 12.5-12.75 GHz (other bands also available)
EIRP: 60 dBW (8 W SSPA)
Figure of merit (G/T): 30 dB/K (11 OK LNC)

Physical characteristics
Container dimensions (L × W × H): 240 × 96 × 102 in (610 × 244 × 259 cm)
Weight: 6,000 kg (shipping configuration)
Antenna diameter: 4.5 m (other sizes also available)
Antenna pointing: 5-85° continuous (elevation); 120° sector (azimuth)

Environmental
Temperature
Operating: −25 to +40°C
Storage: −35 to +50°C
Windspeed: 120 km/h (operating); 175 km/h (survival)
Ice: 25 mm with 150 km/h wind (survival)
Shock and vibration: as encountered during shipment by commercial air, rail or truck

Status

Several terminals have been purchased by a variety of unspecified European customers.

Contractor

BAE Systems, Christchurch, Dorset.

VERIFIED

....

Dagger satellite terminal

Dagger is a modular satellite communications system designed for rapid deployment and ease of use in difficult operational domains with minimal manpower requirements. It is available in a number of variants to meet differing needs. It can provide long-range communications from forward positions back to a main headquarters, or can provide a solution wherever a bearer is needed between an installation and a base station, providing an instant temporary fix for damaged communications links. The data capability allows Ethernet interconnection, giving the possibility of use as a mobile outstation for C4I terminals, such as mobile, short-range radar.

The military tactical headquarters terminal Dagger is built on a standard commercial Land Rover 110 hardtop vehicle, although Dagger can be hosted on a wide range of military and commercial vehicles which are able to accommodate the necessary internal and roof-mounted equipment. The vehicle contains the satellite communications, power generation and environmental conditioning equipment. Specific equipment configurations can allow up to 96 telephone or data circuits to be connected through the satellite network. The telephones can typically be distributed around a tented headquarters or throughout a building.

The Ground Station Module (GSM) network hub Dagger installation can be used to expand the area covered by communications. It houses a GSM hub that allows

Dagger satellite communications system, roof-mounted on a Land Rover 110
(BAE Systems)   0528084

the use of a number of GSM mobile phones. This configuration provides a 20 km island of coverage for mobile communications with a rear link via satellite communications. Alternatively, a TETRA base station or hub can be used in conjunction with the satellite terminal capability within the Dagger vehicle to achieve similar functionality.

The 1.5 m motorised antenna can be aligned to any satellite chosen from the built-in database via a PC interface. The GPS, fluxgate compass and inclinometer give full position and orientation information, allowing speedy satellite acquisition, typically three to four minutes, yielding claimed 'on the road' to 'in traffic' times of less than ten minutes.

The terminal G/T is typically 20.7 dB/K with an EIRP of 51.5 dBW. The current in-service configuration has a maximum bandwidth of 512 Kbits, but this can be enhanced to 2 Mbits. Although the system currently uses Ku-band, military frequencies at X -band are accessible with simple off -the-shelf modifications. 27 in racking holds user equipment, including a multiplexer, interface equipments and cryptographic equipment. A signal interface vault at the rear of the vehicle, fitted with copper or fibre bulkhead connectors, provides a distribution interface for voice and data subscribers. The in-service configuration has 16 voice users and 64 Kbits data on a 128 Kbits satellite channel. This is scaleable up to the 512 Kbits bandwidth and beyond.

The vehicle has its own 4.5 K V A diesel generator. While both the vehicle and generator are diesel powered, the generator uses a separate fuel system from that of the vehicle. All service items for the generator are accessible via a panel from within the vehicle, negating the need for generator removal in the normal course of events and allowing routine servicing within 15 minutes. Dagger can be powered from an alternative external generator if required. Automatic UPS facilities within the vehicle provide power back-up for a minimum of 30 minutes.

To meet an environmental specification of −25 to +55°C, an air conditioner/ heater is mounted above the driver's compartment.

Status
In February 2002 BAE Systems were awarded a contract by the UK DCSA to lease two Dagger systems for at least 12 months, and in January 2003 a contract for a further two systems was announced. The system is in service with the British Army in the Balkans. A total of five other systems are in service with other unspecified NATO countries.

Contractor
BAE Systems, Christchurch, Dorset.

NEW ENTRY

··

Driveaway mobile satellite communications terminal

The Driveaway terminal has been designed to provide tactical and mobile satellite communications that can be quickly deployed using a minimum of manpower. The system has been designed to interface with a wide range of military vehicles and can fulfil a number of roles from highly mobile command post to theatre broadcast studio and incident response vehicle. The system is platform independent and

Driveaway mobile satellite communications terminal (Astrium)   0528497

consists of two principal elements, a roof-mounted pod antenna and an RF element in, typically, two mission cases.

Short duration operations can be conducted by a one-person crew in one vehicle. The antenna housed in the pod deploys and acquires designated satellites automatically. Local or remote equipment monitoring is available.

Specifications
Antenna type: Gregorian-type, dual offset aperture: 1.2 m × 1.5 m shaped construction: one-piece elliptical main reflector
Tx frequency: 7,900 - 8,400 MHz
Radiated power: 53 dBW min
Rx frequency: 7,250 - 7,750 MHz
G/T at 20° elevation and 25° (clear sky): 16.5 dB/K min at 7.5 GHz

Status
The Driveaway is in quantity production for an unspecified NATO country.

Contractor
Astrium, Stevenage.

NEW ENTRY

··

Manpack

The Manpack is a complete lightweight SHF tactical satcom terminal. It is suitable for providing secure global communications for military or government personnel operating in remote or hostile environments.

The terminal can be carried in a rucksack and is designed to survive harsh transportation and operating environments. A single operator can deploy Manpack and establish communications within 5 minutes. The terminal is capable of supporting full duplex voice, data and facsimile communications. Satellite access can be through pre-assigned or Demand Assignment Multiple access (DAMA) carriers. Autonomous operation is from a range of batteries, or alternatively, an external vehicle or mains supply. The modular design allows a variety of configurations and performance options.

Specifications
(a) Frequencies: transmit 7.9-8.4 GHz, receive 7.25-7.75 GHz.
(b) EIRP: 37 minimum dBW
(c) G/T: 6 dB/K
(d) Antenna: 0.6 m cassegrain

Manpack (Astrium)   0101498

Manpack in operational use (Astrium)   0101499

(e) Traffic rates: Data/voice up to 19.2 kbps
(f) Stand-alone operation: > 60 minutes (subject to battery type)
(g) Weight: 10 kg

Status
Manpack is a third-generation man-portable terminal, developed from the man-portable terminals currently in service with French, Italian, Spanish and UK armed forces.

Contractor
Astrium UK, Stevenage, Hertfordshire.

VERIFIED

Military Off-the-Shelf Satellite Terminal (MOST)

The Military Off-the-Shelf Satellite Terminal (MOST) has been designed in response to a UK MOD requirement for a semi-tactical terminal based on commercial components. The requirement demanded a quick delivery at minimum cost for a satellite communications terminal which would be deployed for days rather than hours. The Astrium MOST solution is based on a commercial antenna modified for operation at X-band. The system employs a 90 W SSPA and Low Noise Amplifier mounted on the antenna for maximum efficiency. Up and down converters, antenna control, modems, power distribution and multiplexer modules are all housed in four protective flight cases. The remaining units are housed in man portable containers. Data rates of up to 2 Mbps are achievable and the whole station can be deployed ready for operation in under one hour.

The MOST is suitable for semi-tactical and static hub applications.

Status
MOST terminals are currently used by the British Army and the Royal Air Force for operational roles that do not require full tactical deployment.

Contractor
Astrium, Stevenage, Hertfordshire.

NEW ENTRY

The Military Off-the-Shelf Satellite Terminal (MOST) (Astrium)   0101587

SCOT shipboard satcom terminal

SCOT is the world's most widely used family of SHF satellite communications terminals for naval applications. The first-generation terminals entered service in 1974 to meet the Royal Navy's requirement for reliable, high capacity beyond line of sight communications for its widely deployed forces. SCOT has since been fitted to many types of ship from corvette through to aircraft carriers.

Although originally designed for X-band service exclusively, other versions have now been deployed operating at C-band and other frequency bands can be provided. Multibanded systems, such as SHF/EHF (Ka band) are being offered to satisfy a number of requirements. Antenna sizes range from 1.2 m to 2.7 m and are supplied complete with an all-weather radome for full environmental protection with minimal signal loss. The electronics can be containerised or rack-mounted for installation in a ship's communication suite. The systems are configured to the particular needs of the customer and operation can be via the ship's integrated command and control system or stand-alone.

Terminals are tailored to the individual requirements of each ship and the customer navy can choose from the following options:
(a) Antenna fit: one or two antennas
(b) Antenna size: 1.2 m, 1.5 m, 1.75 m, 2.2 m or 2.7 m
(c) Frequency: C and X-band with S, Ku and Ka options
(d) Output power: 150W SSPA to 2kW TWT
(e) Traffic throughput: up to 8 Mbps on four 2Mbps channels
(f) Baseband: CDMA, FDMA or frequency hopping to meet individual customer requirements
(g) Control: Independent HCI, as part of ship's ICS or via Over-the-Air Control
(h) Installation: containerised or integrated
Terminals for smaller ships are in the final stages of development.

Astrium pays particular significance to the degree of EMC achievable with each of the ship's weapon, sensor and communication systems. SCOT is designed and engineered to provide uninterrupted operation in all sea states with or without the ship stabilisation being activated.

Status
In addition to service with the Royal Navy for which SCOT has recently been selected for the new Type 45 destroyer, the system has been selected for and supplied to the Brazilian, Canadian, German, Italian, Netherlands, Portuguese, Spanish and Turkish navies. Astrium, in conjunction with the Harris Corporation, also manufactures C and X-band subsystems for supply to the United States Navy. In total over 120 ships have been fitted with SCOT. The systems have been used in combat in the Adriatic, the Gulf and the South Atlantic.

The system is now in its fourth generation but many of the original units supplied in the late 70s and early 80s are still in operation. Upgrade packages are available and lease packages are available to new or temporary users.

Contractor
Astrium UK, Stevenage, Hertfordshire.

UPDATED

SCOT terminal
0101503

Skynet satellite communications system

The Skynet system was conceived in 1966 when the USA agreed to launch a synchronous communications satellite for the UK (Skynet 1) which, once in orbit, would be operated by and under the command and control of the UK.

Skynet 1 - 4
Skynet 1, built by Philco-Ford, was launched in November 1969. A second standby satellite was launched in August 1970 but failed to go into orbit. These satellites

SATCOM ground terminal equipment

*RAF Oakhanger
terminal UKFSC 660*

were scheduled to have a three to five year life and to be replaced by two other satellites, Skynet 2A and 2B. Marconi Space and Defence Systems (MSDS), now known as Astrium, built them for launch at the end of 1973. Skynet 2B was successfully placed into orbit at the end of 1974.

A number of contingency plans towards Skynet 3 were considered but, following a UK Ministry of Defence (MoD) review, Skynet 3 was cancelled and the UK requirement was satisfied by arrangements with NATO. However, the UK forces, particularly the Royal Navy, had major plans for increased use of satellite communications. It was thought that the existing NATO and UK DSCS satellites would be unable to provide the UK with the space segment capacity it needed. As a result, in 1981 the UK MoD issued a tender for two new satellites to be known as Skynet 4A and 4B. Initially in different teaming arrangements, the MoD asked British Aerospace Space Systems (BAeSS) and MSDS (now both consolidated within Astrium) to be the joint supplier for the UK satellite communications system. The contract, worth £70 million, was awarded in July 1981 for a project definition phase with full development commencing in December 1981. BAeSS acted as prime contractor and MSDS as payload contractor.

Following a review of requirements in May 1985 the MoD ordered a third satellite, known as Skynet 4C. The Skynet 4 series of satellites was initially designed for launch with the US Shuttle in mid-1986. However, following the Shuttle disaster in January 1986, the satellites were quickly reconfigured for launch on expendable vehicles.

With the Skynet 4, Stage 1 satellites 4A, 4B and 4C in operation, in 1994 the MoD tasked BaeSS and Matra Marconi Space (now both Astrium) to develop the Stage 2 system. The subsequent 'delivery-in-orbit' contract included the launch of the spacecraft and also provided for operational control and in-orbit testing before handover of each satellite to the customer. The Stage 2 satellites are enhanced versions of the Stage 1 and the two similar NATO IV satellites, but with a communications payload which incorporates steerable antennas for SHF spot beam communications and provides increased power and a superior anti-jamming capability. Also, at UHF, a fully tuneable system and new helical antenna offer increased flexibility.

A variety of spot and global beams enable the Skynet 4, Stage 2 satellites to serve an extensive inventory of Earth stations on land, sea and in the air. These range from small manpack sets and aircraft terminals to those on widely dispersed naval vessels - including submarines - and large anchor stations on land. The system utilises signal processing and anti-jamming features to continue to provide

*Skynet 4D in EMC
Chamber with UHF
antenna deployed
0055045*

communications in the face of harsh electronic warfare environments. Improvements to the Stage 2 communications payload over Stage 1 include:

(a) tuneable UHF
(b) power of each of the two UHF transponders increased to 50 W
(c) each UHF transponder serving one channel of 25 kHz
(d) new Earth cover helical UHF antenna
(e) steerable spot SHF transmit and receive antenna
(f) steerable European beam SHF antenna
(g) earth cover SHF antennas
(h) power of each of the three SHF transponders increased to 50 W
(i) reconfigured SHF channel bandwidths to between 60 and 125 MHz.

The Spacecraft Operations Facility is a computer-controlled centre with built-in redundancy, providing RAF operators with full-time and cryptographically protected access to all spacecraft telemetry data and use of all telecommand functions, in both S- and X-band frequencies, via local and remote radio heads. The computers are programmed for ephemeris manoeuvre calculations and fault analysis.

As part of the original Skynet programme a number of ground terminals were produced which included: the Skynet anchor station at RAF Oakhanger; three Marconi-built 12 m terminals (Type 1); MMS shipborne terminals (SCOT) for almost every ship in the Royal Navy; and MMS transportable stations operated in Bahrain, Cyprus, Gan, Hong Kong and Singapore. Siemens Plessey Systems (now BAE Systems) provided a £5 million facility for the Berlin telecommunications office.

In 1982, Siemens Plessey Systems (now BAE Systems) were contracted to build and commission a new ground station complex to operate with the Skynet 4 satellites. Further ground stations for the Skynet system include two Astrium TSC 648 transportable systems, procured under the auspices of NATO for the RAF in 1983. The terminal, with a 15 year life, has undergone over 100 moves and is still in operation today. In 1986, Matra Marconi Space (now Astrium) provided a radome-protected air defence terminal for Saxa Vord in the Shetland Islands linked into the Skynet system. This terminal was subsequently dismantled by the Astrium field support team within two weeks, transported south and reinstalled at Colerne in full working order.

In 1988, MMS (now Astrium) provided a fixed terminal in Birgelen, Germany for routeing STARRNET traffic to RAF Oakhanger and the rest of the UK Military Satellite Communications System (UKMSCS). The project involved construction of a new ground station, and network control centre improvements to RAF Oakhanger and another ground station. The project also included interfaces to the TSC-666 Transportable Satellite Ground Terminals (TSGTs).

In 1992, MMS (now Astrium) replaced one of its original Type 1 antennas with the new UKFSC 660 terminal. The 7.3 m terminal project provides: an antenna and tracking subsystem with a unique motorised control mechanism; RF and IF and baseband subsystems; a control and monitor alarm system; and complete civil works management. The system was completed and commissioned early in 1994.

With the closure of the Birgelen site, Astrium was contracted to supply a new Satellite Ground Terminal (SGT) at Rheindahlen to take on traffic from STARRNET. The SGT is remotely controlled from RAF Oakhanger.

Status

Skynet 4B was launched on Ariane 4 in December 1988 and was retired from service in May 1998. Skynet 4A was launched on a Titan in January 1990 and Skynet 4C on Ariane 4 in August 1990. Skynet 4D was launched on a Boeing Delta 2 on 10 January 1998 and Skynet 4E on Ariane 4 in February 1999. Skynet 4F was launched on Ariane 4 on 7 February 2001. The remaining pair from Stage 1, 4A and 4C, still have some useful life and will continue to serve alongside the Stage 2 satellites in the short term.

Skynet 5

In parallel with the Skynet 4 stage 2 programme, the UK MoD issued Matra Marconi Space and BAE Systems contracts to study the next-generation Skynet 5. This was followed by a two-year study, which began in 1999, into the development of Skynet 5 as a Public Private Partnership under the umbrella of the UK's Private Finance Initiative (PFI)). In July 2000 it was announced that the Skynet 5 programme had

*Skynet 4F launch
February 2001*
0101504

been given approval to proceed as a PFI. By that time the UK Defence Procurement Agency (DPA) had issued invitations-to-negotiate to Astrium, whose Paradigm Secure Communications Team comprised Logica, Nortel, SERCO, Motorola, TRW, BAE Systems and SEA, and to a consortium of Lockheed Martin, British Telecom and BAE Systems. The preliminary UK requirement called for two satellites plus one ground spare. Bids for Skynet 5 were submitted by both teams in January 2001. In February 2002 it was announced that Paradigm was the successful bidder. Paradigm became a wholly owned subsidiary of EADS in early 2003. In October 2003 Paradigm was awarded the contract to provide secure satellite communications services for the UK armed forces, worth an estimated £2.5 billion to 2018. The Paradigm Secure Communications team includes EADS Astrium, Paradigm Services Limited, Cable & Wireless, Cogent, General Dynamics, LogicaCMG, Serco and Stratos.

Paradigm took over the management of the Skynet 4 satellites in mid-2003. Two new Astrium-designed and -produced satellites are expected to be launched: Skynet 5A in late 2006 and Skynet 5B in late 2007. Each of these will provide UHF and X-band (SHF) services, the requirement for EHF having been dropped from the programme and feature enhanced survivability and anti-jam capabilities against the Skynet 4 satellites. The new satellites are expected to provide communications coverage from the east coast of the US to the west coast of Australia. The contract also covers the following terminal upgrades:

Ground Segment
(a) RAF Colerne
    2 additional 9 m X-band heads
    IF/RF upgrades
    New satellite control facility (SCC)
    New Network Management Centre (NCC)
(b) RAF Oakhangar
    IF/RF upgrades
    DCSA Corsham
    New System Operation Centre
(c) System
    Introduction of PMS (Paradigm Modem System) modem
    Addition of Skynet 5 baseband system
    Upgrade of Management System to meet Skynet 5/Paradigm needs

Remote Ground Segment
(a) Land Transportable Terminals - Project REACHER
    36 REACHER Medium
    6 REACHER Large
    2 REACHER RM Variant
    15 Interim REACHER (Talon) (see separate entry)
(b) Shipborne Terminals - SCOT Upgrade (SCUG) (see separate entry)
    Replacement of existing SCOT Terminals with SCOT 3 for 28 ships (Includes Training and Reference System)
    Ship Baseband Improvements - S5MBB
    Skynet 5 Maritime Baseband to be fitted to 28 ships plus a further 8 ships
    Installation of these systems is due to be completed by 2008. Full operating capability of the Skynet 5 system will be achieved by March 2008.

In July 2002 BAE Systems was contracted by Paradigm to supply 12 Talon rapid deployment satcom terminals by the end of 2002 with a further three in 2003. These will initially be used with Skynet 4 but will subsequently be used with Skynet 5. In Oct 2002 it was announced that the first batch of 6 Talon terminals had been delivered. At the end of 2002, 12 terminals had been supplied to Astrium with a further three to be supplied in August 2003.

The UK MoD will have an assured capacity of no less than 50 per cent of PSC's Skynet 5 services over the period of the PFI deal. Remaining capacity will be available for use by the UK or third-party users, such as NATO, which is expected to

make a decision on a new satellite communications system before year end. The French, Italian and UK MoDs have submitted a joint bid to meet alliance requirements, using their respective Sicral, Syracuse and Skynet systems.

Contractors
Paradigm Secure Communications, Stevenage.

UPDATED

Tactical satellite earth terminal

The tactical satellite earth terminal is a mobile, self-contained satellite communications station designed to be carried in a variety of military vehicles. It can be deployed from a Landrover or similar vehicle.

Designed to operate in the SHF band, this equipment consists of a four-petal antenna with the associated RF equipment housed in four easily stowed transit cases. The equipment can be set up within a claimed 15 minutes of arrival at site and typically provides two channels of secure communication, which can be either voice or data. Typical data rates of 128 kbits are available over a range of satellites.

Specifications
Antenna type: 1.7 m diameter nominal circularly polarised
Tx Frequency: 7.90 - 8.40 GHz (X-band)
Rx Frequency: 7.25 - 7.75 GHz
G/T: > 14 dB/K
EIRP: > 86 dBm Status

Status
Variants of this terminal have been deployed in Bosnia, with NATO and it is in service with the, Brazilian, Spanish and UK armed forces.

Contractor
Astrium UK, Stevenage, Hertfordshire.

UPDATED

Tactical satellite earth terminal (Astrium)   **NEW**/0528499

Talon satcom terminal

Talon is a lightweight deployable satellite terminal designed for both tactical and strategic deployments. It is essentially a COTS product packaged to provide a terminal suitable for the military and quasi-military role and in its original COTS form over 400 Talon terminals are in use with the newsgathering industry.

The terminal can operate in C, X, Ku or Ka-Bands, with a simple change of the feed arm and some key electronic components to effect the transition between bands. It is certified for use with many of the commercial satellite operators, who typically use Ku-Band, or can be used with dedicated X-Band military links. To help operation with military satellites, Talon has a tracking system, incorporating separate motors and gearboxes to drive in the azimuth and elevation axes. Replacement of the tracking motors with crank handles will allow manual satellite tracking if required.

The Talon antenna can be supplied as a 1.2 m, 1.9 m or 2.4 m coaxially fed dish, allowing for optimum tailoring of the RF performance to the specified communications need. When used at X-Band with the 1.9 m antenna, the terminal can deliver a G/T of better than 21 dB/km and, using the current generation of single thread SSPA, output powers of better than 61dBW are possible. For higher output powers, modem ruggedised TWTA technology is employed. The antenna architecture uses a simple box support structure of carbon fibre to provide a wide and low stable plinth. The manufacturer claims that, with no additional measures,

The Talon man-portable satellite terminal (BAE Systems)   ***NEW**/0531672*

the 1.9 m terminal can be set up in wind speeds of up to 33 mph, provide guaranteed performance at wind speeds up to 36 mph, and survive in wind speeds of up to 60 mph.

The standard terminal includes all the functionality required to support a communications link, including antenna control and tracking, frequency conversion, and the requisite satellite modem. The standard terminal has four modular cases engineered for outdoor use, which can include both heaters and cooling elements should the deployment environment demand it. A fifth case, housing the main operator interfaces, is normally deployed into a building or shelter.

Status

Three Talon terminals were leased by the UK MOD in 2002 for use by HQ ARRC. At the end of 2002, 12 terminals had been supplied to Astrium for use with the Skynet programme (see separate entry), with a further three to be supplied in August 2003.

In January 2003 a contract for a further two terminals for British forces was announced, with delivery by March 2003. A total of 19 Talon systems will then be in use in UK and NATO forces.

Contractor

BAE Systems, Christchurch.

UPDATED

Towed antenna system

The Towed Antenna System is a satellite communications terminal with a 2.2 m antenna mounted on a trailer. The terminal is supplied with intelligent automatic satellite tracking to aid rapid deployment. The average time for deployment from arrival at site to full operation is within 40 minutes. Separate Baseband equipment is housed in an additional electronics shelter and this arrangement offers interoperability with an operator's own designated satellite. Lightweight materials are used for the antenna which, together with the feed and filter assemblies, provides low, passive intermodulation performance.

Trailer-mounted towed antenna   0055054

The antenna is driven by linear actuators providing ±45° continuous azimuth coverage and 0 to 90° coverage in elevation. This enables flexible antenna tracking mobility following acquisition of the satellite beacon. The rugged, two-wheeled, single axle trailer provides three-point jacking for stability, with overrun braking to aid the towing vehicle. The terminal is transportable by aircraft such as the Hercules C-130.

Specifications

Antenna: 2.2 m Cassegrain antenna
Weight: 1.5 tonnes
Tx Frequency: 7.90 GHz - 8.40 GHz
Rx Frequency: 7.25 GHz - 7.75 GHz
G/T: >20.4 dB/K (for an 85K LNA)
EIRP:>63.5 dBW (for 120 W SSPA)
Data rate: 2.4 kbps to 2 Mbps

Status

In service with several unspecified overseas armed forces, believed to include Spain and Italy.

Contractor

Astrium UK, Stevenage, Hertfordshire.

UPDATED

UK/FSC 658 Satellite Ground Terminal (SGT)

The UK Fixed Satellite Communications (FSC 658) terminal is located at the Joint Headquarters in Rheindahalen, Germany. The system provides voice and data communications links between the British Forces in Germany and the United Kingdom using the Skynet-based UK Military Satellite Communications System (UKMSCS). The contract included the design, supply, installation and commissioning of a fully NBC-protected Satellite Ground Terminal. The system comprises a 9 m antenna with transmit and receive subsystems, Baseband subsystem and terrestrial links to the Ptarmigan and Rodin networks.

The station is unmanned with control provided by a remote management system capable of controlling the terminal and up to five other remote terminals from either of the two UK-based satellite network control centres.

Status

The system was installed in 1999 and handed over to the UK Ministry of Defence in 2000.

Contractor

Astrium UK, Stevenage, Hertfordshire.

UPDATED

The UK FSC 658 Satellite Ground Terminal (Astrium)   ***NEW**/0528501*

UK/FSC 660 satellite ground terminal

The UK Fixed Satellite Communications (FSC660) terminal was originally handed over to the UK Ministry of Defence in 1992 and is located at RAF Oakhanger. The system has recently been upgraded with the GEMS computer-based Control and Monitoring facility. The system was originally supplied to replace one of the original Type One terminals in order to enhance the capabilities of the RAF Oakhanger facilities.

The UK/FSC 660 satellite ground terminal (Astrium)   0528500

The contract included the design, supply, installation and commissioning of a low intermodulation Satellite Ground Terminal (SGT) with a specially designed environmentally protected building adjacent to the antenna. All external equipment and cross-site interfaces are similarly protected and filtered. The system comprises a 7.3 m antenna and RF subsystem for operation at X-band.

The antenna is designed to be capable of withstanding winds in excess of 320 km/h. The antenna provides 180° of travel in azimuth with a single jack drive mechanism and is capable of maintaining pointing and tracking accuracy requirements over the full environmental specification. The transmit power is provided by a 2 kW dual redundant High Power Amplifier (HPA) which is situated some distance from the antenna. The required EIRP is achieved through a complex underground cross-site waveguide run.

Status

The system has recently been enhanced under a contract from the UK MOD to provide computer-based control and monitoring to ensure compatibility with FSC 658 and FSC 692 (see separate entries). The enhancements were handed over in 2001.

Contractor

Astrium UK, Stevenage, Hertfordshire.

UPDATED

UK/FSC 692 satellite ground terminal

UK/FSC 692 is a fixed satellite facility supplied to the UK Ministry of Defence and installed at RAF Oakhanger in the UK.

This system has enhanced the capabilities of the Anchor Station Complex by expanding the existing X-band and S-band Telemetry, Telecommand and Control (TT&C) facilities. It has a fully NBC protected low-intermodulation Satellite Ground Terminal (SGT) with predictive step-track capability selectable for both receive bands. It comprises an 11 m antenna and RF subsystems for simultaneous operation at X-band and S-band and interfaces with existing Satellite Ground Systems (SGS) facilities at IF level. Both X- and S-band subsystems are interfaced with the Satellite Operations Facility (SOF) equipment.

An integrated control and monitoring system containing a hierarchical screen system giving a complete picture of the current health and status of the FSC 692 system is also provided. The operator can control the X- and S-band equipment and will get immediate notification of any equipment faults.

Status

Installation is complete.

Contractor

Astrium UK, Stevenage, Hertfordshire.

VERIFIED

UK Satellite Anchor Station System (UKSASS)

Acting as prime contractor and system design authority, the then Siemens Plessey (now BAE Systems) supplied the £40 million FSC 646 satellite anchor station at RAF Oakhanger. This station forms the hub and central control point of the UK Military Satellite Communications System (UKMSCS). Under a turnkey contract from the MoD, the company was responsible for construction of the hardened building, as well as for the supply of all the electronic and supporting equipment. The station was handed over in 1986, but in 1990 Siemens Plessey Systems was

1595 groundstation

awarded a further £10 million contract to provide significant expansion of the FSC 646 facilities.

Subsequently, a £60 million contract to expand and diversify the UK ground segment was completed. Project 1595 involved the construction of a second large anchor station and a new Satellite Network Control Centre, improvements to RAF Oakhanger and another existing ground station, and provision of two large transportable terminals with network access points. These facilities are linked by terrestrial bearer systems to provide an integrated, survivable ground segment network which is transparent to the users.

Contractor

BAE Systems, Christchurch, Dorset.

VERIFIED

UK/TSC 503 satellite communication system

The UK Ministry of Defence selected BAE Systems to supply the UK/TSC 503 satellite communications system after an 18-month competition. The contract, worth in the region of £15 million, requires the delivery of 15 Satellite Field Terminals (SFTs), together with supporting facilities, for use by the UK armed forces to provide high-capacity, secure communications in support of operations worldwide. A double hop relay station function is included.

The modular design of the SFTs will allow military users to deploy the equipment in a variety of configurations to best suit mission needs and is transported in a number of rugged transit cases. The two main configurations are the Rapid Deployment Terminal (RDT) and the Full Capability Terminal (FCT). The RDT provides a very rapid, 'first in' capacity at rates up to 512 kbps and uses a 2 m antenna. The FCT is intended for deployment during the later phases of capability build-up, supporting multiple trunks through the 4 m antenna. The RDT can be dispersed around a deployment base, with network interface baseband containers located with users up to 4 km from the radio equipment, which is co-located with the antenna. The antenna may be located up to 50 m away from the radio electronics as required.

The 2 m UK/TSC 503 antenna
0093713

The 4 m UK/TSC 503 Antenna NEW/0093712

The antennas use snap-on, lightweight composite reflector panels mounted on tripods which pack into carry cases and provide automatic satellite tracking. They can be quickly assembled with no special tools by two-person RDT and four-person FCT crews in a claimed 30 minutes and 2 hours respectively. Assembly of the 4 metre reflector is helped by a telescoping pedestal. A lightweight fabric radome enables full performance of the 4 m antenna in extreme conditions, such as high and low temperatures and hurricane force winds.

Status

The first four Interim Operating Capability (IOC) terminals were delivered in 2001. Further terminals for other users are likely to be added to the initial order quantity and it is anticipated that the fleet could eventually number around 20. The UK/TSC 503 award builds on BAE Systems' development of the X-Band Flyaway Terminal (X-Flyte), which served as a demonstrator for the programme. Upgrade to 8Mbit/s capacity was planned for 2003.

Contractor

BAE Systems, Christchurch, Dorset.

NEW ENTRY

UK/VSC501 vehicle-borne satellite communications station

The SC2630 (UK/VSC501) is designed to meet military requirements for a land-mobile tactical SATCOM system.

The SC2630 operates in the military SHF SATCOM frequency band of 7.25 to 8.4 GHz via a geosynchronous satellite. Tuning is in 1 kHz steps.

Terminated on the vehicle (or externally) and with a number of modem options, the system provides full-duplex anti-jam digital speech or data circuits operating at

SC2630 transportable SATCOM

2.4, 9.6 or 16 kbits/s or high-speed data at 64 or 128 kbits/s. There are additionally, four telegraph circuits capable of operating at 50, 75 or 100 baud. The aggregate data rate is 144.4 kbits/s. The system will also support one EUROCOM trunk circuit operating at 256 or 512 kbit/s.

The SC2630 is primarily intended for installation in a standard FFR 3/4 tonne Land Rover and trailer, although other vehicle and trailer combinations can be adopted.

The normal manning level for an SC2630 station is a crew of two and the system can be set up to provide communications within 15 minutes. However, in emergencies, the SC2630 can be set up and operated by one person.

Being battery powered from the vehicle equipment batteries, all equipment can remain switched on and therefore warmed up, en route to a new site. Frequency selection and channel configuration can also be actioned prior to departure from a previous location, thereby reducing set up time.

An antenna auto-tracking facility used in conjunction with the 1.9 m dish is capable of tracking geosynchronous military satellites unaided and continuously. Erecting the dish involves coupling together the four segmented petals and attaching them to the highly stable quadrupod base. A flexible waveguide then connects the antenna to the vehicle.

Power handling is in excess of 120 W and maximum power requirement is approximately 1 kW.

Status

The SC2630 is in use with UK and NATO forces. The UK version (VSC 501) has been the workhorse of UK military SATCOM terminals for some years. It is carried in both Land Rover (Army) and BV 206 All Terrain Vehicles (Royal Marines). An update package completed in late 1999 will extend the life of the system until replaced by Project Reacher, part of the Skynet 5 programme.

Manufacturer

Thales Communications UK, Wells, Somerset.

UPDATED

X-Flyte X-band flyaway terminal

X-Flyte is a portable satellite communications terminal for use in conjunction with military X-band satellites such as Skynet, DSCS, NATO and Syracuse. The terminal equipment is housed in 'flyaway' packaging, enabling it to be rapidly deployed to remote or inaccessible locations using a variety of transport means. The entire terminal, including cables and accessories, is packaged within four transit cases. Suitable tie-down points are provided to allow transport by road, rail, sea and air.

The complete terminal takes less than 20 minutes to assemble and initialise. Once communications have been established, the only operator intervention needed is the occasional fine adjustment of the antenna to track the satellite. Even this can be eliminated if the optional motorisation and tracking facilities are fitted to the terminal, in which case the antenna automatically tracks the satellite once it has been acquired. For operation, the antenna mount assembly is attached to the top of the equipment case. This acts as a stable pedestal for the antenna and provides ground clearance. The transit case contains the HPA control panel, the LNA power supply, the frequency converters, the RF modem, satellite beacon receiver and optional Antenna Control Unit (ACU). Tie-down points are provided on each top corner of the equipment case to allow the deployment of ground anchors. The antenna mount is also equipped with a levelling indicator and adjustable feet are fitted to the case.

The antenna is a 1.8 m diameter Cassegrain configuration with the main reflector formed from eight panels arranged around a central hub. The subreflector and feed are combined in a single assembly which locates to the centre of the hub. This connection is the only waveguide joint that the operator needs to make when assembling the terminal. The hub attaches to the antenna mount assembly which comprises an elevation-over-azimuth positioner, a transmit High Power Amplifier (HPA), a receive Low Noise Amplifier (LNA) and the transmit and receive waveguide filters. The HPA and LNA are designed for external mounting and will operate in a wide range of environmental conditions. Adjustment is available over the range 5 to 85° in elevation and ±45° in azimuth. Fine azimuth adjustment is provided over a range ±5° about an initial coarse setting. A motorisation kit is available as an optional replacement for the manual adjustment and this provides pointing over a sectored range of azimuth and elevation via an ACU mounted in the equipment case.

The equipment case contains a number of 19 in rack-mounting units in a shock-mounted subframe. The modem supplied offers data rates of 64 kbit/s and 256 kbit/s. This modem is also equipped with a codec and a multiplexer which enable operator selection of two multiplexed voice channels to be aggregated with the data channel. An engineering services channel is also provided. X-Flyte is equipped with two stages of frequency conversion with the facility to support modems at 70 MHz and/or 700 MHz for communication at X-band. Components are provided to allow simultaneous operation with both of these intermediate

X-Flyte   0010466

frequencies. The equipment rack will accommodate up to 14U of 19 in rack-mounting equipment. In its basic configuration, without the optional ACU, 2U of equipment rack space is vacant.

In addition to the motorisation and tracking upgrades, other options include the provision of a rugged laptop computer and software to enable full remote control of the terminal and the supply of a portable generator to power the terminal. The specifications shown below give typical performance characteristics of the basic X-Flyte system. The system can, however, be reconfigured within this basic package to meet most customer requirements.

Specifications
(a) Azimuth coverage: ±45°
(b) Elevation coverage: 5-85°
(c) Deployment time: 20 min, 2 persons
(d) Transmit frequency band: 7.9-8.4 GHz
(e) Receive frequency bend: 7.25-7.75 GHz
(f) EIRP: 58 dBW
(g) G/T: 17 dB/K
(h) Beamwidth (3 dB): 1.30° at 7.25 GHz; 1.60° at 7.9 GHz
(i) Polarisation: right-hand circular (Tx); left-hand circular (Rx)
(j) Radiation pattern: 1st sidelobe >12 dB below main lobe
(k) Temperature: −10 to +40°C (operating); −30 to +70°C (storage)
(l) Windspeed: 40 mph (operating); 70 mph (survival)
(m) Shock and vibration: as encountered during shipment by commercial or military air, rail or truck
(n) Case dimensions (L × W × H):
   main equipment 720 × 580 × 750 mm;
   antenna petals 760 × 760 × 510 mm;
   feed and mount 930 × 540 × 560 mm;
   hub and accessories 760 × 760 × 510 mm
(o) Weight: main equipment 70 kg; antenna petals 56 kg; feed and mount 73 kg; hub and accessories 60 kg

Status
The terminal was developed as a demonstrator for the UK/TSC-503 (see separate entry), the contract for which the then Siemens Plessey Systems (now BAE Systems) was awarded in 1997. It is uncertain whether there have been any other customers.

Contractor
BAE Systems, Christchurch, Dorset.

UPDATED

United States

2.4 m Tactical Satellite Terminal (TST)

The 2.4 m Tactical SATCOM Terminal (TST) is one of a family of three ground terminals which utilise a common set of electronic modules. These modules are packaged in a high-performance enclosure providing flexibility of options and the considerable transportability. The SATCOM terminal can be mounted on a trailer, HMMWV, free standing on jacks or fixed site installation.

The terminal is offered in a 400 W tri-band core configuration consisting of quick change (< 5 min), separate X-band, Ku-band, and C-band feeds. This configuration consists of a tri-band, single carrier, non-redundant system (modem through antenna) with the following options:
(a) Redundancy (manual or automatic)
(b) Multiple carriers (C-band and Ku-band)
(c) Remote operation
(d) Reed Solomon CODEC
(e) Special test equipment
(f) Baseband electronics
(g) Ancillary equipment (HMMWV or trailer, mounting kit, generator, COMSEC, air conditioner)
The design approach simplifies training, maintenance, repair, and logistics. Two large equipment blowers are capable of cooling the electronics or an external air conditioner is available for systems operating in extreme environments.

Specifications

	C-band	X-band	Ku-band
Frequency range (MHz) Receive	3625-4200	7250-7750	10950-12750
Transmit	5850-6425	7900-8400	14000-14500
Receive figure of merit (G/T) @ 7.5°EL	17.5 dBk	22.5 dBk	25.5 dBk
Transmit EIRP (dBW)	64.0 dBW	67.0 dBW	72.0 dBW
Polarisation	Dual Linear Dual Circular	Circular	Dual Linear
Certification	G	DSCS	E1
Carrier frequency tuneability		1kHz steps	
Long-term stability		1×10^{-11} (with GPS option)	
Built-in-test	Yes		
Operator remoting	1,000 ft		
Set up/Tear down time	30 min		

Status
In production.

Contractor
L-3 Communications-West, Salt Lake City.

VERIFIED

....

AFSATCOM system

The UHF Air Force SATellite COMmunications AFSATCOM system provides reliable, worldwide, C2 communications. These communications are used by designated US Single Integrated Operational Plan (SIOP)/nuclear capable users for Emergency Action Message (EAM) dissemination, JCS-CINC internetting, force direction, and force report-back. AFSATCOM capacity is also provided for a limited number of high-priority non-SIOP users for operational missions, contingency/crisis operations, exercise support and technical/operator training.

The AFSATCOM system is made up of a space segment which consists of UHF transponders aboard several spacecraft and a terminal segment. The terminal segment consists of standard AFSATCOM ground/airborne, manpack, and special communications system terminals. The space segment is air force managed transponders of varying capability and capacity. They are carried aboard the fleet satellite communications (FLTSATCOM), leased satellite communications (LEASATCOM), Satellite Data Systems (SDS), Packages B and C, DSCS III and Lincoln Experimental Satellites (LES) 8 and 9.

VERIFIED

....

AN/FSC-9 SATCOM terminal

AN/FSC-9 is a large 18.2 m parabolic antenna satellite terminal of which there were two installed in the USA: one at Fort Dix in New Jersey and the other at Camp Roberts, California.

It employs a TWT driver and klystron power amplifier, the 70 MHz input signal from the terminal equipment being converted to X-band and amplified to provide a power output of up to 20 kW.

It is equipped with both FM and spread-spectrum modulation equipment. The multiplex equipment can accept 12 incoming user voice channels for operation in the FM mode. Four voice channels and a digital signal of up to 4,800 bits can be provided.

The AN/FSC-9 Terminal at Camp Roberts, California is now the only terminal in operation and provides communications via satellite from the Pacific area to the continental United States.

Specifications
(a) Type of service: 500000KF9W (multilink, multivoice and data)
(b) Frequency range:
   transmit: 7.9 to 8.4 GHz
   receive: 7.25 to 7.75 GHz

(c) Bandwidth:
   down converter: 50 MHz
   parametric amplifier: 500 MHz
(d) Planning range: 16,093 km (10,000 mi)
(e) Power input: 115/230 V AC, 50 to 60 Hz
(f) Power source: Any appropriate AC power source
(g) Power output: Up to 20 kW
(h) Antenna system: Paraboloid surface, 18.29 m (60 ft) in diameter, weight 172.3 MT (190 t)

Contractor
Lockheed Martin Space Systems, Sunnyvale, California.

VERIFIED

....

AN/FSC-78 SATCOM terminal

The AN/ FSC-78 (V) is a fixed SHF SATCOM heavy SGT operating in the X-band frequency range. The terminal consists of six subsystems, including antenna tracking, transmitter, receiver, frequency reference, control and monitoring. The antenna is a 60 ft diameter, high-efficiency, parabolic reflector providing an antenna gain-to-noise temperature ratio (G/T) of 39 dB/K. The reflector is mounted on an elevation-over-azimuth-configured pedestal. Cryogenically cooled, parametric amplifiers provide 30 dB of gain and an antenna G/T ratio of 39 dB/K. The antenna terminal equipment has a tracking converter, 15 down-converters, and 9 up-converters. Only 10 of the down-converters are normally active at one time; the remaining five are in hot standby. The output signals from the up-converters are fed to a 5 kW TWTA, providing a radiated antenna signal of 500 MHz bandwidth at an EIRP of 124 dB referenced to 1 W (dBW). A redundant 5 kW power amplifier can be operated in parallel with the primary power amplifier to provide an output equivalent to 10 kW at an EIRP of 127 dBW. The down-converters translate the receive signal of 7.25 to 7.75 GHz to 70 MHz IF (40 MHz bandwidth) or a 700 MHz IF (125 MHz bandwidth). The up-converters translate the 70 or 700 MHz IF input signal, with bandwidths of 40 or 125 MHz, to the transmit frequency of 7.9 to 8.4 GHz.

Harris Corporation was awarded a firm, fixed-price contract for the heavy terminal/medium terminal (HT/MT) Modernisation Programme which covered the AN/FSC-78, AN/FSC-79 and AN/GSC-39 terminals on 27 March 1992 following a formal competitive source selection. The contract was a one-year basic contract with four option years. This allowed the flexibility to procure equipment as needed and as directed or dictated by schedule and fiscal year funding constraints. The contract included all the hardware, software, logistical support, and a time and materials effort for the contractor installations, antenna refurbishment and on-call engineering support, depot support and post deployment software support. The HT/MT MOD fielding was in two phases: a contractor installation phase and a government installation phase. The contractor installation phase, completed in December 1995, proved the equipment in operational sites and trained government installers. The government installation phase began in January 1996.

Specifications

Antenna type		60 ft Cassegrain
Feed		Five horn pseudomonopulse
Pedestal type		EL/AZ kingpost
Polarization	Transmit	Right-hand circular
	(b) Receive	Left-hand circular
Frequency	(a) Transmit	7.8. to 8.4 GHz
	(b) Receive	7.25 to 7.75 GHz
Beacon carriers		Three maximum receive, used one at a time
G/T		39 dB/K
IF Bandwidths	(a) 70 MHz	40 MHz
	(b) 700 MHz	125 MHz
Amplitude Response	(a) 70 MHz	1 dB/10 MHz
	(b) 700 MHz	2 d8140 MHz
		1 dB/60 MHz
		2 dB/125 MHz
Phase linearity	(a) 70 MHz	0.1 radian/10 MHz
	(b) 700 MHz	0.25 radiant/40 MHz
		0.15 radian/60 MHz
		0.4 radian/125 MHz
EIRP	(a) Normal Mode	124 dBW
	(b) Combined Mode	127 dBW
Simultaneous RF Carriers	(a) Transmit	Up to 9
	(b) Reserve	Up try 15
Tunability		500 MHz in 1 kHz increments
Redundancy		All subsystems except antenna
Fault location		Automatic
Frequency control		Synthesiser, referenced to an atomic (cesium) standard
Carrier power level control		Automatic. or manual
Multiple access capability		FDMA, SSMA. TDMA

Two AN/FSC-78 SATCOM terminals (centre and left)

The HT/MT Modernisation Programme took the existing terminals and replaced the ageing Radio Frequency (RF) electronics with new hardware. This effort was performed in order to extend the SATCOM terminal's operational life by an additional 15 years while reducing the Operational and Maintenance (O&M) costs to the User Command. Some of the RF electronics replaced were the Up and Down Converters, High Power Amplifiers (HPAs), Low Noise Amplifiers (LNAs) and Cesium Standards.

In 1996 Harris was awarded a US$14 million contract to upgrade the AN/FSC-78, AN/FSC-79 and AN/GSC-39 antenna) and it was completed at the end of 1999. A subsequent modernisation programme of the AN/GSC-52 terminal (see separate entry), the contract for which was awarded in May 1998, which renewed the Control Monitor and Alarm (CMA) subsystem was also applied to the AN/FSC-79.

Status
In service with the US Army. Some 22 terminals are believed to be operational worldwide.

Contractor
Harris Corporation (modernisation contract), Melbourne, Florida.

UPDATED

....

AN/FSC-79 SATCOM terminal

The AN/ FSC-79 Fleet Broadcast Terminal is a fixed SHF SATCOM terminal capable of one transmit channel and one receive beacon channel, designed specifically to support the Navy Fleet Satellite Broadcast. It is housed in a permanent facility and uses a 60 ft diameter, high-efficiency parabolic reflector antenna mounted on an elevation over azimuth-configured pedestal. The terminal operates on a single channel, tuneable in 1 kHz increments over a transmitting frequency range of 7.9 to 8.4 GHz, at a maximum output of 10,000 W. It is essentially a modified AN/FSC-78 (see separate entry).

Harris Corporation was awarded a firm, fixed-price contract for the heavy terminal/medium terminal (HT/MT) Modernisation Programme which covered the AN/FSC-78, AN/FSC-79 and AN/GSC-39 terminals (see separate entries) on 27 March 1992 following a formal competitive source selection. The contract was a one-year basic contract with four option years. This allowed the flexibility to procure equipment as needed and as directed or dictated by schedule and fiscal year funding constraints. The contract included all the hardware, software, logistical support, and a time and materials effort for the contractor installations, antenna refurbishment, and on-call engineering support, depot support, and post deployment software support. The HT/MT MOD fielding was in two phases: a contractor installation phase and a government installation phase. The contractor installation phase, completed in December 1995, proved the equipment in operational sites and trained government installers. The government installation phase began in January 1996.

The HT/MT Modernisation Programme took the existing satellite communication (SATCOM) terminals and replaced the aging Radio Frequency (RF) electronics with new hardware. This effort was performed in order to extend the SATCOM terminals operational life an additional 15 years while reducing the Operational and Maintenance (O&M) costs to the User Command. Some of the RF electronics replaced were the Up and Down Converters, High Power Amplifiers (HPAs), Low Noise Amplifiers (LNAs) and Cesium Standards.

In 1996 Harris was awarded a US$14 million contract to upgrade the AN/FSC-78, AN/FSC-79 and AN/GSC-39 antenna. A subsequent modernisation programme of the AN/GSC-52 terminal (see separate entry), the contract for which was awarded in May 1998, which renewed the Control Monitor and Alarm (CMA) subsystem was also applied to the AN/FSC-79.

Status
In service with the US Navy, probably in five locations.

Contractor
Harris Corporation (modernisation contract), Melbourne, Florida.

UPDATED

AN/GSC-39 SATCOM terminal

The AN/GSC-39 Satcom terminal is part of the ground segment of the US Defense Satellite Communications System and is identical in capability to the AN/FSC-78, but it has a 38 ft dish. V(1) is a fixed site terminal; V(2) is transportable.

See AN/FSC-78 entry for details of modernisation programmes.

Status
In service with US forces.

Contractor
Harris Corporation (modernisation contract), Melbourne, Florida.

NEW ENTRY

AN/GSC-49(V) Jam-Resistant Secure Communications (JRSC)

The JRSC terminal (AN/GSC-49(V)) is a ground-based SATCOM terminal deployed within the Defense Satellite Communication System (DSCS).

The 32 terminals in the JRSC network have been modified to improve the performance of their survivability, availability and supportability. This was achieved by upgrading the HEMP resistance, reducing the HPA downtime and replacing obsolete equipment. The programme simplified the terminal design by reducing the quantity and types of LRUs, thus improving reliability and maintainability and, hence, the terminal availability.

The use of COTS equipment increased the kit modularity.

Specifications
(a) Type of service: 50000KF9W (single-channel voice and FSK)
(b) Frequency range:
    Transmit: 7.29 to 8.4 GHz
    Receive: 7.25 to 7.75 GHz
(c) Planning range: 6,093 km (10,000 mi)
(d) Power input: 120/208 V AC, 50/60 Hz
(e) Power source: Two generators (30 kW)
(f) Power output: 3.2 kW max
(g) Antenna system: Quick-reaction, 8-ft parabolic; 20-ft parabolic that can transmit multiple carriers
(h) Set-up time: Approximately 1 hour to satellite acquisition with small antenna; within 12 hours for large dish.

Status
The terminals were due to reach the end of their natural life in 2003 and had been declared unsupportable thereafter.

Contractor
Harris Corporation, Government Communications Systems Division, Melbourne, Florida.

UPDATED

AN/GSC-49 terminal

AN/GSC-52 SATCOM terminal

The AN/GSC-52 is a high-capacity, SHF SATCOM terminal which functions as an integral part of the US Defense Satellite Communications System (DSCS). The terminals have been in operation since the 1980s. They are capable of simultaneous transmission/reception of up to 18 transmit and receive carriers. Each carrier can accommodate CW, pulse modulated (PM), FM, FDMA or SSMA signals. The terminals are provided in either a fixed or mobile configuration. It is a high-capacity, high-altitude electromagnetic pulse (HEMP) protected terminal that uses pseudo-monopulse scanning for operator-selectable manual tracking, memory tracking, or acquisition/ auto tracking of the satellite.

AN/GSC-52 SAMT

The terminal consists of an antenna subsystem, a receive subsystem, a transmitter subsystem, and tracking/servo subsystem. The antenna subsystem has a Cassegrain feed, 38 ft parabolic-reflector antenna, an elevation over azimuth pedestal, and a servo drive mechanism. Modems provide a 70 or 700 MHz IF to the up-converters whose RF outputs are combined into a single RF signal in the 7.9 to 8.4 GHz range with a bandwidth as wide as 500 MHz. The composite signal is amplified by the TWTA and fed, via waveguides, to the antenna subsystem. On the receive side, the antenna receives an RF signal at 7.25 to 7.75 GHz, amplifies the signal using LNAs and the interfacility amplifiers, and passes the signal to down-converters which provide a 70 or 700 MHz IF output to the modem. The terminal uses 12 up-and down-converters.

The AN/GSC-52(V) ground terminal is capable of manned or unmanned operations through a centralised control, monitor, and alarm subsystem that provides computer-aided configuration for control, status and performance monitoring, equipment calibration, fault isolation, and automatic switching of redundant equipment to replace a faulty unit.

The AN/GSC-52 terminals are presently undergoing a modernisation programme. The primary objective of this programme is to extend the life of 39 AN/GSC-52 Satellite Terminals, transportable and fixed sites, for another 15 years. Other major programme objectives include improved performance, enhanced control and monitoring, reduced life cycle costs, increased availability and increased commonality with the other Defense Satellite Communications System (DSCS) strategic terminals. The AN/GSC-52 Mod terminals will be fabricated in two configurations. The fixed configuration consists of the Antenna Group with an Elevated Equipment Room (EER) containing a Radio Frequency Amplifier Assembly and the four each High Power Amplifiers and an Electronic Equipment Building (EEB) housing the balance of the subsystems. The mobile configuration is the same as the fixed except the EEB is replaced by a transportable operations van and a supply/maintenance van. The modernisation of 39 AN/GSC-52 terminals will be accomplished by system upgrade. The contractor (Harris Corporation) will install MWO kits for the first three terminals and the Army will install the remaining 36 MWO.

The modernisation programme will include: the replacement uplink and downlink converters; provision of a PC-based control, monitor and alarm system; provision of an integrated LAN with multiple access points for flexible terminal control options.

Specifications
(a) Frequency Range:
    Transmit: 7.9 to 8.4 GHz
    Receive: 7.25 to 7.75 GHz
(b) Power input: 120/208 V AC 50/60 Hz
(c) Power Source: Commercial or two 50 kW backup generators using UPS technology
(d) Power output: 1 kW
(e) Antenna System: Parabolic 11.6 m (38 ft) in diameter OE-371/G Antenna

Status
There are 39 AN/GSC-52 systems in operation with the US armed forces as part of the Defense Satellite Communications System (DSCS). The modernisation of the Fort Gordon First Article Test (FAT) installation, which supports training of multiservice satellite operators and maintainers, was completed in July 2001. Four other FAT site upgrades - at Fort Belvoir, Virginia; Fort Meade, Maryland; Fort Monmouth, New Jersey; and Fort Bragg, North Carolina - were scheduled for completion by October 2001. The modernisation programme is due to be completed by 2006.

Contractor
Harris Corporation (for modernisation programme), Melbourne, Florida.

UPDATED

AN/MSC-46 satellite terminal

AN/MSC-46 is a 12 m antenna satellite earth station designed for operation with the Defense Satellite Communications System (DSCS). It is designed to operate worldwide with the antenna protected in a geodesic dome. Three mobile vans carry communications and support equipment; a fourth provides additional storage space. Three diesel generators provide emergency power.

AN/MSC-46 satellite terminal

The terminal transmits in the 7.9 to 8.4 GHz band and receives in the 7.25 to 7.75 GHz band. It provides 12 full-duplex voice frequency traffic channels, two full-duplex teletype circuits and a voice order circuit.

Status

The first terminal was built in 1960 for the US Army Satellite Communications Agency (USASATCOMA) by Hughes (now Raytheon) as a mobile ground link terminal within the DSCS. Between 1966 and 1968, 14 were delivered to USASATCOMA. The terminals were refurbished to operate with the new generation DSCS III satellites. The first to be refurbished was the station located in Guam. A US$3.1 million support contract was placed by the US Army in October 1980, followed by a further US$3.6 million support contract in November 1982.

The terminals are now being phased out (see DSCS entry).

Contractor

Raytheon Company, Lexington, Massachusetts.

UPDATED

AN/PSC-3/AN/VSC-7 UHF SATCOM/LOS transceivers

The AN/PSC-3/AN/VSC-7 transceivers provide satellite or direct Line Of Sight (LOS) communications in 300, 1,200 and 2,400 bit/s digital data, plain text voice and FM voice or 16 kbit/s secure voice via an interface with the TSEC/KY-57. Both selective and conference call capabilities are provided. The transceivers operate in the UHF band from 225 to 400 MHz in 5 kHz increments (satellite) or 25 kHz

AN/VSC-7 radio

AN/PSC-3 radio

increments (LOS). Transmit power output is 35 W for satellite operation and 2 W for LOS.

An internal modem provides BPSK and DPSK operation for data and burst rates. Plain text analogue voice is at ±8 kHz deviation and 16 kbit/s cipher text is at ±5.6 kHz. The modem also employs the 300 bit/s data format for call operation. The transceivers receive and differentiate between selective and conference call, and transmit conference call signals.

A microprocessor, in conjunction with a six-pinlite display, provides an electronic control of such functions as frequency, mode of operation and receive offset frequency. Two spring-loaded momentary switches accomplish frequency control. A whip antenna is used for LOS operation or call reception while in motion. An ancillary medium-gain (6 dB minimum) helical antenna is provided for the satellite mode. This crossed-dipole, over-ground-plane, circularly polarised antenna can be deployed in 2 minutes and is collapsible for storage or transport.

Specifications
(a) Modes: voice; 300, 1,200 and 2,400 bit/s data; retransmit; call; X-mode
(b) Transceiver frequency range: 225-399.995 MHz (SAT); 225-399.975 MHz (LOS)
(c) Channel spacing: 5 kHz (SAT); 25 kHz (LOS)
(d) Receiver offset frequency: 4 internally preset spacing from 0 to ±174.995 kHz (SAT); no offset for LOS
(e) Ringing signals: respond to selective or conference calls; transmit conference calls
(f) Retransmit capability: compatible with AN/PRC-70, AN/VRC-12, AN/PRC-77, AN/VRC-43 to -49, AN/PRC-8, -9 and 25, AN/PRT-4, AN/PRR-9, AN/GRC-3 to -8, AN/ARC-54 and -131, URC-101, WSC-3, and LST-5B
(g) Noise figure: 3 dB typical
(h) Transmitter power output: 35 W ±2 dB into 1.5:1 VSWR (SAT); 2 W ±2 dB into 1.5:1 VSWR (LOS)
(i) Power supply: 24 V DC, 110/220 V AC with applique
(j) Battery life: 12 h (min) with 9:1 Rx:Tx duty cycle
(k) Accessories: 6 dB gain helical antenna (SAT); whip antenna (LOS); OA-8990()/P Digital Message Device Group (DMDG)
(l) Height: 79 mm
(m) Width: 178 mm
(n) Depth: 292 mm
(o) Weight: 5.2 kg (excl battery); 11.3 kg (incl accessories)

Status
The AN/PSC-3 manpack satellite radio and its AN/VSC-7 vehicle version have been developed from the AN/PSC-1. The AN/PSC-1 was completed under a contract from the US Army Satellite Communications Agency, Fort Monmouth, New Jersey. First tested at Cincinnati's main plant in Evendale in June 1977, the radios communicated via a stationary Marisat satellite above the Atlantic Ocean. A total of US$3.1 million was requested in FY81 to procure 70 AN/PSC-1 transceivers. In October 1981 the US Army placed an order for AN/PSC-3 and AN/VSC-7 radios worth US$8.6 million. Contracts placed to mid-1991 totalled approximately US$40 million.

Now being replaced by the AN/PSC-5 (see separate entry).

Contractor
CMC Electronics Cincinnati , Mason, Ohio.

VERIFIED

AN/PSC-11 Single Channel Anti-jam Man-Portable (SCAMP) terminal

Description

SCAMP uses the Milstar system to provide worldwide secure, jam-resistant, covert voice, data and imagery communication. It operates at the Low Data Rate (LDR) with up to four full-duplex connections simultaneously. SCAMP is interoperable with Milstar and all satellites with EHF capability that satisfy the LDR satellite data link standards (SDLS). The terminal will automatically acquire the satellite and then establish, maintain and control communication links.

The SCAMP is a self-contained terminal packaged in two sturdy, lightweight, man-portable cases for easy transport, One case holds all communication components while the other contains all deliverable accessory devices. One person can easily deploy and employ a terminal in less than 10 minutes.

The SCAMP terminal can use all LDR satellite beams, depending on mission requirements and the operating environment. It supports up to four simultaneous 2,400 bit/s voice or data channels. The SCAMP software provides the flexibility for performing multimission roles. The program contains system operating parameters that automate, monitor and control overall terminal management functions.

The terminal consists of only three major functional components: the receiver/transmitter (R/T), a hand-held control device and the Interface Unit (IU). The R/T, with its built-in collapsible reflector, contains all the signal processing, waveform generation/detection, terminal control and signalling conditioning hardware. It will establish, maintain and exercise continuous connectivity with the Milstar or other compatible satellites. Ports are available for connecting to secure devices such as STU-111, MMT-1500 and others. Two 24 V batteries permit stand-alone operation.

The SCAMP is controlled from a small hand-held display/keyboard device. In its most simplistic application, the operator merely selects connectivity parameters from a menu of mission presets that were previously programmed using the Collins AN/CYZ-10 Automatic Network Control Device (see associated equipment). These presets include five acquisitions, 20 networks, 20 point-to-point and one automated start-up. Experienced operators have the flexibility to perform diagnostics and edit terminal control functions as mission needs require.

The SCAMP system includes an IU for connecting to clear and secure mission input/output devices, including fax, PC, STU III, imagery and video. The unit has one red and three black baseband ports, allowing for up to four simultaneous communication links.

The SCAMP Radio provides four full-duplex user channels of worldwide secure, survivable, covert, voice and data communication via the Milstar network. The radio will support up to four simultaneous Vocoded Voice (VV) channels on the Narrow Spot Beam (NSB), up to two simultaneous VV on Wide Spot Beam (WSB) and up to four simultaneous 75 bits/TTY channels on the Agile Beam (AB) with full margin. With clear conditions, it will support four VV channels on both Spot Beams and the AB or four TTY channels on the Earth Coverage (EC) antenna. The terminal is packed for storage or transport in two lightweight cases. Case 1 holds everything necessary for voice or data communications. Case 2 houses all deliverable accessories.

Specifications

Radio frequency
uplink frequency 44.0 GHz
uplink bandwidth 2.0 GHz
downlink frequency 20.0 GHz
downlink bandwidth 1.0 GHz
data Rates 75 - 2,400 bit/s
voice (ANDVT compatible) 2,400 bit/s

Mechanical
volume of each case 25 × 13.5 × 11 in
weight of self-contained terminal with case <37 lb
weight of packed accessories case < 34 lb

Environmental
20 mph winds with 30 mph gusts
survive 2 in/h rain
operate −32 to +49° C

Power
internal battery 24 V DC
external DC 20-33 V DC
external AC 110/220

Status

SCAMP is being procured by the US Army. The initial contract is for the first article qualification testing and production of 120 terminals and is valued at US$26 million. The contract options extend the programme through 2002 and provide for the production and support of 512 terminals and support with a total value of US$55 million. Fielding the terminals began in 1999. In 2000 the US Air Force was operational with 29 sets in US Strategic Command (USSTRATCOM). These are designed to replace the obsolete GWEN system.

As at mid-2002 no further detailed information was available on the programme, but it is intended that the equipment will evolve to SCAMP Block II, which will be manpackable and deployed to headquarters at echelons Corps and below.

Contractor

Rockwell Collins Government Systems, Cedar Rapids.

VERIFIED

AN/TSC-85 and AN/TSC-93 Ground Mobile Forces (GMF) tactical SHF satellite terminals

This small satellite terminal family is a tactical group of military communications terminals designed to provide point-to-point or multipoint trunking facilities. Working through the Defense Satellite Communications System (DSCS) satellites, either short- or long-range communications can be established quickly without mid-point repeaters or extensive site preparation.

Each terminal is completely self-contained and has been designed to provide full communications capability within 20 minutes of arrival on site. The 2.4 m parabolic antenna assembly can be erected quickly and provides accurate automatic satellite tracking in wind gusts to 168 km/h.

Each terminal type is housed in an S-250 shelter with the main terminal differences being the degree of redundancy and the baseband communications facilities provided. A companion trailer provides a mobile platform for transport and operation of two prime power motor generators and for stowage. Maximum interchangeability has been provided between all power amplifiers, up- and down-converters, antenna assemblies and other units which are identical and interchangeable from terminal to terminal.

Built-in maintenance facilities are provided throughout to maximise terminal availability. This simple built-in test equipment allows failure diagnosis to the lowest replaceable unit without the need for any additional test equipment and provides a mean time to repair for each terminal type of less than 15 minutes.

The satellite communications terminal AN/TSC-85 is redundant in the RF, modem and tactical satellite signal processor sections and is used as a nodal and non-nodal terminal on tactical trunking networks. The terminal provides capability to transmit up to 96 secure or non-secure voice channels of PCM digital data and order-wire via a single SHF carrier. The terminal can receive up to four SHF carriers, demodulating the carriers and supplying PCM digital data to the multiplexer van.

The satellite communications terminal AN/TSC-93 is non-redundant in both the RF and baseband sections and is used as a non-nodal or point-to-point terminal in tactical trunking networks. The terminal provides the capability to transmit 6/12/24 channels of voice traffic and order-wire utilising self-contained multiplexer equipment. An interleaver provides for substitution of 16/32 kbit/s of data in each of the voice channels. An additional capability to interface on a digital basis with an

AN/PSC-11 SCAMP
0055015

AN/TSC-85B with OE-361 quick react antenna

external 16 kbit secure voice channel or up to 96 channels of secure or non-secure PCM voice traffic is provided. Planned product improvements include a baseband improvement modification and anti-jam control modems.

Specifications
(a) Frequency range
transmit: 7.9-8.4 GHz in 100 kHz spacings
receive: 7.25-7.75 GHz in 100 kHz spacings
(b) Antenna
reflector: 2.4 m
(c) Polarisation
transmit: right-hand circular
receive: left-hand circular
tracking: automatic
(d) Transmit power: 500 W
(e) System noise temperature: 300 K
(f) Amplitude response
transmit: ±1 dB, any 10 MHz b/w
receive:
(AN/TSC-85(V)2) ±1.5 dB, any 40 MHz b/w
(AN/TSC-93) ±1 dB, any 10 MHz b/w
(g) Number of down converters
AN/TSC-85: 4
AN/TSC-93: 1
(h) Phase linearity
transmit: ±15°, any 10 MHz b/w (also ±20°, any 40 MHz for AN/TSC-85)
receive: ±15°, any 10 MHz b/w (also ±20°, any 40 MHz for AN/TSC-85)
(i) Transmission modes on transmit
PCM 48 kbit/channel:
(AN/TSC-85) 6, 12, 24, 48 and 96 channels
(AN/TSC-93) 6, 12 or 24 channels
single-channel digital input: 16 kbit/s
order-wire: FM
(j) Transmission modes on receive
PCM 48 kbit/channel:
(AN/TSC-85) 6, 12 or 24 channels digitally combined, decombined
(AN/TSC-93) 6, 12 or 24 channels digitally decombined
single-channel digital output: 16 kbit/s
order-wire: FM
(k) Modem capability: all rates to 5 Mbit/s

Status
Developed by RCA. First production contracts were placed in 1976 by the US Army for seven AN/TSC-85 and 18 AN/TSC-93 terminals, together with six AN/TSC-94 terminals. Value of these contracts and additional add-ons was US$37 million. In 1979 Harris won a US$79 million production contract, in competition with RCA, for a total of 226 small terminals.

Budget funding for the AN/TSC-85 for FY81 and FY82 was US$17 million and US$23 million respectively. The budget FY81 for US Army procurement of the AN/TSC-93 was US$11.8 million.

In the last quarter of 1980 Harris beat RCA in a competitive bid winning a contract worth US$12.3 million for AN/TSC-85 and AN/TSC-93 terminals. Various add-on contracts were subsequently announced and by mid-1988, 165 terminals had been ordered by the army, 25 by the Marine Corps, 22 by the Joint Communication Support Element and 50 by NATO. Also used by US forces in South Korea.

Deployed in Operation Desert Shield/Desert Storm.

Upgrades to modems and convertors to bring equipment to 'C' status were carried out in 2001.

In 2003 a US$38 million Service Life Extension Programme was begun to upgrade the terminals to 'D' status and to enable them to continue in service until 2012, believed to be caused by delays to STRAT-T fielding. This programme is

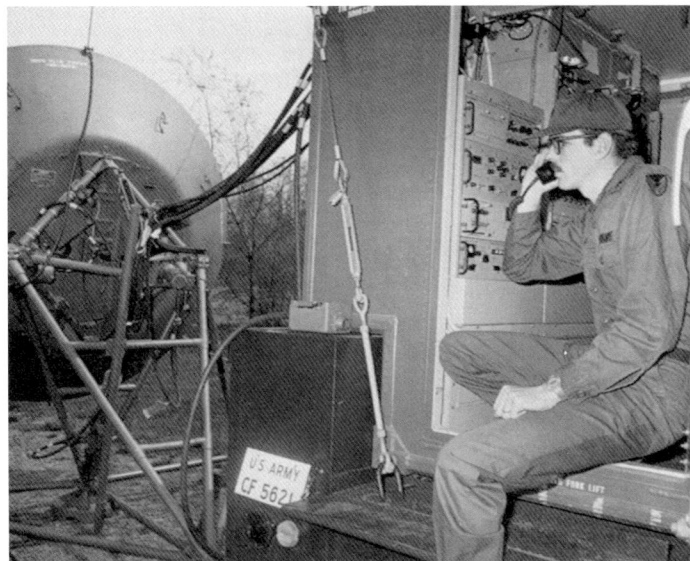

AN/TSC-86 SATCOM terminal under test

believed to include modem, convertor and antenna upgrade, FM order-wire and enhanced tactical satellite signal processors.

Contractor
Harris Corporation, Government Communications Systems Division, Palm Bay, Florida.

UPDATED

AN/TSC-86 vehicle-mounted SATCOM terminal

The AN/TSC-86 is a vehicle-mounted SHF SATCOM terminal and is one of a number of transportable earth stations used by the US Department of Defense to transmit and receive voice messages to and from any point on the globe. The self-contained unit has a 2.4 m diameter antenna that automatically tracks one of the defense communications satellites. An alternative 6 m diameter antenna (high-gain) is also included as part of the AN/TSC-86 package.

The 4,082 kg terminal can be transported by truck or airlifted by helicopter and can be made operational within 30 minutes. It is powered by redundant 30 kW diesel generators.

Status
The terminal was developed by the US Army Satellite Communications Agency for its own use and for the US Air Force. It was designed by RCA. The first of six systems ordered by the US Army was delivered in August 1981. Four systems were procured by the Air Force in FY81 and these were upgraded to AN/TSC-86(A) standard in 1999.

VERIFIED

AN/TSC-94A mobile tactical satellite terminal

The AN/TSC-94A is an SHF mobile communications terminal designed for use with the DSCS III satellite network. Configured as a point-to-point or non-nodal terminal for tactical networks and usually operating as a spoke in a hub and spoke arrangement with the TSC-100 (see separate entry), there are two versions, designated AN/TSC-94A(V1) and AN/TSC-94A(V2). The main difference between the versions is the input power capabilities: the (V)2 is capable of operating on either 400 Hz or 50/60 Hz while the (V)1 is 50/60 Hz only.

All electronic units, except the antenna assembly with its mounted electronic equipment, are housed in an S-250 shelter. The terminal is EMP and EMI hardened. The terminals are designed to operate in a communications jamming environment using a ground-mobile force AJ control modem that uses data rates from 75 bit/s to 32 kbit/s.

The V1 terminal is pallet loaded for transportability. It is equipped with M-720 mobilisers and operates from a three-phase 120/230 V AC power supply. Capable of providing 24 channels for local subscribers, inputs can be either analogue or digital, with 16 or 32 kbit/s CVSD analogue to digital conversion or digital to analogue conversion as required. Digital data rates from 44.5 bit/s to 50 kbit/s can be used, provided the Low Rate Multiplexer (LRM) maximum output rate of 256 kbit/s is not exceeded.

The V2 version is equipped with three 10.5 kW power converters. For local subscribers 12 channels are provided by one LRM. Simultaneous transmit and receive modulation is provided by an all-rate modem with independently separate rates from 16 to 4,999 kbit/s.

Status
In operational use by the USAF in 2003.

Contractor
L3 Communications (East), Camden, New Jersey.

UPDATED

AN/TSC-100A satellite terminal

The AN/TSC-100A is an SHF satellite ground station designed to be transported on an M832 mobiliser and to operate with the DSCS III satellite system. Intended as a nodal or mesh terminal for tactical trunking networks, there are two versions designated the AN/TSC-100A(V1) and AN/TSC-100A(V2). All electronic units are rack-mounted as an integral part of an S-280 shelter, except the antenna assembly with its mounted electronic equipment. The terminals are EMP and EMI hardened.

The terminals are designed to operate in a communications jamming environment using a ground-mobile force AJ control modem. The AJ modem uses data rates from 75 bit/s to 32 kbit/s and can accommodate up to four Low-Rate Multiplexers (LRM).

The V1 terminal is capable of operating from a 120/230 V AC three-phase power source. Provided with six LRMs, the V1 can provide 72 channels for local subscribers. Inputs can be analogue or digital, with 16 or 32 CVSD analogue/

AN/TSC-100A satellite ground station

digital conversion as required. Digital data rates between 44.5 bit/s and 50 kbit/s can be used, provided the maximum output rate of each LRM does not exceed 256 kbit/s.

The V2 terminal is capable of providing 60 channels for local subscribers via five LRMs. Simultaneous transmit and receive modulation is provided by an all-rate modem at independently separate rates from 16 to 4,999 kbit/s.

Status

A contract to build four AN/TSC-100A terminals worth US$11 million was awarded to RCA in January 1980 by the US Air Force (RCA became part of General Electric, now part of Lockheed Martin). In August 1982 the US Air Force placed a US$106.7 million contract which included orders for 19 AN/TSC-100A(V1) and 24 AN/TSC-100A(V2) terminals together with 24 AN/TSC-94A(V1) and 44 AN/TSC-94(V2) terminals. A US$7 million contract for spares for the AN/TSC-94(A) and AN/TSC-100(A) was placed in April 1983.

In operational use by the USAF.

Contractor

Lockheed Martin, Cherry Hill, New Jersey.

UPDATED

AN/TYQ-40 Communications Control Set (CCS) module

The CCS, AN/TYQ-40 (V) 2 or AN/TYQ-63 for Echelon Above Corps (EAC), is the communication centre for the All Source Analysis System (ASAS) (see separate entry) in the Analysis and Control Element (ACE).

Description

The CCS can operate at both the Sensitive Compartmented Information (SCI) and collateral levels. It provides a communication interface that allows the All Source Enclave (ASE) and the Single Source Enclave (SSE) to link into communications networks from a variety of communications systems. The CCS supports communications from remote sensors into the ACE and allows data and voice communications with higher, lower, and adjacent units.

The CCS provides the means to communicate, review, prioritise, and retransmit various types of messages, and provides automatic message routing. It handles numerous links simultaneously using radio, wireline, telephone line, and ethernet LAN. It provides:

- two VHF encrypted datalinks;
- one UHF encrypted datalink;
- two MSE auto dial voice/datalinks;
- four encrypted wireline datalinks, Tri-Tac;
- four 802.3 ethernet links;
- one non-secure voicelink;
- one secure voicelink, STU-III.

The ASAS Block I is equipped with two AN/TYQ-40(V) 2, CCS. It supports collateral and SCI level communications processing, and relay and interfaces with ACUS, CNR, and special purpose intelligence communications systems. The CCS provides secure voice and data communications through MSE, SINCGARS, and the JTT. The CCS equipment provides capabilities for automatic message routing, operator message review, and manual message routing. The CCS consists of the following major systems:

Communications Processing Subsystem (CPS)

The CPS performs message protocol translation, message processing, and detailed auditing of system activity. It has a variety of tools to help operators distribute message traffic automatically. The CPS retains messages on disk packs for temporary storage. The system is capable of processing a number of communications protocols. This capability establishes the basic ASAS compatibility and interoperability with other systems. All data handling internal to the ASAS uses Full Duplex Message Protocol (FDMP)/Digital Data Communications Message Protocol (DDCMP). Outgoing message traffic is translated from FDMP/DDCMP; incoming traffic is translated to it. The CPS provides protocol translation for Automatic Digital Network (AUTODIN), digital communications terminal (DCT), net radio protocol (NRP), and External Digital Data Communications Message Protocol (XDDCMP).

Computer Operator Terminal (COT)

The COT allows the CCS operator to initialise and control the CPS.

TSEC/KY-68 Terminal and Data Adapter

The CCS is equipped with six TSEC/KY-68 Digital Subscriber Voice Terminals (DSVTs) and data adapters for communication into the MSE network. It provides non-secure voice, secure voice, and secure data communications within the MSE network. The data adapter is a carry-in, microprocessor-based communications controller capable of protocol tasks.

AN/ARC-164A

The CCS has one AN/ARC-164A Ultra High Frequency (UHF) radio. It provides secure voice or data communications when used with the TSEC/KY-57 for voice, AN/PSC-2 and TSEC/KY-57 for data, and the TSEC/KG-84A for NRP. The combination of the system's RT-1288A with NRP, a datalink processor, and encryption device provide data communications with NRP-capable sensors and relays such as the AN/TSQ-138 TRAILBLAZER, AN/TRQ-32(V)2 TEAMMATE, and AN/TSQ-175 TIGER.

AN/PSC-2

The CCS is equipped with one AN/PSC-2, DCT. The AN/PSC-2 DCT is used to prepare, send, receive and display reformatted IEW Character Oriented Message Catalogue (COMCAT) messages and free-text messages. The CNR systems in the ASAS Block I CCS support secure data communications when used with the AN/PSC-2. It supports the exchange of SCI tasking and reporting messages between the ACE and AN/PSC-2 equipped IEW assets. The ACE also uses the system to exchange collateral messages with CI teams, interrogation teams, and long-range surveillance teams.

AN/VRC-92A

The CCS has four AN/VRC-92A, SINCGARS, Very High Frequency (VHF) radios that are Frequency Modulation (FM) with Integrated COMSEC Module (ICOM). Operated in the secure non-hopping mode, these systems provide secure voice and data communications.

Data Processor Set (DPS)

The DPS, AN/TYQ-36(V)3 is a mobile, self-monitoring, unmanned data processing station for the Block I ASAS-ASW. Each ASAS Block I has two DPS. The ASAS-ASW applications software and databases reside within the DPS. They provide the communications connectivity between the CCS and ASAS-ASW. The shelter provides environmental control, intrusion protection, fire protection, and secure storage for the ASAS main processors.

Status

The CCS is presently being fielded as part of the ASAS programme. There was a significant upgrade in 1999 in order to overcome Y2K issues.

Contractor

Lockheed Martin Mission Systems, Gaithersburg, Maryland, USA.

VERIFIED

AN/WSC-3 UHF SATCOM/LOS transceiver

The AN/WSC-3 (known as Whiskey-3) is the US Navy's standard UHF satellite terminal and Line Of Sight (LOS) transceiver. It was developed to serve as the US Navy's new generation ship/submarine terminal for the Fleet Satellite Communications System. Whiskey-3 provides a minimum 100 W output in FM Link 11 or data modes. Internal data modulation and detection is 75 bit/s FSK and 75 bit/s to 9.6 kbit/s PSK. A 70 MHz interface capability provides for expansion with a variety of external modems. In 1986 a fleet reliability assessment programme conducted by the US Navy recorded an MTBF of over 15,200 hours for the equipment.

Specifications

(a) Frequency range: 225-399.975 MHz
(b) Number of channels: 7,000 in 25 kHz steps
(c) Preset channels: 20, remote or locally selectable
(d) Transmit modes: AM (wideband and narrowband), FM (wideband and narrowband), FSK, PSK, Link 11.
(e) Also available with programmable modem for higher data throughputs via QPSK and MSK modulation
(f) Power output: 30 W AM, 100 W FM and data mode (FM and data levels adjustable down to <1 W)
(g) Power supply: 115/230 V AC, 60 Hz, SP; 1.4 kVA at 0.8 pf max
(h) MTBF: >15,000 h
(i) Dimensions: 311 × 483 × 588 mm (H × W (rackmount) × D)
(j) Weight: 67.1 kg

Status

Over 11,000 AN/WSC-3(V) units were calculated to be in service in mid-1994 and production continues. The AN/WSC-3(V) is currently in service with all US armed services and with the armed forces of Australia, Canada, Denmark, Egypt, Germany, Indonesia, Japan, South Korea, Morocco, Netherlands, New Zealand, Norway, Portugal, Saudi Arabia, Spain, Turkey and the United Kingdom. In addition

Capabilities and configurations of AN/WSC-3 variants

	Link 11wb/nb AM/FM	Remote parallel control	Remote serial control	Satcom built-in modem	25 kHz channel spacing	5 kHz channel spacing	Single audio system	ECCM mode
AN/WSC-3(V)2	X	X		X[8]	X		Std[6]	
AN/WSC-3(V)3	X	X		X[8]	X		X[6]	
AN/WSC-3(V)6	X	X[11]			X		Std[6]	X[12]
AN/WSC-3(V)7	X	X[11]			X		X[6]	X[12]
AN/WSC-3(V)9	X	X		X	X	X[9]	X[6]	
AN/WSC-3(V)11	X	X[11]			X		X[6]	X[2]
AN/WSC-3(V)14	X		X[4]		X		X[6]	
AN/WSC-3(V)15	X[5]	X	X	X[8]	X		X[6]	
AN/WSC-3(V)17	X[5-7]	X		X[8]	X		X[6]	
AN/WSC-3(V)18	X[7]	X		X	X		X[6]	
RT1217-1	X		X[3]		X		X	X[1]
RT1217-3	X		X[3]		X		X[6]	X[2]
RT1217-4	X		X[3]		X		X[6]	X[10]
RT1244(V)1	X		X[3]		X		X	X[1]
RT1244(V)2	X		X[3]	X[13]	X	X	X[1]	

[1] Allows frequency-hopping controlled externally
[2] Internal AJ
[3] MIL-STD-188
[4] MIL-STD-1553
[5] DAMA compatible
[6] Switchable interface
[7] Includes MTSC

[8] CKA mod kit available
[9] Patrick modification available
[10] Internal AJ, non HQ
[11] Parallel filter IF kit available
[12] HQII mod kit required
[13] PSK mode only

to shipboard installations, Whiskey-3 is also in service installed in vehicles, aircraft and transportable shelters.

Many different LOS and SATCOM versions are deployed. In addition, numerous conversion kits have been developed for upgrading LOS radios to SATCOM, for providing special filter interfaces, for Have-Quick II upgrade, and for meeting particular installation and configuration requirements.

Manufacturer
Raytheon Systems Company, St Petersburg, Florida.

UPDATED

··

AN/WSC-6 SHF SATCOM set

AN/WSC-6 terminals have been provided for the US Navy by both Raytheon (AN/WSC-6(V)1-7) and Harris (AN/WSC-6(V)8-9).

AN/WSC-6(V)1-7
The AN/WSC-6(V) SHF satellite communications set provides a solution for connectivity via DSCS and NATO satellites using the 7.25 to 8.4 GHz X-band. Over 40 of the AN/WSC-6(V) terminals have been fielded. The AN/WSC-6(V) is available with a 122 or 213 cm reflector. The system is configurable for both single- and dual-antenna installations to avoid superstructure blockage during manoeuvres or heavy sea conditions. The terminal operates in either single or multiple carrier configurations and is compatible with common Demand Assign Multiple Access (DAMA), Anti-Jam (AJ), and Single Channel Per Carrier (SCPC) modems. This allows use with current and planned equipment to provide secure digital voice, digital data and facsimile services on a near worldwide coverage basis. The stabilised antennas provide continuous operation on naval platforms with hemispherical coverage up to Sea State 5 conditions.

The AN/WSC-6(V) features single- or dual-antenna operation and DSCS II/III and NATO interoperability. It operates in single or multiple carrier configuration and has centralised terminal control and monitoring with built-in test and RF loopback. Only two below-deck racks are required.

The system is compatible with DAMA, SCPC and AJ modems. It meets MIL-E-16400 requirements and a ruggedised version is also available. Several versions of the power amplifier provide high throughput and an optional VME antenna controller is available.

Specifications
(a) Frequency range: 7.9-8.4 GHz (transmit); 7.25-7.75 Ghz (receive)
(b) Receive g/t: 12.5 dB/K small antenna, 19.0 dB/K large antenna
(c) Transmit EIRP: 59 dBW small antenna (300 W PA), 63 dBW large antenna (300 W PA)
(d) Above decks
    Polarisation: RHCP (transmit); LHCP (receive)
    Axial ratio: <2.0 dB
    VSWR: <1.3:1
    Power handling: 4 kW max large aperture antenna transmit port, 8 kW for small aperture
    Ship's motion (Sinusoidal amplitude/period): 35°/7 s (roll); 12°/5 s (pitch); 8.5°/6 s (yaw); 7.31 m/4.5 s (heave)
    Pointing accuracy: 0.5°
    Size (incl radome): 157 cm diameter × 216 cm height (small aperture); 305 cm diameter × 330 cm height (large aperture)
    Weight: 263.25 kg (small aperture); 495 kg (large aperture)

(e) Below decks
    RF Performance
    spurious performance <60 dBC
    gain stability <1.5 dB per 24 h
    RF tuning steps 1 KHz increments over entire 500 MHz band
    IF performance: centre frequencies 70 MHz, 700 MHz
    Size (below deck): 2 × 483 mm racks, 183 cm high, 76.2 cm deep
    Weight (below deck): 945 kg approx

AN/WSC-6 (V)9
The AN/WSC-6/6A(V)9 shipboard terminal is a COTS/NDI SATCOM system which provides dual-band operation: X-band over DSCS or allied military satellites and C-band over commercial satellites. The design inherently supports upgrade to INMARSAT, Ku-band and simultaneous X/Ka-band operation.

The terminal supports multichannel full-duplex communications at individual channel data rates up to 2.048 megabits per second. For platforms with demanding antenna siting, the terminal can be configured with dual antennas to eliminate superstructure blockage. The system uses interchangeable intermodulation-free feeds and is capable of an upgrade to incorporate L-band (INMARSAT), Ku-band and simultaneous X/Ka-band operation.

The AN/WSC-6/6A(V)9 consists of above-deck and below-deck equipment interconnected by cross-deck cables. The above-deck equipment consists of a single-or dual-antenna system. The antennas provide an INTELSAT/DSCS compliant beam pattern using a 5 ft (1.52 m) reflector mounted on a high-dynamics three-axis pedestal enclosed within a protective radome. The pedestal provides continuous train axis rotation and incorporates inertial elements for stabilisation. The radome and antenna incorporate radar cross-section reduction features to minimise ship observability and identification. The below-deck communications equipment is housed in a single EMI cabinet which contains the modems, upconverters, downconverters, a 2,000 W high-power amplifier, antenna control unit, frequency standard, and supporting equipment and cables. The modems, Harris-built MD-1030(V)8As, utilise advanced signal processing to mitigate the effects of antenna handover at high data rates aboard dual-antenna ships. All equipment is hardened to the naval environment and all control is provided over a LAN via PC-based Operator Interface Units.

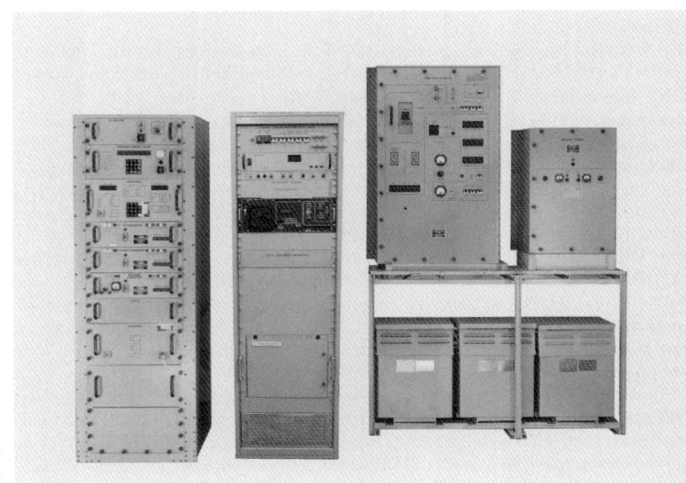

AN/WSC-6 below-deck environment

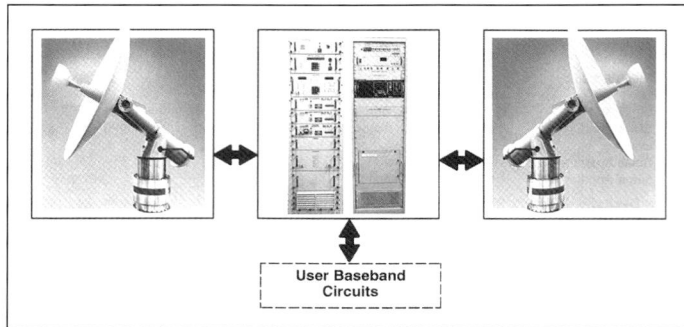

AN/WSC-6 SHF SATCOM set

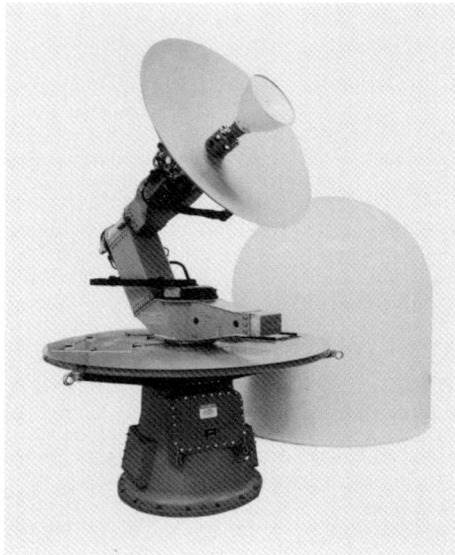

*Optional AN/WSC-6
1.2 m antenna*

Specifications

(a) Satellite operation:
   INTELSAT: Per IESS-601
   DSCS: Per DISA certification
   Up to 12° inclination orbits supported
(b) Transmit frequency:
   C-band 5.850–6.425 GHz
   X-band 7.90–8.40 GHz
(c) Receive frequency:
   C-band 3.70–4.20 GHz
   X-band 7.25–7.75 GHz
(d) EIRP (dBW):
   C-band >58 (max), X-band >63 (linear)
(e) G/T (dBi/K):
   C-band >+12, X-band >+16
(f) Maximum channel data rate: 2.048 Mbytes/s per channel
   Bit count integrity: No loss from antenna handover
(g) Coverage: Full hemispherical
(h) Tracking loss: <0.5 dB RMS
(i) Polarisation: Selectable LHCP/RHCP for C-band operation
(j) Satellite acquisition/reacquisition: <5 min/<90 s
(k) Antenna aperture diameter: 5 ft (1.52 m)
(l) Radome height/diameter: 90.5 in/94.5 in
(m) Out-of-band interference rejection: Up to 100 dB

Status

SHF SATCOM capability is provided to US Navy surface ships and submarines by different WSC-6 variants according to the requirements of those platforms. In early 2000, Surveillance Towed Array Sensor System (SURTASS) platforms were configured with the WSC-6(V)1. The WSC-6(V)2 on numbered Fleet Commander flagships (AGFs/LCCs), and the WSC-6(V)4 on aircraft carriers (CVs/CVNs), flag-capable amphibious ships (LHAs/LHDs), and the Mine Countermeasures Support Ship (USS *Inchon*, MCS-12) were being upgraded to the WSC-6(V)5 in 2002. This upgrade to the below-decks terminal equipment provides a dual-termination capability, enabling the ships to establish and simultaneously maintain their C4I links with Naval Computer and Telecommunications Area Master Stations (NCTAMS) and additional links with an Army, Marine Corps, or Air Force Ground Mobile Force (GMF) SHF terminal ashore in the AOR. The WSC-6(V)7 is a new, single-termination variant beginning in 2002 to be fielded on Aegis cruisers and is also planned for Combat Logistic Force (CLF) ships. All these variants were produced by Raytheon beginning in the late 1990s.

The AN/WSC-6(V)9 is a new, single-termination, dual (C/X) band terminal developed by Harris under a US$111 million contract signed in 2000 to provide wideband, high data rate capability to guided missile destroyers (DDGs) and amphibious ships (LPDs and LSDs). New-construction 'San Antonio' (LPD-17) class amphibious ships are also planned for an SHF SATCOM terminal variant installation.

The Royal Netherlands Navy has procured a number of AN/WSC-6(V) Systems with an extended capability. They feature a 7 ft antenna capable of operating with DSCS, NATO IV and Skynet.

Contractor

Raytheon Systems Company, Marlborough, Massachusetts.
Harris Corporation, Government Communications System Division, Melbourne, Florida.

UPDATED

Chariot S-band Tactical Manportable Terminal

The Chariot S-band Tactical Manportable Terminal (STMT) provides a means to disseminate data, such as secondary imagery products, to field users. Capable of tracking and receiving signals from HEO, GEO and LEO satellites, the terminal processes standard SGLS and DMSP waveforms, converting the received RF energy into user bit streams. An optional transmit capability can provide valuable acknowledgement and mission-related user requests to the transmit source.

The Chariot is a terminal capable of receiving data rates from 1.2 to 1,200 kbit/s. It is lightweight, rapidly deployable and operates from conventional AC power sources. System elements include a prime focus 1.2 m parabolic reflector/feed assembly, RF electronics for receive and transmit, a DSP-based digital transceiver, and a system controller operating as the terminal control processor.

The system components are integrated in a modular manner to support rapid deployment and storage. Each modular section has a soft case for short-distance transport as well as a shipping case for long-distance rugged transport.

The transceiver is a VME-compatible four-card set and is offered in tactical, commercial and rack-mount packaging to accommodate a variety of deployment configurations. The transceiver has an analogue dual conversion tuner followed by digital demodulators. Its function is determined by stored configuration files which define data rate, waveform, modulation, data format/coding and frequency. Mission configurations are stored on disk for rapid set up.

Terminal control is managed via the System Control Processor (SCP), which is either a UNIX Sun OS-based computer or a DOS-based notebook computer. The SCP enables user configuration of the modem with a two-command operation. Configuration of the antenna is reduced to an automatic initialisation, entry of geographic location, file entry of ephemeris from manual or disk and two steps to command tracking. The UNIX SCP also provides image manipulation software.

The antenna servo-control is processor-based, implementing program tracking for satellite tracking.

Optional enhancements include embedded encryption, integrated/embedded GPS receiver, integrated imagery processing software and low-power battery operations.

The unit weighs 110 kg (shipping) and can be set up and operated in 30 minutes.

Status

In service with the US Army. Currently it is deployed supporting TENCAP systems including MITT, FAST and DTES (see separate entries). In January 2003 *Jane's International Defense Review* reported that Chariot was to be replaced in the TENCAP role by a Low-cost S-Band Receiver.

Contractor

Harris Corporation, Government Communications Systems Division, Melbourne, Florida.

UPDATED

STMT

Defense Satellite Communications System (DSCS)

The Defense Satellite Communications System (DSCS) has been designed and configured for several purposes, including presidential communications. It supports the US Global Command and Control System (GCCS), by providing communication services between the National Command Authorities (NCA)/Defense Information Systems Agency (DISA), the unified and specified commands and the general war combat forces and, from peripheral early warning sites and critical intelligence sites. It provides a high-capacity, reliable, independent communications capability in support of contingency and limited war operations and restores primary Defense Communication System (DCS) transmission subsystems that may have become inoperative due to natural causes, sabotage or direct enemy action. It augments the DCS with: a transmission subsystem capable of providing the wideband channels required to handle high-quality secure voice, high-speed data between automated command and control centres; high-resolution graphics and imagery; and rapid transmission of sensor data, together with providing DCS communications service to remote locations not adequately served by other means. The DSCS also provides support for the following: Navy ship-to-shore communications and other authorised users; the voice channel requirements of the Ground-Mobile Forces (GMF); and voice and data requirements of the Diplomatic Telecommunications System (DTS), the UK and NATO.

The Phase I DSCS became operational in July 1967. It comprised 26 operational satellites launched between 1966 and 1968. In 1969 a facility was developed for selected terminals to transmit high-resolution photographic data from Vietnam to Washington.

Phase II of the programme was approved in 1968 with the first satellites launched in the early 1970s.

The Phase II DSCS network comprised a space subsystem, an earth terminal subsystem and a control subsystem. The network accommodated critical requirements which could not be satisfied by other communications systems because of lack of service, geopolitical constraints or the inability of other facilities to provide dependable service under general war conditions. The DSCS II satellites positioned in geostationary orbit were maintained within ±1° of their nominal orbital positions. Each could be repositioned at least once during its operational life to any other equatorial point at the maximum rate of 15° per day. More than one repositioning in the operational lifetime was possible, however, if a less than 15° per day drift were acceptable. The total number of moves depended on the amount of fuel remaining after completing the previous accumulation of moves.

The Phase II satellite transponder (TRW Systems) consisted of a multichannel repeater with cross-linked channels, a receive and transmit Earth Coverage (EC) antenna, a steerable Narrow Coverage (NC) antenna and a steerable Area Coverage (AC) antenna. Each NC and AC antenna was capable of receiving and transmitting simultaneously. This arrangement provided four different channels of operation: Earth Coverage to Earth Coverage (EC-EC), Earth Coverage to Narrow Coverage/Area Coverage (EC-NC/AC), Narrow Coverage/Area Coverage to Earth Coverage (NC/AC-EC), and Narrow Coverage/Area Coverage to Narrow Coverage/Area Coverage (NC/AC-NC/AC).

The DSCS earth terminals are of several types and sizes; the two existing AN/FSC-9 terminals at Lakehurst, New Jersey and Camp Roberts, California, consist of an 18 m diameter paraboloid reflector, matching subreflector and Cassegrain feed system. A superstructure behind the reflector acts as a counterweight and houses the electronic equipment. Antenna weight is approximately 193 tons. Of the 13 AN/MSC-46 earth terminals, 12 are deployed as DSCS terminals and one is used for training. The terminal antenna, normally housed in a radome, consists of a 12 m diameter Cassegrain reflector with a four-horn monopulse feed. The feed and subreflector assembly constitute a single Dielguide (dielectric cone) unit. The remainder of the terminal is housed in vans. Up to 10 kW transmitter power is available.

The AN/TSC-54 is a transportable terminal using a Cassegrain-type antenna composed of four 3 m parabolas in a cloverleaf arrangement. This terminal is air-transportable and requires only a few hours for assembly. Up to 5 kW transmitter power is available. There are nine AN/TSC-54 earth terminals deployed as DSCS terminals, two are Joint Chiefs of Staff (JCS) controlled contingency terminals and one is used for training.

There are three SCT-21s, non-militarised portable earth terminals, deployed in the DSCS and operated and maintained by contractors. The 6 m antenna has automatic tracking and low-noise parametric amplifiers. Up to 5 kW transmitter power is available.

The AN/FSC-78, a fixed terminal, was designed specifically for DSCS Phase II and follow-on compatibility. The antenna subsystem consists of a high-efficiency 18 m solid surface main reflector and a five-horn monopulse feed system supported on a pedestal structure. The receive signal path is divided into two segments: a wideband RF segment provides signal gain over the full 7.25 to 7.75 GHz receive band and the narrowband segment comprises down conversion equipment, which translates 40 MHz bandwidth slots from the downlink band to 125 MHz bandwidth slots. Up to 10 kW transmitter power is available.

The DSCS provides communication services to components of the DoD, the National Security Agency and special authorised users including the DTS, the White House Communications Agency (WHCA), the DCA, NATO and the UK. In addition to serving DoD components, the DSCS directly supports the WWMCCS. This system provides high-priority communications to the JCS, unified and specified commands for the direction and control of forces and for special intelligence and warning.

The DSCS provides analogue and digital transmission paths for virtually every type of telecommunications application. Its configuration can be readily changed to meet contingency requirements. Both strategic and tactical communications needs are met through the global DSCS. Telecommunications services to virtually every geographical area in the world can be established in the time required to deploy a transportable earth terminal. These capabilities make the DSCS an essential subsystem of the DCS for US telecommunications needs.

The DoD military satellite communications (MILSATCOM) systems support three user communities: the wideband, high data rate users, the mobile users and the nuclear capable forces. The DoD MILSATCOM systems include those already in existence as well as future systems for which procurement methods have been decided. The DISA management structure for the DSCS includes the MILSATCOM architect function within DISA and integrates the DSCS into the total MILSATCOM structure. Design and programming of DSCS components is directed by DISA while the responsibility for the acquisition of DSCS earth terminals and control equipment has been assigned to the Army. The Navy acquires shipborne terminals and the Air Force has the responsibility for acquiring the communications satellites and airborne terminals. The DISA through the DISA Operational Control Complex (DOCC) provides direction to the operating ground terminals of the DSCS which are operated and maintained by the military services and owners of non-DoD earth terminals.

All DSCS earth terminals are operated and maintained by the military services controlling the base complex where the earth terminals are located. The Air Force is responsible for procurement, launch and station maintenance of the DSCS satellites.

There are two basic control functions. Satellite communications (SATCOM) control involves the technical monitoring and management of the radio frequency accesses to each DSCS satellite with the objective of achieving the most efficient use of satellite resources. This control function is accomplished by DCA through the DOCC. The network controllers direct SATCOM control functions over terrestrial teletype circuits in each earth terminal. Each earth terminal operator is primarily responsible to the network controllers for maintaining on-frequency carriers and stable uplink power and for monitoring downlink power and the quality of reception. Within the earth terminal complex, two DSCS earth terminals in each satellite area are designated Network Control Terminals (NCT) and each is equipped with RF spectrum analyser equipment to facilitate SATCOM functions. Special DSCS users, the UK, NATO, the DTS and the US GMF, are allocated a certain percentage of satellite power and bandwidth for operation and provide their own NCTs for the control of their respective earth terminals.

Satellite control involves positioning, tracking, monitoring and commanding the satellite during the course of its operational lifetime. The Air Force is assigned on-orbit satellite tracking, telemetry and control service. The Air Force Space Command is the agency through which the Air Force exercises satellite control which involves the manipulative control and monitoring of onboard subsystems or components of a satellite, including those affecting position and attitude as well as the adjustments and switching of subsystems and components. Actual control is accomplished by the Air Force Satellite Control Facility (AFSCF), an element of the Space Command. The AFSCF consists of a worldwide network of Remote Tracking Stations (RTS) used to track, receive telemetry and command DSCS operational satellites. On completion of all testing, operational satellite control over DSCS satellites is exercised by DCA through the AFSCF. The Satellite Test Center (STC), the control point for AFSCF operations is at Falcon AFB, Colorado Springs, Colorado.

The DSCS users have the following priority: Presidential and national command authorities; JCS; unified and specified commands; DCS; other DoD; non-DoD national; NATO and allied governments as specified by international agreements.

DSCS digital communications subsystems are replacing analogue communications subsystems. This transition includes the phase-out of the AN/TSC-54 and some AN/MSC-46 earth terminals with replacement AN/MSC-61 and AN/TSC-86 earth terminals.

The AN/MSC-61 earth terminal (Harris) procured for the Phase II system consists of an 11.5 m antenna subsystem and electronics identical to the AN/FSC-78 electronic equipment. The van-mounted electronic system characteristics are the same as previously described for the AN/FSC-78. The AN/TSC-86 light transportable (LT) terminal (RCA) is a small DSCS earth terminal. The AN/TSC-86 terminal consists of S-280 shelter-mounted electronics, two trailer-mounted 30 kW generators and a 21/2 ton truck for transporting the terminal. The larger, nominally 6 m, antenna system procured for the AN/TSC-86 will be ground-mounted. All electronic equipment units except the antenna-mounted units use drawer and slide construction and are installed in racks mounted in the S-280 shelter. Essentially, all terminal subsystems except the antenna are redundant. However, automatic switch over to redundant units is not provided; this function is performed manually. Space has been provided in the S-280 shelter for the addition of three racks of modulation and multiplex equipments.

Users of the system range from airborne terminals with 838 mm diameter antennas to fixed installations with 18 m diameter antennas and elaborate data processing equipment. Mobile terminals supporting ground and naval operations will communicate with each other and the command chain through the satellite.

On the DSCS III satellites, a six-channel communications transponder with each channel operating through its RF amplifier, serves the users. This allows compatible grouping of users for efficient use of the frequency spectrum and transponder power. Signals are received and transmitted through an interconnected set of multibeam antennas which can spatially distribute receiver pattern gain and the transmission power according to user requirements. Transmitter power can be concentrated on small isolated terminals or distributed optimally over wide areas.

The single channel transponder on DSCS III supplements dedicated AFSATCOM spacecraft for command and control communications from the national command authorities and appropriate commanders to the nuclear capable support forces. The regenerative transponder receives and transmits UHF signals using AFSATCOM I modulation. AFSATCOM II signals are also processed and afford anti-jam protection, receiving and transmitting at either UHF or SHF.

The DSCS III satellites, with latest developments in hardening techniques to ensure survivability, are capable of launch on a Titan IV or Atlas/Centaur. A constellation of five satellites with five reserves is in synchronous orbit over the East and West Atlantic, East and West Pacific, and Indian oceans.

Status

The first DSCS III satellite was launched in October 1982. A contract was awarded that year for initial production of two DSCS III satellites. Among improvements incorporated were anti-jam command capability, improved communications security equipment and adjustable beacon and solid-state amplifiers to replace travelling wave tubes. Procurement of the third and fourth production satellites began in FY83.

TRW received US$5.2 million additional funding from the US Air Force in March 1983 for DSCS II. General Electric was awarded an US$18.2 million contract from the US Air Force in June 1983 to modify the DSCS III for Shuttle capability. The FY84 plan was to continue development of production satellite improvements. In addition to the four DSCS III satellites in the first contract, seven additional satellites were ordered.

In FY85, research, development, test and engineering was completed on two DSCS III preplanned product improvements: solid-state amplifiers and frequency band filters. Two production satellites were launched in late 1985.

In early 1987, Standford Telecommunications Inc won a US$7 million contract to supply operational support services for the system and Magnavox was awarded an increase of US$16.2 million to an earlier US$60 million contract for the AN/USC-28(V) pseudo-noise spread spectrum modulation device which is part of DSCS. The following year, General Electric Astro Space won a US$64 million contract for the DSCS III apogee boost subsystem. In mid-1991, General Electric was awarded a US$8.4 million contract for launch vehicle integration for DSCS III missions 5 to 8. A fourth DSCS III was launched in 1989. In 1992, two DSCS IIIs were launched using the apogee boost subsystem and another two in 1993. In 1994 six satellites remain to be launched. Martin Marietta (now Lockheed Martin) purchased GE Astro Space in April 1993 and is now responsible for DSCS III satellite design, test and fabrication.

The last DSCS II satellite was moved into supersynchronous orbit in September 1998, concluding the 27-year programme. DSCS III now operates with five primary satellites and five reserves. B11 (DSCS III SLEP), launched in October 2000 and now operating in the Eastern Atlantic, is the newest DSCS satellite on orbit. There have been 14 DSCS III satellites built and two remain to be launched.

A DSCS III A3 satellite is due to be launched in February 2003. This will take over from one of the current primary satellites, which will move into reserve.

Contractor

Lockheed Martin Space Systems, Sunnyvale, California.

UPDATED

Fleet Satellite Communications System (FLTSATCOM)

FLTSATCOM is designed to provide multichannel UHF communications for the US Navy. It also supports US Air Force bombers and launch control centres, all airborne command posts and some US Army nuclear capable force elements.

Development

The FLTSATCOM programme was approved in 1971 with the US Navy executive agent for systems development and operation and the US Air Force assigned responsibility for spacecraft development.

The original specification called for four satellites in geostationary orbit to provide global coverage. First launch was planned for 1975 but delays in development postponed it until 1978. Five satellites were launched but the fifth, intended as an in-orbit spare, was damaged during lift-off in August 1981 and is not operational.

In June 1983, the US Air Force placed a US$181.1 million contract with TRW for three new satellites. Two of these carry an EHF communication package serving as a demonstration for initial operational capability for the fast-hopping anti-jam communications package for Milstar. UHF receiving subsystems for the satellites were supplied by E-Systems under a US$30.1 million subcontract to TRW.

Description

FLTSATCOM provides an SHF/UHF anti-jam protected fleet broadcast service to all US Navy ships, as well as providing command and control links for computer-to-computer exchange of digital data among shore stations, fleet ballistic missile submarines, aircraft carriers, cruisers, selected aircraft and other ships and submarines.

Each FLTSATCOM spacecraft has 23 channels in the 244 to 400 MHz range; nine 25 kHz wideband channels (seven low power, two high) for Navy relay communications; 12 5 kHz narrowband channels used as part of the AFSATCOM system; one 500 kHz wideband channel used by the National Command Authorities; and one 25 kHz channel (SHF up, UHF down) for fleet broadcast.

Status

The first EHF-bearing satellite was launched in December 1986. A FLTSATCOM was destroyed on launch in March 1987. The second EHF satellite was launched in September 1989. Now replaced by the UHF Follow-On (UFO) satellite constellation (see separate entry).

Specifications

Configuration: 2 major parts: payload module (incl antennas) and spacecraft module with solar array. A third module was added for the EHF payload on 2 spacecraft.
Launch vehicle: Atlas-Centaur
Operational orbit: geosynchronous (19,000 n mile circular)
Design life: 5 years
Channels: >30 voice and 12 teletype

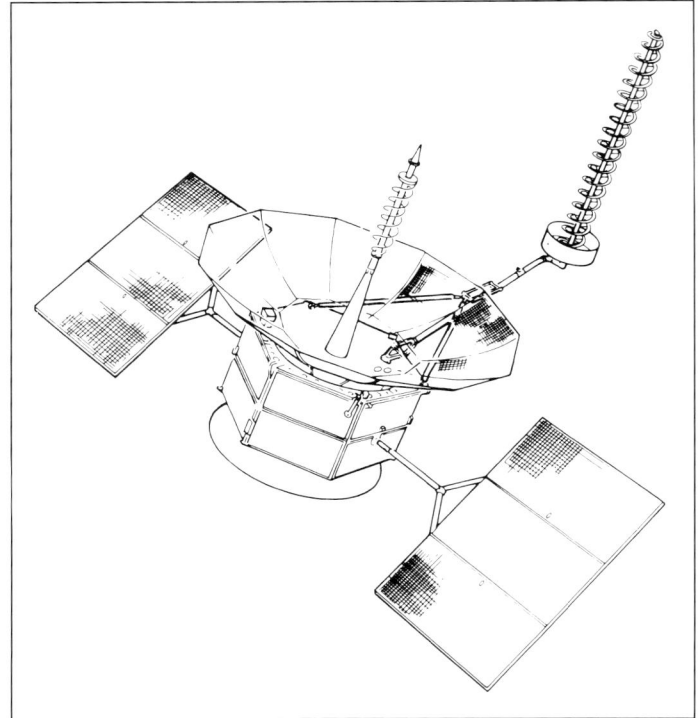

FLTSATCOM satellite principal characteristics

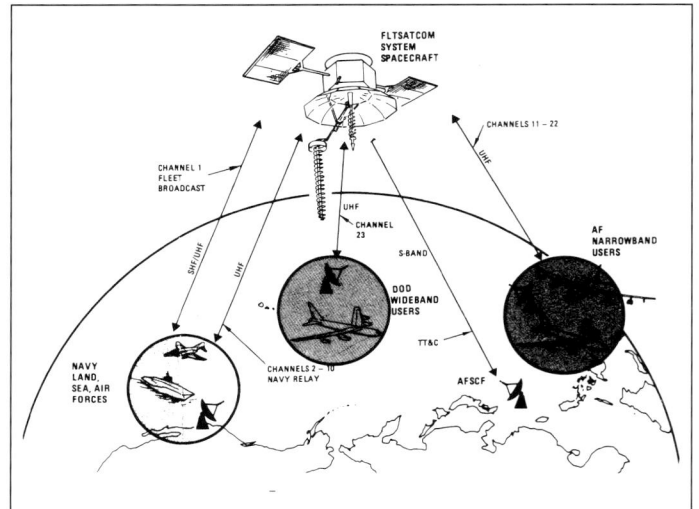

FLTSATCOM communications links

Weight: 1,859 kg nominal at launch, 912 kg in orbit (added 112 kg for EHF payload on 2 spacecraft)
Terminals developed specifically for FLTSATCOM:
AN/FSC-79: SHF uplink broadcast transmitter
AN/SSC-6: SHF terminal
AN/SSR-1: UHF downlink receive-only terminal
AN/SSR-2: SHF terminal
AN/SSR-3: SHF terminal
AN/UCA-2: modem
AN/WSC-1: UHF terminal
AN/WSC-2: SHF terminal
AN/WSC-3: UHF terminal
AN/WSC-5: UHF terminal

Contractor

Northrop Grumman Space Technology, Redondo Beach, California.

UPDATED

Leasat communications satellite system

In September 1978, the US Navy announced a contract award to Hughes Communications Services Inc, a Hughes Aircraft Company subsidiary, to provide a worldwide communications satellite service to the Department of Defense for at least five years at each of four orbital locations. The navy would act as executive agent on behalf of the DoD. The new satellite, known as Leasat, began service in 1984. Users include mobile air, surface, subsurface and fixed earth stations of the US Navy, Marine Corps, Air Force and Army. Hughes Space and Communications Group built the satellites.

Leasat communications satellite

The agreement called for Hughes to design, build, launch and operate a complete communications satellite system. Included were five satellites, one of which was a spare, as well as associated ground facilities, including an operational control centre, a network of four fixed ground stations and two movable stations. The satellites occupy geosynchronous positions over the USA and also the Atlantic, Pacific and Indian oceans.

Leasat is 4.26 m in diameter and 6.17 m high with its UHF and omnidirectional antennas deployed. With its antennas stowed in the launch configuration, Leasat is 4.29 m high. Total payload weight (including launch cradle) in the Shuttle is 7,711 kg. Weight after separation from the Shuttle is 6,895 kg and the satellite's weight on station at the beginning of life is 1,388 kg.

The satellites are spin-stabilised, with the spun portion containing the solar array and sun and earth sensors for attitude determination and earth pointing reference, batteries for eclipse operation and all propulsion and attitude control hardware. The despun platform contains earth-pointing communication antennas, communication repeaters and the majority of the Telemetry, Tracking and Command (TTC) equipment.

Solar panel output is 1,238 W after seven years in orbit. Three Ni/Cd batteries for eclipse operation are designed for a nominal maximum 45 per cent discharge. Redundancy resulting from the three-battery system permits full load support with loss of one of the batteries.

Two large helical UHF antennas provide receive and transmit capability in the UHF band (240 to 400 MHz). Telemetry, command and the Fleet Broadcast uplink and beacon are in the 'exclusive' portions of the SHF band (7,250 to 7,500 MHz and 7,975 to 8,025 MHz). The main communications capability is provided by 12 UHF repeaters.

The principal Navy Fleet Broadcast function includes an SHF uplink, and both SHF and UHF downlinks. The additional antennas for this channel are the SHF uplink and downlink earth coverage horns, which support the uplink and acquisition/timing function, respectively. The UHF downlink for Fleet Broadcast is multiplexed onto the UHF transmit helix.

Status

The Leasat spacecraft were designed exclusively for launch on NASA's Space Shuttle. Leasat F1 was scheduled for launch in June 1984, but the Shuttle mission was aborted only seconds before lift-off. Leasat F2 became the first in the series to be launched, on 30 August 1984. Leasat F1 followed on 8 November 1984. Leasat F3 was launched on the Shuttle on 12 April 1985, but did not achieve orbit when the satellite failed to start. Four months later, NASA and Hughes mounted a salvage attempt during the 27 August 1985 Shuttle mission on which Leasat F4 was launched. After attaching special electronics assemblies to Leasat 3 during two days of space walks, astronauts manually launched the satellite again. The electronics allowed ground controllers to turn on the satellite and, at the end of October, fire its perigee rocket and send Leasat 3 into orbit. Leasat 4 successfully obtained orbit and was undergoing tests about a week after launch when its UHF downlink failed. The satellite was declared a loss. The fifth and last Leasat, which was built as a spare, was successfully launched in January 1990.

In 1997 Leasat 5 ceased operations for the US Navy, having been replaced by the UHF Follow On (FO) programme (see separate entry), and under an agreement with the then Hughes Global Services (now Boeing Satellite Systems) and PanAmSat Corporation (the satellite owners) was leased to the Australian Defence Force for 5 years for use by the Royal Australian Navy. It began limited operation in support of the Australians in October 1997, and in March 1998 was repositioned to a new orbit at 156°E, providing full support from 7 May.

Contractor

Boeing Satellite Systems, El Segundo, California.

Milstar satellite communications system

Milstar is a joint service satellite communications system that provides worldwide secure, jam resistant and low probability of detection nuclear-event resistant communications for all forces across the spectrum of conflict. The multi-satellite constellation will link command authorities with a wide variety of resources, including ships, submarines, aircraft, land vehicles and manportable systems. The objective of the Milstar program was to create a secure, nuclear survivable, space-based communication system and was considered as a top national priority during the Reagan Administration in the 1980's. Milstar is designed to perform all communication processing and network routing onboard, essentially functioning as a 'switchboard in the sky'. This eliminates the use of vulnerable land-based relay stations and reduces the chances of communications being intercepted on the ground.

The operational Milstar satellite constellation will consist of four satellites positioned around the Earth in geosynchronous orbits, each weighing approximately 10,000 lb (4,536 kg) and having a design life of 10 years. The Milstar satellite serves as a smart switchboard in space by directing traffic from terminal to terminal anywhere on the Earth. Since the satellite actually processes the communications signal and can link with other Milstar satellites through crosslinks, the requirement for ground controlled switching is significantly reduced. The satellite establishes, maintains, reconfigures and disassembles required communications circuits as directed by the users. Milstar terminals provide encrypted voice, data, teletype or facsimile communications. A key goal of Milstar is to provide interoperable communications among the users of Army, Navy, and Air Force Milstar terminals.

The Milstar system is composed of three segments: space (the satellites), terminal (the users), and mission control. The US Air Force Space Command's Space and Missile Systems Center (SMC) at Los Angeles AFB, California, is responsible for development and acquisition of the Milstar space and mission control segments. The Electronics Systems Center (ESC) at Hanscom AFB, Massachusetts, is responsible for the Air Force portion of the terminal segment development and acquisition. The 4th Space Operations Squadron at Schriever AFB, Colorado, is the front line organisation providing real-time satellite platform control and communications payload management. Geographically dispersed mobile and fixed control stations provide survivable and enduring operational command and control for the Milstar constellation.

Specifications
(a) Weight: About 10,000 lb
(b) Orbit altitude: 22,250 n miles (inclined geostationary orbit)
(c) Power plant: Solar panels generating 8,000 W
(d) Payload:
Low data rate communications (voice, data, teletype and facsimile) at 75 to 2,400 bps (192 channels) (All satellites)
Medium data rate communications (voice, data, teletype, facsimile) at 4.8 Kbps to 1.544 Mbps (32 channels) (Satellites 3 to 5 only)
(e) Antennas:
LDR: Earth Coverage (uplink and downlink) Agile Beams (5 uplink, 1 downlink) 3 Spot Beams (2 Narrow, 1 Wide) UHF (uplink and downlink)
MDR: 6 distributed users coverage (DUCAs), 2 nulling

Status

The first Milstar satellite was launched 7 February, 1994 aboard a Titan IV expendable launch vehicle and was positioned to provide coverage for Pacific Ocean forces. The second was launched 5 November, 1995 and provides constellation coverage of the Atlantic area. Together, these first two Milstar satellites will provide crosslinked coverage that extends from most of the Middle East, Africa, Europe, Mediterranean and the continental US to beyond the Hawaiian Islands.

The third Milstar satellite (flight-3M) was to be the first of the Milstar II satellites and incorporated a Medium Data Rate (MDR) payload but a failure in the launch

VERIFIED | *Artist's impression of Milstar satellites in orbit*   0001016

programme resulted in its complete loss in May 1999. Milstar 4, with the Medium Data Rate (MDR) payload, was launched successfully on 27 February 2001.

Milstar 5 was successfully launched in January 2002, completing the constellation and providing global coverage. Milstar 6 was scheduled to be launched in January 2003, and will be kept in reserve.

Contractor
Lockheed Martin Missiles & Space, Sunnyvale, California.

UPDATED

Modular Interoperable Surface Terminal (MIST)

The Modular Interoperable Surface Terminal (MIST) is designed for use as a ground datalink terminal system. This modular tactical mobile datalink system consists of the remote equipment group and the surface processing facility. The remote equipment is composed of the antenna, the antenna RF assembly, and an enclosure which contains the tracker controller, motor control unit, enclosure communications and fibre optic subsystem for antenna remoting. The Surface Processing Facility comprises three major equipment groups: Operator Equipment, Link Equipment, and Ancillary Equipment. The system is partitioned to the lowest testable assembly (LTA) around features which can be added or removed to provide an optimum equipment configuration for each application. Some of the mission control features and capabilities are: ranging, situation display, simplex or duplex operation, digital and analogue intercoms, antenna remoting, multiple links, multiple antennas with switching provisions, mission recording, time of day subsystem, GPS and WWV receivers, and UHF radio. SATCOM capability may also be added. The MIST datalink terminal can be used in a tent, building, or integrated in a small vehicle such as an HMMWV. Antennas may be tower mounted, trailer mounted or mounted on a tactical vehicle or a small tripod and the antenna equipment may be remoted up to 7 km from the surface processing facility. Each datalink may consist of up to 4 antennas, which are electronically switchable for redundancy, availability and survivability.

Specifications
(a) Frequency: variable, L through EHF (standard: Dual band; X and Ku)
(b) Tunable in 5 MHz and 100 KHz steps
(c) Integrated 2-way secure voice channel
(d) Interface to standard military encryption devices
(e) Direct sequence spread spectrum (forward link)
(f) Error correction coding and interleaving
(g) Antenna remoting to 7 km
(h) Built in test
(i) Output Power: variable (standard: 50 W or 10 W)
(j) Antenna Gain: variable (standard: 6 ft dish; 43 dB at X-band, 44 dB at Ku-band)
(k) Modulation: Command link: BPSK; Return link: O-QPSK
(l) Forward Link Data Rates: variable, 600 b/s - 200 Kb/s (standard: 200 Kb/s)
(m) Return Link Data Rates: variable, 16 Kb/s - 274 Mb/s (standard: 10.71/137/274 Mb/s)

Status
In use with the US Army. Part of the TES Forward (Full IMINT) configuration (see separate entry). No longer in production.

Contractor
L-3 Communications-West, Salt Lake City, Utah.

UPDATED

Multimission Advanced Tactical Terminal (MATT)

The Multimission Advanced Tactical Terminal (MATT) was developed by the Naval Research Laboratory for joint service use in airborne and ground platforms. MATT is a miniaturised UHF receiver providing near-realtime over-the-horizon threat data for situational awareness and assessment, threat avoidance, targeting, mission planning and communications.

MATT includes modules for embedded decryption, message processing, tactical data processing and multiple external interfaces. A correlation module provides the ability to associate and track known and unknown moving targets. MATT can simultaneously receive and process intelligence reports for the Tactical Receive Applications (TRAP), Tactical Data Exchange System Broadcast (TADIXS-B) and the Tactical Information Broadcast Service (TIBS).

MATT consists of user-configurable SEM-E modules mounted in a standard ¾ ATR chassis. The modules are connected through a common backplane with isolated red and black signal paths. Some special modules, such as the Antenna Splitter Module, are housed in a different mechanical package format to facilitate unique interface requirements. MATT's modular and highly integrated design approach is oriented towards satisfying end-user requirements for tactical link processing functions on platforms where size and space are prime considerations.

Features:
(a) four simultaneous receive channels (TRAP, TADIXS-B, TIBS)
(b) two antenna inputs switchable to any channel

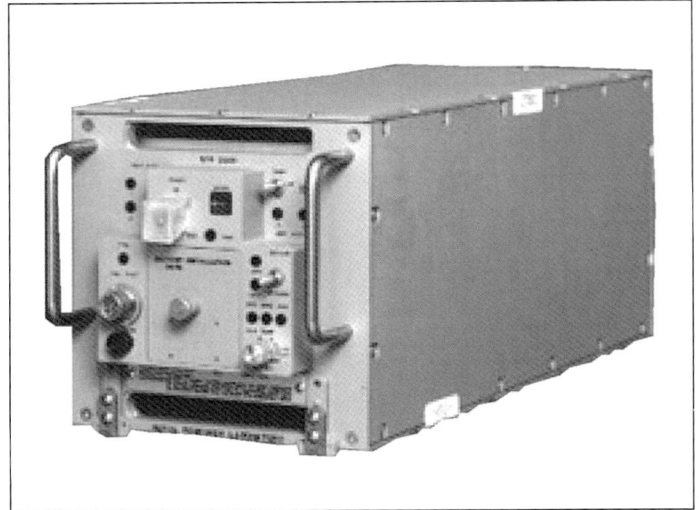

Multimission Advanced Tactical Terminal   0055034

(c) MIL-STD-1553B, RS-232, and RS-422 user inter-faces
(d) SEM-E modules in ¾ ATR chassis
(e) 53 lb/283 W (full capability)
(f) MIL-E-5400; –54 to +71 °C; 0 to 70,000 ft

Planned enhancements include: Transmit VHF/UHF voice, secure voice, and data and six-channel simultaneous receive (TRAP, TADIXS-B, TIBS, TRIXS, OTCIXS, NB-SV, WBSV, DAMA, IMAGE, HSFB).

Status
The MATT has been procured for all three US armed forces and is currently in full-scale production. It is due to be replaced by the JTT (see separate entry).

Contractor
Assurance Technology Corporation, Carlisle, Massachusetts.

UPDATED

NAVSTAR global positioning system

The NAVSTAR Global Positioning System is managed by the NAVSTAR GPS Joint Program Office at the Space and Missile Systems Center, Los Angeles Air Force Base, California.

The GPS consists of three major segments: Space, Control and User. The Space segment consists of 24 operational satellites in six orbital planes (four satellites in each plane). The satellites operate in circular 20,200 km (10,900 n mile) orbits at an inclination angle of 55° and with a 12 h period. The position is therefore the same at the same sidereal time each day, that is, the satellites appear 4 minutes earlier each day. The Control segment consists of five monitor stations (Hawaii, Kwajalein, Ascension Island, Diego Garcia, Colorado Springs), three ground antennas (Ascension Island, Diego Garcia, Kwajalein), and a Master Control Station (MCS) located at Schriever Air Force Base in Colorado. The monitor stations passively track all satellites in view, accumulating ranging data. This information is processed at the MCS to determine satellite orbits and to update each satellite's navigation message. Updated information is transmitted to each satellite via the ground antennas. The User segment consists of antennas and receiver-processors that provide positioning, velocity and precise timing to the user.

GPS provides the following:
(a) 24-hour, worldwide service
(b) Extremely accurate, three-dimensional location information (providing latitude, longitude and altitude readings)
(c) Extremely accurate velocity information
(d) Precise timing services
(e) A worldwide common grid that is easily converted to any local grid
(f) Continuous real-time information
(g) Accessibility to an unlimited number of worldwide users
(h) Civilian user support at a slightly less accurate level

The satellites transmit on two L-band frequencies: L1 equals 1,575.42 MHz, and L2 equals 1,227.6 MHz. Three Pseudo-Random Noise (PRN) ranging codes are in use.
(a) The coarse/acquisition code (C/A-code) has a 1.023 MHz chip rate and a period of 1 millisecond (ms); it is used primarily to acquire the P-code.
(b) The precision-code (P-code) has a 10.23 MHz rate and a period of 7 days; it is the principal navigation ranging code.
(c) The Y-code is used instead of the P-code whenever the Anti-Spoofing (A-S) mode of operation is activated.

The C/A code is available on the L1 frequency, and the P-code is available on both L1 and L2. The various satellites all transmit on the same frequencies, L1 and L2, but with individual code assignments. Owing to the spread spectrum characteristic of the signals, the system provides a large margin of resistance to interference. Each satellite transmits a navigation message containing its orbital elements, clock behaviour, system time and status messages. In addition, an almanac is provided which gives the approximate data for each active satellite. This allows the user set to find all satellites once the first has been acquired.

GPS provides two levels of service, Standard Positioning Service and Precise Positioning Service. Standard Positioning Service (SPS) is a positioning and timing service which is available to all GPS users on a continuous, worldwide basis, with no direct charge. SPS is provided on the GPS L1 frequency and provides a predictable positioning accuracy of 100 m (95 per cent) horizontally and 156 m (95 per cent) vertically, as well as time transfer accuracy to UTC within 340 ns (95 per cent). Precise Positioning Service (PPS) is a highly accurate military positioning, velocity and timing service which is available on a continuous, worldwide basis to users authorised by the US. P(Y)-code-capable military-user equipment provides a predictable positioning accuracy of at least 22 m (95 per cent) horizontally and 27.7 m vertically, as well as time transfer accuracy to UTC within 200 ns (95 per cent). PPS is the data transmitted on the GPS L1 and L2 frequencies. PPS was designed primarily for US military use and is denied to unauthorised users by the use of cryptography.

There are four generations of the GPS satellite: the Block I, the Block II/IIA, the Block IIR and the Block IIF. Block I satellites were used to test the principles of the Global Positioning System, and lessons learned from these 11 satellites were incorporated into later blocks. Block II and IIA satellites make up the current constellation. A total of 28 were put on contract, with the last four tagged as replacements for earlier satellites reaching the end of their service life.

Block IIR satellites boast dramatic improvements over the previous blocks of satellites. They have the ability to determine their own position by performing inter-satellite ranging with other IIR vehicles, reprogrammable satellite processors, enabling problem fixes and upgrades in flight, and increased satellite autonomy and radiation hardness.

Additionally, the Block IIR has the ability to be launched into any of the required GPS orbits at any time with a 60-day advanced notice and requires many fewer ground contacts to maintain the constellation. All these improvements result in increased accuracy for GPS users at a cost of 33 per cent less per satellite than the previous generation of Block IIA satellites. Block IIR satellites will replace Block II/IIA satellites as they reach the end of their service life.

As at June 2003 the GPS constellation consisted of 1 Block II and 17 Block IIA satellites built by Boeing, and 9 Block IIR satellites built by Lockheed Martin. Block IIRs began replacing older Block II/IIAs on 22 July 1997 and twelve remain to be launched. Eight are being modified to radiate the new military (M-Code) signal on both the L1 and L2 channels as well as the more robust civil signal (L2C) on the L2 channel. The M-Code signal is a more robust and capable signal architecture. The first modified Block IIR (designated as the IIR-M) is planned for launch in 2004.

Block IIF satellites are the next generation of GPS Space Vehicles. Block IIF provides all the capabilities of the previous blocks with some additional benefits as well. Improvements include an extended design life of 12 years, faster processors with more memory and a new civil signal on a third frequency. The first Block IIF satellite is scheduled to launch in 2006.

Development History
BLOCK I (Rockwell)
    1974 Contract for eight Block I satellites
    1978 Contract for three Block I satellites
BLOCK II/IIA (Rockwell)
    1981 Contract for qualification satellite (GPS12)
    1983 Contract for 28 Block II/IIA satellites
BLOCK IIR (Lockheed Martin)
    1989 Contract for 21 Block IIR satellites
BLOCK IIF (Boeing-North American)
    1996 Contract and options for 30 Block IIF satellites

Specifications
(a) Block IIA Satellite Characteristics
    Weight (in orbit): 2,175 lb
    Orbit altitude: 10,988 n miles
    Power source: Solar panels generating 700 W
    Launch vehicle: Delta II
    Dimensions: 5 ft wide, 17.5 ft long (including wing span)
    Design life: 7.5 years
(b) Block IIR Satellite Characteristics
    Weight (in orbit): 2,370 lb
    Orbit altitude: 10,988 n miles
    Power source: Solar panels generating 1,136 W
    Launch vehicle: Delta II
    Dimensions: 5 ft wide, 6.33 ft in diameter, 6.25 ft high (38.025 ft wide including wing span)
    Design life: 10 years
(c) Block IIF Satellite Characteristics
    Weight (in orbit): 3,758 lb (accounts for 196 lb for payload adapter and 885 lb for RAP and Flex)
    Orbit altitude: 10,988 n miles
    Power source: Solar panels generating up to 2,900 W (BOL) and 2,440 W (EOL)
    Launch vehicle: EELV (Delta IV and Atlas V)
    Dimensions: 8 ft × 6.47 ft (stowed); 70.42 ft × 12 ft (deployed 4-panel solar arrays)
    Design life: 15 years

Status
The GPS/NAVSTAR system is fully operational.

UPDATED

Predator UAV satellite data terminal

The Predator Satellite Data Terminal (SDT) nominally consists of a commercial SATCOM antenna system and a terminal. The antenna system contains all the components necessary to receive, amplify, and convert the signal to an Intermediate Frequency (IF) feed to the terminal. The terminal, housed in a Personal Computer, contains the demodulator, demultiplexer, and decompression capability needed to output data. Data is available on the PC screen as either a video image or a Synthetic Aperture Radar (SAR) Waterfall display, depending on the Predator's sensor output. The terminal automatically detects the imagery content and compression mode and adjusts its output to ensure an accurate display. Electrical signal data is also available out of the PC terminal in the form of video (RS-170, for input to an external monitor or an annotation station), parallel 8-bit SAR data (for input to a SAR processor), and RS-422 serial telemetry data (for processing by an external Payload/Pilot Operating station).

Key features:
COTS Personal Computer Host
Compatible with 2.4 or larger satellite antennas
Video or SAR Waterfall Windowed Display on PC
Electrical outputs for Video, SAR, and Telemetry

Specifications
Physical characteristics:
Antenna system: Commercial 2.4 m dish standard, options for larger apertures available, or customer may provide a 70 MHz IF feed from their own antenna

Performance characteristics:
Input power: 115 V AC, 60 Hz
Terminal compatibility: Predator T1 return link data transmission format; adaptable for 2T, and ATC radio configurations
Video output: RS-170
Telemetry: RS-422 Serial (38.4 Ku-band)
SAR: 8-bit parallel interface to SAR processor (RS-422)

Special features:
SATCOM Remote Video Terminal (RVT)
Data availability anywhere in satellite beam
Personal computer architecture - office and lab environments
Industrial grade or military PCs - field applications
Real-time video
SAR output
External telemetry port allows correlation of video/SAR with position, heading, altitude

Status
Available. There is at present no confirmation whether the equipment is in operational use.

Contractor
L-3 Communications-West, Salt Lake City.

VERIFIED

Secure Mobile Anti-jam Reliable Tactical Terminal (SMART-T) (AN/TSC-154)

The Secure, Mobile, Anti-jam, Reliable, Tactical Terminal (SMART-T), a Military Strategic and Tactical Relay (MILSTAR) HMMWV (High Mobility Multipurpose Wheeled Vehicle) mounted EHF satellite communications transmit and receive terminal, is a core element of the US Joint Service ground terminal segment of the MILSTAR satellite system. The primary SMART-T mission is multichannel, near global extended range connectivity for the US Army's Mobile Subscriber Equipment, which is the primary tactical communications equipment for corps and division operations, to units beyond line of sight, so allowing communications support of widely dispersed forces. The SMART-T supports the transmission requirements of an MSE node consisting of four trunk groups each containing 16, 32 or 64 channels. An internal GPS receiver provides the accurate positioning and timing required for satellite acquisition. Operating at both the MILSTAR low (75-2,400 bits/s) and medium (up to 1.544 Mbits/s) data rates, it is designed to provide tactical commanders with secure, jam-resistant, extended-range, two-way, point-to-point and network voice, data, and video communications. In addition to overcoming the limitations of terrain masking and distance, SMART-T is designed to operate and survive in severe electronic warfare and nuclear, biological, and chemical environments. The terminal is also capable of stand-alone operation removed from the HMMWV set up or, the manufacturer claims, a single soldier can accomplish tear down in 20 minutes. A self-contained diesel generator supplies the prime power. The terminal can be operated unmanned after set up.

Features:
(a) operates over Milstar I, Milstar II, FEP, and UFO Communications satellites
(b) integrated GPS
(c) self-erecting/stowing antenna
(d) data rates to TI via Milstar satellite
(e) C-130 or helicopter sling transportable

The AN/TSC-154 Secure Mobile Anti-jam Reliable Tactical Terminal (SMART-T)
(Raytheon)   **NEW**/0089128

(f) simultaneous LDR/MDR communications
(g) multiservice interoperability
(h) supports non-linear battlefield
(i) DAMA option for efficient satellite resource utilisation
(j) solid-state GaAs PHEMT EHF amplifier
(k) compact, offset-fed Gregorian antenna with rugged 4.5 ft wide composite main reflector
(l) all steel pallet mounts on/off HMMWV for easy drive in/out or tow capability
(m) militarised hand-held (lap-top) computer for easy operation both local and remote (provided by Miltope Group)
(n) operates with 1.5 kW onboard diesel generator
(o) all terminal electronics mounted in single compact unit.

Specifications
(a) Uplink frequency: 44 GHz
(b) Uplink bandwidth: 2 GHz
(c) Downlink frequency: 20 GHz
(d) Downlink bandwidth: 1 GHz
(e) Set up time: 20 min
(f) Max wind speed: 60 mph
(g) Data rates
     (low data rate) 75-2,400 bit/s
     (medium data rate) 4.8-1,544 kbit/s
(h) Waveforms: Milstar compatible (LDR and MDR)
(i) Reliability: 800 h 80% LCL

Status
The SMART-T is in full-rate production with a total of 313 systems being procured at an average unit cost of US$2.4 million. Of these, 209 are for the US Army, 25 for the USMC, 73 for the USAF and 6 for US Defense Agencies. The SMART-T will replace the AN/TSC-85 and AN/TSC-93 (see separate entries) at Corps level and below, with the displaced terminals moving to support Echelons Above Corps (EAC).

In 2000, it was reported that during Initial Operational Test and Evaluation (IOTE), the equipment failed to achieve required levels of reliability, but in 2001 it was reported that Full OTE had demonstrated these shortfalls had been overcome. In 2001 Raytheon received a three-year, $49 million contract to develop, test and validate an advanced extremely high frequency (AEHF) retrofit kit.

Contractor
Raytheon Company, Marlborough, Massachusetts.

UPDATED

SHF Portable Terminal System (PTS)

The PTS can transmit or receive data and voice messages using existing SHF satellites like DSCS, NATO, SKYNET and HISPASAT. The system includes lightweight man-portable terminals (PT) and a system network control applique (NCA). The NCA is installed at the satellite ground station and uses Demand Assignment Multiple Access (DAMA) protocol to minimise the waiting time for all users. The NCA provides channel assignments and serves as a relay station to transfer voice calls and data messages between remote PT users, using only 500 Khz of the satellite's operating bandwidth. This supports 19.2 Kbps data packet service with up to six simultaneous half-duplex voice circuits.

The PT is a completely self-contained communications unit providing radio, modem, digital voice encoding and user control functions. PT equipment is

SHF Portable Terminal System (PTS)   0010467

packaged in a single environmentally sealed enclosure weighing less than 33 pounds and can also be used with an external encryption device to provide secure communications. The terminal can be set up in a claimed four minutes. Optional capabilities include 64 Kbps data rates, public telephone network interface, C and Ku frequencies, PC file transfer interface and alternate packaging.

Status
Fourteen terminals were delivered to NATO in August 1996 for use in Bosnia.

Contractor
L-3 Communications-East , Camden, New Jersey.

UPDATED

SHF triband advanced range extension terminal (STAR-T)

The SHF triband tactical terminal (STAR-T) is a multiband earth station capable of providing quick reaction communication via satellite. Data rates of 9.6 kbit/s to over 8 Mbit/s are supported. The system is entirely self-contained with integrated enclosures. The basic pallet can be mounted directly to a HMMWV or stand-alone trailer. An optional system is available in a transit case configuration.

STAR-T is capable of up to four links per terminal at group rates up to T1/E1 per link. There are two versions, a switch version and a standard version. The switch version has an embedded commercial circuit switch, compression, router and asynchronous-transfer-mode capability.

The two versions of the switch will support either 140 subscribers (LS version) or 280 subscribers (HS version). The LS version will consist of two heavy HMMWVs. One carries the system and the second carries the power generation and subscriber terminal equipment. The HS version has an additional trailer to carry the equipment required to terminate the additional subscribers.

The standard version consists of two HMMWVs, one to carry the system and the second the power generation and support equipment.

System features
(a) C, X and Ku-band operation
(b) up to 8 Mbit/s data capacity
(c) certified for use on US DSCS satellites
(d) redundant 750 W TWTs
(e) easily configurable, two uplink and height downlink carriers
(f) high efficiency two-axis offset Gregorian antenna
(g) low passive intermod antenna for multiple RF carriers
(h) commercial ISDN switch (option)
(i) commercial ATM switch and ATM router (option)
(j) able to operate with QRSA or LWHG external 20 ft antenna
(k) self-erecting 2.4 m antenna
(l) motorised automatic antenna positioning
(m) programmable and/or beacon tracking
(n) operational in sustained winds to 45 mph, gusting to 60 mph
(o) MTBCF > 6,400 h
(p) payload weight < 4,400 lb
(q) removable pallet with fully integrated antenna

(r) helicopter, rail or C-130/C-141 transportable
(s) set up time less than 17 minutes.

Specifications

(a) Uplink/transmit subsystem
   C-band: 5.850-6.425 GHz
   X-band: 7.9-8.4 GHz
   Ku-band: 14.0-14.5 GHz
(b) Downlink/receive subsystem
   C-band: 3.625-4.2 GHz
   X-band: 7.25-7.75 GHz
   Ku-band: 10.95-12.75 GHz
(c) EIRP
   C-band: 59 dBW
   X-band: 67 dBW
   Ku-band: 66.5 dBW
(d) G/T
   C-band: > 15.5 dBk
   X-band: > 21.2 dBk
   Ku-band: > 25.0 dBk
(e) Modulator/demodulator
   IF operating frequency: 70 ± 20 MHz
   Tuning step size: 2.5 KHz
   Modulation: QPSK or BPSK
   Data rate: variable 9.6 kbit/s - 8 Mbit/s
   Forward error correction: convolutional encoding with soft decision decoding
   Data scrambling: CCITT v3.5

Status

STAR-T commenced operational test and evaluation with the US Army in March 1999. However, the programme was cancelled in 2001, although the equipment remains in Raytheon's product list.

Contractor

Raytheon Systems Company, Marlborough, Massachusetts.

UPDATED

Tri-band transportable earth terminals

The 6.2 m tri-band transportable earth terminal is available in two models: a Tri-band Field Terminal-Consolidated (TFT-C) and a Tri-band SATCOM Subsystem (TSS). Each model can be packed and transported on a single, towable trailer, set up in the field and operated locally or remotely with user-friendly Control, Monitor and Alarm (CMA) software loaded on a laptop PC. Both models incorporate L-3 Communications' patented single-point tri-band feed which allows quick configuration between C-band, X-band and Ku-bands. The primary differences between the two models are trailer length, towing speed and set up time.
   The terminals have the following features:
(a) Transportable on one C-130, C-141, C-17 or C-5 aircraft
(b) INTELSAT, DISA and PANAMSAT type certified
(c) Several satellite tracking modes - Step Track, Program Track, Memory Track and Manual
(d) Built-In Test capabilities

(e) Patented, single point tri-band (C-, X- and Ku-bands) feed
(f) Simultaneous duplex operation (C-band and Ku-band)
(g) Intuitive GUI control, monitor and alarm software
(h) Remote or local terminal control and monitoring capabilities

Status

In production. The TSS will provide communications support for US Army TENCAP facilities, particularly the TES Main (see separate entry).

Contractor

L-3 Communications-West, Salt Lake City.

VERIFIED

Tri-band transportable medium earth terminal

The Tri-band transportable medium earth terminal (37 ft diameter dish) operates at C-band, X-band and Ku-bands, and is transportable on two C-130 aircraft. Its performance rivals that of a heavy (60 ft diameter dish) earth terminal and the manufacturer claims it to be the only one of its kind. This terminal is fully DISA certified.

Key features:
Transportable on two C-130s
Set up time approximately 12 hours
Easily reconfigurable for Tri-band operations
Multiple simultaneous carriers
Fibre optics remote operations 10 km from antenna

Specifications

System parameter	C-band	X-band	Ku-band
System operating frequency range (MHz)	Rx: 3625-4200 Tx: 5850-6425	Rx: 7250-7750 Tx: 7900-8400	Rx: 10950-12750 Tx: 14000-14500
G/T (System) 11.3 Meter Terminal	>31.7 dBk	>37.5 dBk	>37.0 dBk
EIRP (System) 11.3 Meter Terminal	>84 dBW	>88 dBW	>90 dBW
TX & RX Polarisations	Dual CP orDual Linear	CP	Linear
Tuning step size	1 kHz steps		
Tracking modes	Manual, Autotrack, and Program Tracking modes		
Earth station control	Fully remote monitoring and control capability		

Status

In production.

Contractor

L-3 Communications-West, Salt Lake City.

VERIFIED

Specifications: Tri-band transportable earth terminals

Specification	C-band	X-band	Ku-band
Frequency range (GHz)	RX: 3.625-4.200 TX: 5.850-6.425	RX: 7.250-7.750 TX: 7.900-8.400	RX: 10.950-12.750 TX: 14.000-14.500
Tuning: Step Size	1.0 kHz Steps $<2 \times 10^{-8}$		
Accuracy Stability	6×10^{-11} /month		
Receive Figure of merit (G/T @ 7.5° El	>25 at 4.0 GHz	>29 at 7.25 GHz	>31 at 11.0 GHz
Minimum antenna gain - receive	3.625 GHz - 45.30 dBi	7.25 GHz - 51.31 dBi	10.95 GHz - 54.89 dBi
LNA Noise	70.0 dBW	72.6 dBW	77.6 dBW
Temperature saturated EIRP (400 W TWTA)	35K	50K	75K
Minimum antenna gain - Transmit	5.85 GHz - 49.44 dBi	7.9 GHz - 52.05 dBi	14.0 GHz - 57.02 dBi
Transmitter RF power	>300 W	>380 W	>300 W
Polarisation	4 Port CP or Linear	RX: LHCP TX: RHCP	4 Port Linear
3 dB Bw (midband)	RX: .96° TX: .60°	RX: .48° TX: .44°	RX: .32° TX: .25°
Radiation pattern envelope	Meets IESS-207-F1	Meets Mil-Std-188-164	Meets IESS-208-E2
Power handling	2 kW	2 kW	1 kW
Axial Ratio / Crosspol	CP <.75 dB, Linear >30 dB Meets IESS-207-F1	<2.0 dB max. Meets Mil-Std-188-164	Linear crosspol >30 dB min Meets IESS-208-E2
Approved certification	Intelsat F1	DISA	Intelsat E2
Environmental	Best commercial practices		
Exercised set up time	6.2 m standard version - 2.5 h 6.2 m compact version - 4 h		
Weight	Less than 14,500 lbs		

Tri-Tac joint tactical communications system

In 1971, the Department of Defense established the Joint Tactical Communication (Tri-Tac) to design, develop and acquire tactical switched communications equipment for support of all US services. The Tri-Service Tactical (Tri-Tac) signal system is a tactical command, control and communications programme. It was a joint service effort to develop and field advanced tactical and multichannel, switched communications equipment. The programme was conceived to achieve interoperability between service, tactical communications systems, establish interoperability with strategic communications systems, take advantage of advances in technology and eliminate duplication in service acquisitions. Each component of the programme was assigned to one of the services to develop and acquire for the entire US defence community. In the case of the US Army, to accommodate the need for a tactical communications system at Echelon Above Corps (EAC), it was decided that heavy Tri-Tac components would operate at those command levels. This EAC structure is the essential link that facilitates and ensures timely, secure, two-way communications between strategic command authorities and their tactical fighting forces, as the counterpart to MSE (see separate entry) which fulfills this function at Corps and below. Tri-Tac is hybrid in design to accommodate the gradual transition to the objective of a fully digital and automatic system. It can be divided into five main areas: terminals, switching, control, transmission and combining. Tri-Tac and MSE together form the Area Common User System (ACUS), which will ultimately be replaced by the Warfighter Information Network - Terrestrial (WIN-T), still in development.

Terminals

There are six Tri-Tac terminals: the Digital Subscriber Voice Terminal (DSVT), the Digital Non-secure Voice Terminal (DNVT), the Advanced Narrowband Digital Voice Terminal (ANDVT), the Single Subscriber Terminal (SST), the Tactical Digital Facsimile (TDF) and the Lightweight Digital Facsimile (LDF).

The DSVT is a telephone instrument that digitises voice at either 16 or 32 kbit/s. The latter rate is used where analogue-to-digital conversion is required between the analogue and digital elements of some of the transmission links. As these links become predominantly digital, only the 16 kbit/s rate will be used. DSVTs can be connected to Tri-Tac switches or directly interconnected without intervening circuit switches to provide a non-switched back-to-back (sole-user) service. The terminal normally operates in a full-duplex mode but a half-duplex, press-to-talk mode is available for communication with combat net radios using Tri-Tac's net integration facilities. The DSVT can accommodate extension telephones or data terminal equipment such as the TDF or SST. An integral encryption device is used to secure traffic. Power supply can be from the switch common battery or through use of an auxiliary power unit, local battery or AC power.

The ANDVT is a subscriber terminal for use in aircraft, ships or vehicles. It provides transmission of secure digital voice and data traffic over narrowband voice frequency channels. The terminal is used, on a half-duplex basis, in point-to-point or netted circuits that employ HF/VHF/UHF radios or UHF satellites. It digitises voice at 2.4 kbit/s and can be used with a comsec module and other optional modules. The ANDVT can interface with the circuit switched network through a secure digital radio net interface unit.

The SST is a microprocessor-based record/data traffic terminal consisting of a militarised video display unit and a 25-line plasma display. It can utilise an external AN/LGC-74 printer. The SST can be used to compose, edit, store, display, refile, transmit, receive and monitor/record data traffic. It operates in a variety of modes, ASCII and Baudot codes and message formats at rates from 45.5 bit/s to 16 kbit/s.

The TDF provides a facsimile capability with up to 16 shades of grey and can operate on either full- or half-duplex. It interoperates with Tri-Tac and inventory comsec devices at transmission rates from 1.2 to 32 kbit/s and conforms to STANAG 5000.

The LDF is a non-development item which has the capability to provide transmission/reception of black and white graphical material. It is capable of transmitting and receiving in a half-duplex mode via 1.2 to 32 kbit/s wire and radio communication channels through the use of existing comsec and modem equipment. It is compatible with other facsimile devices that are compliant with MIL-STD-188-161 and STANAG 5000 where a digital interface is permitted by the mode of operation.

There are two types of net radio interface. The Basic Net Radio Interface Device (BNRID) provides a semi-automatic, analogue, half-duplex voice-band interface between single-channel radios and circuit system subscribers by a four-wire analogue loop. The Secure Digital Net Radio Interface Unit (SDNRIU) provides comparable service on a digital, secure basis. Both net radio interfaces are interoperable with a variety of comsec devices used with net radios.

Switches

There are three Tri-Tac circuit switches: AN/TTC-39, AN/TTC-42 and SB-3865. The latter two are members of the Unit-Level Circuit Switch (ULCS) family. In addition there are two Tri-Tac message switches: AN/TYC-39 and AN/GYC-7, the latter referred to as the Unit-Level Message Switch (ULMS). The US Army is developing the AN/TTC-39A nodal control circuit switch, a modification to the AN/TTC-39, that expands the switch termination and adds minimum essential control/management functions.

Control facilities

A family of tactical communications control facilities provides flexibility in organisation and near-realtime management of deployed Tri-Tac systems and subsystems.

The Communications Equipment Support Element (CESE) includes circuitry, integral to Tri-Tac circuit and message switches, combining multiplexing and transmission equipment. The built-in test equipment monitors system

performance, generates alarms on performance thresholds and transmits switch and traffic processing performance data to a controlling Communications Nodal Control Element (CNCE).

The CNCE provides centralised management and control for a communications node. It provides the interface among analogue and digital nodal switches, common-user and dedicated subscribers and the inter-nodal radio and cable transmission networks. The CNCE contains a caesium timing standard and acts as the nodal master clock for all digital circuits. The CNCE is a single shelter facility to be used by the US Air Force in controlling its communications networks.

The Communication Systems Control Element's (CSCE) primary tasks are: the exercise of near-realtime control over the allocation and use of system resources within its assigned portion of the deployed tactical communications network; and to establish, maintain and update the database needed for its functioning.

The US Army uses the AN/TTC-39A at nodal (company) level in conjunction with an AN/MSC-32 communications operations central shelter, suitably modified to house facilities of the CSCE. Together they form a complementary pair of shelter facilities for nodal control and management. The management direction and control decisions emanate from the AN/MSC-32(MOD). The dial-up access to the AN/TTC-39A processor permits the nodal manager direct creation or modification of the database resident in the AN/TTC-39A. This is accomplished via directives received at the nodal CSCE (AN/MSC-32(MOD)) from the battalion brigade level CSCE. These directives are translated automatically by the nodal CSCE into the screen image formats required by the AN/TTC-39A and passed over the dial-up link to the AN/TTC-39A processor.

Transmission

Line of sight, tropospheric scatter (troposcatter) and tactical satellite links are used as transmission media within Tri-Tac. The AN/TRC-170 digital troposcatter radio set was the only new transmission equipment developed for Tri-Tac. The army inventory of combining/multiplexing equipment has been adopted, with some modifications, to serve Tri-Tac transmission needs. These are designated AN/TRC-173 and AN/TRC-174.

The AN/TRC-170 radio sets are a family of digital troposcatter radios which operate in the 4.4 to 5 GHz band. The AN/TRC-175, also an army modified unit containing a modified AN/GRC-103 radio, operates at a switching node to provide an up-the-hill link to the radio park. The AN/GRC-103A, located at the radio park, also contains a modified AN/GRC-144 radio. It can operate as a radio/cable terminal, split terminal or radio relay. In the radio/cable terminal role it can terminate up to eight extension systems, up to four internodal systems and a down-the-hill link to a switching node. As a split terminal it can be used in conjunction with the radio/cable terminal to provide two high-capacity internodal links. In a radio repeater role it can extend an internodal radio link.

Tactical satellite terminals, AN/TSC-85A, AN/TSC-93A, AN/TSC-94A and AN/TSC-100A are equipped with compatible combining equipment to interoperate with the Tri-Tac system.

Combining

The Digital Group Multiplexer (DGM) family provides loop transmission access for static subscribers. It consists of shelter-mounted and field-exposed digital multiplexers/modems, cable driver modems, pulse restorers and order-wire control equipment. The AN/GRC-103 modem is also a member of the DGM family and is an integral component of the army's AN/TRC-173 and AN/TRC-174.

Three shelter-mounted multiplexers are associated with the DGM family: the Loop Group Multiplexer (LGM), the Trunk Group Multiplexer (TGM) and the Master Group Multiplexer (MGM). The LGM accepts 7, 8, 15 or 16 channels at 16 or 32 kbit/s and synchronously time-division multiplexes them into a single group. The LGM is used in the AN/TRC-170. The TGM accepts up to four group signals and combines them into one supergroup signal. It is used in the AN/TRC-170 and the AN/TRC-173. The MGM accepts up to 12 group/supergroup signals at rates from 72 to 4,915.2 kbit/s and asynchronously multiplexes them into a single mastergroup signal. It is used in the CNCE as well as in the AN/TRC-138A and AN/TTC-175.

There are two field-exposed multiplexers associated with the DGM family: the Remote Loop Group Multiplexer (RLGM) and the Remote Multiplexer Combiner (RMC). The RLGM accepts up to four conditioned diphase channels at 16 or 32 kbit/s over field wire and multiplexes them into a single conditioned diphase group for transmission over coaxial cable. The RMC accepts up to eight conditioned diphase channels and one conditioned diphase group from an RLGM or another RMC and combines them into a single conditioned diphase group of 128 to 576 kbit/s.

Modems associated with the DGM family are the group modem, the cable driver modems and the cable driver that is used with the remote loop group multiplexer. Associated with the DGM family are two cable order-wire units.

Status

Each of the US services is responsible for equipment and system elements as follows:

US Army: message switch, circuit switch, nodal control circuit switch, BNRID, DGM, SST, CSCE, MSE and LDF.
US Air Force: CNCE, digital tropo radio, TDF and DNVT.
US Navy: ANDVT.
US Marine Corps: ULCs and ULMs.
US National Security Agency (NSA): DSVT, SDNRIU and COMSEC.

Initial Production Delivery (IPD) of the 50-line AN/TYC-39 message switch began in late 1982. IPD for the 300-line AN/TTC-39 circuit switch and basic net radio interface device began in July 1983 and March 1984 respectively. IPD for the DSVT was concurrent with the fielding of the AN/TTC-39. The AN/TTC-39 circuit switch and the AN/TYC-39 message switch were developed by GTE Government Systems Group. In FY80 the army received US$4.5 million for the procurement of two AN/

TTC-39(V)3 systems and, in FY81 the army requested US$11.2 million for three AN/TTC-39(V)1 and US$42.8 million for 15 AN/TTC-39(V)3 systems. Between FY82 and FY85 approximately US$204 million was requested for further procurement.

In FY80 the Army received US$29.5 million to procure 12 AN/TYC-39 systems and in FY81 US$12.9 million was requested for a further four systems.

The digital multiplexer was developed by Raytheon. The Technical Communication Control Facility (TCCF), developed by Martin Marietta, was delivered in 1979. The DNVT TA-954 has been developed by ECI for use by the Air Force; two switches, AN/TTC-42 (150-line) and SB-3865 (30-line), by ITT; the digital troposcatter terminal AN/TRC-170, by Raytheon; and the ANDVT is under development by ITT and TI. In March 1982 Raytheon received a, US$106 million three year contract for DGMs. Other contractors include Litton, Delco Electronics, Collins, Communications Satellite Corporation and Magnavox. In March 1984, Martin Marietta was awarded a US$19.6 million increment to its existing contract to supply the CNCE. In May 1984, GTE was awarded a US$39.3 million increase to an existing contract to develop the AN/TCC-39A circuit switch, followed by a further contract worth US$44.7 million in August 1984. In June of that year, ITT was awarded a US$4.1 million contract for continued research and development on the unit level circuit switch. Other contract increases included US$38 million to GTE for the AN/TTC-39A in August 1984 and US$116 million, US$85.2 million and US$47 million to Martin Marietta for the CNCE in August and November 1984 and August 1987.

By mid-1991, GTE had won Tri-Tac contracts worth US$754 million and had delivered 117 circuit switches and 38 message switches.

In 2002 planned upgrades to the ACUS were to provide an increased capability to support voice, data and video requirements. These upgrades will insert new technologies (Brigade Subscriber Node (BSN), battlefield videoteleconferencing, wireless LAN, and Network Operations Center Vehicles (NOC-V)) in to the US Army's Interim Brigade Combat Teams (IBCT).

UPDATED

Artist's impression of UFO satellite in orbit

UHF Follow-On (UFO) communications satellite programme

The US Navy began replacing and upgrading its Ultra-High Frequency (UHF) satellite communications network during the 1990s with a constellation of customised satellites built by Hughes Space and Communications Company, which is now Boeing Satellite Systems, Inc. Known as the UHF Follow-On (UFO) series, these 601 model satellites support the US Navy's global communications network, serving ships at sea and a variety of other US military fixed and mobile terminals. They are compatible with ground- and sea-based terminals already in service. The UHF Follow-On satellites replace the Fleet Satellite Communications (FLTSATCOM) and the Hughes-built Leasat spacecraft.

In July 1988, the company won the competition for a fixed-price contract awarded by the Program Executive Office for Space, Communications and Sensors in Washington, DC. The initial agreement called for the company to build and launch one satellite, with options for nine more. Options for spacecraft 2 and 3 were exercised in May 1990; for 4, 5, and 6 in November 1990; and for 7, 8, and 9 in November 1991. In January 1994, the contract was extended by ordering a 10th satellite and launch services, bringing the total value to US$1.7 billion. In November 1999, the Space and Naval Warfare Systems Command's Communications Satellite Program Office added an 11th satellite to the contract funded initially at US$27 million, but with options this could reach US$213 million. In January 2001, Boeing was authorised to begin production on that 11th spacecraft, which is scheduled to launch in 2003. This will help sustain the constellation into the latter part of the decade.

In March 1996, under a contract modification for US$150 million, the navy ordered a high-power, high-speed Global Broadcast Service (GBS) payload to be incorporated onto F-8 through F-10. This GBS package is revolutionising

UFO F-5 under construction

communications for the full range of the Defense Department's high-capacity requirements, from intelligence dissemination to quality-of-life programming. The first GBS payload was put into service in 1998 and the final one was launched in November 1999.

The UHF F/O spacecraft has proven to be a very flexible platform for the efficient evolution of critical, advanced DoD communications services. The satellites are versions of the Hughes body-stabilised, three-axis HS 601 model. The spacecraft was introduced in 1987 to meet anticipated requirements for high-power, multiple-payload satellites for such applications as UHF F/O, direct television broadcasting to very small terminals, private business networks, and mobile communications.

The HS 601 body is composed of two main modules. The bus module houses the bus electronics, propulsion subsystem, and battery packs. The payload module contains the communications equipment and antennas. The UHF Follow-On contract calls for a minimum 10-year mission.

The UHF Follow-On satellites replace the Fleet Satellite Communications (FLTSATCOM) and the Hughes-built Leasat spacecraft currently supporting the Navy's global communications network, serving ships at sea and a variety of other US military fixed and mobile terminals. They are compatible with ground- and sea-based terminals already in service.

The UFO satellites are manufactured in El Segundo, California. Using a building-block approach, Boeing and the navy enhanced the constellation's capabilities in stages. Satellites F-1 through F-3 carry UHF and SHF (super-high frequency) payloads to provide mobile communications and fleet broadcast services. Starting with F-4, an additional EHF (extremely high frequency) payload was added to provide protected communications. F-7 introduced an enhancement to the EHF package that essentially doubles capacity. The SHF payload is replaced by the high data rate GBS package on F-8 through F-10. F-11 carries the enhanced EHF package and an upgraded UHF payload as well.

The UFO satellites offer increased communications channel capacity over the same frequency spectrum used by previous systems. Each spacecraft has 11 solid-state UHF amplifiers and 39 UHF channels with a total 555 kHz bandwidth. The UHF payload comprises 21 narrowband channels at 5 kHz each and 17 relay channels at 25 kHz. In comparison, FLTSATCOM offers 22 channels. The F-1 through F-7 spacecraft include an SHF subsystem, which provides command and ranging capabilities when the satellite is on station as well as the secure uplink for Fleet Broadcast service, which is downlinked at UHF.

The Navy added an extremely high frequency communications package beginning with the fourth spacecraft. This addition includes 11 EHF channels distributed between an earth coverage beam and a steerable 5° spot beam and is compatible with Milstar ground terminals. The EHF subsystem provides enhanced antijam telemetry, command, broadcast, and fleet interconnectivity communications, using advanced signal processing techniques. The EHF Fleet Broadcast capability supersedes the need for the SHF fleet uplink. Beginning with UFO F-7, the EHF package was enhanced to provide 20 channels through the use of advanced digital integrated circuit technology.

The GBS payload replaced the SHF payload on spacecraft F-8, 9, and 10. This new package includes four 130 W, 24 megabits-per-second (Mbps) military Ka-band (30/20 GHz) transponders with three steerable downlink spot beam antennas (2 at 500 nmi and 1 at 2,000 nmi) as well as one steerable and one fixed uplink antenna. This modification resulted in a 96 Mbps capability per satellite. Three spacecraft give the US DOD near-global coverage.

F-11 will be most similar to F-7, providing UHF and enhanced EHF communications. The UHF payload incorporates a new UHF digital receiver, providing two additional UHF channels and greater flexibility in configuring communication services.

The Atlas rocket series was chosen to provide the launches from Cape Canaveral, Florida. The Atlas I rocket was used for the F-1 through F-3 satellites. The Atlas II was chosen for F-4 through F-8 and Atlas IIA for F-9 and F-10. An Atlas III will launch F-11.

Status

UHF F-2 was the first in the series to go into service, after its successful launch on 3 September 1993. UHF F-3 was launched on 24 June 1994. Three UHF spacecraft were orbited in 1995: F-4 on 28 January, F-5 on 31 May and F-6 on 22 October. F-7 was launched on 25 July 1996. F-8 was launched on 16 March 1998 and F-9 on 20 October 1998. F-10 was launched on 22 November 1999. F-11 was scheduled for launch in December 2003.

In July 2002 it was announced that F-2, which was being used as an on-orbit spare, had been brought into operation to provide nine additional voice and data channels to support US operations in Afghanistan.

Contractor

Boeing Space & Communications, Seal Beach, California.

UPDATED

INTELLIGENCE SYSTEMS

Surveillance and reconnaissance
Direction-finding
Analysis and support
Signals intelligence
Imagery intelligence

SURVEILLANCE AND RECONNAISSANCE

Brazil

EDT-FILA anti-aircraft fire-control system

The EDT-FILA (Fighting Intruders at Low Altitude) has been developed by Avibras to meet the requirements of the Brazilian Army for a mobile fire-control system for use with 35 or 40 mm anti-aircraft guns and with surface-to-air missile systems. The first unit was handed over to the Brazilian Army in September 1985.

The basic technology for FILA was obtained from the Skyguard AA fire-control system, developed by the Swiss company Contraves, but with a number of improvements such as an additional Ka-band tracking radar, laser range-finder, circular polarisation feeder for the search antenna, a redesigned computer and search radar data extractor. The system retains the basic Skyguard X-band radars for search and tracking, IFF and TV tracker.

The pulse-Doppler X-band search radar has high detection capability, high clutter suppression, automatic target alarm, threat evaluation and integrated operation with the IFF system.

The Ka- and X-band monopulse Doppler tracking radar system has fast PRF and frequency change, automated target exchange, ASM detection and alarm, as well as passive tracking, to ensure a high degree of immunity from noise and jamming. The laser range-finder is fully integrated and an infra-red system automatically measures parallax data for the guns.

A third-generation computer, the C-2001, allows real-time data processing and has ample storage capacity for several types of software including self-test, diagnosis, training and combat programs.

The control console has: a PPI tactical situation display that presents both Doppler and raw video, as well as symbols and markers; an ECCM control panel; a TV tracking system monitor; a joystick for manual tracker control and a matrix panel for data input and output.

The complete system, including an integrated power supply unit, is installed in a four-wheel trailer built of fire-resistant reinforced glass fibre polyester and the cabin is fully air conditioned. It is also provided with an automatic levelling system.

Status

In operational service with the Brazilian Army since 1985. During 1998, AVIBRAS delivered a number of additional EDT-FILA systems to the Brazilian Army. As of April 2001, total EDT-FILA production for the Brazilian Army was reported as being 13 units.

Contractor

Avibras Indústria Aeroespacial SA, São José dos Campos SP.

VERIFIED

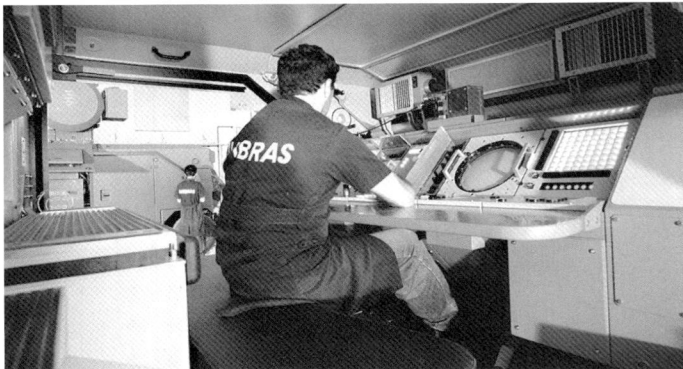

Interior view of EDT-FILA

EDT-FILA anti-aircraft and anti-missile fire-control unit

Canada

Reconnaissance Vehicle Surveillance System (RVSS) sensor suite

The Surveillance System is a suite of remote sensors integrated with the reconnaissance vehicle to provide all-weather monitoring of personnel and vehicle movement at ranges up to and beyond 10 km. Full operation is carried out from the closed-down and concealed vehicles. Two configurations are provided: a vehicle with an extendable 9.5 m mast for rapid deployment and a version with ground-deployed sensors for additional screening. Either configuration is operated by one soldier from the RVSS Operator's Console (see separate entry), which is installed in the reconnaissance vehicle or can be removed and installed in a building/shelter for more permanent surveillance tasks.

All system components are common to both the mast and remote configurations and provide equivalent surveillance capability. The sensors are all non-developmental, latest generation devices tailored for remote application. The system components are:

(a) a lightweight manpack battlefield surveillance radar, all-solid-state, low-power/low-probability of intercept design, menu operation, flat-panel display, with automatic alarms, variable scan and dwell modes, built-in test and able to detect targets up to 24 km away

(b) an 8 to 12 μm FLIR with TV output, low audible noise, integral cooler, variable magnification/field of view, fast cool-down and built-in test which provides all weather recognition/identification of targets to 12 km

(c) a high-resolution visible spectrum camera incorporating a charge-coupled device sensor, a usable image from wide range of lighting conditions including near-twilight, variable zoom/magnification, variable focus, automatic exposure and providing detection/day/night recognition/identification to 18 km

(d) the current standard US Army Eyesafe hand-held laser rangefinder with adaptation for remote operation and range output, 10 km range (±5 m), a 1 mrad beam, fast charge/recharge, false return rejection, built-in test

(e) a fast, precision motorised azimuth and elevation platform with smooth low-speed motion, joystick control, level compensation and boresighting adaptors for sensors.

To accommodate sensors developed/improved during the system life, the surveillance system design is modular and standardised at all interfaces and internal control/communication is configured by software. Additional capacity is provided in the operator's console for advanced features, for example, onboard image enhancement, still or video image capture/compression/transmission, additional sensors and enhanced operator assistance. External access to recorded surveillance imagery is included for transmission to higher echelons.

Specifications
Visible spectrum day/night TV camera
Continuous zoom
NFOV 31 × 23 mrads
WFOV 296 × 222 mrads
High resolution

Thermal imager
8-12 micron spectral range
NFOV 50 × 37 mrads
WFOV 200 × 150 mrads

RVSS battlefield radar and control panel in remote configuration

RVSS sensor suite in remote configuration

Laser range finder
Erbium : glass, eye-safe laser
Range up to 10,000 m, with ±5 m accuracy
Azimuth/Elevation platform
Remote control pan/tilt capability for 360° surveillance
Precise control for LRF targeting at long range

Surveillance radar
Man-portable, battery powered
Ku-band, Doppler radar
Designed primarily for the detection of personnel and vehicles
Radar operational modes
Surveillance - 18°/s scan rate, 180° max swath, 4 W peak output
Acquisition - 1 km × 1 km area 10 m resolution B scope (9°/s scan rate)
Fall of shot - 1 km × 1 km area artillery impact (B scope) (45°/s)

Hybrid cable
Copper wire power transmission
Optical fibre video transmission
Built-in video/data to light converter for no-loss noise free transmission

Operator station
Provides all system control and monitoring
VME computer with 100% growth
Hi-8 VCR
High resolution monochrome monitor
Power conditioning and power control for entire system
Rugged aluminium chassis

Environmentals
−40 to +44°C operating
−46 to +71°C storage - SMS, SLS
Vibration: MIL-STD-810
Wheeled vehicle, Table 514.4 - All
Shock: MIL-STD-810 functional, bench handling, transit

Status
In service with the Canadian Armed Forces in the Coyote reconnaissance vehicle.

Contractor
General Dynamics Canada, Ottawa, Ontario.

VERIFIED

France

Multisensor image interpretation and dissemination system (MINDS)

The multisensor image interpretation and dissemination system (MINDS) is a field deployable system for tactical and operational level intelligence, surveillance and reconnaissance. It is designed to help image interpreters generate image intelligence from raw multisource data and to disseminate the results of the interpretation process to the end users to support decision making, targeting and BDA, mission planning and intelligence collection management. The system has the following features:
(a) Production of IMINT from reconnaissance and ground surveillance missions, involving all types of airborne and spaceborne vehicles.
(b) Real-time digital acquisition and processing of raw data from all types of imaging sensors: electro-optical, infra-red, SAR, MTI, video and film. Automatic selection of an area of interest. Automatic target detection tools to assist operators.
(c) Link-16 capable for quick-reaction targeting.
(d) Network shared management of mission and support data (geographic, intelligence and support databases).
(e) Made up of ruggedised modules which are transportable by two people.
(f) Complies with current and future NATO standards for acquisition and dissemination of images and reports.
(g) The system architecture (multistation network) can be configured to adapt to a variety of intelligence organisations.
(h) Fully integrated GIS.
(i) Data can be exchanged between several stations of a MINDS system through a fibre optic LAN.
(j) Dissemination of intelligence data through WANs (X 25, ISDN, or specific).

Functions
(a) Acquisition, visualisation and selection in real time of all digital multisensor data.
(b) Acquisition, visualisation and selection in deferred time by reading media recorded on board the platform or by digitising wet films with a light table.
(c) Management of dedicated databases (mission data, selected data, geographic data, intelligence files, support data).
(d) Geometric and radiometric corrections
(e) Automatic or assisted geo registration
(f) Use of support data: vector or raster maps, DTM, reference files, image data.
(g) Generation of standardised intelligence reports and target files.

Status
Currently deployed in the French Air Force, Navy (including on board the CV *Charles de Gaulle*), Army and unspecified "certain joint services". Used in support of operations in Kosovo in 1999 and in Afghanistan in 2002. Claimed by Thales to have been used in operations in Iraq in 2003, but by which nation is uncertain.

Contractor
Thales Communications, Gennevilliers.

UPDATED

Stentor battlefield radar

Stentor is a Doppler radar used for ground surveillance of critical areas, such as battlefields and frontiers. A tactical data communication system which enables a command post to simultaneously receive and process the data issued from one to 10 Stentor radar stations is optionally available.

It is claimed that Stentor has the longest range available for ground surface surveillance radars, making possible surveillance of remoter areas and, compared to medium-range radar, giving a significant increase in warning time to aid the interception of hostile units. At shorter ranges the benefits of a powerful signal are evident for high detection probability against very small or intermittent targets, or for minimising the absorption effects of adverse weather conditions.

The Stentor ranges are: 30 to 40 km for pedestrian targets; 50 to 60 km for targets such as a jeep, tank, truck; and 20 to 60 km for boats (depending on sea conditions) and helicopters (a major threat with low-flying at low speeds since they are not detectable by orthodox air defence or low-flying aircraft detection radar).

Description
Stentor is a transportable system comprising the radar head and an operator's console which can be remotely sited several hundred metres from the radar head.

Stentor battlefield surveillance radar

The radar head can be located on a tower, or on a naturally elevated site, to increase the surveyed area and thus, distant detection.

The operating frequency is in I/J-band, tunable over 200 MHz with pulse transmission, ground clutter cancellation being by coherent detection and Doppler filtering; rain cancellation devices provide for all-weather operation. Moving targets are displayed on a daylight CRT with a digital memory. The surveyed sector can be instantaneously adjusted in aperture and in azimuth.

Stentor is simple to operate (even by a single inexperienced operator) and to maintain, with high mobility, fast deployment and effective ECCM capability. Operator training simulators and third echelon bench test units are optionally available.

Specifications
Frequency: I/J-band, tunable
Peak power: 60 kW
Polarisation: vertical or circular
Antenna span: 1.6 m
Range: up to 60 km on tanks and large vehicles
Accuracy: 20 m (range); 4.4 mrad (0.25°) (azimuth)
Weight: 370 kg (radar head and operator's unit)
Power: 1.5 kVA, 220 V, 50/60 Hz
MTBF: >1,000 h

Status
No longer in production. Believed to be in service in the Indian Army. May be in service in other countries.

Contractor
Thales Airborne Systems, Elancourt.

UPDATED

Germany

DO-SAR SWORD radar

The DOrnier Synthetic Aperture Radar Standoff all-Weather Observation and Reconnaissance Drone (DO-SAR SWORD) is an ultralight and compact Synthetic Aperture Radar/Moving Target Indicator (SAR/MTI) sensor developed as a direct interchangeable sensor package for the existing electro-optical sensor of the CL289 drone system. No modification of the CL289 is necessary, the real-time radar raw data being transmitted to the ground station via the existing datalink of the drone system. Image processing is accomplished in real-time on the ground.

Specifications
(a) Radar principle: pulse-Doppler
(b) Pulse modulation: digital linear FM chirp
(c) Image processing: digital, real-time on ground
(d) Frequency: Ku-band
(e) Power (peak): 10 W
(f) Power source: solid-state
(g) Onboard datalink: CL289 compatible
(h) Resolution: 1.4 m
(i) Swath width: 1,800 m

Status
Developed in co-operation with Thales, two demonstrators were operational in December 1997. SWORD is expected to be introduced into service by 2005 as part of a mid-life upgrade of the CL289. For further information see *Jane's Unmanned Aerial Vehicles and Targets*.

Contractor
EADS Dornier GmbH, Friedrichshafen.

VERIFIED

International

COBRA CounterBattery Radar

COBRA (COunterBattery RAdar) is a new, highly mobile weapon location radar being developed for France, Germany and the UK by the EuroArt consortium comprising Lockheed Martin, EADS (Germany), and Thales (France and UK).

Meeting all NATO operating requirements, the system is designed to detect small cross-section targets over the whole battlefield and can classify ammunition types and firing modes (rockets, swarms, salvos and so on). Its high tracking accuracy in conjunction with its precise self-positioning system enables the location of hostile battery positions with the accuracy required for counterbattery fire. The manufacturers claim that a typical mission can be completed in approximately 40 minutes, and that COBRA could perform up to 15 missions at different sites during a typical battlefield day.

The COBRA weapon-locating radar with antenna raised; the PPU is located between the operations cabin and the vehicle cab

The key technical feature of COBRA is its solid-state active modular antenna. This antenna contains about 3,000 low power transmitter/receiver modules (GaAs devices). It has an elliptical aperture distribution, employing phase/phase steering and containing quadrapacks, capable of optimising on demand the aerial patterns for sidelobe level gain. Couplers at the dipoles feed a similar feed network to allow automatic array calibration. For blowers and heat exchangers provide quadrapack cooling and the antenna array is protected by a low-loss RF radome. An inertial navigation unit mounted in the array provides accurate antenna position.

Signal Processor
The signal processor: performs digital pulse compression, the doppler filtering (finite impulse response: FIR), the monopulse error calculation and the false alarm regulation; provides range and doppler interpolation; sets flags indicating if there is a single target, a salvo or a clear target in a swarm; provides jamming analysis; and supports antenna calibration. It is capable of managing a large number of waveforms in order to optimise the detection performance as a function of the ranges requested for the mission. There are four operating modes for the signal processor: passive listening. terrain mask determination. search and track. The data generated for each mode is sent to the radar processing software where it is used to form track files or to provide criteria for further processing.

Doppler correction compensates for the radial velocity of the tracked targets, providing better projectile location and an improved range resolution. The non-linear frequency modulation used for the transmit pulse produces very low-range sidelobes at the output of the matched filter.

The signal processor selects the doppler filtering coefficients depending on the clutter environment. The frequency response for each filter is tailored for increased rejection. where the clutter is stronger close to zero doppler. The false alarm rate regulation is designed to minimise sensitivity losses when there is no clutter and other targets in the vicinity of the target of interest. Range and doppler interpolations are performed to provide target position accuracy finer than the quantisation which is already more precise than a resolution cell.

During passive listening, the signal processor provides an accurate estimate of the direction of any jamming. Prior to each dwell, when the system is radiating. an analysis is performed to determine if no jamming, mainlobe jamming or sidelobe jamming is occurring.

The signal processor supports antenna calibration by measuring the phase and amplitude of the calibration pulses.

Data Processor
The general purpose computer used in COBRA incorporates COTS boards capable of the rapid performance of all processing requirements. It is based on the proven virtual memory system (V1\ 1S) bus architecture. The data processor contains two single board computers (SBC), based on RISC processor technology, each with 1Mb or more of L2 cache memory , up to 256Mb of random access memory (RAM), communication devices to support various external interfaces and a removable 9Gb disk.

Each SBC has its own UN1X operating system. Both SBCs communicate over the VME bus. The man-machine-interface (MMI) is based on OSF/Motif. The mission software is largely written in ADA. The main computer is connected to one or two external terminals, each with their own CPU and memory. The terminal

COBRA with antenna stowed (EuroArt)   0525382

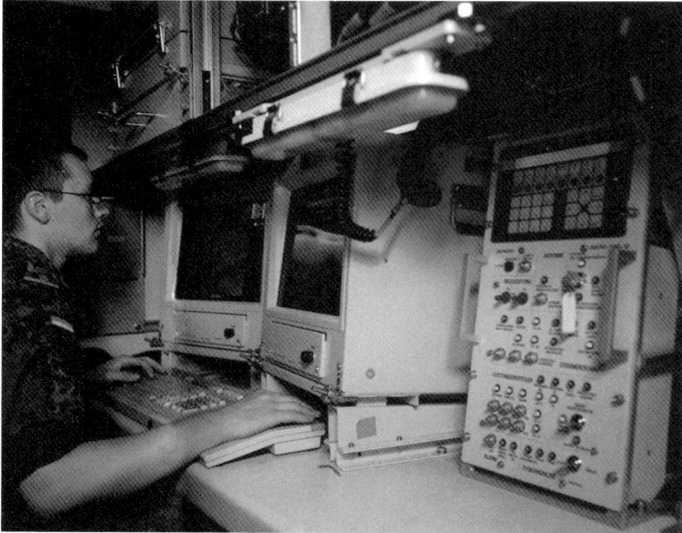

A COBRA operator's workstation showing graphic terminal to the left and alphanumeric to the right, with the trackball under the operator's right hand (EuroArt)   0132920

processors are capable of performing up to 3 Xmarks. There are 32Mb of RAM per terminal and software is held in a flash EPROM.

The mission software which is run on the machine is divided into three computer software configuration items (CSCI). All the real time tasks are performed by the radar processing CSCI, which resides on its own SBC. The other two CSC1s, mission processing and message processing, share a separate SBC.

Radar processing has been assigned all the functions associated with directly controlling and collecting data from the radar. Proper control is essential in scheduling radar dwells to ensure proper energy management of the radar's timeline limitations. The output data from each beam dwell produced by the signal processor is analysed by radar processing to form track files. The reports generated are used by mission processing in performing ballistic calculations to determine battery association, width and orientation and classifying the type of weapon located.

Mission processing is the primary control for all aspects of setting up and carrying out a mission. It determines the validity and compatibility of the mission setup data. It informs the operator of any conflicts detected and, if required, waits for clarification before proceeding with the mission. On command it will proceed with a valid mission and, using data received, optimise the radar's performance. It will respond to user requests by compiling and sending reports based on the data gathered. Mission setup data and results are stored by Mission Processing, ready for later retrieval and display by the operator.

Message processing interfaces with the C3 network within which the system is operating (eg BATES, ADLER or ATLAS) and with the equipment used by the COBRA operator. The C3 interface receives command messages from and transmits data messages automatically to the C3 network. The format of the messages is adapted to the formats used by the C3 network. The operator has access to all data exchanged with the C3 network. The operator interface, based on windows technique, encompasses all actions which are required to control the radar and to display the data defining a mission and the weapons located. The located weapons are displayed on a dedicated screen, which allows the simultaneous display of an electronic map and the areas related to the defined mission in the background.

The data processor has a memory capacity which, in the worst case, exceeds the participating nations' requirements by more than 20 per, and a processor power which exceeds them by greater than 30 per cent. In addition, it contains board space to increase the internal memory by another 30 per cent.

Cabin

The system is contained in a single cross-country truck/operations cabin with the antenna mounted on top. The cabin, which is 6.1 m × 2.4 m × 2.1 m, contains the receiver/processor and command and control equipment, radar operator console, C3 operator console and vehicle intercom. The walls of the cabin, like the antenna, contain Kevlar® to provide some protection against shrapnel and small-arms fire, and the cabin is provided with NBC and EMP protection. AC and DC power is provided by a Prime Power Unit (PPU), but if this should fail, 350 W DC battery power is available to continue NBC filtering and provide lighting inside the cabin. When removed from the prime mover and palletised, the system can be transported by C130 or Transall C160.

The system can be run either entirely from an alphanumeric display and keyboard or in concert with a separate graphic display controlled by a trackball. It is planned for one-person operation for the French Army and for two-person operation in the UK and German forces, enabling the system to operate as a command post.

Survivability of COBRA is afforded by a very short transmit time, wideband frequency agility and various other ECCM modes. Built-in test at the lowest replaceable unit level is provided in addition to online radar status monitoring.

Status

Full-scale production started in March 1998 and includes the requirements from all three participating nations. Under the terms of a fixed-price contract, 29 COBRA

systems will be built; 10 systems for France, 12 for Germany and seven for the UK. The first production COBRA systems were completed in late 2001 and nations will, in general, receive their systems in turn. Deliveries are scheduled to take place from 2003 - 2006.

Contractors

EuroArt consortium comprising:
Lockheed Martin, Moorestown, New Jersey, USA.
EADS, Unterschleissheim, Germany.
Thales Air Defence, Bagneux, France.
Thales Defence Ltd, Crawley, West Sussex, UK.

UPDATED

Israel

ICE (Integrative Component-based Exploitation system)

ICE is a system designed to deliver an integrated solution to the operational cycle of digital imagery, though a process of real-time image processing, use of a digital imagery archive, and image exploitation. ICE imagery data derived from a wide variety of sources can be integrated and interpreted to allow analysis, exploitation and further cueing of sensors. The three basic ICE components are the Generic Geo-Location Gateway (G3), the Intelligence Information Warehouse (IIW) and Digital Image Exploitation workstations.

The Generic Geo-Location Gateway's main function is the real-time processing of the sensors' raw data into geopositioned images. The G3 is a processing unit based on high-end, scalable, multiprocessor PCs. The design incorporates a strong emphasis on:
(a) Software-based, high-rate (hundreds of Mbps) processing capability of incoming, compressed streams of raw data
(b) Automatic, real-time, high precision geo-positioning of the images
(c) Use of photogrammetric algorithms to create a set of 7 zoom level images (RRDS).

The Intelligence Information Warehouse enables the sensors' data, images and intelligence aids to be handled in an integrative, intelligent manner. The IIW's proprietary integration technologies and management work flow result in powerful features and capabilities, including:

Integrative Component-based Exploitation (ICE) twin-monitor workstation (Elbit)
NEW/0593043

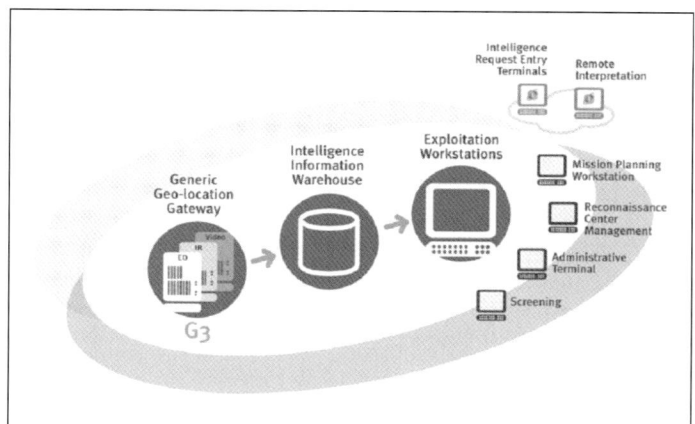

Elbit's Integrative Component-based Exploitation (ICE) architecture (Elbit)
NEW/0593044

(a) Archiving and integration of images from various VISINT sensors, and information, mapping and intelligence aids from multiple sources
(b) Data management capabilities including automatic indexing and storage of the data and images in a geo-positioned manner
(c) Smart searching/query tools for fast retrieval
(d) Oline and off-line multi gigabytes storage
(e) Interactive knowledge management.

The heart of the digital imagery cycle, ICE's exploitation process, comprises two complementary sub-systems:

Task Management

The central managerial and collaboration tool of the reconnaissance centre manager enables the manager to:
(a) Control and monitor all electronic and human resources
(b) Define user's properties and system's status
(c) Manage and monitor the exploitation process by prioritising and balancing tasks among interpreters
(d) Monitor tasks' work flow, processing and interpreters' load balance
(e) Review, authorise and route intelligence reports and products.

Real-time soft copy interpretation

The main features of this are:
(a) Microsoft Windows-based workstations with two synchronized 21 in screens
(b) A unified, generic interpretation desktop - supports interpretation of various imagery data types: EO,IR, SAR and Video
(c) State-of-the-art photogrammetric tools and GIS capabilities
(d) Full stereoscopic 3-D interpretation capabilities
(e) Accurate target generation process
(f) History management capabilities
(g) Automatic, spatially driven, generation of interpretation reports
(h) Full collaboration of data and information among interpreters
(i) High-level personalisation and intuitive user interface. Over 100 image enhancement functions and algorithms
(j) Semi-automatic capability to generate intelligence aids from processed images (for example, orthophotos, DTMs, Target pages, 3-D scenes)
(k) Semi-automatic manual restitution process for highly accurate geo-positioning.

Status

In use in the Israel Defence Force. Also procured by several unspecified customers in both Europe and Asia.

Contractor

Elbit Systems Ltd, Haifa, Israel.

UPDATED

Sweden

ARTHUR Weapon Locating System (WLS)

ARTHUR (ARTillery HUnting Radar) is a highly mobile, medium-range Weapon Locating System (WLS) containing a radar, a data processing unit, operator workstations, communication equipment and a navigation unit. Two carrier alternatives are available. One is the rear cabin of a Hägglunds Bv 206 tracked vehicle with the power generator and the communications equipment in the front car. The other alternative is a 13 ft ISO corner container carried by any cross-country truck (payload ≥5 tonne) with either a towed or integrated power generator.

ARTHUR operates in C-band (NATO G/H-band) with an air-cooled TWT as transmitter. The most prominent component is the large, electrically controlled antenna, phase controlled in azimuth and frequency controlled in elevation.

The ARTHUR WLS in the Bv 206 version for the Norwegian and Swedish artillery
0055185

The ARTHUR WLS as delivered to the Danish Army, mounted on a Unimog 2150L38 all-terrain truck (Ericsson) **NEW**/0593358

The system searches for projectiles along the horizon and tracks detected targets for a few seconds. A number of targets can be tracked simultaneously while searching continues. After tracking, the trajectory is calculated and the enemy launching position is determined. Data on located batteries can then be passed to the artillery Fire Direction Centre for counterfire. The system also provides Fall-Of-Shot (FOS) observation for own artillery. In this mode, projectiles are tracked through the final part of their trajectory and the impact point is determined. Weapon locating and FOS can be carried out simultaneously.

As ARTHUR can also be used in peacekeeping operations to detect and register artillery fire in violation of peace accords, the system includes functions for the registration of all primary data for all events detected.

Status

ARTHUR has been developed for the Norwegian and Swedish armies. In 1996, a series contract was placed by the two countries with delivery due to start in early 1999. The first series unit commenced verification trials in late 1998. In 1997, Denmark placed an order for the truck version to be delivered from the end of 1999. In 1998 a fourth, as yet undisclosed, customer ordered the system and discussions are under way with a number of potential customers. In February 2002 the Greek Army ordered an undisclosed number of systems in a contract worth US$43.8 million. In August 2002 the UK ordered 4 systems for use by Royal Artillery units supporting 3 Commando Brigade RM and 16 Air Assault Brigade, for delivery in two years, with an option to buy an additional four. Before this order the UK was leasing an undisclosed number of systems.

Contractor

Ericsson Microwave Systems AB, Mölndal.

UPDATED

....

Erieye airborne early warning radar

The Erieye airborne early warning radar was originally developed by Ericsson for the Swedish Defence Material Administration. The programme had its roots in studies in the 1960s and 70s, and following the development of phased array technology the concept evolved in the early 1980s. In 1985 the decision was taken to develop an airborne test model. Trials with an airborne demonstrator mounted on a Fairchild Metro III aircraft began in 1990 and in 1992 the Swedish government decided to proceed with development and procurement, with the SAAB 340B selected as the aircraft platform.

Erieye has been designed for both military and non-military missions including:
(a) airborne early warning
(b) intercept control
(c) airspace management
(d) surveillance and control of national borders and exclusive economic zones
(e) detection of illegal activities, for example, drug shipments by air
(f) search and rescue co-ordination.

The system detects and tracks air and sea targets out to the horizon and beyond. Instrumented range is 450 km. Typical detection range against a fighter size target is in excess of 350 km. It features a frequency-agile, phased-array, S-band, pulse-Doppler radar. The antenna is fixed and the radar beam is electronically scanned through 360°. The beam is controlled by an intelligent and automatic energy measurement system which has the ability to transmit in any direction from pulse to pulse. It optimises the beam position providing for quicker detection verification, increased range and improved tracking, compared with a rotodome solution. The single, dual-sided, dorsal-mounted fixed antenna unit weighs 1,000 kg and places

Erieye AEW & C system on the Embraer EMB-145 aircraft   0055189

The Erieye display at Farnborough Airshow in 2002 (Patrick Allen/Jane's) 0523993

much less demand on aircraft size than was previously possible for high-performance AEW&C systems. Ericsson claims it is the world's first high-performance AEW&C system especially designed for commuter-type aircraft and that it drastically reduces acquisition as well as operational costs, with a reported cost per hour of only 10 per cent of that of AWACS .

The onboard command and control system is built with commercial-off-the-shelf (COTS) computers. All operator consoles are equipped with colour displays and user-friendly man/machine interface, and are normally mounted on standard seat rails. Operator workload is minimised by the high degree of automation and computer assistance. In the Swedish AEW role, tracked target data is downlinked (VHF/UHF) to the existing STRIC C2 system (see separate entry), with no onboard operators. The radar is controlled remotely from the STRIC CRC.

Status
Six Erieye systems entered service with the Swedish Air Force between 1996 and 1999 with the SAAB 340B as carrier, with the systems designated PS-890. The total cost of the programme, including integration into STRIC, is believed to have been US$340 million. The Swedish platforms are capable of conducting patrols for 5 to 7 hours at altitudes between 2000 and 6000 m. Ground support is provided by a 'Radar Group' of pilots and technicians; there are two Radar Groups in the Swedish AF, and each can maintain continuous operations for up to one month in separate geographic areas. Swedish development considerations include upgrading the current concept, that is AEW; installing 2 to 3 operators on board, providing AEW&C; and installing 3 to 5 operators on board, providing AEW&C2. Although these latter options represent a fundamental shift from the current remote-control concept, they would enable Sweden to contribute high-value assets in deployed coalition operations. They would also overcome the problem of skill fade in CRC operators, which is understood to be of concern.

Erieye has been selected as the airborne surveillance system in the Brazilian Amazonia surveillance project, SIstema de Vigilancia da Amazonia (SIVAM), using the EMBRAER Emb-145 as the platform. Flight tests with Erieye on the Emb-145 started in mid-1999. Five AEW&C systems are on order, two of which have been operational since mid-2002.

In 1999 Greece ordered four AEW&C systems for the Hellenic Air Force at a cost of US$600 million, using the Emb-145 as the platform with side-facing operator consoles. The first two, mounted on SAAB 340 as an interim solution, were delivered in Oct 2001, enabling Greece to monitor its airspace on a 24 hour basis.

One AEW&C system has also been bought by Mexico, also using Emb-145 with forward-facing consoles.

Contractor
Ericsson Microwave Systems AB, Mölndal.

UPDATED

GIRAFFE air defence radar system

GIRAFFE is the family designation for a series of combined G band pulse Doppler search radars and combat control centres that are designed for mobile and static, short- and medium-range C3I air defence applications. Within such applications, GIRAFFE's primary purpose is the detection of all altitude targets in conditions of severe clutter and electronic jamming. Radars within the family employ broadband travelling wave tube final amplifiers and G band operation has been selected as the optimum for an equipment that is designed to provide low-level coverage against small radar cross-section targets operating in a clutter/jamming environment, in combination with high mobility. Countermeasures proofing is provided via frequency agility and the use of low sidelobe antennas, digital pulse compression, adaptive transmission modes and 'instantaneous' unjammed frequency selection. Digital Doppler processing (with constant false alarm rates) is used to automatically detect, extract and display targets of interest. For communication with associated firing units, the GIRAFFE C3I capability makes use of manpack display units that provide each firing unit with information such as target bearing and range relative to the firing site, target speed and course, cross-distance, detection time limits and target priorities. Five GIRAFFE variants have been identified, the known details of which are as follows:

GIRAFFE 40
GIRAFFE 40 is a short-range (40 km instrumented range) air defence radar with C3I capability that is designed for use with short-range missile and anti-aircraft gun systems. The equipment employs a folding antenna mast which, when fully extended, gives a height of 13 m. GIRAFFE 40 can be integrated with an Identification Friend-or-Foe (IFF) subsystem (including Mk XII) and its detection envelope is given as being from ground level to an altitude of 10 km. In Swedish service, the radar has been designated as both the PS-70 and the PS-701 and a 60 kW power output version has been developed under the designation PS-707 (Sweden)/Super GIRAFFE (export). GIRAFFE 40 is no longer in production.

GIRAFFE 50 A T
With an instrumented range of 50 km, the GIRAFFE 50 A T has been developed for the Norwegian Army's Low-Level Air Defence System (NALLADS) and is installed on a tracked chassis (BV 206) that is capable of operating over 'severe terrain'. The variant's mast-mounted antenna has an operating height of 8 m and other features include:
(a) detection envelope that stretches from ground level to an altitude of 7 km
(b) fully automatic combat control functions (track initiation, tracking (Kalman filters), target identification, target classification and designation, threat evaluation and 'pop-up' target handling)
(c) the ability to exercise tactical control over up to 20 firing units
(d) a radar-to-firing unit target information datalink
(e) automatic hovering helicopter detection and threat evaluation functions
(f) the ability to exchange data with GIRAFFE 75 and AMB systems to facilitate radar co-operation and the compilation of a local air picture
(g) the ability to be integrated with an IFF subsystem (including Mk XII)

GIRAFFE 75
GIRAFFE 75 features a 'complete' C3I capability and is designed for use in medium-range (GIRAFFE 75 has an instrumented range of 75 km) air defence systems or short-range air defence applications where the emphasis is on a high level of electronic counter-countermeasures and C3I performance. Other system features include:
(a) a detection envelope that stretches from ground level to an altitude of 10 km
(b) automatic hovering helicopter detection
(c) a folding antenna mast which, when fully extended, gives an array height of 13 m

The GIRAFFE 75 C3I radar for tactical air defence systems

The Ericsson GIRAFFE 50 AT search radar and C3I system has been procured for use in the Norwegian Army Low-Level Air Defence System (NALLADS)

The Giraffe AMB radar system is designed for use with medium-range anti-aircraft systems such as the RBS 23 BAMSE air defence missile 0006878

(d) fully automatic combat control functions (track initiation, tracking (Kalman filters), target identification, target classification and designation, threat evaluation and 'pop-up' target handling)
(e) the ability to exercise tactical control over up to 20 firing units
(f) a radar-to-firing unit target information datalink
(g) the ability to exchange data with GIRAFFE 50 AT and AMB systems to facilitate radar co-operation
(h) the ability to be integrated with an IFF subsystem (including Mk XII)
(i) an optional add-on unit to optimise the radar for coastal surveillance duties.
In Swedish service, GIRAFFE 75 is designated as the PS-90.

GIRAFFE AMB

Developed for use with the Swedish RBS 23 BAMSE and RBS 97 (Swedish I-HAWK PIP III) air defence missile systems, GIRAFFE AMB is equipped with a phased-array antenna and offers a full 3-D capability. An associated, 'next-generation' C3I capability optimised for use with short- to medium-range air defence systems is included. Other system features include:
(a) multibeam technology
(b) digital beam shaping
(c) a detection envelope that stretches from ground level to an altitude of more than 20 km
(d) automatic hovering helicopter detection and threat evaluation functions
(e) instrumented range of 100 km
(f) a folding antenna mast which, when fully deployed, gives an array height of 13 m
(g) fully automatic combat control functions (track initiation, tracking (Kalman filters), target identification, target classification and designation, threat evaluation and 'pop-up' target handling)
(h) a radar-to-firing unit target information datalink
(i) the ability to exchange data with other GIRAFFE AMB, GIRAFFE 50 AT and 75 systems to facilitate radar co-operation and the automatic compilation of a local air picture
(j) the ability to be integrated with an IFF subsystem (including Mk XII)
(k) Processing architecture based on ruggedised commercial standard equipment.
In Swedish service, GIRAFFE AMB is designated as UndE 23.

GIRAFFE S

GIRAFFE S is a low-level 'gap filler' and coastal surveillance radar that is designed primarily for remotely controlled applications. System features include:
(a) a ground level to 1 km altitude detection envelope
(b) a 180 km instrumented range
(c) automatic hovering helicopter detection
(d) fully automatic system control functions (track initiation, tracking (Kalman filters), target identification, target classification and designation
(e) simultaneous supervision of air- and sea-space
(f) fixed-site, transportable and mobile installations options
(g) a radar-to-firing unit target information datalink.
When used for fixed site operations, GIRAFFE S can be configured for unmanned operation with track and plot data being transmitted to regional or national control centres via narrow-band radio link or landline. System control and status

monitoring are executed via the control centre and the radar's antenna is mast mounted (up to 30 m high). In mobile applications, GIRAFFE S makes use of a folding 8 m high antenna mast that is integrated with an air conditioned, nuclear/biological/chemical protected, cross-country vehicle or truck-mounted equipment/operator cabin. So configured, GIRAFFE S is fully autonomous, incorporates IFF and communications subsystems and is equipped with two operator workstations. Deployment time is given as being less than 10 minutes.

Status

Approximately 450 GIRAFFE radar/C3I systems have been ordered by customers around the world since 1978. According to *Jane's* sources, GIRAFFE 40 systems have been supplied to the armed forces of Singapore, Sweden (PS-70 for use with the RBS 70 missile and PS-707 for use with the RBS 90 system) and Thailand. GIRAFFE 50 AT has been supplied to Brazil and Norway (NALLADS programme), while GIRAFFE 75 customers are understood to include Sweden (PS-90) and Venezuela (mounted on MAN LX90 6 × 6 cross-country vehicles). As of this edition, GIRAFFE AMB was in series production for the Swedish Army (under the designation UndE 23) and was on order (September 1999) for an unnamed export customer. In March 2000 it was on order for the French Air Force. A 2-D variant of GIRAFFE AMB (mounted on a MOWAG armoured command and control vehicle and designated as KAPRIS) is also understood to have been ordered during 1999 for use by Sweden's coastal artillery arm. GIRAFFE S is believed to have been acquired by Finland as an air defence 'gap-filler'. Other GIRAFFE (variant unknown) customers are reported to include Greece and Yugoslavia and Ericsson notes that 'basic' GIRAFFE systems have been supplied to 'more than' 10 countries worldwide. In March 2002 it was announced that Ericsson had secured a contract for two GIRAFFE AMB systems for the US Army Aviation and Missile Command, for delivery in 2004.
In early 2003 Ericsson noted that the AMB generation of GIRAFFE radars were contracted for serial deliveries to eight military services, including land-based and naval applications.

Contractor

Ericsson Microwave Systems AB, Mölndal.

UPDATED

Turkey

Modular-V Armoured Reconnaissance and Surveillance Vehicle (MARS-V)

MARS-V is the combination of advanced sensor technologies and computer controlled infrastructure into a single platform. Mounted on 4 × 4 light-armoured wheeled vehicle, this system provides multispectral surveillance using optical and radar sensors. MARS-V has elevated sensors for extended coverage and is designed for continuous monitoring of the battlefield, detecting possible targets

Aselsan Modular Armoured Reconnaissance and Surveillance Vehicle (MARS-V)
0092761

and transmitting the collected information to command centres and reaction forces.

MARS-V uses ground surveillance radar and a second-generation thermal imager for long-range target detection. Two sensor systems can perform surveillance independently on non-coinciding sectors. When a target is detected by one of the sensors, the other sensor can be automatically repositioned to the target area in order to improve target recognition and classification capability. Target identification can be performed using the Doppler tone generated by the radar, supported by the thermal imager and Day TV in the appropriate conditions. Target information is collected by the sensors and evaluated/classified in the information processing software. It can then be transmitted to the command and monitoring centres using various communication equipment integrated in the vehicle. Target information, including co-ordinate and type, are gathered in formatted reports prepared using the software. This can be transmitted securely via the frequency hopping radio, whereas full-motion target imagery can be sent through microwave video link. Still images can be incorporated in the formatted reports.

The main system functions, including control of peripheral equipment and electro-optic sensor system, are controlled by the System Control Unit, which is operated by the System Operator.

Specifications
(a) Sensors:
    ASELSAN ground surveillance radar
    2nd generation thermal imager
    Day TV
    Target co-ordinate determination system
(b) Communication equipment:
    9600 VHF frequency hopping radio
    SK-4000 digital encrypted radio
    Video transmission system (MW Link)
    Field telephone
    Data terminal
    GPS receiver.
(c) Command and control software:
    Geographical information system
    Database
    Formatted message capability.
(d) Target detection range (with radar):
    38 km (vehicle convoy)
    15 km (personnel)

Status
In operational service and in production.

Manufacturer
ASELSAN Military Electronic Industries Inc, Ankara.

VERIFIED

United Kingdom

Airborne STand Off Radar (ASTOR)

ASTOR is a ground surveillance system designed to provide information regarding the deployment and movement of enemy forces. It uses MTI and SAR technology to obtain high-resolution imagery of static features and to identify and track moving vehicles. It is based on a modified Global Express airframe carrying the radar, datalinks and DAS, which will transmit near-realtime imagery to a network of distributed ground stations. The ground stations will be deployed with the front line forces and will display, analyse and interpret the imagery.

The Raytheon designed system will incorporate technologies developed for Raytheon's HISAR radar, the SAR integrated with the Teledyne Ryan Global Hawk Unmanned Air Vehicle (UAV) and in the Lockheed Martin U2's ASARS-2 improvement programme which recently entered flight tests. The ASTOR system is based on a 4.6 m (15 ft)-long passive antenna, electronically scanned in azimuth with a mechanical elevation scan. The antenna has a single high-power transmitter, which feeds the arrays, each of which has a phase shifter to steer the radar beam, whereas an active antenna has a lower power transmitter at each array. ASTOR 'looks' close to the horizon, negating the need for rapid switching of the radar beam. A passive antenna is also lighter, cheaper and needs less cooling, reducing the airframe and power requirements.

The radar has two modes of operation, Moving Target Indication (MTI) and Synthetic Aperture (SAR). The SAR has two sub modes, SWATH and a spotlight mode. Although the radar performs only a single function at any one time, interleaving between the functions allows continual updating of each radar mode. Mission system operators will sit at Sun Microsystems workstations identical to those in the ground stations. The mission crew will control the radar's data collection, and search patterns and will also perform some data processing to aid this process. High bandwidth datalinks enable secure transmission of radar data, negating the requirement for large mission crews and allow the data to be managed by the ground stations and integrated into a Common Operational Picture (COP). A smaller mission crew allows the ASTOR to be mounted in a business jet which in turn allows the radar sensor to he carried to higher altitudes than would be possible with a converted airliner giving a greater stand-off range. At an operating altitude of 51,000 ft (15,600 m) and a 2° grazing angle, ASTOR will have a stand-off range of more than 300 km (160 n miles).

Radar data will be broadcast to ground stations via two datalinks. Ultra Electronics and Cubic will provide the two-way broadcast link and L-3 Communications will supply the wideband equivalent. As well as 'tethering' the aircraft to ground stations, the datalink will also allow data transfer to and from naval assets, attack helicopters, UAVs, offensive aircraft, ground systems such as long-range artillery and reconnaissance aircraft. A satcom link will allow 'off-tether', non-line-of-sight data transmission.

The contract is for five Bombardier Global Express airborne platforms and 10 Motorola-built ground stations, six of which will be highly mobile tactical units able to keep pace with Army brigade level headquarters as they move forward.

Marshall's Special Vehicles will mount the tactical ground stations on customised Steyr Pinzgauer 6 × 6 all-terrain vehicles, with final integration carried out by Raytheon Systems. A tactical ground station consists of three trucks - workstations, communications and support, and two towed generator trailers and a data link trailer. The larger operational level ground stations will consist of two transportable 7 m containers, one for the operator workstations and the other dedicated to communications. Thales will provide the mission planning system.

Status
In November 2002 it was announced that the first ground station vehicle had been delivered for integration. ASTOR is due to enter service in 2005 with delivery of the full system completed by 2007.

Contractor
Raytheon Systems Ltd (Prime), Harlow, Essex.

UPDATED

CLASSIC and CLASSIC 2000(RGS 2740)

CLASSIC
RGS 2740 Covert Local Area Sensor System for Intrusion Classification (CLASSIC) is an evolutionary extension of the (Project LASS) Local Area Sensor System carried out by the Royal Signals Research Establishment and Racal Ltd (now Thales). It is a system that detects, classifies and remotely displays 'target' information on personnel, wheeled and tracked vehicles by means of a variety of remote ground sensors.

The basic system consists of two main units; the sensor and the monitor. Up to eight sensors can be used with each monitor, the former are designed to be hand-emplaced in suitable tactical locations. Each sensor is coupled to a transducer, either a geophone, magnetic or an infra-red detector. The unit contains signal processing circuitry which classifies the input and broadcasts a tone-coded message by means of a built-in VHF/FM transmitter. The monitor unit receives this signal, decodes the data and presents the information on an LED display to show sensor identification, type and frequency of intrusion. To extend the range of the sensor transmission, a relay unit is available.

CLASSIC monitor unit gives audio/visual warning and indicates which sensors have been activated

CLASSIC is a modular system with a range of optional accessories, which include: alternative antennas and battery units; transducer/pressure pad switches; and a hard-copy printer, to meet a wide range of ground sensor applications. Standard sensors are the TA 2741 used with the MA 2743 seismic transducer, the MA 2770 magnetic detector or the MA 2744 infra-red transducer. The TA 2741 has a switch to select High, Medium or Low Personnel seismic sensitivity and another switch to select the classification code.

The RTA 2746 monitor comprises a VHF receiver, a tone decoder and an LED display. On receipt of a transmission from a sensor unit, an audible alarm alerts the operator. The sensor's identification and alarm mode setting is displayed on a matrix of three LEDs for each of the eight possible sensors. The display is inhibited after approximately 14 seconds to avoid excessive battery drain and a push-button is provided to enable viewing on demand if required. A user's map panel provides for drawing of a diagram of a tactical deployment.

CLASSIC 2000

Thales developed a new system known as CLASSIC 2000 which has replaced the existing CLASSIC (covert local area sensor system for intruder classification) system. This is a small, lightweight, easily deployed, hand-emplaced route or critical point ground sensor surveillance system, which can detect and classify personnel, wheeled and tracked vehicles. The target data is reported in a short covert radio message either directly or via a relay to a distant monitor.

A CLASSIC 2000 system comprises four main elements: the RA 4310 monitor unit, the TA 4312 seismic/PIR sensor, the TA 4314 magnetic/PIR sensor (with built-in transducer) and the RTA 4311 relay unit. Associated with the sensor units are a number of different transducer types which permit detection, classification of target and indication of direction. Each sensor has a non-volatile memory to store data programmed from the monitor or MA 4340 fill software.

A single monitor can receive intrusion data from up to 99 sensors. The monitor can be programmed to display alarms from a limited number of sensors from that range and appropriate to the operational scenario. The monitor is used not only to receive and display information to the operator, but as a fill device, to download information to the sensors and relays. A PC using the CLASSIC 2000 fill software programs the monitor itself.

The TA 4312 seismic/PIR sensor is a versatile unit which can operate with various transducer types, including geophone, piezo cable, short-, medium- or

CLASSIC 2000 is hand emplaced (Thales)   *NEW*/0111112

long-range infra-red (PIR) detectors and contact closure devices such as trip wires and pressure pads. The sensor contains all the algorithms necessary for very effective detection and classification from alarm signals and identifies the transducer type by connection. The algorithms are held in software, which permits future field upgrades, as further algorithms are developed.

The seismic algorithms contained within the sensor to provide the detection and classification process the seismic signals generated by the MA 4320 geophone. As the nature of the ground in which the geophone is buried can affect performance, the sensor allows the operator to adjust the seismic gain through several levels. The algorithm retains an adaptive threshold feature for minimising nuisance alarms due to background noise such as rain. Piezoelectric cable is supplied in various lengths from 25 to 1,000 m and is similar in function to the geophone.

The PIR heads used are sophisticated dual-beam pyrometers with high accuracy optics. The signals from these are processed by algorithms, which adapt to background conditions and employ data fusion to look for correlated activity from the twin detection zones. As a result, alarms can provide directional information. The algorithms help to eliminate nuisance alarms.

The TA 4314 magnetic sensor PIR has an integral magnetic transducer. Its principle function is the detection of slow-moving vehicles on made surfaces and armed personnel. Like the TA 4312 sensor it will also accept the range of PI R heads on contact closure devices.

The RTA 4311 relay unit is used to increase the range from the sensor to the monitor or to overcome problems of radio-dead ground conditions. The relay unit also has a non-volatile memory to store data programmed from the monitor or fill software.

To increase the overall operational flexibility of the CLASSIC 2000 system, all sensors, relays and monitors are programmable. The fill software is a Windows based application programme running on an IBM-compatible PC. All fill options (except certain user functions) are set in the fill software and downloaded to the monitor. The monitor is then used as a fillgun to clone the information to the relative sensors and relay units. Sensors and relays can also be filled directly from the PC. A major advantage of this facility is that operator intervention is minimised.

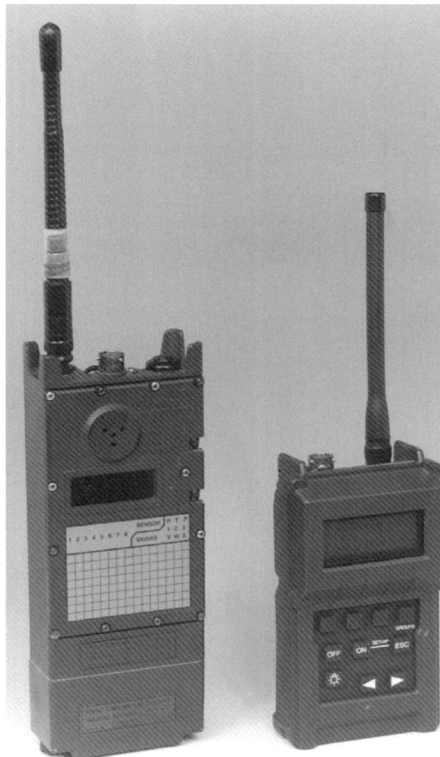

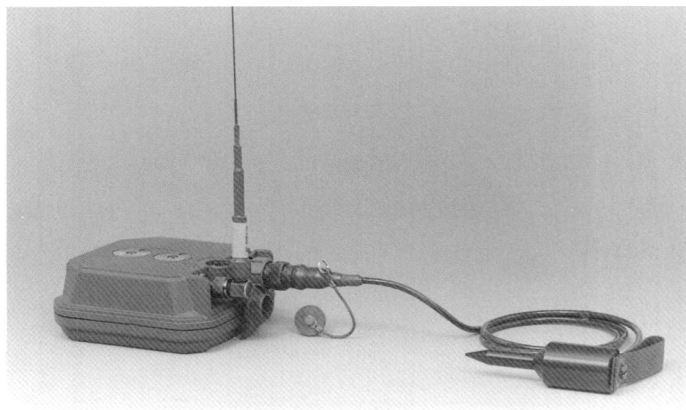

CLASSIC sensor and geophone (Thales)   *NEW*/0053964

CLASSIC (R) and CLASSIC 2000 (L) monitors
NEW/0559365

An additional sensor which can be used with this system is the Miniature Intrusion Sensor (MIS) (see separate entry).

Status

In service with the armed forces of many countries including the British Army. In 1987, Australia placed a contract worth £820,000 for CLASSIC. Thales claims that CLASSIC and CLASSIC 2000 are in service in 42 countries worldwide, including at least 12 NATO countries, in military and para-military (police, customs and border security) organisations.

Manufacturer

Thales Defence Communications, Bracknell, Berkshire.

UPDATED

HALO Hostile Artillery Locator

HALO is a low-cost, passive, acoustic weapon locating system, which is lightweight, mobile and easy to deploy. It provides 360° coverage and is highly accurate, with a typical CEP of 50 m at a range of 15 km, and is operationally available 24 hours a day. The system is resilient to battle damage since failure of one or a small number of sensors will only result in slight performance degradation. This provides the facility in a moving battle to move sensors without losing the capability to locate weapons. In addition, the system provides a high degree of battlefield monitoring of high-energy sound sources, such as tank main armaments and mine explosions, and will not saturate even when there is simultaneous firing from a number of locations. Individual gun positions up to a firing rate of 8 per second can be computed over the area of surveillance; at higher rates of fire, centres of gun batteries are identified. The system can either be used to direct counter-battery fire or to detect cease-fire violations.

The system uses a distributed array of up to twelve sensor posts and a HALO Command Post (HCP). The sensor post consists of a cluster of three microphones, a processor, a solid state meteorological station and a radio working in burst transmission mode. Clusters are nominally deployed between 2 and 4 km apart, their exact location not being critical. If the deployment of a cluster is constrained, the system will still function with the separation of clusters being as little as 500 m or as much as 10 km apart. The microphones are arranged in a small triangular configuration and detect the passing acoustic (pressure) wave generated by gun or mortar fire. Together with real time meteorological data the detection data is passed to the HCP. Further meteorological data is provided from the meteorological station at the HCP.

The HCP consists of a Data Fusion Processor that carries out location calculations, using a digital terrain database, and a Communications Processor that manages the whole communications infrastructure of the HALO system. Received information from each SP is processed and presented to the operator on a screen with a map background, together with system management information. If required target information can then be passed to artillery systems for counter battery fire, and subsequent "fire adjust" information provided" Data can be stored

The HALO meteorological sensor (Patrick Allen/Jane's)
0526711

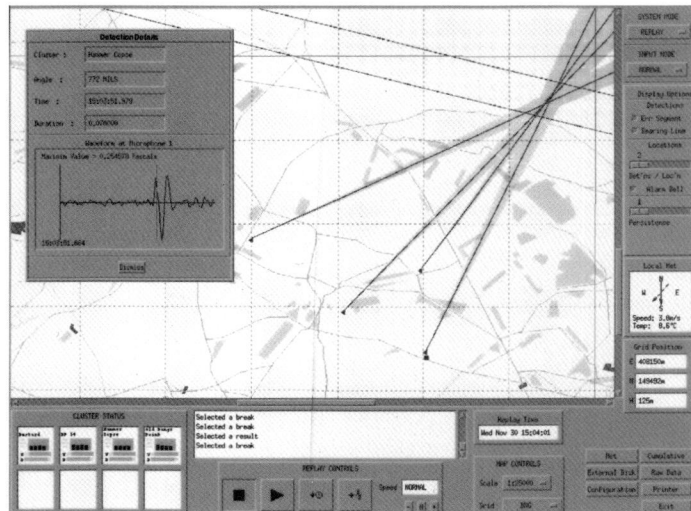

HALO screen   0001026

HALO acoustic sensors (Patrick Allen/Jane's)   0526710

HALO command post   0001027

The HALO central processor unit with attached VHF transceiver (Patrick Allen/Jane's)
0526712

for later more detailed analysis if required. The system features an intuitive, interactive user interface which has been specifically designed to minimise the workload on the operator. The HCP is fully automatic in operation and, by selecting areas of interest (for example, sectors within the ground defence area, an alarm facility is automatically activated. The system does not, therefore, require continuous monitoring.

Status

HALO Mk1 has been in service with the British Army since 1994 and although procured for use in the counter battery role was successfully used in Bosnia and Kosovo in the detection of cease-fire violations. It is claimed to be in use in over twenty armies worldwide. The Mk2 version, which is more accurate, has been selected by the UK for its Advanced Soundranging Programme (ASP); acceptance trials took place in 2002 and the system was accepted into service in 2003. Japan has, and Bahrain is reported to have, acquired the Mk 2.

Contractor

BAE Systems, Christchurch, Dorset.

UPDATED

....

Miniature Intrusion Sensor (MIS)

The MIS is a buried passive ground sensor which provides increased range and capability over the Classic 2000 sensor (see separate entry), with which it retains compatibility. It has an internal seismic transducer and it will accept external transducers such as magnetic, PIR, piezoelectric cable or make/break devices. Detection ranges: seismic 20 to 60 m; piezoelectric cable 20 m; external magnetic 15 to 40 m; external PIR 1 to 100 m (personnel), 1 to 120 m (vehicles). It contains a low-power UHF (433/458 MHz) transmitter which typically provides 0.5 km transmission range, with data being transmitted to a local gateway or a handheld monitor. Further communications relay is possible via Classic 2000, conventional VHF radio, satcom relay or a fixed infrastructure such as GSM or Tetra. The internal battery has a life of 30 days.

There are no external controls and MIS automatically adapts to the background environment. Data fusion takes place internally. In using more than one transducer type, MIS significantly improves target detection and discrimination; further vehicle and aircraft classification algorithms are planned as upgrades. The equipment allows the screening out of unwanted targets and has a low false alarm rate of less than one target per 7 days.

Status

Entering service in several NATO countries. Likely to replace Classic 2000 in many inventories.

Contractor

Thales Defence Communications, Bracknell.

UPDATED

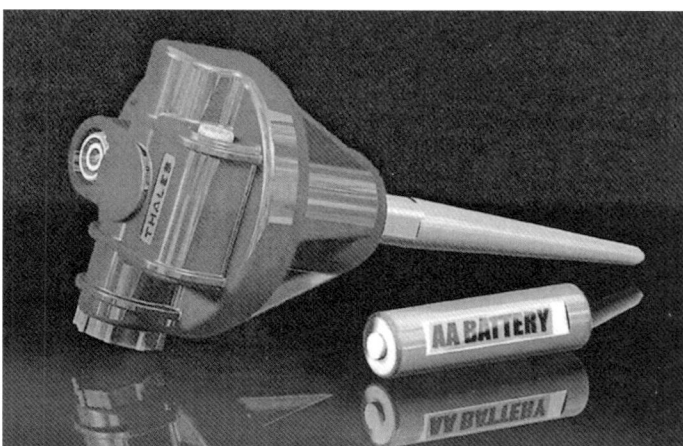

Miniature Intrusion Sensor (Thales)   0096023

United States

Advanced Electronic Processing and Dissemination System (AEPDS)

The Advanced Electronic Processing and Dissemination System (AEPDS) is a C-17/C-5 transportable Tactical Exploitation of National Capabilities (TENCAP) system that supports Corps and Echelon Above Corps (EAC) commanders by receiving and processing intelligence data collected by national, theatre and corps sensors. Following correlation and analysis of this data it produces intelligence reports and imagery products which are then disseminated. The AEPDS has an

external analytical capability and dual function displays for current and future operations. Mounted in a container on a trailer, the system takes 1 hour to set up or tear down, although the associated Radome takes longer.

A variety of products can be produced, including intelligence reports, target data, annotated battlefield imagery and a near-realtime picture. The connectivity options include SUCCESS UHF satcom, MSE, DSCS, TIBS, AUTODIN, STU III, landline and SIPRNET.

The architecture includes VME UNIX based Sun Sparc 20 and a mini data acquisition system. There are four analyst workstations using COTS hardware.

Status

AEPDS entered US Army service in 1997. There are reported to be 10 systems in operation, one with each Corps and the 201st, 501st, 502st, 504th, 513th, and 525th MI Bdes. The USAF also has the AEPDS in two different configurations: the Tactical Data Processing Suite (TDPS) that has the individual components in portable transit cases; and the Enhanced Tactical Data Processing Suite (ETDPS), a similar configuration to that of the Army.

NEW ENTRY

....

AN/FPS-115 Pave Paws radar

Used for SLBM detection and warning and satellite tracking, Pave Paws is the name given to a system of large phased-array AN/FPS-115 UHF solid-state radar to replace an earlier system for ballistic detection and warning. Two of the four original sites are in full operation: at Otis AFB, Massachusetts and Beale AFB, California. Eldorado AFB, Texas and Robins AFB, Georgia have been closed. In addition, versions of the AN/FPS-115 have been used to update the BMEWS sites at Thule, Greenland; Fylingdales, UK; and Clear, Alaska.

The two US-based Pave Paws systems are operated and maintained by the USAF Space Command. The primary mission of detection and warning of SLBM attack also involves the provision of attack characterisation to the NORAD Cheyenne Mountain complex, the Strategic Command and the National Command Authorities. The system's secondary role, in support of the USAF Spacetrack system, feeds in positional and velocity data for display of all earth satellites in orbit. Capable of multitarget tracking, it simultaneously detects and discriminates many objects while providing early warning data, launch, impact, position and velocity information as required.

Automated features of the system include detection, tracking initiation and mission decisions. The two standard computers, CYBER 170/865s, which generally serve as the CPU, are programmed for beam steering and the storage and display of data, as well as performing post-mission data reduction and analysis.

Pave Paws consists of a pair of circular planar phased-arrays about 30 m in diameter, each consisting of nearly 2,000 elements. The arrays are inclined from the vertical by 20° and mounted in adjacent sides of a building measuring about 32 m high, forming sloping walls on the seaward side of the structure. Combined coverage of the electronic beams of the two arrays is 85° in elevation and 240° in azimuth. Range is estimated at about 4,800 km. A version of the AN/FPS-115 system installed at Fylingdales Moor differs from the others in that it has three faces, each with over 2,500 elements, which will provide warning and tracking capabilities over 360° in azimuth.

Specifications
(a) Frequency: 420-450 MHz
(b) Transmitter: solid-state
(c) Module peak power: 322 W
(d) Array type: corporate feed, density tapered
(e) Number of sub-arrays: 56
(f) Antenna gain: 38.4 (directive gain)
(g) Beamwidth (transmit/receive): 2°/2.2° at boresight
(h) Polarisation (transmit/receive): right hand/left hand circular
(i) Array diameter: 22 m (utilised)

Pave Paws phased-array radar installation at Beale AFB, California

(j) Face tilt: 20°
(k) Azimuth: ±60°; 240° (with two faces); 360° (with three faces)
(l) Elevation: 3 to 85°

Status

The Otis AFB Pave Paws was declared operational in 1980, and the Beale AFB system in the following year. The Robins AFB site became operational in 1986 and the Eldorado AFB system in 1987. In March 1988, Raytheon received a US$71 million contract from the USAF to upgrade the two earlier sites, giving them the same enhanced data processing capabilities as the two later systems, and has also received contracts to upgrade two of the three BMEWS sites.

The BMEWS site at Thule was upgraded with a version of the AN/FPS-115 and became operational in 1987. The Fylingdales Moor site was upgraded with a 'three-face' system under a US$167 million contract from the USAF and became operational in 1992.

In late 2001 the AN/FPS-115 was reported as being considered by Taiwan for a ballistic missile early warning system.

Contractors

Raytheon Space and Airborne Systems (prime), El Segundo, California.

UPDATED

Defense Support Program (DSP)

The Defense Support Program (DSP) satellites have been the spaceborne segment of NORAD's Tactical Warning and Attack Assessment system since 1970. Using infra-red detectors that sense the heat from missile plumes against the earth background, the orbiting sentries detect, characterise and report missile and space launches; they also see nuclear detonations.

DSP has repeatedly proven its reliability and potential for growth: the satellites have exceeded their specified design lives by some 30 per cent through five upgrade programmes. These upgrades have allowed DSP to provide accurate, reliable data in the face of changing requirements (for example greater numbers, smaller targets and advanced countermeasures) with no interruption in service. Evolutionary growth has improved satellite capability, survivability and life expectancy without major redesign.

DSP sensor undergoing centre of gravity test

Artist's impression of a DSP satellite in space

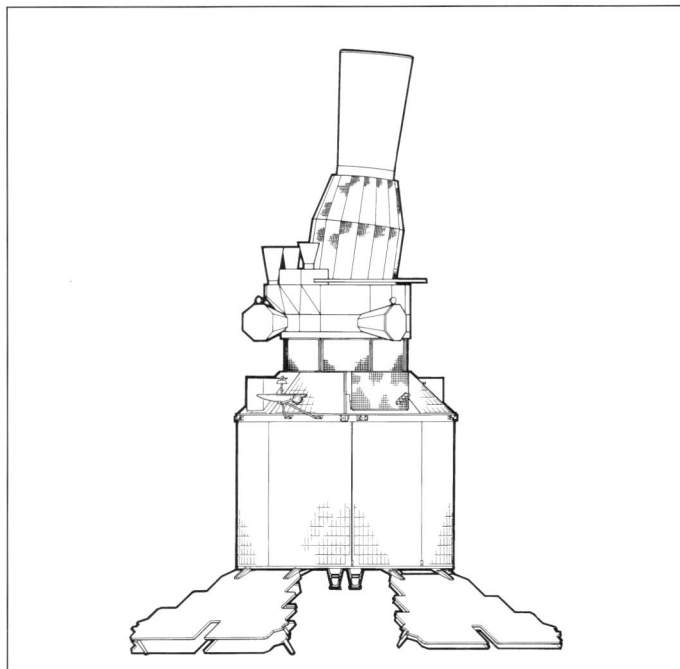

DSP Phase II upgrade - Sensor Evolutionary Development (SED)

DSP performed well in Operation Desert Storm during the 1990-91 Gulf-War, safeguarding coalition military forces and civilian populations by providing a warning of every SCUD missile attack in the conflict. In the case of Desert Storm, US Space Command in Colorado Springs, Colorado, routed data from the satellite system to coalition forces, including Patriot missile batteries, in the Gulf region.

The original DSP satellite weighed 953 kg (2,100 lb), had 400 W of power, 2,000 detectors and a design life of less than 2 years. In the 1970s, as mission requirements changed, the satellite was upgraded to meet the increasingly complex national needs. As part of this upgrade, the weight grew to 1,674 kg (3,690 lb), the power to 680 W, the number of detectors increased by threefold to 6,000, and the design life was three years aiming at a goal of five years. In the 1990s, TRW (now Northrop Grumman) was building a DSP spacecraft that accommodates 6,000 detectors and incorporates significant improvements in survivability. Today's DSP satellite weighs 2,359 kg (5,200 lb) and uses 1,274 W of power.

In 1995 the system was augmented with the Attack and Launch Early Reporting to Theater (ALERT) ground system element for the processing of tactical missile reports (see separate entry).

For more details see *Jane's Space Directory*.

Status

The DSP is operational. Over the last 29 years, there have been 18 satellite launches with five major design changes. These 'blocks' of satellites are:
Block 1: Phase I, 1970-1973, 4 satellites
Block 2: Phase II, 1975-1977, 3 satellites
Block 3: Multi-Orbit Satellite Performance Improvement Modification (MOS/PIM), 1979-1984, 4 satellites
Block 4: Phase II Upgrade, 1984-1987, 2 satellites
Block 5: DSP-1, 1989 - present, 8 satellites launched to date

UPDATED

E-2C Hawkeye Airborne Early Warning aircraft (AEW)

The E-2C Hawkeye is the latest in a series of E-2 Airborne Early Warning aircraft (AEW) designed to provide airborne surveillance and interceptor control at the outermost region of a naval task force's layered defence zone. Operating either from a land base or from the deck of an aircraft carrier, the E-2C can also provide strike and traffic control, search and rescue control, automatic tactical data and communications relay. In 1987, the US Navy had 86 E-2Cs in service and the mission success of these aircraft was reflected in the Navy's intention to procure a total of 160 E-2Cs at an average of six per year well into the 1990s.

Northrop Grumman restarted its Hawkeye production line in St. Augustine, Florida in 1994, after the Navy ordered the first four of an expected 36 new Group II E-2Cs. The Group II AN/APS-145 radar system provides fully automatic overland targeting and tracking capability, an improved IFF system, a 40 per cent increase in radar and IFF ranges, expanded processing capacity and new high-target-capacity colour displays. Additional upgrades include JTIDS for improved secure anti-jam voice and data communications and a Global Positioning System for highly accurate navigation.

The long-range, high-resolution radar, working with Identification-Friend-or-Foe (IFF) and passive detection systems through associated computers, not only develops a picture of the operating environment, but also provides real-time information to air defence centres where command decisions are made. The E-2C

E-2C Hawkeye early warning aircraft

JSTARS showing phased-array radome   0055196

also controls friendly aircraft for pinpoint interceptions through high-speed datalinks. The system can maintain more than 2,000 tracks simultaneously. Track data includes course, speed, altitude and identification of all radar, IFF and passive targets in the computer files.

The newest version is the Hawkeye 2000, which features a new, more powerful, smaller and lighter mission computer, new workstations and other enhancements. The Hawkeye 2000 programme improves the Hawkeye's capabilities in detection, processing, identification, communications, and navigation. Key among the advances is the Mission Computer Upgrade (MCU) with its advanced control indicator set workstations. MCU is a smaller, lighter, more powerful mission computer that allows even more capabilities such as the Co-operative Engagement Center (CEC) upgrade. CEC will enable the Hawkeye to serve as the fleet's information hub, fusing and distributing information from sources such as satellite and shipborne radar. The mission computer upgrade test aircraft flew for the first time in early 1997 and was sent to the Patuxent River (Maryland) Naval Air Test Center for testing. Low-rate initial production of the new mission computer was declared in July 1997.

The Group II E-2C flight crew comprises two pilots, a radar operator, an air control operator and a combat information center operator. The three system operators work independently in all operational roles: sensor utilisation, monitoring and control of the tactical situation and relay of tactical information to key battle group participants.

Powered by twin Allison turboprop engines, the Hawkeye combines fuel economy with short take-off capability. Current production aircraft are equipped with T56-A-427 engines, which provide longer mission duration, greater range and higher altitude capabilities. In its present configuration, the E-2C can cruise on station for more than 4 hours, 200 miles from its base.

Status

The US Navy took delivery of the first E-2A Hawkeye, in 1964, and 59 were delivered by 1967. They flew in Vietnam combat from the USS *Kitty Hawk* and USS *Ranger*. E-2As were modified to E-2Bs with a new programmable, high-speed digital computer. The E-2C programme began in 1968. The E-2C prototype made its first flight in 1971 and the first 11 operational aircraft were delivered to the Navy two years later. Since then, deliveries have totalled more than 140 for the Navy and more than 30 for overseas customers. The US Navy has ordered 21 Hawkeye 2000, with production due to take place until 2006. Hawkeyes are also operated by the air forces of Egypt, France, Israel, Japan, Singapore and Taiwan. The Israeli aircraft are no longer in service, while some from Egypt, France, Japan and Taiwan are to be upgraded to Hawkeye 2000 standard. The first of the six aircraft to be upgraded for Egypt (total programme cost US$174 million) was delivered in March 2003, with the second due in early 2004. See *Jane's All the World's Aircraft* for comprehensive details.

Contractor

Northrop Grumman Corporation.

UPDATED

Joint Surveillance and Target Attack Radar System (JSTARS)

The Joint Surveillance Target Attack Radar System (JSTARS) is a long-range, air-to-ground surveillance system designed to locate, classify and track ground targets in all weather conditions. The JSTARS system is designed to detect, locate and track moving and stationary ground equipment targets.

Description

JSTARS consists of two elements: the E-8C airborne platform with the AN/APY-3 high-performance, multimode, airborne radar system and US Army mobile Ground Station Modules (GSMs).

The E-8C, a modified Boeing 707, carries in a 26 ft canoe-shaped radome under the forward part of the fuselage a side-looking phased-array that is electronically scanned in azimuth and mechanically scanned in elevation. The radar is capable of providing targeting and battle management data to all JSTARS operators, both in the aircraft and in the ground station modules, over a field of view of over 40,000 sq km. The aircraft Operations and Control subsystem has 17 operator workstations, one navigator/operator workstation and an extensive communications suite. The standard crew is 21, but an augmented crew of up to 34 can be carried.

The radar's fundamental operating modes are Moving Target Indicator (MTI) (both Wide Area Surveillance and Sector Search) and Synthetic Aperture Radar. MTI is designed to detect, locate and identify slow-moving targets. Through advanced signal processing, JSTARS can differentiate between wheeled and tracked vehicles. In the MTI mode operators can select the desired radar beam revisit rate, size of the coverage area and resolution. They can also perform history playback, track targets and perform target/position predictions. History playback consists of time compression and time integration functions. In time compression MTI data is recorded over time and the frames are then fast forwarded in a manner similar to a video cassette recorder. This makes it easy to track the target's start point, route of movement and end point. Time integration is the overlaying of successive frames on top of each other for a selected period of time and displaying the frames all at one time assisting, for example, the identification of lines of communications and movement patterns. Since vehicular movement is associated with most military activities, MTI data can be an excellent cue for operators on where and when to collect a high-resolution SAR image.

Synthetic Aperture Radar/Fixed Target Indicator (SAR/FTI) produces a photographic-like image or map of selected geographic regions. SAR data maps contain precise locations of critical non-moving targets such as bridges, harbours, airports, buildings, or stopped vehicles. The FTI display is available while operating in the SAR mode to identify and locate fixed targets within the SAR area. The SAR and FTI capability, used in conjunction with MTI and MTI history display, allows post-attack assessments to be made by onboard or ground operators following a weapon attack on hostile targets.

The Operations and Control (O&C) subsystem provides onboard exploitation and control. It consists of a real-time, VAX-based distributed processing architecture and includes individual DEC ALPHA-based digital processors at each of the 17 operator workstations and the one navigator/operator workstation. The O&C subsystem's computing architecture, combined with its mass memory and high-resolution colour graphic displays, provides all workstations with simultaneous access to the radar's products and the system's exploitation tool kit. Simultaneous access makes it possible to assign workstations in any combination of functionality, ensuring maximum mission flexibility. The tool kit permits the easy and rapid exploitation of the radar's many products. The individual workstations, each with more than 500 Mips of processing power, manage the exchange of radar data and provide displays of easy to exploit radar products. MTI data is displayed as actionable moving imagery that shows moving targets as dots of light while SAR data is displayed as a high-resolution still photo-like image that shows stationary vehicles and surrounding terrain.

The JSTARS Ground Station Module (GSM) is a Mobile Multisensor Imagery Intelligence (IMINT) tactical data processing and evaluation centre. The GSM processes data from the JSTARS aircraft Commanders Tactical Terminals (CTT), Joint Tactical Terminal (JTT), and Unmanned Aerial Vehicles (UAV) and disseminates intelligence, battle management and targeting data to Army Command, Control, Communications and Intelligence (C³I) nodes via LAN, wire or radio. This enables integrated battle management, surveillance, targeting and interdiction plans to be developed/executed using near-realtime data.

Two separate GSM configurations exist. The Medium GSM (MGSM) is housed in a Standard S280 shelter and mounted on a 5 ton truck. A lightweight, rapidly deployable variant, the Light GSM (LGSM) (see separate entry) is housed in a Lightweight Multipurpose Shelter (LMS) and mounted on a High Mobility Multipurpose Wheeled Vehicle (HMMWV). These are being replaced by the Common Ground Station (see separate entry).

Potential future JSTARS GSM improvements were evaluated in JWID-97. As part of a combined demonstration plan, the Time Critical Targeting Aid (TCTA) and the JSTARS Imagery Geolocational Improvement (JIGI) software seek to enhance the value of JSTARS data within the AOC and other users for Theater Missile Defense (TMD) and the application of standoff weapons. TCTA, an intelligence and targeting operations tool, combines in a single display DSP, SIGINT and other intelligence information with MTI SAR data to allow rapid targeting within TMD timelines. TCTA also provides historical database and analysis tools for the identification of traffic patterns, loading points, hide sites and other areas of intelligence value. JIGI, a modified version of the Multiimage Exploitation Tool (MET), produces targeting-quality geolocational information by registering imagery collected by JSTARS' onboard sensor with archived national imagery (or any other precise datum). Once registered, the geolocational accuracy of the national image is transferred to the JSTARS image, improving the accuracy by up to an order of magnitude over a JSTARS-only product. Locations of mobile/relocatable targets not present when national imagery was taken can be obtained with sufficient accuracy to provide a Desired Mean Point of Impact (DMPI) for Joint Stand-Off Weapon (JSOW) and Wind-Corrected Munitions Dispenser (WCMD).

The Joint STARS program office has planned a series of upgrades for the E-8C, both block upgrades and modifications to improve the supportability of the airplane. The Block upgrades planned are:

Block 10 consists primarily of Tactical Digital Information Link-Joint (TADIL-J) upgrade and Y2K compatibility.

Block 20 consists primarily of the Computer Replacement Program (CRP), which replaces the current five-computer system with two Commercial-Off-The-Shelf (COTS) computers, and facilitates upgrading the E-8's computers in parallel with industry.

Block 30 includes the integration of SATellite COMmunications (SATCOM), the incorporation of additional TADIL-J messages, and the integration of the Improved Data Modem. Only SATCOM integration is currently funded.

Block 40 consists of the Radar Technology Insertion Program, which replaces JSTARS' radar, adding several significant improvements to both the SAR and moving target indicator radar modes.

The Radar Technology Insertion Program (RTIP) is a Pre-Planned Product Improvement (P3I) effort where the contractor will be required to design, develop, install, test, and integrate advanced radar systems in the Joint STARS system. The RTIP Engineering Manufacturing Development (EMD) program will design, integrate and test an advanced RTIP sensor subsystem for the E-8C, sufficient to enable a production decision, and transition into a production, retrofit program. The program will explore wide band datalink and anti-tamper security considerations. The E-8C Joint STARS system baseline resulting from the Computer Replacement Program and TADIL-J upgrade will be the starting baseline. On 30 November 1998 Northrop Grumman Corporation's Electronic Sensors and Systems Sector (ES3) and Raytheon Systems Company announced an agreement calling for an equal '50-50' workshare on the radar sensor portion of RTIP. The Northrop Grumman Integrated Systems and Aerostructures (ISA) sector will continue as the RTIP prime contractor with Raytheon as a subcontractor to Northrop Grumman ES3. Northrop Grumman's Integrated Systems and Aerostructures Sector, which is the RTIP prime contractor, will design, develop, install, test and integrate advanced radar systems into Joint STARS at its Airborne Surveillance and Battle Management Systems unit in Melbourne, Florida.

The Improved Data Modem (IDM) connected to four radios and the concurrent installation of a SINCGARS radio with the IDM and an additional SINCGARS hot spare provides an interoperable, full duplex, direct targeting support datalink to the US Army's Army Aviation Command and Control System (A2C2S) and Apache attack helicopters. This installation will be accessible from any E-8 Operator Work Station (OWS), will be fully logistically supportable and includes associated C2 and attack support messages. Joint STARS E-8 does not have the capability to provide direct datalink targeting information to A2C2S and Apaches. The E-8 communications suite does not have fully compatible and interoperable VHF voice and data-capable radios with US Army aviation and ground forces. This capability emulates the initiative to provide Joint STARS data to fighter aircraft. This vastly improves targeting support primarily to army aviation by providing a fully interoperable datalink to C2 and attack helicopters. This capability will also increase Army Joint STARS Common Ground Station (CGS) communications capabilities through the E-8 to Army aviation when helicopters are Beyond-Line-Of-Sight (BLOS) of the CGS. This also has potential to support USMC aviation and USAF IDM-equipped fighters and to decrease fratricide. The IDM can also pass Apache sensor information back to the E-8, increasing situational awareness and improving target correlation capability.

The JT3D-7 engine upgrade modification will allow the E-8 to operate between FL 340 to FL 420 with a climb to altitude within one hour. It will increase the capability of flying a ten hour sortie without air refuelling (2 hour transit time, 8 hours in tactical orbit). It will also increase the capability of flying a 20-hour sortie with an air refuelling (2 hours transit time, 1 hour for air refuelling, 17 hours in the tactical orbit). Low-engine thrust limits deployment of Joint STARS to only those airfields with long runways. The E-8 can operate at maximum gross weight on only the longest runways (10,000 ft) under optimum weather conditions. Any crosswind, gust, or wet runway conditions severely limits take-off gross weight. Low-thrust

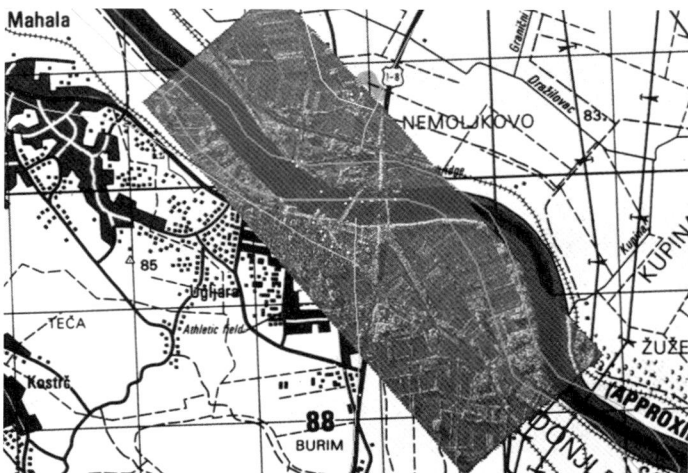

JSTARS Moving Target Imagery superimposed over a Synthetic Aperture Radar image with a map underlay (Northrop Grumman)   0525559

engines limit capability to meet required operating altitudes between FL 340 and FL 420. At heavy gross weights the E-8 cannot meet climb and on-station requirements. Higher thrust engines will provide faster climb to higher operational altitude. This improvement increases sensor coverage, on orbit time, communications reach, survivability, and decreases sensor screening. E-8 engines are being purchased from available commercial stocks which include many JT3D-7 engines. The JT3D-7 engines will have to be down scoped to match the existing JT3D-3B engines if this upgrade is not funded. The total requirement of 80 operational engines plus spares is not reflected in the schedule and cost because some JT3D-7 engines have already been purchased but not yet down scoped.

The Programmable Signal Processor (PSP) replacement replaces four PSPs with two COTS processors with five additional SHARC processor cards for a total of eight. The solution replaces the four PSPs with two COTS processors with five additional SHARC processor cards, replaces the existing PSP/GPC LAN with a fibre ring, redesigns the PSP code from microcode to HOL, redesigns the PSP rack configuration from a four rack to a two rack design to include installation, power, cooling, and cabling and improved diagnostics and Shop Replaceable Unit (SRU) maintainability. The aircraft currently uses four PSPs, which work at maximum processing capacity providing adequate mission support. The E-8 has little growth capability for increased processing required for sensor upgrades. Current processors incorporate inefficient sensor idle time when processing Synthetic Aperture Radar (SAR). The current PSPs also have a high potential for becoming a Diminishing Manufacturing Source (DMS). Currently there is no sensor or processing growth potential and no open architecture capability. This upgrade provides improved radar timeline by eliminating the sensor idle time. It is required to provide growth processing and memory capacity for sensor upgrades (ESAR, ISAR, ATR). It improves supportability of both hardware and software components of the PSPs, and provides an open architecture base and limited weight and space reductions.

The Enhanced Synthetic Aperture Radar (ESAR) and Inverse Synthetic Aperture Radar (ISAR) upgrades allow for target classification and identification through a six-fold enhancement of current SAR resolution with ESAR and the ability to image moving targets and perform mensuration with ISAR. This upgrade assumes the PSP replacement is already implemented. The upgrade also increases both range and azimuth resolution. ESAR and ISAR are concurrent upgrades to reduce cost of Non-Reoccurring Engineering (NRE) and testing. ESAR requires 27.5 kbytes Software Lines Of Code (SLOC) and 34.5 kbytes SLOC for ISAR. The E-8 SAR resolution does not provide for classification or identification. The E-8 SAR resolution provides some target situational awareness and terrain mapping. ESAR and ISAR will contribute to more accurate targeting data and support potential growth to Automatic Target Recognition. ISAR also supports maritime potential by using the translational motion of the targets. The primary applications support Theater Missile Defense (TMD) identification of high-value mobile targets such as SCUD Transporter-Erector-Launchers (TELs). This capability also increases targeting capability, location and identification accuracy, and the potential for fratricide reduction.

The SAR management upgrade is a software modification that allows for the storing of a nominal mission's worth of SAR images in a centralised retrievable database. The estimated Software Lines Of Code (SLOC) count for this is 6 kbytes. The E-8 Operator Work Station (OWS) can only hold 16 SAR images in the local memory. This is basically a screen store and recall capability. This upgrade would provide the capability to store all SAR imagery collected during a nominal mission. The system will have a master SAR file with all the SARs saved as well as individual save files with OWS unique entries (a subset of the master file). When the Radar Management Officer (RMO) receives a request for SAR, the system automatically searches the SAR imagery database to determine if images have already been taken of the area. If imagery exists, the RMO will be notified and also be provided the option of satisfying the Radar Service Requests (RSR) with the existing imagery rather than tasking the sensor again. The database will be accessible from all OWS, include date/time and position data for each image, a search engine capability to recall images, and provide a SAR-to-SAR comparison capability.

There are four phases to the Joint STARS Link 16 Upgrade programmes: Current Capability, TADIL J Upgrade (TJU), Theater Missile Defense (TMD), and Attack Support Upgrade. When added to the current capability, TJU provides a basic, rudimentary Link 16 capability for passing ground tracks to link participants. TJU was operational by the fourth quarter of FY99. TMD will add three messages and part of another message to provide Joint STARS with the capability to identify, monitor and report Transporter Erector Launchers (TELs), TEL reload locations, and TEL hide locations. The ASU adds 25 Link 16 messages to the Joint STARS database. The upgrades allow Joint STARS to realise its attack support role by passing sensor to shooter information for target assignment, target sorting, target/track correlation, and various command and platform management taskings. The implementation of these messages gives Joint STARS a robust, command and control, full up battle management capability. Software development and implementation will occur concurrent with each program software annual release. The Link 16 upgrades will provide Joint STARS with the capability to contribute heavily to TMD, interdiction, SEAD, and CAS mission areas. The current Joint STARS Link 16 capability is very limited. The E-8 can transmit and receive airborne link participant location and identification (PPLI) messages, receive air track and track management messages, and transmit part of a ground track message. Without this upgrade, Joint STARS can not effectively contribute to its attack support mission as called for in the Joint STARS ORD, CONOPS and theatre employment documents. Primary communications between Joint STARS and fighter aircraft will remain voice radio, and without the upgrades, Joint STARS will realise only a small portion of its potential as a sensor to shooter platform for the air force. Concurrent rather than sequential development and implementation of these upgrades will provide substantial cost savings. This upgrade directly enhances TMD targeting mission execution. TJU is funded (except for US$3.8

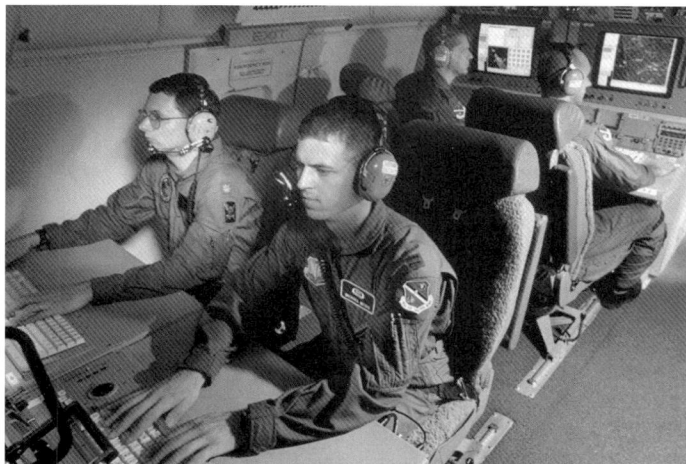

Operators at the Operations & Control workstations in the E-8C JSTARS
(Northrop Grumman)   0525558

million for ground support). TMD, ASU and future enhancements are totally unfunded.

Joint STARS Intelligence Broadcast System initially provides receive-only capability of the TRAP, TADIX, and TIBS broadcast nets. These nets provide near-realtime updates from multiple intelligence sources at the SECRET level to support situational awareness, intelligence preparation of the battlefield, cross-cueing, radar scope interpretation assistance, battle management and mission planning. This upgrade improves Theater Missile Defense (TMD) support, Order of Battle (OB) databases, self-defence awareness, situation assessment and attack planning. The broadcast information would be integrated into all the E-8 workstations to allow the individual operators to conduct overlay and comparison of Joint STARS sensor data and broadcast system data.

Joint STARS Automatic Target Recognition (ATR) provides Joint STARS operators with automated surface target recognition/identification. This enhances operator efficiency in high-density situations and exponentially increases current capabilities for surface target identification. ATR provides higher mission crew situational awareness and increases support to battle management and attack support. In support of TMD, the system will be able to locate, track, and identify missile Transporter Erector Launchers (TELs) vehicles upon cueing from offboard sensors/sources. The ATR concept is based upon algorithms using processed radar data (ESAR and ISAR) and applying Radar Cross Section (RCS) or templating techniques to classify/identify ground and maritime targets. ATR is a computational technique which compares the SAR imagery with imagery templates of high-value targets to quickly identify and locate those targets in the image. This requires a large detailed database of potential target image templates and the processing capability to perform comparisons with Joint STARS sensor data. This capability includes integration into all the E-8 operator workstations and assumes implementation of first the Programmable Signal Processor (PSP) and then secondly the ESAR and ISAR upgrades. The initial effort is to develop and demonstrate an ATR capability on Joint STARS. ATR is not yet at a stage for insertion into Joint STARS production models or retrofit of existing aircraft.

There is no automated target recognition or identification capability on the E-8. Currently, mission crew must cognitively fuse offboard sensor information, current situation awareness, and onboard sensors to make any type of recognition call. This type of analysis produces a low confidence level and requires a level of training which is not provided to operational mission crew members. ATR would allow for more timely and accurate target support and battle management decisions. The PSP replacement and ESAR/ISAR upgrades must be accomplished before development of the ATR capability. JSTARS Tagging is an unfunded requirement to develop and implement a Joint STARS Radar Responsive (R2) Tag System comprising two types of R2 Tags and corresponding functionality on board Joint STARS aircraft. The Joint STARS radar has two primary modes of operation, Moving Target Indicator (MTI) and Synthetic Aperture Radar (SAR). The R2 Tags are designed to work with the rapid revisit, MTI mode of the radar, providing positive identification of targets equipped with the tag. The tagged targets will be visible to the radar operator whether they are moving or stationary, as long as they are located within the radar field of view. The R2 Tags are also designed to interface with Unattended Ground Sensors (UGS) to provide data collected by the UGS to Joint STARS operators on board the E-8 aircraft. This data could be a frame of video image, time/date of acoustic sensor activation, or other data provided by a UGS sensor suite. The Radar Responsive Tag will allow the Joint STARS aircraft to positively identify any tagged vehicle, person, or structure. This capability addresses the need for wide area surveillance capability to detect, locate, track and identify time critical targets and correlate and fuse data.

Enhanced Joint STARS ATR is an unfunded requirement to provide an enhanced means of targeting critical mobile ground targets. The technical objective is to accelerate the transition of targeting enhancements to the Joint STARS system. These enhancements enable more effective targeting against Time Critical Targets (TCTs). There are two primary goals: (1) demonstrate the robustness of using Hi-Resolution Synthetic Aperture Radar (SAR)-based Automatic Target Recognition (ATR) technology for improved identification of Time Critical Targets (TCTs) and; (2) demonstrate the effectiveness of using Hi-Resolution Moving Target Indication (MTI) sorting of targets in a scan mode to pick out target areas of interest in a non-co-operative mode. The approach is to upgrade software capability on board Joint STARS to take advantage of available hardware

and processing to allow for Enhanced Synthetic Aperture Radar (ESAR) and High-Range Resolution (HRR) MTI. The ESAR capability will provide a resolution 6 × the baseline resolution. The HRR/MTI provides target vehicle range extents and integrating this with a tracker capability provides vehicle target length measurements. Conceptually, the HRR/MTI is being used as a cueing mechanism for the detection of TCTs by length measurements. This cue is handed off to the ESAR algorithm for imaging the long length vehicles for the ATR to perform non-co-operative target ID of Tactical Erector Launchers (TELs). The current system performs ATR of TEL-type targets in the clear (no obscurations of the target).

Status

The E-8C programme is for a total of 15 aircraft at a unit cost of US$648.6 million. By February 2003 15 aircraft had been delivered, the most recent five in Block 20 configuration. Upgrading of the first 10 aircraft to Block 20 configuration was also in progress, with two completed and two in progress.

JSTARS was used operationally in support of NATO forces operating in former Republic of Yugoslavia and in support of operations in Afghanistan. It also supports operations over Iraq. All E-8C aircraft are based at Robins Air Force Base and are operated by the 116th Air Control Wing of the Georgia Air National Guard, a mixed unit of both active USAF and Air National Guard personnel.

Contractor

Northrop Grumman Integrated Systems (Prime contractor), Melbourne, Florida.

UPDATED

Relocatable Over-The-Horizon Radar (ROTHR) (AN/TPS-71)

The AN/TPS-71 Relocatable Over-The-Horizon Radar (ROTHR) is a tactical land-based, bistatic ionospheric backscatter radar system originally designed to provide wide-area surveillance as part of US national defences during the Cold War, but now principally used to track aircraft movements in support of the US national Counter Drug (CD) mission. At the operator's option the system can be used for overall surveillance and tracking within the coverage area, spotlighting specific regions to handle targets of interest, or the assessment of the size of a target complex.

The ROTHR system was designed to be relocatable to previously prepared sites in support of its original mission of tactical surveillance. As a CD asset, however, it is a permanent installation and the data is remoted to the main Operations Control Centre (OCC) in Chesapeake, Virginia, from each sensor via a high-quality telephone line.

The ROTHR system depends on the ionosphere to see over the horizon. The changing conditions of the ionosphere must be monitored frequently to ensure accurate tracking performance. This monitoring function is usually performed by remote downrange sounding radars. However, the ROTHR provides this function from the sensor location using a Quasi-Vertical Incidence (QVI) sounder and a backscatter sounder to accurately model the ionosphere using Propagation Management and Assessment (PMA) algorithms.

Description

ROTHR enables the detection of aircraft at ranges of up to 2,000 miles in an arc in excess of 80°. The system consists of three distinct elements: a transmit site, a receive site and an Operations Control Centre (OCC). The transmit and receive sites are separated by 50 to 100 miles. The control centre can be collocated with the receive site but is usually remoted using a high-quality telephone line as the communications medium. The transmit site provides radar illumination in accordance with commands from the OCC. It also provides the transmissions for the QVI and the backscatter sounders used in the PMA function. In the receiver element, the returned beams are formed digitally in the signal processor which also

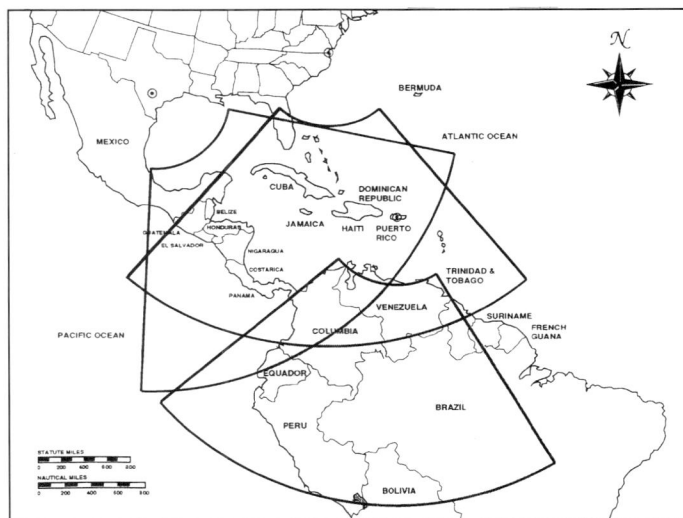

ROTHR Virginia, Texas and Puerto Rico coverage

ROTHR transmit antennas at the Virginia site

carries out the range and Doppler processing and extracts the target detections. The detections, together with the raw processed data from which they were extracted and signals from the sounder transmissions, are passed to the OCC for further processing.

The OCC is the nerve centre of the system and carries out the processing of the returned target information. Within the centre the operators interface with the system via stand-alone displays driven by a data processing subsystem. The system is largely automatic in operation but can be overridden if necessary.

Status

Raytheon/TRW team received a contract in 1984 to develop the ROTHR prototype system. It was tested in 1988 in Virginia and later redeployed to Amchitka, Alaska, where it was further tested and became operational. Based on the results of these tests, Raytheon was awarded a contract in 1989 to build three systems. These systems have been completed. The first system was made operational at the site of the prototype in Chesapeake, Virginia, in March of 1993; the second radar is installed near Corpus Christi, Texas, and became operational in June 1995; the third system was installed in Puerto Rico and became operational in 1999.

The system is being continuously upgraded with new hardware and software under contracts to Raytheon and others. In 2003 Raytheon was awarded a US$24.8 million contract for operation and maintenance services. Work was expected to be completed by September 2004.

Contractor

Raytheon Integrated Defense Systems, Tewksbury, Massachusetts.

UPDATED

REMBASS II

The Remotely Monitored Battlefield Sensor System (REMBASS) is an unattended ground sensor system that detects, classifies and determines direction of movement of intruding personnel and vehicles. It provides worldwide deployable, day/night, all-weather early warning surveillance and target classification. The earlier version, Improved REMBAS (IREMBASS), has now been upgraded and is offered as REMBASS-II, which is half the weight and volume of its predecessor. It is

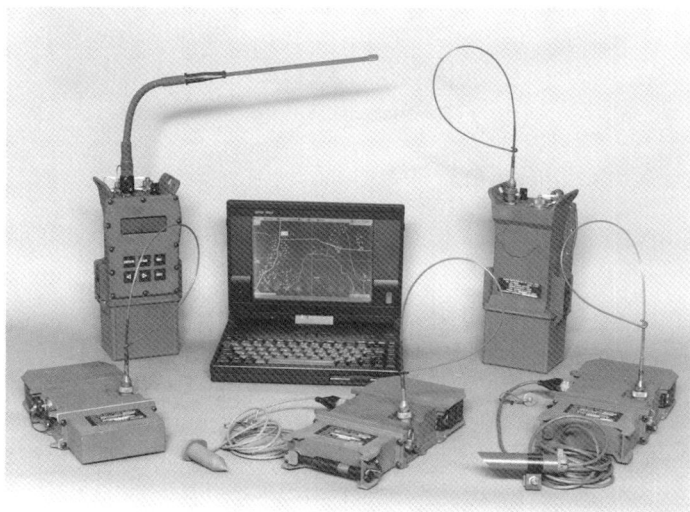

IREMBASS   0011527

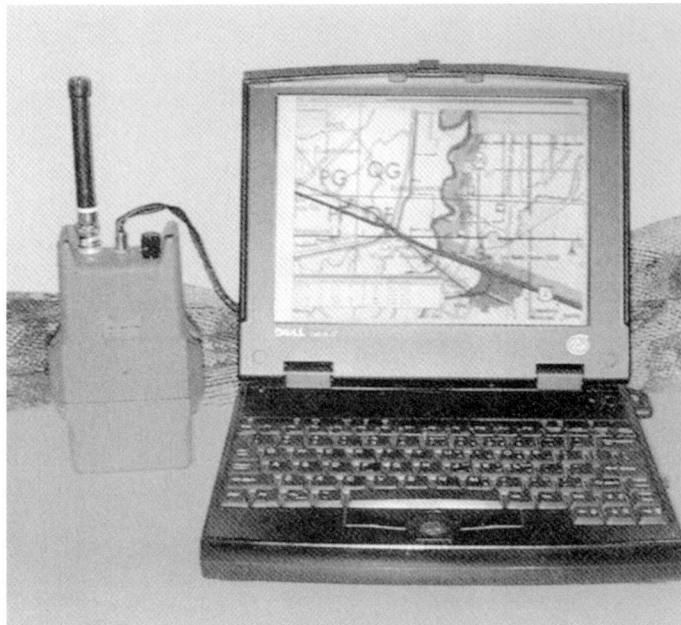

REMBASS-II Situational Awareness Display (L-3 Communications)   0109971

the US Army's type standard unattended ground sensor system and is being procured as the Platoon Early Warning Device-II.

The system works by using remotely monitored sensors placed along likely approaches. These sensors can respond to a wide variety of influences: infra-red, acoustic, seismic, mechanical energy, magnetic field changes and others (REMBASS-II will accept inputs from SIGINT, NBC or meteorological sensors. Data from the latter will interface with IMETS (see separate entry)). The PEWD-II kit contains three Seismic/acoustic and three microphone/geophone sensors. The sensors communicate target data messages up to 15 km using LPI/LPD burst transmissions in the VHF band. Optional relay devices can extend this range by 15 km using a REMBASS-II repeater, 150 km using a UAV relay, or worldwide using a Field Processor Unit (FPU) SATCOM relay.

The data from the sensors is processed by operator display software and can be used to determine the type, number and direction of targets and estimate location and speed. Detections are received and displayed on a small hand-held monitor which can be connected to a laptop via an RS-232 port to produce a graphical depiction of target activity using digital mapping products.

Specifications
REMBASS II

(a) Detection range:
    Seismic/Acoustic: 75 m personnel; 500 m vehicles; 750 m tracked vehicles. 3 sensors give a frontage of 450 m for detection of personnel.

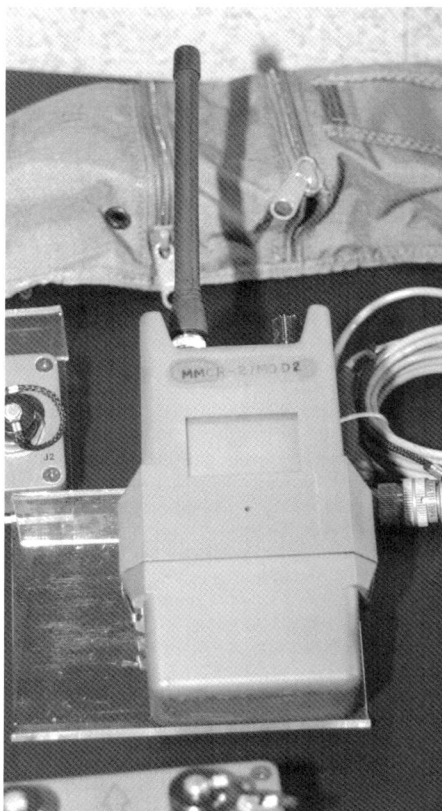

REMBASS-II hand-held monitor
(Patrick Allen/Jane's)
NEW/1029481

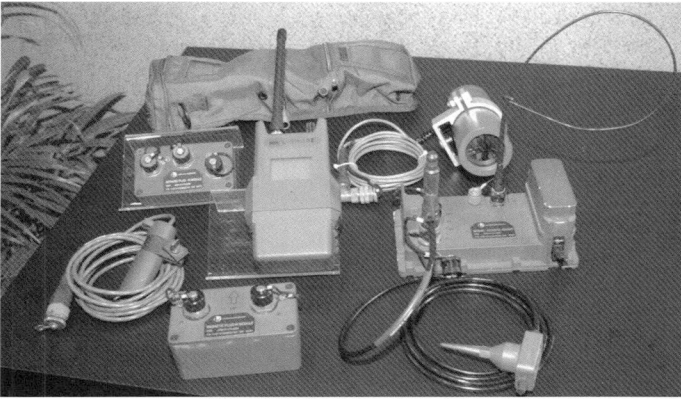

REMBASS-II. Front left, IR sensor; centre, hand-held monitor; front right, seismic sensor; centre right, transmitter; back right, magnetic sensor (Patrick Allen/Jane's)
***NEW**/1029495*

Infra-red: 30 m personnel; 75 m vehicles; 75 m tracked vehicles
Magnetic: 3 m personnel; 15 m vehicles; 25 m tracked vehicles

(b) Frequency: 138-153 MHz, 599 channels at 25 kHz spacing, LPI/LPD burst transmission
(c) Power: 2 W transmitter
(d) Transmitter range: 15 km LOS; 150 km airborne
(e) Endurance (dependent on temperature):
Sensors 30-45 days @ 1,000 activations/day
Supplemental battery box available giving additional 60 days' mission life
Monitor 7-12 days @ 4,000 activations/day
Repeater 25 days @ 4,000 activations/day

Status

IREMBASS is currently in service with the US Army, US Air Force, Israel and Kuwait armed forces, and has been used operationally in Iraq and Afghanistan. Spain and New Zealand armed forces units were delivered in 1998. Over 6,000 units have been delivered to users to date.

167 REMBASS-II (as PEWD-II) were procured by the US Army for two IBCTs in 2002. The planned scale of issue of PEWD-II is one per infantry/engineer platoon and two per MP platoon, but it is a currently unfunded programme.

After the attacks of 11 September 2001 REMBASS sensors were used in conjunction with other surveillance devices to assist in securing the nuclear power stations on the Delaware river in New Jersey.

Contractor

L-3 Communications-East, Camden, New Jersey.

UPDATED

SPAce Detection And Tracking System (SPADATS)

SPADATS is the North American Aerospace Defense (NORAD) Command's worldwide system for the detection, tracking and identification of all objects in space. It is composed of large radar, optical and radio-metric sensors located around the globe and its control centre maintains a catalogue of all objects detected. The US Space Command co-ordinates the contributions from the US Air Force (Spacetrack), Navy (NAVSPASUR) and Army to SPADATS.

Spacetrack is the codename for System 496L, the US Air Force's input to SPADATS and is a mix of radar and optical sensors.

Main radar sensors of this system are the AN/FPS-85 system at Eglin Air Force Base (AFB), Florida; the Cobra Dane radar at Shemya AFB, Alaska and the Pave Paws radars at Otis AFB, Massachusetts, Beale AFB, California, Robins AFB, Georgia and Eldorado AFB, Texas. These radars are supported by the Ballistic Missile Early Warning System (BMEWS) installations at Thule, Greenland; Clear, Alaska and Fylingdales, UK.

The current Spacetrack optical system is a four-site Baker-Nunn camera system with sites at San Vito, Italy; Sand Island in the Pacific; Mount John, New Zealand and Edwards AFB, California. It is capable of tracking and identifying satellites out to synchronous orbit altitudes of 36,800 km.

VERIFIED

Terrain Commander

Terrain Commander is an integrated remotely monitored covert surveillance system that allows a number of surveillance sites to be monitored from a central location. The system combines communications and mapping technologies with a supporting array of acoustic, seismic, magnetic, electro-optical and passive infra-red sensors. Depending upon its configuration, Terrain Commander can detect activity, capture and process images of the activity, and transmit subject data and images to the Command and Control Station (CCS) in near-realtime.

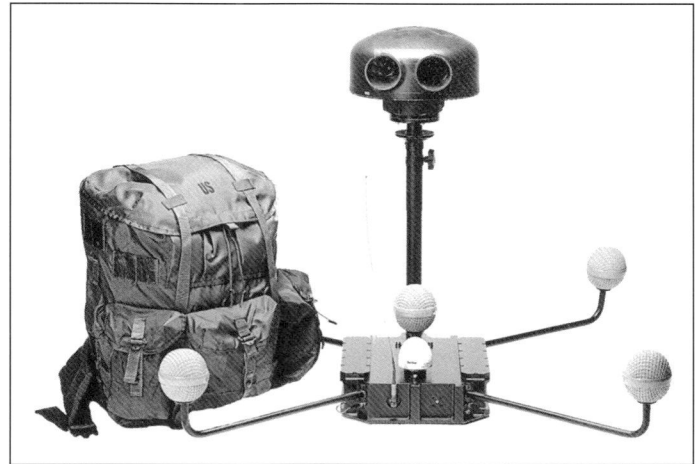

The Terrain Commander OASIS. The electro-optic sensor is on the mast, with acoustic sensors on each of the arms. The signal processing and communications package is in the case in the centre (Textron Systems) 0137198

Description

The Terrain Commander system consists of two major components: the deployed sensor package and the CCS. The sensor package is provided in two forms, the OASIS (Optical Acoustic Satcom Integrated Sensor) which is hand emplaced and the ADAS (Air Deliverable Acoustic Sensor) which can be delivered by a variety of means including in conjunction with various lethal and non-lethal munitions.

OASIS

The main unit comprises extended-range acoustic sensors and signal processing, day/night electro-optics, and satellite-based global communication capabilities. The acoustic sensors detect and classify a variety of intruders including personnel, ground vehicles, watercraft, and rotary and fixed-wing aircraft. The acoustic sensors combined with the signal processing capabilities enable OASIS to identify acoustic signatures and differentiate between predetermined threats and unimportant activity at the surveillance site. Upon acoustic discrimination of a threat, the electro-optical system automatically pans to the target bearing. It then captures a series of images which are processed and compressed using proprietary technology, and transmitted via satellite to the CCS for review. The OASIS unit also functions as the receiving and central processing unit for an integrated array of ancillary devices. Seismic, magnetic, piezoelectric, and passive infra-red sensors can be customised to meet the needs of specific surveillance operations. These supporting sensors also detect and classify intrusions, including personnel, in the surveillance area and transmit data to OASIS for analysis. If OASIS identifies the intrusion as a threat, it captures images, transmits them, with additional corresponding data, to the CCS. OASIS can be further customised by integrating a variety of special use sensors including meteorological, nuclear, chemical and biological detectors, and has recently been integrated with Thales's Miniature Intrusion Sensor (MIS) (see separate entry).

OASIS is one-man portable and all its components fit into an oversized backpack. It can be assembled by one person and is claimed to be fully operational within minutes.

ADAS

An alternative configuration of Terrain Commander field deployable equipment uses ADAS. ADAS features the same range of acoustic sensor and signal processing as OASIS, without the electro-optic component. Used in clusters of three or four, ADAS units are typically networked and are specifically designed for longer range precision tracking of air and ground vehicles in remote or hostile territory. When ADAS identifies a possible threat, acoustic data from multiple sensors is used to track precisely and classify the target. Local area network

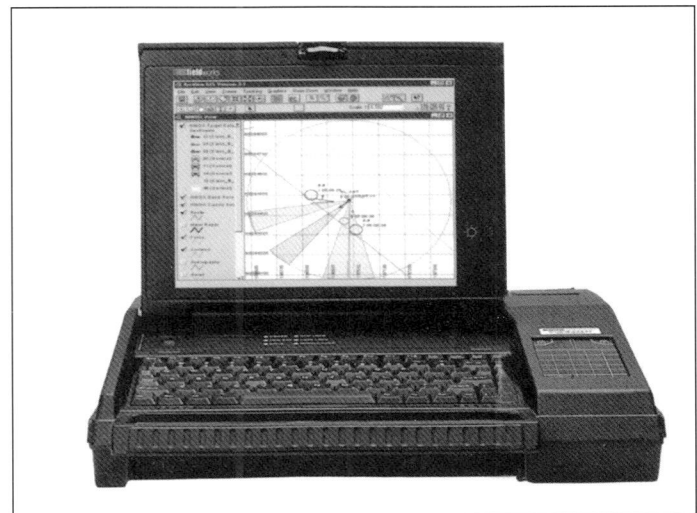

The Terrain Commander Command and Control Station (Textron Systems) 0137199

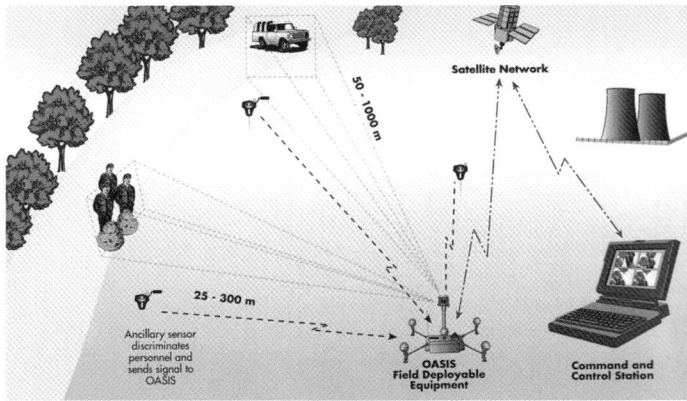

The Terrain Commander system (Textron Systems)   0137200

capabilities enable ADAS nodes to talk to each other and share real-time information. The manufacturer claims that networks have been particularly effective in tracking jets, Unmanned Aerial Vehicles (UAVs) and helicopters at extended distances. The network can also locate the source of acoustic impulses such as artillery firing. ADAS nodes can be air dropped into remote or hostile areas. They are self-righting and self-mapping/orienting and are controlled remotely. Like OASIS, the ADAS can be integrated with ancillary seismic, magnetic, and passive infra-red sensors.

Command and Control Station

The CCS includes a field rugged laptop computer and printer as well as long haul communication equipment designed to interface with the OASIS or ADAS equipment deployed in the field. Installed on the laptop is the integrated software suite necessary to run a Terrain Commander surveillance operation. The proprietary Terrain Commander™ software is used to monitor the surveillance operation. It is Windows-based, user-friendly and easy to operate with pull down menus for easy access to program features. The CCS can display maps of the local area based on any digitised terrain database or satellite imagery. Sensor reports are displayed and the operator can review the history of recent reports. The visual pictures from the OASIS can also be displayed. Imagery transmitted to the CCS is displayed on the monitor as three consecutive still photographs of the area and a three-frame movie. The multiframe motion detection feature enables the viewer to identify distant or obscure images that would be invisible in a static display.

Specifications

	ADAS	OASIS
Weight	13 kg	11 kg (base)
		6 kg (EO)
Size	39 cm h × 20 cm dia	45 × 33 × 11 cm
Deployed	122 cm dia	122 cm dia; up to 2 m/h
Battery life	Up to 90 days	Up to 90 days
Tamper detection	Yes	Yes
Acoustic detection range*		
Light truck	500 m	500 m
Tank	2,500 m	2,500 m
Helicopter	10,000 m	10,000 m
Electro-Optics recognition ranges* (day or night)		
Personnel	NA	150 m
Vehicles	NA	500 m

* *Varies with target and environment*

Status

In production. Selected by the Australian Army, which is believed to have purchased 70 systems as part of Project NINOX. Also believed to have been purchased by Singapore.

Contractor

Textron Systems, Wilmington.

VERIFIED

DIRECTION-FINDING

Australia

WD-3000 Direction Finding system

The WD-3000 WinRadio Direction Finding system relies on the pseudo-Doppler method, combined with statistical signal processing. The main user interface of the system is designed as a 'plug-in' for the virtual control panel of a WinRadio receiver, making it possible to select a frequency and operate the system like a conventional communications receiver.

The direction finding control panel contains four displays: a circular azimuth indicator, a 'waterfall' display showing time progress of the signal bearing, a histogram and a digital display of the calculated bearing. The circular display can also work in a 'polar mode', indicating the relative strength of the signal by the trace length. The waterfall display shows bearing variations against time and can employ either a linear or exponential averaging method, with user-adjustable parameters. The angular histogram makes it easier to assess the direction in cases of random varying multipath propagation conditions or a strong interference between the antenna rotation tone and spectral components of the received signal modulation. The peak of the displayed histogram represents the direction to transmitter with the greatest probability and the width of the base indicates the bearing accuracy.

The number of displayed past sweeps is user selectable. The most recent indication has the greatest intensity and the oldest one is the least visible, in a 'radar like' display. The control panel makes it possible to set alarms which are activated if a measured signal falls inside (or outside) a stipulated bearing range. A choice of rotation speeds exists to minimise interference with the spectral components of the measured signal.

The WD-3000 system is contained in a ruggedised, transportable enclosure with a 15 in TFT LCD display and standard computer facilities. The unit contains one or more WinRadio card receivers with a frequency range up to 4 GHz. The Direction Finding system comes with a variety of antennas covering the range 100 to 3,000 MHz, suitable both for mobile and stationary use.

Specifications
(a) Computer: ruggedised transportable IBM PC compatible
(b) Receiver: WinRadio 3000 Series
    0.15- 1500 MHz (WR-3150i-DSP)
    0.15 - 2600 MHz (WR-3500i-DSP)
    0.15 - 4000 MHz (WR-3700i-DSP)
(c) DF method: pseudo-Doppler
(d) DF accuracy min 2° RMS: (in reflection-free environment)
(e) DF sensitivity: typical 1-10uV/m (depending on antenna and frequency)
(f) Optional antenna arrays for mobile and stationary applications for various frequency ranges are available.

Status
Supplied to at least one unspecified customer.

Manufacturer
WinRadio Communications, Oakleigh, Australia.

NEW ENTRY

The WinRadioWD-3000 Direction Finding system showing the different elements of the display (Patrick Allen/Jane's) **NEW**/1027204

France

TRC 197 HF interceptor/DF

The TRC 197 Direction-Finder (DF) is a compact interception/monitoring direction-finding workstation able to process very short duration signals.

It is based on the interferometry principle and uses a simple, modular broadbase antenna array (an array where the ratio distance between antennas/wavelength is as high as possible). This enables the effects of multiple paths, a source of high bearing error, to be minimised and the azimuth and elevation of the received signal to be determined with very high accuracy.

The TRC 1956 crossed loop antenna ensures excellent sensitivity even for nearby transmissions received from a high elevation angle and regardless of polarisation. Operation is via a PC, including operation of the receiver, making the equipment user-friendly and open ended. Various optional software enables the systems to access new functions (sorting by sector, offline sorting and so on) or to be tailored to any operational mission.

The TRC 197 direction-finding function can work:
(a) either autonomously if the single station localisation option is implemented
(b) or in a system involving several radio direction-finders if the triangulation localisation option is implemented.

The operation of the TRC 197 can be fully remote-controlled using standard interfaces, RS-232 or IEEE488.

Status
The TRC 197 has been supplied to French and other defence forces. No new information received.

Contractor
Thales Communications, Colombes.

VERIFIED

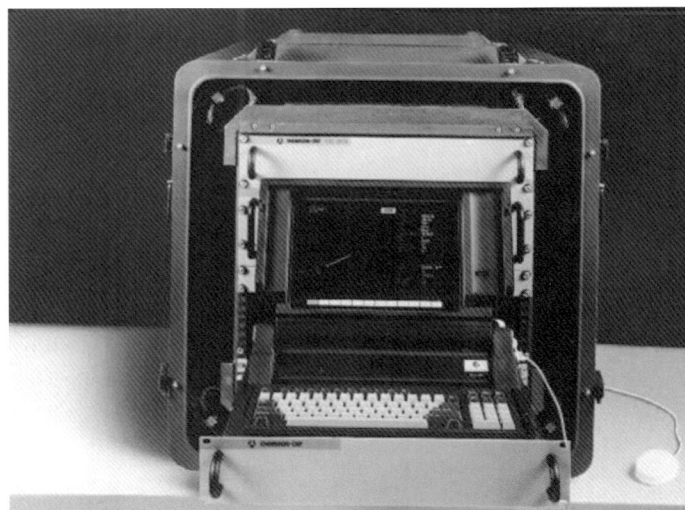

TRC 197 HF interceptor/DF

TRC 297 V/UHF interceptor/radio DF

The TRC 297 is a very compact interception/monitoring and direction-finding workstation able to process very short signals. Based on the interferometry principle, it consists of broad-based antennas. This enables the effects of multipaths, a source of high bearing errors, to be minimised and the azimuth of the received signals to be determined with very high accuracy. Operation by means of a PC, including operation of the receiver, makes the equipment user-friendly and open-ended. Various options of software enable the radio direction-finder to access new functions (location, remote control, offline sorting and so on) or to be tailored to any operational mission. The TRC 297 can work in a system involving several direction-finders if the triangulation localisation option is implemented. It can be fully remote-controlled using standard RS-232 or IEEE488 interfaces.

For details of the latest updates to *Jane's C4I Systems* online and to discover the additional information available exclusively to online subscribers please visit
jc4i.janes.com

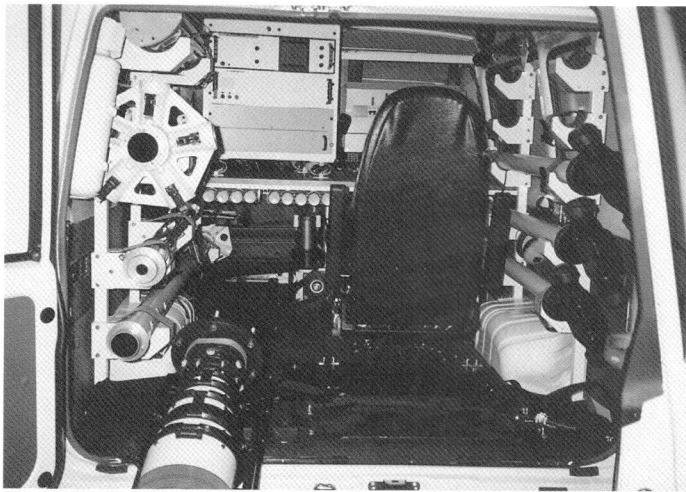

TRC 297 V/UHF DF system in a vehicle

Status
Believed to have been supplied to French and other defence forces. Production status uncertain.

Manufacturer
Thales Communications, Colombes.

UPDATED

TRC 297D lightweight antenna mobile DF

The TRC 297D is a version of the TRC 297, equipped with a lightweight antenna and also using the measurement by interferometry principle to provide a high degree of accuracy.

Designed for spectrum control missions and radio monitoring, it is especially intended to be fitted on board any land-based vehicle, ship or helicopter. It may also be used in a fixed station, when a non-compromising antenna network is required.

Specific processing of the adequate signal allows the various types of transmission used in the V/UHF range to be taken into account, even in severe noise conditions. It can be locally operated, or remotely controlled through standard interfaces for integration into systems. It is equipped with performance BITE which allows immediate and accurate fault diagnosis and maintenance.

Status
Supplied to French and other defence forces in the past. No new information.

Manufacturer
Thales Communications, Colombes.

UPDATED

TRC 297D fitted to a Toyota with concealed antenna and radome

TRC 6100 wideband digital direction-finder

The TRC 6100 is a new generation multirange digital direction-finder. It performs the parallel processing of several channels by digital processing techniques and provides the direction of targets with the best algorithms, interferometry or 'super resolution'.

Designed for electronic warfare systems applications, the TRC 6100 covers the frequency range from 0.3 to 3,000 MHz and has been designed to counter modern military transmissions such as frequency hopping, free channel search and burst transmissions.

TRC 6100 wideband digital direction-finder   0011528

Its modular design allows the TRC 6100 to be adapted for fixed installation, semi-mobile or fully mobile applications in air, ground and sea environments. To fulfil different operational requirements, the TRC 6100 can be used with various types of antenna arrays (tactical and integrated antennas) and various types of DF algorithms (Watson-Watt, interferometer and correlative matrix).

The TRC 6100 is controlled via an external PC running Windows NT. The operating software controls DF functions and supplies various displays (polar histogram, frequency versus azimuth, waterfall) to simplify reconnaissance of targets.

Specifications
(a) Continuous high speed scanning
(b) Scanning synchronisation with GPS
(c) Automatic interception and direction finding
(d) Automatic classification: FF, FH, Bursts
(e) Five parallel channels
(f) PC Windows NT™ environment
(g) HF Direction Finder:
       0.5 - 30 MHz
       300Hz/20 kHz Bandwidth
       40 MHz/s scanning speed
(h) V/ UHF Direction Finder:
       20 - 3000 MHz
       300 kHz Bandwidth
       2 GHz/s scanning speed
(i) DF accuracy: 0.5° RMS

Manufacturer
Thales Communications, Colombes.

UPDATED

Germany

APF 1050 remote-control direction-finding system

The APF 1050 system covers the frequency range from 10 kHz to 30 MHz. Each DF station consists of eight direction-finding receivers which can be operated at the same time and independently of each other. The APF system can be installed either

APF 1050 remote-control direction-finding system (Plath GmbH)   **NEW**/0594248

as an independent DF network or as an extension to existing DF networks and can be used in both semi-mobile and in stationary applications. The remote-control feature of the equipment enables unattended operation in remote areas. The system is designed to enable easy relocation.

Status
In service since 1992.

Manufacturer
Plath GmbH, Hamburg.

UPDATED

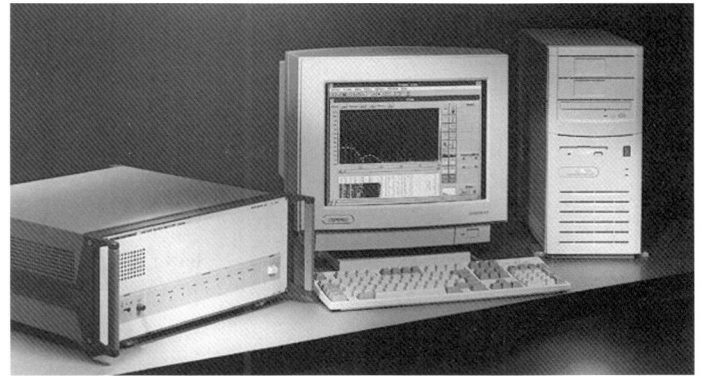

DDF 0xS system

DDF0xM series direction-finders

DDF0xM series equipments cover the 0.3 to 3,000 MHz frequency range and can be operated in Watson-Watt or correlative interferometer modes depending on the type of direction-finding antenna used. A wide range (including units from manufacturers other than Rohde and Schwarz) of such antennas is available for both mobile and stationary applications. The equipments can intercept signals of very short duration (>10 μs in Watson-Watt mode, >500 μs for correlative interferometer) and are capable of operating in scan mode where they make use of the Fast Fourier Transform technique (25 kHz real-time bandwidth for the HF band and 200kHz for the VHF/UHF bands). DDF0xM direction-finders are computer controlled (external or internal as an available option) and can use AC or DC power supplies. The equipments can also be remotely controlled when networked. The series includes the following models:

(a) DDF01M. This covers the 0.3 to 30 MHz band and when operated in the correlative interferometer mode (using a nine crossed-loop or monopole antenna array with a diameter of 50 m), can deliver elevation and azimuth data and thereby provide a single station location capability. For mobile applications, a 1 to 30 MHz, 1.1 m diameter antenna can be used.

(b) DDF05M. This covers the 20 to 1,300 MHz or 20 to 3,000 MHz bands and utilises a range of Watson-Watt or correlative interferometer antenna arrays. It can be used in fixed-site and mobile applications and has a typical scanning speed in interferometer mode of 45 MHz/s with 25 kHz resolution. Provision is made for the interception of GSM signals.

(c) DDF06M. This covers the 0.3 to 1,300 MHz band (extendable to 3,000 MHz) in a single equipment.

Status
In production and service since 1996.

Manufacturer
Rohde & Schwarz GmbH, Munich.

UPDATED

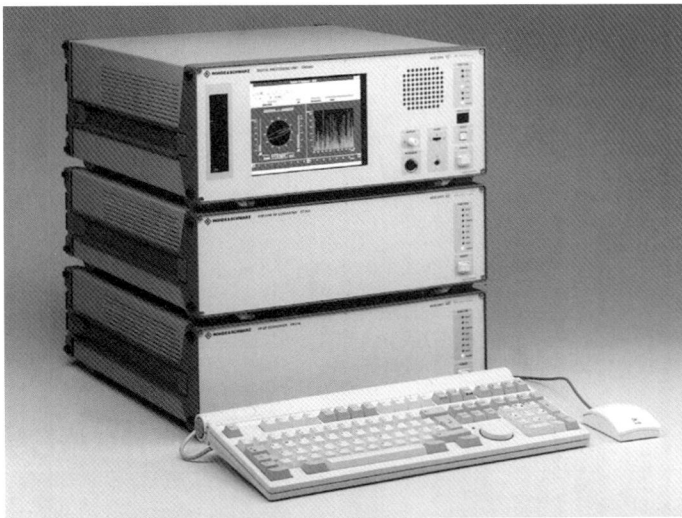

The 0.3 to 1,300 MHz DDF06M digital direction-finder   *NEW*/0550061

DDF 0xS HF/VHF-UHF scanning DF

As scenarios with rapidly changing situations (for instance by using frequency-hopping transmitters or burst transmissions) are more frequently encountered, there is a requirement for new concepts of electronic reconnaissance systems. If separate units are used for detection and direction/position finding, it may often happen that a newly discovered frequency activity cannot be passed on to the DF system because of the short signal duration and thus the signal cannot be processed by the DF system. This means that advanced systems should be capable of performing measurements of frequency, level and bearing of such

signals simultaneously. This concept is realised with the Scanning Direction-Finder DDF 0xS which incorporates a fast-scanning DF receiver that can examine very wide frequency bands within a short time.

The system makes extensive use of digital signal processing, especially of Fast Fourier Transform (FFT), which allows simultaneous analysis of a wide frequency band with selectable channel resolutions within the analyser bandwidth.

Among the applications of DDF 0xS are: automatic position-finding systems with high interception probability; interception and direction-finding of frequency hopping and burst signals; optimised use in automatic interception systems through data reduction, so that results are limited to sources of interest; and Single Station Location (SSL) systems by additional determination of elevation angle using interferometric evaluation in the HF band.

The system is highly flexible both in stationary and mobile applications (vehicles, vessels and aircraft) by selectable use of Watson-Watt or interferometric evaluation algorithms and various antenna configurations, especially those with wide aperture characteristics.

The DDF 0xS covers 0.5 to 1,300 MHz with an analysis window of 200 kHz.

Status
No longer in production, but still in use. Provides the DF element of the early version of the ScanLoc DF system (see separate entry).

Manufacturer
Rohde & Schwarz GmbH, Munich.

UPDATED

DDF195 digital direction-finder

The DDF195 digital direction-finder operates in the 0.5 MHz to 3,000 MHz frequency range, with options available for this to be restricted to 0.5 to 30 MHz, 20 to 1,300 MHz or 1,300 to 3,000 MHz. It consists of three separate DF antennas (one for each band) which can be operated simultaneously using an automatic antenna selector, plus the EBD195 DF processor; a monitoring receiver is also required and it is designed to be connected to the IF output (10.7 or 21.4 MHz) of an external receiver such as the ESMB (see separate entry). The equipment can be operated either using the Watson-Watt DF method (HF band) or the correlative interferometer DF method (VHF/UHF band). The system will locate signals with any modulation; minimum signal duration is 10 ms and bearing results may be displayed in histogram form if required.

The system can be used for both mobile and stationary applications and can be remotely controlled via an RS-232 interface. An optional electronic compass for automatic direction-finding is available for use in mobile applications and a hand-held remote display unit is also available. The power supply can be either AC or DC.

Status
Believed to be available as a COTS item.

Manufacturer
Rohde & Schwarz, Munich.

NEW ENTRY

DFP 5050 digital radio direction-finder

The narrowband radio direction-finder DFP 5050 offers a coherent real-time bandwidth of 20 kHz in the frequency range from 0.3 MHz to 30 Mhz and is designed for use within modern radio reconnaissance systems. Optionally, the frequency range may be extended to reach from 10 kHz to 30 MHz. A spectral resolution of up to 31.25 Hz enables the accurate measurement of individual signals as well as the separation of different transmitters.

The DFP 5050 operates on the three-channel Watson-Watt principle. The DF results are determined concurrently in the time and the frequency range. Thus, in addition to the conventional result data from a narrowband direction-finder (bearing ellipses), also bearing and amplitude values of an FFT direction-finder (frequency spectra) are available for further evaluation. Spectral data, bearing

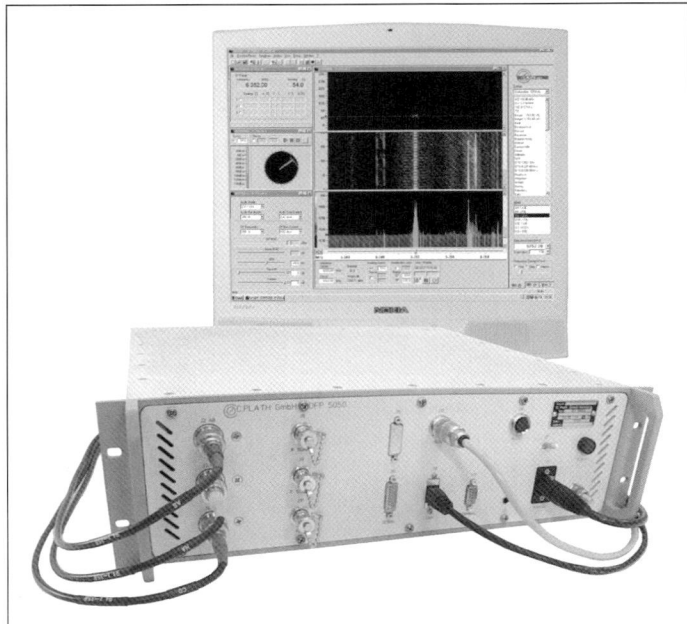

DFP 5050 digital radio direction-finder (Plath GmbH) **NEW**/0594251

ellipses and audio are displayed synchronously by means of the WinDF operating and display software for manual analysis of individual signals. The measurement of signal parameters can be done in pause status or during replay.

Due to its compact design and high environmental specification, the unit is suited to both stationary and mobile operation. In both cases the DFP 5050 may be used as a stand-alone reconnaissance direction-finder or within a radio reconnaissance system.

Specifications
(a) Frequency range: 0.3-30MHz (0.01 MHz to 30 MHz optional)
(b) Frequency step size: 10 Hz
(c) Real-time bandwidth (coherent bandwidth): 20 kHz
(d) Frequency stability: ≤ 1 * 10 -7 (optional: external synchronisation)
(e) Input impedance: 50 W
(f) Input voltage standing - wave ratio (VSWR): 2:1
(g) Noise factor: 14 dB
(h) Minimum discernable signal (MDS): -135 dBm (at 31.25 Hz and S/N=10 dB)
(i) Receiving dynamic range (controlled): 165 dB
(j) DF accuracy: < 0.5°
(k) DF-capable types of modulation: DF-capability independent from type of modulation

Status
In service since 2000.

Manufacturer
Plath GmbH, Hamburg.

UPDATED

DFP 5300 DF and analysing equipment

As a broadband direction-finding equipment, the DFP 5300 offers fast signal acquisition of a claimed 100,000 bearings per second over the frequency range of 0.3 to 30 MHz. Other features are real-time processing, data reduction by sector mode, replay capability and the detection and display of bearings of all emissions.

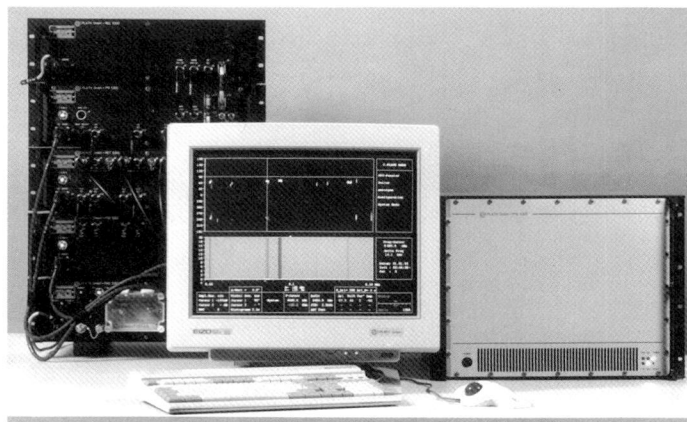

DFP 5300 DF and analysing equipment (Plath GmbH) 0594252

The DFP 5300 has an independent monitoring receiver which can be controlled within a surveillance window of up to 1.6 MHz by means of a cursor. This provides detection and bearing of all transmitters including conventional, frequency hopping, spread-spectrum and chirpsounder. Display modes include frequency over amplitude, frequency over time, bearing over frequency, bearing over time and histogram indication. As a manual analysis tool the DFP 5300 can be integrated into existing direction-finding nets via conventional telephone links.

Status
In service since 1994.

Manufacturer
Plath GmbH, Hamburg.

UPDATED

DFP 5400 broadband radio direction-finder

The broadband radio direction-finder DFP 5400 was developed specifically for automatic HF search and location systems. In a frequency range of 1 to 30 MHz it has a coherent bandwidth of 2 MHz which enables the detection of a burst of 100 ms length and two hits within a frequency range of 10 MHz with theoretically 100 per cent probability. It has a claimed high detection probability due to low noise figure, high dynamics and fast scanning. The low false alarm rate is based on a new mixer concept and excellent values for IP2 and IP3.

The DFP 5400 is reduced both in volume and in power consumption compared to the previous model (DFP 5300, see separate entry), making it suitable for deployment in mobile and semi-mobile systems in which only limited energy and space are available. The equipment assembly consists of the DF tuner, the FFT bearing processor together with the DF-data analyser and a power supply unit.

Specifications
(a) Frequency range: 1-30 MHz
(b) Direction-finding principle according to 3-channel Watson-Watt
(c) Coherent bandwidth: 2 MHz (analogue, can be limited to 200 kHz)
(d) Frequency resolution: 125 Hz
(e) Time resolution: 1 ms
(f) Scanning speed: 200 MHz/s
(g) Noise figure: 14 dB

Status
In service since 2003.

Manufacturer
Plath GmbH, Hamburg.

UPDATED

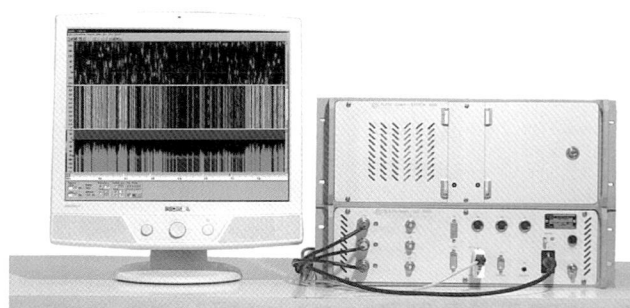

DFP 5400 broadband radio direction-finder (Plath GmbH) **NEW**/0594249

DFP 7107 broadband radio direction-finder

The DFP 7107 is a compact broadband direction-finding receiver developed especially for mobile applications with multichannel time-frequency transformation and DF-evaluation using the Watson-Watt or Interferometer DF-principle. With the DFP 7107 it is possible to monitor broad frequency bands with high transmitter congestion in a short time. The equipment provides detection and bearing of all conventional transmitter types as well as frequency hopping and spread spectrum transmissions.

Specifications
(a) Frequency range: 20-1,350 MHz (optional to 3 GHz)
(b) DF-principle: Watson-Watt or interferometer
(c) FFT window: 1 MHz/3 dB
(d) Resolution: 500; 250; 40 lines

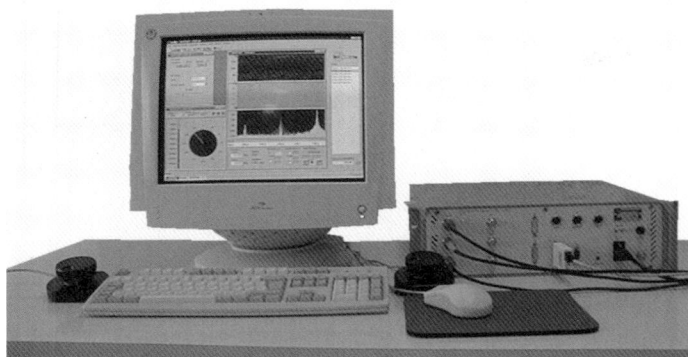

DFP 7107 broadband radio direction-finder (C Plath GmbH)   **NEW**/0594250

(e) Analogue bandwidth: 1 MHz
(f) Analysis bandwidth: 50 kHz-128 MHz, adjustable
(g) Frequency incremental width: 1 kHz
(h) Frequency stability: $\leq 1*10^{-7}$ (optional external synchronisation)
(i) Input impedance: 50 Ω
(j) Precision: ≤ 1° (Watson-Watt), =2° (Interferometer)
(k) Scanning speed: ≥100 MHz/s (at 250 lines resolution)
(l) Acquisition time: ≤10ms
(m) DF-accuracy: up to 500 MHz: ±20° above 500 MHz: ±3°
(n) Image rejection: 70 dB
(o) IF suppression: 70 dB
(p) Dynamic range: 70 dBAGC range: 110 dB
(q) Audio: demodulators - AM (10 kHz), FM (25 kHz), CW
(r) Noise figure: direction-finding - $\leq 10kT_0$
(s) Interfaces: serial - RS-232/ RS-422, fast data interface LAN
(t) Power supply: 21.5-30 V DC, 4A (24 V DC)
(u) Temperature range: operation −10 to +50°C; storage −25 to +75°C
(v) Dimensions: 19 in 3 HU
(w) Weight: approx 19 kg

Status
In service since 2000.

Manufacturer
Plath GmbH, Hamburg.

UPDATED

DIPEC 2000 digital bearing centre

DIPEC 2000 is a modular software/hardware system for creating integrated and combined DF nets, within the frequency ranges from HF up to UHF. It integrates individual DF equipment already existing and manufactured by EADS (from Type Telegon 8 onwards).

The system provides various DF information on a central display with additional user-friendly operating functions to control the DF system, locally or remotely, by one operator. In its basic version DIPEC consists of a workstation computer (PC compatible) that displays on a high-resolution screen the actual DF situation in various shapes.

Real-time display of DF information (bearing ellipse or amplitude histogram), azimuth histogram in polar/Cartesian form, or combined azimuth/elevation histogram in interferometer mode, can be displayed. Waterfall display, replay mode and sector evaluation are also feasible.

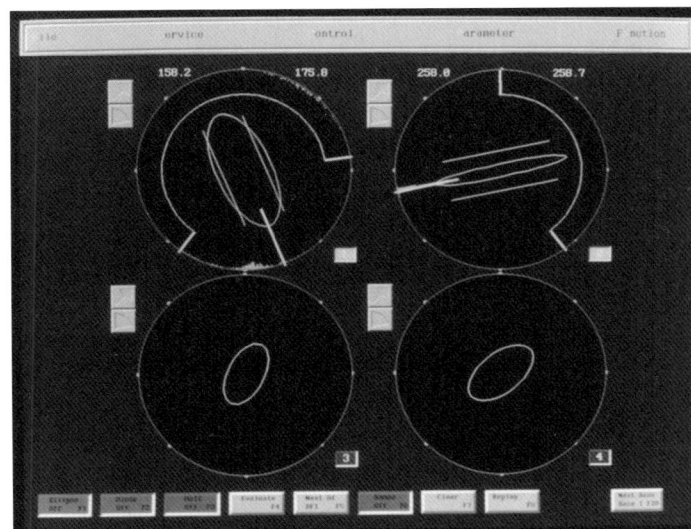

DIPEC digital bearing centre

Up to four DF stations can be remotely commanded and controlled by one central station. This shows the simultaneous actual bearing information of each DF station as well as the calculation of emitter location superimposed on a map display.

Status
No longer in production, but may still be in service with European customers.

Contractor
MRCM
MRCM is a joint co-operation under full ownership of:
EADS Ewation, Ulm, Germany.
Grintek, Pretoria, South Africa.
Herley Industries Inc, Lancaster, USA.
Sysdel, Pretoria, South Africa.
TRL Technology, Tewkesbury, Glos, UK.

UPDATED

GSP3601 HF/VHF DF

The is an HF/VHF Direction-Finder (DF) for installation in APCs with integrated power generators. Designed for mobile use in forward battle areas, the system has a foldable antenna positioned on a hydraulically operated mast 10 m above ground. Mast and antenna can be erected from inside the vehicle in under 2 minutes.

The GSP3601 is based on three independently operating Telegon 10 direction-finders. The Telegon 10 is a Watson-Watt equipment covering the 10 kHz to 1 GHz frequency range with a resolution of 10 Hz and a typical equipment error of less than 1°. The modular DF consumes approximately 150 W and has an eight-digit CRT display. Manual operation can be performed in parallel to automatic.

Status
No longer in production. Has been, and may still be, in service with German forces and an unspecified NATO country.

Contractor
MRCM
MRCM is a joint co-operation under full ownership of:
EADS Ewation, Ulm, Germany.
Grintek, Pretoria, South Africa.
Herley Industries Inc, Lancaster, USA.
Sysdel, Pretoria, South Africa.
TRL Technology, Tewkesbury, Glos, UK

UPDATED

GSP3601 in APC

MapView Geographic Information Software (GIS)

MapView is a digital map display for radio monitoring systems providing:
(a) Fast online display of results on a digitised map
(b) Offline display of results in combination with external databases
(c) Use in direction-finding and radio location systems
(d) Graphical situation display
(e) Special functions for mobile systems
(f) Import of digital maps in various formats
(g) Generating and digitising new maps
(h) Editing and adapting digital maps.

In military applications, this display is used to support operational and tactical analysis. In the case of civil applications, DF evaluation is supported by transmitter-site displays. The GIS MapView is used as a display software for geographical data on digital vector and raster maps. It was primarily designed for radio monitoring and radio location applications, providing fast online result display and features optimised for this task. A range of tools is available for use with the digital maps, including:

(a) Fast map zooming
(b) Measurement of distances and directions
(c) Direct selection of map objects and direction-finding and radio location results.

MapView has functions that make it suitable for use in conjunction with mobile systems (for example in homing DF vehicles):

(a) Optimised for use with small TFT displays
(b) Automatic map navigation on the basis of the current vehicle heading and position (vector maps can be rotated)
(c) Length of the DF beam is proportional to the current RF receive level of the DF signal - evaluation of the DF results is therefore possible by monitoring level variations
(d) RF receive level display immediately next to the DF symbol on the map
(e) Routing function in conjunction with the map and guide option (route display)
(f) Scaling as a function of vehicle speed
(g) Indication of heading (in conjunction with an electronic compass or GPS).

With the aid of graphical elements, MapView makes it possible to display the results obtained from the analysis of the full range of information provided by a radio monitoring system. The situation display is generated using several map layers and can be stored. Besides standard functions such as text entry and line drawing, symbols can be selected with the mouse and positioned on the map. Symbols can be user-generated and saved to various libraries.

In radio monitoring systems such as RAMON (see separate entry), MapView communicates with other software applications from Rohde & Schwarz via a TCP/IP interface. These applications can be the control software for the direction-finders, the location software and database applications. Interacting with these software modules, the current locations of the direction-finders will be automatically displayed on the map and are continuously updated while the direction-finders are moving. In this case, the current direction of the vehicle is also indicated on the map by the DF symbol. DF results are displayed on the map as DF beams; location results as circles. RAMON's location software can be used to record the DF results for more detailed analysis at a later date. This also makes it possible to take running fixes and so locate radio signals with just one direction-finder (direction-finding from different locations). It is also possible to use the offline result-display mode in conjunction with the system software packages. DF and location results, stored in a database, can be displayed on the map. MapView can also be linked to other customer-specific applications via an open TCP/IP interface.

The usefulness of geographic display software depends on the quality of the available maps. MapView uses a special data format to ensure optimal compliance with the requirements of radio monitoring systems. There are various ways of obtaining maps:

(a) The MapEdit option is for generating, converting and maintaining (and so subsequently modifying) user-specific maps.
(b) As an option, maps from other manufacturers can be directly opened in MapView and used without being converted. These are:

Vector maps in map and guide format (road maps with routing information) with the Map and Guide server option

Raster maps of the German Bundeswehr Geographic Office with the CMRG (Compressed Milgeo Raster Graphics) server option

Raster maps in LS telcom format with the LS telcom server option.

MapEdit

MapEdit is for generating and editing user-specific digital maps with many projection modes supported for map generation. There are several ways of generating maps:

(a) Paper maps can be scanned with the MEBASIC basic module, imported into MapEdit in the form of a raster map and georeferenced.
(b) Using a digitiser, paper maps can also be manually digitised and georeferenced. In this case, a vector map for MapView is generated.
(c) Vector or raster maps from an existing GIS can be imported using the ME-VECT option.

MapEdit distinguishes between digitised vector maps and scanned raster maps.

Contractor

Rohde and Schwarz, Munich.

VERIFIED

MCS 3000 Mobile COMINT system

The MCS 3000 system consists of several mobile sensor stations which can operate either independently as stand-alone units or as a direction-finder network to allow location of transmitters. The system provides a fast spectrum-scanning capability together with an integrated DF data analyser.

All stations are identically equipped with VHF/UHF direction-finding and monitoring equipment covering the frequency range from 20 to 3,000 MHz. An extension for the HF frequency range is available as an option. For networked operation each of the sensor stations can serve as a command and control centre.

MCS 3000 Mobile COMINT system (Plath GmbH)   **NEW**/0594253

The mobile sensor stations mainly consist of the following subsystems:

(a) Broadband VHF/UHF DF system
(b) GSM DF extension
(c) VHF/UHF monitoring system
(d) Radio location system with geographic map display
(e) Results database
(f) Communication equipment
(g) Broadband HF DF extension (option)
(h) HF monitoring extension (option)

Status

In service from 2003.

Manufacturer

Plath GmbH, Hamburg, Germany.

UPDATED

MonLoc DF and Location System

The Rohde & Schwarz DF and Location System MonLoc combines two or more digital direction-finders into a DF and location network and enables them to be remotely controlled. Depending on the direction-finder used, conventional radio signals, short-duration signals or broadband emissions can be located. DF and radio location results are displayed on a digital map.

MonLoc allows all the main parameters of the local and remote direction-finders to be displayed on a central PC. The remote direction-finders are connected to the central PC via datalinks using the TCP/IP protocol. Connection may be established via analogue or digital PSTNs (dialled or leased lines), a GSM network or a microwave link. Optimal use of the bandwidth of the datalink allows full use of the system capabilities and a high detection probability is claimed even at low data rates (up to 9.6 kbps). The audio signal from a remote direction-finder can also be simultaneously transmitted on the same communication channel.

The flexible system concept allows an unlimited number of control stations to be combined with an unlimited number of DF stations. Any kind of radio location system from small mobile stations up to nationwide radio location systems can thus be set up and operated with a minimum of personnel. In systems with several networked operator positions, MonLoc also supports the automatic processing of position-finding requests from several operators. In this case the radio location workstation may be unattended.

The system supports fixed frequency (FFM), search-and-scan operating modes. In addition to FFM permitting direction-finding and radio location on single frequencies, the search mode is used for detecting unknown and new radio signals or for monitoring signals stored in a frequency list. Any setting made on the local direction-finder is automatically transferred to the remote DF stations. The scan mode is used for detecting unknown conventional radio signals and especially for detecting and taking bearings of short-duration and broadband emissions. Bearings of such emissions are obtained manually by the operator for all networked direction-finders with the aid of measurement rulers in the azimuth or histogram window of the direction-finder. All bearings on the measurement line are automatically displayed on a digital map and the location is calculated. In these circumstances, a higher transmission rate is required on the datalink (for example 64 kbits/s). For fully automatic location of short-duration and broadband signals, the Rohde & Schwarz ScanLoc system can be used (see separate entry).

Options include the evaluation and storage package DF-EVAL. With the aid of this software option, bearing results can be recorded either continuously, on operator request or by means of preset filter functions, allowing subsequent evaluation. When single mobile DF stations are used, a running fix may be performed: the transmitter location can be calculated from several bearings taken at different positions. Stored bearings and location results can be recalled for subsequent digital display on the Rohde & Schwarz MapView (see separate entry) and for evaluation. This also allows mobile transmitters to be located and their movement traced.

A remote-control interface allows networked direction-finders to be accessed by external software. Via this interface, the direction-finders can be simultaneously controlled and bearing results queried.

Contractor

Rohde & Schwarz, Munich.

NEW ENTRY

MRD30W3/n wideband HF direction-finder

The MRCM products range includes the Polygon MRD 30W3/n wideband FFT DF System. It is intended for wideband spectrum surveillance and DF of fixed frequency and LPI (Low Probability of Intercept) transmissions, such as frequency hoppers, burst, chirp and DSSS (Direct Sequence Spread Spectrum). It provides high probability of intercept with accurate direction finding results achieved by sensitive large aperture interferometer antenna systems and parallel signal processing.

Specifications

(a) Frequency range: 1.5-30 MHz with Interferometer antenna
(b) Instrumental accuracy: less than 1° RMS
(c) Sensitivity (3 kHz 10 dB (S+N)/N): −113 dBm
(d) Instantaneous bandwidth: 500 kHz
(e) Dynamic range for 500KHz bandwidth: 80 dB
(f) Frequency resolution: 625 Hz, 312.5 Hz, 156.25 Hz, 78.125 Hz.
(g) Noise figure: <13 dB
(h) DF Algorithm: Interferometer; Watson-Watt
(i) Search scanning rate: 150 MHz/s (without DF)
(j) DF rate:
    50 MHz/s for 625 Hz channel bandwidth
    19 MHz/s for 312.5 Hz channel bandwidth
    6.5 MHz/s for 156.25 Hz channel bandwidth
    1.9 MHz/s for 78.125 Hz channel bandwidth
(k) Minimum signal duration: 2 ms
(l) Polyphase filters: filter factors 1, 2, 4 and 8
(m) Zoom: zoom factor 1, 2 ,4 and 8
(n) Display Windows: scan and zoom windows with time vs freq waterfall; azimuth vs freq waterfall; 'Time-DF' waterfall display
(o) User interface: Windows NT® XGA colour, 1,024 × 768
(p) Dimensions:
    RF unit including calibration unit: 19 in × 5U
    DSP unit: 19 in × 4U
    MRR 2010 LH1: 19 in × 3U

Status

In production and in service.

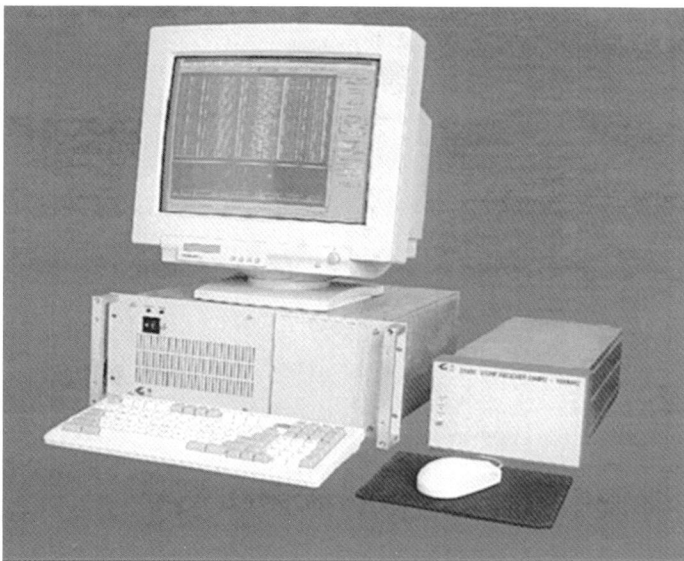

MRD30W3/n hardware   0098332

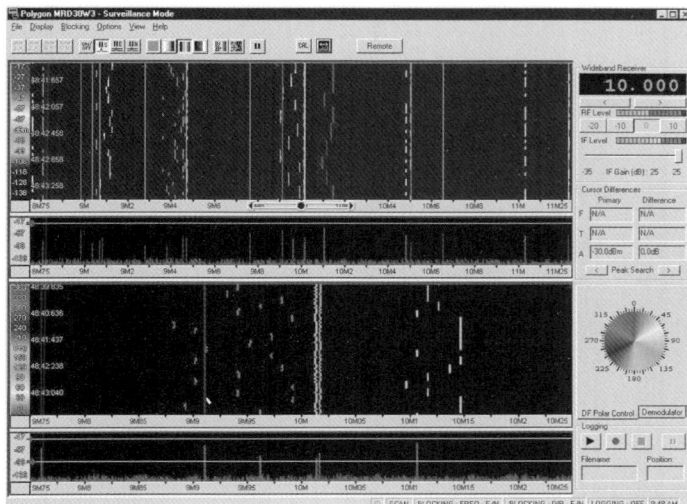

MRD30W3/n MMI   0098331

Contractor

MRCM
MRCM is a joint co-operation under full ownership of:
EADS Ewation, Ulm, Germany.
Grintek, Pretoria, South Africa.
Herley, Lancaster, USA.
Sysdel, Pretoria, South Africa.

UPDATED

OPAL radio location system

The OPAL radio location system consists of multiple (up to eight) HF direction-finders which enables the operator to identify the location of up to three HF radio transmitters on one frequency simultaneously. The direction-finders are interconnected with a location position (of which a number can be deployed) via a central network node (remote communication) computer and are simultaneously controlled via the location position using different parameters such as frequency, bandwidth and class of emission. The digitised direction-finding data (direction-finding ellipses) is evaluated at the location position and the calculated direction-finding angles and location results are displayed on a map. The calculation of locations takes into consideration the time-sorted histograms by means of automatic propagation-time compensation The user can edit, correct, save and print out the results.

Status

In service since 2000.

Manufacturer

Plath GmbH, Hamburg.

UPDATED

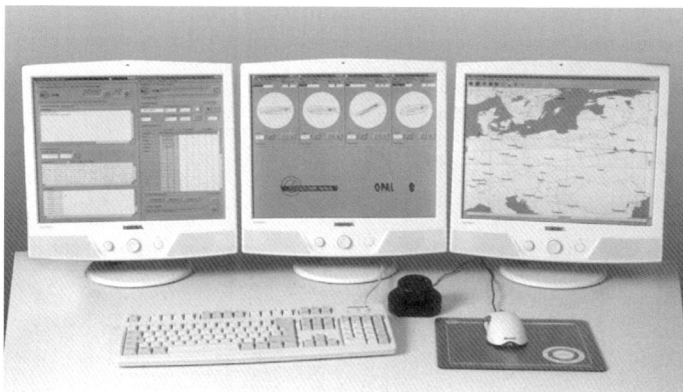

OPAL radio location system (Plath GmbH)   *NEW*/0594257

PA 1555 mobile VHF/UHF DF

The PA 1555 is a compact direction-finder with a foldable antenna system, covering the frequency range 20 to 1,000 MHz. Modes are AM, FM and CW. Bearing accuracy is 2° RMS, sensitivity approximately 2 to 10 µV/m and a minimum signal duration of 50 ms is required. Correlation of bearings is by parameters such as frequency, time and direction. Bearings can be presented in the form of a histogram that allows the identification of different stations involved in a radio communication network. Searches can be conducted within specified frequency limits for activities by using the frequency scan mode and, similarly, up to 100 preprogrammed frequency channels can be scanned.

For use in vehicles, a DF antenna is available which covers the entire frequency range and which is housed in a round plastic radome 1.1 m in diameter and approximately 25 cm high. The DF measures 152 × 254 × 268 mm and weighs approximately 7.5 kg. It has a low consumption DC power supply (10 to 30 V).

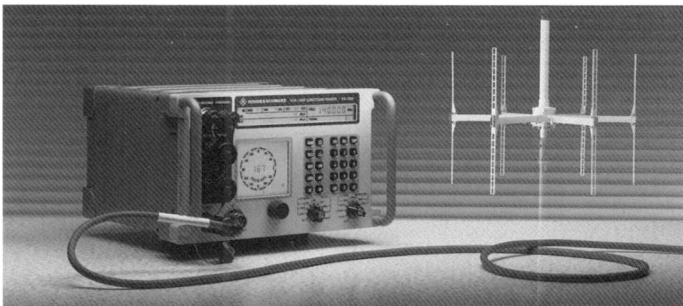

Mobile DF PA 1555 with 20 to 200 MHz DF antenna

The system can be integrated into radio location systems using the WinLoc software package (see separate entry) and can be operated remotely using a detachable display unit.

Manufacturer
Rohde & Schwarz GmbH, Munich.

UPDATED

PSI 2000 HF interferometer DF system

The PSI 2000 HF interferometer is a large aperture, digital Direction-Finding (DF) system, providing high-accuracy bearings by means of parallel digital signal processing in five DF channels. In addition to the advantages of the interferometer DF method, the digital signal processing and filtering provides absolute synchronism of the DF channels in phase and amplitude as a precondition for reliable interception and bearing of the shortest signals (for example, burst transmissions).

A PC-compatible workstation with high-resolution colour monitor serves as a control and bearing display computer. Bearing results are displayed in a polar co-ordinate system as an azimuth/elevation histogram. Map display and Single Station Location (SSL) capability are available optionally.

Specifications
(a) Frequency range: 1.6-30 MHz
(b) DF accuracy: 1° RMS typical
(c) Min signal duration: 2 ms
(d) Power supply: 115/230 V AC
(e) Dimensions: 19 in unit, 3U height

Status
No longer in production but has been, and may still be, in service with several customers.

Contractor
MRCM
MRCM is a joint co-operation under full ownership of:
EADS Ewation, Ulm, Germany.
Grintek, Pretoria, South Africa.
Herley Industries Inc, Lancaster, USA.
Sysdel, Pretoria, South Africa.
TRL Technology, Tewkesbury, Glos, UK.

UPDATED

PSI 2000 HF interferometer DF   0011531

SCANLOC DF and location system

SCANLOC is an HF/VHF/UHF scanning radio location system, designed for short duration and frequency agile emissions (bursts and frequency hoppers). An earlier version was used with the DDF 0XS Direction-Finder (see separate entry); the latest version uses up to four synchronised DDF0XA/E direction-finders (see separate entry). An accurately synchronised scan by all direction-finders ensures a high probability of acquisition and accuracy of results in the location of short-term emitters and frequency-agile transmitters, and permits the real-time display of detected emissions on digitised maps. Time synchronisation is achieved by means of GPS receivers.

Detected emissions are processed at the site of the detached direction-finder and stored with timestamps. These data can automatically be called by the central

DF station. With the use of powerful algorithms for data compression, all key information can be transmitted between direction-finder and DF centre at relatively narrow transmission bandwidths. Results are graphically processed in the DF centre and displayed on two screens. On one screen signal activities are displayed online in the form of different diagrams: signal activity versus time and frequency (waterfall); azimuth versus frequency, and level versus frequency (RF spectrum).

On the second screen the detected signals are displayed on a digital map after processing. If desired, different bearings of a transmitter are automatically combined into plots containing all main signal parameters, for example the detected single frequencies of a hopping emitter. The movement of mobile transmitters can automatically be traced and stored. The system automatically recognises known signals that are in a signal library and marks them so that new emissions can easily be identified.

The system can be extended using the RAMON system software (see separate entry).

Status
The DDF 0XS version is operational. The DDF0XA/E version was scheduled to be available by the end of 2003.

Manufacturer
Rohde & Schwarz GmbH, Munich.

UPDATED

SELOS Broadband Search and Location system

The SELOS system is used for detection, location and analysis of any emitter within the monitored frequency band. It consists of three or more direction-finding sites and the location server, which are connected over a wide area network (WAN). The number of direction-finding sites and of the direction-finders at any single site can be scaled as necessary. The system has been developed to overcome the detection shortcomings of narrowband receivers and is designed for the detection of both standard and of LPI-signals (chirps, DSSS, bursts and frequency hoppers) within the HF range. Although the focus of the system is on the fast search of signals via broadband technology, narrowband direction-finders can be integrated for the separation of neighbouring signals and the detection of weak signals that are hidden within the sideband of strong signals.

After automatic noise and interference suppression and a segmentation of the frequency band, a set of descriptive parameters for each transmitter is available: direction, location, frequency, bandwidth, amplitude, transmitter type, start and stop time as well as keying. Compared to the original bearing data, the data volume of the results gained from the automatic DF bearing data analysis is reduced by a factor of approximately 1000:1. This allows transmission of the direction-finding results to the location server for steerage and to generate the location results for all detected transmitters. The lack of data filtering provides an overview of the signal scenario without limitation.

Manufacturer
Plath GmbH, Hamburg.

UPDATED

SFP 5200 radio direction-finder

The radio direction-finder SFP 5200 operates over the frequency range of 0.01 to 30 MHz. Its claimed strength is its particularly high acquisition rate. A claimed maximum of 8 ms is required from the start of the DF command to the indication of the digitally evaluated bearing together with relevant quality criteria. This time also includes the setting time for the synthesiser at 2 ms and channel tuning with a new system at 1 ms. With a suitable drive this DF receiver is able to detect frequency hoppers and to determine their location within a DF network.

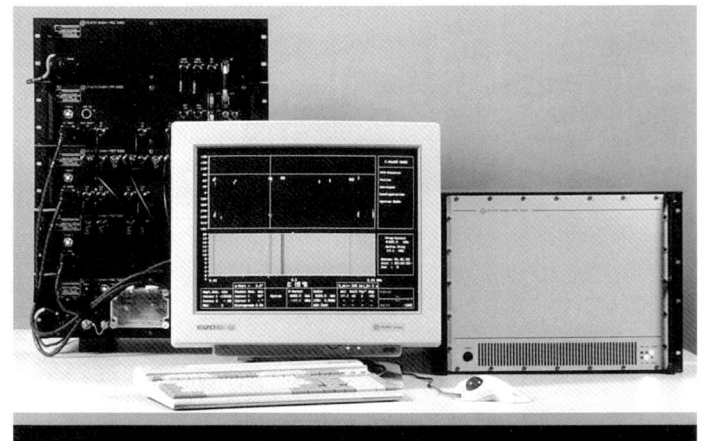

SFP 5200 radio direction-finder (Plath GmbH)   *NEW*/0594255

Different bandwidths can be selected for direction-finding and audio information. The complete contents of a message can be received using the narrow bandwidths which are most favourable for direction-finding. The digital filters that are used also ensure the synchronisation of the receiver during the transient period of tuning or tuning to the filter edge.

Status

No longer in production. May still be in service, but this cannot be confirmed.

Manufacturer

Plath GmbH, Hamburg.

UPDATED

Telegon 10 DF system

The Telegon 10 (Type PGS 1720) communications Direction-Finder (DF) covers the range 10 kHz to 1,000 MHz. The modular construction of the system allows it to be configured to cover smaller frequency ranges.

Telegon 10 employs the Watson-Watt method and is able to obtain a bearing on signals with a duration as low as 1 ms. The system consists of a P 1720 DF receiver, BP 1620 control unit and an SG 1620 CRT display unit.

The main features of the system are:
(a) microprocessor-controlled DF procedure
(b) low signal duration required for DF enables direction-finding of burst signals and frequency hoppers when configured appropriately
(c) real-time DF display enables resolution of non-coherent co-channel interference
(d) integrated automatic bearing processor
(e) 100 memory channels for the DF settings.

Telegon 10 can be employed in stationary and mobile systems and, due to its built in 'intelligence', is designed for integration into EW systems.

Specifications

(a) Frequency range: 10 kHz to 1,000 MHz
(b) DF method: Watson-Watt
(c) Frequency resolution: 10 Hz
(d) Equipment error: =1°
(e) Operational modes: A1A, A2A, A3E, F2A, F3E, J3E

Status

No longer in production, but remains in service with many national and international customers.

Contractor

MRCM
MRCM is a joint co-operation under full ownership of:
EADS Ewation, Ulm, Germany.
Grintek, Pretoria, South Africa.
Herley Industries Inc, Lancaster, USA.
Sysdel, Pretoria, South Africa.
TRL Technology, Tewkesbury, Glos, UK.

UPDATED

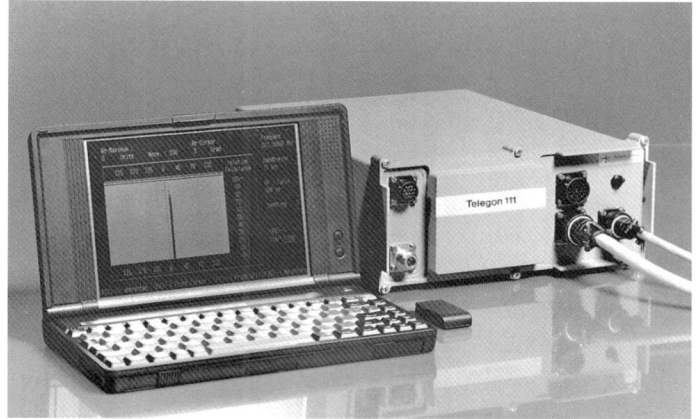

Telegon 10 communications DF

Telegon 111 VHF/UHF DF

The Telegon 111 is a general purpose Direction-Finder (DF) equipment covering the frequency range 25 to 1,000 MHz (1,999 MHz). It is controlled by a PC-compatible computer (notebook or laptop computer).

The Telegon 111 is highly suited for mobile, portable or stationary applications because of its small dimensions and low weight. A serial data interface facilitates the integration into DF systems for spectrum monitoring and security tasks.

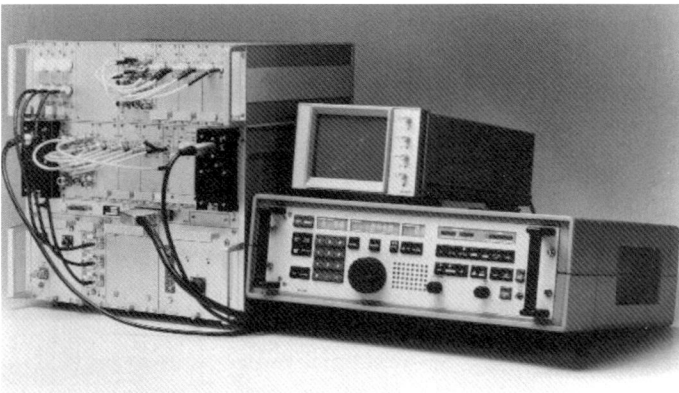

Telegon 111 radio communications DF

Specifications

(a) Frequency range: 25-1,000 MHz (1,999 MHz)
(b) Frequency resolution: 100 Hz
(c) DF method: ADF with electronic rotating goniometer
(d) Demodulation modes: AM, FM, SSB
(e) Min signal duration: 50 ms
(f) Power supply: 12-24 V DC, mains power supply as an option

Status

No longer in production, but has been, and may still be, in service with several customers.

Contractor

MRCM
MRCM is a joint co-operation under full ownership of:
EADS Ewation, Ulm, Germany.
Grintek, Pretoria, South Africa.
Herley Industries Inc, Lancaster, USA.
Sysdel, Pretoria, South Africa.
TRL Technology, Tewkesbury, Glos, UK.

UPDATED

Telegon 112 VHF/UHF DF system

Telegon 112 is a compact communications band DF system with a receiving module which covers the 20-3,000 MHz frequency band and incorporates a standard PC with which to control the system's functions and display data. The system's DF module is based on the digital E 2000 VU receiver (see separate entry) and the equipment as a whole is suitable for both fixed-site and mobile applications.

Specifications

(a) Frequency: 20-3,000 MHz
(b) DF method: ADF with electronic rotating goniometer
(c) Min signal duration: 50 ms
(d) Power supply: 115/230 V AC; optional 12 V or 24 V DC

Status

No longer in production, but has been in service with several customers.

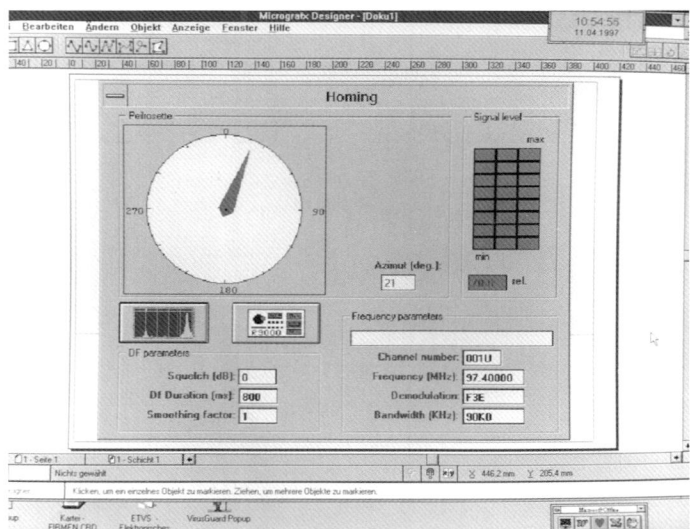

Telegon 112 DF display

0011532

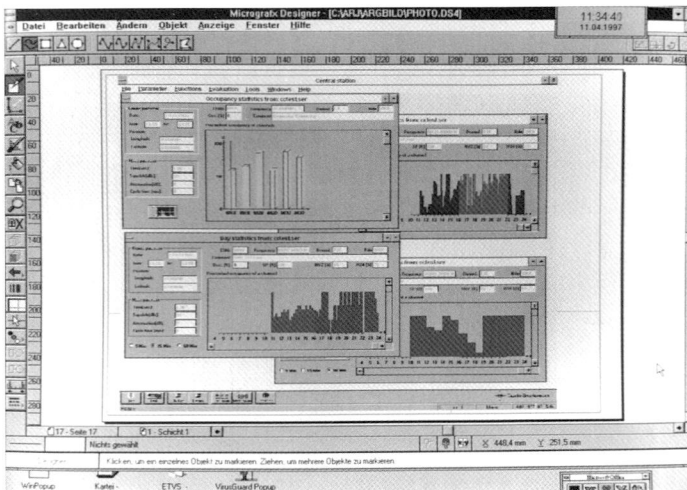

Telegon 112 display of evaluation statistics   0011533

Contractor

MRCM
MRCM is a joint co-operation under full ownership of:
EADS Ewation, Ulm, Germany.
Grintek, Pretoria, South Africa.
Herley Industries Inc, Lancaster, USA.
Sysdel, Pretoria, South Africa.
TRL Technology, Tewkesbury, Glos, UK.

UPDATED

TMSLoc

TMSLoc is a mobile tactical interception and location-finding system which uses several Rohde & Schwarz TMSR systems (see separate entry). It provides detection, identification, location-finding and recording of radio communication signals in the frequency range 20 to 1,300 MHz. Additional extensions are available for the 0.5 to 20 MHz and 1,300 to 3,000 MHz ranges. The system is fully transportable and runs on either DC or AC power supply.

Configurations can consist of one TMSR acting as a master station and up to three TMSR running as slave DF stations. The following types of communication links between the stations can be supplied:
(a) Line communication using analog or digital PSTN or leased lines
(b) Wireless communication using GSM networks
(c) Wireless communication using either military or civilian VHF radios.

Status

Available.

Manufacturer

Rohde & Schwarz, Munich.

NEW ENTRY

TMSR

TMSR is a mobile tactical interception and direction-finding system which provides detection, identification, direction-finding and recording of radio communication signals in the frequency range 20 to 1,300 MHz; extensions are available for both 0.5 to 20 MHz and 1,300 to 3,000 MHz. The system is transportable and runs on either DC or AC power supply and there is a battery-powered manpack version which can be carried by one person. System features include:
(a) Display of high-resolution and high-speed RF and IF spectrum and 2-D waterfall for recognition of radio signals
(b) Recording, replay and statistical evaluation of frequency spectrum data
(c) Display of COMINT results plus additional (for example, tactical) information on electronic maps
(d) Recording of digital audio signals
(e) Option for linking several TMSR within a communications network to form a location system TMSLoc (see separate entry).

Specifications

(a) Frequency range: 20-1,300 MHz, extendible to 0.5-3,000 MHz
(b) DF accuracy: 2° RMS
(c) Min. signal duration for DF: 10 ms
(d) Scanning speed of receiver: 1.3 GHz/s
(e) Power supply: 100-240 V AC, 11-32 V DC, 200 W
(f) Weight (equipment box): approx 45 kg

Status

Available as a COTS item. "Several" systems are operational with unspecified military customers.

Manufacturer

Rohde & Schwarz, Munich.

NEW ENTRY

VKP 4000 wideband HF DF system

The VKP 4000 DF system provides the continuous wideband monitoring and the simultaneous surveillance of all radio activity within the relevant frequency band. It allows the reliable interception of Low Probability Intercept (LPI) transmissions such as burst signals, frequency hoppers, chirp signals, and the detection and analysis of unknown signals, as well as the most efficient search and retrieval of known communications networks after changes of frequency and call signs.

The system is modularly configured and starts with the simultaneous surveillance of an 800 kHz band (real-time bandwidth) within the 0.5 to 30 MHz frequency range. Extensions of the real-time bandwidth to 1.6 MHz or 2.4 MHz are always realisable.

Within this real-time bandwidth, the VKP 4000 system is able to intercept all active transmitters. Signal detection as well as direction-finding of all intercepted emissions is performed in parallel. (that is, 19200 FFT subchannels each 125 Hz wide in a 2.4 MHz real-time band). Surveillance of larger frequency bands can be achieved by an additional scanning facility with a scanning speed of up to 150 MHz/s. The modular system design provides customised solutions (real bandwidth from 400 kHz up to 2.4 MHz). The effective bandwidth is therefore 2.4 MHz modularly divided into three 800 kHz sub-bands for simultaneous surveillance. The signal buffer provides a video buffer for the replay of the last 60 secs (optional 700 secs) of display.

Status

The VKP 4000 is no longer in production, but has been in service with a number of customers since 1997.

Contractor

MRCM
MRCM is a joint co-operation under full ownership of:
EADS Ewation, Ulm, Germany.
Grintek, Pretoria, South Africa.
Herley Industries Inc, Lancaster, USA.
Sysdel, Pretoria, South Africa.
TRL Technology, Tewkesbury, Glos, UK.

UPDATED

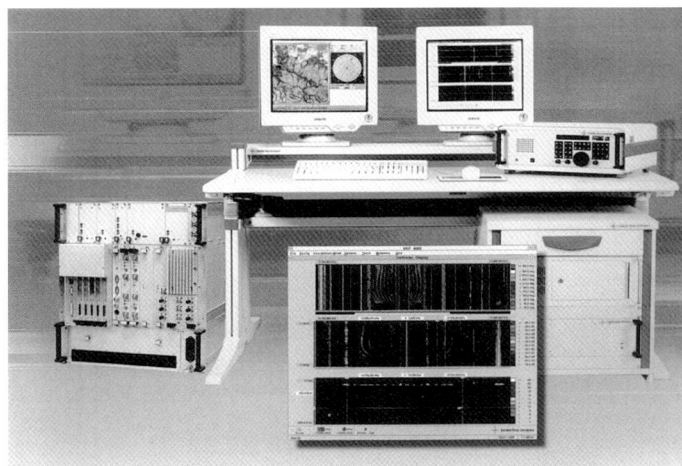

VKP 4000C wideband DF system HF Comint position   0055207

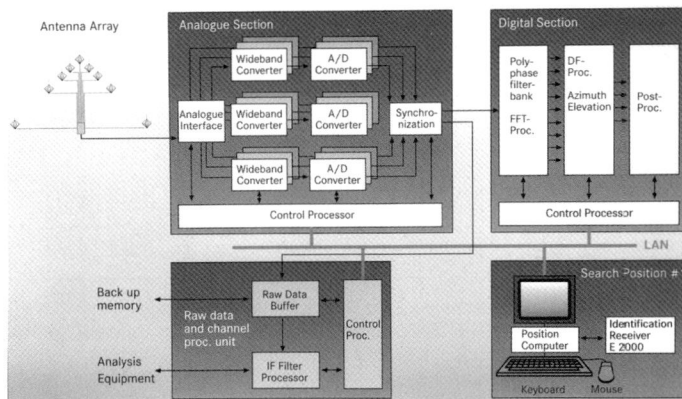

VKP 4000 block diagram   0011535

Hungary

IFR505 HF DF station

The IFR505 radio direction-finder is a modern device operating in the HF band. It is designed for integration into radio surveillance and reconnaissance systems, although it can also be operated in an autonomous mode. The equipment can be integrated into TCP/IP networks.

Features of the IFR505 include: phasometry DF; processing of both sky and ground waves; six-element antenna system; six receiver channels with DSP; large instantaneous DF bandwidth; calculation of DF results for the complete spectrum through the use of FFT techniques; time stamping of all records using a GPS time receiver; storage of raw data records for 60 minutes minimum; NT OS; TCP/IP network compatibility; single beam providing both azimuth and elevation (multibeam algorithm under development in 2002); and SSL algorithm.

Manufacturer

VIDEOTON-MECHLABOR Manufacturing and Development Ltd, Budapest.

NEW ENTRY

International

DFS2000 Man-Machine Interface (MMI) for direction-finders

The DFS2000 MMI provides an integrated and modular environment with a similar user interface for direction-finding applications, and is already in use with the following different direction-finder types:
(a) MRD 1920
(b) MRD 2000 (HF, interferometer with Single Station Location)
(c) MRD 30W3
(d) MRD 3000W5
Other direction-finders can be interfaced to the DFS2000 MMI without major changes. The open system can be integrated with customer equipment, all communications being established through standard interfaces such as TCP/IP.

DFS2000 is a modular MMI running under Windows NT, which lets the user construct their environment. Each element of the MMI (map, histogram display, tune DF) are displayed in separate windows, most of them being freely resizeable and moveable to the user convenience. Although DFS2000 requires a 1,280 × 1,024 screen resolution for a comfortable usage, there is no upper limit for the display area. The MMI can use the complete display area allowed by the graphic card itself: multiple monitors can be used allowing for a more comfortable usage. The DFS2000 MMI does not require any special graphic driver to be run. All the graphic information that has to be displayed goes through the standard Windows NT graphic API. DFS2000 can be run on a single workstation with a single direction-finder, or as part of a network of workstations with a direction-finder coupled or not to each. Several configurations are available:
(a) Stand-alone - single workstation/single DF
(b) Master-slave configuration (up to five slaves)
(c) Multimaster configuration
(d) No predefined master in the networked system. Workstations can exchange result lists, and can work in pairs or larger groups, each being part of a subnet as well.
DFS2000 MMI functions:
(a) Histogram
(b) Memory scan and frequency scan

(c) Zoom function: the operator has the opportunity to zoom in and out for a particular frequency range
(d) Hand-off function: a frequency of interest can be handed off to a monitoring receiver, a memory list or to the DF
(e) Map
(f) Record/replay of all scenario data (audio data and DF results)
(g) Result list management (save, recall, exchange with other stations)
(h) Tracking of mobile emitters
(i) DFS2000 can receive external DF commands with higher priority, sent by non DFS systems. The current activity is stopped for the duration of the higher priority command and is then resumed
(j) DF results can be automatically redirected to external output
(k) GPS input (so that all the systems in the network are synchronised, and the station location is known)
(l) Course input (for shipborne systems)
(m) Optional connection to monitoring or search receivers

Status
In service.

Contractor
MRCM
MRCM is a joint co-operation under full ownership of:
EADS Ewation, Ulm, Germany.
Grintek, Pretoria, South Africa.
Herley Industries Inc, Lancaster, USA.
Sysdel, Pretoria, South Africa.

UPDATED

··

Epsilon

Epsilon is a system which provides automatic detection, location and classification in real time of all HF emitters, including LPI emissions, originating in predefined geographic areas. Key features of the system are:
(a) Simultaneous surveillance and direction-finding of up to 19,200 channels within a total bandwidth of up to 2.4 MHz (subchannel-raster 125 Hz)
(b) Determination and real-time processing of up to 2.4 Mio DF data/s with an instrumental accuracy of 0.5°
(c) Additional scanning facility for surveillance of larger frequency bands (scanning speed up to 300 MHz/s)
(d) Automatic classification of new detected signals
(e) Reliable interception of the start of an emission (First bit /sign)
(f) Demodulation, classification, decoding and production of relevant emission's content
(g) Simultaneous operation by several operators
(h) Use of various DF antennas, such as Adcock, interferometer antenna, Wullenwever antenna
(i) Location of emissions either by Single-Station-Location (SSL) algorithm or by a DF base.
The system performs the tasks automatically by integrating wideband direction-finders, wide and narrowband receivers and a classification subsystem.

Specifications
(a) Frequency range: 0.5 to 30 MHz
(b) Wideband direction finder: Type MRD 4000
(c) Real-time bandwidth: up to 2.4 MHz
(d) Scanning speed up: to 300 MHz/s
(e) Bearing accuracy: 0.7° rms (instrumental). < 1 ° rms with DF antenna

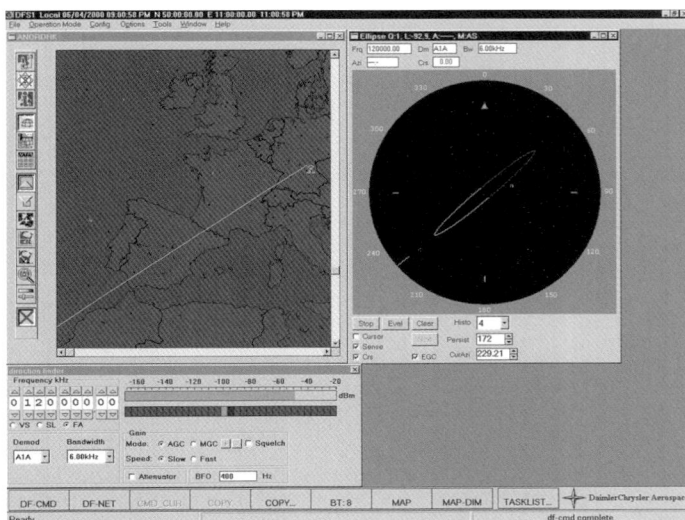

Watson Watt version of DFS2000 0098311

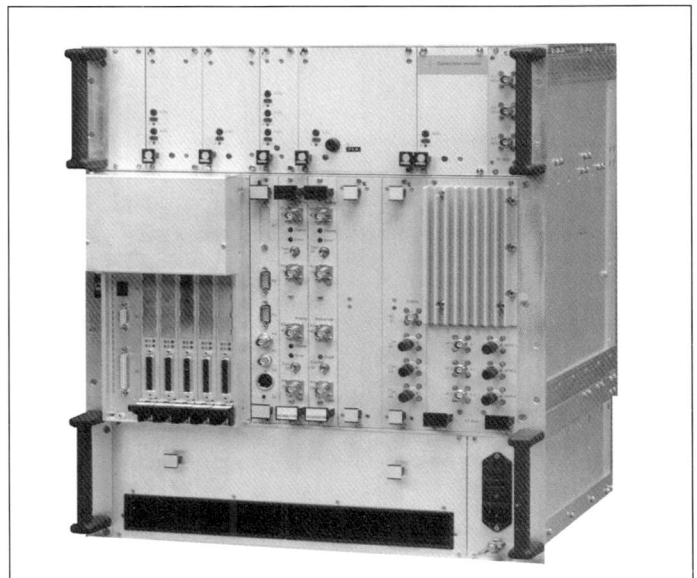

Polygon MRD 4008 wideband direction-finder 0098319

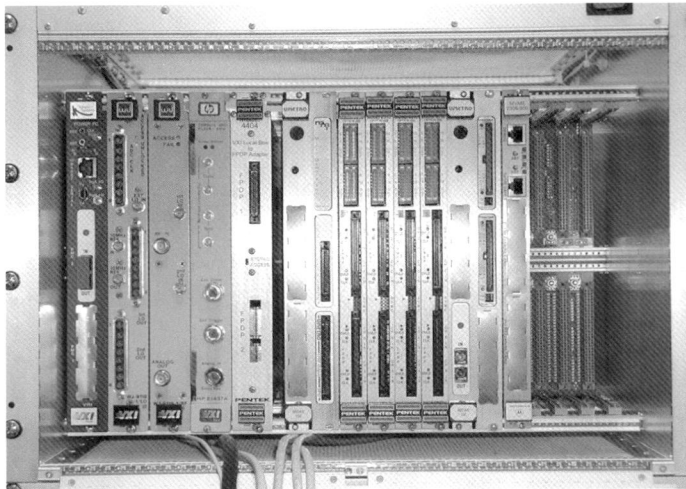

Polygon MRR 4000W wideband receiver   0098321

MCL 4000 signal classifier system
0098320

(f)   Minimum signal duration: 1 ms
(g)   Filterbank resolution: programmable 125 Hz, 250 Hz, 500 Hz, 1 kHz
(h)   DF-mode: Interferometer (small and wide aperture mode) or Watson-Watt
(i)   Wideband receiver: Type MRR 4000 W
(j)   Real-time bandwidth: 2 or 6 MHz, scaleable up to 30 MHz
(k)   Signal filtering: Digital Down Converters (DDCs)
(l)   Signal buffer for pre-detection recording
(m)   Classifier: Type MCL 4000

Status
In service in 2000 with the German armed forces.

Contractor
MRCM
MRCM is a joint co-operation under full ownership of:
EADS Ewation, Ulm, Germany.
Grintek, Pretoria, South Africa.
Herley Industries Inc, Lancaster, USA.
Sysdel, Pretoria, South Africa.

UPDATED

MRD2000 LH HF interferometer direction-finder system

The MRD200OLH HF Interferometer is a high-precision direction-finding system for the frequency range 1.6 (0.5) to 30 MHz. The system consists of:
(a)   A five-channel digital OF receiver, MRR2010 LH5-0F, a member of the MRR 2010 receiver family

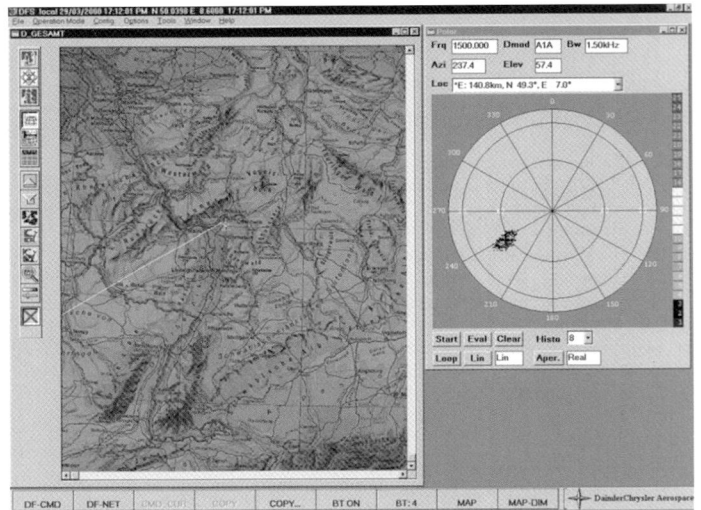

MRD2000LH hardware   0098313

(b)   A large aperture antenna system of 7 or 9 antennas and the antenna switch including cables
(c)   A control and bearing display computer with SW modules for equipment control, bearing display and SSL calculation, with an optional separate manual control unit MRR BFls

The five-channel MRR2010 LH5-DF receiver provides simultaneous processing of five antenna signals, thus enabling the reliable interception and bearing of short signals such as burst signals and frequency hoppers. The digital signal processing and filtering provides synchronisation of the five DF channels in phase and amplitude and provides the necessary facilities to eliminate interfering signals. The DF receiver has a compact design which requires only three height units of a 19 in chassis for the five DF channels, including control and post-processing devices.

The MRD 2000 LH system is available with two different antenna systems, the MRA 2000 and the MRA 2004. The MRA 2000 system is made up of combined active linear/loop antennas, and the MRA 2004 system is made up of passive linear antennas. In both cases, the system consists of seven/nine antenna elements, which are erected in two rows of three/four antennas each in a right angle, and a common reference antenna. The number of elements and the aperture are variable and may be matched to the available site.

The control and bearing display MMI is provided by DFS 2000 (see separate entry).

In the case of skywave propagation the transmitter location can be fixed with only one DF station over distances of "some 1,000 kilometres"; this is known as SSL (Single Station Location) calculation. The distance is computed from the elevation of the incident wave and the height of the ionosphere. The location results from azimuth and distance. The height of the ionosphere, prevailing at any particular time, is obtained from the prediction program "IONCAP", which is valid worldwide for the whole year. The result is displayed in UTM or geographical co-ordinates and also on a scanned electronic map.

Specifications
(a)   No of DF channels: 5
(b)   Frequency range: 1.6 (0.5) to 30 MHz
(c)   Types of demodulation: AM, FM, CW, SSB, FSK
(d)   Intermodulation: IPIP 2: +70 dBm; IPIP 3: +40 dBm
(e)   DF time (min signal duration): 2 ms
(f)   DF sensitivity: -15 dB μV/m
(g)   DF accuracy: $<1°$ rms for groundwaves (typ $0.5°$); $1°/\cos e$ for sky waves
(h)   Claimed SSL accuracy:
        For distances up to 100 km: 15 km rms
        Distances from > 100 km: 10% rms typical

Status
In production and in service.

Contractor
MRCM
MRCM is a joint co-operation under full ownership of:
EADS Ewation, Ulm, Germany.
Grintek, Pretoria, South Africa.
Herley Industries Inc, Lancaster, Pennsylvania, USA.
Sysdel, Pretoria, South Africa.

UPDATED

MRD3000w5 wideband V-UHF direction finder

The Polygon MRD3000w5 is a FFT wideband DF system which is intended for wideband spectrum surveillance and wideband Direction-Finding (DF) of fixed-frequency and frequency agile transmissions in the V/UHF frequency band (up to 3 GHz). It combines an utmost of interception probability with most accurate

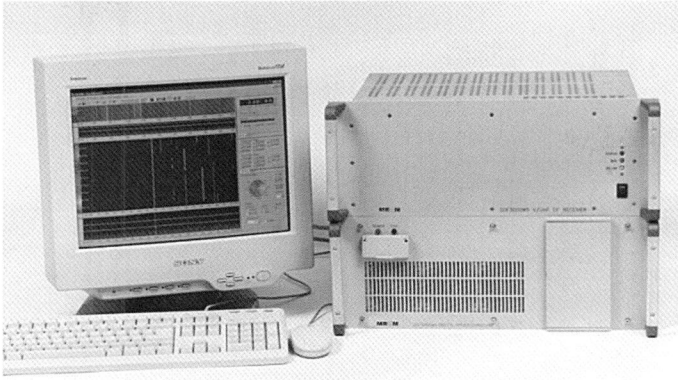

MRD3000w5 hardware 0098324

MRD3000w5 mounted in a Land Rover with antenna erected (MRCM) 0525546

Mobile SIGMA 5000 unit **NEW**/1031949

SIGMA 5000 interior **NEW**/0525299

direction-finding achieved by sensitive large aperture interferometer antenna systems and parallel signal processing in five DF channels. Different system configurations are available to suit fixed, semi-mobile and mobile operations in land-based, shipborne and airborne applications.

Specifications
(a) Frequency range: 20-3,000 MHz
(b) Number of DF channels: 5
(c) Instantaneous bandwidth: 10/1.25 MHz selectable
(d) Frequency resolution: 12.5 kHz, 6.125 kHz, 3.063 kHz, 1.531 kHz
(e) Noise figure: 10 dB
(f) Spurious free dynamic range: 80 dB
(g) Scanning speed: up to 45 GHz/s (search), up to 10 GHz/s (DF)
(h) Signal acquisition time: 80 µs
(i) DF algorithms: Interferometer, Watson-Watt
(j) Typical bearing accuracy for fixed applications
  20 to 1,000 MHz: <1°rms (typ 0.5°)
  1 to 3 GHz: <1.5°rms (typ 1°)
(k) Typical bearing sensitivity for fixed applications
  20 to 1,000 MHz: 1.0 µV/m
  1 to 3 GHz: 1,5 µV/m
(l) Operating system for man/machine interface: Windows 2000

Status
In production and in service.

Manufacturer
MRCM
MRCM is a joint co-operation under full ownership of:
EADS Ewation GmbH, Ulm, Germany.
Grintek Ewation, Pretoria, South Africa.
Herley, Farmingdale USA.
Indra, Madrid, Spain.
Sysdel, Pretoria, South Africa.

VERIFIED

SIGMA 5000: System for Interception, Goniometry, Monitoring and Analysis

SIGMA 5000 is a self-contained dual-operator system which includes a mast-mounted DF antenna for mobile and semi-mobile use. It provides wideband spectrum surveillance and DF in the 20 to 3,000 MHz range and narrowband fixed frequency monitoring in the 10 kHz to 3,000 MHz range. Data decoding and

analysis is available and automatic signal modulation classification is an optional extension. All elements are integrated in a commercial van which contains operator consoles, A/C and power generation.

Vehicles can either be autonomous or in a master/slave configuration. A station can perform either role and can multitask, that is, can be master to one station and slave to another.

Specifications
(a) Scanning frequency range: 20-3,000 MHz Search and DF)
(b) Intercept frequency range: 20-3,000 MHz (monitor); 10 kHz-30 MHz (optional)
(c) DF accuracy
    20-100 MHz: <2.5°rms
    20-1,000 MHz: <1.5°rms
    1-3 GHz: <1°rms
(d) Instantaneous bandwidth: 10 MHz. Zoom: 5MHz/2.5MHz/1.25 MHz
(e) Frequency resolution: 12.5 kHz. Zoom: 6.125 kHz/3.063 kHz/1.531 kHz
(f) Signal acquisition scan rate: up to 45 GHz/s
(g) User interface: Windows NT XGA colour, 1024 × 768
(h) Effective DF scan rate: Up to 1 GHz/s
(i) No of DFs/s: 10,000

Status
In production and in service.

Contractor
MRCM
MRCM is a joint co-operation under full ownership of:
EADS Ewation, Ulm, Germany.
Grintek, Pretoria, South Africa.
Herley Industries Inc, Lancaster, USA.
Sysdel, Pretoria, South Africa.

UPDATED

Israel

TDF 1200 MultiChannel Wide Band DF System

The TDF 1200 Multi-Channel Wide Band Direction-Finding System operates in the VHF/UHF frequency bands. The system is capable of measuring a wide range of DOAs - including frequency hoppers and burst transmissions. The system comprises:
(a) 4 coherent channels receiver
(b) Dedicated Digital Signal Processing (DSP) and controller cards (embedded in system computer)
(c) DF antenna array
(d) DF system computer (PC-compatible)
The TDF 1200 can be operated either as a stand-alone for automatic DOA measurement, or as part of a larger DF Network for Emitter Location (master/slave). The system has a large selection of displays for presenting measured and highly detailed results, including graphic map displays and simultaneous displays of various categories. These include: Direction vs Frequency; Direction vs Time; Frequency vs Time; Polar Displays; DOA Histograms.

The wideband receiver has four coherent channels; high measurement accuracy is provided by the wide aperture antenna using the interferometric principle. The TDF 1200 uses parallel processing techniques for signals within a 4 MHz frequency bandwidth.

Specifications
Frequency range: 20 to 1000 MHz; 20 to 3000 MHz optional
Channel resolution: 25 kHz
Scan speed: 40,000 channels/s
Instantaneous bandwidth: 4 MHz
Azimuth coverage: 360°
Elevation coverage: 10° RMS
Polarisation: vertical
Accuracy: <1.5° typical, <1° instrumental
Response time: fast DF mode:1ms; histogram DF mode: 1, 5, 10, 50 ms
Dedicated signal modulations: analogue and digital modulations

Contractor
Tadiran Electronic Systems Ltd, Holon.

UPDATED

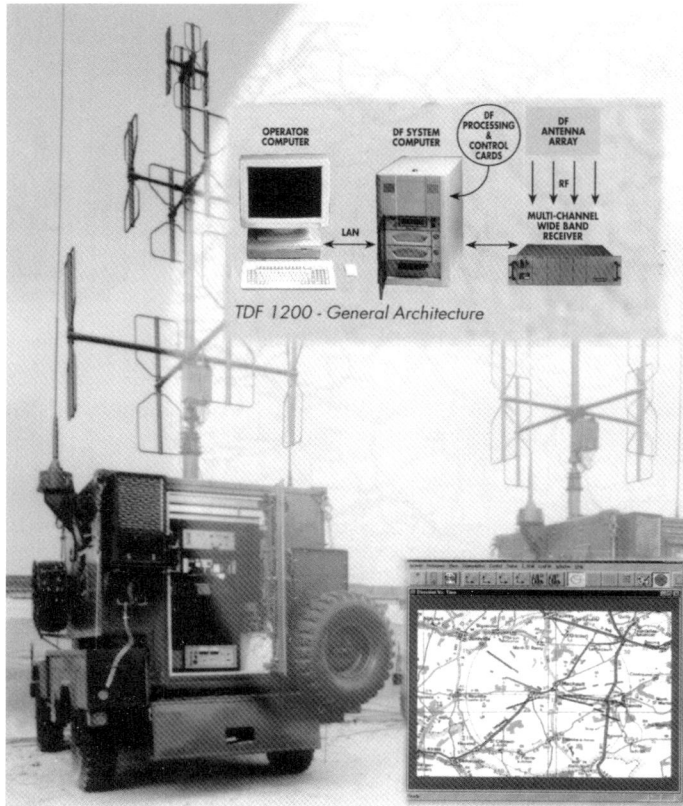

Tadiran TDF 1200 MultiChannel WideBand Direction-Finding System 0084489

TDF-2020 HF/VHF/UHF Direction-Finding System

The TDF-2020 operates in the HF, VHF and UHF frequency bands and is designed for spectrum monitoring applications. It is capable of measuring direction to fixed frequency, frequency hopping, CDMA and burst transmissions, and is designed for easy integration into computerised communication, intelligence and spectrum monitoring systems. The system is based on a weighted beam-forming DF technique that improves system performance by using wide aperture antennas and having algorithms with higher immunity to interference. The weighted beam-forming DF technique is claimed to be superior to other techniques (such as Adcock and interferometer) as it uses both the phase and the amplitude in a combined manner, tailored to each frequency range, to exploit the antenna array to the maximum.

The TDF-2020 system comprises a DF antenna array, an RFDU-2020 RF unit, two TSR-2020 (see separate entry) coherent receiver boards and an RFM-2020 DF processor and reference board. All the boards are installed in a PC compatible computer. The system is suitable for fixed (ground-based), sheltered mobile (light vehicle) including homing, and transportable (semi-mobile) platforms, with an associated family of DF antennas. Location fixing is achieved by combining several TDF-2020 into a net, and the system can be integrated into larger spectrum monitoring systems

Specifications
Frequency range: V/UHF: 20 MHz to 3,000 MHz; HF: 10 kHz to 30 MHz
Azimuth coverage: 360°
Instrumental accuracy: 0.8° RMS
Bearing resolution: 0.1°
Field bearing accuracy (ITU Class A): V/UHF: 1° RMS (Fixed/Sheltered), 2.5° RMS (Mobile);HF: 1.5° RMS (Fixed/Sheltered). 3° RMS (Mobile)
Sensitivity: 1 to 20 m V/m (frequency and antenna dependant)
Minimum signal duration: 5 ms
Scan speed with direction: HF: up to 200 channels/s; V/UHF: up to 1,000 channels/s
Receiver performance: according to TSR-2020 performance

Contractor
Tadiran Electronic Systems Ltd, Holon.

VERIFIED

South Africa

Telegon MRD 1920 multimode direction-finder

The MRD 1920 multimode direction-finder can handle several direction-finding methods (Watson Watt, Interferometer) in the entire communications frequency range from 10 kHz to 1 GHz. The MRD 1920 comes in three different versions:
(a) MRD 1920 LH: 10 kHz-30 MHz
(b) MRD 1920 VU: 20 MHz-1 GHz
(c) MRD 1920 LU: 10 kHz-1 GHz
The MRD 1920 has analogue/digital receivers coupled to a digital signal processing unit that runs direction-finding algorithms.

The dynamic range, sensitivity and selectivity of the MRD 1920 enable it to detect the direction of weak signals almost buried in the ambient noise (or strong nearby emitters producing intermodulation). The algorithms evaluating the direction of arrival ensure, even for shipborne operation, a claimed bearing error of less than a degree on CW and more complex signals like hoppers. For the latter type of signal, the MRD 1920 offers special DF modes for pulse or burst signals including pulse analysis. The MRD 1920 can be fully remote controlled (via modem, radio links, LAN).

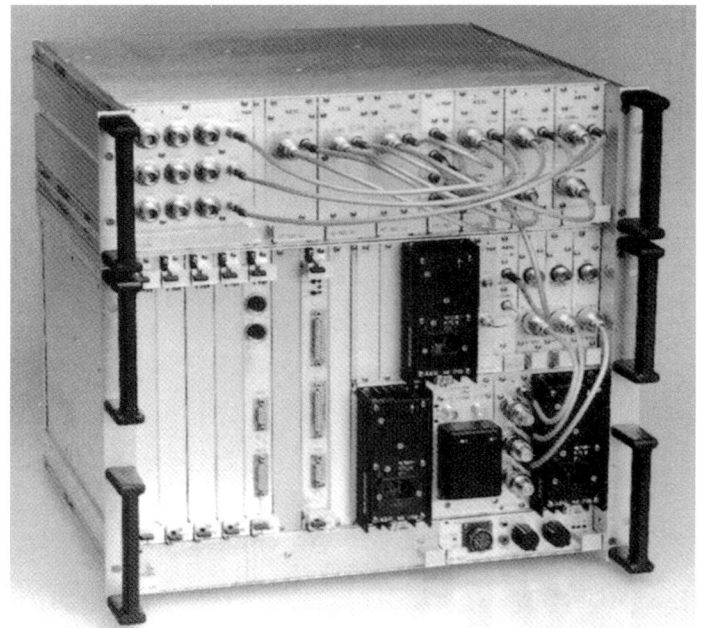

MRD 1920 hardware 0098325

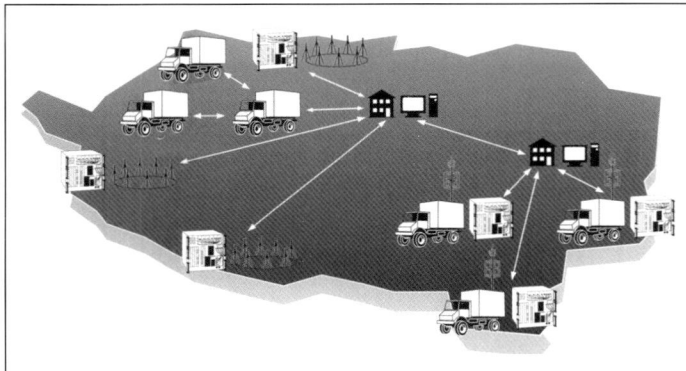

MRD 1920 in operation within a network of fixed and mobile stations   0098326

Specifications

(a) Frequency range: 0.01-30 MHz, 20-1,000 MHz, 0.01-1,000 MHz
(b) Number of channels: 3
(c) Bearing accuracy: 1°
(d) DF algorithm: Watson-Watt, Interferometric (depending on frequency range and antenna system)
(e) Frequency scan linear: 10 MHz/s (channel spacing 25 kHz)
(f) Frequency scan adaptive: 1.5 GHz/s
(g) Memo scan: 400 channels/s BW=30 kHz
(h) Minimum signal duration: 1 ms
(i) Sensitivity (1° bearing fluctuation, integration time T=200 ms, BW=15 kHz): 1 µV/m (with MRA 1228)
(j) Sensitivity: 3 kHz; S/N=10 dB-115 dBm
(k) Dynamic range (BW=0.6 kHz): > 92 dB without gain control; > 140 dB with gain control
(l) Demodulation: CW, AM, SSB, FM, FSK
(m) Noise figure: 10-13 dB depending on frequency range
(n) Temperature:
Operating: −25 to +55°C
Storage: −40 to +70°C
(o) User interface: DFS2000 (see separate entry) Windows NT 1,280 × 1,024 pixels

Status

The MRD 1920 is claimed to be in service in unspecified mobile, fixed and shipborne installations.

Contractor

MRCM
MRCM is a joint co-operation under full ownership of:
EADS Ewation, Ulm, Germany.
Grintek , Pretoria, South Africa.
Herley Industries Inc, Lancaster, Pennsylvania, USA.
Sysdel, Pretoria, South Africa.

UPDATED

Turkey

Armoured Tactical Direction Finding (DF) Vehicle

The ASELSAN Armoured Tactical Direction Finding (DF) Vehicle has been developed to intercept, DF, monitor and record target communications under battlefield conditions that require armoured protection and high mobility. Mounted on the AKREP Light Armoured Vehicle, this system can operate under harsh environmental conditions and can be set up and operational in a short time. The system is based on the ASELSAN DFINT-3T Communications DF System (see

Aselsan Armoured Tactical DF Vehicle
0092760

separate entry). This is married with ASELSAN's command and control software which includes GIS, DBMS and formatted message functions. A comprehensive communications package is also installed. Equipment includes the ASELSAN 9600 VHF frequency hopping radio, ASELSAN SK-4000 digital encrypted radio, ASELSAN data terminal and an audio tape recorder.

Status

The Armoured Tactical DF Vehicle is in production and in operational use.

Manufacturer

ASELSAN Military Electronic Industries Inc, Ankara.

VERIFIED

DFINT-3 series tactical communications DF and intelligence systems

The DFINT-3 series of tactical communications DF and intelligence systems operate in the VHF/UHF frequency band to intercept, DF and monitor target communications. Three different models are available: the DFINT-3A, the DFINT-3A2 and the DFINT-3T. All have similar characteristics but are packaged differently. The systems offer these features:

(a) Simultaneous real-time Line-of-Bearings (LOB) versus frequency, and amplitude versus frequency graphics in DF SCAN mode
(b) On-screen digital panoramic spectrum and polar LOB plot for monitoring or fast scanning of targets in DF STEP mode
(c) Location fixing on a digital map.

All DFINT-3 systems can be integrated through wireless or wired nets.

Common features

(a) Interferometric DF principle
(b) Ability to intercept and DF state-of-the-art frequency-hopping and burst emissions

The DFINT-3A mounted on a Land Rover
0100386

The DFINT-3T   0100390

(c) Netted operation for location fixing
(d) Two-channel receiving and recording facility
(e) Computer controlled modular system architecture
(f) High mobility on rough terrain with a 4 × 4 land vehicle
(g) Immunity to harsh environmental conditions
(h) Ability to operate with external AC and internal DC sources
(i) Built-in GPS receiver and electronic compass
(j) Built-in self-test (BIT) capability, ease of maintenance
(k) Quick set-up and tear-down by two people
(l) Colour graphical and list format data presentations
(m) Secure voice and data communication via wired and wireless nets.

DFINT-3A

The DFINT-3A has a DF scan speed of 500 MHz/sec and functions as an intelligence station that can be operated by one person. It is integrated in a 4 × 4 vehicle that tows a single-axle trailer carrying the AC power generator. Additionally, a built-in battery bank is kept ready for silent missions. A simple keypad, including soft keys, a cursor driver and a high-resolution colour monitor provide menu-driven operation.

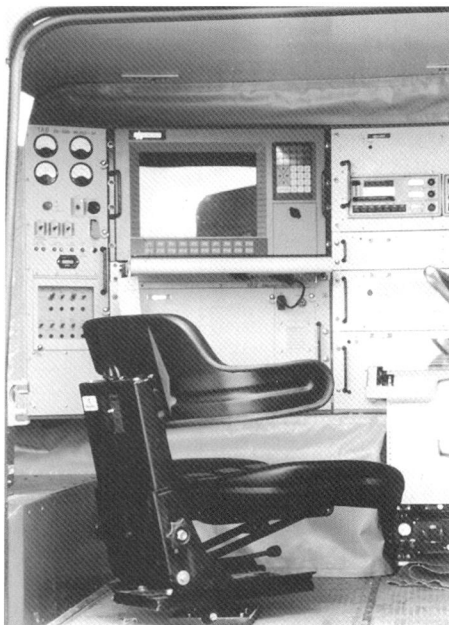

The DFINT-3A interior

The DFINT-3T deployed in the man-portable role (ASELSAN)
0083197

DFINT-3A2

The DFINT-3A2 has a DF scan speed of 1,000 MHz/sec and, like the DFINT-3A, functions as an intelligence station that can be operated by one person, although monitoring receivers and digital voice recorders can be used by a second operator if needed. Both the system shelter and the quiet AC power generator are carried by a 5 ton capacity 4 × 4 truck. A built-in battery bank is provided for uninterrupted operation and silent missions. The antennas of DFINT-3A2 can be located up to 50 m away from the vehicle if required. The DFINT-3A2 shelter has been designed to accommodate two operators. It provides menu-driven operation with a high-resolution 18 in colour monitor, a full keyboard and a cursor driver on each console. All the system resources are accessible from the main console. Monitoring receivers and a digital recording unit can also be operated from the second console.

Additional features

(a) Up to 8 V/UHF monitoring receivers
(b) Operational multichannel demodulators
(c) Up to 32-channel digital recording.

DFINT-3T

The DFINT-3T system has been developed to fulfil the same functions in a variety of environments and has been designed for use on land, in the air and on naval platforms. It has a DF scan speed of 1,000 MHz/sec. It can be deployed with the standard high-gain/lightweight antenna, or subbands can also be deployed individually for further ease of use. Several other antennas are also available for different applications such as low-profile disguised antenna, airborne and heliborne antennas. In land-based transportable applications, the DFINT-3T system is man-portable by two personnel where vehicle access is difficult.

Status

In production and in service with the Turkish forces. In May 2000, *Jane's International Defence Review* reported that the DFINT-3T was being actively marketed abroad.

Manufacturer

ASELSAN Military Electronic Industries Inc, Ankara.

VERIFIED

United Kingdom

MA1122 DF unit

The MA1122 is a Direction-Finding (DF) processing and display unit for use in tactical EW systems. It is suitable for installation in tracked or wheeled vehicles. It forms part of the Thales commutated Adcock DF system operating in the frequency range 20 to 1,000 MHz, with the Thales RA1794 or RA1796 receivers and the AE3020 series of VHF/UHF DF antennas. It can be either manually operated or remotely controlled as part of an automated DF network. An optional MA4233 miniature printer can be used to record results. Alternatively, results can be stored in a PC or another electronic database.

The MA1122 can also be used in a hunting DF role, in which case the frequency range can be extended down to 2 MHz and special low-profile DF antennas are fitted to a suitable vehicle. Covert, mobile DF operations to locate unauthorised transmitters are then possible.

In operation, microprocessors compute a rapid series of bearing data which are integrated and displayed as a four digit number readout. They also calculate a quality factor. A memory facility stores bearing results.

Status

Used in the Seeker III system (see separate entry).

Contractor

Thales Sensors, Crawley, Sussex.

UPDATED

MA1122 DF processor and display unit

Seeker III system

The Seeker III system provides a combined intercept and Direction-Finding (DF) capability in one mobile vehicle installation. It incorporates intercept and DF receivers, a DF processor, MA1122 and communications facilities.

DF is provided over the frequency range 2 MHz to 1 GHz. When three or more Seekers are deployed as a network, target position fixing becomes possible by normal triangulation methods. Antenna masts, generators and other ancillary equipment are transported in an accompanying trailer.

Seeker III is essentially a basic manual system for armies setting up an EW capability. It can, however, provide automated remote control of the DF function, thereby allowing one station to control others in the network. This is achieved by the inclusion of appropriate tactical computers and provision of datalinks over combat radio nets.

Status

A number of Communications EW Systems under the product name of 'Seeker' were delivered to the Australian Defence Forces by Thales Communications UK (previously Racal Defence Electronics Ltd) in the year 2000. The communications EW technology in the system is believed to be related to the UK Odette system (see separate entry), but no further details are available.

Contractor

Thales Defence Communications, Wells, Somerset.

UPDATED

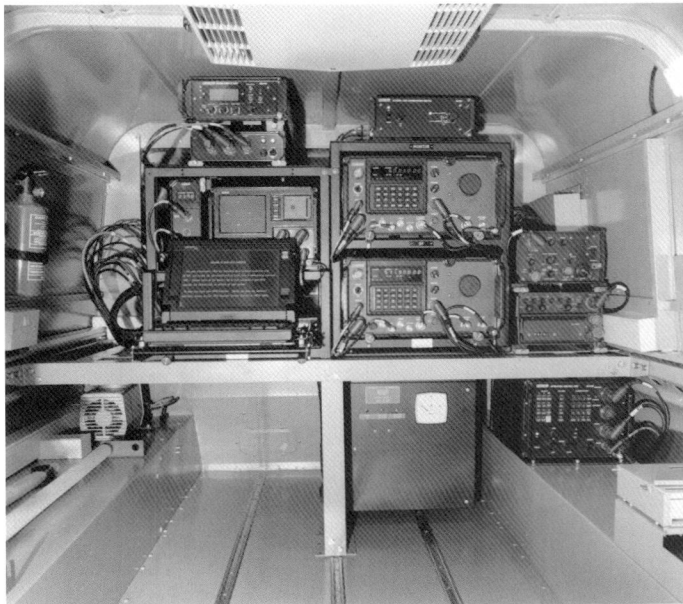

Seeker III intercept and DF system

United States

802 HF DF/SSL systems

The 802 family of HF Direction-Finder/Single Station Location (DF/SSL) systems covers the 1.5 (0.3 optional) to 30 MHz range with a variety of configurations, all based on the TCI 8060 digital DF processor. A specific system configuration is designated 802-N-n where 'N' is the number of elements of the DF antenna array and 'n' is the number of DF receiver channels.

The antenna array is selected to meet requirements of signal intercept, frequency coverage and deployment. Monopole (whip) and crossed-loop elements may be used in linear or circular arrays. The 8060 processor can also accommodate complex circular arrays.

The 8060 processor permits multichannel DF receiver operation without the need for perfect receiver matching. This is achieved through a technique of calibration and software-based compensation for RF path mismatch. When using one DF receiver channel per antenna array element (Nn), no RF switching is necessary and a major source of DF errors is claimed to be eliminated. The TCI 9080 HF receiver has been developed specifically for 802 systems but a wide range of good quality receivers can be used.

The 8060 digital DF processor uses the IBM PC as the basic computer platform. A modified interferometry technique is used for simple arrays and Wave Front Analysis (WFA) for more complex arrays. FFT processing, using modern DSP techniques and components, replaces traditional I-Q demodulators. Signals are processed with a single IF filter (3 kHz bandwidth). The operator may then isolate a signal effected by co-channel interference by reducing bandwidth with software-controlled function keys. A technique of high-speed data sampling (cuts) and cluster analysis is used. The operator can adjust the DF dwell-time, or a number of successful cuts, to handle low-duty cycle signals, such as Morse or SSB.

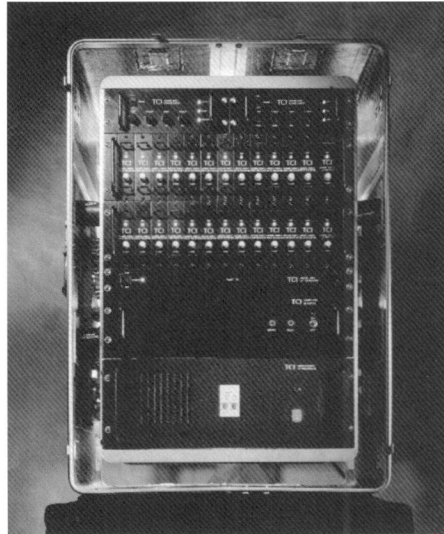

The 802 24-channel receiver with 8060 processor

All 802 systems have SSL capabilities as both azimuth and elevation angles-of-arrival are measured. The TCI 820 vertical ionosphere sounder is used to obtain real-time ionospheric height data necessary for SSL range calculations.

Installations for fixed and mobile stations are available.

Type 802 systems are fully automatic and are controlled remotely. A net of DF stations can provide emitter location results by triangulation or SSL directly to a signal surveillance and monitoring system. Operational displays (with colour VGA) include: polar, with azimuth and elevation angles of arrival; histograms; tabular DF/SSL summaries; signal spectrum and digital map.

Status

In production. Systems fielded and several variants are operational. These include: the 802-24-24 with 820 sounder, a fixed station installation with a 24-monopole circular array and a 24-channel DF receiver section under 8060 control; the 802-9-9 with 820 sounder, a fixed station with an antenna array of nine whip elements and a nine-channel DF receiver; and the 802-2-9, a two-channel DF receiver operating with a nine-whip-element antenna array. The latter is part of the AN/TRD-27 HF DF system (see separate entry), which is a sub system of the US Army AN/TSQ-152(V) TRACKWOLF and AN/TSQ-199 Enhanced TRACKWOLF (see separate entry).

Contractor

TCI, Fremont, California.

UPDATED

8070 monitor operator workstation

The 8070 is one element in a family of modular building blocks that make up the TCI SIGINT Collection, DF/SSL and Reporting System. This collection position is based on the PC/AT architecture, and it can be tailored for transportable, mobile or fixed-station applications. The PC/AT platform can be supplied in a variety of configurations, from rack-mounted rugged commercial to laptop military specification. Other TCI positions compatible with the 8070 and based on the same architecture are the DF/SSL Operator Position (8013) and the Collection Supervisor Position (8014).

The 8070 monitor operator workstation allows the operator to monitor signals, perform automatic direction-finding (DF) and view DF and emitter location results. It can task 8013 DF operator workstations for help in resolving DF on particularly difficult signals. It controls up to four receivers and displays the results of two receivers simultaneously to optimise operator efficiency. The operator initiates all tasks via keyboard command of a Windows-based graphical user interface. The DF Monitor Tasking Screen provides 6 kHz passband pan displays for two receivers that help the operator tune them and initiate requests for automatic, netted or assisted DF. DF results are presented in easy-to-understand formats on multiple high-resolution displays, including a Map Screen, which gives the operator full zooming and panning capabilities on selected areas. Results are saved locally and at the TCI 8014 supervisor workstation if one is on the LAN. A Status Screen displays the 8070's set-up and priority level and the real-time status of other workstations on the LAN.

The 8070 consists of commercially available components that can be tailored to work with customer-provided receivers and PCs. TCI standard architecture is a PC/AT-compatible processor, a hard drive, a CD drive, a diskette drive, a LAN interface, a colour monitor, a keyboard, a mouse and up to four high-performance TCI Model 8074/8174 HF monitor receivers (option). The 8070 can be ruggedised; configured for fixed, mobile and transportable applications; and tailored to satisfy customer specifications.

Contractor

TCI, Fremont, California.

UPDATED

AN/ALQ-151 (V)2 Quick Fix

The AN/ALQ-151 (V)2 special purpose electronic countermeasure system, Quick Fix, is a heliborne electronic warfare system. The system incorporates the EH-60A helicopter, AN/TLQ-17A (V)2 jammer, Electronic CounterMeasure (ECM) group, Electronic Support Measure (ESM) equipment for active ECM, Airborne Radio Direction Finder (ARDF) data processing and a suite of Aircraft Survivability Equipment (ASE). Voice and datalink communications between other airborne Quick Fix systems and select ground systems are provided via secure communications. Quick Fix systems net with each other and interoperate with Trailblazer (see separate entry) in a netted configuration for DF purposes.

Specifications
Target signals:
Frequency: HF/VHF
Distance range: line of sight
Signal types: AM, FM, CW, SSB
Bandwidth: 8, 30 or 50 kHz
Input sensitivity: −110 dB nominal

Communications links:
Types: encrypted voice and data
Frequency: UHF/VHF
Distance range: line of sight
Signal type: FM/AM

Physical characteristics:
Humidity: 0 to 98%
Temperature: −40 to +55°C (−40 to +131°F)

Status
The equipment was developed and produced by TRW Systems, now Northrop Grumman Mission Systems. Production is complete. A total of 66 systems were procured. AN/ALQ-151 (V)3, Advanced Quick Fix, was an upgrade to be applied to 32 of the 66 platforms as part of the overall Intelligence and Electronic Warfare Common Sensor (IEWCS) programme. This programme was cancelled in mid-1999, and replaced by Prophet (see separate entry). Jane's sources indicate that the equipment is due to be withdrawn from service by 2005.

Contractor
Northrop Grumman Mission Systems, Reston, Virginia.

UPDATED

Quick Fix equipment is contained within a modified Sikorsky UH-60A Black Hawk helicopter. The system is characterised by four dipole antennas mounted in pairs on either side of the rear fuselage and a deployable whip antenna beneath the fuselage

AN/TRD-27 HF DF system

The AN/TRD-27 HF Direction-Finding (DF) system is a transportable shelter-based Direction-Finder/Single Station Location (DF/SSL) system incorporating an ionospheric Chirpsounder. A net of AN/TRD-27 stations, with associated communications support (AN/TRQ-41), forms part of the US Army TRACKWOLF AN/TSQ-152(V) (see separate entry).

The AN/TRD-27 is a development of the Trackfinder programme incorporating the TCI 802 HF DF/SSL system and IBM PC/AT-compatible operator workstations.

The AN/TRD-27 forms part of the TRACKWOLF System   0085278

Status
Operational, part of US Army TRACKWOLF AN/TSQ-152(V).

Contractor
TCI, Fremont, California.

UPDATED

AN/TRQ-32A(V)2 Radio Receiving and DF system TEAMMATE

The TEAMMATE AN/TRQ-32A(V)2 Radio Receiving Set is a mobile, ground-based communications intercept and direction-finding system for support of US Army tactical forces. The system, mounted on an M-1097 Heavy HMMVW, provides HF, VHF and UHF communications intercept and VHF direction-finding. The system provides two operator positions to perform all necessary functions. An interface to other similar IEW sensor systems and to control and analysis centres is provided through net radio protocol by a host interface unit.

Status
In operational service within the US Army intelligence and electronic warfare community. It is being replaced by the AN/MLQ-40 PROPHET (see separate entry).

Contractor
Raytheon Systems Company, Fort Wayne, Indiana.

UPDATED

AN/TRQ-32A signal collection and direction-finding system
0085277

AN/TSQ-138 TRAILBLAZER

TRAILBLAZER is a high-capacity ground-based communications intercept, processing, and direction finding system. The system is used to search for, intercept, record, identify, locate, and report on radio signals in the HF/VHF/UHF frequency ranges. The system operates in a netted configuration and interoperates with the airborne QUICKFIX system for direction finding. The original version was the AN/TSQ-114, which was subsequently improved and fielded as the AN/TSQ-138.

Description

The AN/TSQ-138 master control station (TRAILBLAZER) is a tactical ground-based HF/VHF/UHF signal intercept system mounted in a ballistically protected kevlar shelter with a 50 ft quick-erect hydraulic/pneumatic antenna mast assembly mounted on the roof. The system also provides direction-finding (DF) line of bearing (LOB) for intercepted VHF emitters. When used in conjunction with at least two other AN/TSQ-138, it can compute VHF emitter fixed locations through LOB data transfer via a high-speed data link. The data link between other master control sets within a designated network is controlled by the AN/TSQ-138.

It is transported on a 5 ton drop-side truck 6 × 6 with a 5 ton trailer, although there may be a trailer-mounted version towed by a HMMWV, and is manned by 6 operators, some of whom will be linguists.

In 1997 it was announced that dual BARCO MPRD 9651 19 in displays would be integrated into the system as part of an upgrade.

Status

In service with the US Army. Originally allocated with 5 master control sets for each heavy division EW Company, plus further sets for training at Ft Huachuca. Possibly also allocated to some MI battalions. In support of the division it is deployed to cover the divisional front in a loose W formation in sites providing good line of sight to potential targets.

In January 2003 *Jane's Defence Weekly* reported that Trailblazer had been acquired by Taiwan, either in 2001 or possibly in the late 1990s.

Contractor

TRW, Sunnyvale, California.

UPDATED

AN/TSQ-164 tactical HF DF/SSL system

The AN/TSQ-164(V) system is a mobile, automated, ground-based HF skywave and ground wave emitter intercept, collection, analysis, direction-finding, position location and reporting system. The ruggedised system is housed in an S-280 shelter with three identical operator positions. Each position is capable of performing all, or any combination of, the system functions. The direction-finding and position location capabilities are based on the SKYLOC™ architecture. The system can be operated netted or stand-alone as a Single-Station Location (SSL) system using its low-power vertical incidence ionospheric sounder.

The AN/TSQ-164(V) uses VME processors, an Ethernet Local Area Network (LAN), X-Windows, ORACLE relational databases, and the SPOX Operating System™.

Specifications

(a) Frequency range: 0.5-30 MHz
(b) Modulation types: OOK, CW, MCW, ICW, USB or LSB voice, AM, FSK and PSK
(c) Sensitivity: -110 dB for 10 dB S/N in 4 kHz bandwidth; receiver tunable in 10 Hz increments
(d) DF integration time: 6.25-50 ms in steps of 6.25, 12.5, 25 and 50 ms
(e) DF acquisition speed: 8-25 frames/s over range of 50 to 6.25 ms
(f) DF azimuth resolution: 0.1°
(g) DF instrumental accuracy: <=0.5°
(h) External comms interface: RS-232/422
(i) Long-Term DF/SSL Performance

Range	Azimuth (mean)	Range error
Skywave		
40-200 km	<=3°	<=15%
200-700 km	<=1°	<=10%
700-1,000 km	<=1°	<=10%
1,000+ km	<=1°	N/A
Ground wave		
20-400 km	<=1°	N/A

Status

The AN/TSQ-164(V)1 (Dragonfix II) has been deployed with US forces and was used operationally during the first Gulf War; the AN/TSQ-164(V)2 (Palantir II) has been deployed with the Canadian Forces.

Manufacturer

Originally offered by Andrew SciComm, a subsidiary of the Andrew Corporation .

UPDATED

WJ-8991/SYS Independent Collection Equipment (ICE) tactical manpack DF system

The WJ-8991 ICE system is a complete manpack Intercept and Direction-Finding (DF) system. The basic system can be transported by a single person and weighs less than 22.7 kg. It provides DF and intercept capabilities from 20 to 1,200 MHz and can operate either as a single intercept post or as part of a DF network. A variety of options allow customisation to particular requirements. The basic system consists of: the WJ-8996-1 receiver/DF processor; the WJ-9887 DF antenna; the WJ-8996/HHC hand-held controller; tripod, headset and compass and system cables.

The WJ-8991 operates as a stand-alone intercept/DF system or as an element in a DF net intended to conduct emitter location operations. Accuracy is 5° RMS with the antenna on a tripod and 3° RMS with the antenna on a mast. It is ruggedised to MIL-STD-810C.

Status

Acquired by UK for use by Y Troop, 3 Commando Brigade RM during the 1990s. Being replaced by Project SCARUS (see separate entry).

Contractor

BAE Systems North America, Gaithersburg, Maryland.

UPDATED

The WJ-8991 system

ZS-1015 communication signal intercept/DF system

The ZS-1015 communication signal intercept and Direction-Finding (DF) system is designed for tactical or strategic applications. Its software-intensive design provides for the integration of multiple functions within a single workstation. Functions available include signal search and detection, DF, emitter location with digital geographic map display, signal monitoring, parameter measurement and analysis, Digital Audio Recording and Playback (DARP) and audio signal analysis, with all collected data entered and stored in an integrated relational database.

Various configurations of the ZS-1015 system are available covering different frequency bands within the range 1.5 to 1,000 MHz. The system is designed for use in fixed-site stations, mobile installations in equipment shelters for tactical operation in the field, for installation in disguised vehicles and on board ship.

The ZS-1015 comprises a ruggedised PC-compatible system computer mounted in a standard 483 mm rack, together with a precision dual-channel DF receiver specifically designed for making high-performance DF measurements. Various antenna arrays are available for fixed-site, transportable and mobile applications, including both omnidirectional and high-gain directional DF antenna arrays.

ZS-1015 DF vehicles

ZS-1015 DF system

The system achieves frequency channel scanning speeds of up to 3,000 channels/s through the use of an ultra-fast receiver and a high-speed DF processor. The system uses an advanced Correlative Interferometry technique for DF measurement utilising DSP, providing DF measurements which are independent of signal modulation. The system is claimed to achieve high-accuracy DF results even with low-level signals at or below the ambient noise level. Key features in the achievement of this performance are the provision of a software-based correlation quality filter and the use of a proprietary SNR-dependent integration algorithm. DF system accuracy is between 4° and 0.5° RMS, depending on the selected DF antenna array.

With a colour display screen and standard PC keyboard, the system MMI uses Microsoft's Windows NT operating system for real-time multitasking operation. Operation of the system is either fully automatic, semi-automatic or manual, and remote control of all system functions is provided.

Signal search and detection is performed in either continuous scan (band search) or Signal Of Interest (SOI) scanning mode, with the detected signals presented in both tabular listing and panoramic display formats. The monitor mode provides for operation at a single frequency, with audio monitoring of the received signal and a polar histogram display to permit the separate identification of multiple emitters operating at the same frequency. Operation of multiple ZS-1015 systems in a fully automatic DF network is possible with the use of HF, VHF or UHF radio, landline, satellite or telephone communication links, with any station designated as the control station. A time-tagging technique is used to permit the unambiguous calculation of the correct location of each emitter, with all emitter locations displayed on a full colour geographic map on the display screen. All collected data on the intercepted emitters is entered into a relational database that provides the operator with an information analysis tool to develop the electronic order of battle.

A precision software-based signal measurement module is available, providing instantaneous measurement of emitter frequency, signal amplitude, bandwidth and modulation parameters for signal analysis purposes. The DARP module records up to eight audio signals in digital format on the computer hard disk. Simultaneous record and playback, and instantaneous playback of any part of a record, are DARP features. All DARP audio files are stored in the database and can be retrieved together with all other stored data on any signal of interest.

A comprehensive software-based Built-In-Test (BIT) capability is included.

Status
Possibly in service with the USAF and US Army, the Colombian Air Force, Italian Navy and the armed forces of Taiwan and Thailand.

Contractor
Zeta (a division of Sierra Networks Inc), San Jose, California.

UPDATED

ZS-3015/4015 Ultrascan Intercept/DF system

The ZS-3015/4015 Ultrascan Intercept/DF systems are designed for tactical and strategic signal collection in local or remote fixed site, vehicle, airborne or shipboard applications, providing intercept and DF in the 20-300 MHz frequency range. The systems, available in two or five channel configuration, use Zeta's proprietary correlative interferometry DF technique to produce accurate DF measurements. The systems achieve very high spectrum scan speeds through the use of wideband Fast Fourier Transform processing using parallel Digital Signal Processors. The Graphic User Interface is Windows NT. Geographic map displays are used to plot lines-of-bearing and emitter locations. Other system features include Digital audio Record and Playback of received audio signals together with operator comments correlated with signal intercept events.

The systems provide a 'plug and play' capability, with client-server architecture to support multiple operator positions and data fusion from multiple sites, with access to information for each operator in the network. System hardware in 6U VME plug-in modules includes the digital receiver channels, common LO synthesizer, DF/DSP processor, DARP unit, Pentium® series PC computer and system interface modules. System control is via a standard keyboard with integrated trackball, with operator screens presented on a conventional screen or flat screen monitor, allowing the operator to determine system operating modes for signal intercept, direction finding, signal parameter measurement, emitter location and signal database management.

All major system functional elements are housed in an RFI shielded ruggedised enclosure. For systems on moving platforms, position and heading information is supplied by a GPS receiver and magnetic north sensor, or from a ship's or aircraft's navigation system.

Status
Available.

Contractor
Zeta (a division of Sierra Networks Inc, San Jose, California.

UPDATED

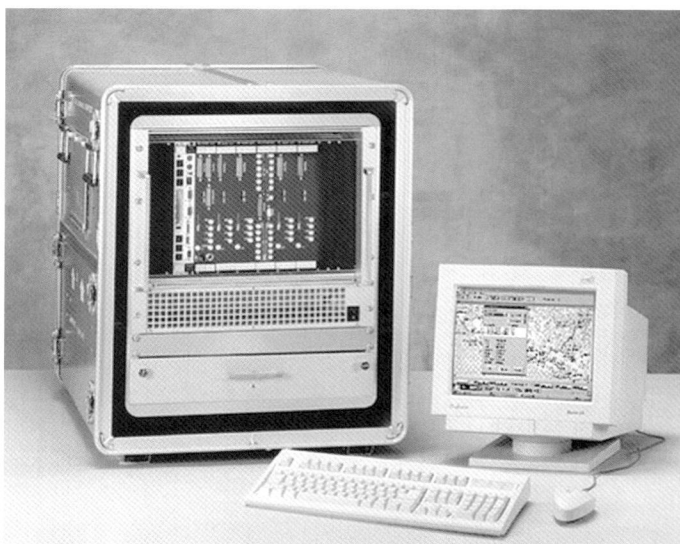

ZS4015 System   0097952

ANALYSIS AND SUPPORT

Denmark

NBC-ANALYSIS

NBC-ANALYSIS is a risk management system for NBC hazards, running on Windows 95, 98, NT, 2000, ME or XP. Integrated versions will also run on UNIX and LINUX. It has been developed as a Commercial-Off-The-Shelf (COTS) product, either as a stand-alone version or integrated into C4I systems. The stand-alone version includes extensive mapping functions, message handling, a communications module and the capability of handling unit positions (these are displayed using the NATO standard APP6-A). When integrated, functions such as mapping and communications are usually provided by the parent system. NBC-ANALYSIS can be customised to meet national requirements. Examples of this include BRACIS-NT in the UK (see separate entry) and NBC-ANALYSIS for JWARN (Joint Warning and Reporting Network) in the US.

Reports of an NBC attack or a release of radiological or Toxic Industrial Materials (TIM) can be entered into NBC-ANALYSIS manually, or automatically using data received from detectors and downloaded onto the platform via a variety of communications channels. This can include data from radiological, chemical and meteorological sensors and GPS, and drivers exist for detectors such as CAM, ACADA and GID-3. The software automatically calculates the hazard area using meteorological data and displays it on a map of the area of interest. Units at risk are also identified and, and a risk assessment timetable can be calculated. Maps in vector or raster format, or aerial photos, can be displayed, and maps can be imported using relevant NATO and US standards.

The data communication module handles the transmission of NBC reports. Communication can be over military or public telephone lines, radio links, LAN or WAN. Standard NATO formatted NBC messages are automatically generated in AdatP-3 format for NATO versions, but the system can be customised to meet other requirements, such as the Joint Variable Message Format for the US Army. Interoperability between nations is based on compliance with ATP-45/AEP-45 and AdatP-3.

NBC-ANALYSIS in operational use following a chlorine spill (Bruhn)   0134232

NBC-ANALYSIS, an NBC hazard prediction, warning and reporting system, contains all types of NBC messages   0055166

NBC-ANALYSIS forms part of the JWARN system (Bruhn)   0134231

The system has been successfully integrated into some CIS. Software is available for integration into MCS/P, MCS Block III, MCS Block IV, FCBCB2 and JMCIS. It is possible that the UK version, BRACIS, will form the basis for the NBC BISA in the UK's Digitisation Stage 2 programme.

For further details see *Jane's NBC Protection Equipment*.

Status

In service with Belgium, Canada, Czech Republic, Denmark, Finland, France, Germany, Hungary, Italy, Luxembourg, Netherlands, Norway, Poland, Slovak Republic, Spain, Sweden, Turkey and in NATO CIS, in the US as part of JWARN and in the UK as BRACIS NT.

Contractor

Bruhn NewTech, Herlev, Denmark; Winterbourne Gunner, UK; and Columbia, Maryland, USA.

UPDATED

United Kingdom

Biological, Radiological And Chemical Information System (BRACIS NT)

The Biological, Radiological And Chemical Information System (BRACIS NT) software program has been developed under a UK Ministry of Defence contract for the UK Armed Forces as a customised version of NBC-ANALYSIS (see separate entry). It is similar to the Joint Warning and Reporting Network (JWARN) used in the USA.

Status

BRACIS NT is in operational use with all three services of the UK forces.

The UK MoD tri-service Biological, Radiological and Chemical Information System (BRACIS)

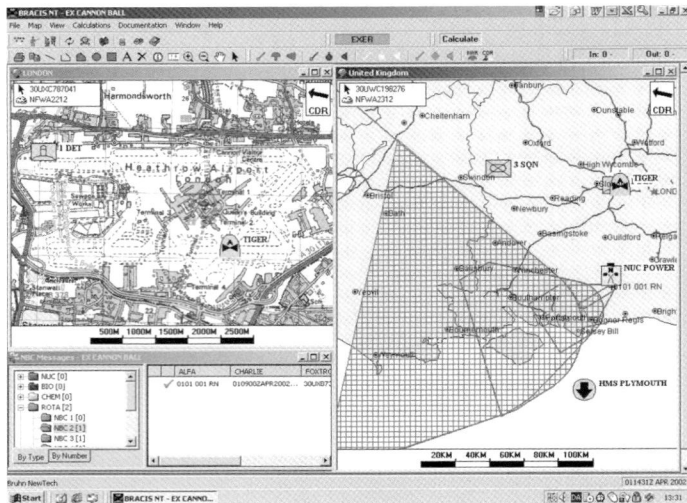

BRACIS NT handling prediction on radioactive material released into the atmosphere from a nuclear facility (Bruhn)   0134230

Contractor
Bruhn NewTech Ltd, Winterbourne Gunner.

VERIFIED

United States

All-Source Analysis System (ASAS)

The All-Source Analysis System is the Intelligence and Electronic Warfare element of the Army Tactical Command and Control System (ATCCS - see separate entry). It is a mobile, tactically deployable, computer-assisted IEW processing, analysis, fusion, dissemination and presentation system designed to support management of IEW operations and target development in battalions, brigades, Armoured Cavalry Regiments (ACR), separate brigades, divisions, corps, and Echelons Above Corps (EAC). It is employed in peacetime and wartime operational environments and is capable of continuous operation 24 hours a day, 7 days a week for extended periods of time. Its hardware and software modularity allows it to be tactically tailored to provide intelligence support to early entry forces as part of a Deployable Intelligence Support Element (DISE) or operate in a split-based mode of operations. ASAS operates in the 'system high' security mode of operation and processes both collateral and Sensitive Compartmented Information (SCI). It interfaces with standard US Army communications systems as well as IEW special purpose communications systems for example, Joint Worldwide Intelligence Communications System (JWICS), and TROJAN Special Purpose Integrated Remote Intelligence Terminal (SPIRIT). ASAS also provides the capability to process GENeral SERvice (GENSER) and Defense Special Security Communications System (DSSCS) record message traffic; and simultaneously maintain both SCI and collateral interfaces. The objective system (Block III) will be capable of operating in the 'multi-security level' mode of operation and support direct computer-to-computer data exchanges across the Defense Information Systems Network (DISN), Army Common User System (ACUS) and Intelligence special user communications at both the collateral and SCI levels.

The ASAS is designed to provide a seamless intelligence architecture between and across echelons. The architecture can be broken down into three major groups: sensors, processors and communications systems. The systems within each group support simultaneous demands for intelligence and targeting information at multiple echelons. They form a seamless intelligence system that supports commanders from tactical through strategic levels anywhere across the range of military operations.

The ASAS is an evolutionary system. Its development and fielding support the near-term needs of units, exploit emerging technology and comply with the standards of the Army Battle Command System (ABCS). The Army is developing the system in several stages, or Blocks. The US Army estimates the cost to develop, procure, and operate the system over its 20-year life cycle at approximately US$5 billion.

ASAS Block I

The ASAS Block I, the initial system, provides ASAS capability to corps and divisions. The ASAS Block I began development in 1984 and provided initial limited interim capabilities. The Jet Propulsion Lab (JPL) was the Block I prime contractor and was responsible for systems integration of the various Block I hardware and software components. The system is housed in trucks and truck-mounted shelters and includes towed electrical generators. The Army spent US$1.4 billion on Block I, most of which was for research and development. Block I procurement was US$345 million for 11 sets to be fielded at corps and divisions, plus one set for training. The ASAS system receives battlefield information and intelligence reports through the Communications Control System (CCS) (see separate entry), stores the information in the Data Processing System (DPS-D) and fuses information in the All Source Correlated Data Base (ASCDB). This fused intelligence is produced in a much-reduced time compared to the old manual systems.

ASAS Block I Components
ASAS-All Source (ASAS-AS) workstation
The ASAS-AS is a component of the ASAS Block I that provides a suite of six AS workstations to the Army Division and Corps Analysis and Control Elements (ACE). The ASAS-AS receives Sensitive Compartmented Information (SCI) level multi-disciplined information and processes it into intelligence products. The AS also assists analysts with Intelligence Preparation of the Battlefield (IPB), maintenance of the enemy situation and targeting.

The ASAS-Single Source (ASAS-SS) workstation
The ASAS-SS Workstation provides a suite of six SS workstations to the Army Division and Corps ACE. ASAS-SS receives SCI level SIGnals INTelligence (SIGINT) information and processes it into multi-discipline intelligence products. ASAS-SS receives Tacrep and Tacelint of reports directly from Joint battlefield and theatre COMmunications INTelligence (COMINT) and ELectronics INTelligence (ELINT) sensors. The SS workstation may be task organised to provide additional workstations and analysts to IMagery INTelligence (IMINT) and SIGINT. The JDISS tool set has been integrated into ASAS-SS. This tool set includes (E-mail, chat, word processing, FTP and imagery).

The Single Source Processor - SIGINT (Product Improvement) (SSP-S PI)
SSP-S PI provides theatre commanders with a high capacity interactive SIGINT processing system, in lightweight transit cases, capable of rapid contingency deployments. The SSP-S PI interfaces with national, other theatre and Corps intelligence systems via record traffic and file transfer data circuits and accommodates rapid exchange and processing of large volumes of SIGINT data.

Remote Workstation (RWS)
The RWS supports collateral intelligence processing at other manoeuvre units below division level. The system is networked to the G2 TOC and is housed in the Lightweight, Multi-purpose Shelter (LMS). ASAS-RWS provides the G2 (S2) with the means to integrate IEW into the ABCS. These workstations provide the G2 (S2) and the ACE the ability to efficiently and effectively process high volumes of perishable combat information and multidiscipline intelligence. This ability in turn supports timely, relevant, accurate and predictive reporting and dissemination of a common threat picture to other battlefield functional areas.

ASAS-Collateral Workstation (ASAS-CWS)
ASAS-CWS is a software package that provides a collateral intelligence processing capability to the G2s of Army divisions and corps. Doctrinally, each unit employees two workstations in the main Tactical Operations Centre. It is the Intelligence and Electronic Warfare component of the Army Tactical Command and Control System (ATCCS) and provides the interface between ATCCS and the ACE. Principle functionality of the ASAS-CWS are Intelligence Preparation of the Battlefield, current enemy and friendly situation, imagery, maps and graphics and analyst tools.   In initial testing, the Block I configuration did not have the performance, reliability, availability and deployability needed to support the Army's operations.

In 1991, elements of the original Block I were merged with a development project called Hawkeye, sponsored by the Army Intelligence School. The Intelligence School initiated the Hawkeye effort because it was dissatisfied with the large, cumbersome equipment being developed for Block I.

US Army Europe (USAREUR) continued to develop Hawkeye and deploy additional intelligence data processing capabilities in a system called Warrior, which cost about another US$15 million. Warrior development continued in a new effort called Warlord, which was initially deployed in March 1994. The Army continued development of Warlord as a rapid prototyping program by agreement among USAREUR, the Army Intelligence School, and the Army program acquisition executive office for ASAS. The ASAS programme manager objected to Warlord because it does not have the automatic features of the JPL equipment in Block I. However, Warlord development products will be retrofitted into Block I, providing Warrior/Warlord capabilities to units not receiving Block I and will be integrated into the concurrent ASAS Block II development as appropriate.

ASAS Block II
The ASAS Block II development programme builds upon and expands the capabilities and functionality developed and produced in the ASAS Block I System. This includes conversion to the Army Tactical Command and Control System (ATCCS) Common Hardware/Software Open architecture and the OSD directed Common Operating Environment (COE) and Modernised Integrated Database (MIDB). ASAS Block II will have greatly improved software that meets baseline system requirements. The ASAS Block II development contract was awarded in October 1993. Block II consists of three subsystems: the Analysis and Control Element (ACE); G2 Tactical Operations Centre (G2-TOC) and the Remote Workstation (RWS) described below.

ASA Block II Components
Analysis and Control Element (ACE)
ACE performance is improved through automatic sanitation and automatic collateral message release. Substantial improvements in communications include four additional channels as well as satellite communications. Making the system smaller has enhanced deployability. Specific improvements to intelligence processing include secondary imagery dissemination with receipt, display and storage capabilities.

G2 Tactical Operations Centre (G2-TOC)
G2-TOC applications provide the interactive tools and automated processes to analyse the mission and provide enemy situation and Intelligence Preparation of the Battlefield products.

Remote Workstation (RWS)

The ASAS-RWS supports collateral intelligence processing at other manoeuvre units below division level. The WARRIOR workstation developed initially for the US Army, Europe (USAREUR) is an example of ASAS prototyping efforts that have supported forces in the field and contributed to improvements to the ASAS-RWS. The RWS is replacing the ASAS Warlord platforms wherever such platforms exist.

ASAS-Single Source (ASAS-SS) Workstation

The ASAS-SS workstation may be task organised to provide additional workstations and analysts to Counterintelligence and Human Intelligence (CI/HUMINT) under ASAS Block II.

Analysis Control Team Enclave (ACT-E)

The ACT-E, a Warfighter Rapid Acquisition Program initiative, is a shelter mounted on a High Mobility Multipurpose Wheeled Vehicle (HMMWV) that integrates the ASAS Remote Workstations used by the brigade ACT with networking capabilities, radios, and other supporting equipment. The integrated shelter helps set-up/tear down, integration of information, and provides environmental protection for the computer equipment and a work area for the operators. The ACT-E is the integrating focal point for intelligence surveillance and reconnaissance management within the manoeuvre brigade. The ACT-E completed a test programme that resulted in a successful In-Process Review in September 2000 for acquisition and fielding. The test programme included Factory Acceptance Tests and a logistics demonstration at Vint Hill Farms, technical tests at the Central Technical Support Facility and the Aberdeen Test Center, and a Limited User test at the Pinon Canyon Maneuver Area.

ASAS Light

The ASAS Light, a laptop providing a sub-set of Remote Workstation functionality to the intelligence sections at the manoeuvre battalions, completed developmental testing in May 2000. The two-phase operational test programme included a controlled event in August 2000 at the Central Technical Support Facility focusing on ASAS Light functionality and a field training exercise in October 2000 focusing on the operational integration and contributions of the ASAS Light to the battalion intelligence staff. Until the fielding of the Force XXI Battle Command Brigade and Below (FBCB2) System (see separate entry), the ASAS Light is the primary interface between the digital environment (ASAS and the rest of ABCS subsystems) and the analogue environment. Until FBCB2 is fully fielded, the Battalion S2 sections must maintain the capability to operate in an analogue, digital, and mixed environment.

The Tactical Imagery Processing System (TIPS)

TIPS is a Block II initiative to allow graphical products to move from the SCI level to the collateral environment. TIPS provides a function similar to the CCS, except the CCS is message text based and TIPS will allow the movement of overlays and other graphical products.

ASAS Block III

The objective system, ASAS Block III, will expand upon Block II capabilities for operational, environmental and performance requirements. In addition to these programmed materiel acquisition versions of ASAS, technology insertion and prototyping efforts will incrementally enhance ASAS and support the rapid distribution of ASAS capabilities to units in the field.

ASAS-Extended (ASAS-E)

ASAS-E is a software package that provides ASAS Block I (ASAS-AS, ASAS-SS and ASAS-RWS) and Block II intelligence processing capabilities on non-standard hardware to selected active and reserve units not scheduled to receive Block I. In March 1994, the Army was directed to accelerate fielding the ASAS capability across the force (including all Military Intelligence reserve units and National Guard brigades) by FY99. The ASAS-Extended program accomplished this through reuse of proven Block I software, use of relatively low cost NDI equipment and tailoring the existing training and maintenance support structure.

In addition to these programmed materiel acquisition versions of ASAS, technology insertion and prototyping efforts will incrementally enhance ASAS and support the rapid distribution of ASAS capabilities to units in the field. The ASAS is upgraded as new technology becomes available. An example of technology insertion is the use of the commercial Alpha Reduced Instruction Set Computing (RISC) processor. The Alpha RISC operates 30 times faster than the initially fielded processor in the Block I ASAS-ASW. It eliminates the need for the two AN/TYQ-36(V) 3 Data Processing Sets (DPSs) in the ASAS Block I. Some units equipped with ASAS Block I will be upgraded with the Alpha RISC processor. The Alpha RISC processor is also used in the ASAS-Extended provided to selected units not receiving the ASAS Block I hardware.

Status

ASAS Block I is fielded to priority divisions and corps, together with ASAS-E. The ASAS Block II is currently under development. 79 ACE, 1031 RWS, 1477 ASAS Light, and 82 ACT-E are currently planned. In November 1999 the ASAS-RWS was first issued to the US 4th Infantry Division at Fort Hood, Texas and further issues are now taking place.

The ACT-E began fielding in September 2000 and the ASAS-Light began fielding in 2001.

Contractor

Lockheed Martin Mission Systems, Gaithersburg, Maryland.

UPDATED

Counterintelligence and Human Intelligence (CI/HUMINT) Automated Tools Set (CHATS) (AN/PYQ-3(V)2, AN/PYQ-3A(V)2, AN/PYQ-3B(V)2)

CHATS is an information management system designed for US Army Counterintelligence and Human Intelligence (CI/HUMINT) teams operating in the field. It is accredited for secret level operation, uses commercial software applications for common tasks and Government developed software for CI/HUMINT tasks. The system provides the team leader with the capability to collect, process and disseminate information obtained through investigations, operations, collections, interrogations, debriefings and document exploitation. CHATS also provides the capability for team level management with case, tasking, intelligence oversight and source management tools. Major components include:

(a) Processor (varies): Pentium 233 to 600 MHz/128 MB RAM, two 3.0GB or 6.01GB removable hard drives, 13.3in Active matrix screen, 24C CD ROM, 3½ in floppy disk drive, slots for three type II or one type III PC cards
(b) Peripherals: colour ink jet printer, colour scanner, digital camera, 520 IB PCMCIA Type III hardcard, 12Mbyte flashcard for transferring pictures, zip drive, precision lightweight GPS receiver
(c) Communications: 56K V.90 and 10Base T/2 combination fax/modem and ethernet PMCIA card, STU III Model 1910 data encryption device, Serial PCMCIA tactical communications interface modem for SINCGARS, KY-68 communications cable
(d) Cases: two wheeled hard cases with retractable handles, one soft case.
(e) Accessories: power and phone adaptors, cables, AC power supplies, operator manuals.

The laptop computer and STU III are packaged in the first hard case. Other larger peripherals such as the printer, scanner, and camera are carried in the second case. Smaller peripheral items such as the Zip Cartridge, CD-ROMs, PCMCIA Cards, spare hard drive, and the technical manuals are distributed between the two. Both cases contain foam cutouts for components. Small accessory items are carried in the soft case. An uninterrupted power supply and power distribution system is built into the first case. CHATS can draw power from both commercial and military AC and DC power sources.

A variety of COTS/GOTS software is employed. This includes MS Windows 95/MS Windows 2000, MS Office 97 Professional with Bookshelf, Sentry Software Hardlock Security (Win 95 only), Norton Anti-Virus, Tactical Packet Network (TPN) Tools, Crime Link, LCON Language Translation SW (some systems), Paint Shop Pro, WinZip, Internet Explorer, Outlook Express, WS-FTP LE (Remote File Access), ProComm, Adobe Reader. Government developed software provides features to set system security and administrative options, and provides the following specific functions:

(a) Reporting Functions:
Perform data analysis, receive/parse incoming messages and reports
Prepare and disseminate results of collection, investigations, operations, and screenings
Conduct local database queries
Transmit messages and reports over a variety of communications paths
Implement DCII/COE Common Message Processor (CMP), link analysis
Provide tools for managing: investigative cases, interrogation serial numbers, intelligence oversight, intelligence contingency funds
(b) Mapping Functions:
Plot MIL STD-2525 symbols
Read ADRG/CADRG/NIFT CD ROMS or off the hard drive
Register scanned maps, position tracking, overlaying of reference lines
DTED Level 1 and 2 information display
Co-ordinate conversion, linear distance and azimuth tools
Self locate and position tracking on local digital map via GPS.

The PCMCIA Ethernet card allows CHATS users to communicate over a wide variety of Local Area and Wide Area Networks (LAN/WAN). Specifically, the system can operate over the MSE Tactical Packet Network and LAN, the Trojan Spirit Network (collateral LAN ONLY) and collateral garrison LANs. CHATS can also communicate over the Single Channel Ground and Airborne Radio System (SINCGARS) with use of the provided SP-TCIM card and/or serial cable. The system can connect to a KY-68 to communicate over MSE wire line. With the use of the STU-111 191 0, secure communications can be established over commercial systems such as analogue telephone networks or INMARSAT. The fax/modem PCMCIA card can be used for unclassified communications over commercial systems.

Status

In operational use in the US Army and US Marine Corps.

NEW ENTRY

....

Digital Topographic Support System (DTSS)

The Digital Topographic Support System (DTSS) is an automated battlefield system that provides geospatial data in digital format for use on ABCS systems. Topographic Terrain Teams located at brigade, division, corps, and EAC use the DTSS capability to perform automated terrain analysis and prepare geospatial products within the timeframes required to support tactical combat operations. The DTSS provides a higher resolution of battlefield terrain visualisation through advanced computing, printing, and scanning of geospatial products and a means of producing a variety of tactical decision aids using terrain analysis models and high-resolution imagery. The system also provides the capability to produce multiple, full colour hardcopy products of the battlefield terrain. Maps not otherwise available in digital format may be scanned in full colour. DTSS products are disseminated to the battalion using the Global Broadcasting System (GBS), the Common Tactical Picture application (depending on bandwidth), a web site, or hard copy (disk, overlay or geospatial printout). There are six distinct configurations that are capable of providing direct topographic support to the Army commanders and staff deployed at key manoeuvre echelons from Theatre down to Brigade. These configurations are the: Digital Topographic Support System-Heavy (DTSS-H), DTSS-Light (DTSS-L), DTSS-Deployable (DTSS-D), DTSS-Base (DTSS-B), DTSS-High Volume Map Production (HVMP) system, and DTSS-Survey (DTSS-S). All of the DTSS configurations are necessary to serve as a complete replacement for the Topographic Support System (TSS).

The DTSS provides a combination of the capabilities of three previous developments: the terrain analysis capabilities of the legacy TSS, the reproduction capabilities of the Quick Response Multicolour Printer (QRMP), and the image processing capabilities of the Multispectral Image Processing System (MSIP). The DTSS has upgraded these capabilities with newer, more capable commercial-off-the-shelf (COTS) technology in UNIX-based computer workstations, printers, scanners and supporting peripherals. The DTSS uses the ERDAS Imagine image processing and ESRI's ArcInfo (see separate entry) geographic information system (GIS) COTS software packages combined with a customised user interface and enhanced terrain analysis software.

Key Capabilities

The DTSS is capable of receiving, formatting, creating, manipulating, merging, updating, storing, retrieving and managing digital topographic data, then processing these data into hardcopy and softcopy topographic products. It accepts topographic and multispectral imagery as input. System functional capabilities include creation of a variety of custom tactical decision aids (TDAs) including visibility, mobility and data query analyses. TDAs generated on the DTSS can be output as map products that include all applicable marginalia. In addition to custom TDA generation, the DTSS provides access to the full capabilities of the image processing and GIS software packages. ERDAS Imagine software can be used to perform ERDAS Imagine image processing. It is being used to process digital imagery in order to perform imagery rectification, image map generation, thematic layer generation, limited digital database creation and 3-D terrain perspective viewing. The DTSS is configured for communications with other Army Battle Command Systems (ABCS) over the ABCS LAN using either fibre optic or copper wire, and includes secure and commercial voice telephone capabilities.

DTSS-Heavy (DTSS-H)

The DTSS-H is housed in a 20 ft ISO shelter and mounted on an army standard 5 ton truck. The legacy DTSS was also mounted on a 5 ton truck, so DTSS-H provides an updated terrain analysis and graphic reproduction capability on a single platform, while preserving the army's investment in 5 ton systems. Equipment includes two Army Common Hardware Software 2 (CHS2) Sun Ultra-2 workstations with 19 in CRT displays, four Hewlett Packard 750C plotters, either two 126 Gbyte Sun RAIDS or 256 Gbyte Winchester RAIDS and other associated peripheral devices. It is tactically mobile and designed for worldwide transport by air, road, rail and sea.

DTSS-Light (DTSS-L)

The DTSS-L is housed in an army standard S-788 Lightweight Multipurpose Shelter (LMS) and mounted on an army High Mobility Multipurpose Wheeled Vehicle – Extended Capacity Version (HMMWV-ECV). The DTSS-L defines the objective configuration of the DTSS, providing the army with a degree of flexibility, transportability and mobility not found in the 5 ton truck-borne DTSS or DTSS-H. The DTSS-L will replace the DTSS-H as part of a five-year cyclic upgrade. The first 20 DTSS-Ls use the same software and workstations as the DTSS-H. The later production systems are using Windows NT-based workstations and will use the same version of the DTSS software as the DTSS-Deployable.

The first 20 DTSS-L mission support equipment includes an Army Common Hardware Software 2 (CHS 2) Sun Ultra-2 workstation with a 20 in flat panel display, a ruggedised portable Sun workstation, two Hewlett Packard 755C plotters, a 256 Gbyte RAID storage device and various associated peripheral devices. The second and third production year systems use CHS 2 Windows NT workstation with dual processors, HP 1000 series printer/plotters, an IDEAL scanner, a 512 Gbyte RAID, with a second RAID set up with a Map Server function. The same LMS shelter is being used throughout production. The DTSS-L is tactically mobile and designed for worldwide transport by air, road, rail and sea.

DTSS-Deployable (DTSS-D)

The DTSS-D was originally acquired at the direction of the US Army Chief of Staff to address an immediate need to provide image maps and terrain data to support army missions. Then, as is still the case, standard map products did not exist or were out of date for a large portion of the earth's surface. The DTSS-D provided a commander with a capability to produce map products quickly from multispectral imagery when standard products were unavailable or unsuitable for reasons of content or currency. The original procurement took place in FY94-FY95. Over the years the software capabilities have been improved to encompass the full range of DTSS capabilities, including terrain analysis, terrain visualisation and terrain data management.

The DTSS-D is configured either for use in a garrison environment, or for field use in a tent or similar enclosure, and does not include a tactical shelter or vehicle. It has undergone a cyclic upgrade, improving workstation processing power, product throughput, and system usability, as well as adding new terrain analysis capabilities. This upgrade was completed in FY01.

The upgraded DTSS-D includes updated versions of the software packages, an improved user interface, ERDAS Virtual GIS for dynamic terrain visualisation, and additional terrain analysis capabilities all hosted on a Windows NT based platform. The upgraded DTSS-D consists of an Army Common Hardware Software 2 (CHS 2) Versatile Computer Unit (VCU) with dual 933 Mhz processors, 2 Gbytes RAM, 3-D stereo graphics video card, DVD RAM drive, JAZ drive, 8 mm tape drive, LS 120 floppy drive, PCMCIA card reader, 3-D stereo graphics monitor, IBM Thinkpad T21 laptop computer with 512 Mbytes RAM, CHS2 1,000 W UPS, 500 Gbytes RAID, CDRW, CISCO switch, 36 in large format scanner and two 36 in large format HP 1055 plotters. All of this equipment has been integrated into ruggedised transit cases for ease of system transport and set up. The upgraded DTSS-D comes equipped with the necessary communications capabilities to operate stand-alone, or in conjunction with a DTSS-H or DTSS-L.

DTSS-Base (DTSS-B)

The DTSS-B (formerly known as the Topographic Imagery Integration Prototype (TIIP)) is the result of a USAEUR initiative to develop the capability to generate terrain information over sparsely mapped areas to support training and mission rehearsal/contingency operations. The need for this capability was highlighted by experiences in Desert Shield/Desert Storm and the former Yugoslavia.

The DTSS-B consists of multiple workstations designed to operate at theatre level in a base-production, garrison environment. The system has increased data production capabilities over the other DTSS configurations, as well as enhanced feature and elevation data extraction tools. It is augmented with a direct link to imagery sources and also has increased data storage, management, and distribution/dissemination tools.

The DTSS-B is designed to augment NIMA capabilities at the theatre level by providing quick-response special purpose mapping, terrain analysis, and terrain-related Intelligence Preparation of the Battlefield (IPB) in areas of obsolete/outdated information or data. The system can produce products such as Digital Terrain Elevation Data (DTED), image map substitutes, 3-D terrain visualisation, and operation/decision graphics. It can also update and produce Topographic Line Maps (TLMs) and create IPB products. Product response time varies from 10 minutes to 36 hours depending on the complexity of the product. Products range from simple image maps through contoured orthophoto maps to extensive 3-D terrain fly-throughs.

DTSS-High Volume Map Production (DTSS-HVMP)

The DTSS-HVMP system is being developed in a effort to modernise the Reproduction Section of the Topographic Support System (TSS). The current TSS Reproduction Section consists of large, heavy, offset lithographic presses with a camera and/or plate maker used to produce the colour-separated negative plates. This equipment is housed in seven 30 ft vans. The current reproduction process using the TSS is a time consuming, labour intensive process.

The DTSS-HVMP will provide a tactical capability to reproduce rapidly large volumes of graphic material including maps, charts and situation overlays. It will be capable of reproducing information from softcopy via a direct digital interface and will interoperate directly, or via the Command Post LAN, with the DTSS systems and/or other Army Battle Command Systems (ABCS) to provide the capability to receive and print their digital products. The system will consist of commercial, full-colour, large format printer(s) and a high-speed paper cutter mounted in a single 20 ft ISO shelter on an army standard 5 ton truck. It will also have a workstation for queuing and submitting print jobs to the printers. The DTSS-HVMP will be capable of printing 2,500 full-colour, large format (22.5 in × 29.5 in) water resistant copies in a 24-hour period and will be capable of producing its first copy in five minutes. Standard NIMA digital products, such as ADRG, can also be printed on the system. The DTSS-HVMP will have no hardcopy scanning capability, but will accept digital data from the Command Post LAN.

DTSS-Survey (DTSS-S)

The DTSS-S is to be developed as a replacement to the Survey Section of the Topographic Support System (TSS). The DTSS-S will continue to provide the capability to perform first to third order topographic survey missions in a tactical environment, providing horizontal and vertical control, and azimuth data in support of weapons systems and mapping operations. The DTSS-S precision survey assets will provide support across the battlefield, supporting field artillery, air defence artillery, aviation, intelligence, communications and construction control points. It will also provide unique capabilities for feature data collection and field data editing that can augment the data collection requirements of other DTSS configurations.

Currently the TSS-Survey is housed in an 18 wheel, 30 ft trailer pulled by a 5 ton standard army tractor. It is envisioned that the TSS-Survey will be downsized to a HMMWV configuration (four HMMWVs) with standard army shelter, similar to DTSS-L. Assets to be included with the DTSS-S are the Automated Integrated Survey Instrument (AISI), Global Positioning System-Survey (GPS-S), a digital level, a laptop/docking station computer to support survey computations and a large-format, low-volume plotter.

Status

Five DTSS-Hs were procured in FY97-98 and four in FY98-99. The system was fielded over an 18-month period, fielding the final systems to topographical units in Germany in December 1999. There are now five systems located in CONUS. There are no plans for further procurements of the DTSS-H, but software updates and improvements will continue for five years, when the DTSS-Hs are scheduled to be replaced with DTSS-Ls.

Twenty DTSS-Ls were fielded in FY00 with 32 more in FY01-FY02. The first 5 DTSS-Ls were issued to 4 Infantry Division. Current plans call for the production of 83 DTSS-Ls so a follow-on production contract for 31 more DTSS-Ls (16 and an option for 15) was awarded to Sechan Electronics Inc in early 2002. Software updates and enhancements will be made over the life of the system and hardware upgrades will occur on a 5-year cycle. A further 16 DTSS-Ls were projected to be procured and fielded in late FY 2003.

Eighty-three DTSS-Ds were purchased in FY00 and FY01. All 83 DTSS-Ds were fielded to topographic units worldwide in FY01. These systems will allow for multiple workstations at some terrain teams and further outfit Reserve Component units.

There are three DTSS-B locations. The DTSS-B received a hardware and software upgrade in FY02 which standardised a mix of hardware and software. The custom DTSS functionality resident on the DTSS-H, DTSS-L and DTSS-D was also added.

In December 1999, Advanced Concept and Technology II (ACT II) funds were awarded to TASC to develop a single DTSS-HVMP prototype. This prototype supported the Joint Contingency Forces Advanced Warfighter Experiment (JCF AWE) in Fort Polk, Los Angeles, during August 2000. In February 2001, a delivery order was awarded to TASC to design and develop two preproduction DTSS-HVMP systems. Operational testing took place at Fort Hood, Texas in early FY 2003.

Development and production of DTSS-S are not yet scheduled.

UPDATED

Integrated Meteorological System (IMETS)

The Integrated Meteorological System (IMETS) provides high-resolution current and prognostic meteorological data and weather effects. It is designed to display and analyse weather products and provide general weather forecasting, weather warnings and weather effects analysis. It provides the Integrated Weather Effects Decision Aid (IWEDA) client to AFATDS, ASAS, CSSCS, AMDPCS and MCS for determining and displaying weather impacts on any of 71 weapons systems over space and time. IMETS is the division meteorological component of the intelligence and electronic warfare (IEW) subelement of the ABCS. It provides commanders at all echelons down to manoeuvre battalions with an automated weather system to receive, process and disseminate weather observations and forecasts. IMETS provides first-in weather support to contingency forces, tailored weather information for deep fires and precision munitions, and weather environment affects decision aids for the planning and execution of manoeuvre and support.

IMETS consists of three basic configurations manned by Air Force Combat Weather Command (AFCWC) Combat Weather Teams (CWT):
(a) Command Post (CP) configuration for fixed facilities at Echelon Above Corps (EAC) level where the IMETS is permanently integrated into the local area network, so a tactical IMETS is not required.
(b) Vehicle-mounted configuration for tactical operations where the supported echelon moves frequently.
(c) Light configuration for task-organised elements of a supported echelon, integrated into a small task force, where lightweight, easily deployed core weather functions can be performed without having its own vehicle, shelter and power source.

The vehicle system is mounted on a High Mobility Multipurpose Wheeled Vehicle (HMMWV). It receives weather information from polar-orbiting civilian and defence meteorological satellites, Air Force Global Weather Central, artillery meteorological and remote sensors, and civilian forecast centres. It processes and collates forecasts, observations and climatological data to produce timely and accurate weather products tailored to specific needs. The most significant weather and environmental support functions are the automated tactical decision aids. These graphics display the impact of the weather on current or planned operations for both friendly and enemy forces.

The IMETS contains a standard configuration of Common Hardware/ Software Version-2 (CHS-2) inside a Standard Integrated Command Post (SICPS) Rigid Wall Shelter (RWS), which is mounted on a -Heavy Variant. The IMETS shelter houses the Weather Effects Workstation (WEW), communications equipment and several Air Force-provided weather collection and processing systems. A towed 10 kW generator issued with the system provides power.

The WEW is a Solaris-based, CHS-2 workstation used to host the IMETS WEW software. The software receives and processes incoming weather data from multiple sources, creates and maintains databases holding the processed weather information and provides methods for the CWT to produce and disseminate weather products tailored for combat operations. Applications residing on the WEW include:
(a) Battlescale Forecasting Model (BFM). This evaluates current weather data, produces forecast weather conditions and creates map overlays depicting current and forecast weather conditions for the area of operations.
(b) Integrated Weather Effects Decision Aid (IWEDA). This displays a Red/Amber/Green 'Go/No-Go' chart and map overlays detailing the effects of weather conditions on specific combat operations and weapons systems.
(c) Night Vision Goggles (NVG). These are used to calculate the effectiveness of night vision equipment for any given date, time and weather condition.
(d) Web Kit. This allows the CWT to post text and graphical weather products on the IMETS Web Page for viewing by other users on the TOC Local Area Network (LAN).

(e) Common Message Processor (CMP). This provides the capability to receive, create and transmit US Message Text Format (USMTF) messages.

In addition to the WEW, there are three standard automated weather forecasting systems onboard the IMETS:
(a) Small Tactical Terminal (STT), which is a Solaris-based, one-way, receive-only system that receives satellite imagery from civilian and defense meteorological satellites. The shelter integration will support either the STT-Enhanced (STT-E) or the STT-Light (STT-L). Software, antenna arrays and capabilities are identical, but hardware components vary slightly.
(b) Tactical Very Small Aperture Terminal (TVSAT), which is an MS Windows NT-based, one-way, receive-only system that receives weather data from AFWA. Data received by the TVSAT is passed to both the WEW and the NTFS via the internal LAN for processing and generation of products.
(c) New Tactical Forecast System (NTFS), which is an MS Windows NT-based workstation that hosts a suite of weather forecasting tools used by the IMETS operators to display incoming products and to generate local and area forecasts.

The STT, TVSAT and NTFS are USAF-owned, USAF-supported systems designed to meet Air Force worldwide forecasting requirements.

Status

Procurement has taken place in two blocks. Planned upgrades to the 15 Block 1 IMETS which were fielded in 1996 were accelerated to resolve Year 2000 (Y2K) problems and were completed in March 1999. Subsequently at least a further 12 systems were procured. IMETS Light, the laptop version of the weather effects workstation, was due for initial fielding in FY02. Development of a Command Post configuration of IMETS was due in the same period, with fielding in FY 03. Fielding to US Army Reserves and the National Guard is projected for FY 04.

UPDATED

MapObjects

ESRI's MapObjects software is a collection of mapping and Geographic Information Systems (GIS) components for application developers. They consist of an ActiveX Control (ACX) and more than 45 programmable ActiveX automation objects that let developers add mapping and GIS capabilities to applications. MapObjects can be used with common desktop software products and can make use of already existing code or other available components. They can be used in standard development environments like Visual Basic, Delphi, Visual C++ and others.

MapObjects can pan and zoom through multiple map layers display data using classifications, graduated symbols and dot density; perform spatial analysis and query; use relational databases and SQL queries; perform address matching and geocoding; and track real-time events with global positioning systems (GPSs).

MapObjects uses ESRI shapefiles, ArcInfo coverages and ArcSDE layers, and displays a wide variety of image formats. Also included are sample applications with code and online help including examples which can be pasted into an application.

The MapObjects Internet Map Server (IMS) extension provides a solution for developers looking for tools to build web mapping and GIS applications.

Contractor
ESRI (Inc), Redlands, California.

VERIFIED

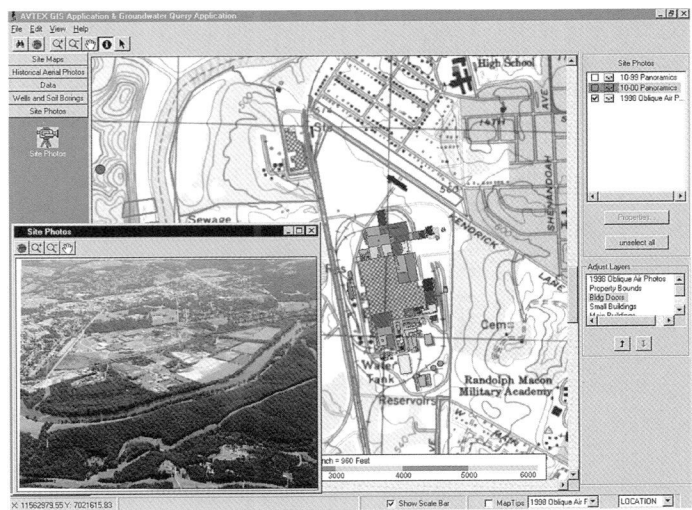

MapObjects Screenshot (ESRI)   0120206

SIGNALS INTELLIGENCE

Canada

AN/SLQ-501 CANEWS EW systems

Specifically designed to perform real-time detection, analysis, identification, classification and warning of both hostile and friendly platforms and missiles, CANEWS architecture comprises four major subsystems:

(a) an IFM-based antenna and receiver subsystem which covers the desired frequency bands and provides a very high probability of intercept
(b) a data transfer unit
(c) Two AN/UYK-505 computers that carry out the real-time analysis and supervisory function and which provide an interface with the ship command and control system
(d) a display console which provides the operator interface.
    Built-in test equipment is also provided.

In operation a main library holds data on numerous radar modes and signatures. Analysis of the detected RF signals identifies the radar type and platform and provides a confidence level indication. In addition, the operator can build a tactical library of threat radar signatures to meet immediate mission requirements. While processing many automatic functions to reduce operator workload and fatigue, the AN/SLQ-501 provides facilities for manual control if required. See *Jane's Radar and Electronic Warfare* for further details.

Status

The AN/SLQ-501 is fitted to all Canadian frigates and destroyers including the Tribal Class Update and Modernization Programme (TRUMP) ships and the Canadian Patrol Frigates. The latest version, CANEWS-2, is under development, and Lockheed Martin (Canada) awarded a contract to Software Kinetics (now xwave) in late 1998. The new system will incorporate an improved processor, modern display technology and new interfaces. CANEWS-2, as currently implemented, forms a functioning ESM prototype consisting of a multiprocessor hardware suite and signal processing software.

Manufacturer

Lockheed Martin Canada Ltd, Kanata, Ontario.

UPDATED

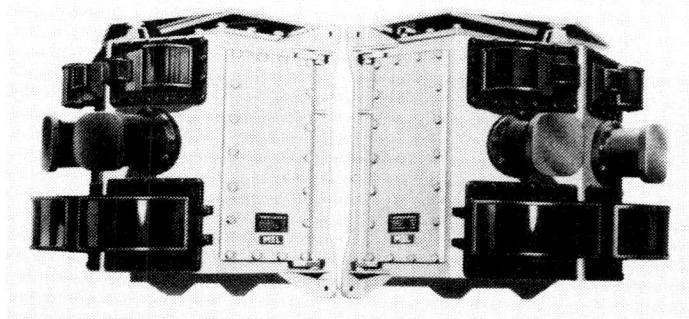

An antenna unit for CANEWS

Chile

Itata airborne ELINT system

The Itata ELINT system is a development by DTS and is a high-sensitivity electronic intelligence gathering system that can detect, locate and measure the parameters of emissions from search, acquisition and fire-control radars. Itata consists of a fully programmable superheterodyne receiver, a digital pulse analyser and a high-gain, wideband, rotating dish antenna which provides 360° coverage and bearing information to within a few degrees accuracy. Although intended primarily for light transport type aircraft, the equipment can also be installed in shipborne or ground vehicle configurations.

The receiver operates over a wide frequency range of 30 MHz to 18 GHz in six bands. It can be used in either a wide open mode over the complete frequency range or in a selective mode over a single band. After detection of a transmission of interest, the receiver locks onto it automatically and measures its frequency and other parameters. Digitised data of each intercepted signal can be recorded automatically for subsequent analysis.

Specifications

(a) Frequency range: 0.03-18 GHz in 6 bands
(b) Sweep modes:
(c) Multiband: total or independent programmable sub-bands

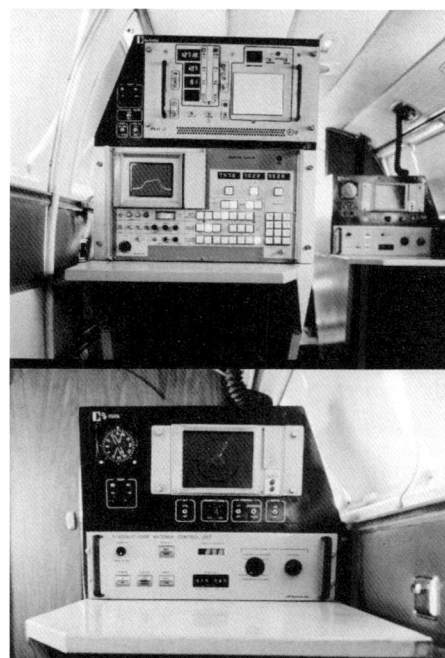

Itata airborne ELINT system

(d) Single band: total or independent programmable sub-bands; manual
(e) Sensitivity: −83 dBm
(f) Dynamic range: 70 dB
(g) Pulsewidth range: 0.05-24 µs
(h) PRF resolution: 0.25 µs
(i) PRF range: 0.092-12.83 kHz
(j) Azimuth coverage: 360°
(k) Polarisation: circular
(l) Azimuth beam-widths: 8° E/F-band; 1.8° J-band

Status

In 2002, the Itata system was reported as installed in three locally modified Beech A99 aircraft operated by the Chilean Air Force's Escuadrilla de Guerra Electronica/2nd Grupo de Aviacion.

Manufacturer

DTS Ltd, Santiago.

UPDATED

China

921-A ESM equipment

The 921-A is a wideband pulse radar direction-finding receiver. It is installed on submarines to detect emitters of airborne, shipborne and shore-based radars. It provides coarse measurement of azimuth, frequency band and operational state of the hostile emitters.

921-A ESM equipment

Specifications

(a) Frequency: 2-18 GHz in 4 bands
(b) Sensitivity: 1.5 - 10^{-2} to 10^{-4} W/M²
(c) Bearing accuracy: better than ±30°
(d) Dimensions and weights
(e) Antenna unit: 560 mm diameter × 515 mm height; 80 kg
(f) Receiver and display: 450 × 468 × 124 mm; 40 kg
(g) Distribution unit: 145 × 214 × 291 mm; 6 kg

Status

In service with the navy of the People's Liberation Army (PLA-N) on 'Xia', 'Han', 'Song', 'Romeo' and some 'Ming' class and North Korean 'Romeo' class submarines.

Contractor

China National Electronics Import and Export Corporation, Beijing.

UPDATED

BM/KZ 8608 ELINT system

The BM/KZ 8608 is an airborne ELINT system developed by the Southwest China Research Institute. It is designed to detect, identify, analyse and locate land-based or shipborne radar emitters with a high probability of intercept and with high sensitivity and accurate measurement of parameters. It consists of five main parts; antennas, superheterodyne receiver, instantaneous frequency measuring receiver, processor and display unit.

Although very few technical details have been released, it is claimed to have a wide frequency coverage, long range, ability to operate in dense electromagnetic environments, automatic signal identification and an emitter fixing capability. No information is available on the size of the processor library.

Specifications

(a) Frequency range: 1-18 GHz
(b) Frequency accuracy: 5 MHz
(c) Azimuth coverage: 360°
(d) Bearing accuracy: 5° (1-8 GHz); 3° (8-18 GHz)
(e) Sensitivity: −100 dBW
(f) Dynamic range: 50 dB
(g) Signal density: 200,000 pps
(h) PRF range: 100 Hz-20 kHz
(i) PRF accuracy: ±1% (100 Hz-2 kHz); ±2% (2-20 kHz)
(j) Pulsewidth range: 0.1-99.9 µs
(k) Power supply: 28 V DC; 115 V 400 Hz AC

Status

BM/KZ 8608 is reported to have been installed on at least one Xian Y-8 (An-12) aircraft of the Air Force of the People's Liberation Army. During the mid-1990s, Jane's sources were suggesting that BM/KZ 8608 is a derivative of an Israeli system.

Manufacturer

Southwest China Research Institute of Electronic Equipment, Chengdu, Sichuan.

UPDATED

Denmark

RX4010 HF ISB/SSB intercept and communications receiver

The RX4010 receiver operates over the frequency range of 10 kHz to 30 MHz in 10 Hz steps and has operating modes of ISB, USB, LSB, RTTY and CW. As an option FM can also be supplied. The RX4010 has a high-frequency stability, but can also be connected to an external standard. The receiver offers extensive scanning and sweeping facilities and can be controlled from either the RC4010 (a maximum of 30 receivers) via a single telephone line or from a computer via RS-232C, RS-422,

RX4010 receiver

or RS-485 databuses. The exciter uses the same remote protocol as the company's receivers.

An ultra-fast synthesiser can be supplied for frequency-hopping and used in computer-controlled DF systems.

Status

Introduced May 1988 and in use by a number of armed forces.

Contractor

Dansk Radio Comm. ApS, Ringsted.

UPDATED

France

ALTESSE Shipborne EW System

ALerT and Surface Ship Evaluation (ALTESSE), is a shipborne alert and awareness monitoring EW system designed for surface warships and submarines. ALTESSE provides the following functions: alert on communication signals, tactical situation development and communication intelligence. Using a high-frequency scanning rate and a short DF bearing measurement time, the system includes new functions to cope with modern threats including: dense environment, frequency hopping signals, burst and so on. ALTESSE may be interfaced with the ship combat system and utilises a user-friendly MMI based on Windows NT. The system covers a frequency range from 20 to 500 MHz with an extension down to 1 Mhz and up to 3000 MHz, providing an Elint capability with short-range radar. It is compatible with the Thales Airborne Systems DR 3000 and may be associated with countermeasure systems such as the TRC 274 HF/VF/UHF digital communication jammer.

Specifications

(a) Frequency range: 20-500 MHz (extension down to 1 MHz and up to 3000 MHz)
(b) Operational bearing accuracy: 1° rmr
(c) Interface with combat system: RS 232 C, Ethernet
(d) Antenna type: ANT 184-A (20-3000 MHz), ANT 206 (1-3000 MHz)
(e) Dimension of the antenna: 1.3 m high × 0.5 m wide (ANT 184-A); 2 m high × 1.5 m wide (ANT 206)

Compact naval ESM antenna for ALTESSE
0055181

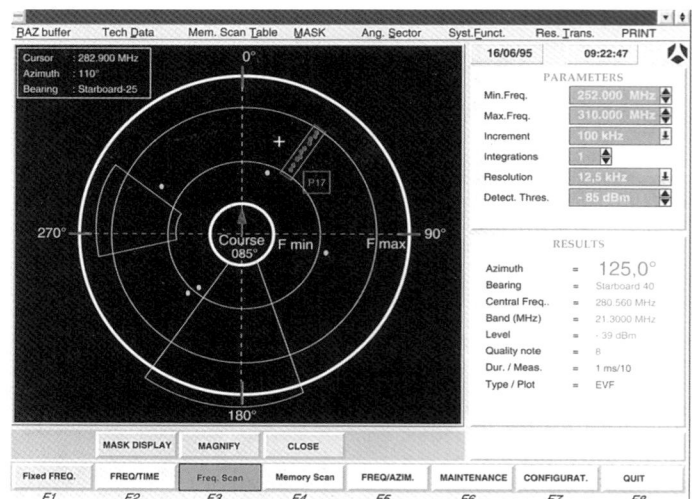

Windows NT MMI for ALTESSE ESM System
0055182

(f) Associated countermeasures: TRC 274 HF/VF/UHF digital communication jammer

Status

In service in the Saudi Arabian 'Al Riyadh' class. Likely to be fitted to the proposed UAE 'Baynunah' class of corvettes. Described as "sea-proven by the French Navy".

Manufacturer

Thales Communications, Gennevilliers.

UPDATED

C-160 Gabriel SIGINT aircraft

ASTAC ELINT system

ASTAC (Analyseur de Signaux TACtiques) is a tactical airborne electronic reconnaissance system consisting of an internally or pod-mounted sensor package and a ground processing station. It is intended to perform detection, identification and location of any radar type in a very dense environment. A datalink between the pod and the ground station enables a very rapid build-up of the electronic order of battle of the observed area.

The main characteristics of the system are a very wide frequency coverage, wide instantaneous bandwidth, high sensitivity, high-discriminating power and high-direction measurement accuracy by interferometer. The system is fully automatic, fully reprogrammable and possesses a very high-speed processing capability of up to 20 radars/s. It can process pulse modulated radar with pulse repetition internal diversity or agility, or radio frequency diversity or agility, as well as pulse compression, Continuous Wave (CW) and interrupted CW systems.

ASTAC uses two wideband compressive surface acoustic wave receivers. One receiver is used to obtain a very precise measurement of the radar frequency and the two together can handle frequency-agile emitters. The system uses interferometer phase-measuring antenna arrays to determine azimuth of any threat emitter operating within the frequency bands that are covered (0.5 to 18 GHz with 18 to 40 GHz as an option). The data is stored in a recording system in the pod for post-flight analysis. The pod also contains the UHF datalink for air-to-ground transmission of data. The onboard control unit for a two-seat aircraft incorporates tabular and liquid crystal displays and a keyboard for situation assessment and targeting.

The lightweight pod weighs 400 kg, with dimensions of 396 cm long and 40.6 cm diameter. It can be easily adapted to any combat aircraft or light transport equipped with powerful navigation equipment. ASTAC LRUs may be installed on board the aircraft and provide the same functions.

Specifications

Frequency coverage: 0.5-40 GHz
Bearing accuracy: 1° (azimuth of threat emitters)
Pod dimensions: 396 cm long × 40.6 cm d
Pod weight: 400 kg

Status

The ASTAC pod system was reported as having been installed on French Air Force C160 Gabriel (internal), DC-8 SARIGUE (internal) and Mirage F1-CR (pod) aircraft, together with Japanese Air Self Defence Force RF-4EJ reconnaissance aircraft (pod). For the Japanese programme, Mitsubishi is the lead contractor. The French Air Force is known to have used ASTAC operationally over Bosnia Herzegovina and in combat missions during the war in Kosovo in 1999. Thales Airborne Systems notes that the system was demonstrated on a NATO F-16 aircraft during 1993 and it is reported that an ASTAC pod is integrated on Mirage 2000-5. Also believed to have been procured by Taiwan for use on Mirage 2000.

Contractor

Thales Airborne Systems, Elancourt.

UPDATED

The ASTAC ELINT pod-mounted on a French Air Force Mirage F1-CR reconnaissance aircraft

Gabriel airborne SIGINT system

Thales Airborne Systems has developed complete SIGINT electronic intelligence systems for integration on board aircraft such as the DC-8, Transall and C-130. One of these systems, known as Gabriel, configures an ELINT subsystem for detection, analysis, identification and localisation of radar emissions (provided by Thales

Airborne Systems and based on ASTAC-type technology (see separate entry)) and a COMINT subsystem for detection, interception, classification, listen-in, analysis and localisation of radio communications stations (provided by Thales Airborne Systems and based on TRC series technology).

The system offers a high degree of automation to assist the operator in accomplishing all types of operational missions.

For further details see *Jane's Electronic Mission Aircraft* and *Radar and Electronic Warfare*.

Status

Two Gabriel C-160 SIGINT aircraft are in service with the French Air Force's 54e Escadre Electronique Tactique based at Metz-Frescaty. Gabriel was used operationally during the 1990-91 Gulf War, has been active in support of UN/NATO operations in Bosnia, in the NATO operation ALLIED FORCE in Kosovo, and reportedly in support of operations in Afghanistan.

Contractor

Thales Airborne Systems, Elancourt.

VERIFIED

Phalanger ESM/ELINT system

Phalanger is a new-generation ESM/ELINT payload for airborne platforms such as unmanned aerial vehicles, helicopters or light multipurpose aircraft. A pod version can be installed on combat aircraft. Based on phase interferometry and compressive receiver techniques, Phalanger offers high performance while featuring minimal weight, volume and power consumption. Aimed at detecting, identifying and locating ground-based radars, it either delivers radar tracks for real-time display and analysis, or uses its high-density recording capability for post-flight analysis. The battlefield tactical situation and electromagnetic order of battle can be displayed either on board on the platform or in a ground-based processing station collecting the data from the payload via datalink. Key features of Phalanger are:

(a) Very high direction-finding accuracy via interferometric antennas and receiver
(b) Very high sensitivity and long-range detection, parameters measurement and very short acquisition time using compressive receiver technology
(c) Target identification, threat library management and localisation by triangulation processed either by a stand-alone computer or by a ground station
(d) Display of the battlefield tactical situation including maps with signal analysis histograms (as an option).

Specifications

Weight: <20 kg
Frequency coverage: E- to J-band (C- to K-band optional)
Field of view: 360°
DOA measurement accuracy: <1°
Power consumption: <300 VA

Status

Phalanger is reported to be in preproduction and has been validated by the French MoD during ground and flight tests.

Contractor

Thales Airborne Systems, Elancourt.

UPDATED

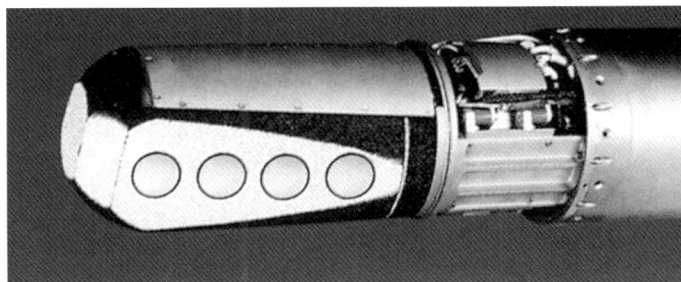

A close-up of the antenna head used in the Phalanger ESM/ELINT system 0009392

Sarigue SIGINT system

Thales Airborne Systems has developed complete SIGINT electronic intelligence systems for integration on board aircraft such as the DC-8, Transall, Business Jet and C-130. One such system, known as Sarigue, configures two subsystems. The first is an ELINT subsystem for detection, analysis, identification and localisation of radar emissions and is provided by Thales Airborne Systems. The other is a COMINT subsystem for detection, interception, classification, listen-in, analysis and localisation of radio communications stations provided by Thales Communications.

The latest version of the overall system is known as Sarigue NG (New Generation). It is based on the ASTAC equipment (see separate entry) and the Thales Airborne Systems TRC 290/600 series communications equipment (see separate entries). The TRC 290 receivers operate in the VHF/UHF band while the TRC 600 series is a range of receivers, analysers and direction-finding equipment covering the frequency range 0.1 to 1,350 MHz. The COMINT function can be extended as far as the 20 GHz region via the use of the ELINT subsystem. For further details see *Jane's Electronic Mission Aircraft*

Status

A single DC-8-55F Sarigue SIGINT is reported to be in service with the French Air Force's 51e Escadron Electronique based at Evreux. Sarigue was used operationally during the 1990-91 Gulf War and has been active in support of UN/NATO operations in Bosnia and Kosovo. In May 1993, Thales Airborne Systems was awarded prime contractorship on the Sarigue NG programme. Sarigue NG utilised an existing French Air Force DC-8 airframe and entered service in 2001.

Contractor

Thales Airborne Systems, Elancourt.

UPDATED

Spectrum Airborne Surveillance (SAS) system

The Spectrum Airborne Surveillance (SAS) system is a communication intelligence system designed to be installed on board any type of aircraft from small single-engine aircraft to business jets. It processes, in real time, all types of signals from voice communication to digital data transmissions.

The main functions are to provide interception, direction-finding, homing, emitter location, listening, analysis, decoding and recording of signals.

SAS is suitable for use on small aircraft   0011539

Spectrum Airborne Surveillance (SAS) system workstation   0011538

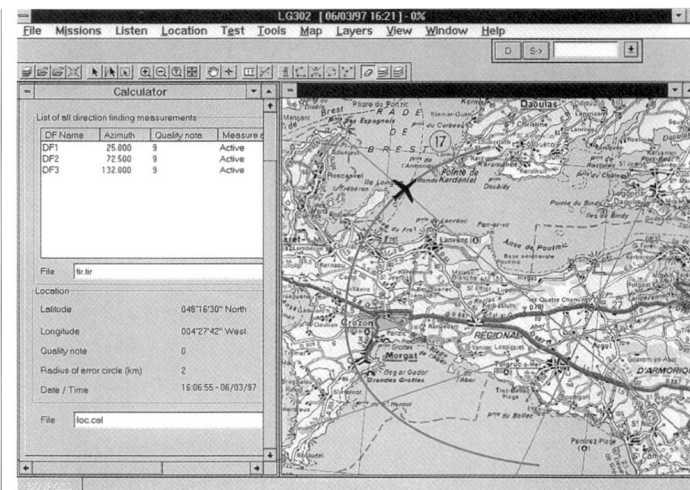

SAS screen   0011540

Housed in one or more operator workstations aboard the aircraft and offering a variety of demodulation and decoding schemes, the SAS collect signals from either a suite of sabre or circular array direction-finding antennas mounted inside a radome.

Principal features include:
(a) digital technology and FFT processing
(b) high scan rate to process mobile radios and cellular radio messaging
(c) man/machine interfaces running on a computer under Windows NT
(d) system programming according to type of mission.

Designed as an open-ended system, SAS can also integrate and merge other data from optical and infra-red sensors. Moreover, SAS can record all data during the mission for debriefings and further investigation.

Specifications

(a) Frequency range: 20-3,000 MHz (DF); 300 kHz to 3,000 MHz (monitoring)
(b) Precision of DF antenna: <1.5° RMS
(c) Scanning rate: 40-2,000 MHz
(d) Demodulation : A1A, A1B, A2A, A2B, A3E, J2A, J2B, J3E, J7B, F1A, F1B, F1C, F3E, F7B, H3E, R3E, B8E
(e) Operational sensitivity: 3 µV/m typical (10 mV at 180 km)
(f) Dimensions: 3U rack standard 19 in + compact PC; 134 × 485 × 520 mm
(g) Weight: 45 kg
(h) Power supply: 115V/230 V AC
(i) Power consumption: 300 W

Status

In late 2003 there was no new information on this system.

Manufacturer

Thales Communications, Colombes.

VERIFIED

STRATEGIE electronic surveillance system

STRATEGIE/AMES is a vehicle-mounted tactical ESM/ELINT system that consists of a network of highly mobile stations designed for deployment in the battlefield area to intercept, identify and localise signals of both airborne and surface radars. The system is operational within 15 minutes of deployment.

Parameters measured and processed in the baseline version include accurate bearing, frequency, signal level, pulse repetition frequency, pulse-width agilities and antenna scan period. ELINT functions are available to allow the detailed analysis of pulse data, signal modulation and recording of these parameters.

Strategie/AMES stations

Automatic identification of the target radar is accomplished by comparison with the threat library, which can be updated by the user. Each station consists of:

(a) a fixed antenna array with a wide field of view, mounted at the top of an erectable mast
(b) a highly sensitive broadband receiver with high scan speed
(c) a high-speed processor for signal sorting and accurate measurement of the direction of arrival of signals
(d) a device for analysing complex radar parameters in real time
(e) a computer for the classification and identification of targets, threat library management and triangulation
(f) an operator's console with a high-resolution screen for tactical situation monitoring, mapping and radar interception analysis like histograms
(g) a system for the interstation transmission of data by radio link
(h) a printer
(i) Global Positioning System (GPS) to gain precise threat location.

Specifications

(a) Frequency band: 0.5-20 GHz
(b) Accuracy: 3 MHz; 0.5° RMS
(c) Field of view: 360°
(d) GUI: Windows NT

Status

The first example of the STRATEGIE system was delivered to the French Army in June 1995 and the system is reported as having been deployed to Bosnia in support of NATO operations. In French service the equipment is known as STAIR. Has also been marketed under the name SURICATE, but as at late 2003 no new information was available.

Contractors

Thales Airborne Systems, Elancourt.

VERIFIED

Syrel ELINT pod

Syrel is a fully automatic electronic reconnaissance system pod which is pylon mounted under the fuselage of the fighter. It is designed primarily for tactical penetration missions at medium or low altitudes. It acquires and records automatically data relating to the identification and location of ground-based electronic systems. It is intended to provide reliable information of early warning, search and acquisition, ground controlled interception and fire-control radars associated with anti-aircraft missiles or artillery.

In normal operations, the aircraft will carry out direction-finding and position fixing of transmitters by flying along a path and taking a series of bearings to establish the position of the emitter.

The pod is 3.57 m long, has a diameter of 0.42 m and weighs 265 kg. It has antennas at both front and rear, with receiver units, amplifier and recorders in the centre section. A downlink enables real-time information to be fed to a ground station. The pylon houses a cooling system which has a ram-air intake in the leading-edge. The system's high-processing speed is assisted by thick and thin film microwave circuit assemblies on ceramic substrates. Thales also produces an associated ground station.

Status

Syrel ELINT systems were reported to be operational on Mirage III, F1 and 2000 aircraft. Sources suggest that one user is the Spanish Air Force. No longer in production.

Contractor

Thales Airborne Systems, Elancourt.

VERIFIED

Syrel ELINT pod-mounted on Mirage F1

TRC 641 HF/VHF/UHF technical analysis equipment

The TRC 641 technical analysis equipment is designed to handle digitised voice and data transmissions within the 300 Hz to 3 GHz frequency range with bandwidths up to 100 kHz and, according to Thales, utilises 'new' processing techniques. The equipment's analysis capability is based on the digital exploitation

TRC 641 HF/VHF/UHF technical analysis equipment   0011541

of received data using high-speed processing algorithms (fast Fourier transform, digital filtering and demodulation) running on fast, specialised processors. Derived parametric data (including centre frequency, modulation type/rate, frequency shifts, encoding, bandwidth, start time and transmission status) is displayed to the operator via a display/control PC using Windows NT. The system will provide automatic modulation recognition; automatic parameters measurement; automatic code recognition; demodulation, decoding and access to content. It can be operated as a stand-alone item or as part of a network.

TRC 641 analysers are quoted as being able to operate with a wide range of receiver types, as being able to be integrated with 'all' types of electronic warfare and electronic support measures systems and as being suitable for use in both fixed-site and mobile stations. Thales also states that the TRC 641 series is suitable for civilian and paramilitary applications as well as strategic and tactical military applications.

Status

In production. Part of the suite of equipment fitted in the Sarigue and Gabrielle SIGINT aircraft (see separate entries).

Manufacturer

Thales Communications, Colombes.

UPDATED

TRC 2000 VLF/HF/VHF/UHF digital receiver

The TRC 2000 is a VLF/HF, VHF/UHF or VLF/HF/VHF/UHF receiver. Exploiting digital technology, the TRC 2000 is claimed to offer new capabilities (easy addition of channel filters, notch and passband filters and so on) which are not available with analogue technologies.

The TRC 2000 system can be used for general search operations, surveillance and monitoring applications in severe operational environments. Designed to be integrated in radio surveillance and intelligence systems, it can be controlled from the front panel and/or an external PC running Windows NT. Several TRC 2000 receivers can be controlled from a single front panel, a PC or a system software. Four channels in a 3U 19 in rack may be controlled by the same computer.

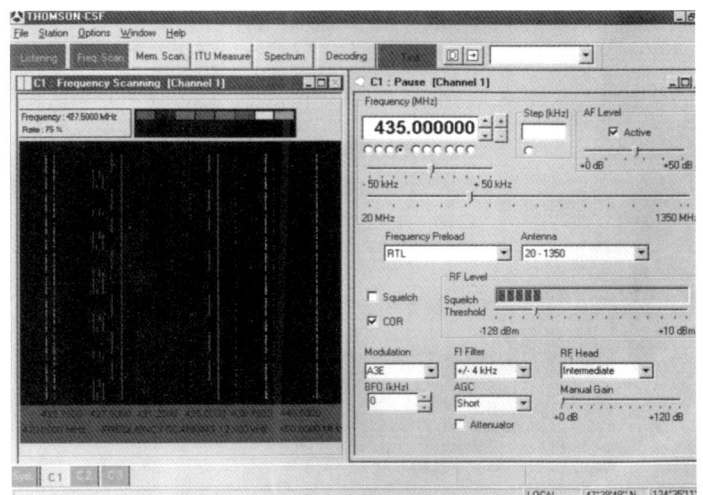

TRC 2000 can be controlled from an external PC running Windows NT   0011543

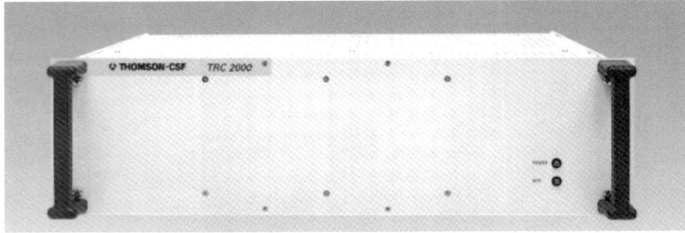

TRC 2000 VLF/HF/VHF/UHF digital receiver   0011542

For all types of operation, receivers can be configured in several modes: listening or technical analysis of digital transmissions; scanning functions allowing the operator to monitor the activity of selected frequencies (up to 100 tables of 100 frequencies) or sub-bands (up to 20). Using frequency masks for up to 800 frequencies and 100 sub-bands makes it possible to monitor only the ranges that are relevant from an operational standpoint.

Status

In production. Believed to be installed in the French Sarigue SIGINT aircraft (see separate entry).

Manufacturer

Thales Communications, Gennevilliers.

UPDATED

Germany

Automatic Modular Monitoring of Signals (AMMOS)

The AMMOS COMINT system is designed for automatic and manual interception, monitoring and analysis of radio communications in the frequency range from 300 Hz - 3.6 GHz for tactical and strategic applications.

The 'sensorgroup' VXI-based sensor equipment (modern VXI-receivers and signal processing boards installed in a VXI-mainframe) provides high modularity and configurability. Signal processing functionality includes interception of voice signals; demodulation, decoding of digital transmissions and Morse signals; classification support; digital narrow and wideband IF recording and replay.

Digital direction-finders (such as the DDF0xM (see separate entry) from the same manufacturer) can be integrated in the AMMOS system functionality. The open system architecture allows the customer to develop his own decoders (in the HF frequency range) or to communicate directly with the AMMOS sensors using the sensorgroup's object-oriented CORBA programming interface.

Status

The AMMOS base system has been available as a COTS item since 2001 and has been delivered to several unspecified customers.

Contractor

Rohde & Schwarz GmbH, Munich.

UPDATED

E 2000 series digital receivers

The E 2000 receiver series is a suite of digital receiver channels and additional attachments. They may be configured in various ways, with up to five separate receiving channels in a single three-unit-high 48 cm (19 in) rack-mounting unit, depending on whether they are being used as general-communication or special-purpose/special-analysis receivers. Furthermore the E 2000 receiver series is subdivided into three frequency-dependent receiving categories:
(a) E 2000 LH communications band receiver covering the frequency range from 0.3 kHz to 30 MHz. In addition to the common features mentioned below,

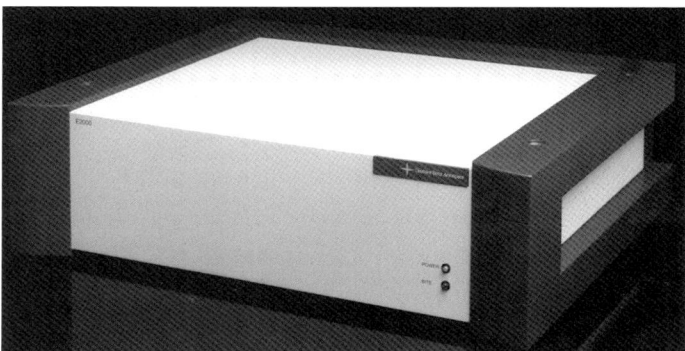

The E 2000 desktop version   0011544

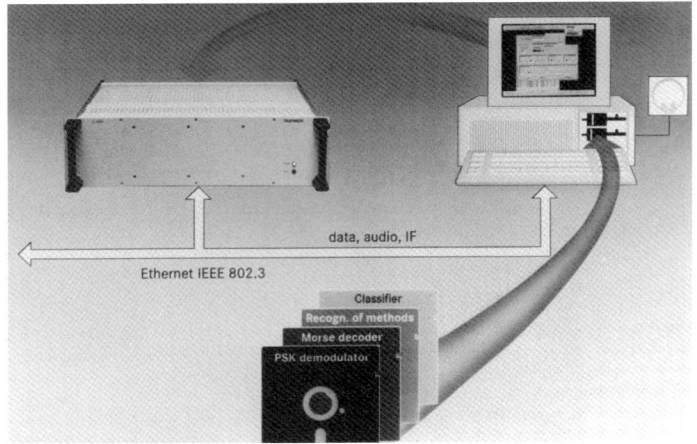

The E 2000 LH system

E 2000 HF receivers set up priorities in universal demodulator functions. Capabilities include ASK, FSK, PSK, AM, FM, FSK/PSK decode including methods and codes, Morse decode and auxiliary functions.
(b) E 2000 VU communications band receiver covering the frequency range from 20 to 3,000 MHz. The system exploits DSP technology to provide the same technical performance as the E 2000 LH receiver.
(c) E 2000 LU communications band receiver covering the frequency range from 10 kHz to 3,000 MHz.

Some of the major features of these HF and VHF/UHF receivers are:
(a) Digital Signal Processing (DSP) and filtering for optimum noise suppression
(b) numerous IF filters
(c) notch filter and passband tuning for the elimination of interference
(d) ancillary slot card processors with software modules for phase/frequency shift keying demodulation, signal classification or recognition of procedures in data transmissions, measuring units for spectrum monitoring systems (following ITU regulations)
(e) computer-controlled operator interface or, optionally, BF 2000 control panel for manual control. This latter is a manual tuning and function control module. Any operation functions and displays which are typically used for the control of a receiver can be set and are shown on an LED display. Frequency tuning is by means of a control knob.

An optional extra is the PI 2000 panoramic display, which is a plug-in module to display received signal spectra or modulation parameters on a screen. Two display functions are available: PB 2000 wide band display (IF = 42 MHz, B = 5 MHz) and PS 2000 narrow band analysis (B = 10 kHz at HF, 100 kHz at V/UHF). The panoramic display has dual-channel display and works with different display modes, such as:
(a) detail, maximum, average or frozen panorama
(b) simultaneous display of two spectra
(c) Phase Shift Keying (PSK) signals, displayed on vector diagram, constellation, eye pattern or spectrum.

ADAS 2000

Developed in conjunction with the E 2000 series, ADAS 2000 is a user-programmable system for the automated acquisition and analysis of radio data signals, especially burst transmissions. It is divided into three subsystems with the following functions:
(a) acquisition subsystem - automated online acquisition of radio data signals relating to interception and decoding of known radio data signals (bitstream) and interception of unknown radio data signals for subsequent fine analysis. The status data set contains, for example, date/time, frequency, type of modulation, signal level, line spacing, bit/band rate, name of recognised transmission protocol and coding procedures.
(b) bitstream analysis subsystem - support at offline analysis of unknown stored signals. The operator can generate an extraction program for the acquisition subsystem supported by the programming subsystem.
(c) programming subsystem - by means of the programming subsystem the user converts the results of an analysis of a radio data signal in a program that describes the signal for the additional hardware modules of the E 2000 receiver.

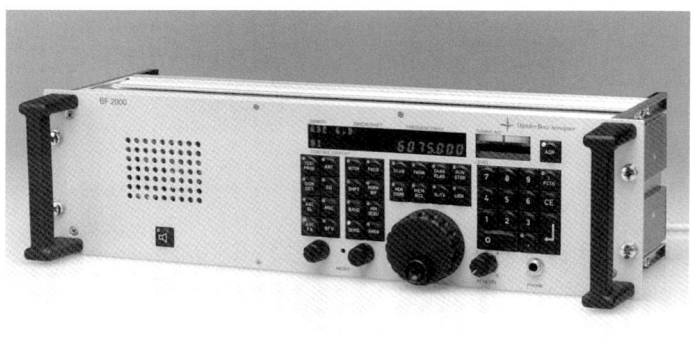

The BF 2000 control panel for E 2000 receivers   0011545

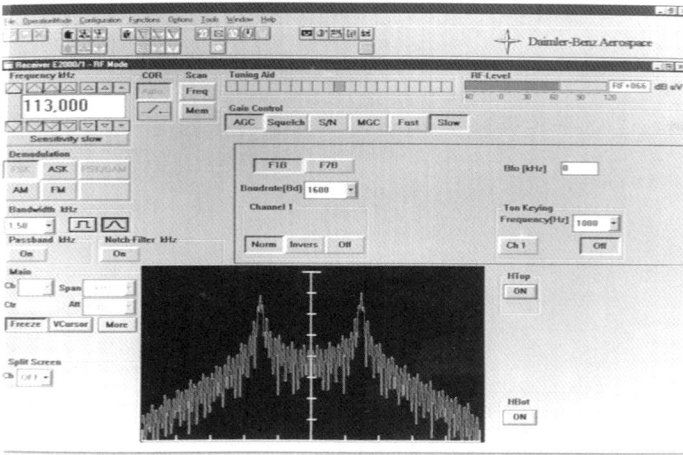

The PI 2000 panoramic display 0011546

Specifications
E 2000 LH

Frequency range: 0.3 kHz to 30 MHz
Frequency increments: min 1 Hz
No of receiving channels per 48 cm unit: 1 to 5
Setting time: typically 20 ms (optional 1 ms)
Notch filters (10 Hz res): 100, 200, 400 Hz
Passband tuning: asymmetrical setting 5 kHz
Types of demodulation: all types used in HF range
Interfaces: RS-232C, RS-485, Ethernet

E 2000 VU

Frequency range: 20 MHz to 3 GHz
No of receiving channels per 48 cm unit: 1 to 4
Setting time: typically 8 ms (optional 100 µs)
Notch filters: 100, 200, 400, 800, 1,600, 3,200 Hz
Passband tuning: ±½ of the selected bandwidth (B ≤15 kHz)
Modes: standard A1A, A1B, A2A, J2A, J2B, A3E, R3E, H3E, J3E, J7B, F3E, G3E; (with UD 2000) 2/4/8/16 PSK, 2/4/8/16 DPSK, F1A, F1B, F1C, F7B (optional)
Interfaces: RS-232, RS-485, Ethernet

E 2000 LU

Frequency range: 10 kHz to 3 GHz
No of receiving channels per 48 cm unit: up to 2
Other technical data: refer to E 2000 LH E 2000 VU respectively

ADAS 2000

Demodulation: FSK, PSK, QAM
    F1B: 1 to 4,800 Bd automatic
    F7B: 1 to 600 Bd automatic separately for the two channels
    PSK, QAM: 50 to 4,800 Bd automatic within a window around setting rating
Line spacing:
    F1B: 20 to 8,000 Hz
Measurement range:
    F7B: 40 to 2,400 Hz

Code/bit structures

Library: scope unlimited; max 100 code/bit structures may be loaded simultaneously; approximately 24 standard code/bit structures are supplied with the equipment
Standard codes and code tables: Baudot 1.5/2.0, CCITT 2; ASCII 1.0/2.0, CCITT 5; SITOR-A/B, Sitor 7; FEC-A, CCITT 2; HNG-FEC, CCITT 2; RUM-FEC, RUM 16; SI-

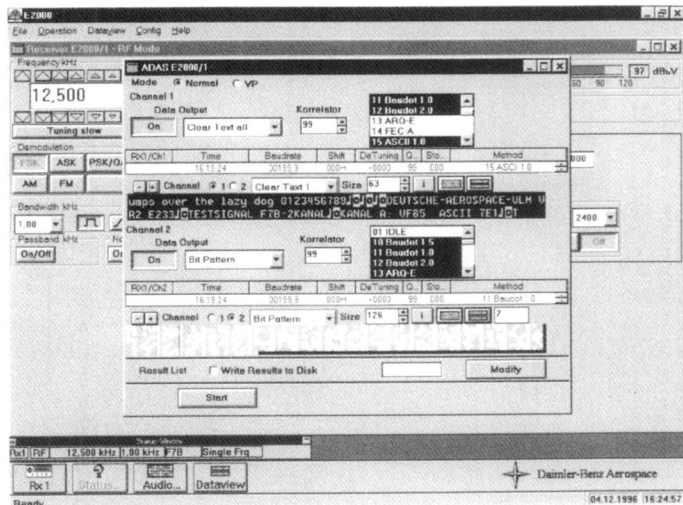

Display of clear text and bit patterns on E 2000 GUI 0055188

FEC, CCITT 3; ARQ-E/-E3/-M2/-M4/-242/-342, CCITT 3; POL-ARQ, Sitor 7; SI-ARQ, CCITT 3; CIS-11/14/27, M2; AUTOSPEC, Bauer; SPREAD 11/21/51, Bauer; AX 25, CCITT 5; IDLE 2/7/8/28/56

Status
The E 2000 series is no longer in production, but has been in service with various customers.

Contractor
MRCM
MRCM is a joint co-operation under full ownership of:
EADS Ewation, Ulm, Germany.
Grintek, Pretoria, South Africa.
Herley Industries Inc, Lancaster, USA.
Sysdel, Pretoria, South Africa.
TRL Technology, Tewkesbury, Glos, UK.

UPDATED

ACOS

The ACOS automatic COMINT system is a scalable broadband radio reconnaissance system. It is designed to acquire, classify and locate any radio signal (fixed-frequency, frequency hoppers, chirp or burst) within the selected frequency range. The overall bandwidth and detection probability can be scaled to the particular user's application with the incorporation of parallel sensors. Narrowband components can be integrated into ACOS in order to monitor individual frequency channels.

Fully automatic analysis of standard signals is designed to allow the operator to concentrate on the registration and analysis of unknown signals. One special feature of the ACOS system is automatic multidimensional DF data compression allowing a user-defined output of the relevant results from the total data available. This simplifies the monitoring task while at the same time offering an almost simultaneous overview of the complete signal scenario. All data are saved on suitable databases and can be post-processed manually.

A system consists of three subsystems - one each for signal registration, search and location - as well as control and evaluation. Signals identified as relevant to the mission undergo detailed analysis and evaluation in the signal registration system. The search and location subsystems provide an overview of the transmission scenario. The control and evaluation part controls the whole system, manages all results and supplies the relevant information to the user in the required format.

In ACOS, the individual emitters are described by their technical parameters. If the modulation and decoding characteristics are known, the individual emission can be analysed down to the message content.

Status
In production in 2003.

Contractor
Plath GmbH , Hamburg.

NEW ENTRY

EB200 MiniPort receiver

The EB200 Miniport Receiver is a portable, battery operated, digital receiver for radio monitoring in the 10 kHz to 3,000 MHz frequency range. The high-speed RF panorama and the internal FFT IF panorama enable the equipment to generate fast spectra and parametric readouts, useful for detecting frequency-agile signals such as bursts and hoppers. All settings as well as the measurement results are presented on the internal graphic display. With the aid of sub menus the receiver settings can be optimised for different tasks. All commands, the measurement results and the demodulated signal (digital audio) can be transmitted via LAN.

Together with the hand-held HE200 antenna, the receiver is appropriate for use in areas where normal receivers cannot be used due to weight and dimensions. For close-range direction-finding, sometimes known as Homing (find target) applications, the EB200/HE200 combination provides specific features, such as a level dependent tone function and a compass on the HE200.

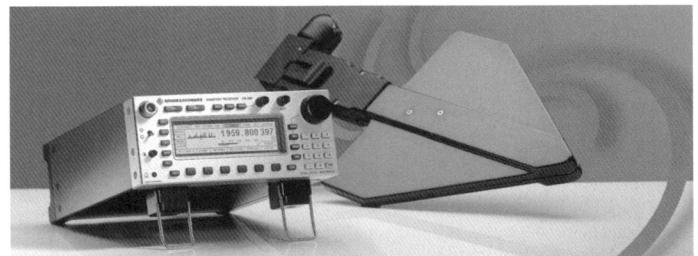

The EB200 Miniport receiver with the HE200 hand-held antenna behind (Rohde & Schwarz GmbH) *NEW*/0592991

Specifications

(a) Up to 2 GHz/s scan speed
(b) Scan modes: RF panorama, frequency scan, memory scan
(c) Internal FFT IF panorama up to 1 MHz span
(d) 12 digital IF filter, 150 Hz to 150 kHz
(e) Demodulation modes FM, AM, Pulse, CW, USB, LSB, IQ LAN or RS-232 interface
(f) remote-controllable via RS-232C
(g) Battery operation time approx 6 h
(h) AC operation (90 V to 264 V) with external power supply (standard accessory)
(i) Weight: 4 kg, 5.5 kg with battery pack
(j) Dimensions: 88 × 210 × 270 mm

Status

EB200/HE200 are in production and available. Deliveries have been made to unspecified customers.

Manufacturer

Rohde & Schwarz GmbH, Munich.

UPDATED

ESMB monitoring receiver

The ESMB monitoring receiver is a high-performance digital receiver for radio monitoring in the 9 kHz to 3,000 MHz frequency range. The equipment is fitted with separate tuners for HF and VHF/UHF in order to achieve high linearity and sensitivity in all frequency ranges. The high-speed RF panorama and the internal FFT IF panorama enables the equipment to generate fast spectra and parametric readouts. The receiver can also measure the modulation index, deviation, offset and bandwidth of the emissions. All settings, as well as the measurement results, are presented on the internal graphic display and with the aid of sub menus the receiver settings can be optimised for different tasks. The equipment can be used in both static and mobile configurations and can be incorporated into larger systems. Both measurement results and the demodulated signal (digital audio) can be transmitted via LAN.

Specifications

(a) Preselection, tracking and sub-octave
(b) Digital IF processing
(c) Fast scan speed, up to 3 GHz/s
(d) Scan modes, RF panorama, frequency scan, memory scan
(e) Internal FFT IF panorama up to 1 MHz span
(f) 18 digital IF filter, 150 Hz to 150 kHz, up to 1 MHz for measurements
(g) Demodulation modes: AM, FM, Pulse, CW, PM, USB, LSB, ISB, Ia LAN or RS-232 interface
(h) Operation voltage: AC 90 V to 264 V, DC 10 V to 32 V
(i) Weight: 8 kg
(j) Dimensions: 132 × 210 × 460 mm (rack model)

Status

Available. Has been supplied to unspecified customers.

Manufacturer

Rohde & Schwarz, Munich.

NEW ENTRY

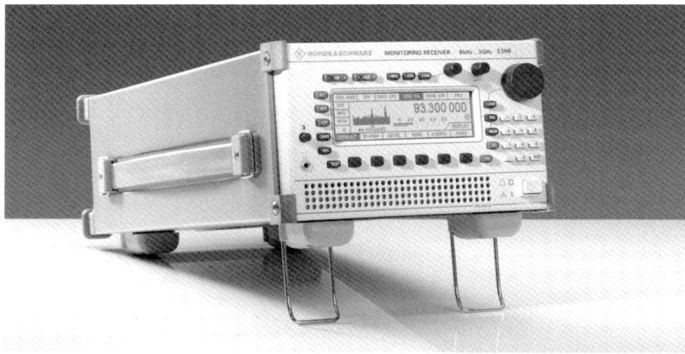

The ESMB monitoring receiver (Rohde & Schwarz)   ***NEW***/0592986

ESMC compact COMINT receiver

The ESMC receiver is half the size of a customary 483 mm multipurpose receiver of three height units, but its performance is claimed to be superior. The receiver has been optimised for radio monitoring applications over the frequency range 20 to 650 MHz, extendible with additional tuners down to 500 kHz and up to 3,000 MHz. It has the following features: operation via direct keys, menu buttons and graphical display; ability to receive any signal in the frequency range; a claimed wide dynamic range and high overload capacity; 1 Hz frequency resolution; low phase noise;

ESMC compact COMINT receiver (Rohde & Schwarz)   ***NEW***/0592983

offset display for channel frequency and internal spectrum display; AC/DC supply without exchange of power supply unit; and full remote control via IEE 488 or RS-232C using the SCPI protocol.

The operating concept of ESMC is designed to meet the requirements of modern radio monitoring receivers. All main functions such as type of demodulation or bandwidth can be set directly via labelled keys. A hot key allows the operator to return to the main menu from any sub-menu. Menu control is organised in priority levels so that signal processing is not interrupted by menu changes.

Specifications

(a) IF Bandwith up to 8 MHz for demodulation
(b) Scan speed up to 13 GHz/s with optional Analogue Scan (ASCAN)
(c) Scan modes: analogue scan, frequency scan, memory scan
(d) Internal IF panorama :1: 100 kHz
(e) 5 IF filters (5 out of 13 between 500 Hz and 8 MHz)
(f) Demodulation modes AM, FM, LOG, PULSE, CW, SSB
(g) IF outputs 21.4 MHz, wideband, wideband controlled, narrowband
(h) Video output
(i) IEEE or RS-232 interface
(j) AC operation (110/220 V) or DC 10 V to 32 V
(k) Weight: 12 kg
(l) Dimensions: 210 × 132 × 460 mm (W × H × D) (rack model)

Status

In production and in operational service since 1995. Reported to form part of the PIT-developed Polish Navy Srokosz shelterised COMINT/ESM system, the first of which was delivered in late 2001.

Manufacturer

Rohde & Schwarz GmbH, Munich.

UPDATED

MCP 8000 monitoring control position

The MCP 8000 monitoring control position is designed for operating up to four monitoring receivers independently. All receivers can be controlled by the MCP in all their functions, including the antenna selection. A significant benefit of the system is that it enables a radio direction-finding net to be commanded via the MCP.

MCP 8000 monitoring control position (Plath GmbH)   ***NEW***/0594235

Each MCP working position includes its own database. Frequencies and location results are stored including receiver settings and additional information. Audio information can be recorded and will be linked to the database. Additional frequency lists can be generated from the database, such as frequency range, type of modulation or target area. Scanning within a given frequency range, with programmed dwell time and frequency step and scanning through frequency lists, are both possible and preprogrammed lists of frequencies can be called up automatically to certain time slots.

The DF net can be remote controlled via conventional telephone lines or datalinks, and the capability of monitoring stations can be increased by the combination of several MCP positions.

Status

In service since 1998.

Manufacturer

Plath GmbH, Hamburg.

UPDATED

RAMON COMINT system

RAMON COMINT systems are customised solutions for radiomonitoring and radiolocation in the 10 kHz to 40 GHz frequency range. The systems are suitable for land-based, naval and airborne applications. Configurations range from small portable single-operator and mobile applications to complex semi-mobile or fixed monitoring stations with dedicated operator positions. These can be supported by remote subsystems through various kinds of wired or wireless communications. The modular concept of RAMON using standardised hardware and software components enables a system to be tailored to meet customer requirements.

Different system modules provide the following tasks:
(a) Creation and exchange of operational and technical orders and reports
(b) Interception of radio signals including:
Search
Direction-finding
Position fix
Recording/playback of audio and IF signals
Classification
(c) Technical analysis of radio signals
(d) Evaluation of generated reports, analysis of the tactical situation and creation of condensed summary reports as a contribution for an electronic order of battle (EOB).

Status

Described as an "available, COTS system". A number of systems are claimed to have been delivered and are in operation with unspecified military and law enforcement customers.

Manufacturer

Rohde & Schwarz GmbH, Munich.

UPDATED

*RAMON COMINT
system*
0011548

Hungary

FCM-MS baseband FDM channel monitor

The FCM-MS equipment is designed for parallel monitoring of the multichannel baseband or video signals of FDM/FM analogue transmissions. The monitor covers the 0.3 to 1,400 kHz frequency range for 2,700 speech (USB or LSB), data or pilot channels in channel increments of 4 kHz. The modular system can be assembled in different versions according to particular requirements and the number of independent channels ranges from one up to 28.

In the 28-channel version, the FCM-MS-28 contains an 8-input to 28- output remote-controlled video matrix, and can be used in a fully computer-controlled configuration for real-time processing of baseband signals from up to eight different microwave receivers.

The independent channels and the matrix are remote-controllable through an RS-485 interface. The nominal input level is –28 dBm/channel, and the output level is 0 dBm/600 Ω in the 0.3 to 3,400 Hz frequency range. Intermodulation suppression is more than 60 dB at 2 X –8 dB input signals. The FCM-MS is powered from 220 V AC single phase or 24 V DC. The dimensions of the 28-channel version are 490 × 430 × 1,380 mm and weight is approximately 95 kg.

Manufacturer

VIDEOTON-MECHLABOR Manufacturing and Development Ltd, Budapest.

NEW ENTRY

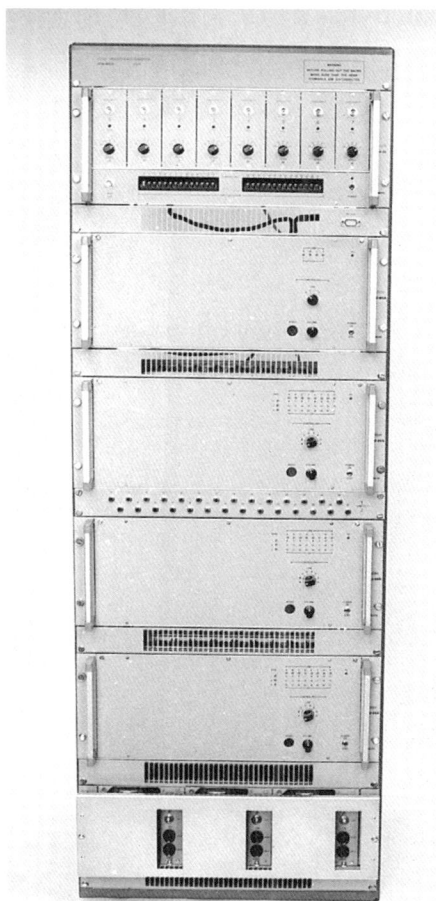

FCM-MS monitor

FH-1 frequency-hopping interceptor

The FH-1 frequency-hopping interceptor is a modern reconnaissance system based on parallel signal processing technology.

The equipment has 160 independent receiving channels covering a 4 MHz wide IF band with 25 kHz channel spacing, 60 dB channel selection and 60 dB intermodulation suppression. The 4 MHz wide IF band is the IF output of a special high-speed front-end receiver which has a 20 to 1,000 MHz frequency range.

The digitised output signals of the channels are multiplexed and fed as 1 Mbits/s data to a fast dedicated signal processing computer. As the processing time of the 160 channels is 200 μs with the front-end receiver 4 MHz frequency setting time, the processing speed of this interceptor is 4 MHz/200 μs or 20 GHz/s. This high speed makes it possible to process the whole 30 to 80 MHz ground-to-ground VHF band within a 2.5 ms time slot. The system's processing algorithm filters out noise spikes and stationary transmissions and in this way hopping transmissions can be classified either in the traditional frequency versus amplitude mode or in a waterfall-

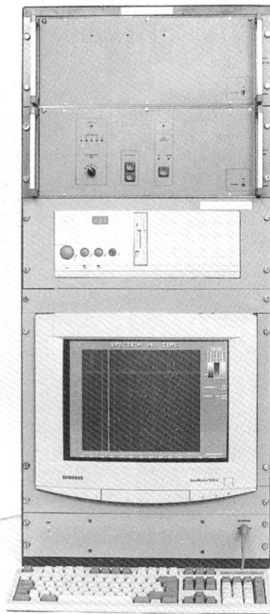

FH-1 interceptor

like frequency versus time display mode. Optional software modules are available for direction-finding the FH transmission and for controlling a remote follower/jammer.

Manufacturer
VIDEOTON-MECHLABOR Manufacturing and Development Ltd, Budapest.

NEW ENTRY

HF-21 IBM PC-mount HF frequency converter

The HF-21 is an HF to IF frequency converter with built-in frequency synthesiser especially designed for operating inside IBM-PC compatible computers. Using appropriate controller software, the HF-21 is able to survey rapidly the whole HF range. A programmable gain function allows it to handle signals of wide dynamic range.

Main features include: 0.1 to 30 MHz frequency range; wide dynamic range; 30 kHz bandwidth; 5 kHz frequency tuning step; less than 1 ms tuning speed; and housing in one PC-ISA slot.

Specifications
(a) Frequency tuning
  Frequency range: 0.1-30 MHz, 0.1-1.6 MHz with reduced parameters
  mixing: double
  tuning step: 5 kHz
  settling time: 1 ms max; 500 µs, typical
  internal frequency stability: 2.5×10-6 (-10 to 60°C)
  external reference input/output: frequency 10 MHz, level 0 to +3 dBm, impedance 50 Ω
(b) Amplitude
  RF input: 50 Ω impedance, asymmetrical, 3:1 max VSWR
  gain (RF in, IF max): 90 dB
  gain control range: 90 dB/in 10 dB steps
  RF-input protection: 2 +30 dBm, min
  dynamic range (at max gain): 15 dBm 3rd order input intercept point min, +40 dBm 2nd order input intercept point
  min sensitivity: -110 dBm max with amplifier, -90 dBm max without amplifier
  IF filter shape factor 3 dB/60 dB: 1.6 max
  analogue IF output: 455 kHz centre frequency, 15 kHz/30 kHz bandwidth (3 dB), 75 Ω impedance

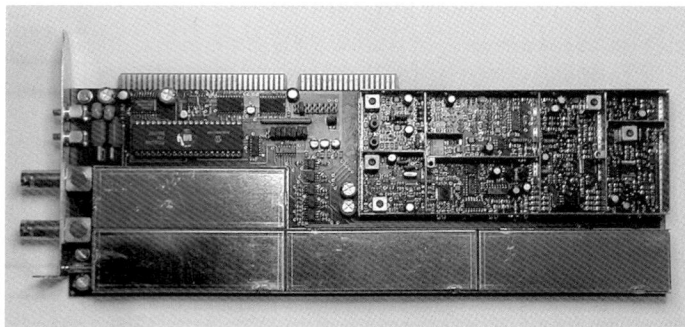

HF-21 board   0098857

(c) Interference protection
  mirror frequency rejection: 90 dB min
  IF rejection: 90 dB min
  oscillator emission: -100 dBm max
  Reciprocal mixing (in 10 kHz bandwidth, typical): offset 30 kHz -70 dB, offset 50 kHz -6 dB, offset 100 kHz -95 dB
(d) Temperature range: 0 to +50°C operating, -40 to +70°C storage
(e) Power consumption: current 65 mA +5 V, 630 mA +12 V, 30 mA +12 V; total power 8.3 W
(f) Size: long PC board

Manufacturer
VIDEOTON-MECHLABOR Manufacturing and Development Ltd, Budapest.

NEW ENTRY

PR-351 panoramic analyser

The PR-351 is a narrowband panoramic analyser designed for applications with HF radio receivers. Connected to the panoramic output of a receiver, it allows monitoring of the received frequency and its immediate neighbourhood, as the narrower bandwidth of the receiver's IF stages does not reduce the bandwidth of the panoramic output. The panoramic analyser is also suitable for: frequency spectrum analysis of received signals; checking the correct tuning of receivers; frequency spectrum evaluation and comparison.

The analyser features microprocessor control and has a 12-key keyboard for manual control. It can be connected to the outputs of two receivers at the same time, with push-button selection of whichever receiver is to be monitored. The band to be scanned can also be selected by push-button control. After A/D conversion, the analyser provides digital data storage and display of signals.

The PR-351 employs a resolution-scan 229 mm CRT display. Analysis is aided by a precision frequency scale and a movable marker with the particular frequency data. Also displayed on the CRT are the actual graph (display A or B), the actual receiver output under analysis (input 1 or 2), and the code for scanning bands of the actual and stored displays (L for large, S for small). It is possible to store a whole display and make comparisons between the stored display and the actual one using two half-pictures. There is a connection facility for a printer to record the stored display.

The analyser has both manual and automatic level controls. In its rack-mounted form, it measures 177 × 483 × 410 mm and weighs approximately 18 kg.

Specifications
(a) Input signal: 200 kHz
(b) Input impedance: 50 and 75 Ω
(c) Sensitivity: better than 10 µV input signal level required for a display of half-screen height
(d) Scanning bands: 12.7 kHz/2.54 kHz
(e) Resolution: 400 Hz/80 Hz
(f) Scanning time: 0.35/2.8 s
(g) Gain control: 11 × 5 dB manual (MGC), or input (AGC)
(h) Input AGC: max 6 dB output level increase for 60 dB input level increase; loading time 1 ms, recovery time 0.1 to 1 s
(i) Horizontal resolution: 128 spectrum line

Manufacturer
VIDEOTON-MECHLABOR Manufacturing and Development Ltd, Budapest.

NEW ENTRY

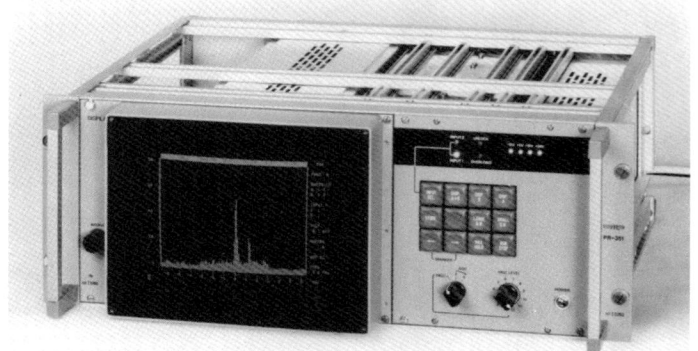

PR-351 analyser

PU-351 panoramic analyser

The PU-351 is a manually or remotely controllable narrowband panoramic analyser designed mainly for applications in UHF radio receivers. Connected to the wideband panoramic output of a receiver, it allows the monitoring of the received frequency and its immediate neighbourhood. The equipment is suitable for the

PU-351 analyser

following tasks: frequency spectrum analysis of received signal; checking the correct tuning of receivers; application in a high-level intelligence reconnaissance system supported by computer; BIT-map reconnaissance operation; indication of the relative amplitude changes of a selected spectrum line in time; and look-through with control of a co-operating jamming transmitter.

The panoramic analyser has two inputs so it may be connected to two receivers simultaneously and the signal of the receiver of interest can be chosen by pressing a selector push-button. The equipment has both manual and automatic gain control. Synthesiser tuning of the panoramic analyser is designed to provide fast and accurate frequency determination. After A/D conversion the analyser provides digital data storage and display of signals in the form of vertical bar graph lines. These can be switched to the linear or logarithmic amplitude scale. Synthesiser, sampling and display are controlled by a microprocessor.

The PU-351 employs a raster scan 229 mm CRT display. Analysis of the displayed spectrum is aided by a precise frequency scale and movable markers for both frequency and amplitude. It is possible to stop scanning, and enter the MAX-HOLD mode, as well as store a whole picture for display at any time. The stored picture and the real-time panorama picture can then be displayed simultaneously in half-size in order to make comparisons. The main settings can be read in the status field under the spectrum.

There is a connection facility for an external printer to print stored display. In rack-mount configuration, the PU-351 measures 177 × 483 × 410 mm and weighs around 21 kg.

Specifications
(a) Centre frequency of input signal: 21.4 MHz
(b) Number of inputs: 2
(c) Input impedance: 50 Ω ±20%
(d) Sensitivity: 30 (45)V input signal level required for a display of half-screen height
(e) Scanning bands: 4 MHz, 320 kHz, 32 kHz
(f) Number of displayed channels: 321
(g) Resolution: 50 kHz, 5 kHz, 1 kHz
(h) Scanning time: 102.4 ms, 80 ms, 200 ms
(i) BIT-map operation mode (FAST):
scanning time: 25.6 ms
simultaneously displayed period: 84 × 25.6 = 2,150 ms
(j) Time mode:
number of picture elements of the time axis: 5 × 640 = 3,200
length of the time axis: 0.80, 0.24, 0.8, 8 s
(k) Manual gain control: max 6 dB signal size increase for 55 dB input level increase

Manufacturer
VIDEOTON-MECHLABOR Manufacturing and Development Ltd, Budapest.

NEW ENTRY

PV-351 panoramic analyser

The PV-351 is a narrowband panoramic analyser designed for applications with VHF radio receivers. Functionally and physically the PV-351 is similar to the PR-351 (see separate entry).

Specifications
(a) Input signal: 10.7 MHz
(b) Input impedance: 50 Ω, ±20%
(c) Sensitivity: better than 30 µV input signal level required for a display of half-screen height
(d) Scanning bands: 128 kHz, 32 kHz, 8 kHz
(e) Resolution: 9 kHz, 3 kHz, 1 kHz
(f) Scanning time: 102.4 ms
(g) Gain control: 11 × 5 dB manual (MGC), or input AGC
(h) Input AGC: max 6 dB output level increase for 60 dB input level increase; leading time 1 ms, recovery time 0.1 to 1 s
(i) Horizontal resolution: 128 spectrum line

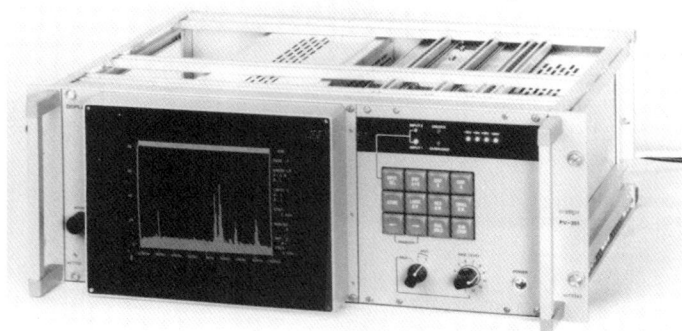

PV-351 analyser

Manufacturer
VIDEOTON-MECHLABOR Manufacturing and Development Ltd, Budapest.

NEW ENTRY

VREV-P wideband panoramic receiver

VREV-P is a wideband panoramic receiver for surveying 10 MHz portions of the 20 to 100 MHz frequency range. Speed of scanning is 78.125 MHz/s quasi-continuous and dynamic selectivity is 30 dB minimum at three-channel off-tune. The receiver is designed for computer-aided intelligence reconnaissance.

Intercepted signals are forwarded via a logarithmic IF module to an analogue/digital converter. The digitised signals are stored in a 'new signals' memory. Stored signals are processed and the newly intercepted frequencies forwarded via an operator post controller to the communication receiver in order to determine the exact parameters of the intercepted emission. Simultaneously, the intercepted signals can be displayed on the CRT display of the device.

The 229 mm CRT displays a 10 MHz wide frequency band in five lines, each representing 2 MHz. The displayed start and end frequencies are shown at the beginning and the end of the lines in digital form. A moving marker also indicates, digitally, the frequency related to the position of the marker in 12.5 kHz steps to facilitate the determination of the frequency of a selected signal.

The VREV-P measures 444 × 192 × 410 mm in cabinet form and weighs approximately 18 kg.

Manufacturer
VIDEOTON-MECHLABOR Manufacturing and Development Ltd, Budapest.

NEW ENTRY

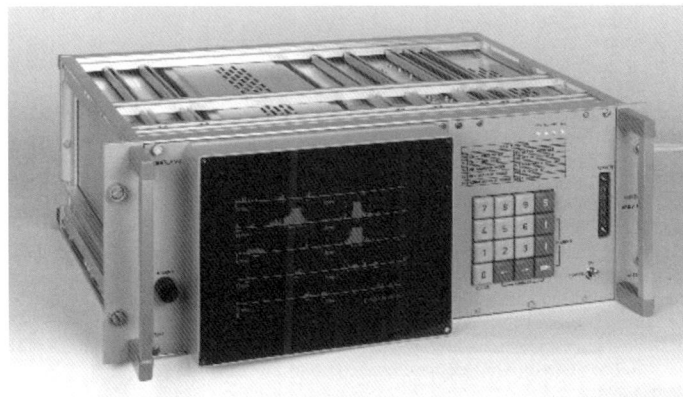

VREV-P receiver

VUREV and VUREV-G VHF/UHF receivers

The VUREV and VUREV-G are receivers which can be used for any task related to reconnaissance of the UHF frequency range including search, scan and monitoring. The VUREV covers 20 to 499.9999 MHz, and the VUREV-G the 20 to 999.9999 MHz range.

The receiver is suitable for both remote-controlled and manual operation. Searching for active emissions can be done in three ways: tuning quasi-continuously in 100 Hz steps by the rotating tuning knob; searching automatically between two adjustable frequency limits in variable frequency steps; and continuously checking those stations whose frequencies have been written into the SCAN memory.

Duration of the investigation of one station is about 25 ms. In SCAN mode, scanning can be stopped for a predetermined period in order for the operator to listen to the emission and to determine its most important characteristics. Another possibility with the SCAN mode is to compile traffic statistics of the transmitters whose frequencies have been stored in the memory. Non-volatile memories are used to store 100 prohibited and 100 task stations' data such as frequency, operation mode and bandwidth.

Remote control and reporting make the receiver suitable for application in computer-controlled systems.

In rack configuration the receivers measure 177 × 483 × 410 mm and weigh 22 kg.

Manufacturer

VIDEOTON-MECHLABOR Manufacturing and Development Ltd, Budapest.

NEW ENTRY

VUREV-P and VUREV-PG wideband panoramic receivers

The VUREV-P and VUREV-PG VHF/UHF receivers are used for surveying an 80 MHz frequency section of the 20 to 500 and 20 to 1,000 MHz frequency ranges respectively and are suitable for use in computer-aided reconnaissance systems.

Intercepted signals are forwarded via a logarithmic IF module to an A/D converter. The digitised signals are stored in the 'new signals' memory. They are subsequently processed and newly intercepted frequencies will be forwarded via the operator post controller to the communication receiver in order to determine the exact parameters of the intercepted emission. Simultaneously, the intercepted signals are also displayed on the CRT of the device.

The 229 mm CRT display shows the 80 MHz wide frequency band in five lines, each representing 16 MHz. It is also possible to display only a selected 20 MHz sub-band of the scanned 80 MHz frequency band. The 20 MHz sub-band is also divided into five lines, each representing 4 MHz. The displayed start and end frequencies are shown at the beginning and the end of the lines in digital form. A movable marker shows the frequency related to the position of the marker in 100 kHz or 25 kHz steps. The CRT can also present a 25-line alphanumeric display without interrupting the operation of the panoramic receiver. When used for panoramic display, a digital message of 49 characters can be written into the bottom line of the screen.

In rack configuration, the receivers measure 221 × 485 × 410 mm and weigh approximately 27 kg.

Manufacturer

VIDEOTON-MECHLABOR Manufacturing and Development Ltd, Budapest.

NEW ENTRY

International

MAIGRET 5000 - Combined Naval CESM and RESM system

The MAIGRET 5000 combined communications/radar band ES system is the latest version of the MAIGRET range, which was first developed in the mid-1990s. It consists of the MAIGRET C 5000 communications band ESM/COMINT subsystem, the MAIGRET R 5000 radar ESM/ELINT subsystem and the MAIGRET S 5000 command and evaluation element and is described as being able to fuse acquired communications and radar data into a single-tactical electronic warfare database in real time. Situational awareness information can be displayed as either a tactical polar diagram and/or as a map. MAIGRET 5000 is understood to be automated, as incorporating automatic threat warning filters (threat/non-threat classification) and a data recording facility and is supported by a shore-based, post-mission analysis tool for offline, multivessel data evaluation. The baseline MAIGRET 5000 ES capability can be extended into the communications and electronic intelligence(COMINT/ELINT) fields and the system employs a single-mast antenna solution that is claimed to offer 'outstanding sensitivity and bearing accuracy'. The antenna assembly used is noted as being 'operationally proven' in a naval environment and the baseline MAIGRET 5000 electronics fit in a 19 in (48 cm) rack to facilitate their use aboard 'compact' surface ships and submarines. The system's software is configured to enable its operation from the host vessel's combat management system multifunction consoles or one or more dedicated operator workstations.

MAIGRET C 5000 Comms ESM/COMINT subsystem features are:
(a) azimuth selective wideband search and occupancy detection
(b) a digital map display
(c) automatic emitter activity detection
(d) data reduction via emitter tracking
(e) automatic emitter classification
(f) a manual signal analysis facility
(g) message content from analogue and digital transmissions.

MAIGRET R 5000 radar ESM/ELINT subsystem features are:
(a) 'high-sensitive', omni-directional, wide-open
(b) direction-finding and reception (−65 dBm)
(c) electronic counter-countermeasures proofing via the division of its frequency coverage into four sub-bands
(d) cartesian, polar and activity displays
(e) automatic emitter classification

A graphic showing the elements that make up the MAIGRET 5000 combined communications/radar ES system. The mast shows the different elements of the combined antenna (MRCM)   ***NEW**/0525253*

(f) the ability to track up to 512 emitters simultaneously
(g) mode and threat libraries (10,000 emitters)
(h) an optional ELINT mode with high sensitivity (better - 80 dBm) for the reception and classification of low probability-of-intercept signals.

MAIGRET S 5000 command and evaluation element features are:
(a) radar and communications data fusion
(b) a mission-planning capability
(c) a reporting/situation display
(d) a platform library
(e) a multi-user capability
(f) remote terminals
(g) signals intelligence applications.

Specifications

Frequency range:
0.01-3,000 MHz (MAIGRET C 5000, monitoring)
20-3,000 MHz (MAIGRET C 5000, DF - extendable to include 1.5-30 MHz range)
2-18 GHz (MAIGRET R 5000, extendable to 0.5-40 GHz)

Status

An early version of MAIGRET is in service with the RNZN. Other versions are possibly in service with the German Navy. The MAIGRET 5000 combined communications and radar ES system is reported as being in service and production ('under permanent design improvement').

Contractors

MRCM
MRCM is a joint co-operation under full ownership of:
EADS Ewation, Ulm, Germany.
Grintek, Pretoria, South Africa.
Herley Industries Inc, Lancaster, USA.
Sysdel, Pretoria, South Africa.

UPDATED

Israel

AES-210-E

The AES-210/E is a family of airborne ESM/ELINT systems designed for installation on various manned and unmanned airborne platforms. The system can be employed for maritime and overland surveillance, elint information gathering, locating and targeting hostile radars, and platform self-protection. The common concept for all installations is based on modular, plug-in Line Replaceable Units (LRUs), minimal effect on platform aircraft performance, provision for integration with onboard systems (RS-422, RS-232, 1553B or ARINC interface), and co-

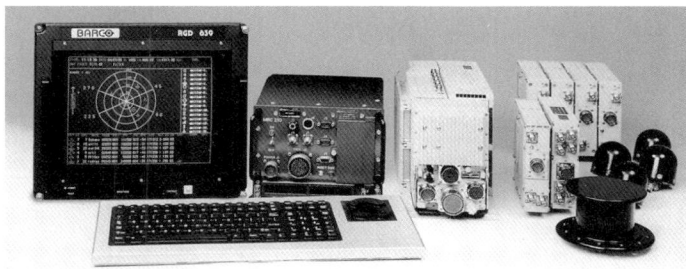

The elements of the AES-210/E (Elisra)   0062851

existence with onboard radars. The system automatically detects, measures and identifies radar emissions from ground-based, shipborne and airborne weapon systems and calculates their location. Intercepted emitters are presented to the operator on an interactive colour graphic situation display and logged on magnetic media and an optional printer for further processing. An updateable identification 'library' of more than 1,000 emitters is a main feature of the system. Long-range detection and a high probability of intercept are enabled by a combination of narrow and wideband receivers. Accurate direction-finding is implemented by amplitude comparison and differential time of arrival (DTOA) techniques. Self-protection capability can include an RWR display. Data can be recorded and played back in flight or on the ground. Optional add-on capabilities include a datalink, for real-time downlink of received electromagnetic signals; and a ground analysis station for enhanced signal analysis of the received threats.

Specifications
(a) Weight: 40 kg typical
(b) Frequency: 0.5-18 GHz
(c) Coverage: 360° azimuth
(d) Accuracy: 7° RMS coarse, 3° RMS fine
(e) Library: >1,000 emitters
(f) Interface: RS-232, RS-422, MIL-STD-1553B, ARINC
(g) Environmental: MIL-STD-5400T
(h) Power: 500 W

Status
Believed to be installed in Royal Australian Navy and possibly Indian Navy helicopters. Also reported as being installed in Argentine Navy MPA.

Contractor
Elisra Electronic Systems Ltd, Bene Baraq.

VERIFIED

....

CR-2740 mobile ELINT system

The CR-2740 is a computer-controlled automatic mobile ELectronic INTelligence (ELINT) system integrating field-proven equipment and operational experience.

Automatic search and acquisition of radar signals is performed with high-gain antennas and superheterodyne receivers which scan the azimuth sectors and frequency ranges. Very high sensitivity is achieved by combining high antenna gain and high receiver sensitivity. The signals are sorted and their parameters are measured. Processed signals are presented on a colour graphic situation display. For manual analysis purposes, time domain and frequency domain real-time displays are provided.

Standard RS-232 and Ethernet interfaces enable data communication with a remote computer. Antennas and receivers can be installed at a considerable distance from the operator's position and operated via a radio link or telephone line. The system interfaces with blanking input and provides log video, linear video and audio output.

The CR-2740 consists of an antenna trailer with receivers for the 0.5 to 4 GHz (VHF/L/S-bands - NATO B/C/D/E/F-bands) range and a shelter with a roof-mounted antenna assembly and receivers for the 4 to 18 GHz (C/X/Ku-bands -

CR-2740 ELINT system

CR-2740 workstation

NATO G/H/I/J-bands) range. Easily deployed and disassembled, the system is transportable by truck, helicopter or cargo plane.

Specifications
(a) Frequency coverage: 0.5-18 GHz by superheterodyne receivers
(b) Azimuth coverage: 360° by rotating dish antennas
(c) Frequency accuracy: better than 300 kHz
(d) Frequency resolution: 100 kHz
(e) DF accuracy: typically better than 1°
(f) Weight: shelter <4,500 kg; trailer (equipment and antenna) <5,000 kg

Status
In operational service.

Contractor
Elisra Electronic Systems Ltd, Bene Beraq.

VERIFIED

....

Electronic Warfare Integrated System (EWIS)

EWIS is an integrated system operating within a broad range of frequencies (0.15 MHz to 18 GHz) which is designed to cope with diverse threats. The system can be adapted to different levels of warfare (tactical, operational or strategic) and draws together the product or capabilities of a wide range of platforms in all environments, comprising COMINT, ELINT, DF, ECM and EW command and control elements. The key element is the latter, which enables the integration of a range of Tadiran's products which perform the various functions. This is compatible with a wide variety of communications media including tactical radio (HF, VHF, UHF), data links, fixed infrastructures, satellite and cellular. The operating system is Windows NT™, and the system provides advanced processing and display capabilities, a central database that can be shared at different command levels, multi-operator functionality due to the system's modularity, and internal LAN communications.

Status
Fully operational, with many customers claimed worldwide.

Contractor
Tadiran Electronic Systems Ltd, Holon.

UPDATED

....

EL/K-1250 VHF/UHF COMINT receiver

The EL/K-1250 is a synthesised receiver operating in the 20 to 510 MHz band. Its volume, weight and power consumption make it a basic building block for larger COMINT or EW systems.

The EL/K-1250 demodulates AM, FM, CW and SSB signals, employing up to four selectable IF filters. RF preselection by voltage tracking filters improves intermodulation protection and rejection of spurious signals. A fast tuning synthesiser settles within 500 μs between any selected channels. Operation is under digital remote control.

Specifications
(a) Modes: AM, USB, LSB, CW, FM
(b) Frequency range: 20-510 MHz
(c) Frequency resolution: 10Hz
(d) Frequency accuracy and stability: ±1 PPM
(e) IF bandwidth: 4 of following: 10 kHz, 20 kHz, 50 kHz, 100 kHz, 300 kHz, 600 kHz, 1 MHz
(f) Noise figure: 12 dB 20-180 MHz; 11 dB 180-510 MHz
(g) Power supply: +5 V/3A, +15 V/1.5A, +30 V/30 mA, −15 V/100 mA
(h) Temperature range
    operating: −20 to +60°C
    storage: −40 to +85°C
(i) Relative humidity: up to 95%
(j) Altitude: 10,000 m
(k) Shock: designed to meet ½ sine of 15 *g*, 11 ms
(l) HxWxD: 193 × 57 × 497 mm
(m) Weight: 4.99 kg

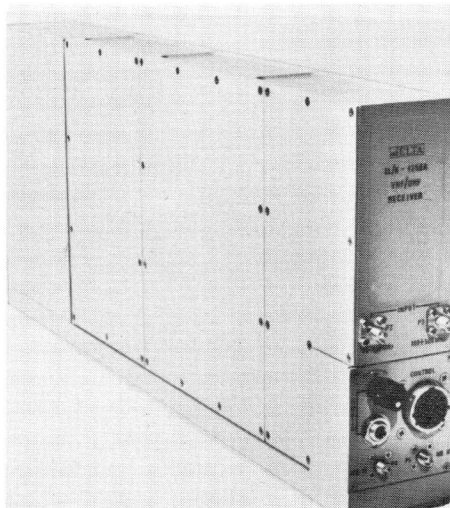

EL/K-1250 COMINT receiver

Status

In production for various countries.

Manufacturer

ELTA Electronics Industries Ltd(a subsidiary of Israel Aircraft Industries Ltd), Ashdod.

UPDATED

EL/K-1250T VHF/UHF COMINT receiver

The EL/K-1250T receiver covers the 20 to 510 MHz frequency range with 1 kHz resolution. Designed as a basic building block for larger COMINT, DF or EW systems, it can demodulate AM and FM signals employing up to four selectable IF bandwidths. RF preselection by voltage tracking filters is designed to provide improved intermodulation protection as well as rejection of spurious signals. A frequency synthesiser is common to both channels and is capable of tuning in less than 500 μs between selected channels. Frequency stability and accuracy is ±1 PPM.

The EL/K-1250T measures 193 × 114 × 356 mm and weighs 3.15 kg. It is powered from +5 V, +15 V, +30 V or −15 V AC and meets MIL-STD-810C.

Status

Remains in the manufacturer's catalogue, but may have been superseded by the EL/K-1250 (see separate entry).

Manufacturer

ELTA Electronics Industries Ltd(a subsidiary of Israel Aircraft Industries Ltd), Ashdod.

VERIFIED

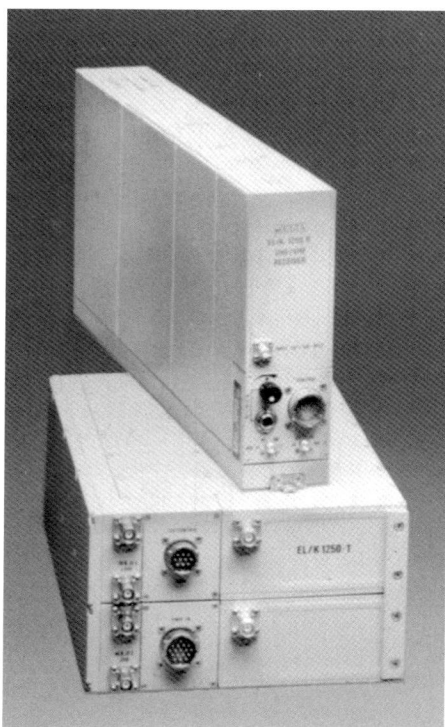

EL/K-1250T receiver (bottom)

EL/K-7035 COMINT system

The EL/K-7035 is a COMINT system which is designed to intercept, monitor, locate, analyse and report radio communication signals in the 20 to 500 MHz range. The system is based on a multitask COMINT workstation and can be configured for land, sea or air applications.

The standard single workstation for all-platform COMINT operation includes two to four receivers, two to four controllers, two to four dual-channel tape recorders, one ruggedised computer, an optional IF panoramic display and an optional time code generator/reader. All units are mounted in a 19 in standard rack. Such a console could be operated as a stand-alone system. More elaborate configurations could include a supervisor console; 2-5 operator consoles; a system controller and mass storage; a remotely controlled DF system with a set of up to eight antennas; a plotter position; and a communications datalink.

Features

(a) General

Each operator's position has independent automatic search and scan capability and access to direction-finding equipment.

Each receiver can be manually operated.

The supervisor can assign tasks to other operators and can monitor and update functional parameters of any operator's station.

(b) Search and Acquisition: The system performs fast (up to 500 channels/sec) search of the frequency range excluding protected channels. The detected signals are automatically monitored and evaluated.

(c) Preset Tasks Monitoring: The system performs scanning of up to 64 preset tasks. The detected signals are automatically monitored and evaluated. Additional receivers may be used for high-priority tasks monitoring and recording.

(d) Localising: Automatic or manual direction finding on targets, utilised for subsequent emitter location analysis.

(e) Analysis: A history of signal activities and DF results can be recorded on hard disk for off-line analysis and study, using special software. Additional, optional equipment can retranscribe selected communications for later analysis, study or editing.

(f) Data Transfer: An optional microwave datalink can transfer received data to remote monitoring stations.

(g) Training: By a simple change of software, the system can be switched over from operational use to simulation. The simulation software provides "realistic" conditions for COMINT trainees.

Specifications

(a) Frequency range: 20-500 MHz
(b) Frequency resolution: 1 kHz
(c) Demodulation: AM, FM
(d) Scan rate: up to 500 steps/s
(e) Preset channels: 64 per receiver
(f) Protected channels: 96 per receiver
(g) DF Bearing display resolution: 0.1°

EL/K-7035 COMINT system

Status

Available. May be in service.

Contractor

ELTA Electronics Industries Ltd (a subsidiary of Israel Aircraft Industries Ltd), Ashdod.

VERIFIED

EL/K-7036 COMINT system

The EL/K-7036 COMINT system has been designed to meet the operational challenge of dense communication environments. It monitors emissions over a wide frequency range characterised by short transmission and frequent changes in network parameters. Claimed to be easy to operate and maintain, the EL/K-7036 is the basic building block of ELTA's new generation of COMINT systems and can be installed on airborne, shipborne and ground-based platforms. It is suitable for mobile or fixed installation. The workstation can operate as a stand-alone system or may be integrated into a larger COMINT/EW system.

Through an automatic three-level screening process, the EL/K-7036 focuses system resources on target signals that meet predefined parameters, drawing the operator's attention to networks of potential importance. Operating in real time, the database provides the operator with all available information required for "fast and accurate" identification of intercepted signals. Thus the decision on whether to monitor the network can be based on accumulated data as well as real-time parameters.

The EL/K-7036 operates in the 2-500 MHz range and optional expansions to give coverage of 0.5 to 2 MHz and 500 to 1,000 MHz are available. The system has a high probability of intercept and has filtering algorithms in the frequency and geographic domains to ensure that only relevant signals are queued for further processing.

The system has a menu-driven man/machine interface using a mouse or keyboard function keys. Among the system's functions are:

(a) search for activity in predefined frequency bands and geographic regions
(b) automatic reject of recognised signals of little or no interest
(c) "very fast" scan of known signals of interest, ensuring a high probability of detection even for short transmissions
(d) modes for accurate DF results and network analysis
(e) audio recording and data input for data accumulation and analysis
(f) reports to a central site or deployed forces
(g) the lines of bearing are displayed on a raster geographical map
(h) an auxiliary monitoring receiver can be added as an option.

Specifications

(a) Interception
Frequency range: 2-30 MHz (HF); 20-500 MHz (V/UHF); optional extensions to 0.5-2 MHz and 500-1,200 MHz, respectively
Modulation types: AM, FM, CW, SSB, PSK, FSK, pulse
IP3: +20 dBm, (HF); +3 dBm (V/UHF)
IF bandwidth: 0.5, 3.2, 6, 16 kHz (HF); 10, 20, 50, 100 kHz (V/UHF)
IF rejection: 100 dB (HF); 80 dB (V/UHF)
Frequency step: 10 Hz (HF); 10-1,000 Hz (V/UHF)
Image rejection: 100 dB (HF); 80 dB (V/UHF)
Fast scan (freq cell/s): 100 (HF); 500 (V/UHF)
(b) Direction-finding
Frequency range: 2-30 MHz (HF); 20-500 MHz (V/UHF); optional extensions to 0.5-2 MHz and 500-1,200 MHz, respectively
Response time: <500 ms (HF) <200 ms (V/UHF)
Accuracy: 2° RMS (HF); 1.5° RMS (V/UHF)

Status

In service.

Contractor

ELTA Electronics Industries Ltd (a subsidiary of Israel Aircraft Industries Ltd), Ashdod.

VERIFIED

EL/L-8300 airborne signals intelligence system

The EL/L-8300 designation describes both an airborne strategic SIGINT suite and a maritime patrol ES system. In its strategic SIGINT guise, the baseline L-8300 is a multi-operator system that probably incorporates the EL/L-8312A ELINT, EL/K-7032 COMINT and EL/L-8350 control and analysis subsystems. The EL/L-8351 ELINT training simulator, EL/L-8352 ELINT data analysis facility and EL/L-8353

tactical ground station support the airborne segment. The SIGINT EL/L-8300 covers the 0.5 to 18 GHz (0.03 to 40 GHz as an option) frequency range, has a detection range of 450 km when being flown at a 'typical operating altitude' aboard a Boeing 707 host platform and offers "high" probability of intercept in dense environments, instantaneous frequency measurement and an electronic order of battle analysis capability. The COMINT subsystem is an acquisition, exploration and monitoring capability that covers the 20 to 1,000 MHz (2 to 1,500 MHz as an option) frequency band and features both wide and narrow band direction-finding. The equipment's airborne 'command station' provides data integration, report generation/mission support, threat warning and air-to-surface communications facilities.

Status

An upgraded version of the SIGINT EL/L-8300 architecture may form the basis of the electromagnetic spectrum hardware used in the mission suite that is installed in Spain's Boeing 707 'SCAPA' SIGINT and electro-optic surveillance platform. As an ES system, EL/L-8300 variants have been selected for use on Australian Lockheed Martin AP-3C Orion (designated as EL/L-8300 AU(?)/ALR-2001 Odyssey) and Singaporean Fokker Enforcer Mk 2 maritime patrol aircraft (EL/L-8300 MPA?). The Australian programme is a joint effort between Elta and Australian industry and involves equipment being installed in 19 aircraft. In 2001 it was announced that Elta had been awarded a $60M contract to provide ESM/ELINT systems for the RAAF Project Wedgetail AEW&C aircraft. It is likely that this will be a similar system. A further variant (designated as EL/L-8300 UK)has been selected as the ES system for the Royal Air Force's Nimrod MRA Mk4 maritime patrol aircraft.

Manufacturer

ELTA Electronics Industries Ltd (a subsidiary of Israel Aircraft Industries Ltd), Ashdod.

UPDATED

EL/L-8300 airborne SIGINT system

EL/L-8300 airborne SIGINT system

For details of the latest updates to *Jane's C4I Systems* online and to discover the additional information available exclusively to online subscribers please visit
jc4i.janes.com

ELTA/IAI SIGINT aircraft programmes

ELTA, working with the elements of its parent company Israel Aircraft Industries (IAI), has made something of a speciality of producing tactical and strategic SIGINT aircraft, usually converted from existing transport airframes, for a range of customers around the world. Most recently, this work has been expanded to include airborne early warning in the form of the Phalcon system.

Status
The ELTA/IAI combine is understood to have supplied Argentina (L-188 WAVE aircraft - see separate entry), Chile (ELINT and COMINT subsystems for the Phalcon/Condor airborne early warning and surveillance aircraft - see separate entry), Spain (SCAPA Boeing 707 aircraft) and Thailand (three SIGINT-configured Arava aircraft operated by the Thai Air Force's No 605 Squadron based at Don Muang) with either complete SIGINT system/aircraft packages, SIGINT subsystems and/or SIGINT system integration services.

Contractor
ELTA Electronics Industries Ltd (a subsidiary of Israel Aircraft Industries Ltd), Ashdod.

VERIFIED

An Arava transport aircraft reconfigured for electronic reconnaissance and countermeasures tasks is typical of ELTA/IAI's work in this field

GES-210E ground ESM/ELINT system

The GES-210E is the ground version of Elisra's AES-210E airborne ESM/ELINT system. The GES-210E provides accurate DF utilising a high-gain spinning parabolic antenna. The DF accuracy is 1° RMS. An advanced combined IFM and superheterodyne receiver enable analysis of all types of signals in the 0.5 to 18 GHz frequency range.

The system enables intra- and inter-pulse analysis.

Contractor
Elisra Electronic Systems, Bene Beraq.

VERIFIED

NATACS 2000

NATACS 2000 is a naval tactical COMINT and DF system operating in the HF and V/UHF frequency bands with particularly fast scanning and activity detection rates. The basic system uses client server architecture and comprises three subsystems. The interception and monitoring subsystem is equipped with active HF and V/UHF receiving antennas, with wide band receivers for fast search, scan and activity detection; signals are classified and recorded. The interferometer-based DF subsystem has a V/UHF wideband DF system as standard with HF DF as optional. The communications subsystem includes HF and VHF radios and a communications controller. Two personnel operate the basic system, although the system can be expanded and further operating positions added. NATACS 2000 can be integrated with the ship's C2, communications and navigation systems.

Specifications
COMINT
Frequency range: HF: 0.3-30 MHz; V/UHF: 20-1,000 MHz (20-1,500 MHz optional)
Scan rate: HF: up to 100 channels/s, 100 MHz/s (optional); V/UHF: 1 GHz/s
Resolution: HF: 1 Hz; V/UHF: 10 Hz
DF
Frequency range: HF: 1.5-30 MHz; V/UHF: 20-1,000 MHz (20-1,500 MHz optional)
Scan rate: HF: 100 MHz/s; V/UHF: 1 GHz/s
DF accuracy: Sea: 1.5° RMS typical; instrumental: <1° RMS

Status
In service in the Israel Navy in the 'Eilat' *(Saar 5)* class corvettes.

Contractor
Tadiran Electronic Systems Ltd, Holon.

UPDATED

NS-9003A-V2 naval ESM system

The NS-9003A-V2 naval ESM system operates in a dense environment carrying out electronic intelligence and is able to cope with frequency agility, frequency hopping and staggered or wobbulated PRI emitters. The system automatically receives, analyses and identifies over-the-horizon radar signals with a 100 per cent probability of intercept over the frequency range from 2 to 18 GHz (0.5 to 40 GHz optional). The direction of arrival is instantaneously measured per pulse, using an accurate multibeam antenna array. This instantaneous direction-finding enables the system to locate any 'exotic' radar signal type in the reception range within a fraction of a second.

The analysed information is presented on a 19 in graphic situation display in several operator-selected modes for various EW tasks. The system has built-in power management provisions for ECM and chaff/flare activation.

The NS-9003A-V2 antenna assembly includes a multibeam static antenna array with associated RF video and digital DF processing hardware and an omnidirectional antenna array. Other units in the system are receiving and processing equipment and an operator console. Interface to radar blankers, chaff/flare equipment and a satellite navigation system can also be provided. An integrated EW suite consisting of the NS-9003A-V2 and the NS-9005 jamming and deception system, known as the NS-9003A-V2/9005 is available.

Specifications
Frequency range: 2-18 GHz (0.5-40 GHz optional)
DF accuracy: 1° I- and J-bands; 2° E/F- and G/H-bands
Sensitivity: −70 dBm
Frequency measurement: instantaneous 2 MHz resolution
MMI: Windows NT
Status
In service with the Israel Navy and other navies throughout the world.

Contractor
Elisra Electronic Systems, Bene Beraq.

VERIFIED

NS-9003A-V2 naval ESM system   0011549

NS-9003A-V2/U submarine ESM/ELINT system

The NS-9003A-V2/U is a compact ESM/ELINT/RWR system for submarines. It features instantaneous high-sensitivity frequency measurement combined with accurate direction-finding. It will carry out accurate and rapid detection and analysis in dense electromagnetic environments.

The system uses an ESM mast incorporating spiral and horn arrays covering the 2 to 18 GHz frequency range. Additional antenna arrays can also be mounted on the periscopic mast.

A variety of data display facilities are available.

The NS-9003A-V2/U can be installed in new submarines or retrofitted.

Status
In production.

Contractor
Elisra Electronic Systems, Bene Beraq.

VERIFIED

Phalcon airborne early warning system

The long-range, high-performance, multisensor Phalcon AEW system provides airborne early warning, tactical surveillance of airborne and surface targets, and the gathering of signals intelligence.

The Phalcon four sensors are: radar, IFF, ESM/ELINT and CSM/COMINT. A unique fusion technique continuously cross-correlates data generated by all sensors and this data is combined with an automatically initiated active search by one sensor for specific targets detected by other sensors.

The Phalcon radar concept replaces the conventional rotodome radar. The radar has several panels of phased radiating elements. They are mounted on the fuselage of the aircraft and can provide up to full 360° coverage. Radar beams can be pointed in any direction in space at any time, with the beam's parameters fully controlled by the radar computer. The radar employs a flexible time-space energy management technique that makes the most of the following capabilities:

(a) surveillance can be limited to the battle zone and other areas of interest. The scan rate in these selected ground stabilised areas is thus much higher than for search

(b) a special mode for manoeuvring and high-value targets employs a high scan rate and beam shapes optimised for each target to ensure tracking performance

(c) verification beams sent at specific, individual, newly detected targets eliminate false alarms. Moreover, track initiation is achieved in 2 to 4 seconds as compared to 20 to 40 seconds with a rotodome radar

(d) by transmitting extra long dwells in selected sectors, an extended detection range is achieved

(e) the system uses distributed, solid-state transmitting and receiving elements. Each element is weighted in phase and amplitude.

The Phalcon IFF system, employing solid-state phased-array technology, implements interrogation, decoding, target detection and tracking using the standard modes. Azimuth measurement is carried out by monopulse processing. The IFF antennas are incorporated in the primary radar array and are co-ordinated with the array to avoid mutual interference. Similar antenna elements and transmit/receive modules are used for both the radar and the IFF.

The Phalcon's ESM/ELINT system is fully integrated with the radar and other sensors, serving as one of the most critical elements of the identification process. It is designed to operate in the densest signal environment, providing simultaneous coverage in all directions. The system uses narrowband superheterodyne receivers and wideband Instantaneous Frequency Measurement (IFM) techniques to provide very high accuracy and probability of intercept of airborne and surface emitters. Very high bearing accuracy for all received signals is achieved through Differential Time Of Arrival (DTOA) measurements. The system also collects and analyses ELINT data.

The Phalcon's CSM/COMINT receives in UHF, VHF and HF Selected radio nets can be monitored for signal activity. A DF capability locates targets. Detected signals can be assigned to monitoring receivers instantaneously. The system makes extensive use of computers to reduce the load on operators.

Features of the computer system include:

(a) loosely coupled distributed processing system composed of multiple computers connected by three separate databusses

(b) each computer consists of several processors, plus a built-in back-up processor

(c) back-up computer can take over the tasks of any computer if a problem cannot be resolved by built-in back-up processor

(d) each sensor system has its own processing facilities

(e) programmable signal processors used in all sensors

(f) radar signal processor features processing power of 1,200 Mflop/s.

Phalcon systems can be installed on a variety of platforms, such as Boeing 707, 747, 767, Airbus and C-130.

A typical configuration includes up to 11 operator consoles. Each operator position has two high-resolution, full-colour graphic displays with keyboard and control facilities and a communications panel. A separate section of the cabin can be equipped as an airborne command post, with a rear projection, large-screen display for presentation of the tactical situation to the entire command staff. Data may also be easily communicated to other users, as the aircraft's communications include long-distance voice and datalinked to other aircraft, naval and ground units, as well as voice and data relay facilities.

For further details see *Jane's Electronic Mission Aircraft*.

Status

One Phalcon-equipped 707 has been procured by Chile, where it is known as the Condor. It is assigned to Grupo de Aviación (Aviation Group)10 based at Pudahuel/Arturo Benitez International Airport, Santiago.

Contractor

ELTA Electronic Industries Ltd (a subsidiary of Israel Aircraft Industries Ltd), Ashdod.

UPDATED

A Phalcon-equipped Boeing 707

SES-210E naval ESM/ELINT system

The SES-210E is the naval version of the AES-210E airborne ESM/ELINT system. The system performs detection and analysis of all types of microwave signals by the use of modern DF antenna arrays, combined IFM and superheterodyne receivers, multiprocessors and so forth.

The compact structure of the SES-210E makes it suitable for installation on small vessels where space and weight are of prominent concern.

Contractor

Elisra Electronic Systems, Bene Beraq.

VERIFIED

STRATUS surveillance system

Rafael has developed STRATUS, a balloonborne COMINT system, that provides an alternative solution to signal collection requirements. It is based on field-proven techniques and subsystems and Rafael's long-term experience with balloonborne systems.

STRATUS consists of an aerostat equipped with a special purpose payload designed to customer specifications, a tether and a ground support system including mooring equipment, computers and operators' consoles. Once the balloon reaches the designated altitude, data are collected and passed to the ground station where they are processed and stored in an information bank. The system can also be used as an over-the-horizon relay. The maximum range is claimed to be 270 km, and the maximum endurance is 30 days.

Status

Reported to be in service.

Contractor

Rafael, Haifa.

VERIFIED

STRATUS aerostat and ground station

Timnex II ELINT/ESM system

TIMNEX ES/ELINT systems are described as being designed for the detection, location, identification and analysis of radar emitters. Such equipments are claimed to offer short response times and a 'high', omnidirectional, probability of detection in 'dense' environments. The 2 to 18 GHz band TIMNEX II subsurface configuration is described as being a fully integrated or stand-alone ES/ELINT system that comprises ES and threat warning antenna arrays, a mast interface processor, a radio frequency test unit, a channelised receiver, a Direction-Finding (DF) processor, a pulse processor, a parallel interleaver digital signal processor unit, a host computer and an interface to the host boat's combat system data bus. As such, the architecture is described as providing:

(a) automatic functionality and threat analysis

(b) 100 per cent probability of intercept

(c) 'short' response times

(d) instantaneous DF and frequency measurement (IDF/IFM)

(e) 'high' angular accuracy for tactical applications

(f) 'high sensitivity' IDF and IFM channels

(g) effective signals analysis and identification in dense environments

(h) 'very' accurate parameter measurement in the ELINT role

(i) automatic ELINT signal identification (by means of library correlation)

(j) operator or computer control in the ELINT role

(k) video and digital data recording

(l) raw or processed data logging for offline analysis.

The TIMNEX II's ES antenna array on a dedicated mast includes the following modules: Omni antenna for the 2-18 GHz band; DF antenna array for the 2-8 GHz band; DF antenna array for the 8-18 Ghz band. All are designed to withstand hydrostatic pressures in excess of 50 bars. Other system features include multiple digital signal processors within the architecture's parallel interleaver processing unit to ensure analysis of the entire frequency spectrum when operating in dense electro-magnetic environments. The system's DF processor can handle both phase and amplitude comparison data, while its channelised receiver incorporates a multiple narrow frequency band blocking capability and the ability to handle co-pulse and low probability of intercept (option) emitters.

Operation is via a two-screen console controlled with full keyboard and trackball. One screen is the Activity Monitor, for the display of a tactical picture of the detected source. The other is the Work Monitor, an alphanumeric data monitor for the display of a picture summary.

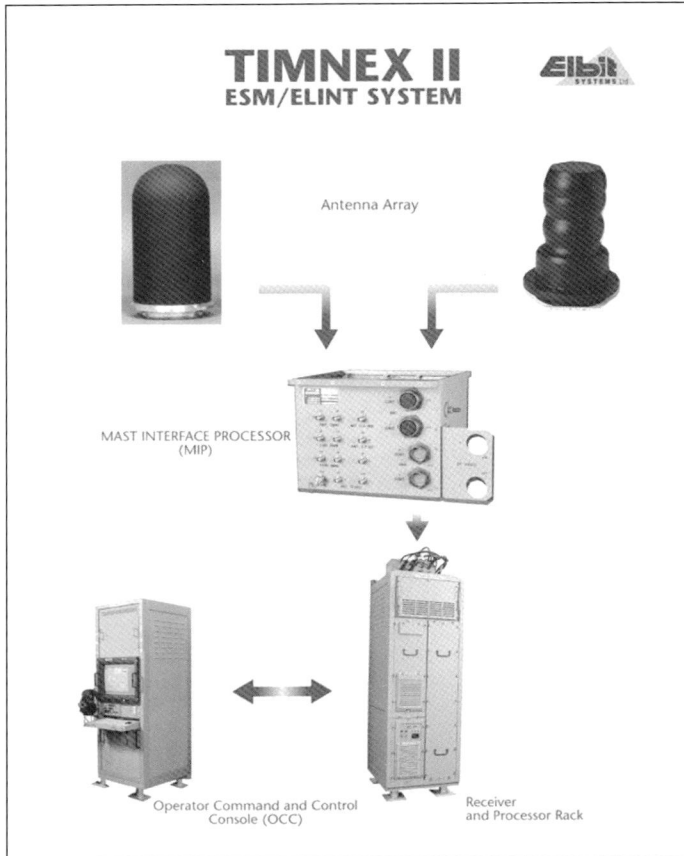

The components of the Timnex II System (Elbit Systems)   0130753

A close-up of the sail of the Israeli 'Dolphin' class submarine Leviathan showing the dedicated ES mast for its TIMNEX II ES/ELINT system at its forward end (right hand side of the picture) (IDR/Michael Nitz)   **NEW**/0049511

Specifications
(a) Frequency coverage: 2-18 GHz (0.5-40 GHz option)
(b) Frequency accuracy: 1.5 MHz (over complete frequency range)
(c) Sensitivity: –68 to –70 dBm
(d) Bearing accuracy: better than 2° (phase comparison antenna array)
(e) Real-time processing capacity: up to 256 emitters
(f) Hydrostatic pressure: in excess of 50 bars

Status
Installed on 'Dolphin' class submarines of the Israel Navy,' Katsonis' class submarines of the Hellenic Navy and 'Hai Lung' submarines of the Taiwanese Navy.

Contractor
Elbit Systems Ltd, Haifa.

VERIFIED

TSR-2020 HF/VHF/UHF Surveillance Receiver

The TSR-2020 is a high-performance computer controlled DSP (Digital Signal Processor) receiver suitable for surveillance and monitoring of communication signals, including signal classification and measurement capabilities. The TSR-2020 covers the 100 kHz to 3,000 MHz frequency range in one PC board. It is internally subdivided into several receivers, each covering part of the large frequency range and has 18 software-programmable, digital IF filters from 100 Hz to 340 kHz. The final IF filtering and demodulation are accomplished by digital signal processing.

The system has up to three RF inputs and two analogue audio outputs and signal samples, I/Q and FFT results can be transferred to the host PC. It has built-in signal measurement and classification capabilities.

With a suitable control interface, the TSR-2020 is part of the TDF-2020 DF system (see separate entry).

Specifications
Frequency range:
    HF: 100 kHz to 30 MHz
    V/UHF: 20 to 3,000 MHz
Demodulation: AM, FM, SSB, CW (PM optional)
IF bandwidth:
    Analogue IF bandwidth: 340 kHz
    Digital IF bandwidth: 18 digital IF filters (0.1 to 340 kHz)
Sensitivity:
    AM (1 kHz, 50%): 1µV for SNR of 10 dB @ 6 kHz BW
    FM (1 kHz, 5 kHz deviation): 1µV for SNR of 17 dB @ 15 kHz BW
    CW: 0.3µV for SNR of 10 dB @ 500 kHz BW
Noise figure:
    HF: 14 dB
    V/UHF: 12 dB (20-1,200 MHz); 13 dB (1,200-3,000 MHz)
Spectrum scanning rate:
    HF: 1,000 channels/s
    V/UHF: 3,000 channels/s

Contractor
Tadiran Electronic Systems Ltd, Holon.

UPDATED

Tadiran TSR-2020 wideband synthesised receiver   0084492

Italy

ELT/888 data collection system

The ELT/888 is an electronic data collection system produced in land-based, naval and airborne versions that feature a range of antenna configurations and platform interfaces. The ELT/888 performs:
(a) intercept of radar and ECM emissions
(b) simultaneous analysis of multiple emissions
(c) real-time emitter classification and identification
(d) emitter direction-finding.
(e) digital recording and printing of the measurement and analysis results.

ELT/888 ELINT system operator console

All functions of the ELT/888 are fully automatic, thereby reducing the operator's workload and enabling him to concentrate on the evaluation of the overall scenario. The provision of high sensitivity receivers ensures high intercept probability, while measurement accuracy is ensured by statistical processing of the measured data.

Data collected by the ELT/888 can be transferred to an Electronic Warfare Analysis Centre (EWAC), also produced by Elettronica, for further evaluation and for integration with intelligence originating from other sources, so as to generate a database of the overall situation.

Status
Fully developed. Further variants may have been developed. As at late 2003 no new information was available.

Contractor
Elettronica SpA, Rome.

UPDATED

....

RQN-5C naval ESM/ELINT system

The RQN-5C is the latest identified variant of Elettronica's RQN-5 shipboard ES/ELINT architecture. System features include:
(a) A compact antenna group that, on ships of a suitable size, can incorporate an optional fine direction-finding antenna (E-mode unit) for 'very' high accuracy direction-finding and 'superior' sensitivity within steerable azimuth sectors

ESM system multifunction console (Elettronica SpA)
0137499

(b) Instantaneous and omnidirectional coverage of the C- through J-band (5 to 20 GHz) frequency range with an upward extension to K-band (20 to 40 GHz) as an available option
(c) 'High' sensitivity receivers for both signal analysis and direction-finding
(d) A 'high-accuracy', full band, Instantaneous Frequency Measurement (IFM) facility
(e) An optional IFM receiver module to enhance frequency measurement accuracy
(f) Fully automated ESM functionality (including automatic real-time extraction and analysis and tracking of all intercepted known and unknown emitters)
(g) Claimed near 100 per cent probability of extraction on a single scan
(h) Automatic warning and identification (by programmable library comparison) of known threats and emitters
(i) Automatic warning of unknown lock-ons and suspected continuous wave threats
(j) Computer aided ELINT-type analysis (including all types of MOP) of individual emitters (with data recording)
(k) An automatic 'fingerprinting' facility on previously analysed emitters
(l) The ability to drive active and passive countermeasures systems automatically
(m) Standard interfaces to facilitate integration with a wide range of peripherals
(n) Configuration modularity to aid installation aboard ships ranging in size from offshore patrol vessels to frigates, and system growth.

Status
Variants of the RQN-5 systems are claimed to be in service with several navies. As at late 2003 no new information was available.

Contractor
Elettronica SpA, Rome.

UPDATED

_____

Netherlands

Sphinx naval ESM

Sphinx (system for passive handling of intercepted transmissions) is a shipborne ESM system for the interception, analysis and identification of all forms of pulsed radar emissions. It is entirely passive and operates autonomously. The system features: a high probability of interception by an instantaneous field of view of 360° and a continuous coverage of the entire frequency range, high sensitivity, high pulse density capability. It is capable of the following functions:
(a) omnidirectional interception of signals in the range 1 to 18 GHz and determination of their frequency and bearing
(b) lock-on detection, that is, automatic detection and warning of any radar locked on to 'own ship'
(c) automatic analysis of selected signals in terms of level, pulse, duration, frequency (maximum, minimum), PRF (maximum, minimum), scan period
(d) automatic tracking in bearing of selected signals
(e) blanking of selected signals tracked in bearing
(f) area blanking
(g) identification of signals by comparison with library
(h) data exchange with ECM and data handling system
(i) data logging on cassette of information acquired.

The basic configuration of Sphinx consists of a frequency receiver and an eight-sector bearing receiver. Both are untuned and wide open in all directions so providing simultaneous coverage of 360° in azimuth and 40° in elevation (nominal). Interception probability approaches 100 per cent. Signals of linear as well as circular polarisation can be intercepted.

The operator's console has three displays:
(a) a situation display for presentation of bearing/frequency plots of detected signals and tell-back data from the ECM system
(b) an alphanumeric display for analyser results
(c) an alphanumeric display for identification results.

Sphinx has provisions for interconnection with a data handling system and an ECM system such as RAMSES. One operator can then take care of both ESM and ECM functions on board the ship. Sphinx has built-in test facilities for checking both before and during operation. The bearing array is designed to form part of the load-bearing structure of a mast. A WM20 series combined antenna system can be easily mounted on top.

Optional enhancements include: a K-band reception package (33 to 40 GHz) with alarm; a threat alarm, sets of parameters and a CW detection package.

Status
No longer in production. Incorporated into the SEWACO VI combat system (see separate entry).

Contractor
Thales Nederland, Hengelo.

UPDATED

Russian Federation

Interception and location systems

A number of intercept and Direction-Finder (DF) systems are reported to have been fielded with the forces of the Russian Federation. Many were in service with the Soviet forces; inclusion here does not signify current operational use. Included are:

The SR-53-V (System A) intercept system covering the 3 to 30 MHz range in AM, CW, MCW and voice modes. Receiver sensitivity is −105 dBm and gain is 15 dBm. A rhombic antenna is used.

The SR-52-V (System B) and SR-51-V (System B-1) are VHF/UHF intercept systems. Operating in FM and voice modes over 30 to 300 MHz, they have a receiver sensitivity of −110 dBm. The SR-52-V, which uses a log-periodic antenna, has a gain of 10 dBm. Corresponding value for the SR-51-V is 1 dBm. The SR-51-V has a whip antenna.

The SR-50-M (System C) is a 30 to 450 MHz, FM and voice intercept system using a whip antenna, with a receiver sensitivity of −110 dBm and a gain of 1 dBm.

The SR-54-V (System D) is a relay intercept equipment operating in voice or TTY modes in the 30 to 300 MHz range. It uses a dish or log-periodic antenna and has receiver sensitivity and gain of −110 dBm and 40 or 15 dBm respectively.

The SR-20-V (System 1) is an AM DF covering 3 to 25 MHz and using an Adcock antenna. Receiver sensitivity is −90 dBm and gain is 10 dBm.

The SR-19-V (System 2) and SR-25-V (System 3) are FM voice DFs covering 30 to 300 MHz with similar performance characteristics to the SR-20-V.

SP2 is a radio DF used by special forces.

Twin Box is a VHF DF system used at division and army level.

Fix-6 and Fix-8 are VHF DF systems.

UPDATED

··

VEGA 85V6-A ELINT system

The VEGA 85V6-A ELINT system is designed to operate within electronic warfare, air defence and other army units. The system can be used within early warning and air traffic control systems and to identify and locate jamming sources as well as an ESM asset. The system is capable of detecting, identifying and tracking up to 100 ground, naval surface and air targets.

A typical 85V6-A system would consist of three 85V6-A ORION detection, location and identification stations (see below) and an 85V6-A control post (CP). Typically, the ORION stations are located up to 30 km from each other with the control post being near one of them. DF and signal parameter data from the ORION stations are transmitted through the data-link channels to the CP, where target positions and tracks are determined and displayed on an electronic map of the area of interest. There is provision for the recording of tracks and for signal monitoring.

85V6-A ORION ELINT station

The ORION station is designed for operation both within the 85V6-A radio-reconnaissance system and independently as a part of electronic warfare and air defence units and divisions. It detects, locates, identifies and classifies ground, naval surface and air targets. It has a high operating speed, achieved through using

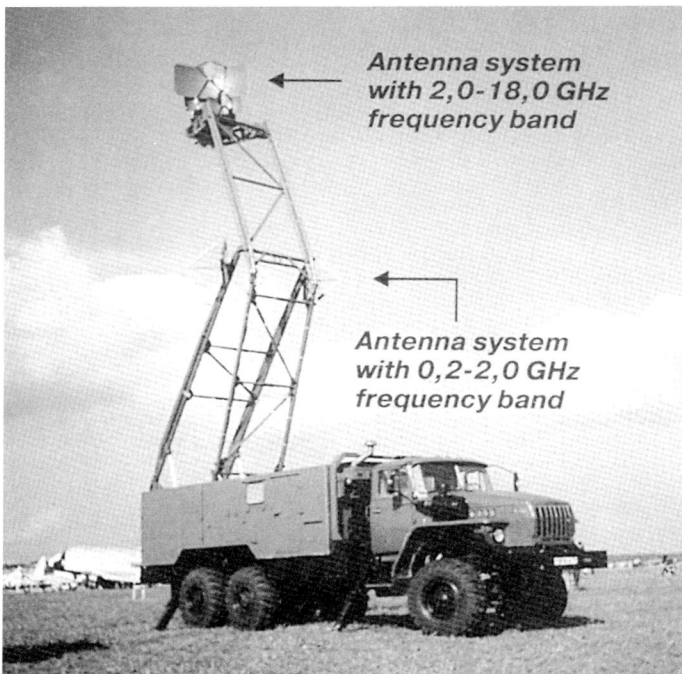

The 85V6-A ORION ELINT station with raised mast, showing the two antenna systems (FSUE)   0143779

single-pulse direction finding and a wide-band acoustic-electronic (compression) Fourier transform processor in a signal processing channel. It is also effective against frequency-hopping equipment. Identification of detected signals is achieved through comparison with the database. Detection, classification and line-of-bearing is achieved and passed to the CP and other users within 6-10 seconds. Manual direction and subsequent automatic tracking is also possible.

Specifications
(a) Operating frequency band 0.2-18 GHz (optional extension to 40 GHz)
(b) Snap detection band 500 MHz
(c) Frequency resolution 1 MHz
(d) Pulse duration measuring accuracy 0.1 µs
(e) Pulse repetition period measuring accuracy 1.0 µs
(f) Azimuth measuring accuracy:
    0.2-2.0 GHz 1-2°
    2-18 GHz 0.2°
(g) Maximum scanning speed 180°/s.

Status
In production.

Contractor
Spetz-Radio RPE CJSC , Belgorod, Russia.

UPDATED

────────────────────────────────────

Spain

SCAPA SIGINT system

The Spanish Ministry of Defence established a requirement for an advanced and computer-controlled SIGINT system which would cover both the communication and radar frequency bands. This system is required to improve the capability of the Spanish defence forces to collect data over the electromagnetic spectrum in zones of interest.

In 1989, the Boeing 707 was selected as the most appropriate airborne platform for the programme and a suitable airframe was purchased in February 1991 from Commodore Aviation, a subsidiary of Israel Aircraft Industries (IAI). The resulting system is known as SCAPA and development was authorised by the Spanish government in December 1991 to the tune of Pta9.9 billion (US$70 million) which makes it probably the largest single Spanish electronic warfare programme. The contractor is Indra, which is responsible for the system design and integration, operational software and data fusion development. Israel's Elta Electronics has supplied the communication and radar sensors (it is understood that the sensors are an upgraded version of those of the EL/L-8300 SIGINT suite - see separate entry) and Taman supplied the onboard long-range stabilised electro-optical observation system integrated by Indra with the other sensors.

The system's main features are:
(a) proven sensors
(b) a friendly man/machine interface. Each operator console has trackball and dedicated soft keys
(c) powerful and ruggedised SPARC workstations in an Ethernet local area network operating under UNIX
(d) extensive use of commercially available software
(e) high capability to record, reproduce and analyse audio and video signals, both in aircraft and ground operational support centres.

Status
SCAPA development and flight testing is reported to have been completed. Understood to be operational. See *Jane's Electronic Mission Aircraft* for further details.

Contractor
Indra EWS, Madrid.

VERIFIED

··

SOCCAM COMINT system

Indra has developed the modular, platform adaptive, communication observation and control system known as SOCCAM on the basis of growth capabilities, commencing with an initial configuration and ending with an advanced version incorporating a multistation. The system is configured for land-based, shipborne and airborne applications. Basically, SOCCAM is a COMINT system for tactical and strategic missions, spectrum control, illegal transmissions control and maintaining territorial integrity.

SOCCAM provides functions for scanning, searching and detection of active transmissions over the 20 to 500 MHz frequency band. The system detects the activity in a series of discrete bands considered to be containing threats within the context of the mission. Analysis of the signal is carried out by the operator who can examine the reception parameters to determine the transmission characteristics

SOCCAM console

and record them for future detailed analysis. A direction-finding capability to establish the bearing and position of a transmission is available but is not included in the initial configuration. There are two different operational modes available: operation in an unknown scenario where the system will search for transmissions of interest in the mission area and known scenario operation where the operator can monitor and locate potential threats.

Specifications
(a) Frequency range: 20-500 MHz
(b) Operation: continuous and discrete scan
(c) Sensitivity: –100 dBm
(d) Modes: AM, FM, CW, PLS
(e) DF accuracy: 4° RMS (instrumental)
(f) Interfaces: RS-422; IEEE-488

Status
SOCCAM is reported to be operational in an airborne application.

Manufacturer
Indra EWS, Madrid.

VERIFIED

Ukraine

Kolchuga

The 'updated' Kolchuga is a passive detection system primarily designed for use in Air Defence. It is intended to:
(a) Detect the take-off and formation of aircraft groups at ranges beyond those of current radar
(b) Determine the course and speed of targets and designate them for Air Defence systems
(c) Identify targets through their emissions
(d) Identify the mode of aircraft weapons control systems.

The station detects, analyses and identifies a wide range of signals in the 0.1 - 18 GHz band. The manufacturer claims that 'practically all currently known' emitters can be identified. Mounted on a 6-wheeled truck, the station consists of four antenna assemblies in the VHF, UHF and SHF wave bands, with both narrow beam (long range) and wide beam (close range) monitoring capability, together with a parallel receiver for analysis, identification and processing equipment, including DF in conjunction with other stations; display and recording equipment; communications equipment including data transmission; power supply. The environmental specifications are claimed to be sufficient to enable the station to operate in conditions ranging from –50°C to +50°C.

The Kolchuga station would normally operate in a complex of three together with a command vehicle, to provide accurate triangulation for location. In this configuration, with the individual vehicles 60 km apart, it is claimed that the system can detect emitters over a front of 1,000 km at up to 600 km range (narrow beam) or up to 200 km (wide beam). The wide dispersal of vehicles aids the elimination of screened areas.

Status
It is believed a total of 76 stations have been produced. In April 2002 *Jane's Intelligence Digest* reported that Western agencies believed a complete system (that is 3 sensor vehicles and a command vehicle) might have been sold to Iraq, although this has not been subsequently confirmed despite much intense and high-level investigation. The system is apparently made by the Joint Stock Holding Company Topaz in Donetsk, but is marketed by FTF "Progress", Kiev.

Contractor
Specialised Foreign Trade Firm "Progress", Kiev.

UPDATED

The Kolchuga station sensor vehicle with raised antenna array (Jane's/IDR) 0083266

The Kolchuga sensor vehicle operator's position (UKRSPETSEXPORT)   0143793

United Kingdom

Barracuda ESM/DF System

Barracuda is a versatile, commercial-off-the-shelf (COTS) tactical communications electronic warfare system combining intercept and direction-finding in one small vehicle.

Description

The system comprises two RA3725 dual receivers, a personal computer and Thales commutated Adcock DF antennas (Type AE3007 HF and AE3020 VHF/UHF). The system covers the frequency range 20 to 1,000 MHz. The HF option extends the frequency range down to 1.6 MHz.

Barracuda uses digital signal processing techniques and provides a number of operator controlled functions so that it can be used against a wide variety of signals. The PC can be deployed remotely from the receiver and antennas. The RDF3725 can be used as part of an automated position fixing system showing target locations on a map display.

The RA3725 receiver can be mounted in a standard 483 mm (19 in) rack and, together with a notebook or desk-mounted PC, is suitable for installation in wheeled vehicles, shelters or fixed sites. It can be either AC or DC powered.

Status

Eight systems were supplied to the British Army in the mid-1990s.

Contractor

Thales Communications, Bracknell, Berkshire.

UPDATED

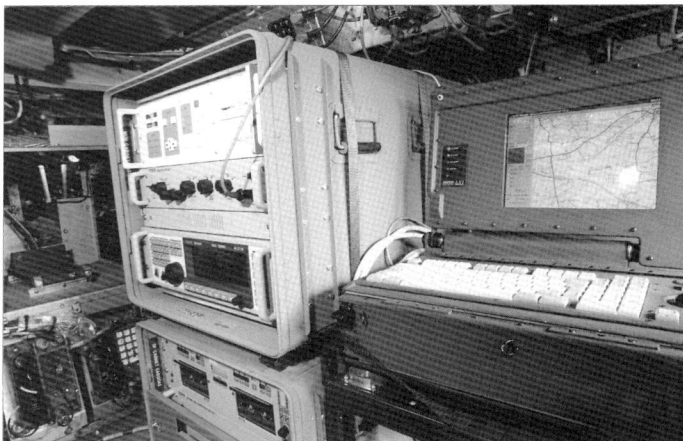

Barracuda ESM/DF 0001028

CORVUS EW products

CORVUS is a family of channelised passive microwave sensor equipment produced by Thales DIS, designed for application to both Strategic and Tactical ELINT signal collection in the radar frequency bands. The product range includes microwave front-end tuners, channelised receivers, pulse analysers, data recording and data analysis software. The standard product covers the frequency range from 0.5 GHz to 18 GHz with options available to extend coverage to 40 GHz, with specialised extensions for coverage beyond. The systems incorporate advanced signal measurement and analysis facilities, including the acquisition of detailed intrapulse characterisation for emitter fingerprinting purposes.

The system is expandable to handle multichannel (antenna) signal inputs, such as would be necessary for DF (direction finding) or signal polarisation measurement, and has been used as the basis for a number of EW sensor systems. For example, CORVUS III and IV configurations are designed for tactical ELINT collection purposes and incorporate advanced automated pulse train deinterleaving and analysis.

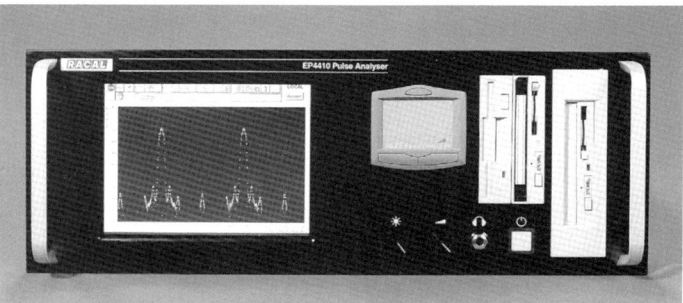

EP4410 pulse analyser *0001025*

Channelised Receivers

The company manufactures a range of channelised receivers for tactical and strategic ELINT collection applications. The EP4220B Receiver offers channelised ELINT intercept capability over a frequency range of 0.5 to 18 GHz and has an instantaneous frequency acquisition bandwidth of 1 GHz, whilst still producing ELINT quality parametric data. Its wide bandwidth (1 GHz compared with the much narrower bandwidths of swept superheterodyne receivers) enables full examination of frequency-agile transmitters. Within each channel, the detection bandwidth is sufficiently narrow to give measurements comparable with narrowband ELINT receivers.The EP4220B has been designed for use in association with the EP4410 Pulse Analyser Unit. Operation to 40 GHz is offered in conjunction with the Corvus IV system configuration.

Pulse Analysis

The EP4410 Pulse Analyser is a new generation signal acquisition and analysis product and is designed to interface with conventional superheterodyne receivers as well as with the EP4220 Channelised Receiver to form the heart of an advanced ELINT system. It is capable of producing high-quality pulse descriptors simultaneously with intrapulse signal parameter detail. The EP4410 incorporates rapid access digital data recording, a colour graphics display, and can process and store up to 64K sequential pulses without the need to transfer to recording media. Modulated IF inputs from receivers at 160 MHz, 1 GHz or special purpose IF frequencies can be accepted via optional interfaces. Eight qualification windows, each set in frequency, amplitude and pulse width, allow selected signals to be extracted from dense scenarios. Data reduction facilities allow high PRI emitters to be captured without excessive use of memory. Signals can be analysed using the built-in facilities of the EP4410 or recorded on recordable media for off-line analysis. Whilst the EP4410 Pulse Analyser offers a pulse-by-pulse intrapulse analysis facility which is fully integrated with measurement of conventional parameters, additional ELINT capability exists in the form of direct digitally sampled systems. These are being incorporated in most of Thales's current advanced strategic ELINT systems in order to provide extended fine grain analysis capabilities.

Features

(a) Analyses and displays up to 64K pulses
(b) Measures: amplitude, frequency, pulse width, PRI, bearing, end frequency, chirp
(c) Tabular display of all parameters on a pulse-by-pulse basis
(d) Colour analysis displays
(e) Intrapulse measurements up to 200 MHz sampling rate
(f) Correlated intrapulse and pulse descriptor measurements
(g) Data reduction modes
(h) Integral high-resolution colour display and simple to operate controls
(i) Continuous recording of pulse descriptors
(j) Remote operation available via Ethernet interface
(k) Modular construction.

Options

(a) 160 MHz input interface
(b) 1 GHz input interface
(c) Integral digital data recorder
(d) Bearing interface
(e) Additional video channels
(f) IEEE 488 remote control (Ethernet is standard)
(g) Precision MOP demodulator (restricted availability)
(h) Digital IF Receiver Interface
(i) Additional data analysis software
(j) Intrapulse Analysis.

CORDAS

Corvus Data Analysis Software (CORDAS), is an ELINT analysis software tool-set, designed to be used by an analyst to process signals that have been collected by the CORVUS ELINT system. Intended to be used in a large screen, multi-window environment, CORDAS allows the analyst to view simultaneously the recorded data collected by the ELINT system in both graphical and tabular formats to ease detailed pulse-by-pulse analysis.

Status

The original CORVUS project is believed to have been developed for an unspecified Gulf state. CORVUS ELINT systems are claimed to have been delivered into naval, ground-based and airborne applications in other unspecified countries. As at late 2003 this was believed still to be a live product.

Manufacturer

Thales Sensors, Crawley.

UPDATED

Cutlass ESM equipment

The Cutlass series of equipment consists of a range of advanced computer-controlled ESM systems, primarily intended for use aboard ships, but capable of deployment on land platforms. Designed for operation in very dense signal environments the equipment receives signals in the 0.6 to 18 GHz frequency range, measures their parameters, compares these with those in a preprogrammed radar

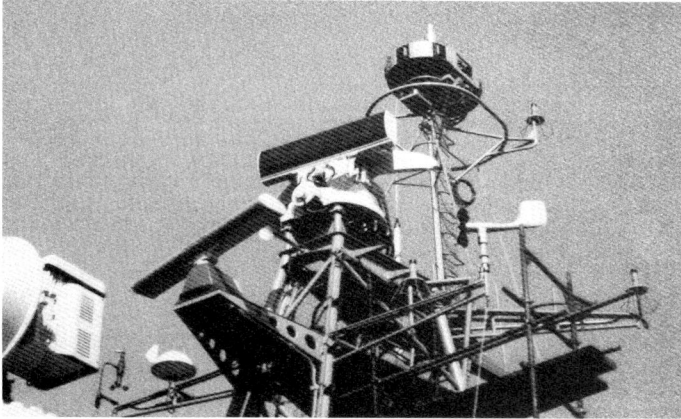

Cutlass antenna system mounted at masthead

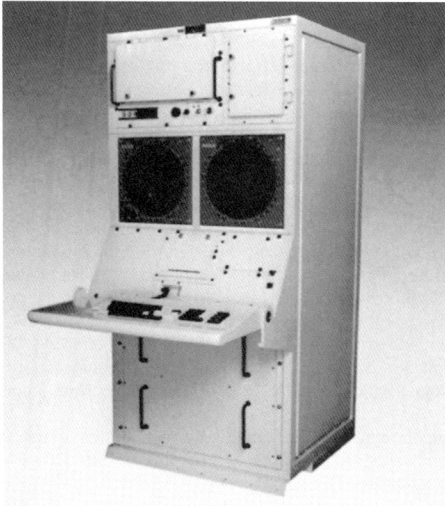

Cutlass console

library and displays the information within 1 second. The EW operator is presented with a tabular display for threat identity and threat evaluation and a tactical display giving a pictorial representation of the RF environment. Selected digital outputs can be sent to other local systems and hard-copy printout of the intercepted radar is also available. The tabular display can indicate 200 intercepts, in the order of priority. The Cutlass central processor is very advanced with a library containing the parameters of up to 2,000 radars.

Cutlass is wide open in both bearing and frequency (that is it does not employ sweep techniques) giving a very high intercept probability (nearly 100 per cent).

Cutlass is modular and can be configured to suit fittings in various classes of ships. The two main variants are Cutlass E and Cutlass B1; both use the advanced Cutlass processor which can be integrated with a variety of man/machine interfaces and both use an Instantaneous Frequency Measurement (IFM) receiver. Cutlass E has a six-port antenna array for bearing measurement using amplitude comparison techniques and a separate omnidirectional antenna to provide RF for the IFM. Cutlass B1 has a 32-element antenna array to provide bearing measurement by phase analysis techniques. This antenna also provides RF for the IFM. In both systems the processor is provided with fast and accurate information on incoming pulses.

For an integrated EW system, Cutlass can be employed in conjunction with the Cygnus or Scorpion jammers. Cutlass will readily integrate with other ships' systems, such as a tactical data system.

Specifications

Cutlass B1
Frequency range: 2-18 GHz
Sensitivity: –60 dBmi
Dynamic range: 60 dB
Accuracy: 2° RMS in bearing
Coverage: 360° azimuth; +40 to –10° elevation
Pulse density capability: 500,000/s
Radar store capability: 2,000 modes
Processing time: <1 s

Status

In production since 1979 and in service with several overseas navies. 22 Cutlass B1 ESM systems and Scorpion jammers have been supplied for the German Navy's S148 and S143B patrol boats.

Variants of Cutlass have been developed: Sabre for surface ships and Sealion for submarines. Sabre is fitted to the Royal Danish Navy's Stan Flex 300 multirole vessels. Sealion is in service in the navies of Denmark, Norway and Turkey.

Contractor

Thales Sensors, Crawley, West Sussex.

UPDATED

INCE

INCE (Interim Non-Comms ESM) has been developed as an interim measure until the UK Soothsayer project enters service, probably in 2006. It is replacing the current Pinemarten and Beady-Eye equipments. It will provide long-range, all weather passive monitoring of radar activity, giving early warning, identification and location, and contributing to EW databases. The system consists of three sensor stations and a control station and will provide a very high probability of intercept over the 0.5 GHz to 18 GHz frequency range. Mounted in Pinzgauer vehicles, it is C130-transportable, and typically can be brought into and out of action in less than 10 minutes.

Much of the system is based on the Thales naval Sealion system, including the rotating dish antenna, the Receiver Digitizer Unit and the Pulse Train Analyser Unit.. The system has a rapid reaction time and a bearing accuracy of between 1° and 2°. It provides fully automatic ESM capability, and automatic and manual ELINT analysis. It has a high gain DF antenna, and an omni antenna can be provided as an option, as can a dedicated ELINT receiver for intrapulse and fine grain analysis. Budget constraints have prevented these options being fitted to the UK version. Each sensor can be operated by one man. The workstation provides multiple views of data, has a manual pulse capture feature which allows the parameter space to be designated by 'rubber-banding' and pulse data files to be saved to the hard disk, and has removable hard drives for the loading of libraries and the recording of intercept activity. The system also provides PC-based analysis tools for ESM and ELINT. Either emitters of interest or a particular space can be selected and data snapshots can be collected in real time or loaded from disk for analysis. The system can operate in stand-alone mode, as master-slave or by remote control. Separate networks for voice and data communications are provided using the Thales Panther V radio.

Marketed as Meerkat-S. Further details can be found in *Jane's Radar and Electronic Warfare.*

Status

In October 2001 it was announced that Thales had been awarded a £6 million contract for the supply of INCE for the UK's 14 Signals Regiment. Believed to have entered service in 2002, although this has not been confirmed.

Manufacturer

Thales Sensors, Crawley, West Sussex.

UPDATED

INCE ESM System Sensor Vehicle (Thales)   0096065

Kestrel airborne ESM

The Kestrel airborne ESM system provides ELectronic INTelligence (ELINT) during all phases from peace to active wartime operations. The system enables staff to gather information on possible hostile threats, their deployment and movements, at long range without the hazards associated with other methods of reconnaissance.

Description

Kestrel receives and processes radar emissions over C- to J-bands. A multiport, amplitude comparison bearing measurement system is used providing instantaneous digital bearing over 360° in azimuth. At the same time a frequency measurement receiver provides instantaneous digital frequency. The pulse-by-pulse digitised information is then passed to the processor which de-interleaves the overlapping pulse trains from the different radars, deriving their pulse repetition, frequency agility, scan type and identification. The most likely identification of the radar is determined by comparison of the measured and derived parameters with those stored in the library of known emitters.

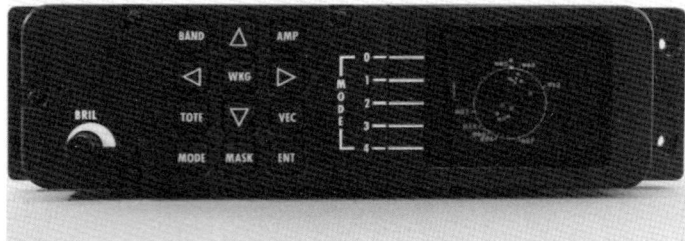

Kestrel cockpit display and control unit

The Kestrel system is based on the Thales Sadie advanced signal processor and uses six RF heads that occupy the same space as the MIR-2. The threat library can contain details of a minimum of 2,000 emitters.

Full information about the radar signal environment is presented to the operator on an ordered tabular or tactical display within 1 second of intercept. Alternatively, the information can be displayed on a remote display via a standard data highway to allow the use of common avionic displays. For the installation in the Royal Navy's Merlin helicopters, the outputs of Kestrel will be shown on the display screens of the central command system. The measured information includes bearing, frequency, frequency agility, PRI, PRI agility, pulse-width, scan type and scan period.

Specifications
Frequency: 0.6-18 GHz (optionally to 40 GHz)
Bearing accuracy: ±3.5° RMS
Frequency accuracy: 5 MHz RMS
Azimuth coverage: 360°
Elevation: 45°
Polarisation: All linear and one hand of circular
Processing time: 1 s max
Receiver types: Log video amplifier, DLVA, IFM and CW Superhet
Warning/identification: All pulse types including CW, ICW, Pulse Doppler, LPI 3-D, jitter/stagger/agile, pulse compression, frequency agile radars and unknown emitters
Emitter library storage: Storage accommodation of at least 2,000 modes
Display: High brightness colour CRT with computer controlled symbology programmable by software
Emitter displays: Up to 400 in pages of 20, each showing tote format parametric data and identity
Range bearing display: 25 highest priority emitters showing lethality, type and track identification
Expanded track information: Showing full parametric data and ranges for a single track
Additional features:
True or relative bearing display presentation;
audio alarm under processor control;
operator-initiated emitter handoff;
specialised computer modes under operator command;
ELINT data recording interface
Weight: 55 kg (Depending on display configuration)
Power consumption: 1 KVA
Interfaces: Serial and parallel interface for jammers, chaff/flare systems, displays, telemetry links, recorders.
Weight: 55 kg (depending on system configuration)

Status
In production. Supplied to the Royal Danish Navy for its fleet of Lynx helicopters. Under the designation Orange Reaper, has been supplied for the Royal Navy's Merlin (EH 101) helicopter fleet.

Contractor
Thales Sensors, Crawley, West Sussex.

UPDATED

Lightweight Emitter Acquisition, Recording and Analysis System (LEARAS)

LEARAS is a software-based Communications Intelligence (COMINT) gathering system, which is COTS equipment within a customised system. It provides wide band monitoring over a frequency range of 2 MHz to 2000 MHz, supported by digital recording and analysis tools. A PC/PDA control interface provides easy operation. Two versions are currently available, a portable system (LEARAS) and a fixed site system (FEARAS). The portable unit has a very low power requirement for sustained battery operations.

LEARAS is configured around a DF4400S, the Operator Control and DF unit (OCDFU). This unit provides the hardware interface between the operator, the receiver and the associated direction-finding system and other essential components. The system includes a full fast acquisition 'staring' spectrum display capability. This technique uses analogue to digital processors and, combined with DSP techniques, allows extremely short duration signals to be reliably identified.

The facility, called Wideband Acquisition and Spectrum Processor (WASP) in this configuration but essentially the Hoka Code300-32 (see separate entry), is provided within the standard LEARAS. The WASP system is able to identify the presence and frequency of signals with a duration of less than 2 ms, to a resolution of 12.5 kHz. It allows the operator to identify frequencies or Signals of Interest (SOI), and hand off the wanted frequency to the receiving system. Once handed off, the frequency and signal is available for further analysis. This includes direction-finding, and within LEARAS a wideband automatic direction finding system (DFS) is provided. In conjunction with portable and low profile antennas, the DFS can provide bearing information on SOI within a short duration, and these may be automatically logged for further analysis. The DFS employs non-commutated techniques, in which all antenna elements are active, to ensure maximum accuracy and sensitivity on short duration signals. DF antennas provide coverage from HF up to UHF. The SOI may be further handed off to an independent monitor receiver. This receiver allows for long term monitoring, thus releasing the WASP and DFS receiver for further acquisition tasks. Up to three receivers may be supplied. The receiver(s) audio signal can be routed directly to the associated PC system, allowing continuous recording of a frequency or SOI.

The main operator interface to LEARAS is via a rugged laptop computer, the System Display and Processor Unit (SDPU). At FOC this will be a Panasonic Toughbook with touchscreen display. The SDPU is equipped with a number of applications, providing the required system capability. It displays the output of the WASP system, including hopping signal detection, fading emitter indication and the ability to log and record activity to the associated hard disk. Visual Radio software provides the operator graphical interface to the DFS, the monitoring receiver, and a configurable emitter or SOI database. The database is of special importance, as it allows SOI activity to be recorded and attached to the technical parameters associated with a transmission. A sophisticated audio system is provided. This allows signals to be recorded for later transcription and analysis. Recordings are automatically time stamped and associated with the recorded frequency. Transcription tools, built into the standard MS Word word processor, allows replay of selected recordings without switching between applications. Decoders and multibit DSP audio systems within the SDPU allow manipulation, analysis and decoding of many of the in-service commercial and military data systems. The emitter's database is linked to a GIS from MapInfo. This allows DF bearings and associated locations to be directly mapped and displayed and provides mission planning tools, such as point to point visibility and propagation analysis.

A Panasonic personal digital assistant (PDA) provides the operator interface when on the move or at brief halts. System headphones are from Sonic and there is a COTS GPS. Also provided is a handheld DF which is stand-alone equipment and, while not accurate enough to conduct detailed triangulation, is intended for use in identifying the direction of signals at close range in, for example, built-up areas.

Included in the system is a "near-universal" modem from ViaSat which will operate over all expected modes of communication and enables automatic network establishment between stations.

Power is provided by a battery developed especially for the system, to meet a mission profile of 10 days without resupply. These are lithium thionyl-chloride batteries with an 80 A/h output which will provide 48 hours continuous operation and replace 5 Clansman batteries. Also provided is a blanket for solar power.

The intercept antenna is mounted on a mast made of a carbon-fibre tube which can be opened longitudinally and rolled up, thus dramatically reducing the mast carriage problem. It has standard Clark mast fittings at top and bottom.

Specifications
(a) Frequency range: 2-2,000 MHz, 3 antennas
(b) DF accuracy: 5° RMS, depending on frequency and deployment
(c) Resolution: 12.5 kHz
(d) WASP bandwidth: 10 MHz (15 MHz with reduced accuracy)
(e) Dynamic range: 80 dB
(f) Minimum signal duration: 2 ms/−110 dBuV
(g) Receive modes: AM, FM (wide), FM (narrow), SSB, CW
(h) Weight: less than 15 kg
(i) Temperature: −20 to +50°C
(j) Power: 10-32 V DC; <16 W PDA mode; <30 W PC mode

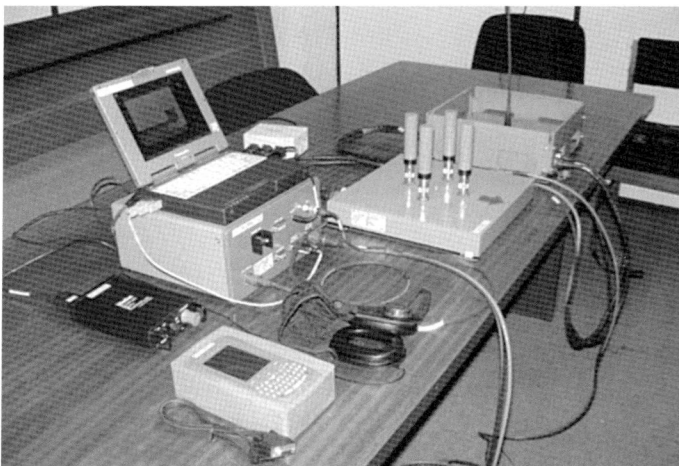

The LEARAS equipment supplied for Project SCARUS IOC, showing SPDA, PDU, modem and antennas (MOD)   **NEW**/0547246

The SCARUS handheld DF (MOD)   *NEW*/0547259

Status

In Jan 2003 it was announced that a £2.4 million contract had been awarded for the supply of twelve manpack communications Electronic Support Systems (ESM), to be known as Project SCARUS. The contract was to replace in-service manpack equipment. The SCARUS system will be carried by between four and six men and will be required to perform On The Move Intercept and Static Automated DF and Intercept. The equipment will provide an early entry capability to the UK's 3 Cdo Bde RM, 16 (AA) Bde and other operations where light forces require EW support. The users will be those elements of 14 Signal Regiment supporting 16 (AA) Bde, and the Radio Reconnaissance Teams from Y Troop of 3 Commando Brigade RM, where it will replace the ICE system (see separate entry).

The Interim Operational Capability (IOC) delivery date was mid-February 2003, with Full Operational Capability by July 2003. At IOC the various software packages had not been integrated, but this will be achieved for FOC, together with an additional signal-processing card. Also at FOC the SPDU will have a "docking station", which will reduce the number of cables attached to it.

The equipment was deployed to Iraq with Y Troop, where it was reported to have been used successfully.

Contractor

Falcon Protec, Tewkesbury, Gloucestershire.

UPDATED

MS336X range of microwave receiver analysers

The MS336X series comprises integrated swept superheterodyne microwave receivers, enabling time and frequency domain measurements for use in technical and tactical ELINT. These units are in operation worldwide and can include an integral pulse analyser capable of producing up to 50 kbits Pulse Descriptor Words (PDWs) detailing captured signals, with functions such as qualification and statistical data for immediate sorting. The PDWs are available via rear panel interfaces for further detailed analysis, and can easily be extended to include angle of arrival and GPS data.

Flexibility is enhanced by the use of modular construction allowing, for example, simple remoting of the display.

Features

(a) Swept superheterodyne receiver operating over the frequency range 0.5 to 18 GHz
(b) A large, flat-screen electro-luminescent display enables the operator to perform complex analysis of signals in real time
(c) Simple operator control provides fast access to a range of collection and analysis modes
(d) 10 receiver settings stored in the MS336X memory for instant recall
(e) Remote control via the IEEE 488 interface. This includes data extraction for external analysis
(f) Can be operated either as a stand-alone receiver, or integrated into larger acquisition systems
(g) Ancillary equipments include frequency extension modules, video recorders, video and IF switching matrices, antennas, analysis and emitter library facility
(h) Options include pulse analyser, integral IF-Tape converter, dual narrow and wide band demodulator, audio compression, ethernet interface and 28 V DC version.

Contractor

TMD Technologies Ltd (TMD), Hayes, Middlesex.

VERIFIED

Odette Intercept and DF System

Odette is a HF/VHF/UHF signals intercept and DF system. It incorporates COTS software and receivers and is designed for installation in a range of platform types. These include Land Rovers, armoured fighting vehicles and the tracked Häggland BV206 all-terrain vehicle. The manufacturer emphasises the system's ability to locate frequency-hopping emitters. It is equipped with a communications subsystem built around Thales Defence Communications' Panther enhanced digital radio.

A BV206-mounted Odette detachment showing the antenna elevated to its full height (MOD)
NEW/0547260

The BV 206-mounted Odette at HQ 3 Commando Brigade RM, at Bagram Air Base, Afghanistan in June 2002 (Patrick Allen)
0129678

An AFV-432 series Odette detachment, probably from 14 Signal Regiment (MOD)
NEW/0547248

An internal shot of the Odette system, showing operator positions   0049377

Porpoise operator's console unit

Status

Thales Defence Communications was awarded a contract in November 1998 to deliver an undisclosed number of Odette systems to the British Army and the UK Royal Marines, following competitive tender in UK, Europe and the USA. The system is now fully in service. An export version, SEEKER-S, is also available, developed using Odette experience.

Contractor

Thales Sensors, Crawley, West Sussex.

UPDATED

Outfits UAP(1), (2) and (3) ESM systems

The UAP family of equipment satisfies the radar ESM requirements of the Royal Navy Submarine Flotilla. UAP(1) *Swiftsure* and *Trafalgar* classes and UAP(3) *Vanguard* class are the current generation of Thales submarine ESM systems procured by UK Ministry of Defence. Both of these equipments have successfully completed all the UK (MoD) mandated trials and are fully accepted into service. UAP(1) and (3) which are 88 per cent common by part, benefit from Thales's EWAM (Extended Window Addressable Memory) based SADIE signal analysis processor. Additionally, computer-assisted manual analysis and a number of other advanced features have been incorporated, including a fully representative onboard training unit which forms an integral part of the ESM operator's console.

UAP(1)'s current performance has been specified for the *Astute* class submarines being built under the prime contractorship of BAE Systems. UAP(2) formerly known as UAC(1) is an earlier Racal (now Thales) submarine radar ESM system which has been removed from Royal Navy service having been superseded by UAP(1), but was fitted to the *Upholder* class of submarine, now sold to Canada.. All variants of the UAP system employ a multifunction periscope-mounted suite of antennas feeding a flexible arrangement of receivers, processors and displays.

Status

UAP(1) and (3) supplied and in service. UAP(1) is the subject of a proposed mid-life update (MLU) programme, which will address the system enhancements necessary to satisfy the evolving role of the SSN. UAP(2) supplied but now removed from Royal Navy service. The 'Astute' class will receive UAP(4).

Contractor

Thales Sensors, Crawley, West Sussex.

UPDATED

Porpoise submarine ESM equipment

A submarine version of the Cutlass ESM equipment (see separate entry), Porpoise is a fully automatic ESM system operating throughout 360° of azimuth. The equipment receives signals in the 2 to 18 GHz frequency range, measures their parameters and compares these with those contained in a preprogrammed radar threat library. The operator is presented with information about the radar signal environment on an ordered tabular display or on a graphic situation display. Up to

2,000 radar modes are held in the library and the operator can feed in data on up to 100 other emitters.

Intercepted signals are preamplified in the mast unit before being passed to the processing and analysis equipment inside the hull. Bearing data is extracted using amplitude comparison. Amplitude, pulse-width, frequency and time are combined in a single digital word before being passed to the processor unit in the operator's console. There the pulse trains of the different radars are de-interleaved and identified from library information.

The system is capable of integration with the vessel's fire-control and communications systems and may also be integrated with periscope-mounted radar warning equipments. Porpoise also has the ability to give an alert warning when prime threats, such as helicopter or maritime surveillance radars, reach a preprogrammed amplitude danger level.

The Porpoise antenna is a compact six port system giving good bearing accuracy and may be mounted on either hull penetrating or non-hull penetrating masts. It is built of titanium to reduce weight and overcome corrosion and is pressure resistant to 60 bar.

Status

In production and in service with several navies, including possibly as an option in the Turkish 'Preveze' and 'Atilay' classes.

Contractor

Thales Sensors, Crawley, West Sussex.

UPDATED

Sceptre ESM system

Sceptre is a shipborne family of modular ESM systems designed for specific operational requirements within any prevailing space and weight limitations. The system provides automatic intercept, analysis, classification and identification of all radar emissions.

A wide range of antenna and receiver options is offered to increase frequency coverage and sensitivity, and to enhance bearing and frequency resolution and passive targeting capabilities. Optional digital processing modules are available to extend the pulse density and library handling capabilities. Output data can be stored either on magnetic tape or in hard-copy format and the system can be interfaced with any central data handling system.

Using 'wide-open' ESM antenna technology, Sceptre provides 360° coverage and 100 per cent probability of intercept. Intercepted signals are automatically analysed and identified by reference to a comprehensive library of known radar types. Processed data is immediately displayed to the ESM operator in both cartesian and alphanumeric formats on two interchangeable colour displays.

Sceptre works on the principle of a central management computer surrounded by subsystems, each containing advanced microprocessors which carry out local processing. This modular approach ensures that sub-units and units can be built into a cohesive system to meet operational requirements while matching the available space and power limitations. In addition to the operational flexibility, the modular approach offers lower procurement and through-life costs and simplifies the fitting and cost of any future options or upgrades.

There are three variants of Sceptre available:
(a) Sceptre O which is a radar threat warner for medium-sized fast patrol boats and MCM vessels
(b) Sceptre X which provides full ESM for corvettes and frigates
(c) Sceptre XL for large warships.

Specifications

Frequency range: 2-18 GHz (optional extensions above and below this range are available)
Dynamic range: 60 dB
Azimuth coverage: 360° instantaneous
Elevation coverage: −10 to +30°
Signal polarisation: horizontal, vertical, and either LH or RH circular
Bearing accuracy: 4.5° RMS (optionally 2° RMS)
PRF range: 100-300 kHz
Pulsewidth range: 50 ns-100 μs
Scan types: circular, sector, steady (simple and complex)
Radar types: simple pulse, frequency agile, PRF agile, CW high-duty cycle, FMOP and PMOP flags
Mission library: 2,000 emitter modes, expandable in multiples of 1,000 modes
Threat warner: 24 emitters with 6 modes each (144 modes)
Operator library: 100 emitter modes

Status

Sceptre X is in service with the RAN and RNZN in their 'ANZAC' frigates.

Contractor

Thales Sensors, Crawley, West Sussex.

UPDATED

Weasel ESM and ELINT systems

Weasel is a combined ESM and ELINT system operating over the frequency range 0.5 to 40 GHz. It is an advanced automatic system designed to operate in a dense radar environment and to achieve a low workload for two-man operation.

The ESM part of the system acts as a search receiver with an instantaneous bandwidth equal to the full input range, thereby providing a virtual 100 per cent intercept probability. As a stand-alone equipment it can provide accurate direction-finding and identification of intercepted emissions. The latter is achieved by measuring the emitter parameters and comparing them with data stored in a library of known emitter characteristics. This is a fully automatic process and the operator is presented with a tabular display of identification and threat significance. Selected digital information can be recorded, hard-copied and transmitted to external stations. An operator can select emitters for close attention and hand over this task to the narrow bandwidth tuneable analysis receiver for attention by the analysis operator. The ESM thereby acts as a filter to reduce the workload on the analysis receiver.

The analysis receiver accepts digital data on any nominated emitter and is automatically tuned to the correct frequency. The higher sensitivity and accuracy of the analysis receiver then permits detailed analysis and direction-finding of the signal. The data obtained can be used to update the library, create new entries and/or be recorded and transmitted to external stations. Both search and analysis receivers are capable of fully independent operation.

The system is designed for installation in a vehicle and for use at both fixed and temporary sites. The equipment is transported and operated in 483 mm (19 in) special transport housings and is capable of withstanding rough handling and harsh environments.

TAC-Weasel

The latest equipment in the Weasel range is TAC-Weasel which has been designed to meet a requirement for a lightweight, mobile, reliable and cost-effective ESM for use with ground forces, particularly for special task forces, border patrols and special missions. The system covers the frequency range 0.7 to 18 GHz and combines the functions of surveillance and analysis receivers. Frequency coverage to 40 GHz is available.

The TAC-Weasel system can be installed in a small vehicle, such as a Land Rover, with a support trailer used to carry the power supply unit. The complete

TAC-Weasel installed in a Land Rover (antenna extended)

Truck-mounted version of Weasel

system is air-transportable and can be underslung from helicopters. Although designed for vehicle installation the system can also be used in fixed or temporary sites.

Status

May have been the basis for the British Army 'Pinemarten' equipment. Believed to be still available, although it may now have been replaced by Meercat-S, a version of which has been sold to the UK as INCE (see separate entry).

Contractor

Thales Sensors, Crawley, West Sussex.

UPDATED

United States

Airborne Digital Automatic Collection System (ADACS)

The Airborne Digital Automatic Collection System (ADACS), an ESM and data collection suite, is a variant of the AN/ALR-66 series. It encompasses integrated ESM capabilities in conjunction with precision emitter parameter measurement and collection. The system handles the modern threat including CW, interrupted CW, pulse stagger/agile, pulse compression and frequency-agile radar and performs precision measurements on all intercepted emitters. The design is flexible for easy installation in both fixed- and rotary-wing aircraft.

The system provides full sensitivity and over-the-horizon capability, covering a full 360° in azimuth and ±45° in elevation in the C- to J-bands in high-density EW scenarios. ADACS has automatic operation of both ESM and data collection with precision parameter measurement and recording of all intercepted emitters. The virtually infinite memory capacity permits full mission recording. A removable data record and library module ensure complete data security.

See *Jane's Radar and Electronic Warfare* for full details of the AN/ALR-66 series.

Status

ADACS is operational on the Lockheed P-3C built for a foreign customer (possibly South Korea). Flight tests were successfully concluded in March 1994. Eight systems have been delivered and associated spares and support have been contracted.

Manufacturer

Northrop Grumman Electronic Systems, Rolling Meadows, Illinois.

UPDATED

Airborne Reconnaissance Low (ARL)

The Airborne Reconnaissance Low (ARL) comprises Communications Intelligence (COMINT) and Imagery Intelligence (IMINT) sensors mounted in the DeHavilland-7 aircraft. ARL is composed of two configurations. ARL-C (COMINT) aircraft contain a complete Communications Intelligence (COMINT) sensor capable of intercepting and locating radio emissions and providing reports to appropriate commanders and intelligence processing centres on the ground. The more capable ARL-M (Multi-INT) configuration combines both a COMINT and imagery (EO/IR and SAR/MTI) capability in one aircraft. The Wide Area Moving Target Indicator (WAMTI) mode scans a 10,000 km^2 area in less than a minute, detecting ground movers, which are depicted on a map of the area. The depicted symbols provide target direction and location information. The SAR spot mode provides 1.8 m resolution imagery of a 10 km^2 area. The WAMTI mode detects movers and provides a cue to invoke the spot mode for a SAR image of the same area.

Status
There are seven ARL systems in service with an eighth currently in production.

NEW ENTRY

AN/ALR-60 communications intercept and analysis system

The AN/ALR-60 communications intercept and analysis system is carried on some US Navy EP-3E electronic reconnaissance aircraft and is used primarily for tracking and identifying warships. It has multiple operator positions equipped with terminals that allow the operator randomly to access raw audio data and processed text data. Digitally controlled receivers are operated on a priority basis.

Status
It is believed only seven systems were built. In operational service.

Contractor
General Dynamics Information Systems & Technology, Mountain View, California.

UPDATED

AN/ALR-73 Passive Detection System (PDS)

The AN/ALR-73 Passive Detection System (PDS) is an airborne ESM unit developed for the US Navy's E-2C Airborne Early Warning (AEW) aircraft. It is the improved version of and successor to the AN/ALR-59 and, because of the extensive update of that system, the ALR-73 was given its own AN designation. The primary operational roles assigned to the E-2C consist of surveillance of both airborne and surface, hostile and friendly forces; early warning of hostile aircraft in order to protect the fleet and the exercise of real-time control of the carrier tactical aircraft. The ALR-73 is intended to augment the AEW, surface, subsurface and command and control role of the E-2C by enhancing the threat detection and identification performed by the aircraft. It is a completely automatic, computer-controlled, superheterodyne receiver/processing system that communicates directly with the E-2C aircraft command and control central processor.

The design of the system was motivated by four major considerations:
(a) very high probability of intercept in dense environments;
(b) automatic system operation;
(c) high reliability;
(d) ease of maintenance.

Description
Features of the ALR-73 which are related to its intercept probability performance include: four quadrant 360° antenna coverage; four independently controlled receivers; dual-processor channels and digital closed-loop rapid-tuned local oscillator. Other features concerned with automated system operation are: low false alarm report rate, automatic overload logic, AYK-14 computer which adaptively controls hardware, and degraded mode operation.

The PDS system can measure DOA, frequency, pulse-width and amplitude and PRI simultaneously. Scan rate information is also available if called for by the central processor. Special emitter tags can also be provided. The PDS detects and analyses electromagnetic radiations within the microwave spectrum. It sends emitter data reports (pulse-width, PRI, DOA, frequency, pulse amplitude and special tags) to the E-2C's central processor via the PDS data processor. The central processor performs the identification function. The PDS immediately reports new emitters to the central processor. It eliminates redundant data on emitters for a programmable period of time, thus significantly reducing the data rate to the central processor. The PDS - a multiband, parallel scan, mission programmable, superheterodyne receiving system - covers the frequency range in four bands through step sweeping. Programmable frequency bands and dwell-

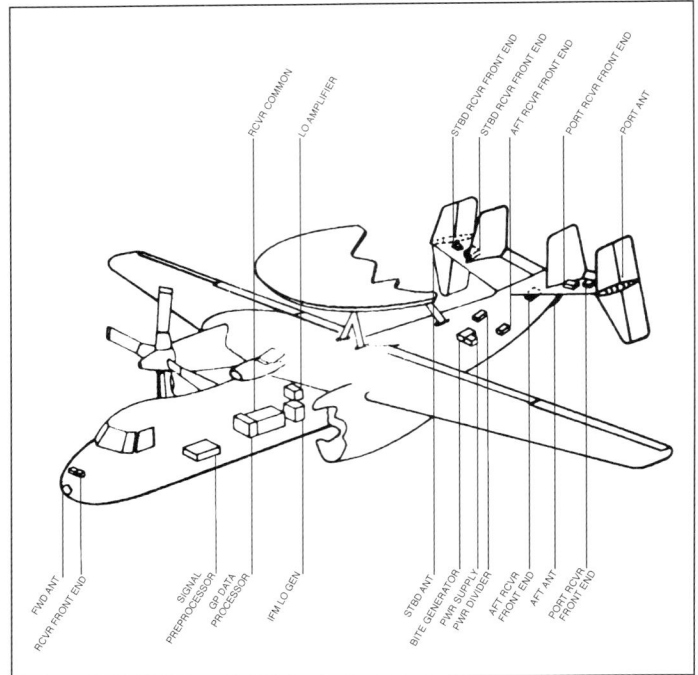

Installation of AN/ALR-73 equipment on E-2C aircraft

time permit very rapid surveillance of priority threat bands. Non-priority bands are also monitored, but at a reduced rate. Probability of intercept is increased without sacrificing sensitivity through the detection of both real and image sidebands.

Installation of the ALR-73 on an existing aircraft required a functional and mechanical modularity in the system partitioning to facilitate installation on the aircraft. Antenna packages located in the four quadrants of the aircraft provide 360° azimuth coverage. The large phase tolerance of the binary beam minimises the sensitivity of the DOA system to radome distortion and aircraft reflection. All four bands in the normal mode of operation scan through their respective frequency limits independently and simultaneously.

Activity indications may be obtained on any of the four bands. Following an indication of activity, permitted dwell-time is increased and processing of the intercepted signal is started. Dual signal-processing circuits allow intercepted signals in any two of the four bands to be processed simultaneously.

The signal processor is a special purpose logic processor which performs pulse train separation, DF correlation, band tuning and timing, and Built-In Test Equipment (BITE) logic functions. The signal processor also contains the I/O circuitry necessary for the computer to communicate with both the signal processor and the aircraft central processor. The AYK-14 provides the system control function, data storage and formatting. Variable frequency coverage, along with variable dwell-times and processing times, provides a means for optimising probability to intercept for any given theatre.

The local oscillator system consists of three units: the IFM/LO generator, LO amplifier and LO power divider. The LO amplifier has been recently upgraded to replace travelling wave tube amplifiers with solid-state amplifiers. The system is unique in that it is extremely fast and accurate. The key unit in the LO system is the Instantaneous Frequency Measurement (IFM) receiver, which samples the LO frequency and converts it to a digital tuning command. This closed loop system permits frequency measurement accuracy, while being largely insensitive to environmental variations. The local oscillator signals from the IFM/LO are amplified by the LO amplifiers to overcome the long LO cable losses and are then routed, via the power divider, to the various receiver front ends.

Status
More than 170 systems have been produced. In service in US Navy E2C and the forces of France, Japan, Singapore and Taiwan. The PDS has been successfully installed on C-130 aircraft.

Contractor
Northrop Grumman Electronic Systems, Baltimore, Maryland.

VERIFIED

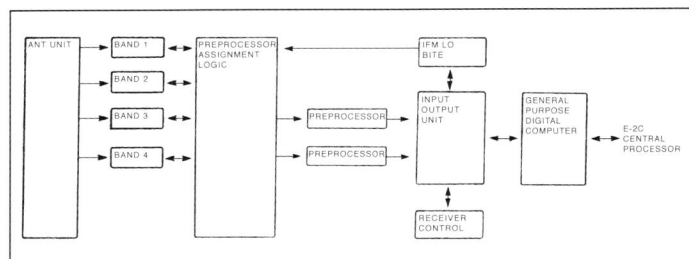

AN/ALR-73 passive detection system block diagram

AN/MLQ-40(V)3 Prophet

In mid-1999, both versions of the Ground-Based Common Sensor (GBCS Heavy and Light) as well as the Advanced Quickfix (AQF) and Intelligence and Electronic Warfare Common Sensor (IEWCS), programmes were replaced by the Prophet System. Prophet will be the division's main Signals Intelligence (SIGINT) and Electronic Warfare (EW) system. It will give the divisional commander a comprehensive near-realtime picture of enemy electronic emitters on the battlefield and provide the ability to detect, identify, geolocate, track, and electronically attack (jam) selected emitters.

Either mounted on a Heavy HMMWV, M1097,or dismounted as a manpackable capability, Prophet's primary mission is to provide 24-hour Force Protection (FP) to

The Prophet HMMWV, shown here with the 6 m telescoping antenna mast extended (Thales)
0118381

the manoeuvre brigade. It will be the sole organic Signals Intelligence/Electronic Warfare (SIGINT/EW), Measurement and Signature Intelligence (MASINT) and Ground Surveillance capability for the tactical commander at divisional level and below. Prophet will operate in direct support (DS) to the manoeuvre brigade at Division, Brigade Combat Team (BCT), Armoured Cavalry Regiment (ACR) and Separate Infantry Brigade (SIB) throughout the spectrum of conflict. Prophet will operate in three configurations: mounted, mobile, and dismounted. The system's electronic support (ES) and ground surveillance capability will provide early warning of potential threats to supported forces in the Brigade Area of Operations, and its stationary and OTM Electronic Attack (EA) capability will provide close-in electronic signals jamming. To prosecute the growing battlefield and future contingencies, Prophet will be a tailorable system able to integrate or interoperate with specific mission and contingency related equipment. It will also have the capability to insert Special Purpose Built System (SPBS) components that are theatre or contingency specific for a particular signal environment. These SPBSs currently are in both NSA and INSCOM's Quick Reaction Capability (QRC) inventories and will be used when required to prosecute advanced modulations or signal types. The SPBS will either plug into the Prophet vehicle or work co-operatively and complement the core Prophet system with a separate collection support system.

Block I System Description

The AN/MLQ-40(V)3 system includes the HMMWV-mounted pallet and the man-portable AN/PRD-13(V)2 (see separate entry). The system operates in the HF, VHF and UHF bands and provides intercept and lines-of-bearing (LOB) data on push-to-talk (PTT) communications. The ES system has three receivers: one designated DF receiver and two monitor receivers. All receivers can be used in directed search (channel scan) and monitor (fix-tuned) modes. In addition, the DF receiver provides general search (band sweep), DF operations and panning operations (manually tuning a signal) and the system can demodulate and collect AM, FM, CW and SSB signals in the specified frequency ranges. An onboard Precision Lightweight GPS Receiver (PLGR) and KVH Tactical Navigation system interface with Prophet to provide accurate worldwide self-positioning locational data to within 10 m and a north sensing device to indicate the heading of the vehicle when on the move. An automatic 6 m telescoping antenna mast allows the Prophet operators the flexibility to switch rapidly between stationary and mobile operations in approximately 90 seconds. The AN/VRC-92A SINCGARS (see separate entry) provides secure voice communications over two communications nets simultaneously. The vehicle has a load-carrying capability for four personnel with their mission essential and personal gear for a 72 hour mission duration (including fuel). The Prophet intercept receiver can also be configured to operate as a manpack system in support of forced entry airborne or air assault operations.

The pallet housed in the modified four-seat M-1097 HMMWV consists of a base, the motorised telescoping mast, integrated MA-723 VHF/UHF Monitor Antenna and MA-458 VHF/UHF DF Antenna, Lighting Protection Unit (LPU) and three distinct manned stations: the Operator Station, the Linguist Station and the Navigator/Team Commander Station.

The Operator Station comprises the human/machine interface (HMI) on a rugged notebook computer, the Control System Interface Unit (CSIU), and System Control Unit(SCU). The wireless-based HMI provides network operation of the MD-405A Receiver/Processor controls on the laptop, status of Prophet controls on the laptop, map displays with LOB overlays, databases of information collected by the system, and interfaces with the communications system to send or receive data to/from command posts or to network with other platforms. The CSIU houses the power circuits and controls. The SCU provides mast control, DF bearing control (true or relative), lamp and BIT control, displays system status and provides audio control, including jacks for the stereo headphones and recorder.

The Linguist Station consists of the MD-405A Receiver/Processor and Processor System Interface Unit (PSIU). The MD-405A is the core component of the man-portable system and provides three receivers (one DF and two monitors) to be used in directed search and monitor modes. The PSIU provides a message exchanger and process monitor and provides control and/or performance monitoring for multiple interfaces (NAV,MD-405A, mast). Additional computer

interface ports and auxiliary audio interfaces are located on this panel with capabilities for selective channel signal monitoring.

The Navigator/Team Commander Station consists of the TACNAV and communications system. The TACNAV system provides dynamic vehicle location information and system orientation for navigation while on the move. SINCGARS is used as the communications system but other communication systems that have voice and tactical data transfer capability can be integrated into the AN/MLQ-40(V)3 system.

In order to overcome the problem of information overload at the Brigade level the product from Prophet can be taken from the system and displayed on a laptop in the Brigade HQ. This concept is called Prophet control.

Operational concept

Prophet in a mounted configuration can deploy close to the brigade Forward Line of Troops (FLOT). Its small size as compared to the legacy systems, as well as its deployability (it will be C130 transportable) will allow for early entry into the contingency area to support force protection missions. The system's mobile collection and DF capability enables Prophet to maintain pace with a supported manoeuvre force. During mobile operations, the Prophet system will provide overwatch and operate within communications range of the supported mobile force. It will normally operate in a stop and go fashion or 'shorthalt', when moving cross-country, which will allow for more precise LOBs. Manoeuvre brigade intelligence and operations officers will have command and control for planning and executing Prophet missions, using the All Source Analysis System (ASAS - see separate entry), while analysts at divisional level will provide Prophet the SIGINT/MASINT technical data required to execute their EW mission, through direct digital interfaces. All SIGINT/MASINT tasking, reporting, and cross-correlation of databases will be accomplished within that framework. The Division Analysis and Control Element (ACE) will process the Prophet intelligence feed and disseminate the information into the joint net using the division communication architecture. In addition to the digital link, Prophet will maintain near continuous voice communications with the manoeuvre brigade's operations and intelligence elements.

Status

Prophet is being procured using a five-block acquisition strategy which is intended to allow tactical SIGINT to keep pace with technology and complex signal environments. There will be six Prophet systems fielded to each division (two per Manoeuvre brigade), four per ACR, three per BCT, two per SIB and five for the Training and Doctrine Command (TRADOC). Prophet will replace the TRAIL-BLAZER, TEAMMATE, TRAFFICJAM, and MANPACK legacy systems (see separate entries).

Block I

The Prophet Block I System entered into the Full Rate Production phase in March 2001 based upon the successful completion of Developmental Testing (DT) and Initial Operational Test & Evaluation (IOT&E). A competitive Source Selection Process to choose a Block I Production contractor culminated in a contract award to Titan Systems, Signal Products Division in June 2001. Titan will initially produce six First Article Test/Production systems, delivery of which started in June 2002. FUE for Prophet Block I will occur in September 2002 with the fielding of the first Prophet Production Systems to the Initial BCT units at Fort Lewis, Washington. Overall, 83 Prophet Block I Production Systems will be fielded to the US Army by the end of 2004. A development version of the system was deployed to Afghanistan on Operation ENDURING FREEDOM, where Jane's sources reported that it worked well, using unspecified different receivers.

Block II/III

Prophet Block II/III will expand the frequency range and signal types addressed by Prophet to include Low Probability of Intercept (LPI) and Modern Signals. An integrated Electronic Attack (EA) capability is also planned. Other key features of the Prophet Block II/III effort include: system netting, signal remoting and beyond LOS communications. In May 2002 it was announced that Thales Defence Information Systems would be teamed with Titan Systems for Block II/III. However, in March 2003 it was announced that the US Army Communications-Electronics Command had selected a team led by General Dynamics Decision Systems to act as the prime contractor and systems integrator for a US$19.9 million programme to upgrade the system to AN/MLQ-40(V)4 Block II/III standard. Raytheon will contribute its expertise in electronic support, with Rockwell Collins providing the off-the-shelf electronic attack systems. Other subcontractors include Antin

The manpack version of Prophet (Titan Systems)   0127377

Engineering, QinetiQ, Spectrum Certification Solutions, Austin Info Systems and Tobyhanna Army Depot.

Block IV

Prophet Block IV will add a MASINT capability on a separate vehicle to the Prophet System. Combined with an upgraded Prophet Block III ES platform, Prophet Block IV will provide a Multi-Spectral Sensor System. Prophet's MASINT component features mobile-attended platform-based sensors along with Unattended Ground Sensors (UGS) deployed as a distributed and networked multiple sensor array. Fusion of these SIGINT and MASINT capabilities onto the future combat vehicle will be achieved when available for integration.

Block V

Prophet Block V is the final planned Block and will incorporate Micro-sensors and Robotic platforms into Prophet. This highly scalable and configurable equipment will have SIGINT/EW (to include DF) and MASINT, allowing the commanders to tailor the collection to the changing mission requirements. Prophet will have the capability to transport, employ, control and relay the collected information from these remote receivers.

Division TUAV SIGINT Payload (DTSP)

Also part of the Prophet programme is the Division Tactical UAV SIGINT Payload (DTSP), formerly known as Prophet Air. This is intended to be a package of ES and EA capabilities operating in the HF, VHF, UHF and SHF bands mounted on a TUAV. Successful trials took place in early 2001 with a HUNTER TUAV and a payload of L3 Communications System - West's Skyhawk 2000 system.

Contractor

Titan Systems Corporation (Prime Block 1), San Diego, California.
General Dynamics Decision Systems(Prime Block II/III), Scottsdale, Arizona.

UPDATED

••

AN/MSQ-103 TEAMPACK ESM system

The AN/MSQ-103 TEAMPACK ESM system is a ground-based transportable unit designed to collect, identify and provide the direction of signals from ground-based transmitters. It operates in the 500 MHz to 40 GHz frequency range and is intended for use in the field at division level. The original system AN/MSQ-103, introduced in 1983, was mounted on a Jeep. Subsequent tracked versions were the AN/

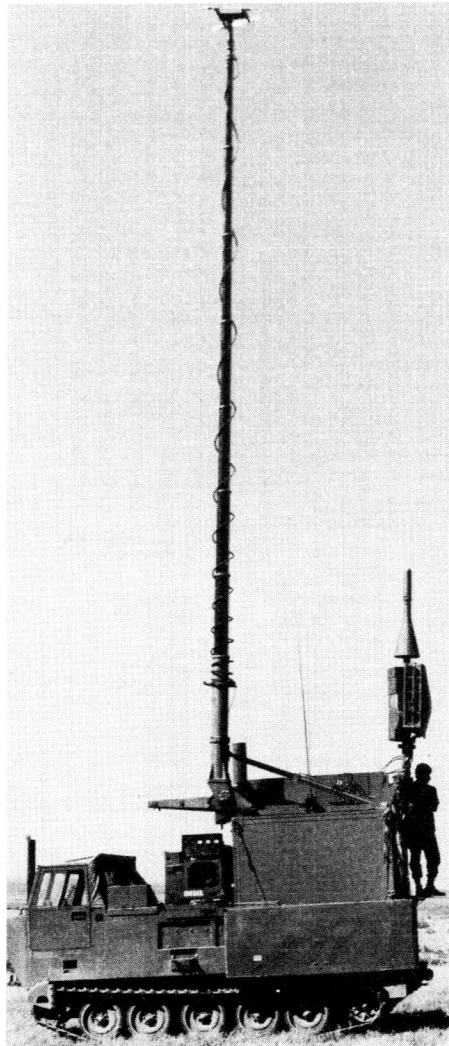

Netted AN/MSQ-103 TEAMPACK system

MSQ-103A and the AN/MSQ-103C mounted on an M-1015; the other version was the lightweight commercial utility vehicle-mounted AN/MSQ-103B. TEAMPACK replaced the US Army AN/MLQ-24 as the basic countermeasures receiving unit and includes built-in computer processing from a ruggedised mini-computer installed in the vehicle. The system also contains a secure voice and wideband datalink to a forward control and analysis centre.

A contract for engineering development and enhancement was completed by Emerson. The enhancement programme, included improvements to reliability, performance and crew protection, as well as the development of a netted system.

Status

In service with the US Army. The AN/MLQ-40 PROPHET (see separate entry) is replacing these systems.

Manufacturer

Systems & Electronics Inc, St Louis, Missouri.

Editor's Note: In 1990, Southwest Mobile Systems and Electronics & Space Corp, became subsidiaries of ESCO Electronics Corporation, a newly formed corporation as a result of a divestiture from Emerson Electric Co. In 1995, Southwest Mobile Systems Corporation and Electronics & Space Corp, combined to create Systems & Electronics Inc(SEI). In 1999 SEI was acquired by Engineered Support Systems, Inc.

UPDATED

••

AN/PRD-13(V) Manportable SIGINT system

The AN/PRD-13(V) is a manportable communications intercept and DF system of which there are two versions, (V)1 and (V)2.

The (V)1 system can support directed search, general search, or a combination of searches. For directed searches (Channel Scan), the system allows the user to programme a list of up to 400 channels with selectable priority. The system's general search (Band Sweep) capability allows for the monitoring of up to 9 bands with automatic, semi-automatic and new energy search strategies supported. Cataloguing the signals detected, the system can maintain an active signals list of 400 signals measuring centre frequency, bandwidth, signal-to-noise ratio, line-of-bearing, number of times seen and visited to provide a duty cycle as a calculated percentage, time first and last seen. In addition, a frequency pass list of up to 400 signals can be maintained. An optional recorder is available to record and playback intercepted audio and signal parameters. The DF subsystem is operable by a single person and the manufacturer claims it can be set up in under five minutes. The system, cables, and components fit in a single pack.

Key system components include the MD-403A Receiver/Processor; MA-445C HF/VHF/UHF Intercept/DF Antenna (2-2000 MHz); MA-713 HF/VHF/UHF Monitor Antenna (1-1400 MHz); and the MA-308 Hand held DF Antenna (2-500 MHz).

The heart of the AN/PRD-13(V)1 is the MD-403A Receiver/Processor. This unit includes all system interfaces, DF/intercept receiver, processor, plus associated display and man/machine/interface (MMI) functions. The processor uses Titan's patented single channel interferometer DF technique. The vector DF mode is used for mobile applications where signal strength and directions can change rapidly. The lightweight, low-profile MA-445C HF/VHF/UHF antenna provides DF coverage from 2 to 2000 MHz. The MA-308 is a hand-held DF antenna provided to support localisation of nearby transmitters during on-the-move (OTM) manpack missions such as the location of a downed pilot's beacon. For monitoring, the MA-713 antenna is a small active whip that can be used alone or with the MA-308 DF antenna. Other antennas may be used with the system to optimise it for ground vehicle, maritime, or fixed-site applications. The system can be operated using an internally mounted battery and an optional solar panel can be provided to power the system for extended operations or to recharge a NiCd battery. Additionally, other power sources can be utilised with use of the provided accessory kit including local power at 110 or 220 V AC at 50 or 60 Hz and DC power, such as vehicle batteries, from 10 to 16 V DC.

The AN/PRD-13(V)2 forms part of the AN/MLQ-40(V)3 Prophet system (see separate entry). The basic characteristics are the same as the (V)1. It supports simultaneous DF and monitor capability as well as simultaneous search and scan capability. A total of 20 priority channels are provided in the search channel list. Changes to the system components in (V)2 are:
(a) The MD-405A Receiver/Processor, which includes three receivers (1 DF/Intercept and 2 Monitor Receivers; manual independent receiver control by user), and dual microcontrollers.
(b) The MA-715A HF/VHF/UHF Monitor Antenna (2-2000 MHz), broadband dual whip antennas which extend the monitoring capability to 2000 MHz.
(c) Direct external power can be accepted up to 28 V DC.

Specifications
(a) Frequency coverage
    DF and Intercept: (MA-445C) 2-2000 MHz
    (V)1 Monitor: (MA-713) 1-1400 MHz
    (V)2 Monitor: (MA-715A) 2-2000 MHz
    DF (MA-308) 2-500 MHz
(b) Demodulators
    (V)1: FMw, FMn, AM, USB, LSB, CW
    (V)2:
    DF/Intercept Receiver FM (200 kHz, 50 kHz, 15 kHz), AM (15 kHz, 6 kHz), SSB (6 kHz, 3 kHz), CW (3 kHz, 0.5 kHz)

Monitor Receivers FM (15 kHz), AM (15 kHz, 6 kHz), SSB (6 kHz, 3 kHz), CW (3 kHz, 0.5 kHz)
(c) DF accuracy: 3° RMS Typical (antenna and location dependent)
(d) DF coverage: 360° azimuth, 0° to +60° elevation
(e) Graphical displays:
PAN (50 kHz, 200 kHz, 900 kHz, 5 MHz spans)
DF (Histogram and Vector)
(f) Remote interface RS-232C
(g) Receiver operating modes:
Directed search (channel scan): 400 normal channels ((V)1) selectable priority; (V)2 20 priority channels)
General search (band sweep): automatic, semi-automatic, new energy modes
Bands: selectable, 9 bands
Signal list: log up to 400: centre frequency, BW, time statistics
Pass list: avoid up to 400 channels
(h) Physical characteristics:
System weight: 18 kg (36.7 lbs) (V)1; 19.5 kg (43 lbs) including Mb-5700 NiCd battery and all field accessories (V)2
(i) Power: 3.9 W max (V)1; 9.5 W max (V)2
(j) Power Input: 10 to 16 V DC (V)1; 10 to 28 V DC (V)2

Status

(V)1 has been acquired by the United States and Australia. (V)2 is the man-portable element of the AN/MLQ-40(V)3 Prophet system.

Contractor

Titan Systems Corporation, San Diego.

NEW ENTRY

AN/TSQ-112 TACELIS ESM system

The AN/TSQ-112 is a location and identification ESM system operating in the frequency range 500 kHz to 500 MHz. The system is fully automatic and performs communication collection, transmitter location and processing functions. It is designed to combat hostile tactical radio communications by generating early warning, target and decision information and jamming.

TACELIS (Tactical Automated Communications Emitter Location and Identification System) is a major component of the US Army TACOM-EWS (Tactical Communications Electronic Warfare System) and consists of two remote master stations and four slave stations. It is deployed with each operations company forward in one 10 ton truck, three 6 ton vans and a truck tractor. Remote master stations are deployed in three 5 ton trucks and each remote slave station is carried by one M113. The component elements of TACELIS include the various stations plus one AN/UYK-7 computer, 13 AN/UYK-19 mini-computers and AN/ULR-17 receivers. The remote slave stations have a direction-finding capability only, while each master station has 14 receivers and two search/acquisition receivers. The TACELIS system is interoperable with forward control and analysis centres, the AN/MLQ-34 TACJAM via the AN/GRC-103 communications link directly out of the remote master stations and the Cefly Lancer aircraft.

Status

This equipment is highly classified and was deployed on Operation Desert Storm with the US Army. No recent contracts have been announced, which is an indication of the programme being complete. It is uncertain whether the equipment is still in service.

Contractor

General Dynamics, Advanced Information Systems, Mountain View, California.

UPDATED

AN/TSQ-152(V) TRACKWOLF and AN/TSQ-199 Enhanced TRACKWOLF

The AN/TSQ-152(V) TRACKWOLF is a mobile, tactical system contained in 205-280 shelters, configured as five distinctive groups and transported by 5 ton trucks. The system is deployed at four sites with signal intercept and collection stations, collection analysis and processing centres, command and control capabilities and a DF net control station with three remote DF stations.

The AN/TSQ-152(V) signal collection/processing subsystem includes extensive signal intercept, classification and monitoring facilities together with data processing and analysis capabilities for elaborating reports to the superior echelon. A dedicated command and control station, AN/TSY-1, controls overall system operation.

The AN/TRD-27 HF direction-finding subsystem of AN/TSQ-152(V) has Single Station Location (SSL) capabilities with ionospheric sounding for improved range determination. The DF voice and data network operates with an HF/UHF 'flashnet' system.

All elements of the TRACKWOLF are provided with secure data and voice communications for command and control of operations. The signal collection/processing and C2 subsystem elements are provided with fibre optic LAN

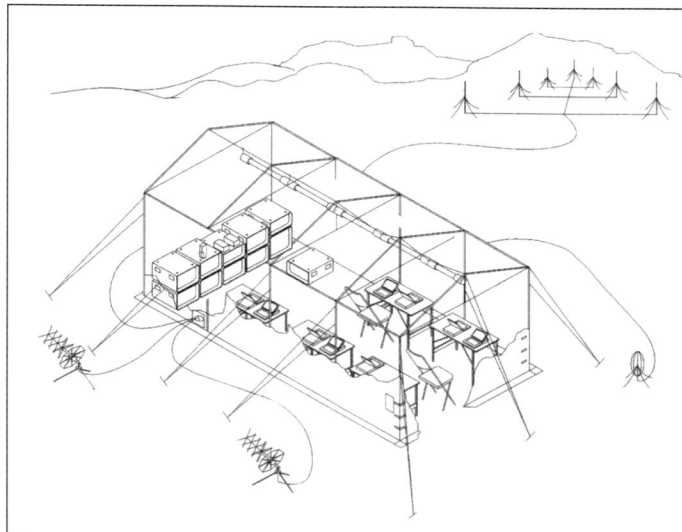

Artist's impression of Enhanced TRACKWOLF   0055190

connectivity. The system interfaces with the All Source Analysis System (ASAS) (see separate entry).

AN/TSQ-199 Enhanced TRACKWOLF (ET) is an Echelon Above Corps (EAC) ground-based, man-transportable transit case high-frequency Direction-Finding (DF) and intercept system. The programme was directed by Congress in FY93 as a result of Desert Shield/Desert Storm, during which the current AN/TSQ-152(V) TRACKWOLF system proved too large and cumbersome for rapid deployment. ET reportedly requires less than 10 per cent of the airlift capacity of the previous system, and in addition to transportability advantages it incorporated several capabilities that allow intercept of modern modulations.

ET consists of three automated AN/TSQ-205 stations capable of netted or stand-alone operations. Each AN/TSQ-205 is a rapid deployment station capable of conventional and LPI COMINT collection and direction-finding. The AN/TSQ-205 is a versatile station that can be configured to meet changing operational needs. A wideband digital system is the heart of each AN/TSQ-205 and each station can have up to nine operator positions to perform any mix of management, analysis, collection and DF functions. ET with all three stations can be configured for leapfrog movement, with one station assuming the duties of net control, while another station is in the process of relocating to another site. ET can communicate externally through SINCGARS LPI VHF radio, AUTODIN/DSSCS, TROJAN SPIRIT II, Single Source Processor-SIGINT, CROSSHAIR, UHF MIL SATCOM Transceiver (DF Flashnet).

Projected future activities include procuring and fielding a satellite communications capability for the DF subsystem outstations, as well as fielding a capability that provides interconnectivity between TRACKWOLF, Navy and National Security Agency direction-finding nets. In mid-2003 there is no confirmation of the progress of these intentions.

Status

TRACKWOLF entered service with the US Army in the 1990s. In mid-1994, E-Systems (now Raytheon) was awarded a US$41 million contract to provide Enhanced TRACKWOLF. In April 1996, E-Systems won a US$7.5 million contract for the production of additional Enhanced TRACKWOLF equipment. Enhanced TRACKWOLF entered service with the US Army in 1997.

Contractors

AN/TSQ-152(V) TRACKWOLF:
TCI , Fremont, California.
AN/TSQ-199 Enhanced TRACKWOLF:
Raytheon Systems Company, Falls Church, Virginia.

UPDATED

AN/TSQ-190(V) TROJAN Special Purpose Intelligence Remote Integrated Terminal (TROJAN SPIRIT II)

TROJAN SPIRIT II is an intelligence dissemination satellite terminal that provides access for intelligence processing and dissemination systems. Connection to the All Source Analysis system (ASAS) (see separate entry) enables further dissemination. The system consists of secure voice, data, facsimile, video and secondary imagery dissemination capabilities. It will receive, display and transmit digital imagery, weather and terrain products, templates, graphics and text between CONUS/OCONUS bases and deployed US forces.

The SPIRIT II programme, which is a development of SPIRIT I, provides a much more robust terminal and increased capability. TS II is a near-term fix for high-capacity imagery data communications capability.

The system's SATCOM system supports up to 14 circuits (8 SCI and 6 collateral) using variable baud rates from 4.8 to 512 kbps per channel on C, Ku, or X frequency bands. System connectivity capability includes DSNET1 and DSNET3, MSE, and

Tactical Packet Network (TPN) interfaces, as well as LAN connectivity. The TROJAN SPIRIT II is shelter mounted on two HMMWVs. It ties into TDN as a mobile switch extension from tactical. The system's two workstations also allow the operators to receive and disseminate secondary imagery, SIGINT databases and reports and UAV video. This capability allows the TROJAN SPIRIT to serve as a temporary communications set for the ACE during redeployment or split-based operations.

The TROJAN SPIRIT II combines the Trojan Data Network with mobile switch extensions to offer a worldwide, forward-deployed, quick-reaction reporting and analysis link. This corps and division asset provides dedicated intelligence communications that is intended to augment EAC and ECB in-theatre communications. It will conduct split-based, inter- and intra-theatre operations through the range of military operations.

The TROJAN Data Network (TDN) is a router, TCP or IP based network. It is overlaid on the communications network that links the AN/FSQ-144(V) TROJAN CLASSIC central operating facilities and switch extensions at various US bases with remote collection facilities worldwide. The TDN is subdivided into three electronically and physically separated networks that correspond to the three security levels required of the system. As with the TROJAN CLASSIC architecture, the TDN has a TROJAN Network Control Centre in the TROJAN Switch Centre at Fort Belvoir, Virginia to provide configuration control and network management. The three networks of the TDN are:

TROJAN Data Network-1 (TDN-1). The TDN-1 operates at the SECRET security level and is the gateway to DSNET1. It provides data exchange between TROJAN Classic facilities, switch extensions, and Special Purpose Intelligence Remote Integrated Terminals (SPIRITs).

TROJAN Data Network-2 (TDN-2). The TDN-2 operates at the TOP SECRET/SCI level. It provides data exchange between selected TROJAN sites requiring access to the NSA network.

TROJAN Data Network-3 (TDN-3). The TDN-3 operates at the TOP SECRET/SCI security level and is the gateway to JWICS. It provides data exchange between TROJAN CLASSIC facilities, switch extensions and SPIRITs.

The AN/TSQ-226(V) TROJAN SPIRIT LITE is a functional equivalent to the TROJAN SPIRIT II and was fielded to the Stryker Brigade Combat Teams (SBCT) in lieu of the TROJAN SPIRIT II, which is out of production. There are three versions: (V)1 a commercial off-the-shelf version in a transit case configuration used to augment Military Intelligence dissemination and communications requirements; (V)2 SBCT (pallet, ECV, trailer); (V)3 SBCT (pallet, shelter, ECV, trailer). (V)3 is similar to (V)2 but adds an additional shelter and workstation. The latter two were designed especially for the SBCTs.

Features
(a) totally nonblocking architecture
(b) up to 248 universal user ports
(c) serves as a nonsecure interconnect system
(d) command and control feature set
(e) emergency response feature set
(f) unlimited conferencing
  (i) one-button access to preset conference calls
  (ii) radio net conferencing and monitor
  (iii) progressive conferencing
  (iv) layered conferencing
(g) English and foreign language voice menus
(h) multimedia broadcast
(i) multilevel security
(j) multilevel precedence and pre-emption
(k) ruthless circuit/radio seizure
(l) multiple trunk hunt groups
(m) fully redundant fail-safe architecture
(n) system speed dial (99 entries)
(o) adjustable 2 second digital radio delay
(p) alarm/monitor input capability

Components
Antenna control system
(a) controls Raytheon E-Systems 2.4 m pop-up antenna azimuth over elevation
(b) controls Raytheon E-Systems 6.1 m transportable antenna x-elevation over elevation
(c) motorised polarisation control
(d) tracking capability for Intelsat and domestic satellites at Ku-band and C-band
(e) automated satellite acquisition.

Cryptographic equipment
(a) KIV-7 Embeddable KG-84 COMSEC module
(b) KY-68 digital secure voice terminal

Mission equipment
(a) 9030T TEMPEST printer
(b) Raytheon E-Systems digital voice compression telephone
(c) DNE 2048AT-16 multiplexer
(d) reconnaissance video compression and dissemination equipment
  (i) parallax graphics PVC SBus board
  (ii) Paradise Systems Uniflix/Simplicity software
  (iii) RGB Spectrum 600-3 VME video tiling board
  (iv) applied Integration VME video switch matrix board
  (v) performance technologies VME RS449 comm board
  (vi) UconX RS449 interface software
  (vii) RGB Spectrum 1600U RGB/NTSC video scan converter
(e) workstation video equipment

TROJAN SPIRIT vehicle

  (i) Sun X486A camera
  (ii) telex intercom/speaker/microphone/headset
  (iii) Sun speaker
(f) TSP-9100A TEMPEST facsimile (2)
  Cyberchron laptop computer:
  (i) 300 Mbyte hard disk
  (ii) 16 Mbyte RAM
(g) HP 1200 C/PS colour postscript printer
  (i) LAN accessible
(h) FORCE workstation
  (i) 64 Mbyte memory
  (ii) 4 Gbyte internal hard disk (removable)
  (iii) 1,280 × 1,024 pixel resolution
  (iv) 4 mm digital audio tape
  CD-ROM
(i) CISCO Systems 4000 Router (2).

Specifications
(a) Ku-band down converter
  Frequency: 10.95-12.75 GHz
  Dimensions: 19 in W × 22 in L (max) × 1.75 in H (partial band) or 3.5 in H (full band) (rack-mountable)
  Synthesised: 1 kHz steps over 1.8 GHz bandwidth
(b) Ku-band up converter
  Frequency: 14.0-14.5 GHz
  Dimensions: 19 in W × 22 in L (max) × 1.75 in H (rack-mountable)
  Synthesised: 1 kHz steps over 500 MHz bandwidth
(c) Ku-band solid-state power amplifier
  Power: 16 W SSPA to meet required EIRP of 59 dBW
  Dimensions: 10 in W × 26 in L × 6 in H (feed-mounted)
(d) Ku-band low-noise amplifier
  Frequency: 10.95-12.75 GHz
  Noise temp: 75 K to meet G/T of 25 dB/K
  Gain: 50 dB (min)
(e) X-band down converter
  Frequency: 7.25-7.75 GHz
  Dimensions: 19 in W × 22 in L (max) × 1.75 in H (rack-mountable)
  Synthesised: 1 kHz steps over 500 MHz bandwidth
(f) X-band up converter
  Frequency: 7.90-8.4 GHz
  Dimensions: 19 in W × 22 in L (max) × 1.75 in H (rack-mountable)
  Synthesised: 1 kHz steps over 500 MHz bandwidth
(g) X-band traveling wave tube amplifier
  Power: 350 W TWTA to meet required EIRP of 72 dBW
  Dimensions: 21 in W × 24 in L × 10.6 in H (feed-mounted)
(h) X-band low-noise amplifier
  Frequency: 7.25-7.75 GHz
  Noise Temp: 55 K to meet G/T of 28.5 dB/K
  Gain: 60 dB (min)
(i) C-band down converter
  Frequency: 3.625-4.2 GHz
  Dimensions: 19 in W × 22 in L (max) × 1.75 in H (rack-mountable)
  Synthesised: 1 kHz steps over 575 MHz bandwidth
(j) C-band up converter
  Frequency: 5.85 to 6.425 GHz
  Dimensions: 19 in W × 22 in L (max) × 1.75 in H (rack-mountable)
  Synthesised: 1-kHz steps over 575-MHz bandwidth
(k) C-band solid-state power amplifier
  Power: 50 W SSPA to meet required EIRP of 56 dBW
  Dimensions: 10 in W × 26 in L × 6 in H (feed-mounted)
(l) C-band low-noise amplifier
  Frequency: 3.625-4.2 GHz
  Noise temp: 40 K to meet G/T of 16.5 dB/K
  Gain: 50 dB (min)

Status

The TROJAN system was originally fielded in the early 1990s, when a limited number of prototypes were deployed in the First Gulf War. Following their success, formal production funding for the TROJAN SPIRIT II was later approved with fielding in the mid-1990s.

A total of 52 TROJAN SPIRIT IIs were built, 38 for the US Army, 8 for the US Marine Corps, and 6 for the US Air Force Predator unmanned aerial vehicle (UAV) satellite communications. The Predator programme eventually selected a different SATCOM platform for production and the six USAF TROJAN systems were refielded to the Army, giving them 44 systems.

Although the original TROJAN SPIRIT systems were fielded in Military Intelligence (MI) units at division, corps and EAC levels, SBCT structures placed three TROJAN SPIRIT LITE systems - two (V)2 and one (V)3 - in the brigade's signal companies with MI operators and analysts attached to those companies.

Contractor

Raytheon Systems Company, Falls Church, Virginia.

UPDATED

AN/WLQ-4 SIGINT system

The AN/WLQ-4(V) is an automated, modular signal collection system which allows for the identification of the nature and sources of unknown radar emitter and communications signals. It incorporates a network of mini-computers and microprocessors and data from these computers is correlated with information received from satellite sensors. The system is part of Sea Nymph, a highly classified US Navy NAVELEX programme.

The AN/WLQ-4 system has a number of key features, including:
(a) automatic search, acquisition and signal processing
(b) automatic logging, bookkeeping and reporting
(c) semi-automatic correlation of real-time measured data with input from an external system
(d) 400,000 lines of AN/UYK-20 source code
(e) 50,000 lines of executable code in 40 microprocessors
(f) a significant growth capability to handle new threats.

Status

Believed to be installed in 'Seawolf', 'Sturgeon' and 'Virginia' classes of US submarines.

Contractor

General Dynamics Advanced Information System, Mountain View, California

VERIFIED

AN/WLQ-4(V) skeletal system in a US Navy SSN-637 submarine

AN/WLR-8 tactical EW receiver system

The AN/WLR-8 is a tactical electronic warfare and surveillance receiver designed for fitting in both surface ships and submarines of the US Navy. The system is of modular construction and provisions are made for operation in conjunction with numerous types of direction-finding or omni-antennas and a wide range of optional

peripheral equipment to provide comprehensive ESM facilities. The WLR-8 is compatible with the Navy Tactical Data System (NTDS) and similar action information automation systems. The system can be expanded in frequency or signal handling capability by means of simple additions and/or software changes. Three versions are available: V(1) for submarines, V(2) for 'SSN-688' class submarines and V(5) for Trident submarines.

Two digital computers are incorporated: a Sylvania PSP-300 for system control, automatic signal acquisition and analysis and file processing and a GTE PSP-200 microcomputer for hardware-level control functions. Digital techniques are employed throughout the WLR-8 system, which is all-solid-state.

Operational facilities provided include:
(a) automatic measurement of signal direction of arrival
(b) signal classification and recognition
(c) sequential or simultaneous scanning over a wide frequency range
(d) signal activity detection for threat warning
(e) analysis of signal parameters such as frequency, PRF, modulation, pulse-width, amplitude and scan rate
(f) logging of signal parameters for display to operator(s) and printout of hard-copy to teletype or printer
(g) extensive built-in test equipment
(h) directed priority searches of specific frequency segments.

Direct reporting to onboard computers, such as NTDS, permits response times in the millisecond range with minimal operator involvement. A two-trace CRT is provided for display purposes and this can be supplemented by an optional five-trace panoramic display for presentation of signal activity data. Another CRT display is incorporated if the WLR-8 is used with automatic or manual DF antenna systems.

Status

Operational with US Navy and possibly other navies. It is possible that more than 2,700 sets have been produced. The system is in use on board 'Los Angeles' class attack submarines and the V(5) version is fitted to 'Ohio' class ballistic missile submarines.

Contractor

General Dynamics Information Systems and Technology, Mountain View, California.

VERIFIED

EW-1017 surveillance system

As an airborne electronic surveillance system, EW-1017 automatically acquires and identifies emissions within the C- to J-bands (0.5 to 20 GHz). The system is also designed to receive and identify all those emissions illuminating the aircraft (including short bursts) particularly when it is operating in very dense signal environments. The warning of possible danger is given both visually and aurally on a display unit while preferential scan is used to ensure immediate recognition of possible lethal threats.

EW-1017 consists of antenna arrays, receiver, a processor system, an electronic support measures operator interactive display subsystem and a pilot's display. Broadband spiral antennas are used to provide omnidirectional coverage which, together with their separate multiband receivers, are mounted in pods on each wingtip. This location drastically limits aircraft 'shadowing' and the proximity of the receiver cuts signal losses to a minimum. Angular bearing of the emissions is determined by using selected pairs of antennas.

The hybrid superheterodyne receiver offers high probability of acquisition, high sensitivity, frequency accuracy and a high degree of frequency selection and selectivity. A broad bandwidth is used in the acquisition mode to obtain the initial intercept with a narrow bandwidth then being used for accurate bearing measurement and analysis. To ensure processing capability in highly dense signal conditions, a high-speed digital computer performs the data processing functions, supplemented by microprocessors. This enables the receivers to scan the frequency band continuously on a reprogrammable basis, so that conventional, continuous wave and agile signals are processed for identification. For special applications, the receiver can be interfaced via a smart post processor unit to a centralised tactical display and control system.

An interactive display subsystem provides a full range of operator facilities to manage and optimise collection. It also provides a readily accessible real-time emitter and platform library storage and analysis capability, facilitated by a modern data management system. A control/display unit allows the operator to monitor and control the automatic surveillance function to resolve possible ambiguities and evaluate and use, to the best advantage, the data displayed.

Status

As of October 2001, EW-1017 was reported as being in service with the German Naval Air Arm (Breguet Atlantic maritime patrol aircraft) and the Royal Air Force (Nimrod Mk 2 maritime patrol, Nimrod 'R' SIGINT and Boeing E-3-D Sentry AEW Mk 1 airborne warning and control system aircraft). In UK service, the system is designated as ARI.18240 'Yellowgate'. During April 1999, Racal Defence Electronics (now Thales Sensors) announced that it had been awarded a contract for approximately £5 million, that covered the upgrading of the 'Yellowgate' Electronic Support (ES) systems installed aboard the UK's Sentry AEW Mk 1 fleet. Jane's sources reported that the upgrade was to focus on the enhancement of the equipment's reliability/maintainability, together with the introduction of Racal's commercial-off-the-shelf 'Melinda' signal processor. The technology used is described as being similar to that employed in the company's Outfit UAT naval ES system and the 'Yellowgate' upgrade was expected to remain operationally viable until 2025.

Contractor
BAE Systems North America Information & Electronic Warfare Systems, Nashua, New Hampshire.

UPDATED

Guardian Star shipborne EW system

Guardian Star is a family of EW systems designed to cover requirements from threat detection to ELectronic INTelligence (ELINT). Various configurations allow the system to be configured to the specific requirements of surface ships and submarines.

The Mk 1 system offers early warning to small surface craft or patrol boats, with an average bearing accuracy of ±10° provided by octave frequency measurements. The Mk 2 system is designed to meet the basic ESM needs of surface ships and submarines. It provides YIG tuned frequency measurement and emitter average bearing accuracy of ±5°. The Mk 3 is an ELINT system for surface ships and submarines. It carries out instant threat warning and accurate frequency measurement via IFM devices in the receiver front end. Average bearing accuracy of less than ±5° is achieved.

Description
The systems consist basically of an antenna assembly, an RF/Digital Interface Unit (RFDIU) and a Display/Controller Unit (DCU). The antenna assembly comprises an omnidirectional and six spiral DF antennas covering the frequency range from 2 to 18 GHz. All the necessary preamplifiers and preprocessing components are included in the assembly. The RFDIU processes all signals before transfer to the main digital processor in the DCU. All signal conversion electronics, auxiliary outputs and power supplies are contained in this RFDIU. The DCU includes the input/output section, main processor, magnetic tape unit, keypad and all required operator interfaces. The operational program and library file data are loaded into the processor memory via the magnetic tape unit.

In the surveillance mode the system operation is broadband from 2 to 18 GHz. This mode is entirely automatic and the operator has only to view the situation summary display. This display will provide data on up to 50 emitters (10 per page). The displayed emitter characteristics are frequency, PRF/PRI, pulse-width, pulse amplitude, true bearing (instantaneous and averaged), scan rate, emitter name (if matched in the library) and threat level (if matched in the library); the library file handles up to 2,000 sets of emitter parameters. The operator can store previously known emitter data together with pre-assigned threat levels in the file. The system matches incoming emitter data with the file and issues an immediate alert on high interest threats. Library data can be changed or entered via the display/controller unit.

Specifications
Simultaneous tracking of 250 emitters
Emitter identification: less than 1s
Display update: 50 emitter/s
Library capacity: 2,200 emitters
Operator's scratch library: 200 emitters
Major emitter information displayed:
(a) bearing
(b) bearing confidence
(c) frequency
(d) Pulse Repetition Frequency (PRF)
(e) Pulse-Width (PW)
(f) Pulse Amplitude (PA)
(g) number of Updates
(h) Scan Rate (SR)
(i) name (with library match)
(j) threat level (with library match)
(k) platform type (with library match)
Bearing accuracy specification: 1-18 GHZ, within 5° RMS
Tunable RF bandpass and band reject filters for high and low bands
20 digital filters for 'look through' blanking
Probability of Intercept: 100% within 1 s. A minimum of four consecutive pulses (default value) received at the wideband (1-18 GHZ) omni-antenna are used to establish and identify an emitter through library matching. Operator controls allow selection of the number of pulses from 100-200 to establish an emitter.
Processing speed: 1,000,000 pulses/s
Ship Parameters:
(a) heading
(b) speed (optional)
(c) periscope position (optional)
(d) latitude/longitude (optional)
(e) date/time

Status
Installed in Canadian 'Victoria' class (ex-UK 'Upholder' class).

Contractor
Northrop Grumman Electronic Systems, Charlottesville, Virginia.

Editor's note: The Sperry and Burroughs Corporations merged to form UNISYS in 1986. Sperry Marine was subsequently acquired by Tenneco in 1987. In 1993 the JF Lehman investment group purchased Sperry Marine from Tenneco. In early 1996 Litton purchased Sperry Marine. In 2001 Litton was purchased by Northrop Grumman.

UPDATED

Guardrail Common Sensor (GR/CS) SIGINT System

The Guardrail Common Sensor (GR/CS) is a Corps Level Airborne Signal Intelligence (SIGINT) collection/location system that integrates the Improved GUARDRAIL V (IGR V), Communication High Accuracy Airborne Location System (CHAALS), and the Advanced QUICKLOOK (AQL) into the same SIGINT platform, the RC-12K/N/P/Q aircraft. Key features include integrated COMINT and ELINT reporting, enhanced signal classification and recognition, fast Direction-Finding (DF), precision emitter location, and an advanced integrated aircraft cockpit. Preplanned product improvements include frequency extension, computer assisted online sensor management, upgraded datalinks and the capability to exploit a wider range of signals.

GR/CS provides near-realtime SIGINT and targeting information to Tactical Commanders throughout the corps area with emphasis on Deep Battle and Follow-on Forces Attack support. It collects selected low-, mid-, and high-band radio signals, identifies/classifies them, determines locations of their sources, and provides near-realtime reporting to tactical commanders. The system uses an integrated processing facility (IPF) which is the control, data processing and message centre for the overall system.

Each system consists nominally of 12 aircraft which normally fly operational missions in sets of three. Up to three airborne relay facilities (ARF) aircraft intercept communications, non-communications emitter transmissions and gather LOB and TDOA data. They then transmit this data to the IPF. The ARF aircraft also serve as the relay platforms for communications between the IPF and the supported commands. The typical system configuration uses one Integrated Processing Facility (IPF), two or three Airborne Relay Facilities (ARFs), approximately nine (up to a maximum of 32) Commanders Tactical Terminals (CTTs), and an Auxiliary Ground Equipment (AGE) van. Special Purpose Equipment (STE) vans are included for maintenance and troubleshooting.

The system incorporates the Communication High Accuracy Airborne Location System (CHAALS) to achieve target locations for its COMINT system, and CHALS-X, which is a continuation of the project which developed the CHAALS precision location subsystem currently in GR/CS systems 4 and 1. The CHALS-X system provides the targeting capability required to support the Division Commander's requirements to locate and kill the enemy by providing for precise location of High-Value Targets (HVTs). Airborne systems mixed with ground based systems will be capable of precisely locating enemy weapon systems and units (regardless of whether the enemy uses conventional or modern radios) producing target locations sufficiently accurate for first round fire for effect by organic artillery. It utilises the previously developed Time-Difference-Of-Arrival/Differential Doppler (TDOA/DD) techniques and incorporates advances in electronics state-of-the-art and distributed processing to provide for improved capabilities; increases frequency range, adds frequency-hopping radios to the target set, and decreases size/weight/power requirements of processing subsystems (3 racks of computer equipment now reduced to two boxes which fit into a standard 19 in rack). The continued evolution of Target Accuracy Geolocation capability using TDOA/DD is a technology advantage over any other country and has been restricted from release to foreign countries.

Guardrail Integrated Processing Facility
0085291

The RC-12K mission aircraft used in the USD-9 (B) GRCS #4 system

GR/CS Targeting accuracy is also provided by the ELINT system. Ground-to-ground (including CTT) communications links also provide an interface with fixed locations and tactical users. Automated addressing to CTT field terminals provides automated message distribution to tactical commanders in near-realtime. Planned improvements include expanded COMINT/ELINT collection, LPI capability, embedded training, CTT (3 channel) retrofit, and automated reporting.

The Radio Remote Receiving Set (AN/ARW-83) is commonly referred to as the Airborne Relay Facility (ARF). The ARF consists of equipment installed in a modified Beechcraft Super King Air aircraft with a military designation of RC-12. The ARFs are manned only by the pilots during a mission. ARF mission equipment is remotely controlled by operators in the Integrated Processing Facility (IPF). The Guardrail systems currently in service include the Guardrail V (RU-21H aircraft), the Guardrail Common Sensor Minus (RC-12H aircraft), and the Guardrail Common Sensor (RC-12K/N/P aircraft). Guardrail Common Sensor (GRCS) combines the Improved Guardrail V (IGRV) Communication Intelligence (COMINT) sensor package with the Advanced Quicklook electronics signals (ELINT) intercept, classification, and direction-finding capability and a Communication High-Accuracy Airborne Location System (CHAALS). GRCS shares technology with the Ground-Based Common Sensor, Airborne Reconnaissance Low, and other airborne systems.

GRCS comprises a series of special purpose detecting systems - AN/USD 9B to E. The GRCS systems are tactical, remotely controlled, airborne mission equipment, and ground-based intercept and emitter location systems. They have an external near-realtime reporting capability that can be operated in six modes (local, isolated, remote, interoperable, training, or maintenance/calibration). These systems are assigned to a B company, military intelligence battalion, aerial exploitation, as part of a corps military intelligence brigade.

The GRCS System 1, AN/USD-9C, and System 2, AN-USD-9E, are the latest addition to this family. They have the additional capability to operate worldwide via the GRCS Tethered Medium Earth Terminal (TMET) and the Direct Air to Satellite Relay (DASR) Aircraft (RC-12Q). Other major system improvements are:

(a) New UNIX-based work stations
(b) Faster (Micro 5) mainframe computers
(c) The fibre-optics distributed data interface (FDDI) local area network (LAN)
(d) The GRCS Data Distribution System (DDS), elementary special signals processing
(e) The GRCS Integrated Processing Facility (IPF) rapid deployment capability (two vans minimum vice four)
(f) The entire system (less aircraft) will be C-130 transportable

Information is processed and reported to joint consumers via TRIXS broadcast primarily over the Joint Tactical Terminal (JTT) which is a subsystem of the GRCS DDS.

A typical mission requires the aircraft to orbit to the rear and parallel to the Forward Line of Own Troops (FLOT). The IPF sends commands to and receives information from the Airborne Relay Facility (ARF) through a secure datalink. The operators in the IPF process the collected information and report the intelligence to the tactical commanders and other possible joint consumers via the JTT relay on board the aircraft.

GUARDRAIL Integrated Processing Facility (IPF)

GUARDRAIL ground processing is conducted in the Integrated Processing Facility (IPF), which consists of four 40-foot SeaLand trailers, interconnected to provide an advanced ground processing centre for the Guardrail mission operations. The Surveillance Information Processing Facility (AN-TSQ-105 (V)), commonly referred to as the Integrated Processing Facility (IPF), is a manned, ground based, control, data processing and message centre. The AN/TSQ-176 Surveillance Integrated Processing Facility, consisting of four 8 × 40 ft electronic shelters, is part of the Guardrail/Common Sensor System-1 (GRCS-1) AN/USD-9C integrated processing facility. The vans are intended to observe, analyse and process information relayed from the Airborne Relay Facility (ARF), AN/ARW-83(V). Interoperable DataLinks (IDL) provide microwave connectivity between the airborne elements and the IPF. Reporting is accomplished via Commander's Tactical Terminals (CTT). Key features include integrated COMINT and ELINT reporting, enhanced signal classification and recognition, fast Direction-Finding (DF), and precision emitter

location. The GRCS Integrated Processing Facility (IPF) rapid deployment capability consists of two vans minimum versus the four of earlier configurations.

In September 2001 it was announced that TRW (now Northrop Grumman Mission Systems) had been awarded a US$13 million contract to develop and field a manned, mobile ground-based Guardrail Information Node (GRIFN) to replace the GRCS IPF.

The Flightline Test Set AN/ARM-163 (V) is commonly referred to as the Auxiliary Ground Equipment (AGE) van. The AGE van is a five ton step van that has been modified to contain automated test equipment used in the preflight checks and maintenance of the Airborne Relay Facillty (ARF).

Information is processed and reported to joint consumers via TRIXS broadcast primarily over the Joint Tactical Terminal (JTT) which is a subsystem of the GRCS Data Distribution System (DDS). A typical mission requires the aircraft to orbit to the rear and parallel to the Forward Line of Own Troops (FLOT). The IPF sends commands to and receives information from the Airborne Relay Facility (ARF) through a secure data link. The operators in the IPF process the collected information and report the intelligence to the tactical commanders and other possible joint consumers via the JTT relay on board the aircraft.

The Precision SIGINT Targeting System (PSTS) is an ACTD under the sponsorship of the Deputy Under Secretary of Defence for Advanced Technology. PSTS is a Joint Service and Defence Agency effort to develop and demonstrate a near-realtime, precision targeting, sensor-to-shooter capability using existing national and tactical SIGINT assets. The third PSTS demonstration, conducted in April 1996 was held in conjunction with an Army Interdiction Counter Fire Exercise (ICE 96-1) at Fort Stewart, Georgia. This demonstration featured processing of both national and tactical SIGINT data at a tactical processing facility, the GUARDRAIL Integrated Processing Facility (IPF) located at Hunter Army Airfield in Savannah, Georgia.

Aerial Common Sensor (ACS)

The Aerial Common Sensor (ACS) system will evolve from the current GUARDRAIL/Common Sensor (GR/CS) and Airborne Reconnaissance Low (ARL) systems. Aerial Common Sensor will combine and evolve all the requirements and capabilities of these systems into a synergistic system which is rapidly deployable, tailorable, and scalable to meet the needs of the land force commander. Based on approved product improvements to the GR/CS and ARL SIGINT payloads, the evolved ACS, SIGINT subsystem will be fully JASA compliant and capable of providing precision targeting locations across the entire spectrum of conflict. The evolved IMINT/MASINT subsystems will be capable of supporting the full range of missions from disaster relief to mid-intensity conflicts (standoff sensors). In addition, both the ground and airborne subsystems will be capable of controlling and exploiting UAV mission payloads, thus facilitating the land force commanders' real-time exploitation, of overflight IMINT, during mid- to high-intensity conflicts. The system will be composed of manned airborne collection platforms, ground based exploitation facilities, wide band datalinks, a robust/tailorable communications capability and a scalable satellite remote capability.

Status

One GR/CS system is authorised per Aerial Exploitation Battalion (AEB) in the MI Brigade at each Corps. Guardrail provided collection coverage along the inter-German border from 1972 to 1990, in Korea from 1974 to the present, and in Central America from 1983 to 1994. Two systems deployed to Southwest Asia during Operations DESERT SHIELD and DESERT STORM. GRCS (Minus) was fielded to Korea in 1988. The first GRCS system was fielded to Europe in 1991, and

Guardrail RC-12 aircraft interior
0085290

For details of the latest updates to *Jane's C4I Systems* online and to discover the additional information available exclusively to online subscribers please visit
jc4i.janes.com

the second was fielded to XVIII Corps in 1994 with a remote relay capability that allowed forward deployment of aircraft while the ground processing facility remains in CONUS. As of mid-2002, one system remained in Korea, one system was in Europe, one was with the XVIII Airborne Corps and one was with III Corps.

Contractor
Northrop Grumman Mission Systems, Reston, Virginia.

UPDATED

LR-100 lightweight ESM system

The LR-100 is a lightweight receiver that provides a fully automatic ESM capability from 2 to 18 GHz. Optional upgrades are available for low and millimetric wavebands. Receiver bandwidths are included for frequency agile CW and complex emitter types. LR-100 is designed for Unmanned Aerial Vehicles (UAVs) and can provide automatic identification, precision direction-finding, cueing and target location. The receiver is connected to a family of available interferometer antennas, which can be tailored in form and fit to a wide variety of ground and airborne applications. A direction-finding accuracy of 0.7° is available from a 254 mm array.

Lightweight is obtained by the integration of ASIC and miniaturised microwave technologies. At 9.8 m^3 and 80 W of power consumption, the low-cost LR-100 is suitable for not only UAV platforms, but any application requiring precision direction-finding capabilities or add-on functions. The LR-100 has automatic self-test and radar identification data can be reprogrammed from any MS-DOS machine with an IEEE 422 interface.

Status
The system was initially designed with the unmanned aerial vehicle (UAV) mission in mind and has since been purchased for the Advanced Seal Delivery System submarine, the Kaman SH-2G Seasprite naval helicopter, notably the SH-2G's purchased by New Zealand, the Sikorsky S-70 Seahawk, and a classified DoD application. The system has been extensively flight tested on the IAI/TRW Hunter UAV and has also demonstrated excellent performance in a mast mounted coastal surveillance application and onboard high-speed coastal patrol boats.

Contractor
Northrop Grumman Electronic Systems, Charlottesville, Virginia.

VERIFIED

IMAGERY INTELLIGENCE

Israel

EL/S-8821 Data and Image Distribution Unit (DIDU)

The EL/S-8821 is a compact Data and Image Distribution Unit (DIDU), a tool for distribution of images to battlefield commanders and users. The images, which are frozen, may be supplied from any standard RS170/CCIR video source or external applications, derived from a wide range of sources, including Unmanned Aerial Vehicles, Imaging Sensor Systems and Unmanned Remote Sensor Systems.

The system's main features include: Alphanumeric/Graphic overlay, Image Enhancement and Image Compression, and Data Reduction for low bandwidth transmission. The system enables the compressed image to be distributed over a variety of tactical field communication systems, digital and analogue radio-telephone systems, and through standard external modems. The images are reconstructed at the Receiving Unit.

The system:
(a) Digitises and displays the frozen images
(b) Alphanumeric/Graphic annotation on frozen or realtime video
(c) Image enhancement: edge enhancement, noise reduction, histogram hyperbolisation, image rotation, image enlargement
(d) Image compression up to a ratio of 1:60
(e) Area of interest enabling definition of two compression ratio areas
(f) Storage up to 64 compressed images in battery-backed memory
(g) Distribution of images or other digital data over a variety of communication systems, including frequency hoppers
(h) Distribution capabilities up to 128 end-users on 16 communication networks
(i) Forward error correction providing perfect data reception, even in noisy and unreliable channels
(j) Signal protocols providing end-to-end positive hand shake, over radio network and multiple relays.

The EL/S-8821 is operated by a remote-control unit and comprises two major modules:

The Video Module handles the input video signal. It displays real time or frozen images and enables the operator to process the frozen image and add an alphanumeric overlay on the real-time or frozen image. This module enables the frozen image, with its overlays, to be compressed at a ratio of 1:60, with high quality decompressed images.

The Communication Module handles the signal protocol, enabling dissemination of the image through the radio network. It includes an intelligent modem which encodes error correction codes and handles the physical interfaces of various communication systems.

Tactical field radio connections can be made through: SINCGARS, VRC-12 family (16 kbps), Digital Radio-Telephones (CDP up to 32 kbps). Analogue Radio-Telephones (DQPSK up to 2400 bps) and standard external modems (up to 64 kbps).

The DIDU can be supplied in vehicular and battery-powered man-pack versions. It can be controlled by a workstation or PC/laptop that implements four communication layers in ISO/OSI model, which can also be used as an expansion storage for more compressed images.

Specifications
(a) Size (H × W × D): 3.2 × 4.7 × 11 in
(b) Weight : 3.3 kg (vehicular version)

Contractor
ELTA Electronics Industries Ltd (a subsidiary of Israel Aircraft Industries Ltd), Ashdod.

NEW ENTRY

IN-TACT Reconnaissance and Surveillance System

IN-TACT is a ruggedised, transportable intelligence workstation that receives and displays real-time video and telemetry data from UAVs and other sensors and payloads. A variety of picture and data manipulation tools is provided to assist the intelligence officer in converting raw data into graphical intelligence overlay on a digital map. The processed intelligence data can then be distributed over tactical radio networks either as compressed video stills or as graphic map overlays and can be stored in graphic, video, static picture and slide form. The system is capable of monitoring the camera footprint onto the video monitor and digital map. It features advanced electronic map manipulation (both 2-D and 3-D) and can interface to a wide variety of RF data links.

The UAV's video and telemetry data are received by means of a Remote Antenna Terminal (RAT), that automatically tracks the UAV's RF signal, receives the RF signal and converts it into base band video and telemetry data. For better reception, the RAT's antenna assembly (rotator, directional and omni), are attached to top of the vehicle or top of the shelter. An automatic mechanism stabilises any inclination. The RAT decodes the down-link information transmitted

by the UAV and passes it on the processing unit for continued processing and display.

IN-TACT has an open architecture and is designed to allow for future expansion.

Specifications
Hardware
(a) Microprocessors:
    Intel Pentium III 800 MHz System Processor
    Video Graphic Processor
    Video Processing Accelerator
(b) Disks:
    20 Gbyte hard disk
(c) Displays:
    15 in high-resolution colour VGA monitor (1,280 × 1,024)
    9 in B/W video monitor (colour option)
Interfaces
(d) Video:
    Input: EIA RS-170, CCIR
    Output: EIA PG-170 CCIR (up to 600 m cable connection)
(e) Serial:
    IBM PC/AT compatible, COM RS-422A/RS-232/parallel
    Front access to LAN communication (10 base-T)

Retrieval of intelligence in real time using IN-TACT 0055194

IN-TACT in a portable container (Tadiran Electronic Systems Ltd) 0113598

(f) Downlink Channels. Biphase modulated signal at selectable rates (L/S/C option x band)

Electrical

(g) Power consumption: 150 W

(h) Power input: 18 to 32 V DC or 220 V 50 Hz AC

Physical

(i) Dimensions: (W × H × D) 535 × 625 × 680 mm (front and rear covers included)

(j) Weight: approx 50 kg

Status

In production. Probably in service with the Israel Defence Force.

Contractor

Tadiran Electronic Systems Ltd, Holon.

VERIFIED

Screenshot from the MRS. In the small window is a picture being transmitted from a UAV or camera, and overlaid on the map is the UAV/camera footprint together with target information and tactical symbology (Tadiran)   0132914

MRS-2000 Manpack receiving system

The MRS is a ruggedised, manpack transportable intelligence computer that receives and displays real-time video and telemetry data from UAVs or other sensors and payloads. The unit comprises a backpack which houses the computer, video receiver and antenna, as well as a lightweight hand-held TFT display unit for the display of video, telemetry, and map data. A directional antenna capability is provided and the system is adaptable to various image standards and datalinks.

A variety of picture and data manipulation tools are provided to assist the intelligence officer in converting raw data into a graphical intelligence overlay on a digital map. Real-time video information is displayed along with platform telemetry data. An electronic map capability and integral GPS receiver provide the operator with a means for system operation, mission management, orientation and a 3-D view of the battlefield terrain. Overlaid on the map are the UAV/Camera footprint along with target information and other standard tactical symbols that help build the complete situation awareness picture.

Applications include: C4I for air, ground and naval forces, mission planning, artillery targeting and ranging, border surveillance and site and sensitive facility security. Still images can be stored in the hard disk for off-line processing. Several users can access the database through a LAN port.

Features

(a) reception and display of real-time AV/UAV imagery and telemetry

(b) advanced electronic map display (2-D and 3-D)

(c) interface to a wide variety of RF datalinks

(d) image processing and video improvement capabilities

(e) tools for building graphic overlays for real-time targets and situation awareness picture

(f) graphic overlays and frozen image storage facility

(g) dissemination of intelligence information over wireless networks

(h) automatic target report generation and artillery correction capabilities

(i) integral GPS receiver for own location

(j) optional up-link for camera control.

Specifications

(a) Functional/capabilities

Reception: C-Band 4.4 to 5.1GHz, L/S band. Optional Ku-band

Range:

C-Band: Up to 10 km with Omni, 40 km with directional antenna

L/S Band: Up to 15 km with Omni, 50 km with directional antenna

Video input standards: RS-170A/NTSC/CCIR/PAL

UAV telemetry support: Pioneer, Ranger, Searcher, UMASS and others

System processor: Pentium III 500 MHz MMX processor with 20 Gbyte hard disk

Hand-held display: 10.4 in TFT 800 × 600 colour, sunlight-readable display monitor with touch panel

Imagery/map overlays: Tactical symbols, text, lines, graphics

MMI display: Digital map and resizeable real-time imagery (full, ½ and ¼) windows

Map raster coverage: Raster and vector maps of all scales

3-D DTM coverage: DTED 1 or 2 for map coverage

Image archive capacity: Limited only by disk capacity

Operating time: Approx 4 h

Batteries: 4 lithium-ion commercial off-the-shelf battery packs

(b) System deployment configurations

Mobile mode: One-person carry, mounted on soldier's back with Omni and GPS antennas for reception on the move

Tripod mode: Unit placed on tripod with solder positioned up to 7 m from unit. System has Omni/directional antenna reception capabilities.

(c) Environmental

Temperature:

Operating: 0 to 50°C

Storage: −20 to 60°C

Shock and Vibration: Per MIL-STD: 81OD/E ground mobile

Water sealing: Rainproof: IP65

(d) External Interfaces

Power input: 18 to 32 V DC

LAN: 10/100BaseTx TCP/IP connectivity

USB: 2 ports

PS2: mouse and keyboard

COM: 2 ports

(e) Options

UAV telemetry support: Multiplatform protocol

Data dissemination: Via tactical radio link for e-mail, tactical situation overlays, frozen imagery using internal proprietary DSP-driven tactical communications controller

Video line driver: Up to 100 m coax composite NTSC/PAL/RS-170A/CCIR.

Auto track antenna pedestal.

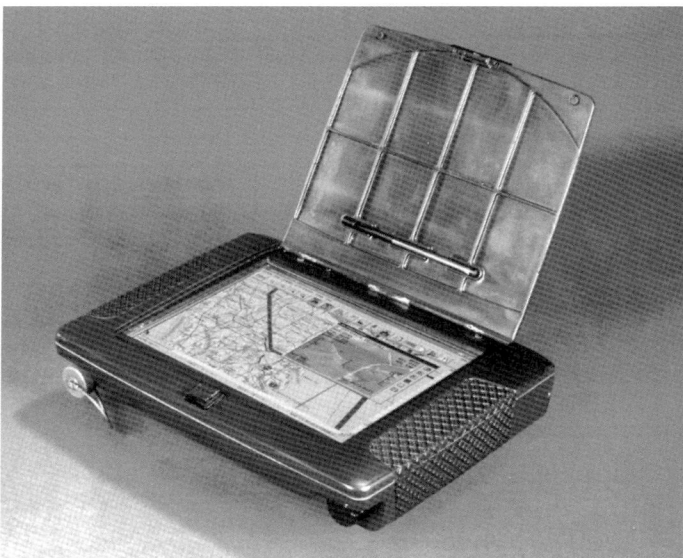

Tadiran MRS Manpack receiving system   0085293

MRS in the manpack configuration (Tadiran Electronic Systems Ltd)
0143767

Status

In service with Israeli forces. Ordered by the USMC, with first shipments scheduled for July 2002.

In June 2003 it was announced that acceptance tests of four systems by the USMC had been successfully completed.

Contractor

Tadiran Electronic Systems Ltd (a member of the Elisra Group), Holon.

UPDATED

Shipborne Receiving Station (SRST)

The SRST is a naval version of the technology employed in Tadiran's Man-pack Receiving System and IN-TACT (see separate entries), enabling the operator to view live images transmitted by an airborne surveillance capability and then overlay them on a map to be fused with other information and analysed. The system receives telemetry and video images from airborne vehicles, including UAVs and fixed-wing and rotary-wing aircraft, and displays them on a B/W video display monitor. Downlinked information transmitted by the AV is decoded and enhanced with relevant telemetry information for better understanding of the displayed image. Frozen images can subsequently be processed offline and displayed on a high-resolution colour graphics monitor (OGD). The OGD displays digital maps correlated to real-time video images. The SRST provides tools for evaluating and analysing the received image and telemetry information.

The downlinked data is received by means of a pair of stabilised Antenna Subsystems (AS), thereby overcoming the problems of coverage caused by screening by the ship's superstructure. The AS auto-tracks the AV's RF signal, receives the RF signal and then converts it into baseboard video and telemetry data. The stabilised AS comprises a rotator, an RF receiver, a dish and omni-antennas, all covered by a radome for environmental protection. Received video may be enhanced and manipulated offline by using functions such as video freezing, archiving and image processing. Digital maps provide battlefield terrain orientation by displaying the 'plots' of the airborne sensors' location. Maps are stored in the system at several scales (1:100,000, 1:50,000 and 1:25,000 or others) in a local magnetic medium and cover an area of 100 × 100 km (configurable according to a customer's requirement). New maps may be loaded during operations by means of a LAN interface from an external digital map source such as a 'Map Loader' Laptop computer. Data input is performed using a trackball and a keyboard, and other elements of the MMI include direct-access functions, switches, displayed buttons and simple menus. The SRST interfaces with the ship's gyrocompass (inertial navigational system) and GPS systems. The interface with the gyrocompass enables stabilisation of the AS by compensating for the ship's attitude in various sea states. Interfacing with the ship's GPS enables the SRST automatically to point the AS to the airborne sensor and pinpoint the ship's position, both while stationary and while under way. The system can operate in conditions up to sea state 3. Up to two SRSTs can be installed on any one platform, and each is capable of receiving data from one airborne sensor, which can be simultaneous without mutual interference.

Specifications

Technical

(a) Reception: C-Band 4.4 to 5.1 GHz; optional L/S, Ku

(b) Range: Up to 75 km

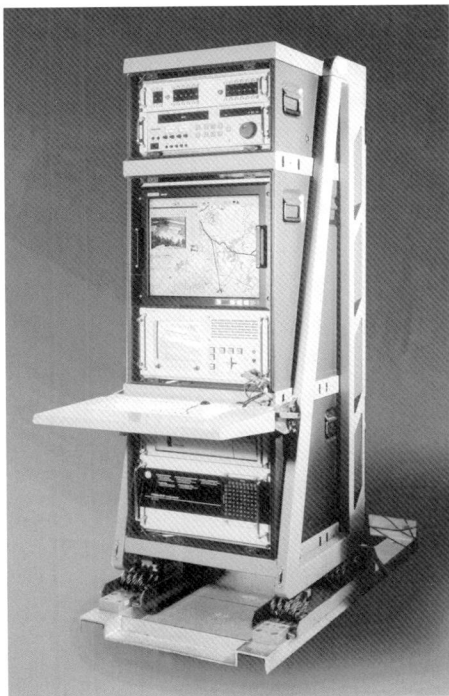

The Shipborne Receiving Station (SRST) showing, from the top, VCR colour graphics monitor and B/W video monitor (Tadiran Electronic Systems Ltd)
0143768

(c) Remote Antenna Terminal: Auto-track RF front-end deployed remotely via F/O interface

(d) Video Input Standards: RS-170A/N

(e) +TSC

(f) Telemetry Support: Standard protocols

(g) System Processor: Pentium III 800 MHz with 20 Gbyte hard disk

(h) Video/Telemetry Recorder: Integral DV compact video recorder for mission recording/interrogation

(i) MMI/Map Display: 20 in TFT 1,280 × 1,024 colours

(j) Video Display: 9 in TFT 800 × 600 B/W

(k) Imagery/Map Overlays: Tactical symbols, text, line graphics and so on

(l) Map Raster Coverage: Multiple scales

(m) DTM Coverage: DTED 1 or 2 for map coverage

(n) Image Archive Capacity: Limited by disk size only

(o) Imagery/Map Overlays: Tactical symbols, text, line, graphics

Environmental

(a) Temperature: Operating 0 to 50°C (optional −20 to +50°C); storage −20 to +60°C

(b) Shock and vibration: Per Mil-Std: 810D/E ground-mobile

(c) Water sealing: Rainproof: IP65

(d) Sea state: Up to Level 3

External Interfaces

(a) Power input: 110/220 V AC

(b) LAN: 10/100 BaseTx TCP/IP connectivity

Options

(a) Telemetry Support: Multiplatform protocol

(b) Data dissemination: Via tactical radio link for e-mail, tactical situation overlays, frozen imagery using internal proprietary DSP-driven PC-COM communications controller

(c) Video input standards: CCIR/PAL

(d) Video line driver: up to 600 m coax composite NTSC/PAL/RS-170A/CCIR

(e) Swappable MAP/video: Two 14 in displays with exchangeable position (lower screen/upper screen)

Status

Claimed by the manufacturer to be in use in several navies.

Contractor

Tadiran Electronic Systems Ltd (a member of the Elisra Group), Holon.

UPDATED

Saudi Arabia

Reconnaissance Mobile Exploitation Facility (RMEF)

The Reconnaissance Mobile Exploitation Facility (RMEF) is a variant of the UK's Ground Exploitation Equipment (GEE) (now replaced by TREF (see separate entry)) and provides the ground elements of the reconnaissance cycle.

In 1987, Computing Devices (now General Dynamics (UK)) and the then British Aerospace (now BAE Systems) defined a ground exploitation facility to support the reconnaissance Tornado aircraft ordered by the Royal Saudi Air Force under Project Al Yamamah.

Although based strongly on the GEE system, the RMEF differs from the RAF's system because the RSAF's planned usage requires mobility and the optional ability to operate as two smaller facilities instead of a single central one. To meet the former requirement, the system is housed in four air-transportable cabins and to satisfy the latter, additions and modifications to the operating software allow the system to be operated as two smaller two-cabin facilities. The cabin installations have inevitably necessitated additional mechanical redesign of the workstation equipment, together with some ruggedisation to meet the air/road transportability requirements. Support facilities for the RSAF include a separate map preparation

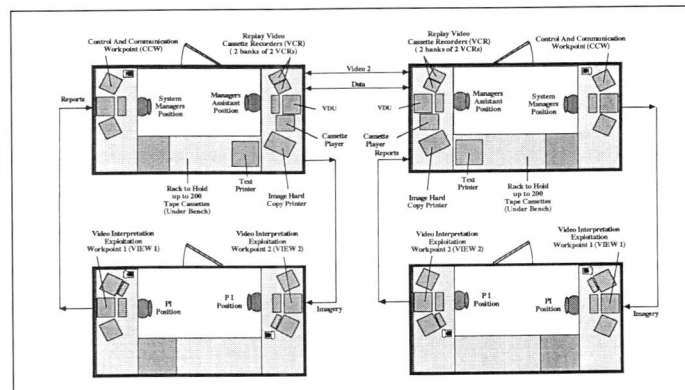

RMEF workstation configuration

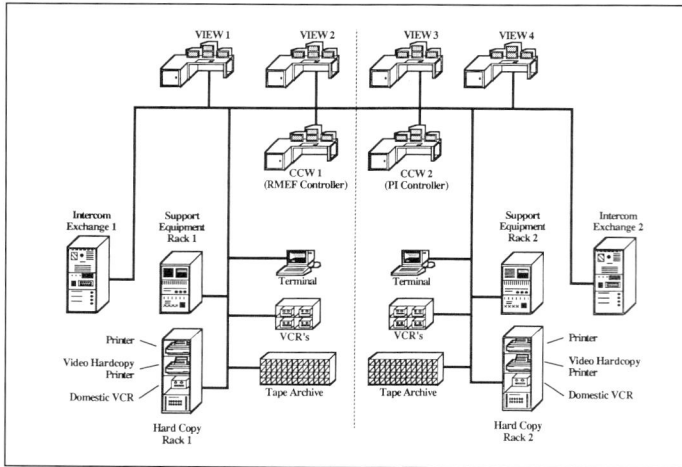

RMEF functional architecture

station (to allow generation of disks for the video map generators in each workstation) and a maintenance rig.

Status
Delivery of the first system took place in early 1992.

Contractor
General Dynamics (UK), St Leonards-on-Sea, East Sussex, UK.

UPDATED

United Kingdom

Ground Imagery Exploitation System

The Thales Ground Imagery Exploitation System (GIES) supports the analysis, exploitation and interpretation of airborne imagery from the Thales electro-optical reconnaissance system, full details of which can be found in *Jane's Electro-Optic Systems*. It is a soft copy system designed for use by image analysts operating in the tactical environment.

The GIES is based on a modular architecture and comprises an Image Interpretation Workstation (IIW), a Map Display/Report-Writing Workstation and a Recce Report Workstation (RRW), driven by Silicon Graphics processors. The prime purpose of the IIW is target location and selection. Having downloaded cassette imagery on to hard disk, the operator can view the dynamic image (with a selected area × 8 zoom facility) with basic image manipulation (brightness, contrast) and insert a single georeference location tag, by using aircraft navigation and sensor data streams which have been continuously recorded during the mission. It can be printed out via a Seikosha VP-4500 thermal screen printer.

The map display workstation generates a moving map display that shows the location of the images and aircraft flight path, while displaying a frozen georeferenced image. It is used for basic and enhanced image manipulation and initial report writing, and is connected to a Paragon ELT 7000 light table for further annotation. It connects to a laser, thermal or photo-quality printer.

The RRW is used to create formal reports and is equipped with Microsoft Office software plus Adobe Photoshop. High-quality reports can therefore be created and transferred to a variety of softcopy formats for dissemination. These formats include floppy/super floppy disks, ZIP, JAZ, CD-ROM or DVD. Alternatively they can be passed electronically via a CIS. In the UK's case, the system is compatible with RAFCCIS and JOCS (see separate entries).

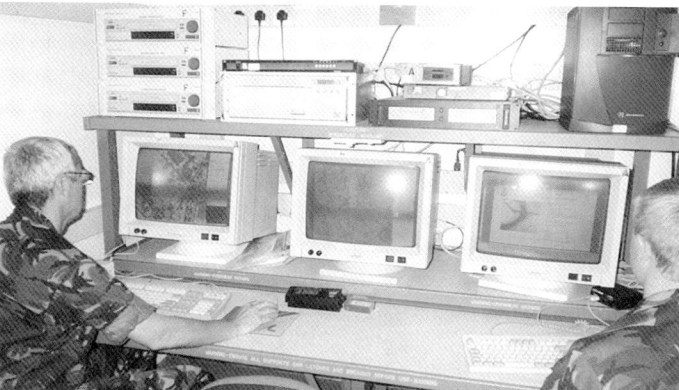

The three-screen workstation of the Ground Imagery Exploitation System installed in a deployable cabin. From the left are the Image Interpretation Workstation, the Map Display/Report-Writing Workstation and the Recce Report Workstation (Jane's/Michael J Gething) **NEW**/0557218

The ground station can either be installed in a vehicle-borne container or dismounted into a building, providing air conditioning is available.

Status
6 GIES are in operational service with the UK. The system is due to be installed in UK CVS. Also in operational service with the Belgian Air Force.

Manufacturer
Thales Optronics , Bury St Edmunds, Suffolk.

UPDATED

Transportable Reconnaissance Exploitation Facility (TREF)

The Transportable Reconnaissance Exploitation Facility (TREF) is a scalable, transportable, multisensor, multiple user exploitation facility which provides both primary (near-realtime) and secondary (off-line) image exploitation functions, collateral data and system management functions, and a full C3I suite for tasking and for voice, text and imagery dissemination. TREF downloads imagery in a variety of formats to digital computer disks, enabling it to be instantly accessible. A high-resolution digital moving map display provides the analyst with correlated geographical and target information. The TREF is fully scaleable from a single workstation up to a full Tempest and NBC proof complex with one to eight 20 × 8 × 8 ft cabins connected using a fibre-optic LAN. Cabins may be towed on integral running gear, loaded on trucks, underslung from Chinook helicopters, or deployed within C-130 Hercules aircraft.

TREF utilises the STANAG 7023 reconnaissance data format to ensure interoperability within NATO. For the transmission and reception of images, the STANAG 4545 standard has been adopted.

A small, portable version, the Real-time Access Portable Imagery Display (RAPID), now known as the X-Lap, has also been developed. X-Lap is supplied as either a single or dual laptop system. The second laptop, although not essential, provides a dual display allowing simultaneous Imagery and Map display and supports additional applications such as the Recognition Material Cell guides from JSPI, and Microsoft Office.

Applications
(a) Image storage and retrieval
(b) Secondary image dissemination (STANAG 4545)
(c) Assisted report writing
(d) E-mail and web browser
(e) Display orientation: north up, head up and 90° to track scales: 1:500,000, 1:250,000, 1:100,000 and 1:50,000
(f) Dynamic overlays: aircraft position, track history and sensor field of view
(g) Multiple map data formats: raster (ASRP 2)

Communications
(a) TCP/IP is used as the communications transport and switching protocol
(b) Point-to-point communications: serial interface for connection to Brent phones and other data modems
(c) Networking communications: 100BaseT/10BaseT auto-sensing Ethernet interface for connection to DLAN and Bowman LAS type infrastructures
(d) A variety of communications interface options:
SATCOM Interface
RTTS interface
Improved Data Modem (IDM) for VHF radio interface
MIDS terminal interface for access to Link-16
High-speed serial interface
ATM interface
ISDN interface
Wireless LAN
Compatibility with the UK's Joint Operations Command System (JOCS) (see separate entry) is provided in a modified TREF
(e) Display orientation: north up, head up and 90° to track scales: 1:500,000, 1:250,000, 1:100,000 and 1:50,000
(f) Dynamic overlays: aircraft position, track history and sensor field of view
(g) Multiple map data formats: raster (ASRP 2)

Exploration preparation
(a) Historic imagery and map review
(b) Historic sensor coverage display
(c) Display orientation: north up, head up and 90° to track scales: 1:500,000, 1:250,000, 1:100,000 and 1:50,000
(d) Dynamic overlays: aircraft position, track history and sensor field of view
(e) Multiple map data formats: raster (ASRP 2)

Imagery exploitation
(a) Comprehensive find and search facilities including find from alternative sensor, map co-ordinates, time, event
(b) Multiple image display modes: waterfall, framing, panoramic roll across
(c) 360 Gbyte high-speed random access image buffer
(d) Magnification and zoom (× 1 to × 16)
(e) Dynamic image enhancement tools: gamma correction; image inversion; edge enhancement; crisp; instantaneous contrast stretch

(f) Dynamic image geometric correction: rectilinearisation; V/H correction
(g) Image analyst event marking
(h) Frozen image exploitation and enhancement: rotate; histogram equalisation; image stretch; image comparison; contrast stretch; zoom; mensuration; target marking
(i) Dynamic map:
Moves in synchronisation with aircraft navigation data
Display orientation: north up, head up and 90° to track scales: 1:500,000, 1:250,000, 1:100,000 and 1:50,000
Dynamic overlays: aircraft position, track history and sensor field of view
Multiple map data formats: raster (ASRP 2)

Status
TREF entered RAF service in mid-1997 in support of Tornado GR1 and GR4a squadrons. A system based on TREF has been developed by GD (UK) Ltd to meet a contract by Lockheed Martin for the USAF/Air National Guard Squadron Ground Station (SGS) in support of the F-16 Theatre Airborne Reconnaissance System (TARS).

Manufacturer
General Dynamics (UK) Ltd, St Leonards-on-Sea, East Sussex.

UPDATED

United States

Forward Area Support Terminal (FAST)

The Forward Area Support Terminal (FAST) is a transportable, modular, survivable, stand-alone UNIX Tactical Exploitation of National Capabilities (TENCAP) system, which was specifically designed for the support of independent brigades. The FAST receives, correlates, integrates and disseminates multidisciplined information, essentially providing the functionality of the MITT (see separate entry) in a box. It can be deployed in a variety of configurations. FAST has an open system architecture using VME UNIX-based SunSparc 20 processors, has an expandable LAN and has a wide range of communications interfaces. The Chariot S-band satcom terminal (see separate entry) is also part of the system.

Its products include intelligence reports; annotated National Imagery Transmission Format (NITF) Secondary Imagery Dissemination (SID) (for example classified satellite imagery); annotated battlefield imagery; target data; a near-realtime picture; and web page reports.

The FAST-Ultra (FAST-U) is a downsized version of the FAST based on a UNIX laptop. It is about one third the size of a normal FAST with the same basic capabilities.

Status
In service with the US Army.

NEW ENTRY

Mobile Integrated Tactical Terminal (MITT)

The Mobile Integrated Tactical Terminal (MITT) is a small, compact, highly mobile, self-sufficient system that provides divisional commanders with Tactical Exploitation of National Capabilities (TENCAP) facilities. The MITT receives, processes and disseminates imagery intelligence (imint) and signal intelligence (sigint) information. It provides some of the functionality of the AEPDS (see separate entry) in a smaller and more mobile configuration. While in the travel configuration and moving, the MITT can continue to receive UHF broadcast and imagery data. Its ability to receive and process national/theatre level data, coupled with its mobility and small size, make it particularly appropriate for use in early entry operations. It is mounted on a HMMWV and can be carried by C-130.

It is installed in two S788 shelters (102 in long × 84 in wide × 67 in high), mounted on two M1097 HMMWV, heavy variant (total weight 4,200 lbs). Power is provided by a specially fabricated 15 kW generator mounted on a modified M116A2 trailer. The MITT comprises three workstations: one communications workstation and two intelligence analyst workstations, configurable as either TENCAP electronic intelligence (elint) or imagery analyst workstations. Any of the three workstations can be remoted outside the shelters. The MITT provides the capability to receive multiple-source imint/sigint data and messages from external sources, analyse and integrate data into correlated databases, and rapidly disseminate intelligence data, secondary imagery and other products to tactical users. The MITT provides the tactical commander targeting information, the current enemy situation, and terrain analysis data.

The E-MITT is an enhanced version of the MITT. The enhancements include the capability to bring in weather data using a modified Chariot antenna (see separate entry) with a MiDAS receiver. The E-MITT receives weather data from the Defense Meteorological Satellite Program (DMSP), Geostationary Operational Environmental Satellites (GOES) and the National Oceanic and Atmospheric Administration to provide weather information. It also has the DAMA SUCCESS radio, which allows both voice and data to be transmitted over the DAMA network using both 5 and 25 KHz channels.

Status
In service with the US Army. The MITT is organic to the MI company in the armoured cavalry regiment, separate brigade, and Interim Brigade Combat Teams; MI battalions in the heavy, light, airborne and air assault divisions; and operations battalions at corps and echelons above corps MI brigade. Due to be replaced by elements of the Tactical Exploitation System (TES) (see separate entry).

NEW ENTRY

Mobile Processing, Exploitation & Dissemination (MoPED II)

MoPED II is a self-contained, C-130 transportable system which processes, exploits and disseminates airborne and satellite-derived commercial hyperspectral and multispectral data, and integrates any remotely sensed digital image data applications. The product is either tactical reports or image products, and because

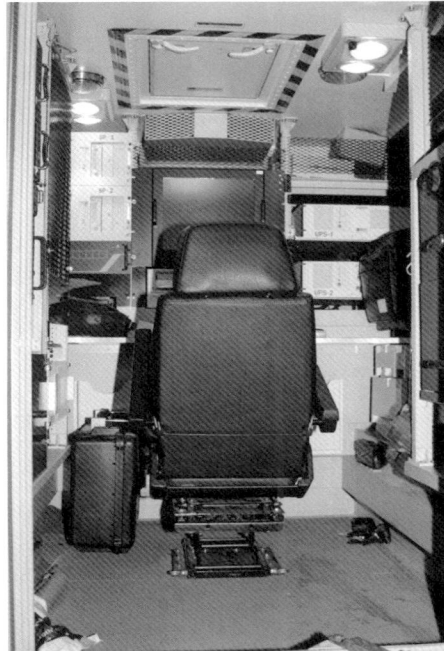

Internal shot of the MoPED II vehicle (Patrick Allen/Jane's)
NEW/0526708

The MoPED II vehicle and tent, showing the CHARIOT satcom aerial (Patrick Allen/Jane's)
NEW/0526722

MoPED II workstations with the 'lunchbox' on the left (Patrick Allen/Jane's)
NEW/0526780

it is derived from commercial sources it can be made available to all members of any coalition of which the US is a member.

MoPED II can support up to 25 workstations in a spectral operations cell. Also available is a light version, known colloquially as a 'lunchbox' which, together with a communications package and a plotter, provides the facilities and capabilities of the heavier version forward to tactical units.

Operational status

The system has been operational since January 2000. Based at Peterson Air Force Base, detachments are reported to have supported operations in both Iraq and Afghanistan, as well as most recent major exercises. Support has been provided to EUCOM, CENTCOM and SOCOM as well as to civil agencies.

NEW ENTRY

Softcopy Search

Softcopy Search is an integrated software package developed to enable the intelligence analyst to search large quantities of imagery for significant changes, replacing the process of searching hard copy film on a light table, in response to a requirement to deal with an expected dramatic increase in data collected by sensor systems. It is built on the Microsoft Windows® operating system with standard Intel-based hardware, to take advantage of the latest specification PC-based hardware using high-performance yet widely available graphics accelerators and disk subsystems. The imagery analyst workflows are brought together into one fully integrated COTS package delivering the functionality for data ingest, management, exploitation, production and dissemination. All imagery analysis workflows are Image Chain Analysis (ICA) approved.

Typically the system will be configured with a high-resolution colour monitor for data fusion (currently 9.2 Mega-pixels (3840 × 2400)). This is used for imagery management, fused display of mapping, imagery and intelligence data, and general reporting tasks. The second monitor is used for detailed imagery analysis. The system will handle multiple images simultaneously, each of which can be multigigabyte in size, performing on-the-fly rectification. To minimise phosphor latency, give the smoothest roam, and provide the optimal spectral response for multibit imagery a high-specification ultra-sharp monochrome monitor is used. Collaboration between analysts can be supported with the addition of a large format, wall-mounted, flat panel display. Extensions for targeting applications, the integration of geo-referenced video, and a stereo module for 3-D (x,y,z) mensuration are currently under development.

Softcopy Search consists of three integrated COTS software components which perform different roles:

(a) Data management is performed by Zeiss/Intergraph (Z/I) TerraShare and OrthoManager software which can ingest and manage terabytes of imagery and other collateral data. Configuration, management and control of search tasks is performed via a standard Windows® Explorer TerraShare interface. The management software provides a single physical storage location for individual files, a logical view of files which permits configuration to suit search task requirements, and file access across distributed networks. It minimises the impact on network loading, provides browse image data right in the Windows® Explorer interface with basic view controls, and will undertake initial imagery checks using geographic coverage searches and metadata, with fast access to imagery.

(b) Data Fusion is performed by GeoMedia Professional Data Fusion (incorporating Paragon ELT5500), which provides geospatial data search capabilities to find, assess, and select imagery for virtual mosaic broad area search. Its open architecture enables on-the-fly data fusion of imagery with raster and vector map data, terrain information and spatially referenced intelligence data, and it provides ease of connection and access to multiple data and database sources. It also provides the ability to control search

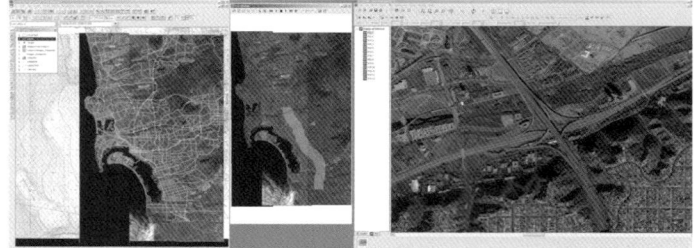

The screenshot illustrates an example Softcopy Search display. It includes three windows that typically would be used in a dual monitor configuration (the data management window is not shown). On the left is the Data Fusion Window, in the middle a Pan Overview Window for controlling the Softcopy Search, and on the right the Image Exploitation Window. The Data Fusion and Pan Overview Windows would typically be viewed in the left monitor. The Data Fusion screenshot illustrates fusion of raster map data with image coverage overviews, Intelligence Database points of interest (simulated), and topographic vector data. The road highlighted in green has been selected as a linear search constraint. The Pan Overview Window is used to control the softcopy search providing video-like controls for start, stop, resume and speed of roam in the Image Window. It provides a record of the area that has been reviewed by the analyst using a 'snail trail' (see green swath in Pan Window). This can be stored for later review or can be retrieved by another analyst on an incoming shift. The Image Exploitation Window is used to view the specified imagery. In this illustration the route selected in the Data Fusion window is being tracked and searched for points of interest (Intergraph)   *NEW*/0529686

patterns with geospatial data (for example, reference point searches; area searches using geographic constraints; or linear searches). For instance, the system can automatically track the imagery along a selected road allowing the operator to roam and search for points of interest in the imagery. Any spatially referenced data may be selected as a search constraint including, for example: line features - rivers, railways, boundaries; point features; designated area features; or other spatially referenced defence/ intelligence data. (See illustration.)

(c) Image Exploitation. In the Softcopy Search mode the image window is primarily used to roam and search imagery. However the window also provides support for a full suite of ELT functionality. Paragon ELT5500 software is incorporated in GeoMedia Image Professional. Its range of functions includes maintenance of precise image quality; support for over 25 image formats; georegister of images to maps, executing change detection using vertical or horizontal wipe, blend or flicker modes; an extensive set of image exploitation and manipulation features, geo features, annotation features, and display tools; and smooth pan performance for softcopy Broad Area Search. It is NITF and NSIF Certified on Windows 2000 ,and DII/COE compliant.

A typical Softcopy Search display in a dual monitor configuration will include three windows (not including the data management window), the Data Fusion Window, a Pan Overview Window for controlling the Softcopy Search, and the Image Exploitation Window. The Data Fusion and Pan Overview Windows would typically be viewed in the left monitor. A screenshot illustrating this accompanies this entry. The rectified virtual mosaic is created on-the-fly from imagery selected in Data Fusion Window. Points of interest can be identified, recorded and saved for further analysis. Approved mensuration tools are fully integrated and there is a full suite of RLT functionality.

The system has considerable interoperability. It provides native read support for standard imagery and raster map formats including: GeoTIFF, NITF (with RPCs), NSIF, TFRD, USGS DOQ, CIB, TIFF (with World files), BMP, PCX, JFIF, GIF, Intergraph, MrSID, ADRG, CADRG, ASRP, KMRG. Read/write support for NITF 2.0 and 2.1, NSIF 1.0, JFIF/JPEG, Raw Image File, Sun Raster File, TIFF, GeoTIFF, and Windows Bitmap. It provides direct read of military standard vector and terrain

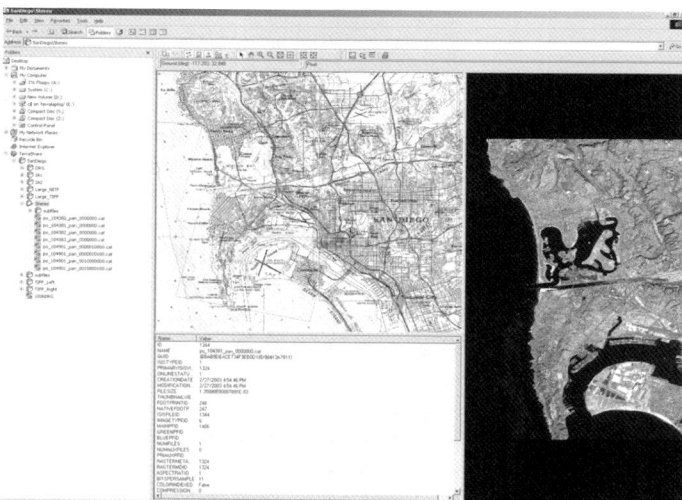

Screenshot showing a typical TerraShare Windows® Explorer view (Intergraph)   *NEW*/0529683

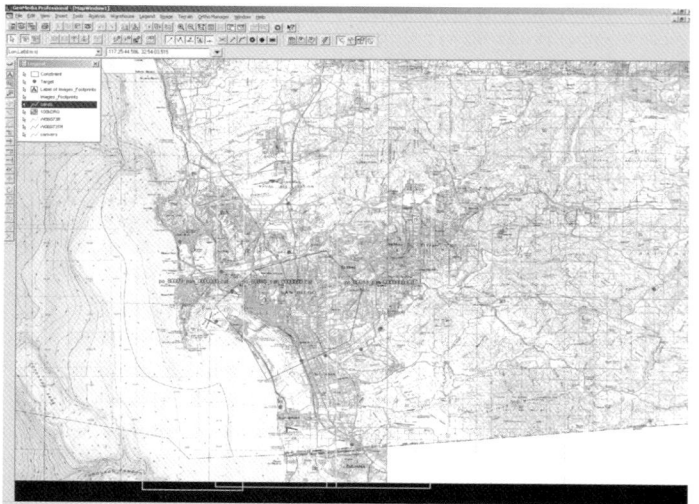

SoftCopy Search screenshot showing the GeoMedia Professional Data Fusion Window illustrating fusion of: USGS DLG raster maps; IKONOS imagery coverages; ArcInfo vector data; and simulated collateral Intelligence data (showing points of interest) (Intergraph)   *NEW*/0529684

An Image Exploitation screenshot (Intergraph)   **NEW**/0529685

matrix formats including VPF and DTED. A large variety of vector formats can be read directly including Shapefiles, ArcInfo, MapInfo (MID/MIF), MGE, MicroStation and AutoCAD. Using the FME data server (available from Safe Software) a considerable number of other formats can be supported. The system supports many OGC standards including the capability to read/ write OGC compliant GML 2.1 data. It can directly read and write to a variety of databases including Oracle, SQL Server, IBM DB2 and Microsoft Access.

Status
As at early 2003 the software had been adopted for integration in major US DoD programmes, and is also in use in Germany and other NATO countries.

Manufacturer
Intergraph Corporation, Huntsville, Alabama.

NEW ENTRY

....

Tactical Exploitation System (TES)

The Tactical Exploitation System (TES) will replace current Tactical Exploitation of National Capabilities (TENCAP) systems, including the AEPDS (see separate entry). It will receive data from satellite and aircraft sensors via direct downlinks and from other ground stations, and process, exploit and disseminate it, acting as the interface between US national systems and in-theatre tactical forces. It will combine all TENCAP functionality into a single, integrated scalable system specifically designed for split base operations, consisting of Forward (a HMMWV-based element) and Main (a large container-based element). TES is designed to provide maximum flexibility ranging from a single HMMWV to a fully co-located facility with up to 40 operator workstations, from which multiple IMINT, SIGINT, CrossINT or dissemination functions can be performed. (CrossINT operations provide a 'layered' view of the data products in TES via CrossINT filter software.) All the configurations are intended to have C-130 drive-on/drive-off capability.

TES will have extensive communications capabilities, including UHF, S, X, C and Ku-band and will simultaneously receive multiple TRAP and TIBS broadcasts. It interfaces with and serves as the preprocessor for ASAS, CGS, DTSS, (see separate entries) and has a digital interface with AFATDS (see separate entry) for high-priority mobile and fleeting targets. It complies with the Defense Information Infrastructure COE and is an integral part of the Distributed Common Ground Station.

The TES is composed of two operational nodes, TES Forward and TES Main. Each node has essentially identical functional capabilities but is carried in different vehicles (TES Forward is HMMWV mounted, while TES Main is in Wolfcoach trucks); have different numbers of workstations; have different antennas (Forward normally has a Modular Interoperable Surface Antenna (MIST) (see separate entry), Main normally a Tri-band Satellite Communication Subsystem (TSS)); and have different mobility capabilities.

General Capabilities
Communications
(a) UHF Satcom and LOS (DAMA SUCCESS)
(b) MiDAS S-band
(c) Interface with Chariot S-band
(d) TSS High Rate channels (X, Ku, C)
(e) Interface to TROJAN SPIRIT
(f) IP network access (Garrison, JWICS, SIPRNET)
(g) MIST/CDL
(h) Secure Telecom (Voice and Data)

SIGINT
(a) Pulse level processing
(b) Correlation and integration of all received data

(c) Multiple input pathways: S-band, UHF, autodin/DMS, IP networks
(d) Access to CrossINT data

IMINT
(a) Direct access to theatre tactical imagery
(b) Complete tactical mission planning and monitoring of U2 and UAVs
(c) Dynamic retasking of U2 sensors
(d) Integration of the common imagery processor
(e) Receipt of national imagery
(f) First phase exploitation and SID production
(g) Integration of NIMA products for exploitation management
(h) Access to CrossINT data

TES Forward
There are five inherent configurations within a TES Forward.
(a) The TES Forward (Basic) consists of two HMMWV mounted shelters providing a communications package which can include SUCCESS radio for broadcast SIGINT and UHF communications, JWICS and SIPRNET; Generic Area Limitation Environment (GALE) for SIGINT analysis; Secondary Imagery Dissemination (SID) functionality; a workstation and a generator. It can be transported on a single C-130 and can be driven on or off without any special handling equipment.
(b) The TES Forward (Full SIGINT) adds a further HMMWV shelter containing the Miniaturised Data Acquisition System (MiDAS) to the two from the Basic configuration, together with additional multipurpose workstations, an additional generator and the MiDAS antenna. It can be transported on two C-130s or a single C-141. It provides the same capabilities of the Basic plus additional SIGINT capabilities provided by MiDAS.
(c) TES Forward (National IMINT) adds to the Basic configuration a HMMWV shelter containing an additional imagery analysis facility, providing the capability to handle 'US national imagery' - a euphemism which usually refers to classified satellite imagery. This configuration includes additional workstations, a generator and an interface to Trojan Spirit (see separate entry), which has to be provided separately by the unit. It can be transported on two C-130s or one C-141.
(d) TES Forward (Full IMINT) consists of the Basic configuration plus three additional shelters which provide the subsystems to receive, process and exploit tactical and national imagery. This configuration includes workstations, a generator, a MIST antenna, and an M1085 vehicle to tow the MIST. It can be transported on four C-130s or two C-141s.
(e) The complete TES (Forward) consists of 6 HMMWVs and a cargo truck. This provides the combined functionality of the different configurations. It can be transported on six C-130s, three C-141s, two C-17s or one C-5A.

TES Main
TES Main is rack-mounted in Wolfcoach trucks. It consists of a main tactical mission vehicle, a main communications vehicle, a TSS vehicle and a main generator vehicle, plus the TSS antenna and a 5-ton truck to tow it. TES Main can be transported on seven C-130s, four C-141s, three C-17s, or two C-5As.

Distributive TES
Distributive TES (DTES) is being procured to replace MITT (see separate entry). It is technically and functionally equivalent to a TES Forward (Basic), but the term DTES is used to indicate the TES configuration that will be delivered as a separate, stand-alone system to Division level, to distinguish it from the TES Forward (Basic) that is an inherent configuration within the larger TES Forward at Corps level, although it can be expended if required. It can be deployed on a single C-130.

The DTES is contained in two HMMWVs, plus a trailer-mounted 30 kW generator. The Forward Communications Vehicle (FCV) contains the system server and database, the UHF SUCCESS radio, public server (secret collateral level), a small Imagery Product Library (IPL) server, tactical communications system processor and the Imagery Support Server Environment Guard (ISSEG). The Communications Support Vehicle (CSV) contains the Sensitive Compartmented Information (SCI) public server, a small IPL server and a multifunction workstation.

DTES provides a broad range of communications connectivity and can operate in a collateral-only environment (that is without using SCI) if necessary.

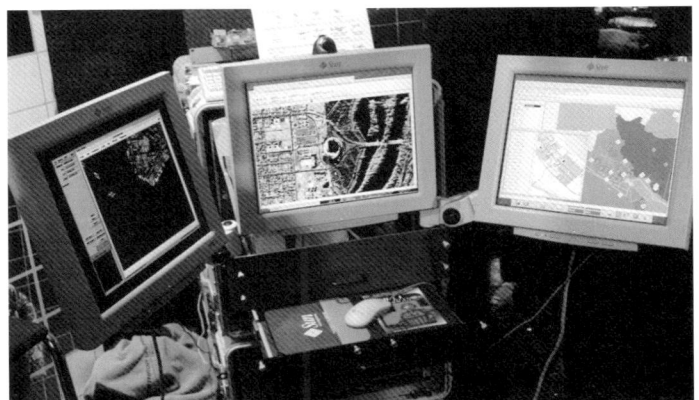

A TES demonstrator, possibly TES-Light, at the AUSA 02 Exhibition
(Patrick Allen/Jane's)   0526709

TES-Light

TES-Light is the TENCAP system intended to provide support at Brigade/Regiment level and to Special Operations Forces. It will be the replacement system for FAST systems (see separate entry), and is intended to provide TENCAP facilities in the smallest possible footprint. It will allow either stand-alone or integrated receipt, processing, analysis and dissemination of selected national, theatre and tactical imagery; SIGINT; and MASINT.

Synthesised UHF Computer Controlled Sub-System (SUCCESS)

SUCCESS is a fully-automated, microprocessor-based, computer-controlled UHF-band radio, which handles data over one transmit and three receive channels simultaneously. The built-in Tactical Receive Equipment (TRE) processor accepts up to three simultaneous TDDS or TADIXS-B broadcasts and will transition to the Integrated Broadcast Service (IBS). A recent improvement has been the incorporation of DAMA capability.

Imagery Support Server Environment (ISSE) Guard

The ISSE Guard system provides a trusted interface for the high-speed, bi-directional, digital transfer of intelligence information including e-mail, imagery, graphics, text and composite products between dissimilar security domains. It provides a link between sensitive processing enclaves and customers' collateral, common use environments. The system consists of two components, the Common Guard Interface (CGI) and the Guard. The CGI provides high and low side users with the ability to select, filter, validate and securely transfer files from the workstation to the Guard. The Guard software, which runs on Sun platforms under the Trusted Solaris multilevel secure B1 compatible OS, serves as the boundary control device. It accepts information from the CGI software and provides the capability securely to connect, verify, downgrade/upgrade and transfer data across the security boundary. The application was upgraded in 1999 from the Harris Night Hawk based platform to an application that can run on any Unix workstation supporting the Trusted Solaris OS. Further upgrades are planned.

Status

The first TES (Forward) was delivered to XVIII Airborne Corps in October 1999, following a 32-month 'contract-to-fielding' schedule. In the spring of 2000, the XVIII Airborne Corps' TES received Synthetic Aperture Radar (SAR) and electro-optical/infra-red (EO/IR) data directly from a U-2 platform, and also was able to receive, process and exploit SAR imagery from the Global Hawk unmanned aerial vehicle via a direct downlink. In June, the same deployed HMMWV-based TES system also demonstrated direct receipt of IR imagery from a US Navy F-14 aircraft. Imagery and SIGINT targeting products were then transmitted to the F-14 cockpit for real-time targeting. A second van-based unit was delivered to Fort Bragg in the summer of 2000. In October 2000 it was announced that Northrop Grumman Corporation's Electronic Sensors and Systems Sector had been awarded a US$122.8 million contract for up to 12 TES in various configurations (TES Main, TES Forward and Distributive TES). DTES was initially fielded to 82 AB Div in July 2001. It is intended to develop and field an initial prototype TES-Light system by mid-2004, with 20-plus systems fielded between 2005 and 2007.

TES is incorporated into the US Navy's Naval Fires Network (rack-mounted on board ship) and the Littoral Surveillance System (HMMWV mounted), and portions of its functionality support dynamic battle management in the USAF's Air Operations Centres.

Contractor

Northrop Grumman Electronic Systems (Prime), Baltimore, Maryland.

NEW ENTRY

COMPUTING

Terminals and workstations
Software

TERMINALS AND WORKSTATIONS

Belgium

Altium 3000

Altium 3000 is Barco's rugged portable computer, designed particularly for use in vehicles and helicopters, and intended for battlefield management, forward observer and fire control applications. The HMI includes 12.1″ TFT display, 12 user-definable function keys, touch screen, backlit keyboard (optional) and a wide range of operational controls (including alarms, voice control and monitoring LEDs). A wide range of I/O options are available, which allows for project-specific customisation and configuration. Up to 8 high-density rugged connectors are available. USB, VGA, serial, parallel, SCSI and antenna can be provided. The open and modular architecture provides spare slots to insert standard (PC 104 / PC 104+ or PCMCIA) cards or custom boards. There are two chassis versions: the one with increased depth provides extra capacity to integrate CD-ROM, DVD, floppy drive, custom-designed boards or GPS.

Specifications
(a) Standard operating system: Windows 2000. Windows NT as an option.
(b) Standard processor is Celeron 400 MHz (low power). Celeron 700 MHz, Pentium 400 MHz or 700 MHz as options.
(c) Standard HDD: 20 Gbytes. 40 Gbytes as option.
(d) Dimensions(W × H × D): 12.6 × 10.4 × 3.9 in/4.9 in deep chassis.

Status
Entered production in 2001 and sold to at least one unspecified customer.

Contractor
Barco, Kortrijk, Belgium.

VERIFIED

Altium 3000 rugged computer (BarcoView)　　　0143774

Altium 4000

The Altium 4000 is a compact multifunctional rugged pentium-based computer designed for use in vehicle management, tactical battle management systems or as a communications server. The unit's design provides a range of customisation options, with room for functionality expansion and technology upgrades (for example PC 104 or PCM/CIA cards). It drives Barco displays (VGA to SXGA, 10 in to 21 in) and the modular concept offers a wide range of I/O options for project-specific configuration.

Manufacturer
Barco, Kortrijk, Belgium.

NEW ENTRY

Barco flat panel displays

Barco have produced a range of flat panel displays in a variety of sizes. See below for specifications table. They are incorporated into a wide spectrum of C4I systems.

Modular Rugged Flat Displays (MRFD)

The MRFD range is designed for applications which require the viewing of graphics, video and other RGB data on a single display and combines a Display Module (DM) with a Display Control Module, connected by a single cable, the Barco Intermodule Link (BIL). This allows the DM to be remoted from the DCM, taking up less space. The DCM can also be attached to the DM to provide a single installation. The main features are:
(a) maximum of 4 inputs (up to 3 simultaneously displayed) from:
　　up to 2 analogue (RGB) inputs
　　up to 2 digital (DVI) inputs
　　up to 2 video inputs (including NTSC, PAL, RGB)
(b) LCD display modules from 17 to 24 in
(c) Resolution from SXGA to WUXGA
(d) Onboard video mixing
(e) Optional LCD flicker compensation
(f) Two levels of ruggedisation.
The DM displays camera images, radar data and graphics, which have been digitally mixed by the DCM. The DCM provides the interface between the DM and the sensor. The specific configuration required can be defined by the selection of appropriate boards for the DCM.

Network-Centric Display Stations

These can be used in Thin Client applications (independent working positions with limited functionality) or as Display Stations providing local computing applications with network interconnectivity. The station can be made application-specific both for input/output, through a customisable panel, and for processing applications. Expansion slots allow mission-specific customisation, future functionality expansion and technology upgrades; a PCI bridge supports the use of various cards; and multiple types of disk drives are available. For increased security the display station has an optional smart card reader which can be integrated into the bezel of the LCD display or the keyboard. The seamless cursor option allows a single user to run network applications on any network-centric display from his working position: the user has an active cursor which can be moved to another display station and can initiate all applications of that station, allowing control of any working position by another if needed. Two variants are available, the 10.4 in TID and the 20.1 in Rugged Display Station.

10.4 in TID 126

The 10.4 in TID (Terminal Input Device) 126 consists of a rugged active matrix TFT colour LCD, incorporating a user-programmable Pentium microprocessor. The TID can be configured as an X-terminal or used as an operator working position for network-centric applications. The TID 126 series can be equipped with a flash disk or hard disk, has a keyboard and mouse interface and also includes a network interface. Options include PCB coating (Humiseal), push buttons on the front

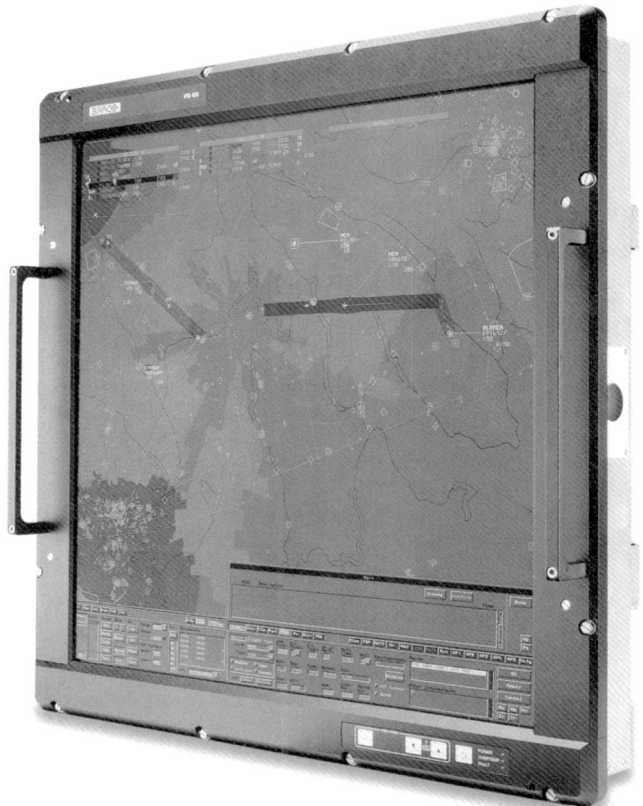

28 in FD 471 flat display (Barco)　　　0143770

Three examples of the MRFD range (Barco)   0143769

panel, Touch Screen input and an external floppy drive. PCMCIA and SCSI interfaces are available on request. Features:
(a) Pentium Processor
(b) Hard or flash drive
(c) Ethernet interface
(d) Optional touchscreen
(e) Operates with most Intel-based operating systems. Tested with Windows 2000, Windows NT and Solaris.
(f) Parallel interface
(g) Serial interface
(h) VGA or SVGA resolutions
(i) PS/2 mouse interface
(j) PC
(k) ASES, Application-Specific Expansion Slot (PC/104)

20.1 in Rugged Display Station
The DS 251 and RDS 251, 20.1 in rugged Display Stations have been developed for operation in extreme environmental conditions. The DS 251 is suited for use in difficult environments, whereas the RDS 251 is a MIL-tailored workstation with greater ruggedisation for use in extremely harsh conditions. The display can be used in 19 in racks, mounted in consoles or used in stand-alone applications. Features:
(a) Front bonded optical stack with integrated heater
(b) 19 in rack-mountable
(c) Drip-proof over angles up to 60°
(d) User-replaceable backlight tray
(e) 20.1 in rugged LCD
(f) Pentium Processor
(g) Hard or flash drive
(h) 2 USB ports
(i) 2 Ethernet interfaces
(j) Optional touchscreen, smart card reader
(k) Optional LCD flicker compensation (patented)
(l) Operates with most Intel-based operating systems. Tested with Windows 2000, Windows NT and Solaris
(m) ASES, Application-Specific Expansion Slot (PC/104)

Integrated Displays
MPRD 126 HB and MPRD 138 HB
The MPRD 126 HB (High-Bright) and MPRD 138 HB are Barco's 10.4 in and 15.0 in MIL-standard rugged flat panel displays. With the Scaler (an image processor which provides the display with a multistandard decoder to accept a wide range of video sources) option, video windows can be inserted, together with a wide range of image processing techniques. Text overlay with user definable characters is possible with On-Screen Display (OSD). The displays have up to two RGB inputs. Functional soft keys, automatic light control (ALC), touchscreen, loop through for 1 RGB input and PCB coating (Humiseal) are optional. The MPRD 126 HB and 138 HB can optionally be configured for NVIS-B compatibility. The 138 HB also comes with a DVI input for direct digital interconnection.

FD 251 and RFD 251
The 20.1 rugged flat panel displays have been designed to meet a wide variety of environmental specifications (airborne, shipboard and land-based). Two levels of ruggedisation are available. The FD series has been designed for use in difficult environments, whereas the RFD series has extended environmental specifications to cope with more extreme conditions. The FD 251 rugged flat panel is protected by a solid mechanical structure. The Active Matrix colour LCD supports different resolutions from VGA to SXGA through upscaling. This direct 19 in rack-mountable series includes an internal database for multiple scanning sets. The RFD series offers MIL-tailored rugged flat panel displays with greater ruggedisation (front bonded optical stack, LCD and backlight heating). The displays are direct 19 in rack-mountable, have a shop-replaceable backlight tray and can be remotely controlled via a serial link. The Automatic Phase Adjust (APA) function locks the sampling clock to drifting graphics generator clocks. A special airborne version (RFD 251A, extended vibration specifications) has been designed to cope with the airborne conditions.

FD 471
The 28 in FD 471 has four field-replaceable backlight trays for consistent image quality and reduced life cycle cost.

Specifications

Type	Size	Panel Resolution	Special Features
Modular Displays			
(R)DM 243	17 in	1,280 × 1,024	MRFD Product Family Features (for all models):
(R)DM 246	18 in	1,280 × 1,024	Analog and/or Digital Inputs
(R)DM 354	21.1 in	1,600 × 1,200	Up to (2) Video Inputs
DM 361	24 in	1,920 × 1,200	Mixing of Video Inputs with Primary Input (PIP) APA, LFC, TS
Network-Centric Display Stations			
TID 126	10.4 in	640 × 480/ 800 × 600	PC, TS, ASES, Network Interface
(R)DS 251	20.1 in	1,280 × 1,024	PC, TS, Smartcard, ASES, Network Interface
Integrated Displays			
MPRD 126 HB/NVIS	10.4 in	640 × 480/ 800 × 600	HB, NVIS, PIP
MPRD 138HB/NVIS*	15.0 in	1,024 × 768	HB, NVIS, Scaler
FD 251	20.1 in	1,280 × 1,024	APA, TS, ED, Removable Backlight Tray
RFD 251	20.1 in	1,280 × 1,024	LFC, APA, TS, ED, Removable Backlight Tray
RFD 251A (Airborne)	20.1 in	1,280 × 1,024	LFC, APA, TS, ED, Removable Backlight Tray
FD 471	28 in	2,048 × 2,048	APA, Removable Backlight Tray, Digital/Analogue input

APA = Automatic Phase Adjust
ASES = Application-Specific Expansion Slot
ED = Extended Dimming Range
HB = High Brightness
LFC = LCD Flicker Compensation
NVIS = Night Vision Imaging System
PC = Computer
PIP = Picture in Picture
TS = Touch Screen

Status
In use in a very wide range of applications, including: Italian and German U212 submarine consoles, EADS SAMOC, NASAMS (RFD 251); DRS Opus 2 console for European minehunter programme (believed to be Turkey); USN Trident submarine, US Mobile Approach Control System, AN/YQ-70 consoles (RFD 251(S)); UK SeaKing Mk 7 helicopter, US E2-C Hawkeye, P-3C Orion upgrade, (RFD 251(A)); USAF RC-135 Rivet Joint, Airborne Laser Programme, US Trailblazer; UK ADVISOR vehicle, MARGOT thermal imaging surveillance system, ITT MACS programme (FD 251); German KZO UAV project (DM 246), German MLC-70 Bridgelayer. See separate entries in most cases.

Manufacturer
Barco, Kortrijk, Belgium.

UPDATED

Barco Vector® Flat Panel Display Systems

The Barco Vector® product line has been designed for observation and display on board armoured vehicles. It comprises a range of rugged LCD panels (7 in, 10 in and 12 in), control modules and HMI software. The display system combines video camera input with standard text On-Screen Display, as well as graphics overlay created by a symbol generator. There are multiple video and infra-red display possibilities and both day and night cameras can be connected.

10 and 12 in Display Systems
The system features a Video Control Module (VCM) and one or more Panel Modules (PM). These modules, in a light-weight composite housing, can be installed separately or mounted together, making for easy integration into confined spaces and consoles. They are connected by means of a single cable, the VectorLink, and can be installed up to 5 m from each other. Mapping and tactical information can be displayed through overlay over camera images or via computer-generated RGB inputs. The system has two PAL-NTSC video inputs, which can also be configured to most thermal imaging standards (Stanag B and C). FLIR images can be displayed in full greyscale mode or can be edited by the Look-Up Tables. The system has a light-weight housing, with extensive use of composite materials, that has been ruggedised in accordance with MIL-STD-810.

The VCM processes all inputs from video cameras and other sensors, provides a communication interface with an external computer or sensor, and transmits the processed commands over a serial communication bus. It directly interfaces with the PM, processes all user inputs from function buttons, touchscreen and serial communication, and translates them into specific control actions (for example, display of OSD messages, Symbol Generator graphics). The VCM also provides optional video recording: the images actually shown on the screen (video, OSD and graphics overlay) are taped on a VCR (PAL or NTSC). The VCM can also be replaced by a Multi-Head Control Box, launched in 2001, which provides the same

The DM118(l) and PVM118(r) 7 in displays (Barco)   0143772

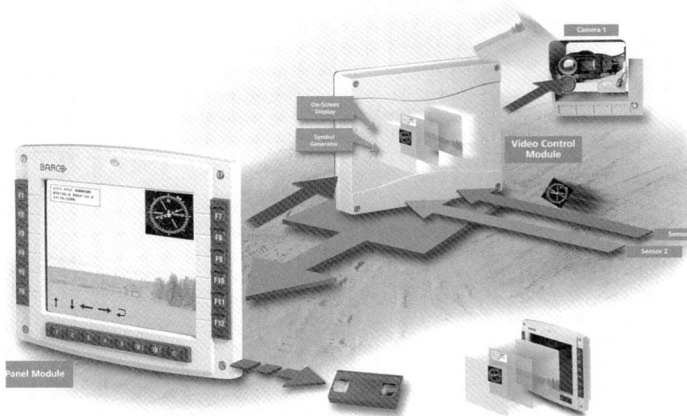

The BarcoVector® display system showing the relationship between the components (Barco)   0143771

functionality but allows the connection of multiple displays, catering for a complete AFV crew. Up to four PMs can be controlled: three have complete functionality while the fourth functions as a slave display (analogue input for video image display) for rear view driver images, for example. Up to 8 camera systems and 3 RGB sources can be connected.

The PM displays camera images (video input) or RGB graphics, generated by the VCM. The camera image can be overlaid with text and a full dynamic graphics overlay from an optional Symbol Generator. Available screen sizes are 10 in and 12 in, with SVGA resolution and an Ambient Light Control function automatically adapts the panel's contrast and brightness settings to changing light conditions. The PM also functions as the operator interface; its 20 function buttons are user-configurable and programmable and have optional illumination and status indication, and a touchscreen can be added as an option. The PM can also be used as a handheld terminal.

7 in Display Systems
The Vector® product family also has two 7in display systems with 16:9 aspect ratio: the DM 118 has video processing, Symbol Generator and On-Screen Display integrated into its housing, whereas the PVM 118 incorporates video processing functionality. The LCD display modules are completely sealed and ruggedised for use in hostile environments and have bonded front glass for optimum optical performance. The 16:9 aspect ratio of the DM 118 and PVM 118 is particularly suitable for armoured vehicles as the image format complies with the format of most direct optical viewing systems such as vision blocks or periscopes. Both display systems can also be used in 4:3 aspect ratio through image scaling and feature image mirroring for use in rear-view applications.

PVM 118
The PVM 118, with dual video input functionality for observation purposes, accepts a wide range of video standards, including PAL/NTSC and thermal imaging timings (STANAG B and STANAG C). The auxiliary display's 16:9 landscape format is compatible with many direct viewing systems and displays images in W-VGA resolution (854 × 480 pixels). The 7 in display system is equipped with two serial communication ports and features auto scan detection (of PAL/NTSC or STANAG).

DM 118
Designed for use as information and observation display, the DM 118 displays a wide range of camera and/or thermal images. The display system simultaneously presents information from several sources. Camera images (video input) can be overlaid with text On-Screen Display and full dynamic graphics overlay from the optional Symbol Generator. The system is equipped with user-configurable and programmable function buttons (operator interface). The DM 118 also features a communication interface with an external computer or sensor and transmits the processed commands via a serial communication bus.

ActEv toolbox
ActEv is a software development tool that enables the implementation of particular functionality in the Barco Vector display system by assigning actions to events. It allows system integrators to generate customised software applications while avoiding time-consuming code programming.

Status
In production and in service in a variety of programmes. The 10 in displays have been selected for the UK Titan and Trojan Armoured Engineer Vehicles. The 16:9 ratio system has been selected for the European MRAV project.

Contractor
Barco, Kortrijk, Belgium.

UPDATED

FlexiVision Series digital VME video mixers

The FlexiVision Series VME board-level video mixers provide the simultaneous presentation of separate multisensor video sources on a single display with any standard RGB video source from 640 × 480 to 1,600 × 1,200 resolution.

They offer a totally digital implementation, which eliminates artefacts and provides unique capabilities when processing data from a wide range of video signals.

For critical military surveillance applications, the FlexiVision video mixers provide both a digital radar input capability and a transparency feature. For missions requiring night vision capability, the FlexiVision products offer a wide selection of video inputs including FLIR monochrome sources.

Status
Selected for the USAF Airborne Laser programme. In use in the UK RAF Nimrod Mk4 Maritime Reconnaissance and Attack aircraft.

Contractor
Barco, Kortrijk, Belgium

UPDATED

FlexiVision VME multisensor video mixer board displays video sources from 620 × 480 to 1,600 × 1,200 pixels   0002975

VISTA Consoles

Vista 1000 Rugged Flat Panel Workstation
The Vista 1000 is a modular, user-configurable flat panel-based rugged workstation for use in a wide variety of C4ISR applications. The console has a preselected set of computing cards and mass storage devices. Users can also choose to integrate their own cards and mass storage devices, select indicators for the front panel and configure the connector panel on the back. The console has a lightweight composite housing and consists of an integrated enclosed Display Head, Computing Enclosure and Operator Desk. The unit is designed for application in hostile environments and can be bulkhead or desktop mounted. The main features are:
(a) Wide choice of rugged mass storage devices: floppy drive, hard drive, optical drive, CD-ROM drive
(b) Operator desk display available with touch input device, video camera input or RGB
(c) Operator desk up/downward tiltable
(d) Several keyboards and trackball options available

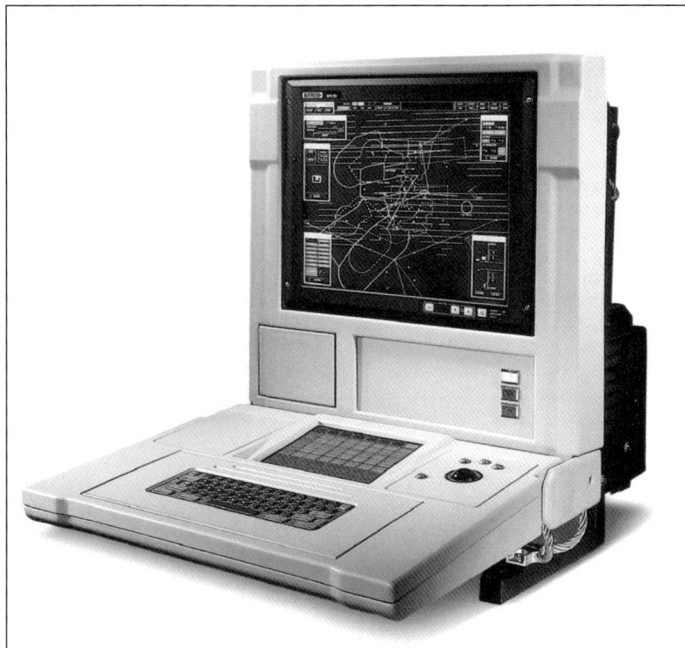

Vista 1000 (Barco)   0116762

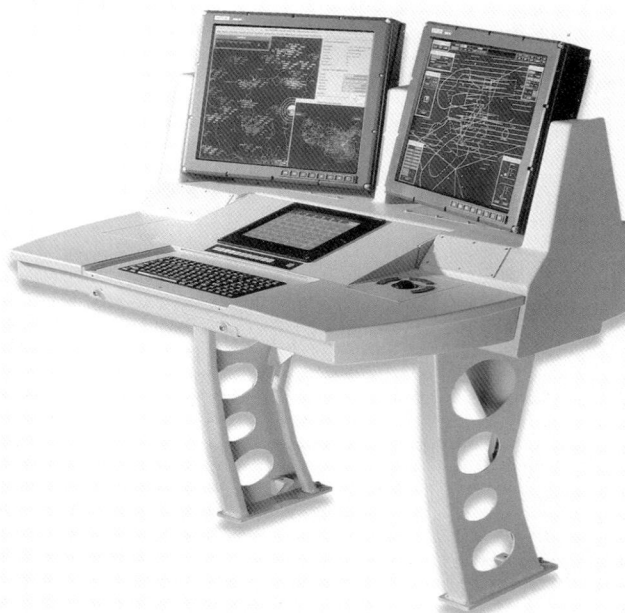

Vista 4000 with horizontal dual screen configuration and no cabinet (Barco)
0137253

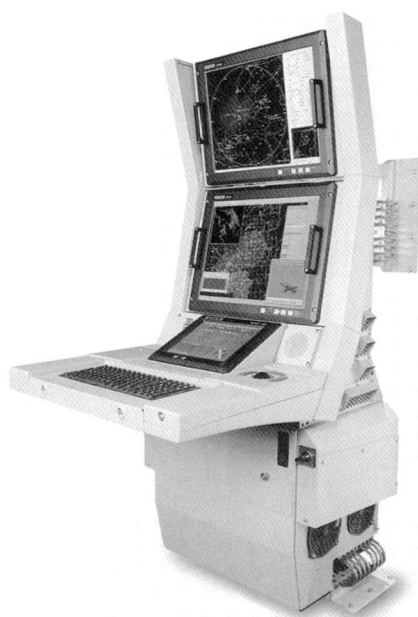

*Vista 2000 with dual
head and touchscreen
input display* (Barco)
0143775

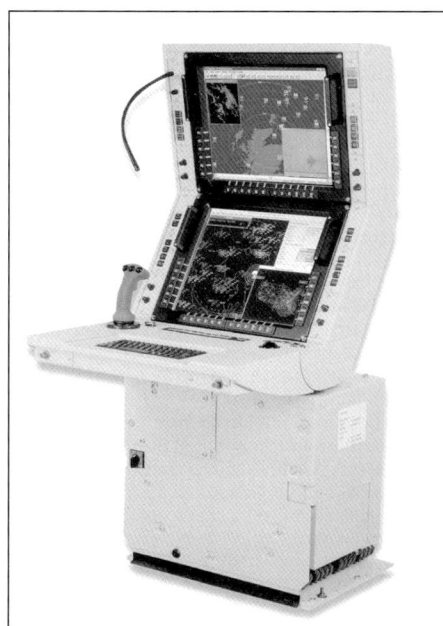

*Vista 4000 with vertical
dual screen
configuration and
cabinet* (Barco)
0137252

(e) User-selectable display: 18.1 in (FD 246 - RFD 246), 20.1in (FD 251 - RFD 251) or 21.3 in (MRFD 354). (See separate entry for display specifications)
(f) Display can be tilted.
(g) Rugged PCI (CPU, 4PCI + 4 ISA slots) or VME (5 slots) based cardcage. Cardcage is on slides and is accessible from the front.
(h) Indicator panel light switches can be customised
(i) Dimensions (H x W x D): 35 × 25.6 × 35.9 in (open); 35 × 25.6 × 20 in (closed)

Vista 2000 Integrated COTS Console

The Vista 2000 (formerly known as C5) integrates Barco's key components (rugged LCD displays and graphics controllers) into an ergonomically designed rugged chassis. The fully modular approach allows users to select their preferred configuration from a wide range of options: single- or dual-head console, VME or PCI rack, touch screen input, trackball position, joystick, UPS input, configurable keys, fixed mount configuration and others. An integrated thermal conditioning system (with controller, anti-condensation heater and cooling fans) protects the electronics from extreme temperatures and quickly renders the console operational on startup. All Line-Replaceable Units (LRU's) can be accessed from the front and rapidly replaced. The Vista 2000 has a MTTR (Mean Time To Repair) of less than 30 minutes. The rugged PCI or VME-based cardcage on slides is accessible from the front. The cardcage has 14 slots which can be filled with Barco's processor boards, I/O cards, graphics controllers, frame grabbers, radar scan converters and other PCI/VME boards. There is extensive BIT functionality, but if necessary the BIT internal diagnostics can be turned off by means of a battle-override switch.

In its standard configuration the console has a tiltable desktop with keyboard and trackball; a backlit keyboard and/or joystick and programmable 10 in touch input display are available as options. The system is operated by means of trackball and joystick and the console can be configured for left-handed people (trackball left of keyboard).To maximise desktop space in normal operational use, the

keyboard can be covered. Customers can select their preferred panel (or a combination of several screens) from Barco's full range of LCD displays (18 or 20 in) and configure the operator desk display for their application. Displays can be automatically tilted and adapted to the operator's viewing angle.

Vista 4000 COTS Console family

The Vista 4000 console's modular, split design combines a Human Machine Interface with Base Unit. This split design provides flexibility and can be customised to user interfacing requirements. Designed for operations in airborne, landbased and naval applications, the Vista 4000 series provides a wide range of configuration options, including various environmental specifications, vertical or horizontal display arrangement and integrated or remoted processing. The three basic versions are: the complete cabinet with integrated VME or PCI processing functionality; the simplified base allowing the integration of an alternative workstation; and the pedestal only, for use where no local processing is needed. HMI options include: single-head or dual-head; side-by-side or top-down configuration; flat panel displays from 18 to 21 in; touchscreen input; trackball; keyboard; joystick; configurable function keys.

Status

All versions are in widespread use in a number of programmes, including:
Vista 100: ADVISOR (UK), ASTROS II (Brazil);
Vista 2000: Programmes with Elbit Systems and IAI (Israel); Boeing UCAV programme;
Vista 4000: Saudi Arabian Navy SAWARI II frigate programme.

Contractor

Barco, Kortrijk, Belgium.

UPDATED

Canada

M60 Improved Fire Control System (IFCS) Ballistic Computer

General Dynamics Canada Limited (previously Computing Devices Canada) supplies the Ballistic Computer System (BCS) to Raytheon Systems Limited for the M60A3 Integrated Fire Control System (IFCS). The BCS is made up of two components, the Fire Control Computer (FCC) and Computer Control Panel (CCP), both of which are proven systems. Included in the system is an inherent capacity for further performance growth, particularly the incorporation of Battlefield Management Systems. The FCC, in conjunction with the CCP, enables the M60A3 Crew Commander (Cdr) and Gunner (Gnr) to control the M60A3 gun. The FCC computes ballistics data based on sensors and/or input by crew members via the CCP. With correct data and error correction calculations, the intent is to achieve 'first-round' hits on a selected target.

Fire Control Computer (FCC)

The FCC incorporates a Motorola MC68000 series microprocessor with a Motorola MC 68800 series co-processor. The three Circuit Card Assemblies in the FCC consist of the:

Processor CCA;

1553 Data Bus/Power Conditioner CCA;

Analog/Digital I/O CCA.

The FCC provides the following functions:

(a) Control of the weapon system databus
(b) Ballistic calculations
(c) Mode selection and implementation
(d) Control of weapon system calibration routines
(e) Built In Test (BIT) and fault logging
(f) Generation of the fire enables control signal
(g) Sensing of ammo select and generation of ammo display outputs
(h) Generation of rate, position, display, mode, laser, test and control signals for the Gun Sight Electronics Unit (GSEU) and Gun Turret Drive Electronics Unit (GTDE).

Computer Control Panel

The Computer Control Panel (CCP), in conjunction with the FCC, allows the weapons system crew to control the M60A3 Integrated Fire Control System (IFCS) and to observe various system conditions through the use of lamps and an LED display. A secondary feature permits maintainers to monitor IFCS serviceability and diagnose faults. This CCP was chosen as the primary system interface for the gunner, because it is extremely robust, simple to use, and also battle proven.

The CCP is responsible for the following functions:

(a) Entry and display of ammo temperature, barometric pressure, air temperature, ammo subtype selection and tube wear.
(b) Display of sensor values of cross-wind, cant, lead and range.
(c) Manual override of sensor values of cross-wind, cant, lead and range
(d) Entry and display of boresight, plumb/sync and zero calibration data
(e) Display of input/output to/from the FCC
(f) Display of software version
(g) Display of various internal software parameters
(h) Initiation of Built In Test (BIT)
(i) Display of BIT results.

Status

The FCC is based on the system designed by Computing Devices for the Challenger 2 Main Battle Tank, which is currently in service with the British Army. This computer system was also selected for the US Army's M8 Armored Gun System (AGS) and is fully qualified on both vehicles. The CCP is in service in the US Army's M1/M1A1 Abrams Main Battle Tank fleet, the Korean K1 MBT, over 1,500 upgraded M48 tanks and on a lesser number of MBTs, all using General Dynamics Ballistic Computer Systems.

The IFCS ballistic Computer is in service in the Raytheon Technical Services Ltd system installed in 50 M60A3 MBT of the Royal Jordanian Army. A production order for an additional 50 systems was received in 2003.

Contractor

General Dynamics Canada, Ottawa, Ontario.

UPDATED

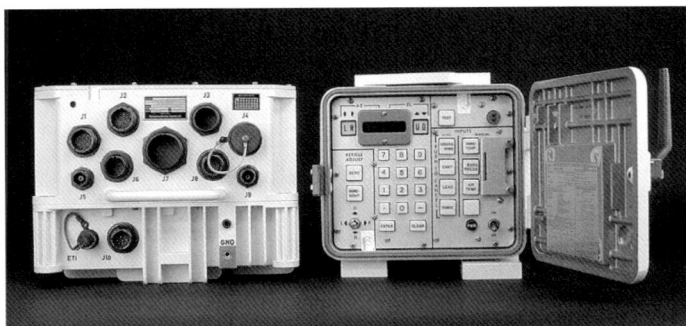

IFCS Ballistic Computer (General Dynamics Canada)   0120245

Vulcan PC6100 Tactical Computer System

The Vulcan PC6100 Tactical Computer System has been developed to operate both fire control and situational awareness applications on combat platforms. The PC6100 consists of a Central Processor Unit (CPU), Flat Panel Display and Keyboard. The core element is the IEEE 1101.2 conduction-cooled Single Board Computer (SBC) in the CPU. The SBC uses a modular design, specifically to accommodate technology insertion. The Compact PCI 6U form factor card is powered by an Intel Mobile Module Pentium III processor and is capable of running both real-time and non real-time operating systems and applications.

Fire Control

The PC6100 CPU is the basis of the computer system being designed for the US Towed Artillery Digitization programme. It will be fitted to the XM777 Lightweight 155 mm howitzer and will run the onboard fire support software. The system will consist of the PC6100 Mission Management computer system with two 2 in × 4 in electro-luminescent flat panel displays for the Gunner and Assistant Gunner and a 6.4 in full-colour AMLCD Section Chief Control and Display Unit. The CPU will be modified with a 1 Gbyte HDD, six additional serial ports and an SP-TCIM to link the Mission Manager to digital radio, and there will be a number of minor modifications to the display units. Linked into this system will be an Inertial Navigation Unit and Global Positioning System receiver to provide gun position and aiming information, digital communications to the Fire Direction Centre, and the capability to add a Muzzle Velocity radar, meteorology sensor and Direct Fire Sight for enhanced fire control. The PC6100 is also being used to host the fire-control software in the Mobile Gun System which is being developed by the GM Defense/General Dynamics Land Systems joint venture responsible for the development and fielding of the US Army Brigade Combat Team Interim Armored Vehicle suite.

Situational Awareness

The PC6100 has operated the US Army FBCB2 (see separate entry) Canadian Army battleWEB (see separate entry), and US Marine Corps C2PC situational awareness applications. These applications ran under the Solaris, SCO Unix and Windows NT operating systems, respectively. The system is fully qualified, is in production and has been designed for operation on the move on combat platforms. These platforms range from wheeled and tracked armoured fighting vehicles to artillery pieces. The PC6100 CPU is also capable of driving two displays, such that it is being used with a 12.1 in display when installed in the turrets of fighting vehicles and a 20.1 in display when installed in command vehicles. The Single Board Computer is in production and has been installed in M1A2 Abrams SEP tanks to operate the FBCB2 software. In this configuration, the tank commander uses the Commander's Tactical Display (part of the SEP upgrade) as the Soldier-Machine Interface.

Basic Technical Specifications

Central Processor Unit:

(a) Intel Pentium III Mobile Module 700/850 MHz processor, with MMX technology
(b) 1Gbyte of Synchronous DRAM
(c) 512 kbytes PBSRAM L2 Cache
(d) 10 Gbytes (minimum) removable hard disk drive in 2.5 in form factor
(e) Optional Flash Disk (up to 288 Mbytes)
(f) 2 Mbytes video memory
(g) Fully IEEE 1101.2 compliant conduction-cooled CCAs
(h) Up to 15 minutes battery backup time
(i) MIL-STD 1275 28 V DC 6U form factor Single Board Power Supply
(j) Software and O/S:
　　　Windows 95, 98, NT
　　　Solaris × 86
　　　SCO Unix
　　　VxWorks
　　　Linux

Status

All elements of the Vulcan PC6100 series are either in service, in production or about to enter service with the United States Army, United States Marine Corps and the Canadian Army.

Contractor

General Dynamics Canada, Ottawa, Ontario.

UPDATED

Vulcan PC6100 Central Processing Unit (General Dynamics Canada) *NEW*/0120250

Vulcan PC6110 Tactical Computer System

The Vulcan PC6110 Tactical Computer System is the latest tactical computing system from General Dynamics Canada, incorporating their Single Board Computer, the PC6010.

Specifications
Computer
(a) Pentium III 700MHz Low Power Mobile Module (MMC-2) (can be up to 850 MHz)
(b) 256 Kbytes L2 cache on-die
(c) 100 MHz Front Side Bus
(d) Up to 1GB SDRAM Memory with ECC support
(e) 4 MB Integrated Video Memory
(f) Standard VGA Interface
(g) LVDS Display Interface
(h) 2 RS170 Video Inputs
(i) Ethernet 10/100 Base-T (Boot From LAN Support)
(j) Ultra Wide SCSI SE 40 Mbytes/s
(k) 2 EIDE hard drive interfaces
(l) Optional up to 576 Mbytes Flash (DiskOnChip 2000)
(m) 2 USB 1.1 ports
(n) 16 bit Sound Blaster Compatible Stereo Audio Input and Output
(o) 6 Serial Ports Software Configurable Protocols (RS232/423/422/485)
(p) 2 Synchronous Serial Ports (Optional - require +/- 12 V DC)
(q) All Standard PC I/O including:
(r) 2 PS/2, Floppy and Parallel Port
(s) Spare PCI Mezzanine Card (PMC) Slot on SBC
(t) Spare Compact PCI Slot on backplane
(u) 140 W Mil-Std 1275A power supply
(v) Full CPCI 2.0 R2.1 6U Single Slot
(w) IEEE 1101.2 Conduction Cooled
(x) Removable 2.5 in hard drive (disk, 10 Gbytes+; solid state 1.5 Mbytes+)

Display
(a) 12.1 in 800 × 600 (64K colours) SVGA ALMCD flat panel display with integral touchscreen
(b) configurable with larger displays
(c) Ability to drive two displays simultaneously
(d) Integral backlit bezel keys/indicators/controls
(e) Infinite dimming capability
(f) Secure lighting blackout switch

Keyboard
(a) 84 key, full sealed
(b) Backlit QWERTY layout
(c) Integrated pointing device
(d) USB hot pluggable connection

Interfaces
(a) Optional Peripheral Expansion Interface
(b) USB 2.0
(c) Second (Gigabit) LAN Interface
(d) FireWire 1394A
(e) TMDS Display Interface and more
(f) Secondary Display: SXGA Interface (Analog)

Software
Operating Systems Supported:
VxWorks
QNX
Windows 9 × , ME, 2000, NT, NTE, XP

SCO Server 5
Solaris × 86
Linux
Supports running from onboard DiskOnChip

Contractor
General Dynamics Canada, Ottawa.

VERIFIED

Finland

Message Terminal M 85200

Patria Finavitec's M 85200 is a Short Burst Message Terminal. It has been developed from Patria Finavitec's field proven DA 8520 Message Terminal by incorporating several advanced hardware and software features in addition to new functions.

Similar to its predecessor, the M 85200 is a microprocessor-based lap top unit for tactical data communication in defence and Special Forces. It is intended for editing, transmission and reception of messages in hostile environments. Serial interfaces allow data input in digital form from sensors and measurement instruments and encrypted data output to communication equipment, computers or display equipment. Owing to its new modular hardware structure the unit is easy to tailor to various applications with varying interfacing requirements. The M 85200 terminal can be used in simple point-to-point communication using radios or telephone lines or in a sophisticated mobile packet switching message transmission network.

The M 85200 plus its accessories fits in a small canvas bag. Operator interface is via a standard QWERTY keyboard with additional function keys. The operations are quickly and easily selected by means of the function keys and prompts shown on the display. The LED matrix display has adjustable brightness, allowing use in daylight conditions and in total darkness. Four LEDs provide the user with important status information.

The unit is microprocessor-based and is consequently easy to tailor to various requirements and languages. The basic software is stored in EPROM whereas the application programs are on FLASH. The M 85200 can therefore be easily reprogrammed via the computer interface from a PC or another Message Terminal without opening the unit.

The basic functions include entering the encryption keys, addresses and other parameters, message store and forward, edit, send, receive, display, automatic/

M 85200 in use in conjunction with combat net radio (Patria Finavitec) 0143781

M 85200 Message Terminal (Patria Finavitec) 0143782

The Vulcan PC6110 Tactical Computer System (General Dynamics Canada)
NEW/0533828

M 85200, an air surveillance variant. Note the different keyboard configuration to the basic variant. (Patria Finavitec) 0143783

manual print, delete, acknowledge and offline encrypt/decrypt. Online encryption/decryption is automatic on message transmission/reception. A sufficient number of fixed format messages can be stored for all practical applications. Further functions include deletion of the entire memory, battery charge, test functions and so on.

The M 85200 can be connected to any standard PC or other type of computer via its standard serial interfaces. It can be configured and remotely controlled from the PC or used as an encrypting/decrypting modem. The computer interface is active all the time independent of other operations performed with the unit.

The unit has large data (S RAM) and program (EPROM and FLASH) memories. Eight separate messages can be kept concurrently in both the transmission and receive memories. A received message can be transmitted from the receive memory or copied to the message entry buffer for editing.

M 85200 has been designed to minimise Tempest. Burst transmission reduces the on-air time to the minimum. The built-in modem is optimised for HF transmission and in addition, adaptive threshold is provided to compensate automatically for the amplitude distortion of the communication channel caused by the selective fading phenomenon. An efficient error correction algorithm and interleaved transmission allow for reliable reception of data bursts during difficult propagation conditions. A real-time clock provides the anti-spoof function as a standard feature. A powerful digital enciphering algorithm provides online and offline encipher/decipher capability and a double encryption function when needed. M 85200 has the capability to simulate message terminal functions by sending dummy messages in a realistic fashion, and messages can also be sent automatically at a predefined time.

The terminal transmits and receives data at various speeds over standard HF, VHF or UHF radios, satellite links or land lines. It uses three character addresses and has a group addressing and broadcast capability. Messages can be printed out over a standard RS-232 serial line using any standard commercial device. An RS-422 serial interface allows input of information in digital form from various measurement instruments and sensors and a computer or display terminal can be connected either via the RS-232 or the RS-422 interface. The computer and printer interfaces have an individual address and group addresses of their own, and therefore messages can be directly addressed to the printer or computer connected to the receiving terminal.

Specifications
(a) Display:
 32 character LE
 5 × 7 dot matrix
 4 LED indicators
 Optional:
 48 × 480 dot graphic type LCD display
 2 × 80 character LCD
(b) Keyboard:
 55-key silicone rubber
 QWERTY with 9 additional function keys
(c) Memory capacity:
 Static RAM 32 kbytes: Receive memory 12 kbytes, Transmit memory 8 kbytes
 Program memory: EPROM 8 kbytes, FLASH 56 kbytes
(d) Interfaces:
 External power supply
 Serial interface RS-232 (110...9.600 bps)
 Serial interface RS-422 (110...9.600 bps)
 Radio interface 4 wire audio 150...600 bit/s FSK
(e) Power supply:
 Internal battery, 4 × 1.5 V D size alkaline dry cells or rechargeable NiCd cells
 External supply: 10...30 V DC
 Built-in charging circuit for NiCd cells
(f) Dimensions and weight:
 (D × W × H): 300 × 220 × 70 mm
 Weight: 3 kg (with battery)

Contractor
Patria New Technologies Oy, Tampere, Finland.

UPDATED

France

CALISTO family of multifunction consoles

CALISTO is a family of multifunction consoles intended for naval command and control systems and has been designed for a variety of applications (search radar, infra-red (IR) search, weapon control, system supervision and so on).
 Key features of the system include:
(a) multifunction console
(b) real-time computer and graphic engine
(c) smart interfaces with board sensors
(d) high processing power
(e) high-resolution colour display
(f) MMI facilities
(g) standard interfaces
(h) standard operating system
(i) BIT and diagnosis software facilities.
 CALISTO provides display of coastline and areas of interest, display of the tactical situation and CMS-oriented software functions of anti-air self-defence, anti-ship warfare, hard/soft-kill co-ordination and decision aids. A variety of man/machine interfaces is available. The basic CALISTO console is equipped with two multifunction keyboards, one trackerball and push-buttons. Other tools such as joystick, alphanumeric keyboard and so on, are also available. Configuration is defined according to the need. The main operational functions are:
(a) multifunction keyboard management
(b) window management (dimension, position, priority, and so on)
(c) IR display (initialisation displayed zone, wavelength, zoom and so on)
(d) radar display (initialisation, display area, absolute or relative display, persistence, decentring)
(e) 25 in TV display
(f) track representation with NATO symbology (creation, suppression, modification of symbols, status, historicals, attribute selection)
(g) plot representation
(h) graphic representation (circle, ring, rectangle, polygon)
(i) visualisation (plans priority, contrast)
(j) map list management.
The graphical library offers:
(a) graphic engine control
(b) drawing of complex shapes
(c) management of graphical objects/attributes
(d) device management
(e) object tracking
(f) picking
(g) clipping
(h) co-ordinates system change
(i) raster graphic functions
(j) soft copy
(k) fonts.
 The CALISTO console's architecture is modular and includes one or two high-resolution displays and sensor interfaces provided by smart boards connected to a standard VME bus. A video preprocessor, associated with a high-performance graphics engine, provides the necessary power for CMS image display requirements. A real-time processor can be added for real-time functions. This processor is based on HP-PA RISC architecture and real-time UNIX. Fully customised, I/O capabilities contribute to multifunction applications (TE/WA). An intelligent interface to ship sensors enable a variety of standard presentations such as synthetic processing and display, TV video display, radar raw video display, infra-red display and any type of mix. The radar interface has normalised radar video input, azimuth and synchro signals and both TV and IR interfaces have standard CCIR input.
 The graphic engine is based on HP 9000-700 workstation series. Its main features are:
(a) clock frequency: 50 up to 135 MHz
(b) cache memory: 64 kbytes for data and 128 kbytes for addressing
(c) main memory: 8 up to 128 Mbytes
(d) disk memory: 150 Mbytes up to 1 Gbyte
(e) operating system: UNIX
(f) one or two graphic heads.
Real-time computing is based on HP VME boards: the processor is an HP-PA RISC 50, 64 or 98 MHz; cache memory 64 kbytes for data and code; main memory is from 8 up to 256 Mbytes and the real-time operating system is UNIX POSIX 1003.4 and 1003.4a. The set of programming tools include X.11 R5 and PEX software, MOTIF, graphical and tactical libraries. The tactical libraries allow the handling of tactical objects or functions for CMS programs.
 See *Jane's Naval Weapons Systems* for further details.

Status
The latest application of the CALISTO consoles family is that of the combat management system (SENIT 8) for the French aircraft carrier *Charles de Gaulle*. Other smaller applications are in production for export to the navies of Pakistan and Kuwait.

Contractor
EADS Systems & Defence Electronics, Vélizy Villacoublay.

UPDATED

Chameleon datalink server

CHAMELEON is a multiple tactical datalink server that supports TADIL-A (Link 11A), TADIL-B (Link 11B) and TADIL-J (Link 16). It receives, transmits, processes and displays tactical data in accordance with the STANAG 5511 and/or STANAG 5516 message specification. The CHAMELEON has the capability simultaneously to process data provided by two networks (for example, Link 11/Link 16, 2 × Link 11).

Depending upon the nature of the application, the CHAMELEON can be configured as a tactical datalink processor only or as a tactical datalink processor together with a graphic display and a keyboard/trackerball. The tactical datalink processor hosts CHAMELEON software and provides for the necessary external interfaces such as serial RS-232 and RS-422, Ethernet, NTDS/ATDS and MIL-STD-1553B.

The CHAMELEON product family includes equipment designed for aircraft, ground station (fixed/transportable) and shipboard applications. It uses commercial workstations for benign environments, 483 mm ruggedised VME-based equipment for ground, shipboard or airborne environments and ½ ATR VME-based equipment for airborne (helicopter) environments.

Status
In production since 1996. Sold to the French Navy, French Air Force and Royal Navy.

Contractor
Rockwell-Collins France, Blagnac.

UPDATED

Commander Panel PC

The 6410-01 Commander Panel PC from IRTS is a ruggedised PC with a fold-down qwerty keyboard, 2 × 10 key functions keyboard and surface wave touch panel screen, in single-block aluminium packaging. The PC uses ETX technology with PC104 expansion slots. Operating systems are Windows 98®, NT 4.0®, Linux® or Lynx OS®.

Specifications
(a) Display: VGA, SVGA or XGA colour TFT 10.4 in or 12 in
(b) Processor: Pentium II MMX 266 MHz, Pentium III 500/700 MHz or Athlon 1.2 GHz
(c) RAM: 64-256 Mbytes (Pentium II); 64-512 Mbytes (Pentium III, Athlon)
(d) Serial/Parallel Ports:1/1
(e) USB:1
(f) HDD: Extractable IDE 2 in 20-60 Gbytes
(g) Floppy drive and CD-ROM: External on 38999 connector
(h) Ethernet: 10/100Mbytes/s
(i) Expansion slots: 3 × PC104 boards with 8 serial ports; 16I/O opto-isolated points; PCMCIA: 1 × type II and 1 × type II or III with PCMCIA locations

The Commander Panel PC showing the keyboard folded down. The two rows of ten function keys can be seen on either side of the screen (IRTS) 0143788

The supplementary display and dialogue screen for the Commander Panel PC (IRTS) 0143789

(j) Operating temperature: −15°C (optional −30 C) to +60°C
(k) Power: 18/36 V DC
(l) Dimensions (W × H × D): 360 × 260 × 120 mm
(m) Weight: 8 kg

A supplementary display and dialogue screen, the 6310-01 rugged LCD monitor, can also be added to provide a two-screen configuration.

Status
In use with the South African Army.

Contractor
IRTS, Toulon.

VERIFIED

IRTS Compact workstations

IRTS provide a range of ruggedised and transportable workstations which are in use in various parts of the world. They are supplied in adaptable containers equipped with dampers. Key common features are:
(a) Active TFT matrix 17 or 18 in flat screen LCD monitors
(b) Front-face access to peripherals
(c) PICMG®, AT®, VME®, Compact PCI® architecture
(d) Windows, Linux, AIX, Lynx OS, SOLARIS environments
(e) Reinforced aluminium structure
Common options include 19 in rack-mounting kit, watertight container, integrated keyboard, customised power supply.

IFPS 7587 in container (IRTS) 0143786

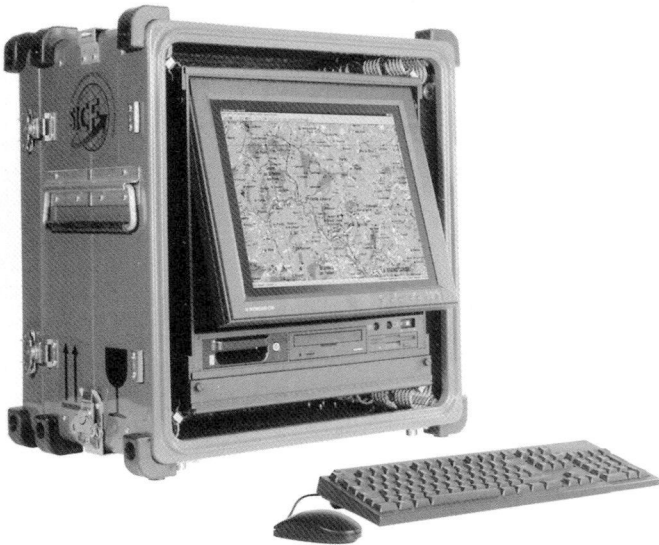

IFPS 7588 workstation in container showing tilted screen (IRTS)   0143784

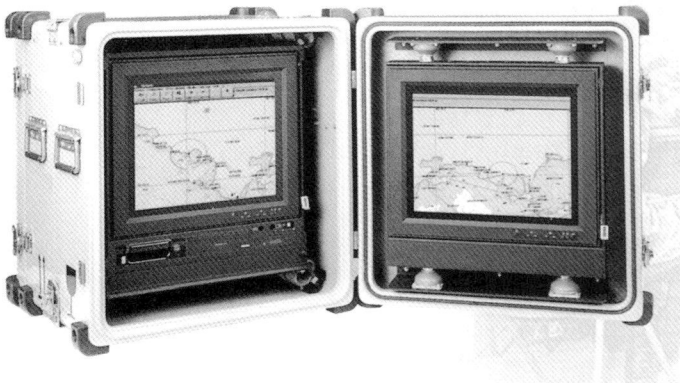

IFPS 7588 dual screen version in container (IRTS)   0143785

Specifications

Variant	IFPS 7588	IFPS7587
Processor	Pentium III 1.26 GHz	Pentium III 1.26 GHz
Memory	Up to 2Gbytes RAM	Up to 2Gbytes RAM
PICMG backplane	6 × PCI; 4 × ISA; 2 × ISA/PCI; 1 × SBC	4 × PCI; 2 × ISA; 1 × SBC
HDD	1 × 3.5 in removable	1 × 3.5 in internal, 1 × 3.5 in removable
Floppy/CD-ROM	3.5 in/IDE or SCSI	3.5 in/IDE or SCSI
Integrated UPS	10 mn	20 mn (basic configuration
Weight	40 kg depending on configuration	50 kg depending on configuration
Closed Dimensions (in container)	625 × 581 × 506 mm	642 × 637 × 506 mm
Display	18 in display can be tilted and locked	

A twin screen version of the IFPS 7588, with the second screen in the inside the front of the container, is also available. There is also a small mobile workstation which is less than 250 mm thick, with dampers made of foam blocks.

Status
The French Army has procured 350 IFPS 7588 for use with the SICF CIS since 1998. The Belgian Army has procured 120 IFPS 7587 since 1999.

Contractor
IRTS, Toulon.

VERIFIED

Rider station

The IRTS rider station is a system of ruggedised packaging to provide a suitable mounting for commercial-off-the-shelf (COTS) processors. The packaging is provided complete with a 17, 18 or 20 in TFT colour monitor and options exist to provide twin or triple monitor configurations. Storage for keyboard, mouse and cables is integrated into the rear cover of the container.

Status
Over 1,500 units have been acquired by all three French armed forces. Used for the SICF system (see separate entry). Also in use in Germany and Denmark.

Contractor
IRTS, Toulon.

VERIFIED

The Rider station in its 3-screen configuration (IRTS)   0143787

SLPRM

Système Local de Préparation et de Restitution des Missions (SLPRM) is a local mission planning and debriefing system which can be used for mission planning for a number of different aircraft including UAV, for the co-ordination of the activities of different aircraft types. Planning functions include track, timing, attack method and weapon; mission analysis and video replay are also provided for debriefing. The system will display the tactical situation including threat locations, with geographical environments provided in 2- or 3-D, including terrain profiles, map backgrounds, recognition images and meteorological data. Once detailed flight and attack planning has been completed on the system, the pilot is provided with navigation, cartographic and tactical data on a data cartridge to be loaded into the aircraft system.

Status
In operational use with the French Air Force and Naval Air Wing, specifically in support of the RAFALE and Mirage 2000.

Contractor
SAGEM SA (Aerospace & Defence Division), Paris

VERIFIED

SLPRM mission planning and debriefing system in support of the Rafale aircraft (SAGEM)   *NEW*/0593087

For details of the latest updates to *Jane's C4I Systems* online and to discover the additional information available exclusively to online subscribers please visit
jc4i.janes.com

Germany

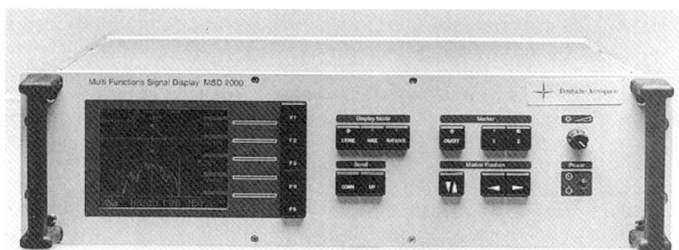

MSD 2000 multifunction signal display

The MSD 2000 is a dual-channel panoramic display unit. It is microprocessor-controlled and, in combination with Daimler-Benz Aerospace receivers, it allows panorama monitoring as well as narrowband analysis. Up to 1 MHz in the HF range or 5 MHz in the VHF/UHF range can be displayed.

Real-time panorama as well as maximum, mean value and 'frozen' panorama displays are shown on a high-resolution electroluminescent screen with controllable brightness. It is highly suited for the presentation of demodulated PSK-specific signals (for instance, by means of general purpose demodulator UD 2000) such as phase vectors or eye patterns.

There are two serial interfaces, RS-232C or IEC/IEEE 488 which facilitate integration into command and control systems.

Specifications
(a) IF inputs: 2 inputs, optionally 10.7 MHz, 21.4 MHz or 42.2 MHz; 2 inputs, optionally 200 kHz or 10 kHz
(b) Frequency resolution: 1 Hz at 250 Hz display width
(c) Display ranges: 250 Hz to 10 kHz (IF 10 kHz, 200 kHz); 1 kHz to 5 MHz (IF 21.4 MHz); 250 Hz to 1 MHz (IF 10.7 MHz, 42-2 MHz)

Status
No longer in production, but has been in service with several customers.

Contractor
MRCM
MRCM is a joint co-operation under full ownership of:
EADS Ewation, Ulm, Germany.
Grintek, Pretoria, South Africa.
Herley Industries Inc, Lancaster, USA.
Sysdel, Pretoria, South Africa.
TRL Technology, Tewkesbury, Glos, UK

UPDATED

MSD 2000 multifunction signal display

India

Spurt Message Alpha-Numeric Radio Terminal (SMART) AS 7306

The Spurt Message Alpha-Numeric Radio Terminal (SMART) AS 7306 is a microprocessor based message processing and transmission equipment for field use. The terminal will provide reliable burst transmission capability to reduce radio channel occupancy time with reduction in probability of interference and jamming. It is suitable for both manpack and vehicular operations. Messages are prepared 'off-air' with the QWERTY alpha-numeric keyboard and can be checked and edited on the 24 character LED display if necessary. Message reception is automatic. The terminal supports both free and formatted text messages and can also be optimised for specific applications. It has a built-in modem for interface to voice links. The equipment uses a powerful FEC and interleaving technique and is capable of operation over poor channel conditions. It also has built-in encryption for message text. The equipment supports networking and both selective call and broadcast call are possible. Automatic acknowledgement is also provided for selective call transmissions.

Specifications
(a) Data rate: 200 bauds (optional 1,200)
(b) Send and receive memory stores: 2,000 characters each
(c) Printer interface: RS-232C/Teleprinter, ASCII/Baudot
(d) Power supply: NiCad battery 12 V DC nominal
(e) Environment: operating −30°C to +55°C, storage −30°C to +70°C
(f) Dimensions: 107 × 335 × 265mm (H × W × D)
(g) Weight: 4.2 kg

Manufacturer
Bharat Electronics Ltd, Bangalore, India.

VERIFIED

Israel

CB-911 Ethernet controller VMEbus

The CB-911 is a member of the VME EL/S-9000 family of military computer systems. It is a high-performance front-end communications processor which connects a VMEbus system to an Ethernet/Cheapernet Local Area Network (LAN).

LAN circuits support the Ethernet/Cheapernet bus. The LAN circuits include an Ethernet controller, Manchester serial encoder and a transceiver. The Ethernet controller gains access to the local memory through its own DMA circuit.

The CB-911 has a VMEbus interface for host computer or global memory connection. The host computer controls the SBIA using a real-time multiprocessor operating system through a communications driver. Twelve SBIA modules are supported by the communications driver, either for message load sharing or for redundancy. The software in use is Intel INA 960, run by the onboard 80386 microprocessor. Communications layers 1 to 4 of the ISO model are supported by the same software. In addition, the board operates in TCP/IP mode as well.

Specifications
General
CPU type: Intel 80386, 20 MHz
Communication controller: Intel 80386, 20 MHz
Address length: 16-, 24- and 32-bit
Data length: 16- and 32-bit
Onboard: VMEbus, master/slave, interrupter
Memory: 256 kbyte EPROM; 512 kbyte SRAM
Serial channel: Ethernet/Cheapernet (IEEE 802.3 base2/10 base 5)
Timer: 3 timers

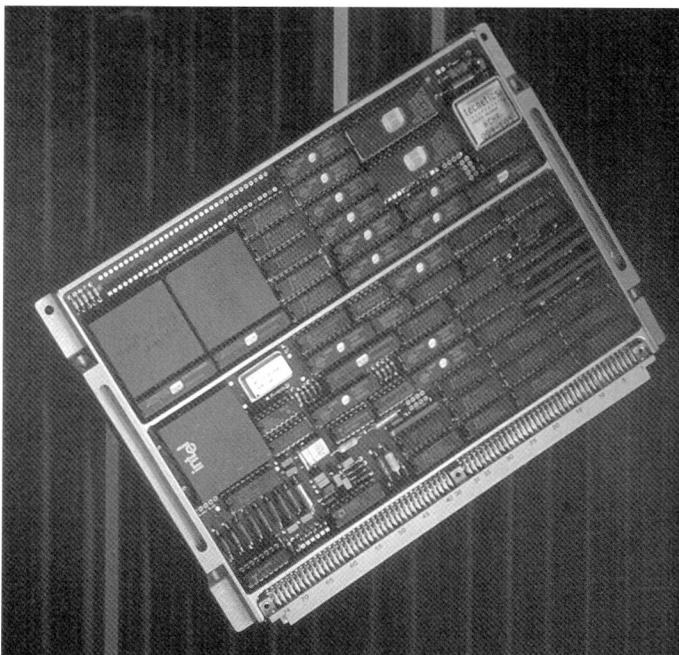

The Spurt Message Alpha-Numeric Radio Terminal (SMART) AS 7306
(Bharat Electronics)   0525557

CB-911 Ethernet controller   0077512

BIT and monitor firmware: Intel INA 960
Communication products: support for 4 ISO layers; ISO and TCP/IP operation modes

Environmental
Temperature
Operational: −20 to +55°C
Storage: −55 to +71°C
Altitude: up to 70,000 ft

Physical
Power: 5 V DC, 4 A (typical)
Physical dimensions: double height (6U) Eurocard, with two 96 pin MIL DIN type-connectors, or 234 × 172.3 mm size card with 222 pin NAFI connector, in accordance with MIL-C-28754.

Contractor
ELTA Electronics Industries Ltd(a subsidiary of Israel Aircraft Industries Ltd), Ashdod.

VERIFIED

CB-912 CPU-40 VMEbus

The CB-912 is a member of the EL/S-9000 family of military, real-time, embedded computer systems.

The module is a 68040 microprocessor-based CPU board especially designed for high-performance multiprocessor applications. The Module supports 32 bit address and data with interfaces to both the VMEbus and the Local Bus Extension (LBE). The LBE provides a high-speed dedicated memory interface to the EL/S-9000 family of memory modules.

The CB-912 module can operate independently in a VMEbus system as a single CPU module, or as part of a VMEbus multiprocessing system. It is designed to operate as a VMEbus master/slave and as a system controller on the VMEbus. The VMEbus interface is implemented by the Cypress VIC068A/VAC068A chipset. The module supports all seven VMEbus interrupt lines and it has the capability to send and receive 'CPU Interrupts' to and from other CPU modules.

The onboard EPROM contains monitor and BIT functions in addition to the user's optional programs such as operating system kernel, system initialisation and extended diagnostics.

Features
(a) Arbiter
(b) Global and local bus time-out
(c) System reset
(d) Block transfer capability
(e) 16 MHz system clock
(f) DMA channel
(g) 32 bit real-time clock
(h) 32 bit watchdog timer
(i) 32 bit user-defined timer
(j) 16 bit user-defined timer
(k) Calendar clock
(l) Double UART
(m) Abort option

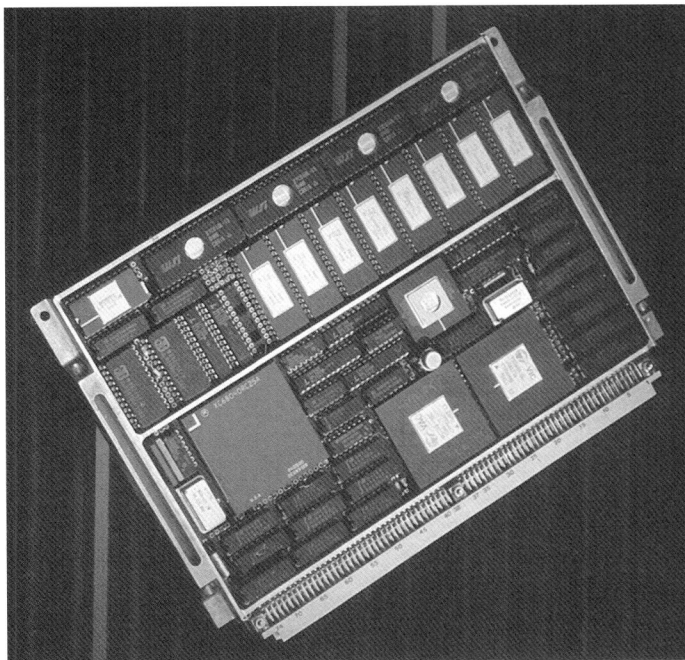

CB-912 68040 microprocessor-based CPU board   0077513

The Module is supported by a choice of multitasking, real-time operating systems such as the MTOS/UX from IPI and the VRTX-SA from MicroTec, an ANSI 'C' compiler, and a wide variety of utilities. Also included is the EL/S-9000 debug firmware package which offers debug, upload/download, on line assembler and disassembler as well as disk bootstrap load functions.

Specifications
Functional
CPU type: Motorola 68040, 25 MHz
External buses: VMEbus, LBE (Local Bus Extension)
Data length: 8/16/32 bit
Address length: 16/24/32 bit
Controllers: VMEbus, interrupt handler, arbiter
Memory
   SRAM: up to 4 Mbytes (eight 32 pin JEDEC)
   EPROM: up to 4 Mbytes (four 32 pin JEDEC sockets)
I/O ports: two serial RS-232/422 ports
Calendar clock: alarm - 99 years
DMAC: based on VIC/VAC068
Block transfer: based on VAC068
Timers: 5
Bit and monitor firmware: EL/S-9000 onboard debug package
Interrupt: multiprocessor capability, up to 16 processors
Battery backup: external battery support

Environmental
Temperature
Operating: −20 to +55°C (−4 to +131°F)
Storage: −55 to +71°C (−67 to +161°F)
Altitude: up to 70,000 ft

Physical
Packaging: double height (6U) Eurocard, with a 222 pin NAFI connector, compatible with MIL-C-28754
Power: 5 V DC, 2.8 A (typical)

Contractor
ELTA Electronics Industries Ltd(a subsidiary of Israel Aircraft Industries Ltd), Ashdod.

VERIFIED

CB-912EV CPU-40 VMEbus

The CB-912 is a member of the EL/S-9000 family of military real-time, embedded computer systems. The Module is a 68040 microprocessor-based CPU board especially designed for high-performance multiprocessor applications. The module supports 32 bit address and data. It also features a built-in Ethernet controller implemented by the Intel 82596 chip, with 10Base-2 and 10Base-5 physical interfaces.

The CB-912 module can operate independently in a VMEbus system as a single CPU module, or as part of a VMEbus multiprocessing system. It is designed to operate as a VMEbus master/slave and as a system controller on the VMEbus. The VMEbus interface is implemented by the Cypress VIC068A/VAC068A chipset. The module supports all seven VMEbus interrupt lines and it has the capability to send and receive 'CPU Interrupts' to and from other CPU modules. The onboard FLASH memory stores the monitor and BIT functions in addition to the user's optional programs such as the operating system kernel, the system initialisation routines and extended diagnostics.

The Module is supported by a choice of multitasking, real-time operating systems such as the Spectra VRTX-Sa, a choice of high-level language compilers such as 'C' and many utilities. The VRTX-Sa board support package includes an Ethernet bridge and a serial bridge, and allows the user to load and debug software over the Ethernet using TCP/IP networking protocols. Also included is the EL/S-9000 debug firmware package which offers debug, upload/download, on line assembler and disassembler as well as disk bootstrap load functions.

Features
(a) Arbiter
(b) Global and local bus time-out
(c) System reset
(d) 16 MHz system clock
(e) Ethernet controller
(f) 32 bit real-time clock
(g) 32 bit Watchdog timer
(h) 32 bit user-defined timer
(i) 16 bit user-defined timer
(j) Calendar clock
(k) Dual UART
(l) Abort option

Specifications
Functional
CPU type: Motorola 68040, 25 MHz
Bus: VMEbus
Data length: 8/16/32 bit

Address length: 16/24/32 bit
VMEbus support: interrupt handler, arbiter
Memory
SRAM: up to 4 Mbytes (2 × 66 pin HEXA) in line socket
FLASH: up to 4 Mbytes (1 × 66 pin HEXA) in line socket
Ethernet: 10Base-5 and 10Base-2
I/O ports: 2 serial RS-232/422 ports
Calendar clock alarm
Timers: 5 timers
Bit and monitor: EL/S-9000 debug
Interrupt: up to 16 processors capability for multiprocessor applications
Battery back-up support: external battery

Environmental
Temperature
Operating: −20 to +55°C (−4 to +131°F)
Storage: −55 to +71°C (−67 to +161°F)
Altitude: up to 70,000 ft

Physical
Dimensions: double height (6U) Eurocard, with two 96 pin MIL DIN type connectors
Power: 5 V DC, 2.8 A (typical)

Contractor
ELTA Electronics Industries Ltd (a subsidiary of Israel Aircraft Industries Ltd), Ashdod.

VERIFIED

CB-914 SBC-20 VMEbus

The CB-914 is a member of the EL/S-9000 family of military computer systems. The module is a 68020 microprocessor-based CPU board, especially designed for high performance multiprocessor applications. The module is generally used as a front-end processor, but may also be used as a stand-alone computer board. It supports 32 bit address and data with interfaces to both the VMEbus and the I/O bus. The I/O bus provides a dedicated 16-bit address/data interface to the EL/S-9000 COM-S/P (serial I/O extension) modules.

The module can operate either independently in a VMEbus system as a single SBC, or as part of a VMEbus multiprocessing system. It is designed to operate as a VMEbus master/slave and as a system controller on the VMEbus. The VMEbus interface is implemented by the Cypress VIC068A/VAC068A chipset. In addition, the module has the capability to send and receive 'CPU Interrupts' to and from other CPU modules. The onboard EPROM contains monitor and bit functions in addition to the user's optional programs such as an operating system kernel, system initialisation and extended diagnostics.

The module is supported by a choice of multitasking, real-time multiprocessor operating system such as MTOS/UX from IPIs and MicroTec's VRTX-SA/Spectra, by high-level language compilers such as 'C' and a wide choice of utilities. Also included is the EL/S-9000 debug firmware package which offers debug, upload/download, on line assembler and disassembler as well as disk bootstrap load functions.

Features
(a) Arbiter
(b) Global and local bus time-out
(c) System reset
(d) 16 MHz system clock
(e) I/O bus
(f) Four dual UART
(g) Five DMA channels
(h) 16 bit parallel port

CB-914 68020 Microprocessor-based CPU board   0077511

(i) 32 bit real-time clock
(j) 16 bit Watchdog timer
(k) 16 bit user-defined timer
(l) 24 bit user-defined timer

Specifications
Functional
CPU: Motorola 68020, 20 MHz
Buses: VMEbus, IOB (I/O Bus)
Data word: 8/16/32 bit
Address width: 16/24/32 bit
VMEbus functions: interrupt controller, arbiter
Memory
SRAM: up to 2 Mbyte (four 32 pin JEDEC)
EPROM: up to 1 Mbyte (one 40 pin socket)
I/O ports: 6 serial RS-232/-422 and two RS-232 ports, a 16 bit parallel port, control ports, 5 DMA channels
Timers: 4
Bit and monitor firmware: EL/S-9000 onboard debug package
Interrupt: multiprocessor applications capability, up to 16 processors
Battery back-up support: external battery

Environmental
Temperature
Operating: −20 to +55°C (−4 to +131°F)
Storage: −55 to +71°C (−67 to +161°F)
Altitude: up to 70,000 ft

Physical
Physical dimensions and packaging: double height (6U) Eurocard, with two 222 pin NAFI type connectors compatible with MIL-C-28754
Power: 5 V DC, 2.8 A (typical)

Contractor
ELTA Electronics Industries Ltd (a subsidiary of Israel Aircraft Industries Ltd), Ashdod.

VERIFIED

CB-914V SBC-20 VMEbus

The CB-912 is a member of the EL/S-9000 family of military real-time, embedded computer systems. The module is a 68040 microprocessor-based CPU board especially designed for high-performance multiprocessor applications. The module supports 32 bit address and data. It also features a built-in Ethernet controller implemented by the Intel 82596 chip, with 10Base-2 and 10Base-5 physical interfaces. The CB-912 module can operate independently in a VMEbus system as a single CPU module, or as part of a VMEbus multiprocessing system. It is designed to operate as a VMEbus master/slave and as a system controller on the VMEbus. The VMEbus interface is implemented by the Cypress VIC068A/VAC068A chipset. The module supports all seven VMEbus interrupt lines and it has the capability to send and receive 'CPU Interrupts' to and from other CPU modules. The onboard FLASH memory stores the monitor and BIT functions in addition to the user's optional programs such as the operating system kernel, the system initialisation routines and extended diagnostics.

The module is supported by a choice of multitasking, real-time operating systems such as the Spectra VRTX-Sa, a choice of high-level language compilers such as 'C' and many utilities. The VRTX-Sa board support package includes an Ethernet bridge and a serial bridge, and allows the user to load and debug software over the Ethernet using TCP/IP networking protocols. Also included is the EL/S-9000 debug firmware package which offers debug, upload/download, on-line assembler and disassembler as well as disk bootstrap load functions.

Features
(a) Arbiter
(b) Global and local bus time-out
(c) System reset
(d) 16 MHz system clock
(e) Ethernet controller
(f) 32 bit real-time clock
(g) 32 bit Watchdog timer
(h) 32 bit user-defined timer
(i) 16 bit user-defined timer
(j) Calendar clock
(k) Dual UART
(l) Abort option.

Specifications
Functional
CPU type: Motorola 68040, 25 MHz
Bus: VMEbus
Data length: 8/16/32 bit
Address length: 16/24/32 bit
VMEbus support: interrupt handler, arbiter
Memory
SRAM: up to 4 Mbytes (2 × 66 pin HEXA) in line socket

FLASH: up to 4 Mbytes (1 × 66 pin HEXA) in line socket
Ethernet: 10Base-5 and 10Base-2
I/O ports: 2 serial RS-232/-422 ports
Calendar clock alarm
Timers: 5 timers
Bit and monitor: EL/S-9000 debug
Interrupt: up to 16 processors capability for multiprocessor applications
Battery back-up support: external battery

Environmental
Temperature
Operating: −20 to +55°C (−4 to +131°F)
Storage: −55 to +71°C (−67 to +161°F)
Altitude: up to 70,000 ft

Physical
Dimensions: double height (6U) Eurocard, with two 96 pin MIL DIN type connectors
Power: 5 V DC, 2.8 A (typical)

Contractor
ELTA Electronics Industries Ltd (a subsidiary of Israel Aircraft Industries Ltd), Ashdod.

VERIFIED

Elbit Hand-Held Computer (HHC)

The Elbit HHC is a ruggedised hand-held computer with communications, mapping and positioning capabilities. The HHC is designed for combat environments and provides rapid update of the operational picture. It features a full colour daylight readable display and makes use of COTS technology.

Specifications
(a) CPU: Pentium 200 MMX
(b) RAM: 64 Mbytes (up to 256 Mbytes)
(c) Cache: 256 kbytes (up to 512 kbytes)
(d) Input: touchscreen and pointing device
(e) Modem: integrated 3 channel
(f) GPS: integrated
(g) I/O: LAN, 2 serial ports, 2 external PCMIA, 2 SCSI-2, external monitor, external keyboard
(h) Size: 9.5 (L) × 8 (W) × 3 in (H)
(i) Weight: 2.0 kg
(j) Power: 24 V DC or 220 V AC

Status
No longer in production, but may still be in use in the Israel Defence Force.

Manufacturer
Elbit Systems Ltd, Haifa.

UPDATED

Elbit hand-held computer   0055158

EL/S-8661E Mil ruggedised computer

The EL/S-8661E targeted market is in the upgrade of systems based on Data General (DG) computers and compatibles. Portable or rack-mountable, the EL/S-8661E is a ruggedised DG/ECLIPSE and Elta's EL/S-861 × compatible computer. It incorporates a Hawk co-processor and I/O add-on card set from Strobe Data Inc, emulating the DG environment. The EL/S-8661E enables systems based on the venerable 16-bit ECLIPSE to achieve a maximum life span by using their existing software on modern PC-based hardware. Based on IBM PC/AT compatible hardware, it utilises commercial-off-the-shelf (COTS) boards and assemblies.

Designed specifically for military and other harsh environments, the main unit is completely shock isolated from the outer shell. All shock and vibration impinging

on the outer surface is safety dampened before it reaches any of the internal components.

The EL/S-8661E also utilises an active temperature control unit claimed by the manufacturer to be unique. The unit maintains the internal temperature within the range that enables COTS boards to work properly in severe environments.

Applications
(a) Hardware upgrade of systems that are based on the Data General ECLIPSE and Elta's EL/S-861 × computers while making full use of the existing application software and operating systems (RDOS, RTOS, AOS, NANOS+).
(b) Use of PC virtual devices to emulate DG peripheral devices, such as disk drive, main console, ALM.
(c) Interoperability between PC Windows™ applications and applications developed in the NANOS+ environment that are compatible with DG computers via shared memory, disk files and interrupts.

Features
(a) MS-Windows™ S/W development environment
(b) Pentium microprocessors
(c) built in 10.4 in LCD-TFT colour display
(d) various COTS I/O units
(e) ECLIPSE I/O bus
(f) front access to removable media
(g) portable or 19 in rack mountable
(h) 28 V DC or 115/230 V AC 50/60 Hz

Options
(a) 19 in rack-mounted adaptor
(b) Available in various temperature ranges
(c) Upgrades: CPU, memory and HDD capacities.

Specifications
Functional
Type: single board computer
Bus: ISA, PCI (option)
Total slots: up to 8 slots
CPU type: Pentium/Pentium II
CPU clock: up to 266 MHz (upgradable upon request)
DRAM: up to 128 Mbytes
EPROM: up to 4 Mbytes FLASH
Battery: support for 128 kbytes or 512 kbytes SRAM
Timer: RTC with battery back-up; Watchdog
BIOS: in FLASH memory, enables system, SVGA and SCSI BIOS updates
I/O ports: 2 RS-232 serial ports, with RS-485 available on COM2; 1 multiple mode parallel port; fast SCSI II; ETHERNET (optional)
FDD: shock-mounted 3.5 in standard
HDD: removable shock mounted up to 10 Gbytes
Keyboard: sealed, detachable with an integrated trackball/mini-mouse
Graphics display: LCD Panel: 10.4 in LCD Colour Active (TFT) matrix
  (Video DRAM) 1 Mbyte
  (Resolution) 640 × 480 VGA
  (Colours) 256 out of 262,144
  (Secondary) 1,024 × 768, 256 colours
  (Video output) 640 × 480 when simultaneous with LCD
Supported operating systems: MS-DOS (standard) Windows 98™, NT SCO UNIX (optional)
RDOS, RTOS, AOS, NANOS+ on the emulated DG ECLIPSE

Physical
Size (portable): (H × W × D) 16.34 × 10.43 × 16.81 in (415 × 265 × 427 mm)
Weight: 22 kg (40 lb) typically, configuration dependent
Power: 28 V DC or 115/220 V AC

Environmental
Temperature
Operating: +25 to +131°F (−5 to +55°C)
Non-operating: −4 to +160°F (−20 to +71°C)
Altitude
Operating: sea level to 15,000 ft
Non-operating: sea level to 40,000 ft

Contractor
ELTA Electronics Industries Ltd (a subsidiary of Israel Aircraft Industries Ltd), Ashdod.

UPDATED

EL/S-8661 Mil-Spec PC compatible computer

The EL/S-8661 is an eight-slot unit based on an ISA or PCI-ISA, passive backplane computer. Six of the slots are available for application expansion. It can be equipped with a Pentium microprocessor and it supports all the software packages and OS supported by any PCs. Designed specifically for military and other harsh environments, the main unit is completely shock isolated from the outer shell. All shock and vibration impinging on the outer surface is safety dampened before it reaches any of the internal components. The EL/S-8661 also utilises what the

EL/S-8661 Mil-Spec PC compatible computer   0077515

manufacturer claims is a unique active temperature control unit. The unit maintains the internal temperature within the range that enables COTS boards to work properly in severe environments.

The EL/S-8661 has a modular structure designed for easy maintainability, which leads to very short MTTR and permits upgrading and easy adaptation for special requirements. It includes a full range of COTS CPUs and peripherals and a full range of COTS PC cards and expansions

Specifications
Functional
Type: single board computer
Buses: ISA, PCI optional
Total slots: up to 8 slots
CPU type: Pentium, Pentium II or Pentium III
CPU clock: up to 750 MHz (upgradable upon request)
DRAM: up to 512 Mbytes
EPROM: up to 4 Mbytes FLASH
Battery: support for 128 kbytes or 512 kbytes SRAM
Timer: RTC with battery back-up; Watchdog
BIOS: in FLASH memory to ease system, SVGA and SCSI updates
I/O ports: 2 RS-232 serial ports with RS-485 available on COM2; 1 multiple mode parallel port; fast SCSI II, ETHERNET (optional)
FDD: shock mounted 3.5 in standard
HDD: removable, shock mounted, up to 10 Gbytes
Keyboard: detachable, sealed with an integrated mini-mouse
Graphics display: LCD panel: 10.4 in LCD Colour Active TFT Matrix
    (Resolution) 640 × 480 VGA
    (Video DRAM) 2 Mbytes
    (Colours) 256 out of 262,144
Secondary: 1,024 × 768, 256 colours
Operating systems: MS-DOS (standard) Windows 98/NT, SCO UNIX (optional) or any of the of-the-shelf PC compatible OS, including Real Time such as pSOS or VRTX

Physical
Size: 16.34 × 10.43 × 16.81 in (415 × 265 × 427 mm) (W × H × D)
Weight: 22 kg (40 lb) typically, configuration dependent
Power: 28 V DC or 115/220 V AC

Environmental
Temperature
    (Operating) −20 to +60°C (−4 to +140°F)
    (Non-operating) −30 to +71°C (−22 to +158°F)
Altitude
    (Operating) sea level to 15,000 ft
    (Non-operating) sea level to 40,000 ft

Contractor
ELTA Electronics Industries Ltd(a subsidiary of Israel Aircraft Industries Ltd), Ashdod.

UPDATED

EL/S-8661P Mil ruggedised computer

The EL/S-8661P targeted market is in the upgrade of systems based on the Compaq 16-bits PDP-11 family of computers and compatibles. Portable or rack-mountable, the EL/S-8661P is a ruggedised Compaq/PDP-11 compatible computer incorporating an Osprey co-processor and I/O add-on card set from Strobe Data Inc emulating the PDP-11 environment. The EL/S-8661P enables systems based on the venerable 16-bit PDP-11 to achieve a maximum life span by using their existing software on modern PC-based hardware. Based on IBM PC/AT compatible hardware, it utilises commercial-off-the-shelf (COTS) boards and assemblies.

Designed specifically for military and other harsh environments, the main unit is completely shock isolated from the outer shell. All shock and vibration impinging on the outer surface is safety dampened before it reaches any of the internal components. The EL/S-8661P also utilises an active temperature control unit claimed by the manufacturer to be unique. The unit maintains the internal temperature within the range that enables COTS boards to work properly in severe environments.

Typical applications are:
(a) upgrade of hardware systems that are based on COMPAQ PDP-11 computers while making full use of existing application software and operating system (RSTS, RSX or RT11).
(b) use of PC virtual devices to emulate the PDP-11 peripheral devices, such as disk drives, main console, serial ports.
(c) interoperability between PC Windows applications and applications developed on the PDP-11 environment.

Features
(a) MS-Windows™ S/W development environment
(b) Pentium microprocessors
(c) built in 10.4 in LCD-TFT colour display
(d) various COTS I/O units
(e) Q-Bus, Uni-BUS
(f) front access to removable media
(g) portable or 19 in rack-mountable
(h) 28 V DC or 115/230 V AC 50/60 Hz

Options
(a) 19 in rack-mounted adaptor
(b) available in various temperature ranges
(c) upgrades: CPU type, clock, memory and HDD capacities are upgradable upon request.

Specifications
Functional
Type: single board computer
Bus: ISA, PCI (option)
Total slots: up to 8
CPU type: Pentium or Pentium II
CPU clock: up to 266 MHz (upgradable upon request)
DRAM: up to 128 Mbytes
EPROM: up to 4 Mbytes FLASH
Battery: support for 128 kbytes or 512 kbytes SRAM
Timer: RTC with battery backup; Watchdog
BIOS updates: in FLASH memory to enable system, SVGA and SCSI BIOS
I/O ports: 2 RS-232 serial ports with RS-485 available on COM2; 1 multiple mode parallel port; fast SCSI II; ETHERNET (optional)
FDD: shock-mounted 3.5 in standard
HDD: removable shock-mounted up to 10 Gbytes
Keyboard: sealed detachable with integrated trackball/mini-mouse
Graphics
    (LCD panel) 10.4 in LCD-TFT Colour
    (Display) active matrix
    (Video) 1 Mbyte DRAM
    (Resolution) 640 × 480 VGA
    (Colours) 256 out of 262, 144
    (Secondary) 1,024 × 768, 256 colours
    (Video output) 640 × 480 when simultaneous with LCD
Supported operating systems: MS-DOS (standard) Windows 98/NT, SCO UNIX (optional),
RSTS, RSX, RT11 (on the emulated PDP-11)

Physical
Size (portable): (H × W × D) 16.34 × 10.43 × 16.81 in (415 × 265 × 427 mm)
Weight: 22 kg (40 lb) typically, configuration dependent
Power: 28 V DC or 115/220 V AC

Environmental
Temperature
Operating: +25 to +131°F (−5 to +55°C)
Non-operating: −4 to +160°F (−20 to +71°C)
Altitude
Operating: sea level to 15,000 ft
Non-operating: sea level to 40,000 ft

Contractor
ELTA Electronics Industries Ltd (a subsidiary of Israel Aircraft Industries Ltd), Ashdod.

UPDATED

EL/S-8663 MIL ruggedised PC-compatible computer

The EL/S-8663 is a member of the EL/S-8660 rugged computer product line. Products in this line withstand harsh environmental conditions even though they are built using COTS PC modules.

The EL/S-8663 is based on a PCI-ISA passive backplane. Featuring a Pentium microprocessor, it supports widely used PC software packages and Operating

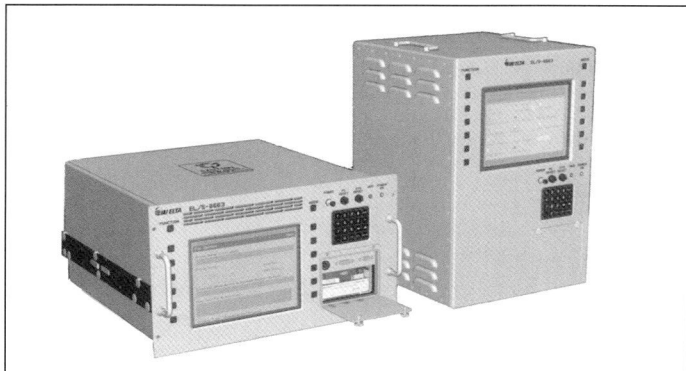

EL/S-8663 MIL ruggedised PC-compatible computer 0055161

Systems such as Windows-95™, Windows-NT™ and SCO-UNIX. Designed specifically for military and other harsh environments, the main unit is completely shock-isolated from the outer shell. All shocks and vibrations impinging on the outer surface are reduced before they reach any of the internal components. The EL/S-8663 is a modular system designed for easy maintainability, with a claimed very short MTTR, and the design also permits easy adaptation for special requirements.

The EL/S-8863 is designed for use in closed vehicles and shelters, commercial and military aircraft and in ships. It is available in two different enclosures, one targeted at rack-mounted applications while the other is for desktop stations. Recommended applications are Standalone PC/Windows applications, C4I as well as control stations for various sensors such as Radar and Sonar. For each application the computer can be equipped with the required processing and I/O boards, using empty slots in the passive backplane.

Features
(a) Availability of a full range of COTS PC modules and peripherals
(b) Easy adaptation for specific applications
(c) Bus selection: ISA, PCI or PCI-ISA
(d) MS-Windows™ software development environment
(e) Pentium and Pentium MMX microprocessor family
(f) Built-in 10.4 in active LCD-TFT colour display
(g) Front access to removable media
(h) Desktop and 19 in rack-mountable

Specifications
Functional
Type: single Board Computer
Bus: PCI-ISA
Total slots: 2 PCI-ISA, 4 PCI and 4 ISA
CPU type: Pentium, Pentium II or Pentium III
CPU clock: 233 MHz (upgradable upon request up to 750 MHz)
DRAM: up to 512 Mbytes
Cache: 256 or 512 Mbytes Flash
Flash disk: up to 4 Mbytes (optional)
Timer: RTC w/battery back-up; dual Watchdog
BIOS: in FLASH memory to ease system, SVGA and SCSI BIOS updates
I/O ports: 2 RS-232 serial ports with RS-485 available on COM2; 1 bidirectional parallel port (LPT1); fast SCSI II; Ethernet (optional)
FDD: shock-mounted 3.5 in standard
HDD: removable shock mounted, 1.6 Gbytes min
Keyboard: sealed, detachable with an integrated mini-mouse
Graphics display
   (LCD panel) 10.4 in LCD Colour Active TFT Matrix
   (Video) 1 Mbyte DRAM
   (Resolution) 640 × 480 VGA
   (Colours) 256 out of 262,144
External secondary video display:
   (Max) 1,280 × 1,024, 256 colours
   (When simultaneous LCD) 640 × 480
Operating systems supported: Windows 95™ (standard), Windows-NT™, SCO UNIX (optional)
Physical
Power: 115/220 V AC, 47-400 Hz
Size and weight (typical, configuration dependent)
   (Desktop) (H × W × D) 13.59 × 9.51 × 13.78 in (345 × 496 × 350 mm), 23 kg (51 lb)
   (Rack mount) (H × W × D) 19 × 10.5 × 19 in (483 × 266 × 483 mm), 22 kg (49 lb)
Environmental
Temperature
Operating: +32 to +122°F (0 to +50°C)
Non-operating: +32 to +149°F (0 to +65°C)
Altitude
Operating: sea level to 15,000 ft
Non-operating: sea level to 40,000 ft

Contractor
ELTA Electronics Industries Ltd (a subsidiary of Israel Aircraft Industries Ltd), Ashdod.

UPDATED

EL/S-8666P ruggedised alpha station

The EL/S-8666P is a ruggedised 19 in rack-mountable computer, based on the COTS COMPAQ Alpha 21164/500 SBC and DMCC family. It will withstand severe shock and vibrations, and operates over wide temperature ranges. All electrical components, including the removable storage units, are isolated against shock and vibrations.

The EL/S-8666P's storage units, such as the hard disk, the tape drive and the CD-ROM drive, are removable and accessible through a front panel door. Each of the standard COTS COMPAQ removable storage units fits into its own slot. Alternatively the workstation can be supplied with a separate storage unit (type EL/S-8668-MSU) for remote installation. The remote unit is available with an independent power supply or it may be powered from the computer. The EL/S-8666P computer is supported by a BIT (Built-In-Test) software package for self-testing during power-up reset.

Main features
(a) Based on COMPAQ modular computing components technology
(b) COMPAQ 21164 ALPHA microprocessor chip at 500 MHz
(c) 32 Mbytes main memory, expandable up to 512 Mbytes
(d) 4.3 Gbyte removable SCSI-2 hard disk drives; expandable up to 6.3 Mbytes
(e) choice of two of the following removable storage units: hard disk, tape drive, CD-ROM
(f) front access to removable storage units
(g) COTS electrical devices
(h) standard I/O ports (SCSI-2, Ethernet, Serial, Parallel and so on)

Specifications
Functional
(a) Type: Alpha 21164 500 MHz SBC-compatible DMCC technology
(b) Buses: 10 PCI, 3 ISA slots
(c) CPU type: Alpha AXP 21164 processor
(d) CPU clock: 500 MHz
(e) CPU performance: 13 SPECint95, 16 SPECfp95
(f) Memory: 1 Mbyte Flash, 32 Mbytes standard and up to 512 Mbytes DRAM, 3 levels of cache: 16 kbytes L1, 96 kbytes L2, 2 Mbytes SRAM L3
(g) Battery: support for TOY and 8 kbytes NVRAM
(h) Timer: TOY w/battery backup; Watchdog
(i) I/O ports:
   2 RS-232 serial ports
   keyboard port
   mouse port
   parallel port
   fast SCSI II
   3 Ethernet ports (2*100BASE TX and 1*10BASE2)
(j) FDD: 3.5 in standard, shock-mounted
(k) HDD: 4.3 Gbytes removable, shock mounted
(l) Graphics: PCI graphics controller
(m) Operating systems supported:
   DEC Open VMS
   VxWORKs for ALPHA
   Windows NT
   DEC OSF/1
Environmental
(a) Temperature:
Operating: 0 to +50°C (+32 to +122°F)
Non-operating: −20 to +71°C (−4 to +160°F)
(b) Vibration (operating): MIL-STD-810E Fig. 514.7(a), Li=0.1 g²/Hz F_1=68 Hz± 5%
(c) Altitude:
Operating: sea level to 10,000 ft
Non-operating: sea level to 40,000 ft
(d) Humidity: up to 90% operating and non-operating (non-condensed)
Options
(a) removable 8 Gbytes SCSI-2 tape drive
(b) removable 4.3 Gbytes SCSI disk drive
(c) DRAM - up to 512 Mbytes

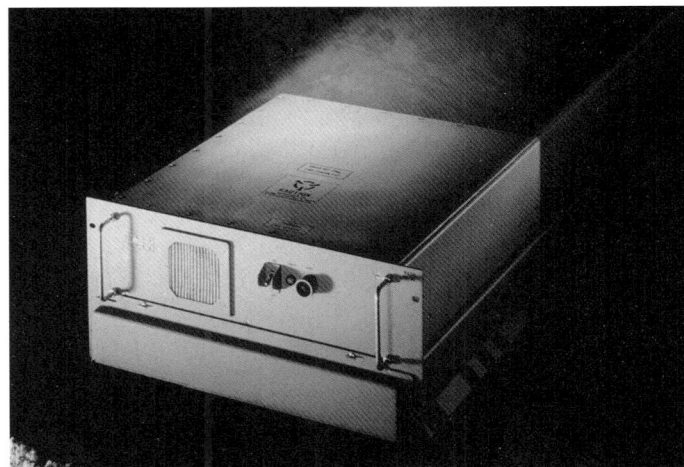

EL/S-8666O ruggedised alpha workstation 0055159

(d) 17 to 21 in ruggedised high-resolution colour monitor
(e) ruggedised keyboard and trackball
(f) various types of graphic boards on PCI

Physical
(a) Size: 19 × 21 × 8.75 in (H × W × D)
(b) Weight: 32 kg (70 lbs)
(c) Power requirements: 90-250 V AC/47-440 Hz single-phase MIL-STD-704D
(d) Power consumption: 320 W (max)

Contractor
ELTA Electronics Industries Ltd(a subsidiary of Israel Aircraft Industries Ltd), Ashdod.

UPDATED

EL/S-8831 Command and Control Operator Console (CCOC)

CCOC is a ruggedised control station based on VME boards. With its subsystems it is a general purpose command and control workstation which provides the capabilities of command, operational control and monitoring of various UAV platforms, payloads, sensors and other systems. Also included in the CCOC is the provision for adaptation to additional generic air vehicles, payloads and sensors. Mounted in a 19 in rack, it includes a mass storage unit (3U), which is large enough to accommodate the installation of various memory storage units such as a hard disk, floppy disk, or CD-ROM. The system includes both a 19 in and a 15 in RGB colour monitor.

The CCOC is built to withstand severe shock and vibration and the electronic components, including the disk drives, are further isolated against shock and vibration. In addition, it is capable of operation over wide ranges of temperature and humidity.

Features
(a) SUN SPARC 5/10/20 workstation compatible
(b) Solaris and VxWorks development environment
(c) Front access to removable media
(d) 2.1 Gbytes hard disk drive
(e) 1.44 Mbytes floppy disk drive
(f) CD-ROM device
(g) Various COTS I/O, Graphic and Video
(h) Options: available in various temperature ranges
(i) Upgrades: CPU type, clock, memory and HDD capacities are upgradable

Specifications
General
Type: operator console
Bus: VME
Total slots: 8 slots

EL/S-8831 Command and Control Operator Console (CCOC)
0009935

CPU type: 1 or 2 SPARC 5/10/20 standard
CPU clock: 110 MHz
Memory: up to 128 Mbytes DRAM; up to 4 Mbytes flash
I/O ports: up to 20 RS-232/422 serial ports; 1 multiple mode parallel port; FAST SCSI II port; ETHERNET port
HDD: 2.1 Gbytes, 3.5 in standard
Keyboard: sealed with integrated trackball
Graphics display: 15 in or 19 in colour monitor; 1 Mbyte DRAM (video); 1,280 × I,024 resolution
External video interfaces: 2 inputs; 4 outputs
Operating systems supported: Solaris and VxWorks

Environmental
Operating temperature: +41 to +122°F (+5 to +50°C)
Non-operating temperature: −13 to +160°F (−25 to +71°C)
Operating altitude: sea level to 15,000 ft
Non-operating altitude: sea level to 35,000 ft

Physical
Size: standard 19 in rack-mounted on shock mounts; 23U, 29U or 30U height; 26 in depth
Weight: configuration dependent
Power: 115/220 V AC

Contractor
ELTA Electronics Industries Ltd(a subsidiary of Israel Aircraft Industries Ltd), Ashdod.

UPDATED

EL/S-8841 Data and Image Distribution Unit - DIDU

Elta's EL/S-8841 is an advanced, compact Data and Image Distribution Unit (DIDU). It is an effective tool for acquiring and distributing high-quality images to battlefield commanders and users via tactical radio and other communication networks. The EL/S-8841 supports exchange of images and data over VHF/UHF Tactical Radio networks. It enables transparent delivery of images and data messages of variable types and lengths. Images can be derived from a wide range of video sources including Unmanned Aerial Vehicles (UAVs) receivers and ground surveillance EO/IR Sensors.

The EL/S-8841 comprises two major modules, the video imagery module and the communication module. The video imagery module handles the video and relevant data received from the video source. It displays real-time colour video or still images and enables the operator to capture, freeze video frames and add alphanumeric and graphic overlay annotations and markings. The module provides compression at a ratio of up to 1:100, while maintaining high-quality of the decompressed images. The communication module handles the communication protocol, enabling dissemination of the image over the tactical radio network. It includes an intelligent modem, which uses error correction codes and handles the physical and link layers of the communication systems.

The EL/S-8841 can be supplied in battery-powered man-pack and vehicular versions.

Features
(a) image handling
(b) display real-time video and relevant data received from the video source
(c) digitises, grabs (freeze) and displays colour images
(d) colour alphanumeric and graphic annotation of images
(e) image enhancement
(f) image compression (up to 1:100) and decompression
(g) designating an Area-Of-Interest, enabling the definition of two compression ratio/image-quality levels
(h) image storage (up to 100 compressed images) and image bank handling
(i) interface to communication controller
(j) data communication
(k) provides interface to external applications
(l) enables transparent transmission of data and images over tactical radio networks
(m) broadcasting a message to several destinations utilising the broadcast capabilities of the network
(n) supports dynamic adaptive routing between stations within the same radio network
(o) network connectivity monitoring to support forwarding
(p) supports end to end acknowledge/retransmit of all messages or specific packets
(q) addressing features
(r) error detection and correction.

Specifications
(a) Interfaces
Tactical radio: SINCGARS, VRC-12 family -16 kbps
Digital radio-telephones: CDP, up to 32 kbps
Analog radio-telephones: DQPSK, up to 2,400 bps
External data modems: up to 64 kbps
Video sources: NTSC (RS-170), PAL (CCIR)

(b) Physical
Power: 28 V DC batteries optional
Size: (H × W × D) 370 × 330 × 90 mm (14.6 × 13 × 3.6 in)
Weight: 8.3 kg (13.97 lb) typically, configuration dependent

(c) Environmental
Temperature
   Operating: −4 to +122°F (−20 to +50°C) (i)
   Non-operating: −4 to +160°F (−20 to +71°C)
Altitude
   Operating: sea level to 15,000 ft
   Non-operating: sea level to 40,000 ft
Humidity: up to 100% operating and non-operating
Notes:
(i) Available in various temperatures ranges.

Status
Remains in the manufacturer's catalogue, but would appear to have been superseded by the EL/S-8821 (see separate entry).

Contractor
ELTA Electronics Industries Ltd (a subsidiary of Israel Aircraft Industries Ltd), Ashdod.

UPDATED

EL/S-8990 Imagery Exploitation Station (IES)

ELTA's Imagery Exploitation Systems (on ground or airborne) provide high-quality, near-realtime IMagery INTelligence (IMINT). They can process imagery data from a variety of day and night, allweather surveillance sensors, including electro optical, infrared and Synthetic Aperture Radar (SAR). The interpretation process is expedited using computerised, softcopy exploitation.

Raw images are geo-referenced processed using proprietary and field proven image registration and image enhancement algorithms. Intelligence quality is augmented by referring to relevant historical data extracted from multiTerabyte archives on a geographical and/or other basis. The image intelligence gathering and dissemination process is managed and prioritised by queuing interpretation requests based on established priorities, automatically distributing incoming image data to available operators. High-speed datalinks accommodate imagery data bandwidth, enable near realtime response and assure data survival. Recording facilities provide archiving and enable further analysis. The advanced EL/S-8990 GES consists of standard UNIX workstations, COTS software packages and libraries of advanced, expandable utilities for image processing and management.

Main building blocks
(a) mission planning and monitoring
(b) softcopy exploitation
(c) management stations
(d) multiTByte size archive servers
(e) highspeed datalinks
(f) dissemination support.

Main features
(a) automatic image georeferencing
(b) automatic target detection (SAR)
(c) automatic change detection (SAR)
(d) automatic cluster identification (SAR)
(e) image enhancements:
   lighting shadowed areas
   sharpening and other convolving filters
(f) GIS and R–DBMS access to archive
(g) hierarchical storage management archive

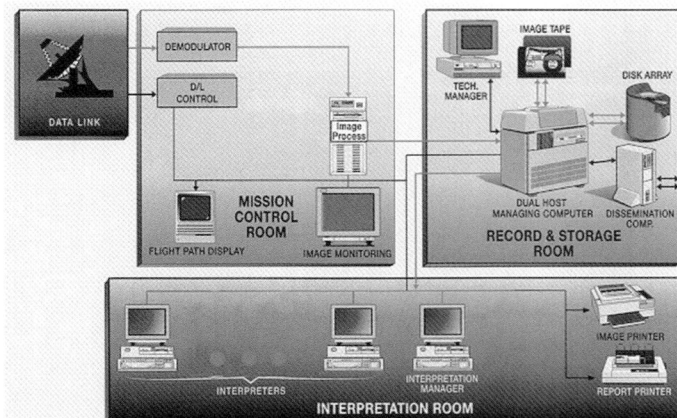

EL/S-8990 block diagram   0077507

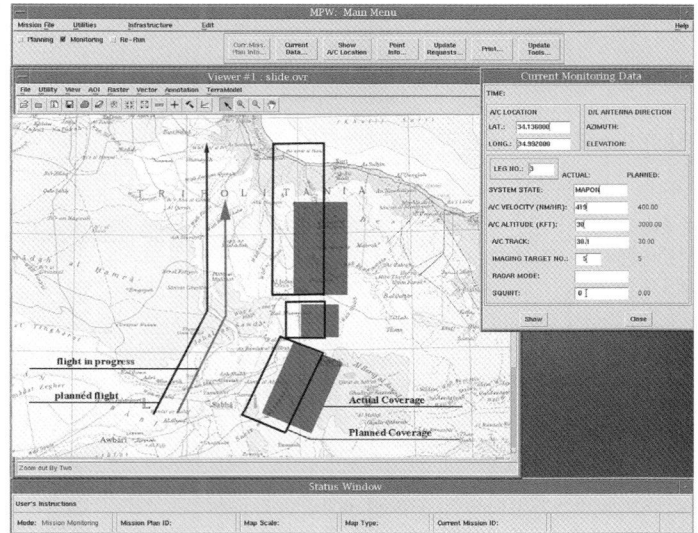

EL/S-8990 Mission Planning Screenshot   0077510

   local (workstation) disk caching
   central RAID
   offline tapes for unused historical data
   offsite tapes for 'disaster recovery'
(h) scaleable system capacity:
   archive capacity
   workstation number
   LAN throughput
   non-blocking 'switch' type LAN
(i) hardware data compression
(j) high-availability and graceful degradation.
The EL/S-8990 GES can be supplied as a 'turnkey' system or can be adapted to specific hardware configurations (number of workstations and so on) and software capabilities.

Contractor
ELTA Electronics Industries Ltd (a subsidiary of Israel Aircraft Industries Ltd), Ashdod.

UPDATED

Enhanced Tactical Computer (ETC)

The ETC 2000 is a versatile rugged tactical PC. It features a sun-readable display and multiple I/O capabilities. It is designed for use as a combat C3I terminal and is able to withstand harsh environmental conditions.

Specifications
(a) CPU: Pentium 933 MHz MMX
(b) RAM: 256 Mbytes (up to 512 Mbytes)
(c) Cache: 256 kbytes (up to 512 kbytes)
(d) HDD: 40 Gbytes
(e) Input: QWERTY illuminated keyboard, touchscreen and pointing device
(f) Display: 10.4 in TFT active SVGA; 120 ft Lamberts (sun-readable)
(g) Modem: integrated 3 channels
(h) GPS: integrated GPS
(i) I/O: LAN, 2 serial ports, 2 external PCMCIA, external monitor (XVGA), external keyboard, parallel port

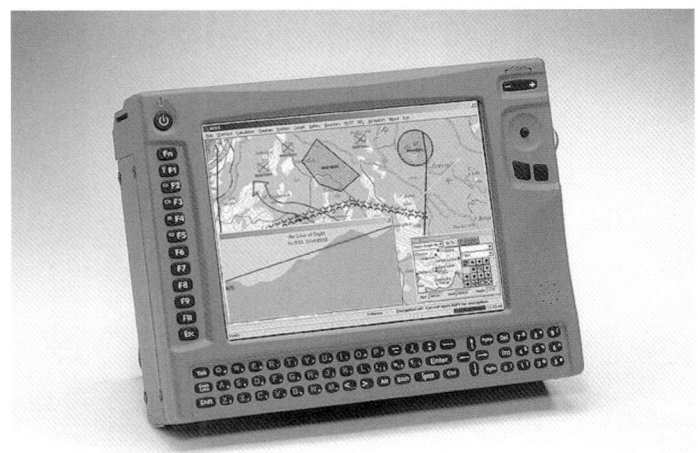

ETC tactical computer   0130752

(j)   Size (L × W × H): 10 × 9 × 3.5 in
(k)   Weight: 5.5 kg
(l)   Power: 24 V DC or 220 V AC

Status

In production. In use in the Israel Defence Force and claimed to be "the choice of many armed forces worldwide". Installed in Merkava and M60 MBT, a variety of artillery systems, and in attack helicopters, as well as in command posts.

Contractor

Elbit Systems Ltd, Haifa.

UPDATED

LTC-800 Tactical Laptop

Tadiran produces a number of Tactical PCs designed for C4I applications, of which the LTC-800 is an example. It is designed to meet the most demanding battlefield requirements and withstand severe environmental conditions. It is available with a wide variety of communication options, including an internal communications controller with repeat and relay and embedded GPS. Potential applications include use as a workstation in various C4I systems at Division, Brigade and Battalion levels, Artillery C4I systems, remote access to large databases, data collection and transmission and map display. The system uses a standard PC architecture, meets full military specifications and is equipped with a vehicle mounting.

Specifications

CPU: Intel Pentium III - 800 MHz
Display: Colour 12.1 in TFT with 1,024 × 786 resolution and external XGA interface
Keyboard: QWERTY, QWERTZ, and external keyboard interface
RAM: 256 Mbyte DRAM, 512 kbyte Second Level Cache
Hard disk: Up to 20 Gbyte (removable)
Pointing device: Pressure pad with 3 buttons and external mouse interface
Serial RS-232 ports
Parallel interface
3.5 in diskette drive
Ethernet 10Base 2/100Base T
CD-ROM

Options:
Internal CD-ROM × 20
2 PCMCIA slots (Type II)
PC-COM 2000, 2 channel communications controller
GPS

Software Operating Systems:
DOS 5.0/6
DOS RMX
Windows 3.11
Windows 95/98
Windows NT
SCO-UNIX
Solaris × 86
Power: 10.8-32 V DC (MIL-STD1275A-AT)
Backup battery: Up to 15 min full operation
Weight: 12 kg max
Dimensions:
(W × L × H): 385 × 435 × 105 mm (closed), 358 mm (opened)

Tadiran LTC-800 Tactical Laptop   0085292

Status

The LTC-800 has been tested and approved by the Swiss Defence Procurement Agency, and there are existing installations for Piranha, M113 APC, Duro and Puch vehicles, but usage cannot be confirmed.

Contractor

Tadiran Electronic Systems Ltd, Holon.

UPDATED

Rugged Personal Digital Assistant (RPDA-88)

The RPDA-88 is a rugged palm top which will provide a situational awareness display and data communications. Four variants are available, providing different PC Card slot configurations.

Specifications
Technical
(a)   Processor: Intel@ StongARM SA1110, 206 MHz
(b)   SDRAM: 64 Mbytes
(c)   ROM: 32 Mbytes Flash
(d)   Operating System: Microsoft Pocket PC 3.0, Microsoft Pocket PC 2002
(e)   Audio: Built-in loudspeaker
(f)   Microphone: Built-in
(g)   Display: Touch-sensitive colour-reflective, TFT LCD, 4,096 colours (12 bit), 240 × 320 (QVGA), 3.77 in diagonal.
(h)   Interfaces: Serial port 115 Kbytes/s, USB, Stereo audio output, Infra-red port 115 Kbytes/s, Internal Bluetooth (optional)
(i)   PC Card slots:
      Type B: 1 Type II PCMCIA slot
      Type C: 2 Type II or 1 Type III PCMCIA slots
      Type D: 1 Type III and 1 Type II PCMCIA slots
(j)   Tactical Modem (optional): Internal Dual Channel Communication Controller, MIL -STD-188-220, TCP/IP.
(k)   Power: Internal Lithium Polymer, BA5800 battery adaptor, AA battery adaptor, AC/DC adaptor.
(l)   GPS: C/A Code GPS.

Physical
(a)   Dimensions:
      Type A: 3.5 × 5.7 × 0.9 in
      Type B: 3.5 × 5.7 × 1.4 in
      Type C: 3.5 × 5.7 × 1.6 in
      Type D: 3.5 × 5.7 × 1.8 in
(b)   Weight:
      Type A: <12.3 oz (380 g)
      Type B: <19.3 oz (550 g)
      Type C: <22.8 oz (650 g)

Environmental
(a)   MIL-STD-810E
(b)   Altitude: Operational 15,000 ft, storage 40,000 ft

The RPDA-88 (Tadiran)   0140069

(c) Temperature: Operating –20 to +55°C; Storage –35 to +65°C
(d) Leakage: 1 m of water

Accessories
(a) AC and DC power supply/battery charger
(b) Serial RS-232 cable
(c) USB cable
(d) Field carrying case
(e) Display protective cover
(f) Wrist strap
(g) Leg strap
(h) GPS antenna and cables

Status
In use in the IDF. Bought in considerable numbers by the US, probably for evaluation and limited operational use. It is a contender for the vehicle maintenance hand-held PC for Stryker.

Incorporated into Innovative Concepts' pocket FAC equipment. Also bought for trials by several unspecified European nations.

In March 2003 it was announced that General Dynamics C4 Systems had placed a more than US$3 million order for RPDA-88; deliveries were to take place over a 6 month period starting in July 2003.

Contractor
Tadiran Communications Ltd, Holon, Israel.

UPDATED

TACTER-31A rugged hand-held computer (RHC)

The RHC (Tadiran designation TACTER-31) is a digital messaging terminal with embedded modem communications that can function as a building block in command and control systems. In addition, navigation and mapping capabilities are supported by an internal GPS receiver. The communication protocols used, such as MIL-STD-188-220 and TCP/IP, provide simultaneous connectivity to two independent tactical radio nets, as well as LANs and other networks.

The TACTER-31 has a range of applications. In a typical combat application, the terminal transmits formatted and free text messages. When used in a C4I system, the GPS capability enables the terminal's position to be displayed on the screen's moving map. This position is transmitted to other units within the net. Forward observers can use the TACTER-31 to determine the exact co-ordinates of targets and reference points on the map and, with the GPS, to locate their positions on the map and transmit them. A configuration connecting the TACTER-31 to a laser range-finder and digital compass creates a Target Dissemination System (TDS). By adding night vision goggles to this configuration, the TACTER-31 functions as a Target Location Designation Hands-off System (TLDHS).

Specifications
(a) Pentium III 500 MHz CPU
(b) 128/256 Mbytes DRAM and 20 Gbytes hard disk
(c) OS: Windows 95/2000/NT (optional Windows XP or 2000)
(d) internal dual-channel communication modem/controller
(e) internal GPS receiver
(f) active LCD VGA colour display 640 × 480
(g) touchscreen operation, folding keyboard, miniature joystick, external keyboard and/or mouse
(h) PC card (PCMCIA) expansion slots (2 type II or 1 type III)
(i) Internal SCSI,USB and LAN interfaces
(j) Physical:
Dimensions 74 × 178 × 229 mm
Weight less than 3.4 kg

Tacter-31 (Tadiran)   0038052

Tacter-31 in use   0038053

(k) Meets MIL-STD-810E for environmental standards
A wide variety of optional extras is available.

Status
In operational use with the IDF, the TACTER-31 has been selected by the US Army and Marine Corps as a preferred system for both dismounted and vehicular applications. Also in use with the US Air Force. It has been purchased by several unspecified countries in Asia, Europe and Latin America.

Contractor
Tadiran Communications Ltd, Holon.

UPDATED

Tactical Pentop Computer (TPT)

The TPT is a full MIL Spec Laptop computer providing considerable computing power, a touchpanel display and a variety of communication options. It has a small foot print and can be used in vehicles on the move. A two channel PC-COM PCMCIA communication controller can be installed to provide data communication capabilities.

Specifications
(a) Hardware:
CPU: Pentium (Up to 233 MHz)
Display: Colour 8.2 in TFT with 640 × 480 resolution with external XGA interface
RAM: 64 Mbyte DRAM
Hard Disk: Up to 12 Gbytes
Pointing Device: Passive pen on touchpanel active colour TFT VGA screen
10 Function keys
External keyboard interface
Serial RS-232 ports
Parallel Interface
Ethernet 10 Base-2/T
4 h internal battery back-up
(b) Software operating systems:
Windows 3.11, 95/98, NT
DOS 6.2
(c) Options:
GPS
2 PCMCIA Slots (Type II)
PC-COM 2000 2 Channel Communications Controller
(d) Environmental:
Temperature range: Operating, –25 to +50 C; Storage, -25 to +65 C
Altitude: Up to 3,000 m above sea level
(e) Physical and electrical:
Power: 9-32 V DC (MIL-STD1275)
Back-up battery: Up to 4 h full operation
Weight : 4.5 kg
Dimensions (W × L × H): 385 × 435 × 105 mm

Status
Possibly in use with the Swiss forces.

Contractor
Tadiran Communications Ltd, Holon.

VERIFIED

Xi ruggedised computer-driven plotters

Xi Information Processing Systems Ltd has developed a family of ruggedised, real-time track plotters for presenting position of tracked objects and targets on standard maps. They come in a number of sizes and configurations to meet a wide range of C3I applications. The largest models are two rack-mounted plotter types, the 76.2 cm (30 in) P-20V and the 101.6 cm (40 in) P-23V, which may be housed in mobile command shelters. The smallest in the range is the 48.26 cm (19 in) rack-mounted P-25V and P-28V that can be transported easily in any military vehicle or carried in the field by a team of two forward observers.

With the Xi plotters, tracked targets and object path are displayed on standard maps, giving users immediate integration of hostile and own troop location and terrain features. System functions enable data presentation and plotting in any co-ordinate system: UTM, geographic, national or any other. Built-in 'intelligence' allows the plotters to handle several maps of different scales simultaneously.

The plotters use standard RS-422C or RS-232C communication protocols so that they can communicate with each other, with the system navigation computer, or with any other C3I installation. A simultaneous display option allows soft copy generation on a graphics terminal.

Specifications
(a) Size:
    (P-25V) 19 in rack mount - 12U height
    (P-20V) 30 in rack mount - 20U height
    (P-23V) 40 in rack mount - 20U height
(b) Resolution: 0.05 mm (0.002 in)
(c) Control unit: PLC-A, a single processor microcomputer with 32-bit floating-point math co-processor.
(d) Operating modes: local, remote (if connected to host computer), track, diagnostic (BIT)
(e) Co-ordinate systems: UTM/geographic or national
(f) Communication: serial RS-232C or RS-422C; parallel 8-, 16-, 32-bits
(g) Power supply: 24-28 V DC; 110/220 V AC, 50 Hz, 60 Hz, 400 Hz

Status
In production.

Manufacturer
Xi Information Processing Systems Ltd, Savyon.

VERIFIED

Italy

MAGICS display system

MAGICS (Modular Architecture for Graphics and Image Console System) combines high-resolution, raster-scan colour graphics with the MARA multiprocessor, the MHIDAS voice-distribution and the MAVID video-distribution systems to provide comprehensive display facilities in a wide range of configurations for land (site and shelter), sea (surface and submarine) and airborne systems applications.

Significant features of the system include:
(a) multisensor interface capability with analogue or digitised video systems (radar, television, infra-red and sonar)
(b) multisensor capability on the same screen (mosaic, for example, Windows with PPI and television displayed simultaneously on the same screen)
(c) contemporary multimode presentation of the same sensor, for example, in a radar application, PPI and A/R of two different radars or PPI and B of the same radar
(d) any combination of sensors and graphic windows can be arranged on the two displays at run time or, by operator control, according to the operational situation. A high graceful degradation capability is achieved through the dual display configuration, the internal architecture and the intrinsic redundancy of the modular electronics.

High-speed geometric operations are supported by a high-performance geometric processor module that allows:
(a) a system area definition with a resolution of 32 bits integer or floating point; 1.1 µs end point two-dimensional transformation; 1.6 µs end point three-dimensional transformation, including perspective; up to 500 k vectors/s transformed and clipped and curves and conics generation
(b) multiple viewports management on the screen
(c) vector writing speed of 50 µs per pixel (20 Mpixel/s), orientation independent
(d) filling/raster symbol writing speed of 20 µs per pixel (50 Mpixel/s)
(e) user programmable graphic or raster fonts
(f) software programmable synchronisation timing and raster format from 580 × 780 interlaced (625-line TV-compatible) to 1,280 × 1,530 - 60 Hz not interlaced
(g) fully programmable mosaic and windows
(h) custom-defined icons.

Note
MARA - Modular Architecture for Real-time processing Application
MHIDAS - Modular High-Integration Distributed Architecture databus System
MAVID - Modular Architecture for Video and Image Distribution system.

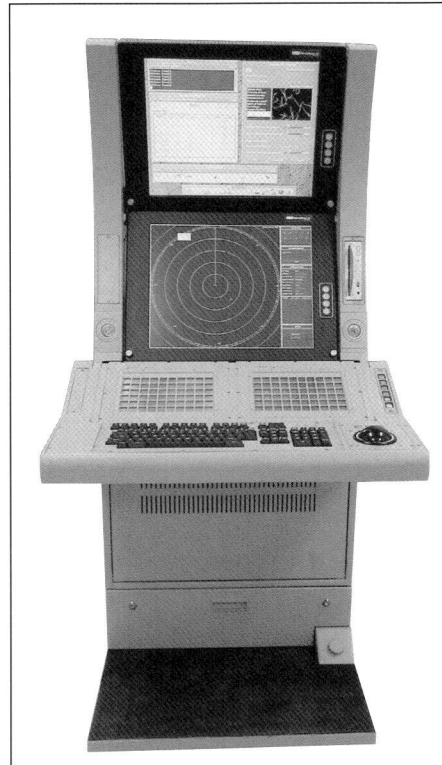

MAGICS-2 bi-monitor multifunction console (AMS)
0132917

Specifications
Monitor:
FPD (AMLCD technology)
20.1 H colour high-resolution (1280 × 1024 pixel)
Ratio 5/4
Angle of view +80°
Triad pitch 0.31 mm
Contrast > 200: 1
Processor Board:
Based on Power-PC CPU family
On board interfaces:
V ME multimasterlslave buses
Serial I/O ports RS-232C
Ethernet port (IEEE 802.3 10 base 5 std)
SCSI II port I
PCI bus

Graphic Board:
PMC
Resolution: user programmable up to 1600 × 1200
8; 16 or 24 bit planes
Up to 4 Mbytes V RAM Frame Buffer

Interface Board:
Ethernet controller
FDDI controller
NDTS A-B-C (MIL-STD-1397 and Stanag 4146)
Serial lines RS-232 and RS-422

Operational Desk:
Configurable with a combination of:
QWERTY keyboard
Multifunction keyboard (60 pushbuttons, 100 pages) with qwerty emulation capability
Track ball
Joystick
High level pushbuttons, protected as necessary
Space available to host external devices (such as communication panel, sonar audio distribution)

Removable Mass Memory Device:
DAT
HD
CD-Rom
Other

Console Software:
Developed according to MIL-STD-498 (DOD STD 2167) using C++ language
AIX Operating Systems (IEEE POSIX 1003 compliant), with X-windows (x 11 R6)/ MOTIF graphic libraries
Framework SW
Main functions:
- Presentation management (radar, TV/IR, range scales, range marks, offset, freeze, zoom)

MAGICS mono-monitor console (AMS) 0132916

- Tactical Synthetic (lines, circles, filled areas)
- Geographic maps
- Data and users synthetic presentation
Auxiliary functions:
- Management of Menus, Alerts, Tabular Data, Graphic Windows
- Category selection
- Track History
- Navigation manoeuvres calculation (collision, CPA/TPA)

Status

In production. In use with the Italian Navy, the Italian Air Force, the Malaysian Navy, as part of the IPN-S C2 system (see separate entry), and with other unspecified navies.

Contractor

AMS, Rome.

UPDATED

Netherlands

Multifunction Operator Console (MOC) Mk 3

The collectively designated Multifunction Operator Console (MOC) Mk 3 is a new generation of operator stations, evolved from Thales Nederland's multifunction operator consoles, MOC Mk 1 and 2, of which more than 200 are now part of 30 combat management systems in ships ranging from FPBs to large frigates. While the MOC Mk 2 generation remains available to fulfil current requirements, the MOC Mk 3 is of a flexible design, including preplanned product improvement for regular technology upgrades far into the future. Because the application of consoles has been extended to functional areas such as command level briefings/debriefings and HQ support, the variety of consoles has been increased as well. Taking into account user efficiency, the MOC Mk 3 design optimises the relationship between the human operator and the computer system. In doing so, it allows a large number of variations in design, functionality and application.

Responding to customer suggestions and an analysis of communication patterns in the naval environment, three styles of communication are identified and supported by the MOC Mk 3 products lines:

(a) PRESENTER for audience-oriented, primarily one-way communication such as briefings and delivery of command operation orders

(b) COMMUNICATOR for interactive human communication such as staff planning and decision analysis which is supported by computer generated information

(c) MEDIATOR for human-to-computer communication when doing routine tasks.

COMMUNICATOR contains two application groups. The CONFERENCE console is a specific derivative, dedicated to close, interactive teamwork in a command team of two to three persons. Besides the normal input devices and display facilities, a large screen is included for a common tactical picture. The COMMAND SEAT or COMMANDER console is designed as a working position for

officers in command. It offers facilities for full scale monitoring, command and control of real-time and non-real-time operations, and tactical situation analysis. This variant is equipped with a complete set of display, control and multimedia facilities. Within each application group a number of versions are available, permitting variation in the number and size of screens and input devices, and/or the type and performance of the processors for the (separated) graphics and system processing tasks.

MEDIATOR comprises the group of workstations covering the whole area of human-computer interactive facilities required for multifunction operator

MOC Mk 3 MEDIATOR operator console as ordered by the Royal Netherlands Navy 0097932

MOC Mk 3 conference console 0052020

MOC Mk 3 desktop mediator console 0052021

workstations. The MEDIATOR group contains versions with horizontally or vertically arranged flat panel screens in various sizes from 18 to 20 in. All versions have the option of a desk-mounted flat screen display, with an integrated hand- or pen- operated touch panel. Each version can be equipped with processors of a customer-preferred type and has scaleable performance.

Status
The first customer was the Turkish Navy which selected the system for the *Barbaros* (Track IIA and Track IIB) class frigates as well as the *Yildiz* class FACS. Commissioned deliveries include systems for Oman and Qatar. Current contracts include systems for the Turkish *Kilic* class and for Indonesia. Based on TACTICOS hardware and software, fire-control systems will be delivered to Korea (*KDX-2* class frigates) and stand-alone Link 11 systems to Portugal (*Joao Belo* class frigates).

Contractors
Thales Nederland, Hengelo.

VERIFIED

Poland

BFC Battlefield Computer

The Battlefield Computer (BFC) is an IBM PC compatible portable Pentium computer. It can be used either as a vehicle based command terminal (with a special shock absorbing mount) or dismounted in a headquarters. It has a choice of two high resolution active matrix displays, a full QWERTY keyboard with additional mouse emulation keys and an optional touchscreen. It has 6 independent serial ports with a choice of RS 232/RS 422/RS 485 configuration. A built-in cable modem enables the computer to be connected to WB Electronics' FONET vehicle digital intercom network from up to 1,500 metres away. Access control is provided by means of an electronic key which is required to activate the keyboard. Typical applications for the BFC computer include fire control and ballistic computer, communication server and CP functions.

Specifications
(a) Processor choice:
Celeron 300, 400, 700 MHz
Pentium III 400, 700 MHz
(b) Main memory: 128 Mbytes max (soldered on mainboard)
(c) Storage:
Choice of 2.5 in solid-state FLASH disks
PCMCIA ATA cards
External FDD 3.5 in 1.44 Mbytes as option
External CD-ROM as option
(d) Display:
10.4 in TFT 800 × 600 pixel. Built-in heater
10.4 in TFT 640 × 480 (high-brightness)
(e) Keyboard:
Integrated, backlit, 102 keys QWERTY + PS/2 mouse emulation keys
8 programmable keys on display side
Waterproof, dustproof
(f) External interfaces:
6 serial (RS-232 / RS-422 / RS-485 choice) or 5 serial + FONET modem
1 parallel printer port (SPP/EPP/ECP)
SVGA monitor
1 USB

The BFC Battlefield Computer from WB Electronics (WB Electronics) 0137505

Ethernet 10BASE-T/100BASE-T connector
PC card in PCMCIA format: 1 × type III or 2 × type II
(g) Power supply:
External 18-48 V DC
Built-in back-up battery 12 V 2.2 Ah with integrated battery charger
(h) Power consumption:
1.5A @24 V (800 × 600 display)
3.0A @24 V (high-brightness 640 × 480 display)
(i) Operating temperature:
−30 to +60 C (800 × 600 display)
−40 to +60 C (640 × 480 display)
(j) EMI emission: meets or exceeds 461D-MIL-STD
(k) Dimensions (L × W × H): 365 × 270 × 85 mm
(l) Weight: 8 kg

Status
In production. In operational use in the Polish Army, notably in artillery fire-control systems. It is installed in the modernised BM-21 from HSW.

Contractor
WB Electronics, Warsaw.

VERIFIED

DD-9620 vehicle-mounted data terminal

The DD-9620 is an IBM-compatible vehicle-mounted data terminal with applications ranging from communication systems controller and fire control system terminal to navigational computer. It has a high contrast colour TFT display and specialised function-numeric keyboard. The standard configuration includes 6 serial ports, external keyboard connector, PCMCIA slot and system lock against unauthorised access. An optional, built-in cable transmission modem allows remote operation (up to 1,500 metres away). The terminal can also be equipped with an external PS/2 type keyboard connector, touch screen, joystick and a set of 8 extra keys located near the display.

Specifications
(a) Processor: Pentium MMX, 166 MHz
(b) Memory: 256 kbytes Flash, 32 Mbytes ECC DRAM
(c) Storage:
ATA FLASH DISK (80-440 Mbytes)
PCMCIA ATA cards
(d) Display:
TFT 6.4 in, 640 × 480 pixel
Optional touchscreen.
Built-in heater
(e) Keyboard: Dedicated 25 keys (functions and numeric)
(f) External connectors:
6 serial ports (RS-232/422/485)
1 PCMCIA slot
Ethernet connector
External PS/2 type keyboard connector (option)
Line connector (optional with internal modem)
Joystick connector (optional)
(g) Power supply:18-48 V DC
(h) Environmental:
High mechanical and environmental resistance
−30 to +60°C operating temperature range

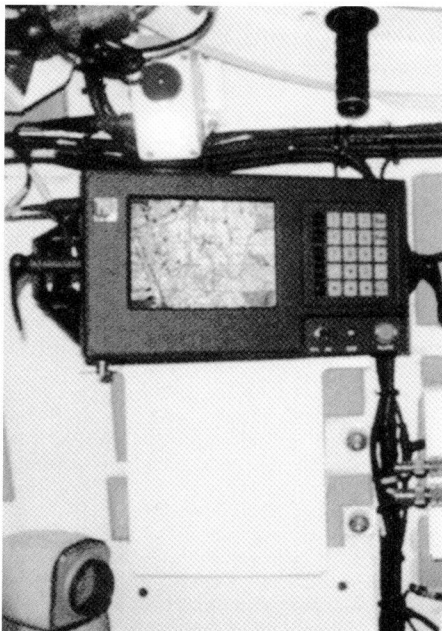

DD 9620 installed in an SP Howitzer
(WB Electronics)
0137501

(i) EMI emission: Meets 461D-MIL STD.
(j) Dimensions (L × W × H): 300 × 165 × 61 mm
(k) Weight: 3 kg

Status
In production. In use in Polish Army.

Contractor
WB Electronics , Warsaw.

VERIFIED

The PC-9600 Hand-Held Terminal (WB Electronics)   0137500

PC-9600 Hand-Held Terminal

PC-9600 is a hand-held computer designed for use at the lower tactical levels which can also be vehicle mounted. It is based on a × 86 architecture and is fully PC compatible. It has 32-bit CPU, large system memory and a high resolution display. Its connectivity facilities allow it to exchange data with up to 5 external devices, including 2 digital radios and a cable transmission modem. It can be automatically configured to operate with transceivers such as Radmor's R3501and Thales's PR4G.

Autonomous, exchangeable battery units allow for continuous 12 hours operation of the terminal. The built-in battery charger and power supply unit can be used with a vehicle supply in the 9-48 V.

Combined in one installation with the R3501 it forms the PCJ-9650 communications set. In this configuration the battery charger charges both radio and terminal batteries.

Specifications
(a) Processor: Intel 386EX, 18-432 MHz
(b) Memory: 5 Mbytes Flash, 3 Mbytes SRAM
(c) Disks:
ROM-DISK with DOS operating system (1 Mbyte)
RAM-DISK with battery back-up (up to 2 Mbytes)
FLASH DISK (4Mbytes).
PCMCIA SRAM/FLASH cards (up to 32 Mbytes).
(d) Display:
LCD 192 × 192 mono - 4 grey levels
LCD 128 × 128 ECB (4 colours)

CD 128 × 128 mono 4 grey levels
LED backlight, heater built-in
(e) Keyboard: Universal, 48 multifunction backlit keys. (IBM PC keyboard emulation)
(f) External connectors:
2 RS-232 or RS-422
2 optional (R × D/T × D only) ports
PCMCIA type I slot
Expansion port for built-in modem or user specific boards
(g) Environmental: Highly resistant to mechanical and environmental conditions −30 to +60°C operating
(h) EMI emission: Meets 461D-MIL-STD
(i) Power supply:
9-48 V external power supply
Up to 12 h operation with built-in battery
Built-in battery charger
(j) Dimensions (L × W × H): 247 × 114 × 64 mm
(k) Weight: 1.4 kg (batteries included)

Status
In production. In use in the Polish Army, particularly for artillery fire control. In July 2003 *Jane's Defence Weekly* reported that new software had been developed for the terminal to enable its use in the fire-control system of the new HSW 98 mm mortar.

Contractor
WB Electronics, Warsaw.

UPDATED

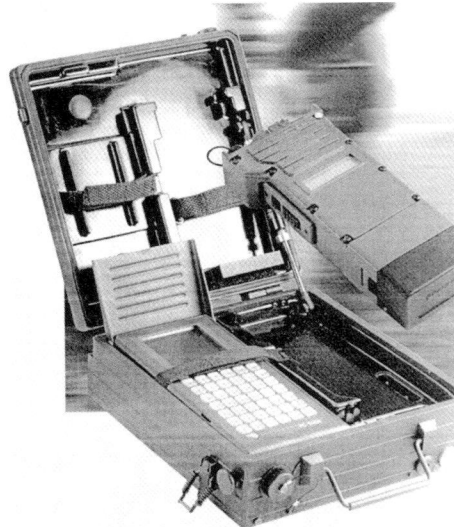

The PCJ-9650 communication set, showing the PC-9600 on the left and the Radmor R-3501 on the right (WB Electronics) 0137498

Sweden

PCQT Rugged monitors

PCQT produces a number of rugged monitors which are in service with Scandinavian forces.

Specifications: PCQT Rugged monitors

Model	QT 9006M	QT9115M	QT9118M	QT 9120M
Display	6.4 in active matrix TFT monitor	15 in active matrix TFT monitor	18.1 in active matrix TFT monitor	20.1 in active matrix TFT monitor
Resolution	640 × 480 to 1024 × 768	640 × 480 to 1024 × 768	640 × 480 to 1280 × 1024	
Brightness	200 cd/m²	250 c/dm²	200 cd/m²	250 cd/m²
Display colours	260K colours	16.7 million colours		
Viewing angle		Left I Right +1- 80°; Up 1 Down +1- 80°		
Keypad	Keypad with 6 programmable function keys			
Pointing device	MS PS/2 mouse compatible with two buttons			
Touchscreen	Optional: Resistive or capacitive	Optional: Resistive, capacitive, infra-red or surface wave		
Dimensions	195 × 170 × 50 mm	370 × 311 × 83 mm	440 × 400 × 120 mm	482 × 429 × 126 mm
Weight	2.5 kg	5.8 kg	12 kg	13 kg
Mounting	VESA 75 mounting or with optional monitor arm	19 in rack with optional mounting kit table top or ceiling with optional stand		
Power requirements	Standard: 28 V DC (18-36 V DC). Optional: AC: External power adaptor, 115-230 V AC, 47- 440 Hz, single phase			

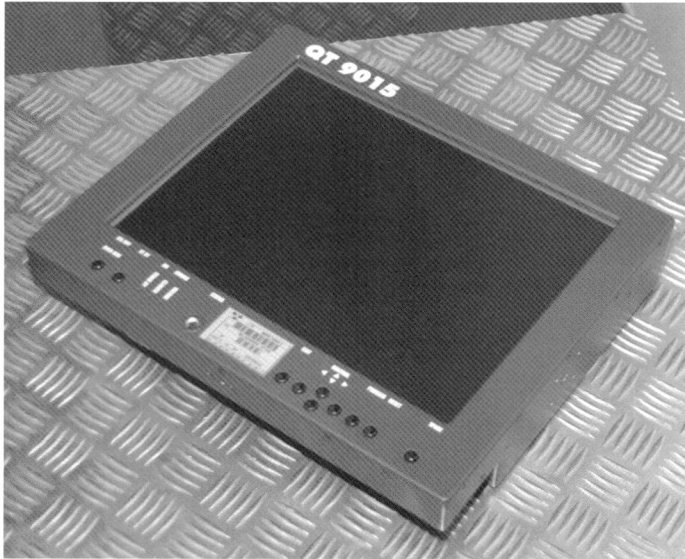

The QT 9115M monitor exhibited at DSEi in 2003 (Patrick Allen/Jane's) **NEW**/1027083

Status

QT 9006M: in service with the Swedish Army, possibly as the monitor for the TCCS (see separate entry).
QT 9115M: in service with the Swedish Navy and the Finnish Army.
QT 9118M: in service with the Swedish Army.
QT 9120M: in service with the Swedish Marines (Coastal Artillery) and the Finnish Air Force

Manufacturer

PCQT, Spånga, Sweden.

NEW ENTRY

QT 9315M

The QT9315M is a rugged 19 in rack-mountable operator's terminal. The unit includes a 15 in LCD monitor, an 83-key keyboard and a pointing device. The monitor folds down when not in use.

Specifications

(a) Display: 15 in active matrix monitor
(b) Resolution: 640 × 480 to 1024 × 768
(c) Brightness: 250 cd/m²
(d) Display colours: 16, 7 million colours
(e) Viewing angle: Left 1 Right +1- 80°; Up 1 Down +1- 80°
(f) Touchscreen: Optional: Resistive, capacitive, infra-red or surface wave
(g) Keyboard: 83-key keyboard with tactile keys
(h) Keyboard layout: English (US and UK), Finnish, French, German, Russian, Spanish, Swedish, Italian and other layouts upon request
(i) Pointing device: MS PS/2 compatible Hulapoint with two buttons
(j) Dimensions: 370 × 311 × 83 mm
(k) Weight: 9 kg
(l) Mounting: 19 in rack with optional mounting kit table top or ceiling with optional stand

The QT9315M rugged monitor and keyboard (Patrick Allen/Jane's) **NEW**/1027085

(m) Power requirements:
Standard: 28 V DC (18- 36 V DC).
Optional: AC: External power adaptor, 115-230 V AC, 47-440 Hz, single phase

Status

In use in the Swedish Army and Navy and the Finnish Army and Navy.

Manufacturer

PCQT, Spånga, Sweden.

NEW ENTRY

Turkey

HT-7243 Hand-held Terminal Unit

The HT-7243 is a lightweight ruggedised portable computer which uses DOS and Windows operating systems. It is used as a data terminal in ASELSAN's BAIKS-2000 artillery fire direction system (see separate entry).

Specifications

Processor	Embedded processor
Display	7.5 in 640 × 480 mono/colour LCD
Pointing device	Micro joystick
Memory	16 Mbytes RAM
Interface	1 × (RS-232/422/485/IRDA Optic interface
	1 × RS-232
	2 × PCMCIA Type 2.1 sealed slots
	External VGA output
	External keyboard/mouse interface
	Parallel port
	2 × SMA RF connector
	(optional GPS, Radio modem, Frame grabber)
Mass data storage	520 Mbytes 1.8 in hard disk
Internal power source	NiMH intelligent battery block
External power source	10.5-32 V DC
	110/220 V AC (with adaptor)
Battery charger	Built-in
Dimensions	240 × 230 × 6U mm
	263 × 230 × 60 mm (with battery block)
Weight	< 25 kg
	< 31 kg (with battery block)
Temperature range	Operating −20 to +50°C
	Storage −30 to +60°C
EMI/RFI	MIL-STD-461/462
Environmental	MIL-STD-810
Options	Up to 32 Mbytes RAM
	2 Gbytes 1.8 in hard disk
	GPS
	Radio modem
	SEMAC modem
	Tactical data/internet communications unit (VIA)
	Frame grabber
	Message encryption
Optional accessories	Li-Ion battery block
	External charger
	110/220 V AC supply adaptor

HT-7243 Hand-held terminal unit. The battery block is the dark strip on the left hand side (ASELSAN)   0137502

Status

In production. In use in Turkish Army.

Manufacturer

ASELSAN Military Electronic Industries Inc, Ankara.

LT-7241 Lap-Top Terminal Unit

The LT-7241 is a ruggedised lap-top computer used in many of ASELSAN's CIS. It uses DOS and Windows operating systems.

Specifications

Processor	K6-II 233 MHz
Display	10.4 in VGA
Pointing device	Micro module
Memory	64 Mbytes RAM
Interface	2 × RS-232
	2 × PCMCIA Type 2.1 slots
	External VGA output
	Parallel port
	1 × SMA RF connector
	2 × ISA internal extension slots
Mass data storage	1 Gbytes 1.8 in hard disk
Other standard hardware	Ethernet (10Base -T)
	Audio (Windows Sound System Compatible)
External power source	18 - 32 V DC
	110/220 V AC (with Adaptor)
Battery block	12 V NiMH
Dimensions	345 × 275 × 90 mm
Weight	< 7.2 kg (with battery block, floppy and CD-ROM)
Temperature range	Operating: -20 to +50°C
	Storage: -40 to +60°C
EMI/RFI	MIL-STD-461/462
Environmental	MIL-STD-810
Options	SVGA TFT LCD
	CD-ROM drive
	Floppy drive
	SEMAC modem
	Frame grabber
	Message encryption hardware
Optional accessories	110/220 V AC supply Adaptor

Status

In use in the Turkish Army.

Manufacturer

ASELSAN Military Electronic Industries Inc, Ankara.

LT-7241 Lap Top Terminal Unit (Aselsan)   0137503

United Kingdom

ICS

ICS is a powerful set of client-server processes providing the foundation services required by C4I systems. These services offer reduced costs across the whole system life cycle, and facilitate interoperability with US and other international systems based on the same foundation.

Core capabilities

The core capability of ICS includes:

(a) messaging and communications services
(b) tracking and correlation
(c) geographic representation
(d) decision and planning aids.

The messaging and communications services provide input and output capabilities to interoperate with a wide variety of systems. Data can be exchanged in many forms including formatted messages, binary data, tactical data links, radars and other sensors, and navigation systems such as GPS. Additional decoders can readily be added to the core capability.

Unit information from each of the input sources is correlated and managed to maintain a synchronised picture across the entire network. Position updates from multiple sources can be associated to improve situation awareness.

ICS users are presented with a geographic representation of units. The chart and mapping service provides a backdrop drawn from a wide variety of map data. Against this map background overlays and tracks can be drawn. Overlays are typically used to represent planning or tactical information, while tracks represent correlated unit contacts.

A wide range of application segments for ICS is also available, including decision aids developed by Logicon, (now Northrop Grumman Mission Systems, formerly Northrop Grumman IT Europe) and third parties. These include:

(a) additional interfaces to tactical data links;
(b) tools to enable ICS to manage the Common Operational Picture;
(c) applications for waterspace management, exercise simulation and others.

The next planned major upgrade of ICS, ICS 3.x, was released in 2003. It reflects the first 'ground-up' redesign for over ten years of the original US system upon which ICS is based. The client side of ICS 3.x is written in Java for NT platforms.

Status

ICS is in operational use as the core foundation software for many joint and maritime systems in NATO and other nations. These include:

NATO Maritime CCIS (MCCIS) - provides the Recognised Maritime Picture and situational awareness capabilities
UK Joint Operational Command System (JOCS) - provides interoperability
UK FOCSLE - situational awareness picture and water space management facilities
UK Command Support System (CSS) - tactical picture
German Navy F124 Frigate CSS - interoperability and command support functions
Italian Navy CSS - interfaces to national sensors and NATO systems
Danish Navy - interoperability and situational awareness

In July 2002 Northrop Grumman announced that it had supplied two ICS 2.2 systems to Mitsubishi Electrical Ltd (MELCO), Kamakura Works, as part of MELCO's contribution to an afloat Command Support System programme for the Japanese Maritime Self Defense Force.

Contractor

Northrop Grumman Mission Systems (formerly Northrop Grumman IT Europe), Southampton.

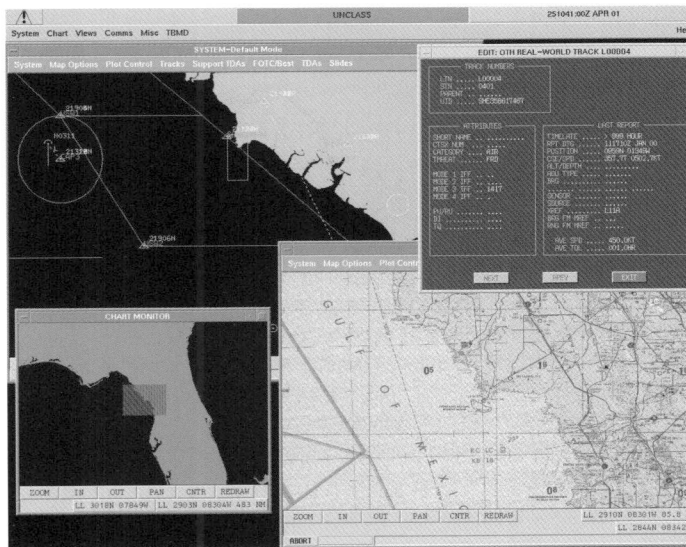

Northrop Grumman IT Europe Ltd's ICS   0109473

LT100 Soldier Digital Assistant

The LT100, shown at DSEi 2003 in London, is an early version of a personal computing system designed for the digitised 'future soldier'. It consists of a CPU, a connection block and a 2.5 in QVGA screen which can be attached to the user's wrist. 3.5 and 6.4 in screens are also available as more traditional hand held devices and optional expansion packs can be added to the CPU to increase memory or to accommodate PCMCIA cards. In the displayed version the components are connected by wires integrated into a vest and the system operated by a simple thumb-and-finger interface, but the intention is to develop a wireless version that will be hands-free and voice activated.

Specifications
(a) Physical:
   Size (mm): CPU 100 × 100 × 25
   Weight: approx 500 g
(b) Electrical interface:
   Power consumption: < 3 W
   Battery: 3 hours standard. Other options available
(c) Interface options: USB, Serial, IRDA, 1394, SPL (Options via 120 way bus)
(d) Additional I/O block options: PCMCIA, extra battery, storage and memory devices
(e) Performance:
   Processor MK1 Superscaler P PC RISC 266-380 MHz with pure dynamic docking for power saving
   Processor M K2 Superscaler P PC RISC 400-760 MHz with pure dynamic docking for power saving
(f) Memory:
   Non volatile 32 Mbytes Flash (expandable)
   Disk on Chip 64Mbytes (expandable)
   SDRAM 32Mbytes (expandable to 1Gbyte)
(g) Operating system: Linux (options: QNX, VX Works, CE*)
(h) Screen options:
   2.5 in QVGA 176 × 220 touch screen fully sunlight readable
   3.5 in VGA 640 × 480 touchscreen fully sunlight readable
   6.4 in VGA 640 × 480 touchscreen fully sunlight readable

Manufacturer
EDO MBM Technology Ltd, Brighton, East Sussex.

NEW ENTRY

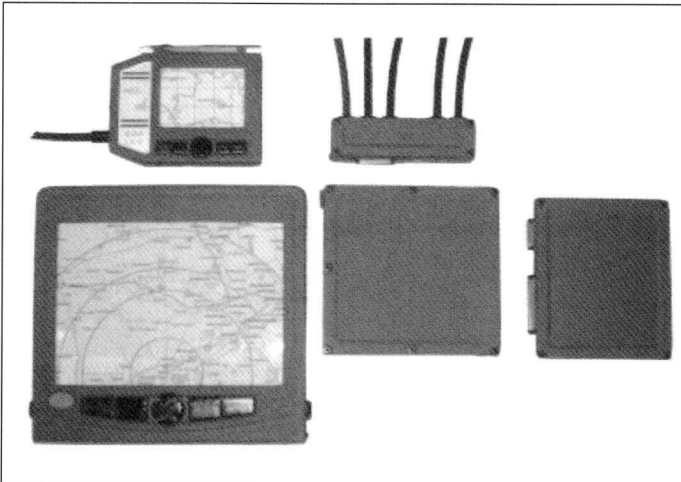

The components of the LT100 (EDO MBM)   *NEW*/0561764

LT450S Termite hand-held computer

The LT450S Termite is a rugged hand-held computer which offers considerable flexibility. The removable keypad allows users to define their own layouts and swap between keyboards to suit specific applications. An extensive range of interconnection options is available through standard or custom connector adaptors which can be changed by the user to suit different applications. These are provided on an interchangeable connector block which fits on the back of the case. The LT450S outer case is available in lightweight magnesium alloys, or aluminium, fully plated and painted with epoxy paints to provide full environmental protection. With a 6.1 in VGA colour TFT LCD with variable backlight and optional viewing hood, Termite can be used in all ambient light conditions, including real-time Video display. With the addition of an inbuilt filter it can be used with Night Vision Goggles (NVG); with this option it is designated the LT450N.

A 5 ampère hour battery pack gives over 2.5 hours of continuous use, and over 5 hours with Windows power-saving modes. Special anti-tamper switches are provided, and user operated purge buttons allow manual purge of data encryption routines; these operate even if the Termite is powered down, or has lost its battery supply. The manufacturer claims that Termite will withstand up to 2 m submersion in water through a dual sealing technique. The seals also ensure that Termite is

LT450S Termite showing battery pack and connector adaptor (Largo Systems)
NEW/0116930

suitable for transport and use at high altitudes, at low pressures and will also maintain EMC compliance.

Variants are the LT455, which has a 1 GHz Transmeta processor and the LT465. The latter was designed to meet the UK Fire Control Battlefield Information System Application (BISA) requirement; the principal differences are the touchscreen interface and the connectors directly mounted onto the case.

Specifications
(a) Processor:
   LT450: 586; 166 or 266 MHz Pentium; 1 GHz Transmeta (LT455)
   LT465: 166/266MHz Pentium user switchable; optional 800 MHz
(b) ROM: 2 Mbytes
(c) RAM: 64 Mbytes expandable to 256 Mbytes
(d) Keyboard: 63 position reprogrammable matrix (12 mm min spacing) with replaceable pad
(e) Display: 6.1 TFT/ LCD with variable backlight. Touchscreen on LT465.
(f) Graphics: VGA 640 × 480 (6.3 in XVGA option)
(g) Mass storage:
   LT450: 10 Gbytes to 25 Gbytes magnetic hard drive (Solid State options 170 Mbytes to 720 Mbytes)
   LT465: Up to 2 Gbytes solid state CF or 4 Gbytes microdrive
(h) Pointing device:
   LT450: Joystick 2 button
   LT465: Touchscreen.
(i) Security features: Covert switch (including display off), purge switches, optional anti-tamper
(j) Connectivity: 102-way connector on rear panel providing connectivity via optional adaptor for up to:
   2 × RS-232: 2 × configurable serial - RS-485 Centronics Parallel Interface; USB interface; additional HDD, FDD, CD ROM via the Centronics or USB interface; ISA
(k) PCMCIA: 2 × Type II or 1 × Type III
(l) Size: 205 × 145 × 72 mm (less battery)
(m) Weight: 2.5 kg
(n) Material/Finish: Aluminium case
(o) Operating system: Windows 98, ME, NT, 2000, XP, Linux
(p) Batteries: Rechargeable Li-Ion
(q) Battery life: 12 h at 30% duty cycle
(r) External DC: Clean, regulated 12 V at 5 amps

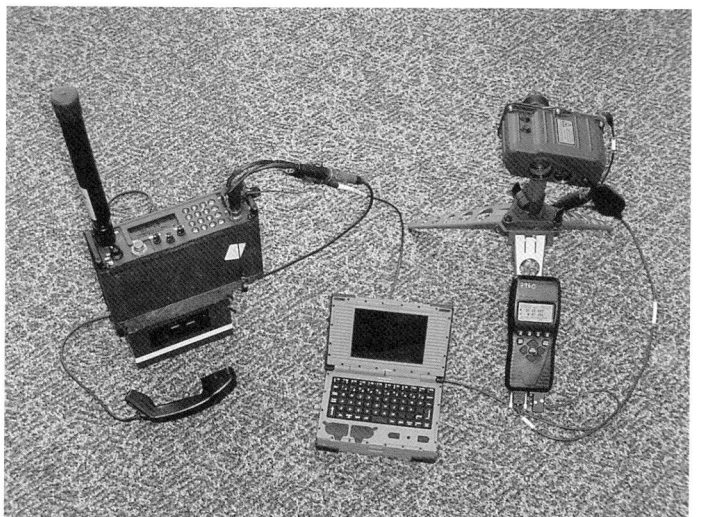

LT450 in QinetiQ's target acquisition suite (QinetiQ)   *NEW*/0111417

(s) External AC: 98/260 V 40/50/400 Hz via adaptor
    Options: Larger battery packs available. 10-l8 V or 16-48 V DC MIL STD 1275 via adaptor
(t) Operating temperature: −30 to +55°C

Status

In production. Used by Qinetiq in the improved target acquisition suite for FACs. Sold to US DoD and other prime contractors. Reported by *Jane's International Defense Review* in June 2002 to be on trial by US Special Operations Command for use with the Special Mission Radio System and the Special Operations Tactical Video System. Used in the airborne and manpack version of the ATRACKS helicopter C2 system (see separate entry).

In August 2002 it was announced that the LT465 had been selected as the handheld computer for the UK FC BISA. The £4.4 million contract, with the prime contractor LogicaCMG, was for the supply of over 600 units to be delivered through 2002 and 2003.

Manufacturer

EDO MBM Technology Ltd, Brighton, East Sussex.

UPDATED

LT500 Modular Tablet Computer

The LT500 Modular Tablet PC is based on MBM's TERMITE™ range of rugged hand-held computers. It is a two-piece design in which the display unit is connected to the main CPU box via a cable. Mounting points allow the LT500 to be vehicle-mounted with the display/touchscreen unit located for the user while the CPU box is stowed separately. Alternatively, an adaptor allows the display to be worn as a kneepad device with the CPU box carried on a belt or, when seated in a vehicle cab or cockpit, stowed nearby.

User input to the LT500 is normally via a touchscreen and virtual keyboard, although USB keyboard/mouse options are also available. A range of connectivity options is available and, of the two PCMCIA slots provided, one is directly user-accessible while the other may be accessed via an external connector. USB 2.0 is available for high-speed video inputs.

The carry case zips together to allow the unit to be carried as a single pack or as a separate display and CPU pack.

Specifications

(a) Processor options: Pentium 266 MHz; Transmeta TM5800 800 MHz; Pentium PIII 400 MHz; Eden Via C3 800 MHz
(b) Size(mm): Display unit 236 × 147 × 27; CPU unit 236 × 147 × 53 (less battery)
(c) Weight (approx): 0.75 kg (display unit); 1.14 kg (CPU unit)
(d) Cache: 512K (Transmeta)
(e) RAM: 32 Mbytes to 256 Mbytes (384 Mbytes Transmeta)
(f) Mass storage: 5 Gbytes hard disk drive with options up to 20 Gbytes. Optional SSD 170 Mbytes to 2 Gbytes.
(g) Display: 6.1/6.5 in LCD transflective with touchscreen. (Portrait mode) NVG option
(h) Interface:
    2 × RS 232 serial port (Options: external mouse, keyboard and monitor connection)
    Two USB 1.1/2.0 ports
    PCMCIA. 1 × factory Type 2; 1 × Type 3 user accessible
    Options: IRDA, PS2, Ethernet
(i) Keyboard: Virtual keyboard via touchscreen (optional external keyboard available)
(j) Power:
    110 V to 250 V AC mains via adaptor cable
    10 V to 32 V DC vehicle power

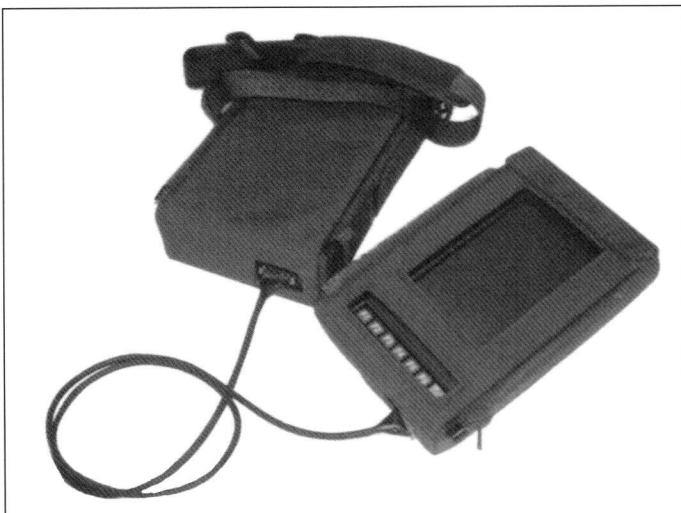

The LT500 is a two-piece design, which can be carried or installed separately, or carried as a single unit (EDO MBM)   **NEW**/0561765

Internal Li-Ion batteries providing up to 24 hours at typical duty cycles
Battery options:- BB2800 or 'Clansman' type.
(k) Operating system: Windows NT4, ME, 2000, XP; Linux

Manufacturer

EDO MBM Technology Ltd, Brighton, East Sussex.

NEW ENTRY

LT600 rugged laptop computer

The LT600 rugged laptop computer is a development of the LT450 Termite (see separate entry). The LT610 variant has a 1 GHz processor.

Specifications

(a) Processor: Pentium II Mobile 266 MHz (166 MHz option). LT610: 1GHz.
(b) Cache: 512 kbytes
(c) RAM: 32 Mbytes to 256 Mbytes
(d) Mass storage: 6 Gbyte hard disk drive with options up to 25 Gbytes. 1.44 Mbyte 3.5 in floppy disk drive. Optional SSD 170 Mbytes to 720 Mbytes. CD ROM drive (24X)
(e) Display: 12.1 in SVGA LCD screen. Transflective option.
(f) Interface:
    2 × RS-232 serial port
    External mouse, keyboard and monitor connection
    Two USB 1.1 ports
    PCMCIA Type III × 1 and 1 × Type II or 3 × Type II
    Centronics parallel printer port
    VGA port
(g) Keyboard: QWERTY 92 key in UK format, with joystick with backlight
(h) Operating system: Windows NT4, ME, 2000, XP; Linux
(i) Dimensions: 100 × 320 × 300 mm (H × W × D)
(j) Weight: 7 kg (approx)
(k) Power:
    110 V to 250 V AC mains via adaptor cable
    10 V to 32 V DC vehicle power
    Internal Li-Ion batteries providing up to 24 h at 30%

Manufacturer

EDO MBM Technology Ltd, Brighton, East Sussex.

UPDATED

LT600 showing backlit keyboard and door concealing 3 × PCMIA slots, 2 × USB ports, PS/2 mouse and keyboard connector (MBN Rugged Systems) **NEW**/0547583

Nauticus

Nauticus is the generic name under which INRI UK Ltd (now Northrop Grumman Mission Systems) delivered foundation Command and Control software based upon that developed by its parent company for the US DoD. This software product is now known as ICS (see separate entry).

Contractor

Northrop Grumman Mission Systems (formerly Northrop Grumman IT Europe), Southampton.

UPDATED

Ultra Compact Console

The UCC (Ultra Compact Console), is a multifunctional twin-screen naval workstation incorporating the latest LCD flat-screen technology. The product developed from studies into the Royal Navy's future requirements combined with Ultra's experience in the development of the Navy's current submarine and surface ship command and control systems (SMCS and SSCS). The UCC provides improved operator facilities, saves space, is simple to install and reduces heat significantly, while improving operational effectiveness. Its extremely low magnetic signature, EMC screening and absence of any degaussing effects make it suitable for many ship classes.

Using packaged COTS technology, the UCC can be mounted on any stable bulkhead, pillar or deck, taking a quarter of the space of CRT based consoles; this enables innovative control room layouts. The folding desk design ensures that when not in use, the UCC does not intrude into operations room space.

The UCC can incorporate communication and control functions managed from a configurable touch input display fitted to the desk, allowing the operator to configure the console for specific tasks. For example, a system tactical picture compiler could log on and have the workstation automatically configure to provide the facilities for his role. This enables the available workstations in the combat system to take on separate distinct or combined roles to enhance operational flexibility.

The LCDs can be fitted with screen diagonals ranging from 12 in to 21.3 in and with resolutions up to 1600 × 1200, which allows for upgrades to accommodate the latest screen technology.

The console is designed to enable technology updating using COTS components. Different processors are readily accommodated to meet specific customer needs. The displays use Windows and Motif techniques with window resizing, and either overlay or tiling are offered. Digitised maps are easily incorporated with superimposed video images from various sensors.

Depending on customer requirements, there is an operational facility to fit an additional flat panel monitor for status display and for the console to be configured as an X-terminal for LAN connection.

The human computer interface is tailored for the military role. The desk area incorporates a space for a touch input display, joystick, trackerball and discrete switches. The keyboard, which is provided primarily for data entry, revolves to provide a clear writing working area. The console can be fitted with display, application and interface processors or any of these may be sited remotely from the console. These processors are commonly linked by a LAN to other parts of the combat system. The reduced power consumption of the flat screens and the facility to put the processors elsewhere means that the need for chilled water in the operational area is eliminated and the installation considerable simplified. With reduced heat and noise (acoustic and electrical) and less space and power, the UCC offers naval forces a significant technology upgrade, particularly where these factors are at a premium.

Status

In service in six unspecified countries.

Contractor

Ultra Electronics Command and Control Systems, High Wycombe, Buckinghamshire.

UPDATED

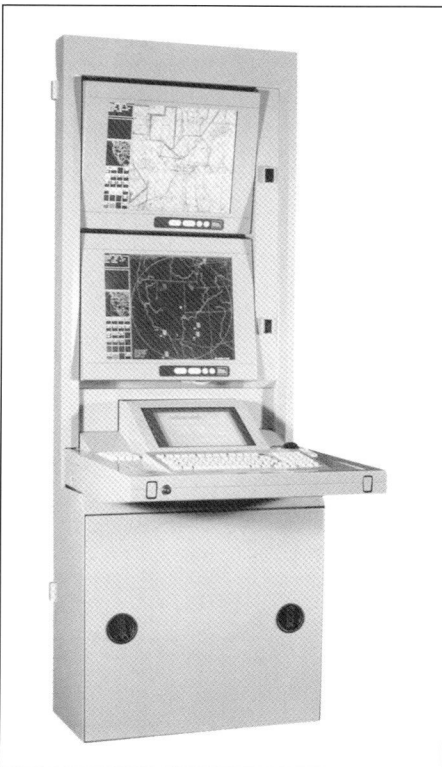

*A typical UCC
installation
(Ultra Electronics)
0109230*

Ultra Electronics MultiFunction Console (MFC)

Ultra Electronics Command and Control Systems, previously Dowty Maritime, is a major supplier for the Royal Navy SubMarine Command and Control System (SMCS) and the Surface Ship Command System (SSCS) for Type 23 frigates. In addition a variant of these systems, known as KDCOM, has been supplied to the Korean Navy for the *Okpo* class destroyer.

For SMCS, Ultra supplies all hardware including cabinets, consoles, processors, displays and interface electronics. It also supplies the local area network system, the device driver software for all interfaces and the foundation and runtime software on which all applications run. In addition to the above, Ultra also provides the test and commissioning software for the system. For SSCS and KDCOM, Ultra supplies a derivative of the above processing environment, device drivers and foundation software.

Using its systems knowledge and experience of the above systems, Ultra has developed the concept of the MultiFunction Console (MFC). The MFC allows customers to procure a flexible, modular and low-cost console system which can be configured to meet their specific role in C^3 and CSS systems.

In situations where space is at a premium, ruggedised flat panel displays, using the latest LCD or plasma technologies can be fitted (known as the Seascape console). These displays offer equal performance to CRT displays, but with reduced weight and cabinet depth requirement, allowing even greater flexibility in design. The processing interfaces, necessary to allow connection to any type of highway and specific application software can be housed in the console or a separate cabinet. This flexibility enables the customer to define the number and type of displays, the MMI, the desktop configuration and the installation layout to suit his specific needs; all of which can be easily reconfigured or have other functions added at a later date with minimal effort.

Specifications

(a) Display capabilities: single or double, 40-53 cm, 1600 × 1200 landscape colour CRT, plasma or LCD displays
(b) Software environment: UNIX operating system, POSIX, X-Windows
(c) Integrated interfaces: standard - data highway, LAN (802.5, FDDI, Ethernet); optional - CCTV, radar, sonar interfaces
(d) Processor performance: RISC-based VME processors (typically SPECint92 - 131.2 and SPECfp92 -153). Processors may be added to meet increased requirements. For example, SPECrateint92 - 8124 and SPECratefp92 - 8906 achievable with four processors
(e) Storage: magnetic/optical disk, CD-ROM, magnetic tape
(f) Standard console dimensions (H × D × W): 1,360 × 1,020 × 550 mm with CRT displays and processor
(g) Seascape console dimensions (typical) (H × D × W): 1,000 × 600 × 650 mm LCD or plasma displays with remote processor
(h) Weight: 220 kg (typical CRT console, including processor); 50 kg (typical LCD or plasma Seascape console with remote processor)
(i) Power requirement: 440 V, 3 phase or 115 V, single phase
(j) Power dissipation: 660 W typical CRT or plasma displays; 450 W typical LCD displays (including processor); either air or chilled water cooled
(k) Environment: operational NES 1004
(l) Reliability: MTBF >1,000 h (Mil Hdbk 217F)

Status

Apart from the use noted above, *Jane's Defence Weekly* reported in July 2001 that the MFC had been selected by the Finnish Navy for use in their six-vessel Squadron 2000 programme. In November 2002 it was announced that Ultra had been awarded a £7.7 million contract for the supply of consoles and associated equipment for the Korean Navy KDX-II destroyer programme.

Contractor

Ultra Electronics Command and Control Systems, Loudwater, High Wycombe, Buckinghamshire.

UPDATED

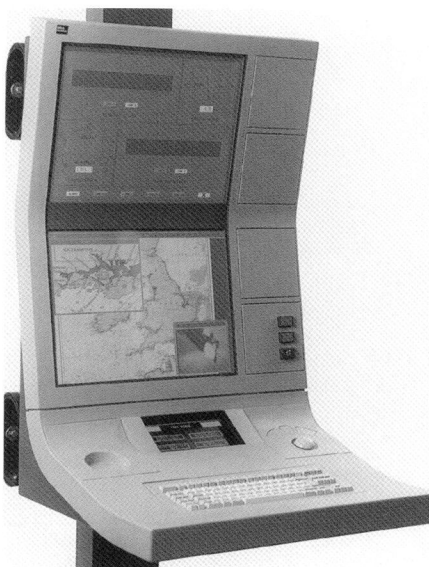

*Seascape MultiFunction
Console (MFC)*

United States

Agama

The Agama rugged handheld computer is the result of a partnership between Raytheon and Compaq (now part of Hewlett Packard (HP)). Raytheon have taken the Compaq iPaq pocket PC and repackaged it to provide a military handheld computer. Agama uses Windows Pocket PC 2002 Operating System with Pocket PC versions of MS applications – Pocket Word, Pocket Excel, Pocket Internet Explorer, and Windows Media Player. It supports map display, GPS display, data message handling and fire control calculations.

Agama has a 206 MHz Intel StrongARM 32-bit RISC processor, with 32 Mbyte ROM, 64 Mbyte RAM, 128 Mbyte SD RAM and an optional 256 Mbyte Secure Digital memory. Expansion modules incorporating CDMA2000 1X, GSM/GPRS modems, GPS, RF LAN and Bluetooth are available.The display is colour reflective Thin Film Transistor (TFT) with analogue resistive touchscreen. Screen resolution is 240 × 320 pixel with 4,096 colours, viewable image size 2.26 × 3.02 ins. User interface is either through handwriting recognition or a virtual software keyboard. A USB or serial connection is provided for communications. Power supply is from a long-life lithium polymer integrated battery, and zinc air or alkaline AAA battery packs are an option for use when AC power is not available. The Agama measures approximately 7 × 3.5 × 1.25 in and weighs 15.5 oz.

A recent development has been the addition of a PCI card slot for a card providing four channels which can receive picture data for display on a moving map.

Status
Available.

Contractor
Raytheon Computer Products, Marlborough, Massachusetts.

UPDATED

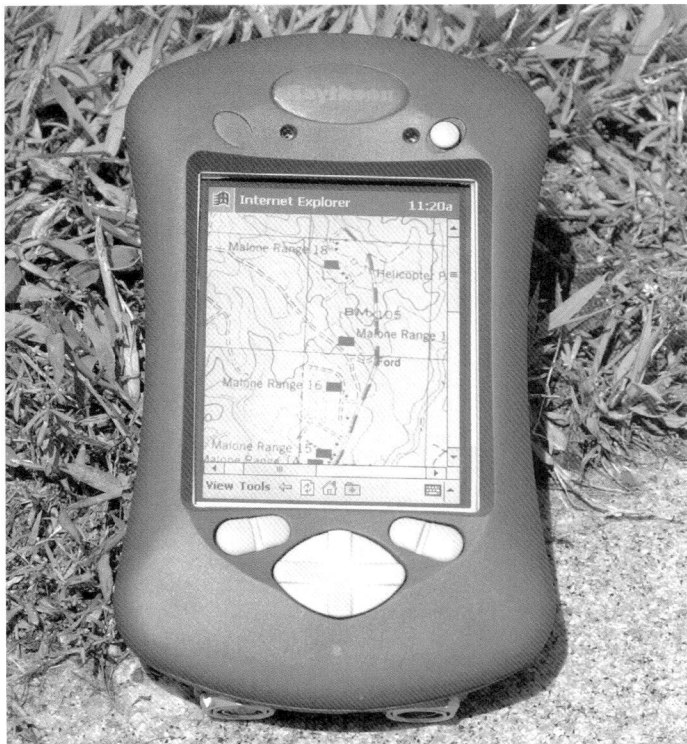

Agama (Raytheon) 0143808

AN/UYQ-70 workstation

One of the first standard combat computer systems implemented with an openly designed architecture, the UYQ-70 supports the common operating environment currently being implemented in US surface, subsurface, land and airborne military platforms to fulfil multiple combat system tasks. The Q-70 provides a single source for a range of solutions for multiple platforms using a variety of configurations. The workstations are configured using COTS modules housed in militarised enclosures and are designed to allow new combat system development or the upgrade of existing systems. The processing families supported by the Q-70 include SPARC, Intel, HP and PowerPC. Operating systems include Solaris, Windows NT, HPUX and VxWorks. The common, open-system COTS architecture enables the use of common components and configurations in the display systems and this design approach has led to savings in logistic support, reduced cross training between different legacy display systems and more flexibility in crew staffing assignments. In addition, the transition to the Q-70's open operating system has not required that legacy architectures be abandoned. The Q-70 programme has introduced many changes to improve the comfort and efficiency of the equipment's operators including ergonomic trackballs, adjustable footrests, redesigned layout for touchscreens and monitors, headset connection and storage improvements.

Status
Q-70 workstations in more than 50 configurations have been and are being widely installed in the USN, including 'Los Angeles' and 'Ohio' class submarines, Aegis cruisers and destroyers, LHAs and E2-C Hawkeye aircraft; approximately 2000 workstations have been supplied. The systems are also planned for retrofit on the 'Ticonderoga', 'Nimitz' and 'San Antonio' classes. Also in service with the navies of Australia, Germany, Japan, Norway and Spain. In May 2002 the first workstation built under licence in Japan by the Oki Electric Industry Co was delivered.

Contractors
Lockheed Martin Naval Electronics & Surveillance Systems(Prime), Clearwater, Florida.
DRS Electronics Systems, Inc, Gaithersburg, Maryland.

UPDATED

Appliqué+ V4 Computer

The Appliqué+ V4 Computer is a versatile, upgradeable solution for situational awareness, command and control and weapons targeting applications and meets the military's digital battlefield requirements. Designed under the Army's Force XXI-Appliqué Programme, it uses lessons learned from the Task Force XXI Advanced Warfighting Experiment. In particular it is used to host FBCB2 applications (see separate entry). The Appliqué+ V4 Computer is a ruggedised computer designed to be hosted in a wide range of wheeled, tracked, or airborne platforms and other demanding tactical applications.

The Appliqué+ V4 Computer has Intel Pentium III® performance. The remote colour display and keyboard facilitate installation in vehicles and modular designs. Expansion features include USB, Ethernet, Compact PCI, PMC and PC-MIP slots, and an AGP bus video and a multi-GB removable hard drive are provided.

Specifications
Functional
(a) Processor: Intel PIII 800+ MHz mobile module
(b) Display: 12.1 in diagonal colour LCD with touch screen, able to be located up to 5m (16.4 ft) from processor unit. 3 additional 5m cables can be added with USB hub spacer adaptor. The video interface from processor to display is all digital high-speed LVDS. 800 by 600 pixels resolution. Sunlight readable at maximum luminance (brightness) with at least 750 units with a contrast ratio of not less than 50:1.
(c) Memory: 512 MB
(d) Mass storage: Removable internal 40 GB hard drive
(e) Keyboard: USB 85-key IBM enhanced keyboard with 101 key functionality able to be located up to 48 in from display unit

The Appliqué+ V4 Computer 0092758

For details of the latest updates to *Jane's C4I Systems* online and to discover the additional information available exclusively to online subscribers please visit
jc4i.janes.com

(f) Pointing device: Keyboard-embedded 2-button force stick type pointing device
(g) Expansion:
     One full compact PCI slot
     Five PC-MIP slots
     One PMC slot
(h) External ports:
     One RS-232C port
     External SVGA display port supporting up to 1,280 × 1,024 pixel resolution, default mode is dual-mode LCD and SVGA at 800 × 600
     Audio Sound Blaster™ compatible audio controller device with speaker outputs and microphone inputs
     One USB ports
     IEEE 802.3 LAN interface 10/100 BaseT
     4-channel RS-422/RS-423
     2 synchronous serial communications port (SYNC-1 and SYNC-2) for Conditioned Di-Phase (CDP) and MIL-STD-188-144 non-return-to-zero communications interfaces

Physical
(a) Weight:
     Processor unit (with HDDC installed): 15.5 lb
     12.1 in display: 8.6 lb
     Keyboard: <2.5 lb
(b) Dimensions (L/H/D):
     Processor unit: 11.5/5.12/10.2 in
     12.1 in display: 12.5/10.2/2.2 in
     Keyboard: 7.25/11.5/1.07in
(c) Electrical - Input voltage:
     28 V DC vehicle power per MIL-STD-1275A(AT)
     Optional AC converter 90-264 V AC, 47-440 Hz

Status
In operational use with the US Army. Originally selected as the main hardware for the FBCB2 programme, but by late 2002 was providing less than half the supply.

Contractor
L-3 Ruggedised Command & Control Solutions , San Diego, California.

UPDATED

DP3 Ultra Workserver

The DP3 Ultra Workserver is based on SUN®'s UltraSPARC™ III Netra™ AX2200 motherboard technology and is powered by two 750 MHz or two 900 MHz 64-bit UltraSPARC™ III processors, each with 8 Mbytes of cache. It also comes with three 33 MHz/32-bit PCI expansion slots, a 66 MHz/64-bit PCI expansion slot, two 120 MHz UPA64S graphics slots and a memory, scaleable to 8 Gbytes. It can be used on the move in both wheeled and tracked vehicles. Configured with a 24-bit graphics board and an audio interface board, the system leaves three PCI expansion slots open for user-defined options. SUN® compatible PCI special purpose boards and mass storage devices can also be used.

Specifications
Computer/Display:
CPUs: (2) SUN® 750 or (2) 900 MHz UltraSPARC™ II
RAM: 2 Gbytes standard, 8 Gbytes maximum DRAM
Storage: up to (8) 5.25 in half-height; (2) 3.5 inch × 1 in devices
Keyboard: SUN® compatible, fully sealed, detachable, 6 USB with integral 1.6 in trackball
Display: 18.1 in colour active matrix TFT, 1280 × 1024 resolution

The DP3 Ultra Workserver (Patrick Allen/Jane's)   **NEW**/0526702

Physical:
Dimensions (H × W × D): 16 × 19 × 25 in
Weight: 90 lbs. (dependent upon configuration)
Interfaces:
Network: 10/100 Base-T IEEE 802.3, via RJ-45
Fibre channel: 1 GB/s FC/AL controller for 2 channels, internal and external
SCSI: 40 MB/s ultra SCSI, one narrow SCSI
Serial: 460.8 Kbaud async port; 384 Kbaud sync port
Firewire: (2) 400 Mbytes/SCC IEEE1394 ports
Parallel IEEE 1284 port
USB: (4) 12 Mbytes/s USB ports
Audio: 16 bit integrated audio
Electrical:
AC: 110+/-10%, VAC 47-63HZ
Optional: 110/220 V AC 40-440 Hz; 21-36 V DC; 5 minute UPS

Contractor
DRS Laurel Technologies, Johnstown, Pennsylvania.

NEW ENTRY

DWT-201e TEMPEST

The DWT-201e TEMPEST is smaller than the traditional PC and can be used vertically or horizontally to make maximum use of available space.

Specifications
(a) Processor: Intel® Pentium® 4 1.7 GHz Processor. Upgrade to 2.2 Ghz
(b) Motherboard: 64 MBytes to 1 GByte Synchronous DRAM for Pentium. 4 Processor. 256 kbytes level 2 cache. Intel. 845 Chipset. 400 MHz Front Side Bus
(c) Video: ATI Rage Pro with 16 Mbytes S DRAM
(d) Input/Output: One each of serial port, DB9; parallel port, DB25; video port, HD15; keyboard port, PS/2 (DB9); mouse port, PS/2 (DB9)
(e) Keyboard/Mouse: 104 key Windows® keyboard, US 3-button mouse, external PS/2
(f) Power: AC to DC Adaptor 100-240 V AC 120 W. Separate DC input
(g) Operating System: Windows® 2000®, NT®, XP®. Other operating systems are available upon request
(h) Sound: Integrated AD 1885 AC '97 Audio
(i) Peripherals: 20 Gbytes Removable Hard Drive; 24 × CD-ROM drive
(j) Physical specifications: 10.5 × 14.0 × 4.6 in (26.7 × 35.5 × 11.6 cm) (H × D × W)
(k) Weight: 15 lbs (6.8 kg)
(l) Operating environment:
     Temperature: Operating 50 to 95°F (10 to 35°C); Storage −40 to +158°F (−40 to +70°C)
     Altitude range: Operating 0 to 10,000 ft (3,048 m); Storage 0 to 30,000 ft (9,144 m)
(m) Options:
     10 Mbps Ethernet-ST fibre connector
     100 Mbps Ethernet-SC fibre connector
     External LCD monitor (15 in and 17 in)
     Ergonomic keyboard
     External mouse
     Two USB ports (DB9)
     Line out
     External + 12 V DC adaptor
     Headset
     Internal USB floppy
     CD/RW
     DVD-ROM
     Removable hard drive options:
     20 Gbytes/40 Gbytes 2.5 in drives
     20 Gbytes/60 Gbytes/120 Gbytes 3.5 in drives
     Additional memory: (128, 256, 512, 1024 Mbytes)
     105 key Windows® keyboard, foreign languages

Status
In use with the Royal Air Force.

Manufacturer
Advanced Programs Inc, Columbia, Maryland.

UPDATED

Explorer MP™

The Explorer MP™ Rugged Portable Workstation is being widely used in digitisation programmes by both the US Army and the US Marine Corps, providing the means to run CIS, fire control, digital mapping and other applications. Developed from the Explorer II, it is capable of supporting a wide range of processors, together with multimedia and graphics systems. Power is provided by

Explorer MP (DRS Technologies) 0134119

an AC/DC power supply augmented by an integral UPS. Options include up to four 5.25 in half-height device bays, removable SCSI hard-disk storage, a variety of I/O options and multi-slot support for an extensive range of PCI expansion cards including the SunPCi™NT coprocessor card. The display is a 15.4 in colour active matrix with 1,280 × 1,024 resolution. A fold-down keyboard and a trackball are provided.

Specifications
(a) CPUs: UltraSPARC™ -IIi (UltraSPARC™ AXi), Intel® Pentium®
(b) Max RAM: 1 Gbyte
(c) Mass storage: four configurable 1/2 height 5.25 in drive bays
(d) Ethernet: Fast Ethernet - 10/100 base-T via RJ-45
(e) I/O: Internal and external ultra-wide SCSI-3, parallel, two RS-232/423, 16-bit audio
(f) Expansion: Four 33MHz/32-bit PCI expansion slots
(g) Display:
  Diagonal area: 15.4 in
  Resolution: up to 1280 × 1024
  Luminance: maximum 200 cd/m2 (Typ.)
  Contrast ratio: 200:1 (Typ.)
  Technology: 24-bit AMLCD, TFT
  Viewing angle: +/−60o(H), +/−50o(V)
  Additional features: EMI shield
(h) Physical
  Size: 14.5 × 22 × 10.5 in (H × W × D)
  Weight: 50 lbs (configuration dependent)
(i) Electrical
  AC: 110 +/−10%, 220 V AC +/−10% V AC, 47-63 Hz
  DC: 20-32 V DC (optional)
  UPS: Internal UPS recharging from AC or DC input, up to 10 min

Status
In widespread use by the US Army for digitisation programmes and by the US Marine Corps. As of August 2003 over 2200 workstations had been ordered.

Contractor
DRS Electronic Systems, Inc, Gaithersburg, Maryland.

UPDATED

Flat Panel Rugged displays

Designed for land, sea, and airborne application, the FPR series has become the accepted standard of the US Army's Common Hardware/Software (CHS-2) programme.

FPR Thin Client computers feature a fully integrated computer workstation within the same thin chassis as the flat panel display. The FPR integrated computer provides Pentium® class processors, fixed or removable hard drives, flash drives, PCMCIA expansion redundant Ethernet, built-in UPS and custom I/O panel configurations. Options for touchscreen, keyboard/trackball, copper/fibre Ethernet converters and a variety of mounting options are available.

FPR Flat Panels function as situational displays and user interfaces. They feature high-resolution, 24-bit AMLCD technology, and options for sunlight readability and touchscreens are available.

Features
(a) Space, weight and power consumption designs for land, sea and airborne applications

Flat Panel Rugged display (DRS Technologies) 0134121

(b) Removable hard drive and PCMCIA expansion options
(c) Built-in redundancy for power, Ethernet and hard drive
(d) Copper/fiber Ethernet conversion (field upgradeable)
(e) Built-in UPS (COTS battery)
(f) 901 Grade A shock qualified
(g) Fully sealed/passive cooling
(h) Fully configured solutions for bulkhead, rackmount and desktop mounting.

Specifications
Computer
(a) CPU: Intel® Pentium® III
(b) Max RAM: 1 Gbyte
(c) Hard drive: Up to 30 Gbytes min EIDE HDD
(d) Flash drive: Up to 1 Gbyte flash
(e) Graphic: Up to 64 Mbytes shared video memory
(f) Bus: ISA, PC/104, PC/104+ and PCI
(g) Ethernet: 10/100 Base-T (dual available)
(h) I/O: PS/2 keyboard/trackball, IDE, parallel, USB, RS-232/422/485.

Physical

Model	Dimensions (H × W × D)	Weight
FPR15	14.1 × 15.6 × 3.8in	14 lb
FPR15TC	14.1 × 15.6 × 4.3 in	22 lb
FPR18	15.8 × 17.8 × 3.8 in	16 lb
FPR18TC	15.8 × 17.8 × 3.8 in	26 lb
FPR20	17.5 × 18.9 × 4.0 in	18 lb
FPR20TC	17.5 × 19.9 × 4.0 in	32 lb
FPR21	17.5 × 19.1 × 5.5 in	20 lb

Electrical
(a) AC: 110/120 V AC (via MIL-C-38999
(b) DC: 20-32 V DC (via MIL-C-38999
(c) UPS (option): 10+ minutes (COTS Ni/Cd batteries).

Environmental
(a) Shock: MIL-STD-901D Grade A
(b) Vibration: MIL-STD-810E/MIL-STD-167
(c) EMI/EMC: MIL-STD-461E
(d) Operating temperature (MIL-STD-810E): Up to −32 to +50°C (−32 to +49°C at 15,000 ft)
(e) Storage temperature: Up to −32 to +66°C
(f) Shock temperature: Up to −18 to +49°C within 5 min
(g) Humidity: Up to 95% RH condensing at +49°C
(h) Explosive atmosphere: MIL-STD-810E, method 516.4, procedure 1
(i) Explosive decompression: 40,000 ft in 0.1 second, MIL-STD-810E, method 511.3.

Status
Examples from this range are in widespread use in both US and other services.

Contractor
DRS Technologies, Gaithersburg, Maryland.

VERIFIED

FlexPac

FlexPac is a rugged portable PC with 14 in XGA TFT display. It has a variety of configurations of CPU and number of expansion slots. An optional wireless mobile display can be used up to 150 ft away.

Specifications

	FlexPAC PIII	FlexPAC P4
Chipset	Intel 815E 133 MHz FSB	Intel 845 400MHz FSB
CPU type	Intel Pentium® III	Intel Pentium® 4
Bus architecture	ISA/PCI	
Memory	PC133 SDRAM 256 Mbytes or 512 Mbytes	PC133 SDRAM 512 Mbytes up to 2 Gbytes
Disk storage	40 Gbytes UDMA-100 minimum capacity unformatted	
Optional HDD	Fixed or removable 2nd HDD	
Floppy disk drive	1.44 Mbytes	
Multimedia drives	CD-ROM or CD-RW option	
Internal display	14.1 in XGA TFT, 8 Mbytes video RAM	14.1 in XGA TFT, 16 Mbytes video RAM
Ext display support	Supports simultaneous external display. 640 × 480, 800 × 600, and 1024 × 768	
I/O ports	Two serial ports (16550 UART compatible); one enhanced parallel port (ECP/EPP/SPP compliant); two USB ports	
Keyboard	AT compatible, 104/105 key with integrated touch-style mouse pad	
Audio option	Onboard audio with built-in speakers	
Operating systems	Microsoft Windows® 2000 Professional	
PCMCIA/CardBus options	Two Type II or One Type III	
Power supply	205 W, 100-240 V AC, 50-400 Hz	300 W, 100-240 V AC, 50-60 Hz

	FlexPAC (4-Slots)	FlexPAC (6-Slots)
Expansion slot options	4 PCI; 1 PCI and 3 ISA; 2 PCI and 2 ISA; 3 PCI and 1 ISA; 4 ISA	5 PCI and 1 Shared PCI/ ISA
Dimensions (W × H × D)	15.78 × 10.18 × 6.72 in (40.08 × 25.8 × 17.07 cm)	16 × 11 × 9.75 in (40.6 × 27.9 × 4.7 cm)
Weight	18 lbs (8.18 kg)	23 lbs (10.55 kg)

Portable Display Specifications

Display	8.4 in – 800 × 600 (SVGA) transflective, daylight-readable colour TFT w/touchscreen
Power	AC/DC adaptor/lithium ion battery pack (1800 mAH/7.4 V). 1.5 to 4 h battery operation depending on brightness selection
Dimensions (W × H × D)	8.7 × 1.2 × 6.4 in (22.20 × 2.95 × 16.20 cm)
Weight	1.5 lbs/710 g

Status

In production. Selected in 2001 as one of the platforms for the US Navy-Marine Corps Intranet (NMCI). Used in the British Army Rapier air defence system.

Manufacturer

Dolch Computer Systems, Inc, Fremont, California.

NEW ENTRY

Fieldworks FW8000 Series rugged workstations

Fieldworks (now Kontron Mobile Computers) FW8000 series workstations are suitable for military applications.

Fieldworks FW8000 Series rugged portable workstation   0084485

Specifications

(a) Processor:
   FW8000 - Intel 500 or 800 MHz Pentium III
   FW8500 - Intel 1.06 GHz Pentium III
(b) Operating system: Windows 98 (std), Windows 2000, Windows XP
(c) Hard drive: 40 Gbytes internal (std), 60 Gbytes Internal (option), 40 or 60 Gbytes dual removable (option)
(d) Drives: Integrated floppy and CD ROM 24X - standard. Integrated CD-RW - optional
(e) Expansion: 3 full-length or 6 half-length PCI/ISA slots
(f) Weight (lbs): 15.8
(g) Size (cm): 38.33 × 44.91 × 10.34 (W × D × H)
(h) Displays:
   12.1 in SVGA Active LCD (800 × 600) - standard on 8000
   12.1 in XGA Active LCD (1024 × 768) - standard on 8500, optional on 8000
   12.1 in Sunlight Readable Transflective - optional

Manufacturer

Kontron Mobile Computers, Eden Prairie.

UPDATED

FW 2000 Series Embedded Vehicle System (EVS)

The FW 2000 series Embedded Vehicle System (EVS) is a rugged, modular PC system built for in-vehicle environments. It includes a server, 10.4 in 800 × 600 SVGA daylight-readable display and back-lit keyboard. The server is separate from the user interface components and can be mounted in a remote, secure location within the vehicle. EVS is a modular system, which is expandable to accommodate up to 270 peripherals and can group together all onboard communications and computing requirements. EVS is built to military specifications.

FW 2000 Series standard features

(a) two open embedded PC card slots (two Type II or one Type III)
(b) 256 kbytes Synchronous Pipeline Burst CACHE
(c) 4 Mbyte video DRAM
(d) three serial ports and one bidirectional (ECP/EPP) parallel port
(e) 10-30 V DC power input
(f) Windows 95 system load
(g) sound
(h) removable hard drive
(i) two USB 1.1 ports.

Specifications

(a) Model FW 2500-III
   Processor: 500 MHz Pentium
   RAM: 128 Mbytes
   Hard drive: 40 Gbytes standard, 60 Gbytes optional.
(b) Model KMC 700-IV
   Processor: 700 MHz Pentium
   RAM: 128 Mbytes
   Hard drive: 40 Gbytes standard, 60 Gbytes optional.

FW 2000 Series Embedded Vehicle System display and keyboard (EVS)   0055163

Status

In use by the US Army. Two separate contracts were awarded, in 2001 (US$4 million) and November 2002 (US$1.6 million). The terminals were for use with FBCB2 (see separate entry) but are also being used with BDI installations (see separate entry).

Manufacturer

Kontron Mobile Computers, Eden Prairie, Minnesota.

UPDATED

Genesis™

The Genesis™ is a multifunctional rugged computer system designed for land, sea or airborne environments. Its internal COTS architecture enables it to address a wide spectrum of application requirements. It has a 10-slot passive backplane enabling it to support a variety of processors (Intel Pentium®, Digital Alpha® or Power PC® processors, or it can be configured with Sun Microsystems UltraSPARC®). Its integrated environmental processor is designed to ensure the correct functioning of the unit in extreme environmental conditions. The power supply will support AC, DC or an internal UPS. The unit weighs 19.4 kg, and can be mounted in a standard 19 in rack mount; an optional tracked vehicle mounting kit is available. A range of associated displays and printers is also available. There are a number of variants based on the Genesis Short Rack (SR).

Genesis™ Commander

Specifications
(a) Computer
  CPU - Intel Pentium®, Digital AlphaPentium®, Power PCPentium®
  Option - Sun UltraSPARC®.
(b) Internal drives
  1 × 9 Gbytes removable HDD
  1 × 3.5 1.44 Mbytes FDD
  1 × PCMCIA
  1 × CD ROM/DVD drive
  1 × back-up tape drive
  1 × 5.25 in spare bay
(c) Interfaces
  SCSI
  RS-232 serial
  Centronics printer
  Keyboard/display
  Video
(d) Options: LAN, WAN, USB, Firewire, RS-422 or Audio.
(e) Expansion slots: Up to 4 × ISA and 2 × PCI expansion slots, dependent on configuration
(f) Display/Keyboard Option: The FPR18 Remote is an integrated 18 in (1,280 × 1,024) rugged flat-panel and detachable keyboard with trackerball. It can be supplied as a single deployable unit or may be mounted in 19 in racks. Optional drive electronics for remote operation of the display/keyboard (up to 50 m). Other display sizes available as standard.
(g) Electrical: Input Voltage:100-240 V AC, 47-63 Hz 24-28V DC to DEF STAN 61-05
(h) Physical
  Weight: 19.4 kg (configuration dependent)
  Dimensions: 480 × 178 × 490 mm (WHD).
(i) Temperature
  Operating: −10 to +40°C
  Non-operating: −20 to +60°C

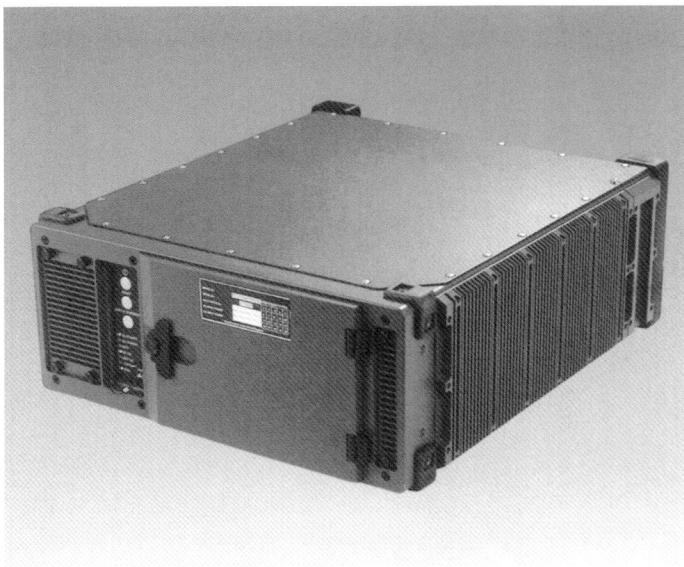

Genesis Commander (DRS Technologies)　　　0110490

Genesis Alpha multiplatform rugged computer (DRS Technologies)　　　0134122

Genesis™ Alpha

Specifications
(a) Computer
  CPU: Digital Alpha up to 500 MHz
  Maximum RAM: 16 Mbytes to 512 Mbytes
  Drives: 2 × 5.25 in half-height, 1 × 3.5 in full height
  Power supply: 110-230 V AC (auto ranging), 47-63 Hz
  Option: 400 Hz; 19-32 V DC; Integral UPS
(b) Physical
  Weight: 19.5 kg, configuration dependent
  Dimensions: 178 × 400 mm.

Genesis™ PowerPC

The Genesis™ PowerPC is based on the same rugged rack-mount enclosure used for the Genesis Ultra and Genesis Alpha, but is configured with the latest PowerPC motherboard. The system combines a full suite of network software with standard PCI components.

Specifications
(a) Computer
  CPU: PowerPC
  Maximum RAM: 32 Mbytes to 512 Mbytes
  Drives: 2 × 5.25 in half height, 1 × 3.5 in full height
  Power supply: 110 V to 230 V AC (auto ranging), 47-63 Hz
  Option: 400 Hz; 19 V-32 V DC; Integral UPS.
(b) Physical
  Weight: 19.5 kg, configuration dependent
  Dimensions: 178 × 400 m.

Genesis™ UltraSPARC®

The Genesis™ UltraSPARC® is based on the same rugged rack-mount enclosure used for the Genesis SR and Genesis Alpha, but is configured with a Sun SPARCengine® Ultra AX motherboard. The system combines 250 MHz UltraSPARC® performance and a full suite of network software with standard PCI components.

Specifications
(a) Computer
  CPU: USI-167, USII-250
  Maximum RAM: 32 Mbytes to 512 Mbytes
  Drives: 2 × 5.25 in half height, 1 × 3.5 in full height
  Power supply: 110-230 V AC (auto ranging), 47-63 Hz
  Option: 400 Hz; 19 V-32 V DC; Integral UPS.
(b) Physical
  Weight: 19.5 kg, configuration dependent
  Dimensions: 178 × 400 mm.

Status

The Genesis™ Commander has been selected as the hardware platform for the British Army WAH-64 Apache Attack Helicopter Ground Support System. The Genesis™ SR is in use with the United Kingdom RAF, the Royal Danish Navy, Royal Australian Navy, Norwegian Army; Swedish Army, an unspecified Middle Eastern navy and an unspecified south eastern Asian army.

Contractor

DRS Electronic Systems Inc, Gaithersburg, Maryland.

UPDATED

Hand-held Terminal Unit (HTU)

The HTU, also known as the Hand-held and Imaging Communications Terminal (HICT) and the Field Data Terminal (FDT), is the US Army's standard, harsh-environment, multi-application computer, supporting battlefield digitisation in programmes for the US Army, US Marine Corps and US Air Force, as well as for allied military forces.

Considered to be a mini-workstation, the HTU is a C4I device suitable for numerous tactical and support applications in the field. The HTU features lightweight hand-held operations, with a high-resolution sunlight-readable colour or monochrome display, mass data storage, multidata ports (dual-channel communications), compatibility with standard tactical radios, and multiple PCMCIA slots for voice, video, GPS and so on.

When used with various tactical software applications, the HTU provides the field user with the capability to rapidly compose, edit, store and display images and messages transmitted and/or received via tactical communications networks. As part of the 'Sure Strike' system linking a ground observer to airborne attack aircraft or helicopters, target information is developed automatically by the terminal through interpretation of positional data and the operator's designation. Target data and position is then automatically linked to a visual display in the aircraft. Subsequent versions provide a GUI for actual target or terrain image transfer between the ground observer and the attack aircraft. The first of these systems was fielded in 1997 and has been used operationally in the Federal Republic of Yugoslavia (now Serbia and Montenegro).

Specifications
(a) Processor: Intel 80586 DX 133 MHz
(b) Mass memory: 32 Mbyte RAM (64 Mbyte option); 260 Mbyte internal hard disk (Type III PCMCIA - 520 Mbyte option)
(c) Multidata ports: dual-channel modem port; 2 × Type III PCMCIA slots (external); parallel and VGA interfaces; 2 serial interfaces; field communications wire binding posts; external keyboard/pointing device port
(d) Display: sunlight-readable, colourised, 640 × 480 pixels (LCD), monochrome or 256 colour option available; 184 mm (7.25 in) diagonal
(e) Size: 246 × 178 × 83.8 mm (9.7 × 7.0 × 3.3 in)
(f) Weight: less than 2 kg (4.5 lb) without battery and accessories
(g) Power options: battery, DC, AC
(h) HUI: thumb controls; backlit, detachable, qwerty programmable keyboard; touchscreen/pen options

Status
In widespread operational use in US service, the HTU has been chosen for such projects as the US Army's Common Hardware Software-2 (CHS-2) programme (GTE prime), the Lightweight Leader Computer and Canada's TCCS-Iris (CDC prime). Originally produced by Northrop Grumman until it sold its ruggedised terminal operation to L-3 communications in 2002.

Manufacturer
L-3 Communications, Ruggedized Command & Control Solutions, San Diego.

UPDATED

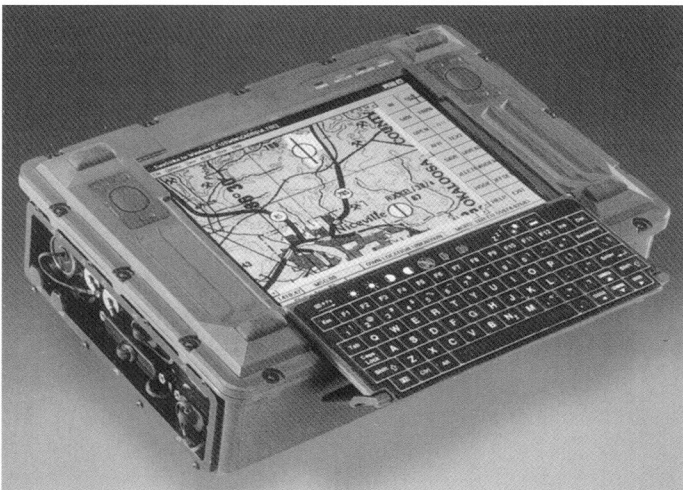

The HTU device provides the processing, storage, display, communications and data handling capabilities needed to perform in tough environments with robust reliability   0055164

KnowledgeBoard

SAIC has developed an advanced data fusion tool called the KnowledgeBoard (KB). The KB provides a web-based distributed electronic 'white board' or 'digital dashboard' which facilitates the display and review of information from many heterogeneous multimedia sources.

Information sources may include video, WWW, relational databases, flat files, custom databases, and so on. The KB user interface is organised according to

cells, each of which is connected to an XML document based on the servicing information source. The software provides the framework for a user to develop a visualisation framework and any required 'drill-down' capabilities for individual cell representations. The KB client and middleware employ a 'plug-in' architecture that allows the end user to extend the graphical capabilities of the KB and the information access components. The system includes a developer's kit for building custom applications and is written in 100 per cent Java and can therefore run on any platform, whether UNIX, Windows, or Macintosh. The KB includes an automatic cell creation capability allowing the end-user to customise the visual interface, including the cell number, size, content, and orientation. The KB is an information portal and enterprise data fusion tool with application in virtually all industries and technology domains.

KB comprises the following components: The Cell Designer component allows the user to create their own views of data and information within the context of the KB. Cell 'wizards' guide the user through the process of creating one or more cells of information. Website Displays allow the user to display web pages on the KB. An automatic refresh time can also be set on the browser to keep the page up to date. View and Cell displays allows the user to monitor multiple events, domains, divisions simultaneously by switching between views comprised of selected cells showing information. By double- clicking on an individual cell, it can be opened to fill the entire KB.

Status
Beta versions of the KnowledgeBoard have undergone extensive testing and early implementations are currently in use in US Department of Defense agencies. Background Work on KB began in 1995 as a result of a Presidential mandate to apply advance technology to deal with the potential of a Chemical Biological Terrorist Threat. In early 1997, working in support of the Marine Corps Chemical Biological Incident Response Force (CBIRF), the first successful demonstration of the KnowledgeBoard then known as the Electronic WatchBoard was achieved. The Defense Advanced Research Project Agency (DARPA) supported this first demonstration of an object technology enterprise data fusion tool. Versions are in use in multiple agencies in the Department of Defense, the National Cancer Institute and the HUBS program, a regional information infrastructure project integrating hospitals, universities, schools and business in Delaware, New Jersey, Maryland and Pennsylvania. It is being integrated into several other DoD programmes, including the Adaptive Courses of Action (ACOA) and Enhanced Consequent Planning and Support System (ENCOMPASS).

Contractor
SAIC, San Diego, California.

VERIFIED

L-3 rugged flat panel displays

L-3 flat panel displays are designed to provide wide viewing angles while maintaining visual clarity even when used in direct sunlight or during night operations. The rugged commercial-off-the-shelf (COTS) module is a low-cost display technology upgrade and retrofit product designed to replace CRTs, accommodate shipboard space, weight and power constraints and to broaden display utility and applications. The display is offered in multiple sizes and with numerous options for flexibility in ground, ground mobile, fixed, airborne and shipboard installations.

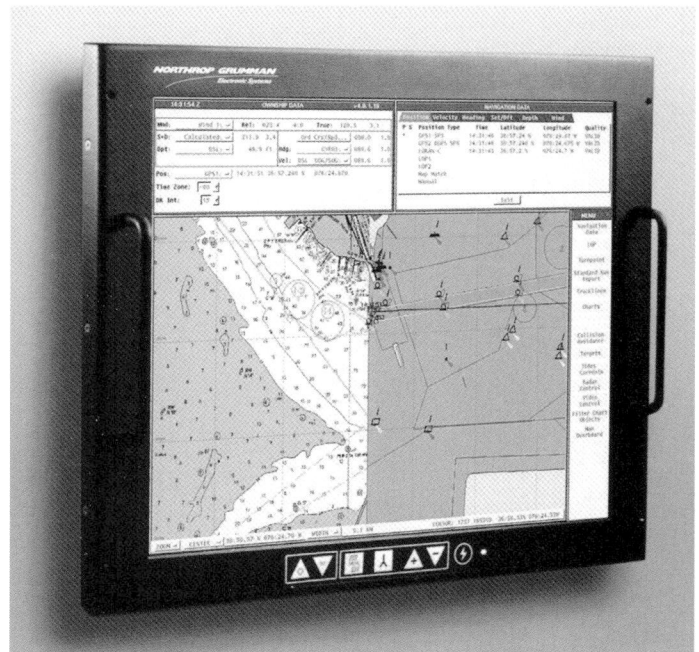

The rugged 20 in flat panel display from L-3 Ruggedized Command and Control Solutions (Northrop Grumman)   0143792

Key features

(a) 18, 20.1, 21.3 in SXGA TFT-LCD
(b) Mil-Spec for shock and vibration
(c) daylight readable and night dimming
(d) EMI protective screen shield;
(e) 1,280 × 1,024 resolution
(f) wide viewing angles
(g) RGB computer input
(h) standard 19 in rack-mountable with hinge-mount option.

Status

Now being offered by the Ruggedized Command and Control Solutions division of L-3 Communications, which was previously part of Northrop Grumman.

Contractor

L-3 Ruggedized Command and Control Solutions, San Diego, California.

UPDATED

LXI-3 rugged portable computer

The LXI-3 is the latest variant of the proven LXI rugged laptop computer range, designed specifically for military applications. Features include Intel™ P3™ processor options up to 1.4 GHz, maximum RAM to 1 Gbyte and a high-capacity removable hard drive. Two PCMCIA slots (one internal and one user-accessible) and one full-length expansion slot are available for user-defined applications, allowing the unit to be adapted to meet customer-specific needs. It can be mounted in a 19 in rack, and the rear I/O panel has been tilted upward to enable the unit to be mounted flush to a vehicle bulkhead. A full-size detachable keyboard is provided, together with an integral trackball designed to be used with a gloved hand, and a 15 in TFT LCD.

Specifications

(a) Computer
   CPU: Intel™ P3™
   RAM: up to 1 Gbyte
   Hard drive: minimum 12.5 Gbytes; removable 2.5 in high-capacity HDD
   Drives: integral CD-ROM/DVD; 3.5 in slimline high-capacity floppy (120 Mbytes); 1 × full-size expansion slot
   Graphic: up to 64 Mbytes shared video memory
   Bus: ISA, PC/104, PC104+ and PCI
   Ethernet: rugged HMFM fibre optic, 10/100 Base-T
   Interfaces: up to 4 × RS 232 serial, 1 × parallel, 4-port USB, integral SCSI (external) and IDE controllers; external video (up to 1,600 × 1,200) 16-bit colour; external audio, Soundblaster™ compatible
(b) Display/Keyboard
   Display: up to 15 in (1,024 × 768) TFT flat panel
   Keyboard: detachable IP65 keyboard with integral tracker ball
(c) Electrical
   AC: autoranging, integral 197 to 260 V AC
   DC: 18 to 32 V DC
   Battery: 2.5 h battery operation (APM support); lithium ion batteries with variable-rate charge suspend-to-disk capability; battery hot swap
(d) Physical
   Weight: 12 kg (configuration dependent)
   Dimensions (W × D × H): 457 × 356 × 150 mm

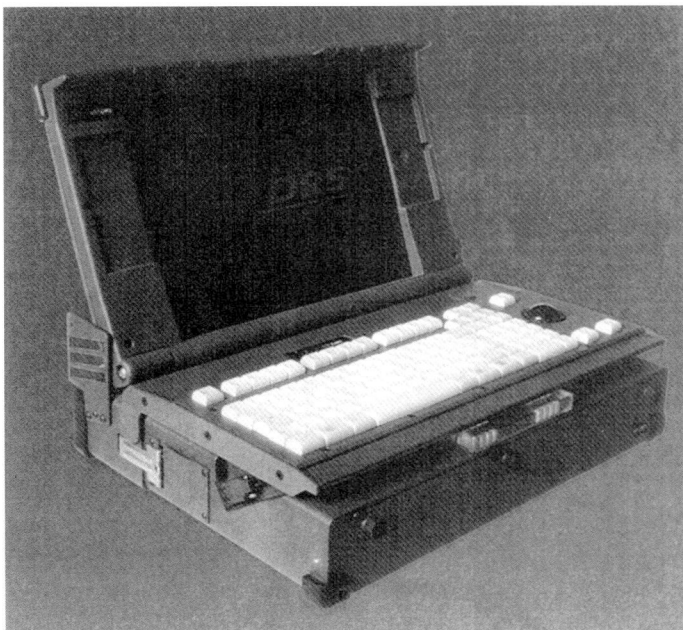

LXI-3 (DRS Technologies) 0130055

LXI-3 Front View (DRS Technologies) 0134123

(e) Environmental
   Operating temperature: 0 to +50°C (option −25 to +50°C)
   Storage temperature: 30 to +71°C (option −40 to +71°C)
   Altitude: operating 10,000 ft; non-operating 40,000 ft

Status

The LXI-3 has been selected for the British Army Tactical Computing System (ATacCS) as the User Data Terminal (see separate entry). Its predecessor, the LXI-2, is in use in several European armies.

Contractor

DRS Electronic Systems, Inc, Gaithersburg, Maryland.

UPDATED

NB3-Te TEMPEST Portable Workstation

The NB3-Te TEMPEST Portable Workstation provides a storage capability comparable to desktops with the ability to secure sensitive information. The 3.5 in floppy drive is interchangeable with a second hard drive. LRU replacement can be achieved in less than 30 minutes.

Specifications

Processor: Intel® Pentium® III Mobile, up to 1 G Hz

Base Memory:
256Kbytes Level 2 Cache RAM
RAM 64 Mbytes
Core Logic Via Chipset
BIOS Phoenix

Video:
LCD panel size:15 in XGA TFT-Zero radiation
LCD panel resolution: XGA (1024 × 768)/SXGA + (1400 × 1050) TFT LCD panel (18 bit)
LCD colour depth: TFT Panel-16 million colours (24 bit)
Maximum resolution of: 1400 × 1050 at 64 kbytes colours
External video resolution and colour depth: 1024 × 768 × 32 bpp
Video chipset: ATI dual display accelerator
Video memory: 16 Mbytes RAM

The NB3-Te TEMPEST Portable Workstation (Advanced Programmes Inc) 0533827

Input/Output:
Keyboard: 102 keys
Pointing device: Touch pad
PCMCIA slots: 2 Type II
External Ports, Standard: one port each of Serial (16550A-compatible); Parallel;
External SVGA; Keyboard/mouse; DC power input

Power:
Battery: Smart Li-Ion, 65 W, 1 h minimum operation (fully charged). Battery life varies depending on system application, power management settings and features.
AC Adaptor: Auto-switching between 100- 240VAC/47.63 Hz

Operating System: Windows® Me, NT 4.0, 98. Other operating systems are available upon request
Physical Specifications:
Dimensions (H × D × W): 3.25 × 13.75 × 15.2 in (7.62 × 34.9 × 38.6 cm)
Weight: 20 lbs (9.0 kg)

Temperature:
Operating: +41 to 95°F (+5 to +35°C)
Storage: -4 to 140°F (-20 to +60°C)

Options:
DVD-ROM
24x CD-ROM
3.5 in floppy
Additional Li-Ion battery
10 mbps ST fibre Ethernet
100 mbps ST fibre Ethernet
Memory: up to 512 Mbytes, using 2 × 144.PIN SODIMM sockets
Removable 10 Gbytes hard drive

Status
In use by the UK Foreign Office, the RAF, the Danish Navy Material Command and widespread amongst security organisations.

Contractor
Advanced Programs Inc, Columbia, Maryland.

Next-Generation Handheld (NGH) Terminal Unit

The second-generation handheld Terminal Unit is an upgraded version of the present HTU and is designed as a 'drop-in' replacement using existing HTU mounting racks, interface cables and support infrastructure. The terminal's tactical modem supports most US military communications protocols, allowing existing H TU application software to run with no new development. The NGH's open architecture design allows easy processor upgrades to 400 MHz Celeron® or 500 MHz Pentium® III. The display is an 8.4 in full-colour sunlight readable SVGA LCD. Built with an open system, non-proprietary architecture, the V3 NGH has the capability to run applications under UNIX (including LINUX, SCO and SOLARIS), MS-DOS, or Windows. An optional touchscreen is available. The NGH accepts all CompactFlash cards and IBM Microdrives®. An optional rugged solid-state hard drive can be offered in lieu of the standard internal 2.5 in hard disk drive, plus interface cables for tactical radios.

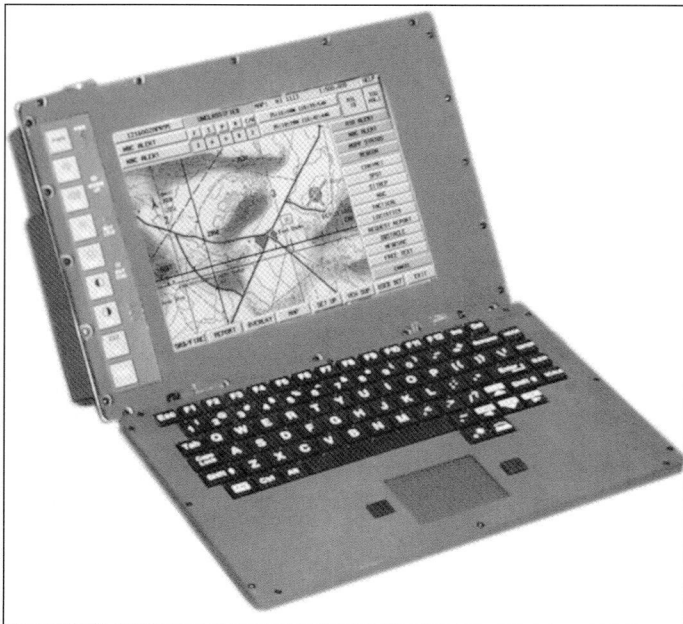

The NGH Terminal Unit (Northrop Grumman)   0130061

Specifications
Technical
(a) Processor: 166 MHz Pentium® MMX, or up to 500 MHz Pentium® III
(b) Display: 8.4 in full-colour transflective sunlight readable SVGA LCD
(c) Memory: 64,128 or 256 Mbytes of DRAM
(d) Mass Storage: 20 Gbyte IBM® Microdrive hard disk drive
(e) Keyboard: Detachable 78-key subset of IBM-PC 101-key
(f) Pointing Device: Embedded two-button touch pad
(g) Expansion:
    Two CompactFlash+ Type II slots
    Two Type II or one Type III PCMCIA slots
(h) External Ports:
    External SVGA display driver supporting up to 1,600 ×1,200 resolution
    Two RS-232 serial
    One parallel
    One audio
    One Ethernet
    One USB 1.1 host
(i) Communications: Tactical dual channel modem
(j) Reliability: 5,000+ hours MTBF (ground mobile)
(k) Operating systems offered: MS-DOS Windows 95/98/2000/NT, UNIX, (SCO, LINUX, SOLARIS)

Environmental
(a) Temperature: Operating; −35° to +52°C;. Non-operating: −40° to +65°C
(b) Altitude: To 15,000 ft

Physical
(a) Weight: 6.4 lb (without battery)
(b) Dimensions: 10.4 × 7.4 × 3.9in (L × W × D)
(c) Electrical:
    28 V DC vehicle power
    Rechargeable NiCad or Lithium Ion battery pack

Options
(a) 110/220 V AC power adaptor (commercial notebook or MIL-SPEC)
(b) External 28 V DC power cable (electronics contained in the NGH; an HTU DC 'Power Stick' is not required with this unit)
(c) Internal MIL-STD battery power
(d) Standard SVGA device interface; serial touchscreen
(e) Various PCMCIA options

Status
In production. Originally produced by Northrop Grumman until it sold its ruggedised terminal operation to L-3 Communications in 2002.

Manufacturer
L-3 Communications, Ruggedized Command & Control Solutions, San Diego.

UPDATED

Operating and Control Console (OCC)

Originally developed for frigates and submarines, the OCC is a programmable multipurpose device which can be used to enhance the flexibility of both existing and new combat systems on surface and subsurface vessels, land-mobile and fixed-site installations, and major airborne applications.

The console may be configured with two 19 in colour monitors or one 19 in colour monitor and two 9 in colour monitors. It also provides an operator desk with keyboard, trackerball and joystick in addition to soft keys, protected and programmable push-buttons and digital potentiometers.

Its modular design permits easy adaptation to a wide variety of requirements. The two graphic engines provide essential redundancy; each engine is driven by three processors operating in parallel giving a capability of 48 million pixels/s rectangular fill. The control processor can also function as back up to the display processor when necessary.

The console accepts both serial and parallel interfaces (RS-232 or RS-422), Ethernet (including MIL-STD-1553B databus), analogue video (RS-170, RS-343 and RS-4112), digital video (40 MHz bandwidth) and language adaptable voice synthesiser. Power consumption is typically 1,200 W.

Contractor
Astronautics Corporation of America, Milwaukee, Wisconsin.

VERIFIED

Opus Rugged Multifunction consoles

The OPUS range of rugged multifunction consoles uses standard COTS architectures to provide computing and display technology for both new and technology insertion programmes. The range has been developed to host applications such as combat systems, fire control, command and control, sonar and other processor and graphically intense tasks, and is designed to meet

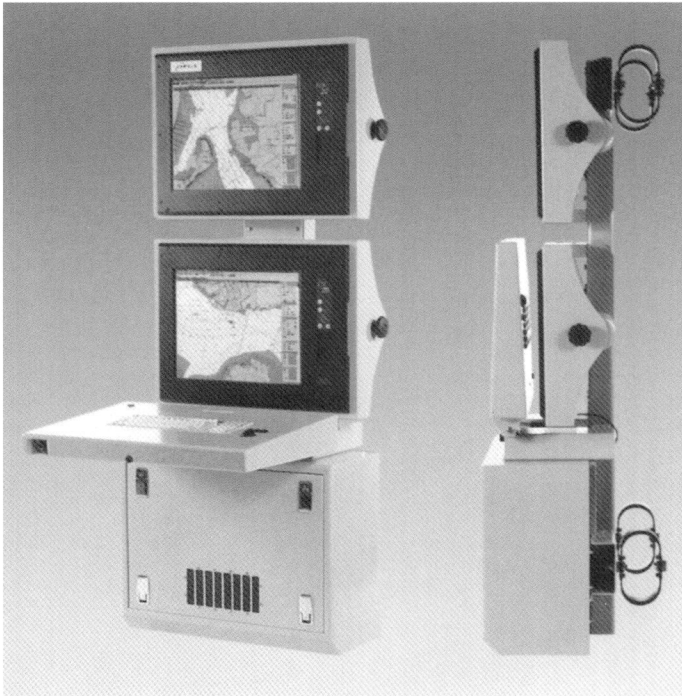

Flat, foldaway OPUS Rugged Multifunction console   0110489

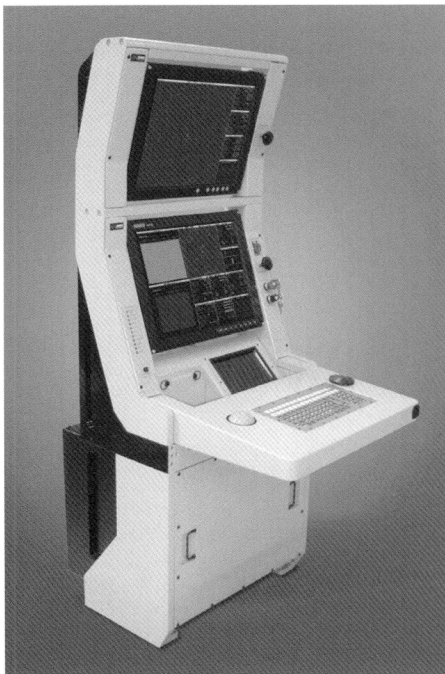

OPUS Rugged Multifunction console (DRS)
***NEW**/0134124*

relevant military requirements for harsh environments. The basic design is capable of supporting a mix of compact PCI and standard VME processors and interface cards, removable hard disks, CD-ROM and associated peripherals. The unit can be configured with one or two high-resolution flat panel displays (15, 18 or 20 in) which are stand-alone devices that can be used either as an integral part of the MFC or as separate desktop or bulkhead mounted display consoles. A dual redundant power supply system is provided.

Status
In December 2001 a £3 million contract was announced for the installation of OPUS 2 consoles in the Royal Navy's Type 23 frigates as part of the Sonar 2087 project. Delivery was expected to start in 2002 and to be complete by 2004. In April 2002 two further contracts were announced. Both were for an unspecified European navy's new minehunter project. One, worth US$2.4 million, was for the display of the AMS NAUTIS 3 C2 system (see separate entry) and the other, worth US$1.6 million, was for the display of Thales's Sonar 2093. In May 2002 a contract in excess of US$1 million was awarded, again by Thales, for the supply of a single display bulkhead-mounted variant for the RN *Hunt* class minehunter mid-life update. In August 2002 a further contract, worth US$3 million, was awarded by AMS for OPUS 2 consoles, for the NAUTIS 3 system for the *Hunt* class. This version will be supplied with integrated Sun® UltraSPARC™ VME electronics and dual 21.3 in, 1,600 × 1,200 pixel high-resolution displays.

Contractor
DRS Technologies, Parsipanny, New Jersey.

UPDATED

Portable User Data Terminal (PUDT)

This is a rugged and lightweight palm-size unit for use with tactical battle management systems. The display is a 6.3 in diagonal colour transflective LCD display for daylight readability with a backlight for viewing in low light levels. The display resolution is 640 × 240 (pixels) with 256 colours. The PUDT is built with an open system, non-proprietary architecture.

Specifications
Functional
(a) Processor: High-performance 206 MHz Intel 32-bit StrongARM SA-1110 RISC processor and the SA-1111 StrongARM companion chip. It will provide performance of 250 (Dhrystone 2.1) MIPS @ 206 MHz
(b) Display: 6.5 in diagonal colour transreflective LCD display for daylight readability with a backlight for viewing in low light levels. The display resolution will be 640 × 240 (pixels) with 256 colours
(c) Memory: 32 Mbytes Flash memory, 64 Mbytes of DRAM memory and 128 kbytes of EPROM as a boot device
(d) Mass storage: CFF2 CompactFlash Card connector; accepts all CompactFlash Cards or 1 Gbyte IBM Microdrive™
(e) Keyboard: Virtual QWERTY onscreen keyboard
(f) Pointing device: Touchscreen that can be used with fingertip or stylus
(g) Function keys: 16 programmable function keys. Each key will power-on device and start application
(h) External ports:
Two RS-232C
One RS-423
USB Device/Slave
USB Host
(i) Reliability: 8,000+ hours MTBF (ground mobile)
(j) Maintainability: Predicted MTTR of 0.2 h

Environmental
(a) Temperature: Operating −20° to +55°C; non-operating −40° to +65°C
(b) Temperature shock: +21° to −25°C and +21° to +65°C, each within 10 min intervals
(c) Shock: 36 in drop per MIL-STD-810 E, Method 516.4, Proc IV and VI
(d) Altitude: To 15,000 ft

Physical
(a) Weight: 2.65 lb (1.2 kg) with standard 8 h battery
(b) Dimensions (L × W × H): 9 × 4.65 × 1.77 in (229 × 118 × 45 mm)
(c) Electrical:
External 11-13 V DC vehicle power supply 110/220 V AC, 47-440 Hz optional power adaptor
Rechargeable battery pack for up to 24 h of operation
8 h rechargeable Lithium Ion battery pack

Operating systems
(a) Microsoft Windows embedded CE3

Optional software (in ROM)
(a) Microsoft Pocket Word, Excel, PowerPoint, and Access
(b) Microsoft Pocket Outlook (Calendar, Tasks, Contacts, Inbox)
(c) Microsoft Pocket Internet Explorer
(d) Omnisolve
(e) Pocket Street
(f) Pocket CAD
(g) Quick View Plus

Status
Selected for the UK Bowman project. Originally produced by Northrop Grumman until it sold its ruggedised terminal operation to L-3 Communications in 2002.

Contractor
L-3 Communications, Ruggedized Command & Control Solutions, San Diego.

UPDATED

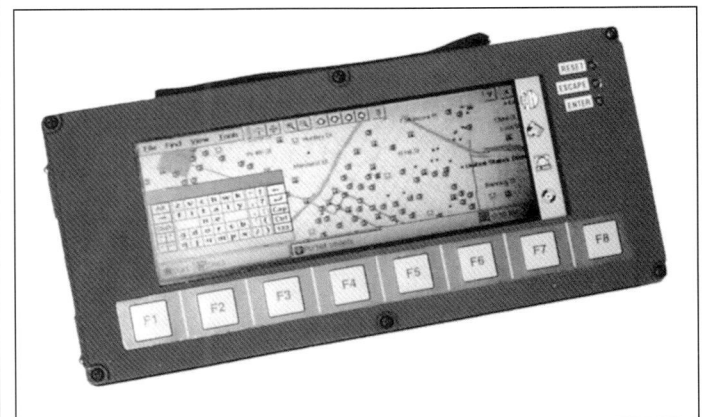

The Portable User Data Terminal (PUDT) (L-3)   0130059

PXI rugged hand-held computer

Intended for low-level tactical use, the PXI is a rugged hand-held computer with processor speeds of up to 700 MHz and a 4 in colour display. A user-accessible Type II PCMCIA slot is available to accommodate industry-standard PC cards, as well as a Type II tactical modem. A QWERTY keyboard plus 15 programmable function keys designed for use in gloves or NBC equipment are provided. Operating systems include Windows 95, 98, NT and XP 2000, as well as LINUX.

Specifications

Computer Features
(a) CPU: Intel® Pentium® III up to 700 MHz processor
(b) Max RAM: 256 Mbytes
(c) Ethernet: Fast Ethernet - 10/100 base-T via military connector
(d) I/O: RS-232, RS-422, PS/2 mouse and keyboard, external video
(e) USB, Ethernet, parallel and audio
(f) Graphic: 2 Mbytes video RAM
(g) PCMCIA: one internal type II PCMCIA and one user-accessible Type II or III
(h) Keyboard: upper 15 programmable function keys; lower QWERTY layout - night readable.

Display Features
(a) Diagonal area: 4 in colour
(b) Resolution: 640 × 480
(c) Contrast ratio: 250:1 (typical)
(d) Technology: low-temperature polysilicon TFT LCD
(e) Colour capability: 256,000 colours
(f) Batteries: four internally rechargeable lithium ion (hot swappable in pairs)
(g) Additional features: pointing device, side-mounted and keyboard-mounted pressure mice.

Physical
(a) Weight: less than 3 lb with batteries
(b) Dimensions (W × D × H): 4 × 2.2 × 7 in.

Environmental
(a) Altitude: operation 15,000 ft, storage 40,000 ft
(b) Operating temperature: —25 to 120°F
(c) Storage temperature: —25 to 150°F.

Status

Selected for a UK mobile satcom programme.

Contractor

DRS Electronic Systems, Inc, Gaithersburg, Maryland.

UPDATED

The Chameleon from Raytheon's mobile computer range (Raytheon)   0143773

Chameleon

The Chameleon contains a CD-ROM, 20 Gbyte removable hard disk drive, 128 Mbytes of RAM (expandable to 256 Mbytes), 3.5 in floppy drive, sealed keyboard and pointing device, Li-Ion battery, 16-bit sound, and provisions for Peripheral Component Interconnect (PCI) docking to the Digital Interface Unit (DIU). The DIU provides expansion capabilities to the Chameleon by incorporating slots for two Industry Standard Architecture (ISA) 1/2 or 3/4 size expansion cards, two PCI expansion cards, or one of each. The DIU powers the Chameleon when docked. Weight 15.5 lbs, dimensions 14 × 4.5 × 12.5 in (W × H × D).

Iguana

The Iguana, derived from the Chameleon, adds a unique one-half size PCI interface card. Both the Chameleon and Iguana are Internet/Intranet ready with a 10/100 Mbps Ethernet port. The Iguana has no docking capability. Weight 17 lbs, dimensions 14 × 5.5 × 12.5 in (W × H × D).

Gecko

The Gecko, also derived from the Chameleon, provides a portable work station with the addition of an 18 in flat panel monitor. The Gecko has the same features as the Chameleon with the exception of the integrated display. Weight 11.5 lbs, dimensions 14 × 3 × 13.5 in (W × H × D).

Status

In use in the USAF and the RAF.

Contractor

Raytheon Computer Products, Marlborough, Massachusetts.

VERIFIED

The PXI rugged hand-held computer (DRS Technologies)   0134125

Raytheon Mobile Computers

Raytheon Company's Chameleon, Iguana and Gecko are compact, lightweight, powerful computers. In addition to desktop computer features, they have an environmental seal, sunlight readable displays and shock isolation of critical components. The standard operating system is Windows 2000, but Windows XP is available as an option. All have a 333 MHz Pentium® II processor with a 256 kbyte Level 2 cache memory and four CardBus (PCMCIA) slots.

RF-6710W wireless message terminal

The RF-6710W wireless message terminal software package integrates Harris-developed data transmission techniques with COTS software. It has the capabilities of the RF-6750W wireless gateway, (see separate entry) but does not interface to LANs. When installed on a PC, the RF 6710W forms a multimedia messaging system that provides transparent relay of e-mail and files over media prone to errors. It also automatically relays messages if the first route is blocked. For HF e-mail messages, the RF-6750W allows one of three protocols/waveforms to be used on the air: STANAG 5066, STANAG 4538 and FED-STD-1052. STANAG 5066 is the second-generation NATO interoperability standard, STANAG 4538 is the third-generation interoperability standard and FED-STD-1 052 is used for interoperability with legacy Harris wireless products.

The RF-6710W-based system extends the home base to remote locations not easily accessed by existing communication lines. To the user, the RF-6710W operates like a standard PC with e-mail, automatically sending messages and data to the final destination, adapting to communications means. For example, messages may be transmitted to a relay point by HF radio and to the final destination by commercial telephone.

In addition to e-mail, the wireless gateway includes the HUITSMail™ imaging application, an Outlook Extension that has many of the capabilities of the Harris

Universal Imaging Transmission Software (HUITS™) (see separate entry). It captures, manipulates and sends high-resolution images directly from the Outlook application and enables transmission of high-resolution digital images using the wireless e-mail system. HUITSMail includes Wavelet compression that provides superior compression to JPEG. It translates between files in several formats. The software supports the industry standard TWAIN interface for acquiring imagery from digital cameras or scanners.

The software runs under the Microsoft Windows 2000 Professional or Windows XP Professional operating systems.

Status

Used by a number of countries during the NATO/PfP Ex COMBINED ENDEAVOUR in 2002.

Manufacturer

Harris Corporation, RF Communications Division, Rochester, New York.

UPDATED

RF-6750W wireless gateway

The RF-6750W software package connects to an Ethernet-TCP/IP local area network and operates as a "radio" mail server. It provides transparent delivery of e-mail and files across multiple transmission media including HF/VHF/UHF radio, LANs, landline, microwave, or satellite and is compatible with e-mail packages that run on a variety of platforms. The software integrates Harris-developed data transmission techniques with COTS software.

When installed on a PC, the RF-6750W forms a multimedia messaging system for transparent relay of e-mail and files over media that are prone to errors. It also automatically relays messages if the first route is blocked. For HF use, the RF-6750W allows one of three protocols/waveforms to be used on the air: STANAG 5066 - the second-generation NATO interoperability standard; STANAG 4538 - the third-generation interoperability standard; and FED-STD-1052 for interoperability with legacy Harris wireless products. Messages are generated and received by terminals that are connected to the LAN or WAN. The sender sends a normal e-mail message and the RF-6750 W automatically routes the message over the proper radio or landline link and sends it to the final destination, selecting the best available channel.

The Wireless Gateway also includes the HUITSMail™ imaging application, an Outlook Extension that has many of the capabilities of the Harris Universal Imaging Transmission Software (HUITS™) (see separate entry). It captures, manipulates and sends high-resolution images directly from the Outlook application. It enables transmission of high-resolution digital images using the wireless e-mail system. HUITSMail includes Wavelet compression that is claimed to provide superior compression to JPEG. It translates between files in several different formats. The software supports the industry standard TWAIN interface for acquiring imagery from digital cameras or scanners.

The software runs under the Microsoft Windows 2000 Professional, Windows 2000 Server or Windows XP Professional operating systems connected to a mail server.

Status

Users include the Japanese Maritime Self Defense Force. Used by a number of countries during the NATO/PfP Ex COMBINED ENDEAVOUR in 2002.

Manufacturer

Harris Corporation , RF Communications Division, Rochester, New York.

UPDATED

RF-6910 Falcon II® C2PC-CNR

The RF-6910 Falcon II® C2PC-CNR Integrated Tactical Command and Control System (ITCCS) combines the tactical IP (Internet Protocol) networking capability and internal GPS (Global Positioning System) of Falcon II radios with Northrop Grumman IT Europe's command and control application C2PC (see separate entry) to provide accurate position information on the battlefield to the user and to friendly forces. The ITCCS receives GPS position reports from radios and immediately updates the digital map of the battlefield to reflect their new positions. It also allows users to pinpoint enemy locations on their map and distribute that and other important information to friendly units who are also using this application. Units can be displayed using either Mil-STD 2525A or NTDS symbology, and a wide range of digital map formats are accepted, including GeoTiff, ADRG, DTED Level 1 and 2, CADRG and ETOP. The IP-based system is flexible enough to meet a variety of requirements, and is easily expandable for additional capability. The commercial IP interface allows other IP applications to share the same channel with no radio configuration changes.

Contractor

Harris Corporation , RF Communications Division, Rochester, New York.

VERIFIED

RHC-500

The RHC-500 is a rugged Palm Pilot with sunlight readable, backlit LCD, rechargeable batteries, sealed circular connectors and is MIL-STD-810 approved.

Specifications

Processor: Motorola 68328 20 MHz Dragonball; Palm OS compatible
Memory: 16 Mbytes DRAM, expandable to 64 Mbytes
Display: 2.8 in sunlight-readable 4 grey level LCD, 160 × 160 resolution backlight
I/O: Sealed circular connector, serial port
User input: Resistive touchscreen
Power: Two (2) AAAs, including rechargeable
Dimensions/Weight: 6 × 3.5 × 1.18 in; 11 oz

Contractor

Paravant Computer Systems, Inc, Palm Bay, Florida.

NEW ENTRY

RHC-1000

The RHC-1000 is a Microsoft Windows Pocket PC, with a sunlight readable colour LCD, USB and PCMCIA interfaces, rechargeable Lithium Ion battery, MIL-STD-810 approved.

Specifications

Processor: Processor Intel StrongArm 206 MHz 32 bit RISC Processor, Pocket PC Win CE
Memory: Memory 64 Mbytes DRAM, expandable to 256 Mbytes
Display: Display ¼ VGA Colour Sunlight Readable LCD, 320 × 240, 95 mm diagonal, backlight
I/O: I/O USB/Serial sealed connector, audio connector, external power connector, PCMCIA slot (optional); IrDA (optional)
User input: Resistive touchscreen ; multifunction keypad; pointing pad
Power: Internal Rechargeable Lithium Ion Primary or External Radio/other Secondary AA Integrated Battery Pack
Dimensions/Weight: 6.5 × 3.375 × 1 in; 11oz with internal battery

Contractor

Paravant Computer Systems, Inc, Palm Bay, Florida.

NEW ENTRY

RHC-2000

The RHC-2000 Rugged Handheld computer was specifically designed as a COTS solution for transferring digital data within the battlefield. First as a Data Transfer Device (DTD) to house the National Security Agency's (NSA) KOV-21 PCMCIA card and CT-3 software used to transfer crypto and other variables needed within the tactical digital battlefield to synchronise and enable communications among the inventory and future communication devices, and also to transfer and access the Signal Operating Instructions (SOI) that define the communication structure among users. It also contains a 1 Gbbyte Hard Disk Drive and USB interface (the Mission Data Loader (MDL)), independent of and separate from the DTD, that is used to transfer files among the US Army's Appliqué platforms to update maps, overlays and such like. A patented cartridge allows easy removal of the DTD NSA card and the MDL hard disk drive for safe/secure storage when not in use. The unit features an Intel StrongArm 206 MHz, 32 bit RISC processor, 16 Mbytes RAM, ½ VGA Sunlight readable display, PCMCIA expansion slots, sealed programmable keyboard and touchscreen.

Specifications

Processor: Intel StrongArm 206 MHz 32 bit RISC processor
Memory: 16 Mbytes DRAM, expandable to 32 Mbytes
Display: ½ VGA sunlight-readable monochrome or colour, 640 × 240 pixels, image size is 6 × 2-5/8 in,
display type is backlit transflective
Expansion: USB mission data loader option, 4, 8, 16, or 32 Mbytes flash ROM, 2 Type I/II PCMCIA ports or 1 Type III PCMCIA slot, USB interfaced hard disk drive option
Keyboard: Sealed, integrated programmable keyboard and touchscreen
Power: 8 AA batteries or BA-5800, over 20 h of life
Dimensions/Weight: 7.5 × 5.25 × 1.8in/ 1 lb 12 oz

Status

In operational use by the US Army. Paravant was awarded a contract worth US$2.8 million as part of the FBCB2 programme (see separate entry). Deliveries began in July 2001 and were completed in February 2002.

Contractor

Paravant Computer Systems Inc, Palm Bay, Florida.

NEW ENTRY

RP8200 rugged portable workstations

These workstations are intended to provide the means to run a variety of applications such as mission planning, command support and fire control. The use of compact PCI as the main system backplane allows the support of different processor architectures, including Intel™ and UltraSparc™. The workstations have a high-resolution display and a wide range of standard peripherals. A variety of options for expansion, drives and interfaces are available for the single portable unit.

Specifications

	RP8200	RP-AGE
CPU	Compact PCI based Processor boards from Intel™	Compact PCI based processor boards from Intel™
Drives	1 × 2.5 in 12 Gbyte + removable HDD 1 × 2.5 in 12 Gbyte + HDD 1 × 3.5 in 1.44 Mbyte FDD 1 × CD-ROM or DVD PCMCIA (option)	1 × 2.5 in 12 Gbyte + removable HDD 1 × 3.5 in 1.44 Mbyte FDD 1 × 3.5 in HDD up to 36 Gbyte (option)
I/O	SCSI, RS-232 serial, keyboard/display, USB, LAN, WAN, RS-422, audio, 2 × 4HPcPCI	SCSI, RS-232 serial, keyboard/display, LAN, WAN, RS422, audio, 2 × 4 HPcPCI, 3 × free compact PCI slots (expansion)
Display	15.1 in 1,024 × 768 TFT LCD	
Keyboard	IP65 and EMC sealed full-size keyboard with integral trackball	
Weight	16 kg	20-24 kg
Dimensions (configuration dependent)	463 (W) × 357 (H) × 510 mm (D) k/b open, 240 mm k/b closed	463 (W) × 400 (H) × 328 mm (D) k/b closed

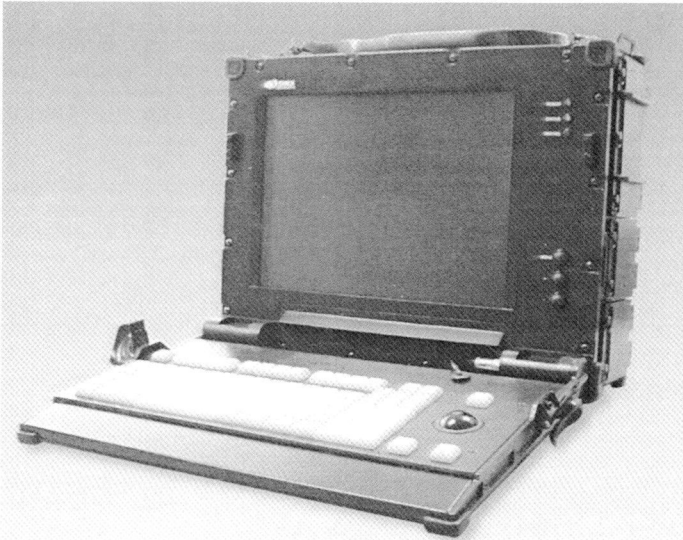

RP8200 rugged portable workstation (DRS)   0143776

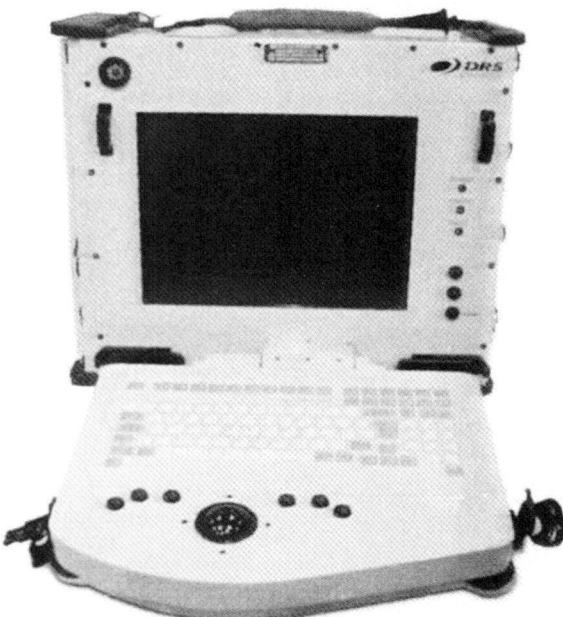

RP-AGE rugged portable workstation (DRS)   0143794

Status

As at late 2003 both variants were available. The RP8200 based on an IBM PC, optimised to support Windows NT, will be used on US typed air stations and surface ships. The RP-AGE is the hardware platform for the database test equipment for the Typhoon Eurofighter.

Contractor

DRS Technologies Electronic Systems Group, Gaithersburg, Maryland.

UPDATED

RVS-250

The RVS-250 is an environmentally sealed rugged vehicle computer with Pentium processor, colour sunlight readable display and a sealed detachable keyboard.

Specifications

Processor: Intel Pentium Processor, 266 MHz or faster
Memory: 64 Mbytes DRAM expandable to 256 Mbytes
Display: Active matrix colour display, 640 × 480, 6.5 in, operator dimmer control and instant black-out switch
Expansion: Two RS-422 ports, one RS-232 port and one parallel port, 2 Mbytes video memory, two Type I/II PCMCIA ports or one Type III PCMCIA port, 10 Gbytes hard disk drive
Keyboard: Sealed, integrated detachable keyboard and pointing device
Power Input:
DC power input: 12 V DC; extended voltage range available
AC adaptor power input: 110 V AC and 220 V AC, 50/60 Hz
Dimensions: 10 × 7 × 3.75 in

Status

In operational use in the US Army in the Movement Tracking System, and has also been used in artillery fire-control displays.

Contractor

Paravant Computer Systems Inc, Palm Bay, Florida.

NEW ENTRY

RVS-330

The RVS-330 is a rugged vehicle computer system which was selected as one of the the Appliqué+ V4 computers used in ground combat systems and vehicles as part of the U.S. Army Force XXI Battle Command Brigade and Below (FBCB2) programme (see separate entry).

Specifications

Processor: Intel Pentium III, 600 MHz or faster. 256 k L2 cache, Socket 370, design also supports Celeron processors
Memory: Up to 512 Megabytes PC-100 SDRAM
Mass storage: Removable internal 10 Gbytes hard drive expandable to 28+ Gbytes
Display: SVGA, 12.1 in diagonal, colour LCD with touchscreen, Active Matrix TFT 800 by 600 pixels resolution, sunlight readable at maximum luminance of 750 nits with a contrast ratio of 2.8:1, display unit interface permits tethered operation up to 25 ft, side-to-side viewing angle +/- 60°, top to bottom maximum viewing angle +10°/-30°
Keyboard: USB 88-key full-travel elastomeric keyboard in QWERTY configuration. 'Hot swappable' while the system is powered
Pointing device: Fully sealed pointing device with right and left pick buttons
Operating system: Compliant with MS-DOS, Windows 95, 98, 2000, Microsoft NT v. 4.0 and Solaris X86 (UNIX) v. 2.6 and 7
Expansion: One PMC slot, Disk On Chip socket, 3U Compact PCI expansion slot, Device Bay DB20 form factor slot
External Ports: External SVGA port supports up to 1280 × 1024 pixel resolution, 18 bit LVDS video, Sound Blaster™ compatible PCI audio input and output, two USB ports, one dedicated to Processor Unit/Display Unit interconnect, IEEE 802.3 LAN interface 10/100BaseT, One RS-232C port dedicated to the Display Unit Touch Screen, 5 RS-422/RS-423 ports, one dedicated to Processor Unit/Display Unit interconnect, 2 synchronous serial communications ports for Conditioned Di-Phase (CDP) and MIL-STD-188-144 Non-Return-to-Zero communications interfaces
Temperature:
    Operating range: -35 to +60°C
    Non-operating range: -35 to +71°C
Weight:
    Processor unit: 18.05 lb.
    Display unit: 7.3 lb.
    Keyboard unit: 2.2 lb.
    Hard disk drive cartridge: 0.8 lb
Dimensions:
    Processor unit: 5.1 × 12.7 × 10.2 in
    Display unit: 2.36 × 13.1 × 9.0 in
    Keyboard unit: 1.0 × 11.5 × 7.25 in
    Hard disk drive cartridge: 0.7 × 3.9 × 5.9 in

Status

The initial year value of the Appliqué contract, awarded to Paravant in December 1999, was over US$13 million, representing 45 per cent of the total shared requirements. After the initial year contract, Paravant was awarded two follow-on contracts. Option Year 1, for US$14.3 million, consisted of a 60 per cent share of the total units to be supplied. Option Year 1 delivery began in June 2001 and was completed within the first quarter of 2002. Option Year 2, awarded in February 2002 for US$24.9 million, consisted of an 85 per cent share of the total units to be supplied. Deliveries are scheduled to begin in the third quarter of 2002 and run through the second quarter of 2003.

Also purchased by the Australian Army for BCSS (see separate entry).

Contractor

Paravant Computer Systems, Inc, Palm Bay, Florida.

NEW ENTRY

Scorpion rugged notebook

The Scorpion rugged notebook computer will be the basis for the majority of the User Data Terminals (UDT) procured for the Bowman programme (see separate entry).

Specifications

(a) Processor: Intel Pentium III, 700 MHz, 256K L2 cache
(b) Memory: 128 Mbytes PC-100 SDRAM, expandable to 512 MbytesMass storage: Removable internal 20 Gbytes hard drive, expandable
(c) Display: SVGA, 12.1 in transflective, 800 × 600 pixel resolution, daylight viewable, XGA, 13.3 in, transmissive, 1024 × 768 pixel resolution, touch screen optional
(d) Keyboard: Sealed, elastomer 88 QWERTY configured keyboard
(e) Pointing device: Fully sealed pointing device with right and left pick buttons
(f) Operating system: Compliant with MS-DOS, Windows 95, 98, 2000, Microsoft NT v. 4.0 and Solaris X86 (UNIX) v. 2.6 and 7.0
(g) Expansion: Two (2) type I/II or one (1) type III PCMCIA ports, additional battery, LS120 floppy disk drive, DVD/CD-ROM drive, custom USB interface device
(h) External Ports: Parallel port, 2 USB ports (1.1 or 2.0), 10/100 ethernet port, external VGA supports up to 1280 × 1024
(i) Communication ports:
Port 1: RS-232, RS-422 or RS-423
Port 2: RS-422 or RS-423
Port 3: RS-422 or RS-423
Port 4: RS-422, RS-423, or internal fax modem
(j) Temperature:
Operating: −20 to +60°C
Non-operating:−40 to +71°C
Optional: −32 to +60°C with heaters
(k) Dimensions: 12 × 10 × 2.5 in
(l) Weight: 12lbs

The Scorpion rugged notebook computer (DRS)　　0533838

(m) Input voltage: 28 V DC vehicle power per MIL-STD-1275A, optional AC converter 90-264 V AC, 47-440 Hz, lithium Ion hot swappable battery.

Status

In August 2002 it was announced that the then Paravant Computer Systems had been awarded a US$6.7 million contract for Bowman terminals. To meet this contract the Scorpion will be configured to meet three UDT variants: the Bowman Management Data Terminal (BMDT) (13.3 in display) a laptop workstation for use in office, CP or command vehicle; the Dismountable UDT (DUDT) (12.1 in touchscreen display) for use in soft-skinned vehicles and dismountable into buildings; and the Vehicle UDT (VUDT), similar to the DUDT except the lid is removed and the touchscreen replaces the keyboard, which is reconfigured in detachable form. The VUDT is therefore a non-dismountable 'tablet' and will normally be mounted vertically, with the keyboard stowed away when not required. Deliveries began in September 2002.

The Scorpion has also been selected by MASS for the ATRACKS helicopter tasking system (see separate entry).

In December 2002 it was announced that the then DRS Technologies had acquired Paravant.

Manufacturer

DRS Tactical Systems, Palm Bay, Florida.

UPDATED

V2A2 Lightweight Computer Unit (LCU)

The V2A2 Lightweight Computer is a third-generation, rugged portable workstation. Interoperable with L3's tactical smart modem, known as the Tactical Communications Interface Module (TCIM), the V2A2 allows continuous communications via e-mail and data-sharing tools.

Two display options are available: a 10.4 in colour active matrix LCD with 256 colours or a 9.4 in monochrome LCD with 16 shades of grey. Built with an open system, non-proprietary architecture, the V2A2 can run applications under UNIX, MS-DOS, or Windows. The V2A2's standard software includes MS-DOS, Microsoft Windows, Diagsoft QA Plus and Century TERM/FT modem communications software.

Specifications

(a) Processor: 200 MHz Intel Pentium P54C processor
(b) Display: active matrix colour 10.4 in VGA and SVGA LCDor 9.4 in monochrome VGA LCD
(c) Memory: 16 Mbytes of RAM expandable to 256 Mbytes
(d) Mass storage: 500 Mbytes, 2.1 Gbytes or 10 Gbytes removable hard drive, internal 1.44 Mbytes floppy drive
(e) Keyboard: PS/2-compatible detachable 85 key subset of IBM-PC 101 key functionality
(f) Pointing device: embedded 3-button trackerball
(g) Expansion:
four full-length ISA slots
one combination PCI/ISA slot
(h) External ports:
one RS-232C serial
centronics parallel
external display supporting up to 1,280 × 1,024 resolution with 16 colours
SCSI-11 (ANSI X3.131 draft dated 11 Nov 91)
external floppy interface
IEEE 802.3 LAN interface
TCIM power

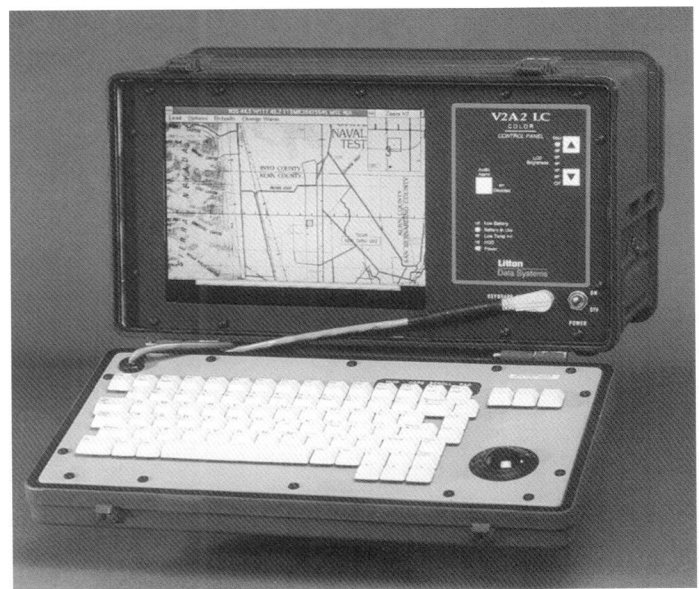

V2A2 Lightweight Computer　　0055175

(i)　Communications: Hayes-compatible V.32 bis 28.8 kbps modem
(j)　Reliability: 2,400 hours MT13F (ground mobile)
(k)　Maintainability: predicted MTTR of 0.25 h
(l)　Physical
　　　Weight: LC: 23 lb (with standard accessories 30.5 lb) (10.5/13.9 kg); External Power Module (EPM): 3.1 lb (1.4 kg)
　　　Dimensions (W × D × H): LC: 16.2 × 11 × 9.5 in; EPM: 10 × 2.1 × 5.5 in
(m)　Electrical:
　　　AC: 110+/−10%, 220 V AC+/−10% V AC, 47-63 Hz
　　　DC: 20-32 V DC (optional)
　　　UPS: Internal UPS recharging from AC or DC input, up to 10 min 110/220 V AC, 47-440 Hz
　　　28 V DC vehicle power per MIL-STD-1275A
　　　rechargeable battery pack for up to 2 h of operation.

Status

Now being offered by the Ruggedized Command and Control Solutions division of L-3 Communications, which was previously part of Northrop Grumman.

Contractor

L-3 Ruggedized Command and Control Solutions, San Diego, California.

UPDATED

V3

The V3 Universal Control Computer is a ruggedised tablet computer.

Specifications

Processor: Intel Pentium class 300 MHz geode ultra low-power processor
Memory: 64 Mbytes DRAM expandable to 128 Mbytes
Display: 8.4 in VGA (640 × 480) transflective colour sunlight readable LCD, adjustable backlight
I/O: 2 sealed I/O interface connectors
2 serial ports configured as 1 RS-232 and 1 RS-422
2 USB ports
1 Ethernet 10/100 port
User input: Five wire touchscreen
Power: 110/220 V AC or 10-32 V DC
Dimensions/Weight: 8.5 × 5.5 × 2.5 in; under 5 lbs

Status

In use as the display for the Joint Service Lightweight Stand-off Chemical Agent Detector (JSLSCAD).

Contractor

Paravant Computer Systems, Inc, Palm Bay, Florida.

NEW ENTRY

SOFTWARE

Canada

MapFusion™

MapFusion™ is COTS software that permits access to geospatial data from diverse sources and formats without requiring complex pre-processing or time-consuming translations. The foundation of the MapFusion product is the Open Geospatial Datastore Interface (OGDI) and the Geographic Library Transfer Protocol (GLTP). Developed in collaboration with defence scientists at the Canadian Department of National Defence, OGDI is an application program interface (API) that enables any software application to access easily heterogeneous geospatial data sources. Developed by Global Geomatics, GLTP provides a unique IP-based data transfer protocol that, when combined with OGDI, allows complete interoperability of distributed - and heterogeneous - geospatial data sources. OGDI is used by MapFusion to access local data sources, while GLTP is used to establish networked exchange and dissemination of geospatial data.

MapFusion merges heterogeneous and distributed geospatial data files into a single map as if the maps were 'fused' together. Multiple vector or image files - regardless of their source co-ordinate systems or formats - can be combined to form a single map. The product uniformly displays tactical overlays on top of geospatial data from any source.

MapFusion supports interoperability between prevalent geospatial military standards including VPF, RPF, ADRG and DTED, as well as among popular GIS business packages from ESRI®, Informix®, Intergraph®, MapInfo®, ObjectFX™, Oracle® and Virtual Prototypes™. It can also retrieve data from proprietary and open GIS file formats. It operates in a broad variety of computing environments including Microsoft® Windows® 98/NT/2000, Sun™ Microsystems Solaris™, and Linux®.

MapFusion components

Comprising interoperable modules, which can be assembled according to customers' needs, MapFusion includes server, workstation and adaptor software components.

MapFusion Server. Geospatial data dissemination and dynamic web map rendering services are handled by MapFusion's server software, which can process requests from both MapFusion Workstations and from standard web browsers. MapFusion Server can also act as a central metadata repository, allowing exchange of geospatial data files. MapFusion Server runs on Microsoft® Windows® NT and Windows® 2000, Sun™ Microsystems Solaris™ and Linux®.

MapFusion Workstation. MapFusion Workstation is user-oriented software that provides access, sharing and visualisation of geospatial data sets. It can search for geospatial data across any local or network file system, hard drive, or CD-ROM, offering an authoring tool for creating maps with features and proportional metrics. The product's rendering engine automatically eliminates duplicate or overlapped annotations and labels. MapFusion Workstation runs on Microsoft® Windows® NT, Windows® 98, and Windows® 2000.

Adaptor software. MapFusion uses adaptor software to normalise geospatial file formats into a uniform data abstraction, which can then be easily manipulated. Adaptors exist for all vector and raster military formats as well as for most popular commercial geospatial software.

Interoperability

MapFusion complies with MIL-STD-2525B and MIL-PRF-89045. The software can directly access all MIL-STD-2407 VPF products including: DCW, VMAP, DTOP and DNC; all MIL-STD-2411 RPF products including: CADRG, and CIB; all ISO8211 raster products such as ADRG, USRP, ASRP; and all levels of DTED. MapFusion can also write to MIL-STD-2407 VPF and MIL-STD-2411 RPF. The product interoperates with a large array of commercial software packages including: ESRI ArcView™ and ArcInfo™, Informix Geodetic Datablade™, Intergraph GeoMedia™, MapInfo Professional®, ObjectFX SpatialFX™, Oracle8i™ spatial formats, and Virtual Prototypes VAPS™ product. It reads data from proprietary and open GIS file

formats including: MrSID, ECW, GeoTIFF, Shapefile, Tab/Map files, GRASS, SDTS and TIGER.

Status

In service with the United States and in the Coyote reconnaissance vehicle in Canada. Selected in February 2002 to provide standard GIS interoperability software for the French Délégation Générale pour l'Armement (DGA).

Manufacturer

Global Geomatics Inc, Outremont, Quebec.

UPDATED

SigNET

SigNET is a frequency spectrum management tool to aid in planning and real-time status monitoring for mobile communication deployments. It aids the planning, frequency assignment and subsequent execution of communications plans. It is a UNIX-based system that operates on PCs or laptops as an integrated component of tactical Combat Net Radio, Command Post communication and Wide Area Networks.

The system provides a full suite of communications management tools including frequency deconfliction, radio path loss, area of coverage, graphical site deployment and net membership. A distributed planning facility enables staff to outline the communication plan while facilities controllers fill in the details for their site. Antenna RF characteristics are entered through forms into the relational database. Radio, antenna, filter, multicoupler and mutual interference characteristics are stored for use in path loss calculations, interference calculations and frequency assignment.

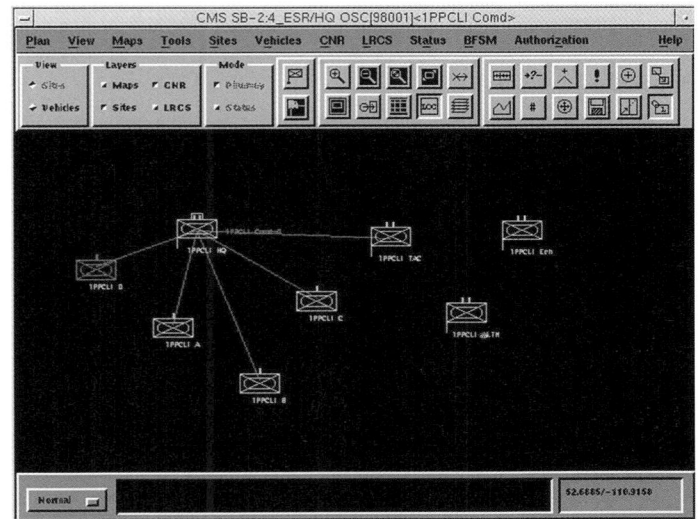

SigNET screenshot with underlying map layer turned off   0055170

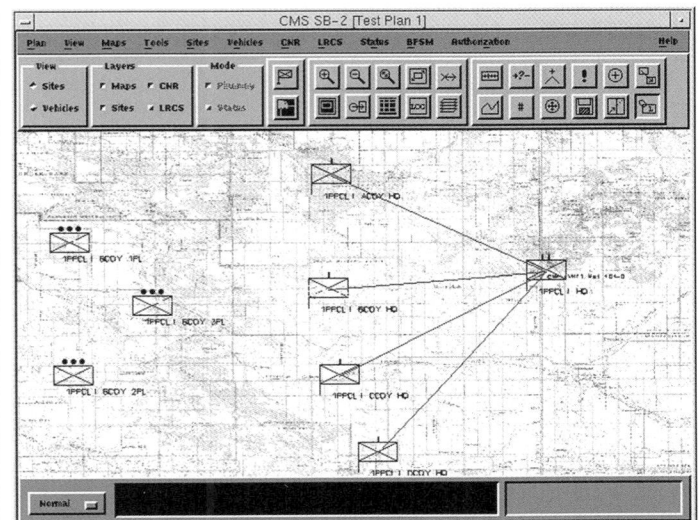

SigNET screen showing site deployment and net membership with map underlay   0055169

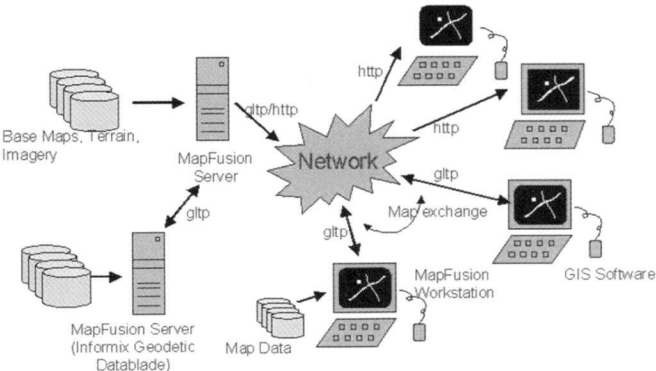

MapFusion system architecture (Global Geomatics)   0132529

The system has the following features:

(a) Battlefield Frequency Spectrum Management (BFSM)

Planning of radio nets, radio relay, and radio rebroadcast elements

Maintaining and assigning frequencies based on allotment, band plans, barred bands and smart re-use algorithms

Visualising the area of coverage and path profile for planning and siting radio assets

Checks on network connectivity, path loss and interference

(b) Graphical Interface System (GIS)

The user is provided with a georeferenced graphical display for deployment planning and management with drag and drop capability to move assets and sites

(c) Distributed planning and assignment

The functionality supports EUROCOM D/1 distributed planning levels to be responsible for increasing level of planning details. High-level planning of links, nets, frequencies and encryption can be performed without involvement in details

Lower-level planners fill in specific details to meet the high-level plan

(d) Relational database

Plans, status and equipment characteristics can be stored to create deployments quickly from previous plans

A backup database can be maintained

(e) Distributed architecture

Workstations can assume new and higher EUROCOM D/1 (ESS) roles as demanded by attrition and equipment loss

Specifications

(a) EUROCOM D/0 (1994) Section 9
(b) Enhanced EUROCOM System Chapter 13
(c) Platform: P3 500 PC or laptop computing platform
(d) RAM: 128 Mbytes minimum
(e) Operating system: UNIX
(f) Hard disk space: 4 Gbytes minimum, 8 Gbytes recommended
(g) Maps: supports DIGEST raster, vector, and DTED mapping data (ADRG, VMAP0, VMAP1, VMAP2, VMAP3, DTED1 and DTED2).

Status

In September 1999, SigNET was fielded with the Canadian Forces at the unit level as the Communication Management System (CMS) portion of the Iris TCCCS System (see separate entry). In May 2002, SigNET commenced field trials at the formation level with the Canadian Forces, which added integrated trunk, satellite, line, STANAG 5040 and STANAG 4206 planning.

Contractor

General Dynamics Canada, Calgary, Alberta.

UPDATED

Denmark

IRIS messaging software

IRIS messaging software provides messaging applications which are included in a wide range of CIS.

IRIS Message Formatting System (IRIS/MFS)

IRIS/MFS is a client-server Message Handling and Message Formatting system. The Message Handling provides facilities for release, reception and distribution of informal and formal messages, and the Message Formatting provides facilities for preparing correctly formatted messages. The software includes tools for working with formatted messages as well as tools for the support of site-specific operational procedures. The Application Programming Interfaces (APIs) available provide the necessary interfaces for integration into CIS. IRIS/MFS supports a number of military message formats, including ACP127, JANAP128, DD173, ACP126m, and e-mail (RFC 822). The software includes functions which enable formatted messages such as USMTF or AdatP-3 to be partly or fully auto-generated from other sources, such as situational databases; and received formatted messages can either automatically update a database or be validated first. Standard procedures for message handling - both incoming and outgoing - can be imposed.

There are four different versions of IRIS/MFS:

(a) IRIS/MFS Lite. A stand-alone product intended for use by single users working independently. Enables the user to produce and validate formatted messages and view incoming messages. Formatted messages can also be sent or received using e-mail through integration with MS Outlook or Lotus Notes. Can be used on Windows platforms.

(b) IRIS/MFS Desktop. Provides the facilities of IRIS/MFS Lite plus additional message drafting facilities and an ODBC V2 database interface to enable manual update of databases and the creation of formatted messages from databases.

(c) IRIS/MFS Professional. Designed for large-scale organisations and integration into CIS developed by a third party. Has a client-server architecture and runs on UNIX. Also supports mixed platforms such as Windows clients running with a UNIX server.

(d) IRIS/MFS Enterprise. Provides a full message-handling solution, with all the features of the other packages plus full functionality for message storage, operational workflow and automatic message profiling and distribution.

IRIS Message text format definition system (IRIS/DEF)

IRIS/DEF supports the definition of unambiguous information exchange formats. It provides a way of defining formatted messages so that the structure, sequence and allowable contents can be exactly defined.

IRIS for Outlook

IRIS for Outlook uses all the built-in features of MS Outlook and Exchange Server while providing all the military messaging functionality of IRIS/MFS. This enables, for example, the use of both non-formal e-mail and formal military messages from the same user interface and the use of standard office automation tools, as well as the production of messages in a variety of formats, workflow control, message profiling and distribution and organisation/role-based messaging.

Status

IRIS applications are embedded in a range of CCIS including TBMCS and DMS in the US; JOCS, CSS, ATacCS, FOCSLE, RAFCCIS, and TAMPA in the UK; SICA, SICF, SITRENS and GRANITE in France; SIACCON in Italy; NORCCIS II and NEC CCIS in Norway; NEC CCIS and FlexCCIS in Denmark; ATHENA and MCOIN III in Canada; MCCIS in NATO; BCSS in Australia (see separate entries for most of these systems).

Contractor

Systematic Software Engineering A/S, Aabyhøj, Denmark.

VERIFIED

France

Circe 2001 family of mission planning systems for air applications

Using the open architecture of its hardware and software, Circe 2001 meets the mission planning requirements for aircraft, helicopters, UAVs and cruise missiles. The range of functionalities enables it to meet the requirements of force, squadron and unit level applications.

Each system can be delivered in a commercial-off-the-shelf (COTS) version or in an easily transportable ruggedised unit. The manufacturer claims that Circe 2001 can be linked to any battle management network in order to transmit or receive orders and exchange intelligence and operational data. In addition to these C3I networks, mission planners can be interconnected, using widely used commercial and military standards, to provide the user with a common tactical situation in real time.

The main available software functions are:

(a) automatic plan generation
(b) flight planning
(c) threat penetration analysis
(d) target analysis
(e) sensor prediction and reconnaissance planning
(f) timing
(g) deconfliction
(h) attack rehearsal
(i) geographical database management
(j) operational data management
(k) intelligence data management
(l) post-flight analysis
(m) three-dimensional fly-through
(n) Mission debrief and replay.

The system is a result of the integration of powerful graphic workstations with dedicated peripherals such as additional memory devices and optical disks.

Circe 2001 for squadron level mission planning

Colour printers and cartridge loaders provide output and data transfer cartridges are loaded through a specific peripheral, which can read the cartridges when the post-flight analysis function is implemented. Navigation log and screen colour hard copies can be used separately or integrated into the combat mission folder which can be easily customised for a given requirement.

Status
Since 1987, SAGEM has provided the French Air Force and Naval Aviation with successive versions of Circe 2001 and it is in service with major commands at force, wing and squadron levels.

Contractor
SAGEM SA, Paris.

VERIFIED

GIPSY

GIPSY is a Geographic Information Production System developed by Sagem. It consists of five modules corresponding to four functions: data acquisition, processing, production and storage of processed geographical data. Each type of data to be processed represents an independent workshop:
(a) Cartography. The cartographic function produces homogeneous mosaics of digital maps of high-precision edge to edge pasting. Geometric accuracy is maintained throughout the process with colour classification, georeference, change of projection and compression without loss.
(b) Imagery. The imagery function processes satellite images from SPOT, LANDSAT, ERS 1 or aerial photos. These images undergo georeference and various radiometric improvements before assembly into mosaics.
(c) Digital Elevation Models. The DEM function produces data with terrain elevation information in 3-D or contour line maps. The data is generated from stereoscopic satellite images or aerial photos, using powerful DEM computation.
(d) Vector. The vector function produces vector data, using most existing vector databases. Data (roads, railway networks, rivers, infrastructures and so on) can be entered once and stored in the database to be merged.
(e) Map production. The map function produces space maps, updated digital maps or hybrid products. The latter are obtained by the fusion of various types of data: map/images, map/vector, image/vector.

Specifications
(a) Acquisition of data: Large size scanners, optical media (CD or DVD)
(b) Production of maps or images: Large variety of printers, including film printers or optical media (CD or DVD)
(c) External exchange formats
Cartography:
Local geodetic systems: (WGS 84, WGS 72, OSGB 36, NAD 83, NTF)
Projection systems: Mercator, Lambert, stereographic
Raster: ERDAS, GEOTIFF, TIFF, SPOT, Landsat, ERS 1, DIMAP, DIGEST
Vector: ARC info, DXF, DFAD, VPF (VMAP), Open GIS, Shape, MapInfo, DWG, VRF, EDIGeo, DX90
DEM: USGS DEM, DLMS/DTED, DIGEST MATRIX

Status
In use in a variety of French Mission Planning Systems and in Digital Mapping Agencies. Also in use in the TACTIS C2 system (see separate entry).

Contractor
SAGEM SA(Aerospace & Defence Division), Paris.
UPDATED

Germany

MilGeo-PCMAP

MilGeo-PCMAP is a geographic information viewer software product based on Dornier's (now EADS) standard software package 'GEOGRID for Windows'. In co-operation with the Bundeswehr Geographic Service AMilGeo, the MilGeo-PCMAP was developed for use within the German armed forces and is available to other military mapping agencies within NATO and the PFP community. MilGeo-PCMAP, as a Military-off-the-Shelf (MOTS) product is permanently upgraded under supervision of AmilGeo and an international user group has been set up by AMilGeo to define further upgrades. As an add-on module EADS offer a conformable military situation package GEOGRID- SitView and programmers' interfaces for integration into other applications are available. The product offers German, English and French HCI.
Standard geographical data packages can be provided by AMilGeo on CDROM:
(a) Germany 1: 50,000 scale and smaller (set of 18 CDROM)
(b) Europe 1: 250,000 scale and smaller (set of 34 CDROM)
(c) World 1: 500,000 scale and smaller aeronautical charts (set of 8 CDROM)
(d) Crisis areas (with additional information such as images.)

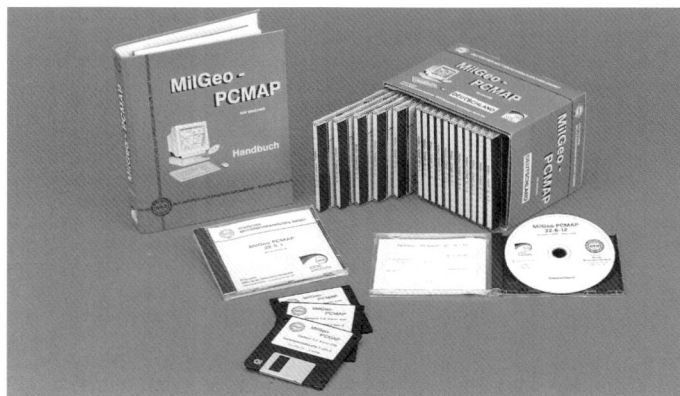
PCMAP   0001032

These data packages include raster map, elevation data and gazetteer. Map vector data can also be presented. The data are stored on the CDROM in KMRG format (STANAG 7151) and other NATO standard formats can be imported (AORG, CAORG, ASRP, CRP, TIFF, GeoTIFF) as well as images, which can be rectified and georeferenced. For the preparation of KMRG data out of standard NATO formats, EADS offers the Map Preparation Software (MAPS NT) package. Data and software are maintained independently based on separate CD-ROM media. Standard software functions are: change of scale, scrolling, panning, zooming, dimming, graphical overlays, indication of coordinate and elevation, large format printing, import and export. The co-ordinate system can be chosen independently of the co-ordinate system of the respective map. MilGeo-PCMAP offers GPS connectivity, route planning, extensive database functionality, inset of images and powerful elevation functions (relief, relief shading, path profile, elevation layers, slope, perspective view, 30 view and line of sight).
MilGeo-PCMAP is designed to run on any PC with MS-Windows (95, 98, NT, 2000) and requires only standard PC performance.

Status
MilGeo-PCMAP has been in operational use by over 5,000 military personnel in the German armed forces for since 1998 and has been introduced so far by the Netherlands, Swiss and Portuguese armed forces. Other unspecified nations are reportedly preparing for its introduction. Version 4.0 has been in use since 2001, version 4.1 was to be available in 2002 although this has not been confirmed. The version for the Swiss armed forces (PCMAP Swissline) comprises both MilGeo-PCMAP and Geogrid-SitView functionality.

Contractor
EADS Defence & Communications Systems, Ulm.
UPDATED

International

Mercury military messaging system

The Mercury military messaging system has been designed to provide secure, flexible and cost-effective military messaging solutions for defence forces. The system can be configured to send and receive large amounts of information - both text and binary data - over a wide range of communications media, conforming to defence messaging protocols. Secure Automated Military Messaging System (SAMMS) (see separate entry) is a variation of Mercury sold separately.
Mercury provides a fully integrated family of products offering:
(a) Conformance to military procedures
(b) Desktop messaging
(c) Interoperability with other message networks
(d) Tactical operation
(e) COTS e-mail interoperability.
The Mercury family comprises a wide range of PC-based Windows products, which fulfil the requirements for any scale of messaging system:
(a) Stand-alone messaging terminals which can be used with virtually any legacy messaging system, in both strategic and tactical environments
(b) Message switches which can undertake all message-switching roles from mininode to large multiport switching, operating with or independently of legacy networks
(c) Message gateways that are fully automatic, simultaneous multiport message conversions between networks using a variety of standards and protocols
(d) LAN-based messaging systems with a client/server architecture which can seamlessly extend military messaging services onto the desktop.
Each of the Mercury products can be readily integrated into legacy message networks, extending their life and capability. Elements of the Mercury family can also be used as 'building blocks' as part of larger communication networks. These networks can be built up simply and quickly, or phased in during the evolution of a system. Mercury networks can operate as:
(a) Ship-based communications systems providing automatic, controlled, accountable links between internal and external ship/shore stations, over LAN, radio and satellite circuits

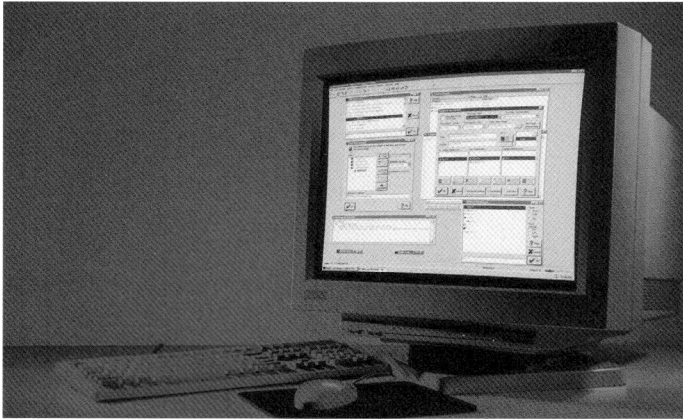

Mercury military messaging system   0062402

(b) HF communications networks with autonomous communication stations operating as part of a dynamic, adaptive HF ALE network

(c) Equipment control - using Mercury elements to control and manage local and remote equipment assets over a wide range of communications media

(d) Command and control systems - extending Mercury communications to provide forms-based messaging services which can be linked into C2 displays, with integrated graphical information systems.

All members of the Mercury family are fully integrated into the Microsoft Windows operating system and can be used in conjunction with most Windows applications. Using Mercury's interoperability features, it can also be accessed from Unix workstations and networks. Mercury products have been designed with security as a key feature. A full audit trail of messages and system events is provided as standard throughout the entire product range.

Status

Elements of the Mercury family are currently in service with the Australian Defence Force. Mercury has also been sold to the UK and in 2000 it was announced that it had been sold to the Hellenic Army as the message-switching element of the Hermes II system (see separate entry). SAMMS has been sold to the Norwegian, Romanian, Turkish and UAE navies.

Contractor

BAE Systems Australia, Canberra.
Aeromaritime Systembau GmbH, Munich.

UPDATED

Israel

MapCore™

MapCore™ is a C2 tool which makes extensive use of graphical display techniques. Using raster or vector maps, sensor data, photos, symbols and graphic overlays, information can be presented and analysed in a variety of ways.

Features

(a) Digital map display with fast zoom, scroll and scale

(b) Full colour moving map with north or heading orientation and continuous smooth movement and rotation

(c) Graphic overlay management. Search, add, display or hide functions can be applied to groups of objects such as symbols, zones, routes and targets

(d) Terrain elevation model. 3-D terrain analysis tools such as line of sight, vertical projection, computed 2-D/3-D shading and perspective terrain view with aerial photo or map overlays

(e) Mission planning tools. ORBAT display, targets and firing positions, course plotting with terrain and threat considerations

(f) Tactical situation management. Unit position and activity plotted manually or received by datalink

(g) Mission debriefing tools. Records tactical situations for future analysis

(h) Navigation system interfaces. Various interfaces to GPS, INS, digital compass

(i) Weapon system and sensor interface. Displays sensor footprint and field of view, and target data

(j) Datalink interface. Can be interfaced to a variety of digital communications systems.

Status

In operational use with the Israel Defence Force.

Contractor

Mitam Advanced Technologies (1999) Ltd (A subsidiary of Elbit Systems Ltd), Petah-Tikva.

UPDATED

Xi vehicle tracking system

Xi Information Processing Systems has developed a vehicle tracking system based on the Global Positioning System (GPS). It includes an in-vehicle unit, a command and control centre and a two-way datalink.

The in-vehicle unit includes an intelligent vehicle Management Control Unit (MCU) with a Vehicle Positioning Module (VPM) based on GPS and optional Dead Reckoning (DR). An interactive in-vehicle display is also available for text messages and optional map display for mobile Geographical Information Systems(GIS).

The command and control centre includes a network communication controller which handles and routes two-way data communication between the vehicles in the field and the command and control centre and a situation station that displays vehicle locations on a moving map. The system is based on a client/server architecture and SQL database.

The tracking system uses Xi's unique 3-D topographic moving map enabling 3-D terrain applications and analysis.

Manufacturer

Xi Information Processing Systems Ltd, Savyon.

VERIFIED

NATO

NATO UHF Frequency Assignment System (NUFAS)

EDS Defence designed and implemented the original NATO UHF Frequency Assignment System (NUFAS) for the Allied Radio Frequency Agency (ARFA) at NATO HQ. It was designed to carry out major reassignment tasks for air-ground-air communications in the military UHF band, and to make assignments on a day-to-day basis for use by NATO nations. NUFAS is also used by the Frequency Management Sub-Committee of the NATO Consultation, Command and Control Board (which succeeded ARFA in 1997) to develop more suitable assignment criteria and strategies, in order to make more efficient use of the available spectrum.

NUFAS was enhanced (NUFAS II) as a result of a necessary port to a UNIX environment due to the replacement of the original IBM mainframe with a secure network of UNIX workstations under the MINERVA project. The enhancements included the use of the Oracle Relational Database, a MINERVA compliant HCI and improved assignment algorithms. NUFAS II was installed in 1993 and provided substantial performance improvements on the original NUFAS.

In 1998 NUFAS 3 was developed. The system replaces NUFAS 2 and the NATO Net Number Assignment System (NANNAS) functionality and interoperates with the Master Radio Frequency List database and the ARFA Computer Aided Data Exchange (ARCADE). The system was installed at NATO HQ in Brussels in 2000 and underwent trials during 2001. It is understood that it is intended to port the system to a Windows environment to enable it to run on the common NATO HQ infrastructure, but no details are available of progress on this.

Contractor

EDS Defence Ltd, Hook, Hampshire, UK.

UPDATED

Netherlands

Code30 and Code300-32 for Windows data analysis and decoding equipment

Code30

Code30 offers military, government and PTT monitoring installations the ability to enhance operators' search and monitor operations in the HF radio spectrum from 9 kHz to 30 MHz. The equipment can be used to analyse and decode a range of digital HF signals, including OOK, FSK, FEK, MFSK, MFEK, PSK, 2DPSK, 4DPSK, OQPSK and QPSK.

Signals are presented on different screens showing, for example, frequency against amplitude, frequency against time, phasor/vector scope and phase against time. Various different bit analysis screens are also available with fully bit-synchronised demodulation. If the intercepted signal requires further post-processing, raw digital bit streams can be demodulated using a fully bit-synchronised demodulator and stored on magnetic media or output as ASCII data.

The Code30 unit itself consists of a totally screened, IBM PC-compatible AT-ISA bus interface card which is installed into the customer's PC installation. The interface requires three BNC coaxial connections for audio signals, two inputs (for optional diversity operation) and one output.

Code30 can be deployed to complement existing or new HF radio monitoring stations. Fully screened against EMI and RFI emissions, it will not compromise TEMPEST installations or degrade existing low noise system performance. As well as the supply of Code30 as an addition to an existing PC installation, turnkey systems can be supplied consisting of an RF screened processor, VDU and keyboard in 483 mm rack size.

Code30's use of digital signal processing and FFT algorithms on all spectrum analysis screens is designed to facilitate immediate reaction to fast operator tuning. More detailed FFT and auto-correlation analysis can be made on emissions of interest enabling experienced operators to 'fingerprint' particular stations.

Code300-32 for Windows

Developed from Code30, Code300-32 is an entirely software-based system which uses the Windows multimedia section and the sound card on a PC or notebook. It can simultaneously process two audio signals from separate sources, and multitasking allows several analysis tasks to be performed simultaneously on the same or different signals. It can be controlled directly by the operator or remotely via a WAN or LAN, with the output being available at any point, and up to four versions can run on the same PC, providing a fully remote-controlled eight channel system. It can also be integrated into bigger systems.

Status

Manufactured in the Netherlands by Hoka Electronik NL since the early 1990s. In use worldwide. Recent customers for Code30 have been the UK MoD, the Pakistan armed forces and the Turkish government. The Code300-32 has been incorporated into LEARAS (Project SCARUS) (see separate entry) for the UK.

Contractor

Hoka Electronics, Pekela, Netherlands.

UPDATED

Turkey

PRC/VRC-9600 network management subsystem

The PRC/VRC-9600 network management system is made up of the KM-9601 network planning unit, the KD-9601 code/key distribution unit and the FG-9601 code/key loading unit.

The KM-9601 is a military personal computer with network planning software allowing 500 different networks to be planned. Planned network data is distributed to other KM-9601 or KD-9601 units at the lower levels, via wired/wireless transmission lines or using diskettes. From this point, network data is distributed and downloaded to users' radios with FG-9601 units.

Contractor

ASELSAN Military Electronics Industries Inc, Ankara.

VERIFIED

United Kingdom

C2PC (Command and Control for the PC)

Command & Control for the PC (C2PC) is a native Windows NT/2000/XP application that can act as a client to an ICS server in a heterogeneous Unix and Windows system. In addition, a PC running C2PC can act as a standalone workstation when fed data from a correlated formatted message stream. It provides COP (Common Operational Picture) visualisation on a map background and track management with related DA/PA tools.

Features provided by C2PC include:
(a) dynamic picture display and management
(b) Vector and Raster map display in multiple formats and projections
(c) live Common Operational Picture display
(d) replication of information across low bandwidth communications.
(e) MIL-STD-2525B/APP-6A Track and Overlay display and creation and comprehensive planning overlay support
(f) Sites, Routes and Formations tools

C2PC provides an open architecture using Microsoft technologies to integrate seamlessly with desktop applications such as Microsoft Word and PowerPoint, as well as any other applications developed using these industry standards. C2PC maps can be quickly and simply copied into briefings or reports generated by such applications.

The C2PC user interface is based on the standard Windows User Interface style providing familiar menus, configurable floating toolbars for frequent operations, uniform help across all features and shortcut keys in common with other Windows applications. This means that new C2PC users are familiar with the style of user interface and are quickly able to navigate between C2PC tools, map windows and help. Units used for common displays within C2PC such as position, distance and bearing are configurable and automatically adopted throughout C2PC.

C2PC Gateway

The C2PC Gateway component provides the interface service between the C2PC client functionality and an external tactical data source. It can receive a live track feed from a number of sources, currently: an ICS server, a correlated OTH-T Gold

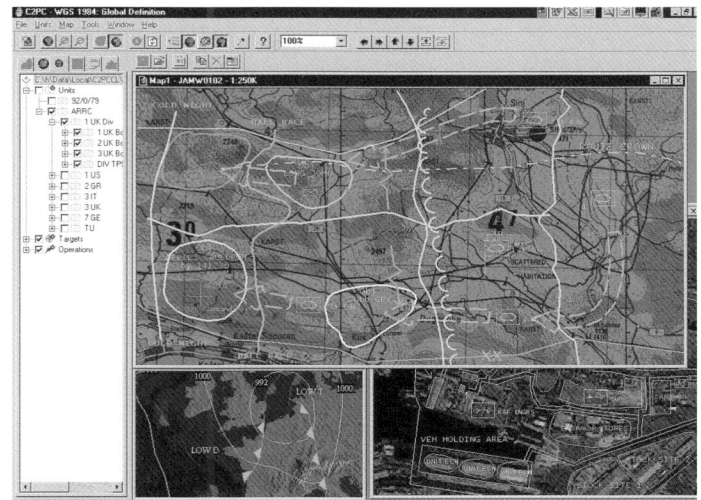

C2PC   0041325

message feed (via a serial interface) or another C2PC Gateway. The Gateway then broadcasts this data to multiple C2PC Clients.

The Gateway can be configured to work on a high bandwidth network (that is EthernetLAN) down to 4800 baud Combat Net Radio. The bandwidth usage can be further reduced with the option of compressing the data and restricting the broadcast to a particular area of interest. The C2PC Gateways can be arranged hierarchically providing a scaleable architecture over multiple bandwidths and areas of interest.

Atlas Display Server

The Atlas component is the C2PC map display server. It provides the management of map windows and co-ordinates the map plug-in components that render digital map data to the map window backdrop. Atlas provides a set of interfaces that client applications use to draw graphical foreground objects onto the map. Support for a number of basic objects (lines, circles, arcs, ellipses, sectors, triangles, symbols, splines, text and clip-art) is provided. In addition to these fundamental line and symbol types C2PC provides extended attribute support for land symbology (MIL-STD-2525/APP6A). Atlas can display any number of Map windows at any time (subject to system resources), with the facility to save maps with areas of interest and plot settings for future recall. These maps can be shared or e-mailed to other users.

The Atlas architecture allows for the extension of map formats, foreground objects, projections and menus. New map data formats are supported by installing additional map plug-ins. These are then automatically available to the Atlas server. Atlas map windows allow object embedding and linking of registered OLE objects such as Microsoft PowerPoint presentations, Internet URLs and sound files. This works in the same way that spreadsheets for example can be embedded into Microsoft Word documents. Equally, Atlas maps can be embedded within other applications such as documents and presentations.

C2PC can display raster and vector map data in the following formats:
(a) World Database II (WDBII) and World Vector Shoreline (WVS)
(b) ARC Digital Raster Graphic (ADRG) *
(c) ARC Standard Raster Product (ASRP) *
(d) Raster Product Format (RPF) including CADRG and CIB
(e) Vector Product Format (VPF)
(f) ESRI Shapefiles
(g) Digital Terrain Elevation Data (DTED) Level 1 and Level 2
(h) GeoTIFF
(i) ETOP *
(* indicates preprocessing before data is displayed)

Most map data can be displayed quickly and easily in the native format. New map formats can be developed as C2PC plug-ins. These are automatically added to the C2PC user interface and appear as an integrated part of C2PC. New projections can also be developed as C2PC plug-ins.

C2PC can display maps in two modes, Flex Draw and Quick Draw. Flex Draw allows the user to display any data in any projection at any scale. Quick Draw limits the map data to the native projection and multiples of the native projection. This provides almost instantaneous rendering of map data and the ability to move quickly between multiple layers of map data.

Navigation around maps is achieved with functions such as zoom in/out, recentre, pan, select area, and map up/down/left/right arrows. In Quick Draw mode, multiple layers of map data can be stepped through using the up/down layer buttons. All of these functions are provided on toolbars (which can float over the map, or be docked into the user interface) as well as menus. C2PC provides the user with a comprehensive set of map management tools. Maps can be centrally stored on a file server and accessed by all C2PC clients on the network, who simply add an entry for the map data location using the C2PC map management tools. Any maps in this location are automatically detected and added to the list of maps available to the user.

Map coverage displays show exactly what areas have map data, distinguishing between remotely stored and locally stored data. Remote map data can be copied to a local machine by highlighting the map coverage on the map display. This is especially useful if a local machine is to be taken off the network containing the map fileserver.

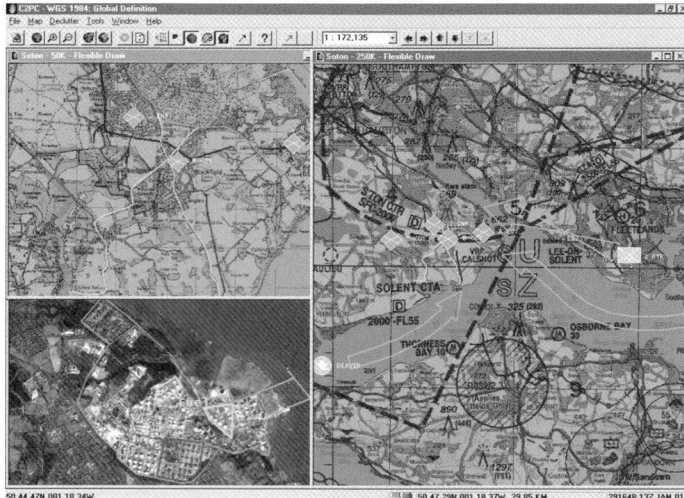

A screenshot of C2PC in operation at JWID 2001, showing overlays and tracks on mapping and imagery (Northrop Grumman)   **NEW**/0569675

Track Management

The Tracks component stores and displays track data received from the Gateway service. Support for new track types is available by installing new plug-ins. Track data is stored locally in an Access-compatible database and is thus readily available for use in many other applications.

In addition to the functionality offered by C2PC, Microsoft Access can be used to view the track data, running user-defined queries and reports. C2PC provides concurrent support for multiple track databases, which can either be updated in real time from an ICS server or maintained by a separate process. Third party client applications can use C2PC TrackPlot interfaces to plot their track databases on the C2PC map. They can add their own options to the C2PC menus and be notified when the user selects their tracks.

Tracks can be displayed using either MIL-STD-2525B/APP-6A or NTDS symbology. With the former, the user has full control over the display of the symbol (that is whether to include frames, icons and fills) and the labels which annotate it. MIL-STD-2525B/APP-6A tracks are easy to create with hierarchical menus for function ID and a palette of common symbols.

Track management facilities include the creation, deletion and editing of platform, unit, acoustic and emitter tracks in any of the databases that are currently being plotted. Plot controls are provided to declutter the display of tracks manually or automatically and to filter the view based upon echelon/threat, category/threat, track type and the MIL-STD-2525B/APP-6A code. Other features include track groups, range circles, plotting of track histories, crossfix and auto panning of the map to a particular track.

Overlays

C2PC supports the creation and manipulation of MIL-STD-2525B/APP-6A overlays. Overlay graphics are inserted, selected, moved and resized easily using the mouse and there is a range of supported graphic styles. Line styles, colours and fill types are selected using floating toolbars. MIL-STD-2525B/APP-6A line styles are supported and OLE objects can also be embedded in an overlay at a georeferenced location. This includes objects such as Microsoft Word documents, video clips and images. These are displayed as an icon that can be viewed by double clicking the object on the map. The graphics within an overlay can be grouped and activated/deactivated on a hierarchical basis. Additionally, overlays can be organised hierarchically into folders and subfolders. Overlays can be imported from an ICS server, and exported back to the ICS server or to other C2PC users. Imported overlays are automatically detected and, if necessary, displayed by the overlays application. This means that dynamic information such as the current weather picture can be centrally collated and regularly distributed to clients. Those clients that have elected to view this information will automatically display the updates as they happen.

Other Features

Other tools provided by C2PC include:
(a) Site creation with display filters
(b) Route planning with CPA calculations
(c) Hierarchical Unit, Operations and Targets Planning
These tools have all been developed using the interfaces exposed by the Atlas Display Server, which means that new tools can be developed using the same public interfaces. Such tools will integrate seamlessly with the existing application, providing the user with a common interface for both core and bespoke tools.

Throughout its development C2PC has been designed to facilitate the integration of additional applications developed both by NGIT and by third parties. Current extensions include:
(a) Air Tasking Order visualisation tool that permits the parsing and display of ATO/TCO generated by TBMCS and ICC
(b) line of sight tool for intervisibility calculations
(c) 3-D visualisation tool that allows the C2PC picture to be exported to a 3-D visualisation application for briefing purposes
(d) oceanographic performance prediction that provides integration with the Wader model from Ocean Acoustic Developments Ltd

(e) Thistle integration layer that enables the Thistle planning tools to be closely integrated with C2PC.

Status

Development of C2PC began as an internal Northrop Grumman (then Logicon) R&D activity in October 1995 and by March 1996 it had been adopted and funded by the US Marine Corps, with whom it is now in operational service. The USMC has established C2PC as the primary component of its single software baseline. In recognition of its reliance on combat net radio for data communications, the USMC has funded much work to engineer C2PC to operate effectively in a combat net radio environment. US version 5.9 was released in 2003 to fulfil requirements sponsored by the USMC, US Navy, Coast Guard and Defence Information Systems Agency (DISA). C2PC is evolving to provide equivalent C4I capability to that currently provided by Unix-based ICS client workstations. In the longer term, development of a Windows-based server also continues.

In operational use in a range of joint and maritime systems in NATO nations. These include:
NATO Maritime CCIS (MCCIS)
UK Joint Operational Command System (JOCS)
UK Command Support System (CSS)
Italian Navy CSS and amphibious C2 system.

Contractor

Northrop Grumman Mission Systems (formerly Northrop Grumman IT Europe), Southampton.

UPDATED

***

EH-101 Merlin Data Collation Device (DCD)

The Merlin DCD is a full mission planning system that supports all roles of the EH-101 Merlin; active and passive anti-submarine operations, active and passive anti-surface operations, naval gunfire support, search and rescue and fleet support.

The DCD automates data preparation, mission planning and post-flight analysis. DCD is capable of reading all the major forms of commercial data storage and can be networked with other systems via a Local Area Network (LAN). Designed to be operated by aircrew the DCD allows data preparation at any time independent of specific sortie. ESM libraries, acoustic signature libraries, system settings and defaults can be prepared and stored for subsequent use. For mission planning the DCD divides a sortie into separate tasks each of which can have different environmental models and targets associated with them. The post-flight analysis function allows mission reconstruction for debriefing purposes and ELINT and acoustic analysis facilities.

A full DCD installation includes a command system interface, a digital map interface and a printer. The Command System Interface (CSI) is connected to the combat system highway in Type 23 frigates and 'Invincible' class aircraft carriers. The CSI enables the DCD to be used for direct access to the host ship's tactical information. The digital map interface reads Compact Disk (CD) and magnetic media and houses the aircraft's data transfer devices. With the exception of the CSI, all components of the system are designed to be removable and portable to allow independent operations from 'green field' sites.

Transfer of information from the DCD to the Merlin aircraft is via four solid-state data transfer devices. The aircraft management computer is loaded with navigation and communication information via its own transfer device as is the mission computer which is loaded with tactical information and equipment details. The digital map has its own system which allows map data to be changed on a sortie-by-sortie basis and the ESM system has a data transfer device for library information.

Status

Delivered to the Royal Navy in late 1997.

Contractor

Thales Defence Information Systems, Wells, Somerset.

VERIFIED

***

ELVIS (Enhanced Linked Virtual Information System)

ELVIS is a Web based segment for the ICS server (see separate entry) that allows a user to browse the Command and Control Picture using just a browser on the client workstation. ELVIS brings the power of the C4I system to the user through adaptive Web technology. Within the ELVIS product range, ELVIS I provides a static picture with access to options for changing the map, interacting with tracks, overlays and Air Tasking Orders (ATOs). The user is provided with plot controls and filtering along with tactical overlay facilities to allow customisation of the Operational Picture. ELVIS II provides a continually updated picture using Java, with dynamic update of tracks, ATOs and participation in collaborative planning sessions.

Status

ELVIS is in operational use in joint and maritime systems in NATO and Allied nations.

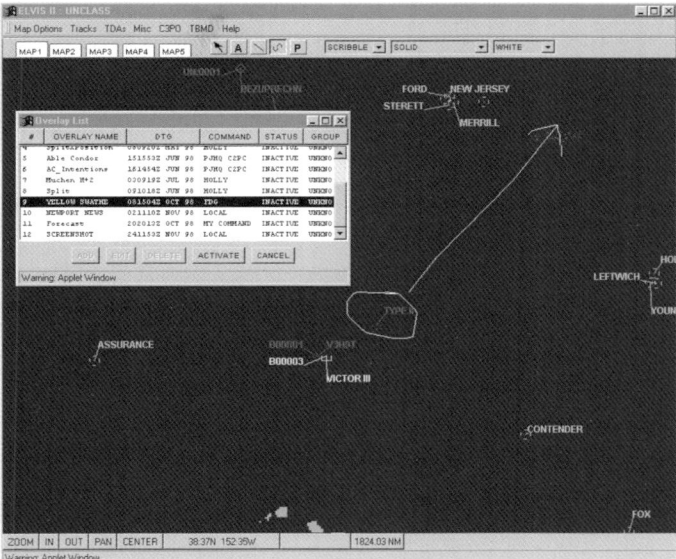

Northrop Grumman IT Europe Ltd's web-based command and control application, ELVIS II   0041326

GP3 in use in HQ ARRC (ARRC)   *NEW*/0111682

Contractor

Northrop Grumman Mission Systems (formally Northrop Grumman IT Europe), Southampton.

UPDATED

Generic Earth Station Management System (GEMS)

Astrium has developed a control and monitoring system called GEMS to control and manage military earth stations both in the UK and overseas. GEMS can be applied to a range of communications architectures and earth station configurations using TCP/IP.

In addition to performing remote-control and monitoring tasks, the system can also be configured to carry out special functions requested by individual customers, usually without the need for code changes. These include functions such as redundancy switching, EIRP calculation and automated test control.

The system can operate with a number of standard computer architectures including PC, Sun and Alpha. Operating systems used to date include Solaris, Solaris × 86 and Digital Unix. GEMS is also available for the Windows™ operating environment. The equipment has the following interfaces: RS-232, RS-485, RS-422, IEEE-488, HDLC and digital I/O. Networking protocols include TCP/IP, Ethernet and X.25.

Status

The first applications of GEMS have been the remote control and monitoring of the UK MOD FSC 658 and FSC 692 earth stations supplied by Astrium in 1999 and 2000 (see separate entries). The first systems were handed over to the UK MOD in 2000. GEMS is being further developed and enhanced for the export market.

Contractor

Astrium UK, Stevenage, Hertfordshire.

UPDATED

GP3

GP3 is an integrated software tool designed to provide information systems support to Army staffs at the four star down to one star levels of command, both in barracks and in the field where elements are deployed to unit level. It provides tools to support the G2 (Intelligence), the G3 (Operations and Plans) and the G6 (Communications and Information) staff cells and forms part of the Digitisation Stage 1 programme for NATO's UK-led Allied Rapid Reaction Corps (ARRC).

GP3 makes use of a client-server architecture and integrates COTS products providing a flexible basis for future CCIS, which in the case of GP3 includes the ability to support Joint/Allied operations and an open integration capability with

other software and systems. In the ARRC it is hosted on IARRCIS and in other British Army headquarters on AtacCS (see separate entries).

Information is presented to the staff on a geographic information system with NATO APP6A military symbology, for which further data is accessible from the database, allowing the monitoring of status direct from the electronic map display. Through the use of automatic replication and distribution by GP3, this information can be shared by a number of different Headquarters/Nodes to provide effective situational awareness.

Among the toolsets provided are those to support: information management; planning, movement, Intelligence Preparation of the Battlefield (IPB), creating ORBATS and TASKORGS, analysing the feasibility of options; creating and distributing orders, directives, intelligence analysis and target analysis.

Status

GP3 originated as part of ACSAS (see separate entry), but has been developed as a separate project for HQ ARRC. Software version 3.1, which included improved replication and distribution of data, enhanced reactivity and faster levels of loading and saving information, was tested in September 2001 and subsequently successfully completed an operational evaluation by the ARRC at the end of the year. In July 2002 it was announced that BAE Systems had been awarded a further contract to produce software V4, to enhance the situational awareness and planning capability, specifically in the area of mapping and symbology, and this was delivered in November of that year.

Contractor

BAE Systems, Christchurch, Dorset.

NEW ENTRY

HUGIN ChartLink System

The HUGIN ChartLink System is a data-fusion system for handling a wide range of different geographic, remote-sensing, hydrographic, oceanographic and meteorological data in an easily assimilated form. The primary purpose of HUGIN is to take all the types of data representing the entire physical environment from across the land, sea and air domains, combine them as layers and produce a Recognised Environmental Picture (REP) for distribution. This is particularly valuable in Joint operations, notably in the littoral.

HUGIN is based upon TENET Defence's own MapLink technology for fast dynamic data fusion and visualisation. The system allows users to combine environmental data from various sources and produce a single-fused product either for viewing separately or within a third-party application. A wide variety of data types are supported including oceanographic, hydrographic, meteorological and geospatial information. Also, 'data object overlays' are included for inclusion of practically any other data source, georeferenced to the background map or chart, such as side scan sonar images and beach gradient diagrams. A facility is provided for the user to build and arrange a product exactly tailored to the requirement by using a structured tree layout to view and arrange the data. More detailed information might be included for areas of particular interest, such as an approach to a specific beach, while maintaining an overview for the entire area of operation. Straightforward updates and additions can be made to the information, as well as

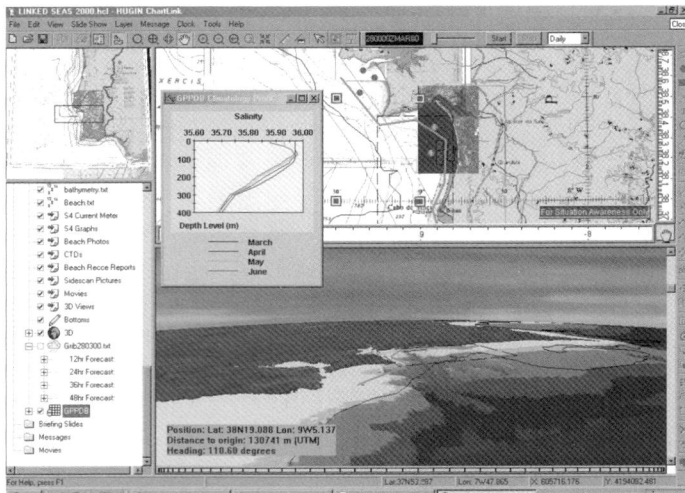

A ChartLink screenshot from Exercise LINKED SEAS showing 3-D effect (TENET Defence)   0533837

allowing the user to explore and assess the data all together in one place. Once a fused product containing all the relevant environmental information has been produced, it can be transmitted to other units. The fused products can then be used to support briefings and mission planning.

The full 3-D HUGIN System is made up of a number of components:

2-D Import Studio

This tool provides an extensive capability to process vector and raster map, chart and image data into a compressed and easy to use form. Data filters included as standard are: VPF, S57 / AML, SHP, ASRP, ARCS, DTED, DBDB-V, CIB, CADRG and multiple image formats including JPEG, TIFF, Geo-TIFF, LAN, GIF, BMP. The Studio is typically only used during expert data preparation, and is not required by most operational users.

3-D Import Studio

3-D Studio data processing environment includes filters for: ADRG, CADRG, CIB, RPF, TIFF, Geo-TIFF, JPEG, BMP, GIF, DTED, DBDB-V, ASRP, and ARCS Charts.

ChartLink

ChartLink is the core component of the HUGIN system and it is where the multiple data sources required for a clear depiction of the physical environment can be fused together as a series of layers for the preparation of fused products such as the REP. Layers included as standard are: Map/Chart, Annotation, Imagery, Object, Met GRIB, SIPS, NATO Overlay 2, GPPDB Query and GPPDB Visualisation. ChartLink then allows the further preparation and analysis of the data through facilities such as contouring, graphing and measurement. ChartLink also handles dynamic changes in source data such as time-charging forecast values and time-stamping of imagery. The fused REP product, along with supporting data, may be saved for sending to other users. Users of ChartLink can generate incremental products and export these as specific update layers for use by other ChartLink users and ChartLink Viewers. Interactive briefings which effectively represent the REP can also be produced and exported for immediate use across the Web.

The 3-D extension gives ChartLink the ability to display a complete representation of the whole earth in a second main-view window. The 2-D and 3-D views are linked together so that any user movement in one view is automatically reflected in the other. The 3-D extension is supplied with low-resolution imagery of the entire earth as standard, and also allows different layers of data, such as maps, imagery, terrain and bathymetry, to be fused together and viewed from absolutely any angle. User annotation and analysis produced in the 2-D view is also displayed in the 3-D view, as well as dynamic information such as temperature or windspeed contours. This gives an advanced mission preview and planning capability - 3-D views and movies can be viewed and integrated into the fused REP product. 3-D ChartLink implements soft pan, zoom and tilt wheels. These allow the user to move from a plan view to a side on view, and to pan around the point of interest.

ChartLink Viewer

This software provides a 'view only' mechanism for access to the core REP data layers as well as the interactive briefings and rehearsals produced with the full ChartLink. Users of ChartLink can produce any number of viewers for unlimited distribution.

Supporting Tools

These include an Imagery Masking Tool and a Symbol Editor which allow the data display to be customised to meet the specific user display requirements.
Interfaces are available for linking the ChartLink display with:
(a) Any GPS signal - for real-time mission monitoring in 2-D and 3-D
(b) Planning overlays, such as shipping lanes or pipelines - at any height or depth
(c) Standard office products - spreadsheets and word documents can be linked to points or objects
(d) Dynamic objects for showing the exact location of all assets in real time
(e) Animated sequencing of weather patterns as they build and decay
(f) Real world models in 3-D such as buildings, vehicles and wrecks
(g) Advanced data configuration, filtering and rendering options.

Standard system requirements:
(a) Personal Computer with Pentium processor
(b) Windows NT 4.0 or higher
(c) 128 Mbytes RAM required; 256 Mbytes RAM or higher recommended
(d) Typically at least 1 Gbyte disk space, depending on size of data sets in use
(e) CD-ROM drive
(f) Mouse or compatible pointing device

Status

The system was trialled by the Royal Navy's survey vessels in 2000, and it was then used during Exercise SAIF SAREEA II in the Oman in 2001, particularly in support of amphibious operations. Further improvements were made as a result of this experience and from use during JWID 02, the latter resulting in the Web export mechanism. The system is still being improved and developed and is now at Version 3.1. It was used by UK forces to support Rapid Environmental Assessment (REA) during operations in Iraq in 2003.

Contractor

TENET Defence Ltd, Horsham, Sussex.

UPDATED

Maritime Acoustic and Navigational Tactical Aid (MANTA)

The MANTA combat support system provides decision support functions for the naval command, enabling them to maximise equipment effectiveness and the warship as a whole. The system is based on ruggedised commercial-off-the-shelf (COTS) equipment and incorporates the latest advanced tactical applications. Originally developed to enhance the existing legacy command system capability on selected UK submarines, the system is currently in service with the UK Royal Navy as DCG(R), a four-user variant of MANTA.

MANTA provides a high-level software environment which is used to integrate and co-ordinate a suite of existing and proven tactical applications which allow the command to optimise functions including:
(a) Tactical picture compilation, display and management
(b) Data fusion, classification and Target Motion Analysis (TMA) tools
(c) Navigation quality charts to enhance the tactical picture
(d) Oceanographic and environmental data analysis, prediction and exploitation
(e) Weapon command tactical aids
(f) Navigation and mission planning facilities
(g) Online operational and engineering information.

MANTA uses a networked architecture of compact, air-cooled, COTS UNIX workstations and flat-panel colour display technology to minimise the impact on platform space and power requirements. All system components are housed in a protective enclosure to meet environmental and electromagnetic compatibility requirements. The displays offer high-performance colour graphics in a compact unit. Combined with the windows, point and click user interaction, this provides a powerful, yet easy to use operating environment. Each display is equipped with a choice of hand-held keypad or alphanumeric keyboard both with an integral trackerball. The system is fully operable from either input device.

MANTA's client-server architecture can be scaled to suit individual requirements. Redundancy is built-in through straightforward network reconfiguration. Through a range of industry standard interfaces and supporting database facilities, MANTA can be configured to interface with other combat system components including command systems, sensors, weapons and communication equipments.

The open environment also supports the use of MANTA as a vehicle for sea trials of new and prototype applications either as stand-alone facilities or integrated functions with access to tactical picture data. This allows end-user feedback to be built into designs, ensures applications are tailored to match their requirements and maximises the investment made in new developments.

The range of functionality provided by MANTA can be tailored and developed to suit specific operational requirements as necessary.

Status

A four-user variant of MANTA is currently in service with the Royal Navy as DCG(R).

Contractor

AMS, Camberley, Surrey.

VERIFIED

Mission Support System (MSS)

The Mission Support System (MSS) is a Command, Control, Communications and Information (C3I) system providing facilities for tasking, management and reporting of air operations including those by maritime reconnaissance, AEW, attack aircraft and helicopters. It provides secure mission planning and mission support for all aspects of UK Maritime Patrol Aircraft (MPA) operations.

The MSS offers connectivity to numerous intelligence and support databases enabling accurate preflight planning and briefing to be conducted at both the main operating base or at deployed sites worldwide. The application provides predictions of sensor performance which, together with tactical applications, fuel and route planning, signature library generation and map generation provide a wide range of information essential for effective MPA operations. Prepared data may be transferred to the aircraft either electronically, via portable transfer devices, or in the form of briefing packs to enable the crews to manually input data to the aircraft systems. MSS is used for crew and command briefing using the workstation/deployable display screens or on large screens or via projectors. Post-flight debriefing, mission tactical reconstruction plus detailed analysis of acoustic and EW information is also conducted using the MSS.

A wide range of terrestrial and satellite communication links can be used, enabling a survivable, geographically dispersed network that supports the operational need to be established quickly. The system provides automated preparation, processing and distribution of e-mail and NATO standard APP4/ADAT P-3/OHT-Gold ATO/ATM messages using ACP127, JANAP 128 and X400 routeing formats. These facilities enable, for example, intelligence data to be automatically entered into the system's database on an event-by-event basis with no operator intervention. The architecture of the system can be tailored to the command structure with data being collated, input and output at the appropriate level. Interfaces can be provided to permit interoperability with other electronic systems, report compilation and Preflight Data Insertion Programmes (PDIP).

Database features enable the storage and management of a wide range of intelligence, signal, asset, parametric, weather, aeronautical, digital map and chart data. Multiple-choice selection windows and data prefiltering options ensure accurate and rapid access to the required data. The map database supports operations anywhere in the world.

Full use is made of colour, graphic and text screens in an OSF/Motif environment. Standard Query Language (SQL) is used for data access by multiple query strings.

Security features allow access to data through a 'privileges matrix' that includes named operator, function to be performed and workstation designation.

The architecture supports the configuration of fixed site, deployable and suitcase systems. The latter is known as the Lightweight MSS (LMSS - see separate entry).

Status

MSS has fixed networked installations at several air stations throughout the UK, including the MPA main operating base at RAF Kinloss in Scotland and a number of command headquarters. It currently supports the Nimrod MR2 aircraft and will support the replacement aircraft, the Nimrod MRA4, when it enters service with the RAF. This version will be known as MS2000. MSS is also used to support main base and deployed operations of other aircraft types including the Sentry E3-D AEW, Nimrod R-1, Tornado, Harrier, and Sea King Mk 3A search and rescue helicopter. MSS will also support the UK MoD's ASTOR aircraft when it enters service.

Contractor

Thales Airborne Systems UK, Crawley, Sussex.

UPDATED

Naval Environmental Command Tactical Aid (NECTA)

The Naval Environmental Command Tactical Aid (NECTA) is a scenario-based sonar performance prediction system suitable for both shallow and deep water environments. It provides a suite of graphical tools for visualising potentially complex underwater environments and, most importantly, tools which provide a rapid assessment and visual presentation of sonar performance.

The key goals of NECTA are to:
(a) provide a simple process for obtaining sonar range predictions quickly, whilst allowing detailed control of the process for more advanced users
(b) perform across the complete range of operational conditions (shallow or deep, LF or HF, passive or active, noise-limited or reverberation-limited.)
(c) provide interactive and intuitive graphical visualisation of data, using both 2-D and 3-D displays
(d) handle the integration of range-dependent environmental data from variable-resolution grids and scattered observations, an especially important feature for data gathering and research use
(e) allow the integration and adoption of different PL models and databases, according to the requirements of individual customers
(f) provide an open architecture, enabling NECTA to run stand-alone or integrated within a command support system or as part of a sonar system.

The main chart display has worldwide coverage and gives the user a geographical context in which to operate. The user can overlay the chart with plan views of various environmental data, including bathymetry, sound speed, temperature, sonic layer, sound channel, CZ potential and the positions of XBT/XSV/CTD drops. NECTA also provides facilities for overlaying tactical information such as other vessels within the task force, target solutions and operating areas, which can be used, for example, to determine the effectiveness of ASW screens.

NECTA contains a suite of Propagation Loss (PL) models, which can be configured to meet different requirements. Supported models include Hodgson, Supersnap, MIMIC, Instant, Mocassin and IFD, and the system architecture allows other models to be "easily" integrated. The system allows batches of PL runs to be executed simultaneously: these may be for different sonars, frequencies, deployment depths or any other variation of input factors, assisting 'what if' planning..

NECTA provides a number of facilities which help promote the user's understanding of the ocean environment, including:
(a) profile viewer - this allows single or multiple profiles to be viewed graphically and in tabular form. The profiles can be annotated with layer depths, channels, cut-off frequencies, critical depths and so on
(b) profile cross section viewer - this allows the user to select two arbitrary points on the chart and display a temperature, sound, speed or salinity cross-section. This display can provide a graphical view of ocean fronts, sound channels and layers
(c) 3-D bathymetry viewer - by selecting two arbitrary corner positions on the chart, the user is able to obtain colour-shaded three-dimensional views of the seabed.

NECTA's data management facilities automate the extraction of range-dependent information from the database. The data management facilities are independent of data source, meaning that NECTA can easily be configured to use different sets of data, according to individual requirements, including:
(a) DBDBS for bathymetry
(b) Levitus for ocean climatology
(c) military geo-acoustic data.

These databases have worldwide coverage but are relatively coarse. NECTA also allows higher-resolution grids to be superimposed on the global backdrop, giving a mosaic of data with different resolutions. This is a common requirement where high-resolution littoral information is available.

NECTA is a suite of modules, some of which are provided as core functionality and others which are supplied as options. In the core NECTA system, sonar detection is calculated on the basis of a sonar Figure-Of-Merit (FOM). In an operational environment this gives instant feedback on the effects on detection range of changing FOM.

The system can also support underwater acoustic research or sonar design activities. The optional Sonar and Platform module allows the user to model sonar geometries and processing characteristics and also to model the acoustics of platforms in terms of target strength, broadband radiated noise and narrowband tonals. Where available, measured noise from the sonar enables directional ambient noise data to be fed into the detection calculations.

In addition to the cross-section views of acoustic parameters which the core NECTA facilities provide, there is a tactical overlay facility which enables plan views of the same results to be superimposed on the NECTA chart. The overlays provide:
(a) PL viewed at any depth
(b) probability of detection viewed at any depth
(c) best evasion depths
(d) best detection depths.
This type of display instantly reveals directions in which detection is better or worse than the norm and clearly highlights topographical effects due to trenches, mounds and so on.

The platform for NECTA is Solaris 2.5 (or later) with a minimum specification of 300 Mbytes of free disk space, 32 Mbytes of memory and a 1,024 × 768 colour monitor, or Windows NT.

A variant is produced specifically for MCM. This adds to the baseline a suite of 3-dimensional graphics facilities which provides:
(a) Three-dimensional visualisation of the underwater terrain
(b) Graphical definition and refining of the waypoints in the survey route
(c) Display and storage of the planned sonar coverage for the survey route
(d) Survey databases of known objects
(e) 3-D 'fly-through' display of the UUV operation, using real-time positional and attitudinal data feeds.

Status

In service with the Royal Swedish and Finnish navies. A derivative is in service with the Royal Navy.

Contractor

AMS, Camberley, Surrey.

UPDATED

Weapons Interfacing Equipment (WIE)

The Weapons Interfacing Equipment in a submarine enables the preparation, launching and control of the submarine's tactical weapons. Two such equipments are used in Royal Navy submarines for this purpose - Outfit DCM which is fitted throughout the 'Vanguard', 'Swiftsure' and 'Trafalgar' classes; and the Submarine Weapon Interface Manager (SWIM) which is to be fitted in the 'Astute' Class. Also supplied by Ultra in association with the weapon handling and discharge system are the Launch Control Processor, in the 'Vanguard' Class, and the Tube Operator Console, in the 'Upholder' now 'Victoria' (Canada) class.

Submarine Weapon Interface Manager

The Submarine Weapon Interface Manager (SWIM) for the Royal Navy 'Astute' class submarine programme is a modified and updated version of Outfit DCM (see below). The new equipment is an evolution of the DCM design, providing the same

The Submarine Weapon Interface Manager (SWIM)
(Ultra Electronics)
0533835

safety, and better reliability and functionality, using commercial off-the-shelf electronics. The re-design has produced a more compact solution, enabled upgrade and expansion capacity and provided a system which is cheaper both to procure and to support.

Outfit DCM
Outfit DCM provides monitoring and data feedback in normal and reversionary modes through the interlinking of the command/fire control system to both the weapon discharge system and the weapons themselves: assuring safe operation of the weapons system. It accepts tactical data and firing and control commands from the command/fire control system via data highways and feeds them, as appropriate, to the weapons discharge system and the weapons. It also feeds tactical data back from the weapons for analysis and action by the command system. The system monitors and reports the weapons loaded, their condition and their readiness state. High reliability is assured by the use of parallel paths and interconnections, and the duplication of critical subsystems. The system is capable of supporting torpedoes such as TIGERFISH and SPEARFISH and submarine to surface guided weapons like SUB-HARPOON. It has recently been modified to enable it to interface with the TOMAHAWK land attack missile and its associated weapon control. Further upgrades are planned.

Launch Control Processor
The Launch Control Processor (LCP) is a part of the 'Vanguard' class weapon handling and discharge system (manufactured by Strachan and Henshaw). The LCP is a microprocessor-based system which monitors and controls the torpedo firing sequences from loading to discharge. It also links the discharge instrumentation system with a comprehensive set of self-diagnostics and provides fault tolerant interfaces to each torpedo tube. The system allows remote control and monitoring of the torpedo tubes from the control room, without operator assistance, and has built in fault recognition to speed diagnosis.

Torpedo Operator Console
The Tube Operator Console (TOC), also a part of the Strachan and Henshaw weapon handling and discharge system, provides similar facilities in the 'Upholder' class, now in service as the 'Victoria' class with the RCN. The TOC also has an operator interface and provides the drive profile for the weapon discharge Air Turbine Pump.

Status
(a) Outfit DCM - In service in all operational RN submarines, with minor variants dependent on the class and command system installed.
(b) SWIM - Evolutionary upgrade of DCM to be fitted in the UK 'Astute' class - completing design proving.
(c) LCP - In service in the 'Vanguard' class.
(d) TOC - In service in the 'Upholder'/'Victoria' class

Contractor
Ultra Electronics Command and Control Systems, Loudwater, High Wycombe, Buckinghamshire.

VERIFIED

United States

ArcGIS™

ArcGIS is a family of software products built on a common architecture that form a complete Geographic Information System (GIS). ArcGIS Desktop refers to a suite of three integrated core applications: ArcMap, ArcCatalog, and ArcToolbox. ArcMap is used for all mapping and editing tasks, as well as map-based analysis. ArcCatalog is the application for managing spatial data holdings, for managing database designs, and for recording and viewing metadata. ArcToolbox simplifies many common GIS data conversion and geoprocessing tasks.

Three different packages, ArcView™, ArcEditor™ and ArcInfo™ comprise the basic ArcGIS Desktop products. ArcView is the basic ArcGIS package, providing core mapping and GIS functionality. It provides geographic data visualisation, query, analysis, and integration capabilities together with the ability to create and edit geographic data. ArcEditor includes all the functionality of ArcView and adds the power to edit topologically integrated features in a geodatabase or coverage. Additional functionality includes support for multi-user editing, versioning, custom feature classes, feature-linked annotation, and dimensioning. ArcEditor allows the user to create and edit all ESRI-supported vector data formats. ArcInfo includes all the functionality of ArcView and ArcEditor and adds further advanced geoprocessing capabilities. ArcGIS uses the following standard software components: Visual Basic® for Applications (VBA) for customisation, commercial database management system (DBMS) for data storage, and TCP/IP and HTTP for networks. The geographic data model supports business logic for versioning and intelligent features.

ArcGIS also supports a series of extensions to its functional capabilities:

ArcGIS 3-D Analyst
ArcGIS 3-D Analyst enables visualisation and analysis of surface data. Using 3-D Analyst, the operator can view a surface from multiple viewpoints, query a surface, determine what is visible from a chosen location on a surface, and create a realistic perspective image draping raster and vector data over a surface. ArcGIS 3-D Analyst features include the following:
(a) Build surface models from any data.
(b) Perform interactive perspective viewing, including pan and zoom, rotate, tilt, and fly-through simulations, for presentation and analysis.
(c) Model real-world surface features such as buildings.
(d) Model subsurface features-wells, mines, groundwater, and underground storage facilities.
(e) Generate three-dimensional surfaces on-the-fly from attributes.
(f) Apply data normalisation and exaggeration on-the-fly.
(g) Drape two-dimensional data on surfaces and view in three dimensions.
(h) Calculate surface area, volume, slope, aspect and hillshade.
(i) Generate contours as two-dimensional or three-dimensional shapes.
(j) Perform view-shed and line-of-sight analysis, spot height interpolation, profiling, and steepest path determination.
(k) Use any data supported in ArcGIS including CAD, shapefiles, ArcInfo coverages, and images
(l) Query three-dimensional data based on attribute or location
(m) Export data for display on the Web using VRML.

ArcGlobe was revealed at DSEi in September 2003. This is a database tool that allows the analyst to import information from any available source or database, irrespective of format, into the visualisation framework. It is to be released in February 2004.

ArcGIS Spatial Analyst
ArcGIS Spatial Analyst provides a broad range of powerful spatial modelling and analysis features. The operator can create, query, map, and analyse cell-based

ESRI ArcGIS editing screen showing an overall map view   0059810

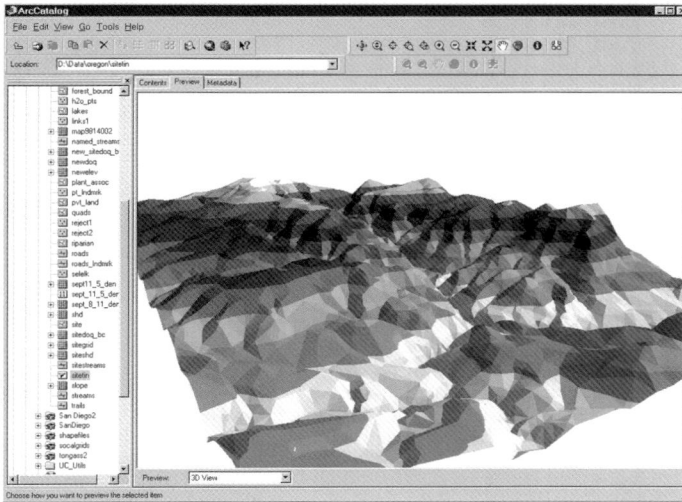

3-D view from ArcCatalog (ESRI)   0120205

raster data; perform integrated raster/vector analysis; derive new information from existing data; query information across multiple data layers; and fully integrate cell-based raster data with traditional vector data sources. ArcGIS Spatial Analyst features include the following:

(a) Convert features (point, line, or polygon) to rasters
(b) Create raster buffers based on distance or proximity from features or rasters
(c) Generate density maps from point features
(d) Create continuous surfaces from point features
(e) Derive contour, slope, view-shed, and aspect maps and hillshades of these surfaces
(f) Perform map algebra-Boolean queries and algebraic calculations
(g) Perform neighborhood and zone analysis
(h) Carry out discrete cell-by-cell analysis
(i) Perform grid classification and display
(j) Use data from standard formats including TIFF, BIL, IMG, USGS DEM, SDTS, DTED, and many others.

ArcGIS Geostatistical Analyst

ArcGIS Geostatistical Analyst provides a powerful suite of tools for spatial data exploration and optimal surface generation using sophisticated statistical methods. Geostatistical Analyst allows the operator to create a surface from data measurements occurring over an area where collecting information for every possible location would be impractical. ArcGIS Geostatistical Analyst features include:

Interpolation techniques
(a) Inverse Distance Weighted (IDW)
(b) Global polynomial
(c) Local polynomial
(d) Radial basis functions include thin plate spline, spline with tension, multiquadratic, inverse multiquadratic, and completely regularised spline*
(e) Kriging * (ordinary, simple, universal, probability, indicator and disjunctive)
(f) Cokriging * (ordinary, simple, universal, probability, indicator and disjunctive)

Spatial analytical tools are available in different combinations for each of the interpolation techniques.
(a) Cross validation and validation
(b) Semivariograms and covariance
(c) Detrending
(d) Declustering
(e) Checking for bivariate normal distributions
(f) Data transformations
(g) Error modeling

Output surfaces
(a) Prediction map
(b) Error of predictions map
(c) Quantile map
(d) Probability map
Other elements of the ArcGIS family include:

ArcSDE

ArcSDE (**S**patial **D**atabase **E**ngine) is the GIS gateway that facilitates managing spatial data in a database management system. ArcSDE allows the operator to manage geographic information in one of four commercial databases: IBM DB2, Informix, Microsoft SQL Server, and Oracle, as well as being able to serve ESRI's file-based data with ArcSDE for Coverages. ArcSDE serves spatial data to the ArcGIS Desktop (ArcView, ArcEditor, and ArcInfo) and through ArcIMS, as well as other applications and it is the key component in managing a multi-user spatial database.

Military Overlay Editor (MOLE)™

MOLE is a symbol generator and editor for military applications. It is used to compose and position unit symbols against a background of geographic data in either raster or vector image formats. It supports a large variety of image formats including CID, JPG, TIFF and CADRG. It can be used with both the Windows and UNIX versions of ArcView GIS.

ArcGIS Military Analyst

ArcGIS Military Analyst is a suite of tools bundled together specifically for the military user. These include Raster and Vector Map tools; a Digital Terrain Elevation Data (DTED) tool; a co-ordinate tool for co-ordinate conversion; 2- and 3-dimensional terrain tools for visibility and threat analysis; and MOLE.

Status

ArcGIS software is in operational use in a wide variety of CCIS all over the world. In July 2002 ESRI was selected as part of a consortium to provide the Commercial Joint Mapping Toolkit, which will be the future standard GIS for US CIS.

Manufacturer

ESRI (Inc), Redlands, California.

UPDATED

CommandVu 4170 display

Tested and field proven in military vehicles, ships and aircraft, the CommandVu 4170 display is extremely flexible, offering multiple sensor inputs and outputs and communication options. Its compact, lightweight design and nominal power dissipation allow its successful integration into any thermal imaging system. It can function as a standard colour computer monitor and offers the ability to add a variety of options including colour video and digital overlays, touchscreen, remote control, and VGA, RGB, SVGA and DVI. It has a 7,500 hour ground-mobile Mean Time Between Failures (MTBF).

The display features 64 true grey shade 8-bit data resolution and is flexible in mounting location. The 10.4 in Active Matrix Liquid Crystal Display has an anti-reflective screen for easy reading in sunlight and at night.

The 4170 has a rugged machined housing and weighs 6.8lbs. The display meets the US military standards for shock, vibration and high impact resistance, providing a high level of crash safety. It also operates over a wide temperature range and is water immersible.

Status

Now being offered by the Ruggedized Command and Control Solutions division of L-3 Communications, which was previously part of Northrop Grumman.

Contractor

L-3 Ruggedized Command and Control Solutions , San Diego, California.

UPDATED

CommandVu 4170 (Northrop Grumman)   0143791

ComputerWall computer-based displays

ComputerWall is a multiscreen display controller designed for real-time, large format graphical presentations. It outputs bright, high-resolution imagery suitable for viewing even under high-ambient light conditions. The system creates wall-sized, computer-based displays and is especially designed to meet the exacting requirements of command centres.

The system displays data, graphics and video on single screens or seamlessly across multiple screens. Data assessment and decision making are enhanced by ComputerWall's display of information from multiple sources, with critical information selected, positioned and scaled.

ComputerWall display

A versatile modular system, it is suitable for both new installations and upgrades of existing facilities. It supports variable size display configurations, video and computer inputs up to 1,280 × 1,024 pixels and a wide variety of projection devices. It connects readily to the RGB analogue outputs of computer workstations. Features include:

(a) Splits computer images into quadrants for display on a 2 × 2 array
(b) Multiple processors can be combined to support larger arrays
(c) Automatically synchronises to input signals from workstations or PCs up to 1,280 × 1,024 pixels
(d) Accepts line doubled NTSC (or PAL) analogue video
(e) Compatible with all software
(f) Advanced interpolation smooths magnified images
(g) Simple hardware installation
(h) Remote control of all functions via RS-232 serial port or front panel push buttons.

Contractor
RGB Spectrum, Alameda, California.

VERIFIED

DatronLINK

DatronLINK is a custom data network management and messaging program. It is designed to automate message and file transfers over radio links, as well as over LANs and through the Internet. It is the successor to Datron's DTS-Gateway.

DatronLINK is completely modular, making customisation for user requirements simple. The program automatically adjusts to the user's Microsoft API-compliant encryption and is compatible with the MIL-STD-IIOA/B HP high-speed modem and the Datron 5300-series of HP modems. DatronLINK is the central control point of a radio-messaging network. It is Windows-based and features automatic station linking (using FED-STD 1 045A for the best channel selection) and error-free communications (ARQ transmission protocol). It allows either point-to-point or station-to-multistation data exchanges, and has high-level adaptive data compression for faster throughput. Once DatronLINK is configured, the exchanges are automatic and require little to no user intervention. Automatic message relaying and alternate path selection are transparent to the user. DatronLINK's mailbox feature provides automatic message forwarding and individual user routing.

DatronLINK runs on any computer that meets the minimum system requirements, and transparently allows access to the user's Internet, LAN, and e-mail services. It is designed to work with the Datron 7000-series or PRCIO99A HP radios, providing system configuration and automatic radio control. Variants of DatronLINK are also offered to satisfy particular user network requirements. These include DatronLINK-Mail, DatronLINK-Light, and DatronLINK-P.

DatronLINK Network Applications.
Some situational data network applications using DatronLINK include the following:

(a) Chat Mode. This allows communication between two users via computer. Once chat mode is initiated, either user enters text by typing on the keyboard. When the user reaches 60 characters or presses the enter key, the data is transmitted over-the-air via the radio.
(b) Relay Mode. Fully automatic message relaying is accomplished by creating alternate paths between stations in a network. Each station sets its own relaying paths. For example, a user at station 1 can set up a path to send the message to station 2 via station 3. Once the alternate relaying paths are defined, no user intervention is required. If conditions prohibit a message from getting through using a direct path, DatronLINK automatically selects an alternate path.
(c) Link Recovery. Automatically resumes interrupted file and message transfers from the point of interruption. Global Relay. Provides for fully automatic message relaying with alternate path selection on a global scale.
(d) LAN Connectivity. DatronLINK allows transparent communications between users on a separate LAN via a radio link. Integration of DatronLINK into a LAN is done with automatic routing through the LAN/WAN. The program supports

Microsoft Exchange, Novell MHS, Lotus Note, and cc:Mail networks and servers.
(e) Multiple Mode. DatronLINK allows seamless integration of HF transmissions with satellite and landline paths (Datron-P). A primary HF path is established between radio stations with handoffs at the end station to a landline or satellite path. If conditions prohibit a message from getting through using a direct path, DatronLINK automatically selects an alternate path.

Specifications
(a) Windows-based, data network programming and messaging program with automatic station linking (ALE) and error-free transmissions (ARQ).
(b) Versions:
DatronLINK: Full-featured package.
DatronLINK- Mail: A network client, operating only to or through a DatronLINK station.
DatronLINK-Light: Point-to-point (radio) only, operating only with each other or a DatronLINK hub.
DatronLINK-P: Operates with VHF/UHF stations in a packet mode.
(c) Data Transfer: Text, binary, and image files.
(d) Data Sources: Keyboard-to-keyboard, radio, e-mail, and fax/imaging.
(e) Data Exchanges: Point-to-point, station-to-multistation, and high-level data compression.
(f) Encryption: Microsoft API-compliant data encryption.
(g) Mailbox: Automatic message forwarding and individual user routing.
(h) ALE: FED-STD-1045A for best channel selection.
(i) Chat and Broadcast Modes: Simultaneous transmission of data to multiple stations.
(j) Radio Control: 7000-series HF transceivers and PRC1099A.
(k) Modems: 5300-series (TW5300 or RT5300), MIL-STD-110A/B (DT110A or E110-series), integrated systems (DT5300 or DT110A).
(l) Computer Requirements:
PC-compatible. Pentium-class. 500 MHz or better, 128 Mbyte RAM. 10 Gbyte HDD, minimum
1 COM port, minimum
Dual, Type-1 PC card (PCMCIA interface for modem and hardware protection key) for E110 operation only
Windows 98, 98SE, NT4, 2000, or XP.

Status
In use in Algeria, Brazil, Morocco, Taiwan, the United States and at least 10 other unspecified countries.

Contractor
Titan Systems Corporation, Vista, California.

VERIFIED

Electronic Warfare Command Station (EWCS)

The Electronic Warfare Command Station (EWCS) is a networked UNIX-C and Java system for naval surface, air and shore-based platforms which accepts stream inputs from passive electromagnetic intercept sensors and fuses this data into a track database available for geographic and parametric displays. Analysis tools are included to identify emitters and perform platform recognition, as well as to determine platform kinematics and threat order of battle. An online, event-by-event, transaction processing environment between EWCS systems on a collection of platforms is supported to produce an integrated battle group electronics warfare capability. Organic track data from this system is exportable to other principal tracking systems.

EWCS allows operators to:
(a) View, edit and manage EW tracks on the EWCS display
(b) Identify the source and threat of an EW track by color and symbol
(c) Enter user-defined EW data directly into the EW track database
(d) Submit EW tracks to the UB TDBM and the ELINT Correlator
(e) Share data among other EWCS platforms in a network using an EWCS Broadcast
(f) Cross-fix analysis on EW tracks from different sources
(g) ELINT Parameter Limits List Queries on EW track data
(h) Manually enter EW intercept reports based on EW data from non-EWCS sources
(i) Tactically collaborate with sanitised data received from sensitive sources

Contractor
Northrop Grumman Information Technology, Reston, Virginia.

UPDATED

EMPSKD unit scheduling program

The process of scheduling US Navy ships, air wings and crews/staffs to meet national commitments is a significant task for fleet schedulers. The Employment Scheduling (EMPSKD) program has been developed to assist fleet schedulers in the preparation and evaluation of plans for present and future naval activities around the world. EMPSKD is PC-based and runs on Microsoft Windows™ workstations. The GCCS-M version of unit scheduling, called NAVSKD, retains all the features inherent in EMPSKD, and features enhancements to support ashore command centre requirements.

Status

Believed to be still in use in the US Navy.

Contractor

Northrop Grumman Information Technology, Reston, Virginia.

UPDATED

FalconView

FalconView for Windows™ is a multimedia mapping application for the PC that displays various types of maps and geographically referenced overlays. It was developed to support flight planning for both the US Air Force and the US Special Operations Command. FalconView's user-friendly interface and its ability to run on a laptop allow pilots to conduct flight planning almost anywhere.

FalconView's moving map display capability coupled with its ability to receive and plot a Global Positioning System (GPS) feed in real time also allow it to be used for tracking vehicles. These GPS trails can be saved into files, displayed and played back for after-action reviews.

FalconView currently permits pilots to develop flight plans by retrieving and displaying the appropriate set of maps on their computer screen. Using the mouse, the pilots can develop the flight plan by pointing and clicking on terrain features along the way. Significant points along the path can be identified or marked using a standard symbol set included in the software. FalconView provides situational awareness through overlays that can track threat locations and provide geospatial visualisation analysis. The resulting flight plans (mission) can be rehearsed in 2-D or 3-D.

The FalconView system is Windows 98 and Windows 2000-based. Map data is provided on CD-ROMs generated by the National Imagery and Mapping Agency (NIMA) or via network connections. Supported data types include CADRG, CIB, GeoTIFF, DTED (1 and 2), V MAP, WOD, WDB II, DNC and Shape Files. The software is expandable by third parties through a rich set of programmer's APIs and a Software Developer's Kit.

For FalconView 3.1.2 or higher SkyView™ is an add-on application that provides a 3-D perspective view of an area, including any overlays opened in FalconView. A special version of FalconView (FalconView for NIMA) was developed for use by US government personnel who fall outside the normal, mission-planning distribution chain. It does not support route planning.

Status

FalconView is widely used as an element of US DoD mission planning systems, notably the Joint Mission Planning System (see separate entry) and the Portable Flight Planning System (PFPS), as well as other federal agencies such as the National Imagery and Mapping Agency, the US Customs Service, and the Federal Aviation Administration. A number of European nations, believed to include the UK and Belgium, are also using an export version of FalconView.

Contractor

Georgia Tech Research Institute, Atlanta, Georgia.

NEW ENTRY

INFOSCENE™

The INFOSCENE™ Situational Awareness Toolkit is a tactical display software package that generates interactive two-dimensional and three-dimensional representations of combat situations. By providing multiple viewpoints, including the Tactical 3D (Tact3D) view, INFOSCENE™ allows users to see the most comprehensive view of a tactical scenario. It also provides the key elements of a tactical information system including a master track panel, track hooking and highlighting, close-control panels, filtering, and multiple icon sets. INFOSCENE™ displays are fully interactive, providing a large number of display manipulations and functions. The Tactical Display Kit is a suite of software tools which provide considerable extensibility. Using the Tactical Display Kit, the end-user can customise and build upon the core product to meet specific tactical display needs.

INFOSCENE™ features a set of powerful tools that can improve the way the user views the battlefield:
(a) The Plan Position Indicator (PPI) view allows the user to hook, pan, zoom, tether, and manipulate the icon set. A 'hooked' or selected track is the only track that can be attacked. Panning and zooming allow the user to get closer and have more detailed views of a specific battlefield area. A tethering feature allows the display to follow a specific track as it moves through the battlefield, simulating a specific track's point of view throughout the combat situation. An adaptable icon set is provided and the use of colour coding allows for clear definition between friendly and hostile tracks.
(b) The Tactical 3D (Tact 3D) Indicator display not only allows the user to see the battlefield from above just like the PPI display, but also from any point in three-

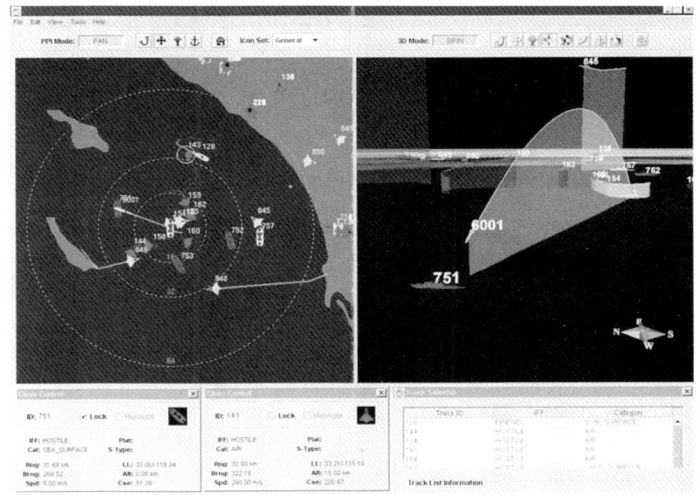

Screenshot from the INFOSCENE™ tactical situation display system with multiple display formats, with 3-D tactical display (R) and a plan position indicator display showing live radar video (L) (Lockheed Martin) **NEW**/0521411

dimensional space. In addition to the PPI display's tools, the Tact 3D display also has the ability to spin, sweep, fly, set the reference point, and spin the earth. The spinning function allows the user to change the display's orientation at a set altitude. The sweeping function changes the view's orientation by moving its vantage point through space. The flying function is a special feature of the Tactical 3D Indicator that displays the simulation through a simulated cockpit windshield. The set reference point function allows the user to change the reference point that the Tact 3D display is centred upon. Spinning the earth is a three-dimensional development of the panning function.
(c) The Range-Height Indicator (RHI) display indicates the altitude of a track as well as its distance from the tracking radar. The aspect of 'track histories' adds an element of time to the graph. In addition to the RHI's unique features it also includes every function of the Plan Position Indicator.

The filter control function allows the user to select which tracks are displayed. The filter can remove individual tracks or groups of tracks based on their altitude, bearing, range, and/or based on their relationship to the 'own ship'.

The track selector window lists all of the tracks input into the INFOSCENE™ application. From within this list, the user can hook a track, hide a track, highlight a track, make a track visible (even overriding the filter), or bring a track to the close control screen. The close control screen is where the user can read specific information regarding a track. The close control function can only be accessed on a track if it is hooked.

The system has been designed to provide a considerable level of customisation. It contains two pieces; a precompiled low-level library (Display Kit); and the display applications, (including all the source code) that are built using components from the library. All of the common functionality is encapsulated in the Display Kit, and only code specific to a particular display is found inside the main application. Although they are fully functional displays and can be used as provided, INFOSCENE™ displays are also intended to be examples of how to implement tactical displays. Because each display comes complete with source code showing exactly how it was put together, it can easily be customised or even rewritten using the sample code as a template, to meet the changing needs of a customer.

Written entirely in Java, the INFOSCENE™ display system will be able to run on any platform that includes the Java Virtual Machine. This includes all versions of Windows, as well as Solaris, Linux, and IRIX. The user interface is provided using Java/Swing, and can easily be modified either by hand or using any commercial Java development package. Because all of the buttons and indicators are software constructed, the application can easily be changed to support new features or an entirely new platform. The system can receive track data from TCP and UDP sockets, Java RMI, or DIS (Distributed Interactive Simulation) broadcasts, and includes a track server engine. It can be purchased as an independent program, or it can be ordered pre-installed and configured.

Contractor

Lockheed Martin, Maritime Systems & Sensors, Moorestown, New Jersey.

UPDATED

InSIGHT™ Advanced Tactical Displays

The InSIGHT™ Advanced Tactical Display (ATD) provides situational awareness with touchscreen user input within a highly ruggedised enclosure. The ATD Thin Client adds computing capabilities within the same thin chassis as the flat panel display and is also capable of functioning as a stand-alone, rugged PC workstation. The ATD Ultra Thin simplifies the client computer to a network appliance and allows more powerful system servers to handle processing functions. Built with Sun Ray™ technology from Sun Microsystems™, the ATD Ultra Thin Client provides simultaneous access to multiple operating environments from a single client node. ATD Thin Clients and Ultra Thin Clients can accommodate copper or fibre optic network systems.

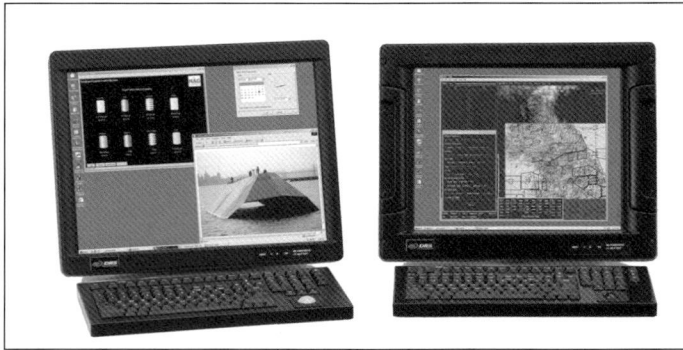

The 21 in Advanced Tactical Displays (DRS Technologies)   0134117

Features

(a) Total system life-cycle cost savings through reduced system administration
(b) Intel® Pentium®III and Sun UltraSPARCTMIII processor
(c) Sun RayTM Integrated Solution enables single access to multiple operating system environments
(d) Windows® XP, 2000, NT, 98, 95, LINUX®, SOLARIS™
(e) Video loop-through connectors
(f) High resolution AMLCD with on-screen display adjustment and retractable handles
(g) Advanced network technology with fully sealed/passive cooling for rugged military environments
(h) ISO-7816 smart card option
(i) Full mounting solutions for bulkhead, rackmount and desktop mounting.

Specifications

Computer
(a) CPU: Intel® Pentium®III
(b) Max RAM: 1 Gbyte
(c) Hard drive: Up to 30 Gbytes min EIDE HDD
(d) Flash drive: Up to 1Gbyte flash
(e) Graphic: Up to 64 Mbytes shared video memory
(f) Bus: ISA, PC/104, PC/104+ and PCI
(g) Ethernet: 10/100 Base-T (dual available)
(h) I/O: PS/2 keyboard/trackball, IDE, parallel, USB, RS-232/-422/-485.

Physical

Model	Height	Width	Depth	Weight
ATD18	16.2 in	17.5 in	3.8 in	25 lb
ATD21	17.5 in	19.7 in	3.8 in	30 lb
ATD18TC/UTC	16.2 in	17.5 in	3.8 in	32/30 lb
ATD21TC/UTC	17.5 in	19.7 in	3.8 in	36/34 lb

Electrical
(a) AC 110/120 V AC (via MIL-C-38999)
(b) DC 20-32 V DC (via MIL-C-38999).

Environmental
(a) Shock: MIL-S-901D Grade
(b) Vibration: MIL-STD-810E / MIL-STD-167
(c) EMI/EMC: MIL-STD-461E
(d) Operating temperature (MIL-STD-810E): 0 to +5C°C
(e) Storage temperature: −18 to +65°C
(f) Humidity: 95% RH condensing
(g) Operating altitude: Sea level to 10,000 ft
(h) Storage altitude: Sea level to 40,000 ft.

Status
Selected for the US Navy Combat DF programme. Incorporated in the DRS SDE-21 workstations in the Royal Australian Navy FFG-7 frigates.

Contractor
DRS Technologies(Electronic Systems Division), Gaithersburg, Maryland.

UPDATED

ISR Warrior

ISR Warrior is an intelligence collection management tool developed from a perceived need by commanders in Kosovo for an intelligence collection picture, to show all collection assets and their activities. The Warrior is a web-based decision system providing a consolidated picture of the collection battlespace. It provides a current picture of ISR assets overlaid with orders of battle, collection plans and planned targets together with tipoff information from SIGINT, MASINT and MTI sources. Also displayed are threat areas and overlays. It therefore assists in making the most effective use of ISR assets and in time-critical and time-sensitive targeting, permitting real-time ad-hoc retasking.

The system is constructed with an open architecture and is expandable, scalable and tailorable to user needs.

Status
A private venture development by Raytheon, the system has been tested on exercise in Korea. In July 2003 it formed part of a demonstration of an Enterprise Expeditionary Strike Warfare Architecture (eESWA), an integrated, open architecture for ISR, time-critical targeting and strike mission planning and execution.

Contractor
Raytheon Company, Garland, Texas.

UPDATED

MSS II and MSS II+ mission support systems

MSS II and MSS II+ are automated mission planning systems developed by Smiths Aerospace to reduce the workload required to plan tactical aircraft missions. MSS II+ is an enhanced performance version of the US Air Force standard MSS II.

MSS II eliminates the need for paper charts and manual calculations of time, distance and fuel during the mission planning process. Instead of paper charts, MSS II uses digitised charts and satellite photographs as well as Digital Terrain Elevation Data (DTED) stored on removable digital optical disks and displayed on a high-resolution colour CRT. Since all of these databases are geographically correlated, the mission planner can select any scale chart (from 1:2 million to 1:250,000) or photograph (Landsat, SPOT or other resource) and display the latitude, longitude and elevation of any point by moving the cursor with the trackerball and pushing a button.

The FLOT, targets, air defence sites, aircraft ground track and way points are displayed as overlays on the digitised chart or photograph backgrounds. When the planning process is complete, an annotated colour combat mission folder can be automatically generated on an optional high-resolution colour plotter. In addition, the MSS II automatically loads aircraft avionics initialisation data into the Data Transfer Cartridge or Data Transfer Module.

The MSS II and MSS II+ hardware is based on the Mission Analysis and Planning System (MAPS) Model 300. For MSS II this comprises:
(a) MicroVAX II computer
(b) 800 Mbyte removable optical disk
(c) 380 Mbyte removable Winchester disk
(d) TK-50 streaming tape drive
(e) IEEE 488 parallel interface
(f) VT-32 data terminal
(g) trackerball with three buttons
(h) 1,280 × 1,024 × 9 bit/pixel colour graphics display.

In addition, a DC 600 streaming tape drive, dual 5.25 in floppy disk drives, 150 Mbyte removable Winchester disk and a laser printer are also provided. An

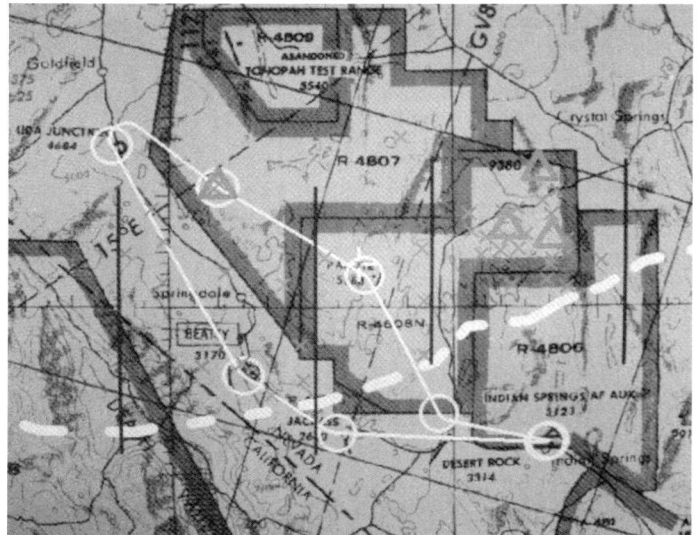

MSS II threat route map

MSS II perspective view

MSS II hardware

interface unit loads data to and retrieves from the Fairchild Data Transfer Cartridge (DTC) and a Smiths Industries Data Transfer Module (DTM). The TEMPEST-proofed system is supplied in two- or four-man-portable, ruggedised enclosures.

MSS II+ MAPS hardware comprises:

(a) MicroVAX II computer
(b) 600 Mbyte removable erasable optical disks
(c) 800 Mbyte WORM optical disk
(d) 760 Mbyte removable Winchester disk
(e) TK-70 streaming tape drive
(f) IEEE 488 parallel interface
(g) VT-320 Data Terminal
(h) trackerball with three buttons
(i) 1,280 × 1,024 × 8 bit/pixel colour graphics display.

A second MSS II+ module, the power distribution/printer enclosure, contains a monochrome photographic quality printer for perspective views and radar predictions, a high-speed high-resolution colour plotter for combat mission folders and a power transformer for 115/230 V AC operation. Both of the ruggedised modules are four-man-portable.

MSS II Aeromap software includes:

(a) host computer operating system (UNIX)
(b) MAPS 300 operating system (MicroVMS)
(c) digital image database management including image retrieval, display, pan and zoom
(d) database correlation with digitised aeronautical charts, digitised satellite photographs, DTED and user-defined overlays
(e) automatic chart updates (CHUM)
(f) DTC/DTM interface driver
(g) built-in test
(h) real beam radar predictions
(i) perspective photographic views
(j) countermeasures employment cueing
(k) threat line of sight overlays
(l) mission planning functions of route editing, route optimisation and automated form 70/691.

Compatible with TAF IMPS software, MSS II is also capable of executing most UNIX and VMS user programs.

MSS II+ software provides the same functions as that for MSS II. However, it also has a terrain-coupled flight path generator and gives automatic generation of annotated colour combat mission folders. Real-time fly-through is also possible with MSS II+.

Status

MSS II is in service with the US Air Force. In 1996 a US$7.5 million contract was awarded for the supply of 16 systems to the Egyptian Air Force (EAF). In July 2003 *Jane's Defence Weekly* reported that a further contract with an undisclosed value had been awarded by the Egyptian Air Force under the US FMS programme for the provision of MSS II to support the EAF AH-64 fleet, with deliveries due in 2004-5.

Manufacturer

Smiths Aerospace, Electronic Systems Division, Germantown, Maryland.

UPDATED

MultiMap

The Smiths Aerospace MultiMap is a software package which turns a computer graphics workstation into a powerful mapping system. Under the direction of the host application software, MultiMap interfaces with geographic databases and application-specific overlay databases to provide comprehensive mapping capabilities to the workstation user.

Smiths Aerospace MultiMap

The backgrounds on the workstation's display are retrieved from the geographic databases, which can include high-resolution, full-colour raster scan reproductions of paper maps, imagery and terrain elevation data. Overlays on the background vary widely among applications. In a battle management application, overlays can represent friendly and enemy forces. In other applications, overlays can represent communication networks, obstacles around airfields and so on.

The databases are all digital and can be stored on any digital media such as magnetic tape, magnetic disk, WORM optical disk and erasable magneto-optical disk. New data can be produced quickly and added to an existing database. Since the databases are all digital, they can be readily shared by sending them across local area and long-haul networks, or by copying them to fixed and removable digital media.

MultiMap is designed to integrate easily with applications-specific software and calls to MultiMap can be in C or Fortran.

The following databases are supported:

(a) paper maps and charts, such as aeronautical charts provided by the Defense Mapping Agency (DMA), National Oceanic and Atmospheric Administration (NOAA) and the US Geographical Survey. Paper maps are raster scanned and stored as pixels ready for display. Multiple map scales are typically used;
(b) ARC Digital Raster Graphics (ADRG) data provided by DMA. This DMA product consists of DMA maps which have been raster scanned and put into the ARC projection. It includes multiple map scales;
(c) digital imagery, from Landsat and SPOT satellites;
(d) film and paper imagery, which can be raster scanned like paper maps;
(e) Digital Terrain Elevation Data (DTED) provided by DMA;
(f) coloured representations of DTED, ready for display;
(g) user-defined overlays consisting of symbols, lines, circles, arcs, ellipses, polygons and text.

MultiMap provides the following functions:

(a) geographic backgrounds; fast retrieval; zoom; recentre;
(b) overlays; georeferenced to the backgrounds; decluttering;
(c) windows; for improved man/machine interface;
(d) cartographic computations; co-ordinates, elevation, range, bearing, co-ordinate transformation;
(e) special functions; line of sight; perspective views; maximum elevation in a polygon; terrain profile;
(f) utilities; geographic database management across network.

MultiMap can be installed on Sun 3, Sun 386i, Sun SPARCStation 1 and VAXstation workstations.

Contractor

Smiths Aerospace, Electronic Systems Division, Germantown, Maryland.

VERIFIED

Naval Integrated Tactical Environmental Subsystem (NITES)

As a primary gateway for environmental products to reach the C4I network, the Naval Integrated Tactical Environmental Subsystem (NITES) collects, displays and disseminates environmental data to navy oceanographers and meteorologists. It also provides real-time weather imagery for mission planners and tacticians throughout ashore and afloat command environments.

NITES offers the tactical user high-performance decision aids to provide surface search radar detection ranges and receives geographical and topographical information that can be used to build contours at various levels. NITES is also known as the Joint Meteorological Segment.

Status

In use in the US Navy.

Contractor

Northrop Grumman Information Technology, Reston, Virginia.

UPDATED

Naval JSTARS Interface (JTI)

JTI gives the operator the capability to view and evaluate Moving Target Indicator (MTI) data (see JSTARS entry) using a GCCS-M (see separate entry) workstation. Using a software only solution, JTI can send MTI raw data to any COE-compliant system used by maritime forces. In the COE, JTI features provide MTI visualisation, MTI replay, track creation, track replay, and track export to the Tactical Database (TDBM).

Contractor

Northrop Grumman Information Technology, Reston, Virginia.

VERIFIED

Naval Modular Automated Communications System II (NAVMACS II)

The Naval Modular Automated Communications System II was developed to replace existing outdated NAVMACS (V) Message Processing Systems onboard US Navy ships. NAVMACS II is designed to provide automatic electronic communications services to multiple users, and is supported by software that performs the communications processing required by all connected systems. This requirement includes providing a user interface, and allows users to perform a variety of tasks, based on security clearance levels, authorisations and needs. The system processes and archives all formatted messages required for maritime operations, including OTX Gold and other Link messages.

The purpose of NAVMACS II is to receive, process, store, distribute and transmit a site's internal and external messages automatically. To achieve this, NAVMACS II provides interfaces to multiple external systems of the Naval Telecommunications System (NTS), including landline and Radio Frequency (RF) communication circuits. This is accomplished by use of the NAVMACS Communication Controller (NCC) which acts as the Front-End Processor (FEP), providing up to 16 front-end off ship interfaces. The NCC provides the serial communications channels needed to interface with crypto devices and radios. Additionally, the system provides interfaces to local systems within a site's communications network.

Basic tasks include reading and sending messages to other user sites. Advanced tasks include configuring system databases, and performing system administration functions. NAVMACS II provides a capability to convert military message formats to the standard e-mail format, for delivery via the Simple Mail Transfer Protocol (SMTP) to local e-mail servers This approach accomplishes direct delivery of message traffic to users who have functional e-mail accounts.

The system architecture consists of two basic hardware configurations hosted on the Tactical Advanced Computer-3 (TAC-3), a Hewlett-Packard 700 series computer. An additional derivative of this architecture employs commercial off-the-shelf (COTS) Personal Computers in place of the TAC-3. All three configurations employ common operating systems and application software.

The AN/SYQ-7A configuration is intended for use on large ships such as CV/CVN, LCC, AGF and LHD platforms that process large daily quantities of messages. The AN/SYQ-7B HP and PC variants are intended for use on smaller ships such as FFG, AOE and AOR platforms which process smaller quantities of messages.

All NAVMACS II systems delivered to date use Uninterruptible Power Supplies (UPS), to provide standby AC power in case of shipboard power loss. The UPS is an important system component because it allows NAVMACS II operators to shutdown the system gracefully, which forces open files of the operating system and running processes to be closed properly. An unexpected system shutdown can lead to data loss and operating system corruption.

AN/SYQ-7A

The AN/SYQ-7A configuration uses the Hewlett-Packard (HP) 750/755 computers as Main Communication Centre processors, and National Computing Device (NCD) X-Terminals as user workstations. In addition, the HP-715/730 computers are used as additional workstations/fileservers acting as the gateway to other shipboard networks. This configuration is fully redundant, employing two MAINCOMM processors and two NAVMACS Communication Controllers (NCC). One system is online, while the second system maintains a backup role should the online system fail. Two internal 2.0 Gbyte hard disk drives and one 2.0 Gbyte removable hard disk drive per computer provide online storage.

AN/SYQ-7B (HP Variant)

The AN/SYQ-7B configuration uses the Hewlett-Packard (HP) 715/730 computers as Main Communication Centre processors. Workstations consist of a HP monitor, keyboard and trackball. The AN/SYQ-7B configuration does not include any additional fileserver/workstations; thus, the interface to other shipboard networks is achieved directly from the Main Communications machine. This configuration is not fully redundant. One processor operates in the MAINCOMM mode, while the other acts as a slave or X-Terminal device. The X-Terminal device has complete functional access to system capabilities. Modes of operation are interchangeable per processor. This configuration includes one NCC, with an 8-channel circuit capability. Online storage is achieved by a shared 10 Gbyte Redundant Array of Independent Disks (RAID) device.

AN/SYQ-7B (PC Variant)

The AN/SYQ-7B (PC Variant) configuration is identical in functionality and operation to the AN/SYQ-7B HP Variant configuration. The PC Variant uses Compaq 1850R Personal Computers for the Main Communication Centre processors.

Status

The first shipboard demonstration of NAVMACS II occurred onboard the USS *America* (CV-66) in January 1992. A second demonstration was completed onboard the USS *John F Kennedy* (CV-67) in March 1993. A third shipboard demonstration, using the TAC-3 computer was completed onboard the USS *America* in August 1993. Developmental Test and Evaluation (DT&E) was conducted onboard the USS *Mount Whitney* (JCC/LCC-20) in February 1995. Operational Evaluation (OPEVAL) was accomplished onboard the USS *Mount Whitney* in September 1995. The first production installation of NAVMACS II was completed onboard the USS *Kitty Hawk* (CV-63) in May 1995. To date, there are over 125 installations of NAVMACS II on US Navy ships. The system is also used in ashore command centres. The current fleet release of software for NAVMACS II HP configurations is 2.1.13.2. The current fleet release for PC configurations is 2.2.2.

A modified version of NAVMACS II, called CMP, has been selected as the Air Force Automated Message Processing Exchange (AFAMPE) system for US Air Force Space Command.

Contractor

Northrop Grumman Information Technologies, Middletown, Rhode Island.

UPDATED

RF-3700H Harris Universal Image Transmission Software (HUITS)

HUITS is a Microsoft Windows 2000/XP compatible software package that integrates image acquisition, manipulation and transmission capability into one application, which is designed for the transmission of high-resolution digital imagery, motion video clips, text and other data over difficult tactical radio communications. A Type II PCMCIA or PCI communications card is included to provide a synchronous digital interface between the computer and the communications equipment.

HUITS supports both analogue and digital camera input and uses image compression and manipulation techniques. Harris' SARQ™ data protocol allows the use of a variety of links, including HF, VHF, UHF, Inmarsat and TACSAT at data rates up to 64kbps. For interoperability purposes, the National Imagery Transmission Format (NITF) and TACO 2 communications protocol are also available. Other features include: the progressive display, which allows the image to be viewed as it is being received; the ability to run on most Windows-based computer platforms; image processing functionality and support for adding annotations, symbol overlays and text file attachments; Harris Prioritised Image Transmission (PIT) for maximising transfer throughput; image compression techniques, including Wavelet; translation between standard image file formats (JPG,BMP, GIF, TIF, CGM, TGA, WVL); a "networking" capability which allows multiple users to use the same communications channel; "burst mode" to limit the amount on-air time; and compatibility with a variety of US and international cryptographic devices.

Status

By the end of 2001 Harris had supplied over 1,500 copies of HUITS to over 20 different countries. These include an unidentified Partnership for Peace nation in Europe (contract announced in mid-1998), a reported order by Polish customers, and New Zealand (in June 2001 Harris announced a deal to supply HUITS to the Ministry of Defence to augment the capabilities of FALCON II radios being delivered under a US$19 million contract).

Manufacturer

Harris Corporation, RF Communications Division, Rochester, New York.

UPDATED

Example of image transmitted using HUITS

0008859

RGB/Videolink video scan converters

The RGB/Videolink series of video scan converters transform high scan rate images from a computer, FLIR sensors or radar to broadcast standard video (NTSC or PAL). The RGB/Videolink is targeted at intelligence, simulation and training applications requiring the recording of a wide variety of computer and other images to video tape. In addition, it is used to overlay video with computer-generated graphics.

The RGB/Videolink allows recording on any video tape recorder and connection to video projectors, teleconferencing systems and composite monitors. The system synchronises to all displays with horizontal scan rates from 20 to 90 kHz, including PCs, workstations from Sun, DEC, HP, IBM, Silicon Graphics and others, FLIR sensors and radar. Both interlaced and non-interlaced signals are accepted. Synchronisation and set up are completely automatic.

All models offer full 24-bit colour processing and real-time operation. Third-generation digital signal processing circuitry eliminates interlace flicker so that horizontal lines appear stable in the output image. The system can map any number of input lines to any number of output lines, allowing all images to be mapped to their ideal screen resolution. For example, 350-line EGA displays can be converted to the full 486 visible lines for NTSC or 575 visible lines for PAL. High-resolution 1,024-line displays can be mapped to NTSC or PAL screens without losing any part of the image. Later models include RS-232 control, a built-in keyer to overlay computer-generated graphics on real-time video and a 31.5 kHz output for projectors and other display equipment accepting higher-than-video rate signals.

RGB/Videolink is available in stand-alone models, with their own enclosure and power supply, or as board level units for integration in VMEbus chassis.

Status

RGB/Videolink is in service with the US Army, Navy, Air Force and Coast Guard as well as the armed forces of Australia, France and Singapore.

Contractor

RGB Spectrum, Alameda, California.

VERIFIED

RGB/Videolink Model 1700 D1   0001030

RGB/View multi-input displays

The RGB/View multi-input processors combine multiple real-time video and computer inputs on a single high-resolution projector or monitor. Inputs may be any combination of NTSC, PAL, S-Video, FLIR or RGB up to 1,600 × 1,200. All image windows can be positioned, scaled from icon size to full screen, overlaid with computer graphics, or overlapped with other windows. Also, each input can be panned and zoomed to emphasise areas of particular interest.. Both standard or task-specific configurations are available. The system is designed for C4I, surveillance, training and simulation applications. There are a variety of systems available:

(a) SuperView 3000 (multi-input, multiple image). This is an advanced display processor that combines up to 12 computer screens and/or video signals on a single monitor or projector.
(b) DualView™ (split screen). This provides a split screen side-by-side display mode (claimed as unique by RGB) for viewing a video and a computer signal, or two computer signals, which can be presented in correct aspect ratio or fit to screen. A picture-in-picture capability is also available.
(c) 4View™ (4 inputs, video quad). The 4View displays four video inputs at full resolution in quads on a 1,280 × 1,024 display. It is compatible with monitors, flat panels, or projection screens, 4:3 or 16:9 aspect ratio.
(d) QuadView® (4 inputs, computer/video quad). The QuadView display processor will accept four computer and video inputs and display them in quadrants or full screen on a monitor or projection screen.
(e) QuadView plus® (Up to 4 input channels; up to 12 switched inputs).The QuadView plus® allows up to four signal sources to be displayed on a single screen. Each input channel offers a choice of high-resolution RGB, S-Video, component or composite video.
(f) SuperView™ 500 (multi-input, multiple image display, real-time applications). The SuperView 500 controller displays up to ten real-time video windows on a

SuperView™ 500 screenshot (RGB)   0101269

high-resolution computer monitor. Inputs may include NTSC (or PAL), S-Video, FLIR, RGB to 1,280 × 1,024. The system was developed for applications requiring the simultaneous display of high-quality video and computer-generated images, for information-critical installations including control rooms, command centres, Tactical Operations Centers (TOCs), simulation systems, and remote surveillance.
(g) RGB/View® 6000 (multi-input, multiple image display, 6U VME format). This has the same features as the SuperView 500 in a 6U VME format.

Status

RGB/View systems are in service with the US Army, Navy, Air Force and Coast Guard as well as the armed forces of Brazil, France and Singapore. In October 2002 it was announced that the systems were being widely used in the Tactical Command Post of III(US) Corps, the first digitised corps.

In January 2003 it was announced that RGB/View 6000 processors were to be used in the dual-display tracking consoles for the South African Navy's MEKO A200S corvettes which will enter service in 2004.

In April 2003 it was announced that the SuperView 3000 had been selected for use in Northrop Grumman's Cyber Warfare Integration Network concept.

Contractor

RGB Spectrum, Alameda, California.

UPDATED

RSC1100 high-performance radar scan converter

The RSC1100 is a high-resolution radar scan converter for workstation based radar operator consoles, based around a 6U board set. It receives radar video, trigger and azimuth signals and processes these to generate a high-resolution raster scan image. Designed to interface with virtually any radar unit, the RSC range provides outstanding radar image conversion for the military or commercial user.

RSC 1100 is a compact, single VME card scan converter which gives a maximum radar resolution of 1Kx1K The radar is scan converted and stored in a dedicated frame buffer. Fade circuits provide smooth decay through 256 levels of intensity to emulate the long persistence phosphors typically found on analogue PPI displays. Both time based and scan based fade modes are supported. Proprietary pixel fill circuits eliminate unwanted spoking and moire patterns. A programmable input look-up table is provided as standard.

Display Reference is available as an option to support ground-referenced displays and True North ship's heading display by processing gyrocompass data. By using RSC 1100 at the heart of a modern high-resolution radar display system, real-time radar display updates are possible without sacrificing graphics performance.

An additional Video Mixer Board can overlay radar images at 1,024 × 1,024 resolution with graphics from any high-resolution source up to 1,600 × 1,200 resolution such as SUN, SGI, PC or other workstations. This approach minimises system integration time, presents no load to the host processor or graphics system and provides a completely platform-independent solution. Overlay, underlay and windowing functions are fully supported. Alternatively, the RSC range can interface directly with Barco-Chromatics IVS4000 series graphics engines.

The RSC range is controlled by an embedded 32-bit microcontroller, which performs all real-time computations required for the scan conversion process. This approach minimises loading on the host VME bus. All embedded software and firmware is stored in flash memory to facilitate upgrades in the field. Control parameters are stored in non-volatile RAM to facilitate automatic configuration upon power-up. When used in a UNIX environment, control is possible either from the VME host using the software control library, or remotely via the RS-232 interface.

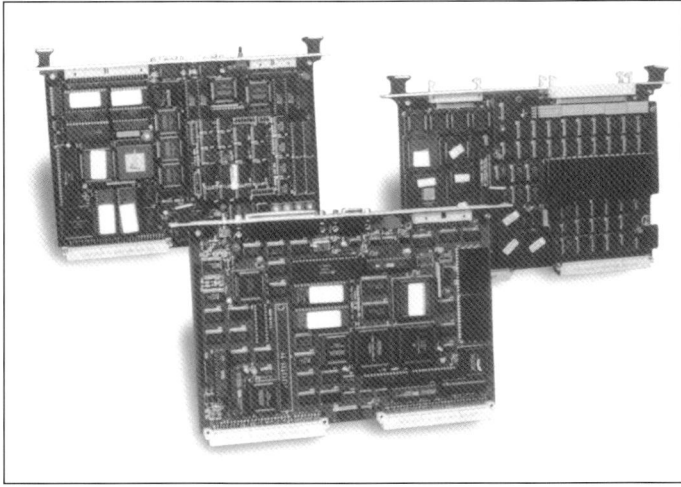

RSC 1100 6U VME modules   0055168

RSC 1100 Key Features include:
(a) Commercial-off-the-shelf single-slot 6U-VME card
(b) Modular radar interface that adapts to any type of radar unit
(c) 40 MHz input sampling rate, 3.75 m sampling resolution
(d) Programmable input look-up table
(e) Frame buffer 1,024 × 1,024 × 8 bits
(f) Origin offset/expanded offset display support modes
(g) Dedicated pixel fill hardware ensures image elements are not skipped
(h) Variable persistence, time or sweep based
(i) Video mixer provides graphics platform independent display of radar/graphics overlays
(j) display resolution up to 2,048 × 2,048 with video mixer card
(k) Built-in test; internal test-pattern generator
(l) Software support from host system via library of C function calls.

Contractor
Folsom Research, Inc., Rancho Cordova, California.

UPDATED

....

Satellite Tool Kit (STK)

Satellite Tool Kit (STK) is commercial-off-the-shelf (COTS) analysis software which enables users to calculate data and display 2-D maps that show a variety of time-dependent information for satellites and other moving vehicles such as launch vehicles, missiles and aircraft. The user can quickly calculate any object's positional attitude over time; evaluate complex visibility relationships among and between land, sea, air and space objects; and calculate satellite or ground-based sensor coverage areas.

All time-dependent information can be viewed in a variety of 2-D map displays, and multiple maps with different styles can be displayed at once. Time can be animated forward, in reverse or in real time to display sensor coverage areas, visibility status, lighting conditions, and star and planet conditions.

STK is available on Windows 2000, Windows NT, Windows XP, LINUX and most major UNIX platforms including SGI, Sun, IBM and HP.

Manufacturer
Analytical Graphics Inc, Malvern, Pennsylvania.

UPDATED

....

SATVUL

The Satellite Vulnerability (SATVUL) program provides Naval commanders with surveillance information indicating when their forces would be vulnerable to detection by hostile reconnaissance satellites and what surveillance might be afforded from friendly satellites. SATVUL produces detailed vulnerability and accessibility reports with supporting graphic displays using orbital data distributed by the Naval Space Surveillance Center (NAVSPASUR). The vulnerability reports calculated by SATVUL can be distributed by GCCS to assist in operations planning. SATVUL can also compute detectability of a ship's radar emissions by hostile satellites. SATVUL graphics support includes plots of a satellite's closest point of approach to a ship's planned route or operating area, and a satellite's ground trace (or footprint) on the world map display.

Status
In service.

Contractor
Northrop Grumman Information Technology, Reston, Virginia.

VERIFIED

Smiths Database Preparation System (DPS)

The Smiths Database Preparation System (DPS) is used to produce outputs in multiple formats by processing the following inputs: charts, paper maps, reconnaissance photographs, SPOT images, LANDSAT images, DTED on CD-ROM, DTED on tapes, ADRG charts on CD-ROM and other paper products. The system is also used to extract elevation data from SPOT stereo pairs, to create and update world catalogues, output data on Write Once Read Many disks (WORMs) or Erasable Optical Disks (EODs) and to prepare multiple copies of the data disks.

Description
The major units used in the system are a drum scanner with controller, a processing station, a verification/mastering station, an elevation extraction workstation, a control point digitising table, Sun workstation, a copy station, a Local Area Network (LAN) and a tape storage rack. Each of these systems is described below and, if the requirements of the user so dictate, can be altered to fit the exact needs of the customer.

A Tangent drum colour scanner and controller with software are integrated into the database preparation system and are used to input paper charts, maps, photographs and so on. The paper product being scanned is placed on the drum and scanned with the output stored on the hard drive of the controller. The controller is used to set up and control the scanner and act as the receptacle for the output of the scanner. It is connected to the network via the LAN to allow the transmission of the scanned image to the processing station. Active chart areas of 103.88 × 165 cm (40.9 × 65 in) can be scanned.

The processing station consists of a VAX processor with hard drives capable of holding several scanned images. The processor is connected to the LAN which allows the station to request the transfer of the scanned image from the scanner controller to the processing station.

At the processing station the image is filed, inspected for colour content, completeness and clarity. The processing station is then used to register the image to a given projection on a representative spheroid. The image is then aligned and rotated so that north is up. During the rotation and alignment process the image density is reduced from 250 to 150 dpi or other user-specified density and transferred to the verification station.

The verification/mastering station consists of a VAX processor with hard drives capable of holding several scanned images. This station also includes a WORM disk drive used to create master disks. The verification station is connected to the local area network to permit the request for transmission of files from the processing station to the verification/mastering station. This station serves as the quality assurance station for the system. The images are inspected at this station prior to being written to the WORM disk and again after the final master has been prepared to ensure that the images are complete and acceptable in accordance with the customer's requirements.

The elevation extraction workstation consists of a Sun computer with a nine track tape deck used to ingest the SPOT images from the CCTs containing the stereo pairs of images for the area of interest, an image processor which is used to facilitate the processing of the data and, optionally, a control point digitising table which is used to input control points into the system. The images are then processed to extract the elevation data using the image processor. The output of the image processor is stored on the high-density disks of the Sun computer. When the elevation data extraction has been completed, the results are processed into either a Digital Elevation Module (DEM) or Digital Terrain Elevation Data (DTED) format and stored on the appropriate media, specified by the user. ORTHO images can be produced at identical scales and projections as the elevation model.

When ground control points are not available from a database source, or from digitised charts, the digitising table is used. A map or chart is placed on the table and known co-ordinates are entered to register the chart. The cursor is then placed over a point on the chart which corresponds to a visible point on the SPOT image and the co-ordinates of the point are entered into the system, thus tying the known co-ordinate of the point to the visible point on the image. This process is repeated for several points on the image and the processor then computes the co-ordinates for all points of the image.

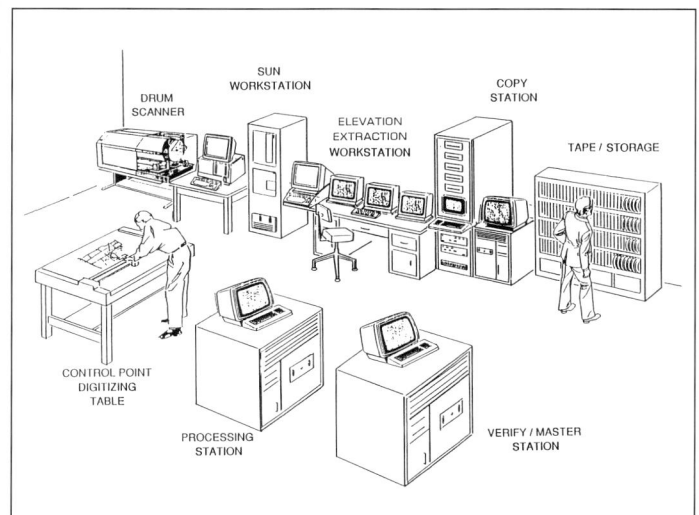

Typical database preparation system

The database workstation consists of a SunSPARC station and the necessary peripherals to meet the requirements of the user. Generally, this consists of a CPU, 2 Gbytes of hard disk space, a 2 in tape drive, SCSI interfaces and a ¼ in tape drive. The workstation is used as the input station for SPOT images, LANDSAT images, DTED data and CD-ROM data. This station processes the data and transmits its output via the LAN to the processing station.

The copy station consists of a VAX processor with a hard drive capable of storing several processed charts, a WORM drive and the necessary controllers to drive up to eight external WORM or EOD drives.

In order to produce multiple copies of the master disk prepared on the verification station, the WORM master is inserted into one of the WORM drives and the software is commanded to process the data onto the target disks, either WORMs or EODs, to meet requirements of the user.

The LAN is the thread that ties all of the above stations together into a network and serves as the transmission media for the transfer of the data between the stations. It consists of a thin-wire LAN where applicable (thick-wire if the distance between the equipments or the quantity of data to be transferred so warrants) and the appropriate transceivers. The software which interconnects the equipment is a combination of TCP/IP and DECNET.

Status

In service with the US Air Force as part of MSS II (see separate entry). One system was part of a larger contract awarded in 1996 to supply MSS II to the Egyptian Air Force.

Contractor

Smiths Aerospace, Electronic Systems Division, Germantown, Maryland.

UPDATED

Tactical Automated Mission Planning Systems (TAMPS)

Description

TAMPS is the US Naval standard unit level aircraft mission planning system. It loads data for the following aviation platforms and subsystems: F/A-18, F-14, E-2C, V-22, C-2, KC-130, AH-1, SH-60, MH-53, HH-60, UH-1, VH-1, P-3C, High-speed Anti Radiation Missile (HARM), Joint StandOff Weapon (JSOW), Joint Directed Attack Munitions (JDAM), Standoff Land Attack Missile (SLAM), Joint Tactical Information and Distribution System (JTIDS), Global Positioning System (GPS), ARC-210, and Forward Area Minefield Planner (FAMP).

TAMPS loads the F/A-18 Data Storage Unit (DSU) with route data (waypoints, sequential steering files), air-to-air radar presets, Tactical Aircraft Navigation Aid (TACAN) and channel identification files. The Data Storage Unit (DSU) in turn provides this TAMPS information to the F/A-18 flight software.

Without the TAMPS load of 'independent overlays' for the aircraft software and bulk files for missile software, weapons such as SLAM, JSOW and JDAM would be unusable. TAMPS is currently the primary means of loading JTIDS data for the F-14D/E-2C. Future systems such as Tactical Aircraft Moving Map Capability (TAMMAC) are planning to use TAMPS for mission planning and data loads. In keeping with the Assistant Secretary of Defense (C³I) direction, TAMPS has been identified as a migration system. Various platform specific aircraft mission planning systems such as Tactical EA-6B Mission Support System (TEAMS), Map Operator and Maintenance Station (MOMS), Common Helicopter Aviation Mission Planning System (CHAMPS), MOMS/AV-8B Maintenance Data System, ES-3 Mission Planning System, Tactical Electronic Reconnaissance Processing and Evaluation System (TERPES) are planned to interface with TAMPS. TAMPS is interoperable with and uses the Global Command and Control System - Maritime (GCCS-M) for data feeds.

Development

TAMPS version 6.2/6.2K was developed and integrated in late 1998. Efforts included the integration of the following modules and functionalities: TAMMAC, H-1 mission planning module, Joint Service Imagery Processing System (JSIPS) interface, Tactical Strike Coordination Module (TSCM), and Tactical Operational Scene (TOPSCENE). This release included improvements to the following modules and functionalities: E-2C module and SLAM module. The inclusion of the following requirements was part of TAMPS version 6.2: full duplex security, Local Area Network (LAN), commercial off-the shelf and operating system upgrades, port to a new hardware suite, intelligence data base in standard extract format and update (MIDB 2.0). System Engineering studies were conducted to identify requirements for various platform specific aircraft mission planning systems (for example CHAMPS, MOMS, H-60, Anti-Submarine Warfare (ASW)) to continue with the execution of the migration plan. The previously planned TAMPS version 6.3 effort has been redefined as a maintenance release 6.2.1 while TAMPS version 6.4 has been redefined as continued maintenance of version 6.2.

TAMPS is programmed to migrate to the Joint Mission Planning System (JMPS), which is currently under development. JMPS flight-planning capabilities are scheduled for release in FY 2002. After the initial JMPS fielding, JMPS will migrate the TAMPS weapons and aircraft data-loading functions, GCCS-M interface and PGM planning functions to an NT-based, DII/COE/JTA-compliant architecture, and this migration is projected to be completed in FY 2003. Strike/force-level planning capability is scheduled for incorporation into the JMPS architecture during FY 2004.

Status

TAMPS is operational as the mission planning system for the US Navy and Marine Corps.

Contractor

Nothrop Grumman Mission Systems, Fairfax, Virginia.

UPDATED

Trusted Exchange 2000

Trusted Exchange 2000 (TREX-2000™) is a 'bolt-on' product to the standard MS Exchange 2000™ Groupware Server that offers centralised functional, intelligence and security capabilities. It is a layer that surrounds the MS Exchange 2000™ and performs active/passive intercepts for tracking and/or additional processing. Its capabilities include:

(a) PKI conversion. TREX-2000™ supports the configuration with a relaxed or no PKI at the user level and stronger encryption services where traffic leaves the local enclave, performing PKI conversion processing for all messages passing from one PKI domain to another

(b) Mail List processing. Secure mass mailing of individual e-mails

(c) Profiling. Automatic content-based key-word distribution (known as profiling). An extensive profiling GUI is provided for creation and maintenance of the key-word 'rules' and associated distribution list

(d) Virus checking for encrypted and unencrypted message contents

(e) 'Fire and Forget' message tracking, providing end-to-end message tracking and notification. TREX-2000™ checks all messages and reports passing through or originating from Exchange 2000™ and an internal audit is created for each message noting key items about the message such as expiry time and intended recipients. It will then continue to track the progress of the message, co-ordinating with other TREX-2000™ servers, until the message is delivered, a delivery error is encountered or the message expires. For delivery failures of any kind, TREX-2000™ will perform automatic user notification actions

(f) The open API allows proprietary modules to be created and bolted onto TREX-2000™ for custom capabilities.

The rules that govern which messages are to be subject to TREX-2000™ value-added processing and which are to flow through untouched are completely configurable. Rule types supported include interception of all messages, messages for particular users, and/or messages containing certain key words. Following TREX-2000™ processing, the message is returned to Microsoft Exchange 2000™ for normal onward processing and distribution.

Status

TREX-2000™ was announced in late 2002. Jane's sources suggest that CommPower is seeking prime contractors who are using an MS2000 Exchange solution in a NATO Message System bid.

Contractor

Communications & Power Engineering Inc (CommPower), Camarillo, California.

NEW ENTRY

Vessel Traffic Service System(VTSS)

The United States Coast Guard (USCG) employs the Vessel Traffic Service System as an integral part of its waterways management efforts to facilitate the safe and efficient transit of vessel traffic and assist in the prevention of collisions, groundings, maritime casualties and ensuing environmental damage. First introduced in 1991, the CGVTS program has undergone numerous changes

The VTS compresses radar, VHF-FM voice and live video data onto commercial leased landlines or microwaves and sends the information to the central VTS Centre (VTC). Vessel data is processed at the VTC and displayed to the operators in a chart-based format that lends itself to early recognition of potentially dangerous situations. A minimum of twelve months of vessel transit data is stored in an Oracle database for later reconstructing collisions, spills or other incidents, as well as a data mining capability used for traffic density and port tonnage studies, and fulfilling Freedom of Information Act requests.

The VTS software is built on DISA's Common Operating Environment (COE). Commercial-off-the-shelf (COTS) and government-off-the-shelf (GOTS) software and hardware is built on top of the COE base. The same base is being used for the majority of USCG systems including the Shipboard Command and Control System (SCCS) for 270', 378' and 210' cutters and the Command and Control PC (C2PC) (see separate entry) for SAR. This commonality allows exchange of information between USCG systems, as well as DOD systems.

VTS features include:

(a) Routine as well as contingency response capability, providing all VTS data simultaneously, and the ability to operate effectively in all waterway environments.

(b) Voice and data communications capabilities via VHF, telephone, and LAN/WAN.

(c) Automatic Identification System (AIS) for traffic surveillance. Automatically associated to vessel in ship database. AIS tracks are easily distinguishable from other tracks.

(d) Independent surveillance via interfaces to several types of radars including SSR. Other radar types are easily accommodated. The system handles multiple radar inputs simultaneously in multiple windows with mixable types and selectable colours.

(e) Estimated Position Tracking, where a track follows a 'Standard Route' (SR) consisting of waypoints, legs, and motions attributes (course, speed). An SR transit can be initiated, updated, or 'parked' (ended).

(f) Automatic association of tracks with vessel data stored in the ship database. The operator-configurable Universal Track Data Cards (UTDC) provide structured queries of vessel and transit data, as well as the ability to define Prospective vessels and start transits. An Editable Track Tag (ETT) function enables customisation of displayed track symbol labels. Operational data (tracks, radar imagery, advisories, alarm areas and so forth) is displayed atop Digital Nautical Charts (DNC), raster (MapTech/BSB) maps, and many other types of electronic charts.

(g) Decision support through immediate access to vessel and non-vessel data, as well as transit data from Oracle database and via PortMaster. An alarms toolkit provides operator-configurable alerts. Advisory data (Safety, Manoeuvring Restrictions, ATON status, Incident and so forth) is stored in the local database, providing the operator with logs and structured queries as needed, and displayed on screen for immediate access.

(h) Colour and low-light closed circuit TV video with touch panel and selection of the various cameras and their controls. Provides time-tagged video with record/playback.

Status

The VTS is installed and operational in four US ports: New York, Puget Sound (Seattle), San Francisco and Houston/Galveston.

Contractor

Northrop Grumman Information Technology, Reston, Virginia.

VERIFIED

Videolink HD scan converter

Designed for critical applications in command and control and training and simulation, the Videolink HD system records computer, radar, sonar or other high-resolution raster scan displays in real time. With HDTV quality video plus two channels of audio, it can provide a definitive record of critical data. Three hours of recording per cassette tape, a small physical package and low cost are significant features.

The Videolink HD converts computer and other signals to the 1125 line HDTV standard. By comparison, ordinary video scan converters transform signals to broadcast standard NTSC or PAL video, resulting in lower image resolution. By using the HDTV standard, the new Videolink HD offers a considerable improvement in resolution.

The Videolink HD is compatible with inputs up to 1,280 × 1,024 pixel resolution. The system works bidirectionally. First, the image is converted to the HDTV standard for recording and storage. Then, the HDTV signal is reconverted to its original non-interlaced format for flicker-free playback on a high scan rate (computer) monitor.

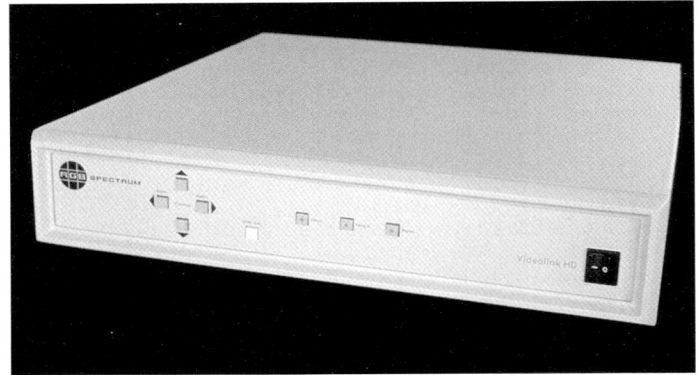

Videolink HD   0018061

Contractor

RGB Spectrum, Alameda, California.

VERIFIED

WaterSpace Management (WSM) route generation and interference avoidance application

The WaterSpace Management (WSM) application has been developed as a tactical decision aid for C4I systems controlling the planning, deconfliction, monitoring and distribution of authorised zones in which a submarine can operate, while under way.

Features include:

(a) Planning tools to construct, edit and display submarine route plans

(b) Interference Algorithm to deconflict WSM routes

(c) Hazard avoidance algorithm and database

(d) Auto generation of tasking messages

(e) Auto parsing of incoming WSM messages

(f) Visualisation of the Under Sea picture

WSM is operated on a UNIX-based LAN of workstations connected to other similar systems via communication links to provide a distributed framework for this critical database of submarine operating areas. Dynamic geographic rendering of currently activated tracks and areas indicates disposition of forces for any interval of time; dynamic track design tools facilitate track construction while controlling deconfliction.

The planning function is referred to as the Track Generation System (TGS) and the deconfliction portion as the Mutual Interference Avoidance (MIA) system.

Status

In use in the US Navy.

Contractor

Northrop Grumman Information Technology, Reston, Virginia.

UPDATED

CONTRACTORS

CONTRACTORS

ADI Limited
Head Office
Level 2, Building 51
Garden Island
New South Wales 2011
Australia
Tel: (+61 2) 95 62 33 33
Fax: (+61 2) 95 62 23 87
Web: http://www.adi-limited.com

Advanced Programmes Inc
7125 Riverwood Drive
Columbia
Maryland 21046
United States
Tel: (+1 410) 312 58 00
Fax: (+1 410) 312 58 50
Web: http://www.drs-ap.com

Advanced Programming Concepts Inc
3300 Duval Road
Suite 200
Austin
Texas 78759
United States
Tel: (+1 512) 327 67 95
Fax: (+1 512) 327 80 43
e-mail: info@apcinc.com
Web: http://www.apcinc.com

Aeromaritime Systembau GmbH
Ludwig-Erhard-Strasse 16
D-85375 Neufahrn b.Freising
Germany
Tel: (+49 8165) 617 10
Fax: (+49 8165) 90 83 89
e-mail: info@aeromaritime.de
Web: http://www.aeromaritime.de

Aeromaritime Systembau GmbH
Hanauer Strasse 105
D-80993 München
Germany
Tel: (+49 89) 14 90 50
Fax: (+49 89) 140 11 10

Aerosystems International Ltd
Alvington
Yeovil
Somerset BA22 8UZ
United Kingdom
Tel: (+44 1935) 44 30 00
Fax: (+44 1935) 44 31 11
e-mail: sales@aeroint.com
Web: http://www.aeroint.com

Alcatel ISR
3 rue Ampère
F-91349 Massy Cedex
France
Tel: (+33 1) 69 76 21 93
Fax: (+33 1) 69 76 21 83

Alcatel Italia SpA
Viale L Bodio 33/79
I-120158 Milan
Italy
Tel: (+39 2) 377 21
Fax: (+39 2) 376 01 18

Alcatel Space
Headquarters
12 rue de la Baume
F-75008 Paris Cedex
France
Tel: (+33 1) 46 52 62 00
Fax: (+33 1) 46 52 62 50
e-mail: alc-spac@pobox.oleane.com
Web: http://www.alcatel.com/space

Alenia Spazio SpA
Headquarters
Via Saccomuro 24
I-00131 Roma
Italy
Tel: (+39 06) 415 11
Fax: (+39 06) 419 06 75
e-mail: communications@roma.alespazio.it
Web: http://www.aleniaspazio.com

Almos Systems BV
Culemborg
PO Box 422
NL-4100 AK
Netherlands
Tel: (+31 345) 54 40 80
Fax: (+31 345) 53 11 46
e-mail: info@almossystems.com
Web: http://www.almossystems.com

Amper Programas SA
Pol Ind. Los Ángeles
Autovía de Andalucía km 12,700
Getafe
E-28905 Madrid
Spain
Tel: (+34 91) 453 24 00
Fax: (+34 91) 453 24 01
e-mail: informacion@amper.es
Web: http://www.amper.es

AMS
Head Office (Italy)
Via Tiburtina Km 12,400
I-00131 Roma
Italy
Tel: (+39 06) 415 01
Fax: (+39 06) 413 10 91
Web: http://www.amsjv.com

AMS
Head Office (UK)
Eastwood House
Glebe Road
Chelmsford
Essex CM1 1QW
United Kingdom
Tel: (+44 1245) 70 27 02
Fax: (+44 1245) 70 27 00
Web: http://www.amsjv.com

Analytical Graphics Inc
40 General Warren Boulevard
Malvern
Pennsylvania 19355
United States
Tel: (+1 610) 578 10 00
Fax: (+1 610) 578 10 01
e-mail: info@stk.com
Web: http://www.stk.com

Anteon Corporation
3211 Jermantown Road
Suite 700
Fairfax
Virginia 22030
United States
Tel: (+1 703) 246 02 00
Fax: (+1 703) 246 02 94
e-mail: rjung@anteon.com
Web: http://www.anteon.com

Arinc Incorporated
2551 Riva Road
Annapolis
Maryland 21401-7465
United States
Tel: (+1 410) 266 41 80
Fax: (+1 410) 266 23 29
e-mail: arincmkt@arinc.com
Web: http://www.arinc.com

Aselsan Electronics Industry Inc
PO Box 101
TR-06172 Yenimahalle
Ankara
Turkey
Tel: (+90 312) 592 10 00
Fax: (+90 312) 354 13 02
Web: http://www.aselsan.com.tr

Astronautics Corporation of America
PO Box 523
Milwaukee
Wisconsin 53201-0523
United States
Tel: (+1 414) 449 40 00
Fax: (+1 414) 447 82 31
e-mail: busdev@astronautics.com
Web: http://www.astronautics.com

Austin Info. Systems Inc
301 Camp Craft Road
Austin
Texas 78746
United States
Tel: (+1 512) 329 66 61
Web: http://www.ausinfo.com

Avibras Indústria Aeroespacial SA
Rodovia dos Tamoios km 14
Jacareí
PO Box 278
12300-000 São Paulo
Brazil
Tel: (+55 12) 39 55 60 00
Fax: (+55 12) 39 51 62 77
e-mail: govsales@avibras.br
Web: http://www.avibras.com.br

Aydin Displays Inc
1 Riga Lane
Birdsboro
Pennsylvania 19508
United States
Tel: (+1 610) 404 74 00
Fax: (+1 610) 404 81 90
e-mail: sales@aydindisplays.com
Web: http://www.aydindisplays.com

Aydin Yazilim ve Elektronik Sanayi AS
Sincan Organize Sanayi Bölgesi
Karamanlilar Cad. No. 7
TR-06935 Sincan
Ankara
Turkey
Tel: (+90 312) 267 27 41
Fax: (+90 312) 267 01 15
e-mail: info@ayesas.com
Web: http://www.ayesas.com

Azdec Ltd
32 Gladstone Road
Sholing
Southampton
Hampshire SO19 8GT
United Kingdom
Tel: (+44 23) 80 44 43 93
Fax: (+44 23) 80 43 20 71
e-mail: information@aztec.ltd.uk
Web: http://www.azdec.co.uk

BAE Systems
Corporate Headquarters
Warwick House
Farnborough Aerospace Centre
PO Box 87
Farnborough
Hampshire GU14 6YU
United Kingdom
Tel: (+44 1252) 37 32 32
Fax: (+44 1252) 38 30 00
Web: http://www.baesystems.com

BAE Systems Australia
DSTO Contractors Area
East Avenue
Edinburgh Parks
South Australia 5112
Australia
Tel: (+61 8) 84 80 88 88
Fax: (+61 8) 84 80 88 00
e-mail: australia@baesystems.com
Web: http://www.baesystems.com/facts/pages/
     business/austral.htm

BAE Systems
C4ISR
Christopher Martin Road
Basildon
Essex SS14 3EL
United Kingdom
Tel: (+44 1268) 52 28 22
Fax: (+44 1268) 88 31 40

BAE Systems
C4ISR
Grange Road
Christchurch
Dorset BH23 4JE
United Kingdom
Tel: (+44 1202) 48 63 44
Fax: (+44 1202) 40 42 21
Web: http://www.baesystems.com

BAE Systems
North America
1215 Jefferson Davis Highway
Suite 1500
Arlington
Virginia 22202-4302
United States
Tel: (+1 703) 416 78 00
Fax: (+1 703) 415 14 59

Barco NV
Theodoor Sevenslaan 106
B-8500 Kortrijk
Belgium
Tel: (+32 56) 23 35 79
Fax: (+32 56) 23 35 80
e-mail: sales.commandcontrol.barcoview@barco.com
Web: http://www.barco.com

Bharat Electronics Ltd
International Marketing
2nd Floor Shankarnarayan Building
25 M G Road
Bangalore
560 001
India
Tel: (+91 80) 226 73 22
Fax: (+91 80) 225 84 10
e-mail: dgmmcco@vsnl.net
Web: http://www.bel-india.com

Boeing Australia Limited
GPO Box 767
Brisbane
Australia
Tel: (+61 7) 33 06 30 00
Fax: (+61 7) 33 06 32 99
Web: http://www.boeing.com.au

Bruhn Newtech A/S
Gladsaxevej 402
DK-2860 Søborg
Denmark
Tel: (+45) 39 55 80 00
Fax: (+45) 39 55 80 80
e-mail: info@newtech.dk
Web: http://www.newtech.dk

Bruhn Newtech Ltd
1 Allenby Road
Winterbourne Gunner
Salisbury
Wiltshire SP4 6HZ
United Kingdom
Tel: (+44 1980) 61 17 76
Fax: (+44 1980) 61 13 30
e-mail: info@bruhn-newtech.co.uk
Web: http://www.bruhn-newtech.co.uk

Bruhn Newtech, Inc
10420 Little Patuxent Parkway
Suite 301
Columbia
Maryland 21044-3636
United States
Tel: (+1 410) 884 17 00
Fax: (+1 410) 884 61 71
e-mail: info@bruhn-newtech.com
Web: http://www.bruhn-newtech.com

CMC Electronics Cincinnati
7500 Innovation Way
Mason
Ohio 45040-9699
United States
Tel: (+1 513) 573 61 00
Fax: (+1 513) 573 62 90
e-mail: sales@cinele.com
Web: http://www.cmccinci.com

CMC Electronics Inc
Executive Office
600 Dr Frederik Philips Boulevard
Ville St-Laurent
Québec H4M 2S9
Canada
Tel: (+1 514) 748 31 48
Fax: (+1 514) 748 31 84
Web: http://www.cmcelectronics.ca

C.MER Industries Ltd
Hazoreff 5
58856 Holon
Israel
Tel: (+972 3) 557 25 55
Fax: (+972 3) 556 79 04
e-mail: homemer@mer.co.il
Web: http://www.cmer.com

Cogent Defence and Security Networks Limited
Meadows Road
Queensway Meadows Industrial Estate
Newport South Wales
NP19 4SS
United Kingdom
Tel: (+44 1633) 29 21 73
Fax: (+44 1633) 29 29 22 48
Web: http://www.cogent-dsn.com

Communications & Power Engineering Inc
1040 Flynn Road
Camarillo
California 93012
United States
Tel: (+1 805) 389 74 14
Fax: (+1 805) 389 74 19
Web: http://www.commpower.com

Computer Sciences Corp
Corporate Headquarters
2100 East Grand Avenue
El Segundo
California 90245
United States
Tel: (+1 310) 615 03 11
Fax: (+1 310) 322 98 05
e-mail: generalinformation@csc.co.com
Web: http://www.csc.com

Comtech Mobile Datacom Corporation
19540 Amaranth Drive
Germantown
Maryland
United States
Tel: (+1 301) 428 21 00
Fax: (+1 301) 428 10 04
Web: http://www.comtechmobile.com

Comtech Systems Inc
2900 Titan Row
Suite 142
Orlando
Florida 32809
United States
Tel: (+1 407) 854 19 50
Fax: (+1 407) 851 69 60
e-mail: csisales@comtechsystems.com
Web: http://www.comtechsystems.com

Concurrent Computer Corporation
4375 River Green Parkway
Duluth
Georgia 30096
United States
Tel: (+1 678) 258 40 00
Fax: (+1 678) 258 41 99
e-mail: press@ccur.com
Web: http://www.ccur.com

Cunning Running Software Ltd
Wykeham House
88 The Hundred
Romsey
Hampshire SO51 8BX
United Kingdom
Tel: (+44 1794) 83 47 50
Fax: (+44 1794) 83 47 51
Web: http://www.cunningrunning.co.uk

Dansk Radio Comm ApS
Jordemodervej 40
DK-4100 Ringsted
Denmark
Tel: (+45 43) 71 60 45
Fax: (+45 43) 71 45 04
e-mail: drac@dansk-radio.com
Web: http://www.dansk-radio.com

Datamat SpA
Via Laurentina 760
I-00143 Roma
Italy
Tel: (+39 6) 50 27 42 79
Fax: (+39 06) 50 27 22 00
e-mail: naval@datamat.it
Web: http://www.datamat.it

DCN International
19/21 rue du Colonel Pierre-Avia
BP 532
F-75725 Paris Cédex 15
France
Tel: (+33 1) 41 08 71 71
Fax: (+33 1) 41 08 00 27
Web: http://www.dcnintl.com

Dolch Computer Systems Inc
Corporate Headquarters
3178 Laurelview Court
Fremont
California 94538
United States
Tel: (+1 510) 661 22 20
Fax: (+1 510) 490 23 60
e-mail: sales@dolch.com
Web: http://www.dolch.com

Dornier GmbH
D-88039 Friedrichshafen
Germany
Tel: (+49 7545) 884 46
Fax: (+49 7545) 844 11
e-mail: simulation.training@dornier.eads.net
Web: http://www.eads.net

DRS Communications Company LLC
1200 East Mermaid Lane
Wyndmoor
Pennsylvania 19038-7695
United States
Tel: (+1 215) 233 41 00
Fax: (+1 215) 233 99 37
Web: http://www.drs.com

DRS Electronic Systems Inc
200 Professional Drive
Gaithersburg
Maryland 20879
United States
Tel: (+1 301) 921 81 00
Fax: (+1 301) 977 61 58
e-mail: info@drs-esg.com
Web: http://www.drs.com

DRS Tactical Systems Inc
3520 US Highway 1
Palm Bay
Florida 32905
United States
Tel: (+1 321) 727 36 72
Fax: (+1 321) 725 04 96
e-mail: sales@drs-ts.com
Web: http://www.paravantcomputersystems.com

DRS Technologies Inc
Corporate Headquarters
5 Sylvan Way
Parsippany
New Jersey 07054
United States
Tel: (+1 973) 898 15 00
Fax: (+1 973) 898 47 30
e-mail: info@drs.com
Web: http://www.drs.com

EADS Astrium
Gunnels Wood Road
Stevenage
Hertfordshire SG1 2AS
United Kingdom
Tel: (+44 1438) 31 34 56
Fax: (+44 1438) 77 30 69
Web: http://www.astrium-space.com

EADS CASA
Headquarters
Avenida de Aragon 404
E-28022 Madrid
Spain
Tel: (+34 915) 85 70 00
Fax: (+34 915) 85 76 68
e-mail: communications@casa.eads.net
Web: http://www.eads.net

EADS Defence & Communication Systems
Headquarters
6 rue Dewoitine
BP 14
F-78143 Vélizy-Villacoublay
France
Tel: (+33 1) 34 63 70 00
Fax: (+33 1) 34 63 70 70
e-mail: webmaster@matra-ms2i.fr
Web: http://www.sysde.eads.com

EADS Defence & Security Systems Division
Headquarters
Willy Messerschmittstraße
85521 Ottobrunn
Postfach 801109
D-81663 München
Germany
Tel: (+49 89) 60 70
Fax: (+49 89) 60 72 64 81
Web: http://www.eads.net

EADS Defence and Security Networks
rue JP Timbaud
BP 26
F-78392 Bois d'Arcy Cedex
France
Tel: (+33 1) 34 60 80 20

EADS Deutschland GmbH
Ulm Office
Wörthstrasse 85
D-89077 Ulm
Germany
Tel: (+49 731) 39 20
Fax: (+49 731) 392 33 93
Web: http://www.eads-nv.com

EDO Communications and Countermeasures
Systems
996 Flower Glen Street
Simi Valley
California 93065
United States
Tel: (+1 805) 584 82 00
Fax: (+1 805) 527 83 32
Web: http://www.nycedo.com/edocorp/pageba13_
edoccs.htm

EDO Corporation, Combat Systems
1801-E Sara Drive
Chesapeake
Virginia 23320
United States
Tel: (+1 757) 424 10 04
Fax: (+1 757) 424 16 02
Web: http://www.edocombat.com

EDO MBM Rugged Systems
Emblem House
Home Farm Business Park
Home Farm Road
Brighton
East Sussex BN1 9HU
United Kingdom
Tel: (+44 1273) 81 05 00
Fax: (+44 1273) 81 06 00
e-mail: sales@edombmrugged.com
Web: http://www.edombmrugged.com

EDS Defence Ltd
1-3 Bartley Wood Business Park
Bartley Way
Hook
Hampshire RG27 9XA
United Kingdom
Tel: (+44 01256) 74 20 00
Fax: (+44 01256) 74 26 12
Web: http://www.eds.co.uk

ESG Elektroniksystem und Logistik GmbH
Einsteinstrasse 174
D-81675 München
Germany
Tel: (+49 89) 92 16 27 45
Fax: (+49 89) 92 16 22 36
e-mail: info@esg.de
Web: http://www.esg.de

Elbit Systems Ltd
PO Box 539
IL-31053 Haifa
Israel
Tel: (+972 4) 831 53 15
Fax: (+972 4) 855 00 02
e-mail: marcom@elbit.co.il
Web: http://www.elbit.co.il

Elettronica SpA
Via Tiburtina Valeria km 13.700
Loc Settecamini
I-00131 Roma
Italy
Tel: (+39 06) 415 40 00
Fax: (+39 06) 415 49 24
e-mail: info@elt.it

Elisra Electronic Systems Ltd
Corporate Headquarters
48 Mivtza Kadesh Street
51203 Bene Baraq
Israel
Tel: (+972 3) 617 55 22
Fax: (+972 3) 617 58 50
e-mail: marketing@elisra.com
Web: http://www.elisra.com

Elta Electronics Industries Ltd
PO Box 330
IL-77102 Ashdod
Israel
Tel: (+972 8) 857 24 10
Fax: (+972 8) 856 18 72
e-mail: market@is.elta.co.il
Web: http://www.elta-iai.com

Environmental Systems Research Institute
380 New York Street
Redlands
California 92373-8100
United States
Tel: (+1 909) 793 28 53
Fax: (+1 909) 793 59 53
e-mail: info@esri.com
Web: http://www.esri.com

Ericsson Microwave Systems AB
Solhusgatan
SE-431 84 Mölndal
Sweden
Tel: (+46 31) 747 00 00
Fax: (+46 31) 747 17 27
e-mail: defenseinfo@emw.ericsson.se
Web: http://www.ericsson.com/microwave

Falcon Protec
Unit 9, Alexandra Way
Tewkesbury
Gloucestershire GL20 8NB
United Kingdom
Tel: (+44 1684) 29 58 07
Fax: (+44 1684) 85 00 11
e-mail: sales@sda-falcon.co.uk
Web: http://www.sda-falcon.co.uk

Folsom Research Inc
11101-A Trade Center Drive
Rancho Cordova
California 95670-6119
United States
Tel: (+1 916) 859 25 00
Fax: (+1 916) 859 25 15
e-mail: sales@folsom.com
Web: http://www.folsom.com

Fujitsu Services
Observatory House
Windsor Road
Slough
Berkshire SL1 2EY
United Kingdom
Tel: (+44 870) 242 79 98
Fax: (+44 870) 242 44 45
Web: http://uk.fujitsu.com

Fujitsu Services
Jays Close
Viables Industrial Estate
Basingstoke
RG22 4BY
United Kingdom
Tel: (+44 80) 234 55 55

General Dynamics Canada
Calgary Division
1020-68th Avenue NE
Calgary
Alberta T2E 8P2
Canada
Tel: (+1 403) 295 67 00
Fax: (+1 403) 730 11 97
e-mail: busdev.calgary@gdcanada.com

General Dynamics Corp
Network Systems
77 A Street
Needham
Massachusetts 02494
United States
Tel: (+1 781) 449 20 00
Fax: (+1 781) 455 52 55
Web: http://www.gd-ns.com

General Dynamics Corp
3190 Fairview Park Drive
Falls Church
Virginia 22042-4523
United States
Tel: (+1 703) 876 30 00
Fax: (+1 703) 876 31 25
Web: http://www.gd.com

General Dynamics Corp
C4 Systems
400 John Quincy Adams Road
Taunton
Massachusetts 02780-1069
United States
Tel: (+1 508) 880 40 00
Fax: (+1 508) 880 48 00
e-mail: info@gdc4s.com
Web: http://www.gdc4s.com

General Dynamics Corp
C4 Systems
8201 East McDowell Road
Scottsdale
Arizona 85257
United States
Tel: (+1 480) 441 30 33
Fax: (+1 480) 726 29 71
e-mail: info@gdds.com
Web: http://www.gdds.com

General Dynamics Corp
Advanced Information Systems
1421 Jefferson Davis Highway
Suite 600
Arlington
Virginia 22202
United States
Tel: (+1 703) 271 73 00
Fax: (+1 703) 271 73 01
e-mail: ais.contact@gd-ais.com
Web: http://www.gd-ais.com

General Dynamics Land Systems
38500 Mound Road
Sterling Heights
Michigan 48310-3200
United States
Tel: (+1 586) 825 40 00
Fax: (+1 586) 825 40 13
Web: http://www.gdls.com

General Dynamics UK Ltd
Bryn Brithdir
Units 3 & 4
Oakdale Business Park
Oakdale, South Wales
NP12 4AA
United Kingdom
Tel: (+44 1424) 85 34 81
Fax: (+44 1424) 85 15 20
e-mail: sales@general dynamics.uk.com
Web: http://www.generaldynamics.uk.com

Geosolutions
11 Elgin Road
Ballsbridge
Dublin 4
Ireland
Tel: (+353 1) 667 76 16
Fax: (+353 1) 667 76 17
e-mail: geosol@iol.ie
Web: http://www.geosolutions.ie

Giat Industries
Headquarters
13 route de la Minière
F-78034 Versailles Cedex
France
Tel: (+33 1) 30 97 35 54
Fax: (+33 1) 30 97 38 97
e-mail: marketing@giat-industries.fr
Web: http://www.giat-industries.fr

Global Geomatics Inc
Head Office
825 Querbes Avenue
Bureau 200
Outremont
Québec H2V 3X1
Canada
Tel: (+1 514) 279 97 79
Fax: (+1 514) 279 02 55
e-mail: info@globalgeo.com
Web: http://www.globalgeo.com

Granit Central Research Institute
3 Gospitalnaya Street
191014 St Petersburg
Russian Federation
Tel: (+7 812) 271 67 56
Fax: (+7 812) 274 63 39
e-mail: cri-granitpeterlink.ru

Grintek Group Ltd
PO Box 8792
0046 Centurion
South Africa
Tel: (+27 12) 672 83 00
Fax: (+27 12) 672 83 01
e-mail: info@grintek.com
Web: http://www.grintek.com

Harris Corporation
Government Communications Systems Division
PO Box 37
Melbourne
Florida 32902-0037
United States
Tel: (+1 321) 727 65 14
Fax: (+1 321) 727 91 00
e-mail: info@harris.com
Web: http://www.govcom.harris.com

Harris Corporation
RF Communications Division
1680 University Avenue
Rochester
New York 14610-1887
United States
Tel: (+1 585) 244 58 30
Fax: (+1 585) 244 29 17
e-mail: rfcomm@harris.com
Web: http://www.rfcomm.harris.com

Hoka Electronics
Flessingsterrein 13
NL-9665 BZ Oude Pekela
Netherlands
Tel: (+31 597) 67 50 40
Fax: (+31 597) 61 26 45
e-mail: hoka@hoka.net
Web: http://www.hoka.net

IBM United Kingdom Ltd
IBM Defence
Meudon House
Meuden Avenue
Farnborough
Hampshire GU14 7NB
United Kingdom
Tel: (+44 1252) 80 55 55
Fax: (+44 1252) 80 55 56
Web: http://www.datasci.co.uk

Indra
Corporate Headquarters
Arroyd de la Vega
Avda de Bruselas 35
Alcobendas
E-28108 Madrid
Spain
Tel: (+34 91) 480 50 00
Fax: (+34 91) 480 50 80
e-mail: indra@indra.es
Web: http://www.indra.es

Indra Sistemas SA
Aranjuez Division
Joaquin Rodrigo 11
E-28300 Aranjuez
Madrid
Spain
Tel: (+34 91) 894 88 00
Fax: (+34 91) 891 80 56
e-mail: magarcia@indra.es
Web: http://www.indra.es

Innovative Concepts Inc
8200 Greensboro Drive
Suite 700
McLean
Virginia 22102
United States
Tel: (+1 703) 893 20 07
Fax: (+1 703) 893 58 90
e-mail: info@innocon.com
Web: http://www.innocon.com

Insys Limited
Reddings Wood
Ampthill
Bedfordshire MK45 2HD
United Kingdom
Tel: (+44 1525) 84 10 00
Fax: (+44 1525) 84 37 04
e-mail: marketing@insys-ltd.co.uk
Web: http://www.insys-ltd.co.uk

Intergraph Corporation
PO Box 240000
Huntsville
Alabama 35824
United States
Tel: (+1 256) 730 20 00
Fax: (+1 205) 730 64 45
e-mail: info@intergraph.com
Web: http://www.intergraph.com

International Data Link Society, United States
e-mail: webmaster@idlsoc.com
Web: http://www.idlsoc.com

IRTS
Bastide de la Giponne
639 boulevard des Armaris
F-83100 Toulon
France
Tel: (+33 4) 94 20 78 00
Fax: (+33 4) 94 20 78 01
e-mail: info@irts-display.com
Web: http://www.irts-display.com

Israel Aircraft Industries Ltd, Headquarters
Ben Gurion International Airport
IL-70100
Israel
Tel: (+972 3) 935 33 43
Fax: (+972 3) 935 82 78
e-mail: corpmkg@iai.co.il
Web: http://www.iai.co.il

ITT Gilfillan
7821 Orion Avenue
PO Box 7713
Van Nuys
California 91409-7713
United States
Tel: (+1 818) 988 26 00
Fax: (+1 818) 901 23 02
Web: http://www.ittgil.com

ITT Industries
Advanced Engineering and Sciences
1761 Business Center Drive
Reston
Virginia 20190-5307
United States
Tel: (+1 703) 438 80 00
Fax: (+1 703) 438 81 12
Web: http://www.aes.itt.com

ITT Industries Inc Aerospace/Communications
1919 West Cook Road
PO Box 3700
Fort Wayne
Indiana 46801
United States
Tel: (+1 260) 451 52 37
Fax: (+1 260) 451 50 92
Web: http://www.acd.itt.com

ITT Industries, Systems Division
PO Box 7463
Colorado Springs
Coloroda 80933
United States
Tel: (+1 719) 591 36 00
Fax: (+1 719) 591 36 94

Kintex
66 James Baucher Street
PO Box 209
1407 Sofiya
Bulgaria
Tel: (+359 2) 266 23 11
Fax: (+359 2) 65 81 01
e-mail: minkin@infotei.bg

Kongsberg Defence & Aerospace AS
Headquarters
PO Box 1003
N-3601 Kongsberg
Norway
Tel: (+47) 32 28 82 00
Fax: (+47) 32 28 82 01
e-mail: office.defence-aerospace@kongsberg.com
Web: http://www.kongsberg.com

Kongsberg Defence Communications A/S
PO Box 87
N-1375 Billingstad
Norway
Tel: (+47) 66 84 24 00
Fax: (+47) 66 84 82 30
e-mail: sales@kdefence.com
Web: http://www.kdefence.com

Kongsberg Gruppen ASA
Headquarters
PO Box 1000
N-3601 Kongsberg
Norway
Tel: (+47) 32 28 82 00
Fax: (+47) 32 28 82 01
e-mail: office@kongsberg.com
Web: http://www.kongsberg.com

Kontron Mobile Computing
7631 Anagram Drive
Eden Prairie
Minnesota 55344
United States
Tel: (+1 952) 974 70 00
Fax: (+1 952) 974 71 99
Web: http://www.kontronmobile.com

L-3 Communication Systems - West
PO Box 16850
Salt Lake City
Utah 84116-0850
United States
Tel: (+1 801) 594 20 00
Fax: (+1 801) 594 21 27
Web: http://www.l-3com.com/csw

L-3 Communications Corp
Systems East
1 Federal Street
Camden
New Jersey 08103
United States
Tel: (+1 856) 338 35 74
Fax: (+1 856) 338 60 14
Web: http://www.l-3com.com/cs-east

L-3 Communications Satellite Networks
125 Kennedy Drive
Hauppauge
New York 11788
United States
Tel: (+1 631) 272 56 00
Fax: (+1 631) 272 55 00
Web: http://www.l-3com.com/snd

Lockheed Martin Canada Inc
Head Office
3001 Solandt Road
Kanata
Ontario K2K 2M8
Canada
Tel: (+1 613) 599 32 70
Fax: (+1 613) 599 32 82
Web: http://www.lockheedmartin.com/canada

Lockheed Martin Information Systems
12506 Lake Underhill Road
Orlando
Florida 32825
United States
Tel: (+1 407) 306 10 00
Fax: (+1 407) 306 11 47
e-mail: info@lmis.com
Web: http://www.lockheedmartin.com/lmis

Lockheed Martin Maritime Systems & Sensors
3333 Pilot Knob Road
Eagan
Minnesota 55121
United States
Tel: (+1 612) 456 22 22
Fax: (+1 612) 456 30 98
Web: http://www.lockheedmartin.com/minn

Lockheed Martin Maritime Systems & Sensors
PO Box 4840
Syracuse
New York 13221-4840
United States
Tel: (+1 315) 456 01 23
Fax: (+1 315) 456 17 93
Web: http://www.lockheedmartin.com/syracuse

Lockheed Martin Maritime Systems & Sensors
199 Borton Landing Road
PO Box 1027
Moorestown
New Jersey 08057-0927
United States
Tel: (+1 609) 722 50 00
Fax: (+1 609) 231 93 63
Web: http://ness.external.lmco.com/ss

Lockheed Martin Mission Systems
700 North Frederick Avenue
Gaithersburg
Maryland 20879
United States
Tel: (+1 301) 240 72 77
Fax: (+1 301) 240 34 16
Web: http://www.missionsystems.external.lmco.com

Lockheed Martin Space Systems
1111 Lockheed Martin Way
Sunnyvale
California 94089-3504
United States
Tel: (+1 408) 742 71 51
Fax: (+1 408) 742 84 84
e-mail: lmms.communications@lmco.com
Web: http://lmms.external.lmco.com

Lockheed Martin Space Systems
Headquarters
12999 Deer Creek Canyon Road
Littleton
Colorado 80127-5146
United States
Tel: (+1 303) 977 30 00
Fax: (+1 303) 897 69 88
Web: http://www.lockheedmartin.com

Logica CMG plc
Milton House
The Office Park
Springfield Drive
Leatherhead
Surrey KT22 7LP
United Kingdom
Tel: (+44 20) 76 37 91 11
Fax: (+44 1932) 83 86 58
e-mail: hackman@logica.com
Web: http://www.logicacmg.com

Maersk Data Defence A/S
Ellegårdvej 25
DK-6400 Sønderborg
Denmark
Tel: (+45 39) 11 30 00
Fax: (+45 39) 11 30 01
e-mail: defence@maerskdata.dk
Web: http://www.maerskdata-defence.dk

Marconi Communications Ltd
New Century Park
PO Box 53
Coventry
West Midlands CV3 1HJ
United Kingdom
Tel: (+44 24) 76 56 20 00
Fax: (+44 24) 76 56 22 47
Web: http://www.marconi.com

Marconi Selenia Communications
Via A. Negrone, 1/A
I-16153 Genova Cornigliano
Italy
Tel: (+39 010) 600 21
Fax: (+39 010) 650 18 97
e-mail: genova.pmpp@marconicomms.com
Web: http://www.marconi.com

Marconi Selenia Communications Ltd
Marconi House
New Street
Chelmsford
Essex CM1 1PL
United Kingdom
Tel: (+44 1245) 35 32 21
Fax: (+44 1245) 28 71 25
Web: http://www.marconi.com

Mass Consultants Ltd
Grove House
Rampley Lane
Little Paxton
St Neots
Cambridgeshire PE19 6EL
United Kingdom
Tel: (+44 1480) 22 26 00
Fax: (+44 1480) 40 73 66
e-mail: systems@mass.co.uk
Web: http://www.mass.co.uk

Mitam Advanced Technologies (1999) Ltd
7 Imber Street
PO Box 3174
IL-49130 Petah Tikva
Israel
Tel: (+972 3) 923 27 23
Fax: (+972 3) 923 27 24
e-mail: mitamtec@mitam.co.il
Web: http://www.mitam.co.il

NATO C3 Agency
Boulevard Leopold III
B-1110 Bruxelles
Belgium
Tel: (+32 2) 707 41 11
Fax: (+32 2) 707 87 70
Web: http://www.nc3a.nato.int

NATO C3 Agency
Oude Waalsdorperweg 61
PO Box 174
NL-2501 CD Gravenhage
Netherlands
Tel: (+31 70) 374 30 00
Fax: (+31 70) 374 32 39
Web: http://www.nc3a.nato.int

For details of the latest updates to *Jane's C4I Systems* online and to discover the additional information available exclusively to online subscribers please visit
jc4i.janes.com

Northrop Grumman, Mission Systems
12011 Sunset Hills Road
Reston
Virginia 20190-3285
United States
Tel: (+1 703) 968 10 00
Web: http://www.ms.northropgrumman.com

Northrop Grumman Corporation
Corporate Headquarters
1840 Century Park East
Los Angeles
California 90067-2199
United States
Tel: (+1 310) 553 62 62
Fax: (+1 310) 201 30 23
e-mail: onewebmaster@ngc.com
Web: http://www.northropgrumman.com

Northrop Grumman Corporation
Electronic Systems
1580-A West Nursery Road
Linthicum
MS A255
Baltimore
Maryland 21090
United States
Tel: (+1 410) 993 24 63
Fax: (+1 410) 981 48 03
Web: http://www.northropgrumman.com

Northrop Grumman Electronic Systems
Navigation Systems Division
21240 Burbank Boulevard
Woodland Hills
California 91367-6675
United States
Tel: (+1 818) 715 40 11
Fax: (+1 818) 715 50 98
e-mail:
customerservice.nsd@northropgrumman.com
Web: http://www.nsd.es.northropgrumman.com
Northrop Grumman Mission Systems
Alpha House
Chilworth Science Park
Southampton
SO16 7NS
United Kingdom
Tel: (+44 2380) 76 04 84
Fax: (+44 2380) 76 04 83
e-mail: sales@northgrum-it.eu.com.
Web: http://www.northgrum-it.eu.com

Northrop Grumman Space Technology
One Space Park
Redondo Beach
California 90278
United States
Tel: (+1 310) 812 43 21
Fax: (+1 310) 814 45 07
Web: http://www.st.northropgrumman.com

Oerlikon Contraves AG
Birchstrasse 155
CH-8050 Zürich
Switzerland
Tel: (+41 1) 316 42 18
Fax: (+41 1) 311 31 54
e-mail: czjag@ocag.ch
Web: http://www.oerlikoncontraves.com

Page Europa SpA
Via Del Serafico 200
I-00142 Rome
Italy
Tel: (+39 06) 50 39 51
Fax: (+39 06) 50 39 52 11
e-mail: e-mail@pageuropa.it
Web: http://www.pageuropa.it

Patria New Technologies Oy
Systems Division
Naulakatu 3
FIN33100 Tampere
Finland
Tel: (+358 20) 46 91
Fax: (+358 20) 469 26 90
e-mail: new.technologies@patria.fi
Web: http://www.patria.fi

PCQT AB
Defence Division
Domnarvsgartan 7
SE-163 53 Spånga
Sweden
Tel: (+46 8) 564 737 30
Fax: (+46 8) 761 29 25
e-mail: info@pcqt.se
Web: http://www.pcqt.se

Plath GmbH,
Nautisch-Elektronische Technik
Gotenstrasse 18
D-20097 Hamburg
Germany
Tel: (+49 40) 23 73 40
Fax: (+49 40) 23 73 41 73
e-mail: c.plath@plath.de
Web: http://www.plath.de

Prodata Systems NV/SA
Leuvensesteenweg 540 Bus 3
B-1930 Zaventem
Belgium
Tel: (+32 2) 722 13 11
Fax: (+32 2) 722 13 99
Web: http://www.prodata-systems.be

Progress
37/41 Artema Street
04053 Kiev
Ukraine
Tel: (+380 44) 490 61 40
Fax: (+380 44) 490 61 82
e-mail: progress@progress.gov.ua/
Web: http://www.progress.gov.ua

Przemyslowy Instytut Telekomunikacji
Poligonowa Street 30
PL-04-051 Warsaw
Poland
Tel: (+48 22) 810 23 81
Fax: (+48 22) 810 23 80
e-mail: office@pit.edu.pl
Web: http://www.pit.edu.pl

Rafael Armament Development Authority Ltd
Corporate Headquarters
PO Box 2250
IL-31021 Haifa
Israel
Tel: (+972 4) 879 40 57
Fax: (+972 4) 879 24 13
Web: http://www.rafael.co.il

Raytheon Company
Intelligence & Information Systems
1200 South Jupiter Road
PO Box 660023
Garland
Texas 75042
United States
Tel: (+1 972) 205 89 84
Fax: (+1 972) 205 45 60
e-mail: bd_igs@raytheon.com
Web: http://www.raytheon.com/businesses/riis

Raytheon Company
Corporate Headquarters
870 Winter Street
Waltham
Massachusetts 02451-1449
United States
Tel: (+1 617) 860 24 14
Fax: (+1 617) 860 25 20
e-mail: corpcom@raytheon.com
Web: http://www.raytheon.com

Raytheon Company
Integrated Defense Systems
50 Apple Hill Drive
Tewksbury
Massachusetts 01876
United States
Tel: (+1 978) 858 52 46
Fax: (+1 978) 858 94 14
e-mail: ask_ids@raytheon.com
Web: http://www.raytheon.com/businesses/rids

Raytheon Company
Space and Airborne Systems
2000 East El Segundo Boulevard
PO Box 902
El Segundo
California 90245
United States
Tel: (+1 310) 647 07 84
Fax: (+1 310) 647 07 85
e-mail: rescprod@notes.west.raytheon.com
Web: http://www.raytheon.com

Raytheon Systems Company
AIS, Falls Church Operations
7700 Arlington Boulevard
Falls Church
Virginia 22046-1572
United States
Tel: (+1 703) 876 19 72
Fax: (+1 703) 849 15 23
Web: http://www.raytheon.com

Raytheon Systems Company
C3I Systems Division
Fort Wayne
1010 Production Road
Fort Wayne
Indiana 46808
United States
Tel: (+1 260) 429 54 16
e-mail: commsys@ftw.rsc.raytheon.com

Raytheon Systems Ltd
80 Park Lane
London
W1K 7TR
United Kingdom
Tel: (+44 20) 75 69 55 00
Fax: (+44 20) 75 69 55 91

Redcom Laboratories Incorporated
One Redcom Center
Victor
New York 14564-0995
United States
Tel: (+1 585) 924 75 50
Fax: (+1 585) 924 65 72
e-mail: info@redcom.com
Web: http://www.redcom.com

Reutech Defence Industries
PO Box 118
3620 Natal
South Africa
Tel: (+27 31) 719 57 11
Fax: (+27 31) 719 58 75
e-mail: info@rdi.co.za
Web: http://www.rdi.co.za

RGB Spectrum
Company Headquarters
950 Marina Village Parkway
Alameda
California 94501
United States
Tel: (+1 510) 814 70 00
Fax: (+1 510) 814 70 26
e-mail: sales@rgb.com
Web: http://www.rgb.com

Rheinmetall Defence Electronics GmbH
Headquarters
Brüggeweg 54
D-28309 Bremen
Germany
Tel: (+49 421) 457 0
Fax: (+49 421) 457 29 00
e-mail: marketing@rheinmetall-de.com
Web: http://www.rheinmetall-de.com

Rockwell Collins France SA
6 avenue Didier Daurat
BP 8
F-31701 Blagnac Cedex
France
Tel: (+33 5) 61 71 77 00
Fax: (+33 5) 61 71 51 69
e-mail: rcf@rockwellcollins.com
Web: http://www.rockwellcollins.com

Rockwell Collins Inc
Headquarters
400 Collins Road NE
Cedar Rapids
Iowa 52498
United States
Tel: (+1 800) 321 22 23
Fax: (+1 319) 295 47 77
e-mail: collins@rockwellcollins.com
Web: http://www.rockwellcollins.com

Rohde & Schwarz GmbH & Co KG
PO Box 801469
D-81614 München
Germany
Tel: (+49 89) 412 90
Fax: (+49 89) 412 91 21 64
e-mail: customersupport@rohde-schwarz.com
Web: http://www.rohde-schwarz.com

Saab Systems Pty Ltd
21 Third Avenue
Technology Park
Mawson Lakes
South Australia 5095
Australia
Tel: (+61 8) 83 43 38 00
Fax: (+61 8) 83 43 37 77
e-mail: sales@saabsystems.com.au
Web: http://www.saabsystems.com.au

Saabtech AB
Nettovägen 6
Jakobsberg
SE-175 88 Jarfalla
Sweden
Tel: (+46 8) 58 08 40 00
Fax: (+46 8) 58 03 22 44
e-mail: info@systems.saab.se
Web: http://www.saabtech.se

SAGEM SA
Head Office
Le Ponant de Paris
27 rue Leblanc
F-75512 Paris Cedex 15
France
Tel: (+33 1) 40 70 63 63
Fax: (+33 1) 40 70 66 40
Web: http://www.sagem.com

Samsung Thales Co Ltd
17th - 20th Floor, Daechi Building
889-11 Daechi 4-Dong
Kangnam-ku Seoul
135-839
South Korea
Tel: (+82 2) 34 58 11 14
Fax: (+82 2) 34 58 11 88
Web: http://www.samsungthales.com

Science Applications International Corporation
East Coast Headquarters
1710 SAIC Drive
M/S 1-14-17
McLean
Virginia 22102
United States
Tel: (+1 703) 676 40 97
Fax: (+1 703) 676 22 69
Web: http://www.saic.com

Science Applications International Corporation
Headquarters
10260 Campus Point Drive
San Diego
California 92121
United States
Tel: (+1 858) 826 60 00
Fax: (+1 858) 546 68 00
Web: http://www.saic.com

Secure Systems & Technologies Ltd
Brunel Court
Waterwells
GL2 2AL
United Kingdom
Tel: (+44 1452) 37 19 99
Fax: (+44 1452) 55 77 00
e-mail: info@sst.ws
Web: http://www.sst.ws

Singapore Technologies Electronics Limited
24 Ang Mo Kio Street 65
Singapore
569061
Singapore
Tel: (+65) 64 13 18 88
Fax: (+65) 64 84 88 40
e-mail: comms.elect@stengg.com
Web: http://www.stengg.com

Sisdef Ltda
Parque Industrial Aconcagua
Camino Con Cón-Quintero Km2
PO Box Casilla 5C-Correo Con Cón
Quintero
Chile
Tel: (+56 32) 81 07 77
Fax: (+56 32) 81 11 90
e-mail: info@sisdef.cl
Web: http://www.sisdef.cl

Smiths Aerospace
Electronic Systems - Germantown
20501 Goldenrod Lane
Germantown
Maryland 20876
United States
Tel: (+1 301) 428 60 00
Fax: (+1 301) 428 69 75
Web: http://www.smiths-aerospace.com

Spetz-Radio Research and Production Enterprise
JSC
4 Promyshlennaya Street
308023 Belgorod
Russian Federation
Tel: (+7 722) 34 22 72
Fax: (+7 722) 34 76 82
e-mail: spetzradio@belgtts.ru

Symetrics Industries LLC
1615 West NASA Boulevard
Melbourne
Florida 32901
United States
Tel: (+1 321) 254 15 00
Fax: (+1 321) 259 41 22
e-mail: symetrics@symetrics.com
Web: http://www.symetrics.com

System Consultancy Services Sdn Bhd
36 Jalan 1/27F
Pusat Bandar Wangsa Maju
53300 Kuala Lumpur
Malaysia
Tel: (+60 3) 41 49 19 19
Fax: (+60 3) 41 49 34 62

Systems & Electronics Inc
201 Evans Lane
St Louis
Missouri 63121-1126
United States
Tel: (+1 314) 553 40 00
Fax: (+1 314) 553 49 49
e-mail: gsevier@seistl.com
Web: http://www.seistl.com

Tadiran Communications Ltd
Marketing Department
26 Hashoftim Street
PO Box 267
IL-58102 Holon
Israel
Tel: (+972 3) 557 46 61
Fax: (+972 3) 557 44 84
e-mail: info@tadcomm.com
Web: http://www.tadcomm.com

Tadiran Electronic Systems Ltd
29 Hamerkava Street
PO Box 150
IL-58101 Holon
Israel
Tel: (+972 3) 557 75 59
Fax: (+972 3) 556 44 96
e-mail: mkt@tadsys.com
Web: http://www.tadsys.com

Technology for Communications International
47300 Kato Road
Fremont
California 94538
United States
Tel: (+1 510) 687 61 00
Fax: (+1 510) 687 61 01
Web: http://www.tcibr.com

Tenet Defence Ltd
North Heath Lane
Horsham
West Sussex RH12 5UX
United Kingdom
Tel: (+44 1403) 27 31 73
Fax: (+44 1403) 27 31 23
e-mail: info@tenetdefence.com
Web: http://www.tenetdefence.com

Terma A/S
Headquarters
Hovmarken 4
DK-8520 Lystrup
Denmark
Tel: (+45) 87 43 60 00
Fax: (+45) 87 43 60 01
e-mail: terma.asy@terma.com
Web: http://www.terma.com

Terma A/S
Naval and Ground Systems
Mårkaervej 2
DK-2630 Tåstrup
Denmark
Tel: (+45) 43 52 15 13
Fax: (+45) 43 52 23 80
e-mail: terma.ncs@terma.com

Textron Systems
Wilmington
201 Lowell Street
Mail Stop 1110
Wilmington
Massachusetts 01887
United States
Tel: (+1 978) 657 21 00
Fax: (+1 978) 657 66 44
Web: http://www.systems.textron.com

Thales Air Defence SA
7/9 rue des Mathurins
F-92221 Bagneux Cedex
France
Tel: (+33 1) 40 84 40 00
Fax: (+33 1) 40 84 33 81
e-mail: info.tad@fr.thalesgroup.com
Web: http://www.thales-airdefence.com

Thales Avionics Ltd
86-88 Bushey Road
Raynes Park
London
SW20 0JH
United Kingdom
Tel: (+44 20) 89 46 80 11
Fax: (+44 20) 89 46 75 30
Web: http://www.thales-avionics.com

Thales Communications SA
Headquarters and General Management
160 boulevard de Valmy
PO Box 82
F-92704 Colombes Cedex
France
Tel: (+33 1) 41 30 30 00
Fax: (+33 1) 41 30 41 86
e-mail: info@tccthomson.fr
Web: http://www.thales-communications.com

Thales Defence Ltd
Communications House
Western Road
PO Box 3621
Bracknell
Berkshire RG2 1WJ
United Kingdom
Tel: (+44 1344) 38 70 00
Fax: (+44 1344) 38 74 03
e-mail: corp_comms@racalgroup.co.uk
Web: http://www.thales-defence.co.uk

Thales Naval SA
7-9 rue des Mauthurins
F-92221 Bagneux Cedex
France
Tel: (+33 1) 39 45 50 00
Fax: (+33 1) 39 45 55 03
Web: http://www.thales-naval.com

Thales Nederland BV
Zuidelijke Havenweg 40
PO Box 42
NL-7550 GD Hengelo
Netherlands
Tel: (+31 74) 248 81 11
Fax: (+31 74) 242 59 36
e-mail: info@nl.thalesgroup.com
Web: http://www.thales-nederland.com

Thales Optronics (Vinten) Ltd
Vicon House
Western Way
Bury St Edmunds
Suffolk IP33 3SP
United Kingdom
Tel: (+44 1284) 75 05 99
Fax: (+44 1284) 75 05 98
e-mail: tovl.info@uk.thalesgroup.com
Web: http://www.wvintenltd.com

Thales Sensors
Manor Royal
Crawley
West Sussex RH10 2PZ
United Kingdom
Tel: (+44 1293) 52 87 87
Fax: (+44 1293) 54 28 18
Web: http://www.thales-defence.co.uk

Thales Underwater Systems
Headquarters
525 route des Dolines
PO Box 157
F-06903 Sophia-Antipolis Cedex
France
Tel: (+33 4) 92 96 30 00
Fax: (+33 4) 92 65 42 77
e-mail: info@tms-ltd.com
Web: http://www.tms-sonar.com

The Boeing Company
Integrated Defense Systems
Headquarters
PO Box 516
St Louis
Missouri 63166-0516
United States
Tel: (+1 314) 232 02 32
Fax: (+1 562) 797 20 20
Web: http://www.boeing.com/ids

Titan Systems Corporation
Linkabit Division
3033 Science Park Road
San Diego
California 92121
United States
Tel: (+1 858) 552 95 00
Fax: (+1 858) 552 96 28
e-mail: ekb@titan.com
Web: http://www.titansystemscorp.com

TMD Technologies Ltd
Swallowfield Way
Hayes
Middlesex UB3 1DQ
United Kingdom
Tel: (+44 20) 85 73 55 55
Fax: (+44 20) 85 69 18 39
Web: http://www.tmd.co.uk

TRL Technology Ltd
Sigma Close
Shannon Way
Tewkesbury
Gloucestershire GL20 8ND
United Kingdom
Tel: (+44 1684) 27 87 00
Fax: (+44 1684) 85 04 06
e-mail: info@trltech.co.uk
Web: http://www.trltech.co.uk

UDS International
525 route des Dolines
BP 157
F-06903 Sophia-Antipolis Cedex
France
Tel: (+33 4) 92 96 34 86
Fax: (+33 4) 92 96 37 69
e-mail: sophia@udsinternational.com
Web: http://www.udsinternational.com

Ultra Electronics Holdings plc
Command and Control Systems
Knaves Beech Business Centre
Loudwater
High Wycombe
Buckinghamshire HP10 9UT
United Kingdom
Tel: (+44 1628) 53 00 00
Fax: (+44 1628) 52 45 57
e-mail: info@ueccs.co.uk
Web: http://www.ultra-electronics.com

Ulyanovsk Mechanical Plant
94 Moskovskoe Road
432008 Ulyanovsk
Russian Federation
Tel: (+7 8422) 32 61 63
Fax: (+7 8422) 32 61 63

United Defense LP
Ground Systems Division
Pennsylvania
PO Box 15512
York
Pennsylvania 17405
United States
Tel: (+1 717) 225 80 04
Fax: (+1 717) 225 81 32

Verint Systems Inc
330 South Service Road
Melville
New York 11747
United States
Tel: (+1 631) 962 96 00
Fax: (+1 631) 962 93 00
e-mail: marketing.lis@verintsystems.com
Web: http://www.verintsystems.com

Viasat Inc
6155 El Camino Real
Carlsbad
California 92009-1699
United States
Tel: (+1 760) 476 22 00
Fax: (+1 760) 929 39 41
Web: http://www.viasat.com

Videoton-Mechlabor Fejleszto es Gyarto Kft
Varosligeti fasor 25
H-1071 Budapest
Hungary
Tel: (+36 1) 342 73 71
Fax: (+36 1) 342 49 34
e-mail: vtmechlab@vtml.videoton.hu

Vista Systems Corp
11460 North Cave Creek Road
Suite 9
Phoenix
Arizona 85020
United States
Tel: (+1 602) 943 57 00
Fax: (+1 602) 943 10 01
e-mail: sales@vistasystems.net
Web: http://www.vistasystems.net

WB Electronics Sp
Filtrowa 63
PL-02-056 Warsaw
Poland
Tel: (+48 22) 825 92 91
Fax: (+48 22) 825 03 49
e-mail: info@wb.com.pl
Web: http://www.wb.com.pl

Wescan Systems Limited
Surveillance Systems
777 Walkers Lane
Burlington
Ontario L7N 2GI
Canada
Tel: (+1 905) 333 60 00
Fax: (+1 905) 333 60 05
e-mail: info@wescanltd.com
Web: http://www.wescanltd.com

Whitehead Alenia Sistemi Subacquei SpA
Via di Levante 48-50
I-57128 Livorno
Italy
Tel: (+39 0586) 84 01 11
Fax: (+39 0586) 85 40 60
e-mail: wassedp@interbusiness.it
Web: http://www.aleniadifesa.finmeccanica.it

Winradio Communications
15 Stamford Road
Oakleigh
Victoria 3166
Australia
Tel: (+61 3) 95 68 25 68
Fax: (+61 3) 95 68 13 77
e-mail: info@winradio.com
Web: http://www.winradio.com

XI Information Processing Systems Ltd
PO Box 60
IL-56915 Savyon
Israel
Tel: (+972 3) 921 58 91
Fax: (+972 3) 534 35 20
e-mail: xi00001@attglobal.net

Zengrange Defence Systems Ltd
Wellfield House
Victoria Road
Morley
Leeds
LS27 7PA
United Kingdom
Tel: (+44 113) 259 75 55
Fax: (+44 113) 259 75 59
e-mail: managing@zengrange.co.uk
Web: http://www.zengrange.co.uk

Zeta, 17680 Butterfield Boulevard
Morgan Hill
California 95037
United States
Tel: (+1 408) 852 08 00
Fax: (+1 408) 852 08 01
e-mail: zeta@zeta-idt.com
Web: http://www.zeta-idt.com

INDEX

Manufacturers' index

Alphabetical index

To help users of this title evaluate the published data, *Jane's Information Group* has divided entries into three categories.

N NEW ENTRY   Information on new equipment and/or systems appearing for the first time in the title.
V VERIFIED   The editor has made a detailed examination of the entry's content and checked its relevancy and accuracy for publication in the new edition to the best of his ability.
U UPDATED   During the verification process, significant changes to content have been made to reflect the latest position known to *Jane's* at the time of publication.
   Items in italics refer to entries which have been deleted from this print edition. These entries are still available in the online version of this product–jc4i.janes.com.